W9-AAZ-795

Donald Knight
Geol 103

Midwest Version

(Illinois, Indiana, Iowa, Kansas, Michigan, Minnesota, Missouri, Nebraska, North Dakota, Ohio, South Dakota, Wisconsin)

Agriculture   Efficient and environmentally responsible farming methods.

Habitat, Wilderness, and Wetlands   Habitat destruction and protection, integrated resource management, logging with mules.

Air Pollution   Setting zero emissions goals, recycling CFCs, reverse commuting.

Water Pollution   Community resource planning, preserving natural wetlands, constructing wetlands for wastewater treatment.

Hazardous Waste   "Back to nature" movement in toxic pollution cleanup, bioremediation.

Solid Waste   Multi-family recycling, yard composting, Illinois' yard waste ban, composting food processing residuals, recycling construction project debris.

Urbanization   Achieving healthy cities and innovative growth management programs.

Northeast Version

(Connecticut, Delaware, District of Columbia, Maine, Massachusetts, Maryland, New Hampshire, New Jersey, New York, Pennsylvania, Rhode Island, Vermont)

Habitat, Wilderness, and Wetlands   Habitat destruction and protection, critical area legislation, protection of non-tidal wetlands.

Air Pollution   Anti-pollution laws, industry's new role in cleaning up the environment.

Water Pollution   Run-off from suburban developments, controls on point and nonpoint sources, the Chesapeake Bay watershed.

Solid Waste   Municipal solid waste crisis, recycling programs, NIMBY issues, private-public partnerships, landfills.

Urbanization and Land Use   Population growth, destruction of ecosystems, urban sprawl, zoning.

South Version

(Alabama, Arkansas, Florida, Georgia, Kentucky, Louisiana, Mississippi, North Carolina, Oklahoma, South Carolina, Tennessee, Texas, Virginia, West Virginia)

Habitat   Habitat and wetland destruction and efforts to restore them.

Air Pollution   Reducing toxic emissions, planting trees, opposing construction of power plants.

Water Pollution   Constructed wetlands, dual water systems, agricultural uses for reclaimed water.

Hazardous Waste   Cleaning up toxic hot spots, preventing improper disposal.

Solid Waste   Recycling, volume-based garbage collection programs, composting, competitive bidding for removing recyclables.

Urbanization   Successful growth management programs, solutions to urban sprawl, "neotraditionalist" architects and town planners.

Environmental Science
Action for a Sustainable Future

Also of interest from the Benjamin/Cummings Series in Life Sciences

General Biology

N. A. Campbell
Biology, Third Edition (1993)

R. J. Kosinski
Fish Farm: Simulation Software (1993)

J. G. Morgan and M. E. B. Carter
Investigating Biology: A Laboratory Manual (1993)

Evolution, Ecology and Behavior

M. Lerman
Marine Biology: Environment, Diversity, and Ecology (1986)

R. Trivers
Social Evolution (1985)

Plant Biology

M. G. Barbour, J. H. Burk, and W. D. Pitts
Terrestrial Plant Ecology, Second Edition (1987)

J. Mauseth
Plant Anatomy (1988)

E. Zeiger and L. Taiz
Plant Physiology (1991)

Animal Biology

H. E. Evans
Insect Biology: A Textbook of Entomology (1984)

E. N. Marieb
Essentials of Human Anatomy and Physiology, Fourth Edition (1994)

E. N. Marieb
Human Anatomy and Physiology, Second Edition (1992)

E. N. Marieb and J. Mallatt
Human Anatomy (1992)

L. G. Mitchell, J. A. Mutchmor, W. D. Dolphin
General Zoology (1988)

A. P. Spence
Basic Human Anatomy, Third Edition (1991)

Molecular Biology and Genetics

F. J. Ayala and J. A. Kiger Jr.
Modern Genetics, Second Edition (1984)

L. E. Hood, I. L. Weissman, W. B. Wood, and J. H. Wilson
Immunology, Second Edition (1984)

R. Schleif
Genetics and Molecular Biology (1986)

J. Watson, N. H. Hopkins, and J. W. Roberts
Molecularbiology of the Gene, Fourth Edition (1987)

Microbiology

I. E. Alcamo
Fundamentals of Microbiology, Third Edition (1991)

R. M. Atlas and R. Bartha
Microbial Ecology: Fundamentals and Applications, Third Edition (1992)

J. Cappuccino and N. Sherman
Microbiology: A Laboratory Manual, Third Edition (1992)

T. R. Johnson and C. L. Case
Laboratory Experiments in Microbiology, Brief Edition, Third Edition (1992)

G. J. Tortora, B. R. Funke, and C. L. Case
Microbiology: An Introduction, Third Edition (1989)

P. J. VanDemark and B. L. Batzing
The Microbes (1987)

Environmental Science
Action for a Sustainable Future

Fourth Edition

Daniel D. Chiras

The Benjamin/Cummings Publishing Company, Inc.
Redwood City, California ■ Menlo Park, California ■ Reading, Massachusetts
New York ■ Don Mills, Ontario ■ Wokingham, U.K. ■ Amsterdam ■ Bonn
Sydney ■ Singapore ■ Tokyo ■ Madrid ■ San Juan

Dedicated to my family

Skyler, Forrest, and Kathleen

whose love and affection and effervescent

smiles brighten my days and make the

struggle seem worthwhile

Executive Editor: Robin Heyden
Assistant Developmental Editor: Kim Viano
Marketing Manager: John Minnick
Art and Design Manager: Michele Carter
Manufacturing Supervisor: Casimira Kostecki
Senior Production Editor: John Walker
Copyeditor: Elizabeth Gehman
Text, Cover Design and Art: Rob Hugel, XXX Design
Photo Research: Kelli d'Angona-West and Sami Iwata
Illustrations: Wayne Clark, Parry Clark, Barbara Haynes,
    Rolin Graphics
Permissions Editor: Marty Granahan
Composition, Page makeup and Film: Jonathan Peck
    Typographers
Printing and Binding: Webcrafters

Photo and text credits appear after the Glossary.

Dan Chiras prepared all drafts of his manuscript on recycled paper and used recycled ribbons in his printer. The text and inserts are printed on recycled paper using soy-based inks. The cover was printed on 100% biodegradable cloth using soy-based inks and the coating is aqueous. The *Action Guide*, which is also printed on recycled paper using soy-based inks, is inserted to avoid shrink-wrapping.

Copyright © 1994 by the Benjamin/Cummings Publishing Company, Inc.

All rights reserved. No part of this publication may be reproduced, stored in a database or retrieval system, distributed, or transmitted in any form or by any means, electronic, mechanical, photocopying, recording, or otherwise without the prior written permission of the publisher. Printed in the United States of America. Published simultaneously in Canada.

Library of Congress Cataloging-in-Publication Data

Chiras, Daniel D.
    Environmental science / Daniel D. Chiras. — 4th ed.
        p.   cm.
    Includes index.
    ISBN 0-8053-4215-X
    1. Environmental science.   2. Sustainable development.   I. Title.
    GE140.C48   1994   93-5239
    363.7—dc20   CIP
Regional Version ISBNs
Northeast: 0–8053–4224–9
South: 0–8053–4225–7
Midwest: 0–8053–4226–5
West: 0–8053–4227–3

1 2 3 4 5 6 7 8 9 10—WC—97 96 95 94 93

The Benjamin/Cummings Publishing Company, Inc.
390 Bridge Parkway
Redwood City, California 94065

Preface

Environmental Science: Action for a Sustainable Future may be one of the most important textbooks you will ever read. Why? Because it deals with one of the most important issues of our time: the environment. More specifically, it details the massive erosion of the life-support systems of the planet upon which our lives and our economies depend.

Like other textbooks on the subject, this one offers an in-depth look at the environmental problems facing the world. It examines the future prospects of human civilization and the environment, intimately bound together, but rarely considered in most human deliberations. It also discusses a variety of positive solutions for businesses, governments, and individuals.

Unlike other texts, this one has changed dramatically in recent years to reflect profound changes in thinking. What prompted those changes?

Twenty years ago when I embarked on a study of the environment, I learned a great many facts about the environmental problems facing humankind. I found that the issues generally fit into three broad categories—population issues, resource issues, and pollution issues. Over the years, I also found that most environmental problems sprung from the same root causes.

I discovered that most governments and businesses were addressing environmental problems in a piecemeal fashion. That is, most policies and practices addressed only small portions of the environmental crisis. Solutions, therefore, typically resulted in marginal adjustments or tackled one portion of a problem, while ignoring or making other problems worse.

Thus, over the years, I've come to realize that many environmental protection efforts, although essential, have missed the mark! In fact, most efforts could be accurately portrayed as stop-gap measures, for many only treat the symptoms and neglect the underlying root causes. Why is it so important to confront root causes?

My studies of environmental issues have made evident that modern society is facing a crisis of epic proportions. Many environmental trends outlined in this text confirm the gravity of the situation and suggest that human society is on a fundamentally unsustainable course. We cannot continue along our current path without risking ecological catastrophe and severe social and economic repercussions. Although that may sound overly pessimistic, even alarmist, I believe it is truthful. As you study the problems and think about the long-term implications of rapid population growth, massive soil erosion, climate change, and a host of other problems, I think most of you will agree that the long-term future is less than rosy. Most of you will probably agree that we must take action to put human civilization onto an ecologically sane path. And, to do so, we must stop treating the symptoms and get down to the heart of the matter.

While adding new material, updating statistics, and polishing the writing, I shifted the emphasis of the text to reflect a pressing need for solutions that confront the root causes of the environmental crisis and create a sustainable human culture, a society that meets its needs without bankrupting the Earth.

In revising this text, then, I looked for sustainable solutions—measures that confront the root causes of the environmental crisis head-on. Such measures can help us create a way of life that coexists in harmony with the environment.

Fortunately, my search paid off handsomely. Studies of the literature and the reasons why natural systems endure revealed a strategy that could lead us away from the quick-fix, end-of-pipe solutions that have characterized the environmental response for so long. You'll learn more about this in Chapter 1.

As in the first three editions, I wanted this text to be user friendly, not bogged down with irrelevant statistics or endless detail. My goal was to continue to present *the* most important facts and concepts in a clear, exciting fashion. As always, I wanted to minimize my own biases by presenting both sides of the issues and by offering Point/Counterpoints on controversial issues. Even though this text presents a strong case for sustainability, it is left to you, the reader, to decide the need and desirability of such an approach. My efforts to make this text as unbiased as possible support my objective of letting students make up their own minds about our predicament and ways to extricate ourselves from it. Critical thinking skills presented in the text also help students learn to analyze issues.

Themes

All textbooks have a central theme or, in some cases, set of themes that shapes the presentation. This text is no different. It is molded by three central ideas: sustainability, critical thinking, and action.

The main theme of this text is that the long-term well-being of this planet and its inhabitants is in jeopardy and that to create an enduring human presence we must make a massive course change. We must make the transition to a sustainable society. A sustainable society seeks balance between human and ecological needs. Its economic systems serve people and the planet. Creating a sustainable society may be our only realistic hope for surviving on a finite planet, but it will not evolve without foresight, planning, and action.

This text also stresses critical thinking skills—skills that teach us to think critically about issues. Critical thinking is essential to the task of creating a sustainable society.

Finally, this text emphasizes an often overlooked point: Building a sustainable future requires actions by all of us. Air pollution is not caused simply by inadequate laws or corporate neglect; it is the result of our own often wasteful life-styles.

Because we are all part of the problem, we must all be part of the solution. Individual action is as essential as responsible corporate and government policies and practices.

Organization

This book is divided into five parts. Part 1 introduces the student to basic ecological principles necessary for understanding environmental issues. These chapters describe the six biological principles of sustainability, useful in revamping modern society. Part 1 also presents an overview of ethics and some basic economic principles necessary for understanding many environmental issues. Part 1 describes basic rules of critical thinking.

Part 2 opens the discussion of environmental issues, dealing with one of the most pressing of all, the population crisis. This part examines the impact of rapid population growth and explores culturally acceptable ways of slowing it down. Part 3 deals with a variety of resource issues, such as wildlife extinction and energy demand, and outlines strategies for solving them sustainably. Part 4 discusses pollution as welll as legal, technical, and personal solutions, including both traditional and sustainable strategies.

Part 5, the capstone of the book, attempts to place the population, resource, and pollution crises in a social context. It reexamines ethics and explores economics and government in more detail.

Special Features

The following special features from the first three editions have been retained to increase student interest and involvement:

Models

This book offers numerous conceptual models that help students understand how the world works. These models are designed to encourage critical thinking and help students organize facts into a solid conceptual framework. Below is a brief description of each model:

Population, Resources, and Pollution Model   This model shows how populations of organisms, like ourselves, affect their environment and how their actions, in turn, affect the populations.

Multiple Causation Model   This model helps students recognize all of the causes of environmental problems.

Impact Analysis Model   This model shows the impacts that humans have on various components of the environment.

Risk Analysis Model   This model presents an overview of the process called risk assessment.

Chapter Supplements

Chapter supplements, found at the end of some chapters, furnish detailed coverage of important topics and provide instructors with an added degree of flexibility. Topics include indoor air pollution, radiation pollution, nuclear war, and environmental law.

Point/Counterpoints and Viewpoints

Complex environmental issues often result in hotly contested debates:

- Is outer space the answer to our population and resource problems?
- Are we responsible to future generations?
- Is population growth good or bad?
- Are we losing the war against cancer?
- Is the spotted owl worth saving?
- Do direct actions (ecotage) to protect the planet help or hinder the environmental movement?

These and many other timely issues are debated in Point/Counterpoints or discussed in Viewpoints by experts such as Norman Myers, Ben Bova, Garrett Hardin, Julian Simon, Amory and L. Hunter Lovins, Frederic Krupp, Lewis Regenstein, Howie Wolke, and others. These essays stimulate individual thinking as well as classroom discussion on complex problems. They're also a perfect tool for developing critical thinking skills.

Color Galleries

Four color galleries are included to emphasize some of the key concepts and issues including: Understanding the Earth, Biomes, Endangered Species, and Resources Misuse.

Case Studies

To give students further insight into timely issues, I've included numerous Case Studies throughout the text. Case Studies discuss important environmental problems and help reinforce material covered in the text. Topics include endangered indigenous peoples, genetic engineering, protection of Antarctica, and problems confronted in Canada's rivers.

Chapter Summaries

Each chapter is followed by a succinct summary of the important concepts and terms. Summaries reinforce the key points and serve as valuable study tools.

Critical Thinking

As pointed out earlier, critical thinking is one of the central themes of this book. Critical thinking enables students to discern fact from fiction. It helps them analyze complex issues and make rational decisions. A number of important critical thinking rules are discussed in Chapter Supplement 1.1. In addition, students are also asked to exercise critical thinking skills after Point/Counterpoints, and many of the discussion questions at the end of each chapter call on students to apply their knowledge and their critical thinking skills to use.

New to This Edition

Updated Coverage

The fourth edition has been thoroughly updated with new discoveries, new concepts, new environmental laws, and the most recent statistics on resources, population, and pollution. New photographs, tables, and line drawings have also been added.

Models of Global Sustainability

To highlight some of the positive steps being taken to build a sustainable society worldwide, I've included numerous examples in Models of Global Sustainability. These essays feature people who are helping to reduce energy demand, to recycle, to tap into renewable energy resources, to restore damaged ecosystems, and to slow down the population growth.

Critical Thinking Exercises

Another feature new to this edition are the Critical Thinking Exercises. Each chapter begins with a brief Critical Thinking Exercise, which asks students to critically analyze an issue, a research finding, or an assertion.

Study Skills

Immediately following the Preface is a list of study skills, another important addition to this edition. The study skills section includes numerous simple but effective tips that help students improve their memory, note-taking skills, reading abilities, test-taking abilities, and much more. Study skills can help all students, even A students, become more efficient learners. Skills learned here will carry over to virtually every other course students will take and will be helpful throughout life.

New Regional Versions to Accompany the Chiras Text

A true understanding of global environmental challenges often starts with a closer look at issues in our own backyards. Based on feedback from hundreds of environmental science professors, we developed four new regional versions designed to complement the core text and introduce students to a variety of specific local problems in the U.S. The results are effective, one-color modules that emphasize sustainable solutions and the necessity of local action in solving global problems. Each regional version begins with a general introduction, outlining the issues pertinent to that particular region. Each issue has a succinct introduction followed by a series of articles and discussion questions. There are four new regional versions of the Chiras text available: Northeast, South, Midwest, and West. Each regional version is available separately or bound with the core text.

Supplements

Environmental Action Guide   Many environmental science instructors are concerned that their students leave this course with a sense of what the individual can do to effect change. To address this need, this edition is published with the newly updated *Environmental Action Guide*. Written by Ann S. Causey of Prescott College, this booklet provides information on environmentally sound products, investments, careers, community action groups, letter writing, and low-impact life-styles.

Instructor's Guide   Ann S. Causey also updated and revised the Instructor's Guide. It includes chapter outlines and test questions. (Transparency acetates for instructors are packaged separately.) In addition, more Case Studies and critical thinking problems are included for further class discussion.

Laboratory Manual   The fourth edition comes with a Laboratory Manual written by Dr. Merle Alexander, Director of Environmental Studies, Baylor University. This manual includes 14 lab exercises, each designed to be conducted in a single class session. Students learn to apply theory through practical applications.

Test Bank   The test questions for each of the 23 chapters are found in the Test Bank. Revised by David Fluharty of the University of Washington, it includes many new questions in multiple-choice, short-answer, and essay formats. Careful attention sought to eliminate ambiguities and confusing wording. Test Bank is available for the IBM PC, IBM PS/2, and the Macintosh computers. This software is available to qualified adopters of *Environmental Science* by contacting the publisher or your local representative.

Acknowledgments

Although I have spent thousands of hours researching and writing, this text is really the product of many people. In truth, it is the product of thousands of researchers in anthropology, biology, chemistry, demography, natural resources, political science, economics, ecology, and dozens of other disciplines. Their findings and their thoughts form the foundation on which this text rests. To them, a world of thanks!

A genuine thanks also to the staff at Benjamin/Cummings. A special thanks to Melinda Adams, my editor throughout much of the revision. Melinda's insights and ideas and her effervescent personality were always appreciated. Thanks also to my developmental editor, Kim Viano, who helped see the book through production. Kim's eagerness to help, cheerful personality, and kind and considerate manner were a godsend in an otherwise hectic schedule. Thanks also to editorial assistant Sami Iwata, who helped with the initial market research and other important tasks. Special thanks to John Walker, who competently guided the book through a rapid production schedule this time around. Also, many thanks to Alyssa Wolf and Kelly Hall, who coordinated the art program. My appreciation to Kelli d'Angona West and Sami Iwata, who procured photographs.

Also a special thanks to my excellent research assistant, Sally Almeria, who helped research and write many of the Models of Global Sustainability. A special word of gratitude to my wife, Kathleen, who helped update the statistics and gather new Point/Counterpoints for this edition. Her persistence and attention to detail were much appreciated. Also, much appreciation to the many people in government agencies and nonprofit groups who shared articles, reports, and data with us. Too numerous to mention here, their part in this project is much appreciated.

Finally, many manuscript reviewers provided helpful and constructive criticism on all editions of this book:

Nancy Bain, Ohio University

Peter Colverson, Mohawk Valley Community College

Sheree E. Cohn, Saint Cloud State University

Craig B. Davis, Ohio State University

David Fluharty, University of Washington

Gary J. Galbreath, Northwestern University

James Grosklags, Northern Illinois University

Zac Hansom, San Diego State University

Pat Hilliard Johnson, Palm Beach Community College

James Hornig, Dartmouth College

John A. Jones, Miami-Dade Community College

Timothy F. Lyons, Ball State University

Paul E. Nowack, University of Michigan

Nancy Ostiguy, California State University, Sacramento

Michael Priano, Westchester Community College

Doris Shoemaker, Dalton College

Steven Solheim, University of Wisconsin, Madison

I am very thankful for their helpful comments.

Daniel D. Chiras
Evergreen, Colorado

About the Author

Dr. Chiras is an adjunct professor of the Environmental Policy and Management Program at the University of Denver, where he teaches "Toward a Sustainable Public Policy." He has been a visiting professor at the University of Washington in Seattle, where he taught introductory environmental science.

Dan Chiras began his teaching career in 1976 after earning his Ph.D. in reproductive physiology from the University of Kansas Medical School, where he was awarded the Latimer Award for his research on ovarian physiology. That year, he accepted a teaching and research position at the University of Colorado, Denver where he taught a variety of courses, including general biology, cell biology, histology, reproductive biology, and endocrinology.

Over the years, Dr. Chiras developed a number of courses on the environment, including a graduate course on pollution, environment, and health as well as several undergraduate science modules on air pollution, nuclear power, noise pollution, impacts from coal development, and strategies for sustainability.

In 1981, Dr. Chiras decided to pursue a full-time writing career. Since that time, he has published over 70 articles and 9 books. A leading advocate of critical thinking and sustainability, Dr. Chiras has published several articles in the *American Biology Teacher* that are helping to reshape environmental education. He recently published a new theory on the roots of the environmental crisis. Currently, Dr. Chiras writes the environment section for *Science Year* published by the World Book Encyclopedia, and serves as coeditor of *Environmental Carcinogenesis and Ecotoxicology Reviews*.

In addition to teaching and writing, Dr. Chiras plays an active role in the environmental movement. Dr. Chiras worked on several EPA projects on the health impacts of chlorinated organics from wastewater treatment and on the impacts of coal mining in the West. He also wrote an assessment of the impacts of oil shale development in Colorado for the U.S. Department of Energy.

Dr. Chiras has served on the Board of Directors of the Colorado Environmental Coalition since 1987, and he was president of this coalition of 40 environmental groups for two years. In 1988, Dr. Chiras cofounded Friends of Curbside Recycling, which was instrumental in convincing the city of Denver to begin a curbside recycling program. In 1989, he cofounded Speakers for a Sustainable Future, an organization that offers slide shows on recycling, water conservation, and sustainability, and that works to establish state chapters. Dr. Chiras cofounded another nonprofit organization in 1993, the Sustainable Future Society (SFS). Its mission is to foster the transition to a global sustainable society by promoting a broader understanding of sustainable principles, policies, and practices and by encouraging adoption of these practices.

When he's not writing, teaching, or working as an activist, Dan Chiras likes to kayak, hike, ski, and bicycle. He lives with his wife, Kathleen, and two children, Skyler and Forrest, in a superefficient passive solar home in the Colorado Rockies where he tries to practice all that he preaches.

Study Skills

College is a demanding time. For many students, term papers, tests, reading assignments, and classes require a new level of commitment to their education. At times, the work load can be overwhelming.

Fortunately, many ways can lighten the load and make time spent in college more profitable. This section offers some helpful tips to enhance your study skills. It teaches you how to improve your memory, how to become a better note taker, and how to get the most out of what you read. It also helps you prepare better for tests and become a better test taker.

Mastering these study skills will require some work, mostly to break old, inefficient habits. In the long run, though, the additional time you spend now learning to become a better learner will pay huge dividends. Over the long haul, improved study skills will save you lots of time and help you improve your knowledge of facts and concepts. That will no doubt lead to better grades and very likely a more fruitful life.

General Study Skills

- Study in a quiet, well-lighted space. Avoid noisy, distracting environments.
- Turn off televisions and radios.
- Work at a desk or table. Don't lie on a couch or bed.
- Establish a specific time each day to study and stick to your schedule.
- Study when you are most alert. Many students find that they retain more if they study in the evening a few hours before bedtime.
- Take frequent breaks—one every hour or so. Exercise or move around during your study breaks to help you stay alert.

- Reward yourself after a study session with a mental pat on the back or a healthy snack.
- Study each subject every day to avoid cramming for tests. Some courses may require more hours than others, so adjust your schedule accordingly.
- Look up new terms or words whose meanings are unclear to you in the glossaries in your textbooks or in a dictionary.

Improving Your Memory

You can improve your memory by following the PMC method. The PMC method involves three simple learning steps: (1) paying attention, (2) making information memorable, and (3) correlating new information with facts you already know.

Step 1 Paying attention means taking an active role in your education—taking your mind out of neutral. Eliminate distractions when you study. Review what you already know and formulate questions about what you are going to learn *before* a lecture or *before* you read a chapter in the text. Reviewing and questioning help prime the mind.

Step 2 Making information memorable means finding ways to help you retain it. Repetition, mnemonics, and rhymes are three helpful tools.

- Repetition can help you remember things. The more you hear or read something, the more likely you are to remember it, especially if you're paying attention. Jot down important ideas and facts while you read or study to help involve all of the senses.
- Mnemonics are useful learning tools to help remember lists of things. I use the mnemonic CARRRP to remember the biological principles of sustainability:

conservation, adaptability, recycling, renewable resources, restoration, and population control.

■ Rhymes and sayings can also be helpful when trying to remember lists of facts.

■ If you're having trouble remembering key terms, look up their roots in the dictionary. This often helps you remember their meaning.

■ You can also draw pictures and diagrams of processes to help remember them.

Step 3 Correlating new information with the facts and concepts you already know helps tie facts together, making sense out of the bits and pieces you are learning.

■ Instead of filling your mind with disjointed facts and figures, try to see how they relate with what you already know. When studying new concepts, spend some time tying information together to get a view of the big picture.

■ After studying your notes or reading your textbook, review the main points. Ask yourself how this new information affects your view of life or critical issues and how you may be able to use it.

Becoming a Better Note Taker

■ Spend five to ten minutes before each lecture reviewing the material you learned in the previous lecture. This is extremely important!

■ Know the topic of each lecture *before* you enter the class. Spend a few minutes reflecting on facts you already know about the subject about to be discussed.

■ If possible, read the text *before* each lecture. If not, at least look over the main headings in the chapter, read the topic sentence of each paragraph, and study the figures. If your chapter has a summary, read it, too.

■ Develop your own shorthand system to facilitate note taking. Symbols such as = (equals), > (greater than), < (less than), w (with), and w/o (without) can save lots of time so you don't miss the main points or key facts.

■ Develop special abbreviations to reduce writing time. E might stand for energy, AP might be used for air pollution, and AR could be used to signify acid rain.

■ Omit vowels and abbreviate words to decrease writing time (for example: omt vwls & abbrvte wrds to dcrs wrtng tme). This will take some practice.

■ Don't take down every word your professor says, but be sure your notes contain the main points, supporting information, and important terms.

■ Watch for signals from your professor indicating important material that might show up on the next test (for example: "This is an extremely important point. . .").

■ If possible, sit near the front of the class to avoid distractions.

■ Review your notes soon after the lecture is over, when they're still fresh in your mind. Be sure to leave room in your notes written during class so you can add material you missed. If you have time, recopy your notes after each lecture.

■ Compare your notes with those of your classmates to be sure you understood everything and did not miss any important information.

■ Attend all lectures.

■ Use a tape recorder if you have trouble catching important points.

■ If your professor talks too quickly, politely ask him or her to slow down.

■ If you are unclear about a point, ask during class. Chances are other students are confused as well. If you are too shy, go up after the lecture and ask, or visit your professor during his or her office hours.

How to Get the Most out of What You Read

■ Before you read a chapter or other assigned readings, preview the material by reading the main headings or chapter outline to see how the material is organized.

■ Pause over each heading and ask a question about it.

■ Next, read the first sentence of each paragraph. When you finish, turn back to the beginning of the chapter and read it thoroughly.

■ Take notes in the margin or on a separate sheet of paper. Underline or highlight key points.

■ Don't skip terms that are confusing to you. Look them up in the glossary or in a dictionary. Make sure you understand each term before you move on.

■ Use the study aids in your textbook, including summaries and end-of-chapter questions. Don't just look over the questions and say, "Yeah, I know that." Write out the answer to each question as if you were turning it in for a grade and save your answers for later study. Look up answers to questions that confuse you. This text has questions that test your understanding of facts and concepts. Critical thinking questions are also included to sharpen your skills.

Preparing for Tests

- Don't fall behind on your reading assignments and review lecture notes as often as possible.

- If you have the time, you may want to outline your notes and assigned readings. Try to prepare the outline with your book and notes closed. Determine weak areas, then go back to your text or class notes to study these areas.

- Space your study to avoid cramming. One week before your exam, go over all of your notes. Study for two nights, then take a day off that subject. Study again for a couple of days. Take another day off from that subject. Then, make one final push before the exam, being sure to study not only the facts and concepts but also how the facts are related. Unlike cramming, which puts a lot of information into your brain for a short time, spacing will help you retain information for the test and for the rest of your life.

- Be certain you can define all terms and give examples of how they are used.

- You may find it useful to write flash cards to review terms and concepts.

- After you have studied your notes and learned the material, look at the big picture—the importance of the knowledge and how the various parts fit together.

- You may want to form a study group to discuss what you are learning and to test one another.

- Attend review sessions offered by your instructor or by your teaching assistant. Study before the session and go to the session with questions.

- See your professor or class teaching assistant with questions as they arise.

- Take advantage of free or low-cost tutoring offered by your school or, if necessary, hire a private tutor to help you through difficult material. Get help quickly. Don't wait until you are hopelessly lost. Remember that learning is a two-way street. A tutor won't help unless you are putting in the time.

- If you are stuck on a concept, it may be that you have missed an important point in earlier material. Look over your notes or ask your tutor or professor what facts might be missing, causing you to be confused.

- If you have time, write and take your own tests. Include all types of questions.

- Study tests from previous years if they are available legally.

- Determine how much of a test will come from notes and how much will come from the textbook.

Taking Tests

- Eat well and get plenty of exercise and sleep before tests.

- Remain calm during the test by deep breathing.

- Arrive at the exam on time or early.

- If you have questions about the wording of a question, ask your professor.

- Skip questions you can't answer right away and come back to them at the end of the session if you have time.

- Read each question carefully and be sure you understand its full meaning before answering it.

- For essay questions and definitions, organize your thoughts first on the back of the test *before* you start writing.

Now, take a few moments to go back over this list. Check off those things you already do. Then, mark the new ideas you want to incorporate into your study habits. Make a separate list, if necessary, and post it by your desk or on the wall and keep track of your progress.

"Study Skills," by Daniel D. Chiras, was originally published in *Human Biology: Health, Homeostasis, and the Environment*. St. Paul: West Publishing Company (1991). Used with permission.

Brief Contents

Detailed Contents

Environmental Science
Action for a Sustainable Future

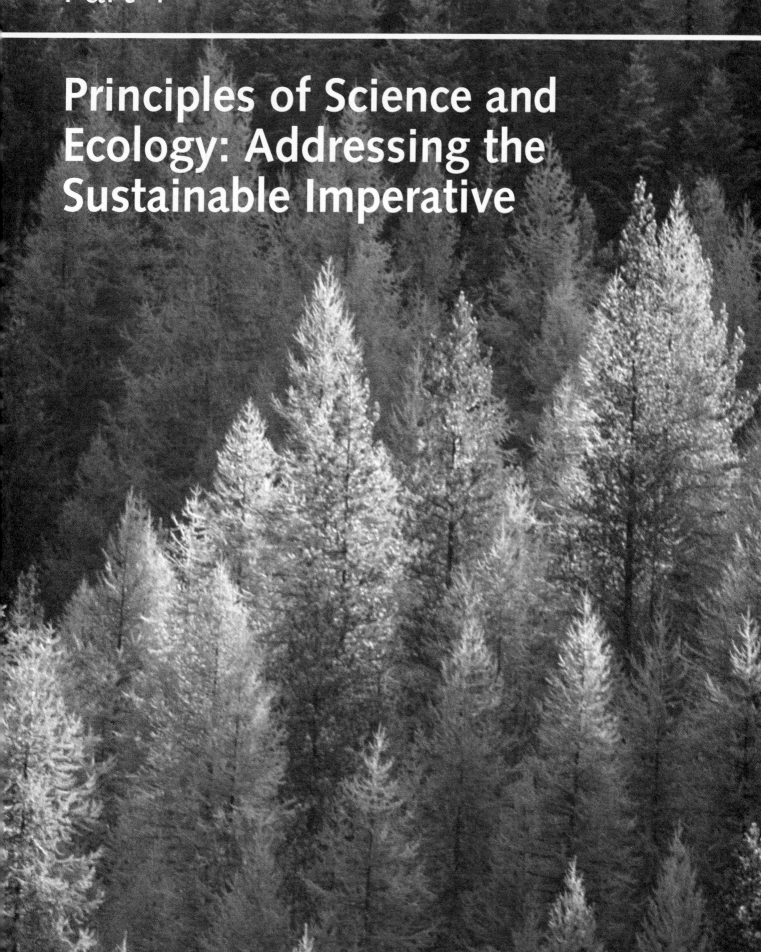

Part 1

Principles of Science and Ecology: Addressing the Sustainable Imperative

Chapter 1

Environmental Science: Meeting the Challenge of Sustainability

There are no passengers on spaceship earth.

We are all crew.

—Marshall McLuhan

Critical Thinking Exercise

A student of mine once remarked that we really don't need to worry about environmental problems as much as most people think. Why? Because technology can solve the environmental problems facing the world. He went on to say that by tightening pollution controls on factories and developing new technologies to provide energy and safely incinerate waste, we can resolve all of our problems. This viewpoint is an extreme case of technological optimism, an outlook prevalent in modern society. Do you see any problems with this logic? What are they?

1.1 The Sustainable Revolution

In 1987, 24 industrial nations met in Montreal, Canada, to sign the Montreal Protocol, a treaty aimed at protecting the **ozone layer**. Found in the upper atmosphere, the ozone layer contains a slightly higher concentration of ozone molecules than the air around it. Ozone molecules, which consist of three oxygen atoms, act as a shield, absorbing harmful ultraviolet light emitted by the sun. Studies performed over the past two decades have shown that this protective shield is being destroyed by chemicals released from aerosol cans, refrigerators, air conditioners, and other sources.

Recognizing the importance of the ozone layer to life on Earth, these **industrial nations**, whose economies rely heavily on manufacturing, committed to steps that would halve the production of ozone-destroying chemicals by the year 2000. In 1990, however, after reviewing new studies that showed ozone depletion was far greater than previously thought, 93 countries, including many nonindustrial nations, agreed to a global ban on ozone-depleting chemicals by the year 2000. In 1992, they accelerated the phaseout.

In 1990, the state of California announced a remarkable new energy policy that could set the stage for an energy revolution in the United States. This plan outlines a strategy to meet future demands in environmentally acceptable ways. According to the plan, approximately one-fourth of the state's demand for new energy will be met by tapping into relatively clean, renewable sources, such as wind, solar, and geothermal. The remaining three-fourths will be met by using existing energy resources much more efficiently. That is, instead of building new power plants, the state hopes to make existing end uses more efficient, freeing up energy to meet projected demand. This policy is part of a global effort to rethink and reshape industrial economies in ways that greatly lessen their damage to the environment.

In other good news, Martin-Marietta's Astronautics Division, located in Denver, Colorado, recently halted use of an ozone-depleting chemical once employed as a cleaning agent for aluminum panels in its Titan missiles. Switching to a relatively nontoxic, recyclable cleaning agent cost the company $270,000 but will save it nearly $1.2 million from 1991 to 1995. Moreover, the new chemical is safer for workers and easier to handle (Figure 1.1).

Finally, at this very moment, in the United States and in several European countries, ordinary citizens are joining together to create more environmentally sound life-styles. Over a six-month period, groups of homeowners will attempt to dramatically reduce the amount of garbage they produce by buying less, recycling, and composting. They will also employ measures to cut water and energy demand in their houses. When they finish this project, sponsored by the nonprofit environmental group called the Global Action Plan for the Earth (GAP), they will have made remarkable changes in their lives. In addition, they will have begun to realize substantial savings. According to GAP, each family will save about $1200 a year for the efforts they take. (See Models of Global Sustainability 1.1.)

From corporate boardrooms to government offices to private living rooms, citizens of the world are rising to a challenge unwitnessed in human history—the task of saving the planet. Of course, the Earth is not in jeopardy; it is the planet's life-support systems that face almost certain ruin. Such efforts are part of a revolution

Figure 1.1 *By switching to a cleaning agent that does not deplete the ozone layer, the Martin-Marietta Astronautics Division will help to protect the environment, saving $1.2 million in the first five years.*

Models of Global Sustainability 1.1

EcoTeams: An International Effort to Create Sustainable Life-Styles

The magnitude of the environmental problems facing us often evokes feelings of helplessness and despair. The problems seem so immense that many wonder if there is anything individuals can do to make a difference. Fortunately, individuals can do a lot to help create a sustainable future. But where does one begin?

One of the best starting points is the Household EcoTeam project, the brain child of the U.S. environmental group Global Action Plan for the Earth (GAP). Designed to assist individuals in taking action, the program consists of three parts: (1) education on the problems and changes necessary to alleviate them, (2) action goals that are easily attainable and measurable, and (3) regular feedback that demonstrates how the combined actions of all individuals add up to real progress.

Those interested in participating join with neighbors who have similar interests to form an ecoteam. Over a six-month period, the team meets to discuss a monthly topic, to explore opportunities for action, and to report on the progress made in the previous month.

The Household EcoTeam Workbook, published by GAP, is the basis of the program. Its six units, one for each month, cover a variety of topics: waste reduction, water efficiency, energy efficiency, transportation efficiency, environmentally conscious purchasing, and empowerment of others to make changes. The workbook includes basic information on environmental issues as well as actions participants can take to attain their goals. It also contains work sheets to record progress and reports that are filed with GAP, which collects and disseminates information on the collective success of the ecoteams.

In the United States, GAP has found that an average household of four that completes the program reduces its water demand by 280,000 liters (73,000 gallons) per year and saves 10 trees each year. In addition, successful completion cuts a family of four's garbage output by over 1400 kilograms (3000 pounds) and reduces gasoline consumption by 2300 liters (600 gallons) per year. Participating households also reduce their annual output of carbon dioxide and of two pollutants that contribute to acid rain by 9 metric tons and 64 kilograms (140 pounds), respectively. GAP estimates that these and other efforts save a family of four about $1200 per year.

Each unit in the workbook suggests several strategies that team members can use in their homes to achieve their goals. Some of these strategies can be implemented quickly at little or no cost, for example, turning down the thermostat, tuning-up the furnace, and plugging air leaks. Turning down the heat during the winter by three degrees reduces the annual heating bill in a house or apartment by 6%. Other actions require larger financial commitments, for example, adding storm windows and doors, buying energy-efficient appliances, insulating ceilings, and installing passive solar heating. The goal of the energy program is to reduce household energy use by 10% in six months and by 30% by the year 2000.

As of February 1992, 277 ecoteams had formed in seven countries, including Norway, Sweden, Holland, the United Kingdom, Germany, and Canada. Half of the ecoteams were in the United States. GAP hopes to have thousands of ecoteams in the United States in the next few years.

Empowering others to change their life-styles is fundamental to the program's success. Thus, household ecoteam participants are encouraged to start programs in their workplaces using the specially written Workplace EcoTeam Program, which is designed to help bring corporations into environmental balance. GAP is also taking steps to involve children. Their Kid's EcoTeam Program includes 24 weekly action plans that children can implement with the help of their families.

Individual action is vital to the task of building a sustainable future. Whether we take actions on our own or with others, most agree individual actions do count. GAP is not only helping to make the change but also helping to show how much small acts contribute to building a better world.

in human society—the **sustainable revolution**. Its purpose is to reshape human society to reverse the Earth-threatening changes now occurring. Its goal is not to revert to antiquated ways but to create a new synthesis—a new way of life that utilizes modern technology and knowledge to protect the Earth's environment from destruction and foster its renewal. How important are these efforts?

Many scientists, activists, and policymakers who have studied the environmental problems facing the world today believe that the threat to the environment is extremely serious. Most agree that immediate action is needed.

All in all, at least 16 major environmental issues confront us, many with dire consequences (Table 1.1). This text will detail these trends and offer solutions to

Table 1.1 Sixteen Major Environmental Trends

Population growth

Species extinction

Deforestation

Destruction of wetlands

Desertification

Soil erosion

Salinization of farmland soils

Farmland conversion

Groundwater contamination

Groundwater depletion

Declining oil supplies

Declining mineral supplies

Surface water shortages

Global warming

Acid deposition

Ozone depletion

them. To understand the gravity of the situation, consider the environmental toll during a single day of human activity.

1.2 A Crisis of Unsustainability

Biologist David Orr points out that if this is an ordinary day on the planet, approximately 140 square miles of tropical rain forest will be leveled to make room for roads, towns, farms, and mines. That's equivalent to a 2-mile-wide by 70-mile-long swath of forest cleared every day to accommodate the human population. Such massive deforestation would not be so devastating if original forests were being restored. Unfortunately, they're not. In fact, in the developing countries where much of the cutting occurs, only 1 tree is planted for every 10 cut; in tropical Africa, only one tree is replaced for every 29 trees felled, according to the United Nations.

By various estimates, 40 to 100 plants, animals, and microorganisms become extinct each day, largely as a result of tropical deforestation.

On this day, 70 square miles of semiarid land will turn to desert primarily because of overgrazing and poor farming practices. That's equivalent to a 1-mile-wide by 70-mile-long strip of land. Tomorrow, another 70 square miles will be lost. Currently, about one-third of the world's cropland is in danger of becoming desert.

On an ordinary day, the industrial nations of the world will pour 15 million tons of carbon dioxide into the atmosphere. Carbon dioxide gas is released by the combustion of oil, gas, and natural gas as well as the burning of forests, garbage, and other organic matter. In the atmosphere, carbon dioxide and other pollutants absorb heat escaping from the Earth's surface and radiate it back to us. Some scientists think that increasing levels of carbon dioxide and other pollutants are responsible for a rise in global temperature, a phenomenon called **global warming**, which could have devastating effects on humans and natural systems.

On this day, industrial nations will also dump 1.5 million tons of toxic by-products of industrial processes, **hazardous waste**, into the air, water, and land. Americans will throw away enough garbage to fill the Superdome in New Orleans two times.

Today, 250,000 new people will join the world population. Each one will require food, water, shelter, and a host of other resources to survive.

At day's end, the Earth's atmosphere will be a little hotter and the water a little more polluted. Crime-ridden and overcrowded cities will be even more crowded. The air in and around them, already choked with pollution, will be a little more filthy. At day's end, the once-rich web of life will be a bit more torn. And, tomorrow, it starts again.

Although these statistics are extremely depressing, my intention is not to discourage you but rather to make three important points. First, the environmental problems facing the world are extremely serious. Second, the environmental difficulties we face are more than a collection of problems. They are signs that human society is on an unsustainable course. In fact, many who have studied the environmental crisis believe that humanity cannot continue as we have been in recent years without destroying the planet's life-support systems. Third, let me hasten to add, this is not to say that humankind and the rich biological world we live in are doomed. Far from it. We have a chance, but because of the severity of the environmental problems and rapidity with which changes are occurring, we have to make profound changes to reverse the trends . . . and soon.

Fortunately, a wealth of information is available to help us meet this important challenge. This text discusses much of that information, but, more importantly, it offers a framework for directing change to create an enduring human presence. To the surprise of many, most of the changes needed to make this shift will not destroy our economy or put millions of people out of work. Far from it. Many changes could actually improve the economy as they improve the long-term prospects for humans and the millions of other species that inhabit the Earth.

Bear in mind as you read this book, small improvements probably won't do; in order to redirect human society onto a sustainable course, we've got to make major changes. Before studying them, let's explore what is meant by a sustainable society.

1.3   What Is a Sustainable Society and What Does It Require of Us?

In 1987, the World Commission on Economics and Development, established by the United Nations, published a ground-breaking book entitled *Our Common Future*. The authors of the book defined a **sustainable society** as one that meets its needs without compromising the ability of future generations to meet theirs. Many others have proposed similar definitions. This text, however, contains a more expansive definition, one which takes into account the requirements of the many other species that share this planet with us. Thus, a sustainable society is defined as one that meets its needs without impairing the ability of future generations *and other species* from meeting theirs. Additional definitions that could deepen your understanding of this notion are shown in Table 1.2.

The mandate of sustainability is to live within the means of the planet and to ensure that future generations and other species can survive and live well, too. As noted above, this will require dramatic changes in society. Above all, it will require a new way of thinking. One of the most important requirements is a change from linear thinking to systems thinking. **Linear thinking** sees events in a straight-line sequence, ignoring complex webs of interaction. **Systems thinking,** on the other hand, entails thought processes that recognize how entire systems function. In the environmental arena, this kind of thinking helps us to see how individual parts work together and how interdependent all life forms are. It also helps us to become mindful of the ways in which individual, corporate, and government actions affect the human and biological systems of which they are a part. By becoming better systems thinkers, we can learn to avoid impacts that threaten the health and well-being of the planet and its organisms.

Because systems thinking encourages us to look at the whole, it will naturally force us to look for the root causes of problems, especially environmental ones. This analysis can help society to identify key leverage points where changes can be made. Thus, our ability to see the "big picture" as well as the connections between various parts is essential to solving the many environmental problems facing the world community and to putting us on a sustainable path.

Building a sustainable society also requires widespread participation with input from rich and poor, conservative and liberal, young and old, and everyone in

---

Table 1.2   Definitions of Sustainability

A sustainable society is one which satisfies its needs without diminishing the prospects of future generations.

> Lester R. Brown, Founder and President
> Worldwatch Institute

Sustainability is equity over time. As a value, it refers to giving equal weight in your decisions to the future as well as the present.

> Robert Gilman, Founder and Director
> Context Institute

Sustainability is living wisely within our resources.

> Prince Charles [paraphrased]

[Sustainability is] the ability to support, provide for, nurture the total life system in our bioregion and all other bioregions on the planet; [and] the ability to supply in perpetuity all life forms with the necessities of life.

> Helen Kolff, *Beyond War*

Actions are sustainable if: (1) There is a balance between resources used and resources regenerated. (2) Resources are as clean (or cleaner) at end use as at beginning. (3) The viability, integrity, and diversity of natural systems are restored and maintained. (4) They lead to enhanced local and regional self-reliance. (5) They help create and maintain community and a culture of place. (6) Each generation preserves the legacies of future generations.

> David McCloskey, Professor of Sociology
> Seattle University

---

between. In fact, the sustainable solutions outlined in this text call for action on the part of large and small businesses, individuals, and governments. You might ask: "Why should individuals be asked to participate? Our part is small compared to governments and businesses, isn't it?"

Individuals are important because each one of us is part of the problem. In other words, many environmental problems result from many small, seemingly insignificant actions. To create a sustainable future, we must all join in. As former Colorado Governor Richard Lamm wrote, "It is not enough for a nation to have a handful of heros. What we need are generations of responsible people." We all have a stake in the outcome of the race to save the planet.

Achieving a sustainable world also requires massive cooperation between citizens and governments to achieve positive solutions. Although signs of this change are underway, cooperation must occur on a much

Understanding the Earth

The late Buckminster Fuller coined the term "Spaceship Earth" to describe our watery planet. The earth, like a spaceship, is a closed system capable of recycling the materials necessary for life. Similar to a spaceship, the earth is vulnerable to disruptions of its life-giving systems.

In the 1970s, the British scientist J. E. Lovelock began to consider the earth as a living entity rather than a spaceship, for the earth in many ways behaves like an organism. It maintains a constant temperature, monitors the chemical composition of its lakes and oceans, and regulates its atmospheric gases in much the same way that an organism maintains its internal constancy.

Although this notion, known as the Gaia Hypothesis, has drawn much criticism, it nonetheless is an elegant metaphor that underscores a key principle of ecology: that all living things operate together. The sequoia and the gazelle, the lion and the rosebush—all are integral parts of a system as interdependent as the cells in your brain and kidneys.

This gallery takes a glimpse at the processes of earth. It shows the active building and tearing down of land, the delicate energy balance, and the oceanic and wind currents that affect life in innumerable ways.

1 The view of earth from the moon initially led to the notion of the earth as a spaceship, capable of recycling its life-sustaining materials. This metaphor has now largely been replaced by the Gaia Hypothesis of the earth as similar to an organism.

2 The earth's water participates in an enormous recycling system—the hydrological cycle. Heated by the sun, water evaporates from the earth's surface into the skies where it forms clouds. Falling back to earth, much of the water evaporates again while the rest flows back to the oceans along rivers and lakes. A small fraction seeps into the earth's surface and migrates slowly as groundwater.

Rain wears at the earth's surface, removing soil and washing it into the sea and lakes. People worsen erosion by stripping the land of its vegetation. Some 20 billion tons of farmland topsoil is washed away each year by wind and rain, often because of poor land management.

The earth's energy balance is as important as the hydrological cycle.

About 30% of the sunlight reaching the earth is reflected by the atmosphere, while approximately 70% is absorbed by the earth's surface and atmosphere and radiated back into space as heat. Water vapor and gases such as carbon dioxide in the atmosphere block much of the outgoing heat and radiate it back to earth, keeping the earth habitable. Unfortunately, modern civilization has added billions of tons of carbon dioxide and other gases to this thermal blanket, causing a gradual rise in global temperature. Some scientists predict that as pollution continues this warming of the earth could eventually melt enough ice to flood nearly 20% of the earth's land surface, with disastrous results for agriculture and our food supply.

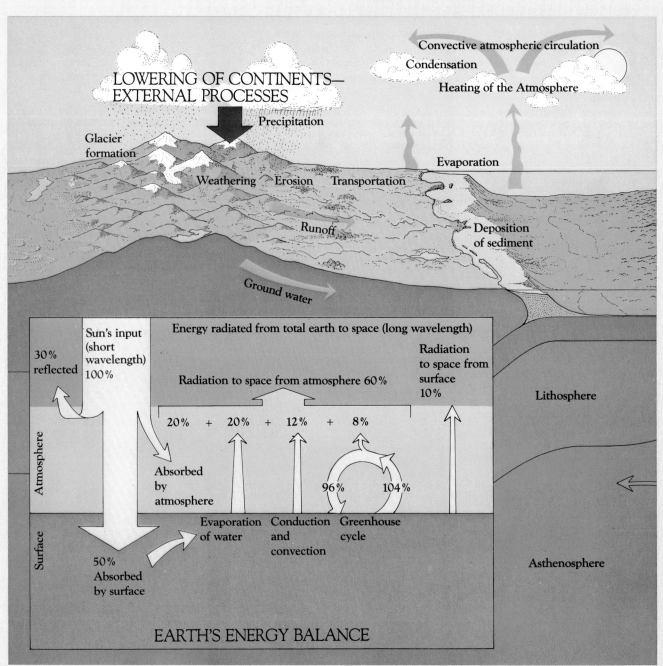

Erosion is to some extent counterbalanced by internal geological processes that build up the land. The action of tectonic plates in this respect is explained on the next page. Lava from volcanoes also adds to the buildup.

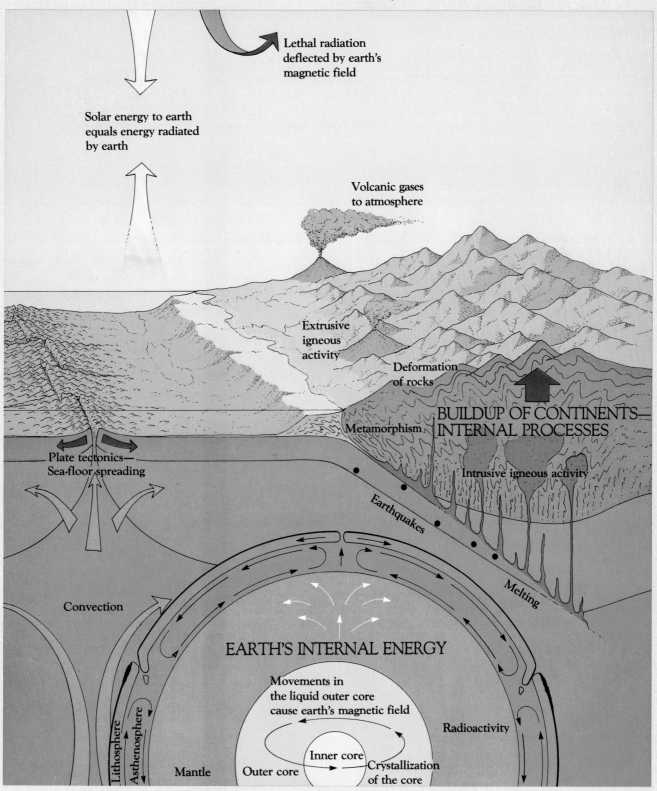

Lethal radiation deflected by earth's magnetic field

Solar energy to earth equals energy radiated by earth

Volcanic gases to atmosphere

Extrusive igneous activity

Deformation of rocks

Metamorphism

BUILDUP OF CONTINENTS—
INTERNAL PROCESSES

Intrusive igneous activity

Plate tectonics—
Sea-floor spreading

Earthquakes

Melting

Convection

EARTH'S INTERNAL ENERGY

Movements in the liquid outer core cause earth's magnetic field

Lithosphere

Asthenosphere

Radioactivity

Inner core

Mantle

Outer core

Crystallization of the core

3 The earth's crust consists of about ten huge movable plates, called tectonic plates, eight of which are shown here. The plates are propelled by convection of the earth's molten interior. Where they pull apart, molten rock can flow upwards and form new ocean floor. Plates that slip under others cause the buildup of continents through mountain formation. Although new crust and new minerals are formed as the earth's crust regenerates itself, the minerals are largely out of reach of modern civilization.

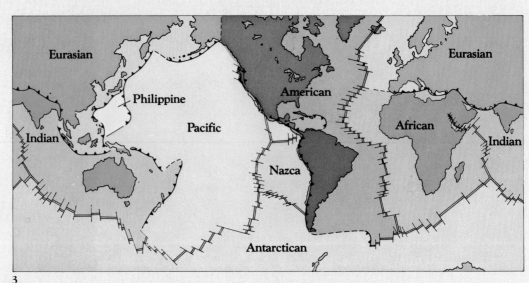

3

4 As the earth orbits the sun, the planet spins on its axis, causing large swirling currents in the earth's atmosphere. These prevailing winds are also generated by the uneven heating of the earth's surface, which leads to differences in atmospheric pressure. Wind occurs when air moves from regions of high pressure to low-pressure zones. For example, hot air rising at the equator causes the low-pressure region known as the doldrums. The northeast and southeast trade winds blow to the equator from the colder high-pressure regions at 30 north and south latitude, respectively. The westerlies blow strongly from the subtropical highs to the so-called subpolar lows at 60 north and south latitude, cooling the climate in those regions.

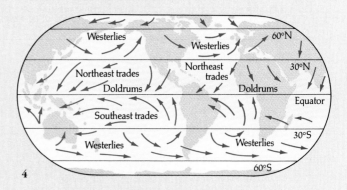

4

5 The major ocean currents are determined by the earth's rotation, by the force of the prevailing winds, and by the location of landmasses that block the currents. The flow of the warm (red) and cold (blue) currents influences regional climates. Because of the presence of the continents the currents move in loops known as gyres. The gyres flow clockwise in the northern hemisphere and counterclockwise in the southern hemisphere, due to the earth's rotation. The Gulf Stream, for example, carries warm water from the equator to the British Isles, keeping them warmer than expected at that latitude. The Humboldt Current, in contrast, brings cold water from the Antarctic to the equator, cooling the western coast of South America.

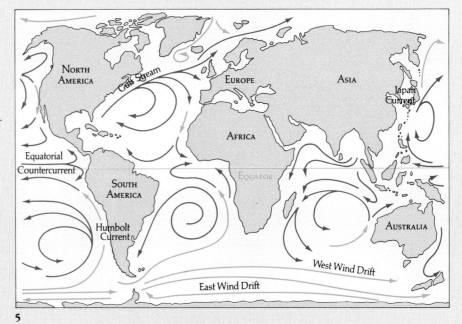

5

grander scale—countries working with one another for the common good. The Montreal Protocol is a good example. Another is the Earth Summit held in Rio de Janeiro in 1992. The Earth Summit produced several global agreements on deforestation, climate change, and species protection (Chapter 23). Participants also published the massive document called *Agenda 21*, which outlines 2500 ways to solve many of the world's environmental problems.

1.4 The Frontier Ethic and the Crisis of Unsustainability

Besides a shift to systems thinking and an increase in participation and cooperation, building a sustainable future requires a new system of values, or **ethics**. The ethical system of a person, religion, group, or even a nation determines how they act. The prevalent system of Earth values in Western culture as well as others is the **frontier ethic**. This ethic embraces a rather narrow view of humans in the environment and an even narrower view of the purpose of nature. The frontier ethic is characterized by three tenets: (1) The Earth has an unlimited supply of resources for exclusive human use—there's always more, and it's all for us; (2) humans are apart from nature and immune to natural laws; and (3) human success derives from the control of nature. In other words, nature is "something to overcome, or conquer." We get ahead by reshaping natural systems to our liking. Let's consider each tenet in detail.

The First Tenet: There's Always More, and It's All for Us

The view of the Earth as an unlimited supply of resources for exclusive human use no doubt evolved in prehistoric time when human numbers were small and the Earth's resources did indeed appear inexhaustible. The massive increase in economic activity and the upsurge in population growth in the last 200 years, however, have brought us face-to-face with the planet's limits. For example, since 1920, dozens of ocean fisheries have been depleted in large part because of overfishing (Figure 1.2). Dozens more are now threatened. In the past 100 years, half of the world's tropical rain forests have been destroyed along with countless species of plants and animals. Within most of our lifetimes, tropical rain forests could be wiped out. Oil supplies are also likely to run out. Several important minerals, including tin and silver, could fall into short supply.

Despite growing evidence that our frontiers have vanished, many people persist in seeing the Earth as an unlimited smorgasbord. In his nomination speech at the Republican National Convention in August 1988, New Jersey Governor Thomas Kean spoke of the dream of a bountiful American future. Senator Phil Graham proclaimed, "There is no limit to the future of the American people." Speaker after speaker echoed similar sentiments. A month earlier, at the Democratic National Convention, the Democrats expressed similar views, all based on a naive assumption of the Earth's inexhaustible wealth, a fundamentally flawed view. Limits are real. By insisting that there's always more, and that it is all for human use, human society may be foreclosing on the Earth's future and the future of our children, not to mention the millions of other species that inhabit the Earth with us.

The Second Tenet: Humans Are Apart from Nature and Immune to Natural Laws

Humankind has sought to position itself outside the realm of nature for thousands of years. Today, many people continue to view humans beings as separate from nature and persist in thinking that we can do what we please without harm. The French philosopher and writer Albert Camus summed up our philosophy of separation best when he wrote: "Man is the only creature that refuses to be what he is." In reality, says ecologist Raymond Dasmann, "a human apart from environment is an abstraction—in reality no such thing exists." Our lives are intricately linked to the natural world. How?

One-half of the oxygen we breathe each year is replenished by plants and algae. All the food we eat comes from plants, soil, water, and air. Even plastic comes from phytoplankton, organisms that lived on the Earth several hundred million years ago. Nature affirms our connections with the environment nearly every day, flooding homes and farms built in floodplains, gobbling up vacation homes on the ever-shifting sands of barrier islands, and filling reservoirs with sediment washed from denuded hillsides. To think of ourselves outside of nature's realm is not only foolish—it is suicidal.

The Third Tenet: Human Success Derives from the Control of Nature

Industrial nations view nature as a force that must be conquered and subjugated. Matthew Arnold, a 19th-century British poet, summarized the modern view best when he wrote: "Nature and man can never be fast friends./Fool, if thou canst not pass her, rest her slave!" That is, if we can't subdue nature, we will become a slave to natural forces.

Believing in the need to reign supreme, humankind has conquered rivers and subdued the wilderness. Today, as evidence of our zealous efforts to control nature, millions of miles of highways cut through the wilderness. Breakwalls restrain storm surges and levees hold back floods. Resource "managers" manipulate wildlife, soils, fisheries, and forests like so many pieces in a board game.

Figure 1.2 *Nearly two dozen ocean fisheries have been depleted as a result of intensive over-fishing. Many other vital resources have been depleted or are rapidly being expended.*

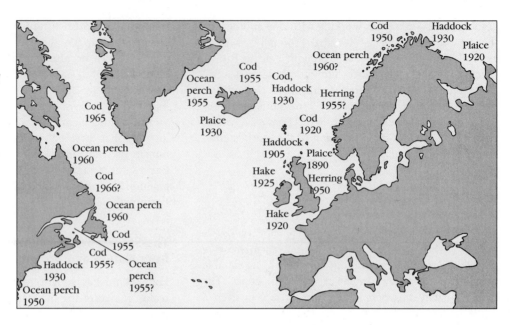

In the past few decades, it has become clear to many that our acts of conquest have been costly (Figure 1.3). In fact, many environmental problems we face today stem from our "success" at controlling and dominating nature. Such impacts are often called **ecological backlashes**.

A good example of ecological backlash can be seen on today's chemical-intensive farms. To increase crop production, many farmers use an assortment of chemical fertilizers and pesticides. **Pesticides** are chemicals used to kill organisms that farmers view as pests. In the past 10 to 15 years, research has shown that pesticide use may be a Pyrrhic victory—one that comes at a terrible cost. Why? Since World War II, chemical pesticide application has increased tenfold, while pest damage has doubled. Today, over 500 insect species are resistant to at least one chemical pesticide and 20 insect pests are resistant to every known pesticide.

Case Study 1.1 offers an example of an ecological backlash, one that resulted from building the Aswan Dam in Egypt. This and other similar efforts call to mind the words of the Greek playwright Euripides: "Chance fights ever on the side of the prudent." For humankind, prudence requires cooperation with nature, not domination.

Some Impacts of Frontier Thinking

The frontier ethic profoundly affects how people act and has led to massive resource depletion and numerous environmental disasters. It also affects our attitudes about the seriousness of environmental problems. Projections of declining fossil fuel supplies, for instance, generally spark little interest among residents of indus-

trial countries because most people are convinced that supplies will never run out, or if they do, that scientists will find substitutes.

The frontier ethic also influences how people solve environmental problems. The solution to impending oil shortages, for example, is to increase the search for new supplies, even if it takes us into one of the few remaining pristine wildernesses. Our faith in unlimited resources prevents many from using resources more efficiently. Why be efficient if resources are unlimited?

The frontier mentality permeates our lives, influencing personal goals and expectations. Most of us tend to make buying decisions based on what we can afford. Very rarely do we ask whether a decision to purchase a product helps or hinders the environment. Nor do we think about the effects on the long-term habitability of the planet. If the world's an unlimited supply of resources, why should we?

Over the years, the unquestioning acceptance of the frontier ethic has spawned a life-style that, while successful in many ways, threatens our future. We've created a hodgepodge of human systems that are out of synch with natural systems that support them.

The term **synergy** is used here to measure how well human systems fit into natural systems. Synergy is an important element of sustainability. Societies that are out of synch with natural systems are called **low-synergy societies**. Such societies tend to grow crops at the expense of soils. They may also destroy wetlands, which are vital habitats for many fish species, which, in turn, are an important human food source. Many other examples of this phenomenon appear in this text. All low-synergy societies are on a path of unsustainability.

In contrast, **high-synergy societies** satisfy their needs for food, shelter, and other resources without destroying

Case Study 1.1

The Aswan Dam: Ecological and Economic Backlash

Africa's Nile River flows from its headwaters in Ethiopia and Uganda, then courses through Sudan and Egypt, eventually emptying into the Mediterranean Sea. For centuries, this great river annually transported 50 million to 100 million metric tons of silt from its headwaters. Flooding its banks in late summer and early fall, year after year, the river replenished nutrients removed from Egyptian farmland by crops. In the sea, the nutrient-rich silt nourished a variety of phytoplankton, such as algae, which, in turn, were food for thriving fish populations.

In the early 1960s, with financial assistance from the former Soviet Union, Egypt built the Aswan Dam on the Nile to provide electricity for the rapidly growing city of Cairo and irrigation water for the lower Nile basin. Government officials who approved the project looked at the economic benefits and some of the potential problems, but felt that the benefits (mostly electricity) more than compensated for the losses, especially possible ecological backlashes.

Not long after the dam was completed and Lake Nassar, the reservoir formed by the dam, began to fill, problems began to appear. First, the periodic flooding that fertilized farmland along the Nile ceased. As a result, farmers had to import fertilizer at an exorbitant cost. Second, the sardine fishery in the eastern Mediterranean collapsed as a result of the loss of nutrient-rich silt. The sardine catch plummeted from 18,000 to 500 metric tons per year in only a few years. Third, the rising waters of Lake Nassar threatened the Ramses Temple at Abu Simbel, built over 3000 years ago. Engineers and construction workers sponsored by the United States, Egypt, and the United Nations dismantled the temple piece by piece and moved it to a site 60 meters (200 feet) above its original level, where it would be safe from the rising waters. The cost of the relocation project was astronomical. Fourth, the incidence of schistosomiasis, a debilitating, sometimes fatal parasitic disease in humans, increased in Egypt as a result of this project. The organism that causes this disease is transmitted by snails. Snails require a constant supply of water, which the lake and the irrigation channels provided. Finally, in recent years, officials have found that the flow of nonsilt-laden water in the lower river is scouring the river's sandy bottom, undermining bridges, levees, and other structures.

This project, although returning sizable economic benefits, also has caused massive environmental and health effects, both of which come at a huge cost. Can you think of ways that the Egyptian society might have supplied the needs for electricity without creating such havoc?

Figure 1.3 *Clear-cutting in the Pacific Northwest leaves ugly scars on the land and may increase soil erosion, which clogs streams, kills fish, and fills reservoirs with silt.*

the natural systems upon which they depend. For example, they grow crops without ruining the soils. They harvest trees while protecting forests and without clogging nearby streams and lakes with sediment eroded from the land. Such communities live in "harmony" with nature.

A high-synergy society does not merely prevent destruction from occurring; it seeks ways to enhance natural systems. A good example is the sustainable farm. Sustainable farming practices tend to increase the organic content of the soil, which makes the land more productive (Chapter 7). Ultimately, creating a sustainable society means fitting human society within the economy of nature.

1.5 Building a Sustainable Society: Ethics and Actions

Creating a high-synergy sustainable society that lives within the Earth's means is, by most accounts, possible only if the world's people adopt a new, sustainable ethic—one that respects limits and seeks to ensure future generations and other species the resources they need to survive. A new value system recognizes our place in the natural order, as one of many millions of species. It is one that favors cooperation over domination.

Principles of the Sustainable Ethic

The **sustainable ethic** contrasts sharply with the frontier ethic. It asserts: (1) The Earth has a limited supply of resources, and they're not all for us; (2) humans are a part of nature, subject to its laws; and (3) success stems from efforts to cooperate with the forces of nature. A fourth tenet reminds us of the importance of the natural world. It is that our future depends on creating and maintaining a healthy, well-functioning global ecosystem. An **ecosystem** consists of a community of organisms and all the interactions among them and their physical environment (Chapter 2).

As you can see, these principles are diametrically opposed to the principles of frontier ethics. Just as the frontier ethic leads us to exploitive behavior, the sustainable ethic could lead to a less exploitive human presence that could endure for thousands of years.

Putting Ethics into Action

The sustainable ethic provides broad guidelines for human behavior. Arne Naess, one of the world's leading environmental philosophers, however, argues that "a philosophy, as articulated wisdom, has to be a synthesis of theory and practice." In other words, a sustainable ethic requires more than lofty goals. It requires actions to manifest those goals. Without actions, the principles of the sustainable ethic become useless philosophy.

Where will the operating principles come from? I believe that answers to the challenge of living sustainably on the planet can be found by studying the workings of nature. Over 4 billion years of evolutionary trial and error, nature has evolved a strategy of sustainability that could be useful in reshaping human society.

Studies of natural ecosystems suggest that at least six biological principles lie at the heart of sustainability: (1) conservation, (2) recycling, (3) renewable resource use, (4) restoration, (5) population control, and (6) adaptability.

Conservation   Natural ecosystems persist because organisms in them use only the resources they need, and often they use them with efficiency. For the purposes of this text, the term **conservation** is used to describe these two related activities.

Waste of the magnitude witnessed in human society is almost unheard of in nature. In fact, the evolution of life in many regions of the Earth hinged on the emergence of structural and functional characteristics that permit organisms to use resources with efficiency (Figure 1.4). Without them, there would be little, if any, desert life. The frozen Arctic would be a barren land, devoid of Arctic foxes and polar bears. In the absence of efficiency, life would probably be restricted to a rather narrow region on Earth where conditions are mild and food abundant.

Recycling   Efficiency is not enough to ensure sustainability, however. Nature persists because it reuses materials over and over, that is, it **recycles** them. The Earth is a **closed system**—one that receives no inputs from the outside except sunlight. Like a terrarium, all the materials necessary for life, such as oxygen, are contained within and are used over and over.

A tear in your eye may have been a tear in Caesar's eye. It may have been in the blood of the first humanlike animal that roamed the savannas of Africa, or it may have been part of the first rain droplet that fell on the newly formed planet 4.5 billion years ago. Without recycling, natural ecosystems would quickly collapse, and life would come to an abrupt halt as the Earth's resources were extinguished.

Renewable Resources   Natural systems also persist because they rely principally on **renewable resources**— resources such as air, water, plants, and animals that regenerate via biological or geological processes. Virtually all life on Earth, including our own, is nourished by plants. Plants, in turn, depend on three renewable resources: air, water, and soil. The energy of life comes from the sun, a vast but nonrenewable resource. Without plants to capture the sun's energy, most life forms could not survive.

Despite the belief that humans are apart from nature, our lives our intimately tied to natural systems.

Figure 1.4 *The kangaroo rat is one of nature's most water-efficient organisms. It excretes dry urine and feces, and can acquire all the water it needs from the breakdown of sugar in its cells.*

Even the energy you expended walking up the stairs this morning came from sunlight energy delivered to you by the cereal you ate. Cereals are made from seeds produced by plants. The minerals contained in your breakfast cereal came from the soil and will return to the environment in human waste. As long as the sun shines and the planet's recycling networks remain intact, life can continue on the planet.

Restoration   Natural ecosystems also endure because they are capable of repairing damage, or **restoration**. Restoration is described in Chapter 3. As you will see, mechanisms that permit the restoration of ecosystems in the **biosphere**, the thin skin of life on the planet, like the system of blood clotting or skin repair in our own bodies, heal "wounds" caused by lightning fires, adverse weather (hurricanes, tornadoes, and floods), geologic disruption (volcanoes), or the activities of life forms themselves (deforestation).

Population Control   Natural ecosystems also persist because they possess mechanisms that control populations within the carrying capacity of the environment. **Carrying capacity** is defined as the number of organisms an ecosystem can support indefinitely. Through a variety of mechanisms, populations living in undisturbed ecosystems are held within the limits imposed by food supply and the availability of other resources. If their demand exceeds resource supplies, numbers are usually quickly adjusted downward to reset the balance. Thus, a natural system of checks and balances keeps organisms from eating themselves out of house and home.

Adaptability   Finally, natural systems persist because of the capacity of organisms within them to change in response to changing environmental conditions, that is, to evolve. **Evolution** is a process that leads to structural, functional, and behavioral changes in species, known as

adaptations. Favorable adaptations increase an organism's chances of survival and reproduction.

Evolutionary change occurs as a result of two factors: genetic variation and natural selection. **Genetic variation** refers to naturally occurring genetic differences in **populations**, groups of genetically similar organisms. Genetic variations result in differences in structure, function, and behavior of organisms. Those that give members of a population an advantage over others in survival and reproduction tend to persist.

Charles Darwin, a 19th-century British naturalist who spent much of his life studying evolution, proposed a **theory of evolution by natural selection**. He described it as a process in which slight variations, if useful, are preserved. Through natural selection, populations of organisms become better adapted to their environment. Organisms endowed with beneficial variations are more likely to survive and reproduce and, therefore, pass to their offspring the genes that gave them an advantage over others. Through this process, the genetic composition of populations of organisms may change over time. This ability to adapt to changing environmental conditions is vital to sustainability. However, for humans, the evolutionary change must be cultural, not biological.

Applying the Biological Principles of Sustainability

The biological principles of sustainability explain why natural systems persist. They offer a set of practical guidelines that help us put the sustainable ethic into effect. In so doing, they could help us revamp economics, ethics, and government. Ultimately, they could help us to reshape systems of waste management, energy supply, housing, transportation, and others, all of which are grossly unsustainable in their current forms. By applying the biological principles of sustainability, human society can build an enduring human presence. These efforts would help humans weave their activities into the web of life, creating a high-synergy society. You can use the guidelines to change your own life-style (Table 1.3).

Applying the biological principles to major sectors of society has many benefits. For example, widespread urban recycling efforts greatly reduce energy demand, urban air pollution, water pollution, habitat destruction, global warming, solid waste disposal, and a host of other environmental problems (Figure 1.5). Efficiency measures, renewable energy use, population control, and other sustainable measures offer similar benefits.

Numerous factors lie at the root of today's crisis of unsustainability. Figure 1.6 lists six of extreme importance: (1) the frontier ethic, (2) inefficiency, (3) overconsumption, (4) linearity (linear thinking and linear systems), (5) fossil fuel dependence, and (6) overpopulation. Interestingly, the ethical and biological principles

---

Table 1.3   The Five Most Important Things You Can Do

---

1. Use only what you need and use all resources efficiently.

2. Recycle all glass, bottles, cans, cardboard, paper products, and plastic. Also, buy products made from recycled materials.

3. Use renewable resources or support government programs to increase knowledge and application of them.

4. Restore damaged areas and support government programs to restore wetlands, forests, grasslands, and others.

5. Limit family size to no more than two children and support government programs to promote family planning at home and abroad.

---

of sustainability outlined earlier confront the root causes of the crisis head-on. Adopting the sustainable ethic, for instance, will help us eliminate the frontier ethic. Applying conservation will help us reverse the massive inefficiency of modern society, which leads to solid waste, pollution, and environmental destruction from mining and timber harvesting. It will also help us reduce our massive consumption of resources. As future chapters point out, vast improvements in recycling will help society reduce its inefficiency as well as its overconsumption. Furthermore, recycling helps us convert linear human systems into cyclic systems, such as those found in natural systems. Converting to clean, inexpensive renewable energy resources will reduce our heavy dependence on fossil fuels, which is the root cause of many problems, among them urban air pollution and global warming. Restoration also addresses linearity—the tendency to use and move on—by rebuilding systems upon which the future depends. Finally, population control measures will help put a lid on continued population growth, which is so dangerous in a finite world (Chapters 5 and 6).

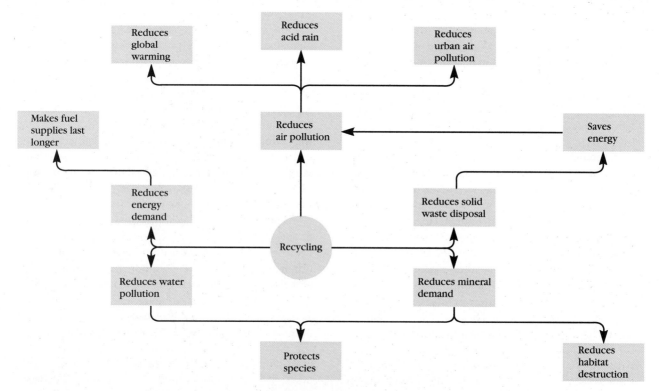

Figure 1.5 *Sustainable solutions, such as recycling, offer many benefits to people and the environment, as illustrated here, because many problems stem from common causes.*

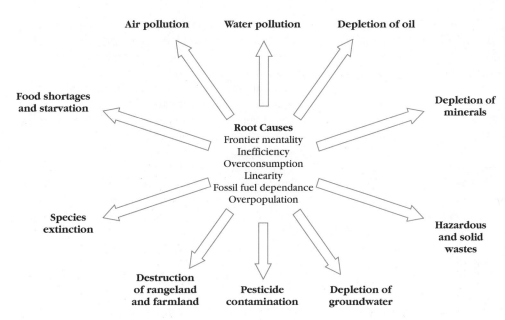

Figure 1.6 *The environmental crisis has many symptoms, including air pollution, water pollution, species extinction, and depletion of groundwater, which result from a common set of root causes.*

This text points out that for many years humans have been treating the symptoms of the environmental crisis, while ignoring the root causes. That is, our approach has been symptomatic, not systemic. As any physician will tell you, to cure a disease, you have to treat the underlying causes. Treating the symptoms, as we've been doing for the last 25 years with the environment, only postpones the day of reckoning.

Creating an enduring human presence on the Earth requires a rebuilding of society according to the pattern laid down by nature. Although many scientists have known this for some time, ecological concepts of sustainability have been ignored, misinterpreted, and sometimes even ridiculed by those entrenched in the present system. Notions of survival by meshing with, rather than transposing our ways onto, the economy of nature seemed outrageous to a society that viewed domination over nature as a key to success and sees humans as immune to ecological law. Moreover, ideas that suggested restraint ran counter to a society convinced that the Earth was a limitless supply of resources for human use.

1.6 Environmental Science: Meeting the Sustainable Challenge

Understanding environmental problems and finding solutions fall in the domain of **environmental science**, a fairly recent arrival on college campuses. Environmental science calls on insights from a number of disciplines, including biology, chemistry, and climatology. It also turns to anthropology, forestry, and agriculture for advice. Yet, solving the highly complex problems of overpopulation, resource depletion, and pollution requires more than a knowledge of science. Because solutions fall into the political and economic arenas, they require a working knowledge of sociology, law, ethics, economics, and, occasionally, psychology. So, as you embark on your study of the environment, be prepared to tap into a broad array of disciplines.

Ambitiously spanning this wide range of specialties, environmental science attempts to piece together an integrated view of the world—a systems view—and our part in it. As such, it is one of our greatest hopes for charting a sustainable future.

Despite numerous victories, many environmental problems persist, and many are growing worse. Dennis Hayes, cofounder of Earth Day, remarked, "How could we have fought so hard, and won so many battles, only to find ourselves on the verge of losing the war?"

One of the most important reasons alluded to earlier is that in many cases society opted to treat the symptoms with quick-fix, Band-Aid measures while ignoring the underlying causes. In effect, many policy and management decisions were designed to reduce but not end the destruction of the Earth's life-support systems.

Recognizing that the symptoms approach isn't enough, many environmental scientists have begun to search for ways to provide systemic solutions. This text offers some guidance on the subject. It also offers some information on critical thinking skills and strategies needed to improve our chances of finding appropriate answers to the problems that lie ahead. (Critical thinking skills are described in Chapter Supplement 1.1.)

The changes required to shift massive industrial and agricultural societies onto a sustainable course are already underway. These changes, which will take generations to complete, will involve arduous work in many fields at all levels of society. The kinds of changes needed will make the moon landing seem like a weekend home-improvement chore. As David Lloyd George reminded us, though, "Don't be afraid to take a big step if one is indicated; you can't cross a chasm in two small jumps."

1.7 Can the Human Race Survive the Human Race?

Some pessimists believe that the human species is doomed, that resource and ozone depletion, global warming, and other problems, alone or together, will wipe humanity off the face of the planet. At the least, they say, things will deteriorate so much that we could lose centuries of technological and economic progress in the next few decades. Our wonderfully diverse biological world, the product of billions of years of evolution, could be eradicated in a fraction of the Earth's history.

Others disagree with the doomsayers. They see a hopeful future where technology or technological fixes save the day. But many "technological optimists" paint an unrealistic picture of the future and frequently fail to distinguish what is technologically possible from what is feasible and affordable. Carried away with technological optimism, they see outer space as a source of new minerals, free energy, and more living space for the world's crowded citizenry. They propose new technologies, such as nuclear fusion, to solve the energy crisis. They call for new pollution control technologies to clean up the waterways and atmosphere. Little thought is given to the cost or impact of their schemes and even less is devoted to simple, cost-effective measures, such as energy conservation, life-style changes, and efforts to slow down or stop population growth.

This text takes a middle position. It contends that the human race can survive itself, but, in order to do so, we must adopt a new, sustainable world view, and we must take concerted steps to create a way of life that protects the planet (Table 1.4). Systems thinking, cooperation, and participation are essential to forge a more hopeful future.

Whether we like it or not, we have become the custodians of an entire planet. The fate of that planet, our home, and the millions of species that share it with us as well as the fate of all future generations lies in our hands.

To cherish what remains of the Earth and to foster its renewal is our only legitimate hope of survival.

—Wendell Berry

❧   ❧   ❧

Critical Thinking Exercise Solution

Perhaps the biggest problem with statements such as the ones made by my student is that they greatly simplify very complex issues, in this case, the entire environmental crisis. Frankly, not all environmental problems lend themselves to technological solutions. For example, technology will not "solve" the problem of rapid population growth. Although we need new contraceptive technologies, they're only part of the solution. As Chapter 22 notes, solving this problem will also require improvements in the status of women, improvements in education, better health care, and changes in value systems. Technology cannot bring back extinct species. Social and political answers are required.

A related problem with this logic is that it assumes that simple adjustments in the ways we do things (technological improvements) are all that's needed. Experience has shown that we could not solve the ozone depletion problem by cutting back slightly on chlorofluorocarbons (CFCs). They had to be banned altogether. By the same token, we can't eliminate global warming by using energy a little more efficiently. We have to use it a lot more efficiently and switch to other energy resources that don't produce pollutants that cause the Earth's atmosphere to heat up.

A third problem with technological optimism is that technological answers often create unanticipated

---

Table 1.4   Summary of Principles and Concepts Essential to Sustainability

New Kinds of Thinking/Acting Required
  Holistic (systems thinking)
  Participatory
  Cooperative

New Values—Sustainable Ethics
  The Earth has a limited supply of resources for all species.
  Humans are a part of nature, subject to its laws.
  Success comes from cooperation with nature.
  All life depends on a healthy, well-functioning ecosystem.

Adapting to Ecological Realities Through
  Conservation
  Recycling
  Renewable resource use
  Restoration
  Population control

problems. Garbage incinerators, for instance, produce a toxic waste that must be dealt with.

The lesson here is that when examining broad generalizations, such as the ones made by my student, stop to think critically about them. You'll almost always find that issues are more complex than they appear.

Summary

1.1 The Sustainable Revolution

■ Numerous efforts are now underway at all levels of society and in most nations to confront the environmental problems facing the world to create a sustainable society.

1.2 A Crisis of Unsustainability

■ A survey of environmental destruction and pollution suggests that the environmental crisis is severe and time is short.

■ To many, environmental problems facing the world's people are a sign that human society is on a fundamentally unsustainable course.

■ Fortunately, there is a wealth of information that could help us create a sustainable future.

1.3 What Is a Sustainable Society and What Does It Require of Us?

■ A **sustainable society** lives within the carrying capacity of the environment. It meets its needs without impairing future generations or other species from meeting theirs.

■ Building a sustainable society will require dramatic changes in the way we live and conduct business.

■ It will necessitate a change from narrow **linear thinking** to holistic **systems thinking**.

■ It will also require participation and cooperation of virtually all people.

1.4 The Frontier Ethic and the Crisis of Unsustainability

■ Creating a sustainable future also requires adoption of a new value system that replaces the frontier ethic, one of the root causes of the crisis of unsustainability.

■ The **frontier ethic** is characterized by three tenets: (1) The Earth has an unlimited supply of resources for exclusive human use; (2) humans are apart from nature and immune to natural laws; and (3) human success stems from the control of nature.

■ The frontier ethic profoundly affects how people think and act. It prevents us from solving our problems and directs us to solutions that are unsustainable in the long run.

■ The frontier ethic creates a **low-synergy, unsustainable society**, one living out of synch with natural systems upon which they depend.

1.5 Building a Sustainable Society: Ethics and Actions

■ The **sustainable ethic** stands in direct contrast to the frontier ethic and asserts: (1) the Earth has a limited supply of resources, not all for us; (2) humans are a part of nature, subject to its laws; (3) success stems from efforts to cooperate with the forces of nature; and (4) all life depends on maintaining a healthy, well-functioning ecosystem.

■ This new ethic could help us establish a **high-synergy, sustainable society** where humans live without ravaging the life-support systems of the environment.

■ Guidance for putting the sustainable ethic into action comes from studies of natural systems, which reveal six **biological principles of sustainability**: conservation, recycling, renewable resources, restoration, population control, and adaptability.

1.6 Environmental Science: Meeting the Sustainable Challenge

■ **Environmental science** offers a broad, multidisciplinary approach to the environmental crisis. It seeks to understand root causes and offers root-level solutions.

■ Its aim is to help foster the transition to a sustainable society.

1.7 Can the Human Race Survive the Human Race?

■ Opinions on the future of the human race range from foolish optimism to gloomy pessimism. The middle position holds that survival is possible but hinges on creating a sustainable way of living and sustainable systems of commerce.

Discussion and Critical Thinking Questions

1. List the things you do in your daily life to help protect the environment. What motivates you to do them?

2. What is your opinion of the environmental crisis? Is it real, or is it greatly exaggerated? How do your values affect this opinion?

3. Analyze the statistics given in Section 1.2. Do you think they suggest that humankind is on an unsustainable course? Why or why not?

4. List and describe the six biological principles of sustainability. How can they be applied to human activities? Can they be applied to your own life?

5. Critically analyze the biological principles of sustainability.

6. Give examples of activities in your community, state, or country that illustrate the biological principles of sustainability in action.

7. One of the principles of sustainability is the ability to evolve. How does this principle apply to human society? Can we evolve biologically to be more tolerant of pollutants?

8. Critically analyze the following statement: "In the United States and other developed countries, the response to the environmental crisis has concentrated on treating symptoms, not the underlying root causes."

9. Define the term *sustainable society*. What new forms of thinking are necessary in order to achieve it?

10. Describe the importance of systems thinking. Can you think of some examples where people who failed to use systems thinking got into trouble?

11. List the basic tenets of the frontier ethic and describe how they affect individual and social behavior.

12. Does your system of values correspond with the frontier or sustainable ethic? Where did you acquire these values?

13. What is meant by the term *high-synergy society*?

Suggested Readings

Brown, L. R., Flavin, C., and Postel, S. (1991). *Saving the Planet: How to Shape an Environmentally Sustainable Economy*. New York: Norton. Contains lots of information and ideas.

Brown, L. R., Flavin, C., and Kane, H. (1992). *Vital Signs 1992: The Trends That Are Shaping Our Future*. New York: Norton. Useful reference.

Brown, L. R. et al. (1992). *State of the World 1992*. New York: Norton. Surveys problems and progress toward building a sustainable society.

Chiras, D. D. (1990). *Beyond the Fray: Reshaping America's Environmental Response*. Boulder, CO: Johnson Books. Describes changes needed to improve our environmental response.

———. (1992). *Lessons From Nature: Learning to Live Sustainably on the Earth*. Washington, D.C.: Island Press. Describes the biological principles of sustainability and how they can be applied to systems of ethics, economics, government, agriculture, industry, energy, etc.

———. (1992). Eco-Logic: Teaching the Biological Principles of Sustainability. *American Biology Teacher* 55(2): 71–76. Describes the biological principles of sustainability and how they can be taught.

Gore, A. (1992). *Earth in the Balance: Ecology and the Human Spirit*. Boston: Houghton Mifflin. Excellent reference.

Meadows, D. H. (1991). *The Global Citizen*. Washington, D.C.: Island Press. Collection of short, insightful essays on a variety of environmental topics.

Meadows, D. H., Meadows, D. L., and Randers, J. (1992). *Beyond the Limits: Confronting Global Collapse, Envisioning a Sustainable Future*. Post Mills, VT: Chelsea Green. Excellent reading.

The Earthworks Group. (1989). *50 Simple Things You Can Do To Save the Earth*. Berkeley, CA: Earthworks Press. Full of important steps you can take to help build a sustainable future.

World Commission on Environment and Development. (1987). *Our Common Future*. Oxford: Oxford University Press. Important reading on sustainable development.

Science, Scientific Method, and Critical Thinking

The term **science** comes from the Latin word *scientia*, which means "to know" or "to discern." Technically, science is defined as knowledge derived from observation, study, and experimentation. Thus, science refers to a body of knowledge and the method in which it is attained.

Many people think of science as an enigmatic endeavor with little relevance to their lives. It may seem dull and uncreative or so difficult that it can be understood only by a few. If the truth be known, science is an exciting endeavor. Paleontologist Robert Bakker, in fact, points out, "Science is fun for the mind!" Good science requires tremendous creativity, insight, and imagination.

Throughout this text, you will learn both fundamental principles of science and many of the scientific facts behind environmental issues. Some of the most important scientific principles come from the science of **ecology**, the study of ecosystems and the interactions that take place among the many components in these complex systems. Thus, a working knowledge of science and scientific method is essential to understanding environmental issues that face the world today and to building a sustainable future.

This supplement outlines some important scientific principles and also discusses **critical thinking skills**, those skills necessary to analyze issues and solutions.

Scientific Method

Scientific study of the world around us is often, but not always, orderly and precise. As a rule, scientific discovery begins with observations and measurements of the subject under study, such as the rate of soil erosion or the effects of a pollutant on tree growth. From observations, scientists often formulate generalizations or **hypotheses**, tentative explanations of what scientists observe. For example, scientists hypothesize that rattlesnakes shake their rattles to warn large animals to stay away. They think the rattle is a protective measure that reduces the chances of a snake being stepped on and killed. This hypothesis is based on observations of rattlesnakes in the wild. Hypotheses based on observation and measurement are derived by inductive reasoning. **Inductive reasoning** occurs when a person uses facts and observations to arrive at general rules or hypotheses.

Once a hypothesis is made, the scientist must determine how valid it is by performing experiments. The results of scientific experiments either support or refute an initial hypothesis. If a hypothesis is refuted, a new one is generally substituted.

Believe it or not, you use the scientific method almost every day of your life. For example, suppose every time you go to a friend's house you become ill. To determine what makes you sick, you'd probably start by making a mental list of all the things you do at your friend's house when you visit. After making the list and weeding out some irrelevant activities, you conclude that the source of the problem is your friend's cat. That's your hypothesis.

To test this hypothesis, you might perform an experiment. For instance, you might ask your friend to keep the cat outdoors and vacuum the rug. Furthermore, you might avoid sitting on the chair where the cat takes its afternoon naps. If you come home feeling sick again, you would have to alter your hypothesis. Maybe it's the new carpet or the new paint in her apartment.

Scientific study is, of course, much more involved than finding reasons why you feel sick every time you visit a friend's house. But, the methodology is the same.

Many scientific experiments are carried out on animals, plants, and organisms. Such studies are designed to test the effects of some form of treatment, for example, the effect of a pollutant on an animal's lung or the effect of a new drug on a certain disease. In these experiments, it is necessary to establish experimental and control groups. The **experimental group** is the one that you "experiment" on. In a study of the effects of air pollution on laboratory mice, for example, the experimental group is exposed to certain amounts of pollution. The **control group** is treated identically, except that it is exposed to purified air. By setting up experiments in this manner, scientists can test the effect of one and only one variable. Thus, any observed differences in the two groups probably result from the treatment.

Studying the effects of pollution on humans is not so easy. Why? First, ethics generally prohibit experimentation on humans that would cause pain and suffering. Second, unlike laboratory mice, which are genetically identical, humans represent a genetically diverse group of organisms. Setting up experimental and control groups that contain large numbers of genetically similar individuals who have been exposed to the same conditions throughout life is impossible.

Consequently, scientists often rely on **epidemiological studies** to understand the effects of chemical pollutants on people. Literally translated, **epidemiology** is the "study of epidemics." Today, however, epidemiology refers to the branch of science that studies the health effects of environmental pollutants, food additives, infectious diseases, and a variety of other agents. Unlike laboratory studies of rats and mice, these studies generally rely on statistical analysis of data of populations that are exposed to various potentially harmful substances.

To understand how these studies are carried out, consider an example. Suppose you want to know the cause of a particular kind of cancer appearing in humans. If you were an epidemiologist, you would generally begin by examining death certificates from local hospitals, searching for individuals who succumbed to the cancer in question. You would then interview their next of kin to learn their personal habits, life-styles, and possible exposures. Next, you would put all the information into your computer and through statistical analysis try to find any commonalities, for example, employment in a chemical factory or smoking.

You would also need to assemble a control group. That group would be as identical to the first group as possible, that is, it would consist of people of the same age, sex, and perhaps occupation. But, these people would have died from causes other than cancer. By making the same statistical analysis of their life histories and comparing results with the cancer group, you might be able to pinpoint the cause of the cancer in your experimental group.

Epidemiological studies also involve comparisons of living subjects. For example, you might compare the health records of workers exposed to a certain pollutant at work to office workers of the same company who are not exposed to the pollutant. This may turn up differences that could be attributed to workplace exposure.

Science requires a fair amount of experimentation to validate various hypotheses. No one study proves anything. By repeating experiments, scientists can gain confidence in their conclusions. Such repetition may seem ridiculous to taxpayers and television commentators, but it is essential to the scientific process.

Scientific knowledge grows with the accumulation of new facts. As facts accumulate, they often lead researchers to formulate **theories**, explanations that account for many different facts, observations, and hypotheses. In other words, theories are broad generalizations about the universe and the way it works. Unlike hypotheses, theories generally cannot be tested by single experiments because they encompass many different pieces of information. Atomic theory, for instance, explains the structure of the atom and fits observations made in many different ways over decades.

The Rise and Fall of Theories

Science is not a static entity. It constantly grows and evolves as new knowledge arises. During this process, scientists sometimes find that new interpretations emerge, replacing entrenched ideas that have prevailed for decades. This sometimes forces scientists to alter or even abandon theories that have enjoyed a faithful following for many years. Consequently, scientists must be open-minded and willing to replace their most cherished theories with new ones as new information renders the former obsolete.

Perhaps the best known example of changing theory is the Copernican revolution in astronomy. The Greek astronomer Ptolemy hypothesized in A.D. 140 that the Earth was the center of the solar system (the geocentric view). The moon, the stars, and the sun, he argued, all revolve around the Earth. In 1543, however, Nicolaus Copernicus showed that the observations were better explained by assuming that the sun was the center of the solar system. Copernicus was not the first to suggest this heliocentric view. Early Greek astronomers had proposed the idea, but it gained little attention until the 16th century.

Supertheories, or Paradigms

The dominant set of assumptions that underlies any branch of science is called a paradigm, a term coined by the philosopher and science historian Thomas Kuhn. A **paradigm** is a basic model of reality in any science. For example, evolution is a paradigm of the biological sciences.

Paradigms govern the way scientists think, form theories, and interpret the results of experiments. They also govern the way nonscientists think. Once a paradigm is accepted, it is rarely questioned. New observations are generally interpreted according to the paradigm; those that are inconsistent with it are often ignored or disputed. In some cases, though, observations that fail to fit the paradigm may amass to a point at which they can no longer be ignored, causing scientists to rethink their most cherished beliefs and, sometimes, toss them aside. This unsettling event is called a **paradigm shift**.

The term *paradigm* is commonly used today in a more general way, referring to the way we "see" the world. Many observers, for instance, speak about the **dominant social paradigm** of modern society when they refer to the set of beliefs that governs our lives. As you learned in Chapter 1, our attitudes and behaviors stem from our beliefs. Moreover, paradigms shape language, thought, and perceptions. Slogans and common sayings that reflect our paradigms are repeated and reinforced day after day. One of the most obvious expressions of the dominant social paradigm is the growth-is-good philosophy. You hear it in the business world every day, where business people proclaim, "A company must grow or die." Politicians echo a similar sentiment, as do most journalists. Donella Meadows wrote in her book *The Global Citizen*, "Your paradigm is so intrinsic to your mental process that you are hardly aware of its existence, until you try to communicate with someone with a different paradigm." Perhaps the most difficult problem is that people get attached to their paradigms and fiercely resist any effort to change them.

Science and Values

This text examines the fundamental attitudes we hold toward nature and the importance of shifting our thinking. It also offers scientific facts that could change people's values, notably how they view themselves and their part in the biosphere.

Nowhere has the influence of science on values been more apparent than in 19th-century England. In British society, Darwin's theory of evolution by natural selection was widely interpreted to involve changes in organisms that led to superior life forms, a view that scientists now reject. In 19th-century England, evolution meant that some life forms were better than others. This view was often used to justify the prosperity of a small elite class of business owners while workers suffered in dangerous factories. Social disparity was seen as an expression of the survival of the fittest. The application of the theory of evolution to social ethics is called **social Darwinism**.

The misapplication of Darwinian evolution points out how important it is for people to understand scientific information and use it correctly. In other words, societal values should be based on accurate knowledge and insights from science.

Environmental science's role in building a sustainable future is, therefore, that of a catalyst for changing values. Sure, you'll learn the scientific facts of environmental problems, but your studies will also help you to see the hidden connections, for example, the role of forests in maintaining atmospheric carbon dioxide levels and global climate. Environmental science will explain not only how endangered many life forms are but will also describe the importance of algae and plants in replenishing global oxygen supplies, which are vital to virtually all life on Earth, including ours. This text and course will illustrate the many ways that natural systems support human life. Knowing how important the planet is to the future, we cannot help but change our attitudes and actions.

Critical Thinking Skills

One of the chief tools a good scientist (and a great many others) rely on is critical thinking. **Critical thinking** is the capacity to distinguish between beliefs (what we think is true) and knowledge (facts supported by accurate observation and valid experimentation). That is, critical thinking helps us separate judgment from facts. Professor Larry Wilson of Miami-Dade Community College calls it the most ordered kind of thinking of which people are capable.

Critical thinking is not just thinking deeply about a subject, although that is necessary. Rather, critical thinking is subjecting facts and conclusions to careful analysis, looking for weaknesses in logic and other errors of reasoning.

Critical thinking skills are essential to analyzing a wide range of environmental problems, issues, and information. Critical thinking has no single formula, but most critical thinkers agree that several key steps are required for this important process. These steps, summarized in Table S1.1, will help you analyze statements you read, arguments, research findings, and issues.

Gather All Information

The first requirement of critical thinking is to gather all the information you can before you make a decision. In order to understand most issues, it is essential to dig deeper. When you do, you will often find a slightly different picture of reality. This is especially true in environmental debates, where advocates tend to simplify issues or present information that supports their case while ignoring contradictory findings. Although you'll never acquire all the information you need, get in the habit of learning all you can before you decide. Don't mistake ignorance for perspective.

Understand All Terms

The second rule of critical thinking is that when analyzing any issue, solving any problem, or judging the accuracy of someone's statements, you must understand all terms. In some cases, you will find that the people presenting a case have inaccurate or incomplete understanding themselves. Look up definitions when in doubt.

Table S1.1   Critical Thinking Rules

Gather all information.
  Dig deeper.
  Learn all you can before you decide.
  Don't mistake ignorance for perspective.

Understand all terms.
  Define all terms you use.
  Be sure you understand terms and concepts others use.

Question how information/facts were derived.
  Were they derived from scientific studies?
  Were the studies well conceived and carried out?
  Were there an adequate number of subjects?
  Was there a control and an experimental group?
  Has the study been repeated successfully?
  Beware of anecdotal information.

Question the source.
  Does the source have an investment in the outcome of
  the issue?
  Is the source biased?
  Do underlying assumptions affect the viewpoint of the
  source?

Question the conclusions.
  Do the facts support the conclusion?
  Correlation does not necessarily mean causation.

Tolerate uncertainty.
  Hard and fast answers aren't always possible.
  Learn to be comfortable with not knowing.

Examine the big picture.
  Study the whole system.
  Look for hidden causes and effects.
  Avoid simplistic thinking.
  Avoid dualistic thinking.

A good example is the term *sustainability*. It's currently being used by people who haven't the foggiest idea what it means. They even use the word to construct oxymorons, such as sustainable economic growth. In a finite system, you will soon see, growth cannot be sustained.

Question the Methods

The third principle of critical thinking is to question the methods by which facts are derived. Were the facts derived from experimentation? Can they be verified? Was the experiment correctly run? Did the experimenters use an adequate number of subjects? Did the experiment have a control group? Were the control and experimental groups treated identically except for the experimental

variable? Have the results been replicated by additional studies?

Beware of **anecdotal information**—stories of isolated incidents that are often made to appear as if they represent the truth. A newscast showing an angry mob in New York, for example, may give the impression that the entire country is in turmoil.

Question the Source

A fourth rule of critical thinking requires us to question the source of the facts. Ask yourself who is giving the information. When the American Tobacco Institute argues that the link between cigarette smoking and lung cancer hasn't been proven, skepticism might be advised. When businesspeople say that pollution from their factories isn't causing any harm, beware. Even environmentalists exhibit faulty reasoning, and resort to exaggeration. And, they're not immune to bias, which taints their interpretation of the problems.

Watch for bias and underlying assumptions. Some very well-entrenched assumptions and myths relevant to the environment are listed in Table S1.2. Study and be aware of them.

Question the Conclusions

The fifth requirement of critical thinking is to question the conclusions derived from facts. Ask yourself, Do the facts support the conclusions? Are other interpretations possible?

Numerous examples of faulty reasoning can be cited. One of the earliest studies on lung cancer, for instance, showed a correlation between lung cancer and sugar consumption. Patients with the highest rate of lung cancer ate the most sugar. A careful reexamination of the patients showed that the experimenters had drawn the wrong conclusion. It turned out that cigarette smokers ate more sugar than nonsmokers. Thus, the real cause of cancer was cigarette smoking, not sugar intake.

This example illustrates a key scientific principle: correlation does not necessarily mean causation. A **correlation** is an apparent connection between two variables. For example, because the economy improves when a certain politician is in office doesn't necessarily mean that he or she had anything to do with it. Because people living near a chemical plant have a high rate of leukemia (cancer of the blood) doesn't mean the factory is causing it.

Tolerate Uncertainty

The sixth rule of critical thinking is to tolerate uncertainty. Although this may seem contradictory at first, it isn't. As noted earlier, science is a dynamic entity. Theories come and go. Scientific knowledge is constantly

Table S1.2   18 Myths and Assumptions of Modern Society

1. People do not shape their future; it happens to them.

2. Individual actions don't count.

3. People care only about themselves and money; they can't be counted on to take action for a good cause unless they'll gain.

4. Conservation is sacrifice.

5. For every problem, there is only one solution; find it, correct it, and all will be well.

6. For every cause, there is one effect.

7. Technology can solve all problems.

8. Environmental protection is bad for the economy.

9. People are apart from nature.

10. The key to success is through the control of nature.

11. The natural world is here to serve our needs.

12. All growth is unqualifiably good.

13. We have no obligation to future generations.

14. Favorable economics justify all actions; if it's economic, it's all right.

15. The systems in place today were always here and will always remain.

16. Happiness stems from material possessions.

17. Results can be measured by the amount of money spent on a problem.

18. Slowing the rate of environmental destruction and pollution solves the problems.

Source: Adapted from Meadows, D. H. (1991). *The Global Citizen.* Washington, D.C.: Island Press, and Chiras, D. D. (1990). *Beyond the Fray: Reshaping America's Environmental Response.* Boulder, CO: Johnson Books.

being refined. What we know, however, is dwarfed by our ignorance. Although knowledge of poisons and hazardous materials is immense, in actuality, very little is known about the toxic effects of chemicals on humans. How do they interact? Are impacts affected by diet? What doses cause cancer or birth defects?

Many current debates on environmental issues involve a fair amount of uncertainty. Hard and fast answers aren't always available. Sometimes we have to accept this reality, learn to be comfortable with not knowing, and find ways to close the gap between knowledge and ignorance.

Examine the Big Picture

The seventh rule of critical thinking is always to examine the big picture. Be a systems thinker. Look for multiple causes and effects. Don't get trapped by simplistic thinking.

Dualistic thinking—black-and-white, right-or-wrong-oriented reasoning—is one of the most common forms of simplistic thinking. It says: "If I'm right, you must be wrong," or "we do it this way or that way." One side thinks this, and the other side thinks that. You have only two choices. In short, dualistic thinking sees two alternatives on every issue and nothing in between.

Humans seem attracted to the simplistic mode of thinking. It makes the choices easier. However, the world is a complicated place, and intermediate positions are often present. Search them out by looking for alternative options.

This book presents several conceptual models or drawings to help you understand the big picture. These learning tools organize a great deal of information into a simple format. Two of special importance are the Population, Resources, and Pollution Model and the Multiple Causation Model (Chapter 4).

Suggested Readings

Chiras, D. D. (1992). Teaching Critical Thinking in the Biology and Environmental Science Classrooms. *American Biology Teacher* 54(8): 464–69.

Cole, K. C. (1985). Is There Such a Thing as Scientific Objectivity? *Discover* 6(9): 98–99. Insightful look at science and the scientific method.

Eckblad, J. W. (1991). How Many Samples Should Be Taken? *Bioscience* 41(5):346–48. Gives examples on how to determine a suitable sample size for biological studies.

Hardin, G. (1985). *Filters against Folly: How to Survive Despite Economists, Ecologists, and the Merely Eloquent.* New York: Viking. Eloquent writing on scientific bias. See Chapters 1–7.

Kelley, D. (1988). *The Art of Reasoning.* New York: Norton. Especially good is Part 4, "Inductive Reasoning."

Klemke, E. D., Hollinger, R., Kline, A., eds. (1988). *Introductory Readings in the Philosophy of Science.* New York: Prometheus Books. An excellent introduction to critical thinking and its application in science.

Kuhn, T. (1970). *The Structure of Scientific Revolutions.* Chicago: University of Chicago Press. The original description of paradigms.

Meadows, D. H. (1991). *The Global Citizen.* Washington, D.C.: Island Press. Excellent discussions of paradigms.

Ecosystems: How They Work and How They Sustain Themselves

Never does nature say one thing, and wisdom another.

—Juvenal

Critical Thinking Exercise

At a meeting of the National Association of County Agricultural Agents in 1991, one of the speakers asserted that environmental problems facing the world are not very serious. He also noted that environmentalists are blowing things out of proportion to scare people. He continued by saying that environmentalists don't really care about people and they really don't care about the environment. They're just using this issue to cripple the economy.

You no doubt have heard similar claims or will hear them in coming years. Do you think the claims about environmentalists are valid? Why or why not? What critical thinking rules helped you to analyze these assertions?

An urban dweller awakens to the buzz of his alarm clock, eats a hurried breakfast, and heads to the concrete, steel, and glass towers of Downtown, U.S.A. Bumper to bumper on the ten-lane highway, encased in his shiny automobile, he listens to the radio in an attempt to counteract the tension. After half an hour on the freeway, he parks his car and rides the elevator to his 15th-floor office.

Sitting back in his chair, gazing at the skyscrapers and the paved arteries that bring more office workers to the city, he entertains only vague notions of his connection with nature. He might even laugh if you were to suggest that he is a part of the natural environment, subject to the rules that govern all organisms.

It is tempting to think of ourselves as apart from, and even superior to, nature. It is equally tempting to think that technology will render us immune to the laws of nature. But, as you learned in Chapter 1, human lives are rooted in the soil and dependent on air, water, plants, and algae. No matter what we think, humans are subject to the laws that govern life on Earth, and our future depends on understanding those laws and obeying them.

This chapter describes basic principles of ecosystem structure and function, information essential to understanding many environmental issues. This material underscores two key principles—recycling and renewable resources—essential for sustaining ecosystems. Finally, this chapter shows how natural systems support our lives, making it clear why the destruction of the life-support system of the planet is so dangerous.

2.1 How Is the Living World Organized?

The study of ecosystems falls within the domain of ecology. **Ecology** seeks to understand the relationships between organisms in ecosystems and their relationships with the environment upon which all life depends. Ecology, therefore, takes the entire living world as its domain in an attempt to understand all organism-environment interactions. It is a branch of science that requires a great deal of systems thinking.

Before we proceed, a word on semantics. The term *ecology* is probably one of the most misused words in the English language. Some environmental advocates argue that "we need to save our ecology." Others warn that our actions are "upsetting our ecology." Why are these uses of the word *ecology* incorrect? As noted above, ecology is a scientific field of inquiry. It is not synonymous with the word *environment*. It does not mean the web of interactions in the environment. In other words, our ecology is not in danger, our environment is. We can save our ecology textbooks, but we cannot save our ecology.

Our study of the science of ecology will begin with a look at life on Earth, starting with the biosphere and successively focusing on smaller and smaller units of organization.

The Biosphere and the Importance of Recycling and Renewable Resources

The part of the Earth that supports life is called the **biosphere,** or **ecosphere.** As shown in Figure 2.1, the biosphere extends from the floor of the ocean, approximately 11,000 meters (36,000 feet) below the surface, to the tops of the highest mountains, about 9000 meters (30,000 feet) above sea level. If the Earth were the size of an apple, the biosphere would be only about as thick as its skin. Within the biosphere, most species are concentrated in a narrow band, extending from less than 200 meters (660 feet) below the surface of the ocean to about 6000 meters (20,000 feet) above sea level.

Life in the biosphere exists at the intersection of land (lithosphere), air (atmosphere), and water (hydrosphere), as shown in Figure 2.2. These vast domains provide the ingredients that make life possible. Soil, for instance, provides minerals; air provides oxygen and carbon dioxide; and water comes from oceans and lakes. As noted in Chapter 1, these three renewable resources are vital to sustaining life on the planet. Also noted in Chapter 1, the biosphere is a closed system, much like a terrarium. In closed systems, all materials are recycled over and over. Only one contribution necessary for life comes from the outside—sunlight. Unlike minerals, water, and other materials that make life possible, sunlight energy cannot be recycled.

One of the first lessons we learn from the study of ecology, then, is that life in the biosphere sustains itself because it depends on renewable resources and because it recycles virtually all matter. Many ecologists believe that human society must pattern itself after nature to become sustainable. That is, we must learn to recycle

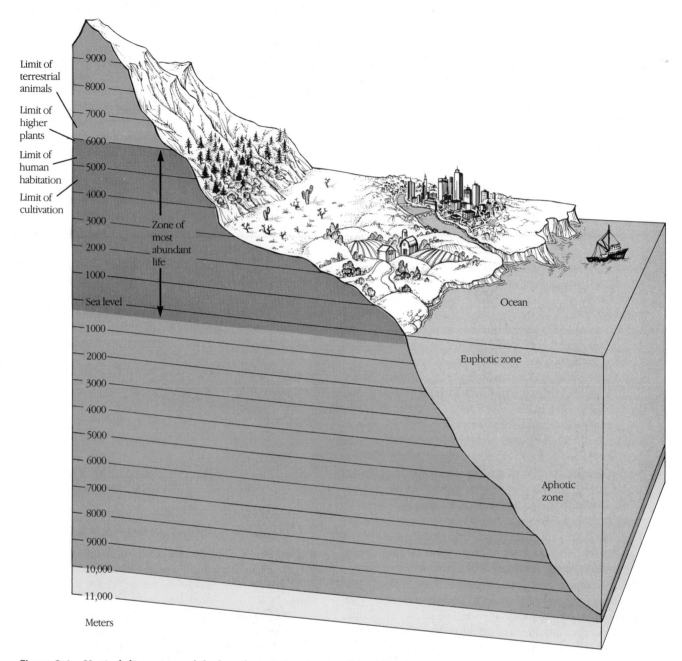

Limit of
terrestrial
animals

Limit of
higher
plants

Limit of
human
habitation

Limit of
cultivation

Zone of
most
abundant
life

Sea level

9000
8000
7000
6000
5000
4000
3000
2000
1000
1000
2000
3000
4000
5000
6000
7000
8000
9000
10,000
11,000

Meters

Ocean

Euphotic zone

Aphotic zone

Figure 2.1 *Vertical dimensions of the biosphere. Life exists in a broad band extending from the highest mountain peaks to the depths of the ocean. However, life at the extremes is rare, and most organisms are restricted to the narrow zone shown here.*

virtually all of the resources needed to support our lives—from wastepaper to automobiles to aluminum cans to plastic milk jugs—and greatly increase our dependence on renewable energy sources, such as wind and solar energy.

Biomes

The terrestrial portion of the biosphere is divided into broad regions characterized by a particular climate and a specific assemblage of plants and animals. These regions are known as **biomes**. A dozen biomes spread over millions of square kilometers of the Earth's surface, sometimes spanning entire continents (Figure S2.1 on page 50). Take a moment to study the biome map to pinpoint the biome in which you live.

Besides varying in plant and animal life, biomes differ with respect to their **climate**, the average weather conditions in a given region. In fact, climate determines the boundaries of biomes and the abundance of plants and animals found in them. Climatic conditions also influence the adaptations found in organisms within a

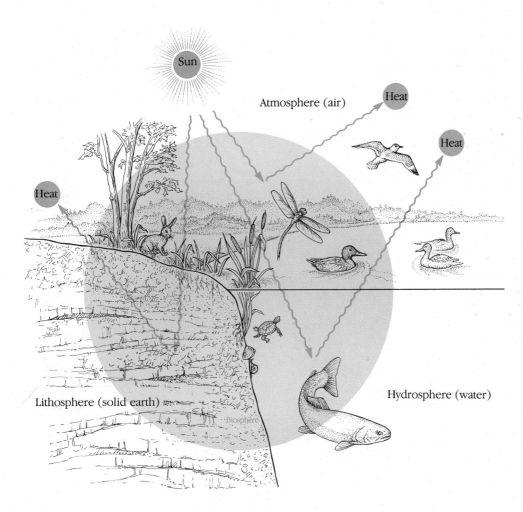

Figure 2.2 *Life exists primarily at the intersection of land (lithosphere), air (atmosphere), and water (hydrosphere). As shown here, the biosphere is energized by sunlight. All materials used in the biosphere are recycled.*

biome. In the desert, for example, where rainfall is scarce, plants generally have thick outer layers that conserve water. In rain forests, where rain falls in abundance, many plants contain leaves that repel water. This reduces the growth of mold. Both adaptations are essential to survival in the unique climate of these biomes.

The most important climatic factors in any biome are precipitation and temperature. As Figure 2.3 illustrates, these two factors combine to produce a variety of conditions that support a diversity of plants and animals, from those unique species of rich tropical rain forests to those that live in arid deserts. (Details of biomes are covered in Chapter Supplement 2.1.)

Aquatic Life Zones

Over 70% of the Earth's surface area is covered by water in the form of lakes, rivers, marshes, and oceans. These vital aquatic regions house many plants, animals, and microorganisms essential to the overall functioning of the Earth's biosphere. Many are essential to human welfare. Each aquatic region can be divided into distinct zones known as **aquatic life zones.** This section looks very briefly at marine life zones. (Freshwater zones present in lakes, as well as some relevant features of rivers,

are discussed in Chapter 16, which examines water pollution.)

The ocean's aquatic life zones include estuaries, coastlines, coral reefs, continental shelves, and the deep ocean. **Estuaries** are the outlets or mouths of rivers where saltwater and freshwater mix. **Coastlines** are the sandy or rocky shores of the world's continents and islands. **Coral reefs** are offshore deposits found in warm waters. They are composed of an animal known as coral, and are home to many fish and other species. **Continental shelves** are the sloping borders of the continents, and the **deep ocean** is the region beyond. Like their terrestrial counterparts, aquatic life zones differ from one another with regard to plant and animal life. The major differences among them can be traced primarily to levels of dissolved nutrients, water temperature, and depth of sunlight penetration.

In the ocean, the biologically richest areas are generally **estuarine zones**—that is, estuaries and **coastal wetlands** (mangrove swamps, salt marshes, and lagoons). In estuarine zones, nutrients from upstream soil erosion support a rich, diverse assemblage of organisms. So rich are these regions that they are often likened to tropical rain forests. The open oceans, in contrast, are relatively barren, and are sometimes likened to deserts.

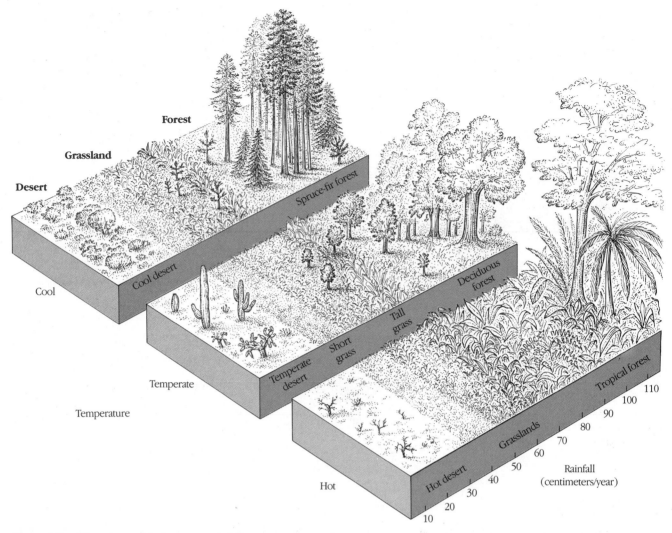

Figure 2.3   *The relationship between rainfall, temperature, and vegetation in terrestrial ecosystems. Rainfall determines the basic type of vegetation, and temperature is responsible for variations of basic types, for example, forests.*

Ecosystems

The term **ecosystem** is an abbreviated form of **ecological system**, which refers to all systems that consist of organisms, their environment, and all of the interactions that exist within. In other words, ecosystems are dynamic, interdependent physical and chemical systems. The biosphere is a planet-wide ecosystem. Because the biosphere is so massive and complex, ecologists generally limit their view to smaller ecosystems. Thus, to simplify one's study, an ecologist might limit his or her study to a pond, a cornfield, a river, a field, a terrarium, or a small clearing in the forest. In a sense, these are human-delineated ecosystems that are all part of the global ecosystem—the biosphere.

The study of ecosystems is invaluable in our pursuit of a sustainable society. The challenge of sustainability outlined in Chapter 1 is essentially a biological one. It calls on us to find a way to sustain a biological species, *Homo sapiens*, in a biological system, the biosphere. Although politics, economics, and ethics play a big part in meeting this challenge, our task is to find a way to fit human culture into the workings of nature in a sustainable way. (For an example, see Models of Global Sustainability 2.1.)

Ecosystems vary considerably in complexity. Some are quite simple, for example, a rock with lichens growing on it. Others, like tropical rain forests, are quite complex. They contain a wide variety of species as well.

Ecosystems aren't delineated by sharp boundaries; in fact, they usually merge with one another. The border of a mountain meadow, for example, is a zone that is half forest and half meadow. **Ecotones**, as these transition zones are called, contain plant and animal species from adjacent regions. In addition, ecotones support many species not found in either bordering ecosystem.

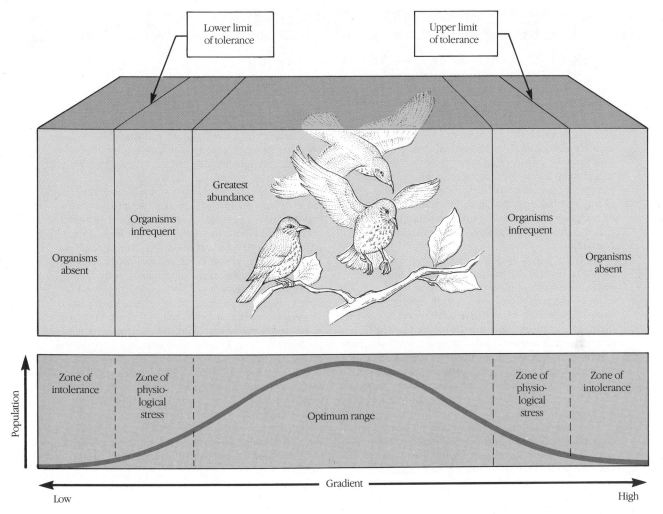

Figure 2.4 *Each organism has a range of tolerance for abiotic factors. It survives within this range but can be killed by shifts below or above the range resulting from human activities or natural phenomena.*

These species are adapted to the conditions in the ecotone. Because ecotones contain both a mix of species and many unique species, they often have a greater number of species than surrounding areas and deserve special protection.

All ecosystems consist of two major factors—abiotic and biotic—which we will study next to help you understand the complex interactions that occur in ecosystems.

Abiotic Factors The nonliving, or **abiotic**, factors are the physical and chemical components of an ecosystem, such as rainfall, temperature, sunlight, air, and nutrient supplies. Each of the Earth's organisms is finely tuned through adaptations to its abiotic environment, and operates within a range of physical and chemical conditions, the **range of tolerance** (Figure 2.4). Although the range of tolerance is sometimes wide, most organisms thrive within a narrow realm of acceptable conditions

known as the **optimum range.** Outside this is the **zone of physiological stress**, where survival is possible but difficult. Outside the zone of physiological stress is the **zone of intolerance**, where an organism will perish. Birds, for example, generally tolerate a narrow range of temperature. If the air cools below the lower limits of their range of tolerance, they will die or escape to warmer climates. If air temperatures exceed their upper limits of tolerance, they may also die or escape. Why is this important?

Human societies alter environmental conditions on the planet locally, regionally, and even globally. Chemical pollution from factories and sewage treatment plants, for example, can alter the chemical composition of streams locally, making them harmful to fish and other life. Overgrazing and poor land management may make large regions hotter and drier. Pollution from automobiles, power plants, and factories may cause a change in the global climate. Such changes alter the conditions in

Models of Global Sustainability 2.1

Think Globally, Act Locally: An Ecological Solution to a Perennial Problem in Texas

The late Buckminster Fuller coined the phrase "Think globally, act locally." He recognized that global environmental problems often result from individual acts and that by taking local action individuals can contribute to the resolution of global problems. For example, when you recycle an aluminum can or walk or ride your bike to work, you're helping to reduce global environmental problems, such as global warming.

In Texas, highway officials have taken Fuller's advice to heart. Fifty years ago, the Lone Star State began sowing wildflower seeds along its nearly 71,000 miles of roadside. Today, the brightly colored blossoms of the state's 5000 native wildflowers adorn the highways. In other states, roadsides are planted in grasses, which require regular mowing, irrigation, fertilization, and applications of insecticides and herbicides—all for roadside lawns as well groomed as a golf course. In Texas, road crews mow a few times a year to help spread the flower seeds, but that's about the extent of it.

Besides beautifying the landscape, the hardy flowers reduce erosion and provide habitat for native species. More importantly, native wildflowers require only the water that falls naturally from the sky, and they need little if any fertilizer and care. Native flowers resist drought, insects, and freezing weather because they inhabit the land where they have evolved over millions of years.

Benign neglect makes good sense economically and environmentally. In 1988, in fact, the Federal Highway Commission took a step that delighted conservationists throughout the nation when it voted to spend 25 cents of every federal dollar invested in roadside landscaping on planting native wildflowers. The wildflower rule was spearheaded in Congress by then-Senator Lloyd Bentsen of Texas, who says that his state's lengthy experience with planting native wildflowers shows that it saves taxpayers' money as well as beautifies highways. Wildflowers seem to deter people from littering as well, thus reducing litter control efforts.

Planting native species reduces water demand, saves streams, reduces energy demand, cuts soil erosion, enhances wildlife, reduces global pollution, and makes the world a more colorful place. Such efforts aimed at cooperating with nature are a key to building a sustainable future. It is a lesson badly needed in the developed as well as in the developing nations of the world.

---

which an organism lives, forcing it to flee or driving it to extinction. As you will see in this text, one of the secrets to living sustainably on the planet is learning to live in ways that retain optimal conditions for all species, including our own.

Although organisms are affected by a variety of chemical and physical factors in their surroundings, one factor, known as the **limiting factor**, usually outweighs the others in determining growth. In other words, the limiting factor is the primary determinant of growth in an ecosystem. The concept of limiting factor was introduced in 1840 by the German scientist Justus von Liebig, who was studying the effects of chemical nutrients on plant growth.

In lakes and reservoirs, phosphorus is a limiting factor. Naturally low levels of phosphorus hold populations of algae and other organisms in check. However, if phosphorus levels rise, for example, by the introduction of sewage rich in phosphates from laundry detergents, algal populations may explode.

From an environmental perspective, limiting factors are important because they can be easily upset by human activities. Another goal of sustainable living, then, is to avoid activities that upset the natural balance. In the example above, low-phosphate or no-phosphate detergents are advisable.

The limiting factor is analogous to the slowest camel in an expedition. The entire party's pace through the desert is set by the slowest camel. In the same fashion, the entire structure of an ecosystem is determined by the limiting factor. If the slowest camel dies, the second-slowest one will set a new pace. The same is true in an ecosystem. An increase in the availability of a limiting factor allows accelerated growth, but another factor invariably sets new limits.

For most terrestrial ecosystems, rainfall is the limiting factor. As shown in Figure 2.3, rainfall determines whether land is covered by forest, grassland, or desert. If annual rainfall exceeds 70 to 80 centimeters (27 to 31 inches), for example, forests develop. Slightly drier climates support grasslands, and the driest regions are always deserts. Figure 2.3 also shows that differences in temperature are responsible for variations on these three basic themes. Consider the wettest ecosystems for a moment. As shown, warm and wet climates support tropical rain forests with huge trees reaching 60 meters (200 feet). Slightly cooler regions are characterized by deciduous forests. In even cooler areas, coniferous forests grow.

Biotic Factors   The **biotic factors** of an ecosystem are its organisms—fungi, plants, animals, and microorganisms. Unbeknownst to many people, each of the Earth's organisms plays an important role in the economy of nature and the sustainability of natural systems. Plants, for instance, trap sunlight and carbon dioxide from the atmosphere, and use them to produce organic food molecules through the process of **photosynthesis**. These molecules nourish most of the living world. Plants also produce oxygen, which is vital to animal life. Animals consume plant material and one another. They use oxygen in their cells to break down food molecules made by plants. In so doing, they liberate some of the sunlight energy that the plants captured to make the molecules in the first place. In this process, however, animals release carbon dioxide, needed by plants to continue the cycle of life. Many microorganisms decompose plant and animal matter, also liberating carbon dioxide and thus also facilitating the recycling of matter essential to sustaining life.

Singer/songwriter Dana Lyons reminds us in his song "Animal" that, despite what many may think, humans are animals. Like other animals, our lives are dependent on the rest of the living world. Everything around us, from the wood we use to build homes to the water we drink to the breakfast cereal we eat, even the gasoline we burn in our cars, comes from other organisms. Even though we humans can travel in space and split atoms, we must not forget that we are just another species, dependent on a vast array of other species for virtually every move we make.

Organisms live in populations. A **population** is a group of the same species occupying a given region. Populations are dynamic groups, however, changing in size, age, and genetic composition as a result of changes in the environment. Because populations do not exist in isolation from one another, they share a common resource base with other populations and contribute to the lives of other organisms in ways described below.

Within ecosystems, populations of different organisms are woven into the complex web of life known as a community. A **community** consists of all of the organisms in a specified region. In Yellowstone National Park in Wyoming and Montana, for instance, the wildlife community consists of grizzly bears, elk, mule deer, coyotes, ravens, and a host of plants and microorganisms.

Within communities, organisms interact in many different ways. One of the most familiar relationships is the predator-prey interaction. **Predation** occurs when one organism kills another for food, for instance, when a robin catches and eats a worm or a grizzly kills and

eats a ground squirrel. Biologists even consider grazing a form of predation. Elk, for example, prey on grass.

Predation obviously benefits predators because it is usually their chief means of acquiring food. Predation also benefits prey populations because predators often kill the weak, sick, and aged members of a population. This helps to reduce prey population size and ensures that the remaining members have a larger share of the available food supply—a benefit especially evident in winter months, when food is in short supply. Predation is, therefore, one form of population control. Destroying predators can have a serious impact on natural systems, as explained in Case Study 2.1.

Another kind of interaction in living systems is **commensalism**, a relationship between two species that is beneficial to one but neutral (neither good nor bad) to the other. Barnacles, for example, attach and grow on the skin of whales. Although the whale provides a substrate for them to grow on and transports them through the water, it probably gets nothing in return.

Mutualism is a relationship that is beneficial to both organisms. A classic example occurs in an organism known as the lichen. **Lichens** are hardy plantlike organisms that typically grow on barren rock, capturing water from rainfall and gathering nutrients dissolved from the surface of the rock. Lichens actually consist of two organisms: fungi and algae. The fungi form the main body of the organism, providing a home for the algae that live inside. Algae are photosynthetic organisms, capable of capturing sunlight and using it to produce organic food molecules from atmospheric carbon dioxide, which are used by the fungus. Thus algae provide food in exchange for a place to live. Both organisms benefit from this relationship.

Some relationships can be harmful to one member. Parasitism is an example. **Parasitism** occurs when one species lives on or even inside another, the **host**. In this relationship, the parasite obtains food by slowly eating the host or by dining on fluids inside the host. A parasite may be a temporary resident, such as a wood tick, or it may set up a long-term relationship, as tapeworms do. When the host is healthy, parasites simply dine on the surplus and have little effect. However, parasitism can kill a host debilitated by disease, old age, and stress. Parasitism, therefore, helps to control populations, especially when populations exceed the carrying capacity of their environment.

The final relationship worth noting is competition. **Competition** occurs any time two organisms vie for the same resources. When it takes place between members of the same species, it is called **intraspecific competition**. Between members of different species, it is referred to as **interspecific competition**. Like predation and parasitism, competition is an important component of population control vital to sustainability of natural systems because it helps to keep numbers within the carrying capacity of the environment.

Human Impact on the Great Barrier Reef

Australia's Great Barrier Reef has been described by some as the eighth great natural wonder of the world. Today, however, many scientists are wondering if this magnificent reef will survive to the year 2000.

For many years, Dr. Robert Endean of the University of Queensland has argued that the Great Barrier Reef was in trouble. He asserted that the delicate life of the coral reef was threatened by a once-rare species of starfish, known as the crown of thorns (Figure CS2.1). This starfish feeds on stationary marine animals called coral polyps, which have given rise to the coral reef over many years. Until recently, starfish populations were so small that they had little effect on coral reefs.

In the 1960s, however, the population of the crown of thorns starfish rose dramatically in the Great Barrier Reef, killing off large sections of the reef. Today, the central third of the reef has been infested with crown of thorns starfish, and scientists now report outbreaks on the northern and southern ends as well. The cause of this outbreak is unknown, but many scientists think that humans have unwittingly destroyed the starfish's natural predators, unleashing a destructive force that could annihilate the Great Barrier Reef within the decade.

One of those predators is a mollusk, the giant triton. Over the years, the triton has been collected in great numbers and sold for food. Once helping to keep starfish populations in check, the triton is becoming rare on the reef. In addition, predatory fish, such as cod and grouper, have been severely overfished in the waters surrounding the reef, contributing to the crown of thorn's dramatic increase in number.

Predators are vital to the sustainability of ecosystems in large part because they help to control populations within them. To help reestablish balance, the

The crown of thorns starfish is wreaking havoc on the Great Barrier Reef.

Queensland government has banned the capture of starfish predators. Bear in mind, though, the Great Barrier Reef is 2000 kilometers long (1200 miles)—roughly the distance from San Francisco to Denver—and lies up to 160 kilometers (100 miles) offshore, making it difficult to patrol.

Destruction of the Great Barrier Reef could have severe economic impacts. Why? The reef attracts scuba divers and sightseers worldwide and thus supports a healthy tourist industry. It also produces many commercially important fish species. Its destruction would affect the world's food supply, which is especially troubling, given the fact that the world population grows by 250,000 people a day! Of course, for the species that live among the coral, its demise would create a tragedy of epic proportion.

Habitat and Niche

If you were to describe yourself to a foreigner, you would probably begin with where you live. Then, you might explain what you do, who your friends and enemies are, how you interact with others, what you eat, and what you do in your spare time. If you ask an ecologist studying armadillos to describe them, she would probably follow a similar approach, first telling you where these armored minitanks live and then what they do. Where an organism lives constitutes its **habitat**. Where it lives and how it fits into the ecosystem are its **ecological niche**, or simply its **niche**.

The niche includes an organism's total functional role in the ecosystem. Therefore, describing an organism's niche requires an understanding of what it eats, what eats it, where it lives, and how it interacts with other living and nonliving components of the ecosystem. A description of an animal's niche would also include answers to questions about its range of tolerance, competitive interactions, shelter requirements, and parasites.

Ecological studies show that no two species can occupy the *same* niche for very long. This is known as the **competitive exclusion principle**. Bear in mind, this does not mean that similar species living in the same habitat cannot coexist. It just means that species with identical requirements cannot coexist for very long. One will drive the other to extinction.

As long as niches differ, though, similar species can coexist in the same habitat. The red-tailed hawk and the great horned owl, for example, occupy the same habitat and prey on mice and other rodents. Although they may seem like direct competitors, an examination of their niches shows marked differences. Owls, for instance, come out at night to feed; hawks prefer daytime hunting. Despite apparent similarities, then, hawks and owls occupy slightly different niches. Because of this, they can coexist in a given habitat without threatening each other's survival.

During evolution, species have evolved to fill many different niches, which minimizes interspecific competition and makes optimal use of the Earth's resources. So why even worry about the competitive exclusion principle if evolution tends to reduce the overlap of niches?

The competitive exclusion principle is important in large part in human-managed systems. Wildlife managers, for instance, have attempted (and continue to attempt) to increase hunting by transplanting foreign species into new regions. If the animals' niches are identical or if they overlap extensively with existing species, trouble begins. The newcomers, called **ecological equivalents**, may conflict with existing species, causing considerable competition. They may perish or may drive their competitor to extinction.

The niche an organism occupies may be narrow and very specific or very broad. Thus, organisms may be defined as generalists or specialists, based on their niches. As a rule, a **generalist** is one that eats a variety of foods or inhabits a variety of different habitats, sometimes both. A **specialist** is one that has a limited diet and habitat. For example, the koala of Australia is a specialist because it lives in a fairly limited habitat and feeds exclusively on eucalyptus leaves (Figure 2.5a). In contrast, the coyote of North America is a versatile generalist that feeds on a wide variety of animals—rodents, rabbits, birds, snakes, house cats, and an occasional sheep (Figure 2.5b). The coyote also occupies a wide range of habitats.

As a rule, generalists are less vulnerable to changes in the abiotic and biotic environment and, therefore, are less apt to become extinct than specialists. If a food source disappears, for instance, a generalist can usually find alternatives, whereas a specialist may perish. Today, the giant panda of China is threatened by the loss of bamboo, its sole source of food. The spotted owl of the Pacific Northwest is another specialist that depends on old-growth forests for food and shelter. If its forests are cut, this bird and an estimated 68 species that live in the same ecosystem will surely perish.

The practical lesson in this discussion is that it is erroneous to think that all species can adapt to changes brought about by human society. A deeper understanding of ecology shows that specialists require very specific abiotic and biotic conditions to survive.

Human beings, like coyotes, are generalists in many ways. We live in diverse climates and eat a wide variety of plants and animals. Although we are generalists,

recent history has been marked by increasing specialization. Today, for example, we depend on a handful of metals, a relatively small number of crops and livestock species, and a few energy sources. If supplies of any of these should be cut off, human society would suffer inordinate hardship. As you will learn in this text, many experts believe that our survival depends in part on reducing the level of specialization, for example, expanding the number of crops we depend on (Chapter 7) and tapping into new sources of renewable energy (Chapter 12).

Another practical application of these ecological principles is that protecting plants, animals, and other species from extinction requires efforts to protect their niches (Chapter 8). Efforts to reintroduce species into the wild also hinge on an understanding of their niches. For instance, in January 1992, two captive-bred California condors were released in the Los Padres National Forest north of Los Angeles. This region was deemed to contain suitable habitat for the condors' survival after extensive field studies of the bird's niche. Only time will tell if the scientists made an accurate assessment.

2.2 How Do Ecosystems Work?

In this section, we turn our attention to ecosystem function. We focus on the ways producers and consumers are related in ecosystems as well as how energy and chemical nutrients flow through them. Human dependence on the natural world will become even more evident. Finally, we examine the importance of recycling and renewable resources to a sustainable future.

Food Chains and Food Webs

In the biological world, you are one of two things—a producer or a consumer. (Only rarely can you be both.) **Producers** support the entire living world through photosynthesis. Plants, algae, and cyanobacteria are the key producers of energy-rich organic materials. Producers are also referred to as **autotrophs** (from the Greek words *auto* meaning "self" and *troph* meaning "to feed") because they literally nourish themselves photosynthetically by using sunlight and atmospheric carbon dioxide to make the food materials they need to survive.

Consumers feed on plants and other organisms and are, therefore, also referred to as **heterotrophs** (from the Greek word *hetero* meaning "other"). Several different kinds of consumers are present. Those that feed exclusively on plants are **herbivores**. Cattle, deer, elk, and tomato hornworms are examples. Those consumers that feed exclusively on other animals, such as mountain lions, are **carnivores**. Finally, those that feed on both plants and animals, such as humans, bears, and raccoons, are **omnivores**.

(a)

(b)

Figure 2.5 *Specialist and generalist.* **(a)** *The koala is a specialist because it eats only the leaves of eucalyptus trees. A loss of this food source would result in its extinction.* **(b)** *The coyote is a generalist. Like humans, it is an opportunist, capable of eating a wide variety of foods. Consequently, its existence is not so precarious.*

As noted earlier in this chapter, all organisms are part of the global recycling mechanism of the biosphere. Their connection is most obvious in food chains and food webs. A **food chain** is a series of organisms, each feeding on the preceding one.

Two basic types of food chains exist in nature: grazer and decomposer. **Grazer food chains** are so named because they start with plants and with grazers, organisms that feed on plants. Figure 2.6 illustrates familiar terrestrial and aquatic grazer food chains.

In **decomposer food chains**, organic waste material, or **detritus**, is the major food source (Figure 2.7). Detritus comes from plants and animals, and is consumed by large scavengers, insects, and microorganisms, such as bacteria. In the grasslands of Africa, for instance, a wildebeest that drowns crossing a river washes up on the riverbank and is consumed by vultures, hyenas, and the larvae of various flies. These organisms are referred to as **detritus feeders** or **macroconsumers**. Decomposition occurs because of **microconsumers**, that is, microscopic bacteria and fungi. All of these organisms derive energy and essential organic building blocks from detritus. In the process, they liberate carbon dioxide, water, and other nutrients needed to continue the cycle of life.

Food chains are conduits by which nutrients and energy flow through ecosystems. As mentioned earlier, the sun's energy is first captured by plants and other autotrophs (algae) and stored in organic molecules, which ultimately pass through grazer and decomposer food chains. In addition, plants incorporate a variety of inorganic materials, such as nitrogen, phosphorus, and magnesium from the soil. These nutrients become part of the plant's living matter. When a plant is consumed,

these nutrients enter the grazer food chain. They are eventually returned to the environment by the decomposer food chain.

This simple system driven by sunlight is the basis of the rich and diverse array of life forms on the planet. It has evolved over billions of years into an elaborate network of which humans are just one part. Protecting this system is essential to protecting all life on Earth.

Classifying Consumers   Biologists are avid namers, happiest when things are categorized, tagged, and dissected. The study of ecology has not been spared this inclination. For instance, ecologists categorize consumers by their position in the food chain (Figure 2.8). In a grazer food chain herbivores are called **primary consumers,** since they are the first organisms to consume the plants. Organisms that feed on primary consumers are **secondary consumers.** Those that feed on secondary consumers are **tertiary consumers,** and so on.

The feeding level an organism occupies in a food chain is called the **trophic level.** The first trophic level marks the beginning of the food chain and is made up of the producers, or autotrophs (self-feeders). Primary consumers occupy the second trophic level. Secondary and tertiary consumers occupy the third and fourth trophic levels. All consumers are heterotrophs. Figure 2.8 shows an example of a food chain broken down into trophic levels.

This discussion of food chains greatly simplifies the structure and function of ecosystems. In truth, food chains exist only on the pages of ecology texts. Food chains are part of more complex networks called

Figure 2.6 *Examples of terrestrial and aquatic grazer food chains. Note that both food chains begin with photosynthetic organisms (grass and phytoplankton), which are consumed by grazers.*

food webs. A **food web** is an interconnected network of food chains that gives a complete picture of the feeding relationships in an ecosystem (Figure 2.9).

Trophic levels can be assigned in food webs just as in food chains; in a food web, however, many species occupy more than one trophic level. As illustrated in Figure 2.10, a grizzly bear feeding on berries and roots is acting as a primary consumer and occupies the second trophic level. When feeding on marmots, animals similar to woodchucks, the grizzly is considered a secondary

Figure 2.7 *The decomposer and grazer food chains. (a) In the decomposer food chain, bacteria and other organisms feed on plant and animal remains and waste products. (b) The grazer and decomposer food chains are linked, that is, all material ultimately comes from the grazer food chain, from producers. And, all material ends up in the decomposer food chain.*

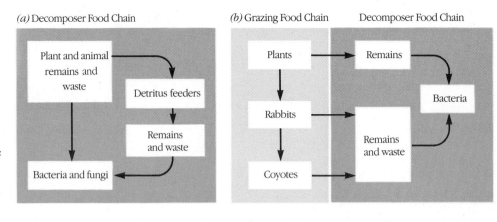

(a) Decomposer Food Chain

(b) Grazing Food Chain | Decomposer Food Chain

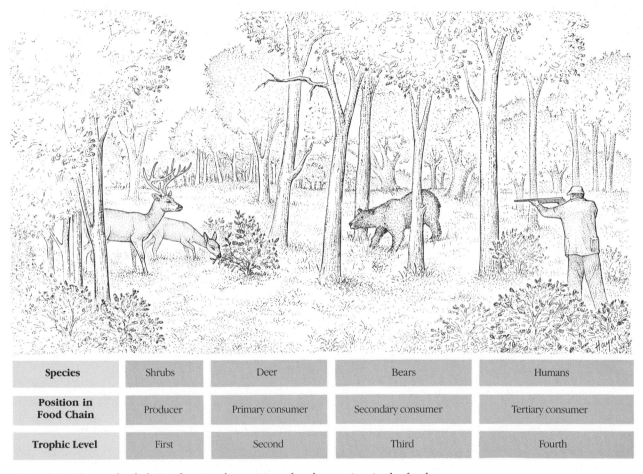

Species	Shrubs	Deer	Bears	Humans
Position in Food Chain	Producer	Primary consumer	Secondary consumer	Tertiary consumer
Trophic Level	First	Second	Third	Fourth

Figure 2.8 *Grazer food chain, showing the position of each organism in the food chain and the trophic level each occupies.*

consumer and occupies the third trophic level. In other instances, a grizzly may feed on insect-eating chipmunks and therefore occupies the fourth trophic level. The grizzly also feeds on animal carcasses, or **carrion**. In such instances, the grizzly is a participant in a decomposer food chain.

As a general rule, the greater the number of channels through which energy and nutrients flow in a food web, the more stable it is. As illustrated in Figure 2.9, the marsh hawk and owl prey on at least six other members of the food web. If one of their prey vanishes, the birds would probably not be affected because they have plenty

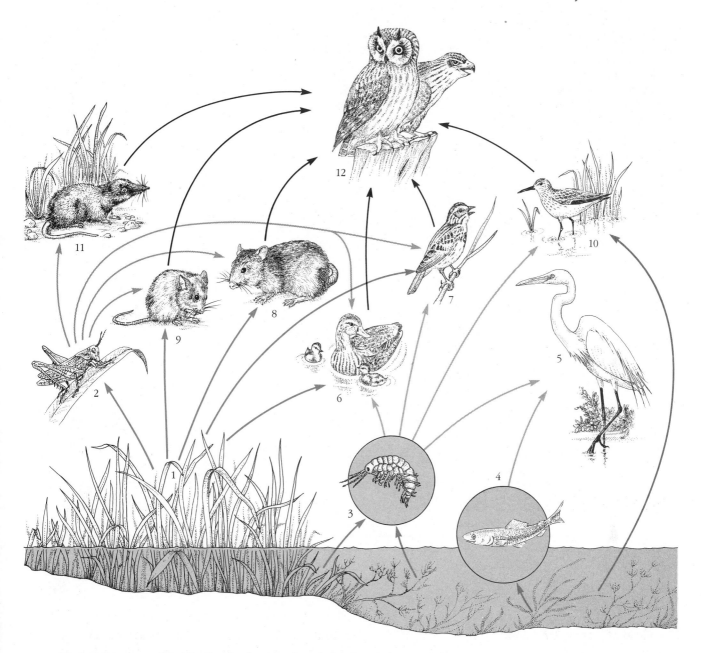

Figure 2.9 *A simplified food web, showing the relationships between producers and consumers. In a salt marsh (San Francisco Bay area), (1) producer organisms, terrestrial and salt marsh plants, are consumed by (2) insects. Marine plants are consumed by (3) herbivorous marine and intertidal invertebrates. (4) Fish, represented by smelt and anchovy, feed on vegetative matter from both terrestrial and marine environments. The fish in turn are eaten by (5) first-level carnivores, represented by the great blue heron and the common egret. Continuing through the food web, we have the following omnivores: (6) mallard duck, (7) savannah and song sparrows, (8) Norway rat, (9) California vole and salt marsh harvest mouse, and (10) sandpipers. (11) The vagrant shrew is a first-level carnivore, while the (12) top carnivores (second level) are the marsh hawk and the short-eared owl.*

of others to satisfy their hunger. However, if there were only one or two sources of food for predatory birds, the loss of one could have a devastating effect. (This explains why specialists are more prone to extinction.) One of the lessons we can learn from this is that ecosystem

stability can be impaired by simplifying ecosystems— that is, by reducing the number of species within their food web. In addition, in many ecosystems, very likely our own as well, species diversity is essential to sustainability.

Figure 2.10 *The grizzly participates in several different food chains in its complex food web and occupies different trophic levels in different food chains.*

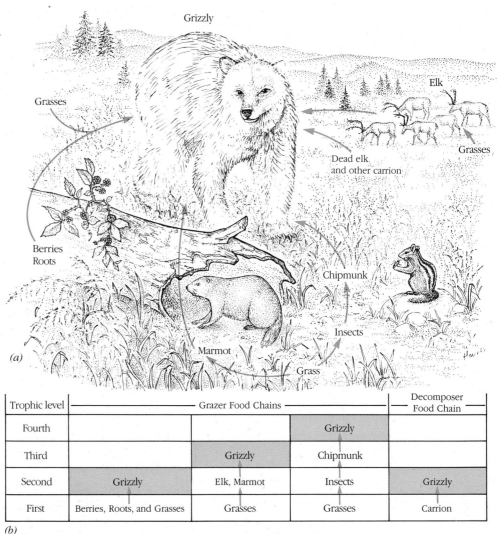

(a)

Trophic level	Grazer Food Chains			Decomposer Food Chain
Fourth			Grizzly	
Third		Grizzly	Chipmunk	
Second	Grizzly	Elk, Marmot	Insects	Grizzly
First	Berries, Roots, and Grasses	Grasses	Grasses	Carrion

(b)

The Flow of Energy and Matter Through Ecosystems

Photosynthetic organisms, such as plants, are the starting point of all food webs, because they alone are capable of tapping into the sun's energy. But, plants capture only 1% or 2% of the sunlight energy that strikes the Earth. The entire living world is built on this small fraction of sunlight. In this section, we turn our attention to the flow of energy and matter through ecosystems, pointing out additional principles vital to understanding the world we live in. Before we begin, let's take a brief look at energy.

What Is Energy?   *Energy* and *love*. We are all familiar with the terms, but we still stumble when trying to define them. One of the reasons is that energy, like love, comes in many forms. Heat, light, sound, and electricity are familiar examples. As different as these forms of energy are, however, they have one thing in common: the capac-

ity to do work. Technically, then, **energy** is defined as the capacity to do work. **Work** is performed when an object—be it a mountain or a molehill—is moved over some distance.

All forms of energy fall into two groups: potential and kinetic. **Potential energy** is stored energy. Coal, oil, and even the food we eat contain potential energy, which, when released, allows us to perform work. **Kinetic energy**, in contrast, is the energy possessed by objects in motion. A falling rock and a swinging hammer can do work and, therefore, possess kinetic energy.

All forms of energy follow basic laws, known as the **laws of thermodynamics**. Understanding these laws will help you better understand ecology, environmental issues, and sustainability.

The First Law   The **first law of thermodynamics** is often called the **law of conservation of energy**. It states that energy can be neither created nor destroyed but can only be transformed from one form to another. The gasoline in your car, for example, contains an enormous amount of potential energy that, when released in combustion, propels your car down the highway. As you drive along, the gas gauge indicates the amount of gasoline your car has burned. Contrary to what you may think, you have not destroyed that energy; you have simply converted it into other forms—electricity to run your radio, heat to defrost your windows, light to show you where you are going, and, of course, mechanical movements that propel your car. If you could measure the amount of energy your car is consuming and the amount of energy being produced in these various forms, you would find the two are equal. In simple terms, energy input is equal to energy output.

Modern society also relies on energy conversions to perform millions of activities every day. Coal, for example, is burned in power plants to generate electricity, which is used to power light bulbs, neon signs, and electric motors. Energy conversions, such as the ones described above, also take place in biological systems where sunlight is trapped by plants and stored in organic food materials during photosynthesis. These molecules are ingested by herbivores and when broken down, provide energy needed to perform a variety of cellular functions.

All forms of energy are eventually converted to heat. Heat is measured in calories. A **calorie** is the amount of heat needed to raise 1 gram of water 1 degree Celsius or 1.8 degrees Fahrenheit. The number of calories contained in a fuel indicates the amount of work that can be performed. All forms of energy can be measured in calories, even the food we eat. Because a calorie is a relatively small unit of measurement, scientists frequently speak about 1000-calorie units, or **kilocalories**, commonly written as Calories (note the capital letter C).

The first law of thermodynamics is also popularly referred to as the no-free-lunch principle—you can't get something for nothing. More specifically, you can't get more energy out of a system than you put into it. Even today, though, people often claim to invent machines that create more energy than they use. Beware of such promises.

Another lesson we can learn from the first law is to balance our energy calculations. When developing new energy sources, for instance, it's imperative to look carefully at the energy required to tap into them (Chapter 11). For example, suppose a company discovered coal in Antarctica. To get the coal to northern markets, the company would have to dig it up, crush it, and then transport it via ship to the north. When the coal arrived, it would have to be unloaded and transported by train to power plants, traveling many more miles. For this to make sense, the energy the company could get out of the coal would have to greatly exceed the energy put into its production. Otherwise, the company would be wasting time, energy, and money.

The Second Law   The first law of thermodynamics is concerned with quantities of energy in energy conversions. The **second law of thermodynamics** explains what happens to energy when it changes from one form to another—it describes changes in energy quality. The second law states that during energy conversions, energy is "degraded." One way of looking at this is that energy goes from a concentrated to a less concentrated form during a transformation. For example, when gasoline is burned in an automobile, it is converted from a very concentrated form to much less concentrated forms, mostly heat. Heat is then dissipated into the air and is dispersed in the random motion of air and water molecules.

Concentrated energy forms, like gasoline, contain a great deal of potential energy. Physicists say such concentrated forms contain a lot of available work. Less concentrated forms, like heat, contain less potential energy and less available work. High-quality energy sources, such as oil and natural gas, can be put to good use, whereas most heat is a low-quality energy source, which is quickly diluted or dispersed. In general, it is a less useful form of energy than concentrated forms.

The second law of thermodynamics has several implications for our lives and ecosystems. It tells us that when we burn fossil fuels, our supply of highly concentrated energy—our finite fossil fuel reserve—shrinks. A word to the wise: Don't waste this precious resource; the supplies we are now tapping are all we have. The second law also tells us that we cannot recycle high-quality energy because when it is burned, it is dissipated into heat and lost into space. It further tells us that no energy conversion is 100% efficient; some is always lost as heat. To maximize energy efficiency, we must reduce the heat loss or capture that heat and put it to good use. For instance, factories that generate their own electricity often produce enormous amounts of waste heat that could be used to heat office buildings or heat water needed for manufacturing. This not only increases the efficiency of energy use but also saves considerable sums of money. As you will see in the energy chapters, building a sustainable society requires methods to reduce the waste of energy by putting waste heat to work for us and by designing more efficient light bulbs, refrigerators, motors, automobiles, and other technologies.

Biomass and Ecological Pyramids   The laws of thermodynamics affect many occurrences in the living world, ranging from the activity of tiny bacteria to the largest mammals, the whales. The following discussion of biomass and ecological pyramids explains one such effect.

In the lexicon of biologists, **biomass** is the dry weight of organic matter created by plants and other photosynthetic organisms that is passed from one consumer to the next in food chains. Because organisms vary considerably in their water content, water is excluded when biologists determine the biomass at any trophic level of

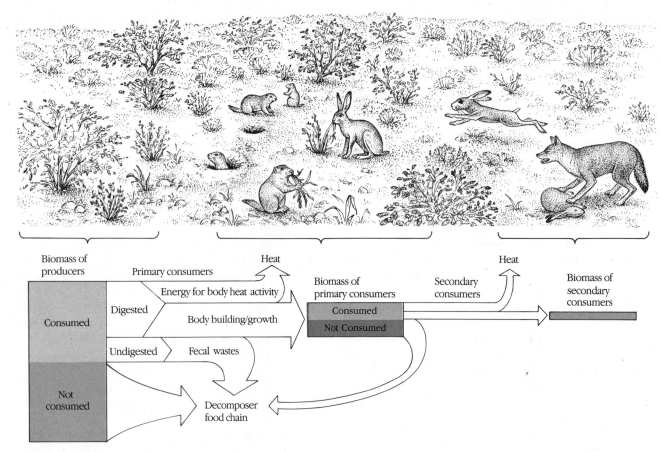

Figure 2.11 *Biomass transfer in the food chain. Note that only small amounts of biomass are transferred from one trophic level to the next. Part of the reason for this is that a large fraction of the biomass of each trophic level is not consumed.*

a food chain. To determine the biomass of plants (the first trophic level), for example, a single square meter of vegetation is removed, roots and all, then dried and weighed.

The biomass at the first trophic level in almost all ecosystems represents a large amount of potential (chemical) energy and tissue-building materials for the second trophic level. The organic molecules that comprise the organisms of the first trophic level, therefore, contain considerable amounts of potential energy that is locked in chemical bonds, a fact substantiated by burning fat, wood, oils, and other organic molecules.

Studies of ecosystems show that although tremendous amounts of biomass are found at the first trophic level of most food chains, only a small fraction is converted into biomass in the second trophic level. Thus, not all plant matter becomes animal matter. Several reasons account for this. First, only a small part of the plant matter in most ecosystems is typically eaten by the herbivores of the second trophic level, as shown in Figure 2.11. Second, not all of the biomass eaten by

the herbivores is digested; some passes through the gastrointestinal tract unchanged and is excreted. Third, most of what is digested is broken down into carbon dioxide and water, which yields energy needed to move about, to breathe, to maintain body temperature, and to carry out other body functions.

Because of these factors, the biomass of the second trophic level is considerably smaller than the biomass at the first trophic level. Ecologists once thought that 10% of the biomass at one trophic level was transferred to the next. They called this the **ten percent rule**. Further research showed, however, that the amount of biomass transferred from one level to the next varies from 5% to 20% in different food chains. Graphically represented, biomass at the different trophic levels forms the pyramid known as the **pyramid of biomass** (Figure 2.12).

Biomass is the substance of which living things are made. As noted earlier, the chemical bonds that hold the organic compounds of biomass together contain stored energy. This energy can be released when organic matter burns or when it is broken down by cells in plants,

Biomes

The biosphere is divided into geographically distinct land parts called biomes, each having its own specialized climate, plants, and animals. These diverse regions, made up of intricately balanced ecosystems, offer us a kaleidoscope of information on evolution and adaptation, which increases our understanding of the way the world works within the constraints of nature.

And yet, whenever we think we understand the workings of a biome (its climate, its topography, its predictability), we experience a Mt. St. Helens—a reminder from nature that life on earth is constantly in flux; changing, adjusting, adapting. Like nature, humans have altered biomes and ecosystems—but unlike nature, many of our changes have become irrevocable.

Through an understanding of the biomes we can learn about structure and function of the environment, which in turn will give us an awareness of how our world works—and, just as importantly, how it doesn't.

1 Atypical of the desert biome, the Great Sand Dunes National Monument in southern Colorado is lifeless and virtually barren.

2 The rim of the Grand Canyon in Arizona shows yet another face of the desert biome.

3 Cactus abound in the desert, blossoming briefly after the spring rains.

4 The tropical rain forest is perhaps the most diverse of all biomes. It is estimated that over 90% of the living organisms in these rain forests remain unclassified. Venezuela.

5 Rain forest in the Olympic Peninsula, Washington. Receiving over 200 inches of rain each year, this area maintains a lush growth of mosses, ferns, epiphytes, and trees.

6 The towering Alaska range overlooks the coniferous forests of the taiga, a biologically rich biome that extends across North America and much of Europe and Asia. Denali National Park, Alaska.

7 The taiga consists of large stands of one or two species of coniferous tree. Meadows occur naturally in mountain valleys, as shown here. Jasper Park, British Columbia.

8 An autumn view of the temperate deciduous forest biome.

9 Golden aspens herald autumn in the Maroon Bells-Snowmass Wilderness area, Colorado. High altitude yields a mix of coniferous and deciduous trees.

4

2

3

5

6

7

8

9

10 Alpine tundra is characterized by a short growing season, prohibiting the growth of trees or shrubs. Wildflowers, like the columbine shown here, grow densely in these areas.

11 Summery wildflowers and lichens add color to arctic tundra in the Soviet Union. Despite minimal precipitation, tundra areas remain damp from the presence of permafrost— permanently frozen soil that resists water absorption.

12 Flint Hills Preserve in Alena, Kansas, is one of the last remaining examples of the tall-grass prairie.

13 The African savannah is richly populated by grazing species that feed off trees, grass, and herbs. Kenya, Masai Mara Game Reserve.

10

12

11

13

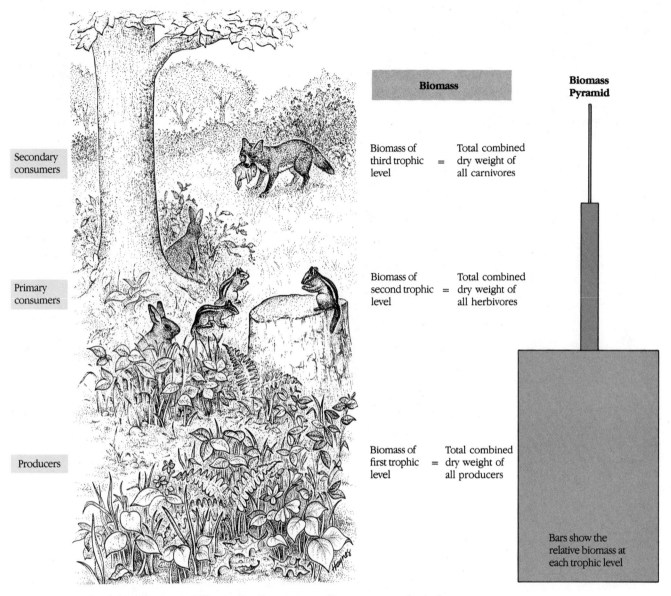

Secondary consumers

Primary consumers

Producers

Biomass

Biomass Pyramid

Biomass of third trophic level = Total combined dry weight of all carnivores

Biomass of second trophic = Total combined dry weight of level all herbivores

Biomass of first trophic = Total combined dry weight of level all producers

Bars show the relative biomass at each trophic level

Figure 2.12 *Biomass pyramid illustrating the amount of biomass at each trophic level in a food chain.*

animals, and microorganisms. If the energy content of biomass is graphed according to trophic level, it also forms a pyramid, a **pyramid of energy** (Figure 2.13).

The loss of energy from one trophic level to the next is determined in part by the second law of thermodynamics. This law, therefore, limits the length of food chains. As a rule, food chains usually have no more than four trophic levels because the amount of biomass (and energy) at the top of the trophic structure is not sufficient to support another level.

Because biomass and energy decrease from the base to the top of a food chain, so does the number of organisms. Thus, when the number of organisms in a food chain is graphed by trophic level, another pyramid

forms. This one is known as a **pyramid of numbers.** Figure 2.14 illustrates a pyramid of numbers for a grassland community in the summer and shows that over twice as many herbivores (trophic level 2) can be supported in a grassland biome as carnivores (trophic levels 3 and 4). Carnivores, in fact, always have the smallest populations in ecosystems. Thus, they are often the organisms in a food web most easily disturbed by human activities that disrupt the lower trophic levels. In a hearing before the Senate in 1991, Susan Weiler, head of the American Society of Limnology and Oceanography, testified that studies in Antarctica have shown that **phytoplankton** (algae and other free-floating photosynthetic organisms) populations decrease about 6% to 12%

Figure 2.13 *Pyramid of energy. Note the rapid decrease in potential energy as one ascends the food chain. Part of the loss conforms to the second law of thermodynamics, which says that energy is degraded as it is converted from one form to another.*

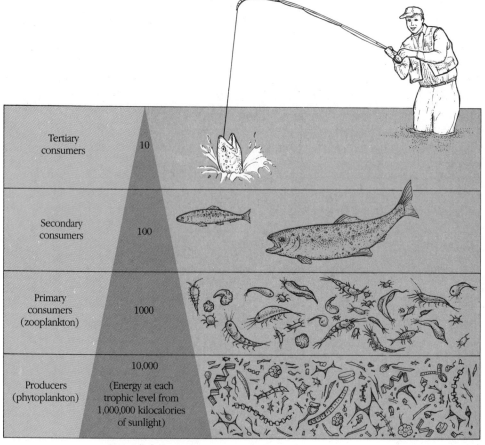

Tertiary consumers	10	
Secondary consumers	100	
Primary consumers (zooplankton)	1000	
Producers (phytoplankton)	10,000 (Energy at each trophic level from 1,000,000 kilocalories of sunlight)	

Pyramid of energy

when ozone concentrations in the upper atmosphere over the region drop by 40%, a common occurrence in the cold Antarctic spring. Since phytoplankton form the base of the aquatic food chain, some scientists are concerned that damage to them could cause widespread problems in fish, penguin, and seal populations in the region.

Another important implication of these ecological pyramids, which result from limits in energy conversion, has to do with feeding the world's people. Since more biomass is available at lower trophic levels, some people believe that efforts to feed the world's people, especially in developing countries, should focus on providing essential nutrients through vegetarian, rather than meat, diets. To illustrate this concept, if 20,000 kilocalories worth of corn were fed to a steer, 2000 kilocalories of beef would be produced (using a 10% conversion). This would feed only one person, assuming that a person can survive on 2000 kilocalories per day. However, if the 20,000 calories of corn were eaten directly, it would feed ten people per day. To improve food supply, then, it makes more sense to increase supplies of grain rather than meat. (Chapters 6 and 7 offer additional solutions to world hunger, among them population control.)

Productivity Economists measure the output of factories and mines in terms of productivity: how much steel is produced and how much coal is mined per hour of labor. In ecosystems, ecologists measure productivity in a similar way. The most common measure is kilocalories of biomass per square meter of surface area per year (kcal/m²/year). In an ecosystem, productivity is the rate at which sunlight energy is converted into the potential energy of biomass. The overall rate of biomass production is called the **gross primary productivity (GPP)**. Like a worker's gross pay, GPP is subject to some deductions. For plants, the chief deduction comes in the form of energy used to meet their own needs. Therefore, by subtracting the biomass broken down to release energy, called **cellular respiration (R)**, ecologists arrive at the **net primary productivity (NPP)**. Much like your net pay, NPP is what's left over after deductions. The simple mathematical equation for net primary productivity is

$$NPP = GPP - R$$

Table 2.1 Net Primary Productivity of Various Ecosystems

Ecosystem	Area (million km²)	Average Net Primary Productivity per Unit Area (g/m²/yr)*	World Net Primary Productivity (billion tons/yr)*
Continental Ecosystems			
Tropical rain forest	17.0	2200	37.4
Tropical seasonal forest	7.5	1600	12.0
Temperate evergreen forest	5.0	1300	6.5
Temperate deciduous forest	7.0	1200	8.4
Boreal forest	12.0	800	9.6
Woodland and shrub land	8.5	700	6.0
Savanna	15.0	900	13.5
Temperate grassland	9.0	600	5.4
Tundra	8.0	140	1.1
Desert and semidesert scrub	18.0	90	1.6
Extreme desert, rock, sand, and ice	24.0	3	0.07
Cultivated land	14.0	650	9.1
Swamp and marsh	2.0	2000	4.0
Lake and stream	2.0	250	0.5
Total continental	**149.0**	**773**	**115.0**
Marine Ecosystems			
Open ocean	332.0	125	41.5
Upwelling zones	0.4	500	0.2
Continental shelf	26.6	360	9.6
Algal beds and reefs	0.6	2500	1.6
Estuaries	1.4	1500	2.1
Total marine	**361.0**	**152**	**55.0**
Total biosphere	**510.0**	**333**	**170.0**

*Units of measure are dry grams and dry metric tons of organic matter.
Source: Adapted from Whittaker (1975). *Communities and Ecosystems* (2d ed.). New York: Macmillian.

Table 2.1 lists the NPP of the Earth's biomes and aquatic life zones. As you can see in column 3, the most productive terrestrial biomes are forests (top four listings), swamps, and marshes. The most productive marine life zones are algal beds, coral reefs, and estuaries.

The importance of each biome to total global biomass production, however, depends on the total area it occupies. Taking into account both productivity and surface area, the largest producers of biomass are the open ocean, tropical forests, and savanna (tropical grasslands).

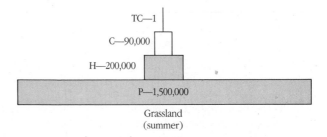

Figure 2.14 *Pyramid of numbers for a grassland community in the summer. The numbers represent individuals per 1000 square meters. P = producer, H = herbivore, C = carnivore, TC = top carnivore (highest consumer in a food chain).*

Viewpoint 2.1

Ecology or Egology? The Role of the Individual in the Environmental Crisis

Peter Russell

The author has been closely involved with the development of the Learning Methods Group, an international organization that helps people make fuller use of their mental potential. His books include The Brain Book, The Creative Manager, The Global Brain, *and* The White Hole in Time.

The environmental sciences usually focus on understanding the intricate relationships and interdependencies that have evolved between the millions of organisms inhabiting the Earth. Rather than looking further into the many facets of these sciences, I wish to explore this question: "Why are such studies necessary in the first place? Why is it that one species out of millions can disrupt the natural balance in so many ways and with such dire consequences?"

The rapid and liberal development of technology is clearly part of the problem. *Homo sapiens* has always been a manipulative species. A million years ago, long before civilization appeared, our opposable thumbs and large brain singled us out as the creature most capable of modifying its environment.

Yet, it is also clear that technological might is not the sole cause of environmental mismanagement. It is ultimately human beings who initiate the actions that result in ecological disturbances. In some cases, disturbances are totally unforeseen. In other cases, impacts are foreseeable, but those concerned ignore the warnings or, worse, seek evidence to the contrary, displaying an apparent lack of care for other species, the biosystem as a whole, and paradoxically, their own long-term welfare. How is it that people can adhere to policies that by nearly all projections look suicidal?

Such behavior stems from individuals not seeing beyond their short-term welfare and from perceiving their interests to be different from the interests of humanity. For most people, an immediate personal fulfillment is more attractive than some distant long-term benefit. Education and social controls may help curb the pull of immediate gratification, but the pull is so strong that these are likely to be insufficient. But, what is it that makes us want to satisfy our short-term needs to the detriment of our long-term welfare? The answer appears to lie deep within our psyche.

In the more developed nations —those responsible for the major environmental problems—the basic needs for food, clothing, shelter, and health care are fairly well attended to. What emerges then is the need for psychological welfare, in particular, the need to be liked. We need people to recognize and reaffirm our worth, and we spend considerable time and effort fulfilling this need. Much of our activity is really a search for personal reinforcement. Some psychologists estimate that as much as 80% of all our actions may be motivated by this search. And when people stand back and listen to themselves, they often find that as much as 90% of their casual conversation is prompted by a need for approval.

One of the most common ways we try to win approval and prestige is through material possessions. We collect many of the various accoutrements of modern living—new cars, fashionable clothes, expensive furniture—not because we need them physically but because we need them psychologically.

This might not be so bad if our various possessions satisfied our psychological needs; but they don't. Our insecurity is rooted far deeper in our psyche. Instead of spotting the obvious flaw, we search for yet more things in the outer world that we hope might fill the inner gap.

Some implications of this for the way we treat the environment are obvious. We gobble up irreplaceable resources, with little regard for the long-term future, partly because the various products they are transformed into briefly satisfy our search for identity. Thus, consumers are as much exploiters of the environment as are corporations.

Consumers are not alone in the quest for security. Ultimately, thousands of people in the world are making decisions to exploit this or that particular resource, initiate industrial processes with severe environmental effects, and set in motion other activities that in one way or another upset the ecological balance. Such people are usually motivated by the need for approval and recognition from their peers and by the mistaken belief that financial security brings inner security. They are caught in the same trap as everyone else. The only difference is that they are more visible.

Another profound consequence of our inner insecurity is a lack of true caring, either for others or for the environment. Each of us, at our core, is a compassionate being capable of deep empathy and caring. If we can get back in contact with this deeper self, we can begin to experience compassion not only for other people but also for the rest of the world.

Letting go of our self-gratifying patterns of behavior, we find not only compassion but humility. We rediscover respect for life on Earth.

Our efforts to solve pressing environmental problems will remain incomplete until they take account of the human psychology that gives rise to them. The only approach that will be successful in the long term is one that attends both to the external environment and to the self.

Nutrient Cycles

The economy of a nation depends on the flow of goods from agricultural and industrial producers to the people, the consumers. The "economy" of nature is also based on the flow of materials from producers to consumers. In ecosystems, biomass and inorganic matter move from plants (producers) to animals (consumers). In nature, these nutrients flow in a cyclical fashion because materials are reused, that is, they are recycled. As pointed out in Chapter 1, recycling is one of the key principles of biological sustainability.

In contrast to nature, in modern industrial societies, materials tend to flow linearly, going from mines to production facilities to stores to our homes. When their useful life is up, most goods are shipped to landfills. During this progression of events, great quantities of waste are generated, very little of which is recycled. Thus, human society attempts to function by linear systems in a distinctly cyclic world, a feature many observers think has limited utility in the long run.

Although nutrients flow in cycles, energy flows unidirectionally through the biosphere from producer to consumer because energy cannot be recycled, as the second law of thermodynamics implies. In other words, all the energy that enters a food chain is eventually lost as heat, which is dissipated into space where it cannot be reused.

The cycles that move nutrients through the biosphere are known as **biogeochemical cycles** or **nutrient cycles**. As shown in Figure 2.15, nutrient cycles involve two general phases: the **environmental phase**, in which chemical nutrients reside in the soil, water, or air, and the **organismic phase**, in which nutrients become part of the living tissue of organisms.

Of the 92 naturally occurring elements, about 40 are essential to life. Six of these elements—carbon, oxygen, hydrogen, nitrogen, phosphorus, and sulfur—form 97% of the mass of all plants, animals, and microorganisms. Elements required in relatively large amounts are known as **macronutrients**. Others, such as iron, copper, zinc, and iodine, are needed in very small amounts and are referred to as **micronutrients**. Both macronutrients and micronutrients "cycle" through the food webs, reaching our dinner plates in foods, such as cereal, vegetables, bread, cheese, and fish.

This section examines three of the most important nutrient cycles: carbon, nitrogen, and phosphorus. The water cycle is discussed in Chapter 10.

The Carbon Cycle
The **carbon cycle** is responsible for recycling the carbon dioxide released by all living things as they break down organic food molecules to liberate energy. The carbon cycle depends in large part

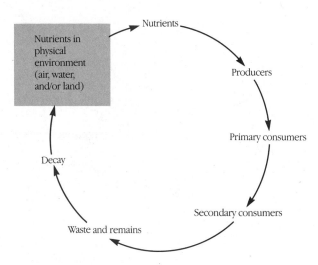

Environmental Phase   **Organismic Phase**

Figure 2.15 *Nutrient cycles have two basic phases: environmental and organismic. Humans are part of the organismic phase and influence both phases in profound ways.*

on photosynthesis. As noted earlier, during photosynthesis carbon dioxide absorbed by plants and algae is converted to food molecules with the aid of solar energy. The chief products of photosynthesis are oxygen and organic molecules, notably glucose, as shown below:

$$\text{carbon dioxide} + \text{water} + \text{sunlight} \rightarrow$$
$$(6\ CO_2) \qquad (6\ H_2O) \quad (\text{energy})$$

$$\text{glucose} + \text{oxygen}$$
$$(C_6H_{12}O_6) \quad (6\ O_2)$$

Oxygen and food molecules made by producers are essential to consumers. In fact, these organisms use oxygen to break down organic food molecules in body cells during **cellular respiration**. In this process, they release carbon dioxide, thus completing the carbon cycle as illustrated below:

$$\text{glucose} + \text{oxygen} \rightarrow$$
$$(C_6H_{12}O_6) \quad (6\ O_2)$$

$$\text{carbon dioxide} + \text{water} + \text{energy}$$
$$(6\ CO_2) \qquad (6\ H_2O)$$

A simplified version of the carbon cycle is shown in Figure 2.16. As illustrated, carbon dioxide in the air is absorbed by plants in terrestrial and aquatic ecosystems. Carbon dioxide molecules combine to form organic molecules such as glucose, which are passed along the strands of the food web from producer to consumer.

Carbon returns to the atmosphere in several ways. In plants and animals, for instance, some of the organic molecules are broken down to generate energy during

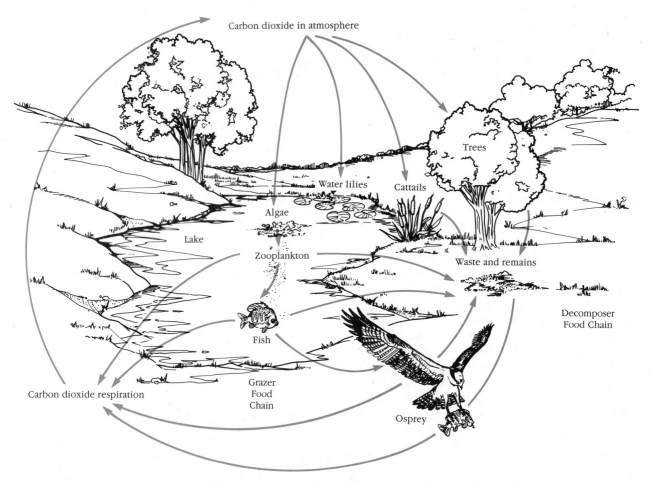

Figure 2.16 *The carbon cycle. Carbon cycles back and forth between the environment (air and water) and organisms. As illustrated here, it is first trapped by plants, then passed to herbivores, carnivores, and decomposers. Carbon dioxide is released by the breakdown of organic materials during energy production in these organisms.*

cellular respiration, which, as shown earlier, is the reverse of photosynthesis. The carbon dioxide gas released during cellular respiration reenters the environmental phase of the cycle for reuse.

Carbon also reenters the atmosphere via the decomposer food chain. As Figure 2.16 shows, the wastes and remains from plants and animals become food for decomposer organisms. These organisms liberate carbon dioxide through cellular respiration. Carbon dioxide is also released when plants are burned by natural causes, such as lightning and forest fires, or as a result of human activities, such as combustion of wood and coal (not shown in Figure 2.16).

Ecology books typically illustrate the details of the carbon cycle as if it were somehow isolated from human life. Nothing could be further from the truth. In fact,

you could easily remove the osprey from Figure 2.16 and insert a drawing of a human being. Nothing would change, except perhaps our appreciation of the importance of this cycle to human life.

Important as it is, the global carbon cycle has been badly altered by deforestation and fossil fuel combustion. By removing forests and vegetation that use atmospheric carbon dioxide to make organic molecules and by liberating carbon dioxide during the combustion of coal, oil, and natural gas—carbon sources that were once locked deep beneath the Earth's surface—humans have increased the global carbon dioxide concentrations by nearly 25% since 1870. As you learned in Chapter 1, further increases could have a devastating effect on global climate, our economy, and all life on Earth (Chapter 16).

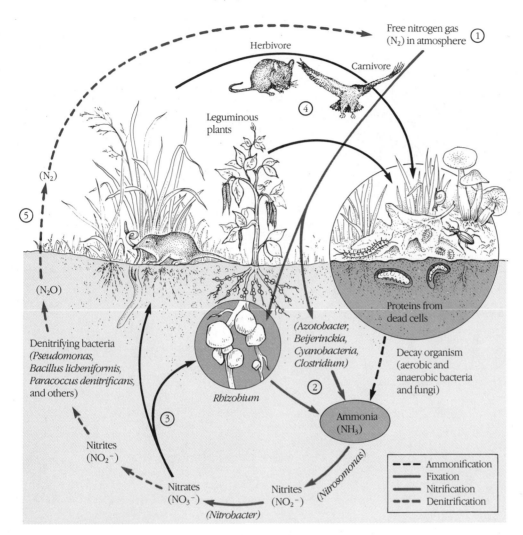

Figure 2.17   *The nitrogen cycle. (1) Nitrogen gas is abundant in the atmosphere, but it can't be used by plants or animals in this form. (2) It must first be converted to ammonia and nitrates by bacteria in the soil and in the roots of certain plants. (3) It can then be incorporated by plants and used to make amino acids. Herbivores acquire the amino acids they need from plants, and carnivores get them from herbivores on which they feed. (4) Nitrogen is returned to the soil in the urine and feces of animals or when dead plant and animal material decays. (5) Some nitrate is converted back into atmospheric nitrogen by denitrifying bacteria.*

The Nitrogen Cycle   Nitrogen atoms are found in many essential organic molecules, such as amino acids (the building blocks of proteins) and the genetic materials RNA and DNA. Although 79% of the air is nitrogen gas (N_2), plants and animals cannot use it in this form. To be usable, it must first be converted into ammonia (NH_3) or nitrate (NO_3).

The conversion of atmospheric nitrogen into ammonia is called **nitrogen fixation**. It occurs primarily in certain bacteria and other microorganisms in the soil and water (Figure 2.17, no. 1). Without these organisms, life as we know it could not exist. One nitrogen-fixing bac-

terium, *Rhizobium*, actually invades the roots of a group of plants known as legumes, which includes beans, peas, alfalfa, clover, and others. The roots respond by forming tiny nodules that serve as sites for nitrogen fixation. Nitrogen compounds can also be formed in the soil. In either case, usable nitrogen compounds are taken into plants where they are used to synthesize amino acids, DNA, and RNA. Animals, in turn, receive the nitrogen they need by eating plants and other animals.

Nitrogen is returned to the soil by the decay of detritus (Figure 2.17, no. 4). Within the soil, certain species of bacteria and fungi decompose the nitrogen-

Figure 2.18 *The phosphorus cycle has two interconnected parts: a terrestrial portion and an aquatic portion.*

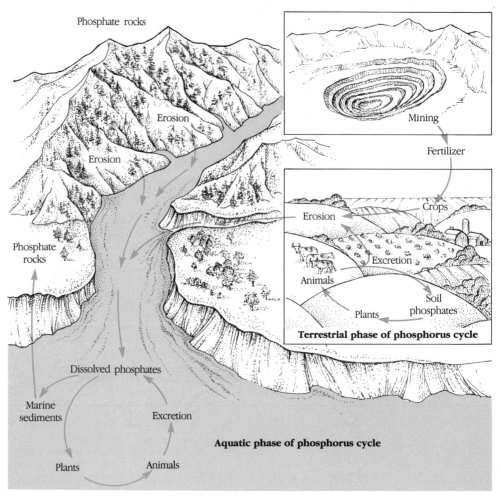

rich wastes from plants and animals. Nitrogen is released from wastes as ammonia and **ammonium salts** (NH^+). Ammonia is further converted into nitrites (NO_2^-), then into nitrates (NO_3^-). Ammonium salts, nitrites, and nitrates can be incorporated by the roots of plants and reused. Additionally, some nitrite is converted into a gas, **nitrous oxide** (N_2O), and released into the atmosphere.

Nitrogen travels from air to soil to plants to animals and then returns to soil and atmosphere in the **nitrogen cycle.** Each nitrogen atom in your body, in fact, came from the plants and animals you consumed, and was once atmospheric nitrogen made available to plants by soil microbes.

The nitrogen cycle, like the carbon cycle, is greatly influenced by human activity. The most significant impacts come from pesticide use, an activity that sometimes destroys soil microbes essential to nitrogen fixation. Nitrogen depletion of the soils may also result from repeated plantings of crops such as corn, which absorb large amounts of nitrogen from the soil. Farmers can replace nitrogen robbed from the soil by applying artificial fertilizers containing nitrates. If applied in excess, though, nitrogen fertilizers may wash into streams, causing serious water pollution (Chapter 16).

The Phosphorus Cycle   Phosphorus (P) is present in living organisms as phosphate (PO_4), which is found in RNA and DNA molecules as well as in fats (phospholipids) in cell membranes.

As illustrated in Figure 2.18, phosphate is slowly dissolved from rocks by rain and melting snow and is carried to waterways. Dissolved phosphates are incorporated into aquatic plants, then passed to aquatic animals that eat them. In terrestrial ecosystems, phosphates also occur in soil and are incorporated directly by plants and passed to terrestrial animals. Phosphorus reenters aquatic and terrestrial environments principally in two manners: (1) some is excreted directly by animals, and (2) some is returned when detritus decays.

Each year, large quantities of phosphate are washed into the oceans, where much of it settles to the bottom and is incorporated into marine sediments. These sediments release some of the phosphate needed by aquatic organisms, but much of it becomes buried and is removed from circulation. New phosphate must come from the sources described above.

In addition to nitrogen, phosphate is a major component of artificial fertilizers. When excess fertilizer is applied to fields, farmers can alter the phosphorus cycle.

Since phosphorus is a limiting factor in lakes and reservoirs, excesses entering from farmland runoff may cause rapid growth of algae and other aquatic plants, as discussed in detail in Chapter 7.

This chapter presented an overview of the structure and function of ecosystems with special emphasis on two of the biological principles of sustainability—recycling and renewable resources. This discussion also showed the importance of ecosystems to our lives. As the end quote of this chapter reminds us, all living things are part of the entire web of life. What we do to it, we do to ourselves. By the same token, planet care is the ultimate form of self-care.

Nothing can survive on the planet unless it is

a cooperative part of larger global life.

—Barry Commoner

☙   ☙   ☙

Critical Thinking Exercise Solution

This book will show that many of the environmental problems facing the world are very serious. Although some proponents do indeed exaggerate and resort to scare tactics, in most cases, they're honest and straightforward. Facts are generally based on good science.

Consider the claim that environmentalists don't care about the environment or don't care about people and that they're simply out to destroy the economy. I suspect that you consider yourself an environmentalist and that you care deeply about the Earth and people. Many other environmentalists you know probably feel the same. You and your acquaintances are, therefore, partial proof that allegations about environmentalists made by the speaker are false.

This brief example shows that it is important to question generalizations regarding entire groups of people. Sure, there are some wacky environmentalists who may not care about the economy or people very much, but they're few and far between. Most environmentalists recognize that a healthy economy is essential to environmental protection. Taking care of the Earth can be a daunting task when an economy is staggering.

In this exercise, critical thinking skills call on us to question the conclusions of the speaker to see if they are supported by the facts. That probably also led you to question the source, whose biases clearly show up here.

Summary

2.1 How Is the Living World Organized?

■ **Ecology** is the study of ecosystems. It concerns itself in large part with the interactions between organisms and their environment.

■ The **biosphere** is that thin zone of life that exists at the interface of air, land, and water. It is a closed system in which all the materials essential to life are recycled over and over.

■ The biosphere is divided into **biomes**, distinct regions each with unique plant and animal life. The boundaries of biomes are determined by the climate, especially precipitation and temperature.

■ Aquatic life zones are the equivalent of biomes but exist in marine and freshwater environments.

■ **Ecosystems** consist of organisms, their environment, and all the interactions that exist between them. All ecosystems consist of two components: **abiotic** (nonliving) and **biotic** (living).

■ All organisms in ecosystems survive within a range of chemical and physical factors, the **range of tolerance**. When the upper or lower limits of the range are exceeded, survival is threatened.

■ In ecosystems, one abiotic factor usually limits growth. It is called a **limiting factor**.

■ Organisms comprise the biotic components of ecosystems. Each organism within a community lives in a specific region, its **habitat**. An organism's **niche** includes its habitat and all its relationships with the abiotic and biotic components of the environment.

■ Organisms may be generalists or specialists. **Generalists** occupy many different habitats and eat a wide variety of foods. **Specialists** generally live in one habitat and consume one or only a few organisms, making them more prone to extinction.

2.2 How Do Ecosystems Work?

■ A **food chain** consists of a series of organisms, each feeding on the preceding one. Food chains are conduits for the flow of energy and nutrients in ecosystems.

■ **Grazer food chains** begin with plants that are consumed by grazers. **Decomposer food chains** begin with the waste and remains of plants and animals, **detritus**.

■ Individual food chains rarely exist in isolation; most are parts of more complex **food webs**.

■ In all ecosystems, organisms are either **producers** or **consumers**. Producers make organic food molecules (**biomass**) that support the entire living world. The energy contained in biomass comes from the sun.

■ Energy is governed by the **laws of thermodynamics**. The first law states that energy is neither created nor destroyed but only converted from one form to another.

■ The second law states that when energy is converted from one form to another, it is degraded, that is, converted from a more concentrated to a less concentrated form.

■ Biomass is transformed from one **trophic level** in food chains and food webs to the next. Graphically represented, biomass at the different trophic levels forms a **pyramid of biomass**.

■ Since biomass contains energy, the amount of potential energy at higher trophic levels also decreases and forms a **pyramid of energy**.

■ The number of organisms also decreases from one trophic level to the next higher one; this forms a **pyramid of numbers**.

■ **Productivity** is the measure of biomass production. Regions with high productivity, such as tropical rain forests and estuaries, are important to humans and other species as food sources.

■ Organic and inorganic matter move within ecosystems in **nutrient cycles**. Examples are the carbon, nitrogen, and phosphorus cycles. These essential cycles can be altered by human activities.

Discussion and Critical Thinking Questions

1. What is ecology?

2. Define the term *biosphere*. Why is the biosphere considered a closed system?

3. Based on your knowledge, do you agree with the following statement? Because the Earth is a closed system, long-term human success depends in large part on recycling virtually all the materials we use.

4. Define the term *biome*. What determines the kind of vegetation in a biome? After studying the biome map in this chapter (Figure S2.1), name the biomes that you have visited.

5. Define the term *ecosystem*.

6. Do you agree with the following statement? Building a sustainable society requires measures to shift to renewable resources. Why or why not?

7. Discuss the concept *range of tolerance*. Draw a graph showing the optimum range, zones of physiological stress, and zones of intolerance. Explain each one.

8. Give some examples of ways in which people alter the abiotic conditions in ecosystems to the detriment of other species and humans.

9. One of the requirements of living sustainably on the planet is minimizing our alteration of the abiotic conditions of the environment. Do you agree with this idea? Why or why not?

10. What is a limiting factor? What are the most important limiting factors in terrestrial ecosystems?

11. How are a habitat and a niche different?

12. In what ways are humans specialists? How are we generalists?

13. What is a food chain? What are the two major types and how are they different? How are they similar?

14. Sketch several simple food chains and indicate all producers and consumers. Also indicate the trophic level of each organism. Can one organism occupy several trophic levels? Give an example.

15. Why does biomass decrease as one ascends the food chain? What are the implications of decreasing biomass in the food chain?

16. Draw the carbon cycle and describe what happens during the various parts of the cycle. How are humans dependent on the carbon cycle?

17. What impacts in the global carbon cycle would you expect from massive tropical deforestation?

18. Why are nitrogen-fixing bacteria vital to life on Earth? In what ways do we alter the nitrogen cycle?

Suggested Readings

Chiras, D. D. (1992). *Lessons from Nature: Learning to Live Sustainably on the Earth*. Washington, D.C.: Island Press. Discusses how to apply the biological principles of sustainability to modern society.

Colinvaux, P. (1986). *Ecology*. New York: Wiley. An excellent textbook of basic ecology.

Ehrlich, P. R. and Roughgarden, J. (1987). *The Science of Ecology*. New York: Macmillan. Higher-level coverage of ecology.

Odum, E. P. (1989). *Ecology and Our Endangered Life-Support Systems*. Sunderland, MA: Sinauer. Excellent introduction to ecology.

Peet, J. (1992). *Energy and the Ecological Economics of Sustainability*. Washington, D.C.: Island Press. A broad-ranging book that examines the current political-economic view and an ecological-sustainable view.

The Biomes

Look out the window. You probably see tree-lined streets, homes or campus buildings, neatly tended lawns, or possibly a parking lot. The vegetation growing around you is probably not "natural" at all. Grasses may have been imported from Kentucky. The trees may have come from Norway or China. Even some of the birds, such as starlings and house sparrows, are aliens, introduced from England. To glimpse natural vegetation, you may have to journey outside your town. There you may find grassland, forest, or desert that resembles the environment before humanity began to reshape it.

Grasslands, deserts, and forests are three of the relatively homogeneous zones called biomes. A biome, as we have seen, is a large region with its own distinctive climate, plant and animal life, and soil type. A dozen biomes spread over millions of square kilometers, spanning entire continents (Figure S2.1). This supplement looks at some of the major world biomes and the way human activities alter them.

Tundra

The **tundra** is a vast, virtually treeless plain on the far northern borders of North America, Europe, and Asia (Gallery 2, Figure 11). It lies between a region of perpetual ice and snow to the north and a band of coniferous forests (the taiga) to the south. It is one of the largest biomes, covering about one-tenth of the Earth's surface.

The tundra receives very little precipitation (less than 25 centimeters, or 10 inches, per year), most of it during the summer. Contrary to popular belief, very little snow falls during the long, cold winter.

The gently rolling tundra is dominated by herbaceous plants (grasses, sedges, rushes, and heather), mosses, lichens, and dwarf willows. Deep-rooted plants, such as trees, cannot grow because much of the subsoil remains frozen year-round, hence the name **permafrost**. Only about 10 centimeters (6 inches) of soil thaw in the summer months, creating a very shallow root zone for plants. Most tundra plants are stunted because of the short growing season, the annual freeze-thaw cycle, which tears and crushes roots and impairs growth, and windblown ice and snow, which abrade vegetation.

The summer days are long and warm, and the land becomes dotted with thousands of shallow lakes, ponds, and bogs because of the permafrost, which hinders drainage. Millions of birds come north to nest and feed on the tundra's abundant insects.

In North America, musk ox, caribou, and a variety of small rodents and mammals inhabit the tundra year-round (Figure S2.2). They share their habitat with snowy owls and ptarmigan (ground-dwelling birds similar to grouse).

The tundra is an extremely fragile biome, containing few species, a condition believed to make it more vulnerable to change (Chapter 4). Vegetation takes decades to regrow after it has been destroyed by vehicles. Large-tired trucks and new vehicles that ride on a layer of air are now being used experimentally and may help to reduce human impact on the tundra.

Taiga

South of the tundra, extending across North America, Europe, and Asia, is a broad band of coniferous (evergreen) forests. This biome is called the **taiga** (Gallery 2, Figures 6 and 7). The average annual precipitation in the taiga is higher than in the tundra, as is the average daily temperature.

With a growing season of 150 days and a complete thawing of the subsoil, the taiga supports an abundance of life forms. Conifers are the dominant form of plant life and are well adapted to the long, cold, and snowy

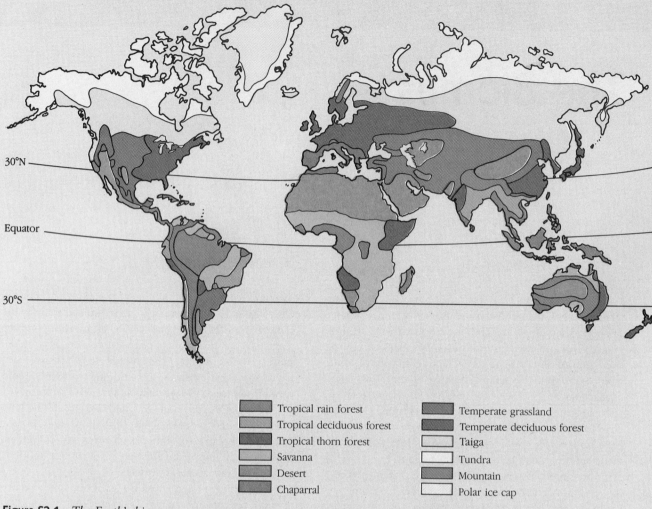

Tropical rain forest	Temperate grassland
Tropical deciduous forest	Temperate deciduous forest
Tropical thorn forest	Taiga
Savanna	Tundra
Desert	Mountain
Chaparral	Polar ice cap

Figure S2.1 *The Earth's biomes.*

Figure S2.2 *The musk ox endures the long, harsh winters on the shelterless arctic tundra.*

winters. Water loss is minimized by waxy coatings on their needles. Reduced evaporation is critical in the winter, when water transport from the root systems halts. The flexible limbs are an adaptation that allows conifers to bend without breaking when covered with snow. Some deciduous trees (ones that shed their leaves each year) inhabit the taiga in localized regions, particularly in areas that are recovering following fires or heavy timber cutting.

Animal life is abundant and diverse. Bears, moose, wolves, lynx, wolverines, martens, porcupines, and numerous small rodents inhabit the forests, along with a variety of birds, such as grouse, ravens, hairy woodpeckers, and great horned owls (Figure S2.3). Insects, such as mosquitoes, also thrive during the warm summer months. Over 50 species of insects that feed on conifers live in the taiga, including the spruce budworm, pine beetle, tussock moth, and pine sawfly. They can cause widespread damage during droughts, when trees' resistance is low.

Numerous lakes, ponds, and bogs spot the landscape. In North America, trappers once traveled by boat into the interior of the taiga on the interconnected waterways in quest of beaver and other fur bearers. The taiga has long been of interest to the logging industry. Through past practices, loggers have stripped large sections of land by clear-cutting forests. Until recently, little or no effort was made to replant denuded areas, resulting in severe soil erosion, destruction of wildlife habitat, and pollution of streams and lakes with sediment.

Figure S2.3 *The moose makes its home in the taiga, feeding off vegetation.*

Temperate Deciduous Forest

The **temperate deciduous forest** biome is located in the eastern United States, Europe, and northeastern China. A region with a warm, mild growing season and abundant rainfall, the temperate deciduous forest supports a wide variety of plants and animals (Gallery 2, Figure 8). The dominant vegetation consists of broad-leafed deciduous trees, such as maple, oak, and hickory. However, the dominant plant species can vary from region to region, depending on the amount of precipitation.

Temperate deciduous forests support a variety of organisms, including white-tailed deer, opossums, raccoons, squirrels, chipmunks, foxes, rabbits, black bears, mice, shrews, wrens, downy woodpeckers, and owls. During the summer, warblers and other bird species migrate into and nest in the rich, green forests.

Deciduous trees act as "nutrient pumps," drawing inorganic chemical nutrients from the subsoil up through the roots to the leaves. When the leaves fall and decay, they add valuable organic and inorganic nutrients to the superficial layers of soil. These nutrients are important for plant growth.

The temperate deciduous forest biome has been extensively exploited by humans. In North America, this biome once extended from the Mississippi River to the Atlantic Coast. Now, however, only about 0.1% of the original forest remains; most has been cleared for farms, orchards, and cities. Early settlers often failed to practice good soil conservation methods on their farms, letting the land erode. Farmers then moved westward into virgin territory, where they continued the same practices. Soil erosion continues today on America's farmland (Chapter 7).

Temperate Grasslands and Savanna

Grasslands exist in both temperate and tropical regions where rainfall is relatively low or uneven. The **temperate grasslands** in North America occupy a region known as the Great Plains, which extends from the Rocky Mountains to the Mississippi River. Moist regions of the Great Basin, lying between the Sierra Nevada and the Rockies, also support grasslands. Temperate grasslands are found in South America, Australia, Europe, and Asia; tropical grasslands, or **savanna**, exist in Africa and South America.

Temperate and tropical grasslands are remarkably similar in appearance (Gallery 2, Figures 12 and 13). Both experience periodic drought and are characterized by flat or slightly rolling terrain. Large grazers, such as the bison in North America and the zebra in Africa, feed off the lush grasses in this biome. The native grasses are well adapted to drought as their roots penetrate deep into the subsoil where they can always find water.

Because their soils are rich in inorganic and organic nutrients, grasslands have been widely used for agriculture. As discussed in Chapter 7, wind and rain take their toll on grassland soils that are farmed improperly.

Desert

Deserts are found throughout the world. The Sahara, the largest desert in the world, stretches across the African continent and is nearly as large as the United States. In North America, deserts lie primarily on the downwind side of mountain ranges. The reason is that warm, moist air flowing toward the mountains is propelled upward when it comes in contact with the mountain range. As the moist air ascends, it is cooled, and water vapor in the air condenses, forming droplets too large to remain in suspension. These droplets form rain, which falls mostly on the windward side, leaving the downwind side dry. When rain does fall in a desert, it is often intense and frequently results in flash floods and severe soil erosion. Water evaporates quickly from the soil because the vegetation is sparse and because of the high temperatures reached on the desert floor in the summer (Gallery 2, Figures 1 and 3).

Contrary to the view held by many people, most deserts are not lifeless lands. A surprising variety of plants and animals have adapted to the desert's unique environmental conditions. Plants include cacti, other succulents, and shrubs (mesquite, acacia, greasewood, and creosote bush). Small, fast-growing annual herbs are also found. Many of these are wildflowers that bloom only in the spring or after drenching rains. Such sudden bursts of growth can turn the desert into a colorful landscape almost overnight (Gallery 2, Figure 3).

Plants have developed a number of adaptations to cope with the dry conditions of the desert. These include: (1) thick, waterproof outer layers, which reduce water loss; (2) an absence of leaves or a reduction in leaf size, both ways to reduce the surface area from which water can evaporate; (3) hairs and thorns, which reflect sunlight and shade the plants; (4) the ability to drop leaves (a strategy used by the ocotillo) when moisture levels are low; (5) extensive, shallow root systems to absorb as much water as possible during cloudbursts; (6) deep taproots, which penetrate into groundwater; (7) succulent water-retaining tissues to store water; (8) recessed pores (stomata), which reduce water loss; (9) wide spacing between plants to reduce competition for available water; and (10) short life spans, which allow plants to develop to maturity quickly following rainstorms.

A surprising number of animals also live in the desert. Like the plants, they have evolved a number of strategies to survive in the harsh climate. Many animals, such as the ring-tailed cat, are active only at night or during the early and late hours of the day, when the heat is less intense. Snakes have thick scales that prevent water loss. The kangaroo rat conserves body moisture by excreting a solid urine containing nitrogen wastes (in the form of uric acid). Some species obtain water from vegetation, and others from the blood and tissue fluids of their prey. Gila monsters combat the heat by remaining underground during the hottest daylight hours.

Large cities have sprung up in the American desert. Farmers have plowed the poor soil to plant their crops, but the desert soil contains little of the organic matter needed to retain moisture and only small amounts of the nitrogen required for plant growth. Consequently, crops are successful only if nitrogen and water are provided. City dwellers and farmers now compete for the limited water in the American West. With a rising population and years of irrigation, water supplies have begun to decline, forcing many farms to shut down.

Human populations have expanded the borders of the world's deserts by allowing livestock to overgraze, deforesting lands, and following poor agricultural practices. The Sahara's southern boundaries are spreading into the Sahel region because of intense grazing in bordering grassland and a shift of rainfall northward, possibly as a result of global warming. Each year millions of acres of land are engulfed by spreading deserts.

Tropical Rain Forest

Tropical rain forests are located near the equator on several continents and islands. With an average annual temperature of approximately 18° C (64° F) and rainfall of 200 to 400 centimeters (80 to 160 inches) per year, the tropical rain forest is one of the richest and most complex biomes (Gallery 2, Figure 4).

The dominant vegetation consists of trees that tower as high as 50 to 60 meters (165 to 200 feet) above the forest floor. The tops of these trees interlace, forming a dense canopy that blocks out much of the sunlight. Smaller trees further reduce light penetration. Only 1% of the sunlight reaches the forest floor, and, therefore, only a few plants adapted to low light can grow on the ground.

The tropical rain forest is well known for its species diversity. As many as 100 tree species are in a single hectare, compared to a dozen or so in a northern coniferous forest of the same size. The rain forest also supports a diverse array of insects, birds, and other animals. There are as many species of butterflies (500 to 600) on a single tropical island as in the entire United States.

Since the ground is bare, many of the animals and insects live in the treetops, where the food is. The treetops are without doubt the most heavily and diversely populated regions of the tropical rain forest.

Woody vines grow in the rain forest, and a large number of plants known as epiphytes live among the branches of the taller trees where sunlight is available.

Epiphytes, like Spanish moss, which grows in the southern United States in temperate deciduous forests, gather moisture and nutrients from the humid air. They have no roots and need no soil. Researchers recently discovered 12 species of trees in the Amazon rain forest whose roots originate in the soil but turn upward to climb up the trunks of other trees. There, it is believed, they scavenge nutrients from rain since the soil is so poor.

Trees of the tropical rain forest have shallow root systems because most of the nutrients are near the surface and because rainfall is abundant, which makes deep root systems unnecessary. Since their root systems are

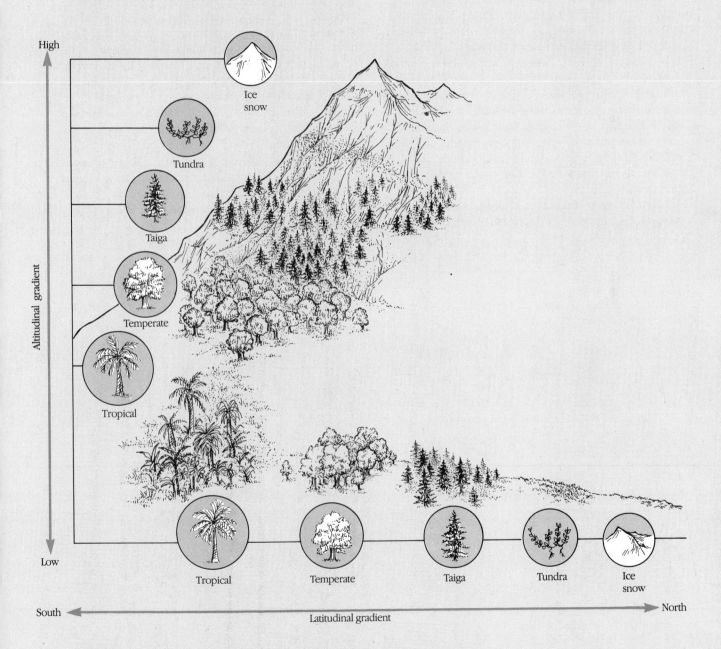

Figure S2.4 *Vegetation varies with altitude in any given region in direct response to climate changes with altitude. Altitudinal gradients mirror latitudinal gradients.*

so shallow, many trees develop wide bases known as buttresses to prevent them from toppling over.

The soils of the tropical rain forests are thin and extremely poor in nutrients. In fact, almost all of the inorganic and organic nutrients found in this biome are tied up in the vegetation. Plant and animal wastes are rapidly decomposed on the forest floor in the decomposer food chain; organic matter does not build up as it does in deciduous forests and grasslands. Minerals returned during decomposition are rapidly absorbed by the roots of trees or are leached from the soil by rain and carried to the groundwater, where they are often unavailable to plants.

Because most are nutrient poor, rain forest soils are unsuitable for conventional agriculture. In addition, many of the soils in tropical rain forests are composed of a red clay known as laterite (from the Latin word *later* meaning "brick"), which forms an impenetrable crust one or two years after the forest is cleared to make farmland. A third problem is that once trees are removed to make room for farms, soils are easily washed away by the frequent rains. Despite these major drawbacks, people throughout the world still clear tropical rain forests for farming, grazing, and timber production. An area the size of the state of Washington is deforested each year.

Destruction of tropical rain forests is so extensive that ecologists project that this biome, which constitutes about half the world's forest, will be gone or at least severely damaged by the year 2000. With it, they assert, thousands of animal species will vanish.

Altitudinal Biomes

Hiking or driving through the Rocky Mountains, the Sierra Nevada, or the Cascade Mountains reveals an interesting shift in plant communities similar to that seen when traveling north from the grasslands through Canada's coniferous forests to the tundra (Figure S2.4). Why does this shift occur? Climbing up a mountain creates the same effect as moving northward; in fact, each 120-meter (400-foot) gain in altitude is like a trip one degree north in latitude. Rainfall and temperature vary with altitude, and, thus, so does the plant and animal life. In the high Rockies, climatic conditions are similar to those of the arctic tundra; so is the plant and animal life. Because of these similarities, high mountain regions are called alpine tundra. As in the arctic tundra, the growing seasons are short, and the winters are cold. Annual precipitation is fairly low. Plant and animal life is similar to that of the far northern tundra. Areas below the tundra resemble the taiga. The forests in these regions are dominated by spruce and fir.

Chapter 3

Principles of Ecology: Self-Sustaining Mechanisms in Ecosystems

And this our life, exempt from public haunt, finds tongues in trees, books in running brooks, sermons in stones, and good in everything.

—Shakespeare

Critical Thinking Exercise

You are working for the Department of Natural Resources in your state and are asked to comment on a proposal that would import a species of deer from Siberia into the state's wildlands. These areas are devoid of large grazers, such as deer, and are also devoid of predators. Currently, the lands are well vegetated and appear to be prime deer habitat. Proponents of this proposal think that it would enhance hunting in the state and would help improve the state's economy. How would you evaluate the proposal? What problems would you anticipate? What critical thinking rules helped you with this exercise?

In 1884, the water hyacinth was introduced into Florida from South America as an ornamental plant. Housed in a private pond, this lovely flowering plant, which grows on the water's surface, eventually escaped into a nearby river. From here, the water hyacinth spread like a cancer throughout the state's many canals and rivers. Aided by a remarkable ability to reproduce—10 plants can multiply to 600,000 in eight months—the hyacinth now chokes waterways throughout Florida, crowding out native plants and making navigation impossible in some areas (Figure 3.1).

Today, the water hyacinth inhabits the waterways of much of the southern United States, occupying nearly 800,000 hectares (2 million acres) of rivers and lakes from Florida to California. In the three states where the plant infestation is the worst, Florida, Louisiana, and Texas, officials spend nearly $11 million a year to keep waterways open.

The story of the water hyacinth shows one way aquatic ecosystems can be thrown out of balance by human action. Fortunately, the plant has some benefits. Today, it is used in commercial wastewater treatment plants to treat human sewage. (For a discussion, see Models of Global Sustainability.)

This chapter discusses how ecosystems respond to various kinds of stress, such as that caused by the water hyacinth. It looks at stresses created by human intervention as well as catastrophic natural events. This discussion provides insights into the operation of ecosystems, information that is vital to protecting ecosystems on which we and millions of other species depend. This chapter also explores the four remaining principles of sustainability—conservation, restoration, population control, and adaptability—which were introduced in Chapter 1. You will see how these factors function to sustain life on Earth and why they are important to our efforts to build a sustainable future.

3.1 Ecosystem Homeostasis

The human body is endowed with a number of physiological mechanisms that help maintain internal con-

Figure 3.1 *The water hyacinth was introduced accidentally into Florida waters and has proliferated at a tremendous rate, choking canals and streams and outcompeting many native species. This pesty species can be put to good use.*

Ecologically Sustainable Sewage Treatment Systems: Mimicking Nature

Residents of eastern Mexico City produce 300 tons of fecal matter every day, much of which is deposited on city streets, vacant lots, and back alleys. When the feces dry, they are often pulverized by cars and trucks and entrained in the city's dust, creating a monumental health hazard encountered in other poor countries as well.

Since the advent of cities and towns, human waste has proved a colossal challenge. Even in rich, industrialized countries, modern sewage treatment practices leave something to be desired. In most cases, solid sludge that remains after treatment is trucked off to landfills. The liquid waste remains are chemically treated and dumped into lakes and streams. In addition, traditional treatment methods waste valuable nutrients and pollute groundwater and surface waters unnecessarily. Sewage treatment is an example of a linear human system designed with disregard for the cyclic nature of the natural world.

Over the years, scientists and others have sought more environmentally compatible—and sustainable—ways to deal with human waste. For instance, ecologist John Todd, who has led this effort, invented a biological system that mimics nature. Todd designed and built his first sunlight-powered sewage treatment plant at Sugarbush ski resort in Vermont. Raw sewage enters a greenhouse, then flows through cylinders where bacteria convert the ammonia in the sewage into nitrate. The effluent then enters special channels where algae dine on the nutrients. Algae are eaten by freshwater shrimp, which are eaten by fish. Snails in the system consume the sludge and also serve as food for fish.

At the far end of the greenhouse is a marsh that filters the effluent before it is released into a nearby stream. The numerous plants in this artificial marsh each play a special role. Cattails and bulrushes, for example, absorb toxic substances. Iris roots secrete a compound that kills disease-causing *Salmonella* bacteria.

In Providence, Rhode Island, a much larger system is being tested by Todd. In this facility, up to 16,000 gallons of raw sewage flow through a greenhouse filled with 1200 aquariums containing organisms that purify the water. This system, which costs one-third as much as an equivalent sewage treatment plant, has negligible environmental impact.

In the mid-70s, while working at NASA, environmental engineer Bill Wolverton began experiments to purify wastewater from NASA facilities using lagoons filled with prolific, nutrient-hungry water hyacinths. One system he designed for a 4000-person NASA facility in Mississippi has saved the agency millions of dollars in sewage fees.

Since he began, Wolverton has designed more than 100 systems in the southern United States. Although most of his systems treat sewage from single-family dwellings, one, in Denham Springs, Louisiana, in use since 1987, treats 3 million gallons of municipal wastewater per day.

These biological treatment systems have met with considerable opposition. One concern is that they will not work in cold climates. To address this concern, Todd built greenhouses, which allow the systems to operate in cold northern climes year-round. His systems also use a variety of species, just as nature does. Some species are active on chilly days, some on hot days, some when it's sunny, some when it's cloudy. This design allows use of the systems in virtually any climate.

A second concern is that the systems cannot be used for large cities. Not so, say proponents. These systems can be scaled up to any size simply by adding more greenhouses or building groups of them. Moreover, biological treatment systems require about the same amount of land as conventional ones.

A third problem is that heavy metals, such as mercury and lead, which are absorbed by plants in these biological treatments plants, need to be landfilled. Todd's colleague, Alan Liss, however, notes: "It's better to have a small amount of highly toxic plants than a huge amount of moderately toxic sludge."

A fourth concern is over malfunction. Proponents of this new technology point out that malfunctions at conventional plants can be serious, sometimes requiring the evacuation of nearby residents to avoid poisonous chlorine gas emitted from them. A malfunction in a solar-aquatic plant may kill off some organisms but poses no threat to the surrounding area.

Biological treatment systems are well suited to developing countries. Because building conventional facilities is expensive, some countries are looking at biological systems that can be added to upgrade existing facilities. They're finding that the necessary upgrades can be made at moderate cost. Moreover, the facilities do not require as much management or use as much energy as conventional systems. They even produce cleaner treated water than their high-tech counterparts.

Because biological systems that mimic nature can be adapted for most situations, their use will undoubtedly increase, welcome news to those interested in building a sustainable future.

ditions essential to our day-to-day survival. Our bodies, for example, possess several mechanisms that maintain a fairly constant body temperature despite fluctuations in the environment. Maintenance of body temperature and other internal conditions within an acceptable range is referred to as **homeostasis** (from the Greek words *homeo*, which means "same," and *stasis*, meaning "standing"). Literally translated, homeostasis means "staying the same."

Homeostasis in our bodies is a **dynamic equilibrium**, a fairly steady state in which internal processes adjust for changes in external conditions. Although some fluctuations do occur, these processes ensure that internal conditions fairly constant. Body temperature, for example, fluctuates only slightly over a 24-hour period, generally falling at night when one sleeps and increasing in the day when one is active. Ultimately, internal conditions must remain within a tolerable range and wide swings detrimental to one's health must be avoided.

Ecosystems have similar mechanisms that help to maintain relatively constant conditions. Predators, for example, help hold prey populations in check. Disease organisms, such as bacteria and parasites, work similarly, helping to maintain the number of species within the carrying capacity of the environment.

Ecosystem homeostasis is readily observable to those who study tropical rain forests and other ecosystems such as coral reefs where conditions remain more or less constant year-round. It is less obvious in temperate climates, such as the deciduous forest biome, where seasonal climate changes result in dramatic shifts in conditions. Even though climatic conditions shift, these systems remain fairly constant from one year to the next. Therefore, if you studied a forest ecosystem near your home each spring for an extended period, you would find that: (1) the total number of species was fairly constant from year to year; (2) the same species were present each year; and (3) the population size of each species was approximately the same from year to year. Such a system is considered to be in a homeostatic condition.

Just as your body temperature varies, so do conditions within an ecosystem. The number of plants growing in a field in Florida or Alabama, for instance, may increase or decrease in response to natural variations in precipitation but remain fairly constant over long periods.

Ecosystems can generally "weather" small variations in environmental conditions as long as the variations occur within the range of tolerance (Chapter 2). When conditions shift out of this range, an ecosystem may begin to deteriorate, losing species that are vital links in food webs. Disruptive shifts result from natural events, such as volcanic eruptions, or some human activities, such as timber cutting, as described later in this chapter.

3.2 What Factors Contribute to Ecosystem Homeostasis?

Ecosystem homeostasis results from the interplay of many factors that influence populations within ecosystems. These factors can be broken down into two groups: **growth factors,** or those that tend to increase population size, and **reduction factors,** or those that tend to decrease it. As shown in Figure 3.2, these factors can be biotic or abiotic.

At any given moment, population size is determined by the interaction of the growth and reduction factors. Since ecosystems contain many species, the balance within an ecosystem can be crudely related to the sum of the individual population balances.

To understand this concept, let's examine growth

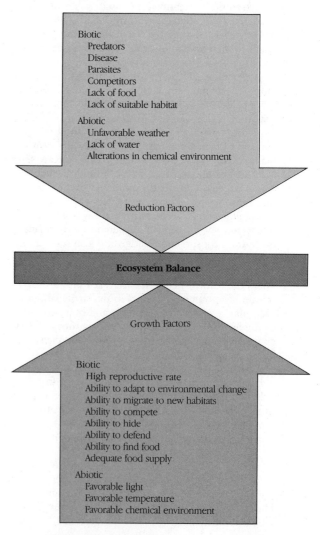

Biotic
  Predators
  Disease
  Parasites
  Competitors
  Lack of food
  Lack of suitable habitat
Abiotic
  Unfavorable weather
  Lack of water
  Alterations in chemical environment

Reduction Factors

Ecosystem Balance

Growth Factors

Biotic
  High reproductive rate
  Ability to adapt to environmental change
  Ability to migrate to new habitats
  Ability to compete
  Ability to hide
  Ability to defend
  Ability to find food
  Adequate food supply
Abiotic
  Favorable light
  Favorable temperature
  Favorable chemical environment

Figure 3.2 *Ecosystem homeostasis is affected both by forces that tend to increase population size and forces that tend to decrease it. Growth and reduction factors consist of biotic and abiotic components.*

and reduction factors in more detail. As shown in Figure 3.2, numerous biotic factors contribute to the growth of populations, for example, the ability to produce many offspring, to defend against enemies, or to blend into the environment. Favorable abiotic conditions also stimulate increases in population size. Ample sunshine, mild temperatures, and an abundance of rain, for example, promote plant growth. Because animals depend on plants, these conditions also promote increases in animal populations. Favorable biotic and abiotic conditions often team up to produce excessive growth. The success of the water hyacinth in southern waterways, for instance, is attributed to its high reproductive potential, its ability to use nutrients dissolved in surface waters, its ability to spread into new territories, and the favorable climate of the South.

In nature, all growth is opposed by a number of biotic and abiotic factors. Ecologists describe these factors collectively as **environmental resistance**. Predators, diseases, parasites, and competition are biotic factors that reduce population size, as do unfavorable weather and lack of food and water, abiotic conditions. When environmental resistance is low, organisms proliferate madly, throwing systems out of balance. The water hyacinth, for instance, faces little environmental resistance in the waters of the southern United States. Even plant-eating native fish are no match for its reproductive success.

The interaction of growth and reduction factors can result in sustainable population fluctuations in natural systems. Consider an example. Living in the grasslands of Kansas is the mouselike rodent known as the prairie vole. Its population size depends on many factors, among them light, rainfall, available food supply, predation, temperature, and disease. In the laboratory, the optimum conditions for reproduction are low temperature (slightly above freezing), long days (14 hours of light a day), and plenty of water and food. Under these conditions, a female will give birth to a litter of seven pups every three weeks for several years. For the captive-raised vole, motherhood is no picnic.

Fortunately, in the wild, optimum conditions never occur simultaneously. In the summer, for instance, when food, water, and day length are optimal, the ambient temperature is warm. Reproduction occurs at a much more moderate pace. In the winter, when the temperature is optimal for breeding, the days are short, and the food supply is low. As a result, reproduction is held and the population is kept partly in check. Hawks and coyotes that prey on the voles also help to keep the population in balance.

Resisting Small Changes

As noted earlier, homeostatic systems are not static but tend to fluctuate within tolerable ranges. In ecosystems, shifts in population occur as a result of normal fluctuations in environmental conditions that influence all of the species in an ecosystem. Among prairie voles, a particularly cold summer might increase reproduction, which increases the population size. As the population of voles increases, populations of hawks and coyotes, which prey on voles, tend to expand. So long as the vole population is high, the predator population will remain high. If normal weather patterns return in subsequent years, however, reproduction in the voles might decrease. The expanded hawk and coyote populations would also begin to reduce the vole population. Eventually, the number of predators would begin to decline as their food supply fell, returning the system to equilibrium. However, if cool summers persist for long periods, the balance might shift to a new equilibrium point.

As this example shows, ecosystems may "bounce back," or recover rapidly, in response to minor changes; this feature is often referred to as **resilience**.

Ecosystems, therefore, consist of an elaborate set of checks and balances that tend to preserve the integrity of the whole. These checks and balances also help to minimize human impact. Sewage dumped into a stream, for example, adds organic and inorganic chemicals to the water (Figure 3.3). The organic molecules are consumed by naturally occurring bacteria whose population is normally low. Because food supplies have increased, the number of bacteria expands. Because bacteria use up oxygen as they consume organic materials, the level of dissolved oxygen in the stream usually drops. This decline may kill fish and other organisms or force them to migrate to new areas. If further spills do not occur, the population of bacteria in the stream will decline as the level of organic pollutants falls. The levels of dissolved oxygen will also return to normal. Fish will return and the stream will be restored to its normal state. This is an example of resilience. (Chapter 17 covers this topic in detail.)

Problems arise when human activities push ecosystems too far. When human activities strain the limits of resilience, severely altering the biotic and abiotic conditions of the environment, damage can become severe. In some cases, damage may be irreversible.

Population Control and Sustainability

The discussion of ecosystem homeostasis illustrates the importance of population control in sustaining natural systems. Population control mechanisms serve one important function: They help to keep populations within an ecosystem's carrying capacity. As shown, in most cases, population control results from extrinsic factors, for example, predation, disease, and adverse weather.

Human populations are also subject to natural controls. Disease epidemics, for instance, can devastate

Point/Counterpoint 3.1

Environmentalism: On Trial

William Tucker

The author, a critic of the environmental movement, has written numerous articles on environmental issues. His book, Progress and Privilege, *is a thought-provoking, controversial discussion of environmentalism.*

Environmentalists Are Subversive To Progress

One of the key realizations of ecology is that the Earth is a kind of living system governed by many self-regulating (homeostatic) mechanisms. The Earth is in a state of equilibrium. If pushed too far in any one direction, the self-regulating mechanisms can become overloaded and break down, resulting in radical changes.

In its scientific aspects, ecology seems to offer an extraordinary broadening of our understanding of life on the planet. Yet, with its transfer into the public domain, it has become little more than a sophisticated way of saying, "We don't want any more progress." Somehow, this exciting discipline has been translated into a very conservative social doctrine. People have often waved the flag of "ecology" as a new way of saying that nature must be preserved and human activity minimized. Ecology is sometimes viewed as "subversive" to technological progress. It supposedly tells us that our ignorance of natural systems is too great for us to proceed any further with human enterprise. Just as nationalistic conservatives always try to throw a veil of reverence around such concepts as *patriotism* and *national tradition*, so environmentalists try to maintain the same indefinable quality around ecosystems.

The lesson environmentalists drive home is that since we do not understand ecosystems in their entirety, and never will, we dare not touch them. Our knowledge is too limited, and nothing should be done until we understand more fully the implications of our actions.

To say that ecology is the science that does not yet grasp the complex interrelationships of organisms is like trying to define medicine as the science that does not yet know how to cure cancer. Environmentalists emphasize the negative parts of the discipline because it fits their concept that we have already had enough technological progress.

The lessons of ecology tell us many things. They tell us that organisms cannot go on reproducing uncontrollably. But, they also tell us that many organisms have developed behavioral systems that keep their populations from exploding. The laws of ecology tell us that we cannot throw things away into the environment without having them come back to haunt us. But, they also tell us that nature evolved intricate ways of recycling wastes long before human beings appeared, and ecosystems are not as fragile as they seem.

In fact, the whole notion of *fragile ecosystem* is somewhat contradictory. If these systems are so fragile, how could they have survived this long? If ecology teaches us anything, it enhances our appreciation of how resilient nature is, and how tenaciously creatures cling to life in the most severe circumstances. This, of course, should not serve as an invitation for us to see how efficiently we can wipe them out. But, it does suggest that the rumors of our powers for destruction may be exaggerated.

The environmentalist interpretation of ecology has been that ecosystems have somehow perfectly evolved and that human intervention always leads to degradation. It should be clear that even if a particular ecosystem did represent biological perfection, that is not reason in and of itself to preserve it at the expense of human utility.

Our ethical position cannot be one of completely detached aesthetic appreciation. We must first be human beings in making our ethical judgments. We cannot be completely on the side of nature.

We are not a group of imbeciles aimlessly poking into the backs of watches or tossing rocks into the gears of Creation. There is purpose to what we do, and it is essentially the same as nature's. We are trying to rearrange the elements of nature for our own survival, comfort, and welfare. We can certainly act stupidly, but we can also act out of wisdom. It is foolish to argue that everything is already perfect and must be left alone. To portray humans as meddling outsiders in an already perfected world is nonsense. In going to this extreme to reaffirm nature, we only deny that we are a part of it.

Environmental writers suggest that we practice an "ecological ethic," extending our moral concerns to other animals, plants, ecosystems, and the entire biosphere. I would accept this proposal, with one important qualification: that is, that our ethical concerns still retain a hierarchy of interest. We should extend our moral concerns to plants, trees, and animals but not at the expense of human beings. Our first obligation is to humanity. We should avoid actions that are destructive to the biosphere, but we must recognize that, at some point, our interests are going to impinge upon other living things.

Daniel D. Chiras

Environmentalists Favor True Progress

Contrary to what some might have you think, the environmental movement is not a homogeneous group of people who uniformly object to all human progress. Founded largely on lessons gleaned from the science of ecology, the environmental movement attracts a following linked by a common interest in protecting the environment.

My experience within the movement, however, shows that environmental protection means many things to many people. People sympathetic with the plight of the natural world range from those who call for "complete preservation," which tolerates little human intervention, to the preserve-it-in-parks folks, who seek to protect small pieces of the environment for future generations to enjoy, while developing most of the rest for human use.

Nowhere are environmentalists' beliefs more varied than on the notion of progress. William Tucker's essay in this Point/Counterpoint casts environmentalists as narrow-minded obstructionists who condemn all human progress and technological development. No doubt, many business people feel the same—and for good reason. Environmentalists have threatened many a proposed dam, highway, and mine. In this narrow vein, few could deny that environmentalism is subversive to progress as some see it.

The truth, however, is not so simple. Progress is not one thing to all people. In fact, I think that much of the environmental debate hinges, quite precariously, on the difference in how people view progress.

To most environmentalists, progress has a much broader meaning than economic growth, faster jets, and new gadgets to make life a little easier. Progress means prospering within the limits of nature, living on Earth without destroying the air, water, and soil on which our lives depend. Progress means finding a way of living sustainably on the planet. We all recognize that environmental protection is a good goal and that planet care is the ultimate form of self-care.

Most environmentalists embrace progress but not blind progress. They think that no human endeavor can be counted a success if it rips apart the Earth's life-support systems. Moreover, those who advocate a sustainable strategy realize that humanity cannot prosper without fundamental changes in the way we go about our day-to-day business.

Tucker's assertion that environmentalists don't want any more progress misses the point. Environmentalists want lots of progress. We want progress in recycling and tapping into renewable resources in a big way. We want progress in using energy and other resources more efficiently. We want progress in restoring ecosystems that we depend on and in controlling human numbers. We can do all this and still have a healthy economy.

As for Tucker's assertion that environmentalists think that nature must be preserved and human activity minimized, my experience in environmental protection shows that most environmentalists believe that restraint and proper design are necessary, but stopping all human activity is not.

Ecology has taught environmentalists to look to the future with their eyes wide open, rather than narrowly focused on material welfare, economic wealth, and convenience. We must be on the side of nature because we are part of nature. In attempting to rearrange the elements of nature to satisfy our needs, we must tread carefully, finding ways to meet our needs without impairing the ability of future generations—and other species—to meet theirs.

As for the claim that environmentalists see society as a stupid meddler whose every activity leads to degradation, let me say this: History teaches us that human activities have had profound, even devastating, effects on the natural world and human society. The dust-bowl days, the great deserts spreading through Africa, and destruction of the once-rich Fertile Crescent are blatant reminders of the severity of our intrusion. More recent studies show that our impact continues on a grand scale, as witnessed by the threats of deforestation, species extinction, global warming, and stratospheric ozone depletion.

Human intervention doesn't always lead to degradation, but it is also not a force to be trivialized. Ecosystems are not as fragile as some would have you believe, but they can be ruined on a massive scale. Witness the widespread loss of wetlands and tropical rain forest throughout the world. Nature may be resilient, but against widespread human intervention, it is often powerless.

In sum, the crux of the disagreement between environmentalists and "developmentalists" lies in each group's view of progress. In my view, many environmentalists have a more sustainable view of progress, which calls for restraint and intelligent-directed action aimed at weaving the human economy back into the economy of nature. Environmentalists recognize the interdependence of all living things and seek to protect other species because they are vital to our own success and survival. Others assert that because nonhuman species have a right to exist, we must protect them.

(continued)

Point/Counterpoint 3.1 (*continued*)

Environmentalism: On Trial

Environmentalism is not subversive. It advocates an alternative form of progress that seeks to foster a sustainable relationship with the planet. Our interests are likely to impinge on the interests of other species, but our challenge is to find ways to minimize these conflicts. Should we extend moral concerns to other species? Absolutely. We have no right to exterminate an-

other species so that we might live better. But, the point is, by adopting the principles of sustainability that I've outlined in Chapter 1, we don't have to live in a win/lose world. We can live well without ripping the planet apart. Let's stop fighting over the environment and find a new way of life and new world view that serve humans and the environment simultaneously.

Critical Thinking

1. Summarize the main points in each essay. Were arguments adequately supported by facts?
2. Use your critical thinking skills to analyze and comment on these views.
3. Whose views most closely represent yours? Why?

Figure 3.3 *The events that follow the dumping of organic wastes into a stream. Note that oxygen levels and fish populations return to normal in the stream, thanks to naturally occurring bacteria that decompose the organic matter in the sewage. Continued dumping or dumping by downstream sewage treatment plant could seriously hinder the stream's recovery.*

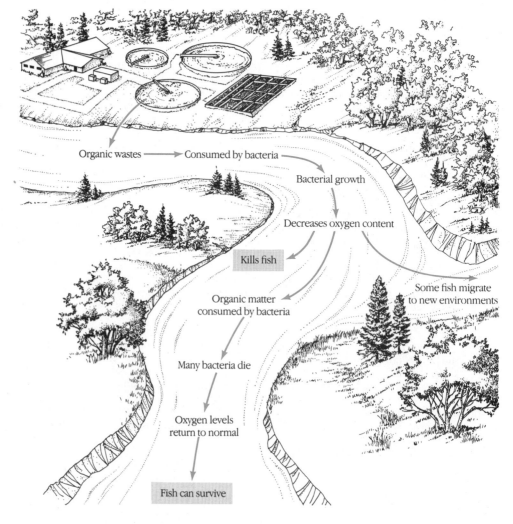

Organic wastes → Consumed by bacteria

Bacterial growth

Decreases oxygen content

Kills fish

Some fish migrate to new environments

Organic matter consumed by bacteria

Many bacteria die

Oxygen levels return to normal

Fish can survive

human populations. Robert Gilman, a leading proponent of sustainability, points out that our continued existence depends on finding ways to consciously control population size. The challenge, he says, is to "set our own limits or have limits disastrously imposed on us." (Chapters 5 and 6 discuss population control in more detail.)

Species Diversity and Stability

Some ecologists assert that ecosystem stability is influenced by species diversity. Roughly speaking, **species diversity** is a measure of the number of species living in a community. The higher the diversity, the greater the stability. Observations that extremely complex ecosystems, such as tropical rain forests, remain unchanged almost indefinitely if undisturbed support this idea. Simpler ecosystems, such as the tundra, are less stable, that is, they experience sudden, drastic shifts in population size. Other simplified ecosystems, such as fields of wheat and corn, are also extremely vulnerable to change, and they deteriorate rapidly if biotic or abiotic factors shift very much.

To understand why ecologists think there may be a connection between stability and diversity, consider the food webs in simple and complex ecosystems shown in Figure 3.4. As illustrated, the number of species in a food web in a mature ecosystem is large. So is the number of interactions among these organisms. In a complex ecosystem, the elimination of one species would probably have little effect on the ecosystem. In sharp contrast, the number of species in the food web of a simple ecosystem is small. The elimination of one species could have repercussions on all other species.

Some ecologists think this is faulty reasoning. They argue that tropical rain forests are stable because their climate is relatively uniform throughout the year. In other words, the stability of the tropical rain forest is not the result of species diversity but rather of constant climate. On the tundra, a relatively simple and somewhat unstable ecosystem, the climate shifts dramatically from season to season. These shifts are responsible for the tundra's relative instability.

Figure 3.5 shows that species diversity among mammals varies with latitude in North and Central America. In the frozen northern regions of Canada and Alaska, for example, species diversity is low. Heading south, diversity increases until one reaches the tropics of Central America, where diversity is highest. The relationship between species diversity and latitude is also found in plants and virtually all other kingdoms. Latitude, therefore, is an important factor affecting species diversity. The connection between latitude and species diversity is climate. Quite clearly, the milder the climate, the more species live there. But, do more species mean a more stable ecosystem?

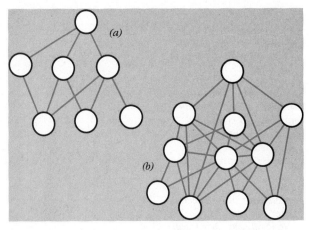

Figure 3.4 (a) *Food web in a simple ecosystem. Circles represent organisms. Note the lack of links in the simplified web.* (b) *Food web in a complex ecosystem. Many ecologists contend that complex ecosystems are more stable because the increased number of links reduces the importance of any one species.*

Truthfully, no one knows for sure whether diversity creates stability. However, we do know that simplifying ecosystems by reducing species diversity makes systems less stable and more vulnerable to outside influences.

Scientists debate the concept of ecosystem stability with vigor because recent research has cast some doubt on the concept of a naturally maintained equilibrium. Some researchers contend that populations and ecosystems rarely, if ever, return to equilibrium once disturbed. In studying the debate, one thing is clear: Return to equilibrium depends in large part on the nature and severity of the disturbance. Evidence suggests that ecosystems can recover from small perturbations, such as changes in rainfall or short-term drought. More severe alterations, such as deforestation of the tropics, may render a system unable to recover.

Unfortunately, the topic is so complex and so easily misconstrued that more confusion than clarity emerges from the debate. At this time, our understanding of ecosystem homeostasis is a long way from complete. As the critical thinking rules outlined in Chapter Supplement 1.1 suggest, sometimes uncertainty is unavoidable. (Species diversity is discussed in more detail in Chapter 8.)

3.3 Correcting Large Imbalances in Ecosystems: Succession

Small shifts in the growth and reduction factors in an ecosystem, whether prompted by natural or human causes, are fairly common and readily correctable. Drastic shifts can seriously upset ecosystems; nevertheless, nature can sometimes restore severe damage. Both abilities are essential for sustaining life on Earth.

Figure 3.5 *The number of mammals (shown here) and most other species varies considerably with latitude. The highest diversity is found in the tropics; the lowest in the tundra. Species diversity may help to create ecosystem stability, but some ecologists think that diversity is a product of climatic stability. Similar maps could be constructed for other species. (After Simpson, C. G. (1964) Species Density of North American Recent Mammals. Syst. Zool. 13: 15–73.)*

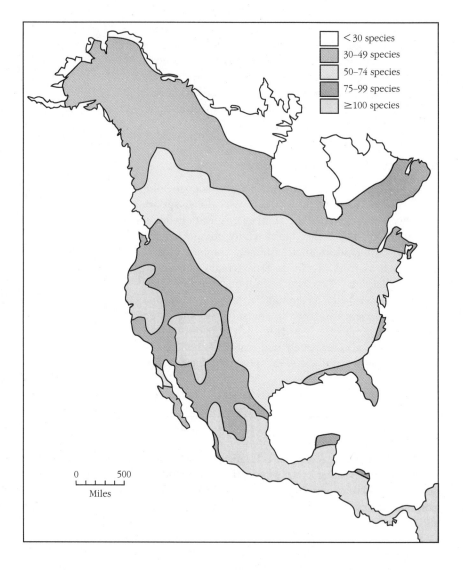

Legend:
< 30 species
30–49 species
50–74 species
75–99 species
≥100 species

0   500
Miles

Walk along a forest path in the aftermath of a volcanic eruption or a forest fire sparked by lightning. Or travel a dusty path over abandoned surface mines or carelessly logged forests. Life sprouts out of the ashes or in the unstable soils, but years or decades must pass before these places return to their predisaster stages. If soil washes away before natural healing occurs, recovery may be impossible.

A biotic community destroyed by natural or human causes often recovers in a series of changes in which one community is gradually replaced by another until the mature, or **climax,** community is reached. This process is called **succession.** The two kinds of succession are primary and secondary.

Primary Succession

Primary succession is the sequential development of a biotic community where none had previously existed. It is not a means of repairing damage but rather of establishing biotic communities in virgin territory. For instance, in North America, when the great glaciers began to retreat 15,000 years ago, large areas of barren land and rock were exposed. The exposed rock first became populated with lichens (Figure 3.6). The lichens, which thrived for a while, were gradually replaced by mosses. The mosses were replaced by small herbs and shrubs, larger shrubs, then trees.

To understand how succession works, let's take a closer look at each biotic community in this example. Lichens, the first inhabitants, cling to the rock's surface, living off moisture from the rain and organic nutrients from photosynthesis. Lichens also secrete a weak acid, carbonic acid, that dissolves rock, liberating nutrients and helping to make soil. Carbonic acid in normal rain also helps to wear down the rock. Tiny insects may join the lichens, forming a **pioneer community,** the first community to become established in a once-barren environment.

The lichens and insects gradually change their environment. Lichens capture windblown dirt particles, promoting further soil development. Dead lichens crumble

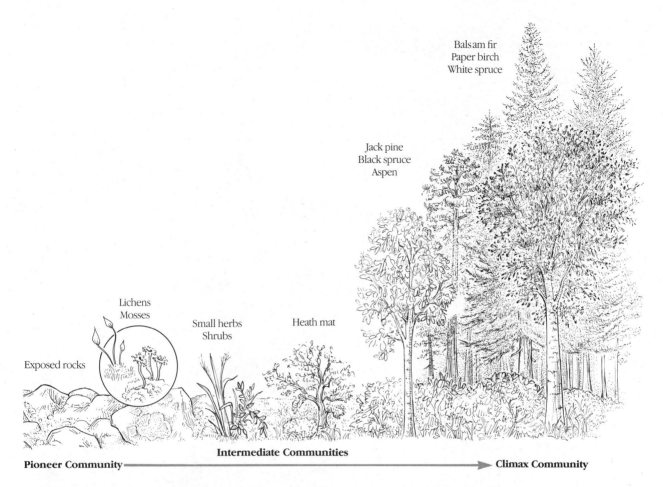

Balsam fir
Paper birch
White spruce

Jack pine
Black spruce
Aspen

Lichens
Mosses

Small herbs
Shrubs

Heath mat

Exposed rocks

Intermediate Communities

Pioneer Community ———————————————→ **Climax Community**

Figure 3.6   *A representation of primary succession on Isle Royale, located in Lake Superior. Rock exposed by the retreat of glaciers is colonized by lichens, then mosses, followed by other plant communities leading to a climax community. One biotic community replaces another until a mature community is formed. During succession, the plants of each community alter their habitat so drastically that conditions become more suitable for other species.*

and become part of the soil, along with the remains of insects, fungi, and bacteria. Over time, enough soil develops for mosses to take root. The mosses, however, shade the lichens and eventually kill them. Mosses, fungi, bacteria, and insects form a new **intermediate community**. This community also brings about changes, which eventually usher in still another community until a forest forms. In this instance, the forest is known as a **climax community**. Climax communities are relatively stable ecosystems.

Another example of primary succession is the establishment of plants on newly formed volcanic islands. The Hawaiian Islands, for example, arose from lava from deep within the Earth. When the lava cooled, plants began to take root. The plants came from seeds deposited in the feces of sea birds that happened upon the islands. Other plants are thought to have come from neighboring continents and islands. Uprooted vegetation, say biologists, may have drifted to the newly formed islands over

many thousands of years, and some took root, turning the once-barren mass of rock into a rich tropical garden.

Secondary Succession: Natural Ecosystem Restoration

Secondary succession is the sequential development of biotic communities after the complete or partial destruction of an existing community. It is a mechanism by which ecosystems restore themselves when severely damaged.

A climax community or intermediate community may be destroyed by natural events, for example, volcanic eruptions, floods, droughts, fires, and storms. The eruption of Mount Saint Helens in 1980 and the devastating fires in Yellowstone in 1988 are modern examples. Established communities are also commonly destroyed by human intervention, such as agriculture,

Figure 3.7 *Secondary succession. Here, abandoned eastern U.S. farmland is gradually replaced by crabgrass, which, in turn, gives way to other herbaceous plants. Trees move in, and over time a mature hardwood ecosystem is formed. Note that many nonplant species (not shown here) also appear in these communities. In the early stages, for example, insects, mice, woodchucks, and seed-eating birds are found. In the later stages, as trees come to dominate, squirrels and chipmunks, which prefer a wooded habitat, invade the new ecosystem.*

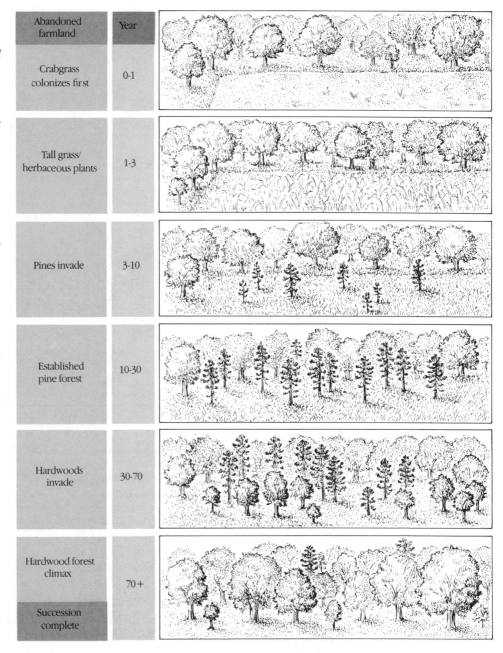

Abandoned farmland	Year
Crabgrass colonizes first	0-1
Tall grass/ herbaceous plants	1-3
Pines invade	3-10
Established pine forest	10-30
Hardwoods invade	30-70
Hardwood forest climax	70+
Succession complete	

intentional flooding, fire, or mining. When secondary succession occurs, it generally takes place more rapidly than primary succession because the long, slow development of soil is unnecessary.

Abandoned farm fields provide an excellent opportunity to observe secondary succession (Figure 3.7). Former farmland is first invaded by hardy species, such as crabgrass or broom sedge, depending on the area. These plants are well adapted to survive in bare, sun-baked soil. In the eastern United States, crabgrass, insects, and mice invade abandoned fields, forming pioneer communities. But, crabgrass is soon joined by tall grasses and other herbaceous plants. The newcomers' shade

eventually eliminates the sun-loving crabgrass. Tall grasses and other herbaceous plants dominate the ecosystem for a few years along with a variety of animals, such as mice, woodchucks, rabbits, insects, and seed-eating birds.

In time, pine seeds settle in the area, and seedlings begin to spring up in the open field. Like crabgrass, the pine trees flourish in the sunny fields. Over the next three decades, pines begin to shade out the grasses and herbs. Animals that feed on grasses, such as woodchucks, move on to more hospitable environments. Squirrels and chipmunks, which prefer a wooded habitat, invade the new ecosystem.

Table 3.1 Characteristics of Mature and Immature Ecosystems

Characteristic	Immature Ecosystem	Mature Ecosystem
Food chains	Linear, predominantly grazer	Weblike, predominantly detritus
Net productivity	High	Low
Species diversity	Low	High
Niche specialization	Broad	Narrow
Nutrient cycles	Open	Closed
Nutrient conservation	Poor	Good
Stability	Low	Higher

Source: Modified from Odum, E. (1969). The Strategy of Ecosystem Development. *Science* 164: 262–270. Copyright 1969 by the American Association for the Advancement of Science.

Shade from the pines gradually creates an inhospitable environment for their own seedlings and a favorable environment for the growth of shade-tolerant hardwood trees, such as maple and oak. As a result, hardwoods take root and over time tower over the pines; their shade gradually kills many of the pines, which had invaded 60 years earlier.

Succession is a kind of biological race to make optimal use of available resources, such as sunlight, soil nutrients, and water. As illustrated, the success of pioneer and intermediate communities is transient. During succession, pioneer species actually create conditions conducive to species that form the intermediate community. These species, in turn, create conditions conducive to species that form the climax community.

During succession, animal populations shift with the changing plant communities. The early stages of succession are characterized by a low species diversity (Table 3.1). Because there are fewer species early on, food webs tend to be simple, and populations tend to be volatile. In contrast, mature biotic communities have a high species diversity and relatively stable populations.

As ecosystems grow increasingly complex, the food chains become woven into more complex food webs. In immature and intermediate stages, grazer food chains account for the bulk of the biomass flow. In climax ecosystems, however, most of the biomass flows through the detritus food chain. In fact, in a mature forest, less than 10% of the net primary productivity is consumed by grazers. Interestingly, in mature ecosystems, nutrients are cycled more efficiently than in immature or intermediate ecosystems, which tend to lose a considerable amount of their nutrients because of erosion and other factors.

In pioneer and intermediate communities, the abiotic and biotic factors that regulate population size are in flux. For example, in an abandoned farm field there is initially little environmental resistance to crabgrass, so there are few limitations to growth. Thus, the growth factors, such as the availability of food, favorable light, and optimum temperature, stimulate rapid population growth.

During succession, each community experiences an increase in environmental resistance, caused in large part by increasing competition from other species (Figure 3.8). In an abandoned farm field, for example, crabgrass is shaded out by taller grass. When environmental resistance reaches a certain level, the crabgrass population begins to decline, and a new community establishes itself.

Climax communities that emerge from undisturbed secondary succession contain organisms that exist in a complex web of life regulated by a variety of growth and reduction factors. These communities tend to stay in relative balance over long periods if undisturbed. These relatively stable assemblages of organisms are sustained for the most part because of the efficient use of resources, recycling of nutrients, and dependency on a reliable renewable resource base. They also sustain themselves because they are capable of restoring damage and maintaining population balance through a variety of population control mechanisms.

Those of you who will go on to study ecology will learn that the view of secondary succession presented above is the classical interpretation. Recent studies have thrown a few key tenets of this theory into question. Research shows that succession varies considerably in different regions and with different types of vegetation. In the Rocky Mountains in New Mexico, Colorado, Wyoming, and Montana, for example, avalanches sometimes destroy stands of conifers. The opening in the forest created by an avalanche first fills with grasses,

Figure 3.8 *A graphic representation of the population growth and decline of a pioneer community (solid black line) and the rise of an intermediate community (colored line) during succession. Note that environmental resistance (dashed line) increases as the pioneer community becomes established, making conditions conducive to the establishment of an intermediate community. Numbers indicate the progression of events.*

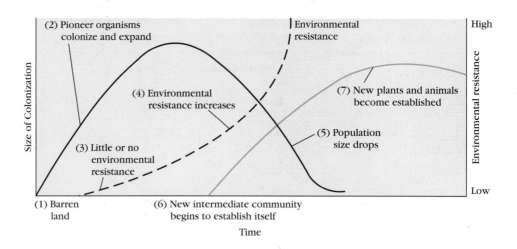

wildflowers, and aspen trees. Over the years, dense stands of aspen grow. Next, shade-tolerant fir and spruce grow up in the forests, eventually reestablishing the climax community. In this system, species diversity in the intermediate community—the aspen forest—actually exceeds that of the mature system. Aspen forests support grasses, herbs, wildflowers, and a great many birds and mammals, such as deer and elk. As the climax community becomes established, species diversity falls because the dense forests inhibit growth of grasses and other sun-loving plants. The needle-covered floors of a fir and spruce forest support relatively few species.

New research also shows that the sequential replacement of one community by another is not always as clear-cut as once thought. In the succession of an eastern farm field to forest, for example, pine trees are often present early in vegetational development. However, because they grow slowly and are often browsed, their presence is not noticeable until later stages.

In summary, this discussion of natural succession shows that ecosystems can recover from damage. This feature, labeled here as *restoration*, is vital to sustaining life on Earth. However, as indicated earlier, not all damaged ecosystems can recover. In a huge region where Tennessee, North Carolina, and Georgia meet, for instance, pollution from a copper smelter, deforestation, and erosion have created a bleak landscape, so badly damaged that plants cannot grow. One journalist described it as a "vast, raw plain, cooking in the summer sun." Remarkably, the scar of this human assault is even visible from outer space. Left on its own, this region might never recover. Today, though, the land is being replanted and coaxed back to life. Similar devastation is occurring in tropical rain forests where millions of hectares are cleared and abandoned each year. Soils are so badly eroded that forests may take many centuries to regrow on their own.

3.4 Evolution: Responding to Change and Maintaining Ecosystem Homeostasis

Chapter 1 noted that life on Earth sustains itself, in part, because populations evolve as conditions change. Evolution has two widely recognized outcomes: (1) In some cases, it results in modifications to existing species, making them better suited to their environment—that is, better able to survive and reproduce. Changes in the coloration of a species of butterfly, for example, might make it better able to avoid being eaten. (2) Evolution also results in the formation of new species.

Most of what scientists know about evolution comes from **fossils**—preserved bones, imprints of organisms captured in rock, or footprints of early animals. By examining these fossils embedded in rock strata whose age can be determined through various means, scientists have been able to construct a biological history of the Earth.

The theory of evolution is often attributed to Charles Darwin, a 19th-century British naturalist. Interestingly, Darwin did not originate the idea; it dates back to ancient Greece, and the theory was already widely discussed in Darwin's time. What Darwin and his colleague, Alfred Russell Wallace, did was propose a mechanism by which evolutionary changes could come about. It is called the **theory of evolution by natural selection**. This theory states that species evolve as a result of **natural selection**, a process described below.

As noted in Chapter 1, some organisms are better adapted to their environment than others, that is, their **adaptations** confer an advantage over others. Adaptations are structural, functional, and behavioral characteristics that occur randomly in populations. Favorable adaptations increase an individual's chances of surviving and reproducing. That is, favorable adaptations tend to persist. Evolutionary biologists say they are selected for. Organisms better adapted to their environment leave more surviving offspring in the population than those without the adaptations. Future generations will therefore have a higher percentage of individuals possessing the favorable adaptations.

Like many other great ideas, natural selection took many years to be understood and appreciated. Not until the 1940s did Darwin's and Wallace's ideas on natural selection become widely accepted, nearly a hundred years after they developed the concept.

Genetic Variation, Mutation, and Natural Selection

Evolution occurs as a result of genetic variation and natural selection. **Genetic variation** refers to naturally occurring differences in the genetic composition of organisms in a population. In sexually reproducing species, genetic variation may result from mutations in the genetic material, or DNA, of the germ cells (ova in females and sperm in males). DNA is a storehouse of information that controls the structure and function of the cells of the body. DNA consists of many segments called **genes**, each of which plays a specific role in regulating cell structure and function. Mutations in the genes may be caused by ultraviolet light or other high-energy radiation, chemicals in the environment, and cosmic rays from the sun. Some may occur randomly. (For more on mutation, see Chapter 14.)

Some germ cell mutations are neutral—that is, they have no effect on the organism. Others are harmful. If not repaired by the cell, they can be passed to an organism's offspring, resulting in birth defects or fetal death. Still others result in favorable adaptations that increase the survivability of an organism's offspring, for example, by improving their ability to escape predators, tolerate cold temperatures, or find food.

Genetic variation may also arise during the production of ova or sperm cells, the **gametes**. During the production of germ cells, genetic material may be transferred from one chromosome, a long strand of DNA with associated protein, to another. This process, called **crossing over**, results in new, favorable genetic combinations.

Another source of new genetic combinations is sexual reproduction. **Sexual reproduction** occurs when offspring are produced by a union of sperm from males and ova from females. Each offspring contains half of each parent's genetic information. Offspring, therefore, represent a new genetic combination that may provide benefits.

Genetic variation within a population provides a broader genetic base manifest by structural, functional, and behavioral variation among individuals. This, in turn, provides a population more leeway in coping with changing environmental conditions. In other words, it results in some members being better adapted to environmental conditions than others. For these reasons, biologists refer to variation as the "raw material of evolution." Organisms with favorable adaptations are said to possess a **selective advantage**.

In instances where conditions are changing or organisms are exposed to a new environment with environmental conditions very different from those in which they evolved, those organisms best adapted to the new or changing conditions persist. This phenomenon, called **directional selection**, can be represented by a simple series of graphs, as shown in Figure 3.9a. The bell-shaped curve in the top panel illustrates cold tolerance in a species of mice. This graph shows that there is a wide range of ability. The shaded area indicates the mice that are best able to withstand cold. If the climate of an area suddenly becomes colder, or the mice are moved to a colder region, the ones best suited to colder temperatures will be more likely to survive and reproduce. The others will die or will reproduce less successfully. Eventually, cold-tolerant members will come to dominate the population.

Sociologist Andrew Schmookler wrote that evolution employs no author but only an extremely patient editor. By that, he meant evolution may appear to move in certain directions, but it is not consciously directed. Variation occurs naturally. Beneficial genetically based adaptations tend to persist because they are selected for by the abiotic and biotic conditions of the environment. In other words, the environment only preserves, or "selects," those organisms in a population with new traits that confer some advantage over the rest.

Ultimately, natural selection results in organisms better adapted to fit their environment. Some think of natural selection as survival of the fittest. **Fitness** is commonly thought of as a measure of strength or survivability. To a biologist, fitness is just a measure of reproductive success and, thus, the genetic influence an individual has on future generations. By definition, the fittest individuals leave the largest number of descendants in subsequent generations. Their influence on the genetic makeup of those generations is, therefore, greater than less fit individuals. In a phrase, then, the fittest individuals are those best adapted to environmental conditions.

One important measure of fitness is the efficiency with which organisms use resources. Especially important are adaptations that render organisms and ecosystems efficient in their use of water and energy. Chapter Supplement 2.1 on biomes, for example, showed that plants and animals in the desert possess adaptations that help conserve water. Numerous other adaptations also

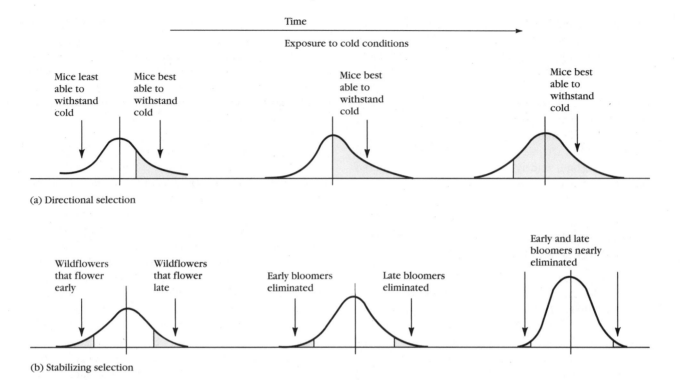

Figure 3.9 (a) *Directional selection occurs when environmental conditions change. Those organisms in a population best adapted to the change have a selective advantage over the other members of the population. Their numbers increase. In time, they may comprise the majority of the population.* (b) *Stabilizing selection occurs when environmental conditions remain the same. In such cases, fringe elements are eliminated, although variation is never completely eliminated.*

contribute to energy conservation in species that live in harsh environments. Sustaining life on Earth, especially in less favorable conditions, depends in part on the presence of such strategies. Human survival will invariably require the evolution of similar, consciously chosen efficiency measures.

Speciation

The foregoing discussion shows why the evolutionary process helps to change and ultimately sustain species in a changing environment. This section discusses how new species arise, a process called **speciation**. Speciation occurs most commonly when members of a species are separated from one another (Figure 3.10). In scientific language, this is referred to as **geographic isolation**. Geographic isolation may result when a new mountain range or a river forms in an organism's habitat. Isolated by impenetrable physical barriers such as these and exposed to different environmental conditions, over time, the two populations may evolve in quite different directions. If the populations are separated long enough, their members may lose their ability to interbreed. This process is referred to as **reproductive isolation**.

When geographic isolation leads to reproductive isolation, new species are formed. Scientists call the formation of new species in different regions **allopatric speciation** (allopatric is derived from the Greek words *allos* and *patra* meaning "other" and "fatherland"). New species may also form without geographical isolation. Common in plants, this is termed **sympatric speciation** ("same fatherland"). Figure 3.10 illustrates allopatric speciation.

Studies of the fossil record suggest that many species form from a common ancestor. As shown in Figure 3.11, evidence suggests that ancient reptiles gave rise to terrestrial, aquatic, and flying reptiles. This process, in which one life form gives rise to many others that occupy different niches, is known as **adaptive radiation**. Ancestral placental mammals may have given rise to a similar array of species. Darwin saw evidence of adaptive radiation on a small scale during his travels to the Galápagos Islands, off the coast of Ecuador. Here he found 14 species of finches that presumably arose from a single species of mainland finch (Figure 3.12). Occupying different islands, the mainland finch evolved into new species, diverging to make best use of the varying food sources.

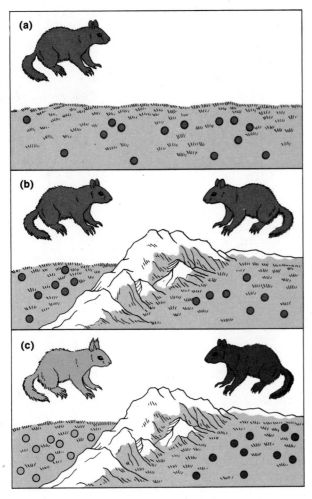

Figure 3.10 *Geographical isolation. When a population splits because of a new geographic barrier, new populations may arise over time because the two groups are subjected to different selective forces. This illustration shows how a mountain range might divide a population of squirrels, over time creating two new species.*

Figure 3.11 also shows that unrelated species may adapt to similar environments in similar ways. In this example, mammals and reptiles both evolved flying forms (in fact, 25% of all mammal species are bats). Mammals and reptiles both gave rise to carnivores, as represented in the figure by the tyrannosaurus and the lion. Evolutionary biologists call this tendency for organisms to develop the same types of adaptations in response to similar environmental conditions **convergent evolution**.

The study of evolution is far from complete. One issue under current debate is the matter of how rapidly a species evolves. For many years, evolutionary scientists believed that such changes occurred gradually over many millions of years. They coined the term **gradualism** to describe this phenomenon. If this were true, paleontologists reasoned, the fossil record should contain many intermediate forms of organisms, a sort of geological recording of the gradual transformation. With few exceptions, however, the fossil record shows few intermediate stages. This gap led two noted paleontologists, Stephen Gould of Harvard University and Niles Eldredge of the American Museum of Natural History, to propose the theory of **punctuated equilibrium**. It states that during evolution of life there are fairly long, quiet periods (equilibrium) that are punctuated by periods of fairly rapid change. This theory accounts for the lack of geological evidence to support gradualism, especially the lack of many intermediate fossil forms.

Evolution Contributes to Ecosystem Homeostasis

The preceding description shows clearly that evolution by natural selection—specifically, directional selection—is the basis of biological change. This process helps sustain life on Earth and accounts for its dynamic nature. In instances in which environmental conditions are unchanging and when a species is already well adapted to a particular environment, natural selection helps to maintain the status quo. In such cases, naturally occurring variation in the population produces adaptations that are not of any benefit under those conditions. New variants are eliminated, and natural selection favors the "average" individual.

This process, referred to as **stabilizing selection**, is illustrated in Figure 3.9b. As shown, stabilizing selection favors organisms close to the average. By eliminating the fringes, it increases the frequency of genes responsible for the fittest phenotype. Consider an example.

In New England, numerous species of wildflowers grow on the forest floor each spring. Most wildflowers emerge in early spring, after the last hard frost but before the leaves of the trees appear. After the trees leaf out, the forest floor is covered in shadow, and is too dark to support flowers. As a result, most wildflowers sprout, flower, and set seed in the brief period between the last frost and full leaf development. Variants in either direction may occur but will generally be selected against. Plants that sprout too early will be eliminated by frosts (unless, of course, they're frost resistant). Plants that sprout too late may not survive to flower and set seed because of inadequate sunlight.

Coevolution—A Mechanism That Contributes to Ecosystem Balance

The discussion of natural selection to this point has focused on environmental conditions as the chief agents of evolutionary change. Numerous studies show that organisms can also serve as selective agents. For instance, barn owls, which hunt for mice at night, may evolve better mechanisms to detect prey. This could profoundly

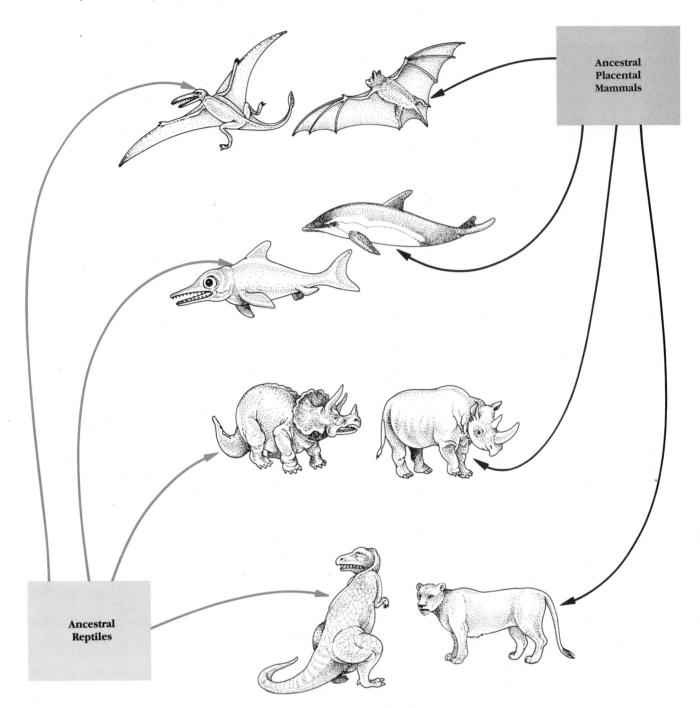

Figure 3.11 *Adaptive radiation occurs as organisms evolve to fill ecological niches, making best use of available resources. The drawing shows the niches first filled by dinosaurs and later in evolution by placental mammals.*

affect the prey population, as the slower mice fall victim to the hungry owl in much the same way that changes in weather might weed out the less fit members of the population. Improvements in predation, however, would eventually result in improvements in avoidance among the mice. Interestingly, changes in the ability of mice to elude the barn owl can act as a selective force on the predator as well.

When members of two species interact in this manner, changes in one species result in changes in the other. Thus, each species can become a selective force that affects the evolution of the other. This process, called **coevolution**, has been likened to an arms race between predator and prey because each improvement in predatory ability is followed by an improvement in the prey's ability to avoid or resist attack. Coevolution

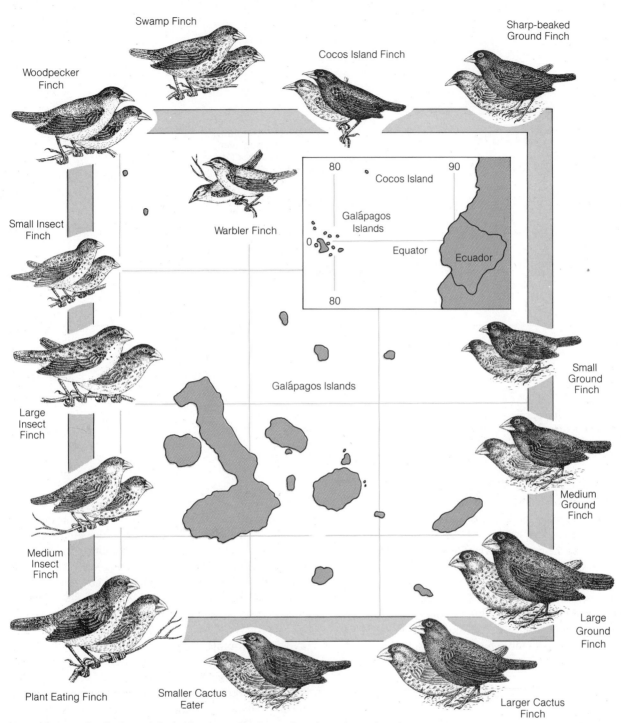

Figure 3.12 *The finches of the Galápagos Islands are found nowhere else on Earth, although other finch species do inhabit the mainland of South America. The Galápagos finches are not uniformly distributed over all the islands. Only the warbler finch is found on all the islands, while the medium insect finch is found only on one island. Adaptation has given rise to different species of finches that fill the ecological niches occupied by hummingbirds, flycatchers, and woodpeckers elsewhere in the world.*

may also occur between plants and the animals that feed on them.

Understanding evolution is important to understanding environmental issues, especially species extinc-

tion. It helps us to broaden our awareness, becoming more competent systems thinkers. For instance, an appreciation of evolution makes us aware of the fact that when a species goes extinct we have ripped a piece

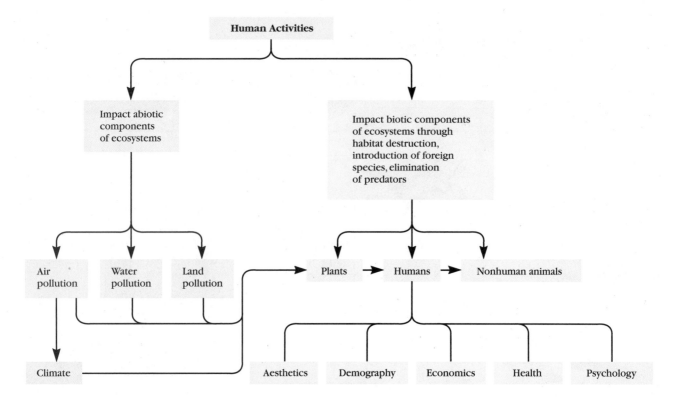

Figure 3.13   *Impact Analysis Model, showing the range of impacts caused by human activities. Note that human activities impact two broad areas, abiotic and biotic conditions. Also note that alterations to the abiotic conditions can impact biotic conditions.*

of the complex ecological network. We may have interrupted important food chains and disrupted coevolutionary processes at work.

Perhaps the most important lesson is that to sustain itself, a species must change in response to changing realities. Organisms generally have no choice; they evolve or they die out. For humans, change is vital at this juncture in our history. But as pointed out in Chapter 1, our change must be cultural in nature and purposefully directed. (More on this in Chapter 4.)

3.5   Human Impact on Ecosystems

Chapter 4 takes you on a journey through human cultural evolution. In it, you will see that throughout history we humans have used much of our knowledge to control the environment—to shape it to our liking and enhance our survival. Recent evidence suggests that many of these efforts have backfired and now threaten our long-term future.

Learning to live sustainably on the planet will require steps to correct past mistakes. It will also require efforts to restructure human institutions and activities to avoid further disruptions of the Earth's life-support systems. As Vice President Al Gore points out in his

book, *Earth in the Balance*, we must ultimately change the entire relationship of humans to the environment.

Building a sustainable world in which humans thrive within the planet's ecosystems without permanent disruption will require systems thinking—an understanding of the whole and anticipation of how our impacts will affect the entire ecological system of the planet. In this section, we will examine one of several tools—the **Impact Analysis Model**—that allows us to anticipate and avoid unfavorable impacts (Figure 3.13).

This model provides an overview of human impact on the environment. As Figure 3.13 shows, human impact on the environment occurs via two avenues: through changes in abiotic and biotic factors in the natural world (Chapter 2). This section outlines some key ways we alter these factors, beginning with abiotic factors. You may find it helpful to refer to the model in Figure 3.13 as you read this material.

Tampering with Abiotic Factors

Many human activities produce pollutants that alter the abiotic factors within ecosystems, creating conditions that organisms cannot tolerate. Chemical disruption is the most common form. As noted in Chapter 2, sewage adds nutrients (nitrates and phosphates) to waterways, which upset the natural chemical balance. Other activities produce pollutants that are totally foreign to natural systems, some of which are quite toxic to life forms. Human pollution can also affect physical conditions of ecosystems. Chlorofluorocarbons, for instance, deplete the ozone layer and increase the amount of ultraviolet light striking the Earth. Power plants release water that changes stream temperature. Carbon dioxide may alter the global climate, resulting in an increase in the Earth's temperature. Thus, all forms of pollution occurring in the air, water, and soil have the potential to affect a wide assortment of life forms, often significantly.

A few examples of ways that humans throw natural systems out of balance by overloading them with chemical pollutants were presented in the nutrient cycle discussion in Chapter 2. Other examples, including climate change, are featured in Chapters 15 through 17 on air and water pollution. This section presents two examples of the introduction of totally foreign chemicals that are toxic to plants, animals, and microorganisms: chlorinated waste from treatment plants and chemical pesticides.

Wastewater treatment plants release a wide assortment of chlorinated compounds into rivers. These chemicals are a by-product of the chlorination process used to sterilize treated wastewater before it is released into streams and lakes (Chapter 17). Unfortunately, many of these chemicals are toxic to fish and other aquatic species.

A second example results from the use of chemical pesticides, substances used to control fungi, viruses, insects, and even weeds. These sometimes highly toxic chemicals may be carried in the air or water to natural ecosystems, where they poison beneficial species, such as honeybees and birds. One of the best-studied pesticides is DDT (dichlorodiphenyltrichloroethane), an insecticide formerly used in the United States and still used elsewhere. DDT contaminated many terrestrial and aquatic ecosystems where it passed through food chains from one organism to another. Stored in body fat, its chief breakdown product, DDE, reached high levels in **top-level consumers**, those animals located at the top of the food chain, such as ospreys, brown pelicans, and peregrine falcons (Figure 3.14).

Interestingly, DDE did not kill the birds outright; rather, it interfered with the deposition of calcium in their eggshells, resulting in eggshell thinning. Because eggs were fragile, they were easily broken. Few embryos survived, and bird populations fell sharply.

The DDT incident illustrates the important phenomenon known as **biological magnification**, or simply **biomagnification**, the accumulation of certain substances in food chains, with increasing amounts found in each trophic level. Because of this phenomenon, fairly low levels of DDT in the environment can result in dangerously high levels in organisms. (Chapter 18 more fully discusses biomagnification.)

The peregrine falcon, which nests on rocky ledges throughout the United States, was nearly destroyed by DDT. In fact, by the time scientists had determined that the decline in the bird's reproductive rate resulted from DDT and DDE, none of the 200 known pairs east of

Figure 3.14 *Peregrine falcon and her chick. The peregrine, once a nearly extinct species in the United States, has made a remarkable comeback with the aid of humans, the species responsible for nearly wiping them out in the first place, as explained in the text.*

Figure 3.15 *Like many coastal cities throughout the world, parts of Miami are built on former marshland, habitat vital to birds and other species but also vital to human food supply.*

the Mississippi River were successfully producing young. Fortunately for the falcon, DDT was banned in the United States (but still manufactured here until 1984), and a determined program was mounted that may well save these birds. By the end of 1991, over 4000 peregrines had been raised in captivity and released into the wild by the Peregrine Fund, now headquartered in Boise, Idaho.

In these and many other cases, human activities create unfavorable abiotic conditions that can reduce or eliminate species and upset the ecological balance. In the Impact Analysis Model, this fact is illustrated by an arrow drawn from the three forms of pollution to the biotic components of ecosystems (Figure 3.13).

Tampering with Biotic Factors

Destroying Natural Systems   Human activities also deplete or destroy resources used by other species, thus impacting them directly. The diversion of mountain streams to supply growing cities, for instance, leaves many streams dry and, thus, kills many aquatic life forms. As another example, in order to build in coastal areas, developers often fill in marshes with dirt (Figure 3.15). Many other examples are cited in this book.

Introducing Competitors   One of the most common impacts we have on the biotic components of ecosystems occurs as a result of introducing foreign species. For instance, African honeybees, commonly known as "killer bees," were introduced to South America in 1956 by geneticist Warwick Kerr. The bees were brought to Venezuela in an attempt to develop a successful stock of honey producers to replace the docile European honeybees, which had fared poorly in the tropical climate.

Kerr hoped that interbreeding the two might yield a more successful tropical strain.

Knowing their aggressiveness, Kerr isolated the bees in screened-in hives. However, in 1957, a visitor unwittingly lifted the screen, allowing 26 queens and their entourages to escape. Trouble soon began. The killer bees quickly spread, moving a remarkable 350 to 500 kilometers (200 to 300 miles) a year, interbreeding with honeybees, and destroying the honey industry in Venezuela and other countries. The bees also assaulted people, horses, and livestock that crossed their paths.

Killer bees arrived in California in the mid-1980s on a load of pipe shipped by freighter from South America. To date, California officials have located at least four colonies and have had to destroy hundreds of commercial beehives to wipe out killer bees that may have mixed with the colonies. Unfortunately, though, the bee has moved into Texas and is expected to move north and east.

If the bee spreads into northern climates and breeds with its tamer cousin, thousands of hives will have to be destroyed to prevent further spread. This could cripple the $140-million-a-year honey industry. But it will be even more devastating to farmers, for each year America's honeybees pollinate 90 major crops, worth an estimated $19 billion.

Many biologists hoped that the northward spread of the killer bee would be halted by the colder climates. A study published in 1988, however, showed that killer bees could survive at 0° C for six months. It is feared that the bees could migrate as far north as Canada, causing widespread damage to the honey industry and to bee-pollinated crops.

Another even more recent example is the zebra mussel accidentally introduced into North America by tankers arriving from Europe. Already prevalent in the

Commonwealth of Independent States, the zebra mussel has invaded many lakes and streams from New York to Minnesota, where it spreads into water pipes from power plants and factories. Here, the mussel proliferates, greatly reducing the flow of water. The mussel also feeds on microscopic phytoplankton, which form the base of the food chain that supports fish populations.

Scientists of the Commonwealth of Independent States have tried many tactics to eliminate the mussel, but none have proved very successful. A group of scientists from Virginia recently proposed introducing blue crabs from Chesapeake Bay in hopes that the crabs would help control zebra mussels without becoming a pest themselves.

The saga of the African honeybee and the zebra mussel are only two of many biological nightmares created by the introduction of a foreign species into a new region. In such cases, foreign species proliferate because there are no competitors or predators to control them. However, not all such introductions have adverse effects. The ring-necked pheasant and chukar partridge, both aliens in this country, have done well in some areas. In other cases, alien species have perished without a trace. Hardy species, such as the killer bee, however, are the ones that demand our attention and remind us of the folly of careless introductions.

Plants such as the prickly pear cactus, water hyacinth, and kudzu can also reproduce uncontrollably in foreign environments. Taking over a new territory, they can destroy native populations that compete for the same habitat. The results can be ecologically and economically disastrous.

Eliminating or Introducing Predators   Predators have never fared well in human societies. Early hunters and gatherers killed them for food and because they viewed them as competition for prey. Modern societies have carried on this dangerous tradition, killing bears, eagles, hawks, wolves, coyotes, and mountain lions with a vengeance, often with serious ecological consequences. (Case Study 2.1 on the Great Barrier Reef illustrates the profound impact of eliminating predators on the rich aquatic life zone off the coast of Australia.)

In some cases, problems arise when predators are introduced into new habitat. The mosquito fish, a native of the southeastern United States, for example, has been introduced into many subtropical regions throughout the world because it eats the larvae of mosquitoes and thus helps to control malaria, a mosquito-borne disease. Unfortunately, the mosquito fish also feeds heavily on zooplankton, single-celled organisms that consume algae. By depleting zooplankton populations, the fish removes environmental resistance that curbs algal growth. This causes algae to proliferate and form thick mats that reduce light penetration and plant growth in aquatic ecosystems.

These examples illustrate that altering ecosystems by introducing or eliminating predators can drastically affect ecosystems as well as human populations.

Introducing Disease Organisms   Organisms that cause disease, **pathogens**, are a natural part of ecosystems, usually held in check by a variety of factors. Unfortunately, humans have unwittingly introduced pathogens into new environments where there are no natural controls. There, they have reproduced at a high rate and caused serious damage.

In the late 1800s, for example, a fungus that infects Chinese chestnuts was introduced accidentally into the United States. It had been carried in with several Chinese chestnut trees brought to the New York Zoological Park. The Chinese chestnut has evolved mechanisms to combat the fungus and is immune to it. But, the American chestnut, once a valuable commercial tree found in much of the eastern United States, had no resistance at all, and was virtually eliminated from this country between 1910 and 1940 (Figure 3.16). (See Case Study 3.1 for a description of the accidental transplantation of a parasite that has devastated the oyster industry of northern France.)

Simplifying Ecosystems

Tampering with abiotic and biotic factors tends to simplify an ecosystem by reducing species diversity. Ecosystem simplification occurs most often when natural ecosystems, such as forests and grasslands, are converted for human use, for example, into farmland. Grasslands contain many species of plants and animals. When plowed under and planted in one crop, called a **monoculture**, the field becomes simplified and vulnerable to insects, disease, drought, wind, and adverse weather.

The reasons for this susceptibility are many. Perhaps one of the most important is that monocultures provide a virtually unlimited food source for insects and plant pathogens, especially viruses and fungi. As crops grow, food supplies increase dramatically, favoring massive growth of pest populations. Viruses, fungi, and insects become major pests because monocultures provide little or no environmental resistance. Protecting monocultures from harm leads to many environmentally harmful practices, notably the application of pesticides with severe environmental impacts, as exemplified by the DDT story cited earlier.

Why Study Impacts?

The Impact Analysis Model allows us to predict impacts on entire systems and thus serves as a preventive tool. It alerts us to potential damage and helps us to seek ways that avoid it, a task essential to building a sustainable society. More commonly in our frontier society,

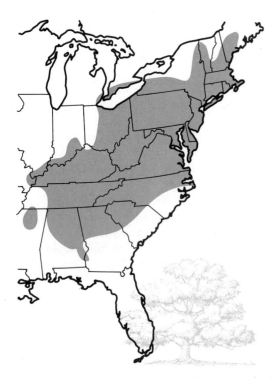

Figure 3.16 *Former range of the American chestnut, a species nearly wiped out by the accidental introduction of a harmful fungus.*

agencies may result in alterations of impact statements, that is, high-ranking officials may change the results of predictions made by agency scientists. This kind of deceit is, according to insiders, more common than one might think. Such activities tend to serve short-term economic interests to the detriment of the long-term future.

Assessing the Probability of Impacts

The critical thinking skills described in Chapter Supplement 1.1 suggest the need to question all conclusions, a rule that should apply to EISs. The discussion on science and scientific method in that section also noted that science doesn't have all the answers. We must expect some uncertainty. In analyzing impact, then, the degree of uncertainty of any given impact is important to know. As critical thinkers, we need to ask about the degree of probability of impact. When assessing the impact of any project, it is also important to note both positive and negative impacts.

impact analysis is used to find ways to **mitigate**, or offset, damage. For example, the loss of recreation opportunities created when a dam is built might be compensated by opening another previously inaccessible stream to recreational users.

In the United States, all projects on federal land or supported by federal dollars require an **environmental impact statement** (**EIS**), which outlines the environmental consequences of proposed actions, as well as the potential social, cultural, and archeological impacts. Many states require similar analyses. An EIS also requires some token review of other options.

After completion, an EIS is subject to public review. Then, the federal agencies responsible for the project either approve or deny it. Although this system may sound good, it is fundamentally flawed. Most EISs focus their attention on only one project and ways to mitigate the impact. To promote sustainable development, say some critics, an EIS should require us to analyze several options thoroughly, then require selection of the option that most promotes desired objectives. This system of **least impact analysis** or **sustainability analysis** might shift our attention away from large environmentally and economically costly projects.

Unfortunately, the science of predicting human impacts on natural systems is imprecise, often requiring those who write these reports to rely on educated guesswork or speculation. Another problem of profound importance is that political pressure in state and federal

3.6 Restoration Ecology: Sustaining the Biosphere

One secret of living sustainably on the planet is to restore the ecosystems we damage. Restoring coastal wetlands throughout the world, for instance, is important not only to protect wild species but also to ensure an adequate food supply for the world's people.

The Birth of a New Science

In 1985, Ed Garbisch began the long and costly process of restoring an ancient freshwater wetland in New Jersey. In a $4 million project, Garbisch and his coworkers seeded marsh grass, cut channels to restore water flow, and built knolls for nesting sites for waterfowl. Despite initial skepticism, Garbisch and his coworkers restored the swamp to its previous condition.

Garbisch and others like him have given rise to a new field of ecology, called **restoration ecology**, the scientific study of how ecosystem recovery occurs and how it can be facilitated. Restoration ecology is one branch of science known as **conservation biology**, which seeks to understand natural ecosystems and ways to protect and restore them.

In many respects, restoration ecology is akin to rehabilitative medicine. Some of the earliest restoration projects were designed to reclaim badly damaged land that had been surface-mined for coal and other minerals. Today, however, ecologists and others are working to restore marshes, tropical rain forests, streams, and prairies throughout the world.

Ecologist John Berger has spearheaded the movement to restore the Earth's many battered ecosystems.

Case Study 3.1

Upsetting the Balance: The Accidental Parasite

Ten years ago, a seemingly harmless event, a shipment of oysters from California to the north coast of France, began the decline of a once-lucrative oyster-growing industry. The California oysters carried with them a parasite that, in the ten years that followed, invaded oyster beds throughout France's northern coast. Today, in this region where 50,000 oyster farmers once thrived, only 10,000 manage to earn a meager income. They succeed primarily because they are growing a Japanese strain that does poorly in the colder waters. European oysters account for just 10% of Brittany's current oyster harvest.

Henri Grizel, an internationally known mollusk pathologist, warned Brittany's oyster growers early in the infestation not to transfer oysters from bay to bay.

He argued that this would facilitate the spread of the disease. Oyster growers typically move their oysters between different waters as many as five times before selling them because, they say, it speeds up growth and improves the flavor. However, the farmers ignored his advice, and through their own actions helped spread the protozoan that is virtually destroying the industry they relied on.

In an attempt to reestablish a healthy industry, researchers are transplanting three- and four-year-old European oysters that have proven immune to the protozoan to regions free of the disease. With careful work and time, the rich oyster beds may someday return to the north of France, providing a sustainable food source.

His book *Restoring the Earth* describes what people are doing in the United States to correct past mistakes. Berger also founded a nonprofit environmental group of the same name, dedicated to restoration of damaged ecosystems. Because of his work and the work of Garbisch, hundreds of marshes have been restored along the East Coast from Maine to Virginia, reestablishing native plants and habitats for many species.

One ambitious project is that of Daniel Janzen, a biologist at the University of Pennsylvania. With the help of others, Janzen hopes to reforest 3900 hectares (9600 acres) of dry tropical forest in northwestern Costa Rica. Janzen's project will take 100 years or so to complete, one-fifth of the time it would take for nature to reestablish forest that has been cut and burned over the years.

Benefits of Restoration

Restoration offers economic, environmental, and aesthetic benefits. For example, protecting an eroding shoreline with concrete or rocks can cost $500 per meter. In contrast, planting a 7-meter- (21-foot-) wide strip of salt marsh to protect the shoreline may cost only $50 to $80 per meter. This can result in a savings of $100,000 or more to owners of shoreline property. Plants can also turn a barren, desolate beachfront into a dense, lush, emerald marsh that attracts a variety of colorful birds.

Creating a marsh costs $5000 to $25,000 per hectare. Although a marsh's value cannot easily be calculated in dollars and cents, biologists in Louisiana estimate that an acre of salt marsh is worth about $85,000, solely for its ability to reduce water pollutants,

such as sediment. Moreover, marshes are one of the world's most productive ecosystems. A single hectare of marsh produces more than 25 tons of organic matter per year, which is more than twice the yield of a corn field and 10 times the yield of coastal waters. Coastal marshes are also an important source of food (Chapter 8). Today, approximately one-half of the salt marshes in the United States have been obliterated. When a salt marsh disappears, shorelines erode at an accelerated rate, fish populations collapse, birds vanish, wildlife retreats, and some of nature's remarkable plant communities are destroyed.

Restoring forests also offers economic, environmental, and aesthetic benefits. One extraordinary example of restoration that illustrates the economic benefits comes from California. In the late 1970s, the Secretary for Natural Resources, Huey Johnson, launched the ambitious program called Investing for Prosperity. Johnson rallied support from industry, banks, agriculture, labor, and environmentalists to convince the state legislature to appropriate $400 million over a four-year period that would be invested in California's economic future.

One of the projects financed by the state was a massive replanting effort on private forestland. At that time, California had 13 million acres of timberland, only about 8 million of which were forested. The remaining acreage needed planting, but for one reason or another, the owners could not afford to borrow the money. The state provided $5 million to private landowners to replant 5 million acres of private land, turning barren hillsides once highly susceptible to erosion into productive forest again.

A study of the project showed that over the next 50 to 75 years, the initial $5 million investment would return more than $400 million in timber sales and $100 million in tax revenues, far in excess of what might be earned by simply investing the money at a 10% rate of return. This project could provide a sustainable source of income to many local communities.

Controversy over Restoration

Some environmentalists worry that restoration legitimizes further environmental destruction. The Sierra Club, for instance, points out that developers who want to build on existing wetlands and other natural ecosystems frequently offer to replace them with wetlands "created" elsewhere. In fact, in Washington and other states, developers are already looking at restoration as a new tool to build on wetlands. If they can make a wetland elsewhere, they ask, why shouldn't they be allowed to destroy an existing one?

A closer look at the issue, essential for critical thinking, suggests several reasons why. First, wetlands are complex ecosystems. A flooded field planted with some swamp vegetation is a far cry from the natural system it replaces. Although it may come to resemble the lost swamp in a decade, it doesn't replace the wetland that was lost. Second, rebuilding wetlands or other ecosystems is a complex, costly task. It requires expert attention and follow-up.

In closing, the end quote in this chapter points out that the fate of the biosphere is in our hands. If so, a knowledge of the ways ecosystems operate is essential to wise stewardship of the Earth's remaining systems. It is essential to foster the restoration of those we have damaged.

For the first time in the whole history of evolution, responsibility for the continued unfolding of evolution has been placed upon the evolutionary material itself. . . . Whether we like it or not, we are now the custodians of the evolutionary process on earth. Within our own hands—or rather, within our own minds—lies the evolutionary future of the planet.

— Peter Russell

Critical Thinking Exercise Solution

To evaluate this proposal, you need to look at the big picture—not just how much money might be brought in by hunting but the systemwide effects of introducing a foreign species. First, you need to know if the deer would adapt to the region into which they're to be introduced. Is their habitat in Siberia similar to their new habitat? If not, would they perish in their new home because environmental conditions are unsuitable?

Second, assuming conditions are appropriate, you need to estimate how the transplanted deer population would grow. Would it grow too fast? Would numbers increase so much that the deer would destroy the habitat? Would hunters be able to keep the population under control to keep it from destroying the range?

Third, you have to study native grazers to see if the introduction of deer would wipe them out. Would the introduced deer bring with them any diseases that could affect native or domestic animals?

Fourth, you need to know whether the deer would remain in the region or if they would likely spread elsewhere, affecting other large grazers, such as white-tailed deer, mule deer, and elk.

As this example shows, many factors must be considered in addition to the possible increase in income from hunting licenses and other incidental expenses. Critical thinking requires a careful look at all possibilities.

Summary

3.1 Ecosystem Homeostasis

■ **Homeostasis** refers to a process by which organisms maintain internal conditions within an acceptable range. Homeostasis is a fairly steady state maintained by internal processes that adjust for changes in internal and external conditions.

■ Ecosystems also contain mechanisms that help to maintain relative constancy under many conditions.

3.2 What Factors Contribute to Ecosystem Homeostasis?

■ Population size within an ecosystem is determined by the interaction of numerous biotic and abiotic factors.

■ Those factors that increase population size are called **growth factors**; those that depress it are called **reduction factors**, which collectively produce **environmental resistance**.

■ Growth and reduction factors interact in ways that tend to promote ecosystem homeostasis.

■ **Species diversity** may also contribute to ecosystem stability, although direct evidence supporting this hypothesis is rare.

3.3 Correcting Large Imbalances in Ecosystems: Succession

■ Small shifts in the biotic or abiotic growth or reduction factors may temporarily tip ecosystem balance, but appropriate responses within the biological community can return the system to normal.

■ Larger shifts may result in a dramatic destabilization of the ecosystem, resulting in its collapse.

■ A biological community destroyed by such large shifts may recover during the process known as **secondary succession,** in which new communities develop sequentially on the remains of the old until a **climax community** is formed.

■ **Primary succession,** on the other hand, is the sequential development of communities where none previously existed.

3.4 Evolution: Responding to Change and Maintaining Ecosystem Homeostasis

■ Life on Earth sustains itself in part because populations evolve in response to change.

■ Evolution results in organisms better adapted to their environment.

■ Evolution depends on two factors: genetic variation and natural selection. **Genetic variation** refers to naturally occurring differences in the genetic make-up of organisms in a population. **Natural selection** is a weeding-out process in which less fit organisms are eliminated from populations.

■ One important measure of **fitness** is the efficiency with which organisms use resources, especially water and energy. Sustaining life on Earth depends in part on the presence of resource-efficient strategies. Human survival may require similar measures that are consciously chosen.

■ Natural selection, particularly directional selection, is responsible for changes in species and the formation of new species, in response to changing environmental conditions.

■ In instances in which environmental conditions are unchanging and when a species is already well adapted to that particular environment, natural selection helps to maintain the status quo by eliminating variants and favoring the "average" individual, a process called **stabilizing selection.**

3.5 Human Impact on Ecosystems

■ Humans cause imbalance in ecosystems by altering abiotic and biotic factors.

■ Pollution severely disrupts abiotic conditions, for instance, by altering the chemical balance in ecosystems or changing physical conditions.

■ Many actions influence biotic factors directly. Introducing or eliminating competitors, predators, and pathogenic organisms, for example, can have devastating effects.

■ Tampering with abiotic or biotic factors in the ecosystem can reduce species diversity and simplify ecosystems, making them susceptible to various forms of stress.

3.6 Restoration Ecology: Sustaining the Biosphere

■ Ecosystem damage continues today, but a growing number of people are finding ways to repair the damage.

■ **Restoration ecology** is a branch of science aimed at studying natural recovery and ways that humans can facilitate the process.

Discussion and Critical Thinking Questions

1. Describe the term *ecosystem homeostasis.* Does it mean that ecosystems stay the same?

2. If you were to examine a mature ecosystem over the course of 30 years at the same time each year, would you expect the number of species and the population size of each species to be the same from year to year? Why or why not?

3. What is environmental resistance? What role does it play in ecosystem balance?

4. Give evidence that species diversity affects ecosystem stability. Is there any evidence to contradict this idea? What is it?

5. What is a mature ecosystem? What are its major features? How does it differ from an immature ecosystem?

6. Do you think that human communities resemble mature or immature ecosystems? If immature, are there any dangers one should be aware of?

7. Describe temporary imbalances caused in ecosystems you are familiar with and how the ecosystems respond to them.

8. The process of secondary succession is nature's way of restoring itself. In some cases, restoration may be impossible. Why?

9. What is the difference between primary and secondary succession? Why does secondary succession generally occur more rapidly than primary succession?

10. Explain why organisms in the pioneer and intermediate communities are replaced by others.

11. Describe how introducing competitors into an ecosystem can affect ecosystem stability. Give some examples.

12. Give some examples of ways humans tamper with abiotic ecosystem components and describe some effects on plants and animals.

13. From an evolutionary standpoint, discuss why simplified ecosystems, or monocultures, are highly susceptible to pests.

14. Define the following terms: *evolution*, *natural selection*, *genetic variation*, *directional selection*, and *stabilizing selection*.

15. Describe how geographical isolation results in speciation.

16. In what ways has the study of ecology broadened your view of life? Has it made you reconsider any of your views?

Suggested Readings

Berger, J. J. (1985). *Restoring the Earth. How Americans are Working to Renew Our Damaged Environment*. New York: Knopf. A delightful book that's a must.

———, ed. (1989). *Environmental Restoration: Science and Strategies for Restoring the Earth*. Washington, D.C.: Island Press. Technical discussion of restoration.

Chiras, D. D. (1992). *Lessons from Nature: Learning to Live Sustainably on the Earth*. Washington, D.C.: Island Press. Shows how we can apply the biological principles of sustainability to modern society.

Ehrlich, P. R. and Roughgarden, J. (1987). *The Science of Ecology*. New York: Macmillan. Higher-level coverage of ecology.

Hudson, W. E., ed. (1991). *Landscape Linkages and Biodiversity*. Washington, D.C.: Island Press. Describes the importance of protecting biological diversity.

Odum, E. P. (1989). *Ecology and Our Endangered Life-Support Systems*. Sunderland, MA: Sinauer. Excellent introduction to ecology.

Ruckelshaus, W. D. (1989). Toward a Sustainable World. *Scientific American* 261(3): 166–75. Excellent introduction to the importance of sustainability.

Smith, R. L. (1992). *Elements of Ecology* (3rd ed.). New York: Harper and Row. Advanced readings on ecology and environmental problems.

Wann, D. (1990). *Biologic: Environmental Protection by Design*. Boulder, CO: Johnson Books. Calls for a revolution in our way of life to include nature-compatible designs.

Chapter 4

Human Ecology: Tracing Our Past, Charting a Sustainable Future

Nobody knows the age of the human race, but everyone agrees that it is old enough to know better.

—Anonymous

Critical Thinking Exercise

Lynn White, professor of history at UCLA, published a paper in 1968 arguing that the emergence of science and technology four generations ago in western Europe and North America spawned an era of massive environmental manipulation. This change was paralleled by an enormous ecological backlash, as evidenced by pollution, extinction, and other damage. Professor White, however, argued that science and technology were greatly influenced by the Bible, which instructs humans to have dominion over the animals and plants and to subdue the Earth. That thinking, he says, is ultimately responsible for the massive ecological transgressions that have occurred over the centuries. Can you see any problems with this logic? What critical thinking rules helped you to analyze this issue?

In 1948, the noted British astronomer Sir Frederick Hoyle predicted that "once a photograph of the Earth, taken from the outside, is available . . . a new idea as powerful as any in history will let loose." It was not too many years later that the first photograph of Earth from outer space came to us, and Hoyle's prediction proved true. In sharp contrast to the darkness of space, the brilliant sphere with its gossamer veil of clouds was breathtaking and yet disturbing, for it showed our home, which we had always seen as inexhaustible, as a tiny, isolated body in a vast universe. Sparkling in the sun's rays, the Earth seemed exquisite, fragile, and, knowing how we have treated it, vulnerable.

In a speech before the United Nations in the 1960s, Adlai Stevenson called our home *Spaceship Earth*, likening the planet to a self-contained spacecraft whose life-support systems recycle all matter necessary for astronauts to survive. This analogy underscored one of the most important lessons of ecology: All life on Earth is dependent on nutrient recycling (Chapter 2). Nutrients are a common thread in the web of life that links organisms to one another and to the environment. Today, the web of life with us in it is threatened.

How did we reach such a precarious place? To answer this question, we will trace human evolution on two levels—biological and cultural. Such a study provides several insights into the modern environmental crisis. This chapter also presents two conceptual models to help you to improve your systems thinking abilities, which are vital to building a sustainable future.

4.1 Human Evolution: An Inquiry into Modern Human Behavior

The humorist Will Cuppy wrote, "All modern men are descended from a wormlike creature, but it shows more in some people." Actually, scientists believe that humans evolved from the tree shrew, an animal that lived 80 million years ago in Africa (Figure 4.1). With handlike paws, the shrew moved about the forest at night, feeding

Figure 4.1 *The tree shrew shown here is thought to have given rise to the first primates.*

(a) *(b)* *(c)*

Figure 4.2 *The first primates were prosimians, which are represented today by (a) the tarsier and (b) the lemur. (c) Monkeys evolved from prosimians. Notice the front-facing eyes, a characteristic of all primates.*

on insects. Over 50 million years ago, tree shrews gave rise to primates similar to modern-day tarsiers and lemurs. These early primates, known as **prosimians**, also lived in trees and eventually gave rise to monkeys (Figure 4.2).

Approximately 20 million years ago, apelike creatures known as **dryopithecines** evolved from the early monkeys. Scientists believe that the great apes—chimpanzees, orangutans, and gibbons—evolved from the dryopithecines. Although fossil evidence is sketchy, scientists believe that dryopithecines were also the predecessor of **hominids**—humans and their fossil relatives (Figure 4.3).

The first hominids to live on Earth belonged to the group called **australopithecines**. The oldest known australopithecine skeleton was unearthed by Donald Johnson, Yves Coppens, and coworkers in Africa, and is about 3.5 million years old. Living in southern Africa, australopithecines were hunters and gatherers who roamed the grasslands, walking upright in search of food.

Roughly 2 million years ago, *Homo habilis* emerged. Possibly an offshoot of australopithecines, *Homo habilis* were the first tool and weapon makers. They eventually spread from Africa to Europe and Asia. About 500,000 years later, a new form arose, *Homo erectus*. With a brain slightly smaller than ours, *Homo erectus* made more sophisticated tools and weapons, such as hand axes and spears. Anthropologists have found evidence in China that *Homo erectus* used fire to cook, to warm their caves, and to frighten away predators. Like their predecessors, they, too, were hunters and gatherers.

Evidence of the emergence of *Homo sapiens*, the species to which we belong, is scanty. Paleontologists believe that our species emerged approximately 400,000 years ago. One of the best-known examples is the European inhabitant, the Neanderthal. Like their predecessors, Neanderthals lived in caves, cooked their food on fires, and hunted animals with tools. They also gathered fruits, berries, grains, and roots. Neanderthals stood fully erect and had slightly larger brains than we do today. They even buried their dead. Paleontologists have found bear skulls at burial sites, along with food and flowers, which they believe are a form of offering left during a ritual performed by clans, the groups in which Neanderthals lived.

Modern humans, or Cro-Magnons, emerged about 40,000 years ago. Originating in Africa, then spreading to Europe and northern Asia, Cro-Magnons rapidly replaced Neanderthals, either killing or interbreeding with them. Characterized by domed heads, smooth eyebrows, and prominent chins, these hunters and gatherers were accomplished stone and bone toolmakers. Living in caves, which they decorated with elaborate art, Cro-Magnons may have had a fully developed language.

At the end of the last great Ice Age, Cro-Magnons spread across Siberia to the New World. Sweeping across North America, they may have been responsible for the extinction of many animal species like the mastodon, saber-toothed tiger, giant beaver, and other large mammals.

The evolution of primates leading to humans is marked by a number of significant developments that are vital to our understanding of the modern human

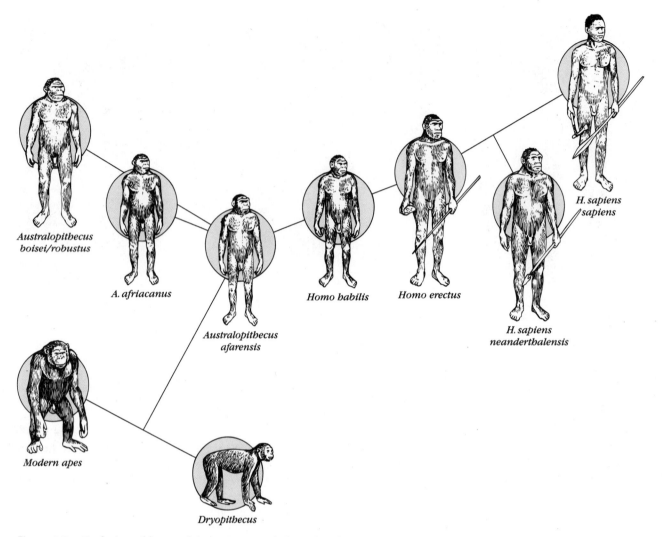

Figure 4.3 *Evolution of human beings. Scientists believe that dryopithecines gave rise to the hominid line.*

predicament. One important change was the development of bipedal (two-legged) locomotion. This development freed the hands to use tools and weapons, and may have facilitated the evolution of manual dexterity—the ability to manipulate objects, which is so essential to the development of technology. In addition, bipedal locomotion requires less energy than quadrupedal (four-legged) locomotion and, thus, probably gave hominid ancestors a competitive advantage over other animals.

A second important development in primate evolution was the increase in brain size. Precise hand-eye coordination afforded by a larger brain probably improved our ability to make tools and to shape the environment, which has markedly influenced the direction of human cultural evolution.

Technological development, resulting from the evolution of our hands and brains, has been a boon to humankind. Most importantly, technology has helped to unleash humankind from environmental resistance. For example, irrigation practiced by early residents of the Tigris-Euphrates Valley reduced the threat of food shortage, which limited population growth. In more recent times, medical technology has eradicated many infectious diseases, such as bubonic plague, that once devastated human populations. Expanded food production and other technological benefits have fueled the massive upsurge in the human population in the past 200 years.

A competitive advantage was also undoubtedly conferred by communication. The ability to create and com-

Table 4.1 Classification of Human Social Systems

Social System	Features
Hunting and gathering	1. The people were nomadic or semipermanent.
	2. They benefited from their intelligence and ability to manipulate tools and weapons.
	3. They were knowledgeable about the environment, and skilled at finding food and water.
	4. On the whole, they were generally exploitive of their resources.
	5. The environmental impact was generally small because of low population density and lack of advanced technology.
	6. They lived healthy lives, were well fed, and experienced low disease rates.
	7. Their widespread use of fire may have caused significant environmental damage in some locations.
Agriculture	1. Farmers were generally either subsistence level or urban based.
	2. They benefited from new technologies to enhance crops and resource acquisition needed for their survival.
	3. They were knowledgeable about domestic crops and animals.
	4. They were highly exploitive of their resources.
	5. The impact of subsistence-level farming was significant, but because population size was small, damage was minimized. The impact of urban-based agriculture was much larger because of new technologies, trade in food products, increasing population, and lack of good land-management practices.
	6. Disease was more common among city dwellers because of increased population density.
	7. Poor agriculture, overgrazing, and excessive timber cutting caused widespread environmental damage.
Industry	1. Industry includes early and advanced forms.
	2. It relies on new technologies, energy, energy-intensive forms of transportation, tremendous input of materials, reduced number of workers, and, recently, biotechnology.
	3. Mass production and modern technology are transferred to the farm.
	4. Industry is highly exploitive, more so than earlier societies; devoted to maximum material output and consumption.
	5. Impact is enormous and includes pollution, species extinction, waste production, dehumanization.
	6. Humans bcome subject to infectious disease and new industrial-age diseases including ulcers, heart disease, and mental illness.
	7. Widespread environmental damage results from industry, agriculture, and population growth.

municate ideas, design technologies, plan for the future, and manipulate objects to our liking gives human society a power unequaled in the biological world. Viewed by many as a means of expanding our capability to alter nature for our own benefit, these and a host of other evolutionary developments have helped to spawn the crisis of unsustainability. In some respects, we've become too smart and too handy for our own good. Our goal today is to learn how to redirect our thinking to create life-styles and technologies that promote a sustainable way of life. To understand the types of changes that are necessary, it is useful to study where we've come from and how we've changed culturally.

4.2 Human Cultural Evolution: Our Changing Relationship with Nature

Anthropologists recognize three major social groupings of human societies: hunting and gathering, agricultural, and industrial (Table 4.1). Although all forms exist today, hunting and gathering societies are present only in isolated regions of tropical rain forests, in parts of Africa, and elsewhere. Many are threatened with extinction, as indicated in Case Study 4.1. Agricultural societies have also largely disappeared, leaving energy- and resource-intensive industrial societies as the dominant form.

Case Study 4.1

The Forgotten People

The term *endangered species* refers to plants and animals destroyed or threatened by habitat alteration. But, certain human populations—tribal peoples—are also threatened with extinction. About 1 of every 25 humans alive today is an Eskimo, Pygmy, Bushman, Indian, aborigine, or some other tribal member. These people live as their ancestors did, as hunters and gatherers or subsistence-level farmers.

Tribal peoples have been uprooted in the name of progress to develop farms, mineral deposits, timber, dams, reservoirs, and wildlife parks. Often driven from their homelands, they are forced into areas unlike those in which they have lived for centuries. In Paraguay, for example, the remnants of the Toba-Maskoy tribe have been moved from their rain forest home to an arid region where their survival is in doubt.

Those native people who are allowed to remain in their homelands are susceptible to new diseases introduced by developers. Brazil's Indian population has shrunk from 6 million to 200,000 since the first Portuguese explorers arrived in the early 1500s. Although war was responsible for some deaths, diseases brought from foreign lands were, and still are, the greatest killers. Barbara Bentley, the director of Survival International (an organization dedicated to the protection of tribal people), says, "The easiest way to dispose of these isolated tribal people is by sneezing."

A government investigation in Brazil revealed that agents of a government bureau charged with protecting the native peoples actually practiced genocide by deliberately introducing smallpox, influenza, tuberculosis, and measles into Indian groups, which wiped out large numbers of native people. The same agents had joined with land speculators and white landowners in systematic murder and robbery of the native peoples.

Some tribes have been "assimilated" into the invading culture with disastrous results. Having been suddenly catapulted two or three centuries in time, they become lost, frightened, and confused by modern technology. Often, they return to their homelands only to find them destroyed. Loss of homeland and traditional values can lead to fatal mental trauma. From Brazil to Australia, alcoholism, severe depression, and poverty take their toll. Once-skilled hunters are often reduced to begging.

The stories continue: Copper mining in Panama threatens thousands of Guaymi Indians; Kalinga tribes in the Philippines fight the construction of hydroelectric dams that would flood their rice terraces; 300,000 Chilean Mapuche Indians have recently been told that their land will be opened for timber cutting; in Peru, the long-isolated Amuesha Indians are threatened by a new highway that would link them to civilization.

The elimination of these cultures will put an end to age-old languages, myths, and social customs and result in the irreversible loss of knowledge, including information on medicinal plants, dyes, and diet. Tribal peoples are responsible for discovering more than 3000 plant species with antifertility properties, a potential boon for birth control research. Some of their plant materials give promising clues to cancer prevention and cure.

The very cultures modern civilization is systematically destroying also offer considerable guidance on sustainable practices. In fact, many goals of the sustainability movement, among them cooperative relationships and holistic thinking, are commonplace in ancient cultures. Thus, the question of whether we can afford to allow these tribal people a continued existence has become this: Can we afford to live without them?

Bushwoman from the Kalahari Desert, in Namibia, resting on her digging stick.

(a)   (b)

Figure 4.4   *Hunting and gathering societies wander the land in search of food and water.* (a) *The men hunt.* (b) *The women are the primary food and fuel gatherers, foraging for wood, roots, berries, bark, and other plant products.*

Each cultural manifestation has used various forms of technology to manipulate the environment, but the impacts vary considerably, depending on the level of development. The most primitive forms of technology, and the least damaging, exist in hunting and gathering societies. The most advanced and Earth-threatening forms are found in industrial societies.

As Vice President Al Gore points out in his book *Earth in the Balance*, during our journey from hunters and gatherers to modern industrialists, our entire relationship to the planet has been dramatically transformed. The following material highlights some important changes in the human-environment relationship over our long cultural evolution.

Hunting and Gathering Societies

Hunting and gathering societies were the dominant form of social organization throughout most of our 3.5-million-year evolutionary history. In fact, 99% percent

of the time spent on Earth, humans have gathered fruits, seeds, and berries and hunted animals. Only in the last few fleeting moments of geological time have they turned to agriculture and industry.

New anthropological evidence suggests that hunters and gatherers may have been much less skilled at hunting large animals than commonly thought. Pat Shipman, an anthropologist at Johns Hopkins University, asserts that hunters and gatherers probably gained a substantial amount of their meat by scavenging, picking the bones of animals that had been killed by others or had died from natural causes.

In any event, studies of present-day hunters and gatherers suggest that members of these societies had a profound knowledge of the environment. Experts in survival, they knew where to find water, edible plants, and animals; how to predict the weather; and what plants had medicinal properties. The Bushmen of southern Africa, for example, can find water in the desert where others fail. The Australian aborigines can locate and catch a variety of lizards, insects, and grubs far better than some of the best-trained field biologists.

Studies of present-day hunters and gatherers also suggest that, contrary to popular conception, in many regions, our early ancestors did not live under the constant threat of starvation and did not spend the greater part of their lives in search of food. Studies further suggest that they were healthy, well nourished, and suffered from few diseases.

Many hunters and gatherers were nomads, wanderers who foraged for plants and captured a variety of animals using only primitive weapons. Because their technology did not give them a great advantage over other species, their populations never grew very large.

Judging from existing hunting-and-gathering societies, anthropologists believe that hunters and gatherers had a deep reverence for the environment and the plants and animals on which they depended. These people understood that they were part of the Earth, dependent on its bounty. But, this is not to say that all hunters and gatherers were environmentally benign. In the Great Plains region of North America, in fact, some Indian tribes ignited grass fires to drive buffalo over cliffs for slaughter, killing many more than they needed. Such a wasteful action does not fit the image of a wise steward of the land. The plains tribes of Canada also depended on intentionally set fires to burn clearings to maintain habitat for deer. Today, Canada is dotted with large, open meadows; ecologists believe that without natural fires and the fires set by native tribes, the land would be covered with unbroken coniferous forests.

Hunters and gatherers fashioned tools from sticks, stones, and animal bones to enhance their survival. On the whole, these humans presented little danger to wild species. Even so, many scientists believe that hunters and gatherers were responsible for the extinction of many species of large animals after the last Ice Age, among them the cave bear, giant sloth, mammoth, giant bison, mastodon, saber-toothed tiger, and giant beaver. These animals may have been killed directly, driven from their preferred habitats, or wiped out as their prey were destroyed.

Some hunters and gatherers developed semipermanent life-styles, setting up homes near rich hunting or fishing grounds that could provide a year-round supply of food. These groups were more likely to cause noticeable damage. New research on hunting and gathering societies also suggests that, for thousands of years and well before the advent of the Agricultural Age, many groups grew their own food and raised animals to feed themselves. Food may have been traded with other groups, thus marking the advent of the first system of commerce. Cave dwellers in Europe, 28,000 to 10,000 years ago, for example, probably participated in extensive networks set up to trade food and other valuable commodities.

On balance, the hunters and gatherers had little impact on the environment and their way of life was generally sustainable, as witnessed by their long history. This form of social structure was sustainable precisely because it did not violate the principles of sustainability. Their numbers were held in check by natural forces. Their demands were small. Because they generally lived nomadic life-styles, the damage they created was easily repaired. Because they lived off the land's renewable resource base and recycled their waste, they fit well within the economy of nature. Moreover, they seem to have exhibited a reverence for the Earth. In short, their life-style and ethics fostered sustainability.

Agricultural Societies

Anthropologists believe that **agricultural societies** emerged between 10,000 and 6000 B.C. The roots of agriculture can be traced to Southeast Asia (Figure 4.5). Here in the moist tropical rain forests, early humans practiced slash-and-burn agriculture. In **slash-and-burn agriculture,** farmers cleared small sections of jungle to plant their crops. They then harvested and planted the same plot for several years, but because the jungle soils were poor in nutrients, crops eventually failed and were abandoned for new clearings. Native species invaded, returning the land to its original state. Damage to the jungle was negligible, and restoration was possible. (For a discussion of this practice in modern times, see Models of Global Sustainability 4.1.)

The early agricultural societies of Southeast Asia also domesticated many animals, such as pigs and fowl. These became vital food sources, greatly supplementing food from crops.

Seed crops originated in a wide region extending from China to eastern Africa (Figure 4.5). The first farmers cleared woodlands known for their rich soils. Agriculture was limited until the development of the plow, which permitted farmers to cultivate fertile grassland soils that had previously been too difficult to plow because of the heavy sod and thick roots of grasses. Invented in the Middle East around 3000 B.C., the first plow was nothing more than a tree limb with a branch that cut through the topsoil. In ensuing years, more elaborate plows were developed and pulled by oxen. This enabled farmers to cultivate grassland soils, dramatically increasing crop production. A variety of domesticated animals, such as goats, sheep, and cows, supplemented the human diet.

The plow gave agricultural societies the means to greatly increase the productivity of the land. As a result, farmers achieved a greater degree of control over their destiny, making an important shift in the human-environment interaction. For the first time, human populations could, by manipulating their environment, expand beyond the limits set by the natural food supply.

Models of Global Sustainability 4.1

Agroforestry: Tapping Ancient Wisdom

The destruction of tropical forests is an issue of worldwide concern. Not only does it wipe out thousands of wild species, it destroys vital economic assets, increases soil erosion, and contributes to global warming.

As Chapter 23 explains, government policies in industrial and nonindustrial nations contribute significantly to rapid deforestation, despite official endorsements of conservation goals. In developing countries, for example, governments seeking to raise foreign exchange earnings or to finance economic development programs often turn to their forests.

Forest destruction also results from rapid population growth in many tropical areas. Migration to deforested regions is often viewed as a means of relieving overcrowding in urban settings and landlessness in agricultural regions.

To curtail rampant deforestation, some experts believe that we need to find ways to use tropical rain forests sustainably. One candidate is the practice called agroforestry. **Agroforestry** is a sustainable management system that combines agriculture and/or livestock with tree crops and/or forest plants. These activities occur on the same parcel of land, either simultaneously or sequentially. Agroforestry maintains or improves the environmental quality of the area, and provides income and food. It can be practiced on forestland that is already degraded and cleared.

Traditional agroforestry systems have evolved over many years. In fact, forests have been prudently managed by indigenous peoples for many generations without any apparent loss of species diversity or deterioration of soil quality.

One form of agroforestry is swidden agriculture. In **swidden agriculture**, people clear small plots of rain forest, where they plant annual food crops and perennial tree crops. When the soil loses its fertility, the plots are abandoned and allowed to regrow. This restores the soil in time, making it useful once again.

The Bora, a native Amazonian tribe of northeastern Peru, practice swidden agriculture. Some scientists believe that large areas of the Amazon forest may have been managed in this manner by considerably larger populations in the past.

Studies indicate that swidden farming is not new to native peoples in the Amazon. The Kenyah people, who live in a remote section of Indonesia, have practiced swidden agroforestry for at least 200 years. The Kenyah have converted much of the virgin forest into a mosaic of secondary forest of different ages, but have retained uncut reserves where they harvest products that are rare or nonexistent in the secondary forest.

Many agroforestry systems studied to date have been those practiced by relatively remote tribes, suggesting that these systems are suitable only for subsistence production. Recent studies, however, prove that they can also be important cash producers near urban centers. In Tanshiqacu, Peru, for example, a nontribal group of 2000 persons of mixed Amazonian and European ancestry produce food, fiber, handicraft materials, and charcoal from the forests. Their products appear in local markets as well as in markets in Lima, the capital, 30 kilometers (18 miles) to the northwest. Most households maintain several agroforestry fields simultaneously and sell a variety of products, a tactic that ensures a year-round income.

In Java, farmers clear forests and plant teak trees, whose wood is exported to many developed nations. Among the teak trees, they plant rice and corn. This system of agroforestry has been modified to include horticultural species, animal food, and fuelwood crops, which provide additional income.

In the Philippines, the rural poor venture into steep, mountainous terrain to clear forests to make room for crops. To prevent widespread destruction of these regions, governmental agencies constructed numerous agricultural demonstration plots during the 1980s to introduce agroforestry practices to the rural people. Since deforestation of steep hillsides leads to soil erosion, the first step in these projects was to stabilize the soil by laying bamboo poles along the contour of the land. The bamboo barriers capture soil, preventing it from washing down the slopes. Fast-growing trees were planted along the contours as well. The trees provide wood for fuel, leaves for animal feed, and soil enrichment. The seeds of these trees are used to produce traditional medicines. Fruit trees, such as mangoes, and food crops, such as pineapples and cassavas, were planted in the fields. Today, the agroforestry plots provide a year-round food and fuel supply for the people. By stabilizing the soil, the plots reduce the need to cut more forest.

As these examples illustrate, agroforestry provides many benefits. It helps protect virgin forest and provides a sustainable supply of food, fuel, and fiber. It also helps reduce soil erosion and land destruction; it benefits the rural poor, providing, in addition to food, income and employment opportunities. Agroforestry is not a panacea; it's just one of many strategies for living sustainably on the planet.

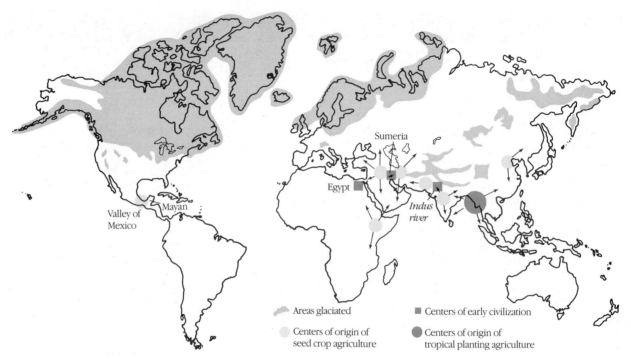

Figure 4.5 *Roots of agriculture. This map shows areas where tropical planting and seed crop agriculture originated.*

Before the development of the plow, most farmers fed themselves and their families. The plow changed all that. At least two significant transformations followed. First, the human population began to swell because more food was available. Second, since fewer people were needed to provide food, many left the farms and moved to villages and cities, where they took up crafts and small-scale manufacturing. Cities became centers of trade, commerce, government, and religion. The face of civilization changed forever.

The plow marks a pivotal point in our cultural evolution—the beginning of the era of modern technology. The growth of population, cities and towns, and small-scale industry placed greater demands on the environment for resources, such as wood, metals, and stone. Heightened exploitation accompanied by poor land management resulted in widespread destruction of the natural environment. Many fertile areas were destroyed by overgrazing, excessive timber cutting, and poor agricultural practices.

The shift to mass-produced food had a potentially more harmful effect: In large part, it severed the link to nature. Unlike hunters and gatherers and early subsistence farmers who depended on a wide variety of plants and animals for food, the new generation of farmers relied on a small number of plants and animals. The profound knowledge of the environment characteristic of their predecessors all but vanished. Agriculture became a way of dominating nature, replacing the cooperation and understanding that had marked earlier cultures.

The environmental impact of more recent agricultural societies and the large urban centers they supported was enormous. Archaeological and historical records show that overgrazing, widespread destruction of forests, and poor farming practices changed many once-productive regions into barren landscapes (Figure 4.6). Ancient civilizations perished as a result, either directly as their crops failed or indirectly as other displaced peoples invaded.

This decline was especially evident in the Middle East, North Africa, and the Mediterranean from 5000 B.C. to A.D. 200. As an example, the Babylonian Empire once occupied most of what is now Iran and Iraq. At the outset, this land was covered with productive forests and grasslands. Huge herds of cattle, goats, and sheep overgrazed the grasslands, however, and eventually destroyed the natural vegetation. Forests were cut to provide timber and create more pasture. The loss of grassland and forest vegetation caused a decrease in rainfall and eventually parched the land. Sediment washed from the barren soils, robbing them of nutrients and filling irrigation canals. These changes and a succession of invading armies eventually destroyed this once-great empire.

This story has been repeated throughout the Mediterranean region. For example, all across Saharan Africa, in what are now dry, uninhabitable regions, remnants of once-wealthy cities can be found buried in the sand.

As this discussion shows, agricultural societies were far more damaging to the environment than hunters and

Figure 4.6 *This ancient city was once surrounded by rich forests and grasslands. Deforestation, overgrazing, and poor agricultural practices turned the land into desert, a trend that continues today at an alarming rate.*

gatherers. This new social form departed significantly from the sustainable practices of their forebears. Because of the development of new technologies, more advanced agriculturalists had a significant impact on grasslands and forests, taxing nature's ability to restore human-caused damage. The emergence of newer technologies also ushered in the age of commerce, which increased the demand for metals and other nonrenewable resources. With rising commerce, material demand increased, and waste from shops and small factories began to accumulate outside cities. With rising demand, human societies began to deplete local resources, a trend that is reaching global proportions today. New, more abundant food sources resulted in an increase in population size, further straining the ties with nature. Finally, as commerce emerged and became established, humans began to see the natural world much differently—more as a source of wealth. As a result, agriculturalists lost the physical and spiritual connection with nature that characterized their hunting and gathering predecessors.

No longer held in check, their numbers began to rise. Their damage became extensive, and in some cases irreversible. They turned more and more to nonrenewable resources to fuel commerce.

Industrial Societies

The **industrial society** is a recent phenomenon in human history. Spawned by the **Industrial Revolution**, a drastic change in manufacturing marked by a shift from small-scale hand production to large-scale machine production, industrial societies emerged first in England in the 1700s and then in the United States in the 1800s. The Industrial Revolution brought about further changes in the human-environment interaction. Residents of the industrial world became even more physically and spiritually disconnected from the Earth, our source of all materials and a repository for all of our wastes.

The invention of coal-powered machines made manufacturing more capital and energy intensive and less labor intensive (Figure 4.7). With the increase in machine labor came a dramatic escalation in energy demand and a need for new means to transport goods to and from the city. Industrialization also led to mass production, which meant that more goods were available to more people. As demand and industries grew, the influx of materials—fuel, food, minerals, and timber—into the city rose sharply.

The shift to machine production changed the working environment, the city, and the surrounding countryside that supplied the resources. The new manufacturing technologies were the fruit of scientific and engineering advances. They were complex and often made work meaningless and boring, and produced large quantities of smoke, ash, and other wastes.

Mechanization also swept the farm (Figure 4.8). Technological advances, such as Jethro Wood's cast-iron plow with interchangeable parts and Cyrus McCormick's reaper, rapidly increased agricultural production. Perhaps one of the most significant advances was the invention of the internal-combustion engine, which made horse-drawn implements obsolete. The motor-powered tractor alone could plow as much land in a week as one of our forebears could work in a lifetime using hand tools. Because of more efficient farming methods, fewer farm workers were needed. Unemployed workers migrated to cities, swelling their populations.

Another significant advance was the development of fertilizers, which allowed an increase in agricultural productivity (output per hectare). Plant breeding, which produced higher-yield crops, also contributed to a rise in output per hectare.

New medicines and better control of infectious disease through improved sanitation during the Industrial Revolution enhanced human survival. As people began to live longer, the human population rapidly increased and demand for resources began to climb.

The agricultural transformation, the development of industry, the growing demand for resources, and population growth had tremendous environmental repercussions. Pollution became more widespread. Agricultural expansion, needed to feed the larger population, destroyed wildlife habitat, depleted soils, and caused severe soil erosion. Sediment from fields polluted waterways. To expand commerce, cities dredged new harbors, destroying wetlands vital to fish and shellfish. Mine waste, city sewage, and industrial discharges pol-

Figure 4.7 *The Industrial Revolution contributed to a massive upsurge in human population, increased the demand on resources, and resulted in widespread pollution and habitat destruction.*

luted rivers and lakes and wiped out native fish populations such as the Atlantic salmon.

The shift to an industrial society further distorted the human relationship with nature as humans sought more control over the environment to ensure their own survival. Industrial people came to view themselves as apart from nature and superior to it. The prevailing attitude was summarized by the 17th-century British philosopher John Locke, who argued that the purpose of government was to allow people the freedom to exercise their power over nature to produce wealth. "The negation of nature," Locke argued, "is the way toward happiness." People must become "emancipated from the bonds of nature." Locke also preached unlimited economic growth and expansion, with the belief that individual wealth was socially important for a harmonious society. These notions have been passed from generation to generation throughout the Industrial Age.

Industrial societies became engaged in a battle with nature. New medicines to combat disease, improved sanitation, an arsenal of chemicals to fight pests, and new technologies to extract resources more efficiently were the key weapons. On the whole, people ate better, lived better, and began to live longer, but environmental deterioration was the price we paid.

Advanced industrial societies arose in the period following World War II. Several major features distinguished this new form of society: (1) a marked rise in production and consumption; (2) a shift toward synthetics, such as plastics, and nonrenewable resources, such as oil and metals; and (3) huge increases in energy demand for farming, industry, and day-to-day living.

Advanced industrial societies are caught in an ever-escalating production-consumption cycle and have only begun to awaken to the costs of environmentally irresponsible and unsustainable behavior. Domination of nature continues as the central theme of modern industrial societies, and economic growth retains its commanding allure, despite evidence that both threaten the long-term future of our planet.

One threat of considerable importance is posed by synthetic substances so common in advanced industrial societies. Among them are plastics, nylon, and chemical pesticides, all derivatives of petroleum. Synthetics create problems in nature because bacteria in soil and water, which decompose natural materials, are frequently unable to break down synthetics. Synthetics may therefore persist in the environment for decades. The persistent insecticide DDT, for example, can accumulate in the fatty tissues of birds and disrupt reproduction. (Chapters 3, 14, and 18 present more detailed discussions of this phenomenon.) Plastic pollution in water has proven to be a problem for some aquatic organisms (see Chapter 17).

Another significant change from the days of hunting and gathering is a dramatic increase in energy use, a

Figure 4.8 *Modern agriculture depends heavily on machinery, energy, and additional resources. Large fields are worked to achieve maximum output.*

trend with serious consequences. In hunting and gathering societies, for instance, individuals require only about 2000 to 5000 kilocalories of energy per day. (Kilocalorie and Calorie, a measure used by dieters, are equivalent.) The 2000 to 5000 kilocalories used by hunters and gatherers provide metabolic energy and additional heat to cook food. Early agricultural societies, however, required twice that amount per person to grow crops. Advanced agricultural societies required twice as much again, or about 20,000 kilocalories per person per day. In industrial societies, per capita energy use climbed again, increasing to 60,000 kilocalories per day per person. Modern advanced industrial societies, with their mechanized farming, energy-intensive industries, and affluent life-styles, doubled that amount, raising energy consumption to about 120,000 kilocalories per person per day. Especially wasteful nations, such as the United States, Canada, and the former Soviet Union, consume, on average, 250,000 kilocalories per person per day.

What's the significance of this? Most forms of energy in use today contribute significantly to environmental decay. They cause most of our air pollution and can result in significant amounts of water pollution, as witnessed by the 1989 oil spill off the coast of Valdez Alaska. Another problem with the energy we use is that it produces carbon dioxide. Carbon dioxide gas released in large quantities now threatens to change the global temperature in ways that alter rainfall patterns, which could disrupt agriculture (Chapter 16).

This survey of human cultural evolution illustrates three points. First, physical and spiritual human ties to the biosphere have been severed over time. Today, in many parts of the world, humanity is living out of touch with the Earth. We are carelessly destroying those very systems upon which we rely. The second key point is that technological development, which has played a

major positive role in our cultural evolution, is also a major determinant in the modern environmental crisis. Third, over time, the human-environment relationship has become increasingly unsustainable. Clearly, the practices of those who live in industrial societies are a far cry from the sustainable ways of early hunters and gatherers. Modern industrial societies use enormous amounts of materials, often inefficiently. They recycle only a fraction of the waste produced in the production-consumption cycle. They rely principally on nonrenewable resources and restore only a tiny portion of the ecosystems they invade. Moreover, their populations have largely been released from natural checks and balances that held numbers in check for most of the 3.5-million-year history of humankind. Attitudes in modern industrial societies neither display respect for nature nor aspire for a peaceful coexistence with the environment. These changes now threaten our existence. To reverse these trends, we need a new sustainable ethic and environmentally sustainable technologies and life-styles based on the principles of biological sustainability. This text shows how this transition might be achieved.

4.3   Understanding the Human-Environment Interaction

Having examined the evolution of human culture, it is important to consider how humans interact with the environment. In Chapter 3, you learned that human activities affect both the abiotic and biotic conditions of the environment. This section shows *how* these impacts arise by presenting another model, the **Population, Resources, and Pollution (PRP) Model** (Figure 4.9).

The PRP Model illustrates several important relationships between humans and their environment. It offers a panoramic view, the big picture, of the human-environment interaction, which is vital to becoming a critical thinker. A view of the whole also helps promote systems thinking. Because an understanding of the big picture is vital to finding systemic solutions, this model also facilitates efforts to build a sustainable society.

The Population, Resources, and Pollution Model

The PRP Model shows that human populations acquire resources from the environment, like all other organisms. The acquisition of resources, for example, coal mining, alters both the abiotic and biotic conditions of the ecosystems. Surface coal mining disrupts wildlife habitat and can lead to pollution; hence the arrow between resource acquisition and pollution in the model. Improper timber cutting in the Pacific Northwest and in the tropics also disrupts habitat and can result in the erosion of tons of sediment into nearby streams.

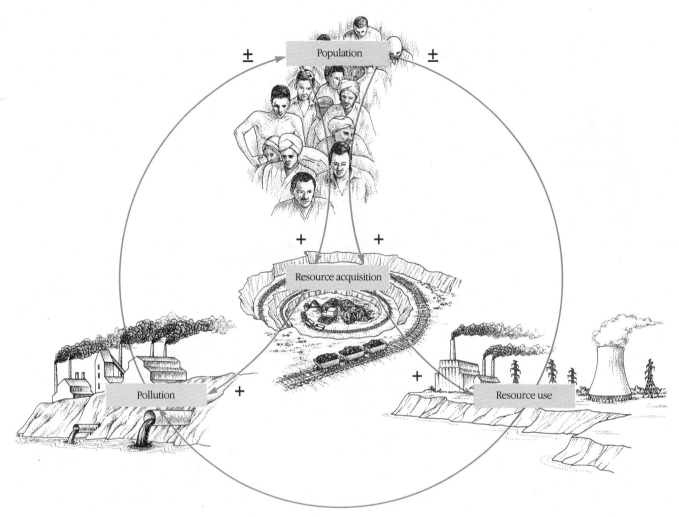

Figure 4.9   *The Population, Resources, and Pollution Model applies to all organisms. It outlines how organisms interact with their environment. The + signs by the arrows indicate a positive feedback loop, where one activity enhances another. The − signs indicate a negative feedback loop, where one activity adversely affects another.*

The resources we acquire are put to use, as indicated by the arrows connecting population with resource use. Coal, for instance, is mined, then burned in power plants. This produces incredible amounts of pollution, as indicated by an arrow connecting resource use to pollution. This model reminds all of us that everything we purchase comes from a mine, forest, field, stream, or the ocean. In acquiring these resources, some physical damage occurs. Nearly everything we buy is also manufactured, which results in pollution of air, water, and soil.

Not all pollutants are created equal. Some are **biodegradable**, that is, broken down by bacteria and other organisms in air, water, and soil. Others are **nonbiodegradable**, not broken down by organisms. Both forms can significantly alter abiotic and biotic conditions of ecosystems.

Scientists also classify pollutants by the medium they contaminate, for example, *air* pollution, *water* pol-

lution, and *land* pollution. In recent years, scientists have found that pollution readily crosses the boundaries between these media, a phenomenon called **cross-media contamination**. Some air pollution, for example, washes from the sky and is deposited in lakes and forests. Some water pollutants evaporate from lakes and streams, entering the atmosphere, only to be deposited downwind in rain and snow. Hazardous wastes deposited in the land drip into groundwater.

The PRP Model says that human populations acquire and use resources from the environment and that these activities have negative impacts on the air, water, and soil. That is, they degrade the environment by altering the abiotic and biotic conditions. Negative impacts of development on the human population are indicated on the PRP Model by minus signs on the arrows that lead from pollution and resource use to population. This response is an example of a **negative-feedback loop**, where one factor leads to a decrease in a second factor.

Negative-feedback loops are the chief mechanism for controlling biological systems and regulating homeostasis (Chapter 3). Many human systems are controlled by negative feedback. A familiar example is the furnace in your home, which is regulated by a thermostat that detects room temperature. When room temperature drops below the setting on the thermostat, the furnace turns on. Heat from the furnace then warms the room air until the air temperature reaches the desired setting on the thermostat. When it does, the furnace shuts off.

Scientists are gravely concerned with the negative feedback loops shown on the PRP Model. The loops could, in fact, have catastrophic effects on human society. Continued soil erosion from the world's farmland, resulting from poor agricultural practices, for example, could cause a marked decline in food production that could have a devastating effect on the global human population.

As the previous discussion of cultural evolution pointed out, resource acquisition and use (especially that permitted by technology) also have positive effects: They enhance our survival and promote population growth. The efficient harvest of food and fiber, for example, has made it possible for large numbers of people to inhabit the Earth's many biomes. This relationship, indicated in the model by the plus signs on the arrows leading from resource use to population, is an example of a **positive-feedback loop**, in which one factor leads to the growth of a second. It, in turn, stimulates the first one in a repetitive cycle. Positive-feedback mechanisms are rare in nature and may backfire in the human-environment system for they can create devastating cycles of depletion and pollution. For example, although fossil fuel energy has increased our capacity to produce food, it could be creating a dangerous rise in global temperature and a massive shift in rainfall that might devastate crop production on much of the existing farmland (Chapter 16).

The PRP Model represents a fundamental ecological relationship true to all living organisms. It's as relevant to humans as it is to black bears. What is more, it provides some insight into the human-environment interaction and allows us to predict the impacts of human actions. Change one variable in the model and all the rest change. Add more people, for example, and resource use is bound to rise. Acquire more resources and pollution is likely to climb. The model also provides insight into sustainable solutions. To solve rising pollution levels, resource demand might be cut. To control resource depletion, controls on population growth might be desirable.

The PRP Model takes on a greater significance when we expand its principal elements. For example, population may be expanded to include factors such as growth rate or distribution of people within a country. Both have important implications for pollution levels. (Table 4.2 lists some additional factors discussed in more detail throughout this text. You might want to study it

Table 4.2   Breakdown of Factors of the Population, Resource, and Pollution Model

Population	Resources	Pollution
Size	Acquisition	Water
Distribution	Use	Land
Density	Supply	Air
Growth rate	Demand	
	Character	

before reading the next section. The models in this section will also prove useful in your study of environmental problems and your quest to find solutions.)

Studying the Interactions: Cross-Impact Analysis

The three elements of the PRP Model interact with one another in so many ways that only a computer could keep track of them. We can get a glimpse of how they affect one another by using the simple technique called **cross-impact analysis**. This technique helps break down complex interactions into simpler ones that are more easily understood. Although it is important to be a systems thinker, it is also important to be able to extract parts of the system and elaborate on them. This model helps you to do that. The technique is sometimes necessary when designing sustainable solutions.

Figure 4.10 is a simplified cross-impact analysis chart. In this chart, the three elements of the PRP Model are lined up in two perpendicular columns. In Box 1, you can describe how population factors (Column A) affect resource factors (Column B), such as supply, demand, acquisition, and use. In Box 2, you can describe how population factors affect pollution, and so on.

Column A			
Population	X	1	2
Resources	3	X	4
Pollution	5	6	X
Column B	Population	Resources	Pollution

Figure 4.10 *The cross-impact analysis chart shows how one factor impacts another. For example, you can analyze how population growth (Column A) affects resource demand (Column B) and pollution (Column B). Box 1 indicate the interaction of population growth and pollution. This chart helps you to systematize potentially complex issues.*

Take some time to fill out the chart. You will find that you already know a great deal about the interactions of population, resources, and pollution. At the end of the course, you may want to repeat the exercise; if you have been conscientious in your study, you will find that your knowledge has grown tremendously and that you have a deeper appreciation of the complexity of modern society. You've become an accomplished systems thinker.

The Root Causes of the Environmental Crisis: The Multiple Causation Model

With this basic understanding of human-environment interaction in mind, we now turn to another useful tool, the **Multiple Causation Model**. This model illustrates an important fact often overlooked in debates over environmental issues or efforts to find solutions: Many factors contribute to environmental problems (Figure 4.11). To create lasting solutions, we must consider the complex web of causation. So, keep this model in mind. You may be amazed how many times it will help you to decipher the complex lines of cause and effect.

In the late 1960s and early 1970s, two noted scientists engaged in a debate over the underlying causes of the environmental crisis. Paul Ehrlich, a biologist at Stanford University, argued that the root cause of all our environmental problems was overpopulation—too many people. His argument was countered by biologist Barry Commoner, who contended that technology was primarily to blame. Their debate, although important, misled the scientific community and the public for years.

On closer examination, it becomes clear that the complex problems of resource depletion, pollution, wildlife extinction, and food shortages are not the result of technology or overpopulation alone but are caused by a variety of factors, as shown in the Multiple Causation Model.

To illustrate the use of this model, we look at wildlife extinction, a problem frequently blamed on habitat destruction and commercial and sport hunting. This view is greatly simplified. To see why, let's examine each element of the model individually.

Population   The human population destroys wildlife habitat in many ways. We build roads through forests once occupied by grizzly bears; we build airports and homes on wetlands that were once home to muskrat, ducks, and a variety of other animals; we pollute streams once home to fish and otters; we strip-mine for coal and gravel, destroying vegetation vital to elk and deer. Each of these activities decreases available habitat.

The extent of environmental damage, however, depends on numerous population factors, such as population size, growth rate, and geographical distribution. In general, the larger the human population, the greater its impact. Likewise, the faster the population grows, the greater its impact.

Per Capita Consumption   The amount of resources used by each member of society—per capita consumption—plays a key role in the destruction of wildlife habitat. For example, in the United States, per capita consumption of water has increased 300% in the past 80 years. Electrical energy consumption has more than doubled in the past 20 years. This rapid increase in resource consumption diminishes resources available to other species, and, in many cases, also pollutes once-healthy ecosystems. The cars we drive pollute the air animals breathe; the oil we use to make the gasoline that powers our automobiles can, if accidentally released, pollute oceans and rivers. Clearly, then, how much a society consumes plays a significant role in wildlife extinction.

Politics and Public Policy   The legal system also has an important role in wildlife extinction. Laws can affect how much habitat is destroyed, how much hunting and poaching occur, and which species will or will not be hunted. Laws can influence population growth, which, in turn, affects wildlife habitat. For instance, some argue that tax laws that give families deductions for each child reduce the financial burden of rearing children and may, therefore, promote population growth. Laws and regulations also affect how we protect the environment from pollution. Strict pollution control laws, for example, can lessen the impact of human society. Lax laws increase it.

Economics   As you will see throughout this text, economics is a key element in all environmental issues, among them animal extinction. One of the driving forces of most economies is profit. The profit motive often dictates shortcuts in production. For example, a company may choose to invest little in pollution control because it makes for higher profit. Such shortcuts obviously have an effect on wildlife habitat.

Economics affects many of our daily business decisions. On the surface, there is nothing wrong with this. However, economics often fails to account for external costs: pollution, habitat destruction, and other environmental impacts that are not entered into the cost of producing goods. Air pollution from factories, cars, and power plants, for example, may cause up to $16 billion worth of damage to crops, health, forests, and buildings in the United States each year. These external costs, or **economic externalities**, are borne by members of the public who suffer ill health, pay a higher price for food, or pay for repairing buildings. If companies invested in pollution control devices, the external costs would be greatly reduced, but consumers would pay a higher price

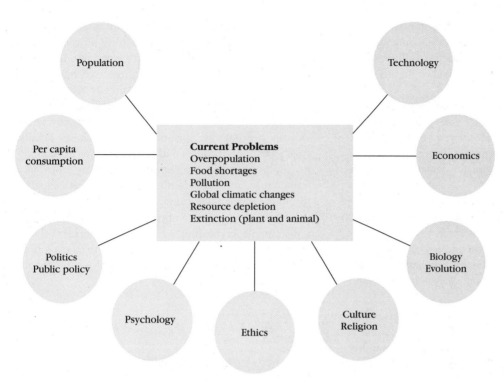

Figure 4.11 *Multiple Causation Model. Environmental problems result from many obvious and not so obvious causes. Critical thinking demands a full understanding of the big picture—the many underlying factors.*

for energy and goods. Thus, a system of economics in which consumers pay to reduce the external cost would diminish environmental pollution and its threat to wildlife habitat.

Our own personal economic decisions are also important determinants in the destruction of the environment and the loss of wildlife habitat. By choosing a cheaper product that may fall apart in a few years and require replacement, we end up consuming more. By making decisions based solely on affordability, ignoring their cumulative impacts on the biosphere, we help fuel a growth-based economic system that is destroying wildlife habitat at a rapid rate.

Psychology, Culture, Religion, and Ethics   How we behave toward the environment is a function of our attitudes and ethics as well as our laws and economic system. The frontier ethic is a major factor in environmental destruction in general and habitat destruction in particular, especially in the industrialized Western world (Chapter 1).

Underlying all of our behavior is a psychology that tends to put immediate human needs before the long-term good of the environment, future generations, and other species. In other words, we worry about our creature comforts but ignore the comfort of other creatures. Although such thinking is shortsighted, it is definitely a biological characteristic. Don't be fooled for a minute; survival in the animal world is dependent on looking out for one's own needs. Being animals ourselves, we tend to think and act with an eye to the immediate.

Because our population is so large and our technologies are so well developed, however, short-term thinking leads to many problems, such as wildlife extinction.

Another crucial psychological/ethical element is the way we perceive our place in the environment. Do we see ourselves as superior to and at odds with the natural world, or are we part of nature, willing to live in accordance with ecological laws? An attitude of superiority and of being separate from nature predominates today. It gives many license to run roughshod over wildlife and plants, contributing to the extinction of many species.

Ethics is dramatically shaped by religion. Christianity, for instance, teaches that humans are apart from nature and that nature is here to serve us. Some argue that this viewpoint, so prevalent today, is one of the root causes of the modern environmental crisis. They note that if animals are revered for cultural or religious reasons, extinction from hunting and habitat destruction may be curbed. (Chapter 20 takes a closer look at the debate and presents some alternative views.)

Technology   Technology is an important factor in the environmental crisis in general and wildlife extinction in particular. Technology refers to instruments or tools of human society used to achieve human needs. At first, primitive technologies helped humankind make better use of available resources and protect itself from threats. More advanced technology has helped us manipulate and control the environment to improve our lives. Technology clearly gives us an advantage over other species in the struggle for survival. As a result of technological developments, we travel freely around the world and through space, inhabit new regions, and acquire resources from distant sites. Technology allows us to prosper in environments where survival might otherwise be impossible.

Our travels, new settlements, and resource acquisitions, however, contribute to the shrinking supply of wildlife habitat. This happens partly because technology enhances population growth. More people need more resources and make a greater demand on wildlife habitat.

Technological advances have had other more direct impacts on wildlife. For instance, improvements in guns used for whaling have allowed humans to deplete whale populations severely. New chemical pesticides have increased agricultural productivity but have also devastated populations of birds and other animals.

Biology and Evolution   Numerous biological features play a role in wildlife extinction (Chapter 8). Some of the most important biological factors are adaptability, number of offspring, and sensitivity to environmental pollutants. Highly specialized species, like the California condor, are unable to adapt to changes brought about by human beings and are more vulnerable than less specialized animals, like the coyote, which seems immune to human presence. The number of offspring a species produces also affects the resistance of the species to human pressures, such as hunting, habitat alteration, and pesticide use. Susceptibility to environmental pollutants also varies considerably among plants and animals.

The Multiple Causation Model, like the PRP Model, can enlarge your understanding of environmental issues, helping make you a systems thinker. It will help you avoid oversimplified diagnosis and treatment of contemporary problems. In short, it helps make you a more critical thinker, better able to find sustainable solutions.

4.4   The Sustainable Revolution: The Cultural Imperative

For most of the evolutionary past, humankind has lived in a sustainable relationship with nature, not necessarily out of choice but out of necessity. If we didn't, we paid the consequences.

In the past few centuries, we have gone astray. Thanks to many developments in the Agricultural and Industrial revolutions, we are now living at odds with the natural world. As author Tim McMillan points out: "For 200 years we've been conquering nature. Now, we're beating it to death."

One thing that this text makes clear is that modern society as we know it cannot continue. Dramatic changes are necessary to create a sustainable future. Part 2 shows that the population growth curve must stop rising and offers solutions to halt the upward trend in population. Part 3 documents how the resources we rely on are reaching or, in some cases, have already reached their limits. It also points out sustainable resource strategies. Part 4 describes pollution and its effects and shows how we can prevent the poisoning of our planet. Finally, Part 5 outlines attitudes and practices of both economics and politics that are outdated, even dangerous, in a finite world, and suggests alternatives that may allow us to build a sustainable future.

Many changes recommended in this text are part of a Sustainable Revolution, which, some observers believe, may be the next phase of the human cultural evolution. These changes are not academic; they're vital to our survival and well being.

You can never plan the future by the past.

—Edmund Burke

Critical Thinking Exercise Solution

One of the critical thinking rules presented in Chapter Supplement 1.1 advises us to consider the source of information and, in doing so, to look for bias. Lynn White, a religious scholar, presents a view of the environmental crisis that is invariably biased by his background and education. White himself acknowledges that environmental destruction has occurred since antiquity, long before Christianity arose.

As this chapter explains, the root causes of the environmental crisis are varied and complex. I think that the problems are rooted in the phenomenon called *biological imperialism,* a natural tendency for populations to expand to make use of existing resources have been facilitated by technology. But, the advent of technology is rooted in the evolution of the human brain and the evolution of our remarkable ability to manipulate our environment. That said, human brain and the evolution of our remarkable ability to manipulate our environment. That said, many other factors still play an important role in environmental destruction, among them Christian teachings. However, to assert that Christianity is *the* root cause is an oversimplification. It fails to look at the bigger picture, another important aspect of critical thinking.

Summary

4.1 Human Evolution: An Inquiry into Modern Human Behavior

■ The origin of the human race can be traced back to tree shrews that lived in Africa 80 million years ago.

■ Over 50 million years ago, tree shrews gave rise to the **prosimians**, primates similar to modern-day tarsiers and lemurs. Prosimians, in turn, gave rise to monkeys, from which the apelike creatures known as the **dryopithecines** evolved about 20 million years ago.

■ Scientists believe that great apes and the first hominids, the **australopithecines**, evolved from the dryopithecines.

■ About 2 million years ago, *Homo habilis* evolved, probably from the australopithecines. These hunters and gatherers, the first toolmakers, originated in Africa and spread to Europe and Asia.

■ About 500,000 years later, *Homo erectus* evolved. With a brain slightly smaller than ours, these hunters and gatherers made sophisticated tools and weapons and used fire to cook, to warm their caves, and to frighten away predators.

■ Paleontologists believe that our species, *Homo sapiens*, emerged approximately 400,000 years ago. One of the best-known examples is the European inhabitant, the **Neanderthal**, who lived in caves, cooked food on fires, and hunted animals with tools.

■ Modern humans, or **Cro-Magnons**, emerged about 40,000 years ago. Arising in Africa, these hunters and gatherers then spread to Europe and northern Asia, and replaced the Neanderthals. Accomplished stone and bone toolmakers, Cro-Magnons may have had a fully developed language.

■ The evolutionary history of primates leading to humans is marked by a number of significant developments, vital to our understanding of the modern human predicament, among them bipedal locomotion, increased brain size, manual dexterity, and communication.

■ These evolutionary developments were essential to the development of technology, an important cultural advancement that helped unleash us from environmental resistance. This, in turn, has spawned massive population growth and considerable environmental damage, especially in the past 200 years.

4.2 Human Cultural Evolution: Our Changing Relationship with Nature

■ Anthropologists recognize three major social groupings of human societies: hunting and gathering, agricultural, and industrial.

■ Although all three forms exist today, hunting and gathering societies are present only in isolated regions.

Agricultural societies have also largely disappeared. Industrial societies are the dominant form today. Key features of each are listed in Table 4.1.

■ Each cultural stage is marked by various forms of technology used to manipulate the environment. The most primitive forms, and the least damaging, exist in hunting and gathering societies. The most advanced and Earth-threatening forms are found in industrial societies.

■ Technological development has resulted in an increasing control over the environment but a growing disconnection from the Earth.

■ Many observers believe that to build a new cultural configuration—one harmonious with natural systems upon which we depend—we need to develop a new ethic, one that fosters physical and spiritual connections.

4.3 Understanding the Human-Environment Interaction

■ The **Population, Resources, and Pollution (PRP) Model** illustrates fundamental ecological relationships in human society. It shows that in the process of acquiring and using resources, we alter both abiotic and biotic conditions of ecosystems.

■ Some pollutants produced by the acquisition and use of resources are **biodegradable**, that is, capable of being broken down by bacteria and other organisms in air, water, and soil. Others are **nonbiodegradable**, not capable of being broken down. Both forms can significantly alter abiotic and biotic conditions of ecosystems.

■ Scientists also classify pollutants by the medium they contaminate, for example, air pollution, water pollution, and land pollution. Pollution readily crosses the boundaries between these media, a phenomenon called **cross-media contamination**.

■ The acquisition and use of resources has both positive and negative impacts on the human population.

■ The PRP Model not only shows how we interact with the environment but also allows us to predict the impacts of human actions and provides insights into solutions.

■ The **Multiple Causation Model** illustrates many factors that contribute to environmental problems, among them population, technology, culture, religion, economics, ethics, biology, psychology, consumption, and public policy.

4.4 The Sustainable Revolution: The Cultural Imperative

■ For most of our history, humankind has lived in a sustainable relationship with nature. It is only in recent times that we have gone astray as a consequence of many technological developments produced during the Agricultural and Industrial revolutions.

■ One thing that this text makes clear is that modern society as we know it cannot continue. Dramatic changes are necessary to create a sustainable future.

Discussion and Critical Thinking Questions

1. What developments occurring during our biological evolution contributed to the modern-day environmental crisis? In your opinion, do these developments mean that we are doomed to continue to live unsustainably?

2. Describe each major form of society—hunting and gathering, agricultural, and industrial. How are they similar, and how do they differ? How did the human-environment interaction shift over time?

3. What general changes are needed to undo the adverse consequences of the Agricultural and Industrial revolutions?

4. Describe the PRP Model. Illustrate through examples some positive and negative feedback loops.

5. Using the Multiple Causation Model, analyze an environmental problem with which you are familiar and give information regarding each of the root causes. After you have analyzed the issue, find ways to address the root causes. Which ones do you think are the most amenable to change?

6. Have the models presented in this chapter helped to develop your thinking on environmental problems and solutions? How? Would their use help to expedite efforts to build a sustainable society?

Suggested Readings

Caufield, C. (1991). *In the Rainforest: Report from a Strange, Beautiful, Imperiled World*. Chicago: University of Chicago Press. Describes efforts to eradicate some indigenous cultures in Latin America.

Chiras, D. D. (1992). An Inquiry into the Root Causes of the Environmental Crisis. *Environmental Carcinogenesis and Ecotoxicology Reviews*. C10(1): 73–119. Detailed look at a network of root causes of the environmental crisis, especially biological and evolutionary roots.

Commoner, B. (1990). *Making Peace with the Planet*. New York: Pantheon. Insightful look into some of the root causes of the environmental crisis.

Isaacs, J. (1980). *Australian Dreaming: 40,000 Years of Aboriginal History*. Sydney, Australia: Lansdown Press. A wonderfully illustrated book on the hunters and gatherers of Australia.

McPhee, J. (1989). *The Control of Nature*. New York: Farrar, Straus & Giroux. Three case studies showing the extent to which modern society attempts to control natural forces.

Meadows, D. H., Meadows, D. L., and Randers, J. (1992). *Beyond the Limits: Confronting Global Collapse, Envisioning a Sustainable Future*. Post Mills, VT: Chelsea Green. Excellent reading.

Shipman, P. (1985). Silent Bones, Broken Stones. *Discover* 6 (8): 66–69. Insightful look at how anthropologists study earlier societies, bias, and some new insights into human evolution.

Part 2

Population

Chapter 5

Population: Measuring Growth and Its Impact

One generation passeth away, and another generation cometh, but the earth abideth forever.

—Ecclesiastes 1:4

Critical Thinking Exercise

An environmentalist speaking to a general audience on the population issue notes that 90% of the current growth in the world population occurs in developing countries. In order to slow growth and protect the environment, he says, something needs to be done about population growth in these countries. His basic philosophy is to attack problems where you can have the most impact. Do you think this singular focus on developing countries is justified? Why or why not? What critical thinking rules can you use to judge this assertion?

Throughout the world, many countries are taking steps to slow their population growth. In Kenya, which has one of the fastest growing populations in the world, large families have been the norm for years. However, because economic conditions are worsening, many parents are worried about supporting their children and are taking action. Between 1984 and 1992, the average number of children a Kenyan woman bears dropped from 7.7 to 6.7, according to statistics from the United Nations. Even though only 27% of women currently practice family planning, 75% of those polled said they would like to.

Even more encouraging news comes from Morocco. In 1980, women in this northern African nation had an average of 7 children. In 1992, that number dropped to 4.5. The use of **contraceptive measures**, devices or techniques that reduce the chance of fertilization, has increased to 40% of all married couples. This progress is even more encouraging when one considers Morocco's social system, where women are relegated to an inferior role, gaining status only by bearing children, especially sons.

Efforts to control the growth of the human population, such as these and the ones discussed in Models of Global Sustainability 5.1, are vital to building a sustainable human presence on the planet. This chapter explains why. It also explains why the human population has exploded and familiarizes you with some basic concepts and terminology needed to understand the population issue in order to make informed decisions about it. It ends with a sobering look at projections of future population growth.

5.1 Dimensions of the Population Crisis

The population crisis can be summarized in six words: too many people reproducing too quickly. That is, the problems posed by the human population are related to its massive size and its rapid growth rate. Let's consider size first.

Too Many People

In 1819, the British poet Shelley wrote, "Hell is a city much like London." If he were alive today, the poet might have put it differently: "Hell is a city much like Cairo, Calcutta, Shanghai, Bangkok, London, Los Angeles, and Mexico City." By the end of the century, at least 22 cities worldwide are expected to have populations in excess of 10 million people. Sixty others will probably exceed the 5-million mark. If present conditions are any indicator of the future, these urban centers will have enormous social, economic, and environmental problems, many linked to their massive human population.

The Plight of Cities In both rich and poor nations, overpopulation brings incredible despair. In Cairo, Egypt, for example, a giant graveyard is home to several hundred thousand people who cannot afford or cannot find housing. They live in makeshift cardboard shanties among the tombstones. Those of Cairo's over 12 million inhabitants who are lucky enough to find housing don't count their blessings. Apartments are tiny. The municipal water system breaks down frequently. One-third of the population lives in housing without toilets.

In Calcutta, India, conditions are even worse. More than 70% of the city's over 10 million people live at or below the poverty level. One water tap supplies, on the average, 25 slum dwellings, and about half of the homes in this city have no indoor toilet. The streets are littered with trash and feces, and an estimated 600,000 people roam the city, homeless.

Shanghai, China, a well-to-do city by comparison, is home to about 12 million people. Housing is provided by the government, but most residents have only 2 to 3 square meters each to call home—about as much space as a medium-sized bathroom. Middle-class Americans have 100 times the living space. The Huangpu River, which smells of raw sewage and industrial waste, runs through this crowded city. The city's air is polluted by cars and trucks powered by diesel fuel.

Cities in the developed world are much better off, but the signs of overpopulation are still evident, especially in the ubiquitous slums in cities such as New York, Chicago, Miami, and Los Angeles, where the poor and disadvantaged dwell. The living conditions in these areas are deplorable. Crime runs rampant, and the streets are strewn with trash. The air in many cities is badly polluted, and the worst conditions are generally encountered in the inner city. In Los Angeles, air pollution from vehicles kills trees in the neighboring mountains, burns the eyes of visitors and residents, and obstructs the remarkable views of the mountains. Rush-hour traffic clogs the highways (Figure 5.1) and raises tension, which

Figure 5.1   *Los Angeles rush-hour traffic: a nightmare on a good day. Overcrowded urban environments exhibit a host of troubling problems, including pollution, water shortages, and crime.*

ing from crowding, which led to a variety of symptoms not uncommon in people: ulcers, hypertension, kidney disease, hardening of the arteries, and increased susceptibility to other diseases. Other researchers argue that additional factors contribute to the stress, such as lack of education, poverty, poor nutrition, and inadequate housing.

Scientists are still unsure how crowding in cities affects humans. But, it is known that urban centers frequently exceed the capacity of the environment to assimilate wastes. Cities throughout the world are centers of intense pollution from automobiles, factories, power plants, and sewage treatment plants. Even pollution from homes contributes to urban air pollution.

Large urban populations also place considerable demands on the outlying countryside for resources, such as fuel, water, and food. And as cities spread, they gobble up farmland and open space, literally devouring the resources they need to survive and prosper.

In 1990, about 45% of the world's people lived in cities; by the year 2000, researchers estimate that over half will reside in urban areas. Without programs to make our cities considerably more sustainable, crowding, and the many problems that result from it, will probably worsen.

Rural Despair   A surplus of people strains rural areas as much as the cities. Rural Bangladesh, Kenya, Ethiopia, Mexico, and other developing nations, for example, reel under the oppressive burden of burgeoning populations. Overpopulation in these countries leads to unsanitary living conditions, water shortages, food shortages, and disease. In Africa, trees around rural villages are cut down for firewood, creating a frighteningly barren landscape. Livestock overgraze grasses, and once-fertile land turns into desert (Figure 5.2).

The problems caused by overpopulation are especially evident in Bangladesh, which lies just east of India along the Bay of Bengal. No larger than Wisconsin, Bangladesh houses ten times as many people, or about 112 million. Eighty-five percent of the population lives outside of cities. Rural and poor, Bangladesh is the most densely populated nation in the world and is also among the fastest growing.

Over the years, cattle have severely overgrazed the land, and hordes of desperate peasants have stripped away the vegetation in search of food, fuel, and shelter. When the rains come, floods follow because denuded hillsides cannot hold back the rain. Soil from these badly abused lands washes into streams. Sediment-choked streams flow to the sea and deposit their silt in river deltas. In search of farmland and a place to live, people flock to the deltas by the tens of thousands. As almost any ecologist will tell you, deltas belong to rivers and the sea, not to people. When the heavy rains and hurricanes arrive, as they do most years, high waters flood

has led to a number of shootings in recent years. Teenage gangs rob, brawl, and vandalize. Drive-by shootings and drug use are widespread. Jobless and homeless men and women rummage for food in dumpsters and garbage cans.

Crowding in urban centers has been implicated in a variety of social, mental, and physical diseases. Many social psychologists assert that social instability, divorce, mental illness, and drug and alcohol abuse are caused, in part, by stress from overcrowding. Prenatal death and rising crime rates may also be attributed to overcrowding. Psychologists call this the **inner-city syndrome**. Research on humans and animals supports their contention. The most notable study was performed by the psychologist John Calhoun. In his study, rats were confined to a specially built room and allowed to breed freely. As the population snowballed, Calhoun observed increased violence and aggression, abnormal sexual behavior, cannibalism of the young, and disruption of maternal behavior. Physiologist Hans Selye performed similar experiments with rats. In these studies, he observed hormonal imbalances induced by stress result-

Models of Global Sustainability 5.1

Thailand's Family-Planning Success Story

Paul and Anne Ehrlich noted in their book *The Population Explosion*, "The population/resource/environment predicament was created by human actions, and it can be solved by human actions. All that is required is the political and social will. The good news is that, when the time is ripe, society can change its attitudes and behavior rapidly." Thailand is a stellar example of what can happen when people decide to take action.

In 1971, Thailand adopted a national population policy. In the next 15 years, the country's population growth rate plummeted from 3.2% to 1.6%. During that period, the use of contraceptives by married couples rose from 15% to 70%. In 1992, Thailand's growth rate was 1.4%; contraceptive use remained about the same.

Thailand's success stems from many factors. The first is <u>religious</u> in nature. Ninety-five percent of the Thai people are Buddhists, and Buddhist scripture warns: "Many children make you poor." In addition, the Buddhist religion embraces family planning. In an effort to support population control, Buddhist monks distribute contraceptives with cards that remind the Thai people: "Many births cause suffering."

Another factor is <u>cultural</u> openness. The Thai people are open to new ideas and relationships between men and women are egalitarian. Consequently, women have equal say in decisions about family matters.

A <u>third</u> reason for the success in family planning is political. The government of Thailand encourages nationwide family planning and offers considerable financial support. Over the years, Thailand's government has made a wide range of contraceptives available to the public. It has also worked cooperatively with an influential nonprofit agency known as the Population and Community Development Association of Thailand (PDA) to elevate family planning to a national goal.

PDA was founded in 1974 by Mechai Viravaidya. Mechai's creative and high-profile methods of promoting family planning in Thailand are considered by some as the single most important factor in the country's remarkable success in population control. Mechai promotes condom use by handing them out in public at any opportunity he can find. He even sponsors a "cops and rubbers" program, during which policemen distribute condoms on New Year's Eve. Today, condoms are affectionately known in Thailand as "mechais."

Mechai started distributing condoms in crowds many years ago. Today, because of his efforts, condoms, birth control pills, IUDs, and spermicidal foams are all available in bus terminals. In addition, his organization has opened vasectomy clinics across the country and celebrates the king's birthday each year

by offering free vasectomies. Sterilization is the most widely used form of contraception in Thailand.

Mechai has also launched an economic development program. As a former government economist, he realized that income-generating alternatives could help change his people's attitude toward large families. Today, widespread commitment of the Thai people to family planning is believed by many to be associated with PDA's economic development programs.

With German financing, PDA introduced a revolving loan scheme for those interested in developing clean drinking water supplies. Loans, however, were only available to those who participated in family-planning programs. Money provided by the fund helped Thai people install thousands of toilets and containers to catch rainwater for drinking. PDA also initiated a program to support agriculture, which offered loans at rates far better than traditionally available. This helped those farmers participating in the family-planning program to avoid heavy debt so prevalent in the country.

The government of Thailand also sponsors a loan program. Previously based on the credit worthiness and character of the applicant and the planned project, the program now gives preference to those practicing family planning. Loan funds, in fact, are apportioned on the basis of a village's use of contraceptives. The total amount of the loan fund increases as the level of contraceptive use goes up. The program is meant to reward contraceptive use, not punish non-use. If the level of contraceptive use falls in the village, money is not taken out of the fund. In addition, participants who discontinue using contraceptives are still allowed to borrow money.

PDA's programs have been so successful in Thailand that they now offer a three-week training course to representatives from other developing countries. Two of the many participants are Bangladesh, where family planning is considered inappropriate, and the Philippines, which is predominately Catholic. The training course is designed to allow participants to develop ways to adapt PDA's strategies for use in their countries.

Thailand's success in family planning has been paralleled by economic success with a doubling of per capita income in the country since 1971. Economic progress has allowed the government to expand its family-planning services and medical facilities throughout the country. On a personal level, economic improvements have meant that families live better and have a wider variety of choices.

Family planning is even helping in the hill tribes of Thailand, where families with ten children have been the norm for many years and where contraceptive use has been held in low regard. The people of the hill

Models of Global Sustainability 5.1 (*continued*)

Thailand's Family-Planning Success Story

tribes still practice slash-and-burn agriculture, but their rising population threatens reforestation and forest conservation programs in northern Thailand. As a result, the Planned Parenthood Association of Thailand began offering Norplant, a contraceptive that is encased in small matchstick-size cylinders implanted under the skin, to tribeswomen. Norplant, which remains effective for five years, has been accepted by the women because it is safe, easily implanted, and

quickly reversible. Although it will take some time to determine Norplant's effectiveness to slow population growth among the tribes, it is an important step forward.

Thailand is still a long way from a stable population, but efforts to date have been remarkable, proving that when people recognize the problems of overpopulation and decide to take action, they can make a big difference.

these low-lying lands, driving people away and destroying their farms and homes. Thousands of lives are lost each year from storms.

In 1985, a hurricane brought a 5-meter (15-foot) storm surge, which crashed inland, devastating makeshift homes and farms on the deltas. The storm left 250,000 people homeless and killed an estimated 4000 to 15,000. Countless livestock died. In 1988, a similar disaster struck. Heavy rains flooded two-thirds of the land, leaving 25 million people homeless. In 1991, the disaster was repeated again, with an estimated 250,000 people killed.

On the surface, the disasters in Bangladesh look like especially severe natural disasters. On closer examination, however, it is clear that much of the blame can be pinned on an underlying human problem—too many people.

In an ironic twist, rural overpopulation feeds the urban crisis. Dismayed peasants and their families, unable to survive on their farms, migrate in large numbers to cities in search of jobs and security. In Mexico City, thousands of rural farmers and their families arrive each year. Nearly every major city in the world serves as a magnet for disheartened rural residents. Unfortunately,

Figure 5.2 *Overpopulation, desertification, and drought have diminished the resources available to these nomadic people in North Africa. Many other rural poor living at or beyond the carrying capacity of their environment face similar problems.*

what awaits them is intense crowding, skyrocketing unemployment, poverty, crime, inadequate food, and pollution.

Reproducing Too Quickly

Hunger, starvation, disease, poverty, illiteracy, pollution, unemployment, and barren landscapes are, to many observers, signs that the human population is already too big for the Earth's resources. Efforts to eliminate these problems have been frustrated by a more immediate problem: rapid population growth. Nowhere is this trend more evident than in Africa.

Fifteen years ago, few African nations admitted the need to control their rapidly growing population. Increasing at a rate of 3%—doubling every 24 years—the African continent fell behind in food production and economic development. Food-exporting African countries became major food importers. Economic development was unable to keep pace with the growth in human numbers. Disease continued to spread. The predicament they and other countries face is described by Paul Ehrlich in his classic book *The Population Bomb*:

> In order to just keep the standard of living at the present inadequate level, the food available for the people must be doubled every 24 years. Every structure and road must be duplicated. The amount of power must be doubled. The capacity of the transport system must be doubled. The number of trained doctors, nurses, teachers, and administrators must be doubled. . . . This would be a fantastically difficult job in the United States—a rich country with a fine agricultural system, immense industries, and access to abundant resources. Think of what it means to a country with none of these.

The world's fastest-growing areas are, in decreasing order, Africa, Latin America, and Asia. In Latin America and Asia, the growth rate has declined in recent years, but further decreases are still necessary to end the mounting environmental destruction and create a sustainable future. In many African nations, however, the rate of growth is still high. In some, it is even increasing, putting the future of this great continent in peril.

The condition of the planet and the health and welfare of all species, including humans, is greatly influenced by the size and growth rate of the human population. As the Population, Resources, and Pollution Model presented in Chapter 4 showed, overpopulation is one of the root causes of the environmental crisis. Robert McNamara, president of the World Bank for 14 years, maintains that "short of nuclear war itself, population growth is the gravest issue the world faces. . . . If we do not act, the problem will be solved by famine, riot, insurrection, and war." Solving the population, pollution,

and resource "trilemma" requires a substantial and immediate downturn in the population growth rate (Chapter 6).

5.2 The Population Explosion

In the 60 seconds it takes you to find this book on your bookshelf and turn to this page, the world population will have increased by 175 people, a rate of about 3 people every second! At this rate of growth, about 1.8 million people join the human population every week, approximately the number of people in Detroit. Overall, the world's population increases by about 93 million people a year. In 1993, world population reached an all-time high of 5.5 billion. By the year 2000, it will very likely exceed 6 billion and may be expanding by 100 million a year. Ninety percent of this growth takes place in poorer nations, where slightly over three-fourths of the world's people currently live.

The rapid growth of world population is a recent phenomenon. Throughout most of human history, our population was quite small (Figure 5.3). Two thousand years ago, for instance, about 200 million people inhabited the Earth. By 1850, though, world population had climbed to 1 billion (Table 5.1). In 80 years, it doubled, reaching 2 billion. Forty-five years later, it doubled again, reaching 4 billion. By 2017, it is expected to reach 8 billion.

In his book, *Earth in the Balance*, Vice President Al Gore puts the population growth in perspective. He points out that it took more than 10,000 generations for the world population to reach the 2-billion mark in 1945, the year of his birth. In his lifetime, about 3 generations, the population could climb to over 9 billion. What caused the rapid upsurge in population growth starting in the 1800s?

The Survival Boom

Throughout most of human history disease, famine, and war controlled the population. These checks on population growth are three of the many reduction factors that constitute environmental resistance (Chapter 3). In the course of human history, though, several important changes occurred, enhancing human survival and reproduction. These include the development of technology (tools and weapons), the Agricultural and Industrial revolutions, and the opening of the New World for settlement (Chapter 4). Each helped to unleash human population from environmental resistance.

Improvements in agriculture, such as the plow and irrigation, which occurred during the Agricultural Revolution, for example, increased food supplies, which spurred population growth. The most dramatic influence

Figure 5.3 *Exponential growth curve depicting world population. Human numbers were held in check by famine, disease, war, and primitive technological development for most of human history. Advances in technology and health care caused a massive increase in human numbers in the last 200 years.*

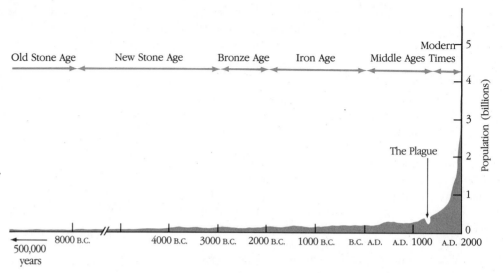

of technology, however, was felt during the Industrial Revolution, when an assortment of farm machinery improved food production once again and permitted the population to swell even more. Modern medicine also enhanced the human condition. New drugs, such as penicillin, dramatically lowered the death rate. Improvements in health care and sanitation decreased infectious diseases and increased survival rates. For example, the pesticide DDT was used in the tropics to combat malaria-carrying mosquitoes.

As a result of these and other developments, death rates in many countries dropped precipitously and life expectancy increased (Figure 5.4). In 1900, for example, the average white American female had a life expectancy of 50 years. By 1992, however, her life expectancy had climbed to 79 years. Life expectancy in white American males born in 1900 was about 47 years, but by 1992, it had increased to 72.

Contrary to what is commonly believed, the increase in **average life expectancy**, the average number of years a person lives, is not because Americans on average are actually living longer. The increase results from the fact that fewer of us are dying early in life. In other words, life expectancy has increased because more Americans are able to survive the very dangerous first year of life (Figure 5.5). Between 1900 and 1978, the life expectancy for an infant increased 26 years, because more were saved in the first year, but the life expectancy of a 60 year old increased by only 5.2 years.

A simplified example will help you to understand this phenomenon. Suppose ten babies were born to an island population, but five of them died in the first year. The rest lived to be 70. The average life expectancy of that group would be a paltry 35 years. Suppose that some time later, a doctor arrived on the island. The next year, ten more babies were born. Because of improved health care, only two died in the first year, whereas the rest lived to be 70. The average life expectancy would leap from 35 to 56 years.

During the Industrial Revolution, as life expectancy increased and death rates dropped, birth rates continued at high levels, at least for a while in many countries. This discrepancy in birth and death rates is responsible for the rapid population growth that continues today in many countries.

Expanding the Earth's Carrying Capacity

Each advance cited above—technology, agriculture, medicine, and others—has helped humans increase the Earth's **carrying capacity**, the number of organisms (in this case, humans) the Earth can support. Carrying capacity of the Earth is determined by at least four factors: food production, living space, waste assimilation, and resource availability. All ecosystems have a specific carrying capacity for each population. Carrying capacity, however, is not a static entity. It fluctuates from year

Table 5.1   World Population Growth and Doubling Time

Population Size	Year	Time Required to Double
1 billion	1850	All of human history
2 billion	1930	80 years
4 billion	1975	45 years
8 billion (projected)	2017	41 years

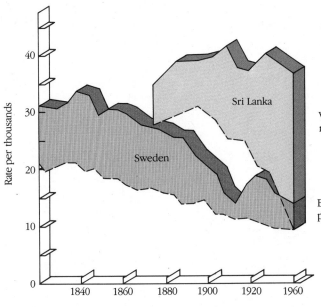

Figure 5.4 *Birth rates (solid lines) and death rates (dashed lines) for two countries. In Sri Lanka, which is typical of many developing nations, a continued high birth rate and a drastic drop in the death rate has resulted in an extremely rapid population growth. In Sweden, which is typical of developed nations, the birth rate declined faster than the death rate, resulting in much slower population growth.*

to year, depending on climate and other factors (Chapter 3). Moreover, carrying capacity can be altered by organisms. As a rule, most species have little effect on the carrying capacity of their surroundings. Humans, however, are an exception to the rule. We can expand the Earth's ability to support our kind through advances in toolmaking, agriculture, industry, and medicine.

Human efforts to expand the Earth's carrying capacity often involve changes in one variable at a time. As biologist Garrett Hardin notes, "Maximizing one [variable] is almost sure to alter the balance in an unfavorable way." Importing water via pipelines to desert areas, for example, increases the regional carrying capacity and results in dramatic population increases that cause sizable increases in pollution. At a certain point, the amount of waste produced cannot be assimilated by natural mechanisms.

Debate over the environment often revolves around arguments of the Earth's carrying capacity. Determining the carrying capacity of the Earth is no easy task. It varies with the standard of living one assumes is possible and with the level of technological development. Current estimates of the Earth's carrying capacity range from 500,000 to 50 billion.

Today, many scientists believe that the current human population of 5.5 billion already exceeds the

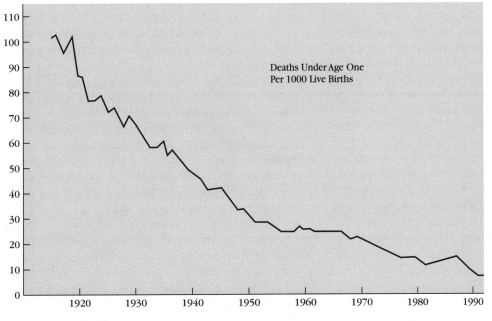

Figure 5.5 *U.S. infant mortality rate, 1915–1992. Note the consistent decline, which has contributed to a marked increase in average life expectancy.*

carrying capacity. They claim that polluted air, depleted ocean fisheries, deforestation, species extinction, denuded landscapes, widespread starvation, poverty, and despair are signs that humanity has pushed beyond the carrying capacity of many regions. We are living on the edge of disaster. If agricultural output should decrease as a result of several years of severe drought or if oil supplies should run out, many people would suffer enormous hardship. Global environmental problems, such as ozone depletion and global warming, suggest that the entire human population may be living beyond the Earth's carrying capacity.

Others are extremely optimistic. They think that the Earth could support many more people. Many of these people, however, forget to factor in the massive amounts of waste that would be produced by further expansion. They forget about the widespread extinction of plants and wildlife, and ignore how dependent human society is on forests, oceans, and other natural ecosystems for the oxygen we breathe and the nutrients we require. (For a debate on population growth, see Point/Counterpoint 5.1.)

Clearly, the cultural evolution of human civilization has resulted in a marked increase in the Earth's carrying capacity. However, this "advance" has been accompanied by depleted resources, increased pollution, and species extinction—all signs that we're living unsustainably. Human civilization is at a crucial point. As this chapter points out, building a sustainable society requires efforts not only to slow the growth of the human population but to stop it. It may also require steps to reduce numbers in a humane way to bring human society back in line with the Earth's carrying capacity.

Exponential Growth

To understand how serious population growth is, one must understand exponential growth. **Exponential growth** occurs when something increases by a fixed percentage every year. For example, a savings account with 5% compounded interest grows exponentially. The remarkable characteristic of all exponential growth is that early growth in absolute numbers is quite slow. Once growth "rounds the bend," however, the item being measured, whether money in a bank account or population, begins to increase more and more rapidly.

To appreciate the true nature of exponential growth, consider an example. Suppose that your parents opened a $1000 savings account in your name on the day you were born. Also suppose that this account earned 10% interest, and all the earned interest was applied to the balance, where it also earned interest. At this rate of growth, your account would double every 7 years. Consequently, when you were 7, your account would be worth $2000. By age 14, it would have increased to

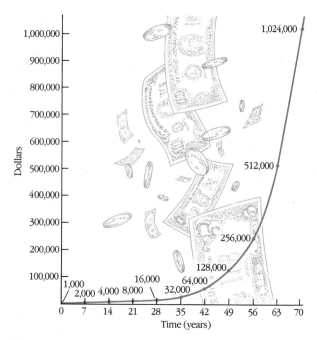

Figure 5.6 *Exponential growth curve of a savings account with an initial deposit of $1000 and an interest rate of 10%.*

$4000. By age 42, you'd have $64,000 (Figure 5.6). If you left the money in a little longer, you'd find it growing faster and faster. At age 49, the account would be worth $128,000; at age 56, you'd have $250,000. In 7 more years, you'd have $500,000. But if you waited until you were 70, your account would be worth over $1 million.

Looking back over your account records, you'd find that during the first 49 years it grew from $1000 to $128,000, but during the last 21 years, it increased by nearly $900,000. The growth rate was constant over the entire period, and the account doubled every 7 years; yet, it was not until the balance in your account "rounded the bend" of the growth curve that the growth seemed to take off.

How did this rapid upturn occur, even though the growth rate was the same? The reason for this is simple: Once the base amount becomes extremely large, even small percentage increases result in substantial increases. For instance, the 1.7% increase in a world of 5.5 billion people results in a net increase of about 93 million people a year! At this seemingly slow rate of growth, world population will double in just 41 years. The problem isn't just that the population is increasing exponentially. It's that population, resource use, pollution, and environmental damage are all increasing at an exponential rate. We've exceeded the Earth's carrying capacity and tems analysts refer to this phenomenon as an **overshoot**. As you will see in Section 5.4, an overshoot will very likely result in the system's collapse.

5.3 Understanding Populations and Population Growth

Newspapers and magazines sometimes present a confusing array of population statistics, making it difficult for the public to understand this global issue. With a little effort, you can learn enough **demography**, or population science, to discern facts from fallacies.

Measuring Population Growth

One of the most important population measurements is the growth rate, which is frequently expressed as a percentage. What does it mean when demographers report that a population is growing at a rate of 3% per year? Is this something to be thankful for or concerned about? To answer this question, let's first see how a population's growth rate is determined.

Growth Rate The **growth rate** of the world population is calculated by subtracting the number of deaths in a population from the number of births in any given year:

Growth rate = (Crude birth rate − Crude death rate)

Crude birth rate is the number of births per 1000 people in a population. **Crude death rate** is the number of deaths per 1000 people. Worldwide, the crude birth rate is currently 26 per 1000 and the crude death rate is 9 per 1000. The difference between the birth and death rates is the population growth rate, in this case 17 per 1000. This means that 17 people join the world population for every 1000 people in the population.

To convert this to a percentage, you simply multiply by 100:

$$\text{Growth rate} = \frac{17}{1000} \times 100 = 1.7\%$$

The relationship between birth rates and death rates determines whether the world's population is growing, shrinking, or staying the same. Some important growth rates are listed in Table 5.2. This table also shows doubling times, which are discussed shortly.

Birth Rates The balance between birth rates and death rates in various regions of the world determines global population growth. Each rate is influenced by a variety of factors. The birth rate in a given population, for instance, depends on: (1) the age at which men and women get married, (2) their educational level, (3) whether a woman works after marriage, (4) whether a couple uses reliable contraceptives, (5) the number of children a couple wants, (6) a couple's religious beliefs, (7) a couple's cultural values, and (8) availability of contraceptives.

Table 5.2 Growth Rate and Doubling Time in 1992

Region	Growth Rate (%)	Doubling Time (years)
World	1.7	41
Developed countries	0.5	148
Developing countries	2.0	34
Africa	3.0	23
Asia	1.8	39
North America	0.8	89
Latin America	2.1	34
Europe	0.2	242
Soviet Union	0.7	104
Oceania	1.2	57

Source: Population Reference Bureau, *World Population Data Sheet*, 1992.

These factors can also be viewed as leverage points for controlling human numbers. For example, efforts to control human numbers rely on improving education as well as changing cultural beliefs, improving economic conditions, and making contraceptives available.

Death Rates Death rates are equally important in determining population growth. A few decades ago, death rates in most developing nations of the world were quite high. Particularly vulnerable were the children. In fact, very few children lived past the first year. Therefore, even though birth rates were elevated, the high death rates kept population growth in check.

Today, modern medicine, pesticides, and better sanitation, among other factors, have greatly reduced the death rate in many developing nations. However, until recently, little attention was given to efforts needed to reduce birth rates. That is to say, the decline in death rates was not paralleled by a decrease in birth rates. Well-intentioned efforts to improve the lives of many through better health care and other measures spawned an unprecedented surge in population growth in the poorer nations.

A similar imbalance between birth and death rates also existed in developed nations, which have, for reasons explained later, successfully lowered their birth rates. In some countries, birth rates and death rates are nearly equal. In others, birth rates have actually fallen below death rates, and their populations are now shrinking.

Point/Counterpoint 5.1

The Population Debate

Julian L. Simon

The author is a professor of business administration at the University of Maryland at College Park. He has written several important books on population and economics, including The Ultimate Resource *and* Population Matters.

The Case for More People

Classical economic theory holds that population growth reduces the standard of living: the more people, the lower the per capita income, all else being equal. However, many statistical studies conclude that population growth does not have a negative effect on economic growth. The most plausible explanation is the positive effect additional people have on productivity by creating and applying new knowledge.

Because technological improvements come from people, it seems reasonable to assume that the amount of improvement depends in large measure on the number of people available. Data for developed countries show clearly that the larger the population, the greater the number of scientists and the larger the amount of scientific knowledge produced.

There is other evidence of the relationship between population increase and long-term economic growth: an industry, or the economy of an entire country, can grow because population is growing, because per capita income is growing, or both. Some industries in some countries grow faster than the same industries in other countries or than other industries in the same country. Comparisons show that in the faster-growing industries the rate of increase of technological practice is higher. This suggests that faster population growth, which causes faster-growing industries, leads to faster growth of productivity.

The phenomenon economists call "economy of scale"—greater efficiency of larger-scale production where the market is larger—is inextricably intertwined with the creation of knowledge and technological change, along with the ability to use larger and more efficient machinery and greater division of labor. A large population implies a bigger market. A bigger market is likely to bring bigger manufacturing plants, which may be more efficient than smaller ones and may produce less expensive foods.

A bigger population also makes profitable many major social investments that would not otherwise be profitable—railroads, irrigation systems, and ports. For instance, if an Australian farmer were to clear a piece of land far from neighboring farms, he might have no way to ship his produce to market. He might also have trouble finding workers and supplies. When more farms are established nearby, however, roads will be built that link him with markets in which to buy and sell.

We often hear that if additional people have a positive effect on per capita income and output, it is offset by negative impacts such as pollution, resource shortages, and other problems. These trends are myths. The only meaningful measure of scarcity is the economic cost of goods. In almost every case, the cost of natural resources has declined throughout human history relative to our income.

Conventional wisdom has it that resources are finite. But there is no support for this view. There is little doubt in my mind that we will continue to find new ore deposits, invent better production methods, and discover new substitutes, bounded only by our imagination and the exercise of educated skills. The only constraint upon our capacity to enjoy unlimited raw materials at acceptable prices is knowledge. People generate that knowledge. The more people there are, the better off the world will be.

Garrett Hardin
The author is a renowned environmentalist, writer, and lecturer. He has taught at several universities, and in 1980, he served as chairman of the board and CEO of the Environmental Fund. The author of numerous books and articles on environmental ethics, he is best known for his article "The Tragedy of the Commons."

Is More Always Better?

To get at the heart of the question "Is more better?" study the daily flow of water over Niagara Falls. You will find that twice as much water flows over the falls during each daylight hour as during each nighttime hour. There, in a nutshell, you have the population problem.

Puzzled? You should be. The connection between Niagara Falls and population is not obvious. Before we can understand it, we need to review a little biology.

For every nonhuman species, there is an upper limit to the size of a population. Near the maximum, individuals are not so well off as they are at lower densities. Starvation appears. Crowded animals often fight among themselves and kill their offspring. Wildlife managers and advocates agree that the maximum is not the optimum.

What about humans? Will we be happiest if our population is the absolute maximum the Earth can support? Few people say so explicitly, but some argue that "more is better"!

Admittedly, we need quite a few people to maintain our complex civilization. A population the size of Monaco's, with about 25,000 people, could never have enough workers for an automobile assembly line. But Sweden, with nearly 9 million people, turns out two excellent automobiles. Nine million is a long way from 5.5 billion, the population of the world today.

Some say, "More people—more geniuses." But is the number of practicing geniuses directly proportional to population size? England today is 12 times as populous as it was in Shakespeare's day, but does it now boast 12 Shakespeares? For that matter, does it have even 1?

Consider Athens in classical times. A city of only 40,000 free inhabitants produced what many regard as the most brilliant roster of intellectuals ever: Solon, Socrates, Plato, Aristotle, Euripides, Sophocles—the list goes on and on. What city of 40,000 in our time produces even a tenth as much brilliance?

Of course, the free populace of Athens was served by ten times as many slaves and other nonfree classes. This left the 40,000 free people to apply themselves to intellectual and artistic matters. Modern peoples are given creative freedom by labor-saving machines, certainly a more desirable form of slavery. But where are our geniuses?

Business economists are keenly aware of "economies of scale," which reduce costs per unit as the number of units manufactured goes up. Communication and transportation, however, suffer from diseconomies of scale. The larger the city, the higher the monthly phone bill. Crimes per capita increase with city size. So do the costs of crime control. All these suggest that more may not be better.

Democracy requires effective communication between citizens and legislators. In 1790, each U.S. senator represented 120,000 people; in 1990, the figure was 2.5 million. At which time was representation closer to the ideal of democracy? To communicate with his or her constituents, each senator now has an average of 60 paid assistants. President Franklin D. Roosevelt had a staff of 37 in 1933, when the population was 125 million. Ronald Reagan had a staff of 1700 in 1981, when the population was 230 million. We have to ask whether democracy can survive unrestrained population growth.

Let's look at another aspect of the more-is-better argument. An animal population is limited by the resources available to it. With humans, a complication arises. Though the quantities of minerals on Earth are fixed, improved technology periodically increases the quantity of resources available to us. In the beginning, copper ores we mined contained 20% copper; now we are using ores with less than 1%. Available copper has increased but not the total amount of copper on Earth.

Let's return to Niagara Falls. Less water flows over the falls at night because more water is diverted to generate electricity when people aren't looking at the falls. It would be possible to use all the water to generate electricity, but then there would be no falls for us to look at. As a compromise, the volume of water "wasted" falling over the falls is reduced only at night. Therefore, the turbines and
(continued)

Point/Counterpoint 5.1 (continued)

The Population Debate

generators are not fully used 24 hours of the day, which means that local electricity costs are just a bit higher.

If the population continues to grow, the day may come when electricity is so scarce and expensive that the public will demand that Niagara Falls be shut down so that all the water can be used to generate electricity. Similar dangers face every aesthetic resource. Wild rivers can be dammed to produce more electricity, and estuaries can be filled in to make more building sites for homes and factories. The maximum is never the optimum.

With human populations, quantity (of people) and quality (of life) are trade-offs. Which should we choose—the maximum or the optimum?

Critical Thinking

1. Is our population now below or above the optimum? To answer this question make two lists. On one, list all the things that you would expect to be better if the population doubled; on the other, list all the things that would be worse.
2. On which list would you put the

availability of wilderness? Noise level? Amount of democracy? Amount of pollution? Per capita cost of pollution control? Availability of parking spaces? Personal freedom?
3. When you are through, compare your list with those of your friends. What value judgments account for the differences? Can these differences be reconciled? How?
4. Outline the main points and supporting evidence given by both authors. Which views do you agree with? Why? Are your reasons based on feelings or facts?

Doubling Time   Another important measurement of population dynamics is the **doubling time**. The following formula is used to determine doubling time:

$$\text{Doubling time} = 70/\text{Growth rate (\%)}$$

In this equation, 70 is a demographic constant. Using this equation, you can quickly convert growth rates into doubling times. The U.S. growth rate of 0.8% in 1992 equals a doubling time of 89 years. Thus, if current trends continue, by the year 2081, there will be 510 million Americans. The world population, which grew at 1.7% in 1992, will double in 41 years if current trends continue.

A Comparison of Growth Rates   The world is crudely divided into the *haves* and the *have-nots*—developed countries and less developed, or developing, countries (Table 5.3). On the whole, **developed countries**, such as the United States, Canada, Great Britain, and Japan, grow fairly slowly, on average about 0.5% per year (doubling time = 148 years). However, some developed countries, such as Canada and the United States, have relatively fast growth rates of about 0.8% per year. A number of other developed countries, such as Austria and Sweden, have extremely slow growth rates, 0.1% and 0.3% per year, respectively. Some, like Bulgaria, have stopped growing altogether. A few others, such as Germany and Hungary, are shrinking or, in the language

of demographers, are experiencing negative growth rates.

In contrast, **developing nations** are growing much more rapidly—at an average of 2.0% per year (doubling time = 34 years). However, this average hides some dangerous trends in countries that are growing well over 3% per year. Kenya, for instance, which currently houses 26 million people, is growing at a rate of 3.7% per year, giving it a doubling time of 19 years.

Fertility and Zero Population Growth

Another important measure and predictor of population dynamics is the **total fertility rate** (**TFR**), the number of children women in a population are expected to have in their lifetime, based on current trends in childbearing. In the United States, for example, the TFR was 2.0 in 1992; this means that each woman in the reproductive age group of 15 to 45 is expected to have 2 children. Canada's TFR is 1.8. In India, a much poorer nation, women have a TFR of 3.9 (Table 5.4).

Many countries have achieved **replacement-level fertility**, the point at which couples produce exactly the number of children needed to replace them. In the United States and other developed countries, replacement-level fertility is a TFR of 2.1 children. That means that 10 women must have 21 children to replace themselves and their husbands. Why the additional child? The reason is

Table 5.3 Comparison of Developed and Developing Countries

Feature	Developed	Developing
Standard of living	High	Low
Per capita food intake	High (3100–3500 cal/day)	Low (1500–2700 cal/day)
Crude birth rate	Low (15/1000 population)	High (33/1000 population)
Crude death rate	Lower (10/1000 population)	Higher (12/1000 population)
Growth rate	Low (0.6%)	High (2.1%)
Doubling time	High (120 years)	Low (33 years)
Infant mortality	Low (20/1000 births)	High (90/1000 births)
Total fertility rate	Replacement level (2.0)	High (4.6)
Life expectancy at birth	High (72 years)	Lower (57 years)
Urban population	High (69%)	Low (26%)
Wealth (per capita GNP) (1985 U.S. dollars)	High ($9,930)	Low ($660)
Industrialization	High	Low
Energy use per capita	High	Low
Illiteracy rate	Low (1%–4%)	High (25%–75%)

Source: Population Reference Bureau.

that in these countries 1 of every 21 children dies before reaching reproductive age. Replacement-level fertility is higher in developing nations because death rates for children are higher.

Today, over two dozen nations (out of 160) have achieved or are near replacement-level fertility. Most of these nations are in Europe. Even though a population reaches replacement-level fertility, it does not mean that it has stopped growing. A population stops growing only when the death rate equals the birth rate and the net migration is zero, that is, the number of people entering the country equals the number leaving it. Demographers refer to this state as **zero population growth (ZPG)**.

In the United States, the TFR fell below replacement level in 1972, thanks in large part to modern contra- ception and widespread public awareness of the popu- lation dilemma. Even though the TFR has remained below replacement-level fertility since that time, the U.S. population has continued to swell. In addition, demog- raphers project that at least 70 years will pass before the U.S. population stops growing. Why?

The U.S. population continues to grow for two rea- sons: First, each year, numerous people move to the United States from other countries. Some people come legally; others arrive illegally. Legal and illegal immi- gration adds 1.2 to 1.5 million people each year to the U.S. population, and is responsible for about 40% to 50% of the annual growth.

The rest of the growth comes from within. To under- stand why, we begin with Figure 5.7, which shows the

Table 5.4   Total Fertility Rate in Mid-1992

Region		Total Fertility Rate	Population Size (billions)
World		3.3	5.420
Developed countries		1.9	1.224
Developing countries		3.8	4.196
Africa		6.1	
Northern	4.8		
Western	6.7		
Eastern	7.0		
Middle	6.1		
Southern	4.6		
Asia		3.2	
Western	4.7		
Southern	4.3		
Southeastern	3.4		
Eastern	2.1		
(incl. China, Japan)			
North America		2.0	
U.S.	1.8		
Canada	2.0		
Latin America		3.4	
Central	4.1		
Caribbean	3.1		
South America	3.2		
Europe		1.6	
Northern	1.9		
Western	1.6		
Eastern	1.9		
Southern	1.5		
Soviet Union		2.2	
Oceania		2.6	

TFR for the United States from 1917. After World War II, the TFR climbed dramatically and remained high for several decades. This increase resulted in the infamous postwar baby boom, a rapid rise in fertility, which, in turn, resulted in a long-term increase in the population. The children of the baby-boom era are something of a tidal wave in the American population. Today, however, many of these individuals are having their own children. Even though they are producing, on average, only 2.0 children each, the number of women giving birth is still on the rise. In other words, because of the postwar baby boom, the reproductive age group (ages 15 to 44) continues to expand. The continued rise in the population, despite its having reached below-replacement-level fertility, results from the fact that more women are having children.

Demographers believe that the U.S. population, which exceeded 255 million in 1992, will increase by about 50 to 60 million before stabilizing. This delayed slowing is called the **lag effect**. Like an oil tanker that glides for a mile or two before coming to a stop, U.S. population will continue to grow long after replacement-level fertility is reached. Stabilization is possible *only* if the number of women entering the reproductive age group levels off, if the TFR remains below replacement level, and if immigration does not climb substantially. Small changes in any of these factors could have drastic effects on U.S. population. For example, Susan Weber, executive director of the nonprofit organization Zero Population Growth, says that if abortion becomes illegal, the TFR could climb to 2.2, slightly above replacement-level fertility. In that case, U.S. population would never stabilize.

Migration

The previous sections dealt primarily with global population, which is determined by the balance of birth rates and death rates. To calculate the growth of individual countries, states, or regions, demographers must take into account the number of people moving out of and into a region as well as the balance, or imbalance, between births and deaths.

Technically speaking, **migration** is the movement of people across boundaries to set up a new residence. The term **immigration** refers to movements into a country; **emigration** refers to movements out. **Net migration** is the difference between immigration and emigration. Population growth in a country will stabilize if the growth rate and net migration are zero.

Immigration is one of the hottest topics today in the United States, in part because legal and illegal immigration into the United States accounts for such a large part of the annual growth. Efforts to stabilize U.S. population will fail unless the country more closely balances immigration and emigration.

Public opinion polls in the United States in the late 1980s showed that a majority favors reducing immigration of all kinds. Accordingly, Congress passed a law in 1986 levying penalties on those who knowingly employ illegal immigrants and requiring job seekers to provide proof of citizenship or legal immigration status. Four years later, Congress passed the 1990 Immigration Act, which will increase the number of legal immigrants by 35%—increasing the number from 530,000 in 1991 to nearly 700,000 per year from 1992 to 1994. After 1994, the number will be reduced to 675,000 per year.

Opponents of the law argue that this generous increase will greatly increase the tax burden on already

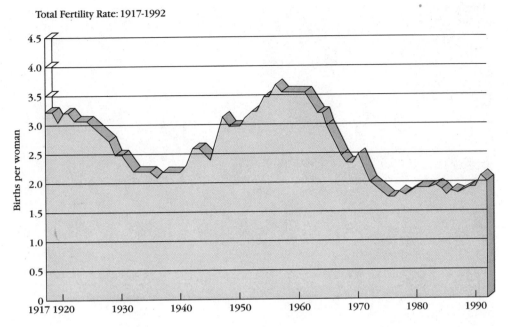

Total Fertility Rate: 1917-1992

Figure 5.7 *Total fertility rate (TFR) is the number of children women are expected to have based on the current age-specific fertility rates (fertility rates in each age group). The TFR in the United States has fluctuated widely with changing economic conditions. In the decade of the Great Depression, women had a low TFR. In the several decades following World War II, the TFR shot up; in 1972, it fell below replacement level, where it has remained. In recent years, it has begun to edge upward.*

struggling cities and will increase crowding and internal strife in racially torn neighborhoods (Figure 5.8). It could also put additional stress on school districts already coping with severe budget problems, and will increase competition for jobs in an economy marked by excessive unemployment.

People also move within the boundaries of a country, often in large numbers. Such migrations can dramatically alter local economies. In the United States, in the 1970s, the primary flow was from the northeastern and north-central regions to the South (mostly Florida) and West (Arizona, Wyoming, Utah, Alaska, Idaho, Colorado, New Mexico, Texas, and California). The migration was so dramatic that for the first time in U.S. history today more than half the population lives in the West and South. In the decade of the 1980s, a similar pattern emerged. Four million people left the North and Midwest, heading for the South and West, which received 3 million and 1.3 million new residents, respectively, according to U.S. Department of Commerce data.

This movement, referred to by many as the "sunning of America," was caused by: (1) expanding industries in the West and Southwest, especially in electronics and energy production; (2) declining industries of the northeastern and north-central states, especially auto and steel manufacturing; (3) a desire for a warmer climate; (4) a desire for a lower cost of living; (5) a preference for abundant recreation; (6) an aspiration for a less hectic, less crowded life-style; and (7) the growth of retirement communities.

The massive migration to the South and West had many important economic benefits for these regions.

Income increased and unemployment fell. New opportunities opened up for builders, restaurant owners, bankers, and others in the service sector of the economy. Ironically, the influx of people into the Sunbelt destroyed many of the values the migrants sought. With growth rates in the range of 3% to 5% per year in the 1970s and early 1980s, many western cities and states were swamped by new people. Air pollution in Sunbelt cities worsened with increasing population; the clean air that many came for disappeared. Traffic became congested. The rapid rise in demand for housing sent the cost of homes skyward. The breakneck speed of growth made it difficult for local governments to provide water, schools, sewage treatment plants, and transportation facilities. Spreading cities engulfed smaller outlying communities and changed the tempo of life.

In the late 1980s and early 1990s, the movement into many regions slowed, and the economic boom of numerous regions died. Many states experienced a substantial rise in unemployment and hard economic times. Housing prices fell, and newcomers left in search of employment. The economic downturn led many government officials to sponsor large and costly development projects in hopes of stimulating these sagging economies. In Denver, Colorado, for example, the city sponsored the construction of a new, larger airport, despite the fact that 25% of the gates at the present airport were unused and despite the fact that the current airport netted about $90 million a year for the city. The new airport, which could cost upward of $5 billion, is viewed by many as governmental make-work—a project supported by federal money and bonds to stimulate the

Figure 5.8 *Immigration offers people a chance to enjoy the wealth and opportunity of the United States, but it does not come without costs, among them social tension, violence, overcrowding, increased tax burden, and unemployment. This scene from the 1992 Los Angeles riots illustrates the rising tensions in poor, crowded urban centers.*

economy. Given cost projections and other problems, the project could end up being a major long-term economic drain.

Internal migrants affect the places they come from as well. Since many of the out-migrants are young, educated, and skilled workers, they often create a significant drain on human resources in areas they leave.

Seeing Is Believing: Population Histograms

Demographers use growth rates, fertility rates, and doubling times to explain the dynamics of the human population—where it is and where it is going. However, few techniques shed as much light as a graph. The old saying "A picture is worth a thousand words" is as true in demography as it is anywhere else.

You have seen a graph of exponential growth and considered the consequences of such growth. What you learned was that "rounding the bend" of the J-shaped exponential growth curve yields a remarkable growth in the population even though the growth rate remains small. Another graphic tool is the **population histogram**. This is a bar graph that offers a profile of a population, that is, it tells the history and the future of a population (Figure 5.9).

The population histogram displays the age and sex composition of a population. Each horizontal bar on the histogram represents the size of a certain age and gender group. As illustrated in Figure 5.10, three general profiles

exist: expansive, constrictive, and stationary. Mexico, a country that is expanding, has a large number of young people. If they produce more offspring than their parents did, the population will continue to expand at the base. If, on the other hand, family size decreases, the base of the histogram will begin to constrict. This is what is happening to the U.S. population. Sweden presents an entirely different picture. For many years, Swedish couples have been having the same number of children as their parents did; as a result, Sweden's population is near stationary.

Population histograms can change over time and, therefore, are not always reliable for making long-term predictions of future growth. Expansive populations can become constrictive, constrictive populations can become expansive, and so on. In essence, then, population histograms give a snapshot view that can be helpful in planning for schools, hospitals, retirement homes, and so on. For example, after studying their histograms, Mexican officials might make special efforts to expand educational facilities and maternity wards. The United States might look for ways to increase retirement homes and services for the elderly.

Shifts in a population histogram can have dire consequences if ignored. In the United States, the postwar baby boom (seen as a bulge in the histogram in Figure 5.9) greatly increased the number of school-age children in the 1950s and 1960s. New schools couldn't be built fast enough to keep up with the pace. Teachers

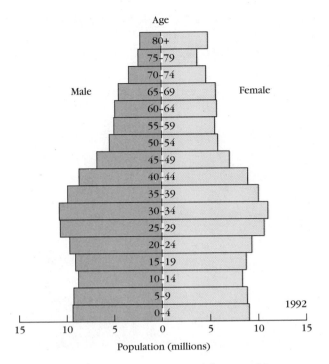

Figure 5.9 *Population histogram of the United States. Males in each age group are shown on the left and females on the right. The bulge in the histogram is the postwar baby boom.*

were in short supply. Housing also became a serious problem as members of this age group married and searched for places to live. Still larger problems loom in the early 21st century, when those people reach retirement age. By the year 2030, individuals 65 and older will make up about 20% of the U.S. population, compared with 11% today. Health costs are almost certain to go up, since people over 65 today account for nearly one-third of the nation's health bill. Gerontologists, new hospital facilities for the aged, and retirement homes will be needed to accommodate this age group (Figure 5.11). Making matters worse, a smaller proportion of the population—including those in college today—will be supporting this large elderly group through the Social Security program.

The aging of the U.S. population also means an aging of the work force, which has positive aspects. It means that competition for entry-level positions will lessen and that the labor force will, on the whole, be more experienced, thus requiring less training. This could increase overall labor productivity, the amount of output per unit of labor.

A population histogram of the world is expansive. In fact, 33% of the world's people today are under the age of 15. If we're going to create a sustainable future, something must be done to curb their childbearing.

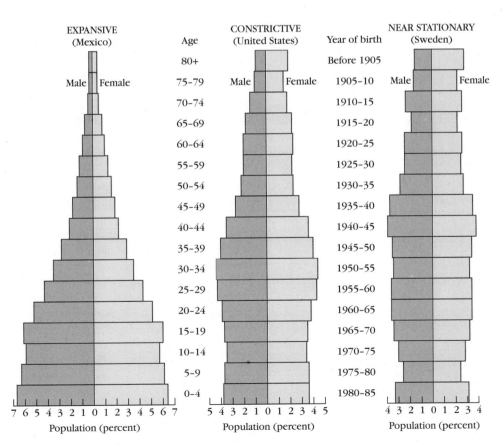

Figure 5.10 *Three general types of population histograms. Expansive populations have a large percentage of young people. Constrictive populations have a tapering base. Stationary or near stationary populations show no expansion or constriction. This results from couples having only the number of children that will replace them. The world population histogram is expansive.*

Figure 5.11 *A glance at the population histogram shows that the population is aging. Within the next few decades, baby boomers will be retiring. To accommodate this shift, private and public sectors will need to provide additional health care, gerontologists, and nursing homes.*

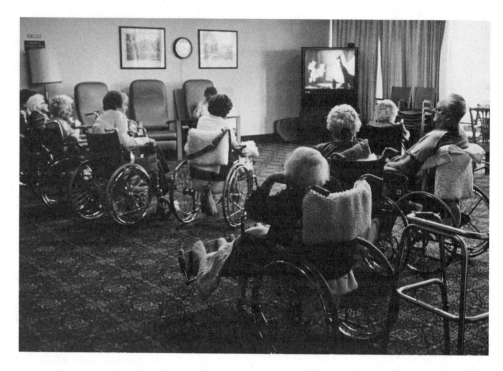

5.4 The Future of World Population: Some Projections

Perhaps, the two questions most often asked of demographers are: "What is the future of the world population? How big will it get?"

Because many factors determine whether a population will grow and how quickly, demographers can only guess about the future of the human population. The factors that determine future population growth, such as birth rates, fertility, and death rates, can change dramatically.

Regardless of the uncertainties involved in making estimates of future growth, many scientists offer growth forecasts. The world population growth rate is currently about 1.7%. If this rate continues for the next three centuries, the world population will reach 1 trillion (1,000,000,000,000) around the year 2300.

Yet, long before then, demographers believe that the population growth curve will level off as a result of dropping birth rates, rising death rates, or both. Figure 5.12 presents several possible scenarios for the United States and the world. As illustrated in Figure 5.12a, a slow reduction in fertility would result in a gradual decrease in world population growth and a stable population by the year 2100. The population size would be about 15 billion, three times what it is today. Still another projection, based on a rapid decline in fertility, shows stabilization around the year 2060 at 8 billion, 2.5 billion more than today. Exactly where the curve levels off will depend on the persistence and success of efforts in family planning and economic development in the developing world (Chapters 6 and 23).

Figure 5.12b shows three scenarios for the U.S. population. The lowest line shows it stabilizing at 268 million if the TFR stays at about 1.8 and net migration is held at 500,000 per year. The middle curve, which experts thought was more likely only a few years ago, given immigration policy and current TFR, shows a continued rise to over 370 million. The third curve shows a more dramatic rise. Because of increased immigration and because immigrants (notably Hispanics and Asians) have a higher fertility rate, this projection is believed more plausible. The U.S. population would equal 509 million in 90 years, up from 255 million today.

Many of the curves in Figures 5.12a and 5.12b suggest a stabilization of population. In these scenarios, the growth curves are **sigmoidal**, or **s-shaped**. However, some demographers predict that shortages and pollution, brought about by exponential growth, will eventually cause the world population to decrease in size, producing a dome-shaped population curve instead of a sigmoidal curve characteristic of a stabilizing population (see Figure 5.13b). The decrease in population size in this scenario, ecologists believe, will occur if populations exceed the Earth's carrying capacity, creating an overshoot. In many regions, human populations have already pushed their numbers beyond the carrying capacity. Further pollution of the air and water; increased shortages of food, water, wood, grazing land, and other natural resources; or continued extinction of plants and animals could cause the population size to decrease over time, or even quite suddenly.

Still another ominous pattern may occur. Instead of leveling off or declining smoothly, the human population might suffer sudden die-offs followed by spurts of growth

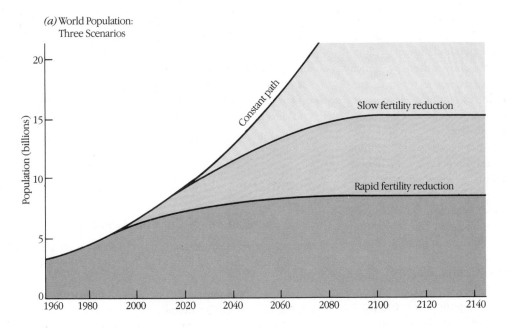

(a) World Population: Three Scenarios

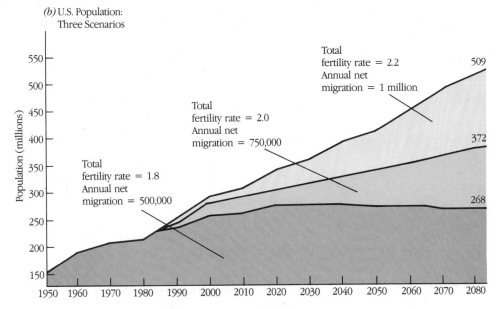

(b) U.S. Population: Three Scenarios

Figure 5.12 *Projected population growth.* (a) *Future of world population. The constant path indicates population growth continuing at the current rate. The middle line indicates the future of world population if fertility drops off to a slow rate. The bottom line shows what might happen with a rapid decrease in fertility. As discussed in the text, the future may lie somewhere between the lower two lines.* (b) *The future of U.S. population depends on the future TFR and annual net migration.*

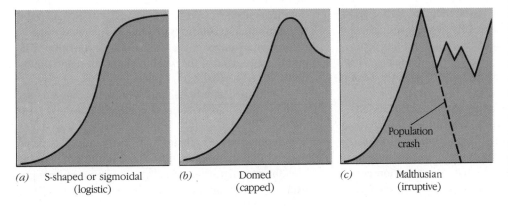

Figure 5.13 *Patterns of population growth.* (a) *Smooth transition to a stable population size.* (b) *Gradual drop-off caused by population exceeding the carrying capacity.* (c) *Population crash caused by irreparable change to the ecosystem. No doubt all three patterns will be seen in individual countries.*

and plunges. This roller-coaster ride creates an unstable, **irruptive** pattern caused by severe ecosystem imbalance (Figure 5.13c). Finally, the human population may plummet in a population crash possibly caused by critical changes in global climate or decreases in resource supplies. This sudden catastrophic die-off may be so severe that the human population becomes extinct or is reduced to only a fraction of its present size.

No one knows what will happen. The only opinion on which experts approach unanimity is that world population cannot grow indefinitely. Some countries will inevitably make a smooth transition to a stable state; others may experience periodic crashes caused by epidemics in crowded urban populations; still others may fall to low population levels because of continued starvation and disease.

H. G. Wells wrote: "Human history becomes more and more a race between education and catastrophe." The current population-resource-pollution bind clearly illustrates this fact. What we decide today will have far-reaching effects that determine the kinds of lives our children and their children will have. What we do and don't do will literally shape the future of the world and determine the future of many species that live on this planet with us.

The doors we open and close each day decide

the lives we live.

—*Flora Whittemore*

ཟ☙   ཟ☙   ཟ☙

Critical Thinking Exercise Solution

Critical thinking skills suggest the importance of examining the big picture, that is, looking at issues in their entirety. As you learned in this chapter, although population growth in developing nations accounts for 90% of the annual increase, residents of the industrial nations actually have far more impact on the environment. In fact, people living in the industrial nations, on average, consume 20% to 40% more resources and produce 20% to 40% more pollution than their counterparts in developing nations. Consequently, the 10 million new residents of the industrial world added to the world population have the impact of 200 to 400 million people in the developing countries. Many experts argue that measures to slow growth in the industrial nations are actually more important than efforts to slow growth in the developing nations.

The critical thinking rules tell us to question conclusions. In this case, and in others, it is necessary to be aware of the fact that people often set priorities and take action without adequate information. They also find it easier to tell others what they need to do and ignore what we ourselves must do. As this chapter points out, to create a sustainable future, it is necessary to control population growth in both developed and developing nations.

Summary

5.1   Dimensions of the Population Crisis

■   Population growth is at the root of virtually all environmental problems, including pollution and resource depletion.

■   Rapid population growth creates difficulties in meeting the basic needs of people.

■   Crowding may contribute to mental illness, drug abuse, and various forms of antisocial behavior.

5.2   The Population Explosion

■   The rapid growth of the world population is a recent phenomenon. Throughout most of human history, population was small because it was controlled by war, disease, and famine.

■   Advances in agriculture and industry, however, allowed humans to expand the Earth's **carrying capacity**, the number of people the Earth can support.

■   Human population is growing exponentially. **Exponential growth** is a fixed-percentage growth that follows a J curve. Initially, growth is slow, but once the curve "rounds the bend," the absolute growth becomes remarkable.

5.3   Understanding Populations and Population Growth

■   World **population growth** is determined by subtracting the **crude death rate** from the **crude birth rate**. These are the number of deaths and births per 1000 people, respectively.

■   **Doubling time** is the time it takes a population to double in size.

■   **Developing countries**, those with low per capita income, inadequate education, and poor nutrition, contain three-fourths of the world's population and double about every 41 years.

■   **Developed countries**, the wealthier, well-fed, and well-educated nations, double on the average every 148 years.

■   Population growth in individual nations is determined by calculating the number of births minus deaths plus the **net migration**, the difference between **immigration** and **emigration**.

■ An important measure of reproductive performance is the **total fertility rate (TFR)**, the average number of children a woman will have. In developed countries, a TFR of 2.1 is **replacement-level fertility**.

■ **Replacement-level fertility** is one of the first steps toward achieving **zero population growth**.

■ **Population histograms** are important tools for studying populations because they show the proportion of males and females in a population in various age groups. The shape of the histogram suggests whether populations will grow, shrink, or remain stable.

5.4 The Future of World Population: Some Projections

■ The future of world population is difficult to determine because many factors, such as total fertility rate, can change.

■ Many demographers believe that world population will stabilize somewhere between 8 and 15 billion and that the exponential growth curve will be converted into a **sigmoidal curve**.

■ However, other possibilities are also likely. The exponential growth curve may be transformed into a **dome-shaped curve** as population levels fall because of overextension of the carrying capacity.

■ The pattern may become **irruptive**, which is characterized by a series of peaks and chasms.

■ Finally, severe damage to the environment caused by exceeding the carrying capacity could result in a devastating crash, virtually eliminating humans from the Earth. The future of the world's population is largely dependent on our efforts.

Discussion and Critical Thinking Questions

1. What is the population of the world? What is the population of the United States?

2. Should the rapid growth in world population concern you? Why or why not? How does population growth in the United States affect your life?

3. Some people flatly deny any connection between environmental problems and population. Do you agree? Why or why not?

4. How many years did it take the world to reach a population of 1 billion people? How quickly did we reach 2, 3, and 4 billion?

5. What factors kept world population in check for so many years? Discuss the advances that have unleashed population growth in the last 200 years.

6. Define the term *exponential growth rate*. Why does growth in absolute numbers increase so dramatically once the bend of the curve is passed?

7. What is a population histogram? Describe the three general profiles. Why are histograms useful?

8. How is the world population growth rate calculated?

9. Define replacement-level fertility and zero population growth.

10. Discuss the pros and cons of a lenient policy toward legal and illegal immigration in the United States. Do you favor strong immigration quotas?

Suggested Readings

Donaldson, P. J. and Tsui, A. O. (1990). The International Family Planning Movement. *Population Bulletin* 45(3). Wonderful overview of worldwide family-planning efforts.

Ehrlich, P. R. and Ehrlich, A. (1990). *The Population Explosion*. New York: Simon & Schuster. Updated version of the 1971 classic *The Population Bomb*.

Frazer, E. (1992). Thailand: A Family Planning Success Story. *In Context* 31: 44–45. Provides additional information on the success of Thailand in promoting family planning.

Jacobsen, J. L. (1991). *Women's Reproductive Health: The Silent Emergency*. Worldwatch Paper 102. Washington, D.C.: Worldwatch Institute. Examines the importance of family planning as a means of protecting women's health.

McFalls, J. A., Jr. (1991). *Population: A Lively Introduction*. Population Bulletin 46 (2). Washington, D.C.: Population Reference Bureau.

National Audubon Society and Population Crisis Committee. (1991). *Why Population Matters. A Handbook for the Environmental Activist*. Washington, D.C.: National Audubon Society. Answers many important questions about population.

PIP. (1992). The Environment and Population Growth: Decade for Action. *Population Reports* Series M(10). Population Information Program. Excellent overview of impacts of human population growth.

PRB. (1992). *1992 World Population Data Sheet and United States Population Data Sheet*. Washington, D.C.: Population Reference Bureau. Published annually with current statistics on U.S. and world population.

Tedesko, S. (1992). Family Planning Media: That's Entertainment! *In Context* 31: 42–43. Interesting look at the promotion of family planning through songs, music videos, and television.

World Commission on Environment and Development. (1987). *Our Common Future*. Oxford: Oxford University Press. Good reference.

Chapter 6

Population Control: Key to a Sustainable Society

*To rebuild our civilization we must first
rebuild ourselves according to the pattern
laid down by life.*

—Alex Carrel

Critical Thinking Exercise

Many people recognize that overpopulation in the developing countries is one of the root causes of the environmental crisis. To slow or perhaps stop the growth of the human population in such places as Kenya, Afghanistan, and Mexico, they suggest that millions of dollars be spent by the United States and other developed countries to finance family planning. Do you agree with the notion that overpopulation is a root cause? Why or why not? If you do, will efforts to slow growth in developing nations be sufficient to create a sustainable human presence? What critical thinking rules can you use in this exercise?

What do a television set and a radio have to do with family planning? Not much, you say? Think again.

In many developing nations, television and radio are helping promote family planning. The idea of using entertainment to encourage family planning originated in Mexico in 1977. The television soap opera "Acompáñame" ("Come Along with Me") is extremely popular in the country and has been credited with encouraging many Mexican women to visit family-planning clinics and helping to decrease Mexico's population growth rate from 3.1% to 2.7% and to increase contraceptive use by 33%.

Popular entertainment provides characters with whom individuals can identify. In India, the television show, "Hun Log," which first aired in 1987, promotes smaller families and equal status for women. In Kenya, television and radio shows support family planning. Music and music videos spread the message in the Philippines and Nigeria. In Brazil, television ads promote vasectomies, and in the Dominican Republic, Turkey, and Egypt, they introduce modern contraceptive methods.

Although criticized as being manipulative, these methods have proven to be an effective tool for social change. They empower people to choose a brighter future for themselves and for the planet.

6.1 Limits to Growth

Most experts are convinced that the current exponential population growth cannot continue indefinitely without serious consequences. This conclusion gained wide acceptance following an extensive computer study of the human future by a research team at the Massachusetts Institute of Technology (MIT) led by Donella Meadows. The team's analysis of population growth, resources, food, pollution, and industrial output, published in 1972 in the book *The Limits to Growth*, showed that the

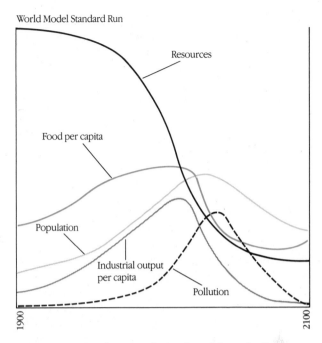

Figure 6.1 *Computer analysis of world trends. Because of natural delays in the system, both population and pollution continue to increase for some time after the peak of industrialization. Resource depletion and declining food supplies, however, cause a dramatic decline in population.*

human population would exceed the planet's carrying capacity within a century if exponential growth continued. As shown in Figure 6.1, world food supplies, industrial output, and population would grow until natural resource supplies began to fall, largely as a result of the expanding population. As the graph shows, the rapid depletion of resources would bring about a precipitous decline in food supplies, industrial output, and population. In short, we'd fall flat on our face.

What would happen if resource supplies were much larger than the researchers estimated? To address this question, the team doubled their estimated available supply of nonrenewable resources. What they found was that the human population would still overshoot the Earth's carrying capacity and crash, just a couple of decades later.

In another scenario, the authors assumed that world resources were unlimited. Under these conditions, population growth was halted by rising levels of pollution. The conclusions of the MIT study were unequivocal. Any way you look at it, if human civilization continues on the same path, we will exceed the carrying capacity of the biosphere with perilous results. Can anything be done to change the course?

Several computer runs in the study showed that a stable world system could be achieved by: (1) immediately achieving zero population growth, (2) employing massive recycling efforts, (3) controlling pollution,

(4) greatly reducing soil erosion, (5) rebuilding soil, and (6) emphasizing food production and services rather than industrial production (Figure 6.2). These actions are prescribed by the biological principles of sustainability discussed in Chapters 1–3.

The Limits to Growth study created a storm of controversy worldwide. Blinded by the frontier notion of unlimited resources, many critics simply couldn't believe its conclusions. Since the study's release, however, 17 independent computer studies have all reached a similar conclusion: infinite growth in a finite system is impossible.

Despite such warnings, some political leaders remain oblivious to the consequences of continued exponential growth in population and industrial output. According to Meadows and her colleagues, "Every day of continued exponential growth brings the world system closer to the ultimate limits of . . . growth. A decision to do nothing is a decision to increase the risk of collapse." Moreover, the longer we allow exponential growth to continue, the smaller the chance of achieving a sustainable future.

In 1992, the MIT team published a sequel to its study. In this book, *Beyond the Limits*, Meadows and her team wrote:

> In 1971, we concluded that the physical limits to human use of materials and energy were somewhere decades ahead. In 1991, when we looked again at the data, the computer model and our own experience of the world, we realized that in spite of the world's improved technologies, greater awareness, [and] stronger environmental policies, many resource and pollution flows had grown beyond their sustainable limits.

The new study still predicts that a sustainable world is possible but only if drastic steps are taken to curb population growth, resource consumption, pollution, and habitat destruction. This chapter examines ways to bring population growth under control.

6.2 How Do We Control Population Growth?

Recognizing that a concerted worldwide effort is needed to deal with the rapid growth in the human population, we need to set some goals.

Setting Our Goals

But, what should our goals be? Basically, the nations of the world have to make decisions regarding two issues: population growth and population size. The more immediate problem is what to do with population growth. Should countries let population growth continue unabated, should they try to reduce it, or should they stop

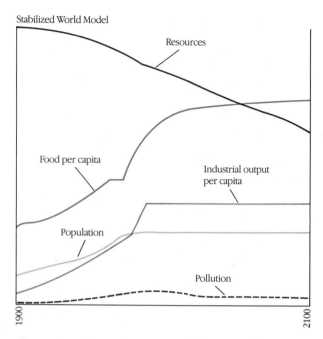

Figure 6.2 *The same computer analysis as in Figure 6.1 shows that a sustainable system can be achieved by global zero population growth, pollution control, soil management, and sustainable agriculture.*

growth altogether? And, what about population size? Should we shoot for 10 billion? Should we level off, then try to decrease population size?

The answers to these questions differ, depending on who is asked. One reason for the difference may be attributed to personal bias. Some people object to family planning, arguing that it violates religious and personal values. Many of these people argue that the world can support an unlimited number of people. A second reason for the difference is ignorance. Many people are woefully uninformed about the condition of the biosphere and the widespread human suffering already occurring in many parts of the world.

Narrowness of view also contributes to the difference of opinion about the optimal population size. Some observers contend that the world can comfortably accommodate only 500,000 people—1 for every 10,000 present today. Others say that a population of 100 billion would be possible. Those who project larger numbers base their assertion on a narrow view of what is necessary to sustain human life. For example, they often look only at food supplies, assuming we can safely convert all of the potential farmland to food production. They project carrying capacity on the basis of this simpleminded assumption, ignoring the problems that would arise from accompanying energy use, habitat destruction, species extinction, and so on. As noted in the previous chapter, this approach, which maximizes one variable, often ignores all of the other necessities for a healthy existence on the planet.

Some experts believe that the human population has already exceeded the Earth's carrying capacity; they contend that growth should be halted as soon as possible (Chapter 5). Let's consider the details of this argument.

Today, about 800 million to 1 billion live in extreme poverty. These individuals are inadequately fed and sheltered. They wander the streets of Calcutta, Bangkok, Cairo, and other cities begging for food or stealing what they can. At night, they sleep in alleyways, under bridges, or in makeshift cardboard shelters. Another 2 billion people live on the edge, with inadequate food and shelter and few amenities. In some places, 4 families live in a two-room apartment and share a water tap with 25 other families. Many have no sewage disposal systems.

All told, nearly three-fifths of the world's population live in dire circumstances. Strenuous efforts to improve the economic condition of the world's poor, in hopes of increasing individual wealth, have failed to keep pace with growth. More and more people fall into poverty each year. Because of these facts and the deteriorating condition of the environment, many observers believe that it is essential to stop human population growth now to reduce further suffering, environmental pollution, and resource depletion.

Some observers believe that the human population should eventually be reduced humanely via attrition—keeping birth rates below death rates. Reducing the size of global human population, they say, may be essential to building a sustainable society.

Stopping growth, reducing population size, and pursuing sustainable economic plans, all discussed below, can help break the vicious cycle of poverty and environmental destruction. However, the road ahead will be long and difficult. Even if we could miraculously reach replacement-level fertility today, the world population would still swell to 8 billion, in large part because of the lag effect described in Chapter 5.

Population Control Strategies

How do we go about controlling population growth? For a number of years, many population experts advocated economic growth as a means of slowing population growth. The equation was simple: By expanding the economies of all nations, more jobs are created. More jobs mean more personal wealth. People can then afford decent housing, food, and education. Poverty and disease would be eliminated. Data throughout the world show that as incomes rise, fertility rates decline.

Demographic Transition The notion that economic growth in the developing nations would solve the population crisis is based on evidence from the history of developed countries. In these countries, population growth unleashed by the Industrial Revolution was brought under control as economic conditions improved. This phenomenon is called the **demographic transition**.

The demographic transition takes place in four stages (Figure 6.3). The industrial nations began in

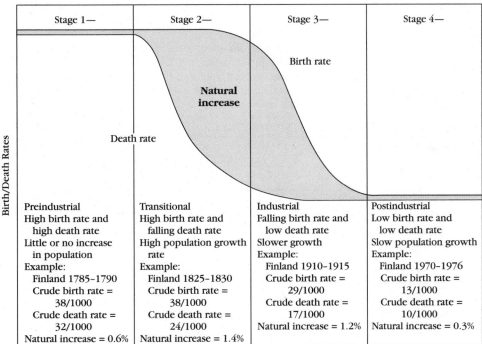

Figure 6.3 *The demographic transition shown is a transition from high birth and death rates to low birth and death rates brought about by industrialization and an overall increase in wealth in a population.*

SOURCE: Haupt, A. and Kane, T. (1978). *Population Handbook.* Population Reference Bureau

Models of Global Sustainability 6.1

Controlling Urban Sprawl: The Other Kind of Population Control

Controlling populations means more than slowing the rate of growth. It also entails efforts to halt the spread of human settlements onto farmland, forest, fields, and wildlife habitat. This goal is as crucial to the developed nations of the world as it is to the developing nations.

A world leader in growth management is the state of Oregon. In the early 1970s, Oregon's state legislature passed a law requiring all cities and towns to establish urban growth boundaries designed to protect open space, forests, wildlife habitat, and farmland. Within the growth zones, city officials are permitted to plan for an annual population growth of 2% over a 20-year period. At the end of this period, they can revisit the boundaries and set new ones, if necessary.

Urban boundaries have had many positive effects in Oregon. They have helped reduce the leapfrogging of subdivisions, a leading cause of the destruction of open space and agricultural land. Furthermore, because urban sprawl usually increases the costs of services, such as roads and sewer lines, urban growth boundaries have reduced taxes and utility costs.

Within urban growth zones, zoning laws can provide a wide range of housing options; for example, single-family dwellings, apartments, duplexes, and condominiums. By concentrating growth in a more confined region, urban growth boundaries have helped to make Portland's light rail system a colossal success. The number of passengers who rode the light rail system in its first year was twice what planners had projected. In 1991, the system carried the same number of passengers that would be carried on two new lanes on every road entering downtown.

Downtown Portland's share of the regional retail market has increased from 7% to nearly 30%, and has added 30,000 jobs without any increase in the number of cars. The number of days that air quality exceeds health standards has dropped from about 100 per year to none.

By greatly slowing the outward expansion through urban growth boundaries, cities help to make all urban systems—from waste management to transportation to energy—more efficient. Urban growth boundaries are possible in virtually all cities because considerable amounts of land are usually available within cities for growth. In 1989, in fact, the amount of vacant and underused land in Portland was estimated to be nine times the space needed to accommodate Portland's projected growth for the next 20 years.

Portland has contained 95% of its growth within urban growth boundaries since the 1970s. Today, the city covers approximately 900 square kilometers (350 square miles). Denver, with approximately the same population size and no urban growth policy, covers 1300 square kilometers (500 square miles) and is expected to sprawl to over 2600 square kilometers (1000 square miles) by the year 2010.

Several states, including Maine, Florida, Vermont, New Jersey, Rhode Island, Washington, Hawaii, and Georgia, have laws similar to Oregon's. Worldwide, many nations and cities have taken steps to control urban growth, among them Belgium, France, the Netherlands, and West Germany.

Several Canadian cities, including Toronto and Vancouver, have taken the lead in providing compact, well-mixed urban centers with outlying, high-density suburban areas. Because of efforts to concentrate housing and business, public transportation became a feasible way to link areas, and thereby decrease automobile usage, energy consumption, and pollution. Canadians walk, cycle, or use local public transport within a given area and rely on rapid light rail or express buses to reach other areas.

Curitaba, Brazil, has made tremendous strides to create a more sustainable design. In the 1960s, Curitaba was a typical automobile-dependent city plagued by traffic congestion and urban sprawl. In the 1970s Mayor Jaime Lerner initiated measures to integrate transportation and land-use policies, including ways to concentrate population near major transport corridors. Each region within the city was also designed to include a mix of jobs, homes, and services, making them more self-reliant. To reduce the outward sprawl of the city, new development was concentrated within city limits. An extensive network of bikepaths was set up to permit commuters to move about by bicycle. The public transport system was designed to consist of main thoroughfares served by express buses and frequent feeder routes that link neighborhoods to the main system. That way, people living in outlying areas could easily access a main route to get into or out of the city.

The effects of this integrated, far-reaching policy in Curitaba have been profound. More than 1.3 million passengers use the bus system daily—that's 50 times more than 20 years ago. Shopping districts once on the decline have made a dramatic comeback. Open space per capita has increased from .5 square meters to 50 square meters. Gasoline use per vehicle is 30% less than in other Brazilian cities its size, and inhabitants of Curitaba spend only 10% of their income on transportation, one of the lowest rates in Brazil.

Urban growth can be controlled by numerous strategies described in Chapter 23, all of which are vital to making cities and towns sustainable.

Stage 1, which is characterized by high birth rates and high death rates. In this phase, the population is stable because high birth rates and death rates cancel each other out. In Stage 2, improvements in health care and sanitation brought about by improving economic conditions cause death rates to begin to fall. However, as Figure 6.3 illustrates, birth rates in Stage 2 tend to remain high. The discrepancy between birth rates and death rates results in a period of rapid population growth. In Stage 3, as countries continue to develop economically, birth rates begin to decrease and population growth slows. Finally, over time, birth rates and death rates come into balance. Population growth stops or slows dramatically in Stage 4.

The decrease in birth rates in Stage 3 can be attributed to several factors. Perhaps the most important is the shift in people's attitudes toward children. Preindustrial farmers view children as an asset because they help on the farm and often support their parents in old age. With industrialization and the inevitable migration of families to cities, however, children become an economic liability. Because of competition for living space, each child means that more money must be devoted to food and housing. If the children do not work, they create an additional financial drain on the family. As a result, smaller families generally prevail.

Virtually all industrial countries have experienced demographic transitions and are in Stage 4, with birth rates at or near death rates. If it worked for these countries, why won't it work for Ghana, China, or India?

When one examines the issue in detail, at least four reasons become apparent. First, the economic resources of many developing countries are too limited to build the kind of industry needed for demographic transition. Second, the classic transition did not occur overnight: It took Finland over 200 years to approach a balance between birth and death rates. Developing countries with rapid doubling times do not have this much time. Third, population growth in many countries outstrips economic growth. Recent studies show that a 1% growth in the labor force requires a 3% economic growth. Think of the economic growth needed to sustain populations growing at 3% and 4% per year. For many countries, it is almost impossible for economic growth to keep pace with population growth. Getting ahead has proven to be a pipe dream. The fourth reason is that the fossil fuel energy sources essential to demographic transition in the developed countries are diminishing and becoming ever more costly. Without the rich mines of England or the great oil deposits of Arabia and North America, developing nations will probably never experience similar economic growth.

Many developing countries have entered Stage 2 of the demographic transition, with high birth rates and low death rates. Because a demographic transition brought on by industrialization is unlikely, the transition to stable populations must come about in other ways, most likely through family planning and small-scale sustainable economic development (Chapter 22).

Family Planning **Family planning** allows couples to determine the number and spacing of offspring. For countries mired in Stage 2, it can greatly accelerate the decline in the birth rate. For family planning to work, information on birth control must be readily available. Birth control must also be accessible and inexpensive. Furthermore, people must be motivated to practice it. **Birth control** includes any method to reduce births. Generally, birth control measures fall into two categories: **contraceptives**, any chemicals, devices, or methods that prevent sperm and egg from uniting, and **induced abortions**, the intentional interruptions of pregnancy through surgical means or drug treatments.

Family-planning programs vary considerably. Some, such as those noted in the chapter introduction, offer education on the importance of birth control and the various methods that are available (Figure 6.4). Others offer education as well as contraceptives or surgical abortions.

Family-planning programs may be privately run or may be state sponsored. Planned Parenthood in the United States, for example, is a private, nonprofit organization with clinics in many large cities. It offers low-cost medical care, contraceptives, and abortions.

State-sponsored family-planning programs lie on a continuum from the purely voluntary to the compulsory. **Voluntary programs**, as a rule, make birth control information and methods available to the public at low cost. The government exerts no pressure; people are free to choose the type of birth control and family size they prefer.

Family-planning programs promoted by governments are referred to as **extended voluntary programs**. In these cases, governmental agencies may distribute information on birth control and sterilization or sponsor posters, newspaper ads, television and radio announcements, and billboards (Figure 6.5). In Egypt, for instance, a song promoting birth control was played on a government-sponsored commercial. The song became so popular that it became a national hit.

In extended voluntary programs, payments or other incentives are often offered by governments to couples who practice birth control or undergo sterilization. Transistor radios, for example, were handed out in rural Egypt to couples who adopted some form of birth control. Special loan programs are used in Thailand (Models of Global Sustainability 5.1). Some government programs offer reminders. In Djakarta, Indonesia, for example, music plays at intersections, and in Bali, church bells ring at 5:00 P.M., reminding women to take their birth control pills.

Figure 6.4 *Family planning in India. A doctor lectures to a group of Indian women on various methods of birth control at a family-planning clinic in New Delhi. These and other clinics throughout the world are vital to efforts to educate women on family planning.*

Government programs often attempt to change people's thinking about family size. Posters in Vietnam, for instance, extol the virtues of a one-child family. Murals and posters in India feature a smiling family of four with the slogan, "two or three children is enough." These efforts strike at one of the root causes of the population problem—people's mind-set. Clearly, as Varinda Vittachi, a writer from Sri Lanka, notes, "The world's population problem will be solved in the mind and not in the uterus."

Information campaigns can also alter stereotypical sex roles, persuading men that masculinity and self-importance are not related to the number of their children or convincing women that the childbearing role is not the only one of value.

Forced family-planning programs involve strict government limitations on family size and punishment for those who exceed quotas. These programs are extremely rare. They may impose sterilization after a family reaches

the allotted size, limit food rations for "excess" children, or tax couples who exceed the allowed number of children.

China, which has what many observers consider a forced family-planning program, adopted a one-child policy to halt its rapid growth. This policy is intended to reduce China's population size to 700 or 800 million, down considerably from the present 1.2 billion. Female workers meet regularly in small groups to discuss birth control. Government workers reportedly urge men and women with one child to undergo sterilization. According to unsubstantiated reports, pregnant women who have already had one child were "forced" to have abortions. The Chinese government also supplements the monthly income of couples who pledge to have only one child. If a couple that has pledged to have one child has a second one, however, the government often requires repayment of the monthly bonus and cuts the monthly salary by 15%.

Many programs do not fit neatly into these categories. The government of Singapore, for instance, uses radio, billboards, and school curricula to affirm that two children per family are enough. These aspects clearly constitute an extended voluntary program. Several coercive measures are also used. For example, the first two children of any family are allowed to attend local schools, but additional children must often be bused elsewhere. Furthermore, hospital costs increase for "surplus" children, and government employees who are not sterilized are given a low priority for public housing.

Small-Scale Sustainable Economic Development

"Family planning cannot exist in a vacuum. You can't just distribute contraceptives and tell people to go ahead and start lowering the birth rate," says Aziz el-Bindary, head of Egypt's Supreme Council for Population and Family Planning. "To have an effective family-planning program," he adds, "you also have to have an effective economy—where jobs are available, where health facilities are adequate."

Throughout the world, governments are finding that in order to promote family planning they must attack the problem on several fronts. Programs must create jobs and education for women, small-scale economic development, better health care, and family planning. To those versed in critical thinking skills, this should come as no surprise. Childbearing is a complex issue; the reasons for having children are many.

Developed Countries—What Can They Do?

Many people think that population control pertains only to the developing countries, where 90% of the present growth occurs. Observers point out, however, that the high per capita consumption of the developed countries places enormous strains on the Earth's environment. The

Figure 6.5 *A familiar poster seen everywhere in India, with a smiling family and the slogan "Two or Three Children . . . Are Enough." India's family-planning program is closely meshed with the government's health services. All doctors, nurses, and other health workers are trained and expected to provide family-planning advice and assistance as part of their regular duties. Similar posters are found in Thailand and other Asian countries.*

environmental impact of a nation is often approximated by the equation:

Environmental impact = Population size × Per capita consumption × Pollution and resource use per unit of consumption

In developed countries, especially the United States and Canada, the per capita consumption, or the amount of resources each person uses, is remarkable. According to several estimates, a single American or Canadian uses 20 to 40 times more resources than a citizen of the developing world.

What does this mean? Each year, approximately 93 million people join the world population, 84 million of which are in the developing nations. The remaining 9 million are born in developed countries. Yet, the latter consume as many resources and produce as much waste as 180 to 360 million residents of developing nations.

Widespread pollution, species extinction, and resource depletion are three signs that many developed countries have exceeded critical ecological thresholds. Although it may be a difficult concept to accept, to foster a sustainable revolution, population growth control is as important, perhaps even more important, in developed nations as it is in developing nations.

There is some good news to report on this matter. As pointed out in Chapter 5, about two dozen European nations have slowed growth considerably. The populations of several European nations, among them Germany, Hungary, and Bulgaria, have stopped growing. Some countries have actually begun to shrink because total fertility rates have fallen below replacement-level fertility.

Despite these improvements, developed countries are expanding at a rate of 0.5% per year, which yields a doubling time of 148 years. In Canada, the United States, Australia, New Zealand, and the Commonwealth of Independent States, population growth is 40% to 50% higher than the industrial world average. Growth in these nations is of great concern for two reasons: First, these nations comprise over half the population of the developed world. Second, the people of these nations, excluding the Commonwealth of Independent States, lead some of the most resource-intensive life-styles on the planet.

In order to lower their environmental impact and become more sustainable, developed nations need to slow their growth rate even more and greatly reduce their demand for resources. Three sustainable options to meet the second goal are conservation, recycling, and renewable energy use. Restoration is also vital to protect the renewable resource base and natural systems upon which they depend.

By using less, many people argue, developed countries can also make possible a more equitable sharing of the Earth's resources. This redistribution could help to reduce world tensions, but sharing resources is controversial. Garrett Hardin, author of the book *Filters Against Folly*, contends that global sharing of resources is not the answer to global stability, as some suggest. The rate of growth and needs of the nearly 4.2 billion residents of the developing nations far exceed our capacity to help, he argues. However, developed countries can assist developing nations by sharing their knowledge of birth control, agriculture, health care, environmentally friendly technology, and sustainable development.

Table 6.1   Population Control Strategies for Developed Countries

Strategy	Rationale	Benefits
Stabilize population growth by restricting immigration, and by spending more money and time on sex education and population awareness in public schools.	High use of resources taxes the environment. Immigrants create serious strain on the economy and create social tension in conditions of high unemployment. Education helps citizens realize the importance of population control.	Limiting resource use leaves more for future generations and developing countries.
Provide financial assistance to developing countries for agriculture and appropriate industry. Aid should come from government and private sources.	Economic growth in developing countries will raise the standard of living and aid in population control.	The rich-poor gap would narrow. A decrease in sociopolitical tension and resource shortages would result.
Provide assistance to population control programs.	Better funded population programs can afford the increased technical assistance and community outreach programs necessary to provide information to the public.	Could result in faster decrease in population growth.
Make trade with less developed countries equitable and freer.	Freer trade will increase per capita income and raise standards of living with little effect on home economy.	A higher standard of living and increased job opportunities could result.
Concentrate research on social, cultural, and psychological aspects of reproduction.	Techniques available today are effective and reliable. What is needed is more motivation for population control, especially among poor countries.	Money will be better spent; research of this nature may help facilitate family planning in less developed countries.

Financial assistance to help achieve a moderate rate of industrialization, using **appropriate technologies**—industries that are labor intensive and use local resources to meet local needs—can also help. Table 6.1 lists some additional suggestions for developed countries.

William and Paul Paddock propose a triage system to determine who gets financial and technical aid. According to this system, countries would be categorized in three groups: (1) those that have an adequate resource base and could survive hard times without aid, (2) those impoverished nations that would probably not survive drought and food shortages even with aid, and (3) those that can be helped. Group 1 needs no assistance. Group 2 must be abandoned as hopeless, for no amount of aid will help. Group 3, if aided, could pull through a drought or other difficult period. Therefore, the Paddocks suggest that developed nations, all faced with limited financial resources, should offer financial aid for population control, food, and sustainable development to Group 3. To some, this seems unethical. Others see it as the only answer to apportioning the limited aid. (See Viewpoint 6.1 for a discussion of why population growth in the developing world should concern us.)

Developing Countries—What Can They Do?

Most developing countries recognize the need for population control. Today, in fact, 93% of the world's population lives in countries that have some form of population control policies. Support for these programs comes from a variety of sources including the developing nations themselves. One of the major players in the international effort to promote family planning in developing nations is the **International Planned Parenthood Federation (IPPF)**, a nonprofit organization established in the 1950s. IPPF disseminates information on family planning and provides assistance to many countries.

Another major source of financial support for family planning is the **United Nations Fund for Population Activities (UNFPA)**. Established in the late 1960s,

Viewpoint 6.1

Population Growth in the Developing World: Why Should We Worry?

Daniel D. Chiras

The developing countries are growing at a remarkable rate. In fact, by the year 2025, the population of the developing nations is expected to double, climbing from 4.2 billion in 1992 to 8.5 billion! This massive growth will create problems of epic proportions, among them deforestation, wildlife extinction, erosion, and poverty in the developing nations. However, many people in the developed world consider population growth in the developing nations an issue of little importance. What difference does it make to a North American resident?

A survey of current impacts indicates that the rapid growth of population in developing nations could have profound effects on the lives of North Americans. Consider some examples.

The rapid growth of the human population in Latin America has resulted in a dramatic increase in the size of the work force. Many Central American workers, however, cannot find jobs at home, and consequently, many cross the borders illegally, hoping to find work. Although this cultural infusion is seen as a positive impact by some, these individuals compete for low-level jobs with U.S. workers. They also strain U.S. schools and hospitals, which are not well equipped to accommodate Spanish-speaking immigrants.

Illegal immigrants also strain our environment. At least 500,000 illegal workers and their families annually enter the United States, each needing water, food, shelter, and other goods and services. Some say the number may be as high as 1.5 million. If this is true, the United States is adding another Los Angeles (excluding the suburbs) to its population every year and a new California every 20 years, placing additional burdens on already stressed land, air, and water.

Population growth abroad also impacts our lives. For instance, 54 million people lived in the Middle East 30 years ago. Today, the population has climbed to nearly 140 million. The rapid growth of population fans religious, ethnic, and political turmoil, which often leads to bloodshed. Regional turmoil threatens access to one of the world's largest oil reserves, on which the United States, Japan, and Europe rely. The 1991 Persian Gulf War is a case in point.

Growing population abroad increases U.S. food production. Although potentially good for farmers, it places stresses on already badly abused soils (Chapter 7). It also means additional pesticide use, energy consumption, and irrigation water—all of which impact our environment.

Expanding population abroad also results in the destruction of forests, fields, and wetlands as well as the many species that live in

them. The loss of species has numerous economic implications (Chapter 8). Many new drugs, for instance, could come from plants in tropical rain forests that are now being cut to make way for more people.

Population growth also indirectly threatens global climate. The loss of tropical rain forests, for example, is partly responsible for rising carbon dioxide levels that contribute to global warming (Chapter 15). Deforestation of rain forests as far away as Africa may have a profound effect on rainfall patterns in Europe.

Rapid population growth in the developing nations contributes to the vicious cycle of poverty gripping many poor nations, making it difficult for them to repay loans to developed countries. Today, billions of dollars in international debt remain unpaid because countries cannot cope with their population growth. When countries fail to repay loans to U.S. banks, consumers suffer. Banking services invariably become more costly. The savings and loan failures in the United States resulted in part because developing nations were unable to meet loan payments.

Why worry? Because, in the highly interconnected world, population growth in developing nations creates hardship and pain everywhere. Few boundaries are immune to this insidious problem.

UNFPA doles out about $170 million a year to developing nations for family planning and related activities, and is largely responsible for much of the progress in global population control in recent years. UNFPA has raised international awareness of the dangers of rapid population growth and has helped many governments develop, finance, and implement family planning programs. Unfortunately, because of the growing interest in population control in developing countries requests for support often exceed available funds. The United States

withdrew its fairly substantial support of UNFPA in the Reagan era because of its alleged support of abortion in China. As of December 1992, the United States had not changed its position.

UNFPA also invests money in economic development programs. In Egypt, for instance, UNFPA recently made substantial investments in clothing factories that will employ Egyptian women. The logic behind these and other such activities is that working women often delay marriage and childbearing and thus have fewer children. (Chapter 22 describes the topic of sustainable economic development in detail.)

Foreign assistance programs sponsored by developed countries are a third source of support. The **U.S. Agency for International Development (AID)**, for instance, was the major sponsor of family-planning programs in the developing world from 1965 to 1980. In 1990, the United States gave $250 million through this and other programs.

Even international lending agencies help out. For example, the World Bank, which is largely financed by the United States, has offered financial aid to developing countries with population control policies since the 1960s, and has recently stepped up its support. All told, approximately $600 million is donated to developing nations from outside sources. Although that may sound like a great deal of money, it is far from sufficient. More money is needed. Those who oppose increased funding, however, note that financial aid for family planning has increased dramatically in the past decade. What they fail to recognize is that because of inflation, overall spending has actually fallen.

Today, although many countries with active family-planning programs provide most of their own financing, few of these governments spend more than 1% of their national budget on family-planning services, which is far below what is needed. The director of the AID family-planning program estimates that the cost of providing family-planning services in developing countries in the year 2010 will reach $9 to $10 billion. Many believe that major international donors, such as the United States, Japan, and Germany, will be unable to meet the projected rise in costs. As a result, many developing countries will have to shoulder an even larger share of the burden.

Despite the obstacles facing the developing countries, family planning makes good sense from both environmental and economic standpoints. Depending on the country and the program, each birth averted by family planning yields a savings of between $15 and $200 per year in social services. Estimates in the United States suggest that $1 invested in family planning saves at least $3 in health and welfare costs in the first year.

Table 6.2 lists population control strategies for less developed countries. Accompanied by appropriate development, these suggestions could improve the lives of many people.

6.3 Making Strategies Work

Almost any venture encounters obstacles. Population control is no exception, especially because it strikes at the heart of what to many people is a cherished personal freedom: the right to have children. Besides economic barriers noted earlier, three major obstacles impede the progress of family planning: (1) psychological inertia, (2) lack of education, and (3) religious and cultural beliefs.

Psychological Barriers

Large families are an asset in developing countries, since children help with the chores and later care for their parents when they retire. Given the high mortality rates in these countries, having many children ensures that some will survive. In spite of a rapid decline in death rates among children, birth rates remain high.

Traditional views of family size often change slowly after a decline in death rates. Nowhere is this more evident than in India. In some economically developed regions, for instance, sons are still held in high esteem because they represent a source of security in their parents' old age. Although the present government promotes small families of two or three children, many parents see the ideal family as two sons and one daughter. In trying to reach this goal, however, couples in 1992 averaged 3.9 children. "One son is no sons," one Indian argues. "To be sterilized is to tempt fate," another proclaims. Until people begin to realize that one son is enough and that that son will probably survive, India's population will continue growing at its current rapid rate.

Having children is an enriching activity for many people. Sociologists report that in many developing nations men and women are admired for the number of children they have, a phenomenon present in developed nations as well. Humorist and educator Bill Cosby pokes fun at this view, saying, "If a chimpanzee can have a baby, the human female [and male, I would add] should realize that the feat is something less than an entry for the *Guinness Book of World Records*."

Unfortunately, social acceptance and other psychological factors result in the birth of many children who will never have adequate food, clothing, shelter, and education. Why do people continue to have large families?

Citizens of developed countries, even those who love children and put stock in family life, tend to view children as an economic drain. Recent studies show that it costs low-income families in the United States about $60,000 to raise a child, including education at a publicly subsidized college. Middle-income families spend about $90,000 per child. In developing countries, however, a child deprives the family of virtually nothing. In fact, since children represent a form of material wealth (free labor and support in old age) and satisfaction, they are often viewed as an asset.

Educational Barriers

Education is a key barrier to smaller families for several reasons. In the United States, for example, studies by the

Table 6.2 Population Control Strategies for Developing Countries

Strategy	Rationale	Benefits
Develop effective national plan to ensure better dissemination of information and availability of contraception and other methods of population control. Do not rely on one type of control.	Each country better understands its people and thus can design better programs to spread population control information and devices.	More effective dissemination of information and, probably, a higher rate of success.
Finance education in rural regions, emphasizing population control and benefits of reduced population growth.	Education can help make population control a reality.	Slower population growth, more effective use of contraceptives, and more incentive.
Seek to change cultural taboos against birth control and cultural incentives for large families.	Changes in culture and psychology may be needed to make population control programs effective.	Such changes will help programs succeed.
Develop appropriate industry and agriculture, especially in rural areas to reduce or eliminate the movement of people from the country to the city.	Appropriate agriculture and industry will create jobs and better economic conditions for families. A higher standard of living could translate into better health care and greater survival of young, thus destroying need for large families.	This will result in higher standard of living, better health care, and impetus for control of family size.
Seek programs of development that attain a maximum spread of wealth among the people.	Development must not just help a select few, because benefits may not trickle down to needy.	Plans of this nature yield good distribution of income and help the needy rather than select few.
Integrate population policy with economic, resource, food, and land-use policy to achieve a stable state.	Finite resources require wise allocation and use; success in the long run depends on attempts to achieve a sustainable future.	Longevity and permanence are attainable if policies are integrated and take into account the requirements of a sustainable society.
Seek funding from the United Nations and developed countries.	Developed countries have a stake in stabilizing world population growth.	Developed countries could provide significant financial support.

Department of Commerce show that the higher the educational level, the lower the TFR (Figure 6.6). The reason for this is fairly simple: Educated women often pursue careers and thus postpone marriage and childbearing. Since the childbearing years are from 15 to 44, a woman who graduates from college at age 21, marries, but delays children until she is 30 has decreased her childbearing years by half.

The relationship between education and fertility is evident in the developing world as well. Men and women who lack educational opportunities or who choose not to go to school generally marry younger and pursue careers that do not interfere with childbearing. Thus, the period of childbearing is much longer than for couples that pursue higher levels of education. A lack of education also makes it more difficult for people to learn about alternatives to childbearing and proper use of contraceptives. It is no wonder, then, that the birth rate is still high in rural India, where 80% to 90% of the women cannot read or write.

Religious Barriers

Religion may also be a powerful force in reproduction. As pointed out in Models of Global Sustainability 5.1, Buddhism actually promotes family planning in Thailand. Other religions, such as Catholicism, forbid all "unnatural" methods of birth control, such as the pill, condom, diaphragm, and abortion. Theoretically, the Catholic church guides the sexual practices of approximately 600 million people. Recent surveys show, however, that the use of contraceptives by Catholic women,

Figure 6.6 *Total fertility rates of U.S. women, by level of education. In the United States and abroad, education opens up numerous employment opportunities for men and women and lowers fertility rates. As such, education is crucial to all efforts to reduce fertility rates needed to build a sustainable world community.*

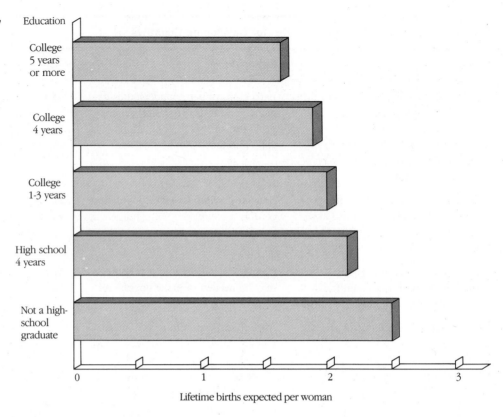

Lifetime births expected per woman

especially in Western nations, is nearly as high as that among non-Catholics. In Latin America, in fact, many priests speak out against the Vatican's official policy.

Birth control is a generally undiscussed subject among other religions. In many Eastern and Mid Eastern religions, which compete with one another for followers, birth control is frowned on. Not surprisingly, TFRs in Iran and Iraq are 6.1 and 7.0, respectively.

Overcoming the Obstacles

The main barriers to widespread adoption of family planning—economic, psychological, educational, and religious—are formidable but not insurmountable. If family-planning programs are going to work, they must address these problems.

First and foremost, all programs must be culturally and religiously acceptable. For example, in many Middle Eastern countries where family planning is culturally unacceptable, programs can be promoted as a way of spacing births to improve maternal and child health, thus bypassing religious or cultural objections.

Second, attitudes toward family and childbearing size must be changed. Especially needed are means to improve the status of women and to give women a larger say in the economic decisions of a village.

Third, because many people most in need of family planning are those least able to pay, contraceptives and advice on family planning must be inexpensive or free and readily accessible. Meaningful work for men and women can change their economic status.

Fourth, because education levels are often low, birth control measures must be easy to understand and simple to use to reduce the risk of failure. Trained individuals must be available to counsel those seeking advice on family planning. Educational programs can also help raise the literacy rate, making it easier for people to understand the benefits of family planning as well as the instructions that accompany contraceptives.

Fifth, new programs must be created to provide better health care and sanitation to lower the death rate and slow the impetus for having numerous children.

In the past two or three decades, family planning was viewed as a measure to reduce population growth, protect the environment, and improve the economic condition of many of the world's people. Today, family planning, while still embracing these goals, is considered in some countries as a human right. New emphasis has been placed on improving the health and welfare of mothers and their children. This new direction appeals to more people and broadens the base of support for family-planning programs. A couple who may not relate to lofty goals of slowing the growth rate of their country can surely understand the benefits family planning brings to them in improved economic conditions and maternal and infant health.

6.4 Ethics of Population Control

No issue in environmental science is so burdened with ethical conflict as population control. To deny the right to reproduce is to deny one of the most basic and important of all human activities. To some, population control is a violation of deep religious beliefs. To others, it is an intrusion into a private matter; for minorities, population control has overtones of genocide. Three important ethical questions discussed in this section serve as an introduction to this complex ethical issue:

1. Is reproduction a personal right?
2. Is it ethical not to control population growth?
3. Should we try to achieve zero population growth?

Is Reproduction a Personal Right?

Some argue that the right to reproduce at will should be curtailed when the rights of the individual interfere with the welfare of society, that is, the **collective rights** of all the people. For example, in 1975, India's government, under the late Prime Minister Indira Gandhi, began a program of forced sterilization. Changes in India's Constitution placed the collective rights of society above the rights of the individual. This short-lived and extreme program illustrates that an important end—the welfare of an overpopulated nation's people—can prompt governments to adopt tyrannical means to achieve an admirable goal.

The opposite view was well stated in 1965 by Pope Paul in a speech to the United Nations: "You must strive to multiply bread so that it suffices for the tables of mankind, and not, rather, favor an artificial control of birth, which would be irrational, in order to diminish the number of guests at the banquet of life."

Individuals who support the right to reproduce freely often argue that denying such rights takes away personal freedom. Ecologist Paul Ehrlich argues that we "must take the side of the hungry billions of living human beings today and tomorrow, not the side of potential human beings. . . . If those potential human beings are born, they will at best lead miserable lives and die young." He argues that we cannot let humanity be destroyed by a doctrine of individual freedoms conceived in isolation from the biological facts of life.

Ecologist Garrett Hardin argues that the integrity of the biosphere should be the guiding principle in the debate over population control. Recognizing that the welfare of the biosphere determines the welfare of all living things, including humankind, and that we have obligations to future generations to protect the biosphere, many environmentalists perceive human population control as a biological and cultural imperative.

Clearly, most countries today favor the environmental viewpoint and have taken steps to humanely reduce population growth. They realize that building a sustainable society requires controls on human numbers. While few countries put strict limitations on family size, most encourage smaller families, leaving the decision to the parents. Many countries that promote family planning are striving to slow the growth rate. Others, who recognize the inevitable consequences of continued growth, are looking for ways to reach zero population growth. Still others, such as China, who recognize that they are already exceeding the carrying capacity, are seeking to reduce human numbers over the long run.

Is It Ethical Not to Control Population Growth?

Another way to look at ethical issues surrounding the population growth controversy is to ask the opposing question: Not is it ethical to control human numbers, but is it ethical not to control population growth?

If we do not control population growth, uncontrolled growth will very likely cause a massive deterioration of the environment and rob future generations of the opportunities many of us now enjoy, say environmentalists. They base their conclusions on studies of population growth, resource supplies, and pollution, such as the Limits to Growth and the Beyond the Limits studies mentioned at the beginning of this chapter. Other research also suggests that continued population growth is an invitation to disaster. Collectively, these studies remind us that humankind comes with essentially the same warranty the dinosaurs had.

Werner Fornos, director of the Population Institute, notes, "National leaders are recognizing that population growth without adequate resources and services results . . . in national disaster." In a nutshell, not controlling population, say many observers, is unethical because it violates our obligation to future generations and to other species.

Critics of this viewpoint argue that scientists will develop technology and substitutes to feed the hungry new mouths and that more people will lead to a better tomorrow. Julian Simon, a leading advocate of this outlook, argues that more people mean more knowledge and that knowledge is the key to a better future. With knowledge and technology, we can overcome all limitations. (For a debate on this topic, see Point/Counterpoint in Chapter 5.)

Should We Try to Achieve Zero Population Growth?

Most of the leading experts who study resources and pollution argue that to build a sustainable human community, we must stop growing altogether, not just slow the rate of growth. Arguing that humanity is already living beyond the Earth's carrying capacity, they say that we must reduce human numbers. Ethical obligations to future human generations and to other species also suggest the importance of zero population growth.

6.5 The Status of Population Control

"The world of the mid-1980s is a world of stark demographic contrasts," writes Lester Brown, president of the Worldwatch Institute. Never have the differences among countries been greater. "Some populations change little in size from year to year or decline slightly," Brown notes, "while others are experiencing the fastest growth ever recorded." The same could be said for the early 1990s. This section discusses demographic contrasts and explores areas of progress and setback in current efforts to guide society onto a sustainable path.

Encouraging Trends

Developing Nations   Most of the world's developing nations recognize the need for programs to control population growth. Fifteen years ago, this was not the case. At that time, the developing nations rallied behind the cry "Development is the best contraceptive." A decade of falling per capita food production and income and rising debt, however, has changed many leaders' minds.

The first step in solving the population problem is awareness. The developing countries are now aware of the problem, and most are trying to do something about it. In fact, for every $1 of foreign aid received to promote family planning, developing nations now spend about $4.

Many developing nations have successful population control programs. The largest reductions in growth rate have occurred in China, Taiwan, Korea, Thailand, Tunisia, Barbados, Hong Kong, Singapore, Costa Rica, and Egypt. China's decline in birth rate in the 1970s is one of the most rapid of any country on record and may be one of family planning's greatest success stories. China's 2.5% annual growth rate in the 1960s (doubling time = 28 years) decreased to 1.3% (doubling time = 54 years) by 1989.

The Republic of Korea runs another extremely successful program. One of the earliest in the developing world, South Korea's family-planning program was a combination of private and government effort aided by an incredible improvement in economic conditions. Between 1962 and 1992, in fact, per capita income in South Korea increased tenfold, skyrocketing from $60 to $5400. The proportion of women completing secondary education also rose dramatically during that period from 25% to 86%. In the 1960s, only one in every ten women was practicing family planning. Today, over 75% of all couples use contraception. The TFR fell from 5.4 in 1962 to 1.6 in 1992.

Developed Nations   In the early to mid-1970s, many developed countries updated their laws and policies governing family planning. Nearly two dozen countries in Europe now have near stationary populations. At least three European countries are shrinking. The Common-

wealth of Independent States and Japan, two of the most populous developed countries, show signs of slowing population growth. According to some demographers, the decline in growth in the developed nations is largely responsible for a drop in the world's growth rate from over 2% in 1974 to 1.7% in 1992.

The United States exemplifies what has happened throughout much of the developed world. In the early 1970s, U.S. abortion laws were liberalized and access to birth control increased in the 1980s and 1990s. Moreover, a growing number of women entered the work force. Educational opportunity for women increased, and the feminist movement articulated a life-style in which motherhood was but one option. Many couples decided to remain childless. In addition, economic troubles in the past two decades, including two recessions, one inflationary period and soaring hospital costs, contributed to a reduction in birth rates. The TFR in the United States dropped to slightly below 2.1 in 1972 and since then has remained below replacement level, although it has edged upward in recent years (Figure 6.6). If abortions are restricted, TFR could increase even more.

Discouraging Trends

Unfortunately, the world population dilemma is far from solved; several discouraging trends are evident. For instance, one-third of the world's population is currently under the age of 15. When this age group marries and begins having children, it could spur a massive increase in world population.

Another discouraging trend is that during the 1970s death rates turned upward in many poorer countries, primarily because of hunger and malnutrition. In addition, food demand now outstrips production, and world grain reserves have dropped from a high of 102 days in 1986 to 66 days in 1992, according to the U.S. Department of Agriculture.

Another discouraging trend is that contraceptive availability is below demand. Studies suggest that although many women of reproductive age throughout the world do not want additional children, about half of them are not using effective methods of birth control. Furthermore, whereas more than 80% of all married women in developing countries have heard of contraception, in some countries, such as Pakistan, only one in ten women has ever used it.

Contraceptive use in Africa and Latin America remains low (Figure 6.7). Expanded efforts are needed in these regions to avert widespread starvation, pollution, poverty, and resource depletion.

Rapid growth is expected to continue in Africa, Latin America, and Asia. Figures 6.8a and 6.8b show the relative growth of developing countries and the developed world. With the rapid expansion of population in the developing world will come a whole host of

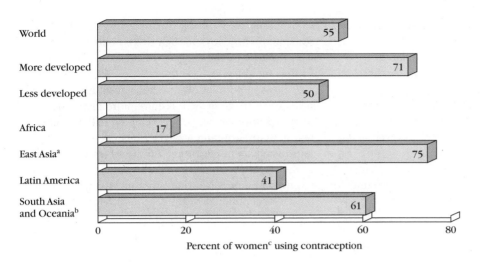

Figure 6.7 *This graph shows the percent of women using contraception in the world, in developed versus developing countries, and various continents.*

[a]Excludes Japan.
[b]Excludes Australia and New Zealand.
[c]Married women of childbearing age.
Note: Extrapolated from 1983 estimates.

SOURCE: UN, (1989). *Levels and Trends of Contraceptive Use as Assessed in 1988.* New York: United Nations, p. 33.

(a)

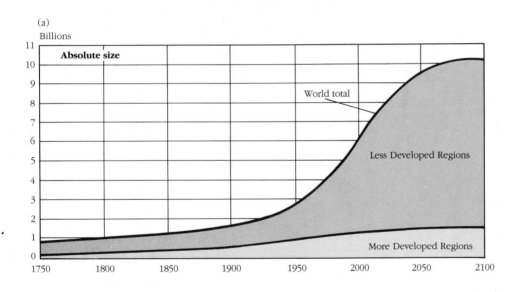

Figure 6.8 (a) *Graph showing that most of the world population growth will occur in the developing countries.* (b) *Another way of showing the same phenomenon, this graph indicates the relative proportion of new residents in the world population in developed countries and developing countries. Both graphs point to a dangerous trend.*

(b)

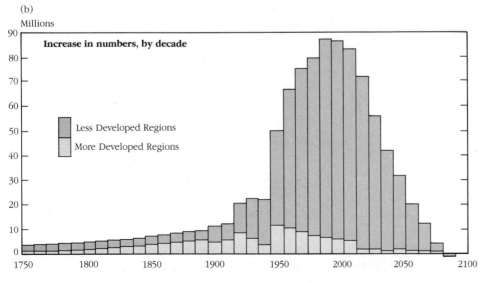

problems, many of which could affect the world in significant ways (see Viewpoint 6.1).

One of the most visible failures in population control is in India, even though a national birth control program has been in effect since 1951. India's program was the first of its kind in the world. The rate of growth slowed from 2.6% in 1969 to 2.0% in 1992. India had a population of over 882 million in 1992, however, and by 2010, it will rise to nearly 1.2 billion.

Population growth is a social and environmental problem. It begins with people as the cause and ends with people as one of the chief victims. Human civilization need not decay in an explosion of people, but, many experts agree, we can improve only if we comply with the ecological mandate to hold our numbers within the carrying capacity of the environment. To many this means stopping the growth of the human population.

In a short-sighted society, in which wars and economic crises frequently divert our attention, we may have difficulty sustaining interest in controlling population growth. Long-term action, lasting to the end of this century and well into the next, is necessary if we are to ensure a balance between human beings and the planet that will allow us to explore our full potential.

I cannot believe that the principal objective of humanity is to establish experimentally how many human beings the planet can just barely sustain. But I can imagine a remarkable world in which a limited population can live in abundance, free to explore the full extent of man's imagination and spirit.

—Philip Handler

ɹ   ɹ   ɹ

Critical Thinking Exercise Solution

As noted in previous chapters, population growth is one of the driving forces of environmental destruction. Despite what some may have you think, more people mean more pollution, more resource use, and more destruction. No matter how carefully we conduct our affairs, human society will always have an impact on the environment. The more of us there are, the greater the impact. It's hard to say exactly what the impact will be because that depends on how many resources people use and the level of technology and the steps we take to control or eliminate pollution. As the criti-

cal thinking rules advise, it is necessary to tolerate some ambiguity. It is encugh to know that population is at the root of the crisis and then to attempt to do something about it.

Slowing population growth is absolutely essential, but reductions in growth require more than family-planning improvements in developing nations. Slowing, perhaps stopping, growth is also dependent on improvements in education and health care as well as in the status of women in the developing world. Also required are measures that promote widespread sustainable economic development in developing countries.

Summary

6.1   Limits to Growth

■   Many experts believe that the growth of population and use of resources cannot continue indefinitely without irreparably damaging the global environment.

■   Computer studies suggest that building a stable, sustainable society is possible but will require global zero population growth, recycling, pollution control, soil erosion control, soil replenishment (restoration), and more emphasis on food production and services rather than on industrial production.

6.2   How Do We Control Population Growth?

■   Family planning allows couples to determine the number and spacing of offspring. Programs may be voluntary, extended voluntary, or forced.

■   Many experts believe that family planning in the developing countries should be part of an integrated program that promotes sustainable economic development and jobs for men and women as well as improvements in health care and education.

■   Developed nations have a role to play in lessening their impact through reductions in population growth and sustainable economic development strategies based on the biological principles of sustainability. They can also assist developing countries through financial aid and information sharing.

6.3   Making Strategies Work

■   To be effective, family-planning programs must address cultural and religious beliefs, psychological factors, and educational barriers.

6.4   Ethics of Population Control

■   No other issue in environmental science creates so much ethical conflict.

■ To some, population control implies denial of one of the most basic and important of all human freedoms—the right to bear children. To others, it violates deep-seated religious beliefs.

■ Some believe that the right to reproduce should be curtailed when the rights of the individual interfere with the welfare of society; these people ask whether it is ethical not to control population.

6.5 The Status of Population Control

■ Many encouraging signs are evident. The world population growth rate has dropped in the past 20 years; some countries have achieved remarkable success in controlling numbers.

■ Unfortunately, the population problem is far from being solved. In 1992, 33% of the world's population was under 15 years of age and will soon enter their reproductive years. In addition, even though family planning is on the rise, many couples cannot afford it or are unable to access it.

Discussion and Critical Thinking Questions

1. Critically evaluate this statement: "The world cannot support the people it currently has at a decent standard of living, so we should help developing nations become industrialized. Population will fall as a result, so population control programs are not necessary."

2. Define *family planning*. Make a list of the three major types of family planning programs. Give examples of each.

3. The United Nations appoints you head of population control programs. Your first assignment is to devise a population control plan for a developing country with rapid population growth, high illiteracy, widespread poverty, and a predominantly rural population. Outline your program in detail, justifying each major feature. What problems might you expect to encounter?

4. Describe ways in which developed countries might aid developing countries in solving the population crisis.

5. Discuss the "value" of children in less developed countries. How do these views differ from those of the developed countries? Are they similar to or different from your views?

6. Discuss reasons why the total fertility rate tends to be lower among more educated women.

7. Discuss general ways to ensure a high rate of success in population control programs.

8. Do we have the right to have as many children as we want? Should that right be curtailed? Explain.

9. Discuss some of the encouraging and discouraging news regarding world population growth. What progress has been made? Where do we need to concentrate our efforts in the near future?

10. Using your critical thinking skills, evaluate the following statement: "Population control should be encouraged only in developing nations. Countries like the United States are not overpopulated."

Suggested Readings

Brown, L. R. et al. (1990). *State of the World 1990*. New York: Norton. See Chapter 8 for a discussion of means to end poverty.

———. (1992). *State of the World 1992*. New York: Norton. Excellent information vital to population control.

Chiras, D. D. (1992). *Lessons from Nature: Learning to Live Sustainably on the Earth*. Washington, D.C.: Island Press. Offers insights into development measures useful in curbing population growth in the developing world.

Cole, H. S. D., Freeman, C., Jahoda, M., and Pavitt, K. L. R. (1973). *Models of Doom: A Critique of the Limits to Growth*. New York: Universe Books. A rebuttal of the computer study.

Gupte, P. (1984). *The Crowded Earth: People and the Politics of Population*. New York: Norton. Superb! Book on family-planning successes.

Jacobson, J. L. (1988). *Environmental Refuges: A Yardstick of Habitability*. Worldwatch Paper 86. Washington, D.C.: Worldwatch. Excellent.

———. (1991). India's Misconceived Family Plan. *World-Watch* 4(6): 18–25. Describes fatal flaws in India's family-planning program.

———. (1992). Thailand's (Qualified) Success Story. *World-Watch* 5(2): 36–37. Outlines some reasons why Thailand's family-planning program has proven successful.

Lowe, M. (1992). City Limits. *World-Watch* 5(1): 18–25. Explains different strategies of controlling urban growth.

Meadows, D. H., Meadows, D. L., Randers, J., and Behrens, W. W. (1974). *The Limits to Growth* (2nd ed.). New York: Universe Books. Excellent study of population, resources, and pollution.

Meadows, D. H., Meadows, D. L., and Randers, J. (1992). *Beyond the Limits*. Post Mills, VT: Chelsea Green. Sequel to the Limits to Growth study.

Misch, A. (1990). Purdah and Overpopulation in the Middle East. *World-Watch* 3(6): 10–11, 34–35. Describes an important cultural factor that impedes population control in the Middle East.

World Resources Institute (1991). *Environmental Almanac*. Boston: Houghton Mifflin. Compilation of information on global environmental problems with important population data.

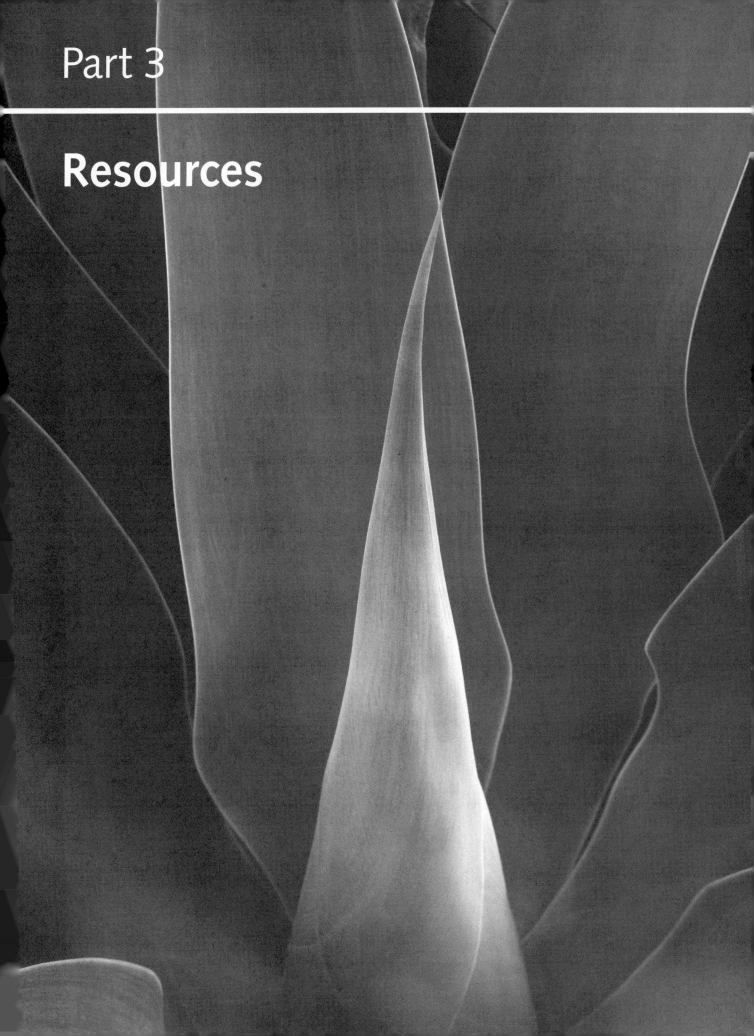

Part 3

Resources

Chapter 7

Feeding the World's People Sustainably: Food and Agriculture

It is not in the stars to hold our destiny but in ourselves.

—*Shakespeare*

Critical Thinking Exercise

Some agricultural interest groups object to the phrase *sustainable agriculture*. They argue that it implies current agricultural practices are somehow not sustainable. In support of their case, they point out that agricultural productivity (the output of food per acre) in many countries, such as Canada, Australia, and the United States, has been rising steadily, thanks to widespread use of insecticides, herbicides, fertilizers, irrigation, and other modern practices. They also point out that farmers in many modern agricultural countries are feeding more and more people every year. What's wrong with this line of reasoning? What critical thinking rules are essential to analyze this issue?

Nebraska farmers typically plant huge fields of corn and sugar beets, sometimes spreading as far as the eye can see. Today, though, some Nebraskan farmers alternate 3-meter (10-foot) -wide strips of corn with strips of soybeans. This planting technique increases the productivity of soybeans by about 11% and increases the productivity of corn by a remarkable 150%. Increased yields result from the fact that the corn protects sugar beets from the drying effects of the wind, and less dense stands of corn result in better sunlight penetration, which stimulates growth. This new practice also reduces the need for chemical pesticides (for reasons explained later).

This is only one example of many efforts aimed at building a system of **sustainable agriculture**, one that produces high-quality foods while maintaining or improving the soil and protecting the global environment—the air, water, soil, and wealth of wild species.

This chapter describes ways to create such a system. It begins with a discussion of hunger and malnutrition, then outlines some of the environmental problems caused by modern agricultural practices. It ends with a discussion of the principles and practices of sustainable agriculture.

The discussion of hunger shows that the current system of agriculture is woefully inadequate. In addition, it describes how hunger and malnutrition contribute to a vicious cycle of poverty and environmental destruction in developing nations. Agriculture is also shown to be a major cause of environmental pollution and habitat destruction. Building a sustainable system of agriculture is, therefore, essential to improving the lives of many people and creating a sustainable human presence.

7.1 Feeding the World's People/ Protecting the Planet

By various estimates, hunger afflicts 17% to 40% of the world's people, mostly in Asia, Africa, and Latin America. These people may suffer from **undernourishment** (hunger), a lack of sufficient number of calories, or **malnourishment**, a lack of proper nutrients and vitamins. Many people in the poorest nations suffer from both, that is, they don't get enough to eat and what they do get fails to provide all of the nutrients and vitamins they need. According to the World Health Organization (WHO), one of every four people alive today (1.5 billion people) is undernourished. These people are too poor to grow or purchase the food they need. The World Bank estimates that about 630 million people receive less than 80% of the calories, nutrients, or both, that they require. Yet hunger and malnutrition are not restricted to developing nations. In the United States, 10% to 15% of the population suffers from hunger. Hunger is most prevalent in Mississippi, Arkansas, Alabama, New Mexico, and the District of Columbia.

Diseases of Malnutrition

Many people die from starvation each year. Making matters worse, hunger and malnutrition increase the risk of contracting **infectious diseases**, which are caused by bacteria or viruses. People weakened by malnutrition are more likely to die from normally nonfatal diseases than those who are well nourished. Although no one knows exactly how many die each year from undernutrition, malnutrition, and nonfatal diseases worsened by poor nutrition, the number is estimated to be about 12 million.

Scientists recognize two types of malnutrition. The first, called **kwashiorkor**, results from a lack of protein. The second, **marasmus**, results from an insufficient intake of protein and calories. Kwashiorkor and marasmus are two extremes of protein-calorie deficiency, and most individuals who are malnourished exhibit symptoms of both.

Kwashiorkor In the rural villages or city streets of Latin America, Asia, or Africa, children lie in their mothers' arms. Their legs and arms are thin, and their abdomens are swollen with fluids (Figure 7.1). They look up with sleepy eyes, moving only occasionally. These children are suffering from kwashiorkor, the protein deficiency most common in children one to three years of age. This disease generally begins when children are weaned (losing the protein-rich milk of their mothers) and are fed a low-protein, starchy diet.

Marasmus Other children are thin and wasted (Figure 7.2). Their ribs stick out through wrinkled skin. They suck on their hands and clothes to appease a gnawing hunger. Unlike victims of kwashiorkor, children suffering from marasmus are alert and active. Marasmus often occurs in infants who are separated from their mothers' milk as a result of maternal death, a failure of milk production (lactation), or the use of milk substitutes. In years past, slick advertising campaigns designed to pro-

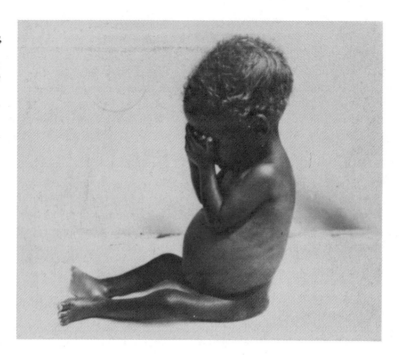

Figure 7.1 *Kwashiorkor, a protein deficiency, leads to swelling of the abdomen. Note the thin arms and legs caused by loss of protein in muscles. Children are physically and mentally stunted, apathetic, and anemic.*

mote powdered milk substitutes persuaded many women in the developing world to bottle-feed their children. After starting their children on these supplements, however, many women found that they could not afford to continue. By then, their breast milk had dried up. To compensate, some women diluted what milk they could afford with water, reducing their children's intake of protein and calories. Because many women used water from contaminated streams, their children developed diarrhea, which further reduced food intake and worsened their malnourishment.

In developing countries, for every clinically diagnosed case of marasmus and kwashiorkor hundreds of other children suffer mild to moderate forms of malnutrition, a condition much more difficult to detect. Like the more severely malnourished children, these children are prone to infectious diseases.

The Cycle of Poverty and Environmental Destruction

Growing evidence shows that malnutrition early in life often leads to mental and physical retardation. The more severe the deficiency, the more severe the impairment. Mental retardation results because 80% of the brain's growth occurs before the age of two.

Malnourished children who do survive to adulthood remain mentally impaired. Typically plagued by malnutrition their entire lives, they are often prone to infectious diseases and provide little hope for improving their lives. Many have large families. Hunger and malnutrition, therefore, contribute to growing poverty, rising population, and worsening environmental conditions.

Declining Food Supplies

From 1950 until 1970, improvements in agricultural production and expansion of the land area under cultivation increased world per capita grain consumption by approximately 30%. This resulted in a substantial improvement in the diet of many people. From 1971 to 1984, however, world grain production barely kept pace with population growth. Between 1984 and 1989, food production per capita fell a surprising 14%. From 1989 to 1993, it fell again by about 3%.

The decline in food production per capita over the past decade results from numerous factors, among them global warming, population growth, soil erosion, and soil deterioration from causes described in the next section. Many experts think the decline will continue throughout the 1990s. If global warming, population growth, soil erosion, and other problems worsen, hunger, poverty, and environmental destruction will become even more widespread.

Because of the declining per capita food production, many countries have lost the ability to feed their people and have become dependent on imports from developed countries, such as Canada, Australia, and the United States. In 1950, grain imports by developing countries amounted to only a few million metric tons per year; today, imports total over 450 million metric tons a year, according to the U.S. Department of Commerce.

Some experts believe that the safety net provided by major food-producing nations is in danger. One of the most serious threats is global warming (Chapter 16). In 1988, after record-high temperatures throughout the Midwest, U.S. grain production fell by 35%, plummeting to 190 million metric tons—barely enough to satisfy

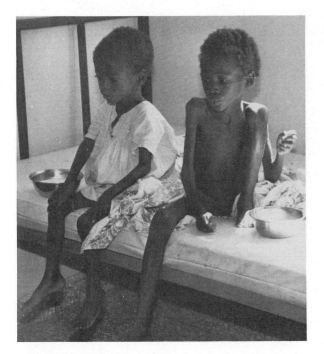

Figure 7.2 *Victims of marasmus, a protein and calorie deficiency, await medical attention at a hospital in Angola. Marasmus victims are thin but alert and active. Survivors of malnutrition may be left with stunted bodies and minds.*

domestic needs, let alone foreign demand. Fortunately, previous surpluses satisfied domestic and foreign demands. Continued hot weather could slow food production and exports considerably, threatening developing nations that rely on outside help.

According to a study by the U.N. Food and Agriculture Organization (FAO), by the year 2000, numerous developing countries will be unable to feed even half of their population, including Kenya, Nigeria, Uganda, Ethiopia, Haiti, Afghanistan, Yemen, and Rwanda.

Long-Term Challenges

Unless decisive, far-reaching steps are taken, widespread starvation is inevitable; millions will perish in the poor countries. The famines in Ethiopia, Chad, Somalia, and Sudan, in which hundreds of thousands of people have died in recent years, may portend what is to come. Many believe that the rich nations, to whom the developing nations are highly indebted, will not be immune to the resulting upheaval.

Three interrelated challenges face the world today. The first is feeding the malnourished and undernourished people who are alive today—the immediate challenge. The second is meeting future needs for food—the long-term challenge. The third is growing food without depleting the soil and water upon which agriculture

depends—the continuing challenge. That is, we have to ensure that current agricultural practices are sustainable. In order to understand the changes needed to make agriculture sustainable, we turn our attention to some of the major problems facing world agriculture.

7.2 Problems Facing World Agriculture

World agriculture can be summed up this way: In many poor nations, food production falls short of demand. Many people cannot afford the food that is available. Moreover, because of intense demand and poor land management, the cropland in many countries is in a state of rapid decline. Once-productive land is washed away by rain or is becoming dry and barren. Good farmland is often used to grow export crops, such as coffee, tea, and fruit, for the developed countries. As a result, many poor nations must turn to the outside for food assistance.

In rich nations, farmers generally produce an abundance of food in most years, enough to feed their own people and anyone else who can afford to buy it. But, food production does not come without a price. Heavy fertilizer and pesticide use is contaminating drinking wells and nearby lakes and streams. Poor land management results in excessive soil loss from erosion.

Soil Erosion

Thomas Jefferson wrote that "civilization itself rests upon the soil." The first towns, early empires, and powerful nations trace their origins to the deliberate use of the soil for agriculture (Chapter 4). But, as agricultural expert R. Neil Sampson wrote, in most places on Earth, "We stand only six inches from desolation, for that is the thickness of the topsoil layer upon which the entire life of the planet depends."

Everywhere one looks soil is being eroded by wind and water. **Soil erosion,** the loss of soil from land, is one of the most critical problems facing agriculture today in the poor developing countries and the industrial nations. Erosion occurs when rock and soil particles are detached by wind or water, transported, and deposited in another location, often in lakes and streams.

Soil erosion is a natural process. **Natural erosion** generally occurs at such a slow rate that new soil is generated fast enough to replace what is lost. However, **accelerated erosion,** which results largely from human activities, such as overgrazing, occurs at a rate that outstrips new soil formation. Accelerated erosion decreases soil fertility and can cause a decline in production (Figure 7.3). Numerous studies on corn and wheat productivity indicate that each 2.5 centimeters (1 inch) of topsoil lost to erosion results in a 6% decline in productivity. Severe soil erosion may result in the formation of gullies that can entirely destroy land.

Figure 7.3 *Soil erosion on farmland. All soil erosion above replacement level bodes poorly for farmers and the world's people. This overgrazed land is suffering from extreme soil erosion that renders it useless.*

Soil erosion is an urgent problem because new soil forms very slowly; 2.5 centimeters (1 inch) of topsoil may take anywhere from 20 to 1200 years to form. Soil erosion also has a number of serious environmental impacts. For instance, soil particles attach to pesticides. Transported to nearby waterways, these pesticide-laden particles may contaminate fish and other aquatic organisms, which, in turn, may be passed to birds and human consumers in the food chain. Sediment deposited in waterways also increases flooding, destroys breeding grounds of fish and other wildlife, and increases the need for dredging harbors and rivers. The World Resources Institute estimates the off-site damage from soil erosion in the United States is over $10 billion a year.

Since 1880, one-third of the topsoil in the United States has been lost to erosion, according to the U.S. Soil Conservation Service. Unfortunately, soil erosion continues today. The U.S. Department of Agriculture estimates about 1 billion metric tons of topsoil were lost from U.S. farmland in 1990. Although that is less than the 1.45 billion metric tons lost in 1985, the current rate of erosion remains far too high. The average rate of erosion on U.S. farmland is approximately seven times greater than soil formation, a situation that is clearly unsustainable. Should erosion continue, the U.S. agricultural system could experience substantial declines in productivity. (Chapter Supplement 7.1 describes soil erosion in more detail.)

Unfortunately, little information is available on soil erosion rates throughout the world. Scientists currently estimate that approximately one-third of the world's cropland topsoil is being eroded faster than it is being regenerated.

Soil erosion is especially rapid in many developing nations. In China, for example, the Yellow River annually transports 1.6 billion metric tons of soil from badly eroded farmland to the sea. In India, the Ganges carries two times that amount. Overall, the Worldwatch Institute estimates that 24 billion metric tons of topsoil are eroded from the world's croplands each year! At this rate, the world loses about 7% of its cropland topsoil every ten years. Another way of looking at it is, if this rate continues, 240 billion metric tons will be lost in the 1990s, which is equivalent to more than half of the topsoil on U.S. farms.

Desertification: Turning Cropland to Desert

The U.N. Environment Programme predicts that by the end of this century one-third of the world's cropland will have become desert. The conversion of range land, pasture, and cropland into desert is known as **desertification**. Such lands experience mild to severe decreases in productivity.

Numerous factors contribute to desertification, which occurs most frequently in semiarid lands—lands that are already fairly dry. One of the leading causes is drought, which may result from natural climatic changes or from human induced changes, among them global warming, overgrazing, and deforestation. Overcropping semiarid lands, that is, planting too frequently, is also a contributing factor. Desertification is just one of the many consequences of overpopulation.

Desertification afflicts numerous countries, including the United States, Africa, Australia, Brazil, Iran,

Afghanistan, China, and India. Worldwide, an area about the size of West Virginia (about 6 million hectares or 15 million acres) becomes barren desert each year. The U.N. Environment Programme estimates that 63% of the world's rangeland and 60% of the rain-fed cropland are threatened. About one-third of the irrigated cropland faces a similar danger.

Desertification is not new to humankind. In the ancient Middle East, for instance, deforestation, overgrazing, and poor agricultural practices reduced local rainfall. Coupled with a long-term regional warming trend, the decline in rainfall turned once-productive pastureland and farmland in much of the Fertile Crescent, where agriculture had its roots, into desert (Chapter 4). In more recent history, the United States found itself immersed in the infamous Dust Bowl era of the 1930s. This environmental catastrophe resulted from prolonged drought combined with fence-post-to-fence-post cultivation of fields, in part to supply Europe with food in the early years of World War II. Because the drought persisted, crops withered and died. Field after field turned into arid tracts of dry dirt. Winds swept the parched topsoil into huge dust storms and carried it away. Only through extensive conservation measures in the postwar years were farmers slowly able to rebuild their soils. Today, however, some of these gains have been lost as farmers attempt to increase their gross earnings, by increasing food production. Mini dust bowls are occurring in southern California and Texas.

Desertification is rampant in parts of Africa, especially in the sub-Saharan region known as the Sahel. Beginning in 1968, a long-term drought in the Sahel, coupled with overpopulation, overgrazing, and poor land management, caused the rapid southward migration of the desert in Ethiopia, Mauritania, Mali, Niger, Chad, and Sudan. The Sahara is also spreading northward. An estimated 100,000 hectares (250,000 acres) of rangeland and cropland are lost in Africa each year.

Desertification and erosion are taking a huge toll on world food production. In Africa, a continent straining under the pressure of more than 650 million people, about 100 million do not have enough food to eat. In Ethiopia, nearly one of every three people is undernourished. In Chad, Mozambique, Somalia, and Uganda, four of every ten people are undernourished. One sure sign of the troubles is the rising death rate. In Madagascar, an island nation off the east coast of Africa, infant mortality, which serves as a fairly accurate indicator of nutritional status, has risen from 75 per 1000 in 1975 to 110 per 1000 in 1989 to 115 per 1000 in 1992.

Latin America is experiencing declining food supplies as well. The number of undernourished preschool children in Peru now stands at nearly 70%. Infant mortality in Brazil continues to rise. The Worldwatch Institute notes that "Latin America's decline in food

production per person will almost certainly continue in the nineties."

Depletion of Soil Nutrients

In Chapter 3, you learned that farming can severely disrupt nutrient cycles. Excess fertilizer, for example, washes into lakes and streams, upsetting the natural balance. Harvesting crops also has a deleterious impact on the environment. Ten percent of the dry weight of plants is mineral matter from the soil. Current farming practices remove most of a crop, including the nutrients. Thus, many farmers are mining the soil, removing minerals faster than they can be replaced. This problem is especially noticeable on lands where the climate permits planting two or three times a year and on farms where land is treated only with artificial fertilizers, which replace only three of the dozens of soil nutrients extracted by plants.

High Energy Costs and Diminishing Supplies

Few relationships are as clear as the one between the cost of food and the cost of fossil fuels, such as oil and natural gas. Oil by-products power farm equipment. Oil and natural gas run food-processing plants and factories that produce farm equipment. Natural gas is converted into nitrate fertilizers. Oil by-products are converted into pesticides that are sprayed onto fields from airplanes and tractors. Trucks, trains, and ships, powered by oil, move food across countries and oceans. Even the stores that sell the food to the public require oil and natural gas for heating and lighting. So high is the energy demand, that modern agricultural societies invest 9 kilocalories of energy for each kilocalorie of food produced. Agriculture's dependency on fossil fuels became evident in the United States and elsewhere when petroleum prices increased from $3 a barrel to about $35 a barrel in the 1970s. Food prices climbed rapidly in tandem. Rising food prices hurt consumers in the developed nations. They also aggravated conditions in developing countries by raising both domestic production costs and import costs.

Oil is quickly being used up. By various estimates, in 25 to 45 years, world oil supplies will be depleted (Chapter 11). By the year 2000 or thereabouts, the demand for oil is expected to exceed supplies, a change that could fuel a worldwide spiral of inflation. World population and food demand are expected to be at least 50% above current levels by 2005. Unless alternative fuels are developed, many more of the world's people will be unable to afford food. World hunger, starvation, and disease are likely to rise dramatically. As pointed out earlier, rising poverty is likely to contribute to the unsustainable deterioration of the planet.

Water Mismanagement

A large percentage of the world's food production is dependent on irrigation. In the United States, for instance, one-eighth of all cropland is irrigated; this land produces approximately one-third of the nation's food. Irrigated farmland in countries such as China, Egypt, India, Israel, Indonesia, North and South Korea, Pakistan, and Peru produces over half of all domestic food production. Globally, about 18% of the world's cropland is irrigated and that farmland produces about 33% of the food.

From 1950 to 1980, irrigated cropland more than doubled, increasing from 93 million hectares (230 million acres) to 211 million hectares (520 million acres). In the 1980s, however, irrigated cropland only increased about half as quickly, by about 24 million hectares (60 million acres). This represents a net decrease in per capita land under irrigation, which may help to explain why food production per capita declined in the mid- to late-1980s. Further increases are likely to be modest. Why?

Irrigated agriculture faces several constraints, most importantly groundwater depletion and intense competition for water supplies. According to the Worldwatch Institute, increases in land under irrigation resulting from new irrigation projects are likely to be offset by declining groundwater supplies in other locations, siltation of reservoirs caused by soil erosion, waterlogging of soil, and salinization, problems discussed in this chapter and in Chapter 10.

Agricultural Groundwater Depletion   One source of irrigation water is **groundwater**, which comes from underground reservoirs known as aquifers. An **aquifer** is a zone of porous material, such as sandstone, that is saturated with water. Aquifers are naturally replenished by water from rain and snow that percolates through the soil and rock from the surface in regions called **aquifer recharge zones**.

Unfortunately, water from many aquifers is being withdrawn faster than it can be replenished (Figure 7.4). Of major concern is the depletion of groundwater in the Ogallala Aquifer, a major prehistoric water deposit. This enormous aquifer lies under the highly productive irrigated land of Nebraska, Kansas, Colorado, Oklahoma, Texas, and New Mexico. To date, more than 150,000 wells have been drilled into the aquifer, mostly for agricultural irrigation. Some parts of the aquifer, which took approximately 25,000 years to fill, have been depleted in only a few decades; water levels are falling 1 meter per year in heavy-use areas, compared with a 1-millimeter replenishment rate, forcing many farmers to abandon their wells or to drill deeper. By the end of this century, some analysts predict that many parts of the aquifer will be completely drained. Irrigated agriculture in these regions will decline sharply.

Serious groundwater overdraft is also occurring in other countries, among them China and India, two of the world's major food producers. Heavy pumping in parts of northern China is causing the **water table**, the uppermost surface of an aquifer, to drop 1 meter per year. In parts of southern India, the water table has fallen 25 to 30 meters in a decade.

Competition for Water   Agriculture uses about 70% of the world's water. However, growing population centers and industry are increasingly coming into conflict over water. In the United States, municipal water demands in Arizona and California have already made deep cuts into agricultural water supplies and put an end to much irrigated farming.

Certain sectors of agriculture also compete for water. Livestock production, which you learned in Chapter 2 feeds fewer people per kilocalorie than crops, demands enormous amounts of water in some regions. In California, for example, approximately 450,000 hectares (over 1 million acres) of grassland are irrigated for cattle production. Grassland irrigation in this relatively dry climate consumes one-seventh of the state's water. But, revenues from the cattle industry are only 1/5000 of the state's total annual earnings. Some critics argue that cattle production is a wasteful drain of water resources that deprives fish and wildlife as well as other more fruitful agricultural uses.

Certain crops also use extraordinarily large amounts of water. California's central valley, for example, receives only 10 centimeters (4 inches) of rain per year but is used to grow water-intensive nonfood crops, such as cotton, which require approximately 100 centimeters (40 inches) of moisture.

In the long run, competition could reduce the amount of land that can be irrigated, resulting in lower food production, higher prices, and more hungry and malnourished people .

Waterlogging and Salinization   Although irrigation has greatly increased food production, in many semiarid regions, it has created serious problems that could significantly reduce food output in coming years. Irrigating poorly drained fields, for example, often raises the water table. Two consequences of this practice are waterlogging and salinization (Figure 7.5).

Waterlogging occurs if the water table rises too near the surface, filling the air spaces in the soil and suffocating the roots of plants. It also makes soil difficult to cultivate. **Salinization** occurs when irrigation water that has accumulated in the soil evaporates, leaving behind salts and minerals once dissolved in it. If not flushed from the soil, enormous quantities can accumulate in a couple of decades, greatly reducing crop production and making some soils impenetrable.

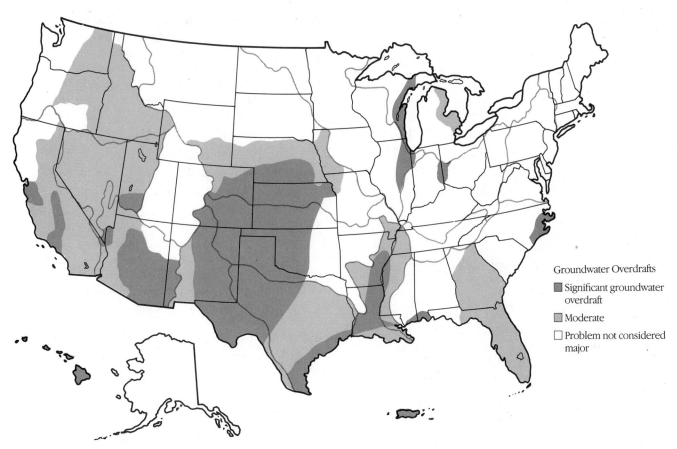

Figure 7.4   *Groundwater overdraft. Shaded areas indicate regions where water is being removed from aquifers by farmers, industries, and cities faster than it can be replenished. As shown, some areas are more severely affected than others. Without serious water conservation measures in many of these areas, severe shortages are bound to develop.*

Groundwater Overdrafts

■ Significant groundwater overdraft

▨ Moderate

☐ Problem not considered major

Worldwide, about one-tenth of the irrigated cropland, an area slightly smaller than Idaho, suffers from waterlogging. As a result, productivity on this 21 million hectares (52.5 million acres) of cropland has fallen by approximately 20%.

Worldwide, about one-fourth of the irrigated farmland suffers from salinization, an area equal to about 45 million hectares (112 million acres). In the United States, salinization occurs to some degree on an estimated 300,000 hectares (800,000 acres). Salinity expert James Rhodes believes that salinization lowers crop productivity by 25% to 30%. It is a growing problem in many other nations, including Pakistan, Iran, Iraq, Mexico, and Argentina. In India, 20 million hectares (45 million acres) of irrigated farmland are salinized to some degree. In Mexico, salinization is believed to be reducing crop output by the equivalent of 1 million metric tons of grain per year, enough to feed 5 million people.

When saline buildup reaches a critical level, soil becomes unproductive and must be abandoned. In India, an estimated 7 million hectares (nearly 18 million acres) have been abandoned because of salinization. One expert estimates that 1 to 1.5 million hectares of land are abandoned worldwide each year.

In summary, waterlogging and salinization reduce grain production and cause land to be lost to production entirely, and, all the while, the world population grows by 250,000 per day.

Conversion to Nonagricultural Uses

In addition to being eroded, waterlogged, and salinized, farmland is being rapidly converted to nonproductive uses, a phenomenon called **farmland conversion**. Expanding cities, new highways, shopping malls, and other nonfarm uses rob millions of hectares of farmland each year in the United States and abroad (Figure 7.6). In the United States, an estimated 1400 hectares (3500 acres) of rural land are lost every day. At this rate, the United States loses 530,000 hectares (1.3 million acres) of farmland, rangeland, and pastureland each year. One-third of rural farmlands will disappear in the next 100

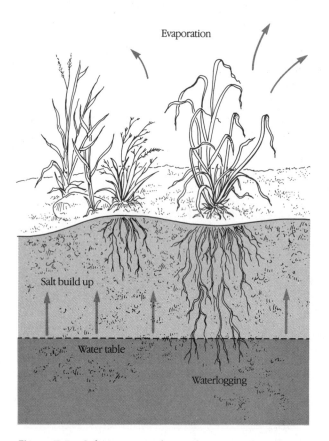

Figure 7.5 *Salinization and waterlogging. Salts and other minerals accumulate in the upper layers of poorly drained soil (salinization) when irrigation waters raise the water table and water begins to evaporate through the surface. The rising water table also saturates the soil and kills plant roots (waterlogging). Lower arrows indicate the movement of water from groundwater into the topsoil. Upper arrows indicate evaporation.*

years. Farmland conversion is a worldwide phenomenon. Former West Germany, for example, loses about 1% of its agricultural land by conversion every 4 years, and France and the United Kingdom lose about 1% every 5 years. Little is known about the rate of agricultural land conversion in the developing world, but it is believed to be substantial.

Conversion of Cropland to Fuel Farms: A Future Problem

Some experts predict that when petroleum supplies begin to fall below demand, around the year 2000, many countries will turn to another form of liquid fuel, ethanol, to power their transportation systems. Ethanol can be produced from sugar in certain crops, such as sugar cane and corn, by microorganisms. This process, called **fermentation**, is basically the same process used to make beer, wine, and other alcoholic beverages. Alcohol can be mixed with gasoline in a ratio of one part of alcohol to nine parts of gasoline, creating the fuel known as **gasohol**. It can also be burned in pure form. In Brazil, for instance, specially designed cars and trucks now burn 100% ethanol.

Ethanol is a renewable, relatively clean-burning energy resource, but tapping this largely ignored energy supply could have serious impacts on world food production. For example, Brazil eventually hopes to achieve complete self-sufficiency in automotive fuel through domestic ethanol production. Such a feat would require Brazil to convert 50% of its farmland to fuel production, greatly reducing food exports. If ethanol were to replace oil worldwide, major food producers, such as Canada,

Figure 7.6 *Urban sprawl, as shown here in Des Moines, Iowa, swallows up farmland at an alarming rate throughout the world. Once houses and other structures are built, the land is forever lost from agricultural production, a trend with serious consequences because of the rapidly growing human population.*

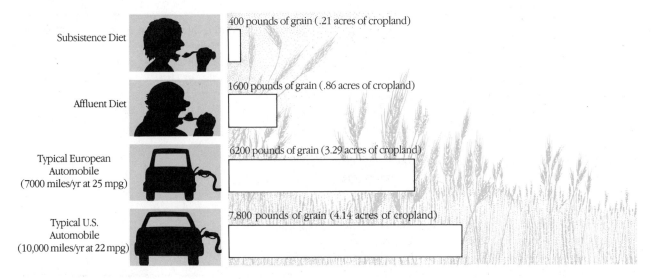

Figure 7.7   *Annual grain consumption—people versus cars. As this illustration shows, ethanol production for automobiles will require enormous amounts of grain and land if Western nations turn to ethanol for liquid fuel. Wealthy car owners will inevitably be in a better position to pay, leaving many additional people hungry.*

the United States, and Australia, would have to divert grain that once fed the world to fuel production. In fact, 20% of the currently exportable corn produced in the United States would be needed to produce 2 billion gallons of ethanol, a fraction of total U.S. liquid fuel needs.

If this scenario became a reality, food prices would very likely skyrocket, affecting all peoples, rich and poor. Today, most countries import some grain, and nearly a dozen countries import more than half of the grain they consume each year. Unless the developing countries become self-sufficient in food production, the world's people will be forced to compete with the automobile users of the world for precious grain (Figure 7.7).

Politics and World Hunger

The world's food problems are not all environmental and economic. Some are governmental. In fact, some agricultural experts maintain that plenty of food is produced each year in the developing countries but that about half of it never reaches the table. One reason is that governments, whether with good intentions or fraudulent ones, often become involved in the production and distribution system, reducing food availability.

In the now defunct Soviet Union, for example, government officials once controlled farming. Officials decided how much fertilizer should be applied by region, rather than individually tailoring fertilizer demands to local needs. Such decisions ultimately reduced productivity. Today, the farms are being turned over to the people, but because so many farm workers have spe-

cialized in only one or a few tasks, an overall view of farm management is lacking.

Political decisions in one country can also have tremendous impacts in other countries. In 1954, for example, Congress enacted legislation known as the **Farm Surplus Disposal Act**, which established what is commonly called the **Food for Peace Program**. This legislation authorized huge surplus grain shipments from the United States to developing countries. Free or inexpensive U.S. grain entered many developing countries and drove many farmers who were unable to compete with cheap imports out of business. Therefore, this program stimulated U.S. production at the expense of production in the countries it was intended to help.

Government policies can also have subtle long-range impacts on agricultural production. The U.S. program called **Payment in Kind** (**PIK**), enacted during President Ronald Reagan's first term, encouraged farmers to curtail excess production in order to reduce grain surpluses that were holding down prices. In this program, farmers who had cultivated a piece of land for more than two years were allowed to retire it from production; to offset farmers' economic losses, the government donated grain from huge federal stockpiles. Grain payments amounted to 80% to 95% of their expected production. Farmers were then free to sell the grain on the market.

Unfortunately, land speculators got into the farm business, bought marginal land, planted it for two years, and then took it out of production so that they could collect the government grain payments. This practice was euphemistically known as "sod busting." Because much of this land was rangeland not well suited to farming and because little care was given to it, the meager topsoil quickly eroded. Thus, through PIK, the government ended up promoting soil erosion on land that could have been used for grazing or crop production in the future in conjunction with good soil management.

Lawmakers throughout the world have unwittingly facilitated the creation of a massive unsustainable system of agriculture characterized by excessive erosion, waterlogging, and salinization, among other problems. Subsidies are an underlying cause. In the United States, the federal government subsidizes farmers through price supports. Farmers receive a guaranteed price for certain crops, such as wheat and soybeans. This inadvertently encourages farmers to plant only one or, at most, a few crops. It also encourages them to plant all of the land they can, even marginal land that may be easily eroded by wind and water. Huge monocultures are susceptible to insects and other pests. Consequently, farmers turn to a legion of insecticides and other chemicals to reduce damage. Many of these chemicals end up in groundwater and in lakes and rivers, where they poison many species.

In Mexico, most credit for irrigation systems and roads is given to farmers who produce cash crops, such as tomatoes and cattle, for export to the United States. Cash crop farms and cattle pastures usurp farmland that was once used to produce crops for domestic consumption.

Countless examples of government interference in agriculture exist. One final example illustrates the complex problem facing Ethiopian farmers, who traditionally left land fallow for seven-year periods so that nutrients from the highly weathered, poor soil could be replenished by natural vegetation. This practice, however, is now condemned by the Ethiopian government, which is interested in increasing farm production. If land is not cultivated within 3 years, it is confiscated. Unfortunately, bypassing the fallow period results in rapid deterioration. Compounding matters, Ethiopia has been engaged in civil war for many years, which deters the flow of food from Ethiopian farms and foreign countries.

Loss of Genetic Diversity

Before the advent of modern agriculture, grains and vegetables existed in thousands of varieties. Now, only a few of these varieties are commonly used (Table 7.1). (The same trend is occurring on ranches throughout the world as ranchers adopt livestock breeds developed for maximum yield.) In Sri Lanka, farmers once planted 2000 varieties of rice; today, only 5 varieties are grown. In India, 30,000 strains of rice were once grown; today, about 75% of the total rice production can be attributed to 10 varieties. Many experts believe that this loss of variety, or **genetic diversity**, could have a significant impact on world agriculture in the long term. Reducing

Table 7.1   Limited Diversity in American Agriculture

Crop	Varieties Available	Major Varieties in Use	Percentage of Total Production
Corn	197	6	71
Wheat	269	10	55
Soybeans	62	6	56
Rice	14	4	65
Potatoes	82	4	72
Peanuts	15	9	95
Peas	50	2	96

Source: Reichart, W. (1982). Agriculture's Diminishing Diversity. *Environment* 24(9): 6–11, 39–44.

genetic diversity makes crops more susceptible to insects and other pests as pointed out in Chapters 3 and 4.

Why is genetic diversity dwindling? First and foremost, the new varieties often have a higher yield on mechanized farms. Between 1903 and 1976, for instance, new varieties of wheat allowed American farmers to double their yield; new varieties of corn allowed them to quadruple output per acre. These varieties were better suited for machine harvesting and responded well to irrigation. Seed companies also benefited economically by concentrating their efforts on a few varieties.

The development of high-yield varieties in the 1960s is part of a worldwide agricultural movement often called the **Green Revolution**. It began in 1944, when the Rockefeller Foundation and the Mexican government established a plant-breeding station in northwestern Mexico. The program was headed by Norman Borlaug, plant geneticist at the University of Minnesota, who developed a high-yield wheat plant for which he was later awarded a Nobel Prize (Figure 7.8). Before the program began, Mexico imported half of the wheat it consumed each year. By 1956, it had become self-sufficient in wheat production, and by 1964, it was exporting 500,000 metric tons.

The success in Mexico led to the establishment of a second plant-breeding center in the Philippines. High-yielding rice strains were developed there and introduced into India in the mid-1960s. Again, the results were spectacular. India more than doubled its wheat and rice production in less than a decade, and has become self-sufficient in wheat production.

Important as it was, the Green Revolution greatly contributed to the decrease in species diversity in cultivated crops. Moreover, new crop varieties were not as resistant to diseases and insects. Local varieties of plants

Case Study 7.1

The Promises and Perils of Genetic Engineering

No technological advance offers as much hope for a better future as genetic engineering, say its proponents. And none promises to revolutionize as much of our society—how we treat diseases and how we grow plants and rear livestock. Consider some of the promises.

Scientists have successfully implanted the gene for the human growth hormone into livestock embryos in hopes of producing marketable cattle much faster than by conventional means.

To help combat pest damage, researchers have isolated a bacterial gene that produces a naturally toxic chemical, and have successfully transplanted that gene into another bacterium that grows on plant roots, giving plants a new way of warding off pests.

To help combat salinization, geneticists have introduced into oats certain genes that allow them to thrive in salty soils. Researchers are also experimenting with genetically engineered bacteria that could help plants absorb nutrients more efficiently, thus increasing crop yields.

On another front, genetic researchers have developed a mutant bacterium that retards frost development on plants. This development promises to save farmers millions of dollars each year.

In an attempt to help protect soils from nutrient depletion, genetic engineers are experimenting with ways to transplant the genes that would allow non-leguminous plants to fix atmospheric nitrogen. If successful, this could help farmers to reduce fertilizer needs and water pollution created by excess fertilizer use.

Through genetic engineering, plant scientists are also developing a bacterium that produces a natural antifungal agent, protecting wheat from the devastating infestation of the take-all fungus, a disease that causes millions of dollars per year in crop damage.

The successes of genetic engineering have fostered extraordinary enthusiasm in the business community. Dozens of new companies have formed in recent years, and billions of dollars have been invested in the fledgling industry.

The Risk Debate

Nevertheless, safety questions remain. Could the genetically engineered bacteria escape into the environment, upsetting ecological balance? Experts agree that, once unleashed, a new form of bacterium or virus would be impossible to retrieve. Controlling it could prove costly and damaging.

Some individuals have criticized genetic engineering as a means of tinkering with the evolutionary process. Deliberate genetic manipulations, such as the transfer of chromosomes from one species to another, are different from anything that ordinarily occurs during evolution. Is it right, critics ask, to interfere with the genetic makeup of living organisms? Will these intrusions alter the evolution of life on Earth?

Recent research suggests that the dangers of genetic engineering have been blown out of proportion and that genetically engineered bacteria are not generally a threat to ecosystem stability. Most scientists agree.

At least two studies now indicate that genetically engineered bacteria applied to seeds, so that the bacteria take up residence on the roots, migrate very little from the site of application in the short term. Critics, however, are concerned with long-term consequences—ecological backlashes that may occur months or even years after application.

Many supporters of genetic engineering oppose attempts to halt its progress. To them, the promises of this revolutionary technology far outweigh any potential damage. The benefit of feeding the world's hungry with a genetically engineered strain of supercattle, they argue, far exceeds any possible threats. Many leading environmental scientists joined genetic engineering critic Jeremy Rifkin and others at the outset of the battle in the 1970s but have since switched sides. Nonetheless, ecologists want to see the industry properly monitored and careful testing conducted before release. Genetically altered species, they say, are analogous to alien species introduced into new environments. The history of such introductions has been fraught with difficulties (see Chapter 8).

Environmental organizations also recommend caution. The federal government, they say, has poured enormous amounts of money into research on genetic engineering but has spent very little to assess its potential risks. In *An Environmental Agenda for the Future*, a list of recommendations for environmental improvements in the United States, ten leading environmental groups called for a no-release policy until the government develops a full understanding of potential effects. They also called for tighter standards on designing and operating genetic engineering facilities.

Genetic engineering stands at a threshold. Still in its infancy, it offers an unparalleled opportunity for improving agriculture, animal husbandry, and medicine. At the same time, this revolutionary science carries with it an unknown potential for environmental harm that worries many of its critics.

Figure 7.8 *Comparison of an old low-yield variety of wheat (left and right) with a new short-stemmed high-yield variety (center). The first of the short-stemmed varieties was developed under the direction of Norman Borlaug.*

are adapted to their environment; natural selection has ensured this. New varieties, on the other hand, often have little resistance to insects and disease.

In addition, as diversity dwindles, huge monocultures become more and more common. Expansive fields of one genetic strain facilitate the spread of disease and insects. The potato famine in Ireland in the 1840s is an infamous example of the effects of reducing crop diversity. At that time, only a few varieties of potatoes were planted in Ireland. When a fungus (*Phytophthora infestans*) began to spread among the plants, nothing could stop it, and there was no backup supply of resistant-seed potatoes. Within a few years, 2 million people perished from hunger and disease, and another 2 million emigrated.

In addition to their susceptibility to disease, high-yield hybrids are generally less resistant to drought and flood. In 1980, the American peanut crop, consisting of two varieties, was almost entirely destroyed by drought and disease.

The loss of genetic diversity among crop species is paralleled by an equally troublesome extinction of wild species throughout the world, especially in the tropics. The loss of species in the tropics could have a devastating effect on modern crops because many modern crops came from tropical regions. Many of their relatives remain there today, growing as they have for centuries; their fate is vitally important to the future of agriculture. These wild relatives, as well as other species, could serve as a source of new genetic information for domestic crops to combat drought, disease, and insects.

The world agricultural system is on an unsustainable course. Much of the world's farmland is being destroyed by erosion, waterlogging, salinization, and farmland conversion. Vital groundwater supplies needed to irrigate farmland are being depleted. If these trends continue, human civilization could conceivably destroy or greatly reduce its ability to provide food for itself, causing a population crash.

By recognizing the problems and taking evasive action on a large scale, humanity can end the destruction of these valuable resources and rebuild its soils. That is, through concerted efforts, we can begin to build a sustainable agricultural system. But, how do we go about this task?

7.3 Building a Sustainable Agricultural System

One of the most important solutions to end famine is population control, that is, reducing demand by slowing the growth of the human population (Chapter 6). Beyond that, efforts must be made to increase food supplies sustainably. Such measures must protect and improve the soil, water, and other resources upon which farming is dependent. Most analysts recommend a multilevel approach that includes efforts to: (1) increase the amount of cultivated land; (2) raise output per hectare, that is, increase productivity; (3) develop alternative foods; (4) reduce food losses to pests; (5) increase the agricultural self-sufficiency of developing nations; and (6) enact legislation and policies to ensure a better distribution of food and to spur domestic food crop production rather than cash crop production.

Increasing the Amount of Agricultural Land

Increasing the amount of cropland and rangeland available to us can be achieved by protecting what we have and expanding the land under cultivation.

Protecting What We Have The best way to maintain and improve food production is to prevent the loss of farmland, rangeland, and pastures from the destructive forces of erosion, desertification, waterlogging, salinization, and urban sprawl.

One of the highest priorities must be to end excessive soil erosion in all countries. To date, farmers in most of the world have done little to reduce erosion. In the developing world, farmers struggle to meet their basic needs and claim they have neither the time nor the means to care properly for the land. Few can see the benefits of soil conservation because the gains tend to materialize slowly and usually take the form of a decrease in losses rather than an increase in food output.

Economics and short-term thinking impair soil erosion control in developed nations as well. Caught between high production costs and low prices for grains, farmers often ignore the long-term effects of soil erosion while synthetic fertilizers artificially help them to maintain yield in the short term.

Governments can promote conservation through a variety of programs and laws, many of which are discussed in Chapter Supplement 7.1. One landmark attempt to end the devastating loss of U.S. topsoil is a 1985 farm bill. This law created a land conservation program that directs the federal government to pay farmers to "retire" their most highly erodible cropland from production for ten years and to plant trees, grasses, or cover crops to stabilize and rebuild the soil. By 1990, farmers had retired nearly 14 million hectares (34 million acres), cutting erosion by an estimated 500 million metric tons per year.

The farm bill also established a federal program that calls on farmers to develop soil erosion plans in exchange for eligibility for federal crop insurance, subsidies, and other benefits. Thus far, 1.5 million farmers have enrolled in the program and have worked with the U.S. Soil Conservation Service to develop plans encompassing 53 million hectares (134 million acres), about 25% of U.S. farmland. These efforts could cut soil erosion on some of the most productive soils in the country by another 360 million metric tons per year by 1995, reducing annual erosion from 1 billion metric tons to 0.56 billion metric tons per year. Although impressive, any loss above natural soil replacement is not sustainable in the long term.

Protecting soil from desertification is also needed to preserve cropland and rangeland. Pollution controls to reduce global warming are vital to this effort (Chapter 16). More direct actions are also possible. In China,

for instance, agricultural officials have embarked on an ambitious program to plant a 6900-kilometer (4300-mile) "green wall" of vegetation to stop the spread of desert in the northern region (see Models of Global Sustainability 7.1). In Australia, huge semicircular soil banks have been created in the windswept plains to catch seeds and encourage regrowth in areas denuded by livestock. Despite these efforts, much more work is needed to end the destruction and foster the renewal of these important lands, so essential to creating a sustainable system of agriculture.

Efforts to prevent urban sprawl, the proliferation of highways, and other nonfarm uses of arable land are also needed. Careful city planning and new zoning laws could help reduce farmland conversion by ensuring that homes, roads, airports, and businesses are not built on agricultural land. (For more on this topic, see Models of Global Sustainability 6.1 and the discussion in Chapter 22.)

Waterlogging and salinization can be reduced by more careful irrigation practices, which could substantially limit the annual loss of farmland. By using computerized sensors that measure soil moisture, farmers can apply only the amount of water needed by crops, reducing both problems. Special drainage systems can be installed to draw off excess water and prevent the buildup of soils and the potential for waterlogging. Government programs could discourage irrigation in soils susceptible to these problems as well.

It is also important to protect water supplies. Without them, many irrigated fields will be abandoned. Farmers can improve irrigation efficiency through many simple, cost-effective measures. Lining irrigation ditches with cement or plastic can cut water losses by 30% to 50% (Figure 7.9a). Transporting water in pipes can result in even greater savings. Farmers can also use drip irrigation systems to deliver water directly to the roots of some crops, even fruit trees (Figure 7.9b). Conventional center-pivot irrigation systems, which spray water into the air with tremendous losses to evaporation, can be modified to spray water downward at a considerable savings. Computer systems, as noted above, can help farmers monitor soil moisture so they apply water only when it is needed and in the amount required by crops. These and other techniques are replacing wasteful practices and will inevitably increase in use as water supplies drop and prices for water rise.

Expanding into Farmland Reserves For years, virtually all nations have solved the problem of rising food demand by opening up new lands to the plow. Today, in most parts of the world, abundant reserves are in short supply. In the United States, for example, farmers currently cultivate 167 million hectares (413 million acres). The cropland reserve, land that could be farmed if necessary, is a paltry 51 million hectares (127 million

(a)

(b)

Figure 7.9 *Increasing the efficiency of irrigation. (a) Elevated pipes reduce water loss by 30% to 50% compared to open (unlined) ditches. (b) Trickle systems deliver water to roots, cutting evaporation losses substantially. Trickle systems can only be used for certain crops, among them fruit trees.*

acres). Projected increases in foreign and domestic demand will require cultivation of this land by the year 2000. In most of the major industrial nations, farmland reserves are also small.

In several other areas, the situation is similar. In Southeast Asia, 92% of the potential agricultural land is being farmed. In southwestern Asia, more land is currently being used than is considered suitable for the rain-fed agriculture. Consequently, per capita food production in Asia has begun to fall as the population continues to grow.

For those countries that have little farmland reserve, population control measures (urban growth management and a reduction in growth rate) and efforts to protect soil already under cultivation offer the greatest hope for meeting future demand.

According to the FAO, Africa and South America have large surpluses of land that could be farmed. In Africa, for example, only 21% of the potentially cultivable land is in use. In South America, only 15% is being farmed. Some experts believe that this land should be developed. However, much of it is currently covered by tropical rain forests, many of whose soils are poor in nutrients. These soils also are prone to erosion in the intense tropical rains and may become hardened when exposed to the sun (Chapter Supplement 3.1). The potential for expansion on these continents is, therefore, grossly overrated.

Although tapping unfarmed land may be an option in some areas, it would severely deplete wildlife populations and could disrupt many of the free services provided by nature, for example, flood control and local climate control. (Chapters 8 and 22 describe the importance of maintaining wildlife populations and natural ecosystems.)

Additional Measures to Increase Food Output

Numerous other strategies that can be employed to increase the world's food supply are discussed in the following paragraphs.

Preserving Genetic Diversity In a forest in Mexico in 1979, four scientists discovered a few tiny patches of a wild, weedy-looking grass called *Zea diploperennis*. A relative of modern corn, whose known population numbers only a few thousand, this species is different from its modern cousin because it is a **perennial**, that is, it grows from the same root structure year after year rather than from seeds. Most modern corn is an annual crop, which must be planted from seeds each year.

To geneticists who create new types of corn, this was an important discovery. Why? Corn, like most other crops, is only as good as the genetic "boosts," or infusions of fresh genes, it receives. These can provide resistance to disease, drought, and insects.

Models of Global Sustainability 7.1

Stopping the Spread of Desert in China

China is a nation in trouble. With a growing population approaching 1.2 billion people, China's land is falling into ruin. Centuries of overgrazing, poor agricultural practices, and deforestation have resulted in severe erosion and rapidly spreading deserts that gobble up the once-productive countryside.

The spread of deserts affects the lives of millions of peasants in China. In the Loess highland of northern China, for example, the land is cut with gullies, some hundreds of meters deep. Erosion from the raw, parched Earth is an astounding 30 to 40 metric tons per hectare (12 to 16 tons of topsoil per acre) per year, far above replacement level. In all, some 1.6 billion metric tons are carried into the Yellow River annually, making it one of the muddiest rivers in the world.

According to one estimate, an area larger than Italy has become desert or semidesert in China in the last 30 years. Although most of the desertification is occurring in northern China, few areas are immune to the rampant destruction of the land.

In 1978, the Chinese government launched a reforestation project to stem the tide. By planting trees, shrubs, and grasses, they hope to form a giant green wall across the northern reaches of the nation. The wall will extend 6900 kilometers (4300 miles) and will be 400 to 1700 kilometers (250 to 1000) miles wide. Moreover, it will return the land to productive use.

By 1985, the first phase of the program was completed. More than 5.9 million hectares (14.6 million acres) of barren land had been replanted with trees and shrubs, many of which have survived and halted the growing desert. The second phase of the project is scheduled to be completed by the end of 1995, at which time government officials hope to have reforested nearly 6.4 million hectares (16 million acres).

The Yulin District is one of China's success stories. Before 1949, more than 400 villages and 6 towns had been invaded or completely covered by sand. Today, four major tree belts have been planted in the area, decreasing the southward push of the desert by 80%. Towering sand dunes now peep through poplar trees, and rice paddies sparkle in the sunshine. Grain production has been replaced by a diversified agricultural system, including animal husbandry, forestry, and crop production.

The trees provide shade and help to reduce the shifting sand dunes. Shrubs and grasses thrive on land once stripped of its rich vegetation. Trees and shrubs grow in gullies and grasses carpet slopes, helping hold the soil in place and reversing the local climate change.

Local residents have built a diversified economy in what was once simply a desert. Juice from the desert cherry tree, which thrives in the desert climate and is extremely rich in vitamin C and amino acids, is now used to produce soft drinks, preserves, and beer. Twigs of the desert willow are used to make wicker baskets and trunks that earn local residents $2 million in U.S. currency every year.

Despite the encouraging signs in China, a report by the Shanghai-based *World Economic Tribune* says that while nearly 10 million hectares (75 million acres) are planted every year, twice that amount is still being lost. In the northern province of Heilong Jiang, home of China's largest concentration of virgin forest, loggers have reduced the tree cover from 50% to 35% in just 30 years. Government pricing policies promote overcutting.

The reforestation project is further plagued by a shortage of money and technical expertise. Because of shortsighted land use practices, some experts believe that China's Yangtze River, the nation's longest, could turn into a second Yellow River. Each year, its tributaries are turning muddier.

Reforestation is needed throughout the world to help reverse centuries of land abuse that have led to the spread of deserts and the gradual deterioration of the earth's soil. Forest replanting can also reduce global warming that now threatens the world climate.

Adapted from: Ming, L. (1988). Fighting China's Sea of Sand. *International Wildlife* 18(6): 38–45.

Today, valuable new genes present in modern strains come from wild plants or early varieties still grown by people in remote corners of the world. The increased vigor they confer on corn results in larger harvests, and because corn accounts for 25% of the world's cereal grains, a large harvest is important.

Genetic boosts have an enormous impact on corn production. In the last 60 years, for instance, corn harvests have more than quadrupled from 20 bushels per hectare to 100 to 250 bushels because of genetic improvements. Consequently, breeders are always looking for the rare varieties that could help produce hardier, more resistant plants. The primitive corn found in Mexico is valuable because it contains genes that could provide resistance to several diseases. Some researchers are developing a resistant perennial hybrid that grows in fields much like grass. This new variety would reduce erosion and save farmers annual costs of plowing, sowing, and cultivating. The potential savings are enormous.

Protecting tropical rain forests and other natural habitats—and their immense genetic diversity—is, therefore, important. By destroying these lands, we severely limit the future of agriculture. (Chapter 8 describes efforts to preserve genetic diversity of plants by storing seeds and cuttings from plants in special repositories.)

Developing Higher-Yield Plant and Animal Varieties

Another way to increase food supplies is to develop new genetic strains. New high-yield varieties of rice and wheat developed during the Green Revolution, for example, can produce three to five times as much grain as their predecessors. New varieties of plants created by breeding closely related plants to combine the best features of the parents are called **hybrids.**

Unfortunately, as new hybrids were introduced into many poor nations, the hopes of the Green Revolution dimmed, for farmers soon realized that the hybrids required large amounts of water and fertilizer, unavailable in many areas. Without these, yields were not much higher than those of local varieties; in some cases, they were lower. The cost of the new varieties prevented many small farmers from buying them. Many were forced out of business as larger farms converted to the high-yield varieties. Also, new plants were often more susceptible to insects and disease.

The Green Revolution, once written off by its critics as a failure, was the first step in a long, tedious process of plant breeding aimed at improving yield. Today, plant breeders throughout the world are developing high-yield crops with a higher nutritional value and greater resistance to drought, insects, disease, and wind. Plants with a higher photosynthetic efficiency are also in the offing. Efforts are even under way to incorporate the nitrogen-fixing capability of legumes into cereal plants such as wheat, a change that would decrease the need for fertilizers and reduce nitrogen depletion (Chapter 2).

One exciting improvement announced in 1988 is a new variety of corn. Corn is a staple for 200 million people worldwide, many of whom are chronically malnourished. Because corn is such an important source of calories and protein, researchers sought to improve it. The result of nearly two decades of work is a product called quality-protein maize (QPM). Studies show that only about 40% of the protein in common corn is ultimately used by humans. In contrast, roughly 90% of QPM's protein can be digested and used. In countries where corn is a staple, such as Africa and Mexico, QPM could help curb malnutrition. It could also help them become more self-sufficient in food production if costs could be lowered to make it affordable.

Some researchers are exploring the use of perennial crops for agriculture. Today, most agricultural crops are annuals. Preliminary research suggests that productivity from perennials may be equal to or slightly lower than conventional annuals, such as wheat, but the benefits from soil conservation, soil-nutrient retention, and energy savings may overwhelmingly favor them.

Just as new varieties of plants help increase yield, so do fast-growing varieties of fowl and livestock. Efforts are being made to improve plants and livestock by **selective breeding** and **genetic engineering**, a complex process involving several steps. In genetic engineering, scientists first identify genes that give plants resistance and other important properties that might increase yield. Next, they isolate the genes and chemically analyze them so they can make copies. Finally, the genes are inserted into seeds, which then sprout and produce new plants whose cells contain the beneficial gene. The genes are then transferred to the plant's offspring.

Genetic engineering may be used in other ways to improve agriculture. A new strain of bacteria developed by scientists at the University of California at Davis inhibits the formation of frost on plants, thus potentially offering farmers a way of reducing crop damage and increasing yield. Other scientists are working on genes that give plants resistance to herbicides used to control weeds. A group of scientists at the Monsanto Company has developed a strain of bacteria that grows on the roots of corn and other plants. When eaten by insects, the bacteria release a toxic protein that kills the pest. Animal geneticists are also trying to improve livestock, combining genes from one species with those of another to increase efficiency of digestion, weight gain, and resistance to disease.

Genetic engineering was once touted as a savior to world agriculture. Today, however, researchers are finding that it is more difficult to apply to agriculture than proponents once thought. According to Lester Brown and John Young of the Worldwatch Institute, "It will not revolutionize agriculture overnight—but it could be an important new weapon in the fight against hunger." (For more on genetic engineering, see Case Study 7.1.)

Soil Enrichment Programs   Soil erosion controls help to preserve farmland, as noted earlier, but also protect soil nutrients vital to soil fertility. As any good farmer will tell you, protecting, even enhancing, the fertility of the soil is vital to increasing its productivity.

Soil fertility can be enhanced by the use of fertilizers and crop rotation, topics discussed in Chapter Supplement 7.1. Soil fertility can also be improved by human wastes from sewage treatment plants, an activity that recycles nutrients originally gained from the land.

Developing New Food Sources: Native Species and Fish Farms   Another approach to meeting current and future demand is to develop new food sources, including native species and fish and other aquatic species grown on fish farms.

Many native plants and animals, not currently eaten by humans, could be tapped to help meet the rising demand for food. The winged bean of the tropics, for example, could become a valuable source of food because the entire plant is edible: its pods are similar to green beans, its leaves taste like spinach, its roots are much like potatoes, and its flowers taste like mushrooms. Food scientists are looking for other plants with similar potential.

Native animals may also provide an important, sustainable food source in years to come. In Africa, for instance, native grazers are far superior to cattle introduced from Europe and the United States. Unlike cattle, native grazers carry genetic resistance to disease and rarely overgraze grasslands. Native grazers also generally convert a higher percentage of the plant biomass into meat and may be cheaper to raise.

Fish provide about 5% of the total animal protein consumed by the world's population. Although three-fourths of the fish catch is consumed by developed nations, fish protein is important to the people in many poorer countries, often supplying 40% of the total animal protein consumed.

During the 1970s, the world catch stabilized between 66 and 74 million metric tons per year. In the 1980s, it began to climb again, peaking at 100 million metric tons in 1989, the latest year for which statistics are available. Although the fish catch rose in the 1980s, the amount of fish available per capita remained more or less constant until the late 1980s, at which time it rose slightly. What does the future hold?

The FAO says that fish catches in 9 of the 19 world fishing zones it monitors are above the lower limit of estimated sustainable yield. Fourteen fish species that account for 20% of the world fish catch have been so seriously overfished that they would take 5 to 20 years to recover even if all fishing stopped. Biologists are, therefore, concerned that efforts to catch more fish will result in a dramatic population decline in many commercially important fish species.

Overfishing, or depletion of stocks, results when commercial fishing interests deplete the breeding stock, so that a natural fishery cannot be maintained. Protecting current stocks so vital to maintaining a sustainable harvest will require global cooperation.

Barring any substantial increases in fish from the sea, many observers believe that fish farms are one of our greatest hopes for increasing fish and shellfish. A **fish farm** is a commercial endeavor in which fish are raised in ponds and natural bodies of water. Fish farms are already common in many parts of the world, and new ones might help to increase protein supplies. Fish farms are forms of aquatic agriculture, called **aquaculture** in fresh water and **mariculture** in brackish or saltwater.

Fish farming employs two basic strategies. In the first, fish are grown in ponds. In some ponds, the population density is kept high by intensive feeding, which is costly and, therefore, not always suited to developing countries (Figure 7.10a). Fish and shellfish can also be maintained in enclosures or ponds, where they feed on algae, zooplankton, and other fish that occur naturally in the aquatic ecosystem (Figure 7.10b). This system requires little food and energy, and is quite suitable for developing countries.

Worldwide, fish farms produce about 4 million metric tons of food per year. Intensified efforts could double or triple this amount, providing additional food for the growing population.

Eating Lower on the Food Chain If everyone ate like North Americans, current food supplies would feed approximately 2.5 billion people—about one-half of the world's population. However, if everyone ate a subsistence diet, getting only as many calories as needed, the annual world food production could supply an estimated 6 billion people, 500 million more than are alive today.

Armed with statistics such as this, some propose that wealthier citizens of the world can contribute to solving world hunger by eating less and eating lower on the food chain. The logic is that by eating less and consuming more grains, vegetables, and fruits—and less meat—citizens of developed nations would free grain for those in developing nations. A 10% decrease in beef consumption in the United States, for instance, would release enough grain to feed 60 million people in the developing world. (Chapter 2, in fact, explained the biological reason why many more people could be sustained on a vegetarian diet.)

The problem with this idea, say critics, is that sacrifices on the part of the wealthy would most likely not translate into gains abroad. Another problem is that land suitable for grazing is often not arable.

This is not to say that a vegetarian or a meat-conservative diet is not desirable. It is. It is healthier, and it consumes far fewer resources. If there is a lesson to be learned from this issue, it is that to feed their people, developing nations should concentrate on grain rather than on meat production.

Reducing Pest Damage and Spoilage

Rats, insects, and birds attack crops in the field, in transit, and in storage. Conservatively, about 30% of all agricultural output is destroyed by pests, spoilage, and diseases. In developing nations, this figure may be much higher, especially in humid climates where crops are grown year-round in conditions optimal for the spread of crop diseases and insects.

Reducing the heavy toll of pests could help increase the global food supply, a topic discussed in Chapter 18. To create a sustainable system of agriculture, however,

(a)

(b)

Figure 7.10 (a) *Commercial catfish farm near Monticello, Arkansas. These feeders release food pellets when a catfish nuzzles an extended rod.* (b) *Fish raised in irrigation ditches and ponds in China and other developing countries supply needed protein.*

pest control measures must be safe and sustainable. Thankfully, numerous strategies for pest management that fit these requirements are available. These measures are detailed in Chapter Supplement 7.1 and Chapter 18.

One measure described here is an improvement in storage and transportation. Inefficient transportation can delay food shipments while rats, insects, birds, or spoilage take their share. Spoilage can be greatly reduced by refrigeration. In the developing world, some scientists believe that the supply of fish could be increased by 40% by improving refrigeration on ships, in transit, and in stores. Grain supplies in the developing nations could be stored in dry silos or sheds to prevent the growth of mold and mildew and reduce rodent problems. Technical and financial assistance from the developed countries could go a long way toward improving food storage and transportation, potentially increasing the world food supply by 10% or more.

Increasing Self-Sufficiency and Sustainability of Farms

Proponents of sustainability argue that one element of sustainability is self-sufficiency—individual, state, and national self-reliance. Nowhere is the need for self-sufficiency more evident than in the food production in developing nations.

Recommending increased self-sufficiency in today's global economy may seem foolish. For many countries, however, agricultural self-sufficiency may be one of their only hopes for survival, for many people in these nations are so poor they cannot afford imported food. Nor is it likely that wealthy agricultural nations, such as the United States and Canada, can continue to feed them.

As Section 7.2 points out, in the very near future, production shortages created by soil erosion, salinization, fuel farms, farmland conversion, global warming, and other problems could severely reduce the exports of the developed countries or eliminate them entirely. For these reasons, developing nations would be well advised to find ways to feed their own people.

Agricultural self-sufficiency, detailed in Chapter 22, could be achieved in part by technical and financial aid from the developed nations that seek to transform current agricultural systems based on the biological principles of sustainability: conservation of soil, energy, and water; recycling of soil nutrients; renewable energy for farm machinery; restoration of damaged lands; and population control. Sustainable agriculture in the developing nations should be based on abundant human labor. Like sustainable farms in the United States and elsewhere, this new system of agriculture would seek to protect the environment while producing high-quality food. (For more insights into sustainable agriculture, see Models of Global Sustainability 2.1.)

Political and Economic Solutions

Solving world hunger will require dramatic changes in government policy worldwide. Laws and policies that promote unsustainable farming practices must be changed or abandoned. To help ensure water supplies, laws that encourage waste should be modified or replaced. In the western United States, for example, farmers or ranchers who reduce the water they are allocated lose the rights to it, eliminating the incentive to be frugal. Simple changes in water laws could free up enormous amounts of water. Some farmers may find it advantageous to abandon crops that use large amounts of irrigation water. Shifting water-thirsty crops, such as cotton and rice, from the desert of California's central valley to more suitable climates could free up enormous amounts of water for more productive uses.

This chapter began by listing three challenges in agriculture: the immediate need to feed malnourished and undernourished people, the long-term need to provide food for future generations, and the continuing need to protect and enhance the soil and the environment. It should be clear that there are many steps required to solve these problems. These ideas must be integrated into a comprehensive policy to build a sustainable agricultural system. Anything short of this is doomed to fail in the long run.

To a man with an empty stomach,

food is God.

—Gandhi

Critical Thinking Exercise Solution

The problem with this line of reasoning is that food production per hectare—or productivity—is only one measure of sustainability. Moreover, several other measures are more relevant to the long-term sustainability of agriculture, including the amount of topsoil on farms, the health of the topsoil, and the supply of water available for irrigation. As this chapter points out, current agricultural practices are often unsustainable. Gains in productivity are likely to be a short-lived phenomenon and not a measure of the future viability of the systems. Taking a broader view of agriculture is essential to assess its sustainability. Focusing on one measure, such as productivity, can be very misleading.

This exercise requires the use of four critical thinking rules. First, it requires that one question the conclusions, notably, that rising productivity means agriculture is sustainable. Second, it requires that we define terms like *sustainability* and *productivity*. This helps us understand what we're talking about and enables us to think critically. Third, it requires one to look for hidden assumptions. In this case, it is assumed that rising productivity signifies a healthy system. Finally, it requires that we examine the big picture—that is, we must look at all of the factors that contribute to the sustainability of farming.

Summary

7.1 Feeding the World's People/Protecting the Planet

■ Three challenges confront world agriculture: (1) feeding the current population, (2) meeting the demands of the rising population, and (3) converting the unsustainable system of agriculture into a sustainable one.

■ **Sustainable agriculture** consists of practices that produce high-quality food and enhance, not destroy, agricultural land, water supplies, and the global environment.

7.2 Problems Facing World Agriculture

■ Many problems stand in the way of the goals listed above: soil erosion, nutrient depletion, desertification, dwindling fossil fuel supplies, groundwater depletion, waterlogging, salinization, declining species diversity, and farmland conversion.

■ If these problems are not confronted, it is doubtful that hunger can be eliminated and that future demands for food will be met.

7.3 Building a Sustainable Agricultural System

■ Numerous strategies must be employed to feed the world's people sustainably now and in the future.

■ Population control is first in importance. Beyond that, we must protect existing agriculture by decreasing soil erosion, reducing the spread of deserts, and slowing farmland conversion.

■ Efforts are also needed to protect water supplies and reduce waterlogging and salinization.

■ Farmland reserves can be tapped in some countries by converting forests and fields but not without serious ecological costs.

■ In conjunction with these efforts, food production can be increased by developing new plant and animal varieties, by soil enrichment, and by improved irrigation.

■ New high-yield plant and animal varieties can be developed through special breeding programs or by genetic engineering.

■ Native species, especially grazers, can provide protein and have many advantages over introduced domestic species.

■ Increasing the fish catch may be possible in the short term but could have serious effects on populations. A better strategy would be to develop commercial fish farms.

■ Cost-effective measures could be devised to reduce the amount of food destroyed by pests during the production-consumption cycle.

■ Developing nations could take steps to become self-sufficient in food production. Most importantly, all nations need to convert currently unsustainable agricultural systems into sustainable ones by applying the biological principles of sustainability.

Discussion and Critical Thinking Questions

1. What percentage of the world's population is malnourished? What are the short- and long-range effects of malnutrition?

2. What is desertification, what factors create it, and how can it be prevented?

3. Critically analyze the statement: "Soil erosion control is too expensive. We can't afford to pay for it because our crops don't bring in enough money."

4. How does the long-term outlook for oil affect world agriculture?

5. Describe waterlogging and salinization of soils. How can they be prevented?

6. Describe the decline in agricultural diversity. How could this trend affect world agriculture? Give some examples.

7. Critical analyze the statement: "Technology can solve all of our food problems, so there is no need to slow population growth."

8. List and discuss the major strategies for solving world food shortages. Which are the most important? How would you implement them in the United States and other countries?

9. Critically analyze the statement: "Simply by practicing better soil conservation and replenishing soil nutrients, we can reduce the need for new farmland."

10. Describe the successes and failures of the Green Revolution. What improvements might be made?

11. Critically analyze the following statement: "The world fish catch climbed nicely during the 1980s, and there is no reason to believe that the ocean won't continue to be a major source of new food."

12. Debate the following statement: "The developing countries should become more self-sufficient in agricultural production."

13. Although you have just been introduced to the concept of agricultural self-sufficiency, can you think of other activities where self-sufficiency may be important? How is self-sufficiency related to sustainability?

14. You have been appointed head of a U.N. task force. Your project is to develop an agricultural system in a poor African nation that imports more than 50% of its grain and still suffers from widespread hunger. Outline your plan, giving general principles you would follow and specific recommendations for achieving self-sufficiency.

Suggested Readings

Brown, L. R. et al. (1989). *State of the World 1989*. New York: Norton. Chapters 1 through 3 provide an excellent overview of modern agriculture problems and some sustainable solutions.

Brown, L. R. and Young, J. E. (1990). Feeding the World in the Nineties. In *State of the World 1991*. New York: Norton. Excellent study of the prospects for world agriculture.

Durning, A. T. and Brough, H. B. (1992). Reforming the Livestock Economy. In *State of the World 1991*. New York: Norton. Important discussion of reforms needed in meat production.

Norse, D. (1992). A New Strategy for Feeding a Crowded Planet. *Environment* 34(5): 6–11, 32–39. Important reading on sustainable agriculture.

Owen, O. S. and Chiras, D. D. (1990). *Natural Resource Conservation*. New York: Macmillan. Excellent sections on agriculture, soils, and land management.

Reganold, J. P., Papendick, R. I., and Parr, J. F. (1990). Sustainable Agriculture. *Scientific American* (June): 112–120. Excellent overview.

Soule, J. D. and Piper, J. K. (1992). *Farming in Nature's Image: An Ecological Approach to Agriculture*. Washington, D.C.: Island Press. Important reading.

Young, J. E. (1990). Bred for the Hungry? *World-Watch* 3(1): 14–22. Interesting look at biotechnology's potential contribution to world agriculture.

Soil and Soil Management: The Cornerstone of Sustainable Agriculture

To build may have to be the slow and laborious task of years. To destroy can be the thoughtless act of a single day.

—Winston Churchill

ða   ða   ða

Soil conservation involves a variety of techniques aimed at preventing soil erosion and the loss of nutrients. This supplement discusses some cost-effective ways of reducing erosion and maintaining, even improving, soil fertility, measures that are cornerstones of sustainable agriculture. To begin, we turn first to some basic information about soils.

What Is Soil?

Soil is a complex mixture of inorganic and organic materials with varying amounts of air and moisture. Clay, silt, sand, gravel, and rocks are the inorganic components of soil. Detritus, organic wastes, and a multitude of living organisms are the organic components. Soils are described according to six general features: texture, structure, acidity, gas content, water content, and biotic composition.

How Is Soil Formed?

Soil formation is a complex, slow process, even under the best of conditions. The time it takes soil to develop depends partly on the type of **parent material**, the underlying substrate from which soil is formed. To form 2.5 centimeters (1 inch) of topsoil from hard rock may take 200 to 1200 years, depending on the climate. Softer parent materials, such as shale, volcanic ash, sandstone, sand dunes, and gravel beds, are converted to soil at a much faster rate (in 20 years or so) under favorable conditions.

Numerous physical processes contribute to soil formation. For example, daily heating and cooling cause the parent rock material to split and fragment, especially in the desert biome where daily temperatures vary widely. Water enters cracks in rocks and expands when it freezes, causing the rock to fragment further. The roots of trees and large plants reach into small cracks and fracture the rock. Rock fragments produced by these processes are slowly pulverized into smaller particles by streams or landslides, by hooves of animals, or by wind and rain.

Soil formation is facilitated by a wide range of organisms. Chapter 3 described how lichens erode the rock surface by secreting carbonic acid. Lichens also capture dust, seeds, excrement, and dead plant matter, which help to form soil. Plant roots serve as nutrient pumps, bringing up inorganic chemicals from deeper soil layers. These nutrients are first used to make leaves and branches, which can fall and decay, thus becoming part of the uppermost layer of soil, the **topsoil**.

Grazing animals drop excrement on the ground, adding to the soil's organic matter. The white rhinoceros, for example, produces about 27 metric tons of manure each year, which is deposited in its habitat. A variety of insects and other creatures, such as earthworms, also participate in soil formation.

The Soil Profile

Soil is often arranged in layers of different color and composition. These layers are called **horizons**. Soil scientists recognize five major horizons (Figure S7.1). The uppermost is the **O-horizon**, or **litter layer**. This relatively thin layer of organic waste from animals and detritus is the zone of decomposition and is characterized

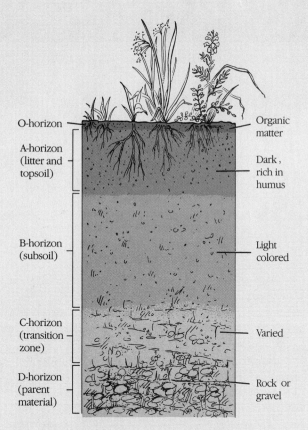

O-horizon

A-horizon
(litter and
topsoil)

B-horizon
(subsoil)

C-horizon
(transition
zone)

D-horizon
(parent
material)

Organic
matter

Dark,
rich in
humus

Light
colored

Varied

Rock or
gravel

Figure S7.1   *Soil profile, showing the five horizons. Not all are found in all locations.*

by a dark, rich color. Plowing mixes it in with the next layer.

The **A-horizon,** or **topsoil,** is the next layer. It varies in thickness from 2.5 centimeters (1 inch) in some regions to 60 centimeters (2 feet) in the rich farmland of Iowa. Topsoil is generally rich in inorganic and organic materials, and is important because it supports crops. It is darker and looser than the deeper layers. The organic matter of topsoil, called **humus,** acts like a sponge, holding moisture.

The **B-horizon,** or **subsoil,** is also known as the "zone of accumulation" because it receives and collects minerals and nutrients leached from above. This layer is lightly colored and much denser than topsoil because it lacks organic matter. The next layer, the **C-horizon,** is a transition zone between the parent material below and the soil layers above. The **D-horizon** is the parent material from which soils are derived. Not all horizons are present in all soils; in some, the layering may be missing altogether.

The soil profile is determined by the climate (especially rainfall and temperature), type of vegetation, parent material, age of the soil, and organisms. Soil profiles tell soil scientists whether land is best for agriculture, wildlife habitat, forestry, pasture, rangeland, or recreation. They also tell how suitable soil might be for vari-

ous other uses, such as home building and highway construction.

Soil Conservation

Sustainable farming depends on soil conservation, which, as mentioned earlier, consists of two basic steps: erosion control and nutrient preservation.

Erosion Control

Erosion can be minimized, even halted, by a variety of simple techniques. This section discusses six major strategies: minimum tillage, contour farming, strip cropping, terracing, gully reclamation, and shelterbelts.

Minimum Tillage   Typically, farmers plow their fields before planting a new crop, then break clumps of soil up with a device called a disk. This makes the soil suitable for sowing seeds. In areas where soil is too wet in the spring, farmers plow and disk in the fall, leaving their land barren and subject to wind erosion during the winter months.

With special implements, however, farmers can forgo these costly and time-consuming steps and plant over the previous year's crop residue (Figure S7.2). This technique is one form of **minimum tillage,** or **conservation tillage,** a strategy that reduces the physical disruption of the soil. According to the U.S. Department of Agriculture, conservation tillage is practiced on about 33% of all U.S. farmland. Their projections suggest that by the year 2000, conservation tillage could be in place on about one-half of all farms in the United States. Unfortunately, this practice is not widely used in other countries (Figure S7.3).

Figure S7.2   *Minimum tillage planter, designed to dig furrows in the presence of crop residue.*

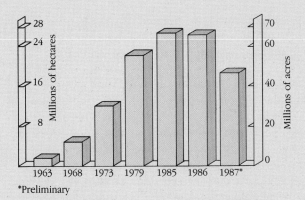

Figure S7.3 *Growth in minimum, or conservation, tillage in the United States. Source: U.S. Department of Agriculture.*

Figure S7.4 *This land is farmed along the contour lines to reduce soil erosion and surface runoff, thus saving both soil and moisture.*

Because fields are protected much of the year by crops or crop residues, soil erosion can be decreased substantially, in some cases, by as much as 90%. Minimum tillage also reduces energy consumption by as much as 80% and conserves soil moisture by reducing evaporation. Crop residues can increase habitat for predatory insects that prey on pests. This may reduce pesticide use and contamination of the environment (Chapter 18).

Despite its benefits, minimum tillage has several drawbacks. For example, **herbicides**, chemicals that kill weeds, are often used in place of mechanical cultivation to control weeds. In addition, crop residues may harbor harmful insects that damage crops. Minimum tillage also requires new and costly farm equipment.

Contour Farming   On hilly terrain, crops can be planted along lines that follow the contour of the land, a technique called **contour farming**. As shown in Figure S7.4, rows are planted across the direction of water flow on hilly terrain, which reduces the rate at which water flows across the land, resulting in a 60% to 80% reduction in erosion in some areas. This technique markedly increases water retention, thus reducing water demand for irrigation water.

Strip Cropping   As Figure S7.5 shows, **strip cropping** is a measure in which farmers plant strips of two or more crops in single fields in alternating bands. Strip cropping reduces wind and water erosion and increases productivity. It can be combined with contour farming to further decrease soil erosion. For example, farmers may alternate row crops, such as corn, with cover crops, such as alfalfa, on hilly land. Water flows more easily through row crops and begins to gain momentum, but when it reaches the cover crop, its flow is nearly stopped.

Terracing   For thousands of years, many people have grown crops in mountainous regions using **terraces**,

small earthen embankments placed across the slope to check water flow and minimize erosion. Terraces have also been used in the United States for over 40 years on land with less-pronounced slope (Figure S7.6). Terraces are expensive to construct and may interfere with the operation of large farm equipment.

In a University of Nebraska study, scientists showed that terracing and contour farming each reduced erosion by 50%. Together, they reduced soil loss by 75%. Minimum tillage was even more effective. By itself, it reduced

Figure S7.5 *Strips of corn protect small grain plants from the effects of wind.*

Figure S7.6 *This Iowa corn is grown on sloping land with the aid of terraces, small earth embankments that reduce water flow across the surface. The corn is planted in the stubble of last year's crop.*

erosion rates by 90%, but when combined with terracing and contour farming, soil erosion rates were reduced by 98%.

Gully Reclamation   Gullies are a danger sign indicating rapid soil erosion. Some gullies can work their way up hills at a rate of 4.5 meters (15 feet) per year. To prevent gullies from forming, farmers must reduce water flow over their land. Contour farming, strip cropping, and terraces all help. Special diversion ditches can be dug in channels where water naturally flows off fields. Seeding already formed gullies with rapidly growing plants, including trees, can prevent further erosion. Small earthen dams can be built across gullies to reduce water flow, retain moisture for plant growth, and capture sediment, which will eventually support vegetation as erosion is reduced. Too often land with severe gullies is abandoned or haphazardly reclaimed, only to suffer worse erosion in time.

Shelterbelts   In 1935, the U.S. government mounted a campaign to prevent the recurrence of the disastrous Dust Bowl era. This program involved planting long rows of trees as windbreaks, or **shelterbelts**, along the margins of farms in the Great Plains to slow the fierce winds that carry soil away (Figure S7.7). Today, thousands of kilometers of shelterbelts have been planted from Texas to North Dakota.

Besides decreasing soil erosion from wind, shelterbelts reduce damage to crops caused by windblown dirt particles. In the winter, they permit snow to accumulate in fields, which increases soil moisture and groundwater supplies, reducing irrigation demand and groundwater overdraft. Shelterbelts can also improve irrigation effi-

ciency by reducing the amount of water carried away from sprinklers. In addition, shelterbelts provide habitat for animals, pest-eating insects, and pollinators. Shelterbelts protect citrus groves from wind that blows fruit from trees. They have the added benefit of saving energy by reducing heat loss from homes and farm buildings.

Despite the importance of soil erosion controls to sustainable agriculture, many farmers are reluctant to invest in these practices. One prominent reason, explained in more detail in Chapter 21, is short-term economics. Because erosion control costs money and could cut into marginal profits, some farmers are reluctant to take steps. Such a view ignores the long-term cost of permitting unsustainable loss of soil. The 1990 farm bill encourages American farmers to practice soil conservation by offering federal crop insurance and other benefits to those who participate in such programs.

Nutrient Preservation

Preventing soil erosion preserves soil nutrients needed to maintain productivity. Several additional methods to prevent soil nutrient depletion, which were mentioned in Chapter 7, are discussed in more detail below.

Organic Fertilizers   Organic fertilizers, such as cow, chicken, and hog manure and human sewage, are excellent soil supplements that replenish organic matter and important soil nutrients, such as nitrogen and phosphorus. Other organic materials may be used to build soil; especially valuable are leguminous plants, such as alfalfa, that are grown during the off-season and plowed under before planting crops. Referred to as "green manure," they add organic matter to the soil, retain moisture, and also reduce soil erosion during fallow periods.

Figure S7.7 *Shelterbelts used to protect farmland in Michigan from the erosive effects of wind.*

Soil enrichment with organic fertilizers (1) improves soil structure, (2) increases water retention, (3) increases fertility and crop yield, (4) provides a good environment for the bacterial growth necessary for nitrogen fixation, (5) helps prevent shifts in the acidity of soil, and (6) tends to prevent the leaching of minerals from soil by rain and snowmelt. In addition, the use of human wastes on farmland could significantly reduce water pollution by waste treatment plants.

Organic wastes have been successfully applied in some countries, such as China and India, but this practice is not without its problems. One of the leading problems is the cost of transporting waste to farms by pipelines or trucks. Although the initial investment may be high, the long-term benefits of applying organic fertilizers could outweigh the costs. Another problem is that waste from municipal sewage treatment plants may be contaminated with pathogenic organisms, such as bacteria, viruses, and parasites. Theoretically, some of these could be taken up by crops and reenter the human food chain. In industrialized nations, toxic heavy metals, such as mercury, cadmium, and lead, may be present in municipal wastes and could enter crops. Better controls at sewage treatment plants or at the factories that produce these materials could alleviate the problem.

Synthetic Fertilizers   In the developed countries, like Canada, Australia, and the United States, farmers apply millions of tons of **synthetic fertilizers** to boost crop production. Without them, world food production would fall 40% or more, according to the Worldwatch Institute.

Synthetic fertilizers are manufactured by chemically combining nitrogen and hydrogen to form ammonia and ammonium salts. Some nitrates and urea are also made. These are combined with phosphorus and potassium.

Artificial fertilizers can be applied directly to the soil as a liquid or a dry, granular mix.

Although they partially restore soil fertility, artificial fertilizers do not replenish organic matter or micronutrients necessary for proper plant growth and human nutrition. As a result, on farms treated only with artificial fertilizer, soil is slowly degraded over the years. Moreover, excess fertilizer may be washed from the land by rains and end up in streams, causing a number of problems discussed more fully in Chapter 17 on water pollution.

To prevent the slow depletion of nutrients and to help develop a sustainable agricultural system, many proponents recommend much wider use of organic fertilizers. In fact, they note that synthetic fertilizers can be eliminated entirely in the developing nations or on small farms located near abundant sources of organic fertilizer.

Combined with programs of soil conservation and organic enrichment, synthetic fertilizers will inevitably play an important role in feeding the world's people. As fossil fuels become scarcer, however, synthetic fertilizers may become more costly.

Crop Rotation   In "modern" agriculture, synthetic fertilizers and pesticides have allowed farmers to grow the same crop year after year on the same plot. This way farmers can concentrate their efforts on one crop that they know well. However profitable this practice may be, it is unsustainable, for it gradually depletes the soil of nutrients. It also tends to increase soil erosion and typically worsens problems with pests and pathogens.

Crop rotation is a practice in which farmers alternate crops they plant in their fields. It helps to build the soil and to reduce pest damage and the need for costly and potentially harmful chemical **pesticides**, substances that kill damaging insects, weeds, and other species. Using the time-tested practice of crop rotation, a farmer may plant a soil-depleting crop, such as corn, for one to two years and then follow it with a cover crop, such as grasses, or various legumes, such as alfalfa. The cover crop reduces soil erosion, and legumes replenish soil nitrogen. Often, cover crops are not harvested but are simply plowed under to replenish organic matter and return nutrients to the soil.

Crop rotation puts conservation, recycling, renewable resource, and restoration principles of sustainability into action. Properly planned and executed, it can produce yields 10% to 15% higher than those achieved through monocultures.

Erosion Control and Beyond: Sustainable Farming

Today, thousands of farmers are running profitable farms either without chemical fertilizers and pesticides or with only small amounts of them. **Organic farming** is a system that excludes synthetic fertilizers, pesticides, growth regulators, and livestock food additives. On these farms, farmers rely on crop rotation, green manure, animal wastes, and off-farm organic wastes to fertilize their fields and maintain soil quality. Organic farmers also rely on alternative, less environmentally harmful means to control pests (Chapter 18). In some cases, organic farmers use crop rotation to reduce pest population and may plant several crops in the same field in alternating strips, creating greater crop diversity, which reduces pest damage. They may spray their fields with natural insecticides or release predatory insects from time to time to control crop pests. Organic farmers also generally pay close attention to their soil and practice erosion control, minimum tillage, contour farming, strip cropping, and terracing, among other measures.

The success of organic farming often baffles those who view modern chemical-intensive agriculture as the only way to produce food. Ecologist Barry Commoner compared organic and conventional farms of similar size and found that the organic farms produced nearly as well. His study showed that organic farms produce about 11% less than a comparable chemical-intensive farm, but the **net income**, or **net profitability**, per hectare, how much money the farmer nets per hectare, was about the same because of lower input costs. For example, organic farms require 60% less fossil fuel energy to produce corn crops with only 3% lower yields.

Organic farming arose from a concern about the chemical-intensive nature of modern agricultural systems, and focused on human health and the health of the biosphere. Since its origins, organic farming has expanded considerably to include numerous sustainable practices aimed at ensuring a long-term supply of food. Consequently, the term *organic farming* is being replaced by **low-impact sustainable agriculture**, or simply **sustainable agriculture**. Sustainable agriculture does not represent a return to preindustrial methods of farming but rather combines traditional conservation-minded farming practices with modern technologies.

Sustainable farms may not perform as well as their modern counterparts in the short term, but they offer long-term benefits of inestimable economic and cultural value. In two adjacent farms near Spokane, Washington, for instance, researchers compared the impacts of sustainable and conventional farming on the soil. Both farms began in 1909. In 1948, one farm adopted the chemical-intensive, erosion-prone new methods of modern agriculture. The other retained the "old" practices, such as crop rotation.

The sustainable farm in this study operates on a three-year cycle, producing winter wheat the first year, then a crop of spring pea the following year. In the third year, the farmer plants Austrian winter pea, which is plowed under to provide nutrients and organic matter for the soil. The conventional farm, in contrast, alternates between crops of winter wheat and spring pea.

Although crop yields are similar for both farms, the sustainable farm produces a cash crop on a given field only two out of every three years. To some that may seem economically foolish, but that's because they view agriculture too narrowly—strictly as a matter of production. A comparison of the two farms reveals the fallacy of such thinking.

Researchers measured the topsoil on the sustainable farm and found that it was, on the average, 15 centimeters (6 inches) thicker than on the neighboring farm. Soil moisture levels and soil structure were also much better on the sustainable farm. When researchers computed the loss of topsoil on the two farms, they found that the conventional farm would be stripped of its topsoil in 50 years, exposing the dense, less fertile clay subsoil, greatly decreasing the land's productivity. Because

topsoil loss on the sustainable farm was negligible, it could produce forever. If one factors in the long-term earnings of the two farms, the sustainable operation would win hands down.

When analyzing agriculture, many experts look solely at the yield per hectare—the output of the land. This is a little like an employer who considers only the amount of work an employee produces each day, without concern for the worker's health or the quality of the product. Critical thinking suggests the need to look more broadly at agriculture, including the long-term prospects of soil productivity, effects of farming on the environment, and effects on the health of farmers, farm workers, and farm families.

To be sure, organic farming is not a panacea. Organic fertilizers may not replace all of the phosphorus and potassium drained from the soil by plants. Legumes may not always replace all of the nitrogen removed by high-yield crops, such as corn, and total production may be lower than on a conventional farm because of land devoted to soil conservation. When shifting to sustainable farming practices, yields and losses to insects and weeds may be higher during a three- to five-year transition period.

Despite these problems, sustainable farming must become the wave of the future. However, the wholesale adoption of organic farming techniques will not come quickly. Farming has become a big business, dominated by corporate giants and an industrial mentality that seeks maximum short-term profit despite the environmental costs.

In developing countries, opportunities to shift to sustainable farming practices are numerous, given the high cost of fossil fuels and their abundant labor pools. Unfortunately, rapid population growth has shifted attention to the immediate problem of hunger, causing people to ignore the continuing problem: how to farm while protecting the soil, the resource that makes all life possible.

Achieving a Global System of Sustainable Agriculture

Soil protection and enhancement are means to build a sustainable system of agriculture. Below are some suggestions for achieving these changes quickly:

1. Governments can develop programs to target areas where topsoil erosion is the most severe. If necessary, governments may need to enact laws that prohibit highly erodible land from being farmed.

2. Governments can expand programs, such as the U.S. Soil Conservation Service, to educate farmers on

ways to farm sustainably, conserve water and energy, minimize chemical input and costs, and tap into renewable energy.

3. Governments can offer tax breaks, low-interest loans, and other incentives to farmers to help defray the cost of soil conservation and other sustainable farming practices, especially during the transition years.

4. Governments can require farmers applying for government loans, crop insurance, or aid to have approved water and soil conservation programs in operation.

5. States can strengthen their role in soil conservation and foster ways to increase cooperation among state, local, and federal agencies involved in soil conservation.

6. Universities can establish programs in sustainable agriculture.

Soil conservation is a cornerstone of sustainable agriculture, but sustainable agriculture is also a cornerstone of a sustainable human society. If farming isn't sustainable, it isn't agriculture; it's anticulture. Unsustainable forms of agriculture are tears in the fabric of human civilization so severe that the whole could come unraveled.

Suggested Readings

Edwards, C. A. et al. (1990). *Sustainable Agriculture.* Ankeny, Iowa: Soil and Water Conservation Society.

National Academy of Sciences. (1989). *Alternative Agriculture.* Washington, D.C.: National Academy Press.

Reganold, J. P., Papendick, R. I., and Parr, J. F. (1990). Sustainable Agriculture. *Scientific American* (June): 112–20.

Chapter 8

Wildlife and Plants: Preserving Biological Diversity

The worst sin toward our fellow creatures is not to hate them, but to be indifferent to them; that's the essence of inhumanity.
—George Bernard Shaw

Critical Thinking Exercise

You own a guest ranch in western Colorado. Besides coming to ride horses, your guests come to view deer, elk, eagles, hawks, and other wildlife. In addition to horses, you raise sheep to supplement your income. One day, a neighbor asks for your help in controlling coyotes, which are killing off his sheep and costing him lots of money. He wants you to join him in spreading Compound 1080, a poison bait, on your land. It's not legal, but he says that he can extract the poison from sheep collars, which are. He would like to inject the poison into scraps of meat that he would distribute on your property and his.

What concerns would you have about this? Make a list of questions you will need to answer before deciding whether to participate with your neighbor. What critical thinking rules can you apply?

In 1986, biologists captured the last known surviving black-footed ferrets living in the wild in the United States. The ferret's numbers had been reduced by habitat destruction, loss of prairie dog populations—the ferret's main food source, disease, and other factors. In captivity, the ferrets flourished; their numbers rose from 18 to over 300 animals by 1991. Because of the animals' remarkable proliferation in captivity, biologists decided to release 49 captive-bred ferrets onto the short-grass prairies of Wyoming in September 1991. Although only 4 to 6 animals lived through the first year, another 60 animals were released in the fall of 1992.

In other good news, in January 1992, two captive-bred California condors were released in the Los Padres National Forest north of Los Angeles (Figure 8.1). This was the first attempt to reintroduce this endangered species into its former habitat. California condors, which were on the brink of extinction only a few years earlier, were hatched in a zoo-breeding program sponsored by the U.S. Fish and Wildlife Service and the National Audubon Society. The last surviving wild condors had been captured near Bakersville, California, in 1987.

These are only two of many success stories that portray humankind's heroic efforts to save species that we have pushed to the brink of extinction. This chapter examines the extinction of plants and animals, and describes the underlying causes. It also looks at the countless benefits of plants and wildlife to human societies, and suggests actions needed throughout the world to protect the organisms that share this planet with us.

8.1 The Vanishing Species

By some estimates, as many as 500 million kinds of plants, animals, and microorganisms have made this planet home since the beginning of time. Today, scientists estimate that the world contains between 10 and 80 million species, although only about 1.5 million have been identified and named. Thus, 420 to 490 million species have become extinct.

All this is to say that extinction is an evolutionary fact of life. But, natural extinction differs considerably from the impending doom now facing the rhino and tens of thousands of plants and animals. Natural extinction differs from the accelerated extinction now taking place for at least two reasons.

First, during the course of millennia, old species evolved into new ones (Figure 8.2). Consequently, many of the millions of extinct species are represented today by their descendants. Modern extinctions, on the other hand, eliminate species entirely. If the rhino vanishes, it will leave no evolutionary legacy. It will be gone forever.

Second, the rate of extinction varies considerably. Even though some species did vanish because of severe climatic changes or other reasons, the rate of natural

Figure 8.1 *The California condor is on the verge of extinction. The bird's future lies in the hands of scientists and zoo personnel. Hunting, habitat destruction, lead poisoning, and a low reproductive rate have resulted in the decimation of the condor population in the United States.*

Figure 8.2 *Stages in the evolutionary history of the horse. Modern horses, like other species, evolved through intermediate stages that are no longer alive today.*

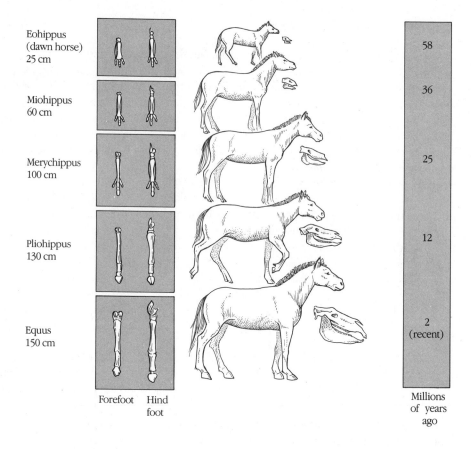

extinction—about 1 species every 1000 years—is slow compared with today's accelerated extinction. Although no one can tell exactly how many species become extinct today, estimates based on loss of habitat and species diversity within these ecosystems suggest that as many as 40 to 100 species go extinct every day. Harvard biologist E. O. Wilson thinks the number may be higher.

Many experts believe that we have entered an era of extinction unparalleled in the Earth's history. John Ryan of the Worldwatch Institute writes, "Difficult as it is to accept, mass extinction has already begun, and the world is irrevocably committed to many further losses." Today, many millions of species are endangered. An **endangered species** is one that is threatened with extinction. Without concerted efforts to protect it, the species will become extinct. A **threatened species** is one that is still abundant but declining so rapidly that it is likely to become extinct. Birds are especially threatened. Three-fourths of the world's bird species are declining in number or threatened with extinction. Moreover, more than two-thirds of the world's 150 species of primates are threatened. A recent study in Malaysia showed that more than half of the 266 fish species that inhabited the rivers before the advent of logging are gone. Today, approximately 25,000 species of plants are threatened with extinction—1 of every 2 plant species on Earth. In the United States, approximately 3000 species of plants are in danger of extinction. More than 700 plant species will probably disappear in the next ten years unless concerted efforts are made to save them.

Scientists worldwide are disturbed by an alarming disappearance of amphibians. Many species of frogs, toads, and salamanders are either experiencing steep declines in population size or are vanishing altogether. Amphibians are disappearing from a wide variety of habitats—from the jungles of Brazil to the suburbs of New York City.

Endangered pandas, blue whales, rhinos, and chimpanzees generally make the headlines because they are the most appealing or visible victims. Interest in the less appealing species is often difficult to stir, but many less conspicuous species are important components of natural systems and contribute significantly to human welfare. An adult frog, for example, can eat its weight in insects every day. In India, sharp declines in the frog populations may be partly responsible for higher rates of insect damage to crops and an increase in malaria, which is transmitted by mosquitoes.

Unless we curb population growth, reduce pollution and habitat destruction, and manage our resources better, vast expanses of forests, wetlands, and grasslands will disappear. As many as 1 million species may vanish between 1980 and 2000. The impact of habitat destruction and mass extinction will be felt worldwide.

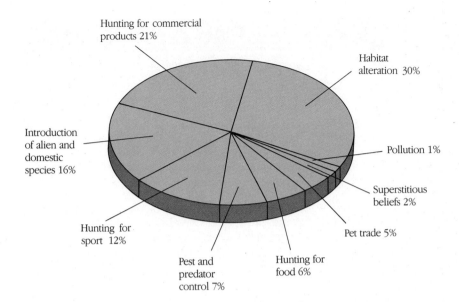

Hunting for commercial products 21%

Habitat alteration 30%

Introduction of alien and domestic species 16%

Pollution 1%

Superstitious beliefs 2%

Pet trade 5%

Hunting for sport 12%

Pest and predator control 7%

Hunting for food 6%

Figure 8.3 *An approximate breakdown of the human activities that lead to extinction. Note that the number one cause of extinction today is habitat alteration. In many cases, two or more activities contribute to species extinction.*

8.2 What Causes Extinction?

Plant and animal extinction, like many other environmental problems, results from many activities (Figure 8.3). The top five, in decreasing order of importance, are: (1) habitat alteration, (2) commercial hunting and harvesting, (3) hunting for sport and subsistence, (4) introduction of alien species, and (5) pest and predator control.

Habitat Alteration

Humans have always altered habitat, but today our numbers and our high demands place inordinate strain on lands vital to many species. Habitat alteration, ranging from moderate to extreme, is *the* most significant factor in extinction.

Habitat damage results from many human activities, among them urban sprawl, farming, mining, road building, and timber cutting. Dams, airports, shopping malls, and housing developments now occupy many once richly endowed ecosystems. Even recreational areas have adverse effects on wild species. Many dams constructed to enhance boating, fishing, and water skiing, for instance, release extremely cold water from the bottom of their reservoirs. The change in temperature may destroy native fish unable to withstand the cold conditions. Dams also block the migration of species, such as salmon, that migrate from the ocean to inland rivers to spawn. Dams on the East Coast have eliminated the Atlantic salmon and those on the West Coast have greatly reduced the Pacific salmon. On the Columbia River, 10 to 15 million salmon once spawned on

thousands of miles of streambed now blocked by dams. Today, the Columbia River system supports only 2.5 million salmon per year. Recent estimates suggest that the number of salmon is far lower and many people think that some species of salmon should be listed as endangered animals. Another recreational impact results from ski resorts, with their attendant condominiums and restaurants, which destroy the winter feeding grounds for deer, elk, and other species.

Nowhere is the loss of habitat more noticeable than in tropical rain forests, which house at least half of the Earth's species, perhaps more. Once covering an area the size of the United States, tropical rain forests have been reduced by about one-half. Countless species have perished as a result. (Chapter 9 discusses tropical rain forests in more detail.)

Coral reefs, wetlands, and estuaries (the mouths of rivers) are other critical habitats already greatly reduced and rapidly declining because of human development. Wetlands and estuaries are home to many species, but are also highly prized by humans. Damage to wetlands has been particularly severe in industrial nations. New Zealand and Australia, for example, have lost more than 90% of their wetlands and countless species. India, Pakistan, and Thailand have lost at least 75% of their mangrove swamps, a type of coastal wetland. Their destruction threatens the future of fish and waterfowl throughout the world (see Chapter Supplement 10.1).

Another ecosystem that has experienced serious losses is the tall grass prairie of North America, which today is virtually nonexistent. Temperate rain forests, such as those in the Pacific Northwest, have also been largely destroyed. Worldwide, 56% of the temperate rain forests have been logged or cleared. In the United States,

Models of Global Sustainability 8.1

Restoring Canada's Troubled Fisheries

Canada is a country of incredible beauty. Vast open plains and rich forests blanket her lands. Wildlife abound in her unspoiled wilderness. But, the rich, natural resource base that supports Canada's 28 million people and pumps billions into the economy is slowly being destroyed. In eastern Canada, for example, 10,000 lakes are so acidic that experts believe they no longer contain fish. In Lake Ontario, Canadian anglers reel in salmon ulcerated by tumors.

In south-central British Columbia, the number of nonresident anglers has dropped by half since 1960. In the Yukon, anglers are catching 28% fewer arctic grayling, pike, and lake trout than in 1975. In southern Manitoba, anglers can land only one-tenth as many walleye as they did 20 years ago.

Canada's freshwater fisheries are in a state of decline. Fish populations are falling. Each year, numerous spawning grounds and rearing areas are lost to urban development and industrialization. In some areas, fish are contaminated with pollutants and they are unfit for human consumption.

The destruction of Canada's once-famous fisheries results from many different causes. Overfishing by commercial interests and recreationists is a leading cause. Air pollution from Canadian and U.S. smoke-stacks and cities contributes, as do water pollutants from Canadian cities and paper mills. Fertilizers and pesticides from farms are also major factors. Unfortunately, recreational fishing management falls under the auspices of provincial governments, which generally have insufficient funds to do the job right.

Many Canadian anglers have decided to take matters into their own hands. In record numbers, they are devising methods to rescue their favorite streams and to influence political leaders to help protect this vanishing resource. They are pressuring the federal government to tackle some of the problems that require national solutions.

Recently, anglers turned out in record numbers to save the Credit River, a trout stream 70 kilometers (40 miles) from Toronto, a city with a population of 3 million. The anglers labored hard to remove dead

trees and other obstructions that have changed the stream flow, making it warmer and unsuitable for trout. And, they have begun to revegetate the river-banks to reduce erosion, which increases the sediment load in streams and destroys spawning sites.

The Credit River is only one of many rivers that have been rescued by these ambitious Canadians. Twenty years ago, the Bow River, a large stream that flows through Calgary, was compared to an open sewer, but local anglers applied pressure to political leaders to ensure minimum-flow dam releases, better waste treatment, and tighter fishing regulations. As a result, the river is now a blue-ribbon trout stream. Brown trout of 50 centimeters (20 inches) or more are common in its waters.

Jim Gourlay, editor of a Canadian fishing magazine, says, "Instead of just griping as in the past, people are out there cleaning and improving streams, planting eggs and fry, pushing for mandatory hook and release and other conservation measures." In Ontario, anglers lobbied to institute fishing licenses. As a result of their successful campaign, Ontario fishing licenses now produce millions of dollars used to protect fish and manage their fisheries better.

Another exciting development is the Community Fisheries Involvement Program (CFIP) in Ontario, a cooperative program between the government and private interest groups. The government provides technical advice and some money; the groups provide labor. Thousands of individuals work on projects such as cleaning up streams, building incubation boxes, and anything else to help improve trout and salmon streams.

Ontario's CFIP is a model program. British Columbia has started a similar program to protect streams. In a single year, 250 fishery projects, involving 8000 volunteers, from all walks of life, are carried out in British Columbia. This idea is spreading fast to other parts of the province, and it is a good example of how people can make a difference in their environment and help to create a sustainable future.

---

no more than 10% of the original old-growth temperate rain forests remain.

Commercial Hunting and Harvesting

Commercial hunting and harvesting of wild species represents the second largest threat to the world's animal species. Whale hunting is one of the most widely publicized examples. Commercial whalers have hunted one

species after another to the brink of extinction. The result has been a severe reduction in whale populations (Table 8.1). Thanks to recent efforts by the International Whaling Commission, commercial whaling has been greatly reduced. In its place is a new industry—whale watching—with annual revenues that exceed those of the commercial whaling industry itself.

Despite bans on whaling, Japan and Iceland continue to hunt whales under the guise of scientific

Table 8.1   Whale Populations—Then and Now

Species	Number Before Commercial Whaling	Current Estimate
Blue	166,000	7,500–15,000
Bowhead	54,680	3,600–4,100
Fin	450,000	105,000–122,000
Gray	15,000–20,000	13,450–19,200
Humpback	119,000	8,900–10,500
Minke	250,000	130,000–150,000
Right	50,000	3,000
Sei (includes Bryde's)	108,000	36,800–54,700
Sperm	1,377,000	982,300

elephants, rhinos, jaguars, tigers, and cheetahs in Africa. In 1973, 130,000 elephants lived in Kenya. In 1992, only about 26,000 were alive—a reduction of 80%. This decrease was largely the result of commercial killing of elephants for their ivory. Recognizing the plight of the African elephant, the Kenyan government banned further export of ivory in 1989. Despite the ban, poachers continue to kill large numbers.

The world's fisheries (fishing grounds) have also been heavily overfished, and many have been severely depleted, as noted in Chapter 7. Worldwide, an estimated 45 fisheries have been abandoned thus far as a result of unsustainable harvest. Many others are currently threatened. In Thailand, the introduction of trawl nets, coupled with the rise in motorized boats, has resulted in a dramatic increase in commercial fishing—so great, in fact, that several natural fisheries may soon be depleted. The Thai fishing industry provides protein to the nation's population and nets over $700 million per year in export sales to countries in Asia, Europe, and North America.

research. Such killing, say critics, is an effort to keep their whaling industry alive. Japan recently tried to reclassify one of its minke whaling ventures as a subsistence hunt, even though 90% of the whale meat was sold on the open market. Iceland businesses have also profited handsomely on whale meat from "research" slaughter, reaping over $8 million in revenue in one year.

Commercial harvesting of wild species is systematically reducing the populations of many other endangered species, causing a drastic decline in gorillas,

Hunting for Sport and Subsistence

Hunting of properly managed, nonendangered species frequently benefits populations by keeping them within the carrying capacity of their habitat. It is two other kinds of hunting that threaten rare and sometimes endangered wildlife: trophy hunting and subsistence hunting (Figure 8.4).

Trophy hunting can, if not properly regulated, wipe out local populations of animals, and trophy hunters, who generally go after the prize animals, can diminish the genetic stock of a population.

Figure 8.4 *Sport hunting can endanger wildlife populations that are not carefully managed.*

Eskimos annually kill a number of endangered bowhead whales. Environmentalists, with the support of scientific evidence, claim that the whale population (about 2300) cannot support further hunting. The Eskimos, however, argue that their own survival depends on the hunt. In a compromise decision, the Eskimos agreed to cut back on their annual kill and to reduce the loss resulting from wounding. Nonetheless, some environmentalists fear that the new quotas could eventually eliminate this species.

Introduction of Alien Species

Foreign, or alien, species introduced accidentally or intentionally into new territories often cause the extinction of native species. The water hyacinth is such an example (Chapter 4). The English sparrow fluttering in the bushes outside your window is another. Deliberately introduced into this country in the 1850s, the sparrow quickly spread throughout the continent. It now competes for nesting sites and food once used by bluebirds, wrens, and swallows.

Islands are especially vulnerable to new species. In Hawaii, for example, human inhabitants and organisms (such as rats) introduced by humans have destroyed 90% of all bird species. In New Zealand, half the native birds are endangered or extinct.

Florida is a showcase of alien species gone wild. Australian pines, introduced as ornamentals, have spread rapidly along coastal beaches. Their shallow roots are so dense that they destroy sandy beaches where many sea turtles lay their eggs. The worst offender is a species called the punk tree. This thirsty plant grows in swamps, creating a dense tangle of vegetation impassable to many animal species.

Pest and Predator Control

Pest control also influences wild populations of plants and animals. DDT and other pesticides have taken a huge toll on American wildlife (Chapter 17). The peregrine falcon disappeared in the eastern United States by the 1960s as a result of DDT-induced reproductive failure. DDT caused eggshell thinning and destroyed the entire falcon population east of the Mississippi River (Chapter 3). Eagles and brown pelicans met a similar fate. Even the California condor suffered from eggshell thinning. DDT poisoning was, however, just one of many factors that spelled doom for this species. Among other factors were lead poisoning (discussed later), loss of habitat from home building and farming, and fire control, which eliminated takeoff and landing areas needed by these giant scavengers, whose wings span up to 3 meters (9 feet). Especially harmful to U.S. migratory birds are the persistent pesticides, such as DDT and

related compounds, that have been banned in the United States but are still used in Latin American countries. Birds in Latin America have also been hard hit.

Predator control, once the cornerstone of wildlife management, has endangered or wiped out populations of wolves, bears, and other animals. Killing off predators creates an ecological backlash as the prey populations are thrown out of balance.

Additional Causes of Extinction

Collecting for Individuals and Research Animals and plants are gathered in the millions throughout the world for zoos, private collectors, pet shops, and researchers. In 1990, more than 250 to 300 million fish, 1 to 1.5 million live reptiles, and 3 to 4 million reptile skins were imported into the United States, according to the World Wildlife Fund (Figure 8.5). Each year, millions of tropical birds are brought to the United States, Canada, and Great Britain. However, for each bird that makes it into someone's home, 10 to 50 may die en route. Survival after reaching an individual's home is also low.

With the influx of legally imported animals come countless members of endangered, threatened, and rare species. Smuggled in, many of these animals go to private collectors. Plants are also high-ticket items. Cacti and orchids, for instance, are in high demand today, and support a growing industry. About 2 to 2.5 million cacti are imported each year into the United States from more than 50 countries. At home, collectors pillage Texas and Arizona deserts in search of salable cacti to adorn the lawns of eager customers. In an area of Texas near Big Bend National Park, 25,000 to 50,000 cacti were uprooted in a single month.

International trade in exotic flowers, especially those originating in the Mediterranean, threatens native populations. Although most flower bulbs sold in the United States are from domestic stock, tens of millions of bulbs are imported from Turkey to the Netherlands, where they are propagated or re-exported.

To reduce the ravaging impact of commercial cactus rustling, Arizona has made it illegal to remove 222 different plant species. With maximum penalties of $1000 and jail sentences of one year, Arizona has taken a small step to protect its native plants. Still, with only seven "cactus cops" to patrol the state, little can be done.

Worldwide, scientists use a variety of primates—monkeys and apes, such as chimpanzees—for research. Taken from their homeland in Africa, as many as five chimpanzees die for every one that enters a laboratory. Although annual primate imports have dropped off rapidly in the United States and elsewhere, some rare and endangered primates are still captured and sent to the United States (Figure 8.6). Some highly endangered mon-

Figure 8.5 *In developing countries, local residents sell birds and reptiles they catch in nearby forests to dealers who import them, sometimes illegally, to industrial nations. Many animals die along the way. These owls were for sale in Katmandu, Nepal.*

keys and apes, particularly chimpanzees, are seriously threatened by AIDS research.

In 1975, the United States banned the importation of all primates as pets but allowed continued importation for zoos and research. Because research animals often do not breed in captivity and because they have a high mortality, continual replenishment from wild populations is likely to continue.

Today, two-thirds of the world's 150 species of primates are threatened with extinction. In the past, researchers exploited many of these species with little concern for their declining population and their survival in the wild. Most people agree that research must continue under humane conditions but believe that it should not be carried out at the cost of extinction. One solution is captive breeding, in which animals are raised and bred

Figure 8.6 *Decline in primates imported by the United States.*

in captivity to supply researchers. Another solution is the use of laboratory tests, for example, cell cultures, that replace live-animal tests. A surprising number of substitutes are already available. Such actions could stop the flow of these animals from their natural habitat and prevent the extinction of many primates.

Pollution   Pollution plays a minor role in species extinction, but it is of growing concern as the human population expands and as global environmental problems worsen. Global warming, acid deposition, and ozone depletion, all caused by pollution, could bring about massive species extinction (Chapter 16). Global warming, for example, is believed to be responsible for a massive die-off of the world's coral reefs. Already besieged by sediment from onshore development, physical damage from ships, and poisoned by toxins and human sewage, numerous coral reefs are dying. Reefs are composed of a primitive animal species, the stony coral, that houses photosynthetic microorganisms known as dinoflagellates. These organisms provide nutrients for the coral in exchange for protection. When deprived of their symbiotic partners, though, coral bleach out, grow much slower, and may die. Recent studies indicate that the symbiotic relationship is being disturbed worldwide, possibly by a gradual rise in water temperatures caused by warming global temperatures. If the waters continue to warm, many coral reefs could be destroyed.

Pollution is also of concern on a local level. In the semiarid farmland of California's San Joaquin Valley, irrigation waters draining from farm fields were diverted to a series of specially built evaporation ponds in the Kesterson National Wildlife Refuge. In 1985, though, biologists found an unusually high incidence of abnormalities in chicks of waterfowl and wading birds at Kesterson. Many dead embryos and chicks without eyes, beaks, wings, and legs were discovered. In natural conditions, biologists expect a 1% deformity rate; at Kesterson, the rate in some species reached 42% (Figure 8.7). Adult birds were also affected. Crayfish, snakes, raccoons, and muskrat that once flourished in the rich biological community also vanished. As one journalist put it, "The Kesterson Refuge had become a place that killed the animals it was supposed to protect."

Researchers found that levels of the toxic metal selenium in the irrigation water, drained from the land to prevent waterlogging and salinization, were a thousand times higher than was considered safe. Selenium and other metals were concentrated by evaporation in the ponds, compelling the Water Resources Control Board to reclassify the refuge as a hazardous waste pit.

To solve the problem, state and federal officials plugged up the inlets, stopping the inflow of polluted irrigation water. Poisonous ponds were allowed to evaporate, then were filled with soil and reseeded, providing upland habitat for other species. Farmers reduced irrigation on thousands of hectares, changed crops, and recycled some of the water. An experimental treatment plant was operated for a short while but had so many problems it was shut down.

Kesterson illustrates how a solution to one problem—in this case, waterlogging and salinization—can lead to another. Unfortunately, it is only one of many examples of the threat that pollution poses to wildlife. Federal officials note that refuges in Utah, Wyoming, Texas, Nevada, Arizona, and other parts of California are also laden with toxic metals from farm drainage. Several of these refuges have confirmed the presence of deformities in chicks.

Ecological Factors That Contribute to Extinction

Because of peculiarities in behavior or limits to their reproductive capacity, some species are more vulnerable to extinction than others. Consider the passenger pigeon. At one time, the passenger pigeon inhabited the eastern half of the United States in flocks so large they darkened the sky. Probably the most abundant bird species to ever live, the passenger pigeon is now extinct, in large part because of widespread commercial hunting and habitat destruction. Between 1860 and 1890, countless pigeons were killed and shipped to cities for food. In 1878, the last colonial nesting site in Michigan was invaded by hunters. When the guns fell silent, over 1 billion birds had been killed. By that time, the passenger pigeon was nearly extinct; only about 2000 remained. Broken into flocks too small to be hunted economically, the birds were finally left alone. However, the number of birds dwindled year after year, until in 1914, the last one, which was held captive in the Cincinnati Zoo, died.

This tragic, oft-told story has a lesson: Some species have a critical population size below which survival may be impossible. The passenger pigeon population dropped below that level. The bird needed large colonies for successful social interaction and propagation of the species. Two thousand were simply not enough. Scientists know very little about the critical population size for many species. Overhunting, habitat destruction, and other activities discussed in this chapter can do irreparable damage before we even realize it.

Organisms can be categorized as specialists or generalists (Chapter 2). Specialists tend to become extinct more readily than generalists, who exploit more food sources and can live in diverse habitats.

Animal size also contributes to extinction. Larger animals, such as rhinos, are easier, and often more desirable, prey for hunters; they are also more likely to compete with humans for desirable resources, such as grazing land. Larger animals generally produce fewer offspring,

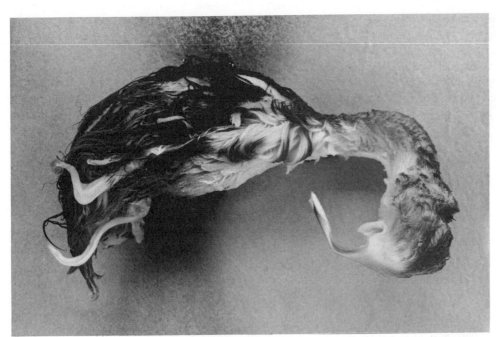

Figure 8.7 *Defective killdeer hatchling from the Tulare Basin, California. Selenium and other toxic metals have resulted in a number of embryonic defects. This embryo has a grossly deformed beak and legs.*

making it more difficult for reduced populations to recover. The California condor, for example, lays a single egg every other year. Young condors remain dependent on their parents for about a year but are not sexually mature until age six or seven. Combined, these factors give the condor little resiliency to adjust to pressures of human populations or natural disasters.

Another factor in extinction is the size of an organism's range. The smaller the range, the more prone the organism is to extinction. Finally, organisms often exhibit varying rates of tolerance to human presence. In North America, bluejays, elk, and coyotes, for example, coexist with humans quite nicely, but grizzly bears move out when humans move into an area. In the Amazon, woolly monkeys avoid roads and clearings.

Keystone Species According to one ecological theory, in some ecosystems the extinction of a critical species, called a **keystone species**, may lead to the collapse of an entire ecosystem. Consider some examples.

The gopher tortoise is believed to be a keystone species of the southeastern United States. This reptile digs long burrows in the sand, which it occupies for decades. Many other species also live in the burrow. For instance, the gray fox and diamondback rattlesnake visit the burrow regularly. The Florida mouse and the gopher frog live exclusively in the gopher tortoise burrow. Mice excavate tiny side tunnels for their own living quarters; opossums and indigo snakes frequent the burrows as well. The gopher tortoise is so important, in fact, that in areas where it has been eliminated, 37 species of invertebrates have also disappeared. The gopher tortoise's

role as a keystone species in Florida prompted wildlife officials to list the animal as a "species of special concern."

Keystone species are also seen in tropical rain forests. Ecologist John Terborgh and a team of researchers from Princeton University have been studying plant communities in Peru for over ten years. They recently found that three-fourths of the birds and mammals of the Amazonian rain forest rely on fruit as their major food source. But, most fruits are available only nine months of every year. During the remaining period, monkeys, peccaries, parrots, and toucans live on figs. If figs are removed from the ecosystem, it could very well collapse.

In the marine ecosystem of the U.S. Pacific is a keystone species known as the sea otter. Sea otters are voracious eaters, feeding on sea urchins, abalone, crabs, and mollusks that inhabit kelp beds. The sea otter helps to control sea urchin populations, which feed on kelp and seaweed. In locations where sea otters have been eliminated, kelp beds have also disappeared. As the sea otters repopulated the regions, urchin populations fell, and the kelp and seaweed recovered. The seaweed provides a habitat for fish and other fish-eating creatures, such as harbor seals. Thus, when the sea otter returns to an area, so do other species.

The existence of keystone species raises concern about habitat destruction. Many little-known animals may be keystone species important to human societies. For instance, bats are a keystone species because they pollinate flowers and disperse seeds of rain forest plants. Many tropical plant species are entirely dependent on bats, including many important food crops, such as avocados, bananas, cloves, cashews, dates, figs, and mangoes.

Ironically, many of the species that receive the lion's share of conservation funds are not keystone. Keystone species, in fact, tend to be unobtrusive, rare, or little-known plants and animals. Their survival, however, may ensure the survival of the more glamorous, well-known species.

8.3   Why Save Endangered Species?

Why should disappearing beetles, plants, or birds concern us? To many of us, the question hardly requires an answer. All species are the unique, irreplaceable products of millions of years of evolution.

Aesthetics

Wildlife expert Norman Myers wrote, "We can marvel at the colors of a butterfly, the grace of a giraffe, the power of an elephant, the delicate structure of a diatom. . . . Every time a species goes extinct, we are irreversibly impoverished." Wildlife and their habitat are a rich aesthetic resource. The sight of a female trumpeter swan gently nudging her offspring into the water for the first swim, the eerie cry of the loon at night, the lumbering grizzly bear on a distant grassy meadow, the sputtering of a pond full of ducks, the playful antics of sea otters, the graceful dive of the humpback whale—these sights enrich our lives in ways no economist could calculate.

Ethics

Preserving endangered plants and animals is an ethical issue as well. What right, critics ask, do we have to drive another species to extinction? Other organisms also have a right to live. Preserving life has become our duty because we have acquired the means to destroy the world. With that ominous power comes enormous responsibility.

Food, Pharmaceuticals, Scientific Information, and Marketable Products

Aesthetics and ethics are adequate answers for many of us, but others require a more compelling reason. What's in it for them? The answer is, "Plenty."

"From morning coffee to evening nightcap," writes Myers, "we benefit in our daily life-styles from the fellow species that share our One Earth home. Without knowing it, we utilize hundreds of products each day that owe their origin to wild animals and plants." Unbeknown to most of us, we depend on numerous plants and animals from grasslands, oceans, lakes, rivers, and forests, among others. From the oceans, we harvest millions of tons of fish; from rain forests, we harvest a wide assortment of edible fruits and nuts and numerous plant products such as rattan, which is used to make wicker furniture and baskets. Tropical rain forest plants also yield medicinal chemicals long used by native tribes. Interestingly, 80% of the world's people, according to the Worldwatch Institute, get most of their medicine from plants, especially those that live in rain forests. As noted in the last chapter, important plant and animal genes needed to improve domestic crops and livestock come from nature. Wildlife provides a wealth of enjoyment for hunters, anglers, and nature lovers, and wild species are a huge source of scientific information that provide valuable insights into our world.

The economic benefits of wild species are huge. By some estimates, for example, 50% of all prescription and nonprescription drugs are made with chemicals that came from wild plants. The commercial value of these drugs is over $20 billion per year in the United States and over $40 billion worldwide. The U.S. Department of Agriculture (USDA) estimates that each year genes bred into commercial crops yield over $1 billion worth of food. Similar gains can be documented for other major agricultural nations. Internationally, rattan is the most widely traded nontimber product, which is estimated to be worth $3 billion a year.

Wild species provide a wealth of scientific information that could be of great practical value. Myers, in fact, asserts: "Wild species rank among the most valuable raw materials with which society can meet the unknown challenges of the future." Many developments from wild species now loom on the horizon and may offer us considerable financial gains and perhaps even healthier lives. For example, the adhesive that barnacles use to adhere to rocks and ships may provide humankind with a new glue to cement fillings into teeth. Or a chemical derived from the skeletons of shrimps, crabs, and lobsters may help prevent fungal infections.

Free Services: Ecosystem Stability and Economic Benefit

Ecosystems provide us with many invaluable services free of charge. For example, natural ecosystems control pests. They produce oxygen and absorb carbon dioxide. They help maintain local climate. They replenish groundwater supplies, reduce flooding, and control erosion. They even help to purify water.

The economic benefit of these "free services" is astronomical. In Seattle, Washington, for example, one study showed that continued clear-cutting in the watershed of the Cedar River, from which the city acquires its drinking water, could increase sediment in the river.

Viewpoint 8.1

Playing God with Nature: Do We Have Any Other Choice?

Norman Myers
The author, a fellow at Oxford University and wildlife expert, has spent 25 years advising governments on park management. He has written The Sinking Ark *and* A Wealth of Wild Species.

I still hurt to recall the first time I went out on an elephant-culling foray. It's hard to forget the screams of terror, the fountains of blood, and the sudden silence, broken only by the clinical talk of the technicians and scientists. In South Africa's Kruger Park, where elephant culling is a fact of life, officials work under the seemingly arrogant notion that only humans can keep a park wild in this human-dominated world. I spent months agonizing that there must be a better way to deal with nature.

Eventually, I came to the conclusion that Kruger's approach is the right one. It is our only choice, but we must do it with great caution. Despite my realization that such management is necessary, it sticks in my throat.

When I first went to Africa almost 30 years ago, human com-munities tended to be islands of settlement among a sea of wildlands. Today, it is the wildlands that are islands. Africa is bursting at the seams with people; population pressures threaten parks from Ethiopia to Zimbabwe, from Kenya to Senegal, making intensive management a bitter but necessary reality. The huge Kruger Park, two and a half times the size of Yellowstone National Park, is no exception. A 950-kilometer (600-mile) fence surrounds the park, turning it into an island in an otherwise crowded land. That's where the trouble begins. Biologists theorize that ecological islands tend to have fewer species than a similar-sized portion of contiguous habitat. They are probably more susceptible to environmental change.

In fenced-off islands such as Kruger Park, wardens and scientists have concluded that they cannot allow the rich diversity of species to dwindle. To ensure diversity, they have chosen to manage the park to the hilt, which is where this essay began.

The thought of the flesh of wild elephants ending up in cans in a supermarket may be repellent. But that's what is happening, and for good reason—to prevent overpopulation and habitat destruction within the park. The alternative to control can be dangerous to parks, as I saw in Kenya. During the early 1970s, the country was hit by drought. The elephants in Tsavo Park, already suffering from overcrowding, started to die in the thousands as food supplies shrank. People were also starving. Yet, park officials refused to allow anyone to touch the meat; they were aghast at the idea that the park's wildlife might be used to meet human needs.

Several years ago I returned to Tsavo and found that the local people were trying to acquire sections of the park for cultivation and grazing. One group of elders told me that their overall aim was to have Tsavo abolished altogether. "That park is an insult to us," one of them said. "We have a score to settle."

For their own survival, Africa's parks must take a lesson from Tsavo. Desperate, hungry people do not take pleasure in the pristine wilderness; they need food in their stomachs. Park managers must realize this. In protecting wildlife from the encroachment of humanity, they can serve dual needs. Culling herds to eliminate overpopulation preserves habitat and ensures a sustainable ecosystem. In the process, they can provide meat, mountains of it for neighboring peoples, reducing the animosity and making them more aware of the benefits of wildlife.

But, let us tread the path toward a human-dominated future very carefully. Once we have accepted such total management of a vast area, how long will it be before we go to extremes? Will we choose to eliminate species that aren't of any direct value to us? Will we begin to seed the savanna with "improved" grasses to increase yield, destroying native species uniquely adapted to the environment?

We established our parks as arks against a rising tide of humanity. Now, we discover that it isn't enough to play Noah; willy-nilly, we are playing God. Let's do it carefully.

To purify water, the city would have to build a $120 million water treatment plant that would cost millions of dollars a year to operate. The service is provided free now by a relatively intact forest ecosystem. Collectively, wetlands, forests, grasslands, and other natural systems provide billions of dollars worth of services, most of which we take for granted.

Many wildlands can sustain significant harvests of natural resources while providing environmental services, for example, providing clean water and erosion control, without threatening species diversity. Mangrove swamps, for instance, are a nursery ground for commercially important fish and shrimp but also reduce coastal flood and storm damage and help filter sediment from waterways. If properly managed, mangroves can provide timber for construction, pulpwood for paper, and charcoal for energy. They can also provide food for livestock, shellfish for human consumption, and a number of other products.

In Matang, Malaysia, well-managed mangroves produce fish and wood products worth more than $1000 per hectare ($400 per acre) per year. One job is provided for every 3 hectares (7.5 acres). If all of Southeast Asia's mangroves were managed sustainably, they would provide about $25 billion annually to the economy and create about 8 million new jobs.

Although a species lost here and there may be of little consequence for ecosystem stability unless it is a keystone species, in the long term the cumulative effect of such losses could threaten our survival. Numerous examples cited in this text show that the continued destruction of nature could place human populations on the endangered species list.

8.4 How Can We Save Endangered Species?

Most efforts to save species over the past two and a half decades involved laws and regulations that seek to protect the most endangered species. A good example is the Endangered Species Act.

The Endangered Species Act

In 1973, in response to the plight of wildlife and plants in the United States and abroad, Congress passed the **Endangered Species Act**. This act requires the U.S. Fish and Wildlife Service to list endangered and threatened species in the country. Currently, endangered and threatened species are classified on the basis of biological criteria, primarily their population size and rate of population decrease. However, some people want eco-

nomic considerations to be included in the listing process. If they are successful, an endangered species will not be listed if its protection costs too much or costs too many jobs.

The Endangered Species Act also requires the establishment of federal programs to protect the habitat of listed species, and provides money to purchase this habitat.

Finally, the law enables the United States to help other nations protect their endangered and threatened species by banning the importation of these species and by lending technical assistance.

Protection formally begins with the listing of an endangered or threatened species. In November 1992, 278 animals and 298 plants were designated as endangered in the United States. An additional 100 animals and 72 plants were listed as threatened. At least 3600 species currently await listing. In the time it has taken officials to consider their case, several species have become extinct.

According to the law, all federally funded or approved projects that could have an impact on an endangered species must be reviewed by appropriate agencies. These agencies can then deny permits in order to protect the species or ask for modifications that remove the danger.

Since the Endangered Species Act went into effect, thousands of projects have been reviewed for their impact on listed species. In most cases, differences have been amicably worked out. The most renowned exception was the case of the snail darter and the Tennessee Valley Authority's (TVA) Tellico Dam on the Little Tennessee River (Figure 8.8). Problems began in 1975 when a federal court order halted construction of a multimillion-dollar dam, already 90% completed, which would flood the fish's only breeding habitat. The order was upheld in the U.S. Supreme Court, and Congress established a committee to review a request for an exemption to the law. In 1979, the committee refused to grant an exemption, saying that the project was of questionable merit. The TVA applied more pressure on Congress, however, and later that year, Congress authorized the completion of the dam. The snail darter was transplanted to several neighboring streams, where it is doing well. Additional populations were discovered in several nearby streams.

Another more recent example is the battle in Oregon and Washington between the logging industry and environmental groups over protection of old-growth forests and the spotted owl, which was listed as a threatened species in 1989 (Chapter 9).

The Endangered Species Act is one of the toughest and most successful environmental laws in the United States. The success of the act, says Bob Davison, a National Wildlife Federation biologist, "is that there are

Figure 8.8   *Measuring only 8 centimeters (3 inches), the snail darter created a controversy between environmentalists and industry. The impending destruction of the snail darter by the TVA's Tellico Dam brought the multimillion-dollar project to a standstill. After years of debate, Congress ordered the dam to be completed.*

Figure 8.9   *Inside a genetic repository laboratory. The lab is sponsored by the U.S. Department of Agriculture to stock seeds of hundreds of thousands of plants from all over the world for future study.*

species around today that would not have survived if the law had not forced agencies to consider the impacts of what they're doing while allowing development to proceed. . . . To a large extent, the law has succeeded in continually juggling those two competing interests."

The Endangered Species Act allows listing of species in other countries as well and forbids their importation. To protect these endangered species, governments throughout the world have joined the United States in an unprecedented legal effort to stop the illegal trade in rare and endangered species. But, in many cases, inadequate funding makes enforcement a joke. Inspectors can be paid off by illegal traffickers of endangered species. Government agents patrol only a small fraction of poachers' range, and the courts have routinely been lenient toward poachers.

Zoos Lend a Hand

Zoos are an important component of the global effort to save species from extinction. For many years, the world's zoos have been breeding endangered species in captivity to prevent them from disappearing from the face of the Earth and to establish populations that could be released into suitable habitat. One recent project was launched by zoos in the mainland United States to rescue three rare species of birds threatened by tree snakes on the island of Guam. Numerous zoos are participating in special breeding programs to rebuild wild populations of the Guam rail, the Micronesian kingfisher, and Mariana crow. As noted in the chapter introduction, the survival of the California condor lies in the hands of biologists and personnel at the San Diego Zoo.

Germ Plasm Repositories

Because much of the habitat of many organisms is bound to be destroyed by bulldozers and chain saws, biologists have been searching through forests and fields to gather seeds for cold storage in genetic repositories (Figure 8.9). Here they can be preserved for future study and possible use. The U.S. Department of Agriculture currently supports the National Germ Plasm System, which has over 290,000 plants in "stock" and plans to add 12,000 to 15,000 per year in the next decade.

In 1985, the developing nations belonging to FAO voted to establish a worldwide system of storing seeds, cuttings, and roots from native plants. This system was created to thwart "genetic imperialism" by the developed nations, which collect plants and seeds from developing nations and extract genes and important medicinal drugs, sometimes reaping huge profits.

However important genetic repositories are to the immediate goal of protecting the world's species, this strategy has some major drawbacks. First, despite storage at low temperature and humidity, many seeds decompose and must be replaced. Others undergo mutations when stored for long periods and are no longer useful. Finally, storage systems do not work for potatoes, fruit trees, and a variety of other plants.

Protecting Endangered Ecosystems: A Global Effort

Species-by-species protection afforded by the Endangered Species Act, zoo programs, and other similar measures now in place throughout the world is important. Unfortunately, efforts to protect endangered species are only stopgap or Band-Aid measures, an approach that typifies our response to many social and environmental problems. What's wrong with them? Frankly, they don't address the long-term challenge of preventing species from becoming endangered in the first place. Moreover, they can't keep pace with runaway population growth, burgeoning agricultural development and urbanization, and global pollution. Additional measures are needed. Accordingly, some U.S. wildlife advocates are calling for the passage of an **Endangered Ecosystems Act**, which would protect endangered ecosystems, such as old-growth forests. Worldwide efforts are also underway to set aside regions of biodiversity. (See Models of Global Sustainability 8.2 for some economic solutions.)

The world map in Figure 8.10 shows numerous high-priority areas in dire need of protection. Those who advocate protection of ecosystems argue that maintaining these regions represents the best economic investment in wildlife protection we can make. As you can see

from the map, most of these regions are located in developing countries who lack financial resources needed to protect them. The wealthy, "biologically poor" developed nations, however, are the ones that will probably benefit the most from preserving these genetic resources. For this reason, many argue that the rich nations should share the cost of preserving the tropics. A 0.1% tax on internationally traded oil, which would net $1 billion a year, would go a long way toward establishing and maintaining large reserves in high-priority areas.

Many countries have taken matters into their own hands. The government of Colombia, for example, has turned over 18 million hectares (45 million acres) of rain forest, an area approximately three-fourths the size of Great Britain, to tribal peoples who have lived on the land sustainably for hundreds of years. The lands will be held in common by the people and cannot be sold unless three-fourths of the adults in the tribal community agree.

The government is also purchasing mountain land where settlers of European descent have lived for almost 200 years, and returning it to indigenous people. Combined with rain forests, some 26 million hectares (65 million acres) are now in Indian hands. The success of this program prompted other developing countries to follow suit, among them Bolivia and Venezuela.

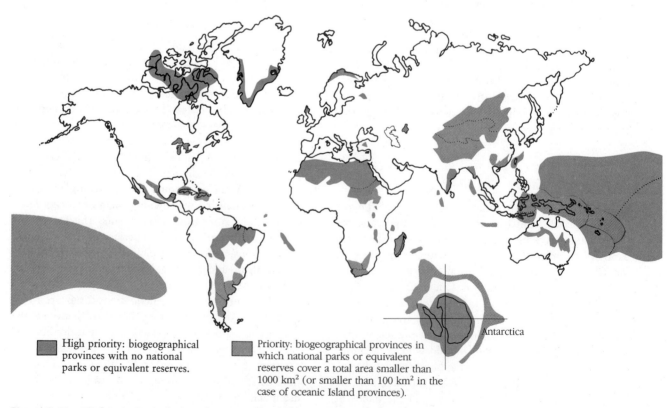

Figure 8.10   *High priority (color) and priority (gray) regions of the world in need of protection to preserve important plants, animals, and other species.*

Models of Global Sustainability 8.2

Debt-for-Nature Swaps

In 1989, the U.N. General Assembly drafted a resolution asserting that poverty and environmental degradation are closely related issues in the developing world and must be dealt with as such. Not surprisingly, developed nations can play an important role in solving these problems.

In 1984, Dr. Thomas E. Lovejoy, then with the World Wildlife Federation, proposed a radical new idea called **debt-for-nature swaps**. It linked environmental protection with reductions in the monstrous debt of developing nations owed to the industrial world. Lovejoy reasoned that by reducing debt countries could free up money badly needed to solve other problems, among them illiteracy, health, and sustainable development.

The money currently used to repay the enormous debt of the developing nations is often derived from the export of natural resources, such as cash crops, timber, minerals, and petroleum. Caught in a vicious cycle, most developing nations are depleting their natural resources just to keep up with interest payments.

In a debt-for-nature swap, a developing country's foreign debt can be exchanged for a commitment to invest some of the debt in local conservation programs. A debt-for-nature swap involves three players: a country in debt, a nation or bank to which the debt is owed, and an entity, usually an international nongovernmental organization (NGO), that purchases the country's debt at a significant discount. In a typical debt-for-nature swap, Country A, which owes a bank or a government a large sum of money, say $10 million, sells its debt to an NGO at a considerable discount, for example, 10 to 20 cents on the dollar. In this scenario, Country A would sell its debt to the NGO, which would pay the creditor $1 to $2 million, and the debt would be forgiven. In exchange for purchasing the debt, the debtor nation would enter into an agreement with the NGO to invest a certain amount of money in wildlife parks or other conservation measures.

Three international environmental NGOs based in Washington, D.C., have played a major role in promoting these swaps: The Nature Conservancy, World Wildlife Fund, and Conservation International. In 1987, Bolivia initiated the world's first debt-for-nature swap. The agreement between Bolivia and Conservation International resulted in a commitment on the part of the Bolivian government to protect an area of biological importance in trade for a small debt reduction. Conservation International purchased $650,000 of debt at an 85% discount. In exchange for the debt, the Bolivian government agreed to provide $250,000 in local currency to help manage the newly established Beni Biosphere Reserve, to create several protected areas nearby, and to ensure that the forest in a vast tract of adjacent land would be managed sustainably.

Implementation of the plan was postponed for two years because of internal opposition resulting from a diplomat's casual remark about Bolivia relinquishing territory and economic rights to foreign interests. In addition, government funds were delayed because of Bolivia's tight financial resources.

Since the first debt-for-nature swaps in Bolivia, similar deals have been struck in Ecuador, Costa Rica, and the Philippines; other developing countries, especially Peru, Brazil, and Mexico, are showing growing interest.

Although debt-for-nature swaps help debtor nations reduce debt payments and allow them to convert foreign debt into local obligations, such deals are not without problems. First, there is concern for the sovereignty of debtor nations. Understandably, many countries do not want outsiders dictating their environmental and monetary policy priorities. Second, some governments and peoples are particularly sensitive to foreign investment in their countries.

Creditors also experience benefits and costs. Commercial banks, for example, receive immediate cash payment for selling a debt. This is money that they may have never seen had the debtor nations foreclosed on their obligations. Some governments give tax credits to the banks for such swaps, helping to reduce their losses. On the downside, however, debtor nations may withhold payments on their debts in hopes of striking a more favorable agreement.

NGOs and environmentalists benefit because they help to reduce debt and poverty and to protect the environment. Some swaps increase NGO funds available for international environmental protection programs.

Debt-for-nature swaps are not a panacea. They represent only one way of protecting biodiversity, but they are a step in the right direction. Existing programs have sparked a great deal of enthusiasm and hope, and infused conservation and environmental efforts with badly needed money. To date, though, the greatest benefit of debt-for-nature swaps may be that they focus attention on the burden that debt places on developing countries and the world's devastating environmental problems.

In 1991, for example, Venezuela gave permanent title to a region of forest about the size of Austria to a native tribe.

Another way of protecting vital habitat in developing nations is by establishing an **extractive reserve**, land set aside for native people to use on a sustainable basis (Figure 8.11). Huge tracts of tropical rain forest, for example, are being preserved for sustainable harvesting of rubber, nuts, fruits, and other products. While providing a sustainable income, the reserves also help to protect native species. (Models of Global Sustainability 9.1 provides more information on extractive reserves.)

Interestingly, several studies show that income from extractive reserves actually exceeds revenues generated from agriculture, grazing, and lumber on the same land. Officials in the African nation of Niger recently halted a project and decided to let the rain forest regrow *after* discovering that the economic and social benefits of a sustainable harvest from the intact forest outstripped those expected from a proposed plantation.

The world's first extractive reserve was established in the Amazon, and since that time, 13 additional reserves have been established, protecting 3 million hectares (7.5 million acres) of tropical rain forest. Proponents eventually hope to establish reserves on 25% of the Amazon's rain forest, about 100 million hectares (250 million acres). Encouraged by Brazil's success, several other developing countries, among them Peru and Guatemala, have established forest and lake reserves.

A word of caution is in order, however. Sustainable harvest of these forests could provide income for people and help to maintain the complex communities that live in them. But, intensive exploitation, such as that proposed by some, could have devastating effects. The secret lies in achieving the proper balance between use and preservation.

Buffer Zones and Wildlife Corridors

The success of protected areas can be enhanced by establishing buffer zones around them. A **buffer zone** is a region in which limited human activity is allowed, for example, timber cutting or cattle grazing.

Another potentially exciting idea in wildlife protection is the **wildlife corridor**, a strip of land that connects isolated patches of habitat set aside to protect species. Ecologists refer to isolated patches of protected habitat as *ecological islands*. Existing amid crops, cities, pastures, towns, and mines, these protected areas have been only marginally successful in protecting biodiversity, much to the surprise of many. American ecologist William Newmark studied the loss of mammal species in national parks and reported an alarming decline in the number of species in all but the very largest parks. Bryce Canyon National Park, which is one of the nation's smallest parks, lost 36% of its species. Yosemite, which

Figure 8.11 *Brazilian rubber tappers harvest rubber from the forest on a sustainable basis at a large extractive reserve. The bark of the trees is cut and latex is collected in buckets. This activity has little effect on biodiversity in tropical rain forests.*

is nearly 20 times larger than Bryce, lost 23% of its species.

The decline of species in ecological islands occurs because the areas often do not contain enough habitat for all of the species that live within them. Wide-ranging animals are especially hard hit. Small islands of habitat also reduce populations, sometimes below levels needed for successful reproduction. Tiny fragments of habitat often lead to interbreeding among members of a small population resulting in inferior offspring, which are less fit.

Wildlife corridors allow species to migrate from one habitat to another, breeding with members of other subpopulations. This, in turn, increases genetic diversity within a species, ensuring a better chance of survival. Connections between islands also open up new habitats. Food shortages encountered in one region can be offset by migrating to another area.

The state of Florida and The Nature Conservancy are currently developing a series of wildlife corridors to protect the endangered Everglades panther, which has been relegated to a few patches of land widely dispersed throughout the state. If efforts to connect the islands are successful, the panther may someday roam much more freely in search of food and mates, living in relative harmony with the human population, which has transformed the state into a patchwork of farms, roads, cities, and towns.

Improving Wildlife Management

Besides establishing buffer zones and connecting corridors, efforts are needed to improve the way we manage fish and wildlife and the ecosystems they inhabit. Most important are efforts to regulate harvests of fish and other commercially important species to avoid depletion. Steps are also needed to restore damaged wildlife habitat. Ideas on proper management of forests and rangelands are presented in the next chapter.

Addressing the Root Causes: The Sustainable Strategy

Confronting the underlying causes of wildlife extinction, among them habitat destruction, overharvesting, and pollution, requires systemic changes in the way we live and conduct business. By implementing the biological principles of sustainability, we can eliminate many of the pressures that endanger our planet and the many species that live on it with us.

The sustainable strategy calls for a worldwide effort to use only the resources we need and to use them with extreme efficiency. Such an effort could greatly reduce habitat destruction caused by timber harvesting, mining, and pollution. Recycling similarly reduces our need for natural resources and landfills, and greatly reduces the amount of pollution society produces. Turning to renewable energy resources, such as sunlight and wind, would have an equally beneficial impact on wild species and natural habitats. It could help to reduce some of our most serious pollution problems, such as global warming, acid deposition, and oil spills. The restoration of natural habitats is also needed to save wild species and to protect the free services nature provides. Finally, and most importantly, serious efforts are needed to stop the spread of human populations through better growth management and by reductions in the human population growth rate.

Personal Solutions

As in all environmental issues, individuals' actions can make a difference. By practicing the principles of sustainability, you can become an important part of the solution. Use only what you need and use all resources efficiently. Recycle and buy recycled products. Use renewable resources and support government programs aimed at increasing their use. Help to restore ecosystems and support groups that take an active role in these efforts. And, most importantly, limit your family size and support private and government efforts throughout the world to provide family planning services and other factors that could help slow the growth of the human population.

You can also help by educating others about protection of the world's biota. You can join groups and spread the word through educational campaigns, lobbying, television ads, posters, books, and pamphlets. Support organizations and politicians who address the problems.

Joining wildlife groups is one of the best ways to learn from dedicated experts with well-developed plans for wildlife protection. (See the *Environmental Action Guide* for a list of organizations and their addresses.) Some organizations, such as The Nature Conservancy and the Trust for Public Lands, purchase habitat for rare and endangered species. Others, such as the National Wildlife Federation, Sierra Club, Audubon Society, and Wilderness Society, concentrate much of their efforts in the legislative arena to promote sound environmental policy.

In closing, progress in protecting the millions of species that inhabit the planet and provide us so much has been considerable. On balance, however, the prospects for plants, animals, and other species are not bright. For most species, the situation is growing worse. The expanding human population and our growing demand for resources threaten to destroy hundreds of thousands of species in our lifetimes. The time for action is now. The next decade is extremely crucial, not only for the species that share this planet with us but for ourselves.

What is civilized in us is not opera or literature but a compassion for all living things and a willingness to do more than simply care.

—*Daniel D. Chiras*

Critical Thinking Exercise Solution

Before joining your neighbor, you should take a look at the big picture. First, you might want to know if poisoning coyotes would decrease wildlife populations on your ranch and affect your future income. Would it increase the number of rabbits, on which coyotes prey? Would this decrease the amount of grass your sheep can eat? Moreover, do you really need to kill all of the coyotes to make the sheep safe? Are there other ways to control coyote predation that are less damaging to the environment and more economically sound?

Another important critical thinking strategy requires examining the basic premise of your neighbor—that is, that coyotes are killing a large number of his sheep. Can you accept his assertion? Is he exaggerating? Is he biased by a hatred for coyotes? What is the true extent of the damage? Is he a reliable source? Do his losses merit the economic and environmental costs of the control measures he is proposing?

Summary

8.1 The Vanishing Species

■ By various estimates, 40 to 100 species become extinct each day. The natural extinction rate is 1 species per 1000 years. Many millions of species are now **threatened** or **endangered**.

8.2 What Causes Extinction?

■ Plant and animal extinction results from many activities. The top five, in decreasing order of importance, are: (1) habitat alteration, (2) commercial hunting and harvesting, (3) hunting for sport and subsistence hunting, (4) introduction of alien species, and (5) pest and predator control.

■ **Habitat destruction**, the single most important factor, is especially severe in tropical rain forests, coral reefs, and wetlands.

■ Numerous ecological factors also play a role in extinction. Some species require a **critical population size** for successful reproduction. If a population is reduced below this level, survival may be impossible.

■ Additional ecological factors include the degree of specialization, an organism's position in the food chain, size of an organism's range, reproductive rate, and tolerance to human presence.

■ According to ecological theory, some species, called **keystone species**, may be critical to the well-being of many others. Their loss may have many ecological repercussions.

8.3 Why Save Endangered Species?

■ Many reasons exist for protecting endangered species.

■ All species are a unique and irreplaceable product of evolution.

■ Species improve the quality of our lives, providing many free services vital to economic and ecological well-being.

■ Plants and animals are also a source of considerable food and wealth.

■ Wild species are of great interest to science and may provide new cures and new products.

8.4 How Can We Save Endangered Species?

■ Many stopgap measures are helping to protect threatened and endangered species, such as the Endangered Species Act, captive breeding programs, germ plasm repositories, biodiversity reserves, buffer zones, connecting corridors, and improvements in wildlife management.

■ Preventive strategies are also needed, notably strategies that put into practice the biological principles of sustainability: conservation, recycling, renewable resource use, restoration, and population control.

Discussion and Critical Thinking Questions

1. Debate the statement: "Extinction is a natural process. Animals and plants become extinct whether or not humans are present. Therefore, we have little to be concerned about."

2. List and describe the factors that contribute directly to animal and plant extinction. Which ones are the most important?

3. Trophy hunters generally try to shoot the dominant males in a population. Natural predators, on the other hand, remove the sick, weak, and aged members of the population. Using your knowledge of ecology and evolution, how do trophy hunting and natural predation differ in their effects on the prey population?

4. Why are islands particularly susceptible to introduced species?

5. Discuss the "ecological" factors that contribute to species extinction.

6. Describe the concept *keystone species*. What are its implications for the modern conservation movement?

7. You are placed in a high government position and must convince your fellow executives of the importance of preserving other species. How would you do this? Outline a general plan for preserving species diversity.

Suggested Readings

Achiron, M. (1988). Making Wildlife Pay Its Way. *International Wildlife* 18 (5): 46–51. Elaboration on Norman Myers's Viewpoint (in this chapter).

Domalain, J. (1977). Confessions of an Animal Trafficker. *Natural History* 87 (5): 54–57. Startling account of illegal practices in the animal trade.

Harris, T. (1991). *Death in the Marsh*. Washington, D.C: Island Press. Tells the story of the Kesterson National Wildlife Refuge.

Hudson, W. E., ed. (1991). *Landscape Linkages and Biodiversity*. Washington, D.C.: Island Press. Discusses innovative conservation strategies to protect biodiversity.

Kohm, K., ed. (1990). *Balancing on the Brink of Extinction. The Endangered Species Act and Lessons for the Future*. Washington, D.C.: Island Press. Important collection of writings.

Laycock, G. (1966). *The Alien Animals*. Garden City, NJ: Natural History Press. A classic account of the troubles created by species introduction.

Owens, M. and Owens, D. (1985). *Cry of the Kalahari*. New York: Houghton Mifflin. Important information on behavior, ecology, and conservation.

Plotkin, M. and Famolare, eds. (1992). *Sustainable Harvest and Marketing of Rain Forest Products*. Washington, D.C.: Island Press. Outlines ways of sustainably harvesting tropical rain forest products while protecting native species.

Reisner, M. (1991). *Game Wars: Undercover Pursuit of Game Poachers*. New York: Viking. Riveting account of the poaching in the United States and efforts to stop it.

Ryan, J. C. (1992). Conserving Biological Diversity. In *State of the World*. New York: Norton. Describes many ways to protect biological diversity.

Sunquist, F. (1988). Zeroing in on Keystone Species. *International Wildlife* 18 (5): 18–23. Good reference on this new concept.

Tudge, C. (1992). *Last Animals at the Zoo: How Mass Extinction Can Be Stopped*. Washington, D.C.: Island Press. Describes how captive breeding programs and restoration of natural habitat can be used to save endangered animals from extinction.

Wilcove, D. (1990). Empty Skies. *The Nature Conservancy Magazine* 40 (1): 4–13. Excellent overview of factors causing the decline of songbirds in the United States.

Wilson, E. O. (1989). Threats to Biodiversity. *Scientific American* 261 (3): 108–16. Graphically illustrated discussion of habitat destruction here and abroad.

Chapter 9

Grasslands, Forests, and Wilderness: Sustainable Management Strategies

Our duty to the whole, including the unborn generations, bids us restrain an unprincipled present-day minority from wasting the heritage of these unborn generations.

—*Theodore Roosevelt*

Critical Thinking Exercise

One of the most dramatic changes on the planet in the last two decades has been the steady march of the world's largest desert, the Sahara. In the 1970s and 1980s, researchers estimated that the desert spread southward at a rate of 5 kilometers (3 miles) per year, which they attributed to drought, overgrazing, and agricultural land abuse in semiarid grasslands bordering the desert. This projection, however, was based primarily on measurements in a few locations, assuming that they represented the entire continent.

Using satellite observations of vegetation, scientists recently found that the Sahara has advanced and retreated in the past 11 years in response to rainfall. From 1980 to 1984, for example, the desert's southern boundary moved 240 kilometers south. Between 1984 and 1985, it moved north by 110 kilometers. In 1987 and 1988, it shifted northward again by 155 kilometers. In 1989 and 1990, however, the desert boundary shifted southward 77 kilometers.

Although the southern border of the desert in 1990 was 130 kilometers farther south than in 1980, some researchers believe that the shift does not reflect a long-term trend but rather differences in year-to-year rainfall.

Critics of global warming use these statistics to argue that desertification caused by climatic shift is not occurring. They say that the shift of the desert is a natural phenomenon. How would you answer this claim? What critical thinking rules can you use?

❧❧❧   ❧❧❧   ❧❧❧

In the early 1970s, scientists pored over satellite photographs of the drought-stricken African Sahel, a band of semiarid land that borders the southern Sahara. One of them noticed an unusually green patch of land amid the desert. On the surface, it appeared much like an oasis. Curious to find out the reason, Norman MacLeod, an American agronomist, flew to the site. There, surrounded by newly formed desert, was a privately owned ranch of 100,000 hectares (250,000 acres). Its grasses grew rich and thick even though vegetation in the surrounding fields had long since died, exposing the sandy soil. Why?

Stretching around the perimeter of the ranch was a fence that held out the cattle of the nomadic tribes, which had overgrazed the surrounding communal property for decades. The ranch was divided into five sections, where a rigidly controlled number of cattle were grazed once every five years—a regime sustainable in the semiarid land.

The photographs of the drought-stricken Sahel show us not only the devastation wrought by mismanagement but also the possibilities created by sustainable management. This chapter examines the state of the world's grasslands and forests—sources of food and fiber and numerous ecological services—and offers ideas on sustainable management. It also looks at wilderness areas, vital sources of recreation, which, if managed sustainably, can also become important means of preserving wild species.

9.1 A Tragedy of the Commons

The desert surrounding the ranch described in the chapter introduction was at one time lush, communal grazing land. Tribesmen of the Sahel grazed their livestock on the land with abandon, eventually causing its decimation. As far back as ancient Greece, Aristotle recognized that property shared freely by many people often deteriorates severely. Early civilizations, for example, clearcut communal forests and overgrazed their cattle on grasslands. History shows that these civilizations paid dearly for their disregard. The skeletons of buildings from ancient cities stand out in deserts that were once rich forest and grassland ecosystems of the Fertile Crescent. Much of Iran and Iraq, now barren desert, once supported cattle, farms, and rich forests (Figure 9.1). Historians believe that the decline of Greece and Rome may be partly attributed to the rampant misuse of their lands.

Economists have debated the fate of other common resources, such as air, water, and land, for decades. It was not until 1968, however, that professor Garrett Hardin exposed the cycle of destruction in a paper entitled "The Tragedy of the Commons."

In England, Hardin noted, cattle growers grazed their livestock freely on fields called **commons**, which fell into ruin as the users became caught in a blind cycle of self-fulfillment. Families found that they could enhance their personal wealth by increasing their herd size. Each additional cow meant more income at little, if any, cost because farmers did not have to buy new land or feed; the commons provided them.

Hardin argues that the cattle growers were rewarded for doing wrong. Although they realized that increasing their herds would lead to overgrazing and deterioration of the pasture, they recognized that the negative effects would be shared by the entire community. Thus, each herdsman arrived at the same conclusion: he had more to gain than to lose by expanding his herd. This shortsighted thinking resulted in a spiraling decay of the commons. As each pursued what was best for himself, the whole was pushed toward disaster. "Freedom in a commons brings ruin to all," wrote Hardin.

The logic that compels people to abuse communal holdings has been with humankind as long as common property. Today, however, the process has reached epic

Figure 9.1 *Aerial view of barren hillsides of Iran. Once covered by rich forests, these lands have been ruined by centuries of abuse, starting with deforestation.*

proportions. The overgrazed communal property of the Sahel is one of the most recent signs. This tragedy has been worsened by unsustainable government policies. In the 1960s, for example, nomadic people obtained loans to drill for water. With a steady supply of water, previously nomadic tribes no longer migrated south during the dry season. The land around the wells and human settlements deteriorated as herds exceeded the carrying capacity.

The tragedy of the commons provides partial explanation for many of the world's problems, even the rapid increase in human population. In rich, industrial nations, as well as in poor, developing countries, fertility is largely driven by personal desire. In India, children provide security in old age or improve the social status for women. In Mexico, childbearing may be motivated by male pride. In the United States, it may be personal satisfaction. Individual self-serving acts of procreation create a tragedy that befalls everyone.

Hardin's analysis of the tragedy of the commons, while important, gives the impression that common resources are alone in their mismanagement. In many places, privately held resources are also recipients of pathetic abuse. Erroneous frontier notions of the Earth as an unlimited supply of resources, ignorance of the impacts of human actions, and a short-term view of economics dictate unsustainable management strategies on a wide variety of privately held resources (Chapter 21).

Today, virtually all lands are gripped by the tragedy of exploitation. Short-term exploitation may have been permissible at one time, when the human population was small in relation to the Earth's resources. Today,

such actions are intolerable. Too many people depend on the biosphere for food, water, and other resources, and the cumulative effect of many small insults is staggering. Local problems have expanded to affect entire regions. Regional issues are spreading to create a global nightmare.

9.2 Rangelands and Range Management: Protecting Grasslands

Cattle and other livestock range over half the Earth's surface, munching on grasses. In much of the world, domesticated animals outnumber humans three to one.

Grasslands on which many of the world's livestock depend are known as **rangelands**. They are a vital component of global food, leather, and wool production. When properly managed, grasslands and the livestock they support can provide useful products indefinitely. Regrettably, the vast majority of the world's grasslands on which livestock graze are mistreated like many other renewable resources and could fall into ruin. This section examines the problems facing grasslands from livestock production and shows how they can be managed sustainably.

Rangeland Deterioration

In the United States, over 90 million hectares (225 million acres) of grassland, much of it public land, have been turned to desert in the last 200 years. This is an area equivalent to one and a half times the size of Texas! The Navajo Indians, for example, live on a 6-million-hectare (15-million-acre) reservation in Arizona, New Mexico, and Utah. Theirs is a sun-parched land, dusty and dry. To feed and clothe their people, the Indians have gradually increased the size of their sheep herds. Today, the herd size exceeds the carrying capacity by at least four times. Baked by the hot summer sun and swept by fierce winter winds, the overgrazed reservation is quickly becoming an arid dust bowl, further adding to problems of unemployment.

Northwest of Albuquerque, New Mexico, is another tract of parched desert land, the Rio Puerco Basin. In the late 1800s, its lush grasslands supported huge cattle herds. A century later, however, the land had deteriorated under the strain of overgrazing. Erosion formed gullies that widen by 15 meters (50 feet) a year. Wind and rain erode the soil five to ten times faster than it can be replenished naturally.

According to the **Bureau of Land Management (BLM)**, which manages most of the federally owned rangeland, 62% of the public rangeland is in fair to poor shape. On rangeland classified as fair, vegetation is 50% to 75% below what the land is capable of supporting. Poor rangeland supports 75% or less. Studies show that private rangeland is in even worse condition (Figure 9.2). The U.S. Soil Conservation Service estimates that 75%

of all U.S. rangeland and pastureland needs better management.

Worldwide, the picture is much the same. The U.N. Environment Programme (UNEP) estimates that nearly 75% of the world's rangeland is at least moderately desertified with a reduction in carrying capacity of at least 25%. The major culprit in this deterioration is overgrazing. The results are severe erosion, desertification, a drop in the water table, loss of wildlife, and invasion of weeds.

Forests are also damaged by livestock. In some regions, branches are cut from trees for animal food, and entire stands are removed to make room for cattle ranches. In India, for instance, grasslands do not provide enough food for the nation's 196 million cattle, which forces people to turn to state forests and other lands. The connection between livestock and forests is elaborated in the next section.

Sustainable Range Management

The prospering island within the Sahel, discussed in the chapter introduction, provides living proof that rangeland can remain productive despite aridity. In fact, proper range management can benefit the land in many

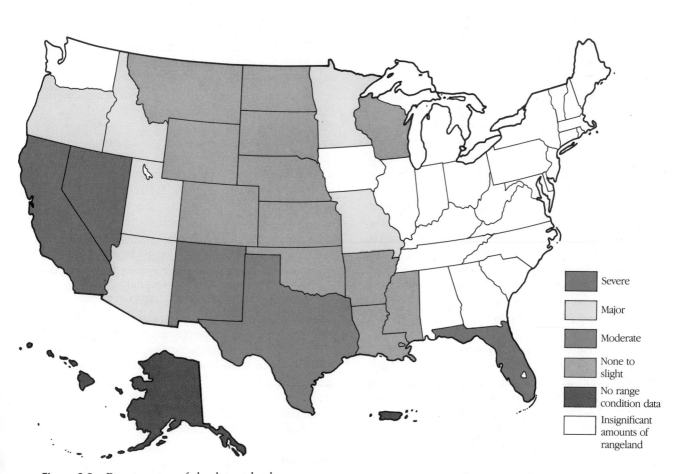

Figure 9.2 *Damage to nonfederal rangeland.*

Figure 9.3 Properly placed water holes help to distribute cattle evenly on the range and avoid overgrazing of select areas. Ranchers can also minimize overgrazing by periodically moving cattle from one section of rangeland to another.

ways. Livestock, for instance, can help to spread seeds and fertilize the soil.

Range management involves two basic techniques often employed simultaneously: sustainable grazing management and range improvement. **Sustainable grazing management,** the first line of defense, requires careful control of the number of animals and the duration of grazing on a piece of land. **Range improvement** means controlling brush, revegetating barren areas, fertilizing impoverished soils, constructing fences and water holes, and similar measures to promote uniform grazing (Figure 9.3).

The key to proper range management is keeping livestock populations within the carrying capacity of the ecosystem. This is often difficult because carrying capacity varies from one year to the next depending on the weather. In dry periods, the carrying capacity may be half of that in a normal year. Truly effective range management requires a willingness to cooperate with nature, benefiting during the good times and cutting back during the bad.

The **Public Rangelands Improvement Act,** passed by Congress in 1978, promotes better range management and calls for improvements on lands managed by the BLM and the U.S. Forest Service. Both agencies have guidelines for proper range management that direct management decisions. However, some observers believe that the BLM's policy, formulated by a rancher advisory board, is not as sound as the U.S. Forest Ser-

vice's, which is conceived by range management specialists and others.

This act also requires the BLM and U.S. Forest Service to reduce grazing where damage is evident, a strategy that is unpopular among ranchers, who either cannot see the benefits of improving range conditions or dispute the claims that they are overgrazing the land. One of the chief weaknesses of the Public Rangelands Improvement Act is that it does not pertain to Indian lands in the West, where grazing reductions are badly needed. As a result of private and federal actions, U.S. rangelands seem to be gradually improving, but much more work is needed.

Tremendous improvements are needed in other countries as well. In Australia's Northern Territory, for example, ranchers who fail to keep a minimum number of cattle on public land are fined by the government. No maximum limits are set. Such policies need to be scrapped and replaced with sustainable ones.

In Africa, government programs rob tribal grazing lands from nomadic cattle growers. Such programs tend to destroy the sustainable practice of migrating with cattle herds in accordance with seasonal rainfall. Programs that reduce pastoralism tend to crowd these people and their livestock onto arid lands that deteriorate under heavy use. International aid to developing countries that support western-style ranching endeavors in place of pastoralism must also be curtailed to support sustainable practices.

Endangered Species

In 1978, the Supreme Court blocked the completion of the nearly finished Tellico Dam in Tennessee in order to save a small, ugly fish called the snail darter.

While many applauded the decision, some asked how we can allow a fish to stand in the way of progress. The answer is twofold: (1) Everytime we destroy a species, we destroy an intricate web of interdependency. The near-extinction of the Guam rail, for instance, has had a dramatic effect on both its prey and its predators. The altered ecosystem struggles to adapt or ceases to exist. (2) We can be certain that plants and animals contain some of the answers to the myriad of problems we have not yet been able to solve. Is the cure for cancer hidden in one of the thousands of yet unclassified plants that are being destroyed at the rate of 100,000 acres per day in the tropical rain forest? Can the rapidly disappearing gorilla teach us what we need to know about our own evolution? We can be sure that, as the diversity of the gene pool shrinks, so do our chances for answering some of human-kind's more serious problems.

With the *Endangered Species Act* of 1973, the government assumed legal responsibility for vanishing plants and animals, but it has done little to fulfill this obligation. For instance, during the Reagan administration, over 3800 species were nominated, but only 36 received protection.

1 The southern sea otter lives off the Pacific coast amongst the kelp beds, which are home to its favorite food—the abalone. Widespread killings by their competitors, abalone fishermen, brought the otter to the verge of extinction. Its recent comeback is a result of legal protection.

1

2 The West Indian Manatee, now numbering only 1000, makes its home in the coastal waters of Florida. Its numbers have been reduced by habitat alteration, overhunting by humans, and injuries caused by collisions with power boats.

3 The largest living primate, the mountain gorilla has been hunted to the brink of extinction. Killed for its meat, collected for zoos, and slaughtered for body parts, the population has dwindled to about 1000.

4 The Panamanian Golden Frog lives in a three-square-mile region in Panama. Attractive to tourists in search of unusual pets, the endangered frog is now protected by law.

5 Once enjoying the largest geographic distribution on this continent, the timber wolf has been reduced to a handful of small packs in Alaska, Canada, and the Soviet Union. Probably one of the most feared and misunderstood of all animals, timber wolves were victims of widespread slaughter by humans who believed they were protecting their livestock. It is now known that wolves kill only the old or infirm in the herds.

6 The nation's symbol, the bald eagle, is endangered or threatened in most of its habitats in the United States. Habitat destruction coupled with pesticide poisonings have brought the birds' numbers down to an alarming 1300 pairs in the lower 48 states.

7 A shy animal cursed with a beautiful pelt, the snow leopard was first photographed in the wilds of Nepal in 1970. Studies indicate that as few as 16 remain today.

8 A sow grizzly will charge anything that threatens her cubs. At 1000 pounds, with a 30-inch neck and a 55-inch waist, it seems unlikely that this animal could run faster than 65 kilometers (40 miles) per hour. But she can—easily catching any human or beast that might unwittingly wander between her and her cubs. Only 800 grizzlies remain in the lower 48 states today.

9 Zoos, once contributing to the decimation of wildlife populations, have come to the assistance of endangered species. Here two zoologists introduce a motherless newborn African elephant to a lactating, and hopefully cooperative, female.

4

2

3

5

6

7

8

9

10 This California condor was hatched in the San Diego Zoo in 1983. Condor eggs are taken from nests in the Los Angeles mountains, and hatched and raised in captivity. Because they are victims of habitat alteration, the baby condors' chances of surviving to sexual maturity are less than 1 in 10.

11 A zooworker feeds a hatchling using a condor puppet. The use of the puppet teaches the baby to recognize its own kind and encourages it to feed normally. Careful avoidance of human contact will help to keep the birds wild and enable them to breed naturally once they are set free.

12 All wild condors were captured in 1986 and 1987 and placed in captivity in southern California. Successful programs at the San Diego Wild Animal Park and the Los Angeles Zoo, have helped to bring the number of California condors up to 38. So far, in 1990, 7 eggs have hatched, bringing the total number of eggs hatched in captivity up to 19. In the fall of 1991, up to four California condors are planned to be released into the wild.

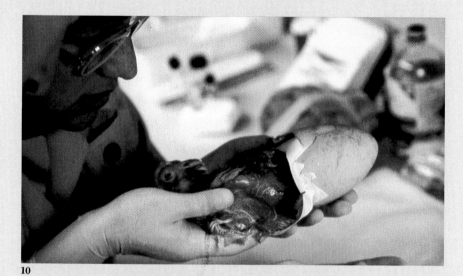

10

12

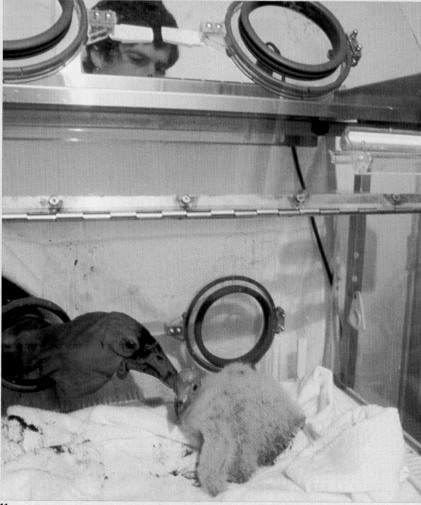

11

Sustainable Livestock Production

In many countries, cattle and other livestock are raised in confined quarters for at least part of their life cycle. Although this does not affect grasslands directly, it does have a tremendous environmental impact. For example, pen-raised cattle produce incredible amounts of manure in limited spaces. At one time, farmers applied the mountains of manure their animals produced to nearby cropland. Today, however, many livestock operations are specialized, that is, not combined with crop production. Thus, their manure creates a huge waste disposal problem. Ironically, farmers who now supply grain to feed cattle use artificial fertilizer on their land. This linear system disrupts one of nature's vital loops—nutrient cycles—and is clearly unsustainable.

Another problem with livestock raised in enclosures is that they require enormous quantities of grain, mostly corn and sorghum. In developing nations, meat primarily feeds the wealthy class. Because livestock are fed grains or are sometimes produced on land that could grow food crops, meat production reduces overall food supplies and makes food more costly for the poor. In Egypt, for example, corn to feed animals is now grown on cropland previously used to grow staple grains, such as wheat and rice. The percentage of that nation's grain fed to livestock has increased from 10% to 36% over the past quarter century. In Mexico, the share of grain fed to livestock has increased from 5% in 1960 to 30% today, despite the fact that 22% of the nation's people are malnourished.

Livestock production on rangeland and in confined spaces can be sustainable, but it must be scaled back, according to the Worldwatch Institute. To downsize this activity, rich countries have to reduce their meat consumption; developing nations have to decrease, not increase, meat consumption. A sustainable system also requires a reintegration of livestock and crop production. Rangelands need to be managed with an ecosystem approach, one that adjusts herd size to the carrying capacity of the land. Efforts are also needed to restore damaged grasslands. Central to this strategy are efforts to slow the growth of the human population.

9.3 Forests and Forest Management

Covering about one-third of the Earth's land surface, forests provide many benefits. The most notable direct benefits are an estimated 5000 commercial products, such as lumber, paper, turpentine, and others, worth billions of dollars a year. Forests also provide refuge from hectic urban life and opportunities for many forms of recreation. In many poorer nations, forests are a source of wood for cooking and heating. Forests are also home to many of the world's species.

Forests benefit us indirectly by protecting watersheds from soil erosion, and keeping rivers and reservoirs relatively free of silt. Forests reduce the severity of floods and facilitate aquifer recharge. Forested lands also assist in the recycling of water, oxygen, nitrogen, carbon, and other nutrients.

The United States has about 300 million hectares (740 million acres) of forestland (Figure 9.4). About two-thirds are commercial timberland, most of which are privately held. Each year U.S. forest products sell for over $30 billion, and forestry employs 1.5 million people, making an important contribution to the economy.

According to estimates of the Worldwatch Institute, world forests cover an area 15 times greater than that in the United States—approximately 4.2 billion hectares (10.6 billion acres). About two-thirds of these forests have already been logged or disturbed in some way by human activities. The economic benefits, job opportunities, and ecological services provided by the world's forests are staggering. International forest products, for example, are worth an estimated $85 billion.

Despite the great benefits of forests, only about 13% of the world's forestland is under any kind of management. In addition, only about 2% of the world's forests is protected in forest reserves.

Worldwide Deforestation—Then and Now

Since the advent of agriculture, about 33% of the world's forests have been cleared and converted to other uses, mostly farms and human settlements. To date, the United States and Africa have both lost about one-third of their forests, while Brazil, the Philippines, and Europe have lost 40%, 50%, and over 70%, respectively. Moreover, deforestation continues virtually everywhere. By one estimate, 17 million hectares (42 million acres) of tropical rain forests—equal to an area the size of the state of Washington—are leveled each year. The World Bank estimates that within a decade the number of tropical countries that export wood will drop from 33 to about 10. In India, forestland is shrinking by 1.5 million hectares (3.75 million acres) per year. At its current rate of harvest, China will lose all of its commercial forests within ten years. In the 1980s, softwood harvests on the West Coast of the United States exceeded sustainable yield by 25% on privately owned land and 61% on national forests.

The heavy use of forests might not be so bad if efforts were made to replant trees at a rate commensurate with cutting. In developing countries, for every 10 trees cut down, only 1 tree is replanted. In Africa, the ratio is 29 to 1.

Besides destroying habitat for many species, deforestation decreases sustainable fuel supplies needed for cooking and home heating. According to estimates of FAO, 100 million people in 26 countries now face acute

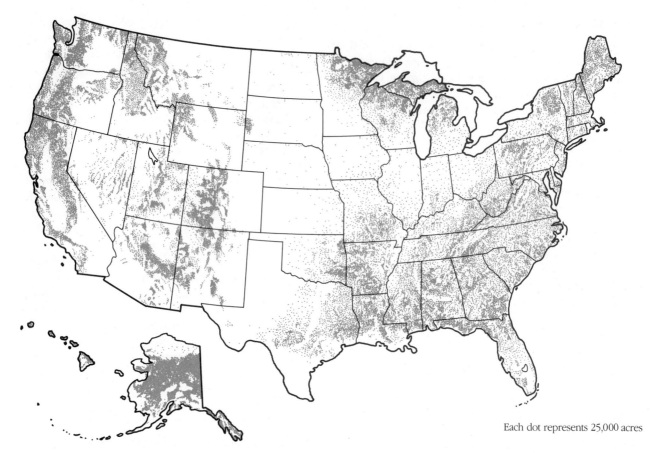

Each dot represents 25,000 acres

Figure 9.4   *Distribution of forestland in the United States.*

firewood shortages. In rural Kenya, shortages mean that some women must spend up to 24 hours per week in search of wood. (Models of Global Sustainability 9.1 discusses deforestation in tropical rain forests.)

The destruction of forests is exemplified by the American experience. When the first colonists arrived, forests covered about half of the nation's land surface. But, settlers cleared much of the land for farms, towns, and roads, and used wood to build ships and homes. White pines were especially hard hit by commercial harvesting, starting in the early 1800s in New England. After depleting eastern forests, companies headed westward, felling trees along their way and finally reaching Minnesota and Wisconsin in 1870. By the early 1900s, the white pine had been reduced to the point that it was no longer commercially profitable to harvest. Lumber companies then moved into the South to cut slash, loblolly, and longleaf pines. Within a few decades, most of the profitable stands were felled.

Fortunately, the southern pines recuperated from heavy harvesting. The trees grow rapidly in the hot, open areas created by previous tree harvests and in abandoned cotton and tobacco fields. Today, under better management, pines support a profitable timber industry in the South.

The cycle of depletion and exploitation of new forests, which has been repeated throughout the world, continues today. Thailand's loggers, for example, have been banned from their own forests because of widespread overcutting and are now actively depleting the world's last great teak forests in nearby Myanmar, formerly known as Burma. Japan, the world's leading importer of tropical wood, imported lumber from the Philippines in the 1960s, Indonesia in the 1970s, and Malaysia in the 1980s, shifting the market as overcutting depleted once-productive forests.

History of Forest Conservation in the United States

Commercial interests in the United States and elsewhere have traditionally taken a narrow economic view of the world's forests, seeking monetary gain with little concern for the future. This rapacious attitude led others to press for forest protection measures. In the United States, such efforts began in earnest in 1891 when President

Models of Global Sustainability 9.1

Saving the World's Tropical Rain Forests

Rain forests are found in the tropics. Forming a thick, lush carpet of vegetation with a stunningly diverse array of species, tropical rain forests of the world once covered an area about the size of the United States. Today, tropical forests have been reduced by half, and logging continues at a feverish pace in many areas. Some experts think that tropical rain forests in all but a few places could be virtually obliterated early in the next century if we do not enact strict measures to protect them.

When I wrote the last edition of this book, in 1989, about 11 million hectares (27 million acres) of tropical forests were being cut down each year—an area about the size of Ohio. More recent studies estimate that the annual rate of deforestation in the tropics is closer to 17 million hectares (43 million acres), a region the size of the state of Washington.

The loss of tropical rain forests is one of the most serious problems facing the world today. It is a major cause of extinction since tropical forests contain about one-half to two-thirds of the world's species.

Al Gentry, a researcher who studies tropical forests, says that the "loss of so many species is not only a tragic squandering of the Earth's evolutionary heritage but also represents depletion of a significant part of the planet's genetic reservoir, a resource of immense economic potential." Genes from tropical rain forests help to boost agricultural production (Chapter 7). Wild species are also a source of new drugs to battle diseases and a host of new products.

Tropical forests play an important role in global recycling of oxygen and carbon dioxide. By various estimates, global deforestation accounts for about 13% of the world's annual increase in carbon dioxide. It may, therefore, be a major factor in global warming, discussed in detail in Chapter 16.

Clear-cutting tropical rain forests exposes the soil to intense rains that wash away the soil, filling nearby streams and rivers with sediment. These problems are commonplace in Pakistan, India, Thailand, and the Amazon Basin of South America. In Thailand, a devastating flood in 1988, caused by massive deforestation by commercial timber companies, ultimately led to a nationwide ban on timber cutting.

Much of the rain falling on rain forests evaporates to become clouds again, only to rain down on the "downwind" forest. When forests are cut, surface runoff increases, and local rainfall decreases by 50% or more in the immediate vicinity. As the area of deforestation spreads, rainfall declines over large regions, and once-lush areas turn into barren tropical deserts. Some tropical soils bake to a hard, bricklike consistency when exposed to sunlight.

Ironically, tropical soils are generally nutrient poor. In fact, approximately 95% of the nutrients are in the biomass and only 5% are in the soil, just the opposite of a temperate forest. Cutting and burning the forests to make room for ranches and farms releases enough plant nutrients to support crops for a few years, but because the soil has little reserve and because nutrients are quickly taken up by crops or washed away, the land quickly falls into ruin.

The International Tropical Timber Organization estimates that less than 0.1% of tropical logging operations is carried out on a sustainable basis.

Because of these problems, many nations have enacted laws to protect their forests. In 1988, Brazil established its first extractive reserve in the Amazon Basin in the state of Acre. Set aside for harvesting rubber, nuts, fruit, oils, and other products, the 40,000-hectare (100,000-acre) reserve will allow people to reap sustainable economic benefits while protecting biodiversity of the forest. Thirteen additional extractive reserves have been in the Amazon, with hopes of many more in the future.

Extractive reserves can be sustainably managed and represent a long-term solution to economic and environmental problems facing developing countries (Chapter 8). To understand why, contrast cattle ranching with the sustainable harvest of rain forest products. In the state of Rondonia, ranchers and farmers have cut down or disturbed at least 35% of the rain forest, creating millions of hectares of wasteland. On most soils, ranchers can graze one cow per hectare for five years. In the next two years, because of declining soil fertility, they can only graze one cow per four hectares. After that, the land is often destroyed.

In contrast, large tracts of rain forest set aside for the sustainable harvest can produce a variety of products, among them latex, nuts, and fruits, indefinitely. The income potential of sustainable harvest of these and other products far exceeds cattle ranching and plantations (Chapter 8).

One of the most important strategies for protecting existing tropical rain forests and providing wood for present and future generations is replanting the land abandoned by timber companies or settlers. Although replanting does not duplicate the diversity of forests, commercial plantations with a modest degree of diversity could be developed to provide wood and other products. Plantations could dramatically reduce pressure on primary forests.

Although positive steps are being taken, rain forests are being destroyed much faster than they can be protected. Individuals can help by joining groups fighting to protect the forests, by avoiding wood products derived from rain forests (teak, disposable chopsticks, parquet floors, and many hardwoods), by recruiting friends, and by writing government representatives.

Benjamin Harrison established the first forest reserve, known as the Yellowstone Timberland Reserve. Theodore Roosevelt added more land to the forest reserve system, and by the end of his term, over 59 million hectares (148 million acres) of forest had been saved from commercial interests.

Many of Roosevelt's actions were at the urging of the noted conservationist and forester Gifford Pinchot. Pinchot and Roosevelt argued that forests could be harvested without permanent damage. Carefully managed, they said, forests could continue to produce valuable timber indefinitely.

In 1905, Roosevelt established the **U.S. Forest Service** as part of the U.S. Department of Agriculture, and forest reserves became known as the **national forests**. Pinchot, who became the first head of the U.S. Forest Service, favored judicious forest development over strict preservation. His ideas laid the foundation for U.S. forest management policy for decades.

Today, the U.S. Forest Service controls about 76 million hectares (187 million acres) of public land in 155 national forests and 19 national grasslands. About half of the U.S. Forest Service land is open for commercial harvesting. U.S. Forest Service lands have many uses besides commercial timber cutting, among them grazing, mining, hiking, skiing, hunting, and camping.

The U.S. Forest Service lands are managed by guidelines set out in the **Multiple Use–Sustained Yield Act** (1960). This law requires that national forests and grasslands be managed to achieve the greatest good for the greatest number of people (multiple use) in the long term (sustained yield). With so many interests competing for the national forests' economic and scenic riches, numerous conflicts have arisen over this mandate. They generally involve environmentalists, on the one hand, and mining interests, timber companies, and resort developers, on the other. Central to the conflict is the question of how particular parcels of land should be used. Should a region remain undeveloped for backpackers, hunters, and wildlife, or should it be leased to a ski resort developer or a timber company?

Sustained yield means that timber cutting should not exceed timber growth and should not destroy long-term productivity of forests. In other words, the U.S. Forest Service's lands should be managed so that future generations can enjoy the same benefits as present generations. This requires erosion control measures and, in some cases, reseeding to help ensure regrowth. Special controls on clear-cutting, described later in this chapter, also increase the likelihood that forests will regrow. Critics of the U.S. Forest Service argue that this agency violates its own policy by allowing overharvesting and timber cutting on steep slopes. In addition, they say that major reforms are needed to manage forest ecosystems to preserve their long-term productivity.

Factors Contributing to Global Deforestation

Before we examine the principles of the "new forestry" and measures to meet future demands sustainably, we must examine the reasons why forests are declining. Deforestation throughout the world stems from many factors. Analysts have traditionally viewed deforestation in developing nations as a natural social response to poverty, unsustainable population growth, and landlessness.

Land ownership in much of the developing world lies in the hands of a select few. In Zimbabwe, Africa, white farmers comprise 1% of the total population but control 39% of all land, much of it prime farmland. In El Salvador, 2% of the landowners holds title to 60% of the land. Poor rural peasants seeking farmland often enter forests and cut down trees to carve out a plot to farm. Many of these people exploit fragile hilly terrain, semiarid grasslands, or rain forests. Norman Myers, a British scientist, estimates that landless peasants clear 60% of the tropical rain forest lost each year. Because they are relegated to marginal lands, their farms often fail. Consequently, they go deeper into the forest and repeat the cycle.

A new study by economist Robert Repetto of the World Resources Institute puts much of the blame for global deforestation on ill-advised government policy that influences how a nation's forests will be used. Even governments that are committed to conservation often have contradictory policies.

Unfortunately, many governments—the United States and Canada included—believe that forest protection can only occur at the expense of economic development. Repetto argues that the misuse of forests actually costs countries billions of dollars per year. Among the hardest hit are the poor countries of the developing world, many of which are saddled with immense foreign debt. For example, the devastating and costly flood that struck Thailand in 1988 was caused in large part by massive deforestation.

Governments worldwide have typically sold timber below market value to logging companies. Studies show that the U.S. Forest Service loses money on most, if not all, of its national forests because it routinely auctions off timber rights below the cost of building roads, surveying, filling out paperwork, and conducting auctions for which the U.S. Forest Service and the taxpayer pay. Below-cost timber sales are also a form of public subsidy to the lumber industry amounting to about $100 million a year. Such an amount obviously discourages conservation by companies, builders, and individuals.

The United States is not alone in selling off timber at below cost. Many other countries, among them Ghana, Indonesia, and the Philippines, let their timber go cheaply.

Economic policies encourage an unsustainable exploitation of forest resources. Many developing countries, for example, restrict the export of raw wood by international companies to create jobs at home, and encourage economic development of domestic wood-processing industries. Bans or heavy taxes that limit raw-wood exports, they think, will increase the export of finished-wood products (for example, furniture), netting higher revenues than the sale of raw wood. Unfortunately, says Repetto, many of the small mills are highly inefficient and use 50% more logs than the industry standard to achieve a given output of milled products. Such policies consequently result in a higher rate of deforestation.

Another problem is short-term contracts. In many developing nations, 35 years or more are required for a stand of trees to recover from logging, but contracts are written for 20 years. Because companies have no long-term interest in their concessions, such contracts discourage them from protecting forests. Longer contracts might encourage them to harvest forests sustainably.

As noted in Chapter 8, heavy borrowing from international banks and industrialized nations has created enormous debt in the developing world. To repay loans, countries often cut as much timber as they can. Forests are also cleared to grow export crops. In addition, export crops are often grown on prime farmland, forcing peasants to turn to forests and other fragile ecosystems to make a living.

Government tax policies further encourage deforestation. For instance, in Brazil, the government once offered income tax credits to investors in cattle ranches for up to 75% of a project's cost and for up to 50% of a company's tax liabilities (the taxes owed in a given year). Cattle ranches were once a leading cause of deforestation in the Amazon. Ironically, many of the projects subsidized by tax credits could not turn a profit without these generous subsidies.

In the developing world, many countries see forests as a safety valve for overpopulated cities. Governments often promote settlement of forests by giving free land or building roads into forests where the urban dispossessed move, clear the forests, and attempt to eke out a living.

The Canadian government has long encouraged deforestation with little regard for the environment. In British Columbia, where forests are falling at a rate far greater than the estimated annual sustainable yield, practically no institutional channels exist by which citizens can influence forest management on public lands. Citizens do not even have the right to sue to stop harmful forest cutting. As a result, many citizens have taken the law into their own hands. To deter timber companies, they have formed human roadblocks or camped in the path of road construction. Local communities are worried about the loss of recreational opportunities; many people who turn out to block the bulldozers are anglers, artists, whale-watching guides, small business owners, and so on. They don't want an end to logging, but they do want a voice in deciding how and where it should occur.

When conservationists argue for controls on deforestation, the timber industry responds with the threat of lost jobs. But, in British Columbia as well as in the Pacific Northwest in the United States, wood products jobs have been steadily declining for years because of automation, while the annual cut has risen sharply. The timber companies wield an incredible amount of power and use the job issue as a smokescreen, say critics. Their power and influence throughout the world are additional factors responsible for the widespread deforestation.

Policies that contribute to deforestation stem in part from the frontier mentality (Chapter 2). To build a sustainable system of forestry requires new attitudes, new policies, and actions on the part of companies and individuals. To understand the kinds of reform necessary, we must first look at the way trees are typically harvested in the United States and other nations.

Forest Harvest and Management

Trees are commercially harvested by three basic methods: clear-cutting, selective cutting, and shelter-wood cutting.

Clear-cutting is a standard practice used primarily for softwoods (conifers), which grow in large stands with relatively few tree species. It is also used in tropical rain forests, which have tremendous species diversity. In U.S. clear-cutting operations, loggers remove all the trees in 16- to 80-hectare (40- to 200-acre) plots, although on U.S. Forest Service land, clear-cuts are now limited by law to 16 hectares (40 acres) with some exceptions. For instance, in California, Oregon, and Washington, Douglas fir stand clear-cuts can reach a maximum size of 25 hectares (60 acres). In the South, clear-cuts can be larger, up to 30 hectares (80 acres). In Alaska, the limit is 40 hectares (100 acres).

On clear-cuts, loggers remove the commercial timber from a plot and often burn the remaining material. This returns nutrients to the soil, facilitates regrowth, and reduces the threat of fires that could damage the regenerating forest. As the new stand grows, trees are thinned to eliminate overcrowding.

Clear-cutting is one of the fastest and cheapest methods to harvest trees. Clear-cutting may also increase **surface runoff**, the flow of water over the ground's surface, which enhances stream flow. This may increase the supply of water to cities, farms, and industry.

Clear-cutting increases suitable habitat for some species, such as deer and elk, which benefits hunters.

However, not all clear-cuts are equal when it comes to elk habitat. A small clearing, for example, is better for elk than a larger one because elk generally avoid open spaces of more than 8 hectares (20 acres). Elk prefer to remain at the edge of meadows so they can escape into nearby forests should a predator arrive. Thus, large square or rectangular blocks are less advantageous than smaller, irregular cuts. Another factor that determines whether a clear-cut increases or decreases elk habitat is the location of the cut. Winter range is a limiting factor in elk populations. Thus, clear-cuts in winter range, which make more food available to elk, are more beneficial than cuts in the more abundant summer range. In Rocky Mountain states, however, clear-cuts are generally made in elk summer range, high in the mountains.

For a long time, researchers thought that clear-cuts were beneficial to deer and elk because they permit herbs, grasses, and bushes to grow, thus providing additional food. However, a Washington State University wildlife biologist, Charles Robins, recently discovered that although plants grow faster in clear-cuts, they contain lower levels of important nutrients. Robins found that the available protein content of huckleberry growing in clear-cuts was less than half of that of huckleberry found in old-growth forests. Plants in clear-cuts also produce more defensive compounds, such as tannins, that lower their nutritional value. Robins hypothesizes that these changes in food value could hinder reproduction in wildlife.

Despite their benefits, clear-cuts create unsightly scars that can take years to heal (Figure 9.5). If not replanted or reseeded naturally, soil erosion may become severe, especially on steep terrain. Eroded sediment fills streams and lakes, destroying fish habitat and greatly increasing the cost of water treatment. Sediment also reduces the water-holding capacity of lakes and streams, which increases flooding, already more likely because of the elevated surface runoff. Erosion in clear-cut areas can deplete the soil of nutrients, thus impairing, even preventing, revegetation. In some instances, clear-cutting decreases runoff from spring snowmelt, which can deplete river water. Large open patches in mountainous terrain, for example, may accelerate a process called **sublimation**, the conversion of snow to water vapor. When this occurs, snowmelts actually decrease. Routine burning in clear-cuts can damage soils by destroying bacteria needed for nutrient cycling. Burning can also volatilize soil nitrogen, robbing nutrients from the soil itself. Burning may also destroy **mychorrizal fungi**, which grow in soils and attach to plant roots, greatly increasing their uptake of water and nutrients.

Clear-cutting fragments wildlife habitat, creating ecological islands. These islands are exposed to wind and dramatic changes in temperature, humidity, and light, which greatly affect indigenous species. In the Pacific Northwest, studies suggest that for each 10 hec-

tares (25 acres) clear-cut, an additional 14 hectares (35 acres) will be degraded.

Finally, clear-cutting destroys habitat and can contribute to the decline of many species, such as ivory-billed and red-cockaded woodpeckers and numerous tropical species. In the Pacific Northwest, heavy cutting of old-growth forests, ancient forests more than 250 years old, with many sections from 500 to 800 years old, threatens the spotted owl and dozens of other species dependent on this habitat (Figure 9.6). Excessive cutting of old-growth forests in the past century has devastated valuable salmon runs in Washington, Oregon, and California.

Canadian old growth is also threatened. One of the hardest hit areas is Vancouver Island in British Columbia. Vancouver Island has remarkable old-growth forests. Western hemlock is one of the main species, but massive red cedars also exist, some with a circumference of 20 meters (60 feet). Sitka spruce can climb to 100 meters (300 feet) or more. A century of feverish logging, though, has taken most of the best trees from the island.

In tropical rain forests, clear-cutting is a prescription for ruin: soils become baked in the sun and too hard to support growth; others wash away in torrential rains. (For more on the effects of clear-cutting in tropical forests, see Chapter Supplement 3.1 and Models of Global Sustainability 9.1.)

New regulations by the U.S. Forest Service are helping to reduce the impact of clear-cutting in national forests, but on private lands timber harvest remains largely unregulated. About 34,000 tree farms in the United States, covering approximately 30 million hectares (75 million acres), are privately owned. Large commercial tree farms operate much like agribusiness. Seedlings are planted, fertilized from airplanes, doused with herbicides to control less desirable species, and sprayed with pesticides to reduce losses. When the trees reach the desirable size, they are cut down, and the cycle begins again.

Arguing for smaller clear-cuts on private and federal lands, E. M. Sterling, an expert on forest management, notes that Austria harvests as much wood from its forests as does the Pacific Northwest. Yet Austrian forests show little evidence of clear-cutting because of strict forestry laws that apply to public as well as to private lands. Austrian law, for instance, forbids clear-cutting on all steep, erodible land. It also limits the size of clear-cuts. A private landowner may cut .6 hectares (1.5 acres) without permission but must obtain a permit for larger clear-cuts. Seldom do clear-cuts exceed 2 hectares. Most clear-cuts are narrow strips that blend in with the terrain and reseed on their own. The lesson from Austria is not that U.S. clear-cutting should be banned but that it should be improved to reduce erosion and the visual impact, for instance, by making cuts smaller and by blending them with the terrain.

(a)

(b)

Figure 9.5 (a) *Clear-cuts in the South Tongass National Forest, Alaska, on steep slopes become an eyesore but may also increase soil erosion, impairing forest regrowth and polluting nearby lakes and streams.* (b) *Clear-cut up close in Kootenai National Forest, Montana.*

Figure 9.6 *The spotted owl is just one of many species that is adapted to old-growth forests. When its habitat is destroyed, the owl disappears.*

Selective cutting, as its name implies, is the removal of a limited number of trees from a forest. The object of selective cutting is to preserve species diversity to protect forests from disease and insects and to reduce visual scarring. In this method, foresters periodically selectively harvest trees of all ages, leaving an equal percentage of each age class. In some cases, though, foresters remove only the best trees. This practice, viewed skeptically by some forest managers, removes the genetically superior trees whose seed is needed to keep the forest healthy. In so doing, a forest may slowly degenerate, producing lower quality wood.

Selective cutting may sound like a good alternative to clear-cutting. Unfortunately, it has several major disadvantages. The most critical are cost and time involvement. Selective cutting in tropical rain forests can also be quite damaging. A study of the eastern Amazon showed that when only 3% of the trees were removed, 54% had been uprooted, crushed, or damaged during the construction of roads and logging. Roads can also accelerate soil erosion and because larger tracts of forest must be harvested to achieve the same output as a clear-cut, more roads need to be built. Selective cutting is also not suited to shade-intolerant tree stands, that is, trees whose seedlings grow in sunny locations.

If properly implemented, however, this technique leaves little scar, causes little or no erosion, and does little damage to wildlife habitat. Despite these benefits, forestry experts point out, it cannot be viewed as a replacement for clear-cutting.

Shelter-wood cutting is an intermediate form of tree harvesting between clear-cutting and selective cutting. In this technique, poor-quality trees are removed first, leaving the healthiest trees intact. These trees reseed the forest and provide shade for their seedlings. Once seedlings become established, loggers remove a portion of the commercially valuable mature trees, leaving enough in place to provide shade for the seedlings. Finally, when the seedlings become saplings, the remaining mature trees are harvested.

Shelter-wood cutting has many of the advantages of selective cutting. It leaves no unvegetated land, except for roads, minimizes erosion, and increases the likelihood that the forest will regenerate. However, it is more costly than either clear-cutting or selective cutting.

Shelter-wood and selective cutting can be economically competitive with clear-cutting in **second-growth forests**, ones that have been cut previously, if logging roads are already present. Even in low-diversity forests, containing only one or two tree species, these techniques can be economically competitive with clear-cutting. Since shelter-wood and selective cutting prevent the scarring of the land, they provide additional economic and aesthetic advantages to regions that rely heavily on tourism.

Creating a Sustainable System of Forestry

With the world population growing and the demand for wood for fuel and wood products rising, global efforts are badly needed to protect, even expand, forests. Measures are also required to create a sustainable system of forestry that optimizes yield while protecting the long-term health and diversity of forest ecosystems. This section describes a number of strategies.

Reducing Demand by Increasing Efficiency
Approximately 50% of the world's commercially harvested wood is cut into lumber for construction. Another 25% is used to make paper, and about 12% becomes plywood and chipboard. All these activities waste rather sizable amounts of wood. Tapping into this waste is, therefore, a key to creating a sustainable system.

According to the U.S. Forest Service, new homes could be built with about 10% less lumber by spacing studs farther apart and other measures that would not compromise the structural integrity of the house.

In Southeast Asia, about 60% of the raw wood entering a plywood mill is wasted. In Japan, the amount wasted is 30%. Reducing waste in these and other plants could reduce deforestation significantly. How can wood be saved?

Thinner saw blades that reduce the kerf (wood removed by the blade), improved machines that grind up logs for plywood more efficiently, and a host of other technologies could dramatically increase the efficiency of mills. In fact, the U.S. Forest Service estimates that wood required to make a sheet of plywood or plank of lumber could be cut by 33% through such measures.

Decreasing Demand by Using Less and Recycling
Conservation, one of the key biological principles of sustainability, means not only using resources more efficiently but also using only what we need. The average American consumes over 272 kilograms (600 pounds) of wood per year in the form of lumber and paper. This is 4.5 times what the average European consumes and 40 times what citizens of less developed countries use.

Individuals can reduce consumption by using less, for example, by avoiding overpackaged items, using the backs of scrap paper for homework and notes, using cloth shopping bags instead of paper, and refusing bags for small items. Individuals can stop the constant bombardment of printed advertising material by writing to companies and asking them to delete their names from mailing lists. Smaller homes, which use 20% to 30% less wood, can be built. Earth-sheltered housing, discussed in Chapter 12, can also reduce the demand for wood while drastically cutting fuel consumption. Many other options are available at work and at home.

Another way of reducing the demand for trees is by recycling paper. Each year, 1250 acres of Canadian forest are cut down just to supply wood pulp for newsprint for the Sunday edition of the *New York Times*. Increased recycling of newsprint could reduce the amount of forest cut down each year. Individuals, companies, colleges, and governments can help by purchasing paper products made from recycled paper (Chapter 19).

Increasing Wood Supply by Protecting Forests and Biodiversity
Protecting forests and the soils that support them is essential to creating a sustainable system of forestry. Clear-cutting, especially in the tropics, should be more carefully regulated on public and private land. Clear-cuts on steep terrain should be eliminated to decrease erosion and its many impacts. Smaller clear-cuts in the tropics would help protect biodiversity and ensure reseeding. In the Amazon, a group of Indians that operates a timber-cutting cooperative clear-cuts narrow strips. This leaves large areas of forest intact, ensures reseeding, and minimizes damage to surrounding trees. In the United States and other countries, reseeding by logging companies should be monitored more carefully, especially on public lands.

In 1986, Congress passed a law that prevents the U.S. Agency for International Development (AID) from funding projects in developing nations, including dams and roadways, that could lead to the destruction of tropical forests. The law also directs AID to help countries find alternatives to forest colonization, and requires it to support preserves and other measures to save forests and promote biological diversity.

Protecting forests from natural hazards, including diseases, insects, fires, droughts, storms, and floods, is also important to creating a sustainable system of forestry (Figure 9.7). Diseases, insects, and fires account for most damage. Sound management that seeks to maintain trees and forests in healthy states can minimize these problems. This may require periodic thinning and efforts to protect the soil and increase water retention. Steps are also needed to protect biodiversity. Insect- and disease-resistant trees could be developed. Finally, all imported trees and lumber should be carefully inspected to avoid accidentally introducing pests.

Fire accounts for 17% of U.S. forest destruction, leveling 0.8 to 2.8 million hectares (2 to 7 million acres) of forest each year. According to the U.S. Forest Service, 85% of all forest fires are accidentally or deliberately started. The remaining 15% are ignited by lightning. However, lightning fires account for about half of the annual forest damage.

To protect watersheds, timber, and recreational opportunities, the U.S. Forest Service and state governments attempt to reduce forest fires by posting fire danger warnings and sponsoring television and radio announcements. Each year, the U.S. Forest Service spends from $450 million to $600 million for fire fighting and surveillance.

Protecting forests from fire began with Gifford Pinchot in the early 1900s and it has no doubt saved billions of dollars' worth of property and timber and countless animals. But, ecologists and foresters now realize that strict fire control can be detrimental to forests. Fires are a natural event, with many benefits to the forest. Minor, periodic fires, for instance, burn dead branches that have accumulated on the ground, returning the nutrients to the forest soil (Figure 9.8). Most animals can escape minor ground fires, and living trees are generally unharmed by them. Periodic ground fires also forestall intense, destructive fires.

In protected forests, the story is quite different. If a fire erupts in a forest that has been protected for long periods, the ample fuel supply may permit it to burn uncontrollably, spreading from treetop to treetop as a **crown fire**. Huge areas are destroyed in firestorms so hot that the soil is charred and wildlife perish in large numbers. Trees may be so severely burned that they die.

Periodic fires protect forests and, in some cases, foster their renewal because many forest species require occasional fires for optimal growth. The cones of the jack pine, for instance, open up and release their seeds during fires as do those of the Douglas fir, sequoia, and

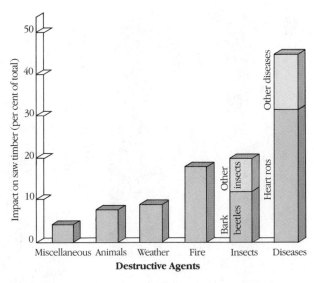

Figure 9.7 *Causes of damage to U.S. forest expressed as a percentage of the total annual damage.*

lodgepole pines. Besides returning nutrients to the soil, fires remove brush that shades seedlings and help reduce disease and insect populations.

Recognizing the benefits of periodic ground fires, forest managers now allow many naturally occurring forest fires to burn. The let-it-burn policy of the U.S. Park Service, begun in 1972, for example, permits fires caused by lightning to burn so long as they do not endanger people, human property, special sites within the park, or endangered wildlife. The U.S. Forest Service, which manages national forests, actually starts hundreds of fires each year (**prescribed fires**) to remove underbrush and litter, reducing the chances of potentially harmful crown fires and improve wildlife habitat, soil fertility, timber production, and livestock forage.

The Park Service's permissive, ecologically correct policy for fires was tested in the summer and fall of 1988 when Yellowstone National Park, one of the oldest and most treasured parks in the nation, erupted in flames. Yellowstone has a history of fires burning in huge blocks of forest, often 40,000 hectares (100,000 acres) at a time, but this one was different. It began outside the park in U.S. Forest Service land that had been protected for many years, making it unstoppable once it crossed into the park.

Critics blame the devastating fire on the let-it-burn policy. However, to do so is to ignore the facts. In 1988, record-breaking heat and drought turned the park into a tinderbox. Almost no rain fell that summer, making it the driest in the park's history. Humidity on the forest floor was a startling 2% or 3%, much lower than kiln-dried wood, which contains 12% moisture.

Many of the fires that started outside the park were fought immediately, but fire fighters could do nothing to stop them. High winds and arid conditions conspired

(a)   *(b)*   *(c)*

Figure 9.8 *Benefits of forest fires.* (a) *Dense undergrowth in an Oregon pine stand results from the control of forest fires.* (b) *Controlled burning removes the undergrowth.* (c) *Periodic burning prevents disastrous fires, returns nutrients to the soil, and increases forage for cattle and wildlife.*

against them. Burning embers jumped entire valleys and fire lines with ease. The fires were simply beyond human control.

Surprisingly, the fires in Yellowstone had remarkably little effect on wildlife. By official estimates, only 250 of the park's 40,000 elk perished. Only 2 grizzly bears and a dozen buffalo of 3000 died. Soil damage was minimal.

The Yellowstone fire, like many before it, created a mosaic—some areas burned while others were passed over by flames that jumped from one location to another. This checkerboard pattern increased biodiversity of the Yellowstone ecosystem and could play a role in controlling fire in Yellowstone in future years because it created patches of young fire-resistant forest with minimal dead wood. According to a widely accepted theory on forest fires, recently burned forests are not susceptible to fire for 200 years. Young forests have long been thought to serve as natural firebreaks. Fires slow to a crawl or die in them.

Saving Primary Forests   Some forestry experts believe that we need to protect remaining untouched or **primary forests**. This will require laws that set aside large tracts of primary forest throughout the world to ensure continued free services the forests provide and also to help protect biodiversity and indigenous people that depend on them.

Forestry experts also argue that we need to retain as much diversity as possible in the world's remaining forests. This will require laws and regulations that reduce the size of clear-cuts and favor selective and shelter-wood cutting.

Intact stands of native trees provide homes to insect predators, such as birds and other insects, and act as physical barriers to the spread of pests. Old-growth forests in the Pacific Northwest, for instance, house 100 times more insect predators than neighboring tree plantations.

Switching to Sustainably Managed Secondary Forests and Plantations   Approximately two-thirds of the remaining forests are second growth. Although they are important to the ecological health of the planet, many could be sustainably harvested with minimum damage because they are close to access roads and are often scattered in small fragments.

Another vital strategy for meeting future demand in a sustainable fashion is to replant cleared forests, creating plantations that support a diversity of species. An estimated 7 million square kilometers (3 million square miles) of tropical rain forests have been cleared, two-thirds of which are available for replanting. If managed sustainably, this land could become a major source of wood and wood products. Planting only 5% of it could nearly double the supply of commercially harvested wood. In China, adding 10 million hectares (25 million acres) of plantations, or about 4% of the area now available for timber harvesting, could double timber production.

Plantations that support a mixture of species and different-age trees can supply wood and nonwood products while providing wildlife habitat. Even in monoculture plantations, diversity can be encouraged by leaving nearby forests intact. During harvesting, diversity can be enhanced by protecting ground vegetation and leaving some trees uncut. In tropical rain forests diversity can be encouraged by planting trees to provide fruit and animal food to support local people.

In sum, the world's forests are being cut and degraded at an unsustainable rate. To create a sustainable system of forestry, efforts are needed to reduce demand, recycle, replant, protect, and promote diversity. Instead of narrowly focusing on yield, the new emphasis on forestry is to find ways to sustain the entire system and the natural systems that make tree growth possible.

9.4 Wilderness and Wilderness Management

In 1987, the Sierra Club, a U.S. environmental group, published a study showing that 34% of the Earth's surface was **wilderness**, land not significantly altered by human activities. One-third of this land was forest, much of it tropical rain forest now under pressure; the rest was mostly tundra and desert. Only about 20% of the world's wilderness was legally protected.

Wilderness, as defined by U.S. law, is "an area where the Earth and its community of life are untrammeled by man, where man is himself a visitor who does not remain." Why is it important? Why should we set aside large tracts of wilderness?

From a strictly human perspective, wilderness provides a temporary escape from modern society (Figure 9.9). Joseph Sax, author of *Mountains Without Handrails*, writes that nature "seems to have a peculiar power to stimulate us to reflectiveness by its awesomeness and grandeur." It helps us to understand ourselves and the world we live in, awakening us to the forgotten interdependence of living things. "Our initial response to nature," Sax writes, "is often awe and wonderment: trees that have survived for millennia; a profusion of flowers in the seeming sterility of the desert. . . . [It] is also a successful model of many things that human communities seek: continuity, stability and sustenance, adaptation, sustained productivity, diversity, and evolutionary change."

Wilderness, especially forested regions, is vital to the Earth's wealth of wild species and the preservation of nature's free services. Many of the reasons for protecting wildlife (Chapter 8) also pertain to wilderness protection.

But, not everyone would agree. To many people, wilderness is simply a playground for the upper-middle-class elite. They protect these lands to the detriment of others who could reap economic benefits from mining

Figure 9.9 *Wilderness restores us. It is a vital resource in our world, offering numerous free ecological services, from climate control to watershed protection.*

Point/Counterpoint 9.1

Controversy over Old-Growth Forests in the Pacific Northwest

by Ralph Saperstein
Ralph Saperstein is vice president of Northwest Forestry Association in Portland, Oregon. He has spent the past ten years tracking the spotted owl/old-growth issue for the forest products industry.

Old Growth, Spotted Owls, and the Economy of the Pacific Northwest

The public debate over national forests in the Pacific Northwest focuses on the volatile passions generated by the images of ancient forests and endangered wildlife. Unfortunately, these images are far from reality. The forests of the Pacific Northwest have always been a mosaic of ages and ecosystems. Natural fires, volcanoes, floods, windstorms, insects, and disease have all played a major role in the evolution of our present forests. These forests were never all old growth, and the wildlife that they support have evolved in a wide variety of age classes and habitat types.

Congress designated national forests to provide the American people a stable source of timber, water, grazing, minerals, recreation, fish, and wildlife. The principle of multiple use has guided the management of national forests for nearly 100 years. Comprehensive

land-use management plans developed under congressional directive have assigned millions of acres of forestland to nontimber uses. In fact, before the listing of the northern spotted owl as a threatened species, only 30% of national forests in the Pacific Northwest were available to grow trees to produce lumber and paper products for millions of people. Protection plans for the spotted owl remove another 8% of the forestland from management.

Restrictions on forest management to protect old growth and the spotted owl have drastically reduced the supply of timber from our national forests. Literally dozens of companies, and the communities they support, were established on the promise of a sustained yield of timber from our national forests. Close to 100 towns in the Pacific Northwest are dependent on a lumber mill for their economic lifeblood. When the forest products industry gets a cold, the local communities catch pneumonia.

Trees are a valuable resource that can be planted and grown to produce lumber and paper products for future generations. Over 40% of the lumber used in the United States to build homes is manufactured in the Pacific Northwest. Stopping the scientific management of our national forest will lead consumers to turn to countries with little or no environmental restrictions for their forest products. All forests harvested in the Pacific Northwest are promptly replanted to grow trees for the future, as they have been for over 50 years. Federal and state laws require that streams are protected, wildlife habitat is created, and productive soils are cared for.

The forest products industry supports protecting old growth and maintaining habitat for the spotted owl. The issue that must be addressed is one of balance. Over 1.6 million hectares (4 million

acres) of old-growth forest are already preserved and will never be harvested. One must recognize that preserving forestland does not assure that it will never be susceptible to fire, wind, insects, and disease. These natural forces and the normal progression of forest growth will eventually drastically alter today's old-growth forests.

The spotted owl has grabbed the attention of the entire nation. This nocturnal bird has served well as a surrogate for the preservation of old-growth forestland. Special interests have used the courts and administrative processes to force massive land set-asides for the owl. The forest products industry has opposed most of the protection schemes proffered to date. All spotted owl protection strategies are based on the simplistic notion that owls need old growth to survive. In the last ten years, the inventory of spotted owls has climbed dramatically. In 1973, there were a couple hundred individuals. Today, federal biologists' estimates exceed 10,000 owls. Most significant, many of the new owl discoveries are taking place in younger, managed forestlands.

There is new information about the wide range of habitat usage by the spotted owl. This information leads the forest products industry to advocate a management plan that relies on existing preserves and forest management to create and maintain habitat for the spotted owl. Following such a strategy would significantly reduce the 50,000 jobs that will be lost as a result of spotted owl protection.

The spotted owl issue has raised important philosophical issues about man's role in the natural environment. The Endangered Species Act requires protection of all threatened or endangered species without regard to the social and economic costs of such protection. What once was a universally acceptable premise is now being scrutinized at all levels of society.

The spotted owl is one of literally thousands of species that threaten to stop development, agriculture, fishing, transportation, and power generation in all 50 states.

The elevation of wildlife above people's needs is a philosophical change that will shake the foundations of modern society. The answer cannot be nature versus

people. The solution to these conflicts must come from the wise management of our natural resources in balance with social and economic values.

by Victor Rozek
Victor Rozek is the general manager of the National Forest Council.

Owls, Lies, and Taxpayer Waste

For years, the tobacco companies would trot out their cigarette-scientists who announced the results of their latest studies "conclusively proving" that cigarette smoking was not linked to lung cancer. Today, against all evidence that logging the public's national forests is an economic and an ecological disaster, the timber industry continues to indulge in its own version of chronic denial. To listen to industry spokespeople, there are no problems in our national forests that more cutting will not solve.

Let's be clear: 95% of the original native forests that once covered most of our nation is gone. The 5% that remains is primarily in the Northwest, much of it badly frag-

mented by clear-cuts—the practice of cutting, then burning every living thing in 40- to 80-acre increments. What is left resides almost exclusively on public lands as part of the system of national forests. These forests belong to present and future generations of Americans, not to the timber industry.

The mere fact that in 150 years—a short time in the life of a forest—we have managed to dispatch all but 5% of this once-dominant ecosystem attests to unsustainable forestry practices. We have, in fact, been hacking down our native forests at twice Brazil's rate. Ironically, everyone seems to be in agreement that Brazil, which still has 80% of its original forests intact, should stop cutting.

Typically, the industry argument is framed in terms of jobs versus owls; a narrow and inaccurate characterization. The two primary reasons for decreased timber employment are automation and exports. Tellingly, in Oregon, during the decade *before* the emergence of the spotted owl, while the total cut from national forests *increased* 15%, employment in the timber industry *decreased* 16%. The industry exports *more timber annually than the entire cut from all national forests*. According to Oregon Congressman Peter DeFazio's office, one-fourth of all the trees cut in the Northwest are exported. If you add minimally processed timber, exports account for up to 60% of all timber cut. It is clear we do not need federal timber for domestic consumption. Like a third-world colony, we export our raw materials—and our jobs—to

foreign nations who then sell us back finished goods. If we simply stopped exports, we could cease logging national forests and experience no timber shortage.

Standing national forests offer much wider and more essential values than timber alone. Forests provide us with clean air and pure water. They abate flooding, moderate the climate, and deter desertification. Forests provide wildlife habitat and abundant fisheries. They are a source of medicines. The bark of the Pacific yew tree contains a chemical called taxol, a potent anticancer substance remarkably effective against ovarian cancer. Tragically, for decades, the yew has been cut and burned as a weed species.

Standing forests also act as a vast carbon storehouse. Once cut, they release enormous amounts of carbon into the atmosphere, hastening global warming. And, of course, forests offer us recreation and inspiration. Yet, incredibly, when preparing timber sales, the government attributes no value to standing trees. In Alaska, the U.S. Forest Service sold 400-year-old trees for $1.48 each!

A congressional study showed that the U.S. Forest Service lost $5.6 billion in direct taxpayer subsidies over the past decade on timber sales. We are asked to pay for the destruction of our own heritage for the sake of temporary employment and short-term profits that will inevitably disappear with the last of our native forests.

As for the spotted owl, the disregard for existing federal laws has been so blatant that in May of
(continued)

Point/Counterpoint (*continued*)

Controversy over Old-Growth Forests in the Pacific Northwest

1991, U.S. District Judge William Dwyer issued a temporary injunction against timber sales in the Northwest. Judge Dwyer observed: "More is involved here than a simple failure by an agency to comply with its governing statute. The most recent violation of the National Forest Management Act exemplifies a deliberate and systematic refusal by the Forest Service and the Fish and Wildlife Service to comply with the laws protecting wildlife."

It is absurd to suggest that endangered species are "threatening" to stop development or that the Endangered Species Act (ESA)

represents an "elevation of wildlife against people's needs." To the contrary, it is an act of last resort. It is precisely a century-long *imbalance* of placing human needs and economics ahead of everything else that has brought the forests to the point of near total ruin. The owl is an indicator species by which scientists judge the health of the entire forest ecosystem. And science tells us that the ecosystem is in deep trouble. As to balance, there are some 1500 pairs of spotted owls and 5.5 billion people. Clearly, humans are not endangered. They can even be restrained, while owls cannot. Are we so impover-

ished that we need to kill the last handful?

The time has long passed for "compromise" or "balance." It's time to stop managing our national forests like a private social welfare program and return them to present and future generations of Americans. We would not think of hiring displaced quarry workers to fill in the Grand Canyon. We would be far more foolish to sacrifice the enormous value of standing national forests for the benefit of wasteful employment that consumes much more than it produces.

and timber harvesting. Historically, wilderness has been viewed as something to exploit for short-term gain. In early Colonial and post-Colonial periods, U.S. lands represented untapped wealth—an unequaled opportunity to sustain a young, growing nation. Some frontierspeople, and particularly early farmers, perceived many natural resources more as obstacles than as assets. Forests needed to be cleared to permit farming. Marshes needed to be drained. The concept of wilderness preservation, had it arisen at that time, would have seemed absurd. Part of the reason for the obstacle mentality of the early settlers was a false perception of the inexhaustibility of the vast resources of the new country.

Preservation: The Wilderness Act

The earliest efforts at wilderness preservation in the United States began in the 1860s. John Muir, founder of the Sierra Club and a longtime wilderness advocate, is credited with much of the early interest in saving wilderness for future generations. Further advances came in the early 1900s under the leadership of Roosevelt and then in the 1930s, when the U.S. Forest Service began to set aside large tracts of forestland, called **primitive areas,** for protection. Between 1930 and 1964, the U.S. Forest Service established over 3.7 million hectares (9.1 million acres) of primitive areas in the national forests.

In 1964, Congress passed the **Wilderness Act,** establishing the **National Wilderness Preservation System.**

The U.S. Forest Service's primitive areas were renamed **wilderness areas.** The Wilderness Act forbids timber cutting, motorized vehicles, motorboats, aircraft landings, and other motorized equipment, except to control fire, insects, and diseases or where their use was already established. Although the Wilderness Act sought to create an "enduring wilderness," many unwildernesslike activities were allowed to continue—notably, livestock grazing and mining for metals and energy fuels if claims were filed before the end of 1983. Wilderness areas throughout the United States are riddled with private inholdings, property owned by individuals and companies who control the mineral and water rights on the property. In Boulder, Colorado, for instance, a lawyer who owns a large section within the Indian Peaks Wilderness Area cut through the public property with bulldozers to repair a dam on his property, causing considerable damage. Another man owns mineral rights in the Maroon Bells Wilderness Area near Aspen, Colorado, and is actively pressuring the U.S. Forest Service to allow him access so he can open a marble quarry along the main path into the area.

The Wilderness Act directed the U.S. Forest Service, the U.S. Fish and Wildlife Service, and the National Park Service to recommend additional land within their jurisdictions for wilderness designation. As of July 1992, 38.6 million hectares (95.4 million acres) of land were protected as wilderness. Sixty percent of the wilderness is in Alaska.

The Wilderness Act did not provide a means of designating wilderness on the 180 million hectares (450 million acres) of land the BLM manages, mostly in the western states and Alaska. In 1976, Congress passed the **Federal Land Policy and Management Act** (1976), which calls on the BLM to submit recommendations on the wilderness suitability of its land. By January 1992, however, only 127,000 hectares (315,000 acres)—less than 1%—of the BLM's land had been designated as wilderness.

Controversy over Wilderness Designation

Nothing stirs controversy in the western United States like wilderness designation. While environmentalists press for more land to be set aside, the mining and timber industries generally oppose it, arguing that wilderness designation locks up resources. In the battles that ensue, environmentalists are accused of costing U.S. society jobs by excluding those who would profit from the Earth's riches. Environmentalists counter that wilderness experiences and ecological benefits provided by wilderness (erosion control, habitat protection, recycling of nutrients, etc.) are enormous. As a rule, many of these benefits cannot be translated into dollars and cents, whereas lumber and mineral ores are easily translated into economic terms. This makes it difficult for nonenvironmentalists to understand the rationale behind wilderness protection.

The U.S. timber industry is one of the nation's leading opponents of wilderness protection. Because only slightly over 16% of all government land (excluding military bases and land on which public buildings are built), or about 4% of all land in the United States, has been afforded wilderness protection, environmentalists argue that the timber industry's claims that wilderness is locking up the Earth's riches are unfounded. Locally, however, wilderness tracts can tie up huge parcels of land, threatening the economic well-being of communities that have historically made a living by timber harvesting and mining. Advocates note that wilderness designation can create a more sustainable economy by promoting businesses in rural areas that service hunters, anglers, cross-country skiers, hikers, backpackers, and the like.

Mining companies also oppose wilderness designation, claiming that valuable mineral deposits are being locked up by such actions. If anything, environmentalists argue, mining interests have been catered to excessively. By allowing some mining in wilderness areas, Congress acted against the best interests of preservation for mining conflicts with wilderness as much as any human activity could.

Should the United States set aside wilderness if it contains oil, natural gas, or minerals that could be used today? Many environmentalists believe that wildlands are more valuable than these finite resources because there are no substitutes for wilderness once it has been destroyed. And, they note, there are many more sustainable ways to meet our demands for resources, such as energy, minerals, and wood (Chapters 12, 13, and 19).

Sustainable Wilderness Management

Lured by the thought of quiet and solitude, backpackers pour into some U.S. wilderness areas only to be dismayed by crowds and special camping restrictions aimed at protecting lakes and streams from pollution. To many, overcrowding signifies the need for more wilderness, especially if the U.S. population is to grow by 50 to 60 million people before it stabilizes, as many demographers now predict (Chapter 5). To others, it is an example of one of the major problems caused by designating an area as wilderness: too many people are attracted, ultimately destroying the wilderness experience for many.

Wilderness crowding and environmental degradation that results from overuse can be reduced or eliminated by better management: (1) educating campers on methods to lessen their impact; (2) restricting access to overused areas; (3) issuing permits to control the number of users; (4) designating campsites; (5) increasing the number of wilderness rangers to monitor use; (6) disseminating information about infrequently used areas to divert campers from overused areas; and (7) improving trails to promote use of underutilized areas.

Globally, interest in wilderness protection is slowly growing. Several countries have established extractive reserves, which, if properly managed, could help protect biodiversity. Although some countries, such as Costa Rica, Brazil, and Colombia, have set aside large parcels of land for protection, many others have little or no protection whatsoever. Where they do exist, protected areas are often understaffed and overused.

Wilderness, grasslands, and forests are vital to our future. They cater to many different needs: eating, shelter, relaxation, and escape. A world without them is unimaginable. A group of scientists peering through the glass of their space shuttle 100 years from now will see the evidence of our actions. Whether they see patches of ancient desert within rich, productive land or just the opposite depends on decisions we make today.

The art of progress is to preserve order

amid change.

—A. N. Whitehead

Critical Thinking Exercise Solution

Critical thinking rules instruct us to question the methods by which information is gained. Early studies on desertification were based on studies that looked at only isolated regions. Nonetheless, the conclusion that the Sahara is expanding may still be correct. Looking at the new evidence suggests that although the desert may ebb and flow, overall it is marching southward (Chapter 7). These studies suggest that heavier precipitation may result in some vegetative recovery from year to year, but the recovery reported in some years may be temporary. Clearly, further research is needed. At this point, it is necessary to exercise one of the most paradoxical of all critical thinking rules—to tolerate uncertainty. Only time will tell if the desert is indeed marching southward as a result of global warming and other problems mentioned in the introduction to this chapter.

Summary

9.1   A Tragedy of the Commons

■ Garrett Hardin described a phenomenon called the **tragedy of the commons**, which applied to communal grazing land where the British once grazed their livestock.

■ Without regulation, he noted, the commons fell into a cycle of decay. Each cattle grower increased his herd for personal gain, bringing ruin to the unmanaged commons.

■ Many other resources are held in common, especially air and water. The misuse of these biospheric commons is the root of considerable environmental degradation. Resource mismanagement also occurs on private lands.

9.2   Rangelands and Range Management: Protecting Grasslands

■ A history of overgrazing has caused the deterioration of both private and public rangelands throughout the world, resulting in permanent loss of vegetation, erosion, desertification, wildlife extinction, invasion of weeds, and a drop in water tables.

■ Sustainable range management helps reduce or eliminate these problems through **grazing management**, controls on the number of animals on a piece of land and the controls on duration of grazing, and **range improvement**, such as fertilizing, reseeding, and techniques that encourage uniform grazing.

■ Government policies of many nations must change to promote sustainable use of grasslands.

■ Since many livestock are grown in confined areas and fed grain that could be used more efficiently to feed people, programs are needed to reverse growing global meat consumption.

9.3   Forests and Forest Management

■ Forests benefit society directly by providing numerous commercially valuable products and opportunities for recreation.

■ Forests also benefit humanity by providing free services: protecting watersheds from soil erosion; reducing surface runoff; recycling water, oxygen, carbon, and other important nutrients; and providing habitat for a diversity of species.

■ Worldwide, millions of hectares of forest have been cut down. At the root of this devastation are overpopulation, poverty, shortsightedness, and ill-advised public policy.

■ In the United States, forest protection began in the late 1800s. In 1905, President Theodore Roosevelt established the U.S. Forest Service. Its first head, Gifford Pinchot, promoted a policy of careful use of forests over one of strict preservation.

■ Pinchot's notion of **multiple and sustained use** has persisted, although critics point out that it focuses on timber production, and tends to ignore the long-term health of the forest ecosystem.

■ Serious efforts are needed to build a sustainable system of forestry that pays attention not only to yield but also to the long-term health and diversity of the forest ecosystems of the world. Some suggestions are: (1) reduce waste, for example, by making mills more efficient; (2) reduce the demand for wood, for example, by building houses more efficiently and by recycling; (3) protect existing forests from damage, such as fire and pollution; (4) halt the rapid depletion of old-growth forests, for example, by setting up forest preserves, and (5) switch to sustainable managed secondary forests and biologically diverse plantations.

9.4   Wilderness and Wilderness Management

■ **Wilderness** is defined by U.S. law as "an area where the Earth and its community of life are untrammeled by man, where man is himself a visitor who does not remain."

■ For some people, wilderness provides an escape from hectic urban life, a chance to watch animals, exercise, or relax. Wilderness is also a vast repository of biological diversity that offers many free services. To others, though, wilderness is an untapped resource with valuable minerals and timber.

■ In 1964, Congress passed the **Wilderness Act**, which established **wilderness areas** within land owned by the National Park Service, U.S. Fish and Wildlife Service, and U.S. Forest Service.

■ The Wilderness Act prohibits timber cutting and motorized vehicles and equipment except in certain instances. Mining can take place in designated wilderness areas if claims were filed before the end of 1983.

■ Even though commercial interests often oppose wilderness preservation, conservationists note that rising population and the fact that more people are turning to outdoor recreation necessitates more wilderness to avoid overcrowding and damage.

■ In addition to setting aside more wilderness, existing areas must also be managed better. Solutions include educating campers on ways to minimize impact, restricting access to overused areas, issuing permits to control the number of users, designating campsites, increasing the number of wilderness rangers to patrol areas and pick up garbage, disseminating information about infrequently used areas, and improving trails to encourage the use of underutilized regions.

Discussion and Critical Thinking Questions

1. Critically analyze the concept *the tragedy of the commons*.

2. How can the biological principles of sustainability discussed in Chapters 1 through 3 help to reshape livestock and forestry systems? Give specific examples of ways these principles can be applied and potential outcomes of such actions.

3. What are the major problems facing the world's rangelands? What suggestions would you make to create a sustainable system of providing meat and other animal products for human society?

4. Define the following terms as they relate to forest management: *sustained yield*, *multiple use*, and *clear-cutting*.

5. What is meant by sustainable forestry? How is it different from the present method of cutting trees based principally on sustained yield?

6. List and discuss strategies to satisfy the growing need for wood and wood products in the coming years. Which of your ideas are the most ecologically sound?

7. How can you reduce paper and wood waste and increase recycling?

8. Critically analyze the statement: "Wilderness is essential to humanity."

Suggested Readings

Brough, H. B. (1991). A New Lay of the Land. *World-Watch* 4(1): 12–19. Shows how land distribution affects poverty and environmental decay and how land reform can reduce deforestation.

Brown, B. (1982). *Mountain in the Clouds*. New York: Simon & Schuster. Classic account of deforestation practices in the Pacific Northwest.

Brown, L. R., Flavin, C., and Postel, S. (1989). Outlining a Global Action Plan. In *State of the World*. New York: Norton. Tells how reforestation can be part of a global strategy to build a sustainable future.

Durning, A. T. and Brough, H. B. (1992). Reforming the Livestock Economy. In *State of the World*. New York: Norton. Outlines ways to create a sustainable livestock industry.

Fearnside, P. M. (1989). A Prescription for Slowing Deforestation in Amazonia. *Environment* 31 (4): 16–20, 39–41. Detailed description of sustainable forest practices.

Goodland, R., ed. (1990). *Race to Save the Tropics. Ecology and Economics for a Sustainable Future*. Washington, D.C.: Island Press. Important writings on solutions to rain forest depletion.

Hardin, G. (1968). The Tragedy of the Commons. *Science* 162: 1243–48. A classic paper.

———. (1980). Second Thoughts on "The Tragedy of the Commons." In *Economics, Ecology, Ethics*, ed. H. E. Daly. San Francisco: Freeman.

Hughes, J. D. and Thirgood, J. V. (1982). Deforestation in Ancient Greece and Rome: A Cause of Collapse. *The Ecologist* 12 (5): 196–207. Detailed paper.

Maser, C. (1989). Life Cycles of the Ancient Forest. *Forest Watch* 9 (8): 11–23. Important for understanding forest ecology.

Monastersky, R. (1988). Lessons from the Flames. *Science News* 134 (20): 314–17. Worthwhile reading on the benefits of the Yellowstone fire.

Myers, N. (1991). Trees by the Billions: A Blueprint for Cooling. *International Wildlife* 21(5): 12–15. Excellent discussion of revegetation potential of tropical lands and benefits.

Owen, O. S. and Chiras, D. D. (1990). *Natural Resource Conservation: An Ecological Approach* (5th ed.). New York: Macmillan. Chapters 11 and 12 cover rangeland and forest management.

Postel, S. and Heise, L. (1988). Reforesting the Earth. In *State of the World*. New York: Norton. Important reading.

Postel, S. and Ryan, J. C. (1991). Reforming Forestry. In *State of the World*. New York: Norton. Superb article.

Robinson, G. (1988). *The Forest and the Trees: A Guide to Excellent Forestry*. Washington, D.C.: Island Press. Excellent source for information on sustainable forestry.

Teitel, M. (1992). *Rain Forest in Your Kitchen*. Washington, D.C.: Island Press. Shows what you can do to help save the rain forest.

Watkins, T. H. (1988). Blueprints for Ruin. *Wilderness* 52 (182): 56–60. Extraordinary article on the fate of old-growth forests in North America.

Chapter 10

Water Resources: Preserving Our Liquid Assets

A river is more than an amenity—it is a treasure.

—*Oliver Wendell Holmes*

Critical Thinking Exercise

Hydropower represents a vast resource in the United States and many other parts of the world, which can be tapped to generate electricity to fuel economic growth, says one energy expert. Moreover, hydropower is a clean, nonpolluting energy source. It won't contribute to global warming and acid rain. As a result, he says, it is foolish not to take advantage of it. Critically analyze this statement. Can you detect any weaknesses or loopholes in the logic? What critical thinking rules did you use to analyze this assertion?

Water. We drink it. We wash with it. We cook with it. We play in, on, and underneath it. We irrigate our crops with it and we use it in our factories. The abundance or lack of water often determines where we live and how well off we are. Despite its importance to human society and other life forms, water is squandered and polluted by industry, agriculture, sewage treatment plants, and municipalities.

This chapter discusses water supply problems and flooding, and presents sustainable solutions. Chapter 16 addresses water pollution. Before looking at the problems of water supply, however, let's examine the water cycle.

10.1 The Hydrological Cycle

Water is part of a global recycling network known as the **hydrological cycle,** or **water cycle.** The hydrological cycle runs day and night, free of charge, collecting, purifying, and distributing water, which serves a multitude of purposes along its path. This all-important cycle is driven by evaporation and precipitation.

Evaporation occurs when water molecules escape from waterways and lakes, from land, and from plants, becoming suspended in air (Figure 10.1). When water molecules depart, they leave behind impurities. Thus, evaporated water is free from contamination until it mixes with pollutants from automobiles, factories, and natural sources.

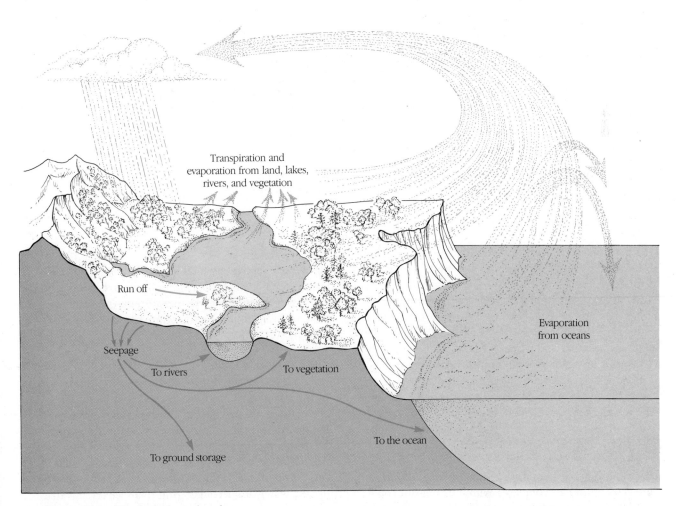

Figure 10.1 *The hydrological cycle.*

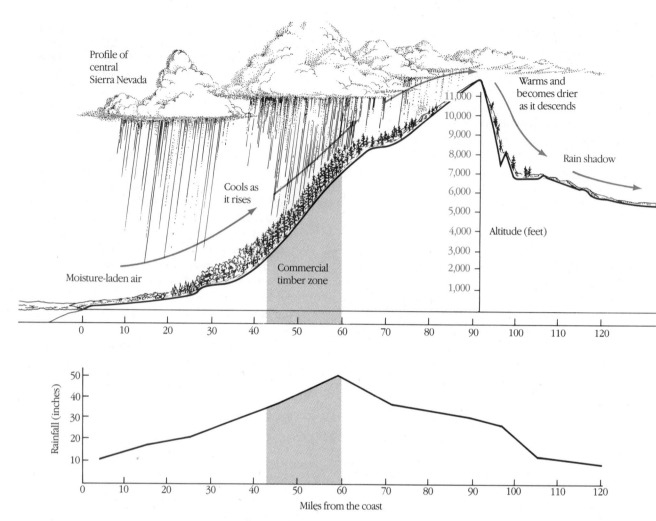

Figure 10.2 (a) *Mountain ranges thrust air upward, causing it to cool. This increases the relative humidity of the air, forming clouds. Moisture in clouds condenses to form raindrops, which fall as precipitation.* (b) *Most of the precipitation falls on the windward side of the mountain. As the air descends on the leeward side, it expands and warms, and the relative humidity decreases. Leeward sides of mountain ranges are often arid.*

In plants, the evaporation of water from leaves helps draw water-dissolved nutrients up from the roots through the stems, much in the way that sucking on a straw draws water from a glass. The evaporative loss of water from leaves is called **transpiration**. The loss of water from the soil and the leaves of plants is known as **evapotranspiration**.

In the atmosphere, water is suspended as fine droplets known as **water vapor**. The amount of moisture, or water vapor, air can hold depends on the temperature. The warmer the air, the more it can hold. Atmospheric moisture content can be expressed as **absolute humidity**—the number of grams of water in a kilogram of dry air—or as **relative humidity**, the more common measurement. Relative humidity measures how much moisture is present in air compared with how much it could hold if fully saturated at a particular temperature. At a

relative humidity of 50%, for example, air has 50% of the water vapor it can hold at that temperature. If the relative humidity is 100%, the air is said to be **saturated**.

When moisture exceeds the saturation point, clouds, mist, and fog form. Clouds form when moisture-laden air comes in contact with cold air, for example, when moist air is raised by mountain ranges or when warm moisture-filled air rises to cooler levels because of ground-level heating (Figure 10.2). For rain to form, air must contain small particles, or **condensation nuclei**, on which water vapor collects. Condensation nuclei may be salts from the sea, dusts, or particulates from factories, power plants, and vehicles. Over a million fine water droplets must come together, or **condense**, to make a single drop of rain. If the air temperature is below freezing, the water droplets may form small ice crystals that coalesce into snowflakes.

Clouds move about by winds, generated by solar energy, and deposit their moisture throughout the globe as rain, drizzle, snow, hail, or sleet. This process, called **precipitation**, returns water to lakes, rivers, oceans, and land. Water that falls on the land may evaporate again or may flow into lakes, rivers, streams, or groundwater, eventually returning to the ocean, from which it may once again evaporate.

On the average, over 15 trillion liters (4 trillion gallons) of precipitation fall on the United States every day. Two-thirds of all this precipitation (10.5 trillion liters) evaporates. Thirty-one percent (4.9 trillion liters) finds its way to streams, lakes, and rivers, and 3% recharges groundwater (Figure 10.3a).

At any single moment, 97% of the Earth's water is in the oceans. The remaining 3% is freshwater. Of this, 99% is locked up in polar ice, glaciers, and in deep, inaccessible aquifers. This leaves a paltry 0.003% available for use by humans, but most of this water is difficult to reach and much too costly to be of any practical value. At present, the entire human population and all terrestrial life forms are maintained by only a tiny fraction of the Earth's water supply, a fact that underscores the importance of treating our freshwater supplies with care. (For a discussion of groundwater, see Chapter 7, Section 7.2.)

10.2 Global Supply and Demand

Water shortages face virtually every nation (Table 10.1). Despite the abundance of sparkling blue water around the Caribbean, for example, the islands are plagued with severe shortages of fresh water. To meet domestic needs, rainwater must be captured on rooftops and used sparingly. Water shortages occur in parts of Peru, Chile, Mexico, Panama, Africa, New Zealand, Australia, Korea, Japan, India, Pakistan, Iran, Italy, Spain, all of the Arab states except Syria, and the United States.

In many developing countries, people spend a large part of their waking hours fetching water, often walking 15 to 25 kilometers (10 to 15 miles) a day to get it, frequently from polluted streams and rivers. According to the World Health Organization, three out of every five people in the developing nations do not have access to clean, disease-free drinking water. Additionally, 80% of all disease in these countries results from the contaminated water that people both drink and bathe in.

The U.N. General Assembly proclaimed the 1980s as the International Drinking Water Supply and Sanitation Decade. Its goal was to provide the world's population with clean drinking water and adequate sanitation by 1990. Unfortunately, when the decade ended, the United Nations fell far short of meeting its goal. Work continues, however, to supply hundreds of millions of people safe, clean drinking water.

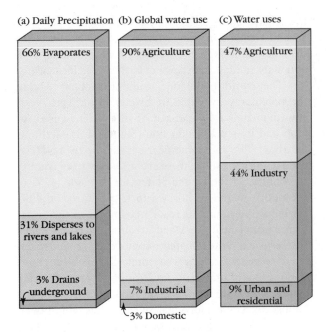

Figure 10.3 (a) *Fate of precipitation in the United States,* (b) *global water use, and* (c) *use of water in the United States.*

To many, it may seem as if the world is running out of water. Far from it. Today's freshwater supply is the same as it was when civilization began thousands of years ago. In fact, enough drinkable water falls as rain each year to flood the land to a depth of 86 centimeters (33 inches), providing sufficient quantities to meet our needs several times over. Water shortages arise from three factors: First, rainfall is not evenly distributed across the face of the Earth. Tropical rain forests are drenched with rain, whereas deserts where many people

Table 10.1   Common Water Resource Problems

Inadequate surface water supply

Overdraft of groundwater

Pollution of surface water and groundwater

Quality of drinking water

Flooding

Erosion and sedimentation

Dredging and disposal of dredged materials

Drainage of wetlands and wet soils

Degradation of bays, estuaries, and coastal waters

live receive under 25 centimeters (10 inches) per year. Second, demand in many cities and towns has exceeded the sustainable supply of water, even in some areas with abundant rainfall. Third, water is often used inefficiently.

The world population withdraws about 9% of the potentially available freshwater runoff each year. Approximately 90% percent of this water is used for crop and livestock production. Industrial use accounts for about 7%, and domestic use makes up the rest (Figure 10.3b). Of course, these averages vary from country to country. In the United States, for example, 47% of the fresh water is withdrawn for agriculture and 44% for industry (Figure 10.3c). Three-fourths of this water comes from lakes, rivers, and streams, collectively called **surface waters**; the remainder comes from **groundwater**.

As a general rule, it is economically feasible for most nations to withdraw 10% to 20% of the annual stream runoff, although some wealthy countries now siphon almost 30%. On individual rivers, water withdrawals may be much higher, causing some to run dry.

Figure 10.4 shows the current water withdrawals throughout the world. The solid horizontal bars on the left of each graph mark 10% and 20% of the runoff. As illustrated, in Africa, Asia, and Europe, people are already withdrawing massive amounts of water. Because of future population increases, water withdrawals in Africa and Asia will likely exceed 10% to 20% of the annual runoff. To meet their needs, people will require water far in excess of **stable runoff**, the amount of water flowing in rivers and streams that can be counted on from year to year. In most countries, stable runoff is about 30% to 50% of the stream-flow average taken over many years. When water demand exceeds the stable runoff, extreme water shortages can develop in dry or even moderately dry years.

The graphs in Figure 10.4 suggest that most continents should be in fairly good shape. Graphs and statistics can be deceiving, however. In this case, continental averages hide local conditions. For instance, the graph for North America, which suggests that enough water is available to meet future demands, does not show that many parts of the United States already tax their water supplies and suffer severe water shortages in dry years. North America also appears well off because the graph includes massive untapped water resources in Canada and Alaska, which will probably never find their way to many water-short areas, such as the desert Southwest. This example illustrates how important it is to critical thinking to delve deeper to understand what statistics and projections really mean.

Drought also contributes to water shortages. A **drought** exists when rainfall is 70% below average for a period of 21 days or longer. A severe drought results in a decrease in stream flow; a drop in the upper surface of the groundwater, the **water table**; a loss of agricultural crops; a loss of wildlife, especially aquatic organisms; a drop in the levels of lakes, streams, and reservoirs; a reduction in range production and stress on livestock; an increasing number of forest fires; and considerable human discomfort.

Whether you live in New York, Alberta, Oklahoma, or California, you've probably felt the impacts of water shortage at least once in your lifetime. In 1987, severe drought struck the southeastern United States, marking the fourth year in a row of below-normal rainfall. Farmers in northern Alabama, Georgia, Tennessee, and the Carolinas were badly hit. Farm production fell drastically and food prices rose throughout the nation. The West, particularly Oregon, California, Washington, and parts of Idaho and Montana, also experienced severe drought in 1987 and 1988, which reduced water supplies for cities and farms. The drought also hurt natural fisheries and the hydropower industry, and caused numerous forest fires. In fact, 1987 had more forest fires than any year in the previous 30 years. Drought conditions persisted in parts of California and other western states into the early 1990s.

Most people think of droughts as natural events. Although this is true, drought conditions can also be caused by human activities. Moreover, droughts are likely to become more common and more devastating if global warming becomes a reality and if humans continue to overcut forests and overgraze grasslands, both of which reduce local and regional rainfall.

Worldwide, irrigated land is expected to double between 1975 and 2000. Domestic use is expected to increase fivefold, and industry's demand is projected to increase twentyfold. By the year 2000, water demand is expected to exceed water supply in at least 30 countries.

10.3 Impacts of Traditional Approaches to Water Shortages

Much of the water taken by humans from rivers, lakes, and aquifers never returns to its source. For example, over 80% of the agricultural water applied to crops evaporates. Consequently, many rivers flow at a fraction of their natural rates, especially during high-use seasons. This reduces populations of aquatic species and recreation activities. Overuse also causes aquifers to dry up completely, driving farmers out of business (Chapter 7).

The traditional approach to water shortages has been to seek new water supplies or to overexploit existing supplies. This frontier strategy results in part from a dogged belief in the notion of unlimited supplies. Economics and politics also contribute to the pursuit of unsustainable solutions. The following material examines some of the impacts of the traditional "supply-side solutions," and suggests the need for a sustainable approach to water management.

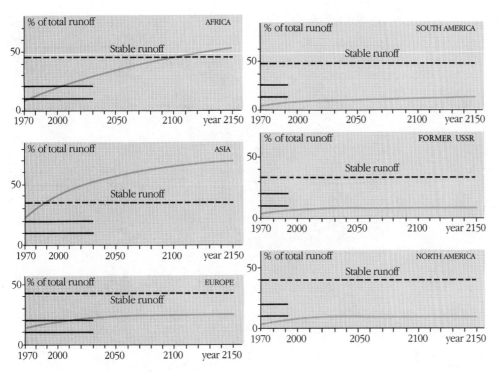

Figure 10.4 *Graphic representation of water demand by continent. The bars on the left indicate 10% to 20% of the total runoff, which most countries can capture without major problems. When a country's demand (indicated by green line) exceeds these levels, it may suffer severe shortages in dry years. Some continents may exceed the stable runoff, that is, the amount of water that can be counted on from year to year. These graphs hide local and even regional shortages.*

Impacts of Excessive Groundwater Withdrawals

Overexploitation of existing water supplies can have severe, long-term implications. In coastal regions, groundwater overdraft can lead to **saltwater intrusion**, the movement of saltwater from marine aquifers into freshwater aquifers (Figure 10.5). When wells remove fresh water faster than it can be replaced, freshwater aquifers shrink. Saltwater from the ocean moves in, contaminating wells.

Ponds, bogs, and streams are sites that mark the intersection of aquifers and land surface. Many think of a pond as "exposed groundwater." Because of this link, groundwater overdraft also drains swamps and ponds, at times drying them up completely (Figure 10.5). Fish, wildlife, and recreation are often affected. Excessive withdrawal of groundwater also threatens the long-term prospects for irrigated agriculture in the United States (Chapter 7). Over 81,000 hectares (200,000 acres) of farmland on which corn had been grown have been phased out of production since 1977 because of groundwater depletion. By the year 2000, Texas may lose half of its irrigated farmland, 1.2 million hectares (3 million acres). Excessive groundwater withdrawals are causing the water table to drop in parts of China and India, the two most populous countries in the world.

Groundwater fills pores in the soil and thus supports the ground above the aquifer. When the water is withdrawn, the soil compacts and sinks, a process called **subsidence**. The most dramatic examples of subsidence have occurred in Florida and other southern states,

where groundwater depletion has created huge **sinkholes** that may measure 100 meters (330 feet) across and 50 meters (165 feet) deep (Figure 10.6). Subsidence has occurred over large areas in the San Joaquin Valley in California, damaging pipelines, railroads, highways, homes, factories, and canals (Figure 10.7). Southeast of

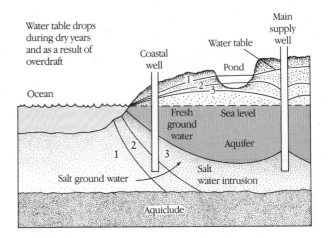

Figure 10.5 *Saltwater intrusion into groundwater. Depletion of groundwater causes freshwater aquifers to retreat, allowing saltwater to penetrate deeper inland underground. The numbers on the left indicate different positions of the water table drawn downward by overdraft from wells. Numbers on top indicate the effects of groundwater overdraft on surface water (pond). Coastal wells are first affected by falling groundwater. Inland wells pick up saltwater only after severe depletion.*

Figure 10.6 *This large sinkhole in Winter Park, Florida, developed quickly, swallowing part of a community swimming pool, parts of two businesses, a house, and several automobiles. It was caused by depletion of groundwater, as explained in the text.*

Phoenix, over 300 square kilometers (120 square miles) of land has subsided more than 2 meters because of groundwater overdrafting. Huge cracks have formed, some 3 meters wide, 3 meters deep, and 300 meters long.

Impacts of Dams and Reservoirs

To meet the growing demand, many municipalities have constructed dams on streams and rivers to hold rainwaters or spring snowmelt. The city of Denver, for example, has constructed a number of dams in the mountains of Colorado to capture snowmelt to meet urban water demands. Dams and reservoirs destroy the habitat of black bears, bighorn sheep, and many other animals. One dam, in Waterton Canyon near Denver, resulted in the death of one-half of the bighorn sheep population. The Denver Water Department had hoped to build another reservoir on the same river. Many times larger, it would have flooded about 60 kilometers (35 miles) of stream that offers excellent kayaking and trout fishing. It would also have reduced the flow of the South Platte River in Nebraska, a stopping-over spot for migrating endangered whooping cranes and tens of thousands of sandhill cranes. Thanks to efforts of local and national environmental groups, the Environmental Protection Agency (EPA) denied the city's permit in 1990.

New dams threaten recreational rivers and place towns that depend on tourist revenues from rafting and kayaking in direct conflict with sprawling cities and sub-

urbs that want their water (Chapter Supplement 10.1). Land developers and energy companies often battle farmers or recreationists over rivers as each group attempts to meet its own needs.

In developing countries, dams built for irrigation, hydropower, and drinking water often flood valuable farmland on which people have lived sustainably for hundreds of years. Pakistan's recently completed Tarbela Dam displaced 85,000 people. China's mammoth Three Gorges dam and reservoir on the Yangtze River will displace 2 million people and destroy nearly 41,000 hectares (100,000 acres) of farmland in the world's most heavily populated and agriculturally productive river valley. People displaced by dams and reservoirs must often move to marginal land that deteriorates rapidly under heavy use.

Dams also interfere with the migration of fish to spawning grounds. Dams built without "ladders," which allow fish to bypass dams, have decimated salmon runs on many streams in the Pacific Northwest and the East Coast. Even when ladders are built, there's no guarantee that fish populations will thrive. Salmon, for example, migrate downstream to the ocean after hatching. Reservoirs eliminate current, which normally helps them to migrate. Many young salmon get lost in reservoirs, never reaching the sea, where they mature and reach full size. In addition, dams generally release the coldest water from the bottom of the reservoir, which can make downstream water exceedingly cold year-round and decrease the spawning of native species, as noted in Chapter 8.

Figure 10.7 *Subsidence caused by groundwater overdrafting damaged this building and driveway in Contra Costa, California. Occurring throughout the world, subsidence can damage highways, buildings, and even farmland.*

The Glen Canyon Dam on the Colorado River, for instance, releases chilly water from the reservoir's bottom, and has converted what was once a warm-water fishery to a rainbow trout fishery. The cold water threatens several native species, such as the humpback chub. To avoid changing stream temperature, some existing dams have been retrofitted with devices that combine warm water from the reservoir's surface with cooler, deep waters. Newer dams are often fitted with multiple gates to ensure proper downstream water temperature.

Because up to 80% of the annual runoff of some streams may be withdrawn, wildlife habitat and good recreational sites may be destroyed by dams. In addition, these projects usually transfer water from rural areas of low population density, often agricultural regions, to cities. This practice creates bitter conflicts between rural and urban residents.

Water withdrawals from rivers and streams have many "downstream" impacts. For example, reduced water flows may allow saltwater to intrude into estuaries during high tide, upsetting the ecological balance in these important life zones. This is a serious problem in Everglades National Park in Florida.

Diminished water flow also reduces the flow of nutrient-rich sediments to estuaries (Case Study 1.1). This interrupts the natural flow of nutrients to floodplains, river deltas, and coastal waters. Good farmland along riverbanks once replenished by periodic flooding must be abandoned or fertilized, often at a high cost.

Sediments that collect in reservoirs slowly fill them, making them useless. In the United States, well over 2000 small reservoirs are totally clogged with sediment. Many larger reservoirs throughout the world face a similar threat. The $1.3 billion Tarbela Dam on the Indus River in Pakistan took nine years to build, but because of upstream soil erosion, the reservoir could fill with sediment in 20 years. Lake Powell on the Colorado River will be filled in 100 to 300 years.

Water withdrawals can also affect downstream water quality. Diversion of clean, high-mountain water in tributaries of the Colorado River, for instance, diminishes the flow and water quality of the river, a source of water for over 20 million people. Because the lower sections of the river carry a heavy burden of sediment and dissolved salts, removing relatively pure water from the spring snowmelt increases the concentration of salts downriver. By the time the river reaches Mexico, its salt concentration is over 800 parts per million (ppm), which violates a U.S.-Mexico treaty, compared with 40 ppm at its headwaters.

Water with a salt concentration over 700 ppm cannot be used for agriculture unless it is diluted or desalinated. Drinking water must have a salt concentration of 500 ppm or less. In 1973, Congress passed a bill that authorized the construction of three desalination plants along the Colorado River to remove about 400,000 metric tons of salt from the river each year. Costing an estimated $350 million, these projects help purify water destined for use in California and Mexico.

Excessive water withdrawal from two rivers that once fed the Aral Sea in the Commonwealth of Independent States has resulted in one of the world's most

Viewpoint 10.1

The Third Stage of Environmentalism

Frederic D. Krupp

The author is the executive director of the Environmental Defense Fund.

I believe that U.S. conservationists have begun to embrace a major shift in tactics that will create a newly constructive third stage in environmentalism's evolution.

The first stage of the conservation movement, represented by President Theodore Roosevelt and the early Sierra Club and National Audubon Society, was a reaction to decades of rapacious exploitation of natural resources, especially in the West. The early focus was on stemming the direct loss of wildlife and forest lands.

A key change occurred in the 1960s as people began to realize that they, too, were becoming victims of environmental abuses, that careless contamination of water, land, and air had sown seeds of destruction in the food chains of both wildlife and humans. The environmental movement's response in this phase was to work to halt abusive pollution.

The Environmental Defense Fund (EDF) was born in the forefront of the second phase, more than 20 years ago, in a victorious effort to stop the use of DDT, which threatened the osprey, bald eagle, and other species with extinction, and had established an alarming presence in mother's milk. The EDF's original vision—to present the evidence of environmental science in a court of law—proved to be an effective strategy to halt and even reverse environmental damage. Lawsuits, lobby-ing, peaceful protests, and other direct efforts became the common expressions of environmental concern in this period.

Most environmental organizations still emphasize this form of reaction, but a new age of environmentalism may be dawning. The "New Environmentalists" say we cannot be effective solely by opposing environmental abuses. We must go the extra mile by finding alternatives to answer the legitimate needs that underlie ill-advised projects, like destructive dams. Otherwise, we are treating only the symptoms of problems that will surface again and again. When we answer the underlying needs, we perform a lasting cure.

If conservationists worry about the impact of a dam, for example, they had better address the water-supply or power-supply problem the dam was proposed to solve. They must concern themselves with the science and economics of environmental protection. Jobs, the rights of stockholders, and the needs of agriculture, industry, and consumers for adequate water and power—all of these issues must become part of the new environmental agenda.

For us to move beyond reactive opposition, to become a positive movement for better alternatives, means facing a difficult challenge. Our organizations will need to keep and recruit people who can envision and persuasively lead the nation toward an environmentally sound economic future.

The EDF's experience in California utility regulation is one of several case studies in this New Environmentalism. In the late 1970s, one of the country's largest utilities, Pacific Gas & Electric (PG&E), had plans to build coal and nuclear power plants worth $20 billion. Forces on both sides were strong, and deeply entrenched. Then, an EDF team—a lawyer, an economist, and a computer analyst—developed an unprecedented package of alternative energy sources and conservation investments and ultimately convinced PG&E to adopt the plan. Why? Because it not only met the same electrical needs but also meant lower prices for consumers, higher returns to PG&E stockholders, and a healthier financial future for the company itself.

The EDF's plan not only blocked construction of the polluting plants, it literally made the plants unnecessary and made obsolete an entire "bigger is better" mind-set among electric power planners.

In the 1970s, society badly needed environmental action groups and public-interest law firms to meet the problems of that era. Today, the need is for these institutions to become well equipped to envision solutions and to assemble new coalitions—even coalitions of former enemies—to bring about answers to environmental problems. But, the third stage of environmentalism is in no sense a move toward compromise, a search for the in-between position. We will still need skillful advocacy—even in court—against narrow institutional vision or vested interest in the status quo. We must, however, become advocates of a new course of action, not mere opponents of the old.

egregious environmental disasters (Figure 10.8). One river is completely dried up; the other is a mere trickle. The sea has dropped 12 meters (40 feet) in 30 years. Two-thirds of its water has evaporated now that it is no longer replenished. Towns once situated on the shore are now 30 miles inland. Because freshwater inflow has all but stopped, the concentration of salts in the sea has risen, killing off 20 of the 24 commercial fish species in the sea. As a result, dozens of fishing towns have been abandoned.

Figure 10.8 *The Aral Sea. This body of saltwater in the former Soviet Union has been reduced by massive withdrawal of water to irrigate farmland. Most species of commercial fish have disappeared, and fishing villages have been abandoned. These fishing vessels have been left high and dry.*

10.4 Meeting Present and Future Demands Sustainably

To avert future water shortages, we need a sustainable water policy and management strategy based on four of the biological principles of sustainability: conservation, recycling, restoration, and population control.

Water Conservation: Using What We Need and Using It Efficiently

U.S. water resource policy was defined in the past by the flow of federal dollars to dams and canals. That era, however, is ending for three reasons: (1) federal support has dwindled; (2) the general public has shown a strong interest in preserving its rivers for recreation and wildlife; and (3) many of the best sites for dams have been used, and those that remain will be more expensive to develop and are often more environmentally objectionable. Therefore, many experts think that water policy will shift from water development to the efficient use of existing facilities and supplies, making water conservation and recycling the wave of the future.

Water conservation measures can be brought online quickly and at costs well below the construction costs of new dams and reservoirs. Conservation measures also offer numerous environmental benefits.

Since agriculture and industry are the biggest water users, efforts to conserve water in these sectors should be given highest priority. Substantial gains in agricultural water efficiency can be achieved by lining open, dirt-lined ditches that deliver water to crops and lose about 40% to 50% of their water. Concrete- or plastic-lined ditches are 80% to 90% efficient. Pipes are even more efficient. Sprinklers, which waste up to half of their water, can be replaced with drip-irrigation systems, which lose only 5%. However, drip irrigation can be used for only a limited number of crops, such as fruit trees, grapevines, and some vegetables. For vast fields of wheat and corn, more conventional methods must be used. Modifications in sprinkler systems for these crops can cut water losses by 30% to 50% (Chapter 7). Since 1970, farmers in the Commonwealth of Independent States have cut water withdrawals per hectare of irrigated land by 30% through improvements in irrigation efficiency.

Water conservation can help offset current shortages and can meet future demands at a far lower cost than dams. In the San Joaquin Valley in California, for example, irrigation water costs farmers about $5 per acre-foot (1 acre of water 1 foot deep). Water is inexpensive to farmers partly because federal and state taxes are used to pay for water projects. In other words, the public subsidizes the private use of water to grow taxpayer-subsidized crops. Water is so cheap because the economic costs of the damage—for example, the loss of fish—created by water projects are not included in the cost.

According to the Environmental Defense Fund (EDF), recycling water and scheduling water application according to crop and soil needs cost about $10 per acre-foot. Switching from irrigated crops, such as cotton, to dryland crops costs farmers about $40 per acre-foot in lost crop yield. Water conservation by more efficient drip and sprinkler systems costs about $175 per acre-foot.

At first glance, farmers have little incentive to conserve water. A critical analysis of the situation shows that this conclusion is false. For instance, by cutting back on water demand, farmers can reduce farm runoff and salinization (Chapters 7 and 8). Conservation also reduces the need for new water projects, which cost taxpayers considerably more—up to $500 per acre-foot—now that federal monies are less available. In the long run, water conservation makes good economic and ecological sense.

There are other reasons as well. Perhaps the most important is that by investing in water conservation, farmers can cut water demand and sell excess water to willing buyers. California cities and industries, for instance, currently purchase water for about $200 per acre-foot per year. If a farmer could invest $50 per acre-foot to save water and sell the excess for $200, he would reap a nice profit. Because farmers would be making a profit on publicly subsidized water, some critics say that it seems only fair that the public should be reimbursed as well. Farmers shouldn't be allowed to reap the full profit of a publicly subsidized resource. No matter what the outcome of the debate, water conservation makes good sense in the long run as well as the short run. A 10% decrease in agricultural water use could double the amount of water available for cities and industries.

Farmers can also save water by improving the organic content of their soil (Chapter 7). Measures that reduce surface runoff (strip-cropping) or the flow of wind across farm fields (shelterbelts) may also result in a substantial reduction in irrigation water demands. In India, farmers plant rows of grass along the contours of hilly terrain. The rows form a natural barrier that retards the flow of water, reduces soil erosion, and increases soil moisture content. Yields can increase by 50%. Although simple, such measures can be extremely cost-effective in large regions and make costly irrigation dams unnecessary. Shifting water-intensive crops, such as cotton, from regions of low moisture (California) to areas of higher precipitation (Alabama) can also result in substantial water savings.

To avoid unsound future water projects, many propose that federal and state governments should require those who benefit from a project to pay its full costs. "Demand for water projects would reflect the discipline of the marketplace, rather than the undisciplined pursuit of subsidies," say the authors of *An Environmental Agenda for the Future*, a collection of recommendations set forth by ten leading environmental groups. Making users pay the full cost, they argue, might encourage them to focus on least-cost alternatives, particularly conservation. The cost of a water project should also include the costs of correcting future environmental problems, a step that would make conservation even more attractive.

In January 1989, two major water supply agencies announced an historic agreement. The Metropolitan Water District of Southern California agreed to pay for water conservation measures in the nearby Imperial Irrigation District. By lining irrigation canals and other measures, farmers will free up substantial amounts of water for use in Los Angeles, San Diego, and neighboring cities at a cost far below that of new dams and diversion projects.

In industry, water conservation can be achieved by redesigning existing processes and facilities to make them more efficient. The use of water by steam-cooled electric power plants, one of the most water-consumptive industries in the world, could be cut one-fourth by using dry cooling towers (Figure 10.9), although these require more energy and are more expensive to operate than wet cooling towers.

Improvements can be made in cities and towns as well. Municipal water systems throughout the United States lose, on average, 12% of their water through leaks. New pipes can reduce water loss enormously, saving streams and the wildlife that depend on them. Such measures also provide important employment opportunities.

You and I can do many things to cut down on water use. Table 10.2 compares typical wasteful habits and shows how simple adjustments in behavior translate into great water savings. One very simple and cost effective solution is a high-efficiency showerhead. Using 70% less water than a traditional showerhead, the new models also use less energy and can save a family of four $200 to $250 per year in energy bills—all for a $10 investment and five minutes of installation time. Individual savings add up. The Colorado Environmental Coalition, which has 1300 individual members, found that if each used a low-flow showerhead, 330,000 liters (85,000 gallons) of water per day could be saved, or about 92 million liters (24 million gallons) per year! Imagine the savings if everyone in the United States used one.

Water Recycling and Groundwater Management

Water recycling is also essential to creating a sustainable system of water supply. Wastewater from cities and factories can be purified and reused over and over, thus greatly reducing the amount of water taken from streams. How can cities convert their existing sewage treatment plants to recycling facilities?

Sewage treatment plants remove many impurities from wastewater from homes and factories. Although the product is hardly potable (drinkable), further treatment could purify the water to make it safe. In Tokyo, for example, Mitsubishi's 60-story office building has a fully automated recycling system that purifies all of the building's wastewater to drinking-water purity.

Another form of recycling that lets nature do the work is the release of wastewater into **aquifer recharge zones**, regions where rain and snowmelt replenish aqui-

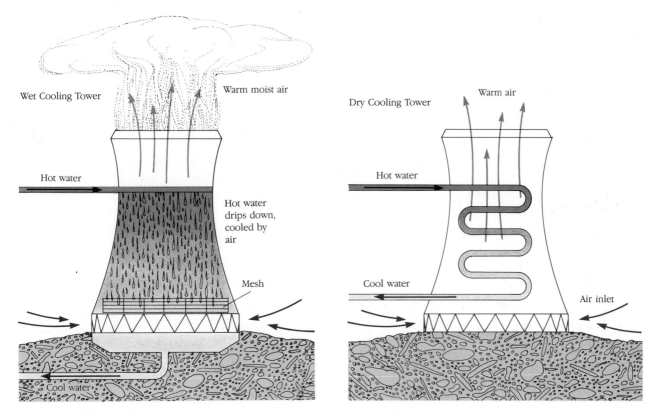

Figure 10.9 *Wet and dry cooling towers. Water from electric power plants is cooled in them before being reused. Dry towers cost more to operate but conserve water.*

fers. Nutrients are taken up by plants and bacteria in the soil. Therefore, as wastewater percolates through the soil, it is cleansed. Effluents from sewage treatment plants, irrigation runoff, storm-water runoff, and cooling water from industry could all be used to replenish aquifers. However, aquifer recharge is costly since many cities are far from aquifer recharge zones.

Other measures to create more natural purification systems are highlighted in Models of Global Sustainability 4.1.

Water recycling could help stabilize water supplies and because it eliminates discharge into streams, help cut water pollution, also vital to a sustainable future. As water shortages worsen and public and governmental awareness increases, water recycling may become more popular. Although building and maintaining large-scale recycling plants can be expensive, they are often environmentally and economically cheaper than new water diversion projects and dams.

Nutrient-rich urban wastewater can also be applied to pastures and cropland (Chapter 7). This practice returns valuable nutrients to the soil and reduces pollution of surface waters. In Israel, where fresh water is limited, 35% of the municipal wastewater is used to irrigate cotton. By the year 2000, Israel hopes to use 80% of its wastewater for irrigation.

Finally, another measure needed to create a sustainable water supply system is to regulate groundwater withdrawal so it does not exceed recharge.

Restoration

Restoring vegetation in watersheds is yet another component of a sustainable system of water supply. Revegetation, especially in the tropics, reduces sedimentation of dams and reservoirs, extending their lifetime. It also helps to reduce surface runoff and enhances the infiltration of water into the soil, increasing groundwater supplies.

Population Control

Given the limited amount of fresh water available to humans and the importance of free-flowing streams to maintaining aquatic life, creating a sustainable system of water supply requires measures to control, even stop, population growth. Especially important are measures to control growth in water-short areas, such as the desert Southwest of the United States, northern Africa, the Middle East, and parts of the Indian subcontinent. Special attention must be paid to the placement of new

Table 10.2   Water Savings Through Conservation

Activity	Typical Use (Liters/Use)	Efficient Use (Liters/Use)	Suggestions
Shower	270	77	Install flow restricters
		46	Shorter showers with flow restricter
Toilet	19	15	Bottles or bricks to displace water in tank
		13	Install low-flush toilet
		4	Install washdown toilet
Faucets	(liters/min.) 12	10	Low-flow faucets
		6	Flow restricters
Brushing teeth	20–40	2	Brushing with faucets off, then rinsing briefly
Shaving	24–48	2	Shaving with faucets off, followed by brief rinse
Washing dishes	60–80	10–20	Rinsing dishes in drainer or turning water on and off each time; using short cycle for dishwashers
Washing clothes	140	80	Front-loading washing machine
		100	Adjusting water level for each load
Outdoor watering	(liters/min.) 40	—	Nighttime watering can cut water demands in half

housing and other development so that it is not built over aquifer recharge zones. (See Case Study 10.1 for more.)

Education: Learning to Use Water Wisely

Massive public education can aid the battle to ensure an adequate water supply. Teaching children throughout the world ways to use this precious resource efficiently may be as important as the math or history they learn.

Legal Solutions That Promote Sustainable Water Use

Governments can have a substantial impact on achieving sustainable water supply systems by encouraging education and promoting water conservation, especially among the large water users—agriculture and industry. One important step is to price water properly, especially agricultural water. Irrigation systems throughout the world are often built with huge public subsidies, and users pay only a fraction of the cost of the water. In developing countries, government revenues from irrigation projects only cover 10% to 20% of the costs. The same holds true in many developed nations. When water prices reflect scarcity, conservation and recycling will become commonplace.

New laws and building codes can promote reductions in household water consumption in both new and existing buildings. Requirements that mandate low-water toilets, low-flow showerheads, and water-efficient lawns could cut water use in new homes by half. Under a newly amended plumbing code, the state of Massachusetts now requires that all new toilets use no more than 6.2 liters (1.6 gallons) of water per flush, about one-half as much as current models and one-third as much water as toilets in older homes. Because toilets use 30% to 40% of the indoor water consumed by a family, water-conserving models could generate significant savings.

Considerable efforts are needed in arid regions to replace water-thirsty lawns with water-thrifty species of grass, trees, and shrubs. In northern Marin County in California, for example, 33% of the county's supply is used to irrigate lawns. One town offers sizable subsidies to homeowners who will **xeriscape** their lawns—remove water-intensive vegetation and replace it with water-conserving species. Each converted lawn saves about 463 liters (120 gallons) of water per day during peak months. Smaller lawns and water-conserving drip and root-zone irrigation systems can be encouraged and planted with low-water grasses.

Government programs can also encourage the installation of water-conserving measures in existing

Case Study 10.1

Ecological Solutions to Flooding and Water Supply Problems in Boston, Massachusetts, and Woodlands, Texas

What do Boston, Massachusetts, and Woodlands, Texas, have in common, besides the distinguishing accents of their residents? The answer is that both regions have made important decisions in land-use planning that will save millions of dollars by reducing flood damage.

Boston boasts a fine park system that stretches from the center of the city into the suburbs. Few realize that the park system with its meandering stream was built in large part to help control flooding. After heavy rains, excess water flows into this basin, where it is slowly released to the sea, reducing flooding and property damage.

The city also purchased large wetlands in an outlying area rather than letting developers drain the wetlands and build on them. The wetlands were set aside to reduce flooding and also provide valuable habitat for fish and wildlife. The project cost one-tenth as much as a dam to control flooding.

In 1971, landscape architect Ian McHarg wrote a landmark book entitled *Design with Nature*. In it, he suggests that builders consider wildlife needs, natural flooding, soil stability, and a half-dozen other factors when constructing homes. By careful site analysis and design with nature, builders can greatly reduce their impact on the land, air, and water.

Texas developer George Mitchell decided to build a new town, called Woodlands, using McHarg's ideas. On his forested tract north of Houston, Mitchell envisioned a city in harmony with the forces of nature. He and his staff of planners first analyzed the region and found that they could allow the natural drainage system to remain as open space, which would carry away water more effectively and more cheaply than a storm sewer system. That step alone saved $14 million in construction costs. Roads were built on high ground and buildings were restricted from aquifer recharge zones, protecting the groundwater that supplies Houston. In 1979, rainwaters drenched the site. The streams swelled by 55%. In neighboring towns, built with little regard for nature, water flows increased 180%. The towns suffered considerable flood damage while Woodlands remained safe.

Woodlands is an attractive community. Most of its trees still stand. The floodplains that were set aside for natural drainage and aquifer recharge harbor numerous birds and mammals, including bobcats and white-tailed deer. This community stands as a testament to the benefits of designing with nature, an approach to development that permits nature to direct the design of human settlement and helps people live in harmony with nature, so vital to building a sustainable future.

homes. Some water departments now offer free water audits, free high-efficiency showerheads, and sizable rebates if a homeowner replaces existing toilets with low-water models. Systems that retain rainfall and used water from showers and faucets, or **gray water**, could be installed. Gray water is suitable for lawn and garden irrigation. Finally, municipal water agencies could adjust water rates, charging customers more for water used when evaporation is greatest (daytime) and less for water used when evaporation is low (evenings). This incentive could help reduce water evaporation from lawns and save huge amounts of water.

Perhaps the most serious deficiency in water resource regulation is the lack of coordination between water resource management and economic development. In many cases, the two are discussed exclusively. Builders erect new housing developments, then ask where the water will come from. City planners scramble to find water, constructing dams and diversion projects that drain rivers and disrupt wildlife habitat. By meeting new demand by using existing supplies more efficiently, devel-

opment could occur without further environmental deterioration.

Tapping New Sources of Water

As a last resort, rising demand can be met by cautiously developing new water supplies.

New Dams   Because the economic and environmental costs of dams and reservoirs are substantial, discretion is advised when undertaking such projects. In developing nations, small-scale irrigation projects, combined with efforts to use water more efficiently, can eliminate the need for large and costly dams. Smaller projects help build local self-reliance, so vital to sustainability.

Tapping into the World's Oceans: Desalination

Another way to increase water supply in communities near the ocean is by **saltwater desalination**, removing salts from seawater (Figure 10.10). The two main methods are distillation and reverse osmosis. In **distillation**,

Figure 10.10 *Desalination plant in Key West, Florida. This plant removes salt from seawater, allowing people to live in areas that lack fresh water. Unfortunately, expanding the carrying capacity of such regions often has many adverse environmental impacts.*

saltwater is heated and evaporated, a process that leaves behind the salts and minerals. The steam is then cooled, and pure water condenses out. In **reverse osmosis**, water is forced through thin membranes whose pores allow the passage of water molecules but not the salts and minerals.

Through these methods, seawater can be purified for drinking and irrigation. Since 97% of the Earth's water is in the oceans, desalination might seem like the best solution to water shortages. Unfortunately, water produced by desalination is four to ten times more expensive than water from conventional sources such as dams. Nevertheless, since 1977, the world's desalination capacity has increased dramatically. Still, desalination produces only a tiny proportion of the fresh water consumed by humans. In the United States, over 100 desalination plants now produce an excess of 1250 million liters (330 million gallons) per day, or about 0.001% of the total freshwater requirement. The majority of the plants are located in California, Texas, Florida, and the Northeast. Desalination plants are also in operation in Saudi Arabia, Israel, Malta, and a few other countries.

Even though costs have recently decreased, energy requirements and construction costs of desalination plants may remain prohibitively high. Desalination seems even more unlikely for U.S. agriculture, since water-short areas are often located far from the sea.

Desalination plants permit a further extension of the Earth's carrying capacity with potentially serious ecological impacts. For example, population growth in the Florida Keys permitted in part by a new desalination

plant threatens coral reefs and wildlife, such as the Key deer. Construction of houses and condominiums causes erosion that pollutes coastal waters. New residents produce an increasing amount of sewage and other pollutants that decrease water quality. Large quantities of salt and minerals from desalination plants could worsen water quality.

10.5 Flooding: Problems and Solutions

Ironically, after shortages, the next major U.S. water problem is flooding. Despite years of flood control work, floods cause damage valued between $2 billion and $3 billion per year.

Causes of Flooding

A deceptively simple correlation can be drawn between floods and their apparent cause: heavy rainfall and snowfall. Closer examination reveals many causes that remain unnoticed by most people. As shown in Figure 10.11, precipitation that does not evaporate must either run off or percolate into the soil. Whether it flows across the land and empties into rivers where it may spill over the banks or sinks quietly into the soil to become groundwater is largely determined by surface features, especially the vegetative cover and ambient temperature. For example, forests and grasses retard water flow and promote percolation. Heavily vegetated watersheds act as sponges. Light vegetation, for instance, in deserts, increases surface runoff and, hence, flooding. In addition, major spring floods often occur when snow melts before the ground has thawed.

The fate of rainfall in many cases rests not so much in the hands of nature as in the hands of farmers, urban planners, developers, and homeowners, who often strip vegetation from the land and increase runoff. Water flowing rapidly over the surface of barren land into streams causes flooding and often transports a substantial amount of soil in the process. In a vicious circle, eroded sediment fills rivers and lakes and reduces their holding capacity. This makes flooding more likely, even after moderate rainfall.

Flooding in urban areas frequently results from highways, airports, shopping centers, office buildings, and homes, which greatly increase the amount of impermeable surface. Instead of soaking into the ground, rainwater washes off the surface in torrents. Figure 10.12 shows regions of the United States that are susceptible to flooding. Many of these areas are located along the Mississippi River and its tributaries. The regions along the banks of rivers naturally subject to flooding, the **floodplains**, are popular sites for cities, towns, and farms. Thus, continued development that increases sur-

Models of Global Sustainability 10.1

Undoing the Damage to Florida's Kissimmee River

Between 1964 and 1970, the Army Corps of Engineers hacked away at the winding Kissimmee River in south-central Florida, hauling up mud and dumping it along the banks. When they finished this huge flood control project, a river that had once lazily meandered nearly 150 kilometers (100 miles) through the Florida marshes was reduced to a canal 65 kilometers (40 miles) long, 60 meters (200 feet) wide, and 9 meters (30 feet) deep. The canal was designed to drain water quickly from the northern reaches of the watershed. On the heels of the huge dredgers came contractors, who threw up concrete and earthen dams and locks every 16 kilometers (10 miles), creating huge reservoirs along the river's previous course. These were designed to control flooding below.

Once a rich habitat for bald eagles, deer, fish, waterfowl, and alligators, the Kissimmee River became a sterile tribute to our tireless efforts to control flooding. Most scientists condemned the channelization as a major environmental catastrophe, which destroyed three-fourths of the original 16,000 hectares (40,000 acres) of marsh, once a major breeding ground and stopping-off place for dozens of species of water birds. Secondary canals built by landowners along the main canal drained another 80,000 hectares (200,000 acres). Soon after the canal's completion, the vast flocks of ducks disappeared. Gone, too, were the wading birds. By Florida Game and Freshwater Fish Commission estimates, 90% of the waterfowl and 75% of the bald eagles vanished from the region, as did the largemouth bass that once attracted anglers from all over the country.

Two years after this enormous project was completed, Florida biologists began noticing changes in Lake Okeechobee, into which the Kissimmee's clean waters once flowed. Dead fish and dying vegetation were the most blatant signs that something was awry in the lake, which provides drinking water for Miami and coastal cities. It didn't take biologists long to determine that the loss of marshlands, which purify waters and hold back sediment, was the reason for Lake Okeechobee's sudden deterioration. The loss of the natural cleansing of the wetlands and a heavy load of pesticides, fertilizer, animal wastes, and sediment from cattle ranches and farms that sprang up along the river's banks created a monumental water-quality problem for the lake.

The waters of the Kissimmee River, which flow south, once fanned out across southern Florida to nourish a huge multimillion-hectare wetlands known as the Everglades. On the southernmost tip of Florida is Everglades National Park. To make room for farms, much of the Everglades has been drained, and the water from Lake Okeechobee and the Kissimmee

River basin has been shunted via canals to the coast. Reduced flows have disrupted the ecology of the Everglades, seriously threatening many species, some already endangered. Reduced water flows also produced an ironic backlash. Because of a lowered water table, farmland once prized because of its rich soil began to sink at a rate that could hinder farming in the region. Lower water flows also resulted in saltwater intrusion into surface water and groundwater.

Ironically, studies made after the canal was completed indicate that it provides little or none of the expected flood control above Lake Kissimmee. Making matters worse, the canal is now seen as a major threat to downstream areas. After heavy rains in central Florida, for instance, a slug of water travels rapidly southward along the canal, wiping out nesting waterfowl and drowning unsuspecting wildlife.

Less than two years after the Army Corps of Engineers finished, a special governor's committee released a report calling on the state to reflood the marshes that it had just drained. The report concluded that channelizing the river had been a big mistake. With a price tag of $30 million, it had been a costly one. Even the Army Corps of Engineers commissioned a study to reevaluate the project and prepare recommendations for returning the river to its original state.

In 1983, Governor Robert Graham took steps to reverse the damage. In 1984, the Kissimmee River restoration began, as a demonstration project aimed at testing the effectiveness of four plans. However, because no federal funds were available to reclaim the river, funding had to come from another source: property taxes collected from residents in southern Florida. The South Florida Water Management District, which was put in charge of the project, built three small dams to divert water from the canal back into the old river channel. This flooded the marshes along 20 kilometers (12 miles) of the river and cost $1.5 million.

These dams helped wetland vegetation, waterfowl, and fish populations recover, but full recovery could take decades, and will require other measures as well, especially steps to restore more normal water flows. As it was, the water management strategy to control flooding resulted in periods of very high flow and long periods of zero flow. These unnatural flow regimes affected "recovering" wetlands tremendously. To help solve this problem, the water district began buying back drained wetlands along the river's banks and letting the system return to its natural state. By the time the state is through purchasing land, it could end up spending $40 to $50 million.

(continued)

Models of Global Sustainability 10.1 (*continued*)

Undoing the Damage to Florida's Kissimmee River

In 1992, after reviewing recommendations from the state and the Army Corps of Engineers, Congress approved legislation that will begin a 15-year restoration project on 22 miles of river. The water district will continue to buy up land along the river and will take an ecological approach to restoration. The goal is not to optimize one or a few valuable species, such as bass, but to restore the ecosystems badly damaged by channelization. The water district is also working upstream to help prevent runoff from farms and dairies.

Restoring the Kissimmee River is part of a major ecological experiment aimed at saving Florida's fast-vanishing wetlands. By the year 2000, the complex wetlands of Florida will function more like they did 100 years ago. But, at this very moment, engineers and construction companies are diligently draining wetlands the world over. One of the largest projects lies along the Nile River. In the home of countless birds and wildlife, huge dredgers are now busily sucking up the mud and straightening the river.

face runoff and our penchant for building in floodplains ensure humankind a future of flooded basements.

Controlling Flooding

Flood prevention measures along riverbanks include dams and **levees**, embankments to hold back the water.

These approaches treat the symptoms, not the underlying causes, of a serious and costly problem. **Watershed management** is a more effective approach. Watershed management is to flood control what preventive medicine is to health care. It includes steps to reduce deforestation and overgrazing in flood-prone regions and to replant trees, shrubs, and grasses on denuded hillsides.

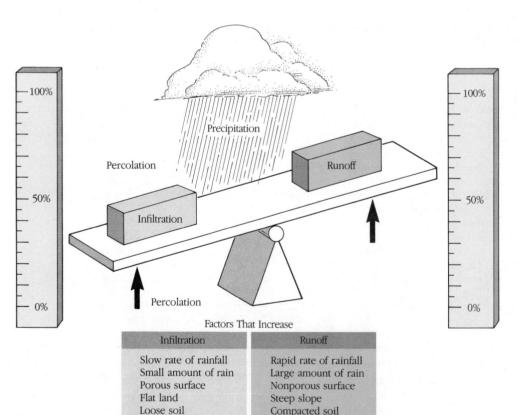

Figure 10.11 *Percolation-runoff ratio. When 75% of the water flows over the surface, 25% percolates into the soil (ignoring evaporation).*

Factors That Increase	
Infiltration	**Runoff**
Slow rate of rainfall	Rapid rate of rainfall
Small amount of rain	Large amount of rain
Porous surface	Nonporous surface
Flat land	Steep slope
Loose soil	Compacted soil
Abundant vegetation: Forests, marshes, grasses	Lack of vegetation: Clearing (clear-cuts), deserts

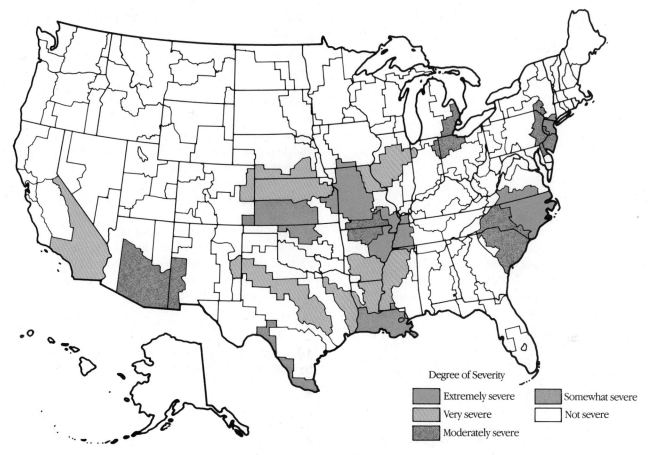

Figure 10.12 *Flood-prone areas in the United States. State boundaries are shown in black.*

It also includes measures to redesign urban environments to slow down the rush of water during a rainstorm, for example, the construction of holding ponds and underground storage tanks that absorb storm water. Lakes in city parks can trap much of the surface flow in a rainstorm, which can then be released slowly to rivers or used to water lawns or recharge aquifers. Storm sewers can also divert water to special holding ponds or water tanks that serve factories. Individual homeowners can divert gutter water to underground holding tanks to be used later for washing cars and watering lawns and gardens. (For an example of ways to avoid flooding, see Case Study 10.1.) Finally, floodplain zoning laws can help restrict building and other human activities in flood-prone areas.

One technique to reduce flooding that is falling out of favor is streambed channelization. **Streambed channelization** is a streamlining of streams and rivers achieved by bulldozing vegetation along stream banks and deepening and straightening river channels. This creates a glorified ditch, which may then be lined with concrete or rock.

Although channelization generally eases flooding in the immediate vicinity, its minor benefit is outweighed by habitat destruction, increased erosion of stream banks, a loss of river recreation, and increased flooding downstream. (For more, see Models of Global Sustainability 10.1.)

Streambed channelization in the United States was begun seriously in 1954, when the **Watershed Protection and Flood Prevention Act** authorized the U.S. Soil Conservation Service to drain wetlands adjacent to rivers to increase farmland and to reduce flooding. To date, 13,000 kilometers (8000 miles) of U.S. streams have been channelized. Another 13,000 kilometers have been targeted for similar "improvements." Critics argue that many projects are of dubious merit and should not be undertaken.

In closing, living sustainably requires efforts to tap into water resources without depleting them. It also requires methods to control flooding. Not surprisingly, our two most pressing problems and their solutions are related. In protecting and restoring ecosystems, we find ways to maintain, even increase, water supplies while reducing flooding, illustrating the fact that planet care is the ultimate form of self-care.

Let him who would enjoy a good future

waste none of his present.

—*Roger Babson*

Critical Thinking Exercise Solution

Hydropower is indeed a vast, untapped resource in the United States. The trouble is most of the untapped potential is in Alaska. In order to tap into this source of energy, large dams would have to be constructed far from human habitation. This would require extensive road building. To transport the electricity to cities and towns where it would be used, extensive power lines would need to be built. Such a project would not only be costly; it would also cause considerable environmental damage. Moreover, it would be inefficient because a great amount of energy is lost in transmission.

On the issue of cleanliness, although hydropower may be relatively pollution free once it is up and running, incredible amounts of energy are required to build roads, dams, and power lines, generating large amounts of pollution. In addition, dams can have significant impacts on fish and wildlife populations, and ruin excellent recreational resources as well.

In analyzing this subject, you probably used at least two critical thinking rules. First, you found that by defining your terms more correctly, the statement that hydropower is an abundant, untapped resource in the United States was true but misleading. Second, upon examining the big picture, you found that although a dam is a clean energy supply once it is functional, it is far from being environmentally benign.

Summary

10.1   The Hydrological Cycle

■   The **hydrological cycle** is a natural system driven by solar energy. Two basic processes are fundamentally important to this cycle, **evaporation** and **precipitation**.

■   At any single moment, only a small portion of the Earth's fresh water is available for human use.

10.2   Global Supply and Demand

■   Water shortages face virtually every nation and arise because: (1) rainfall is unevenly distributed across the face of the Earth, (2) cities and towns have often grown in excess of the sustainable supply, and (3) water is often used inefficiently by all sectors of society.

■   The world population withdraws huge amounts of water. Approximately 90% percent of it is used for crop and livestock production and 7% is used by industry; the rest is for domestic use.

■   In the United States, 75% of the water used by society comes from surface waters and the rest from groundwater.

■   The long-term prospects for meeting water needs for industry, agriculture, and individuals are dim in many parts of the world. Africa and Asia may find it difficult to find adequate water supplies, as may many other regions within countries, even in the developed world.

10.3   Impacts of Traditional Approaches to Water Shortages

■   Water shortages are usually met by increasing supply. This leads to an overexploitation of groundwater and surface water in many parts of the world.

■   Groundwater depletion may also induce **saltwater intrusion** in coastal zones and may cause inland ponds and swamps to dry up. In some regions, **groundwater overdrafting** has caused severe **subsidence**, which results in the collapse of highways, homes, factories, and pipelines.

■   Dams and reservoirs retain snowmelt and rainwater, help control floods, generate electricity, and increase certain forms of recreation. But, dams often inundate wildlife habitat, farmland, and towns. They also reduce stream flow into the ocean, resulting in changes in the salt concentration of receiving waters and reductions in the flow of nutrient-rich sediment to coastal waters, with devastating effects on the aquatic food web.

10.4   Meeting Present and Future Demands Sustainably

■   A sustainable water supply system can be built, but it will require massive efforts in conservation, recycling, restoration, and population control.

■   Since agriculture uses a large percentage of the world's water, measures to reduce irrigation losses are needed. Businesses and homes also need to develop efficiency measures.

■   New laws and building codes can help increase water efficiency in various sectors of society.

■ Improvements can be made in municipal water systems throughout the United States, which lose, on average, 12% of their water.

■ Water recycling by industries, municipalities, and individuals can also help create a sustainable water supply system. Industries can recycle water on-site. Wastewater can be applied to crops or to aquifer recharge zones.

■ Restoration of vegetation in watersheds is essential. Revegetation reduces sedimentation of dams and reservoirs, and enhances the infiltration of water into the soil, increasing groundwater supplies.

■ Measures to control population growth, especially in water-short regions, are important.

■ New sources of water may still be required. In developing countries, small dams, coupled with water conservation measures, can eliminate the need for large, environmentally destructive projects.

■ **Desalination** of saltwater is feasible in some places, but it is much more expensive than conventional projects. In addition, desalination plants produce salts that must be disposed of and encourage population growth in water-short regions.

10.5   Flooding: Problems and Solutions

■ Human activities that increase the surface runoff or decrease percolation can increase flooding. This problem is worsened by a human tendency to inhabit **floodplains**, river valleys that are subject to periodic flooding.

■ **Streambed channelization**, once considered to be an effective measure of reducing floods, destroys wildlife habitat, increases stream bank erosion, diminishes recreational opportunities, alters the aquatic environment, and often increases flooding in downstream sites.

■ Flooding can be prevented by revegetating denuded watersheds and by measures taken in urban environments to reduce surface flow.

Discussion and Critical Thinking Questions

1. What is the hydrological cycle? Draw a diagram showing how the water moves through the cycle. Why is this cycle important to you? How does human society alter the cycle?

2. Define *transpiration, evaporation, relative humidity, absolute humidity, saturation,* and *condensation nuclei.*

3. What sector is the largest water user in the world? In the United States?

4. Define the following terms: *groundwater, water table, saltwater intrusion,* and *subsidence.*

5. Discuss the problems caused by exploitation of groundwater and surface water.

6. You are appointed by the governor of your state to study floods and flood control projects. List reasons why flooding is now severe and sustainable solutions to correct these problems.

7. How can desalination of seawater help solve water shortages? What are the limitations and the problems created by this method?

8. Describe ways by which you and your family can help conserve water. Calculate how much water your efforts will save each day. How much will they save in a year?

9. What are the connections between flooding and water shortages? How can people solve both problems simultaneously?

10. Explain how the biological principles of sustainability can be used as a pattern for a sustainable system of water supply.

Suggested Readings

Brown, L. R. (1991). The Aral Sea: Going, Going . . . *World-Watch* 4(1): 20–27. Gripping story of what can happen when people abuse their water resources.

Dzurik, A. A. (1990). *Water Resources Planning.* Totowa, NJ: Rowman and Littlefield. Comprehensive survey of resource planning and management.

Howe, C. W. (1991). An Evaluation of U.S. Air and Water Policies. *Environment* 33(7): 10–15, 34–36. Skillful analysis.

McPhee, J. (1989). *The Control of Nature.* New York: Farrar, Straus & Giroux. Excellent reading.

Okun, D. (1991). A Water and Sanitation Strategy for the Developing World. *Environment* 33(8): 16–20, 38–43. Describes ways to provide clean drinking water in developing countries.

Postel, S. (1990). Saving Water for Agriculture. In *State of the World 1990.* New York: Norton. Outlines ways to avert water shortages in the 1990s.

Reissner, M. and Bates, S. F. (1989). *Overtapped Oasis: Reform or Revolution for Western Water.* Washington, D.C.: Island Press. Critique of western U.S. water policy with recommendations for change.

Wetlands, Estuaries, Coastlines, and Rivers

From certain vantage points, Chesapeake Bay on the eastern seaboard of the United States resembles a vast ocean. Perhaps fooled by its size, Americans have for years treated it with disrespect. Today the bay is in trouble. The rich abundance of organisms is diminishing, threatened by pollution, overfishing, and other activities.

The bay and its surrounding wetlands are home to a variety of fish and shellfish, including blue crabs, oysters, and striped bass. Properly managed, the bay could provide enough food to feed Japan. Chesapeake Bay is much more than a food source for the 13 million people who live near it. It is also a source of recreation for hunters, anglers, boaters, and nature enthusiasts.

Although it is only 310 kilometers (195 miles) long, the bay's shoreline measures nearly 12,000 kilometers (7000 miles). In many ways, the bay is a symbol of how wetlands and coastlines have been mismanaged for 200 years. Old-timers claim that her waters once contained "wall to wall" oysters. But, oyster populations have declined markedly, falling by 99% since 1870, with most of the decline occurring in the last 30 years. Striped bass populations have also fallen off alarmingly. Various strategies have been tried to reverse the downward trend in the bass population, among them a moratorium on striped bass fishing in January 1985. Because bass populations recovered slightly between 1985 and 1990, limited fishing has been permitted.

EPA studies show that the bay's submerged vegetation, so vital to fish, has dropped by 76% in the last 25 years, partly as a result of large **algal blooms** (bursts of algal growth resulting from certain pollutants) that block sunlight and impair the growth of submerged plants. Further damage to the bay is caused by aerobic (oxygen-requiring) bacteria, which deplete the oxygen supply when they decompose dead algae and organic pollutants in the water.

Population growth, commercial development, and the resulting pollution are the biggest threats to the bay.

The population is expected to reach 16 million by the year 2020. Most of the pollution comes from oil spills, sewage, toxic chemicals, heavy metals, and runoff from the bay's extensive drainage system, an area slightly smaller than Missouri. One of the worst pollutants is nitrogen from commercial duck farms, municipal sewage, and farms.

The story of Chesapeake Bay reminds us that living sustainably on the planet requires much more than protecting human civilization from floods and providing adequate amounts of water. It requires efforts to protect wetlands, estuaries, coastlines, and rivers, the subject of this supplement.

Wetlands

Wetlands, perpetually or periodically flooded lands, fall into two groups. **Inland wetlands** are found along streams, lakes, rivers, and ponds, and include bogs, marshes, swamps, and river overflow lands that are wet at least part of the year. **Coastal wetlands** are wet or flooded regions along coastlines, which include mangrove swamps, salt marshes, bays, and lagoons (Figure S10.1).

The Hidden Value of Wetlands

Wetlands are an extremely valuable and productive habitat for many animal and plant species. Deer, muskrat, mink, beavers, and otters are just a few of the animals that live in or around wetlands. In addition, shellfish, amphibians, reptiles, birds, and fish also call these endangered places home.

Aside from their importance to wildlife, wetlands play an important role in regulating stream flow. A study in Wisconsin showed that wetlands act like sponges, holding back rainwater and reducing natural flooding.

Figure S10.1 *Wetlands in the United States.*

The sponge effect has the added benefit of recharging groundwater supplies. Wetlands also remove sediment from surface runoff and thus reduce sedimentation in streams. In addition, wetland plants absorb nitrogen and phosphorus, two common pollutants washed from heavily fertilized land, thus reducing water pollution. According to another study, an acre of coastal wetland is the equivalent of an $85,000 sewage treatment plant. Coastal wetlands also help protect human settlements by absorbing **storm surges**, high waves accompanying high winds. Wetlands are used to grow certain cash crops, including rice, cranberries, and blueberries. Because their usefulness is not always apparent, they are often filled in or dredged to make way for housing, recreation, and industry.

Declining Wetlands

U.S. wetlands once covered an area the size of Texas (70 million hectares or 170 million acres). Today, less than one-third of America's wetlands remain, and what is left is fast heading toward oblivion as 48,000 to 200,000 hectares (120,000 to 500,000 acres) are destroyed each year.

Some areas have suffered greater losses than others. In California, New Zealand, and Australia, development has claimed more than 90% of their wetlands. Canada contains about one-fourth of the world's wetlands and has lost relatively little except along the Pacific coast where coastal marshes have been reduced by two-thirds.

No matter where you look, wetlands are in trouble. Coastal wetlands fall victim to dredgers that scoop up muck from streams and bays to increase navigability. Dredging also drains adjoining swamps. Cities often fill in swamps to accommodate homes, recreational facilities, roadways, and factories. Farmers fill in swamps to expand their arable land.

Mangrove swamps are one of the most valuable forms of wetlands and the most threatened, suffering heavy losses in Asia, Latin America, and western Africa. In Ecuador, nearly half of the mangrove swamps have been cleared to build shrimp ponds. India, Pakistan, and Thailand have lost three-fourths of their mangroves.

Protecting Wetlands

Concern for the loss of wetlands has stirred many governments into action. For example, Florida passed legislation in 1972 to regulate all wetland development. Strict controls in Nassau County, New York, have also slowed destruction. In 1988, New Jersey passed a law requiring buffer zones of 30 meters (100 feet) around important wetlands. Although a plan does not ensure protection, it is a step in the right direction.

The federal government has also assumed an increasing role in wetland protection through executive orders that prohibit its agencies from supporting construction in wetlands when a practical alternative is available. The federal **Coastal Zone Management Act** (1972) calls on coastal states to develop plans to protect their wetlands. Virtually all such states have either federally approved plans or state plans to safeguard their coastal wetlands. However, only 16 states have measures to protect inland, freshwater wetlands.

The federal government can purchase wetlands and set them aside as part of the **National Wildlife Refuge System**. The system today contains 4 million hectares (10 million acres) of wetlands. The U.S. Fish and Wildlife

Service also buys wetlands for protection. States own additional wetland acreage. All told, a little over one-fourth of existing U.S. wetlands are protected. In 1986, Congress passed a measure to raise more money for the purchase of wetlands by increasing duck stamp fees and charging admission at certain wildlife refuges. Many environmentalists, however, believe that Congress should find additional sources to fund the purchase of wetlands to sharply accelerate the current rate of acquisition.

Another important action aimed at protecting wetlands was the 1985 Farm Bill (Chapter 7). It included important "swampbuster" provisions—rules that deny federal benefits, such as low-interest loans and crop insurance, to farmers who drain and farm their wetlands. In 1987, the swampbuster provisions were tightened because of pressure on Congress from national environmental groups. The new rules could save thousands of hectares of **prairie potholes**, small ponds that dot the landscape in Iowa, Minnesota, Wisconsin, the Dakotas, and parts of Canada, which are now under heavy pressure from farmers.

Unfortunately, many wetland protection laws and regulations are ineffectual. In many instances, little is done to enforce these important laws.

Individuals can write to their representatives in Congress and their state legislature to find out what laws exist and how they are being enforced. Local action groups can help stimulate stronger enforcement when necessary. Personal action is needed more than ever because of deep cuts in the federal budget.

On an international scale, nations can cooperate to preserve wetlands. In 1971, representatives of many nations met in Iran to discuss the plight of wetlands, and agreed to protect those lands within their jurisdiction. In 1986, the United States ratified the agreement. Four U.S. wildlife refuges were added to a list of wetlands of international importance.

Estuaries

Estuaries are the mouths of rivers, where saltwater and fresh water mix. Estuaries, like wetlands, are critical habitat for fish and shellfish. Together, coastal wetlands and estuaries make up the **estuarine zone**. Two-thirds of all fish and shellfish depend on this zone during some part of their life cycle. Because fish are an important source of food, protection of the estuarine zone is vital to human survival and a sustainable existence.

The estuarine zone derives its richness from the land—eroded land. Eroded sediment, rich in nutrients, is carried in streams to the ocean, where it supports an abundance of aquatic organisms, especially algae, the base of the aquatic food web.

Damaging This Important Zone

The estuarine zone is vulnerable to a variety of assaults: pollutants from sewage treatment plants or industries; sediment from erosion that buries rooted estuarine vegetation; oil spills; and dams that cut off the life-giving flow of nutrients from the land. Cities may withdraw so much fresh water upstream that rivers run dry. In Texas, for example, drought and heavy water demands in past years have critically reduced water flow into estuaries. The Mexican delta of the Colorado River is a remnant of its former self. Freshwater inflows are critical to maintaining the proper salt concentration in coastal wetlands where mollusks and other organisms dwell. Salinity may be one of the most important factors determining shellfish productivity.

New research suggests that pollutants from the ocean can also concentrate in estuaries. Plutonium, a radioactive material deposited in the ocean from aboveground nuclear blasts, attaches to particles suspended in seawater that can concentrate in estuaries. Particles can also carry heavy metals, organic pollutants such as PCB, and insecticides such as DDT.

Many organisms inhabit the estuarine zone, but the most sensitive to pollution are clams, oysters, and mussels. Pollution is generally thought to be one of the major factors responsible for the decline in mollusk harvests in the last 25 years. Mollusks concentrate toxic heavy metals, chlorinated hydrocarbons, and many pathogenic organisms, including those that cause typhus and hepatitis. These pathogens may not affect the mollusks' survival, but they make them unsafe for human consumption.

The estuarine zone is a battlefield of sorts, beleaguered by pollution, water loss, sedimentation, dredging, and filling. Compounding the damage is the widespread problem of overharvesting of fish and shellfish. It is generally agreed that the decline in U.S. oyster production after 1950 was largely the result of overharvesting. Clams in the Northeast have likewise been severely overharvested. Chesapeake Bay is certainly a victim of heedless overfishing.

Over 40% of the U.S. estuarine zone has been destroyed. The most severe damage has occurred in California, along the Atlantic coast from North Carolina to Florida, and along the entire Gulf of Mexico. Despite state and federal laws to protect this zone, destruction continues.

Protecting the Estuarine Zone

Protecting estuaries and coastal wetlands is largely a matter of common sense and good resource management, part of an overall sustainable management strategy whose beneficial effects could ripple through the biosphere just as the adverse effects do now. Improved water pollution control (Chapter 17) is a key element of the plan. Erosion control is equally important. Water conservation to preserve vital water flow into estuaries is also needed. Restraint is the final element. Restraint means restricting dredging and filling and ending the overharvesting of fish and shellfish.

Protecting the estuarine zone presents a unique challenge in the United States, because 90% of the coastal land in the 48 conterminous states is privately owned. In 1972, Congress responded to the plight of the estuaries and coastal wetlands by passing the Coastal Zone Management Act mentioned earlier. This law set up a fund to provide the 35 coastal and Great Lakes states with assistance in developing their own laws and programs. It also provided them with money to purchase estuarine zones and estuarylike areas in the Great Lakes states. These regions are to be set aside for scientific study; as of December 1992, 18 national estuarine sanctuaries had been established. Six more are pending. Eventually, 20 to 30 could be established.

The Coastal Zone Management Act also allows for the establishment of national marine sanctuaries. Nine have been set aside off the coasts of California and the state of Washington; six more are pending. The purpose of these sanctuaries is to protect vital habitat for a variety of marine mammals and fish.

Half of the U.S. population lives in counties that are within an hour's drive of a coast, and a majority of the major cities are coastal. Discoveries of offshore oil, gas, and minerals pose new problems that require immediate solutions. An abundant supply of cooling water makes the coastal zones prime candidates for new power plants and oil refineries. Because of current and potential problems, the Coastal Zone Management Act is an important step in preserving U.S. coasts.

Still, according to critics, the act leaves too much discretion to the states. Some states, in fact, either have not adopted programs or enforce their programs poorly, leaving their coastal waters open to misuse.

An additional 15 federal laws have been passed to promote coastal zone management, but they are only as good as their enforcement. Lackadaisical enforcement is almost as bad as no enforcement at all. Without further, more serious efforts to improve coastal zone management, we will almost certainly lose more of this important habitat.

Barrier Islands and Coastlines

The eastern coast of the United States and Mexico is skirted by a chain of **barrier islands**, narrow, sand islands separated from the mainland by lagoons and bays (Figure S10.2). An estimated 250 barrier islands lie along

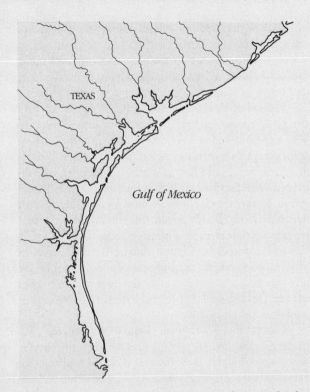

Figure S10.2   *The Texas Gulf Coast is a barrier island coast. Barrier islands are moving bodies of sand that make poor sites for homes and resorts.*

the Atlantic and Gulf coasts. Many of them are popular sites for recreation. Some have been purchased by the federal government and are used for recreation (as national seashores) and wildlife habitat, but most barrier islands are under private ownership.

In the last 40 years, many of these islands have been developed for vacationers. Summer homes, roads, stores, and other structures have invaded the grass-covered dunes. According to estimates of the U.S. National Park Service, in 1950, only 36,000 hectares (90,000 acres) of barrier islands had been developed, but by 1980, 113,000 hectares (280,000 acres) had been developed. In 1992, the U.S. Fish and Wildlife Service estimates put the number at about 170,000 hectares (420,000 acres).

Barrier islands and their beaches are part of a river of sand that migrates down the East Coast. The islands grow and shrink from season to season and year to year in response to two main forces. First, waves, which tend to arrive at an angle, erode the beaches. Waves create **beach drift**, a gradual movement of sand along the beach (Figure S10.3a). In addition, the wind creates **longshore currents** parallel to the land, which move sand along the beach. Combined, beach drift and **longshore drift** (movement of sand by longshore currents) are called **littoral drift**. This causes the barrier islands to move parallel to the main shoreline, shortening on one end and elongating on the other. Homes built on the up-current side of the island may collapse into the sea. Second, winter storms tend to wash over the barrier islands and move the sand closer to the land, destroying houses, roads, and other structures.

Federal actions and various relief programs have, in the past, encouraged development on barrier islands. When erosion and storm damage occurred, the government stepped in with money for disaster relief. Federal flood insurance paid for the damage, and federally subsidized construction projects helped rebuild roads and stabilized the islands. In subsequent years, when storms devastated the islands again, the rebuilding cycle was repeated at considerable expense. Recognizing this cycle, Congress passed the **Coastal Barrier Resources Act** in 1982 to prohibit the expenditure of federal money for highway construction and other development on barrier islands. Despite denial of federal subsidies, development continues on many islands.

Coastal beaches also fall victim to human activities. Understanding the dangers requires a look at the natural processes that affect these beaches. **Coastal beaches**, like barrier islands, are eroded by longshore currents. In fact, U.S. beaches are like great rivers of sand kept in constant motion by the major coastal currents. Sand lost in one area is replaced by sediment carried to the sea by rivers. Consequently, dams that trap sediment diminish the natural replacement of sand on coastal beaches. According to one estimate, dams hold back nearly 40% of the sediment that once reached the mouth of the Santa Clara River north of Los Angeles. This robs California beaches of 15 million metric tons of sand each year. From New Jersey to Texas, the story is the same.

Some communities erect barriers, called **jetties**, to prevent erosion by longshore currents. These structures only slow down the process. Jetties are sometimes built to maintain navigable passageways in coastal harbors. In 1911, for instance, two 300-meter jetties were built on the New Jersey coast north of Cape May to prevent sand from filling in the harbor. While performing admirably in their appointed duty, the jetties had a disastrous effect on down-current beaches. The beaches at Cape May grew thinner and thinner. By the 1920s, the town was actively fighting back by building small jetties to keep the remaining sand from being washed away and to trap the sand flowing in the longshore currents. To the townspeople's dismay, their efforts were fruitless. Beaches retreated by 6 meters (20 feet) per year. Lighthouses fell into the sea. The ocean threatened to swallow the airport. After years of anguish, the town turned to an expensive pumping system that draws sand from above the two large jetties and moves it down to the beaches.

To protect coastlines, the delicate balance between sediment flow and erosion must be maintained. Dams that retain sediment should be avoided or rocky beaches will become common in the future. Beaches must be allowed to grow and shrink with natural cycles. Realizing that it makes more sense to cooperate with nature, many government officials on the Atlantic coast have begun to develop plans that would return developed shorelines to their natural state following destructive hurricanes. This step would begin a retreat from the shorelines, where nature probably intended humankind to be only a visitor.

Wild and Scenic Rivers

More and more people are flocking to rivers to embark on a variety of sports, such as fly-fishing, kayaking, rafting, inner tubing, and canoeing. This great recreational resource, however, is increasingly imperiled by dams and diversion projects. Foreseeing the need to protect rivers for recreation and habitat for fish and other species, Congress passed the **Wild and Scenic Rivers Act** in 1968. It was designed to prevent the construction of dams, water diversion projects, and other forms of undesirable development along the banks of some remaining free-flowing rivers.

Undammed rivers offer much more than excitement for the river runner. Like wilderness, they provide unrivaled scenery and opportunities for relaxation and reflection. For the adventurous kayaker and rafter, they offer a taste of danger and the opportunity to test one's skills and strength. They are home for fish and other forms of wildlife. Their canyons offer opportunity for geological study. Taken together, these opportunities make a river much more than a source of water.

Because rivers exist in varying states of development, Congress established a three-tiered classification scheme in the Wild and Scenic River System: **wild rivers** are relatively inaccessible and "untamed," **scenic rivers** are largely undeveloped and of great scenic value, and **recreational rivers** offer important recreational opportunities despite some development. As of December 1992, 151 river segments had been included in the system, totaling over 17,300 kilometers (10,379 miles).

Wild or scenic river designation is often marked by controversy because so many interests vie for a river's benefits: municipal water consumers, paper manufacturers, farmers, anglers, and white-water boaters. Competing interests make compromise difficult or impossible. A dammed river provides water for a new paper mill, water skiing, and boating but irretrievably floods the kayakers' rapids and the fly-fisherman's favorite pools.

The fight to protect free-flowing U.S. rivers from development continues today. Much of the pressure to dam American rivers, however, has been reduced because of economic and legal forces. After years of paying for questionable dams, Congress found that many projects return only a few cents for every dollar invested. As a result, federally subsidized water projects, often handed out as political favors, have fallen into disfavor.

Further conflicts can be avoided by finding alternative means of supplying water, as discussed in Chapter 10. Many conservationists argue that valuable recreational rivers must be preserved just as endangered species are. When a scenic river gorge is dammed, it is gone forever.

A river is a vital resource, but dammed and diverted to water-hungry, often-wasteful consumers, that river becomes a tragic symbol of poor planning and undisciplined greed. Our goal should be to manage rivers wisely and efficiently, minimizing waste and damage and ensuring future generations the use of treasures we now enjoy and too often take for granted.

Suggested Readings

Horton, T. and Eichbaum, W. M. (1988). *Turning the Tide: Saving the Chesapeake Bay*. Washington, D.C.: Island Press. Outlines efforts to save this valuable resource.

Kusler, J. (1992). Wetlands Delineation: An Issue of Science or Politics. *Environment* 34(2): 6–11, 29–37. Summarizes the recent debate over U.S. wetlands delineation.

Natural Resources Defense Council. (1990). *Coastal Alert: Ecosystems, Energy, and Offshore Oil Drilling*. Washington, D.C.: Island Press. Practical guide for protecting coastal waters.

Platt, R. H., Beatley, T., and Miller, H. C. (1991). The Folly at Folly Beach and Other Failings of U.S. Coastal Erosion Policy. *Environment* 33(9): 6–9, 25–32. Explains how poorly conceived coastal erosion policies can backfire.

World Wildlife Fund. (1992). *Statewide Wetlands Strategies*. Washington, D.C.: Island Press. Practical guide for developing comprehensive statewide wetlands strategies.

Nonrenewable Energy Sources

Our entire economic structure is built from and propelled by fossil fuels. We have invaded the long-silent burial grounds of the Carboniferous Age, appropriating the dead remains of yesteryear for the use of living today.

—Jeremy Rifkin

Critical Thinking Exercise

The oil embargoes of the 1970s and the 1991 Persian Gulf War led many proponents of nuclear energy in the United States to lobby the public and Congress to support nuclear energy to lessen America's dependence on foreign oil. Nuclear power, they argue, has the added benefit of not contributing to global warming. Is this thinking valid? Why or why not? What critical thinking rules did you use in analyzing this issue?

Pumped up from wells or scooped up by the truckload in surface mines, energy in its many forms is the lifeblood of modern industrial societies. At the same time, it is their Achilles' heel. Cut off the supply for even a brief moment, and disaster would strike. Industry would come to a standstill. Agriculture and mining would halt. Millions would be out of work. Automobiles would vanish from city streets.

Many developing nations have pinned their hopes for economic progress on their ability to tap into oil, coal, natural gas, and, to a lesser extent, nuclear power, sources which have fueled the industrial transformation of wealthy nations. China, for example, is hoping to get much of its future energy from coal.

At this juncture in history, it is important to consider the sustainability of current energy strategy. Two vital questions arise: First, is the industrial world's dependence on coal, oil, natural gas, and nuclear energy sustainable? Second, can developing nations achieve success by following in the same footsteps?

This chapter examines **nonrenewable fuels**, those that cannot be regenerated in the foreseeable future by natural forces. It shows the impacts of their use and presents information on projected supplies. It closes with some guidelines on creating a sustainable energy future.

11.1   Energy Use: Our Growing Dependency on Nonrenewable Fuels

U.S. Energy Consumption

The history of energy use in the United States is one of shifting dependency (Figure 11.1). One hundred years ago, Americans had few choices for energy. Wood, a renewable resource, was the chief form of energy. Today, the nation's options are many: coal, oil, natural gas, hydropower, geothermal energy, solar power, nuclear power, and wind (Figure 11.1).

American energy options began to expand in the late 1800s as wood supplies became depleted. Coal began to be used in factories, but coal was a dirty, bulky fuel that was expensive to mine and transport. When oil and natural gas were made available in the early 1900s, coal use began to decline. These new, cleaner-burning fuels were easier to obtain and cheaper to transport.

Today, despite numerous energy options, the United States depends primarily on three nonrenewable fuels: oil, natural gas, and coal. In 1991, oil accounted for slightly more than 40% of total energy consumption (Figure 11.2a). Natural gas provided about 25% and coal provided a little over 22% of the energy. All told, fossil fuels accounted for 87% of U.S. energy use. Nuclear power, another nonrenewable fuel, provided 8%. Renewable sources—solar, geothermal, and hydropower—supplied a measly 4%. Although it does not appear in the energy calculations, conservation also provides us with a great deal of energy. (Figure 11.2b breaks down energy consumption by user.)

With only 6% of the world's population, the United States consumes about 30% of the world's energy. In 1991, the United States consumed 81.5 quadrillion BTUs (quads) of energy, exceeding a previous all-time high of 79 quads in 1978.[1] As Figure 11.2b shows, industry and business consume more than half of the energy. Transportation consumes about 27% and residential use is about 20%.

Global Energy Consumption

Virtually all industrial nations rely principally on non-renewable energy sources, with the mix varying from one country to the next, depending on local resources and access to markets (Figure 11.3). Japan, for example, gets 56% of its energy from oil, 19% from coal, and 11% from nuclear—all of which must be imported.

In the developing countries, renewable fuels, mostly wood and cow dung, provide 35% of the energy needs. Of the nonrenewable energy fuels used in these countries, oil supplies the largest share, on average about 50% (Figure 11.3). Coal, natural gas, and hydroelectric power supply the rest, with nuclear contributing only a tiny fraction.

Worldwide, human civilization consumes nearly 350 quadrillion BTUs of commercially traded fuels, or **primary energy**, excluding wood, peat, and animal wastes, which though important in many countries are unreliably documented. That's equivalent to over 7.8 million metric tons of oil each year. The biggest users of energy are Americans. On a per capita basis, Americans consume more than twice as much energy as the people of Japan and western Europe and about 16

---

[1]A British thermal unit (BTU) is the amount of energy it takes to heat 1 pound of water 1°F. One quadrillion equals 1 million billion (1,000,000,000,000,000) BTUs.

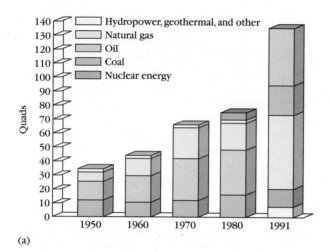

(a)

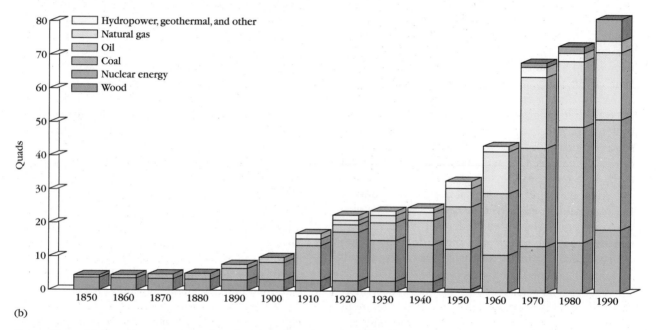

(b)

Figure 11.1   (a) *Energy consumption from 1950 to the present. U.S. energy dependency has shifted over the years. Today, oil, natural gas, and coal are the main sources of energy in the United States.* (b) *Energy consumption in the United States by fuel type from 1850 to the present.*

times more per capita than the people of developing nations (Figure 11.4). As a result, Americans have a disproportionately large impact on the environment.

11.2   Fossil Fuels: Costs and Benefits

Energy is not cheap. In addition to economic costs, society pays a huge environmental price for use of nonrenewable energy, including damage to human health and the environment. Lung disease, acid rain, global warming, and urban air pollution are but a few of the many costs.

Pollution, much of it caused by energy production and consumption, is the topic of Chapters 14 through 18. As these chapters point out, the economic impacts of pollution are generally borne by society and not by the manufacturers directly responsible for them. For many years, the environmental movement has sought to expose the external costs and to reduce them through regulations.

In order to understand the external costs of all fuels, it is necessary to understand where energy comes from and the many steps involved in delivering energy to homes, factories, and gas stations. Figure 11.5 presents a diagram of some of the major steps involved in extracting energy and delivering it to end users. This chain of

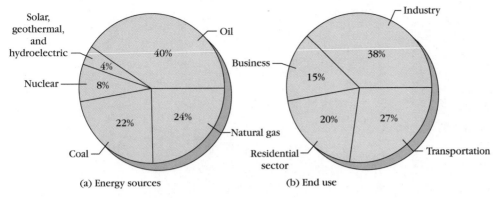

Figure 11.2   (a) *Major energy sources in the United States.* (b) *Breakdown of energy use by sector.*

events constitutes an energy fuel cycle, or an **energy system,** and is composed of five phases: exploration, extraction, processing, distribution, and end use. The most notable environmental impacts occur at the extraction and end-use phases. Keep this diagram in mind as you explore the nonrenewable energy options.

Crude Oil

Petroleum, or **crude oil,** is a thick liquid containing many combustible hydrocarbons. Found in deep deposits in the seafloor and on land, crude oil often occurs with natural gas (Figure 11.6). Geologists locate oil deposits through a careful study of the type of rocks found in various regions and by other fairly low-impact means. However, to explore for oil in remote areas, roads must often be built into them so that exploration crews can enter. (For a discussion of the potential impacts of oil exploration in the Arctic, see Case Study 11.1.)

Once located, crude oil is extracted via wells drilled into deposits. Oil flows into the wells and is pumped up

to the surface. This process is known as **primary recovery.** To increase yield, water can be injected into nearby wells, forcing the thicker crude oil into a central well, where it is pumped to the surface. This process, known as **secondary recovery,** combined with primary recovery, removes about 33% of the crude oil in a deposit. Another 10% can be acquired by injecting steam and carbon dioxide into wells, a process known as **tertiary recovery.** Tertiary recovery yields more oil from existing deposits, but it is costly.

After the crude oil is removed, it is transported by ships, trucks, or pipelines to refineries. Transportation of oil by ships and pipelines is a source of considerable environmental damage. Ships all too often run aground, spilling their contents into lakes and oceans with devastating environmental consequences (Chapter 17). Pipelines may leak or break, causing additional problems.

In a refinery, the oil is heated and distilled, a process that separates out many of the different hydrocarbon molecules in crude oil, including those that make up

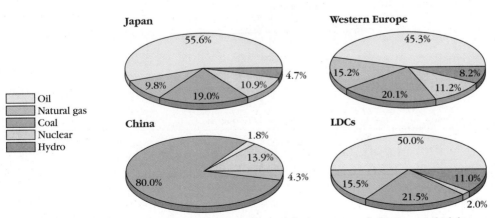

Figure 11.3 *Energy sources in other countries.*

Noteworthy points are coal's pre-eminent position in China and the importance of oil in Japan (which has very limited indigenous production).

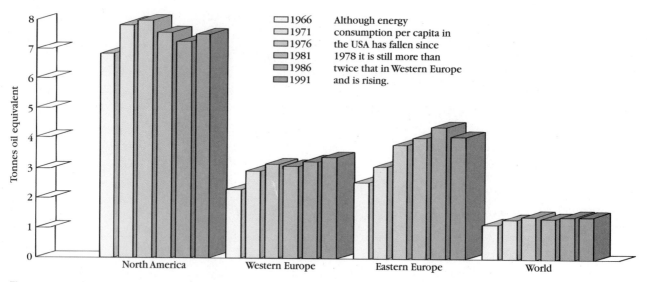

Figure 11.4   *Per capita energy consumption in North America, Western Europe, Eastern Europe, and the world.*

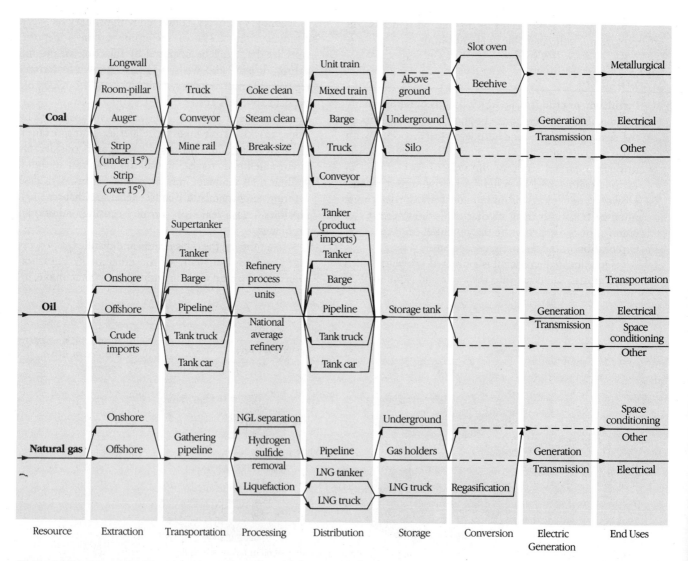

Figure 11.5   *Energy we use comes to us through an elaborate set of steps, the energy fuel cycle. Environmental impacts occur at each stage.*

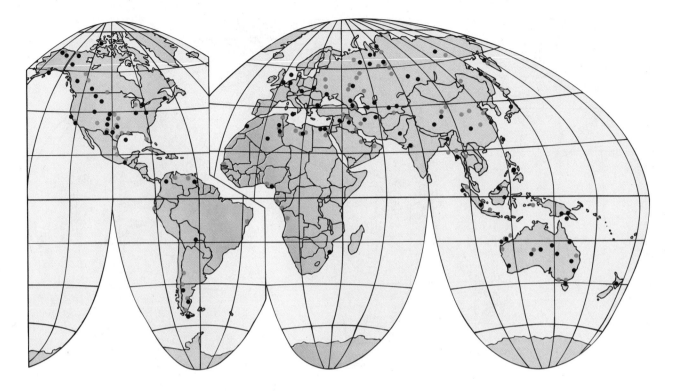

Oil
● Natural gas

Figure 11.6 *Map of global oil and natural gas deposits. Note that two-thirds of the world's oil deposits are located in the politically volatile Middle East.*

gasoline, diesel fuel, heating oil, and asphalt. Many small organic compounds are also extracted during this process. They are used to make a variety of medicines, plastics, paints, and pesticides. Refineries use considerable amounts of energy and are major sources of air and water pollution.

The combustion of crude-oil derivatives, such as gasoline and diesel fuel, produces enormous quantities of carbon dioxide, a greenhouse gas (Chapter 16), and other pollutants, among them sulfur dioxide and nitrogen dioxide, which are converted into acids in the atmosphere. In fact, many of the most important local and global environmental problems cited in this text stem from industrial nations' heavy dependence on oil and its by-products.

Natural Gas

Natural gas is a mixture of low-molecular-weight hydrocarbons, mostly methane. Burned in homes, factories, and electric utilities, natural gas is often described as an ideal fuel because it contains few contaminants and burns cleanly. Like oil, it is easy to transport within countries via pipelines and generally can be transported

safely in the gaseous form. However, to transport it across oceans, it must be liquefied. In the liquid form, natural gas is unstable and highly flammable. A ship containing liquefied natural gas could burn intensely. Natural gas is also fairly economical and requires relatively little energy to extract.

Natural gas comes from wells as deep as 10 kilometers (6 miles), often in association with oil. On land, drilling rigs generally have minimal impact unless they are located in wilderness areas, where roads and noise from heavy machinery and construction camps can disturb wildlife. However, natural gas extraction can cause subsidence in the vicinity of a well. One notable example is in the Los Angeles–Long Beach harbor area, where extensive oil and gas extraction, which began in 1928, has caused the ground to drop 9 meters (30 feet) in some areas.

Coal

Coal is the most abundant fossil fuel, one not likely to be depleted soon. In fact, world coal supplies could last hundreds of years. The United States has about one-third of world reserves (Figure 11.7).

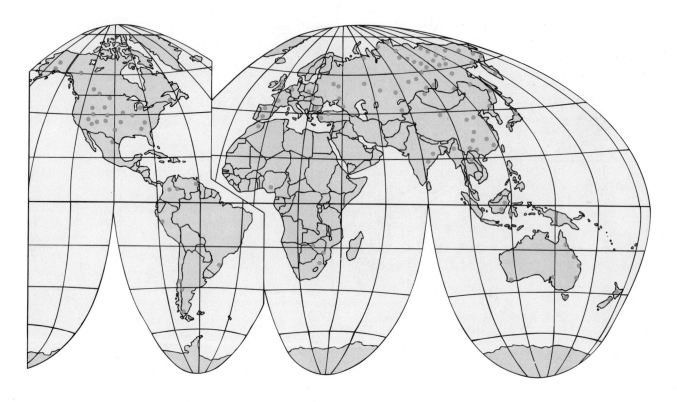

Coal

Figure 11.7 *Map of global coal deposits.*

In the United States, coal is widely used to generate electricity for homes and businesses. Some businesses also burn coal directly to generate heat and steam. Like nuclear energy, coal fits well in the existing electric grid system; in the United States, it costs consumers about 5 to 7 cents per kilowatt-hour (about half the cost of nuclear energy). Energy conservation, cogeneration, and some new wind generating facilities are the only energy sources economically cheaper than coal (Chapter 12).

Coal-fired power plants are an established technology and are relatively inexpensive to build compared with nuclear power plants. Some experts consider coal a promising candidate for the future because it can be converted into a synthetic natural gas and oil, thus providing fuel for transportation and buildings.

On the other hand, coal is dirty and environmentally costly. At virtually every step in the coal energy system, significant impacts may occur. (Table 11.1 summarizes many of these problems.)

Coal exploration once caused enormous damage in the eastern United States because companies used bulldozers to carve mountainsides in search of coal seams. Dirt and debris were often pushed over hillsides, burying trees and vegetation. When coal was found, the bull-

dozers removed more material, creating additional waste that was similarly discarded (Figure 11.8). Soil erosion from these sites was enormous.

Today, these harmful exploration and mining practices have been replaced with less damaging ones. The rock and dirt lying over coal seams, **overburden**, is hauled away rather than pushed over the hillside, and is stored until the coal seam is exhausted. It is then trucked back to the site and used to fill in the scar. Even though practices have changed, coal strip mines still create eyesores and can increase erosion. Roads can also erode, filling streams with sediment.

In the East, much of the coal comes from **underground mines**. Coal mines are notorious for explosions and cave-ins, making underground coal mining the most hazardous of the major occupations in the United States. Since 1900, more than 100,000 Americans have been killed in underground coal mines, and 1 million have been permanently disabled. As a result of stricter safety regulations, death rates have dropped substantially in the last 50 years.

Underground coal mines also cause **black lung disease**, or **pneumoconiosis**, a progressive, debilitating disease caused by breathing coal dust and dirt particles

Table 11.1 Major Environmental Impacts of Fossil Fuels

Fuel	Extraction	Transportation	End Uses
Coal	Destruction of wildlife habitat, soil erosion from roadways and mine sites, sedimentation, aquifer depletion and pollution, acid mine drainage, subsidence, black lung disease, accidental death	Air pollution and noise from diesel trains	Air pollution from power plants and factories—especially acid pollutants and carbon dioxide, thermal pollution of waterways
Oil	Offshore leaks and blowouts causing water pollution and damage to fish, shellfish, birds, and beaches; subsidence near wells	Oil spills from ships or pipelines	Air pollution similar to that from coal
Natural gas	Subsidence and explosions	Explosions, land disturbance from pipelines	Fewer air pollutants than coal and oil, but nitrogen oxides and carbon dioxide

(Figure 11.9). Victims have difficulty getting enough oxygen because the tiny air sacs, or alveoli, in the lungs break down. Exercise becomes difficult, and death is slow and painful. Despite safety improvements, one-third of all U.S. underground coal mines still have conditions conducive to black lung disease, a problem that costs U.S. taxpayers over $1 billion per year in federal worker disability benefits.

Collapsing mines also cause subsidence, a sinking of the surface. Cracks form on the surface, ruining good farmland. In some cases, streams vanish in the fissures. Over 800,000 hectares (2 million acres) of land has subsided in the United States from underground coal mining. For every hectare (2.5 acres) of coal mined in central Appalachia, over 5 hectares (12.5 acres) of surface becomes vulnerable to subsidence.

In the West and Midwest, coal is extracted from **surface mines.** In surface mining operations, the topsoil is first removed by scrapers and is set aside for re-application (Figure 11.10). Next, the overburden is

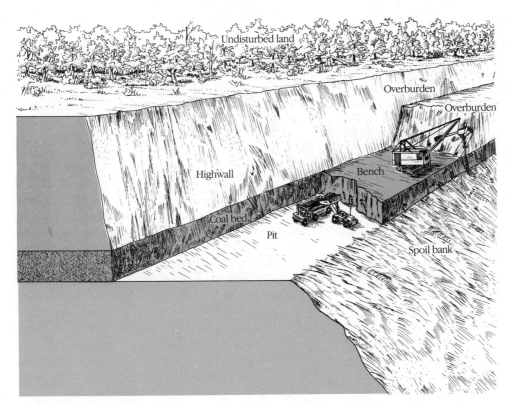

Figure 11.8 *Contour strip mining is common in hilly terrain, for example, in the eastern U.S. coal fields. For many years, overburden was dumped on the down slope, destroying natural vegetation and creating enormous erosion problems. Today, overburden is hauled away and stored. When the coal seam is exhausted, the overburden is replaced, graded, and planted.*

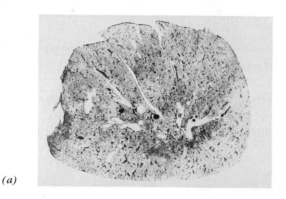

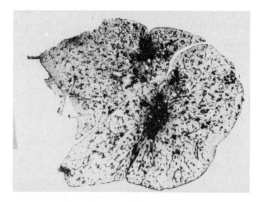

Figure 11.9 (a) *A cross section of a normal lung.* (b) *A cross section of a lung from a retired coal miner with black lung disease. The black material is carbon from coal dust.*

dynamited and removed by huge shovels called **drag-lines** to expose the coal seam. The coal is removed and hauled away, and another parallel strip is cut. The over-burden from the new cut is placed in the previous one, as shown in Figure 11.10. This type of surface mine is often referred to as a **strip mine**. As required by the **Surface Mining Control and Reclamation Act** (1977), the overburden must be regraded to the approximate orig-inal contour, and the topsoil replaced. Seeds are sown, increasing the likelihood that the area will re-vegetate.

Strip mines create eyesores, destroy wildlife habitat and grazing land, and may increase erosion (Fig-ure 11.10). Strip mining in the hilly terrain of Kentucky, for instance, has been shown to increase erosion from 0.4 metric tons per hectare to 2 to nearly 150 metric tons per hectare. Proper reclamation can restore wildlife habitat and grazing land and eliminate the eyesores. However, unless erosion control measures are taken dur-ing mining operations and reclamation is carried out immediately afterward, soil erosion can become a major problem.

Surface mines can also disrupt and pollute ground-water supplies in the West because many aquifers are located near or in coal seams. In Decker, Montana, extraction of coal from an aquifer seam resulted in a drop in the water table of 3 meters (10 feet) or more within a 3-kilometer (2-mile) radius of the mine. Resi-dents who depended on the aquifer were forced to drill deeper wells or find new water supplies.

If not properly constructed, the dirt roads that trans-port workers to mines create enormous problems in hilly eastern regions. During heavy rains, sediment from road-ways washes into nearby streams, reducing their water-carrying capacity and killing fish and other aquatic

organisms. When rain falls, water spills over stream banks, flooding farms and communities.

Many abandoned coal mines in the East leak sulfuric acid into streams. **Acid mine drainage**, as it is called, consists of sulfuric acid formed from water, air, and sulfides (iron pyrite) in the mine. A bacterium (*Thio-baccillus thioxidans*) facilitates the conversion, making the waters highly acidic. Acid mine drainage kills plants and animals and inhibits bacterial decay, allowing large quantities of organic matter to build up in streams. Sul-furic acid also leaches toxic elements, such as aluminum, copper, zinc, and magnesium, from the soil and carries them to streams.

Acid can render water unfit for drinking and swim-ming. Municipal and industrial water must be chemically neutralized before use. Acid also corrodes iron and steel pumps, bridges, locks, barges, and ships, causing damage in the millions of dollars each year.

U.S. mines, most of them abandoned, produce about 2.7 million metric tons of acid per year. Acid mine drain-age pollutes over 12,000 kilometers (7250 miles) of U.S. streams, 90% of which are in Appalachia (Figure 11.11). Cleaning up abandoned mines could take decades and cost billions of dollars; further increases in coal produc-tion could increase acid mine drainage, although aware-ness of the problem has resulted in efforts to reduce it.

Underground mines produce enormous quantities of wastes, which are transported to the surface and dumped around the mouth of the mine. These wastes, called **mine tailings**, contain heavy metals, acids, and other pollutants that wash into streams during rains. Coal-cleaning plants, designed to crush the coal and wash away impurities, also produce enormous quantities of waste. These wastes are often stored in holding ponds that sometimes leak and pollute nearby streams.

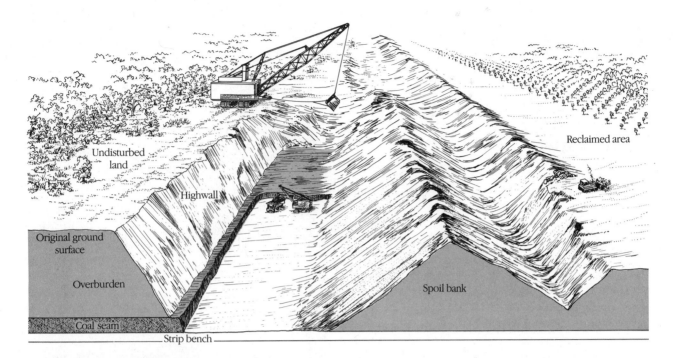

Figure 11.10 *In flat terrain, coal is extracted by strip mining (above). A dragline removes the overburden to expose the coal seam. Overburden is placed on the previously excavated strip and eventually regraded and replanted. Aerial view of a strip mine (left). If land is not carefully reclaimed, it can be permanently ruined.*

After the coal is extracted, it is loaded onto trucks or rail cars and shipped to power plants. Diesel trains, which move most of the nation's coal, pollute the air with a black cloud of particulates and other potentially harmful pollutants, such as sulfur dioxide.

About two-thirds of U.S. coal is burned to generate electricity; the rest is burned by industry to generate heat or to produce steel and other metals. A 1000-megawatt coal-fired power plant, which produces electricity for about 1 million people, burns about 2.7 million metric tons of coal each year. Numerous environmental pollutants are released into the atmosphere during coal combustion. The most notable are particulates, sulfur oxides, nitrogen oxides, carbon monoxide, and carbon dioxide. These pollutants are responsible for many of the world's most threatening environmental problems

and cause billions of dollars in damage to fish, lakes, buildings, and human health (Chapters 15 and 16).

Coal combustion also produces enormous quantities of solid waste. During combustion, the fine dust known as fly ash is produced. **Fly ash** is mineral matter that makes up 10% to 30% of the weight of uncleaned coal. It is carried up the smokestack with the escaping gases and if captured by pollution control devices, becomes a hazardous solid waste. Sulfur dioxide gas is also emitted from the stack, but it can be removed by a pollution control device known as a **smokestack scrubber** (Chapter 15). Scrubbers produce a toxic sludge containing fly ash and sulfur compounds that must be disposed of. Some mineral matter that is too heavy to form fly ash remains at the bottom of the coal-burning furnace as **bottom ash**. It, too, is a hazardous waste that

Figure 11.11 *Streams affected by acid mine drainage in the eastern United States. Acid mine drainage is prevalent in regions where rock contains the sulfur-containing compound known as iron pyrite. Acid mine drainage kills fish and other aquatic organisms.*

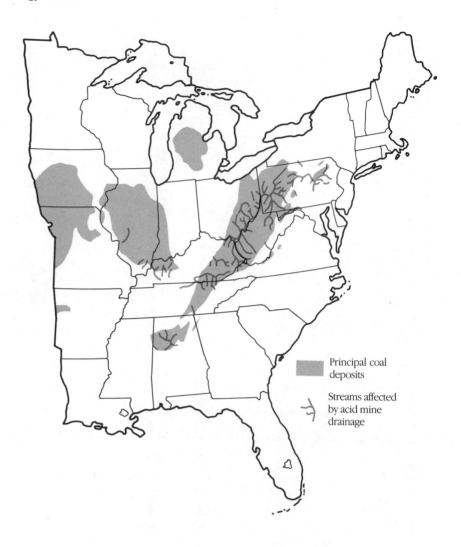

Principal coal deposits

Streams affected by acid mine drainage

must be disposed of. Wastes from coal-burning furnaces and their pollution control devices are generally buried in landfills, but toxic substances from these sites may leak into groundwater supplies, polluting them.

A 1000-megawatt power plant may produce 180,000 to 680,000 metric tons of solid waste each year, including fly ash, sludge from scrubbers, and bottom ash. In 1991, U.S. coal-fired power plants produced about 78 million metric tons of solid waste. Approximately one-fourth of the waste was used to make road surfaces and other products; the rest ended up in landfills.

Understanding the coal fuel cycle and the hidden consequences of energy consumption makes it clear that a simple flick of a light switch by a consumer creates a trail of environmental destruction. It also shows that each of us is partly responsible for the long list of environmental problems facing the world. Knowing that the environmental crisis stems from individual actions stimulates many to take steps to reduce their contribution.

The impacts of the coal production-consumption cycle can be reduced by passing tougher laws and enforcing existing laws. In addition, technological improvements can help us burn coal more efficiently and more cleanly (Chapter 15). Energy efficiency measures can help to reduce demand for this environmentally costly fuel.

Synthetic Fuels from Oil Shale, Tar Sands, and Coal

Oil is essential for transportation, heating, and chemical production, but global supplies are fast declining. Three nonrenewable substitutes are available: oil shale, tar sands, and coal derivatives. Each of these can be converted into combustible liquid and gaseous fuels known as **synthetic fuels,** or **synfuels.**

Case Study 11.1

Controversy over Oil Exploration in the Arctic National Wildlife Refuge

In the far northeastern corner of Alaska lies the Arctic National Wildlife Refuge (ANWR). Set aside to protect wildlife, this refuge encompasses 7.7 million hectares (19 million acres). It has been described as the last great American wilderness. But, if the oil companies have their way, a major portion of this refuge—critical to wildlife—could be opened to oil exploration. If oil is found, the pressure will build to develop widespread drilling in the coastal plain, a 0.6 million-hectare (1.5 million-acre) region—an area almost as large as Yellowstone National Park.

Oil companies want to search for oil in the delicate arctic tundra of the coastal plain, a region 210 kilometers (115 miles) long and 50 kilometers (30 miles) wide. The only section of land along the entire 1800-kilometer (1100-mile) north coast of Alaska that is closed to oil exploration and drilling, the coastal plain is the annual calving ground of several large caribou herds. It also provides habitat for polar bears, grizzlies, wolves, moose, wolverines, and numerous small mammals. Many thousands of birds spend the summer there, raising their young and feeding on insects. It is also the habitat of the musk ox, which was nearly hunted to extinction by 1969. Today, as a result of conservation efforts, the herds number over 400.

In 1991, after public protest over the tragic 1989 oil spill in Prince William Sound subsided (Chap-

ter 17), then-President George Bush and the oil companies introduced legislation that would allow drilling in ANWR. Because of a massive outpouring of public protest, the bill was defeated, but most experts believe that the oil companies may soon reintroduce legislation that would permit them to explore for oil in the region.

If oil is found, the pristine area would very likely turn into a nightmare of roads, oil platforms, waste ponds, buildings, and gravel pits. Four airfields and 50 to 60 oil platforms are proposed. A power plant and several oil processing plants would be built on the tundra, and hundreds of kilometers of roads and pipelines would crisscross this delicate land.

While proponents maintain that oil development and wildlife can coexist in harmony, critics predict a 20% to 40% decline in one of the major caribou herds. They expect over half of the musk oxen to perish. Populations of grizzlies, polar bears, wolverines, and other animals are also likely to decrease substantially. Air pollution, water pollution, and hazardous wastes are expected to have major impacts on wildlife and the long-term ecological health of the region. One exploratory well alone requires 35,000 cubic meters (115,000 cubic feet) of gravel, which would be excavated from streambeds to make pads and roads. Oil spills on the tundra and careless waste disposal, common in nearby Prudhoe Bay, would change this

An oil drilling pad in Prudhoe Bay, Alaska. The delicate Arctic tundra could be ripped apart to satisfy America's insatiable thirst for oil.

(Continued)

Case Study 11.1 (*continued*)

Controversy over Oil Exploration in the Arctic National Wildlife Refuge

delicate landscape, whose short growing seasons and harsh winters greatly impair the natural healing that normally takes place in the wake of human interference.

Proponents argue that we need the oil to reduce reliance on foreign sources and that we need to start exploring in ANWR now because of the 10- to 15-year lead time required to fully "develop" the area. Geologic evidence, some say, indicates a high potential for oil discovery, and oil is needed to help keep the Alaskan economy alive. They say that environmental regulations will ensure minimal impacts on wildlife and the environment.

Those in favor of preserving the refuge intact argue that the environmental costs are too high. Experience in Prudhoe Bay shows that oil companies often ignore environmental regulations. Frequent violations have been cited. The U.S. Department of Interior, which generally favors oil development, has recorded over 17,000 oil spills in the Arctic since 1973. Where the oil saturates the soil, vegetation fails to recover. Opponents of exploration also argue that since the

entire north coast of Alaska is already open to oil exploration, ANWR should not be—now or ever. Furthermore, they point out that slight improvements in automobile gas mileage could easily "provide" as much oil as ANWR could generate over its lifetime—and at a much lower economic and environmental cost. Moreover, critics say that proponents have exaggerated the potential for finding oil that is economically feasible to recover. Based on the oil industry's own reports, they say, there's only a one-in-five chance of finding economically recoverable oil if oil prices were $33 per barrel. Today, oil sells for about $20 per barrel.

Few of us will ever visit ANWR, but most of us take comfort in knowing there are places where wildlife is free from the torment of modern industrial society. We can keep it that way by insisting that the ANWR stay closed, by reducing our own energy consumption, and by convincing government officials to promote energy conservation and renewable energy resources.

Oil Shale   Oil shale is a sedimentary rock containing an organic material known as **kerogen**, which can be separated from the rock by heating. In a liquid state, this thick, oily substance is called **shale oil**. Like crude oil, it can be refined and purified to make gasoline and other by-products.

Oil shale is found in large deposits lying under much of the continental United States. The richest ones are in Colorado, Utah, and Wyoming. Large deposits are also found in Canada, the Commonwealth of Independent States, and China.

The chief advantages of oil shale are its versatility and its large supply. The main disadvantage is cost. Oil shale technology is expensive because it is so energy intensive. In fact, about one-third of a barrel of oil is needed to mine, extract, and purify a barrel of shale oil. **Net energy analysis**, which determines how much energy it takes to extract and process an energy resource, shows that shale oil production creates only about one-eighth as much energy as conventional crude oil production for the same energy investment. Its **net energy yield** (energy

invested − energy produced = net energy yield) is, therefore, rather low.

Oil shale creates serious environmental impacts as well. Strip mining to remove oil shale disturbs large tracts of land, causing erosion and reducing wildlife habitat. Mined shale is then crushed and heated in a large vessel, a process called **surface retorting** (Figure 11.12). Surface retorting also produces enormous amounts of solid waste, called spent shale. A small operation producing 50,000 barrels per day would generate about 19 million metric tons of spent shale per year. Since the shale expands by about 12% on heating, not all of it could be disposed of in the mines from which the raw ore came. Dumped elsewhere, the spent shale may be leached by water, producing an assortment of toxic organic pollutants that could contaminate underground and surface waters.

Oil shale retorts also require large quantities of water, but oil shale deposits are typically found in arid country. Furthermore, retorts produce significant amounts of air pollutants unless carefully controlled.

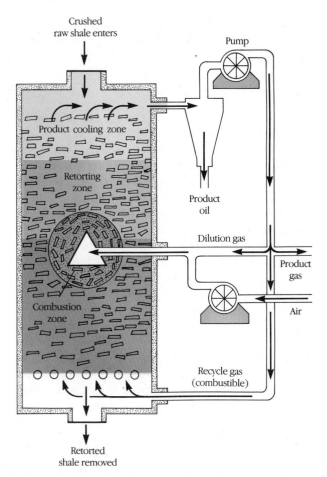

Crushed
raw shale enters

Pump

Product cooling zone

Retorting
zone

Product
oil

Dilution gas

Product
gas

Combustion
zone

Air

Recycle gas
(combustible)

Retorted
shale removed

Figure 11.12 *A surface retort used to extract kerogen from oil shale. Fractured shale is introduced at the top of the retort vessel. Retorted shale is removed at the bottom after being burned. A fire is maintained in the center of the vessel and oil is driven off as a vapor in the hot gas at the top of the retort.*

To bypass the solid waste problem, oil shale companies have experimented with **in situ retorting**. In this process, shale deposits are first fractured by explosives. A fire is then started underground in the oily rock and forced through the deposit. The heat from the fire drives off some of the oil, which is collected and pumped to the surface.

In situ retorts eliminate many impacts caused by surface mining but have not worked well, in part because groundwater often seeps in and extinguishes the fires. Operators have also found it difficult to fracture shale evenly, which is necessary for uniform combustion. In addition, in situ retorts produce more sulfur emissions than surface retorts, raising the cost of shale oil extraction.

A 1980 study by the U.S. Office of Technology Assessment found that huge government subsidies would be needed to help the oil shale industry produce even small amounts of oil. In 1985, the federal government withdrew virtually all of its support from the industry

because of the poor economics. In the early 1990s, the only oil shale plant operating in the United States, which cost $1 billion to build, closed because it was barely covering operating costs and was producing oil at $42 per barrel, two times greater than the cost of conventional crude oil.

Tar Sands   **Tar sands** are sand deposits impregnated with a petroleumlike substance known as **bitumen**. Although they are found throughout the world, the largest deposits are in Alberta, Canada; Venezuela; and the Commonwealth of Independent States. In the United States, six states have commercially attractive deposits.

Tar sands can be strip-mined and treated in a variety of ways to extract the bitumen. Hot-water processing is the only method used commercially. In situ methods similar to those used in oil shale extraction are also being tested.

Tar sands are plagued with many of the problems that face oil shale. They are expensive to mine. The sand sticks to machinery, gums up moving parts, and eats away at tires and conveyor belts. Tar sand expands by 30% after processing, and as with oil shale, its production requires large amounts of water, which is badly polluted in the process.

The most significant barrier to tar-sand development is poor economics. Like oil from oil shale, synthetic crude oil produced from tar sands has a poor energy efficiency: at least 0.6 of a barrel of energy is consumed per barrel produced. Worse, world reserves of tar sand oil are insignificant compared with world oil demands.

Coal Gasification and Liquefaction   The abundance of coal in the United States and abroad has stirred interest in coal gasification and liquefaction. In both technologies, coal is reacted with hydrogen to form synfuels. **Coal gasification** produces combustible gases. **Coal liquefaction** produces an oily substance that can be refined.

Coal gasification creates numerous air pollutants and requires large quantities of water, making its product a dirty alternative to natural gas. The cost of synthetic natural gas is high. For example, building a plant may cost $1.3 billion to $2 billion. An additional problem is the low net energy production. Synthetic gas produced from surface mines is about 1.5 times more expensive than natural gas, and synthetic gas from coal extracted from underground mines is 3.5 times more expensive.

Coal liquefaction, the production of synthetic oil from coal, is similar to coal gasification. Although there are at least four ways to produce a synthetic oil, each involves the same general process: adding hydrogen to the coal. The oil produced by liquefaction must be purified to remove ash and coal particles.

Coal liquefaction could provide liquid fuels, but it is costly. It produces air and water pollutants and requires large amounts of energy. Like coal gasification, it might be preempted by cheaper, cleaner, and renewable energy sources.

11.3   Nuclear Energy: Costs and Benefits

Nuclear energy provides a growing percentage of the world's energy demands. Nuclear reactors in use today in the United States and most other places are fueled by naturally occurring **uranium-235 (U-235)**. This is a form of uranium whose nuclei split, or **fission**, when struck by subatomic particles known as neutrons. When they split, uranium atoms give off enormous amounts of energy (Figure 11.13). In fact, the complete fission of 1 kilogram (2.2 pounds) of U-235 could yield as much energy as 2000 metric tons of coal!

Fission reactors are special devices in which fission reactions can be produced. In a reactor, U-235 is housed in long **fuel rods**. These are located in the **reactor core** (Figure 11.14). U-235 naturally emits neutrons that bombard other nuclei, causing them to split. Heat released during fission is then transferred to water that bathes the fuel rods in the reactor core. As Figure 11.14 shows, the heated water around the reactor core heats water in another closed system. In the latter, hot water is converted to steam, which drives a turbine that generates electricity. The steam is then cooled and used again. Most nuclear plants are cooled by water and are called **light water reactors** (**LWRs**). Other reactors use coolants such as liquid sodium but operate on the same principle.

The splitting of a U-235 nucleus produces two smaller nuclei, called **daughter nuclei** or **fission fragments** (Figure 11.13). Over 400 different fragments can form during uranium fission, many of which are radioactive. (Radioactivity is described in Chapter Supplement 11.1.) A U-235 nuclei fission reaction also releases neutrons, which strike other nuclei in the fuel rods, creating a chain reaction. The chain reaction is controlled to prevent overheating and reactor core meltdown. An atomic explosion in such a case would be unlikely because the fuel is not sufficiently concentrated.

Fission reactions are held in check by the water that bathes the fuel rods. Water absorbs some of the neutrons, thus reducing the rate of fission. Fission reactions are also held in check by the **control rods**, slender rods of neutron-absorbing materials (boron) that lie between the fuel rods. Raising or lowering the control rods regulates the rate of fission. When the control rods are completely lowered, the reactor shuts off.

The entire assemblage of fuel and control rods and their liquid coolant are housed in a 20-centimeter- (8-inch-) thick steel **reactor vessel**. It is surrounded by a huge shield, and the entire unit is encased in a 1.2-meter- (4-foot-) thick cement shell, the **reactor containment building**.

Nuclear Power: Pros and Cons

When nuclear power was first proposed, proponents claimed that its energy would be so cheap that it wouldn't be cost-efficient to meter houses. That dream has failed to materialize. In the United States, for instance, nuclear power costs about 10 to 12 cents per kilowatt-hour, about twice the cost of electricity from coal and wind farms (Chapter 12). Building a nuclear power plant is two to six times more expensive than an equivalent coal-fired power plant. Nevertheless, nuclear power provides approximately 99,000 megawatts of electricity, enough for 99 million people. It fits well into the electric grid system that provides electricity to large numbers of people. Perhaps the most convincing argument for using nuclear power, as opposed to coal and oil, is that it produces very little air pollution. Studies show that the release of radioactive materials into the atmosphere from nuclear power plants is insignificant under normal operating conditions. In fact, a coal-fired power plant may release more radioactivity than a normally operating nuclear power plant. Moreover, nuclear plants do not produce toxic gases, such as sulfur dioxide and nitrogen dioxide, which are converted into acids in the air. Carried back to the Earth in rain, snow, and on particulates, these acids can cause considerable damage to the environment (Chapter 16). Another advantage of nuclear power is that it requires less strip mining than coal because the fuel is a much more concentrated form of energy. This results in less land disturbance and fewer impacts on groundwater, wildlife habitat, and so on. The cost of transporting nuclear fuel is lower than that for an equivalent amount of coal.

Despite its many advantages, nuclear power has substantial drawbacks, including: (1) waste disposal problems, (2) contamination of the environment with long-lasting radioactive materials from accidents at plants and during transportation, (3) thermal pollution from power plants, (4) health impacts, (5) limited supplies of uranium ore, (6) low social acceptability, (7) high construction costs, (8) a lack of support from insurance companies and the financial community, (9) vulnerability to sabotage, (10) proliferation of nuclear weaponry from high-level reactor wastes, and (11) questions about what to do with nuclear plants after their useful life span of 20 to 25 years. This section examines four major areas of concern: reactor safety, waste disposal, social acceptability and cost, and the proliferation of nuclear weapons.

Reactor Safety   An accident at the nuclear power plant at Three Mile Island, Pennsylvania, in 1979,

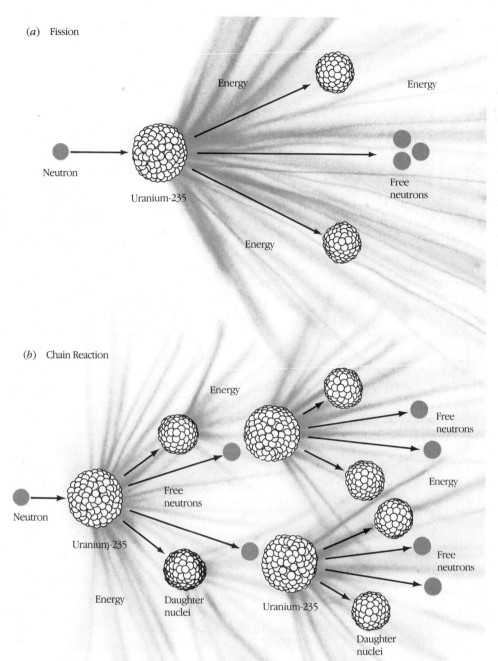

(a) Fission

Energy

Energy

Neutron

Uranium-235

Free neutrons

Energy

(b) Chain Reaction

Energy

Free neutrons

Neutron

Uranium-235

Free neutrons

Energy

Energy

Daughter nuclei

Uranium-235

Free neutrons

Daughter nuclei

Figure 11.13 (a) *In a fission reaction, a U-235 nucleus struck by a neutron splits into two smaller nuclei. Neutrons and enormous amounts of energy are also released.* (**b**) *A chain reaction is produced by placing fissile U-235 in a nuclear reactor. Neutrons liberated during the fission of one nucleus stimulate fission in neighboring nuclei, which, in turn, release more neutrons. Thus, the chain reaction can be sustained.*

alerted many Americans to the dangers inherent in nuclear power. A malfunctioning valve in the cooling system triggered a series of events that led to the worst commercial reactor accident in U.S. history. Radioactive steam poured into the containment building. Pipes in the system burst, releasing radioactive water, which spilled onto the floors of two buildings. Some radiation escaped into the atmosphere, and some was dumped into the Susquehanna River. The accident then took a turn for the worse. Hydrogen gas began to build up inside the reactor vessel, threatening to expose the core and cause a meltdown. The gas bubble was slowly eliminated, but

thousands of area residents had to be evacuated. Photographs of the core showed that a partial meltdown had occurred.

The accident at Three Mile Island had many long-term effects. It cost the utility and its customers many millions of dollars to replace the electricity the plant would have generated. Even more money (over $1 billion) was needed for the cleanup, which still isn't complete.

Although utility officials claim that the accident will cause few cancers, John Gofman and Arthur Tamplin, radiation health experts, contend that the exposure to

Figure 11.14 *Diagram of a nuclear power plant. As shown here, nuclear fission reactions in the reactor core heat up water to generate steam. The steam runs a turbine that generates electricity as in conventional power plants. Note that the water surrounding the reactor core circulates in a closed system, through which it heats up water in the steam generator. This dual water heating system helps prevent the escape of radioactivity in the steam.*

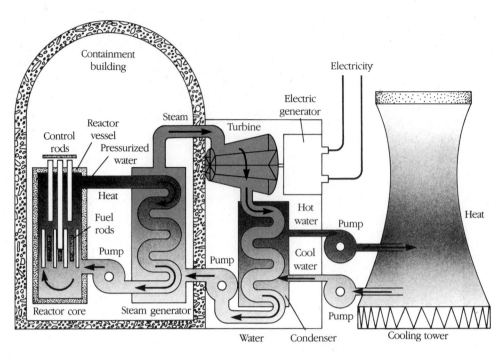

low-level radiation that residents received for 100 hours or longer would cause at least 300, and possibly as many as 900, fatal cases of cancer or leukemia.

A 1975 study on reactor safety, called the Rasmussen Report for its principal author, Norman Rasmussen of MIT, showed that the probability of a major accident at a nuclear power plant was not more than 1 in 10,000 reactor-years. (A **reactor-year** is a nuclear reactor operating for 1 year; for example, 10 reactors operating for 20 years are 200 reactor-years.) If 5000 reactors were operating worldwide, as some advocates propose by the year 2050, we could expect a major accident every other year! Each accident might cause between 825 and 13,000 immediate deaths, depending on the location of the plant. In addition, 7500 to 180,000 cancer deaths would follow in subsequent years. Radiation sickness would afflict 12,000 to 198,000 people, and 5000 to 170,000 genetic defects would occur in infants. Property damage could range from $2.8 billion to $28 billion.

In 1992, 420 nuclear reactors were operating worldwide, and 76 were under construction or soon to be built. If the Rasmussen Report estimate is correct, when all 496 reactors are functional, we can expect a major accident every 20 years.

Because the Rasmussen Report failed to include such possibilities as sabotage and human error, it has been discredited by the Nuclear Regulatory Commission, which argued that it underestimated the real risk. Human error could result in an even higher accident rate. As a case in point, many experts believe that the accident at Three Mile Island was worsened by operator confusion, which turned a small problem into a disaster. Many people contend that misjudgment and perform-

ance errors among personnel could negate technological improvements designed to make plants safer. Human errors and oversight can also occur during construction of nuclear power plants.

The accident at Chernobyl in the Soviet Union in 1986 reinforced concerns about the possibility of human error (Figure 11.15). Releasing large amounts of radiation throughout Europe, on farms and cities, this accident was caused by plant operators who were running tests on the reactor. After reducing power output to test the turbines, the men deactivated several safety systems to simulate an emergency, a direct violation of plant regulations. During the test, the level of the cooling water flowing through the reactor core fell rapidly. Without sufficient coolant, the temperature of the 200 tons of uranium in the reactor's fuel rods soared as high as 2800° C (5000° F), twice the temperature required to melt steel. An enormous steam explosion blew the roof off the building. Flames from 1700 tons of burning graphite, a neutron-absorbing agent in the core, shot 30 meters (100 feet) into the air. While fire fighters risked their lives to contain the disaster by spraying water down on the reactor core from the roofs of nearby buildings, the uranium fuel melted, spewing radioactive isotopes into the atmosphere.

Soon after the accident, 116,000 people living within a 30-kilometer radius of the plant were relocated. In July 1989, the *New York Times* reported that 100,000 additional Soviet citizens would probably have to be evacuated from nearby areas as far as 330 kilometers (200 miles) from Chernobyl. Many of those that lived closest to the nuclear power plant can never return to their homes.

Figure 11.15 *Chernobyl nuclear reactor soon after the 1986 explosion, which blew the top off of the reactor building. This reactor and many others in the Soviet Union were not built with suitable containment.*

Worse than losing their homes and possessions, tens of thousands of people may have been exposed to high levels of radiation. Estimates of human exposure near Chernobyl indicate that whole-body radiation for persons in the immediate area of the plant was 4 to 20 times higher than the allowable exposure for U.S. nuclear workers. Although no one knows for certain, it is likely that about 1 out of 10 people, or about 15,000 people, exposed to radiation from Chernobyl in the 30-kilometer radius will die from cancer. The government of the Ukraine asserts that the accident at Chernobyl has already caused 6000 to 8000 deaths, suggesting that the estimate of 15,000 deaths may be too low. The accident also caused a number of immediate deaths from radiation poisoning—32 within the first four months. Workers who were saved by heroic procedures are likely candidates for cancer.

Radiation from Chernobyl also spread throughout Europe. One study estimates that at least 13,000 Eur-

opeans will die from cancer in the next 50 years. Radiation was greatly diluted by the time it hit the United States, and estimates suggest that only 10 to 20 Americans will die from cancer from the Chernobyl accident in the next 50 years. That may not seem significant, unless you're one of the victims.

In addition to exposing large numbers of people to potentially harmful radiation, the Chernobyl accident threatened crops, farmland, and livestock in the Ukraine, the "breadbasket" of the former Soviet Union. By some estimates, up to 150 square kilometers (60 square miles) of land is so contaminated that it will be unsuitable for farming for decades unless the contaminated topsoil is removed. Agriculture outside the country also suffered from the ill effects of radiation. Soon after the accident, for example, Italian officials turned back 32 freight cars loaded with cattle, sheep, and horses from Austria and Poland because of abnormally high levels of radiation. Radioactive fallout from the accident also settled on Lapland, an expanse of land encompassing northern Sweden, Norway, Finland, and the northwestern part of the former Soviet Union. Lapland is occupied by a seminomadic people who raise reindeer for food. The reindeer feed on lichen, which was contaminated by radiation after the accident. The meat was so heavily contaminated that Laplanders were forced to round up and slaughter their herds.

Critics have always been concerned with unforeseen technical difficulties. The hydrogen bubble at the Three Mile Island plant, for example, took experts by surprise. Numerous backup systems in nuclear power plants are designed to prevent a meltdown and the release of radiation, but as the accident at the Three Mile Island showed, plants are not invulnerable.

Clouding the issue of reactor safety is the possibility of terrorism. In 1975, two French reactors were bombed. Nuclear power plants could become targets of similar attacks. Damage to the cooling system could result in a meltdown, with radiation leakage. Most plants are vulnerable to attack. Protection from ground and air assaults may be impossible. Even though security has been improved at many plants, the threat of well-planned terrorist actions cannot be ignored.

To address the question of reactor safety, the nuclear industry is promoting new, smaller designs that they claim are much safer than present models. Skeptics question whether this is true and point out that making safer plants only addresses one of a dozen problems, among them what to do with the enormous amounts of radioactive waste produced by nuclear power plants.

Waste Disposal   Nuclear power has been in use in the United States for well over 30 years. Despite this fact, the industry still lacks acceptable means of disposing of all of the wastes produced by reactors and **uranium**

mills, where ore is crushed and the uranium is extracted to make nuclear fuel (the enrichment process).

According to one estimate, approximately 125 million metric tons of mill wastes were dumped on or near the facilities, a practice that continued until the late 1970s (Figure 11.16). Nearly 11 million tons were dumped along the banks of the Colorado River and its tributaries. Some wastes were used for fill in construction of homes and buildings. In Grand Junction, Colorado, for example, 4000 homes were built over mill tailings. Residents in these homes are exposed to radiation equivalent to ten chest X rays per week. Not surprisingly, the leukemia rate in Grand Junction is twice that of the rest of Colorado. Complete removal in the area, costing tens of millions of dollars, should be completed by 1994.

Nuclear power plants also produce huge amounts of radioactive wastes. In the United States, nearly 4 million cubic meters of low-level radioactive wastes from reactors and other facilities, such as hospitals and laboratories, have been buried in shallow landfills. These wastes were improperly disposed of for many years. In some instances, they were buried in boxes or steel drums that leaked radioactive materials into groundwater and surface waters. Because of such problems, half of the low-level waste sites in the United States have been closed.

Until the 1960s, low- and medium-level radioactive wastes were often mixed with concrete, poured into barrels, and dumped at sea. Thousands of barrels of waste were dumped into Massachusetts Bay near Boston; some were dumped into the harbor itself. An estimated 47,000 barrels of radioactive wastes were dumped 48 kilometers (30 miles) off the coast of San Francisco near the Farallon Islands (now a marine and bird sanctuary), and numerous barrels were dumped in the ocean near New York City. Although this activity has been banned in U.S. waters since 1970, it still occurs in the Irish Sea where England annually dumps large quantities of radioactive wastes without even placing them in barrels. Officials of the former Soviet Union also dumped huge amounts of radioactive wastes off its shores.

Low-level wastes are hazardous for about 300 years; high-level wastes can be dangerous for tens of thousands of years. The common way of measuring the life span of a radioactive substance is by **half-life,** the time it takes for half of the material to decay. The half-life of one highly radioactive waste product of nuclear reactors, **plutonium-239 (Pu-239),** is 24,000 years. As a rule, it generally takes about eight half-lives for a material to be reduced to 0.1% of its original mass, at which point it is considered safe. For Pu-239, this is about 200,000 years!

Currently, an estimated 23,000 metric tons of high-level radioactive waste from nuclear power plants is being stored on site, awaiting a permanent disposal site. In the 1980s, Congress passed a law requiring the U.S. Department of Energy (DOE) to create a suitable site to

Figure 11.16 *Radioactive mill tailings from a uranium processing plant in Grants, New Mexico. Wastes like these have been dumped throughout the American West, sometimes along riverbanks where they can enter waterways.*

build a disposal facility for such waste. After eliminating a number of candidates, DOE selected a remote site in Nevada (Yucca Mountain) and to date has already spent over $1 billion studying the area, which is scheduled to open in 1998.

Meanwhile, high-level wastes are accumulating at nuclear power plants and weapons facilities throughout the United States. In some locations, wastes have begun to leak. At Richland, Washington, where fissile ores are enriched for atomic bombs, high-level wastes have been stored in large steel tanks since the plant opened during World War II. Since then, the plant has produced several hundred million liters of highly radioactive liquid waste. Gradually, the tanks have deteriorated. Since 1958, at least 2 million liters of waste have leaked out of tanks into the soil and, possibly, the groundwater. Approximately 150 million liters of waste is still held in storage tanks that could leak. Clearly, something needs to be done about these wastes and the thousands of metric tons of waste currently stored at nuclear power plants. (For a discussion of radioactive waste, see Chapter 19.)

Viewpoint 11.1

The Best Energy Buys

Amory B. Lovins and
L. Hunter Lovins

Amory Lovins is a consulting physicist; his wife and colleague, Hunter, is a lawyer, sociologist, and political scientist. They have worked as a team on energy policy in over 20 countries, and are principals of the nonprofit Rocky Mountain Institute (RMI) in Old Snowmass, Colorado, which explores the links between energy, water, agriculture, security, and economic development. Most of RMI's income comes from advising utilities on how to produce electricity more efficiently and more economically.

The energy solution is fairly simple: Use the energy more efficiently whenever doing it is cheaper than new supply (almost always); use renewable sources like sun, wind, falling water, and biomass whenever cost-effective (frequently); and use natural gas (relatively cheap, abundant, and environmentally benign) to make a smooth transition to a completely sustainable energy system. Careful studies in 15 countries have documented that efficiency and renewables can meet essentially all our long-term global needs.

Since 1979, the United States has gotten four and a half times as much energy from conservation as from all net increases from supply put together, and a third of the new supply was renewable. We did this using simple methods: caulk guns, duct tape, plugging steam leaks, and measly car efficiency improvements.

Vast energy saving opportunities remain. Fully implementing the thousand or so best electricity-saving innovations on the market could displace over half the electricity the country now uses. These savings cost under 2 cents per kilowatt-hour, less than the cost of running a nuclear plant. Thus, if a nuclear plant cost nothing to build (it actually costs up to 15 cents per kilowatt-hour to deliver nuclear electricity), was perfectly safe (it's not), and didn't produce radioactive wastes or bomb materials (it does), your utility would save money closing the new plant, never running it, and paying for efficiency instead. Lighting, for example, uses roughly 20% of U.S. electricity. Compact fluorescent lamps use 18 watts to deliver the same illumination as 75-watt incandescent bulbs. Each bulb lasts a dozen times longer (saving enough installation labor and replacement bulbs to more than pay for the lamp). The electricity saving displaces 770 pounds of coal or 62 gallons of carbon dioxide, and 18 pounds of

sulfur dioxide don't get released, reducing global warming and acid rain.

Just the lighting improvements now commercially available can, if fully used, save nearly one-fourth of all electricity used in the United States. That is enough to displace 120 Chernobyl-sized power plants. In contrast, nuclear power has already cost the United States over a trillion dollars but delivers today less energy than wood.

Similar oil savings are available. The U.S. household vehicle fleet now averages 19 miles per gallon (mpg). Improving that to 22 mpg could displace all the oil the United States imported from Iraq and Kuwait before the hostilities of July 1990. Increasing the efficiency of the vehicle fleet average by another 9 mpg would end the need for the United States to import any oil from the Persian Gulf. Basically, we put our kids in 0.56 mpg tanks and 17 feet-per-gallon aircraft carriers because we failed to put them in 32 mpg cars.

Ten auto manufacturers have built and tested attractive, peppy cars that get between 67 and 138 mpg. Some of the prototypes comfortably hold four or five people. Space-age materials and better design are making some of them safer than many of today's models.

With savings from efficient jets, trucks, buses, buildings, and industrial processes, the present U.S. economy could operate on one-fifth the oil we now use. The savings usually cost less than $5 per barrel. Even achieving only 15% of this potential would displace all the oil we now import from the Gulf.

Saving energy also helps communities. In Osage, Iowa, the municipal utility chose to help its customers get more efficient, rather than raise rates to build another power plant. Over nine years, this program to weatherize homes and control electric load saved the utility enough money to enable it to prepay all its debt, build up a nice

(Continued)

Viewpoint 11.1 (*continued*)

The Best Energy Buys

cash surplus, and cut customers' rates by a third. These lower rates attracted new factories to town. Meanwhile, each household saved more than $1000 per year. The extra money in people's pockets helped local businesses, making Main Street noticeably more prosperous than in comparable towns nearby.

Ultimately, the choice is yours. Every time you buy weather stripping instead of electricity—because you can get comfort cheaper that way—you're part of the transition. The United States could eliminate oil imports in the 1990s, before a power plant ordered now can provide any energy whatsoever and at

a tenth of its cost, just by making buildings and cars more efficient. Conversely, each dollar spent on reactors can't be spent on energy efficiency, and, hence, it delays energy independence. Power plants also provide fewer jobs per dollar than any other investment. Thus, every big plant loses the economy, directly and indirectly, about 4000 net jobs, by starving the rest of the economy.

The United States is already a quarter more energy-efficient than it was 15 years ago. We're spending $150 billion a year less for energy. But, we're among the least efficient of Western industrial countries. If we were as efficient as

Sweden, which has one of the world's highest standards of living, we would save another $200 billion per year. That's more than the entire federal deficit. Whether you care about the environment, peace, saving money for yourself, or making America competitive again, energy efficiency is the answer.

Critical Thinking

Make a list of the major points made by the authors. List the supporting evidence for each one. Using your critical thinking skills, critically analyze each point.

Social Acceptability and Cost   Two of the most important factors controlling the future of nuclear energy are its social acceptability and its cost. In 1989, voters in California passed a public referendum calling for the closure of the Rancho Seco nuclear power plant. Internationally, several countries, including Sweden and Italy, have voted to phase out nuclear power. Why is nuclear power so unpopular these days?

Besides safety concerns and waste disposal problems, nuclear power plants are expensive to build. A new nuclear power plant today costs between $2 billion and $8 billion, compared with $500 million to $1 billion for an equivalent coal-fired power plant. Costs are high because of strict building standards, expensive labor, construction delays, and special materials needed to ensure plant safety. Because of the high construction costs and risk of damage, U.S. banks refuse to finance nuclear power plants, and utilities must turn to foreign investors.

The costs of operating nuclear power plants are also high. Repair costs are many times higher than those for conventional coal-fired plants. For example, saltwater corrosion of the cooling system in a reactor owned by Florida Power and Light cost over $100 million to repair and $800,000 per day to compensate for the lost electricity. A similar problem in a coal-fired power plant would cost only a fraction of this amount.

Nuclear power is plagued by costly poor performance. Colorado's only nuclear reactor generated its first electricity in 1976. In its 13-year lifetime, the plant was shut down half of the time because of various mechanical problems. The long "downtime" cost that utility and its customers $20 to $30 million per year. In 1989, the company decided to close the plant permanently.

A fourth factor in the cost equation is retirement costs. According to the Worldwatch Institute, one out of every six reactors already built has been retired. Their average life span was 17 years, compared to an industry estimate of 20 to 30 years. Dismantling these plants may cost utilities about $500,000 to $1 million per megawatt, bringing the cost of decommissioning a 1000-megawatt plant to $500 million to $1 billion. This cost will inevitably be added to electric bills.

A rash of cancellations has also increased costs. Even though utilities halt construction of nuclear plants, they must repay the billions of dollars they borrowed to start construction. Customers inevitably foot most of the bill. Bondholders, who often finance such projects, can be left in financial ruins, as was the case when the Washington Public Power Supply System defaulted on $2.25 billion in bonds (Figure 11.17). Because of the default, electric rates increased dramatically. The region is now saddled with a debt that may climb to $9 billion with interest. Although the utilities assumed the

Figure 11.17 *One of the Washington Public Power Supply System's abandoned nuclear power plants.*

financial responsibility for three canceled reactors, the remaining costs will be passed on to the customers and bondholders.

High construction and maintenance costs, combined with other factors, make nuclear energy the most expensive form of electricity currently available on a large scale. In fact, France, which gets 60% to 70% of its electricity from nuclear power plants, is thought to be paying 35% to 60% more for that electricity than it would for electricity from coal.

A lack of public support, exorbitant costs, and a rash of canceled plants in recent years, among other problems, have slowed the growth of the nuclear industry. In fact, the Worldwatch Institute predicts that unless there is an immediate turnaround in orders, nuclear power may decline by the year 2000.

Proliferation of Nuclear Weapons   At least 21 countries have the materials and the technical competence to build nuclear bombs. Many of these countries are politically unstable or are in volatile regions where war could easily erupt. Six countries—the United States, China, the Soviet Union, Great Britain, France, and India—have already test-fired nuclear weapons. The plutonium used in atomic bombs comes from special nuclear reactors as

well as conventional reactors. Critics argue that the spread of nuclear power plants throughout the world will make bomb-grade plutonium more widely available.

Nuclear fuels could be stolen by terrorist groups and easily fashioned into a crude, but effective atomic bomb that would fit comfortably in the trunk of an automobile. Strategically planted, a terrorist bomb would prove disastrous.

Breeder Reactors

The world's supply of U-235 used in light water reactors will last about 100 years at the current rate of use. Increases in the production of electricity by nuclear power plants, however, could greatly reduce the life span of uranium reserves. Because uranium supplies are limited, the nuclear industry has developed an alternative, the **breeder reactor**. Similar in many respects to the LWR described earlier, a breeder reactor performs an additional function: it converts abundant U-238 into reactor fuel, Pu-239. In the breeder reactor, fast-moving neutrons from the reactor core strike the nonfissionable U-238 placed around the core and convert it into fissile Pu-239 (Figure 11.18). The neutrons come from small amounts of Pu-239 located in the fuel rods of the breeder reactor. Theoretically, for every 100 atoms of Pu-239 consumed in fission reactions, 130 atoms of Pu-239 are produced, hence, the name *breeder*.

The attraction of breeder reactors is that they could be fueled by U-238 found in the wastes of uranium processing plants or in spent fuel from fission reactors. In the United States, the estimated supply of fuel for breeder reactors would last 1000 years or more. In addition to providing a long-lasting supply of electricity, breeder reactors could reduce the need for mining, processing, and milling uranium ore. Fuel prices might remain stable because of the abundance of U-238. The breeder technology does not create chemical air pollution if problems from mining and milling are discounted.

Breeder reactors have been under intensive development in the United States for well over 30 years. The most popular design is the **liquid metal fast breeder reactor**. It uses liquid sodium rather than water as a coolant. Heat produced by nuclear fission in the reactor core is transferred to the liquid sodium coolant, which transfers this heat to water. The water is then converted to steam and used to generate electricity.

On the surface, the breeder reactor sounds like the answer to the world's electric energy needs. Unfortunately, it has numerous problems, some so great that they may make the technology impractical. The most significant problem is that it takes about 30 years for the reactor to break even—that is, to produce as much Pu-239 as it consumes. The fate of breeder reactors hinges on a drastically shortened pay-back period. The second major problem is the cost of construction:

Figure 11.18 *Nuclear reactions in a breeder reactor. Neutrons produced during fission strike nonfissionable "fertile" materials, such as U-238. U-238 is then converted into fissionable Pu-239, which can be used in the reactor as fuel.*

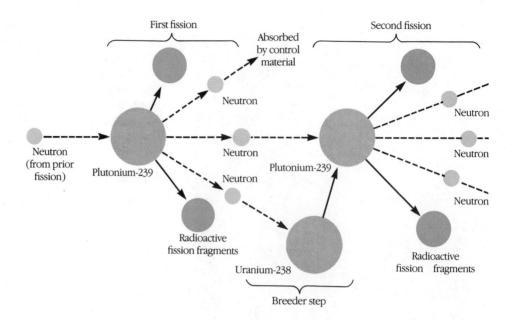

$4 billion to $8 billion. In addition, many of the problems encountered in light water reactors also occur in the breeder reactor.

A third problem is the large quantity of plutonium located at breeder reactors. Plutonium is long lived and extremely toxic if inhaled. It burns when exposed to air. Liquid sodium coolant is also dangerous. It reacts violently with water and burns spontaneously when it is exposed to air. Leaks in the coolant system could trigger a catastrophic accident at a breeder reactor. Should the core melt down, a small nuclear explosion equivalent to several hundred tons of TNT might occur. Rupturing the containment building, the explosion could send a cloud of radioactive gas into the surrounding area.

In 1983, Congress canceled funding for an experimental breeder reactor, the Clinch River project in Tennessee. After having spent nearly $2 billion for planning, Congress decided that further investments were unmerited. The prospects of another breeder reactor are currently being assessed by France now that its commercial breeder reactor, the Superphenix, has been closed down because of technical problems.

Nuclear Fusion

Another form of nuclear energy is nuclear fusion. **Nuclear fusion** is responsible for the production of light and heat in the sun and other stars. It occurs when four hydrogen nuclei *fuse* to form a slightly larger helium nucleus. Fusion requires extremely high temperatures to overcome the mutual electrostatic repulsion of the positively charged nuclei. When hydrogen nuclei fuse, they emit large quantities of energy. Scientists hope to harness these reactions and capture this energy to generate electricity.

If it proves successful, controlled fusion could supply enormous amounts of energy. One source of fuel, **deuterium** (a hydrogen atom with a neutron), is plentiful in the oceans. As illustrated in Figure 11.19, two deuterium nuclei can fuse to form a helium nucleus. Another fuel source, **tritium** (a hydrogen with a neutron and two protons), does not exist naturally but can be made from lithium, which is abundant. Deuterium can fuse with tritium. Energy analyst John Holdren estimates that at current rates of energy consumption in the United States fusion would meet energy needs for up to 10 million years!

Unfortunately, fusion has significant drawbacks that may make it unattainable commercially. The first of these is that fusion reactions occur at temperatures measured in the hundreds of millions of degrees Celsius. The main obstacle, then, is finding a way to contain such an extremely hot reaction. No known alloy can withstand these temperatures; in fact, metals would vaporize. To contain fusion reactions, scientists have devised reactors that suspend tiny amounts of fuel in air within a metal reactor vessel. The most popular technique in experimental fusion reactors is **magnetic confinement** (Figure 11.20). In these reactors, a magnetic field suspends the superheated fuel. The heat that is omitted is used to boil water to make steam. Small-scale experimental reactors of this type have been developed and operated in the United States, Europe, the Commonwealth of Independent States, and Japan. Despite 40 years of research, however, researchers have failed to reach the break-even point. Achieving this goal would be just the first step in a long, costly climb to commercialization.

In the proposed designs, heat released from the fusion reaction would be drawn off by a liquid lithium

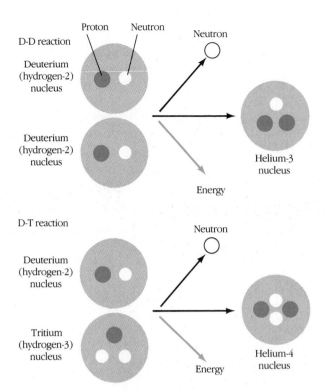

Proton   Neutron
D-D reaction

Deuterium
(hydrogen-2)
nucleus

Deuterium
(hydrogen-2)
nucleus

Neutron

Energy

Helium-3
nucleus

D-T reaction

Deuterium
(hydrogen-2)
nucleus

Tritium
(hydrogen-3)
nucleus

Neutron

Energy

Helium-4
nucleus

Figure 11.19   *Two potentially useful fusion reactions.*

blanket, which transfers the heat to water. The lithium blanket would also capture neutrons, creating tritium, which could be extracted and used as additional fuel.

Scientists estimate that a demonstration fusion reactor could be running by the year 2000, but commercial-scale plants would not be possible until 2020 or 2030 at the earliest. The cost of a commercial fusion reactor cannot be accurately assessed, but it could cost three to five times more than a breeder reactor, about $12 billion to $20 billion.

The deuterium-tritium fusion reactor is the most feasible type of fusion reactor, but tritium is radioactive and difficult to contain. Because of the high temperatures in fusion reactors, tritium can penetrate metals and escape into the environment. Fusion reactors would produce enormous amounts of waste heat, the ramifications of which are discussed in Chapter 17. Fusion reactions would also emit highly energetic neutrons, which would strike the vessel walls and weaken the metal, necessitating replacement every two to ten years. Metal fatigue could lead to the rupture of the vessel and the release of tritium and molten lithium, which burns spontaneously in air. A leak might destroy the reaction vessel and the containment facilities. Neutrons emitted from the fusion reaction would also convert metals in the reactor into radioactive materials. Periodic maintenance and repair of reactor vessels would be a health hazard to workers, and radioactive components removed from the reactor would have to be disposed of properly.

11.4   Fossil Fuel Energy: Supplies and Demands

Since 1860, energy use has grown worldwide at a rate of 5% a year, except for brief respites during the Great Depression (1930–1935) and in the worldwide economic recession of the early 1980s. In 1992, global consumption rose to about 350 quads. Assuming a growth rate in energy use of 4% to 5% per year, world energy consumption could reach 550 to 600 quads by the year 2000. Will there be enough fossil fuel energy to meet these demands?

Answering this question is difficult because no one knows how much economically recoverable fossil fuel lies within the Earth's crust. In addition, future demand is not entirely predictable. Will the Commonwealth of

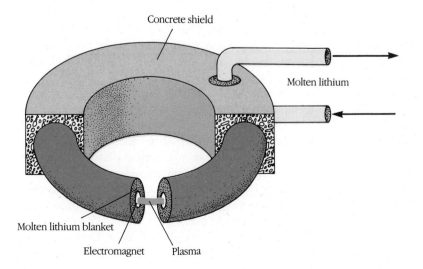

Concrete shield

Molten lithium

Molten lithium blanket

Electromagnet   Plasma

Figure 11.20   *One type of fusion reactor. The fusion reactions occur suspended in an electromagnetic field.*

Independent States' transformation to a capitalistic economic system dramatically increase fossil fuel consumption? Will efforts in developing nations have a similar effect, greatly increasing the depletion of oil and other fossil fuels? Or will nations adopt energy-efficient technologies on a large scale, cutting energy demand? This section examines the supply-and-demand picture for three fossil fuels: oil, natural gas, and coal. When examining the prospects for energy supplies keep in mind the many uncertainties outlined above.

Oil: The End Is Near

Some experts predict that when the last drop of crude oil has been burned, the world's oil fields will have yielded about 2000 billion barrels of oil. This figure is called the **ultimate production**. To date, 675 billion barrels have been produced and consumed globally. Of the remaining 1325 billion barrels, approximately 1000 billion are **proven reserves**, that is, they are known to exist. The remaining 325 billion barrels are **undiscovered reserves**, that is, are thought to exist.

On the surface, it appears as if the world has an abundance of crude oil. At the current rate of consumption, though, the 1000 billion barrels of proven reserves would last about 42 years; the remaining 325 billion barrels, if they exist, would last about 14 years.

Fifty-six years of oil—that's the good news. Now for the bad news: If oil consumption continues to increase at 5% per year, the remaining oil will be gone by the year 2018, 25 years from now. Some oil company geologists believe that the world's ultimate oil production may be as high as 5000 billion barrels. Although that may sound encouraging, 5% per year growth in oil consumption would still deplete the world's reserves by the year 2038, attesting to the powers of exponential growth!

Long before final production, signs of shortage will be evident. You can understand this by studying Figure 11.21. As illustrated, oil production is expected to rise to a peak, then fall as reserves are depleted. As the proven reserves decline, oil companies will work harder and harder to maintain production. In addition, as domestic oil reserves fall, many nations may turn to the politically unstable Middle East, which currently houses about two-thirds of the world's oil reserves (Figure 11.22). The Middle East's share of the oil market is expected to reach 40% by the end of the decade, up from 27% in 1990. Christopher Flavin and Nicholas Lenseen of the Worldwatch Institute, who liken the developed world's dependency on oil to an addiction, note: "Not only is the world addicted to cheap oil, but the largest liquor store is in a very dangerous neighborhood."

Increasing our dependence on fuel from a politically volatile region of the world can have devastating con-

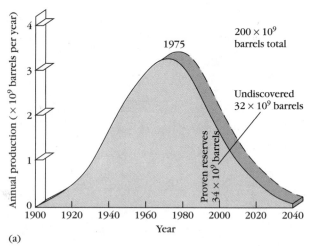
(a)

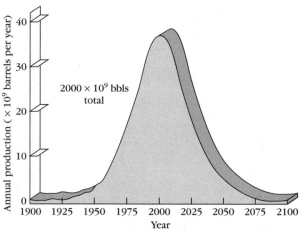
(b)

Figure 11.21 *Petroleum production curves for the United States (a) and the world (b).*

sequences. For example, the 1973 oil embargo imposed by the **Organization of Petroleum Exporting Countries (OPEC)** and a second embargo by Iran in 1979 drove the price of oil from $3 a barrel to $34 in a short period, stimulating rapid inflation that brought the industrial world to its knees. The 1991 Persian Gulf War was another costly consequence. This war cost the developed nations $61 billion, a price paid primarily by the United States, Japan, and Saudi Arabia.

As Middle East supplies begin to decline, continued high demand could outstrip supplies, resulting in worldwide shortages and exorbitant oil prices. Countries that have not developed alternative energy options could experience severe economic turmoil: inflation, economic stagnation, and widespread unemployment.

Global oil production is expected to peak sometime between the years 2000 and 2010. U.S. oil production peaked in 1975. At that time, half of the domestic supplies had been extracted and consumed. Despite greatly increased efforts, the United States produced only 7.4 million barrels of oil per day in 1991, compared

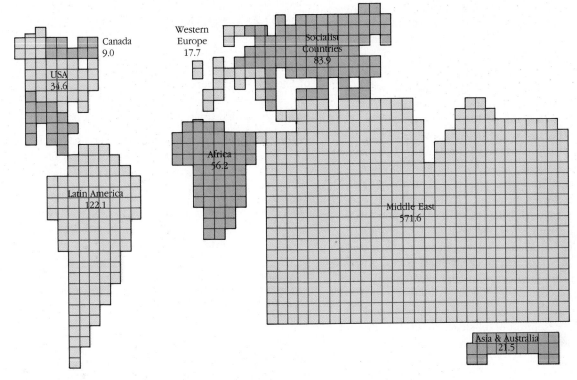

Figure 11.22 *Where will industrial countries get their oil? This "map" of the world shows where the world's oil lies. Each square equals 1 billion barrels of oil in proven reserves. The numbers represent the total billions of barrels.*

with 11 million in 1973. As a measure of how difficult it is to find oil, in 1973, the 11 million barrels came from 497,000 oil wells; in 1991, approximately 590,000 wells were in operation, producing the 7.4 million barrels per day.

Fortunately, energy efficiency measures can rescue the world from the impending shortfalls. To prevent widespread economic turmoil, however, nations must act soon to use oil more efficiently and to find replacements for oil, which powers our transportation system, heats our homes, and provides the raw materials for plastics and other synthetic materials, such as nylon. Some energy options are discussed in Chapter 12.

Natural Gas: A Better Outlook

The outlook for natural gas is considerably brighter than the outlook for oil. Although experts debate the size of our ultimate production, estimates range from 5000 to 12,000 trillion cubic feet. Taking 10,000 trillion cubic feet as a reasonable estimate, let's examine the global outlook. About 2600 trillion cubic feet of natural gas have already been produced and consumed as of the end of 1992, leaving 7400. Of this, 4400 is proven reserve (we know it exists), and 3000 is yet to be discovered.

How long will the global supply of natural gas last? At the current rate of usage, which is about 70 trillion cubic feet per year, proven global reserves would last about 62 years. The undiscovered reserves would last about 42 more years. As with oil, increased consumption could reduce the life span of natural gas considerably, perhaps by as much as one-half. For the time being, it appears that global supplies are adequate.

In the United States, the picture has recently brightened. In the 1980s, the total known reserves of natural gas were about 187 trillion cubic feet. New studies put the total reserves at seven times this number—between 1000 and 1300 trillion cubic feet. At current rates of consumption, this would last about 60 years.

Because natural gas supplies are huge and because it is a clean-burning fuel that can be used in cars, power plants, industry, and homes, natural gas could replace oil and coal. Increasing the world's dependence on natural gas could help wean us from the most damaging fossil fuels, oil and coal, as we develop a sustainable energy system based on conservation and renewable energy sources.

Coal: The Brightest but Dirtiest

Coal is the world's most abundant fossil fuel. World proven reserves, estimated to be 786 billion tons, will last 200 years at the current rate of consumption. Geologists believe that proven reserves are only a fraction of total resources, all the coal in the ground, which are estimated at 12,600 billion metric tons, half of which are believed to be recoverable. However, some energy analysts think that this estimate exaggerates the reserves by five or six times. The recoverable reserves, an estimated 6300 billion tons, would last 1700 years. Growth rates of 4% to 5% would greatly reduce the life span of coal, but the large surplus suggests that coal will last for a long time. If the world's powerful energy companies have their way, coal will probably play a large role in the near term because it can also be used to make liquid fuels to power transportation systems and can be converted into a synthetic natural gas for home heating.

The United States has about 30% of the world's proven coal reserves, or about 225 billion metric tons that can be recovered. This could last the country 100 to 200 years even at increased rates of consumption. It is believed that at least 360 billion additional metric tons of coal can be recovered in the United States, giving Americans a supply of coal that could last several hundred more years.

Unfortunately, coal is not a clean-burning fuel. In fact, with the exception of oil shale, coal is the dirtiest fossil fuel known. New and cleaner ways to burn coal are being developed, but progress has been slow. New technologies to burn coal and pollution control devices may help lessen the environmental damage, although some problems, such as carbon dioxide pollution, are inescapable (Chapters 15 and 16).

11.5 Guidelines for a Sustainable Energy Future

Choosing a sustainable energy future requires us to consider a number of factors by which we can judge energy options. First, energy resources should have a positive net energy yield. That is, energy we obtain from an energy resource must exceed the energy we invest in its exploration, extraction, and transportation. The higher the yield, the better.

Second, energy supplies should match energy demand. Energy comes in a variety of forms. Each of these forms has a different **energy quality**, a measure of the amount of available work one can get out of it (Table 11.2). Oil and natural gas, for instance, are highly concentrated energy resources. When burned, these high-quality energy resources produce large amounts of heat useful for certain tasks, such as melting steel. Sunlight is a lower-quality form. Streaming through the south-facing windows of a house, it produces heat appropriate for warming homes and offices.

Third, options chosen should be as efficient as possible. By increasing efficiency, we cut energy use and reduce environmental impacts.

Fourth, selected options should minimize pollution and maximize public and environmental safety.

Fifth, selected options should be abundant and

Table 11.2 Energy Quality of Different Forms of Energy

Quality of Energy	Form of Energy
Very high	Electricity, nuclear fission, nuclear fusion*
High	Natural gas, synthetic natural gas (from coal gasification), gasoline, petroleum, liquefied natural gas, coal, synthetic oil (from coal liquefaction)
Moderate	Geothermal, hydropower, biomass (wood, crop residues, manure, burnable municipal refuse), oil shale, tar sands

*Workable nuclear fusion reactors do not yet exist. Even if they were technically operable by the end of the first quarter of the twenty-first century, they would probably remain economically unfeasible.

renewable in order to maximize returns on money invested in research, development, and commercialization.

Sixth, selected options should be those most affordable now and in the future.

Establishing Priorities

With energy trends described in this chapter and the guidelines for a sustainable energy strategy listed above, we can establish a list of goals and actions needed to meet future energy demands.

Short-Term Goals (within 20 years)

- ACTION 1 Greatly improve the energy efficiency of all machines, homes, appliances, buildings, factories, motor vehicles, airplanes, and so on. This action will stretch fossil fuel supplies and give us more time to transition to renewable energy resources.

- ACTION 2 Find clean, renewable replacements for oil because supplies are limited. Also find replacements for coal because it is such a filthy fuel.

Long-Term Goal (within 50 years)

- ACTION 3 Find a replacement for natural gas. Its supplies will begin to decline.

To create an energy system that is economically and environmentally sustainable is an enormously complicated task, made more difficult by current controversies and the failure of many proponents to subject their "pet energy sources" to critical analysis. To make wise choices, we must often abandon allegiances and base our decisions on what is socially, economically, and environmentally sustainable. Our goal is to design a sustainable energy system that makes sense given the world's limited resources, one that fits into the economy of nature.

The trouble with our times is that the future

is not what it used to be.

—Paul Valery

࣭   ࣭   ࣭

Critical Thinking Exercise Solution

To begin, you should look at the uses of nuclear power and oil. In other words, be sure to define and understand all terms. As pointed out in this chapter, nuclear power is a source of electricity. In the United States and most other nations oil is used primarily to produce gasoline, diesel fuel, home heating oil, and jet fuel. It is also used to produce chemicals needed to make plastics, medicine, and other products. Very rarely is oil burned to generate electricity.

Increasing our reliance on nuclear power will do little, if anything, to reduce our dependence on oil. Some see this effort, which continues today in an expensive media campaign sponsored by the nuclear industry, as a ploy to dupe the public. Proponents of nuclear power are promoting a desirable goal (energy independence) but offering a false solution (nuclear power). By defining terms more carefully, you can see they're consciously or unconsciously hoodwinking the public.

On the issue of nuclear power and global warming, it is quite clear that what proponents say is true. A nuclear power plant produces no carbon dioxide and does not contribute to global warming. But what about nuclear power's impacts? Are we simply trading one problem for another?

Summary

11.1 Energy Use: Our Growing Dependency on Nonrenewable Fuels

■ Energy is the lifeblood of modern industrial society.

■ Over the years, industrial societies have shifted their dependency on energy sources. Today, the United States and most other developed countries depend primarily on three **nonrenewable fuels**: oil, natural gas, and coal.

11.2 Fossil Fuels: Costs and Benefits

■ The acquisition and use of energy are responsible for many serious environmental problems.

■ Understanding the impacts of various energy sources requires an examination of **energy systems**, the chain of events from exploration to extraction to processing to distribution to end use.

■ **Crude oil** is found in deep deposits in the seafloor and on land. It often occurs with natural gas. Crude oil is distilled in refineries to produce gasoline, diesel fuel, heating oil, asphalt, and many other products.

■ Crude oil is extracted via wells drilled into deposits. Once the crude oil is removed, it must be transported by pipelines, trucks, or ships to refineries. Transportation of oil by ships and pipelines is a source of considerable environmental damage.

■ The combustion of crude oil derivatives produces enormous quantities of carbon dioxide, a greenhouse gas, and other pollutants, among them sulfur dioxide and nitrogen dioxide. The latter two are converted into acids in the atmosphere.

■ **Natural gas** is a clean-burning fuel that is extracted from wells. On land, drilling rigs generally have minimal impact unless they are located in wilderness areas, where roads and noise from heavy machinery and construction camps can disturb wildlife. Natural gas extraction can also cause subsidence within a considerable radius of wells.

■ Natural gas is generally safe to transport in the gaseous form in pipelines. To transport it via ships, it must be liquefied, a process that creates an unstable and highly flammable fuel.

■ Substitutes for oil and natural gas could come from oil shale, tar sand, and coal, which can be converted into liquid and gaseous fuels, or **synfuels**.

■ **Coal**, the most abundant fossil fuel, is widely used to generate electricity, heat, and steam.

■ Compared with nuclear power plants, coal-fired power plants are relatively inexpensive to build. Coal can also be converted into a synthetic natural gas and oil, providing fuel for transportation and buildings.

■ Coal is dirty and environmentally costly. At virtually every step in the coal energy system, significant impacts occur, among them acid mine damage, subsidence, and acid deposition.

■ **Oil shale** contains an organic material known as **kerogen**, which can be extracted by heating. Widespread development of this resource would require mining of extensive regions and large energy inputs, and would result in the production of large quantities of air pollution and solid waste. Currently, the costs of shale oil are about twice the cost of crude oil.

■ **Tar sands** are sand deposits impregnated with a petroleumlike substance called **bitumen**, which can be removed and refined. Only a small proportion of the bitumen can be recovered, however, and global supplies are tiny compared with energy demand.

■   **Coal gasification**, the production of a gaseous fuel, produces large quantities of air pollution and solid waste. Synthetic gas produced in such operations is expensive because of the low net energy yield. **Coal liquefaction**, similar in principle, has many of the same problems.

11.3   Nuclear Energy: Costs and Benefits

■   **Nuclear reactors** are fueled by U-235, whose nuclei split when struck by neutrons. This process, called **fission**, releases enormous amounts of energy that can be trapped to generate electricity.

■   The most common nuclear power reactors, **light water reactors**, produce very little air pollution. Land disturbance and transportation of fuel is lower than for an equivalent number of BTUs of coal.

■   Many problems are associated with the use of nuclear power, including radioactive waste disposal, low social acceptability, high construction costs, questionable reactor safety, vulnerability to sabotage, and their contribution to the proliferation of nuclear weapons.

■   Some consider the **breeder reactor** an answer to the limited supplies of U-235. Besides producing electricity, the breeder reactor makes fissionable plutonium-239 from the abundant uranium-238 available from waste piles, spent fuel rods, and uranium ore.

■   **Liquid metal fast breeder reactors**, the technology of choice today, have many problems. They require about 30 years in order for fuel production to equal consumption. They are also quite costly.

■   In addition, the presence of large quantities of fissionable materials in the core could lead to meltdowns or very low level nuclear explosions during accidents. In addition, they could display all the problems found in light water reactors.

■   Another proposed energy system is fusion power. **Fusion** occurs when two or more small nuclei form a larger one, a process accompanied by the release of energy.

■   The forms of hydrogen needed to fuel the fusion reactor are abundant and could provide energy for millions of years. However, after four decades of research, fusion is still a long way from being commercially available.

■   Fusion reactions occur at extremely high temperatures. Safely containing such reactions is a major challenge.

■   The cost of a commercial fusion reactor could be many times more than that of conventional fission reactors, which currently face financial difficulties.

■   The emission of highly energetic neutrons from the fusion reaction would weaken metals and necessitate their frequent replacement. Metal fatigue might lead to rupture of the vessel and the release of radioactive materials or highly reactive lithium.

11.4   Fossil Fuel Energy: Supplies and Demands

■   If estimates of current oil supplies and future demand are correct, the world oil supply could be depleted somewhere between the years 2018 and 2038. Energy efficiency and substitutes for oil are therefore badly needed.

■   The prospects for natural gas are better, and the prospects for coal are even better. Unfortunately, coal is the dirtiest of the three.

11.5   Guidelines for a Sustainable Energy Future

■   Developing a sustainable energy future requires us to judge options by a number of guidelines. New supplies and technologies should: (1) have a positive net energy yield, (2) match energy quality with end use, (3) be efficient, (4) rely on resources that do minimal environmental damage, (5) rely on abundant, renewable sources, and (6) secure resources that are affordable.

Discussion and Critical Thinking Questions

1.   Critically analyze this statement: "There is enough oil to go around for at least 65 more years."

2.   Describe the term *energy system*. List all of the steps you can think of to produce gasoline from oil, starting with exploration, and discuss some of the environmental impacts associated with each step.

3.   Critically analyze the statement: "Coal is our most abundant fossil fuel and, therefore, should be a major source of energy in the next 100 years."

4.   Describe how a light water fission reactor works. What is the fuel, and how is the chain reaction controlled?

5.   What are the advantages and disadvantages of nuclear power?

6.   Critically analyze the statement: "Nuclear energy is a clean source of energy and should be actively promoted."

7.   What is a breeder reactor? How is it similar to a conventional fission reactor? How is it different? Discuss the advantages and disadvantages of the breeder reactor.

8.   What is nuclear fusion? Discuss the advantages and disadvantages of fusion energy.

9. What is oil shale? Discuss the benefits and risks of oil shale development.

10. Define *coal gasification* and *liquefaction*.

11. You are studying future energy demands for your state. List and discuss the factors that affect how much energy you'll be using in the year 2000.

12. Using your critical thinking skills, debate the pros and cons of oil exploration in the Arctic National Wildlife Refuge.

13. What principles of sustainability pertain to designing a sustainable energy system?

Suggested Readings

Brown, L. R., Flavin, C., and Postel, S. (1991). *Saving the Planet*. New York: Norton. Contains an important overview of the need for an alternative, non-fossil-fuel-based economy.

Chiras, D. D. (1992). *Lessons from Nature: Learning to Live Sustainably on the Planet*. Washington, D.C.: Island Press. Describes why fossil fuel combustion is not a sustainable activity.

Flavin, C. (1987). *Reassessing Nuclear Power: The Fallout from Chernobyl*. Worldwatch Paper 75. Washington, D.C.: Worldwatch Institute. Important look at nuclear power and its future.

———. (1992). Building a Bridge to Sustainable Energy. In *State of the World 1992*. New York: Norton. Describes the role of natural gas as a transition fuel to a sustainable energy system.

Gibbons, J. H., Blair, P. D., and Gwin, H. L. (1989). Strategies for Energy Use. *Scientific American* 261 (3): 136–43. Discusses possible role of several energy sources and conservation in meeting future energy demands without increasing global warming.

Gofman, J. W. and Tamplin, A. R. (1979). *Poisoned Power: The Case against Nuclear Power Plants before and after Three Mile Island*. Emmaus, PA: Rodale Press. Well-written analysis.

Hirsch, R. L. (1987). Impending United States Energy Crisis. *Science* 235 (March 20): 1467–72. Important analysis of energy costs in the coming years.

Hollander, J. M. (1992). *The Energy-Environment Connection*. Washington, D.C.: Island Press. Comprehensive survey of the problems created by energy use.

Jungk, R. (1979). *The New Tyranny*. New York: Warner Books. An important, well-written book.

Kunreuther, H., Desvousges, W. H., and Slovic, P. (1988). Nevada's Predicament: Public Perceptions of Risk from the Proposed Nuclear Waste Repository. *Environment* 30 (8): 16–20, 30–33. Outlines problems with efforts to place a high-level radioactive waste disposal site in Nevada.

Lenssen, M. (1991). Designing a Sustainable Energy System. In *State of the World 1991*. New York: Norton. Contains important information on the future of oil.

———. (1992). Confronting Nuclear Waste. In *State of the World 1992*. New York: Norton. Detailed account of nuclear energy and its problems.

Radiation Pollution

This chapter discusses nuclear energy, one of many potential sources of radiation exposure. This supplement describes radiation—its sources and its effects. To understand radiation, we must understand the atom.

The **atom** is composed of a nucleus and an electron cloud (Figure S11.1). The **nucleus** contains two tiny subatomic particles, protons and neutrons, and constitutes 99.9% of the mass of an atom. Protons are positively charged particles. Neutrons have no charge. The much lighter electrons of an atom orbit around the positively charged nucleus in the comparatively large region known as the **electron cloud**. To give you an idea of the size of the electron cloud and nucleus, consider a simple analogy. If an atom were the size of Mt. Everest (about 8800 meters or 29,000 feet), the nucleus would be about the size of a football. The rest of the atom would be made of the electron cloud.

Atoms of a given element, such as carbon or uranium, all have the same number of protons in their nuclei. But, they may contain slightly different numbers of neutrons. For example, uranium atoms all contain 92 protons. Some uranium atoms may have 146 neutrons, while others have 143. These alternate forms are called **isotopes**. To distinguish them, scientists add up the protons and neutrons and tack the sum onto the name of the element. The isotope of uranium containing 146 neutrons is called uranium-238, or U-238 (92 protons + 146 neutrons = 238). The form containing 143 neutrons is called uranium-235 (U-235).

Excess neutrons in some isotopes sometimes make them unstable. To reach a more stable state, they emit **radiation**, high-energy particles or bursts of energy. Unstable radioactive nuclei are called **radionuclides**. They occur naturally or can be produced by various physical means. For the most part, naturally occurring radionuclides are isotopes of heavy elements, from lead (82 protons in the nucleus) to uranium (92 protons). Physicists have identified three major types of radiation:

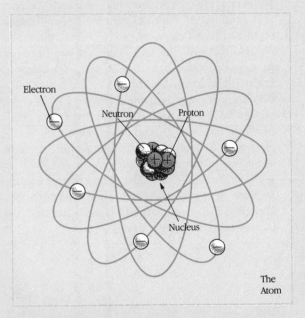

Electron

Neutron   Proton

Nucleus

The Atom

Figure S11.1 *Atoms consist of a dense central region, the nucleus, which contains protons and neutrons. Electrons are found in the area that surrounds the nucleus, the much larger electron cloud.*

alpha particles, beta particles, and gamma rays. X rays, which are also considered in this supplement, are artificially produced.

Alpha particles consist of two protons and two neutrons, the same as a helium nucleus. Alpha particles are, therefore, positively charged. Alpha particles have the largest mass of all forms of radiation. In air, they travel only a few centimeters. They can be stopped by a thick sheet of paper, so it is easy to shield people from them. In the body, alpha particles can travel only about 30 micrometers (about the width of three cells) in tissues. Because they cannot penetrate skin, they are often erroneously assumed to pose little harm to humans.

But, if alpha emitters enter body tissues, say, through inhalation, they can do serious, irreparable damage to nearby cells and their chromosomes.

Beta particles are negatively charged particles that are emitted from nuclei. They are equivalent to electrons found in the electron cloud except they are more energetic. Beta particles arise when neutrons in the nucleus are converted into protons, a process that helps stabilize radionuclides. During this process, a small amount of mass and energy is lost; this is the energetic beta particle that is ejected out of the nucleus.

The beta particle is much lighter than the alpha particle and can travel much farther. It can penetrate a 1-millimeter lead plate and can travel up to 8 meters (27 feet) in air but only 1 centimeter in tissue. Beta particles from some radionuclides have enough energy to penetrate clothing and skin but generally do not reach underlying tissues. They can, however, damage the skin and eyes, causing skin cancer and cataracts.

Gamma rays are a high-energy form of radiation with no mass and no charge, much like visible light but with much more energy. Gamma rays are emitted by nuclei to achieve a lower-energy, more stable state. They are often emitted after a nucleus has ejected an alpha or beta particle because the loss of these particles does not always allow the nucleus to reach its most stable state. Some gamma rays can travel hundreds of meters in the air and can easily penetrate the body. Some can penetrate walls of cement and plaster or a few centimeters of lead.

Unlike the three previously discussed forms, **X rays** do not originate from naturally occurring unstable nuclei. Rather, X rays are produced in X-ray machines when a high voltage is applied between a source of electrons and a tungsten collecting terminal in a vacuum tube (Figure S11.2). When electrons are ejected, they strike the collecting terminal. Colliding with tungsten atoms, they are rapidly brought to rest. The energy they carry in is released in the form of X rays, which behave like gamma rays but have considerably less energy. They cannot penetrate lead.

All the forms of radiation described above are called **ionizing radiation** because they possess enough energy to rip electrons away from atoms, leaving charged ions. Ions are the primary cause of damage to tissues.

How Is Radiation Measured?

Radioactive elements lose mass over time because of the emissions from their nuclei. Each radionuclide gives off radiation at its own rate, called the **radioactive decay rate**, which is measured in disintegrations per second. For example, 1 gram of radium decays at a rate of 37 billion disintegrations per second! The rate of radioactive decay determines the half-life of a radionuclide,

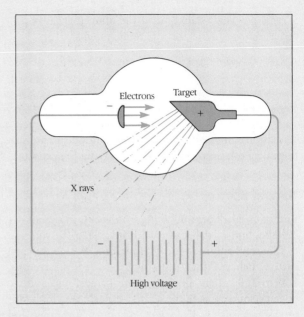

Figure S11.2 *An X-ray machine. Note that electrons strike a tungsten filament, causing it to release high-energy X rays.*

that is, the time it takes for half of a given mass of a given radionuclide to decay into more stable isotopes.

Radiation exposure in humans is expressed in several different ways. One of the most widely used measures is the rad. **Rad** stands for the "radiation absorbed dose" or, simply, the amount of energy that is released in tissue or some other medium when it is irradiated. One rad is equal to 100 ergs (a unit of energy) deposited in 1 gram of tissue.

As radiation travels through tissue, it loses its energy. The rate of energy loss is called the **linear energy transfer (LET)**. Put another way, LET is the amount of energy lost per unit of distance the radiation travels. Energy lost by radiation as it travels through the body is transferred to tissues. Because of their mass, alpha particles travel only short distances through tissue and, therefore, lose their energy rapidly. They are said to have a high LET. X rays, gamma rays, and beta particles travel farther through tissues and lose their energy more slowly; they have low LETs. Consequently, 10 rads of energy from beta particles would do less damage than 10 rads from an alpha particle because the energy from an alpha particle is lost in a shorter distance.

The term **rem** takes into account LET and thus indirectly indicates the damage that a given amount of radiation will cause in tissue. For X rays, gamma rays, and beta particles, 1 rem is essentially equivalent to 1 rad, but for alpha particles, because of their high LET, 1 rad is equivalent to 10 to 20 rems.

As a point of reference, a medical X ray may be equivalent to about 0.1 to 1 rem, depending on the type. The safety standard for workers in the United States is

5 rems per year. Background radiation is measured in thousandths of rems, or millirems (mrems).

Sources of Radiation

Radiation comes from two sources: natural and anthropogenic. Both contribute to our daily radiation exposure.

Natural Sources

Radiation is all around us. It is in rocks, in the air we breathe, and in the water we drink. Even the sun and distant stars bombard us with radiation. For many years, experts believed that Americans received, on average, 160 to 200 mrems per year from natural and anthropogenic sources, with the highest exposures in high-altitude states, such as Colorado, Montana, Wyoming, and others. In 1988, however, the National Council on Radiation Protection and Measurements announced that the average American may be exposed to nearly twice as much radiation each year as previously thought. The revised estimate takes into account the exposure resulting from radon, a radioactive gas found throughout the country. **Radon** is emitted from a naturally occurring radioactive element, radium, found in rock and soil. It seeps into homes and buildings and is breathed into the lungs, where it gives off damaging radiation. (For more on radon, see Chapter Supplement 15.1.)

Table S11.1 shows that radon probably accounts for 55% of the annual radiation exposure. All told, natural sources are responsible for slightly more than 80% of our exposure to radiation. Slightly under 20% of our exposure comes from human sources, such as X rays and consumer products (televisions and luminous-dial watches).

Anthropogenic Sources

Anthropogenic radiation sources are many: (1) medical therapy (X-ray treatment for cancer) and diagnosis (X rays for bone fractures), (2) detonation of nuclear weapons in testing and World War II, (3) nuclear energy, (4) television sets, (5) luminous dials on watches, and (6) air travel.

Effects of Radiation

How Does Radiation Affect Cells?

All forms of radiation ionize and excite biologically important molecules in tissues. Positively charged alpha

Table S11.1   Estimated Radiation Exposure in the U.S.

Sources of Radiation	Exposure and Percent of Annual Dose
Natural	
Radon	200 millirems (55%)
Cosmic	27 millirems (8%)
Rocks and soil	28 millirems (8%)
Internal exposure	40 millirems (11%)
Anthropogenic	
Medical X-rays	39 millirems (11%)
Nuclear medicine	14 millirems (4%)
Consumer products	10 millirems (3%)
Others	Less than 1%

Source: National Council on Radiation Protection and Measurements.

particles, for example, draw electrons away from atoms in body tissues. Negatively charged beta particles in tissues may repel electrons of various atoms, causing them to be expelled from their atoms. They, too, produce positively charged ions. Gamma rays and X rays, on the other hand, are uncharged, but they possess lots of energy, which may be transferred to electrons as these forms of radiation pass through tissue. This energy excites the electrons and may cause their expulsion from the atoms, forming ions. Alternatively, the energy imparted to the electrons may make chemical bonds in molecules unstable and more easily broken.

Ionization of the atoms in water and other molecules in tissues is responsible for much of the damage caused by radiation. Water molecules become positively charged when electrons are ripped from their atoms, as shown in Figure S11.3. Electrons freed from water molecules may combine with uncharged water molecules, forming negatively charged water molecules. Both positively and negatively charged water molecules rapidly break up into highly reactive fragments called **free radicals** (Figure S11.3).

Free radicals react almost instantaneously with biologically important molecules. When they react with oxygen, for example, hydrogen peroxide is formed. This powerful oxidizing agent damages or destroys proteins and other molecules, causing cell death. If it is extensive enough, cellular destruction can kill the organism. In some instances, damage may be quickly repaired by the cells without any long-term effects. In other cases, the damage in the form of mutations and cancer may not be expressed until years after the exposure (Chapter 14).

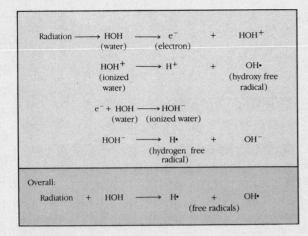

Figure S11.3 *The ionization of atoms in water molecules is responsible for much of the damage caused by radiation.*

Figure S11.4 *For many years, much of what we knew about the effects of radiation came from studies of the people injured and killed by the atomic bombs dropped on Japan at the end of World War II.*

Health Effects of Radiation

The effects of radiation on human health depend on many factors, such as the amount, the exposure time, and the type of radiation. The half-life of the radionuclide, the health and age of the individual, and the part of the body exposed also influence the outcome. Finally, damage depends on whether the exposure is internal or external.

Numerous studies of radiation have revealed some interesting generalizations: First, fetuses are more sensitive to radiation than children, who are, in turn, more sensitive than adults. Second, cells undergoing rapid division appear to be more sensitive to radiation than those that are not. This is especially true in regard to cancer induction. Thus, lymphoid tissues (bone marrow, lymph nodes, and circulating lymphocytes) are the most sensitive of all the body's cells. Epithelial cells, those that line body organs such as the intestines, also undergo frequent division and are highly sensitive to radiation. In sharp contrast, nerve and muscle cells, which do not divide, have a very low sensitivity and rarely become cancerous. Third, most, if not all, forms of cancer can be increased by ionizing radiation.

Health experts divide radiation exposure into two categories. Exposures over 5 to 10 rems per year are considered high level. Below this, the exposure is low level.

Impacts of High-Level Radiation

The most important information on high-level radiation comes from studies of the survivors of the two atomic bombs dropped on Japan at the end of World War II (Figure S11.4). Studies of these groups have led to several important findings. First, the lethal dose for one-half the people within 60 days is about 300 rads. Second, a dose of 650 rads kills all people within a few hours to a few days. Third, sublethal doses, or doses that do not result in immediate death, range from 50 to 250 rads. Individuals exposed to these levels, however, suffer from **radiation sickness**. The first symptoms, which develop immediately, are nausea and vomiting; 2 to 14 days later, diarrhea, hair loss, sore throat, reduction in blood platelets (needed for clotting), hemorrhaging, and bone-marrow damage occur. Fourth, sublethal radiation has many serious delayed effects, including cancer, leukemia, cataracts, sterility, and decreased life span. Fifth, sublethal radiation also has profound effects on reproduction, increasing miscarriages, stillbirths, and early infant deaths.

High-level radiation exposure is rare today. Such exposures usually occur only in individuals working at or living near nuclear power plants or munitions factories where a major accident, such as Three Mile Island or Chernobyl, has occurred. Nuclear war, even on a limited scale, would expose large segments of the human population to dangerously high levels of radiation (Chapter Supplement 16.1). Accidents during the transportation of nuclear fuels and high-level wastes might also result in dangerously high exposures.

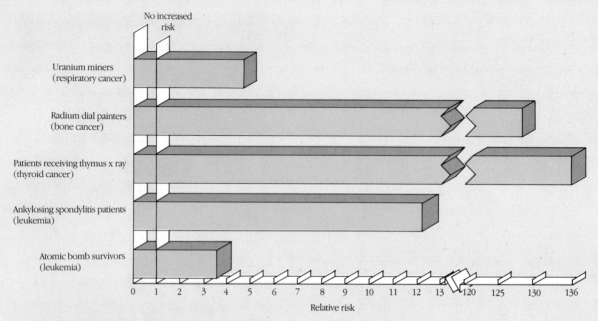

Figure S11.5 *Relative risk of cancer in people exposed to radiation. Relative risk is a measure of the probability of contracting cancer. As shown here, a uranium miner is 4.0 times more likely to develop respiratory tract cancer than someone who is not a miner. Atomic bomb survivors, in general, are 3.5 times more likely to develop leukemia than non-irradiated people.*

Impacts of Low-Level Radiation

The effects of high-level radiation are well established, but as is the case with low-level exposure to any toxic agent, health effects are harder to discern with low-level radiation. A growing body of evidence shows that low-level radiation increases the likelihood of developing cancer (Figure S11.5). For example, studies of individuals who years earlier were treated with radiation for acne, spinal disorders, and even syphilis show elevated levels of cancer and leukemia. In addition, children whose necks have been X-rayed have an elevated rate of thyroid cancer. Lung cancer rates are elevated in uranium and fluorspar (calcium fluoride) miners, both of whom are exposed to radon gas. Factory workers who painted watch dials with radium early in this century developed bone cancer and a serious disease of the bone marrow called aplastic anemia. Several studies show that the rates of leukemia, tumors of the lymphatic system, brain tumors, and other cancers are 50% higher in infants whose mothers have been exposed to ordinary diagnostic X-rays (2 to 3 rads) during pregnancy. One study showed that a 1-rem exposure to fetuses causes an 80% increase in mortality from childhood cancer. According to some estimates, low-level radiation from testing of nuclear weapons in the 1950s and 1960s caused 400,000 deaths in children because of cancer. Studies of workers at a plant in Hanford, Washington, the U.S. government's source of radioactive materials for

nuclear weapons, showed that the death rate from cancer was 7% higher than expected and that workers who died of cancer had received, on the average, only about 2 rems per year, well below the supposed safe level of 5 rems a year. As a final note, a new study of 27,000 Chinese radiologists and X-ray technicians showed that exposure to low levels of radiation increased the risk of developing cancer by 50%. Leukemia (a kind of blood cancer), breast cancer, thyroid cancer, and skin cancers were the prevalent types. This study, which confirms research done in the United States and Europe, is important because of the large number of subjects studied.

Studies suggest that low-level radiation is much more harmful than many scientists thought a decade ago. Noted radiation biologist Dr. Irwin Bross, for example, estimates that low-level exposure is about ten times more harmful than previously calculated. He and others have called for major revisions of the maximum allowable doses for workers.

But, is there a threshold level below which no damage occurs? No one can say for certain. Some health officials believe that no level is safe because the effects of continued low-level exposure are cumulative.

According to radiation experts John Gofman and Arthur Tamplin, exposure to the current public health standard set for anthropogenic sources—0.17 rad per year—will result in a 30-year exposure of 5 rads. This would cause 14 additional cases of cancer each year for every 100,000 people exposed, or about 14,000 additional cancer cases per year in adults over 30 and at least 2000 cases of cancer in individuals under 30 years of age. Studies by the National Academy of Sciences and the National Research Council suggest that a 0.17-rad exposure to anthropogenic radiation will probably increase the rate of cancer by 2% and the incidence of serious genetic diseases by about 1 birth in every 2000. Thus, some experts suggest lowering the health standard to 0.017 rad per year.

→   Some geneticists warn that genetic disorders caused by radiation may be passed from generation to generation and may increase in incidence with subsequent generations. For example, if the incidence of genetic disorders and birth defects with a genetic basis were 1 in 2000 in the first generation, it would be 5 in 2000 in their offspring. A major consideration, then, is what impact radiation exposure today will have on subsequent generations. Are current policies and radiation standards posing a danger to future generations?

→   Low-level effects are small and hard to detect. The long latent period between exposure and disease makes it difficult for researchers to link the cause and the effects. Thus, studies such as those cited above have stirred a considerable amount of controversy. Although the results and conclusions of individual low-level radiation studies are debatable, on the whole, they consistently seem to point to one conclusion: a substantial risk is created by subjecting people to low-level radiation. The questions become: What level of risk is acceptable? At what point do the benefits of X rays, nuclear power, and other uses outweigh the risks?

Bioconcentration and Biological Magnification

Table S11.2 lists some radionuclides emitted from nuclear weapons and nuclear power plants. Some of
→ these radionuclides are absorbed by humans and organisms, and may become concentrated in particular tissues.

For example, iodine-131 (I-131) is released from nuclear power plants, during both normal operations and in accidents. Fallout on the ground may be incorporated in grass eaten by dairy cows. It is then selectively taken up, or bioconcentrated, by the human thyroid gland, where it irradiates cells and may produce tumors. Milk contaminated with I-131 is especially harmful to children.

Strontium-90 is released during atomic bomb blasts. It may also be released from reactors in small amounts under normal operating conditions but in large quantities in accidents. Strontium-90 is readily absorbed by plants and may also be passed to humans through cow's milk. It seeks out bone, where it is deposited like calcium. With a half-life of 28 years, it irradiates the bone and can cause leukemia and bone cancer.

Table S11.2 Radionuclides from Nuclear Weapons and Reactors

Nuclear Weapons	Nuclear Reactors
Strontium-89	Tritium (Hydrogen-3)
Strontium-90	Cobalt-58
Zirconium-95	Cobalt-60
Rubidium-193	Krypton-95
Rubidium-106	Strontium-85
Iodine-131	Strontium-90
Cesium-137	Iodine-130
Cerium-141	Iodine-131
Cerium-144	Xenon-131
	Xenon-133
	Cesium-134
	Cesium-137
	Barium-140

Accumulation of radionuclides within tissues has important implications for human health, as seemingly low levels may become dangerously high in localized regions. Some radionuclides may be biologically magnified in the higher trophic levels of food chains.

Minimizing the Risk

Radiation can be reduced in several ways. Since X rays are the most significant anthropogenic source of exposure, prudence dictates a cautious use of them. High-dose exposures especially warrant thorough discussion with the physician.

Important ways to reduce X-ray exposure include: (1) asking the physician if previous X rays or diagnostic procedures you've had would provide the same information; (2) reducing X-ray exposure in children; (3) informing physicians and dentists that you are pregnant (don't wait to be asked); (4) if you are pregnant, avoiding all X-rays of the pelvis, abdomen, and lower back unless they're absolutely necessary; (5) avoiding mobile X-ray units because they tend to give higher-than-necessary doses; (6) if you are a woman under the age of 50 and have no family history of breast cancer, avoiding routine mammographs; (7) questioning the necessity of preemployment X-rays; (8) if you must be X-rayed, requesting that a full-time radiologist do it; (9) asking if the X-ray machine and facilities have been inspected and set to minimize excess exposure; (10) requesting that a lead apron be placed over your chest and lap for dental X rays and that a thyroid shield be placed around your neck; (11) cooperating with the X-ray technologist (do not breathe or move during the X ray); and (12) making sure the operator exposes only those parts of the body that are necessary.

Radiation exposure from medical diagnosis and treatment is by far the easiest to control on an individual level. Controls on exposures from nuclear weapons testing, possible nuclear war, and possible catastrophic accidents at nuclear power plants may seem out of reach for the average citizen. Nevertheless, individual citizens can have a significant cumulative impact on nuclear policies by educating themselves and others, voting responsibly, becoming involved in political organizations, and writing letters to their representatives in Congress. For an opinion on the influence of such letter writing and for some suggestions on how to write letters on political issues, see the *Environmental Action Guide* published with this text.

Suggested Readings

Gofman, J. W. (1981). *Radiation and Human Health.* San Francisco: Sierra Club Books. Comprehensive survey of the effects of low-level radiation on health.

Harley, N. H. (1991). Toxic Effects of Radiation and Radioactive Materials. In *Toxicology: The Basic Science of Poisons* (3rd ed.), ed. M. O. Amdur, J. Doull, and C. D. Klaassen. New York: Pergamon Press. Excellent review.

Laws, P. W., and Public Citizen Health Research Group. (1983). *The X-Ray Information Book.* New York: Farrar, Straus, & Giroux. General coverage of X rays, their effects, and ways to minimize exposure.

Foundations of Sustainable Energy Strategy: Conservation and Renewable Energy

I cannot say whether things will get better if we change; what I can say is they must change if they are to get better.

—G. C. Lichtenberg

Critical Thinking Exercise

An internationally known expert from the oil industry lectures to a group of congressional representatives that solar energy is not an economically sound option. It's a great idea, he argues, but the economics of solar energy aren't good enough to merit widespread use. He contends that the conventional fuels, such as oil, coal, and nuclear energy, are more desirable. If you were one of the panel members listening to this testimony, what questions would you ask? How would you analyze this information? What critical thinking rules does this examination require?

Michael Reynolds is a maverick in the construction industry. Based in Taos, New Mexico, Reynolds builds homes that embody virtually all of the principles of sustainability. The walls, for example, are constructed of old automobile tires, which otherwise would have ended up in landfills. The tires are stacked on top of one another and packed with dirt from the construction site. Adobe is then applied to the tire wall, creating an appealing design (Figure 12.1a). Reynolds's houses are generally built into the side of hills and with their thick walls are extremely energy efficient. For people who like a living space that is cool in the summer and warm in the winter (and who doesn't?), these houses are just the ticket.

Reynolds's homes are heated by the sun in the winter and can be designed with interior planters that line the south wall, permitting residents to grow a variety of vegetables year-round (Figure 12.1b). Some of his homes generate their own electricity from sunlight and are equipped with the most efficient lighting systems and appliances available.

Reynolds compares the typical American home to a patient in an intensive care unit. Without intensive outside support in the form of food, water, and energy, its occupants could not survive. In many ways, the typical American home is a microcosm of cities and towns. Cut off from food, water, and oil, a city or town would face serious difficulties.

Reynolds is a pioneer in creating sustainable living spaces that supply the needs of the occupants with minimal environmental impact. He is part of a growing legion of people working to help human society make the transition to a sustainable culture, one that supplies human needs like energy while enhancing the Earth's life-support systems.

This chapter examines two major sources of energy needed to create a sustainable global energy system—conservation and renewable energy—and looks at the pros and cons of both approaches. The final section of the chapter shows how these alternative and environmentally sustainable forms of energy could eventually replace oil, coal, natural gas, and nuclear energy.

12.1 Conservation: Foundation of a Sustainable Energy System

The United States needs energy, and needs it badly. But, at least one-half—perhaps as much as three-fourths—of the energy consumed in the United States is wasted. The second law of thermodynamics states that when energy is converted from one form to another, some energy is lost as heat. In other words, no energy conversion is 100% efficient; some waste is inevitable. However, the amount of energy wasted in the United States far exceeds the inevitable loss.

Other countries are also extremely wasteful in their use of energy, among them, Canada, the Commonwealth of Independent States, many Western European countries, and most developing nations. Even the most energy-efficient nations of the world, such as Japan and Germany, have found that they need to make substantial improvements in energy efficiency.

Given our heavy dependence on nonrenewable energy supplies, their economic and environmental costs, and their eventual depletion, waste is economically, environmentally, and socially irresponsible. Writer Bruce Hannon put it best: "A country that runs on energy cannot afford to waste it." You could easily substitute *world* for *country*.

Economic and Environmental Benefits of Energy Conservation

Conservation is one of the biological principles of sustainability and encompasses two related aspects: first, using what we need and, second, using it efficiently. Energy conservation is essential to the sustainable transition now under way and offers numerous economic and environmental benefits. In industry, energy efficiency can reduce the cost of producing goods, giving companies a decided economic advantage in the marketplace. An Alcoa plant in Iowa, for example, found that with a minimum investment, it could cut its $6 million per year energy bill by a third, reducing energy costs by $2 million per year. Energy conservation also results in substantial reductions in pollution. Since many factories capture pollutants in pollution control devices, such as smokestack scrubbers, which produce substantial amounts of waste, energy savings translate into additional economic savings by reducing waste disposal costs.

The economic savings from energy efficiency can be illustrated by comparing the cost of this strategy with

(a)

(b)

Figure 12.1 *A house made out of tires?* (a) *Architect and builder Michael Reynolds builds homes out of tires that would have otherwise been thrown away.* (b) *When walls are covered by adobe, they appear quite attractive. These planters can be used to grow food and flowers year round.*

from coal-fired power plants costs 5 to 7 cents per kilowatt-hour. Electricity from nuclear power plants, on average, costs 10 to 12 cents per kilowatt-hour. What this means is that for every penny a company invests in energy efficiency it will save 4 to 11 cents in electric bills, depending on the cost of energy in the area. Similar calculations can be made for other fuels. According to energy expert Amory Lovin of the Rocky Mountain Institute, oil conservation costs about $2 per barrel, one-tenth of the cost of a new barrel. If you own a factory, every $1 you invest on energy-efficiency measures that cut your heating-oil demand will save you nearly $10 in oil costs. Not a bad investment.

Not surprisingly, the most energy-efficient companies and nations are also the most successful economically (Table 12.1). In the United States, for example, 10 cents of every dollar of the total output of goods and services, the **gross national product** (**GNP**), is spent on energy. Japan, on the other hand, spends about 4 cents per dollar of GNP. This relative inefficiency in fueling industry costs the United States $220 billion per year. In other words, U.S. industries spend over $200 billion per year more than is necessary (by Japanese standards) to produce the same goods and services. The cost differential gives the Japanese a slight economic edge on everything they sell in the United States and in foreign markets.

Like sustainable strategies outlined in previous chapters, energy efficiency simultaneously addresses a number of environmental problems. For example, it helps reduce acid deposition, global warming, and urban air pollution. By cutting demand, it helps reduce the number of oil spills from tankers and pipelines, thus saving aquatic species and birds and mammals that

the cost of supplying energy from conventional sources. For example, measures that improve the efficiency of machines and appliances cost about 1 to 2 cents per kilowatt-hour of energy saved. In contrast, electricity

Table 12.1   Oil Consumed Per Unit of National Output, Selected Countries, 1989/90[1]

Country	Oil Consumed (million barrels per day)	Oil Dependence (million barrels oil per billion dollars GNP)
United States	16.20	1.11
Japan	5.25	0.66
West Germany	2.40	0.59
China	2.28	1.95
India	1.04	1.48
South Korea	0.81	1.41
Brazil	1.20	1.25

[1]Oil dependence figures for industrial and developing countries may not be comparable.
Source: from *State of the World, 1992.*

depend on clean water. In addition, because less fuel is needed, this strategy reduces the destruction of wildlife habitat by decreasing mining and extraction of fossil fuels.

The transition to a sustainable energy system could take a minimum of 30 to 50 years. Thus, energy-efficiency measures rapidly brought on-line could help stretch limited supplies of fossil fuel, especially clean-burning natural gas, which may prove to be vital to the sustainable transition.

Energy-Efficiency Options

The range of energy-efficient options available to businesses and homeowners is enormous. In some cases, efficiency can be achieved through minor changes. The Denver Marriott Hotel, for instance, made substantial cuts in energy demand simply by asking the cleaning staff to turn the heat up in unoccupied rooms during the summer and down during the winter. Around the home, shutting off lights and turning off the television when they're not in use, or lowering the thermostat a little in the winter and raising it in the summer can result in significant reductions in energy demand. For commuters, driving within the speed limit, keeping one's car tuned, and maintaining proper tire pressure could translate into millions of gallons of gasoline saved each year. Such measures require little or no initial investment, and can save individuals enormous sums of money.

Many devices that save energy are readily available

for homes, businesses, and factories. Water-saving devices, especially showerheads, for instance, can save substantial amounts of energy in homes. High-efficiency lighting systems can cut electric demand in homes and businesses. The compact fluorescent light bulb, shown in Figure 12.2, replaces the standard incandescent light, using one-fourth as much energy. Thus, a 75-watt incandescent light bulb can be replaced by an 18-watt compact fluorescent bulb. Although the compact fluorescent costs more initially ($10 to $20), it will outlast 9 to 10 ordinary bulbs, saving $30 to $50 over its lifetime in reduced electric demand. It will also reduce carbon dioxide pollution by about 1 metric ton and will reduce cooling costs.

Substantial energy savings can also be achieved by installing better insulation in homes, businesses, and factories. Doubling the insulation of a new home, for example, adds about 5% to the cost but pays for itself in five years in reduced energy costs. Conservation measures in new office buildings can cut energy for lighting, heating, and cooling by 80% or more! Energy-efficient homes can be built with heating bills of $100 per year even in cold climates!

Planting shade trees around homes can reduce summer cooling costs. One study showed that three trees planted next to light-colored houses in a residential neighborhood in Phoenix cut cooling demand by 18%. In Sacramento and Los Angeles, the savings would be even greater—34% and 44%, respectively. Planting conifers on the north and west sides of homes can cut wind and reduce winter heating costs. Nationwide, tree planting near homes and businesses could cut energy demand by nearly 1 quad, or about one-eightieth of the total energy demand.

Conservation does not mean "freezing in the dark," as some would have us believe. For homeowners, adding attic insulation to older homes reduces heat loss in the winter and retains cool air in the summer, resulting in not only a significant savings on utility bills but also a more comfortable existence. Likewise, installing storm windows saves energy and reduces drafts. Both insulation and storm windows require small investments that are repaid in short periods by reduced energy bills. (For additional suggestions, see Table 12.2.)

One extraordinary means of reducing energy waste is **cogeneration**. In industrial cogeneration, waste heat from one process is captured and used in another, thus reducing waste and improving the energy efficiency of a plant. For example, a large factory that produces electricity on site may capture waste heat that would have otherwise been vented into the atmosphere. Waste heat can then be used to warm office buildings or heat water.

For years, most American industries have produced their own steam on site for various purposes with natural gas or oil, which they purchased, along with electricity, from local utilities. Steam generation in such factories

Figure 12.2 *The compact fluorescent light bulb is one of many cost-effective means of saving energy.*

was 50% to 70% efficient. By generating their own electricity on site with waste heat, the efficiency of the system can be boosted to 80% to 90%. The cost of electricity from cogeneration is about 4 cents per kilowatt-hour, slightly less than from coal and substantially less than from nuclear power.

Because of its favorable economics, cogeneration is an emerging source of energy. In 1989, 2300 cogeneration plants were operating in the United States, producing 27,000 megawatts of energy that otherwise would have been wasted. In 1992, over 3000 facilities generated 40,000 megawatts of energy. Plans to increase cogeneration are under way. Another 60,000 megawatts are under development in the United States alone. By the year 2000, cogeneration could provide 100,000 megawatts; its estimated full potential is seven times greater than this estimate. U.N. estimates put the world potential at 2.7 million megawatts.

In Germany, small cogeneration technologies are being installed in restaurants, hotels, and apartment buildings. They supply space heat, hot water, and electricity at a lower cost than conventional systems. In Chula Vista, California, a McDonald's restaurant produces its own electricity and hot water using a small cogeneration system.

Table 12.2 Energy Conservation Suggestions

1. Water Heating
   Turn down thermostat on water heater.
   Use less hot water (dishwashing, laundry, showers).
   Install flow reducers on faucets.
   Coordinate and concentrate time hot water is used.
   Do full loads of laundry, and use cooler water.
   Hang clothes outside to dry.
   Periodically drain 3 to 4 gallons from water heater.
   Repair leaky faucets.

2. Space Heating
   Lower thermostat setting.
   Insulate ceilings and walls.
   Install storm windows, curtains, or window quilts.
   Caulk cracks and use weatherstripping.
   Use fans to distribute heat.
   Dress more warmly.
   Heat only used areas.
   Humidify the air.
   Install an electronic ignition system in furnace.
   Replace or clean air filters in furnace.
   Have furnace adjusted periodically.

3. Cooling and Air Conditioning
   Increase thermostat setting.
   Use fans.
   Cook at night or outside.
   Dehumidify air.
   Close drapes during the day.
   Open windows at night.

4. Cooking
   Cover pots, and cook one-pot meals.
   Turn off the pilot lights on stove.
   Don't overcook, and don't open oven unnecessarily.
   Double up pots (use one as a lid for the other).
   Boil less water (only the amount you need).
   Use energy-efficient appliances (crock pots).

5. Lighting
   Cut the wattage of bulbs.
   Turn off lights when not in use.
   Use fluorescent bulbs whenever possible.
   Use natural lighting whenever possible.

6. Transportation
   Car-pool, walk, ride a bike, or take the bus to work.
   Use your car only when necessary.
   Group your trips with the car.
   Keep car tuned and tire pressure at recommended level.
   Buy energy-efficient cars.
   Recycle gas guzzlers.
   On long trips take the train or bus (not a jet).

The Potential of Energy Efficiency

The United States and other industrial nations have made some modest improvements in energy conservation, but the conservation potential has hardly been tapped. The World Resources Institute, in fact, estimates that the world could meet 90% of its new energy needs between 1987 and 2020 simply by making more efficient use of the energy we now use. Even though the world population is expected to double between 1980 and 2020, the institute estimates only a 10% increase in energy production would be needed if existing energy-efficiency technologies were effected. California's 1990 energy plan outlines a strategy in which 75% of new demand in the next 40 years would come from using existing sources more efficiently. The remainder would come from renewable energy.

Enormous opportunities to conserve energy exist in transportation, buildings, and industries.

Transportation Savings   Over one-fourth of the energy consumed in the United States is used in transportation. According to energy experts, dramatic improvements in vehicle efficiency are possible. Automobiles, the single largest source of fuel consumption, represent the greatest potential for reductions. Without any major technological breakthroughs, new cars averaging 60 miles per gallon (mpg) could be on the market by 1995. Geo Metro, made by the Japanese and sold by Chevrolet, gets 24 kilometers per liter (58 mpg on the highway). Several auto manufacturers have test models that get 40 kilometers per liter (98 mpg). In contrast, the average new car rolling off the assembly line in the United States in 1992 got only 11.6 kilometers per liter (28 mpg), one of the lowest averages in the developed world. By increasing the average mileage to 20 kilometers per liter (50 mpg), Americans could reduce automobile emissions by half.

Buildings   About one-third of the energy in the United States is consumed in buildings. It is used for heating, lighting, cooling, and operating appliances. Unfortunately, efficiency improvements in buildings lag behind other areas. Why? New houses and commercial buildings are built by contractors seeking to minimize initial expenses. As a result, they often overlook investments in efficient appliances, heating, and lighting systems. In addition, incentives for reducing energy use in existing buildings are rather weak. Today, renters occupy more than 30% of the nation's housing. Because they are largely responsible for paying utility bills, landlords have little incentive to invest in energy efficiency.

One encouraging development is the emergence of energy conservation companies, which provide innovative financing schemes for building owners who want to invest in energy efficiency. Energy conservation compa-

nies, successful in Europe for years, are becoming more common in the United States. Some experts believe that cost-effective energy-efficiency measures could cut total energy consumption in U.S. buildings by as much as 30% over the next two decades despite a 15% to 20% increase in building stock.

Industries   U.S. industries consume more than one-third of the nation's energy. Over the past ten years, industrial energy savings have exceeded many expectations, but industries still have a long way to go. Corporate executives now realize that an investment in energy efficiency is one of the most cost-effective ways of reducing expense and boosting profits.

Recycling is also an important strategy for reducing industrial energy consumption (Chapter 19). Roughly three-fourths of the energy consumed by industry is used to extract and process raw materials. Making metals from recycled scrap uses a fraction of the energy needed to produce them from virgin ore. Recycling 1 metric ton of steel, for instance, uses only 14% of the energy needed to produce a ton of steel from raw material. One metric ton of aluminum from recycled scrap uses only 5% of the energy needed to make this metal from raw ore.

Promoting Energy Efficiency

Given the economic and environmental advantages of improved energy efficiency, it seems logical that countries would be vigorously pursuing energy-efficiency strategies. In some, this is true. To date, at least 24 industrial nations have pledged to stabilize or reduce their carbon dioxide emissions in an effort to control global warming. They will achieve this in large part by becoming more energy efficient.

In other countries, such as the United States, interest in energy conservation has slowed to a crawl since 1985. How can individuals, businesses, and nations be compelled to use energy more efficiently?

At least six avenues are available: (1) education, (2) taxes, (3) "feebate" systems, (4) government-mandated efficiency programs, (5) price changes, and (6) least-cost planning.

One of the most important steps is **education**. Educating people of the economic and environmental benefits of energy conservation is vital to the task at hand. The public must also be made aware of strategies and technologies.

Federal taxes on fossil fuels, especially oil and coal, could help promote efficiency. By making fossil fuels more costly to consumers, governments could play an important role in stimulating energy conservation in all sectors of society.

In the United States, the average consumer pays a tax of about 30 cents per gallon for gasoline. In Denmark, the tax is nearly $3 per gallon. This tax encourages the efficient use of gasoline. Increased taxes on gasoline and other fuels in the United States and Canada could help cut energy consumption and free up money for building efficient mass transit systems, promoting conservation, or for funding sustainable alternative energy sources.

Another innovative measure, proposed by Amory Lovins and others, is the **"feebate" system**. It consists of a fee (tax) paid by those who buy gas-guzzling cars, which would create a disincentive to consumers. A rebate, which would serve as an incentive, would be given to those who purchase energy-efficient autos. The rebate would be paid by the fees. The state of Maryland passed a feebate program that charges purchasers of gas guzzlers $100 extra and offers new car buyers a rebate of $50 for every mpg over 34, but the Bush administration forbade the state to implement the program.

A fourth means of stimulating conservation is through **government-mandated efficiency standards**. The **National Appliance Conservation Act** passed by Congress in 1987, for instance, established energy-efficiency standards for all new appliances. From 1992 on, all major household appliances, such as refrigerators, must consume 20% less electricity than 1987 models. The act will reduce peak U.S. demand for electricity by nearly 22,000 megawatts by the end of the century, the equivalent of 20 large power plants. California passed a similar law, calling for a 50% cut in electric usage by new appliances. Efficiency standards could be applied to all new homes and factories as well. Automobile mileage standards, already in place, could be tightened.

Conservation can also be stimulated by changes in pricing. For example, some utilities now charge customers more for electricity used during peak hours because meeting peak demand is very costly for utilities. It often requires construction of additional power plants, which are needed only a few hours per day. If utility companies can reduce peak demand through pricing, they won't have to build expensive new facilities.

Finally, efficiency can be stimulated by a process called **least-cost planning**. In the United States, individual states regulate utilities. Thus, when a power company decides that it needs to build a new power plant to meet rising demand, it must first receive state approval. Currently, over half of the public utility commissions now require utilities to select the least-expensive means of providing electricity. Utilities forced to perform least-cost analysis frequently find that new plants are far more expensive than conservation strategies. Some cost-saving options include improvements in generating efficiency, peak-pricing schemes, purchasing electricity from other companies, cogeneration, and promoting energy conservation measures in homes, factories, and businesses. These steps often save customers substantial sums of money and prove to be quite profitable for utilities, so much so that some companies offer cash rebates to consumers who buy energy-efficient appliances. Overall, companies have found that these incentives are two to three times cheaper than providing energy by building a new plant; moreover, the payback period of conservation is only 2 or 3 years rather than 20. Other utilities provide free energy audits or low-cost loans to customers who invest in insulation or storm windows.

Roadblocks to Energy Conservation

Despite the many advantages of conservation, it still is not as widely practiced as many would like. One reason is that the United States, Canada, and other developed nations have had abundant energy resources. Today, many suffer under the delusion that fossil fuel energy supplies are vast and, therefore, see little need to conserve energy.

Another reason is that federal programs have subsidized fossil fuels and nuclear energy, giving them an unfair advantage in the energy market. With federal subsidies paid by taxpayers, U.S. oil, which sells for about $20 per barrel, actually costs us about $100 to $200 per barrel, according to estimates of Amory Lovins. At this rate, gasoline really costs us about $5 to $10 per gallon. Lovins also claims that Middle Eastern oil costs about $495 per barrel when one considers the cost of the military escort of tankers moving in and out of the Persian Gulf. This doesn't include the cost of the Persian Gulf War!

Still another reason is that high-efficiency products, such as compact fluorescent light bulbs, sometimes cost more than less energy-efficient ones. Many people neither calculate the long-term savings nor think about the environmental consequences of waste.

In addition, many governments have not been committed to energy conservation programs. In 1987, the U.S. government spent only $160 million on energy conservation, down from $345 million in 1980. In 1992, because of a rising concern for energy efficiency, conservation research, development, and grants from the U.S. Department of Energy amounted to $540 million. Although an impressive turnabout, the current outlay for conservation is equal to about 17 hours of military spending. This poor showing is hard to reconcile when the government investment-to-savings ratio for conservation can be as high as 1 to 1000—$1000 saved for every $1 invested.

The lack of interest in energy conservation in the United States and other countries can further be attributed to the influence of energy companies, a dominant force on the political scene.

Despite these barriers, Christopher Flavin and Nicholas Lenssen recently noted: "Powerful economic, environmental, and social forces are pushing the world toward a very different energy system." One of the cornerstones of that system will be energy efficiency. Its potential is great, and many people, businesses, and nations are beginning to recognize that one of the greatest future sources of energy is our waste.

12.2 Renewable Energy Sources

Imagine a world powered by sunlight, wind, and other clean, renewable forms of energy. Although this may sound like a dream, it could very likely become a reality. Fifty years from now, in fact, you may live in a world powered by a diverse mixture of renewable energy resources. In some regions, energy-efficient houses would be heated and cooled by the sun. Windmills and photovoltaics would provide electricity to power mass transit systems, electric cars, and homes. Hydrogen gas would heat our food and power vehicles. Some liquid fuels might even come from waste wood.

The shift to a renewable energy future has already begun. Brazil, for example, is turning to ethanol produced from sugarcane to power trucks and cars. California is turning to wind and geothermal energy. Israel and other Middle Eastern countries are increasing their dependence on solar energy. Many Pacific rim nations now acquire substantial amounts of energy from geothermal sources and plan to obtain substantially more in the future.

As world oil reserves decline, environmental problems precipitated by fossil fuel use intensify, and population spirals upward, more and more countries will shift to renewable energy resources. Each one will need to find many different clean and abundant locally available renewable fuels. In doing so, they can create more self-reliant economies. This section looks at six options and discusses the pros and cons of each.

Solar Energy: Abundant, Clean, and Profitable

Oil, natural gas, oil shale, and coal all have limits. So does the sun, the origin of solar energy. In the sun's case, though, the supply of energy is expected to last for at least 2 billion years. As a result, most people consider solar energy as a renewable resource.

Each day, about two-billionths of the sun's energy strikes the Earth. Although this is a small amount, it adds up to an impressive total. In fact, if all of the sunlight striking an area the size of Connecticut was captured and converted into useful energy, it could power the entire United States, including all homes, factories, and vehicles. Despite the enormous potential, solar energy provides only a fraction of U.S. energy needs. Contrary to popular misconception, this poor showing is not because solar energy is limited to a few areas. In fact, significant sources of solar radiation are available throughout the world (Figure 12.3).

Four major solar technologies are in use worldwide: passive, active, photovoltaics, and solar thermal electric. Understanding each one can help to assess the potential of this largely untapped energy source.

Passive Solar Heating   Passive solar heating systems are designed to capture solar energy within a building and provide an excellent source of space heat (Figure 12.4). In a passive solar system, sunlight streams through south-facing windows and is absorbed by interior walls and floors of brick, tile, or cement. The heat stored in these structures radiates into the rooms, heating the air day and night. On cloudy days, solar homes are kept warm by residual heat that continues to radiate from heat-absorbent materials (thermal mass) and by backup systems. Properly designed passive solar homes and buildings require good insulation and shutters or heavy curtains to reduce the outflow of heat at night. Overhangs block out the summer sun.

Passive solar energy is sometimes described as a system with only one moving part, the sun. It can be added to an existing home by building on a greenhouse, which, in addition to supplying winter heat, provides a year-round source of food (Figure 12.5).

Well-designed passive systems can provide 100% of a home's space heating. One passive solar home in Canada, built by the Mechanical Engineering Department of the University of Saskatchewan, for instance, has an annual fuel bill of $40, compared with $1400 for an average American home. The house is so airtight and well insulated that heat from sunlight, room lights, appliances, and occupants provides sufficient energy to maintain a comfortable interior temperature. The construction cost of this house was only slightly more than that of a tract home. My superinsulated solar home at 2500 meters (8200 feet) in the Colorado Rockies is heated for about $100 to $120 per year (Figure 12.6).

As a rule of thumb, solar houses cost about 10% more to build than conventional houses of similar size. Rising energy costs, however, could easily offset the slightly higher initial price. In addition, individuals who want a passive solar home can build one that fits their budget by trimming a little unnecessary floorspace. Through careful design, one can maximize the available floorspace, and still have plenty of room and a house that's energy independent. Because used-home sales are primarily based on square footage, passive solar homes are often a good buy. My house cost about the same as a conventional used home, with only a fraction of the utility bills!

Thousands of American homeowners have selected another solar option, the **earth-sheltered house**. Built partly or entirely underground to take advantage of the

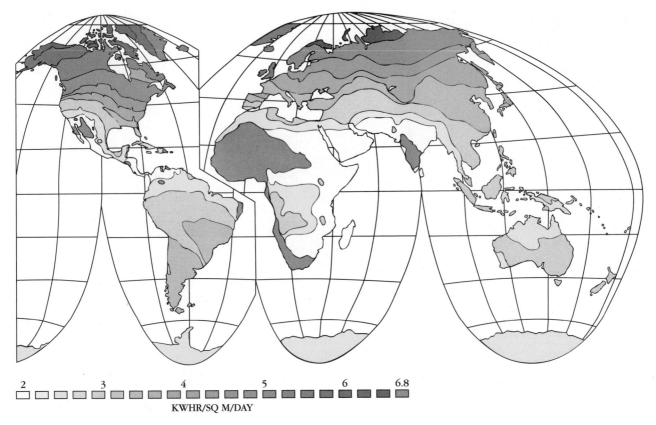

Figure 12.3 *Map of global solar energy availability.*

insulative properties of soil, properly designed earth-sheltered homes are well lighted, dry, and comfortable. They require less external maintenance.

Active Solar Heating   Active heating and cooling systems employ **solar collectors**, generally mounted on rooftops. Most collectors are insulated boxes with a double layer of glass on the sunny side (Figure 12.7). These are called **flat plate collectors**. The inside of the box is painted black to absorb sunlight and convert it into heat. The heat is carried away by water or some other fluid flowing through pipes in the collector or by air blown in by a fan. The heated water or air is then carried to a storage medium, usually water, in a superinsulated water storage tank. After transferring its heat to the storage medium, the water or air is returned to the collectors, thus completing the cycle.

The hot water in the storage tank can be used for showers, baths, and washing dishes, and can also be used to heat homes. In some parts of the country, active solar water and space heating are competitive with electric heating. In Cyprus, Israel, and Jordan, solar panels supply 25% to 65% of the domestic hot water. Israel eventually hopes to heat 90% of its domestic hot water via

rooftop collectors. Specially designed active solar systems can be used in industry to provide hot water and steam, a major consumer of energy.

Photovoltaics   **Photovoltaics** provide a way of generating electricity from sunlight. A photovoltaic cell consists of a thin wafer of silicon or some other material that emits electrons when struck by sunlight. Electrons liberated from the material then flow out of the wafer, forming an electric current (Figure 12.8).

Electricity from photovoltaics currently costs about 27 cents per kilowatt hour, or about three to six times more than electricity from conventional sources. Fortunately, costs have fallen rapidly in the last two decades. Experts predict that improvements in production could make photovoltaics competitive with electricity from nuclear power plants by the year 2000. Photovoltaics could be competitive with coal by 2030.

The best commercially available cells have an efficiency rate of 18% to 20%. In 1988, researchers at Sandia National Laboratories in New Mexico reported an efficiency rate of 31% in a new line of photovoltaics. The improvement, along with more government support, could make photovoltaics cost competitive sooner.

Figure 12.4 *Schematic representation of a passive solar house. Sunlight streams into the house during the winter months when the sun is low in the sky. During the summer months when the sun is higher, the overhang reduces sunlight entering the structure. Interior walls absorb sunlight energy and emit heat, keeping the house warm and cozy at a fraction of the cost of conventional heating systems.*

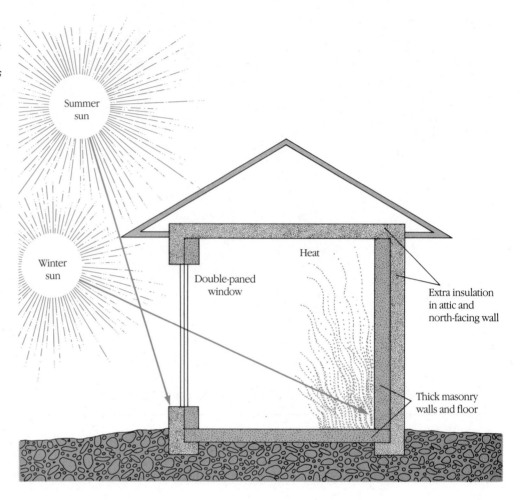

Interestingly, photovoltaics are already in wide use in remote villages in many less developed nations. In such locations, photovoltaics are far cheaper than transmission lines that carry electricity from distant power plants. Today, more than 6000 rural villages in India rely on them. The governments of Sri Lanka and Indonesia have launched ambitious programs to install photovoltaics in remote areas. In the United States, it is cheaper for homeowners to install photovoltaics than it is to pay to string lines if their homes are more than half a mile from power lines.

Solar Thermal Electric   In Southern California, one company has constructed a series of aluminum troughs to reflect sunlight onto small oil-filled tubes. The hot oil heats water, which is turned into steam that drives a turbine to generate electricity. This system, which is 22% efficient, produces enough electricity to supply 170,000 homes at a cost of 8 cents per kilowatt-hour. Engineers are also working on small modular systems consisting of parabolic reflective dishes that track the sun. These systems would focus sunlight onto a single point where the energy is captured fand used to generate electricity.

Pros and Cons of Solar Energy   One of the most notable advantages of solar energy is that the fuel is free. The only costs are for devices to capture and store it. While the construction of solar technologies, such as flat plate collectors and photovoltaics, creates pollution and solid wastes, as does any manufacturing process, once a solar system is operating, it is a very clean form of energy. It does not add to global warming, urban air pollution, and other environmental problems. Over their lifetime, solar systems produce much more energy than is needed to make them. Years of pollution-free operation offset the pollution created by production. Because most solar systems can be integrated with building designs, they do not take up valuable land.

Solar energy offers the advantage of great flexibility. Current systems provide energy for remote weather-sensing stations, single-family dwellings, and commercial operations. Solar energy can be collected to meet the low-temperature heat demands of homes or the intermediate- or high-temperature demands of factories. Solar electricity today provides energy to power radios, lights, watches, road signs, stream-flow monitors, space satellites, automobiles, and industrial motors.

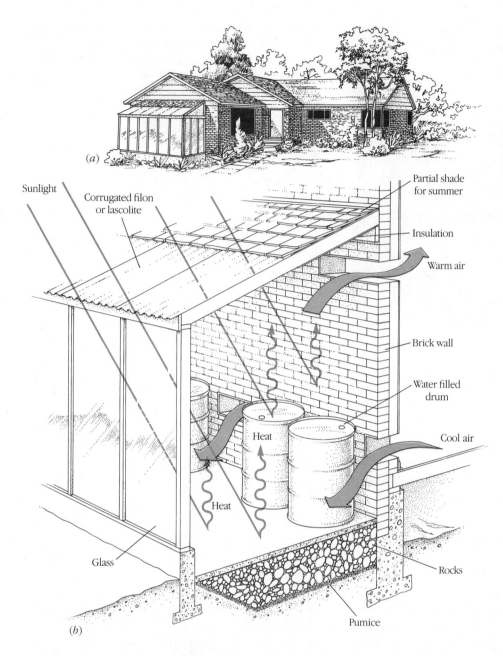

No major technical breakthroughs are required to use many existing solar systems, such as active solar water heating and passive solar space heating. Some improvements in design and costs could enhance the economic appeal of others, such as active solar space heating and cooling and photovoltaics.

Because rising energy prices will take a larger and larger chunk out of family and corporate budgets in the near future, those who invest in solar now could well enjoy an advantage over those who continue to use costly fossil fuels. Over a lifetime, passive solar energy can save a homeowner in cold climates $50,000 to $80,000 at current energy prices. As fossil fuel prices rise, solar homes may be the only affordable homes of the future.

Solar energy is a huge energy resource. Its major limitation is that the source is intermittent: It goes away at night and is blocked on cloudy days. Consequently, solar energy must be collected and stored, but current storage technologies are limited. Photovoltaic systems, for example, require an array of storage batteries. Passive solar stores heat in thermal mass, but most solar users must install a backup system to provide heat during long cloudy periods. The section on hydrogen discusses an innovative way to store solar energy.

Another disadvantage of some forms is cost. Most solar energy technologies do not compete well economically with conventional sources. However, this comparison ignores the massive economic damage caused by

Figure 12.6 *The author's superinsulated, passive solar house at 2500 meters (8200 feet) above sea level in the foothills of the Rocky Mountains costs only about $100 to $120 a year to heat.*

Table 12.3 Land Use of Selected Electricity-Generating Technologies, United States

Technology	Land Occupied (square meters per gigawatt-hour, for 30 years)
Coal[1]	3642
Solar thermal	3561
Photovoltaics	3237
Wind[2]	1335
Geothermal	404

[1]Includes coal mining.
[2]Land actually occupied by turbines and service roads.
Source: From Brown, L. R., Flavin, C., and Postel, S. (1991). *Saving the Planet: How to Shape an Environmentally Sustainable Global Economy.* New York: Norton.

conventional fuels and the huge subsidies that help support them. When these two factors are taken into account, the economics of solar energy are quite good.

Commercial operations that generate electricity via photovoltaics and solar thermal electric technologies require a great deal of land (Table 12.3). When compared with electricity from coal (including mining), however, land requirement of the renewable technologies is somewhat smaller. Therefore, this limitation is minor. Photovoltaics mounted on rooftops would bring sources closer to home, reduce transmission costs, and save valuable land.

Wind

About 2% of the sun's energy striking the Earth is converted into wind. Winds form in two major ways. First, because sunlight falls unevenly on the Earth and its

Figure 12.7 *An active solar heating system. The flat plate collectors shown here circulate a fluid that absorbs heat captured by the black interior of the panels. The heated fluid is then pumped to a storage tank and the heat is transferred to water. The water can be used for space heating or heating water for a variety of uses.*

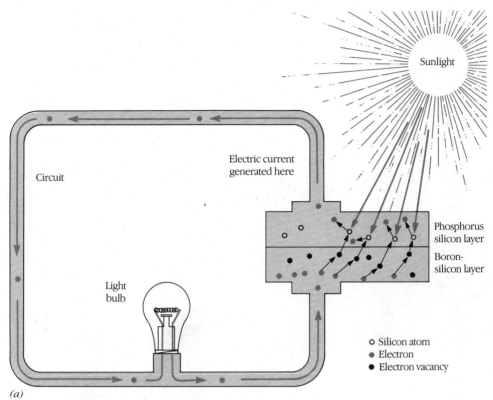

Sunlight

Electric current
generated here

Circuit

Phosphorus
silicon layer

Boron-
silicon layer

Light
bulb

o Silicon atom
• Electron
• Electron vacancy

(a)

Figure 12.8 (a) *Photovoltaic cells made of silicon and other materials. When sunlight strikes the silicon atoms, it causes electrons to be ejected. Electrons can flow out of the photovoltaic cell through electric wires, where they can do useful work. Electron vacancies in the silicon atoms are filled as electrons complete the circuit.* (b) *Array of solar voltaic cells. These cells are being used to power a railroad switching station in Alaska.*

(b)

atmosphere, some areas are heated more than others. The warm air rises, and cooler air flows in from adjacent areas. The Earth's most important circulation pattern develops as warm air near the equator rises, drawing cooler polar air toward the tropics. The Earth's rotation then causes air to circulate counterclockwise in the Northern Hemisphere and clockwise south of the equator (as viewed from the North Pole). The second major wind-flow pattern results from the unequal heating of land and water. Air over the oceans is not heated as much as air over the land. Therefore, cool oceanic air often flows inland to replace warm, rising air.

Figure 12.9 *Windmill generators near Livermore, California. Generators like these produce electricity costing about 5 cents per kilowatt-hour, about the cost of coal-generated electricity without the adverse environmental impacts. The land can also be used for grazing and growing crops.*

Wind can be tapped to generate electricity, pump water, and perform mechanical work (grinding grain, for example). Electricity is produced by mechanical turbines that are driven by propeller blades (Figure 12.9).

Today, large wind farms are being built worldwide to provide electricity. The leading suppliers of wind-generated electricity are the state of California and Denmark. Wind energy now costs on average about 8 cents per kilowatt-hour, although some wind farms in California produce it for about 5 cents, clearly cost competitive with coal.

Pros and Cons of Wind Energy   Wind energy offers many of the advantages of solar energy. It is an enormous resource, abundant in certain parts of every continent. Tapping the globe's windiest spots could provide 13 times the electricity now produced worldwide. The Worldwatch Institute estimates that wind energy could provide 20% to 30% of the electricity needed by many countries.

Wind energy is clean and renewable, uses only a small amount of land, and is safe to operate (Table 12.3). On a typical wind farm, only 10% of the land is used for roads and windmills; the remaining land can be grazed and even planted.

Windmill technology is fairly well developed, and the fuel is free. Wind energy is now cost competitive with many other sources of electricity (Table 12.4). When the environmental benefits are considered, the economics of wind energy, like solar energy, become even more favorable.

Wind systems, of course, have disadvantages. First, the wind does not blow all of the time, so backup systems and storage are needed. Storage is a major weakness. Second, because many states haven't extensively surveyed their winds, businesses and homeowners have difficulty in deciding whether wind energy would be practical. Third is the visual impact. Individual windmills and wind farms can be eyesores. Fourth, large wind generators may be noisy and may impair television reception, although fiberglass blades reduce interference by half. Some generators may also impair the microwave communications used by telephone companies.

Biomass   The organic matter contained in plants is called **biomass**. The energy found in this organic plant matter comes from sunlight. Useful biomass includes wood, wood residues from the timber industry, crop residues, charcoal, manure, urban waste, industrial wastes, and municipal sewage. Some of these fuels can be burned directly; others are converted to methane (a gas) and ethanol (a liquid). The simplest way of getting energy from biomass is to burn it, but, in some countries, it may be more economical to convert it to gaseous and liquid fuels and raw materials for the chemical industry to replace declining oil and natural gas supplies.

Biomass, for example, wood, supplies about 20% of the world's energy. In fact, it is the primary source of energy for about half of the world's population, primarily in developing countries. In sub-Saharan Africa, three-fourths of the energy comes from burning wood. In the United States and other developed countries,

Table 12.4 Present and Estimated Electrical Generation Costs*

Source	Cents per Kilowatt-Hour 1992	2000
Nuclear	8–12	na
Coal	5–7	na
Gas and oil	6–9	na
Hydroelectric	3–6	3–6
Wind	5–8	4–5
Geothermal	4.5–5.5	4–6
Photovoltaic	27	6–18
Solar thermal	8–12	5
Biomass	5	5

*Generation costs are costs to companies. (na = estimates not available)
Sources: Public Citizen Critical Mass Energy Product and Worldwatch Institute.

biomass supplies a smaller portion of energy needs, about 3%.

The U.S. Office of Technology Assessment recently projected that biomass could supply 15% to 18% of the nation's energy needs by the year 2000 with aggressive research and development, government sponsorship, tax breaks, and other incentives (Table 12.5). Fuel farms could add additional energy (Chapter 7). One of the most important contributions from biomass would be **ethanol,** a liquid fuel that can be burned in vehicles (Chapter 8). Wood burned in factories and homes can provide large amounts of energy. Burnable municipal trash could likewise help produce heat, electricity, and steam, but incinerators are a poor substitute for recycling programs (Chapter 19).

Certain nonfood crops could be grown to produce liquid fuel. For example, a desert shrub (*Euphorbia lathyris*) found in Mexico and the southwestern United States produces an oily substance that could be refined to make liquid fuel. In arid climates, the shrub could yield 16 barrels of oil per hectare on a sustainable basis. The copaiba tree of the Amazon yields a substance that can be substituted for diesel fuel without processing. Sunflower oil can also be used in place of diesel. Farmers could convert 10% of their cropland to sunflowers to produce all the diesel fuel needed to run their machinery. Eventually, the entire transportation system could be powered by renewable fuels, which would help reduce global warming.

Pros and Cons of Biomass   Some proponents argue that by using wastes and converting cropland to fuel farms, many countries, such as the United States and Canada, could make biomass a significant, renewable energy resource. Biomass can help reduce our dependence on nonrenewable energy resources, and it offers many other advantages. The most notable advantages are its high net energy efficiency when it is collected and burned close to the source of production and its wide range of applications. Biomass does not pollute the atmosphere with carbon dioxide, long implicated in the greenhouse effect (Chapter 16), providing the plant matter burned equals the plant matter produced each year. Burning some forms of biomass, such as urban refuse, reduces the need for land disposal (Chapter Supplement 18.1).

Biomass is not a panacea. Although the world's reliance on this form of renewable energy will increase in coming years, it will probably not increase as much as some proponents suggest. If global warming materializes, it could seriously alter crop and forest production. In many areas, water shortages will constrain biomass production. Rising food demand may force governments to outlaw the use of farmland to produce ethanol to power transportation systems. Furthermore, removing crop and forestry residues may reduce soil nutrient replenishment. Biomass can create large amounts of air pollution, for example, smoke from wood stoves. Finally, transportation costs for biomass are higher than traditional fossil fuels because biomass has a lower energy content.

Several recent developments could help improve the prospects for biomass. First is the use of gas turbines, similar to those used in jet aircraft. In gas turbines, hot gases from combustion turn the blades and generate electricity at a higher efficiency than conventional systems. At this writing, a small demonstration plant is under construction in Sweden. A second improvement is the development of an enzymatic process that improves the efficiency of ethanol production from wood wastes. This procedure has lowered the cost of ethanol derived from wood from $4.00 per gallon to $1.35 in a decade. Further improvements could decrease costs to 60 cents per gallon by the end of the decade, say researchers. Wood wastes now dumped in landfills, urban tree trimmings, and sustainably managed tree farms could eventually form the base of a sustainable ethanol production that supplements other renewable fuels.

Hydroelectric Power

Humankind has tapped the power of flowing rivers and streams for thousands of years to run flour mills and, more recently, to produce electricity. River flow is made possible by two factors: sunlight energy, which drives

Table 12.5   Estimate of Available Biomass in the U.S.

Type	Total Resources (Million Tons)	Total Energy Potential (Quads)	Recoverable Energy (Quads)
Crop residues	340	5.1	1.0–4.6[1]
Forestry residues	300	4.5	0.9–4.0
Urban refuse	135	1.2	0.3–1.1
Manure	45	0.7	0.1–0.6
Total	820	11.5	2.3–10.3

[1]Recoverable energy probably lies in the middle to lower part of the range given here.
Source: Kendall, H. W. and Nadis, S. J. (1980). *Energy Strategies: Toward a Solar Future*. Cambridge, MA: Ballinger, p. 170.

the hydrologic cycle (Chapter 10), and gravity, which is responsible for movement of water in streambeds.

Hydropower is a renewable resource, usually tapped by damming streams and rivers. Water in the reservoirs behind dams is released through special pipes. As it flows out, it turns the vanes of electric generators, producing electricity.

Brazil, Nepal, China, and many African and South American countries have a large untapped hydroelectric potential. In South America, for instance, hydroelectric-generating potential is estimated at 600,000 megawatts. By comparison, the United States, the world's leader in hydroelectric production, has a present capacity of about 70,000 megawatts and an additional capacity of about 160,000 megawatts.

However, estimates of hydroelectric potential can be deceiving because they include all possible sites, regardless of their economic or environmental costs (Chapter 10). For example, half of the U.S. hydroelectric potential is in Alaska, far from places that need power. The potential for additional large projects in the continental United States is rather small because the most favorable sites have already been developed. In addition, the high cost of constructing large dams and reservoirs has increased the cost of hydroelectric energy by 3 to 20 times since the early 1970s.

Two of the most sensible strategies for increasing hydropower may be to: (1) increase the capacity of existing hydroelectric facilities, that is, add more turbines, and (2) install turbines on the many dams already built for flood control, recreation, and water supply. In appropriate locations, small dams could provide energy needed by farms, small businesses, and small communities. But, all projects must be weighed against impacts on wildlife habitat, stream quality, estuarine destruction, and other adverse environmental effects.

In developing nations, small-scale hydroelectric generation may fit in well with the demand. In China, over 90,000 small hydroelectric generators account for about one-third of the country's electrical output.

Pros and Cons of Hydroelectric Power   Hydropower supplies about one-fifth of the world's electric demand. It creates no air pollution or thermal pollution and is relatively inexpensive. Furthermore, the technology is well developed. However, it is not without its problems. One of the major problems is that reservoirs behind dams often fill with sediment, giving a hydropower facility a typical life span of 50 to 100 years, although large projects may last 200 to 300 years. Thus, even though hydroelectric power is renewable, the dams and reservoirs needed to capture this energy have a limited lifetime (Figure 12.10). Once a good site is destroyed by sediment, it is gone forever. Dams and reservoirs create many additional problems (Chapter 10).

Geothermal Energy

The Earth harbors an enormous amount of heat, or **geothermal energy**, which comes from the decay of naturally occurring radioactive materials in the Earth's crust and from **magma**, molten rock beneath the Earth's surface. Geothermal energy is constantly regenerated, but because the renewal rate is slow, overexploitation could deplete this resource regionally.

Geothermal resources fall into three major categories. The map in Figure 12.11 locates zones in the United States where the two most practical forms of geothermal energy can be tapped.

Hydrothermal convection zones are places where magma penetrates into the Earth's crust and heats rock containing large amounts of groundwater. The heat drives the groundwater to the Earth's surface through fissures, where it may emerge as steam (geysers) or as a liquid (hot springs).

Figure 12.10 *This view of Mono Dam in California shows the reservoir filling with silt eroded from the surrounding watershed. In succeeding years, the reservoir filled completely and gradually was reclaimed by the surrounding forest. Many other dams in use today are slowly filling with sediment and will be taken out of production in coming years.*

Geopressurized zones are aquifers that are trapped by impermeable rock strata and heated by underlying magma. This superheated, pressurized water can be tapped by deep wells. Some geopressurized zones also contain methane gas.

Hot-rock zones, the most widespread but most expensive geothermal resource, are regions where bedrock is heated by magma. To reap the vast amounts of heat, wells are drilled, and the bedrock is fractured with explosives. Water is pumped into the fractured bedrock, heated, and then pumped out.

Geothermal energy is heavily concentrated in a so-called ring of fire encircling the Pacific Ocean and in the great mountain belts stretching from the Alps to China. It is also prevalent around the Mediterranean Sea and in East Africa's Great Rift valley, which extends along the eastern part of the African continent. Within these areas, hydrothermal convection zones are the easiest and least expensive to tap. Hot water or steam from them can heat homes, factories, and greenhouses. In Iceland, for example, 65% of the homes are heated this way. Iceland's geothermally heated greenhouses produce nearly all of its vegetables; the Commonwealth of Independent States and Hungary also heat many greenhouses in this way.

Pros and Cons of Geothermal Energy Steam from geothermal sources can be used to run turbines to produce electricity. Geothermal plants can produce electricity day and night, and can provide electricity when wind or solar systems are not operating. Geothermal energy can also provide electricity in areas without sizable wind or solar resources.

Although still in the early stages of development in most countries, geothermal electric production is growing quickly in the United States, Italy, New Zealand, and Japan. El Salvador in Central America currently generates 40% of its electricity from geothermal sources. Kenya and Nicaragua acquire 11% and 28%, respectively. By the year 2000, some experts believe, the United States could produce 27,000 megawatts of electricity from geothermal energy, enough for 27 million people, over one-tenth of the population.

Hydrothermal convection systems have several drawbacks. The steam and hot water they produce are often laden with minerals, salts, toxic metals, and hydrogen sulfide gas. Many of these chemicals corrode pipes and metal. Steam systems may emit an ear-shattering hiss and release large amounts of heat into the air. Pollution control devices are necessary to reduce air and water pollutants. Engineers have proposed building closed systems that pump the steam or hot water out and then inject it back into the ground to be reheated. Finally, because heat cannot be transported long distances, industries might have to be built at or near the source of energy.

Hydrogen Fuel

Hydrogen is another renewable fuel with enormous benefits. It could help replace oil and natural gas and be used in automobiles, homes, and factories. Hydrogen

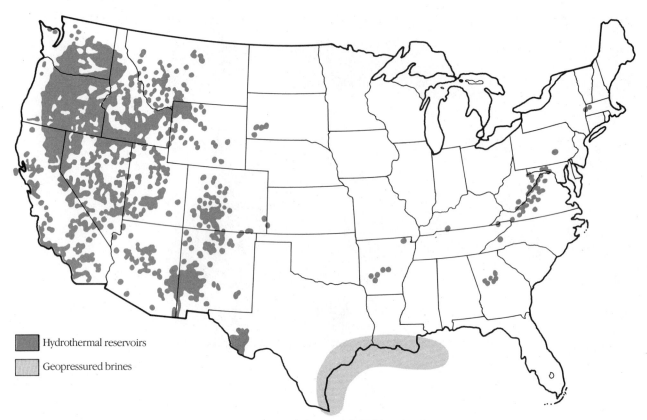

Figure 12.11   *Geothermal resources in the United States.*

gas is produced by heating or passing electricity through water in the presence of a catalyst, a chemical that facilitates the breakdown of water into oxygen and hydrogen without itself being changed.

When hydrogen burns, it produces water, energy, and small amounts of nitrogen oxide, which is formed by the combination of atmospheric oxygen and nitrogen. Nitrogen oxides, which contribute to acid deposition, can be minimized by controlling combustion temperature and special pollution control devices. Unlike fossil fuels, when burned, hydrogen produces no carbon dioxide.

Hydrogen may serve as a way of storing energy from hydroelectric, wind, solar, and other renewable energy sources. For example, when demand from these sources is low, electricity they generate could be used to produce hydrogen from water. It would be stored for later use. When renewable sources are not available, say on a calm day or during the evening, hydrogen could be burned to produce electricity.

Pros and Cons of Hydrogen Fuel   Because hydrogen could be generated from seawater, it is an essentially limitless renewable energy resource. Hydrogen is also easy to transport and has a wide range of uses. Unfortunately, it takes considerable energy to produce hydrogen. This low net energy yield makes it an expensive form of energy.

Although the immediate prospects for hydrogen are not spectacular, efforts to produce hydrogen more efficiently could help to make this a more cost-effective energy source. Some sunlight-, wind-, and water-rich areas could become major sources of hydrogen, piped around the world in existing natural gas pipelines.

The Renewable Energy Potential

Robert L. San Martin, who heads the U.S. government's renewable energy program, estimates that renewable energy sources provide about $18 billion in energy each year, or about one-twelfth of the total energy demand. What is more, he estimates that this is only a tiny fraction of the renewable energy potential.

San Martin examined total energy resources available to the United States. **Total resources** is a measure that includes all energy whether or not it is economically recoverable. In his study, San Martin found that renewable energy sources accounted for about 93% of total resources. Coal, oil, natural gas, and oil shale accounted for the remainder.

Making some reasonable allowances for what is accessible using current technology, San Martin found that renewable energy sources made up 90% of all **accessible resources**. Accessible resources is a measure that looks at technological feasibility but does not consider economics. Therefore, although it is much more realistic than total resources, it is not a true reflection of current circumstances. In a sense, accessible resources is a measure of promise. It indicates energy that will be available as the economic picture changes and as finite resources become depleted.

Accessible renewable energy that is available each year is equivalent to 70 to 80 billion barrels of oil. In contrast, nonrenewable energy resources under U.S. control are equivalent to 9 billion barrels of oil. Because the latter are nonrenewable, they represent a one-time source. Renewable energy, therefore, not only outstrips nonrenewable sources but would be available to the United States in large amounts year after year.

12.3 Is a Sustainable Energy System Possible?

Many observers see coal and nuclear power as the mainstays of the world's energy diet in the coming decades. But, others envision a renewable energy transition in the making. They believe that energy efficiency and solar energy in its many forms could supply the bulk of the world's energy demands. Figure 12.12 shows one potential shift in the energy supply picture.

In the sustainable transition, conservation, active solar heating, passive solar heating, photovoltaics, and windmills could replace huge centralized nuclear- and coal-powered electric plants. Locating energy sources closer to the consumer could be beneficial; knowing where our energy comes from may help us respect its importance and use it more efficiently.

But, how will renewable energy supplant the non-renewables that currently power U.S. society? Where will the fuel come from to power the transportation system? As we make the transition to a sustainable energy system, the first step will be to use existing energy much more efficiently. Drastic cuts in energy waste in every sector of society are absolutely essential. Cars that routinely get 50 to 80 mpg could double or triple the existing oil supply. Mass transit, many times more efficient than conventional automobiles, would further stretch energy supplies.

Renewable fuels will very likely play a growing role as concerns for environmental protection increase, and as fossil fuels, especially oil, begin to decline. Ethanol from crops or, more likely, wastes could power automobiles, trucks, buses, and planes. Electricity from sunlight and wind sources and hydrogen could also provide a significant amount of fuel for the transportation sector.

Low- to intermediate-temperature thermal energy in the sustainable society would come from solar energy. Solar sources could be used for home space heating, water heating, and many industrial processes. Electricity would come from photovoltaics and windmills.

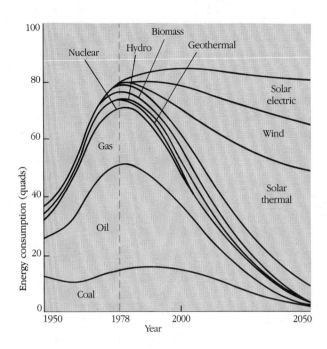

Figure 12.12 *One projection of possible energy resources in the United States. These figures are based on full commitment to renewable resources.*

Table 12.6 gives a further breakdown of energy demand by sector, showing what kind of energy is needed and how it would be used.

12.4 Economic and Employment Potential of the Sustainable Energy Strategy

Contrary to the beliefs of some, a sustainable energy strategy is not a prescription for economic disaster. It is, in fact, a road to economic health. Numerous studies show that conservation and certain renewable technologies produce energy more cheaply than conventional sources and eliminate many costly environmental impacts. They also employ more people than conventional energy-supply strategies. The chief reason for this is that conservation and alternative technologies tend to be more labor intensive (requiring more people) than oil, coal, and nuclear energy, which are more capital intensive (requiring huge investments in machines and fuels). Consider some examples:

A California-style wind farm generates electricity for about 5 cents per kilowatt-hour, one-half of the cost of electricity from a nuclear power plant and without the serious environmental impacts. In addition, a wind farm that produces the same amount of energy as a nuclear power plant will employ over 540 workers, compared to 100 workers in the nuclear facility (Figure 12.13).

Table 12.6   Meeting Energy Needs of a Solar-Powered Society

Demand Sector	Sources	Application	Percentage of Total Energy Use
Residental and Commercial	Passive and active solar systems, district heating systems	Space heating, water heating, air conditioning	20–25
	Active solar heating with concentrating solar collectors		
	Solar thermal, thermochemical, or electrolytic generation	Cooking and drying	~5
	Biomass		
	Photovoltaic, wind, solar, thermal, total energy systems	Lighting, appliances, refrigeration	~10
			Subtotal ~35
Industrial	Active solar heating with flat plate collectors, and tracking solar concentrators	Industrial and agricultural process heat and steam	~7.5
	Tracking, concentrating solar collector systems	Industrial process heat and steam	~17.5
	Solar thermal, thermochemical, or electrolytic generation		
	Solar thermal, photovoltaic, cogeneration, wind systems	Cogeneration, electric, drive, electrolytic, and electrochemical processes	~10
	Biomass residues and wastes	Supply carbon sources to chemical industries	~5
			Subtotal ~40
Transportation	Photovoltaic, wind, solar thermal	Electric vehicles, electric rail	10–20
	Solar thermal, thermochemical, or electrolytic generation	Aircraft fuel, land and water vehicles	
	Biomass residues and wastes	Long-distance land and water vehicles	5–15
			Subtotal ~25
			100

Source: Kendall, H. W. and Nadis, S. J. (1980). *Energy Strategies: Toward a Solar Future.* Cambridge, MA: Ballinger, p. 262.

Energy conservation can achieve similar gains. A study in Alaska found that state expenditures on weatherization—home energy conservation—would create more jobs per dollar investment than the construction of hospitals, highways, or new power plants. Conservation spending, for instance, would create three times as many jobs as highway construction.

Weatherization of all homes in the United States would create 6 million to 7 million job-years, according to the Worldwatch Institute projections. (A job-year is

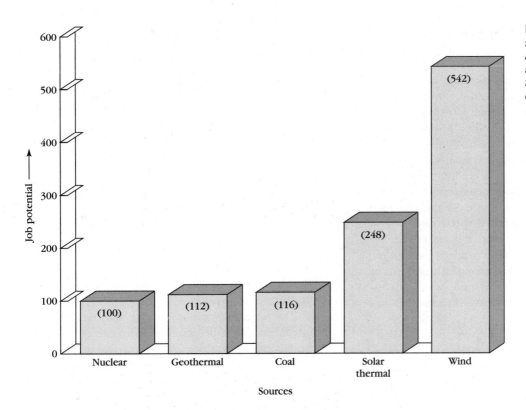

Figure 12.13 *Employment opportunities created by different energy strategies, all producing the same amount of energy per year.*

1 job for 1 year.) Six million job-years is the equivalent of 300,000 jobs over a period of 20 years! Not all jobs would be low wage either. Energy-efficiency companies would be run by high-paid personnel and would employ sales people, managers, accountants, engineers, and installers.

A wise energy future is economically, environmentally, and socially sustainable. The current system fails on all of these criteria. Nonetheless, great barriers lie in the path to a sustainable system. One of the most significant is that industrial nations have been built on fossil fuels. Billions upon billions of dollars have been spent on the present fossil fuel–based system. Huge investments have been made in mining and drilling equipment, transportation networks, processing facilities, and power plants. Although the system is unsustainable, enormous resistance to change comes from many powerful political and economic interests: the large energy companies, mining companies, and the public representatives from energy-producing states, companies that expect to profit from their previous investments.

The easiest steps toward a sustainable energy system are those that will make it much more efficient, but converting to renewable fuels means starting more or less from scratch. It means phasing out many existing facilities and building new ones.

Shifting to renewable energy resources and conservation will require massive investment capital to finance wind farms, geothermal plants, and so on.

Fortunately, some parts of the current energy system

are salvageable. Electric lines, for instance, could be used just as easily to transport solar- and wind-generated electricity as coal-generated electricity. Natural gas pipelines could be used to transport hydrogen gas. Rail lines that carry coal could be used to carry people on high-speed trains.

Although the shift to a sustainable energy strategy could create many new economic opportunities, jobs will be lost in some sectors. If proponents are right, the number of potential jobs exceeds those lost. Not only will there be more jobs, but they will undoubtedly be safer. Some workers will be able to turn their skills to similar activities. Petroleum geologists and oil well crews, for example, might use their expertise to drill for geothermal resources.

The shift to renewable energy resources will take many years, even given the urgency created by global warming. The length of the transition will depend on our political will and our willingness to change for the sake of the planet's future.

There are many ways of going forward,

but only one way of standing still.

—Franklin D. Roosevelt

Critical Thinking Exercise Solution

Critical thinking warns us that broad statements such as these often fall apart on closer examination. When one examines the big picture and defines terms, one frequently sees a different picture of reality.

Solar energy, as you learned in the chapter, consists of at least four different technologies. One of them, solar thermal electric, is currently cost competitive with nuclear power. Other forms, like photovoltaics, are not currently cost competitive with conventional fossil fuel resources in most applications in developed countries. In rural villages in the developing world, however, they're quite cost competitive.

This exercise shows that it is also important to define what is meant by *uneconomic*. Does the term refer only to conventional economics, which ignores externalities like damage to the environment? In this case, the oil industry spokesperson is talking only about the cost of producing conventional fuels and ignoring enormous external costs.

Along that same line, does the cost of production include subsidies? As you learned in the chapter, oil is very heavily subsidized by the federal government. The nuclear and coal industries are also heavily subsidized. Thus, this exercise also shows the important critical thinking concept of considering the source of information, who may be biased by economic self-interest.

Summary

12.1   Conservation: Foundation of a Sustainable Energy System

■ Most countries rely primarily on fossil fuels, which are often used inefficiently. Given our dependency on energy, its economic and environmental costs, and the limited supplies in nonrenewable energy resources, continued waste is foolish.

■ **Conservation** represents a huge, untapped energy resource. Besides reducing pollution and environmental destruction, energy-efficiency measures can save large sums of money. In addition, energy-efficiency measures can be brought on-line very rapidly.

■ The range of **energy-efficient options** available is enormous. In some cases, efficiency can be achieved through minor behavioral changes. In others it can be achieved by installing energy-efficient technologies, such as high-efficiency compact fluorescent light bulbs.

■ Energy efficiency can be increased by fuel taxes, feebate programs, government-mandated efficiency standards, pricing changes, and least-cost planning.

12.2   Renewable Energy Sources

■ **Solar energy** is abundant but currently provides only a fraction of our energy needs.

■ **Passive solar systems** capture sunlight energy and store it within walls and floors, thermal mass; the stored heat is gradually released into the structure.

■ **Active solar systems** rely on collectors that absorb sunlight and convert it into heat.

■ **Photovoltaics** are made of silicon or other materials that emit electrons when struck by sunlight, thus producing electricity.

■ **Solar thermal electric systems** use concentrated sunlight energy to heat fluids that generate steam and electricity.

■ Solar systems provide many advantages over conventional power sources. The fuel is free, nondepletable, and clean.

■ The major limitations are that the source is intermittent, making it necessary to store energy overnight or on cloudy days.

■ Winds can be tapped to generate electricity. **Wind energy** offers many of the advantages and disadvantages of solar energy.

■ **Biomass**, a form of indirect solar energy, has some potential. Useful biomass includes wood, wood residues, crop wastes, industrial wastes, manure, and urban waste.

■ Energy in biomass can be released by burning it or converting it to gaseous and liquid fuels.

■ **Hydroelectric power** is a relatively inexpensive renewable resource that creates no air pollution.

■ The potential for hydroelectric power is limited in developed countries because the best sites have already been developed or are located far from population centers where the energy is needed.

■ In developing nations, sites capable of producing large amounts of energy are available, but high construction costs impede their development.

■ The Earth harbors a great deal of energy (**geothermal energy**) from the decay of naturally occurring radioactive materials in the crust and from magma, molten rock.

■ The most useful geothermal resource is the **hydrothermal convection zone**, where magma penetrates into the crust and heats rock formations containing groundwater. The heat pressurizes the groundwater and drives it to the surface through fissures.

■ Hydrothermal convection zones can be used for space heating and electricity. Such systems, however, produce steam and hot water laden with toxic minerals, salts, metals, and hydrogen sulfide. Noise pollution is also a problem.

■ **Hydrogen fuel** is produced by heating or passing electricity through water in the presence of a catalyst. Water breaks down into hydrogen and oxygen.

■ Hydrogen is a clean-burning fuel that could replace gaseous and liquid fuels. It is easy to transport but explosive. Electricity needed to make hydrogen could be generated from solar energy, wind energy, or hydroelectric facilities.

■ Renewable energy resources now technologically accessible far outstrip nonrenewable energy resources, including coal.

12.3 Is a Sustainable Energy System Possible?

■ Energy experts have shown that we can substitute renewable energy resources, such as solar energy, for nonrenewables, such as oil, natural gas, coal, and nuclear power.

■ A smooth transition can be made into a sustainable future, but it will require an immediate investment in renewable energy resources by governments and individuals.

12.4 Economic and Employment Potential of the Sustainable Energy Strategy

■ Numerous studies show that conservation and several renewable technologies produce energy more cheaply than conventional sources while eliminating many of the environmental impacts of the latter.

■ Because these strategies are more labor intensive, job opportunities will increase.

Discussion and Critical Thinking Questions

1. In your view, is it imperative that we change to a sustainable energy system? Why or why not?

2. Describe the types of solar energy technologies available. How does each one operate? What is it used for?

3. What are the advantages and disadvantages of solar energy? How can technical problems be solved?

4. A person living in the Pacific Northwest argues that renewable energy is a fluke. The sun rarely shines in this part of the country. How would you answer this?

5. Describe the difference between passive and active solar systems. What features are needed in a home to make passive solar energy work?

6. What are photovoltaic cells? Why are they economical to use in rural villages in developing nations?

7. Wind energy is cost competitive with conventional electricity. Should we develop this energy resource in preference to nuclear power, coal, or shale? Why or why not?

8. What is biomass? How can useful energy be acquired from biomass?

9. How is geothermal energy formed? How can it be tapped? Describe the benefits and risks of geothermal energy.

10. Using your critical thinking skills, debate this statement: "Hydroelectric power is an immensely untapped resource in the United States and could provide an enormous amount of energy."

11. What are the major problems facing hydrogen power? How could these be solved?

12. Using your critical thinking skills, discuss this statement: "Conservation is our best and cheapest energy resource."

13. Discuss ways in which you could conserve more energy at home, at work, and in transit. Draw up a reasonable energy conservation plan for you and your family.

14. Using your critical thinking skills, debate this statement: "A sustainable energy strategy won't work. It will cost money and lose jobs."

Suggested Readings

Brown, L. R., Flavin, C., and Postel, S. (1991). *Saving the Planet: How to Shape an Environmentally Sustainable Global Economy*. New York: Norton. Contains important advice on creating a sustainable system of energy, including economic and employment opportunities.

Chiras, D. D. (1992). *Lessons from Nature: Learning to Live Sustainably on the Earth*. Washington, D.C.: Island Press. Outlines key tenets of sustainable energy and transportation systems with numerous examples.

Davidson, J. (1987). *The New Solar Electric Home*. Ann Arbor, MI: aatec publications. Contains lots of practical information on photovoltaics.

Flavin, C. (1990). Slowing Global Warming. In *State of the World 1990*. New York: Norton. Overview of nonfossil fuel options to help reduce global warming.

Flavin, C. and Lenssen, N. (1991). Designing a Sustainable Energy System. In *State of the World 1991*. New York: Norton. Superb overview.

Group, L. (1978). *Solar Houses: 48 Energy-Saving Designs*. New York: Pantheon. Filled with interesting information.

Johansson, T. B., Kelly, H., Reddy, A., and Williams, R. H. (1992). *Renewable Energy: Sources for Fuels and Electricity*. Washington, D.C.: Island Press. In-depth analysis of renewable energy options.

Kendall, H. W. and Nadis, S. J., eds. (1980). *Energy Strategies: Toward a Solar Future*. Cambridge, MA: Ballinger. Detailed survey of energy sources and their prospects for the future. Superb!

Pollock, C. (1986). *Decommissioning: Nuclear Power's Missing Link*. Worldwatch Paper 69. Washington, D.C.: Worldwatch Institute. Authoritative coverage of the costs involved.

Renner, M. (1988). *Rethinking the Role of the Automobile*. Worldwatch Paper 84. Washington, D.C.: Worldwatch Institute. Detailed coverage of the growth in automobile use and alternative transportation systems.

———. (1992). Creating Sustainable Jobs in Industrial Countries. In *State of the World 1992*. New York: Norton. Excellent data on the job potential and costs of renewable energy and energy efficiency.

Reynolds, M. (1990). *Earthship. How to Build Your Own*. Taos, NM: Solar Survival Press. Delightful reading that shows how tire homes are built.

Shea, C. P. (1988). *Renewable Energy: Today's Contribution, Tomorrow's Promise*. Worldwatch Paper 81. Washington, D.C.: Worldwatch Institute. Excellent resource.

Shifting to a Sustainable Transportation System

Americans are engaged in a dangerous love affair with their automobiles, one that many experts think cannot be sustained much longer. Pollution, declining oil supplies, and crowding on urban highways are three forces that could end it.

Today, nearly 30% of the energy Americans consume is used by the transportation sector and much of that powers our automobiles. The American passion for automobiles has spread throughout the world. The global automobile fleet, in fact, has expanded from 50 million in 1950 to over 400 million in 1990. By the year 2000, the world's automobile fleet could rise to 700 million.

Automobile travel accounts for 90% of the motorized passenger transport in the United States and 78% in Europe. Each year, Americans travel nearly 2000 billion miles in their automobiles—the equivalent of more than ten round-trips to the sun, 93 million miles away.

Because oil supplies are declining and global warming is very likely to worsen, the automobile could fall into disfavor in the coming decades. Liquid fuels from coal, oil shale, and fuel farms will probably not be able to save the auto from extinction. Gradually, a sustainable system of transportation is bound to emerge.

Phase 1: The Move Toward Efficient Vehicles and Alternative Fuels

The first step in the transition to a sustainable transportation system, occurring within the 1990s, is a dramatic increase in automobile efficiency. Improving automobile efficiency will prolong the automobile's life span. In 1982, the average new American automobile got about 9 kilometers per liter of gasoline (22 miles per gallon). By 1992, the average fleet mileage had edged up to 11 kilometers per liter (27.8 miles per gallon; 26.9 domestic; 29 foreign), still a long way from what is achievable.

Not surprisingly, the best gas mileage was achieved by a Japanese-made vehicle, the Geo Metro, which averages 24 kilometers per liter (58 miles per gallon) on the highway. The British Leyland, a four-passenger prototype vehicle, leaves the Geo Metro in the dust, getting 34 kilometers per liter (83 miles per gallon). A Japanese prototype currently gets the same mileage.

Improvements in gasoline mileage can be achieved by engine redesign, smaller cars, and new materials, among them new foams and plastics, which make cars lighter and more efficient. Space-age materials and air bags can increase the safety of the smaller, more energy-efficient vehicles. The alleged dangers of smaller cars could also be mitigated by tougher drunk-driving laws, enforcement of speed limits, and better driver education.

Los Angeles, plagued by traffic congestion and air pollution, is attempting to solve some of its problems by requiring "ultraclean" cars, beginning in 1994. Ultraclean cars include electric cars as well as automobiles that run on a mixture of methanol and gasoline. The concept also includes cars with advanced catalytic converters and cars that burn natural gas or propane. By 2000, all new cars sold in Los Angeles will be ultra-clean.

The shift to clean-burning cars is essential but, like improvements in efficiency, only a stop-gap measure. Eventually, large numbers of commuters will have to shift to mass transit. Recognizing this, the program in Los Angeles also calls on employers with 100 or more workers to develop plans to reduce vehicle miles traveled by workers. If they don't, they risk a $25,000-per-day fine. Companies can offer bonuses to employees who commute via car pools or mass transit. Employers may also permit workers to work at home, linked to the office by phone lines, or **telecommute**. Others may shift to four-day workweeks.

Phase 2: From Road to Rails and Buses

No matter how much the automobile fuel economy improves, the car does not compare to bus and train transportation (Table S12.1). In urban centers, buses and trains achieve a fuel efficiency of about 62 passenger kilometers per liter (150 passenger miles per gallon) of fuel. Intercity bus and train transport increases efficiency to 82 passenger kilometers per liter of fuel (200 passenger miles per gallon)—seven times better than the average new car today.

Given the relative efficiency of mass transit, declining fuel supplies, congestion, and pollution, many urban residents may give up their second and third automobiles within the next few decades. They will turn to more efficient and less polluting forms of transportation, among them light rail and buses (Figure S12.1).

Some experts think that as oil supplies decline, nations may be forced to ration their gasoline, giving agriculture and mass transit precedence over automobile use. If this happens, it will cause a further decline in automobile transportation.

In Stockholm, Sweden, the Office of Future Studies has proposed that the city work to phase out the private automobile. It recommends considerable expansion of the existing mass transit system, in part because it is more efficient and cleaner. It also recommends expanding the fleet of rental vehicles for vacations and other special occasions.

The shift to mass transit is inevitable over the coming decades, but cities will have to improve their present systems, making them much more rapid and more convenient. With declining automobile traffic, cities may be able to convert highway lanes to light rail lines. Median strips could be converted to light rail systems serving

Table S12.1 Fuel Efficiency by Passenger Transportation Mode in Western Europe

Mode of Transportation	Kilojoules per Passenger Kilometer/Mile[1]
Van pool	400/640
Rail	400/640
Bus	450/720
Car pool	650/1,040
Automobile	1,800/2,880
Airline	3,800/6,080

[1]A kilojoule is 1000 joules, a unit of energy or work.
Source: Worldwatch Institute.

surrounding suburbs. Fast, efficient buses could carry commuters from their homes to outlying rail stations, where people board high-speed trains that whisk them into urban centers.

To be profitable, high-speed rail requires high participation, high-density population in outlying areas, and a large central business district. In order to achieve this, it may be necessary to "densify" new suburbs—that is, to place houses closer together. Reducing street widths, building houses closer to the street, building smaller homes, and having backyards open onto a large commons will help to facilitate the change. Communal swing sets and communal gardens could be added to create a sense of community and save resources. Existing zoning laws could be changed to allow homeowners to add small apartments in the tradition of the carriage house to existing homes or to convert unused space into apartments for renters.

New housing developments should be as self-contained as possible, that is, they should be within walking or biking distance to shops and stores. Builders can also concentrate new housing developments along major transportation routes—bus routes and new or planned light rail systems—to make mass transit more convenient.

New laws can help contain urban growth by promoting a more efficient use of land and controlling urban sprawl. These efforts will make mass transit more economical. (For a discussion of urban growth management strategies, see Models of Global Sustainability 6.1.)

Urban centers can densify by converting empty parking lots, produced by the decline in the automobile use in cities, to office buildings.

Unfortunately, the economics of mass transit is currently skewed by massive subsidies. According to national statistics, the automobile is subsidized to the

Figure S12.1 *Light rail systems and electric trains like this one can move commuters to and from cities with great speed and at a fraction of the energy cost.*

tune of about $260 billion per year—or about $1500 per car. This includes expenses for police protection, traffic control, city-paid parking, and others. Like solar energy and other sustainable strategies, mass transit has a hard time competing with the automobile because the playing field is tilted heavily in favor of the automobile. Removing the hidden subsidies from oil and automobiles—or providing more support for mass transit— would clearly make mass transit compete more favorably with the auto.

In some locations, the bicycle could supplement buses and high-speed trains. Investments that promote bicycle commuting represent one of the cheapest options available to cities and towns.

For decades, the bicycle has been a major means of transportation in many European and Asian countries. Following suit, some cities in the United States have laid out extensive bike paths for commuters (Figure S12.2). Davis, California, is a leader in promoting bicycle transportation. Today, 30% of all commuter transport within the city is by bicycle. Some streets are closed entirely to automobile transport, and 65 kilometers (40 miles) of bike lanes and paths have been established.

Bicycles won't replace cars, buses, and trains, but they can augment them. Because of vast differences in climate, layout, and topography, the bicycle won't find a place in all of our cities and towns.

Figure S12.2 *The bicycle is an efficient and economical alternative to the automobile for some commuters.*

Economic Changes Accompanying a Shift to Mass Transit

The shift to more efficient forms of transportation is likely to lead to significant shifts in the world's economy. The automobile industry is the world's largest manufacturing industry and supports a number of other economically important industries. Manufacturers of rubber, glass, steel, radios, and numerous automobile parts will also feel the impacts of the shrinking automobile market. So will the service sector: gas stations, automobile dealerships, and repair services.

Today, 20 cents of every dollar spent in the United States is directly or indirectly connected to the automobile industry and its suppliers. Eighteen cents of every tax dollar the federal government collects comes from automobile manufacturers and their suppliers.

Although shifting toward a sustainable transportation system will result in a dramatic shift in the economy, it is feasible. Steel and glass now destined for autos could be used for buses and trains. Automobile workers could find jobs in plants that produce buses and commuter trains. Mechanics could shift as well to service the new fleet of more efficient vehicles. Some workers, however, will inevitably be forced to find employment in new areas. Helping them adjust to the changes will be an important task.

Studies suggest that the employment potential of mass transit, like other sustainable strategies, exceeds that of the current automobile-based economy. A study in Germany showed that spending $1 billion on highways yields 24,000 to 33,000 (direct and indirect) jobs. The same amount spent on mass transit produces 38,000 to 40,000 jobs.

A sustainable transportation system is possible, but it requires a massive restructuring of a system that is largely unsustainable and inevitably bound to collapse. Making that transition will require foresight and considerable political will.

The Earth and Its Mineral Resources

Conservation is humanity caring

for the future.

—*Nancy Newhall*

Critical Thinking Exercise

Detractors of recycling sometimes argue that it doesn't make sense. They often point to the decline in price in recent years for recycled newspaper and other materials as a sign that the system should be scrapped. As a logical extension, that means we should stick with the old way of making paper, which involves cutting down trees to make pulp, which is then made into paper. To strengthen their case against recycling, critics also point out that paper recycling is a pretty dirty process. Using your knowledge of recycling and your critical thinking skills, how would you analyze this point of view?

To a person gazing out on a vast expanse of land—the prairies of Oklahoma or Texas, for instance—the Earth seems stable and permanent. But, churning deep within its interior is molten rock that can spew out in frightening displays, burying villages and farmland, leveling trees, and killing wildlife. Huge land masses fold and twist as they crash together. Others move like pieces of a jigsaw puzzle.

Evidence of the Earth's restlessness is all around us, manifested in earthquakes, volcanoes, and mountain ranges. Nowhere is this more apparent than in California. Consider, for instance, that in 1.5 million years Los Angeles will lie where San Francisco is.

What's happening is that a narrow strip of land, bounded on the east by the San Andreas fault and on the west by the Pacific Ocean, is inching its way northward. It is part of a huge plate in the Earth's crust encompassing nearly the entire Pacific Ocean. On its slow northward path, it is taking along this tiny strip of exposed land—and millions of unsuspecting passengers. The rumbling and earth-shattering quakes that now plague the West Coast give evidence of the Pacific plate grinding against the American plate, which includes North and South America (Figure 13.1).

13.1 The Earth and Its Riches

To understand this process and to understand the important resources we gather from the Earth's restless crust, let us turn back to the time when the Earth began to form. Five billion years ago, the Earth began to cool and the surface of this molten mass gradually formed a thick, rocky crust. As the Earth cooled, water vapor in the atmosphere condensed and rained down, forming oceans, lakes, and rivers. Today 29% of the Earth's surface is land; 71% is water. Beneath the crust is the mantle, and beneath that, the core.

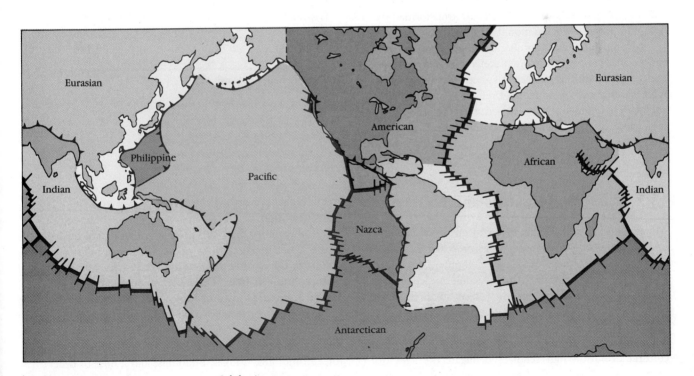

⊢ Regions of spreading ⊿⊿ Subduction zones

Figure 13.1. *The Earth's crust is broken into thin moving plates. Dark black lines indicate regions of spreading. Lines with solid triangles indicate subduction zones where one plate slides under another.*

A Rocky Beginning

After the Earth cooled, its crust consisted of solid rock. Over time, this rock was subject to many altering forces. Wind and rain, for example, created tiny particles that gave rise to soil. Rock was compressed and ripped apart by the Earth's internal forces, creating new types of rock. These and other forces are active today, constantly reshaping the planet.

Despite these dramatic changes, the Earth's crust contains the same inorganic compounds, called **minerals**, that were present during its fiery beginning. Minerals are made up of elements, such as silicon and oxygen. The most abundant elements in the Earth's crust are oxygen, silicon, aluminum, iron, and magnesium. Others, like gold and platinum, are extremely rare.

Minerals are grouped into four basic categories: (1) **metal-yielding minerals**, such as aluminum and copper ore; (2) **industrial minerals**, such as lime; (3) **construction materials**, such as gravel and sand, and (4) **fuel minerals**, including uranium, coal, oil, and natural gas. All are integral to modern economies. This chapter concerns itself primarily with the first three kinds. Fuel minerals were described in Chapter 11.

Minerals occur in **rocks**, solid aggregates that usually contain two or more different types of mineral. Geologists divide rocks into three major classes: (1) **Igneous rocks**, such as basalt and granite, form when molten minerals cool. (2) **Sedimentary rocks**, such as shale and sandstone, form from particles eroded from other types of rock. (3) **Metamorphic rocks**, such as schist, form when igneous or sedimentary rocks are transformed by heat and pressure during mountain building.

Most metal-yielding minerals come from igneous rocks. These minerals are often concentrated in igneous rocks by geological processes. A concentrated deposit of minerals that can be mined and refined economically is called an **ore**. Most ores are mined, then treated to produce their final product, for example, metals such as aluminum and zinc.

Table 13.1   Estimated World Production of Selected Minerals, 1990

Mineral	Production[1] (thousand tons)
Metals	
Pig iron	552,000
Aluminum	18,100
Copper	8,920
Manganese	8,600
Zinc	7,300
Chromium	3,784
Lead	3,350
Nickel	949
Tin	216
Molybdenum	114
Titanium	102
Silver	15
Mercury	6
Platinum-group metals	0.3
Gold	0.2
Nonmetals	
Stone	11,000,000
Sand and gravel	9,000,000
Clays	500,000
Salt	191,000
Phosphate rock	166,350
Lime	135,300
Gypsum	99,000
Soda ash	32,000
Potash	28,125

[1]All data exclude recycling.

13.2   Mineral Resources and Society

Minerals are extremely important to our lives. For example, metals derived from some ores are in many products, among them, buildings, computers, bicycles, glasses, and automobiles. Construction minerals are used to make roadbeds, schools, office buildings, and homes. Industrial minerals are used in fertilizers and concrete.

Minerals are so important to our welfare and our cultural evolution that scholars delineate the ages of human history by the chief minerals in use at the time: Stone, Bronze, and Iron. Although the Industrial Revolution is described in reference to the growing use of fossil fuels, minerals also played a key role in this transformation.

Today, more than 100 nonfuel minerals are traded in the world market. These materials, worth billions of dollars to the world economy, are vital to industry, agriculture, and our lives. The major minerals used in the United States are shown in Table 13.1. As shown, global production of construction materials far exceeds metals.

Among metals, the most important are iron, aluminum, and copper. Iron is largely used to make steel for automobiles, bridges, buildings, and a host of other products. Aluminum is used to build jet aircraft and beverage cans. Copper is mostly used to make electric wire. So important are the metal-yielding minerals that

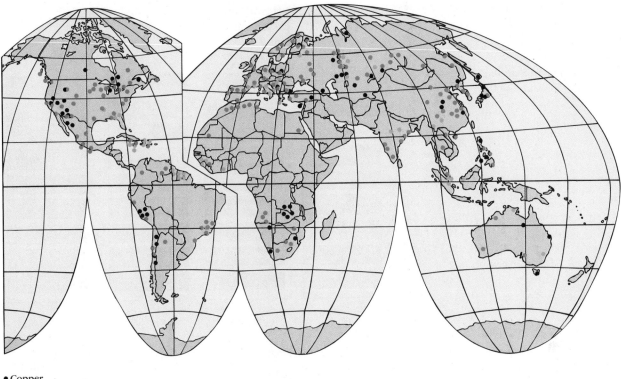

● Copper
● Bauxite
● Iron ore

Figure 13.2 *Map of global mineral reserves.*

if any one of several dozen key minerals were suddenly no longer available at a reasonable price, industry and agriculture would be brought to a standstill.

Who Consumes the World's Minerals?

The developed countries are the major consumers of minerals. With 25% of the world's population, they consume about 75% of its mineral resources. The United States, with only about 6% of the world's population, consumes about 20% of its minerals. Japan, Canada, Europe, and the Commonwealth of Independent States also consume large quantities.

In the past few years, mineral consumption by industrial nations has leveled off and, in some cases, declined. One reason for this is that industrial nations have largely completed their **infrastructure**—bridges, highways, and buildings—and no longer need massive inputs of minerals. Another is that heavy manufacturing is on the decline in many countries while nonmineral-intensive industries, among them high technology and services, are on the rise. Because many developing nations are becoming industrialized, their share of the world's mineral consumption is increasing. In fact, between 1981 and 1989, the latest year for which statistics are available, the developing nations' use

of aluminum grew from 13% to 18% and copper consumption grew from 12% to 18%. Most of this growth in demand is occurring in Mexico, India, Brazil, Thailand, and Korea.

Growing Interdependence and Global Tensions

World mineral production is widely dispersed (Figure 13.2). However, some minerals can be found only in specific locations, making the rest of the world highly dependent on them and consequently fairly vulnerable to disruptions in supply. As shown in Figure 13.3, the United States, Japan, and Europe rely heavily on mineral imports. In contrast, the Commonwealth of Independent States and Eastern bloc nations produce large quantities of minerals domestically. The dependency of many developed countries on foreign markets results primarily from three factors: (1) the depletion of high-grade ores in many industrial nations, (2) the presence of relatively untapped and rich deposits in developing countries, and (3) lower labor costs in the latter.

The vast majority of U.S. mineral resources comes from politically stable countries. What concerns some analysts, however, is that some vital minerals, such as chromium and platinum, are imported from politically volatile regions (Figure 13.4). To protect against embar-

Figure 13.3 *Reliance of the United States, Japan, the European Economic Community (EEC), and the Eastern bloc on foreign and domestic mineral supplies.*

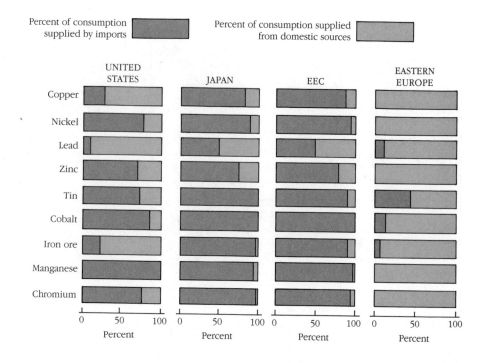

goes or sudden declines in export resulting from political upheaval in such nations, the United States stockpiles a three-year supply of strategic minerals, hoping that political troubles in suppliers will be resolved before the supply runs out.

Industrial nations have for many years exploited deposits in the developing countries, buying huge quantities of minerals at extraordinarily low prices. Developing countries understandably feel cheated by the Western world, which converts the raw materials it acquires at bargain prices into products that reap trillions of dollars per year. To offset this problem, many governments of developing nations have taken steps to increase their share of profits, for example, by charging royalties or taxing raw ore exports. The success of these efforts is reflected by statistics from the U.S. Bureau of Mines, showing that in 1982 the United States imported $4 billion worth of raw minerals and $25 billion worth of finished metals from foreign sources, mostly developing countries. In 1991, the imported material dropped to $2 billion, and the finished metals rose to $34 billion.

Inspired by the actions of OPEC nations, some mineral-exporting countries have united to form OPEC-style cartels to raise the price of several commodities, among them copper, iron ore, and aluminum ore (bauxite). In the future, developed countries could be forced to pay much higher prices for these minerals.

Will There Be Enough?

The long-term outlook for minerals is mixed. Current estimates of world mineral reserves and projections of consumption suggest that about 75% of the economi-cally vital minerals are abundant enough to meet our needs for many years, or if they are not, have adequate substitutes. However, approximately 18 economically essential minerals will fall into short supply, some within a decade or two. Gold, silver, mercury, lead, sulfur, tin, tungsten, and zinc are among them.

Some experts warn us not to count on new discoveries and improved extraction technologies to save the day. Even if we could extract five times the currently known reserves of these materials, the 18 "endangered" minerals will be 80% depleted on or before the year 2040. Combined with declining oil supplies, the depletion of these important minerals could devastate the world economy unless we make changes in consumption soon.

13.3   Environmental Impacts of Mineral Exploitation

Another impetus for changing our ways is the extraordinary impact that mineral exploitation has on the environment. Like fossil fuels, minerals are part of a production-consumption system that involves exploration, mining, processing, transportation, and end use. Many of these activities cause significant environmental damage. In fact, mining and **smelting** (melting ore to extract metal) have created huge environmental troubles worldwide.

Copper production is a case in point. Copper is first extracted from the Earth from **open pit mines**, one type of surface mine. During mining, the rock above the deposit is removed (overburden), and the ore is ex-

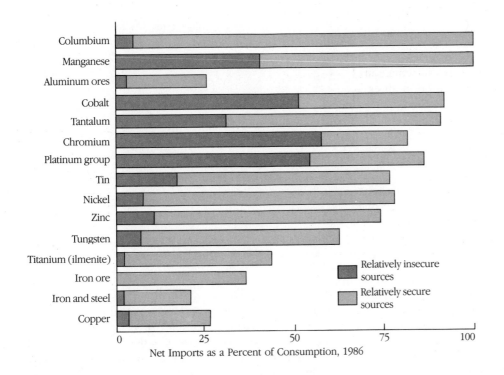

Figure 13.4 *U.S. imports of some major minerals as a percentage of total consumption. The complete bar shows the total imports as a percentage of U.S. consumption; the colored portion indicates the percentage of the minerals consumed that comes from insecure sources. The difference between the total and the colored bar represents the imports from secure sources as a percentage of total consumption.*

Net Imports as a Percent of Consumption, 1986

tracted. It is then crushed and run through a concentrator that removes impurities. The concentrated ore is then heated at high temperatures in a smelter to extract a crude metal, which is later resmelted to produce pure copper.

Each stage in this system produces significant environmental problems, even in the best-regulated countries. For example, mine wastes from open pit mines are typically deposited in piles or ponds near the mine. These wastes, called tailings, can erode into streams, clogging them and causing a number of problems outlined in Chapter 11. Toxic metals in the mine wastes can contaminate nearby streams and reservoirs. Sulfur present in the tailings can combine with water to form sulfuric acid, creating acid mine drainage.

Smelting produces enormous quantities of air pollution. Globally, copper and other nonferrous (noniron) smelters produce about 8% of the world's sulfur dioxide emissions, which contribute to acid deposition (Chapter 16). Smelters also release enormous amounts of toxic metals, among them arsenic, lead, and cadmium. Not surprisingly, smelters have produced some of the most noteworthy environmental disasters—huge "dead zones" where vegetation has perished because of the toxins raining down from the sky in the vicinity. Around the Sudbury smelter in Canada, 10,400 hectares (26,000 acres) have been turned into a barren moonscape. Similar but smaller dead zones have been created in Japan and the United States. A new one is now forming in the Commonwealth of Independent States around one of its smelters.

Finally, transporting ore and finished metals requires enormous amounts of energy, which further adds to global environmental problems.

Other mineral extraction efforts rely on equally devastating techniques. Gold in the Amazon, for example, is extracted by a process known as **hydraulic mining.** Hillsides are blasted with high-pressure streams of water that wash away the soil and rock containing gold. The runoff is directed into sluices that separate out the gold. The sediment runs off into streams, killing fish and other aquatic life.

In North America, gold is sometimes extracted from piles of crushed ore or old tailings by a process called **heap leaching.** In this technique, miners spray a cyanide solution on piles, letting it percolate down through the material. During its sojourn, the cyanide solution dissolves gold. The liquid is then collected and gold extracted. Cyanide collection reservoirs and contaminated piles endanger wildlife and groundwater. In 1990, for instance, 38 million liters (10 million gallons) of cyanide solution broke through a containment dam in South Carolina, flushing into a stream and killing at least 10,000 fish. Ducks and other waterfowl die by the thousands each year when they land in cyanide ponds near mines.

Mineral exploitation creates environmental damage on a scale matched by few other human activities. It is responsible for deforestation, soil erosion, water pollution, and significant air pollution. The environmental impacts are particularly severe in developing countries, which produce a large portion of the world's minerals.

13.4 Meeting Future Needs Sustainably

Two interrelated challenges face the world community: (1) meeting future needs and (2) doing so in an environmentally sustainable fashion.

The Sustainable Path: Recycling, Conservation, and Restoration

Future demands can be met in part by putting into practice the biological principles of sustainability: recycling, population control, conservation, and restoration.

Recycling   Although recycling is discussed in Chapter 19 in detail, it is important to examine a few key points. Instead of being discarded in dumps, valuable minerals can be returned to factories, melted down, and used to manufacture new products. Recycling, therefore, helps increase the time a mineral or metal remains in use, or its **residence time**. Recycling of materials not only helps to stretch limited mineral supplies but also greatly reduces energy demand, cuts pollution, and reduces water use. For example, manufacturing an aluminum can from recycled aluminum uses only 5% of the energy required to make it from aluminum ore (bauxite).

In the United States and many other countries, recycling efforts have fallen short of their full potential, except in a few instances such as the automobile. (Approximately 90% of all American cars are recycled.) Recycling of aluminum, steel, and many other metals could easily be doubled.

Meeting future demands sustainably will require a massive shift to recycling. Important as it is, recycling will not permit our current exponential growth in mineral use to continue indefinitely for several reasons. First, during the production-consumption cycle, some minerals and metals enter into long-term uses, for example, aluminum used for wiring or bronze for statuary. Second, some materials are lost through processing inefficiencies, are lost accidentally, or are purposely discarded. Thus, it is impossible to recycle 100% of any given material. A practical goal would be 60% to 80%. As a result, recycling can slow down the depletion of mineral supplies but cannot stop it. In theory, recycling can double our mineral resource base, but continually rising demand and inevitable losses will eventually deplete reserves.

Conservation   Conservation is another key element of a sustainable system of mineral supply. As in energy supply, conservation is often the cheapest, easiest, and quickest means of stretching mineral resources.

Conserving minerals may mean making smaller automobiles and designing many other products that use less material. It also means making products more durable, so they last longer. Combined with recycling, con-

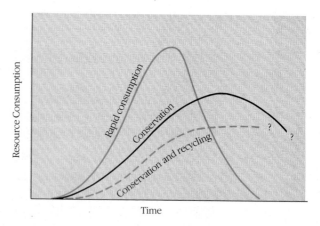

Figure 13.5 *Three possible scenarios on a hypothetical time scale. Which one will we take?*

servation measures can greatly extend the lifetime of many valuable mineral resources. As Figure 13.5 shows, continued exponential growth—the track we have been on until recently—is the fastest route to depletion. Recycling will slow down the depletion. Recycling and conservation measures combined will slow it down even more, giving us more time to develop new mining technologies and find substitutes.

Promoting Conservation and Recycling   Changes in existing laws are needed to stimulate recycling and conservation. For example, the **General Mining Act** (1872) permits miners to purchase federally owned lands for $5 per acre or less. Since the government sells the land to miners and does not collect royalties on minerals extracted from it, mineral prices are artificially low. This discourages conservation and recycling.

U.S. mining companies also receive a sizable tax exemption of about $500 million per year. Known as a **depletion allowance**, this exemption allows companies to deduct 5% to 22% of their gross earnings (depending on the mineral) from their taxes each year, even if the ore deposit is still producing. This law was intended to compensate mining companies for declining ore reserves. What it actually does is make the price of minerals artificially low, discouraging conservation and recycling. Many countries subsidize mineral development through tax laws or even direct investment of government funds (tax dollars) in mineral development projects.

Governments could help promote recycling and conservation by offering tax breaks or other incentives to companies that use recycled materials or offer longer warranties. Deposit and refund systems for products could also stimulate recycling. Changes in these policies and others are badly needed to promote a more sustainable use of the world's mineral resources.

Restoration and Environmental Protection   Meeting mineral demands sustainably requires efforts outlined

Models of Global Sustainability 13.1

Antarctica: Protecting the Last Frontier

Antarctica comprises one-tenth of the Earth's surface. This vast, cold wilderness is home to many animals, among them penguins, seals, and whales. Relatively untouched by humans, Antarctica has been the center of a huge controversy over minerals.

The first real push for Antarctic mineral exploration came after the perilous climb in oil prices in the 1970s. Reports from the U.S. Geological Survey indicated that huge oil reserves might lie beneath Antarctica's continental shelf. But, evidence of oil, many experts say, is sketchy. Similar studies of the Atlantic coast of the United States, for instance, have suggested the presence of oil beneath the continental shelf, but so far, virtually no oil has been discovered there.

Sizable nonfuel mineral deposits may also lie beneath the ice and snow of Antarctica. Traces of minerals have been located in the Pensacola Mountains. Like the early rumors of gold in the American West, these discoveries have accelerated research activity in Antarctica. During the 1980s, many nations began to line up for their share of the envisioned wealth. But, trouble broke out as environmentalists and many scientists projected the environmental and financial costs of mineral development. They argued that Antarctica should be left in a pristine state.

Between 1908 and 1943, Great Britain, France, Norway, Australia, New Zealand, Chile, and Argentina claimed 85% of Antarctica. The United States and the Soviet Union, which had done their share of South Pole exploration and research, did not establish claims. Believing it preferable to retain an interest in the entire continent, the superpowers refused to acknowledge the territorial claims of the others.

In 1959, 16 nations signed a cooperative agreement, the **Antarctic Treaty System**, with two objectives: to maintain the continent for peaceful uses by prohibiting all military activities and weapons testing and to promote freedom for scientific research. The treaty set aside arguments over land claims, permitting the nations who had signed it to carry out research unencumbered. International cooperation for science became the basis for a whole new political order.

Since then, hints of Antarctic minerals and oil have threatened to destabilize the truce, pitting one nation against another. Part of the conflict arose between the haves—the seven claimant nations—and the have-nots—those who signed the treaty and have participated in Antarctic research and management but have no legal claim to the land or resources. The have-nots also include a growing number of developing nations, which have done no research in Antarctica but would like a share in any potential wealth. Added to the growing controversy were several international environmental organizations, such as Greenpeace International and the International Union for the Conservation of Nature and Natural Resources.

In 1988, the race for resources began when the signatories of the treaty agreed to regulations that would permit oil and mineral exploration. If approved, oil and mineral prospecting by seismic testing and other techniques with relatively minor environmental impacts could have begun. Full-scale development, such as mining and drilling, however, would have been barred until a new agreement was reached.

Many critics were concerned that in the rush for wealth, Antarctica's environment would have been decimated. Such concerns were well founded. Offshore oil rigs, for instance, would operate in some of the roughest seas in the world. Huge icebergs could rip apart drilling rigs. Ice floes could trap and crush ships. Mining would require huge amounts of energy to drill through the mile-thick ice cap.

Environmentalists were also concerned about the potential impacts of resource development on the rich marine ecosystem. Oil spills, they said, could be carried ashore, wiping out huge breeding colonies of penguins, birds, and seals. Oil slicks might also kill algae, the base of the aquatic food chain. Caught under the ice and in the chilly waters, oil could persist 100 times longer than in warmer waters. Finally, cleaning up an oil spill would be impaired by the short "warm" season. If a well blew out, winter might set in before workers could plug it up. Oil would collect under the ice for up to nine months with devastating effects.

Nonfuel mineral extraction would require enormous amounts of energy, create pollution, damage the ice pack, and severely disrupt the land surface. Pollution from mining operations would also disrupt scientific research on air pollution. Most damaging to wildlife would be the use of ice-free coastal areas for processing and shipping ores.

Research activity has taken precedence over environmental protection. The activities of researchers and their daily maintenance have created significant air and water pollution. Further occupation and human activity on a massive scale could create an environmental disaster. Should oil and minerals be discovered, this relatively pristine wilderness could be sullied beyond imagination.

In October 1991, after considerable political pressure from environmentalists, 24 industrial nations signed an agreement that bans environmentally damaging oil and mineral exploration in Antarctica for at least 50 years. Korea signed in 1992, but as of December 1992, Japan had not yet signed and is the last active participant in the region not to do so.

Besides prohibiting oil, gas, and mineral development, this treaty includes provisions that will protect the environment from the impacts of human activities. Although the agreement still leaves the possibility of oil and mineral development open, for now Antarctica is saved from the ravages of modern society.

above but also measures to restore the damage to natural systems caused by mining and to minimize environmental disruption and pollution in current operations.

New laws and tighter enforcement of existing laws could improve mining practices and reduce pollution from smelters in industrial and nonindustrial nations. In the United States, mining and smelting operations are often exempt from existing laws. Mine wastes, which contain toxic metals that leach into nearby waterways, are currently not regulated by the **Resource Conservation and Recovery Act** (1976), which among other things, places tight controls on hazardous wastes. This despite the fact that mines are the single largest producer of wastes in the nation. Moreover, smelters are not required to report their emissions to the EPA as other industries are. Efforts are currently under way to close these loopholes, but given the political power of the mining industry, many observers are skeptical about their outcome.

Expanding Our Reserves

Because of the continued growth of the world population and rising expectations of the residents of the developing nations, recycling and conservation will not satisfy 100% of future demands. Some new deposits need to be discovered.

Geologists use the term **reserves** to indicate deposits that they are fairly certain exist and are feasible to mine at current prices. Several factors determine the size of available reserves.

Rising Prices, Rising Supplies   As noted in Chapter 1, the law of supply and demand is the centerpiece of economic thinking. This law says that when demand outstrips supply, prices tend to rise. Rising prices tend to stimulate production, increasing supply. Many people seem to think that this can continue ad infinitum.

This line of reasoning may work in the short term, but in the long term, it is bound to fail. Why? The main reason is that at some point finite resources could simply run out. No amount of money will help at that point.

Using Technology to Expand Reserves   Improvements in technology, such as more efficient means of extracting ore, can help expand reserves. In fact, technological improvements in the mining and processing of ore have increased global reserves of several key minerals since the 1970s (Figure 13.6).

Scientists and engineers are currently developing techniques to improve mining efficiency even more, thus allowing companies to exploit even lower-grade ores. One group, for example, recently found that certain algae bind gold ions; it is believed that these algae could help extract gold from mine wastes or even natural waters.

Although technological advances are needed to increase mining efficiency, it is important to remember that technology is cost driven. When mineral supplies fall below a certain point, costs could conceivably become too great for consumers to bear. Mining would be curtailed. Thus, although technology can help, it is only part of the solution.

Factors That Reduce Supplies   Higher prices and new technologies can expand reserves within limits. However, many factors have an opposite effect. Rising labor costs, for example, tend to reduce reserves. Exploration, mining, and production are also influenced by interest rates. High interest rates on capital needed for exploration and mining may slow these activities. Some economists predict that competition for capital will escalate in coming years, driving up interest rates and slowing the expansion of global reserves.

Energy prices will also play a major role in determining when mining a particular mineral becomes uneconomical. As the concentration of an ore decreases, the amount of energy required for mining and refining is fairly constant up to a certain point, beyond which it increases dramatically (see Figure 13.7 on page 319). At this point, these reserves become too expensive to mine; their costs exceed what the market will bear. For all intents and purposes, these reserves are depleted.

According to the U.S. Department of the Interior, the costs of minerals are expected to increase 5% per year after the year 2000, the same rate at which energy costs are predicted to increase. Rising energy prices combined with declining ore deposits could very well bring an end to some important mineral resources within 30 to 40 years. Runaway inflation could grip the world economy unless something is done.

A final factor is environmental costs. Mining lower-grade ores results in greater environmental damage than higher-grade ores. Why? Larger surface mines are needed to produce the same amount of ore, more material is transported to smelters for processing, more waste is produced at the mines and smelters, and more air and water pollution results. This could cause environmental protection costs to escalate, adding to the cost of production and accelerating the economic depletion of some minerals.

The prospects for expanding mineral reserves are mixed. As noted earlier, reserves for some minerals appear to be quite limited. Although supplies of others may be huge, environmental constraints and energy costs could put a damper on expansion, making conservation and recycling even more attractive.

Minerals from Outer Space and the Sea   The mineral deposits on land are finite, and, in many cases, they have been heavily exploited, leaving little for future gen-

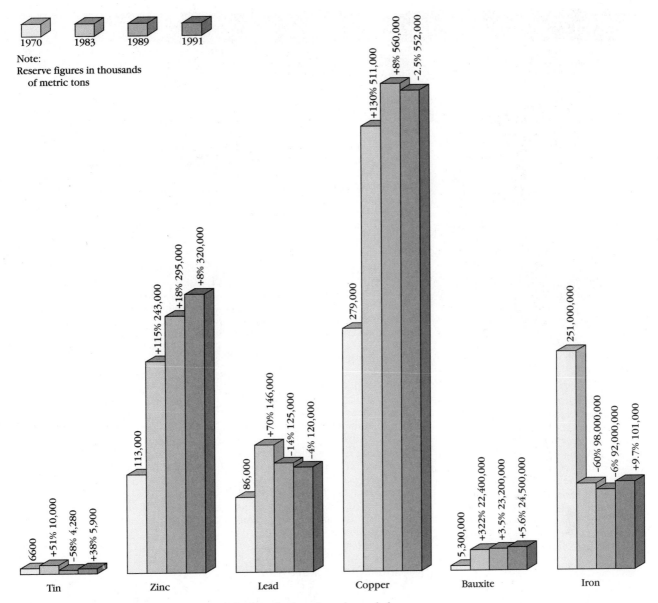

Note:
Reserve figures in thousands
  of metric tons

1970   1983   1989   1991

Tin
6600
+51% 10,000
−58% 4,280
+38% 5,900

Zinc
113,000
+115% 243,000
+18% 295,000
+8% 320,000

Lead
86,000
+70% 146,000
−14% 125,000
−4% 120,000

Copper
279,000
+130% 511,000
+8% 560,000
−2.5% 552,000

Bauxite
5,300,000
+322% 22,400,000
+3.5% 23,200,000
+5.6% 24,500,000

Iron
251,000,000
−60% 98,000,000
−6% 92,000,000
+9.7% 101,000

Figure 13.6 *Changes in world reserves of selected minerals and metals between 1970 and 1991.*

erations. As a result, some companies are looking toward three untapped frontiers—outer space, Antarctica, and the floor of the world's oceans—as potential sources for new minerals. Superficially promising, these options face serious economic, environmental, and social barriers.

The Point/Counterpoint in this chapter discusses the pros and cons of the outer-space option. Models of Global Sustainability 13.1 examines the mineral potential of Antarctica. This section describes seafloor deposits in the ocean.

The ocean is a vast resource of minerals, many of which are dissolved in the water itself. However, the concentrations of most dissolved minerals are generally too low to be of economic importance. More important

are mineral deposits on the seafloor. The most important of these minerals are small lumps called **manganese nodules** (see Figure 13.8 on page 320). Abundant in the Pacific Ocean, manganese nodules contain several vital minerals, notably manganese (24%) and iron (14%), with smaller amounts of copper (1%) and cobalt (0.25%).

Several mining companies have explored the possibility of dredging the seabed to mine nodules. Preliminary studies show that seafloor mining may be technically feasible as well as profitable, but little is known about the environmental impact of such a venture. At this point, the biggest impediment appears to be a political one. To date, the world is in a quandary over the issue of ownership.

Point/Counterpoint 13.1

Is Outer Space the Answer to Our Population and Resource Problems?

Ben Bova

Ben Bova is the author of more than 75 futuristic novels and non-fiction books about science and high technology. Among them are Welcome to Moonbase, *an examination of the economic, social, and scientific benefits to be reaped from a permanent settlement on the moon. He is the former editor of* Omni *and* Analog *magazines, and president emeritus of the National Space Society.*

Toward a New World

We live in a solar system that is incredibly rich in energy and raw materials. In interplanetary space, more wealth than any emperor could dream of is available for every human being alive. Instead of thinking of our world as a finite pie that must be sliced thinner and thinner as population swells, we must go out into space and create a larger pie so that everyone can have bigger and bigger slices of wealth.

What can we gain from the New World that begins a couple of hundred miles above our heads? Of immediate concern, there is en-

ergy in space—enormous energy from unfiltered sunlight. The sun radiates an incomprehensible flood of energy into space—the equivalent of 10 billion megatons of H-bomb explosions every second. The Earth intercepts only a tiny fraction of this energy, less than 0.2 billionths of the total energy given off by the sun.

For the purposes of our generation, a few dozen solar-powered satellites in geostationary orbit above the equator may be sufficient. Each would beam 5 billion watts of energy to us, making a substantial contribution to our energy supplies.

Natural resources are also there in superabundance. Thanks to the Apollo explorations of the moon, we know that lunar rocks and soil contain many valuable elements, such as aluminum, magnesium, titanium, and silicon.

Looking further afield, there are asteroids that sail through the solar system, especially in a region between Mars and Jupiter, called the Asteroid Belt. To a miner or an industrialist, they are a bonanza. An MIT astronomer, Tom McCord, estimates that there are hundreds of millions of billions of tons of nickel-iron asteroids in the belt. The economic potential of these resources is incalculably large. And, mixed in with these, are many other valuable metals and minerals.

This "mother lode" of riches is a long way from Earth—hundreds of millions of miles—but in space, distance is not so important as the amount of energy you must expend to get where you want to go. Like sailing ships, spacecraft do not burn fuel for most of their flight; they simply coast after achieving sufficient speed to reach their destination. Space is not a barrier, it is a highway. The biggest and most difficult step in space flight is getting off the Earth's surface.

The opportunity offered by space is not to export people from

Earth but to import valuable resources to Earth.

We will move outward into space because it is biologically necessary for us to do so. Driven by forces beyond our understanding, we head into space citing all the good and practical and necessary reasons for going. But, in actuality, we go because we are driven.

The rocket pioneer Krafft Ehricke has likened our situation on Earth today to the situation in a mother's womb after nine months of gestation. The baby has been living a sweet life, without exertion, nurtured and fed by the environment in which it has been enclosed. But, the baby gets too big for that environment, and the baby's own waste products are polluting that environment to the point where it becomes unlivable.

Our biological heritage, our historical legacy, our real and pressing economic and social needs are all pointing toward the same conclusion: It's time to come out into the real world. Time to expand into the new world of space.

Daniel Deudney
The author is a Hewlett fellow in Science, Technology, and Society at the Center for Energy and Environmental Studies at Princeton University. He writes on international security, space, and global commons issues.

No Escape from the Population Bomb

Much of the recent writing about humanity's future in space has been dominated by outlandish proposals for large-scale space operations that aim to bypass the Earth's resource limits, either by exporting people from the planet or by importing energy and materials from space. These include space colonies, solar-powered satellites, and asteroid mining operations. At first glance, these massive undertakings have a logical appeal: The Earth is limited, space is infinite.

Even though space contains vastly more "living" room, energy, and materials than the Earth, this abundance cannot be brought to bear meaningfully on the Earth's problems. The Earth's population and resource problems will have to be solved on Earth and soon.

In recent years, space colonies have been advanced as a solution to the Earth's population and environmental problems by Professor Gerard O'Neill, a Princeton University physicist. In his writings, he details plans to build colonies with first 10,000 inhabitants, and later a million. Manufactured out of materials from the Earth, and then the Moon and asteroids, colonies would be made completely self-sufficient by harnessing sunlight energy. Those who favor space colonization as a solution to overcrowding and environmental degradation argue that by the exporting of ever greater numbers of people into these orbiting cities, the wildlife and wilderness on Earth could be protected, perhaps even expanded.

Life in space is envisioned as pastoral, pollution-free, and pluralistic—like floating garden cities. Yet, in reality, space habitation would probably be bleak. Thick metal shielding would be necessary to block the lethal cosmic and solar radiation. Life would be like that in a submarine—cramped and isolated. For at least the next several decades, people will go into space to perform various specialized missions—or perhaps briefly as tourists—but they will not live there in significant numbers.

No scientific laws forbid large space colonies. Structures of the size envisioned are thousands of times larger than anything yet built for space; no doubt there will be unforeseen and even insurmountable technical problems. Ecologist Paul Ehrlich points out that scientists have no idea how to create large, stable ecosystems of the sort that would be needed to make space colonies self-sufficient. The key to such knowledge is, of course, much more study of the ecosystems on Earth, many of which are becoming less diverse and less stable. It could be many decades before scientists know enough to understand—let alone re-create—ecosystems as complex as those now being degraded. There are also unanswered questions of human biology: Can babies be born and grow up in a weightless environment? Would the various forms of cosmic and solar radiation make the mutation and cancer rate unacceptably high?

Space colonies are not even a partial answer to the population and environmental problems of Earth. Simply transporting the world's daily increase of about 250,000 people into space would consume the annual GNP of the United States. Each launch of the space shuttle costs $300 million, and that doesn't even include the $25 billion in research and other initial costs. Maintaining the complex life-support systems in orbit costs thousands of dollars an hour.

Solar-powered satellites are often seen as complements of large-scale space colonization and as a source of energy for Earth. The principal appeal of these satellites is their ability to collect virtually unlimited amounts of solar energy day and night without polluting the atmosphere. The construction of them, however, would be an undertaking of unprecedented size and cost. A NASA and U.S. Department of Energy study estimated that 60 satellites, each as big as Manhattan Island, would be needed to produce the current U.S. electric usage. The cost estimates ranged from $1.5 trillion to $3 trillion. If two satellites were built each year, one heavy-lift rocket—seven times the largest rocket ever built—would have to be launched each day for 30 years to build and service them.

More troubling are planetary-scale environmental risks. Beaming trillions of watts of microwaves through the atmosphere for extended periods is almost certain to alter the composition of gases in unpredictable ways. Launching

(continued)

Point/Counterpoint 13.1 (continued)

Is Outer Space the Answer to Our Population and Resource Problems?

millions of tons of material into orbit would also release large quantities of exhaust gases into the upper atmosphere, perhaps disrupting the ozone layer which screens out ultraviolet light. An operational system big enough to make a difference in the terrestrial energy equation would be a shot-in-the-dark experiment with our atmosphere.

The other mirage of abundance in space that has recently received attention is asteroids. These irregularly shaped rocks, which orbit the sun, range in size from as small as a grain of sand to as big as the state of Texas. Although there is probably enough metal in the asteroid belt to meet world needs for many centuries, getting them to the

Earth's surface would be costly and energy intensive and would risk an accidental collision and ecological disruption on a colossal scale. Long before it becomes feasible or economical to bring rare metals from space, scientists should be able to turn the abundant clay, silicon, alumina, hydrocarbons, and iron in the Earth's crust into materials that can be used to meet global needs.

In summary, large-scale space colonization and industrialization is an unworkable attempt to escape from the problems of the Earth. In the struggle to protect the Earth from overpopulation, ecological degradation, and resource depletion, outer space has a great, largely unfulfilled role to play. It is

valuable not as a source of energy or materials or as a place to house the world's growing population but rather as a tool both for learning more about our planet and for assisting problem solving here on Earth.

Critical Thinking

1. Summarize the views of both authors on using outer space to reduce population pressure, acquire energy, and find mineral resources. Using your critical thinking skills, analyze each argument.
2. Based on your analysis, which author do you most agree with? Why?

Mining companies in the Western world have the wealth and technology to exploit the seafloor and argue that they should be allowed to do so. However, the developing nations contend that the seabed belongs to all nations and believe that they should receive a portion of the proceeds.

At a U.N. Conference on the Law of the Sea, developing countries proposed an international tax on seabed minerals that could provide them with millions of dollars per year for agricultural and economic development. A comprehensive Law of the Sea Treaty, worked out with U.S. negotiators during the Carter administration, would have included this plan, but the United States refused to sign the treaty under Presidents Ronald Reagan and George Bush, as did Great Britain and former West Germany. Today, many people believe that a seabed taxation scheme is fair and necessary. Others see it as an unjust way of cutting into corporate profits. As a result of the growing conflict, progress toward mining the seabed has come to a halt.

Finding Substitutes

Historically, the substitution of one resource for another that has been depleted has been a useful strategy for industrialized nations. Shortages of cotton, wool, and natural rubber, for example, have been eased by synthetic materials made from oil. Synthetic fibers are used in American clothing. Synthetic rubber has replaced most natural rubber from trees.

Substitution will unquestionably play an important role in the future, too. It could help find alternatives to some minerals that fall into short supply. It could help replace materials whose production is environmentally damaging. For example, fiber optics are thin glass fibers made from sand that are currently replacing many uses of copper, a mineral whose extraction and processing are environmentally harmful, even under the best of circumstances. But is substitution a crutch we can lean on forever?

Critics argue that substitutions have created unreasonable faith among the public in the ability of scientists to find new resources to replace those that are being depleted. Some resources may not have substitutes. For instance, it may be impossible to find substitutes for the manganese used in desulfurizing steel, the nickel and chromium used in stainless steel, the tin in solder, the helium used in low-temperature refrigeration, the tungsten used in high-speed tools, or the silver used in photographic papers and films. Scientists also note that

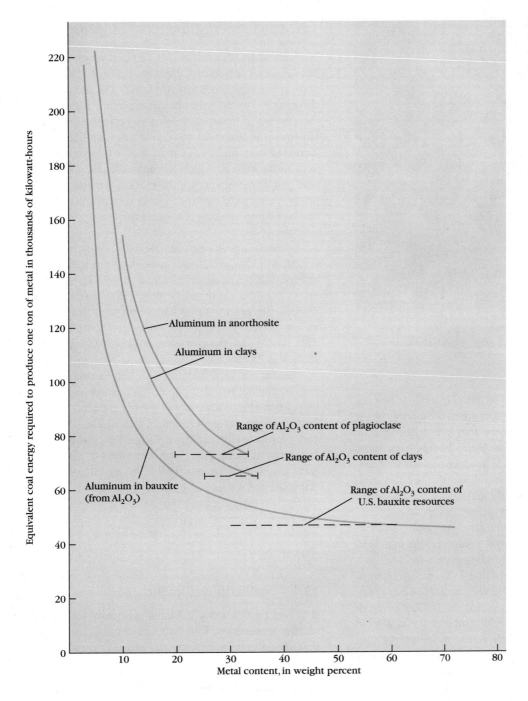

many substitutes have limits themselves. Plastics have replaced many metals, for example, but the oil from which plastics are made is a limited resource.

Finding substitutes is a race against time. Since we have rounded the bend of the exponential curve of demand, the time of economic depletion is fast approaching, perhaps more quickly than substitutes can be found. A wise strategy would be to identify those resources that are nearest to economic depletion and then promote widespread conservation and recycling.

Research to find substitutes should begin immediately, or if research is already under way, it should be greatly accelerated.

Some Personal Actions

Tolstoy wrote: "Everyone thinks of changing the world, but no one thinks of changing himself." Personal actions can greatly help to conserve mineral supplies and can be financially advantageous. By putting into practice the

Figure 13.8 *Manganese nodules on the ocean's floor. These nodules range from the size of peas to the size of oranges.*

principles of sustainability, you can play a role in helping create a sustainable world community. Buy only what you need. Choose containers that are returnable or recyclable. Recycle all metals. When buying something, look for quality: a well-made product outlasts inexpensive imitations and could be well worth the extra cost. In the long run, high-quality materials save resources. Write letters complaining to manufacturers whose products fall apart soon after you get them home. You can also help by supporting legal solutions, for example, local, state, and national recycling programs.

Polls in the 1980s showed that 75% of all Americans questioned favored increased recycling. However, only 14% of U.S. waste was recycled in 1991. "Between saying and doing," writes F. W. Robertson, "is a great distance." Blessed with cheap energy resources and abundant food sources, the United States and other countries are throwing away their future.

Given the need to create a sustainable industrial system, conservation and other measures that reduce waste and minimize environmental destruction are badly needed. They are not a roadblock to progress but a prerequisite for it.

Tomorrow's growth depends on the use we

make of today's materials and experiences.

—Elmer Wheeler

Critical Thinking Exercise Solution

In recent years, the market for recycled newspapers has faltered because of a dramatic upsurge in recycling programs. What has happened is that the rise in recycling in the United States has dramatically increased the amount of newspaper going to recycling mills. However, mill capacity to recycle this material has increased only slightly. That is, there's a huge bottleneck in the system, which is responsible for the current glut of newspaper on the market. Because of this, prices for recycled newsprint have dropped considerably, sometimes making this endeavor rather uneconomical.

However, problems with recycling don't mean the system will always be inadequate. Current trends can change. Gathering more information on the issue, so essential to critical thinking, shows that nearly 70 mills are gearing up to handle recycled paper and cardboard in the United States and Canada. This should ease the glut considerably starting in 1993 and 1994.

Businesses, governments, and individuals can also help by buying recycled paper. Recycled newsprint can be converted into a variety of other products, including insulation, tar paper, mulch, boxes, kitty litter, and gypsum wallboard liner. Citizens can lobby their governments to support businesses that put waste to good use. On the issue of pollution, like almost everything else we do recycling creates pollution. Fortunately, it creates far less than making materials from raw ore or virgin materials.

Summary

13.1 The Earth and Its Riches

■ The Earth's **crust** contains inorganic compounds known as **minerals**. Concentrated deposits of minerals that are economical to mine and process are called **ores**.

13.2 Mineral Resources and Society

■ Minerals are worth billions of dollars in the world market and are vital to industry and agriculture.
■ The developed countries are the major consumers of minerals, but consumption is beginning to rise in developing countries as well.

13.3 Environmental Impacts of Mineral Exploitation

■ Mineral production creates extraordinary environmental impacts at nearly every phase.
■ Mining scars large areas and creates enormous quantities of waste that erode into streams and lakes.

Wastes also contain toxic metals that contaminate waterways.

■ **Smelting** produces huge quantities of air pollution, including carbon dioxide, sulfur dioxide, and toxic metals.

■ Transporting ore and finished metals requires large amounts of energy, further adding to global environmental problems.

■ Mineral exploitation is responsible for deforestation, soil erosion, water pollution, and significant air pollution. The environmental impacts are particularly severe in developing countries, which produce a large portion of the world's minerals.

13.4 Meeting Future Needs Sustainably

■ The challenge facing us today is twofold: meeting future needs and doing so sustainably.

■ Conservation can greatly extend the lifetime of many valuable minerals and give us more time to find substitutes.

■ Recycling can also help stretch limited supplies, cut energy demand, reduce environmental pollution, and create jobs.

■ New reserves will need to be developed in environmentally acceptable ways.

■ Some economists believe that rising prices will stimulate exploration and new technologies, which will increase mineral reserves in a never-ending cycle. As mineral reserves of lower and lower concentration are exploited, however, energy investments in mining and processing will become higher and higher.

■ Sometime after the year 2000, the real cost of minerals may begin to rise by 5% a year, making many minerals uneconomical to mine and process. Competition for capital may raise interest rates, which would make it economically unprofitable to mine and process lower-grade ores. Mining low-grade ores increases environmental damage as well.

■ Minerals from the sea may help expand our reserves, but economic, environmental, and legal questions have yet to be resolved.

■ Substitution of one resource for another that has become economically depleted has been a useful strategy in the past but may be only partially helpful in the future. Substitutes for scarce minerals may have limits themselves. Some minerals have no adequate substitutes. Finally, in some cases, economic depletion may occur before substitutes can be developed.

Discussion and Critical Thinking Questions

1. Why is it difficult to determine the level of future mineral demand? How would you go about calculating the demand for minerals in the year 2000?

2. Critically analyze the statement: "Economic forces will ensure a continual supply of mineral resources. As prices rise, we'll find new resources, develop new technologies, and find substitutes for minerals currently used."

3. What are reserves? Explain how mineral reserves expand and contract.

4. The trend over the past few decades has been increasing reserves of many minerals. Why has this occurred? Will it continue? Why or why not?

5. Outline a plan to meet the future mineral needs of our society sustainably. Describe each element of your plan.

6. What legal/legislative changes would help promote a sustainable system of mineral production and consumption?

7. Debate the statement: "There is no need to worry about running out of minerals; we will find substitutes for them."

8. List the ways in which you can reduce your resource consumption.

Suggested Readings

Bogart, P. (1988). On Thin Ice: Can Antarctica Survive the Gold Rush? *Greenpeace* 13 (5): 7–11. Telling account of the troubles in Antarctica.

EDF. (1988). *Coming Full Circle: Successful Recycling Today*. Washington, D.C.: Environmental Defense Fund. Superb survey of recycling programs.

Holdgate, M. W. (1990). Antarctica: Ice Under Pressure. *Environment* 32(8): 4–9, 30–33. A discussion of the political and environmental considerations affecting the future of Antarctica.

Pollock, C. (1987). *Mining Urban Wastes: The Potential for Recycling*. Worldwatch Paper 76. Washington, D.C.: Worldwatch Institute. Fact-filled reading.

U.S. Bureau of Mines. (1990). *Mineral Facts and Problems*. Washington, D.C.: U.S. Government Printing Office. Superb reference.

———. (1992). *Mineral Commodity Summaries*. Washington, D.C.: U.S. Government Printing Office. Excellent source of information on minerals published every year.

Wilkinson, C. F. (1992) *Crossing the Next Meridian: Land, Water, and the Future of the West*. Washington, D.C.: Island Press. Discusses outmoded laws that govern today's policy.

Young, J. E. (1992). Free-Loading Off Uncle Sam. *World-Watch* 5(1): 34–35. Describes policies that give mining an unfair economic advantage.

———. (1992). Mining the Earth. In *State of the World 1992*. New York: Norton. Excellent source for more detailed information on minerals.

Part 4

Pollution

Chapter 14

Toxic Substances: Principles and Practicalities

Life is a perpetual instruction in cause and effect.

—*Ralph Waldo Emerson*

Critical Thinking Exercise

You are an environmental manager for a major corporation. Your boss tells you that the company needs to cut back on toxic pollutants emitted into the air and water. If you the company doesn't, it could face heavy fines. You turn to your staff and ask them to come up with a plan to reduce pollution. In a week, they report with a proposal that calls for the addition of a $20 million pollution control device. You present the proposal to your boss, who says that the company can't afford it. It would be cheaper to continue releasing toxic pollutants and be fined than comply with the law. What would you do? Would critical thinking skills help you out of this mess?

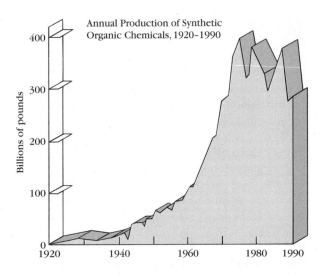

Figure 14.1   *Growth in the production of synthetic organic chemicals in the United States. Note that production has more or less leveled off since 1970.*

Across the globe, many companies are taking a new approach to the problem of hazardous chemical wastes. Instead of dumping them in the ground, releasing them into the air, or pumping them into nearby streams or groundwater, they are finding ways to reduce hazardous chemical waste production. That is, these companies are greatly scaling back their output through a variety of innovative measures. This approach, called **pollution prevention**, resembles energy efficiency, discussed in Chapter 12, in that it is not only good for the environment but also great for the bottom line. This encouraging shift has been stimulated in part by costly regulations and also by responsible industry leaders who realize that industry is partly to blame for the deteriorating condition of the planet and that industry can play a significant role in solving environmental problems.

This chapter lays the groundwork for Part 4 (Chapters 14 through 19) on pollution. It examines principles of the study of toxic chemicals, **toxicology**, and looks at chemicals that adversely affect living organisms, **toxins**. It discusses their effects and means of controlling them. It concludes with an assessment of the risks and benefits of life in a world dependent on technology and an assortment of potentially harmful chemicals.

14.1   Principles of Toxicology

In the United States, 60,000 chemical substances are sold commercially. By various estimates, 8600 food additives, 3400 cosmetic ingredients, and at least 35,000 pesticides are in use. U.S. chemical production and consumption have skyrocketed since World War II (Figure 14.1). Of the commercially important chemicals, however, only a small number—perhaps 2%—are known to be harmful. Nonetheless, this small percentage amounts to hundreds of potentially dangerous chemicals that pose a threat not only to workers but also to the general public. According

to the U.S. Department of Transportation statistics, approximately 6500 accidents occur each year during the transportation of toxic wastes in the United States. Approximately 300 people are killed or injured each year as a result. Many more accidents occur in the workplace and in toxic waste facilities.

Perhaps the biggest problem with toxic substances is our lack of knowledge about their effects. The National Academy of Sciences notes that fewer than 10% of U.S. agricultural chemicals, mostly pesticides, and 5% of food additives have been fully tested to assess long-term health effects. Testing potentially harmful substances is a costly and time-consuming task, made more difficult by the annual introduction of 700 to 1000 new chemicals to the marketplace.

Biological Effects of Toxins

This chapter is primarily concerned with the effects of toxic chemicals on humans. Subsequent chapters take a broader view, showing how toxins affect plants, animals, and microorganisms.

People are exposed to toxic substances at home, at work, or in the out-of-doors. In many cases, people have little control over exposure. Polluted air from nearby power plants or highways, for instance, exposes them to dozens of potentially harmful substances. In some cases, we expose ourselves voluntarily to harmful substances, such as toxic fumes from cleaning agents.

Immediate and Delayed Toxicity   Toxins produce a wide variety of effects. Some stimulate immediate effects. Some effects are subtle, such as a slight cough or headache from urban air pollution. Others can be pronounced, such as the violent convulsions induced by exposure to certain insecticides. As a rule, immediate effects disappear shortly after the exposure ends and are generally caused by fairly high concentrations of chemicals resulting from short-term exposures.

Other toxins produce delayed effects, such as cancer or birth defects. They may occur months to years after exposure and usually persist for years, as in the case of emphysema caused by cigarette smoke or pollution. Delayed effects often result from low-level exposure over long periods, chronic exposures. Short-term exposures may also have delayed effects. A one-time exposure to certain cancer-causing agents, for example, may cause the disease, although the symptoms may not appear for many years.

Reversible and Irreversible Toxic Effects   Some toxic effects are reversible. That is, they alter normal body functions for a while, but the effects are fleeting. The reversibility of any effect depends on how long a substance stays in the body and also on the ability of tissues to repair damage caused by it. Others toxins cause irreversible damage, for example, a chronic disease or death.

Local and Systemic Toxic Effects   Still another important distinction regarding toxic chemicals is their site of action. Some exert local effects, most often at the site of contact. Certain industrial chemicals, for instance, cause skin rashes. Other toxic substances enter the body and circulate to many different sites where they exert their effects, often adversely affecting entire organs or organ systems, such as the brain and spinal cord. These effects, described as **systemic**, are the most common type. Most systemic toxins act on the central nervous system.

How Toxins Work

Toxic substances exert their effects at the cellular level in four major ways: First, some toxins affect **enzymes**, the cellular proteins that regulate many biochemical reactions. A disturbance of enzymatic activity can seriously alter the functioning of an organ or tissue. As examples, mercury and arsenic both bind to certain enzymes and block their activity. Second, some toxins can bind directly to molecules in the body, upsetting the chemical balance. Carbon monoxide, for example, binds to hemoglobin in red blood cells, blocking its ability to bind to oxygen. This decreases the transport of oxygen throughout the body and can lead to death if levels are high enough (Chapter 15). Third, some toxins can cause the release of other naturally occurring substances that adversely affect the function of cells and organs. Carbon tetrachloride, for example, stimulates certain nerve cells to release large quantities of epinephrine (adrenaline), believed to cause liver damage. Fourth, some toxins interact with the genetic material of cells, causing **mutations,** potentially harmful changes in the structure of the DNA (described below).

Factors Affecting the Toxicity of Chemicals

Predicting the harmful effects of chemicals is no easy task. Age, sex, genetic composition, health, and a variety of other elements are all contributing factors. Consider the case of a family of six living near a Canadian lead and zinc smelter, which released large amounts of lead. Although each member was exposed to high levels of lead, the symptoms varied. For example, the father and a four-year-old boy suffered from colic (acute abdominal pain) and pancreatitis. The mother developed a neural disorder. Two other children experienced convulsions, and the last developed diabetes.

Individuals can develop tolerance to certain toxins. In other cases, they can become extremely sensitive to tiny doses, or **hypersensitized**. People exposed to formaldehyde, a chemical released from furniture, plywood, and other products, are often initially tolerant of the chemical at high levels but over time become sensitive to extremely low levels. Numerous factors influence the effects of a given chemical. The four most important are: (1) dose, (2) duration of exposure, (3) biological reactivity of the chemical in question, and (4) route of exposure.

Dose and Duration of Exposure   In general, the higher the amount given—the **dose**—of a toxin, the greater its biological effect. To uncover the effect, toxicologists expose laboratory animals to varying doses, determining the response at each level. The resulting graph is called a **dose-response curve** (Figure 14.2).

To compare one chemical to another, toxicologists often determine the dose that kills half of the test animals—that is, the lethal dose for 50% of the test animals, or **LD_{50}**. By comparing LD_{50} values, scientists can judge the relative toxicity of two chemicals. For example, a chemical with an LD_{50} of 200 milligrams per kilogram of body weight is half as toxic as one with an LD_{50} of 100 milligrams. Thus, the lower the LD_{50}, the more toxic a chemical.

The **duration of exposure** is the amount of time an individual or laboratory animal is exposed to a toxin. Exposures generally fall into two categories: **acute**, or short-term exposures, generally lasting less than 24 hours, and **chronic**, or long-term repeated exposures, lasting more than three months. Obviously, these are very broad categories, and many intermediate exposure possibilities exist. Studies show that for many toxins, acute exposures result in very different effects than those resulting from chronic exposure. An acute exposure to benzene, for instance, may result in a transient bout of depression. Chronic exposure may lead to leukemia, a cancer of the white blood cells.

Biological Activity and Route of Exposure   The toxicity of a chemical is a function of its biological activity, that is, how it reacts with enzymes or other cellular

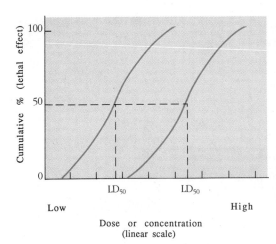

APPROXIMATE ACUTE LD50s
OF A VARIETY OF CHEMICAL AGENTS

AGENT	LD50 (mg/kg)
Ferrous sulfate	1,500
Morphine sulfate	900
Phenobarbital sodium	150
DDT	100
Picrotoxin	5
Strychnine sulfate	2
Nicotine	1
d-Tubocurarine	0.5
Hemicholinium-3	0.2
Tetrodotoxin	0.10
Botulinus toxin	0.00001

Figure 14.2 *Dose-response graph for two chemicals with differing toxicities. The LD_{50} is the amount of chemical that kills one-half of the experimental animals within a given time. The higher the LD_{50} value, the less toxic the chemical.*

components. The more reactive it is, the more effect it has. Inert substances—those that do not chemically react with cellular components—generally are not harmful, although there are notable exceptions, such as asbestos (described below and in Case Study 14.1).

Toxins enter the body by one of several routes. Some may be inhaled, entering the bloodstream via the lungs. Others may be ingested in food and water. Still others are absorbed through the skin. The effect a toxic agent elicits is greatly influenced by its route of entry in the following order: inhalation, ingestion, and dermal absorption. This hierarchy is largely determined by the ease with which a chemical enters the bloodstream.

Other Contributing Factors

Further adding to the complexity, several additional factors come into play.

Age   Although we think of youngsters of all species as resilient creatures, young, growing organisms are generally more susceptible to toxic chemicals than adults. For example, two common air pollutants, ozone and sulfur dioxide, affect young laboratory animals two to three times more severely than they affect adults. Among humans, infants and children are more susceptible to lead and mercury poisoning than adults because their nervous systems are still developing.

Health Status   Health is determined by many factors, among them one's nutrition, level of stress, and personal habits, such as smoking. As a rule, the poorer one's health, the more susceptible he or she is to a toxin.

Chemical Interactions   People live in a sea of potential toxins that may interact in one of several ways. Some chemical substances, for example, team up to produce an **additive response**—that is, an effect that is sim-

ply the sum of the individual responses (for example, 2 + 2 = 4).

Others may produce a **synergistic response**—that is, a response stronger than the sum of the two individual ones (for example, 2 + 2 = 6). One of the most familiar examples of synergism is the combination of barbiturate tranquilizers and alcohol. Although neither taken alone in small amounts is dangerous, the combination can be deadly. Pollutants can also synergize. For instance, sulfur dioxide gas and particulates (minute airborne particles) inhaled together can reduce air flow through the lung's tiny passages. The combined response is much greater than the sum of the individual responses. (Case Study 14.1 discusses the synergistic effect of smoking and asbestos.)

Another fascinating interaction is **potentiation**. This occurs when a chemical with no toxic effect combines with a toxic chemical making the toxin even more harmful. This response can be represented by the equation 0 + 2 = 6. Isopropyl alcohol (rubbing alcohol), for instance, has no effect on the liver, but when combined with carbon tetrachloride, it greatly boosts the toxicity of the latter.

Certain chemicals can also negate each other's effects, a phenomenon called **antagonism**. In these cases, a harmful effect is reduced by certain combinations of potentially toxic chemicals (for example, 2 + 4 = 3). In mice exposed to nitrous oxide gas, mortality is greatly reduced when particulates are present. Scientists are uncertain of the reasons for this phenomenon.

Bioaccumulation and Biological Magnification

Two additional factors that profoundly influence toxicity are bioaccumulation and biological magnification. **Bioaccumulation** is the buildup of certain chemicals within tissues and organs of the body. It results from several factors: a resistance to breakdown, slow excretion rates, and selective absorption. Heavy metals, for example, cannot be broken down in the body, and they

Case Study 14.1

The Dangers of Asbestos

Asbestos is the generic name for several naturally occurring silicate mineral fibers. Asbestos is useful because of its resistance to heat, friction, and acid; its flexibility; and its great tensile strength.

In the United States and other industrial countries, asbestos was once added to cement to make it more weather resistant. Asbestos was also applied as a heat insulator on ceilings and pipes in factories, schools, and other buildings. In addition, asbestos was sprayed on walls and ceilings to make them more soundproof and fireproof. It was also used in the manufacture of brake pads, brake linings, hair driers, patching plaster, and a multitude of other products.

Asbestos fibers can be easily dislodged. Floating in the air, these fine particles may be inhaled into the lungs, where they are neither broken down nor expelled. Remaining in the lungs for life, they produce three disorders: pulmonary fibrosis, lung cancer, and mesothelioma.

Pulmonary fibrosis, or **asbestosis**, is a buildup of scar tissue in the lungs that occurs in people who inhale asbestos on the job or in buildings with exposed asbestos insulation. This disease makes breathing nearly impossible and takes 10 to 20 years to develop.

Exposures to asbestos at low levels, even for short periods, can cause **lung cancer**. The death rate from lung cancer in asbestos insulation workers in the United States is four times the expected rate. Interestingly, the incidence of lung cancer in asbestos workers who smoke is 92 times greater than in asbestos workers who don't, a striking example of synergism.

Asbestos is the only known cause of **mesothelioma**, a cancer that develops in the lining of the lungs. Highly malignant, this cancer spreads rapidly and kills victims within a year from the time of diagnosis. Although the incidence of all asbestos-related diseases is clearly related to dose, mesotheliomas have been observed in individuals with only brief exposure.

Scientists have long wondered how asbestos causes cancer. They have found measurable amounts of DNA in tissue fluids surrounding cells of the body. Research suggests that asbestos fibers may attach to this DNA, then pierce the cell membrane, carrying the DNA inside. What happens to the DNA inside the cell is still unknown. Several possibilities exist. One possibility is that the DNA disrupts or turns off the genes that control a cell's growth. With the control mechanism altered or completely shut down, the cell begins to duplicate wildly. Another possibility is that the DNA may carry cancer-causing genes into the cell. Once inside, the genes become activated, triggering the cell to divide. Still other research suggests that asbestos fibers inside cells may damage chromosomes during cell division. This, in turn, may cause harmful mutations.

An estimated 8 million to 11 million American

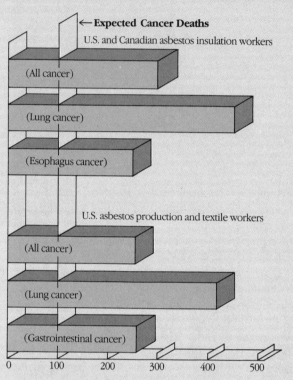

Incidence of cancer in asbestos workers in the United States and Canada. (Ratio of the number of observed to the number of expected deaths times 100.)

workers have been exposed to asbestos since World War II. Studies of these workers show that over one-third developed lung cancer, mesothelioma, or gastrointestinal cancer. The expected death rate in the population for these diseases is about 8%. Between 1990 and 2020, some health experts expect about 2 million people, mostly workers, to die from exposure to asbestos.

Based on these and other statistics, the use of asbestos in the United States for insulation, fireproofing, and decorative purposes was banned in 1974. In 1979, the EPA began to help states and local school districts identify crumbling asbestos from pipes and ceilings. In 1989, the EPA ordered bans on almost all other applications, which will go into effect by 1997 and will, with the previous ban, eliminate almost 95% of all asbestos use in the United States. To protect workers in the meantime, the Occupational Safety and Health Administration toughened its rules.

Companies that produce asbestos argue that cleanups and bans on asbestos will not protect public health. Workers have been the main victims of asbestos-related diseases. Furthermore, as noted earlier, workers who smoke are at a much higher risk.

bind strongly to proteins, reducing their excretion. Chlorinated hydrocarbons, such as DDT, are also resistant to breakdown, and because they are fat soluble, they remain in the body fat permanently.

Scallops and other mollusks that feed on material suspended in water selectively take up certain toxic elements from seawater, such as the heavy metals zinc, copper, cadmium, and chromium. The level of cadmium in scallops in polluted waters, for example, may be 2.3 million times that of seawater. While such concentrations do not always cause problems to the organisms, they can create troubles for those that eat them because of a phenomenon known as biological magnification.

As shown in Figure 14.3, the pesticide DDT in water is absorbed by **zooplankton**, single-celled (nonphotosynthetic) organisms that live in water. Small fish ingest DDT when they feed on zooplankton, and the persistent insecticide is concentrated in their body fat. DDT also accumulates in the fat of fish-eating birds. As illustrated, tissue concentrations increase substantially at each level of the food chain.

This increase in tissue concentrations of a toxin in successive trophic levels of a food chain is called **biological magnification**. Biological magnification occurs because DDT is a fat-soluble chemical that is stored in body fat, not readily broken down, and not excreted. The more fish an osprey eats, the higher its DDT levels become. The concentration of DDT, in fact, may be several million times greater in fish-eating birds than it is in water (Figure 14.3). For humans, magnification that occurs in the food chain may be as much as 75,000 to 150,000.

Biological magnification exposes organisms high on the food chain to potentially dangerous levels of persistent toxins. Synthetic chemicals like DDT, some lead and mercury compounds, and even some radioactive substances are all biomagnified. The presence of this phenomenon is important to consider when judging the risk that a chemical poses to humans and the many species that live among us.

14.2 Mutations, Cancer, and Birth Defects

Much of the concern about toxic chemicals today stems from their effects on the genetic material and two likely offshoots of these effects—cancer and birth defects.

Mutations Agents that cause mutations are called **mutagens**. In general, three types of genetic alteration are possible: (1) changes in the DNA itself, (2) alterations of the chromosome that are visible by microscope (deletion or rearrangement of parts of the chromosome), and (3) missing or extra chromosomes. The term *mutation* is used to describe all three.

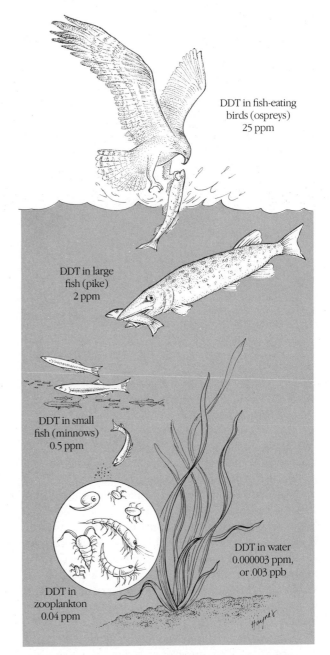

Figure 14.3 *Some toxic metals and fat-soluble organic compounds, such as the pesticide DDT, increase at higher levels in a food chain. This phenomenon is known as biological magnification and results from the fact that fat-soluble molecules tend to be stored in the fat of organisms and not broken down. As a result, they are passed with nearly 100% efficiency from one level of the food chain to the next.*

Mutations can be caused by chemical substances, such as benzene, or physical agents, such as ultraviolet radiation and high-energy radiation (Chapter Supplement 11.1). In humans, mutations can occur in normal body cells, or **somatic cells**, such as skin and bone. Such mutations occur quite frequently, but are usually repaired by cellular enzymes. If a mutation is not

Figure 14.4 *The incidence of Down syndrome caused by abnormal chromosome numbers is related to the mother's age. Note that the incidence of Down syndrome is about 5 in 1000 at age 35 but 40 in 1000 by age 45, about the time of menopause, when women stop producing ova.*

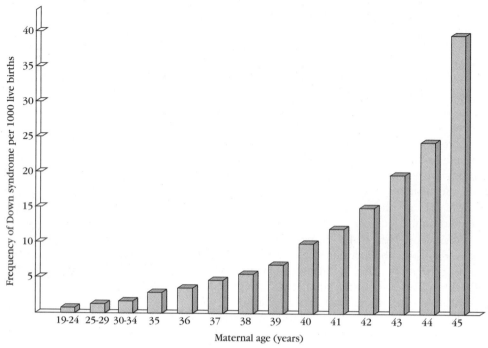

repaired, and it affects important genes, the cell may die or may become cancerous.

The reproductive cells, or **germ cells**, in the male and female gonads are also susceptible to mutagens. Unrepaired germ-cell mutations may be passed to offspring. If a genetically damaged ovum, for example, is fertilized by a normal sperm, the mutation is passed on to the offspring. The defective gene may prove lethal, killing the embryo, or it may manifest itself as a birth defect, a metabolic disease (a biochemical disorder), or childhood cancer. Some germ-cell mutations may not be expressed in the first generation but may appear in the second and third generations. This delayed effect makes it difficult for scientists to pinpoint the causes of some diseases.

Genetic mutations are present in about 2 of every 100 newborns. The causes of mutations in humans are not well understood. Abnormal chromosome numbers, responsible for diseases such as Down's syndrome, are related to maternal age (Figure 14.4). Broken and rearranged chromosomes are also related to maternal age. As women enter their 30s, their chances of having a baby with an abnormal number of chromosomes increase; after age 40, the chances skyrocket. Geneticists once hypothesized that the older a woman is, the greater the chance that she has been exposed to mutagens, hence the greater the chance that her child will have a mutation. New research suggests that the increased incidence of birth defects in babies of older women may result from the fact that older mothers are, for unknown reasons, more likely to carry a defective fetus to term than younger women.

Other diseases, associated with structural defects in the DNA molecule itself, seem to increase in incidence as the father ages and are unrelated to the mother's age (Figure 14.5). These defects may be caused by mutagens.

Cancer   Cancer annually kills 500,000 people in the United States. **Cancer** is an uncontrolled proliferation of cells that forms a mass, or **primary tumor**. Cells may break off from the tumor and travel in the blood and other body fluids. The spread of cancerous cells is called **metastasis**. In distant sites, the cancerous cells may form **secondary tumors**.

Every cancer starts when a single cell goes haywire, a process that occurs most often in tissues undergoing rapid cellular division, for example, the bone marrow, lungs, lining of the intestines, ovaries, testes, and skin. Nondividing cells, such as nerve cells and muscle cells, rarely become cancerous.

Despite years of intensive research, scientists remain uncertain about the causes of many types of cancer. Studies suggest that the formation of a cancerous tumor involves two separate steps: (1) conversion and (2) development and progression (Figure 14.6). **Conversion** occurs when the DNA of a normal cell is mutated by a chemical (benzene), physical (X ray), or biological agent (virus). Interestingly, 90% of all chemicals known to cause cancer also cause mutations in bacterial test systems.

Agents that trigger conversion are known as **cocarcinogens**. When a cocarcinogen causes a mutation in a gene that controls cell division, a cell may be partially or completely released from normal growth control. A

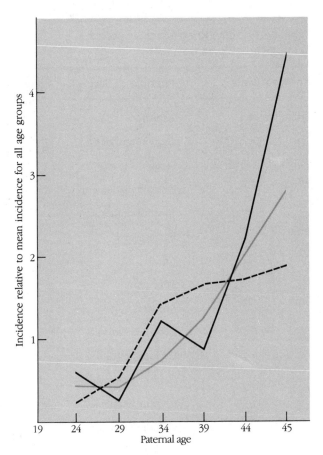

Figure 14.5 *The incidence of several diseases in newborns, caused by DNA damage, is related to the father's age. The dashed line indicates hemophilia; the green line indicates achondroplasia; and the solid black line indicates myositis ossificans. These are diseases of the blood, cartilage, and muscles, respectively.*

mutated cell that is fully released from growth controls may divide uncontrollably, forming a cancer. However, most cells are held in check by neighboring cells, which release chemical substances that prevent the converted cell from dividing.

Partially released cells often remain dormant for long periods. Eventually, they may be spurred on to divide by chemical substances called **promoters** or by errors in DNA replication that unleash cell division. This occurs during the **development** and **progression** phase. Once released, these cells begin to divide uncontrollably.

Chemicals that induce cancer by altering the DNA of cells are generally referred to as **DNA-reactive carcinogens.** A growing body of evidence, however, indicates that some carcinogens do not react with DNA. These substances are called **epigenetic carcinogens.** Some of them are thought to cause hormonal imbalances that lead to cellular proliferation. Others may alter the function of the immune system, resulting in a cancerous proliferation. Still others may cause chronic tissue injury. Asbestos fibers, for instance, may end up in cells where they slice dividing chromosomes.

Seemingly innocuous chemicals can also be converted into dangerous substances in the body. For example, nitrites are fairly benign chemicals that are converted to carcinogenic nitrosamines in liver cells and possibly others. Although the liver usually detoxifies chemical substances and protects us from harm, in this instance, it converts a relatively harmless substance into a carcinogen with potentially lethal consequences.

New studies indicate that emotions may also play an important role in the development of cancer and other

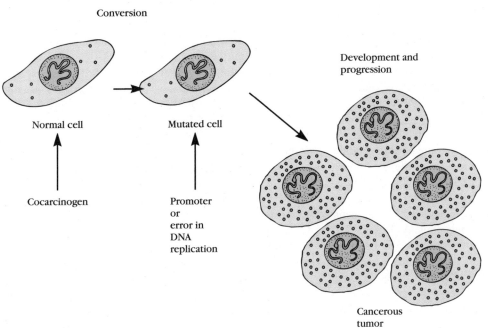

Figure 14.6 *The induction of cancer is thought to require two steps: conversion, a mutation that transforms a normal cell into a potentially cancerous one, and development and progression, the unleashing of the cell that permits it to divide uncontrollably.*

diseases, possibly by acting through the immune system. Researchers at Johns Hopkins University, for example, studied the incidence of cancer in medical students who took a personality test between 1948 and 1964. The research showed that students who suppressed emotions were 16 times more likely to develop cancer later in life than students who vented their emotions. More research is needed to determine if the cause-and-effect relationship between mental health and cancer is real.

According to two scientists from Oxford University, 8000 Americans die each year of cancer caused by environmental factors, such as air pollution. Another 8000 cancer deaths are attributed to food additives and industrial products, such as pesticides used around the house; 16,000 deaths result each year from occupational exposure to harmful substances. The researchers note, by comparison, that tobacco causes at least 142,000 lung cancer deaths each year.

(Point/Counterpoint 19.1 looks at the presumed epidemic of cancer in the United States and Viewpoint 14.1 discusses successes and failures of cancer research and treatment.)

Birth Defects   A **birth defect** is a physical (structural), biochemical, or functional abnormality. The most obvious defects are physical abnormalities, such as cleft palate, lack of limbs, or spina bifida (incomplete development of the spinal cord, often resulting in paralysis). According to some estimates, about 7% of all U.S. newborns have a birth defect. Others believe that the incidence of birth defects may actually be higher, about 10% to 12%. That is, 10 to 12 of every 100 babies may have a birth defect. The discrepancy occurs because many minor defects escape detection at birth, among them mental retardation and certain enzyme deficiencies.

Agents that cause birth defects are called **teratogens**; the study of birth defects is **teratology** (from *teratos*, the Greek word for "monster"). Teratogenic agents include drugs, physical agents such as radiation, or biological agents such as the rubella virus, which causes German measles (Table 14.1). No one knows for sure what percentage of birth defects is caused by chemicals in the environment.

· Embryonic development can be divided into three parts: (1) the period of early development immediately after fertilization, (2) the period when the organs are developing (**organogenesis**), and (3) the growth phase, the period during which organs have formed and the fetus primarily increases in size (Figure 14.7). Teratogens exert their effects during organogenesis. As Figure 14.7 shows, organs are most sensitive early in organogenesis, and it is during this period that the most noticeable effects occur. Thus, the effects of a teratogen are related to the time of exposure.

Teratogenic effects are also determined by the type of chemical involved. Certain chemicals affect certain

Table 14.1   Some Known and Suspected Teratogens in Humans

Known Agents	Possible or Suspected Agents
Progesterone	Aspirin
Thalidomide	Certain antibiotics
Rubella (German measles)	Insulin
	Antitubercular drugs
Alcohol	Antihistamines
Irradiation	Barbiturates
	Iron
	Tobacco
	Antacids
	Excess vitamins A and D
	Certain antitumor drugs
	Certain insecticides
	Certain fungicides
	Certain herbicides
	Dioxin
	Cortisone
	Lead

organs; for example, methyl mercury damages the developing brains of embryos. Other chemicals, such as ethyl alcohol, can affect several systems. Children born to alcoholic mothers exhibit numerous defects, including growth failure, facial disfigurement, heart defects, and skeletal defects. Like most toxins, teratogens usually exhibit a dose-response relationship: the greater the dose, the greater the effect.

Fetal development occurring after organogenesis may also be affected by physical and chemical agents in homes and workplaces. Some chemical toxins, for instance, stunt fetal growth and, in higher quantities, even kill fetuses, resulting in a **stillbirth** or **spontaneous abortion**.

Reproductive Toxicity   Reproduction is a complex process, involving many steps. Chemical and physical agents may interrupt any of these complex processes, interfering with reproduction. The field of study that examines the effects of physical and chemical agents on reproduction is called **reproductive toxicology**.

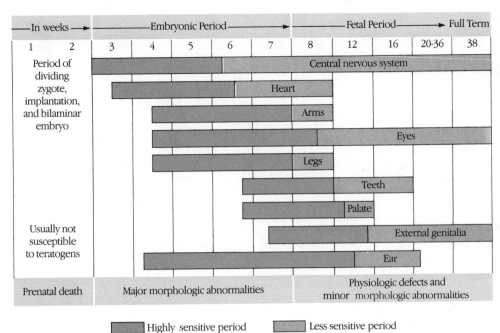

Figure 14.7 *Schematic representation of human development showing when some organ systems develop. Sensitive periods are early in development, indicated by the green color bar. Exposure to teratogens during these times will almost certainly cause birth defects. Exposure to potentially harmful chemicals after organogenesis may result in physiological defects, minor anatomical changes, or death if levels are sufficiently high.*

The effects of drugs and environmental chemicals on reproduction have become a major health concern in recent years. Studies have shown that male factory workers become sterile when exposed on the job to DBCP (1,2-dibromo-3-chloropropane), some permanently. Men who routinely handle various organic solvents often have abnormal sperm, unusually low sperm counts, and varying levels of infertility. A wide number of chemicals, such as borax, cadmium, diethylstilbestrol (DES), methyl mercury, and many cancer drugs are toxic to the reproductive systems of males and females.

Reproductive toxins may exert their effects long after exposure. Two researchers at Laval University in Quebec, for example, examined the records of 386 children who died of cancer before the age of five. Their study showed that at the time the children were conceived many of their fathers had been working in occupations that exposed them to high levels of hydrocarbons. Some were painters exposed to paint thinners; others were mechanics exposed to car exhaust. This study suggests that hydrocarbons had entered the bloodstream of the fathers at work and may have traveled to the testes, where they damaged the germ cell DNA. The resulting mutation was passed to the offspring.

In 1970, seven women (ages 15 to 22) were diagnosed with clear-cell adenocarcinoma of the vagina and cervix, a disease never seen before in women below the age of 30 and usually seen in women over 50. A study of these women showed a correlation between their cancer and maternal ingestion of the drug known as diethylstilbestrol (DES). DES was widely used from the mid-1940s to 1970 in the United States. It was administered to women who either had a history of miscarriages or

had begun to bleed during pregnancy. Since bleeding is an early symptom of miscarriage, DES was given as a preventive measure. (It is now known that DES cannot prevent miscarriage.) Years later, cervical and vaginal cancers and other abnormalities began to appear in the daughters of DES-treated women. Additional research has uncovered reproductive abnormalities in approximately one-fourth of male offspring exposed to DES through their mothers. Problems include small testes, cysts, and low ejaculate volume. No evidence of testicular tumors has been found.

14.3 Controlling Toxic Substances

The United States produces about 125 million metric tons of synthetic chemicals and creates 200 million metric tons of hazardous wastes each year. The need to control such toxic substances has grown dramatically; many laws have been passed to regulate them (Table 14.2).

Federal Control

In 1976, Congress passed the **Toxic Substances Control Act (TSCA)**. TSCA consists of three parts. The first is premanufacture notification, which requires all companies to inform the EPA 90 days before they import or manufacture a chemical substance not currently in commercial use. The agency then has 90 days to approve the chemical and place restrictions on its use if necessary. To do this, scientists at the EPA review existing toxicity data on the chemical or data on chemicals with a similar

Viewpoint 14.1

Are We Losing the War Against Cancer?

**John C. Bailar III
and Elaine M. Smith**
*John Bailar is a physician and
biostatistician at the School of
Public Health at McGill University.
Elaine Smith is a biostatistician at
the University of Iowa Medical
Center. This viewpoint is adapted
from an article published in the*
New England Journal of Medicine,
May 1986.

Since 1950, both private- and
government-sponsored cancer research has grown at a tremendous
rate. But what progress have we
made in the fight against cancer?
Have our efforts to find treatments
for this disease been successful?

In 1962, cancer was the recorded cause of death for 278,562
Americans. In 1986, 24 years later,
469,330 Americans died of cancer,
an increase of 68%. This increase
is somewhat deceptive. During that
period, the U.S. population was
growing. The relative proportion
of people in older-age categories
also increased. When these two
factors are considered, the real
growth in the cancer rate turns out
to be 10.1%.

Mortality data do not tell the
whole story. We might ask not
how many Americans die of cancer
but how many contract the disease? From 1973–1974 to 1985–

1986, the age-adjusted incidence
rate increased by 12.3%. The conclusion: Cancer is on the rise.

But, have we made inroads in
other areas, such as treatment? To
look at the overall effectiveness of
new cancer treatments, we might
focus on long-term survival rates.
Our data show that five-year survival rates for Caucasian patients
with all forms of cancer ran from
50% in 1975 to 51.3% in 1981.
Since then, there has been little
change.

We have witnessed clear progress against some uncommon
forms of cancer, especially those
that strike children and young
adults. Their mortality rates have
decreased sharply, but such improvements have not been seen in
older individuals. If we could wipe
out cancer entirely in the age
groups under 15, the overall death
rate from cancer would decline by
about 0.5%.

Our data show that overall cancer mortality rates have increased
slowly but steadily over several
decades. There is no evidence of a
recent downward trend. There has
been, in our view, little progress in
treating most cancer as reflected in
the survival-rate data. In this sense,
we are losing the war against cancer. Substantial increases in our
understanding of the nature and
properties of cancer have not led

to a corresponding reduction in the
incidence or mortality of this
disease.

These comments are in no way
an argument against the earliest
possible diagnosis and the best
possible treatment of cancer. The
problem, as we see it, is the lack of
any substantial recent improvements in treating the most common forms of cancer.

The main conclusion we draw is
that 40 years of intense effort focused largely on improving treatment is a qualified failure. The
results have not been what they
were intended or expected to be.
But, we think that there could be
much current value in an objective
review of the reasons for this failure. Why were hopes so high?
What went wrong? Can future efforts be built on more realistic expectations? And, why is cancer the
only major cause of death for
which age-adjusted mortality rates
are still increasing?

On the basis of past medical experience with infectious and other
nonmalignant diseases, we suspect
that the most promising route to
cut cancer rates is prevention. Reducing smoking, indoor air pollution, and workplace exposure and
other efforts could pay huge dividends in the long run. History suggests that savings in both lives and
dollars would be great.

---

structure. If the new chemical is believed to pose insignificant risk, it is approved. If it could be hazardous, the
agency usually asks the manufacturer to test its toxicity
and report back. At this time, many manufacturers drop
the candidate because of the cost involved in toxicity
testing.

The second part of TSCA required the EPA to examine chemicals that were in commercial use before the
law passed. Those the agency deemed potentially hazardous were required to undergo toxicity testing.

Finally, the act called for controls on chemicals that
Congress knew were hazardous. The most radical controls were placed on **polychlorinated biphenyls (PCBs)**,
an insulating fluid used in electrical transformers. PCBs

are stable in the environment because they resist biodegradation. They are also widely dispersed in the environment, are subject to biological magnification, and are
fairly toxic to laboratory animals. Because of these factors, in May 1979, Congress banned their manufacture
and distribution except in a few limited cases.

Market Incentives to Control Toxic Chemicals

In 1988, one of the most controversial environmental
laws in the United States went into effect. California's
Safe Drinking Water and Toxics Enforcement Act set in
motion a market strategy aimed at dramatically reducing
the exposure to toxic chemicals in foods and consumer

Table 14.2 Federal Laws and Agencies Regulating Toxic Chemicals

Statute	Year Enacted	Responsible Agency	Sources Covered
Toxic Substances Control Act	1976	EPA	All new chemicals (other than food additives, drugs, pesticides, alcohol, tobacco); existing chemical hazards not covered by other laws
Clean Air Act	1970, amended 1977, 1990	EPA	Hazardous air pollutants
Federal Water Pollution Control Act	1972, amended 1977, 1978, 1987	EPA	Toxic water pollutants
Safe Drinking Water Act	1974, amended 1977	EPA	Drinking water contaminants
Federal Insecticide, Fungicide, and Rodenticide Act	1948, amended 1972, 1973, 1988	EPA	Pesticides
Act of July 22, 1954 (codified as § 346(a) of the Food, Drug and Cosmetic Act)	1954, amended 1972	EPA	Tolerances for pesticide residues in food
Resource Conservation and Recovery Act	1976	EPA	Hazardous wastes
Marine Protection, Research and Sanctuaries Act	1972	EPA	Ocean dumping
Food, Drug and Cosmetic Act	1938	FDA	Basic coverage of food, drugs, and cosmetics
Food additives amendment	1958	FDA	Food additives
Color additive amendments	1960	FDA	Color additives
New drug amendments	1962	FDA	Drugs
New animal drug amendments	1968	FDA	Animal drugs and feed additives
Medical device amendments	1976	FDA	Medical devices
Wholesome Meat Act	1967	USDA	Food, feed, and color additives; pesticide residues in meat, poultry
Wholesome Poultry Products Act	1968		
Occupational Safety and Health Act	1970	OSHA	Workplace toxic chemicals
Federal Hazardous Substances Act	1966	CPSC	Household products
Consumer Product Safety Act	1972	CPSC	Dangerous consumer products
Poison Prevention Packaging Act	1970	CPSC	Packaging of dangerous children's products
Lead Based Paint Poison Prevention Act	1973, amended 1976	CPSC	Use of lead paint in federally assisted housing
Hazardous Materials Transportation Act	1970	DOT (Materials Transportation Bureau)	Transportation of toxic substances generally
Federal Railroad Safety Act	1970	DOT (Federal Railroad Administration)	Railroad safety
Ports and Waterways Safety Act	1972	DOT (Coast Guard)	Shipment of toxic materials by water
Dangerous Cargo Act	1952		

CPSC = Consumer Product Safety Commission
DOT = U.S. Department of Transportation
EPA = U.S. Environmental Protection Agency
FDA = Food and Drug Administration
OSHA = Occupational Safety and Health Administration
USDA = U.S. Department of Agriculture
Source: Council on Environmental Quality

Case Study 14.2

Waterbed Heaters and Power Lines: A Hazard to Our Health?

In 1979, two researchers from the University of Colorado reported a link between high-current electric power lines and the incidence of childhood leukemia. Their study suggested that extremely low-frequency (ELF) magnetic fields produced when electricity flows through wires may be the cause of the increased incidence of cancer in children living nearby. The researchers found that the death rate from cancer in children was twice what was expected in the general public.

ELF magnetic fields are virtually everywhere. Moreover, they easily penetrate building walls and the human body. Magnetic fields are also found around power stations, welding equipment, subways, and movie projectors.

In 1986, researchers from the University of North Carolina announced the results of a study that supports the Colorado research. The new study showed a fivefold increase in childhood cancer, particularly leukemia, in residents living near the highest magnetic fields, 7 to 15 meters (25 to 50 feet) from wires that carry electricity from power substations to neighborhood transformers. Adding to the concern, a researcher from Texas recently found that magnetic fields increase the growth rate of cancer cells. In addition, cancer cells exposed to magnetic fields are 60% to 70% more resistant to the body's naturally occurring killer cells.

In 1986, another group of scientists reported that magnetic fields from by electric heaters in waterbeds and electric blankets increased the likelihood of miscarriage. In the group that used electric heaters in their waterbeds, 61% of the miscarriages occurred from September to the end of January. By comparison, women using neither a waterbed heater nor an electric blanket had a 44% miscarriage rate during this same period.

Although researchers are not sure how the general public could be protected from magnetic fields created by power lines, many manufacturers are designing their beds with thick insulative coverings over the waterbed mattress so that no heaters are necessary.

ELF magnetic fields may also be a cause of birth defects in humans. Scientists know that magnetic fields affect fetal development in pigs, chickens, and rabbits.

The power industry is concerned about the effects of magnetic fields. Leonard Sagan, manager of the Radiation Sciences Program at the Electric Power Research Institute (EPRI) says that the real cause of the increased cancer may be something else. For instance, individuals living near magnetic fields may also be exposed to increased pollution from traffic. Sagan agrees that electric blankets and waterbeds deserve more attention, however. The EPRI is now spending over $2 million per year to study magnetic fields.

Many scientists and industry representatives are skeptical about the link between cancer and magnetic fields. Eleanor Adair, environmental physiologist and senior research scientist at Yale University, notes that of the more than 30 epidemiological studies, about half show a weak correlation between magnetic fields; the other half do not. She finds many serious problems with the design of virtually all epidemiological studies, further shedding doubt on their validity. In most studies, she notes, scientists did not measure the strength of the magnetic fields. Instead, they relied on measures of the proximity of homes to power lines or other sources and measured the size of the wires carrying electric current to homes. She also notes the failure of such studies to take into account other variables, such as smoking and exposure to pollution, that might be responsible for reported cancers. Other scientists, like David Carpenter from the University of Albany School of Public Health, disagree and feel that the studies probably underestimate the risk of cancer.

Given the scientific uncertainty, what should we do? To some, the scientific uncertainty suggests that we should not rush carelessly forward to regulate electricity. Nonetheless, those more convinced about the link between magnetic fields and cancer think that it is time to inform people of the wisdom of avoiding unnecessary exposure.

---

products. Created as a result of a citizen-sponsored initiative, Proposition 65, which was passed overwhelmingly by voters in 1986, this law directs the state to set acceptable levels for potential toxins in various consumer products and foods. The law requires that manufacturers who violate these standards print warnings on their products, noting that the amount of the chemical in the product exceeds the state's acceptable level. Manufacturers are not required to meet standards, only pub-

licize the fact on the packages. The proponents of this law believe that consumers will shun products that do not meet the standard, thus creating a market force that will encourage manufacturers to reduce levels of toxins in their products. California's regulators worked quickly. In a single year, they generated more standards than the EPA has managed to create in over a decade.

Numerous other market-based incentives for controlling toxic pollution are available. Two of the more prominent are **pollution charges**, taxes on toxic chemicals used or released by industries, and **tradable permits**, permits for release of certain levels of pollution that companies can buy and sell. (Other measures are discussed in Chapters 15 through 19 and Chapter 22.)

Market-based policies are designed to produce monetary incentives for reducing pollution and to reduce regulatory burden. They also provide more freedom of choice for businesses to select methods that suit their unique situation. Although market-based actions are gaining popularity among businesses, government regulators, and environmentalists, they still fall far short of reaching their full potential.

The Multimedia Approach to Pollution Control

The EPA typically regulates toxic chemicals by controlling emissions into various media, such as air, water, and soil. Each medium is monitored and regulated by a separate branch of the agency. However, in the past, EPA officials found that companies often disposed of wastes in the least-regulated medium. Wastes from a refinery that were subject to hazardous waste rules, for instance, might end up in the wastewater, which is less strictly regulated. EPA inspectors have traditionally been trained to check for compliance with one medium. As long as companies were obeying the necessary regulations, inspectors were satisfied.

In 1991, William Reilly, then administrator of the EPA, called for a **multimedia approach** to stop this environmentally costly toxic shell game. In the regional offices of the EPA, Reilly set up programs to train inspectors in several media. Their job is to determine if companies are in compliance with all pertinent environmental regulations.

The EPA hopes that multimedia inspections will help the agency identify cross-media contamination and find companies that are illegally or improperly disposing of wastes. Such a tightening in the regulatory approach could promote pollution prevention, say EPA officials, because increased agency awareness of a facility's activities decreases a company's ability to hide pollution and avoid regulation by transferring pollution from one medium to another.

14.4 Determining the Risks

Ralph Waldo Emerson wrote, "As soon as there is life, there is danger." Every day of our lives we face many dangers, some obvious, some hidden. The study of the risks of modern technological societies has become an important policy-making tool, vital to efforts to build a sustainable society in the most cost-effective way possible. This section looks at risks and how they are assessed.

Risks and Hazards: Overlapping Boundaries

Two types of hazard are broadly defined by risk assessors: anthropogenic and natural. **Anthropogenic hazards** are those created by human beings. **Natural hazards** include events such as tornadoes, hurricanes, floods, droughts, volcanoes, and landslides. As you learned in Chapter 5, natural hazards often have a human component. For instance, the damage resulting from floods is in large part the result of our living along floodplains, channelizing streambeds, or changing the vegetative cover. Similarly, earthquake damage can be greatly magnified by overpopulation and bad building practices.

Hazards befalling society exact an enormous price: human lives, human health, economic ruin, social disruption, mental illness, environmental destruction, and animal and plant extinction. Measuring the damage is never easy.

Risk Assessment

Since the mid-1970s, a new and rather imprecise science, called **risk assessment**, has evolved to help us understand and quantify risks posed by technology, life-styles, and personal habits, such as smoking, drinking, and diet.

Risk assessment involves two interlocked steps: hazard identification and estimation of risk. **Hazard identification** involves steps to identify potential and real dangers. **Estimation of risk** generally involves two processes (Figure 14.8). The first is determining the probability of an event or occurrence. This process answers the question "How likely is the event?" The second stage is determining the severity of an event, answering the question "How much damage is caused?" For risk assessment of a toxic chemical, the estimation of risk requires an estimate of the number of people or other organisms exposed, levels of exposure, duration of exposure, and other complicating factors, such as age, sex, health status, personal habits, and chemical interactions.

Estimating risk is fraught with uncertainty, especially in the area of toxic substances because human knowledge of toxicology is far from complete for several reasons. One of the most important is that it is neither practical nor ethical to test toxic chemicals on human beings. As a result, toxicologists must rely on tests on laboratory animals, such as rats, mice, and rabbits, to estimate human toxicity. Unfortunately, the results of experiments on laboratory animals cannot always be extrapolated to humans. As my graduate adviser once reminded me, "Contrary to popular belief, the human is not a large rat." Lab animals often react differently to chemicals than humans do; they may be able to break them down better, or they may not be able to break them down as effectively. Physiological differences between humans and lab animals, therefore, make it difficult to predict if a chemical that's harmful to an animal will be injurious to humans.

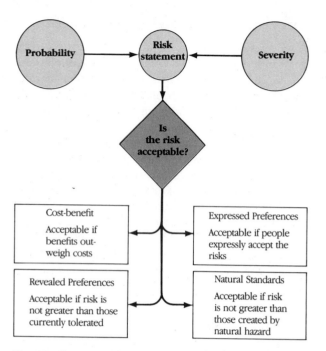

Figure 14.8 *Determining the acceptability of a risk. Cost-benefit analysis is the most common method of determining risk acceptability, but three other methods can also be used, as shown here.*

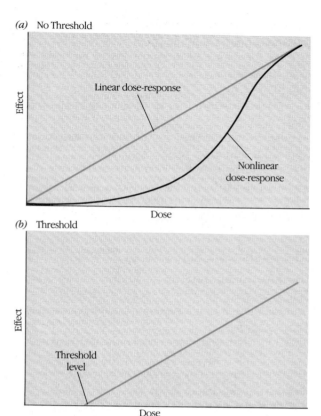

Figure 14.9 (a) *A hypothetical dose-response curve indicating the absence of a threshold level, a level below which no effect occurs. Note that some responses are linear. For every increase in the dose, there is a corresponding increase in the response. In others, the response is nonlinear. That is, there is not a 1 to 1 correlation between dose and response, although the response increases with increasing dose.* (b) *A hypothetical dose-response curve showing a threshold level.*

Ignorance of toxic effects also stems from the fact that humans are often exposed to many potentially harmful chemicals and may be exposed over long periods. For practical reasons, most toxicity tests are performed on one substance at a time. Because of chemical interactions described earlier, extrapolating the results from these tests to the real world can be misleading.

Another problem is that most animal tests of toxicity, especially those for cancer, are performed at high-level exposures. It is assumed that if a chemical is harmful at high levels, it will also be harmful at lower levels. However, the fact that a large dose of a chemical induces cancer in a lab animal does not necessarily mean that the chemical will cause cancer at low doses to which humans are typically exposed. This is because most, if not all, toxins have a **threshold level**, that is, a level below which no effects occur (Figure 14.9). Thus, extremely low levels of certain chemicals may be completely safe. The reason for this may be that at low levels protective mechanisms in the body can cope with the chemical, either inactivating it or excreting it fast enough to prevent damage. Tissues may also be able to repair damage at low levels. At levels above the threshold, these mechanisms may not be able to keep pace and noticeable effects may appear.

Although thresholds may exist for most toxins, there are some exceptions, among them asbestos (Case Study 14.1). Radiation is also thought to have no threshold (Chapter Supplement 11.1). Many scientists believe

that even the smallest exposure has an effect and that damage caused by repeated low-level exposures accumulates over time. However, controversial new studies in humans challenge the no-threshold view for radiation.

If scientists can't make accurate extrapolations from high doses to low doses, why do they perform such experiments? The primary reason is economics. Researchers use high doses to speed up their experiments. As a general rule, the entire process from exposure to manifestation takes about one-eighth of the life span of an animal. In humans, the time required to develop a noticeable cancer is 5 to 30 years after exposure. Any measure that accelerates the process, such as a high dose, helps cut costs. To test for low-level effects, scientists would need very large numbers of experimental animals to generate statistically valid results. High-dose studies, therefore, also reduce the number of lab animals needed, cutting time and costs—a significant factor, since cancer studies can cost $500,000 to $1 million per chemical.

Risk Management: Decisions About Risk Acceptability

Risk assessment is ultimately designed to help society manage its affairs. No matter what we do—whether it is screwing in a light bulb or flying cross-country in a jet—we put ourselves and possibly our environment at risk. Nothing is safe, that is, entirely free from harm. The science of risk assessment recognizes that human life is haunted by hazards. Therefore, rather than talking in terms of safety, which is absolute, the risk assessor speaks in terms of risk, which is relative. Activities that we commonly consider safe are referred to as *low-risk functions*. *Unsafe* activities are better labeled *high-risk functions*.

Knowing the relative risk of a technology is one thing. Knowing whether the risk is acceptable to the general public is another story. **Risk acceptability**, in fact, is one of the trickiest issues facing modern society. Because we are fickle, what appears safe one day becomes suspect the next after a widely publicized accident. Irrational fears crop up and frighten us away from relatively low-risk activities.

The acceptability of risks is also influenced by **perceived benefit**—how much benefit people think they will get from something. In general, the higher the perceived benefit, the greater the risk acceptability (Figure 14.10). As an example, the risks of a new steel mill might be acceptable to a community with high unemployment. Automobile travel provides the most telling example of the way in which perceived benefits affect our decisions. The risk of dying in an automobile accident in the United States is 1 in 5000 in any given year. Over your lifetime, the risk of dying in a car accident is far higher if you don't wear seat belts, about 2 in 100, than if you do, 1 in 100. Meanwhile, substances believed to be far less hazardous than driving are banned from public use because their benefit is not so highly valued.

Perceived harm, the damage people think will occur, also heavily influences risk acceptability. In general, the more harmful a technology and its by-product are perceived to be, the less acceptable they are to society.

Decisions, Decisions

Decisions on modern sources of risk—technologies, personal habits, and pollution—are becoming more and more commonplace. Nowhere was this more evident than in Tacoma, Washington, a city south of Seattle. A copper-smelting plant owned by Asarco annually contributed $20 million to $30 million to the local economy. However, it daily spewed out 765 kilograms (1700 pounds) of arsenic into the air, 23% of the total arsenic emissions in the United States. Arsenic has been linked with lung cancer, skin cancer, and neurological disorders and was found in alarmingly high levels in the blood of

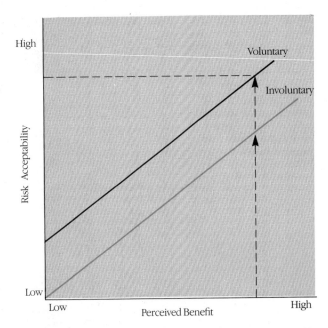

Figure 14.10 *The acceptability of a risk increases as the perceived benefit rises. Risks that people agree to—voluntary risks—are generally more acceptable than risks imposed without consent.*

smelter workers and their children at a nearby school. In 1983, residents nearest the smelter had a lifetime cancer risk of 9 in 100. Even if the best available control technology were instituted, experts predicted that the cancer risk would be about 2 in 100. The company argued that reducing arsenic emissions would force it to close the factory, costing the community millions of dollars in tax revenue and income for its 1300 employees.

In this landmark case, William Ruckelshaus, then head of the EPA, proposed that the residents be brought into risk-benefit analysis (Figure 14.11). This action angered many critics, who accused the agency of asking residents to choose between clean air and jobs. The EPA held extensive hearings in Tacoma to outline the risks. The company hired a public relations firm to outline the benefits of the smelter. By the end of the hearings, dozens of residents were wearing buttons that expressed their sentiment. The buttons said "Both." During the public clamor over the Tacoma smelter, Asarco decided to close the plant because of the declining metals market, which made the business uneconomical.

The Tacoma experiment points out the trade-offs that must sometimes be made when environmental and health risks are weighed against economic risk and employment. Clearly, the choices are not easy. Uncertainty about the effects of arsenic levels made the situation more confusing. One of the most important benefits of a sustainable strategy, including efforts such as energy efficiency, pollution prevention, recycling, and renewable resources, is that it offers many ways of

Figure 14.11 *Asarco parents at a town meeting in Tacoma, Washington. The sign reads "Don't risk our children." Faced with a dilemma of jobs or a clean environment, the people responded that they wanted both. As pointed out in the text, applying sustainable strategies to many problems can make this vision a reality.*

achieving both a clean, healthy environment and economic opportunities. It makes the environment and economics complementary, not antagonistic, forces. But, for now, many decisions must be made about risk acceptability.

Figure 14.8 shows that decisions can be based on one of four different strategies. The most common decision-making tool is the **cost-benefit technique**, the method left to the people of Tacoma. In this technique, one weighs costs against benefits. Although that may sound simple, it isn't. One reason is that the benefits are generally measured easily: financial gain, business opportunities, jobs, and other tangible items. Many of the costs, however, are less tangible. External costs are among the most difficult to quantify (Chapter 21). In addition, it is difficult to assign a monetary value to human health, environmental damage, and lost species. Moreover, the costs may be long term and thus are borne by future generations.

Another problem with cost-benefit analysis is that the benefits, most importantly financial gain, often accrue to a select few who have inordinate power to influence the political system and sway people's opinions through sometimes-less-than-honest media campaigns. American companies that design and build hazardous waste incinerators, for instance, spend extraordinary sums of money trying to convince people in rural areas of the benefits of siting an incinerator near them.

In sum, cost-benefit analysis suffers because many costs are poorly documented, spread out, and unquantifiable, whereas the benefits are often obvious and readily quantifiable. Recent efforts by economists to assign a dollar value to environmental and health costs may help improve the process. In addition, the efforts

of scientists to measure the impacts of technologies and their by-products on wildlife, the environment, recreation, health, and society in general may also help. Because of the lack of information on environmental and health costs, authors of *An Environmental Agenda for the Future* (a list of recommendations made by ten leading environmental groups) suggest that society should adhere to a better-safe-than-sorry policy even if predictable costs purchase benefits that cannot be calculated with certainty. In the long run, we may be better off.

Actual Versus Perceived Risk

The main purpose of risk assessment is to help policymakers formulate laws and regulations that protect human health, the environment, and other organisms. Ideally, good lawmaking requires that the **actual risk**, or the amount of risk a hazard really poses, be equal to the **perceived risk**, the risk perceived by the public. When actual risk and perceived risk are equal, public policy yields cost-effective protection (Figure 14.12).

When the perceived risk is much larger than the actual risk, costly overprotection occurs. In contrast, when the perceived risk is much smaller than the actual risk, underprotection occurs. It, too, can be costly to present and future generations.

The Final Standard: Ethics

Ultimately, environmental decisions are based on the values we hold or, simply, what we view as right and wrong—our ethics. Values that affect decisions come

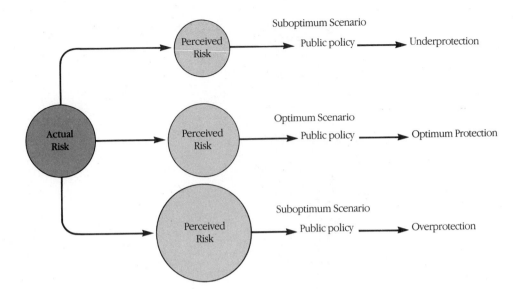

Figure 14.12 *Matching the actual risk and the risk that a society perceives is essential to the formulation of good public policy. But, perceived risk and actual risk do not always match, as shown. Policy decisions may result in overprotection or underprotection when the actual and perceived risk do not match. Optimal protection occurs when actual and perceived risk are equal.*

from parents, relatives, friends, enemies, teachers, religious leaders, and politicians. These values shift over time, sometimes subtly, sometimes dramatically. They often change as we age and as our priorities shift. Although our ethics are often never explicitly stated, they play an important role in our lives. They determine how we think, how we vote, how we treat one another, how we carry on business affairs, how we treat the Earth, and how we treat future generations.

Prioritizing Values Because values influence the way we think in profound ways, they play an important role in decisions regarding risk acceptability. Because benefits and costs are incurred in virtually all of our decisions, even seemingly sustainable solutions, we must always balance costs and benefits. This requires us to prioritize our values. If a new coal mine were to be built near your community, for instance, it would create clear benefits, such as more jobs and a stronger economy. However, certain costs, such as air and water pollution, might be incurred. The decision to open the mine would be influenced by the priority of values.

Prioritizing values requires us to ask what we value the most. What is more important to us in environmental decision making? Economics? Health? Wildlife? Future generations? A sustainable human presence? For example, a new reservoir will bring much-needed water to an area, allowing it to grow and prosper, but it could destroy valuable wildlife habitat and recreation areas. How do we choose? Do we save recreation areas and wildlife habitat and find other solutions to the water shortage, or do we dam the river, destroying the wildlife and the recreation area?

One of the most important values is sustainability. If an action is not sustainable in the long run, that is, if it destroys the life-support systems of the planet, it should be viewed skeptically.

Space-Time Values Further insight into our thinking comes from looking at **space-time values**. These are simply the concerns we have for other people and other living organisms in time (the present and the future) and space (you, your family, community, state, nation, and world). As the hypothetical scatter diagram in Figure 14.13 shows, individual interest can be identified by a single point that denotes one's space and time concerns. Most people's values lie toward the lower end of the scales, tending toward self-interest and immediate concerns. Some people call this selfishness, but it can also be considered a natural biological tendency to be concerned with the self. Among animals, awareness of the needs of others is a feature only of social creatures, like monkeys and lions; however, concern for the uppermost end of the space-time graph is a distinguishing feature of the human animal. The unique human ability to ponder the consequences of actions is fortunate because humans have reached a position of unprecedented power as molders of the biosphere. Our power to change the world to our liking has never been greater; nor has our power to destroy ever attained such heights.

Sound decision making in a sustainable society requires that we know where our priorities lie on the scatter diagram. Three important space-time questions require answers: (1) Is our decision based primarily on self-interest? In other words, are we looking at the issue solely in terms of how we might benefit or be harmed? (2) Is our decision based primarily on concern for others who are alive now who might benefit or be harmed? And, how far does our concern go? Are we concerned with the well-being of stockholders or citizens of the community, state, nation, or world? (In some cases, local actions can have global impact.) (3) Finally, is our decision based on what is good for future generations and other species? Will our decision benefit or harm those who follow? Will it promote sustainability?

Figure 14.13 *Various people's spatial and temporal interests are indicated by the points on the graph. Most individuals tend toward the lower end of the scales, being concerned primarily with self and the present.*

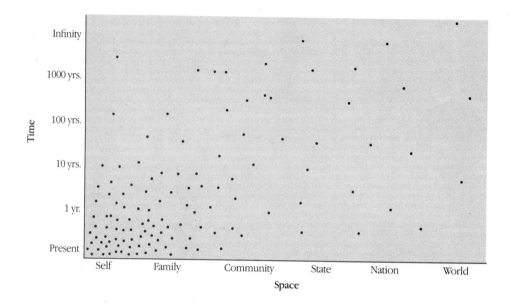

Building a Sustainable Future

The Toxic Substances Control Act and risk assessment are attempts to control toxic substances and protect humans and the many species that make the Earth their home. What we decide today will, in many ways, lay the foundation for the future. If we are reckless, future generations and other species will pay the price. If we minimize the use of toxic substances and use them wisely, future generations will probably be better off.

Slowly, modern society is learning what the American Indians have known for centuries. Unborn generations have a claim to the land, air, and water equal to our own. The need for new, broader ethics that protect the future can no longer be ignored. The need to manifest those ethics through prudent actions is a prerequisite for a sustainable society.

Chance fights ever on the side of the prudent.

—*Euripides*

Critical Thinking Exercise Solution

As a systems thinker with a strong concern for the environment, you would probably begin by investigating to see if adjustments could be made to the existing production system in order to reduce hazardous emissions. In other words, by modifying processes in your factory (which would cost very little compared to their returns), finding substitutes for toxic chemicals, recycling wastes, and a number of other cost-effective measures, you might find that you can comply with

the emissions standards and save your company huge sums of money. All it takes is a look at the bigger picture and a little digging to find some options. This is a good reminder to beware of conventional strategies, which often overlook cost-effective measures that could help save money and build a sustainable society.

Summary

14.1 Principles of Toxicology

■   **Toxins** are chemicals that cause any of a wide number of adverse effects in organisms. **Toxicology** is the study of these effects. People are exposed to toxins at home, at work, and out-of-doors.

■   Toxins produce a wide variety of effects. Some stimulate immediate responses. Immediate effects, which often disappear shortly after the exposure ends, are generally caused by fairly high concentrations of chemicals resulting from short-term exposures.

■   Other toxins produce delayed effects, such as cancer or birth defects, which may occur months to years after exposure and usually persist for years. Delayed effects are often the result of low-level exposure over long periods, or chronic exposures. Nonetheless, short-term exposures may also have delayed effects.

■   Some toxic effects are reversible; others cause irreversible damage.

■   Another important distinction regarding toxic chemicals is their site of action. Some exert local effects, most often at the site of contact. Other toxic substances affect entire organs or organ systems. These effects are described as **systemic** and are most common.

■   Toxic substances exert their effects at the cellular level by: (1) Altering enzyme structure and function;

(2) binding to important biological molecules, thus upsetting important physiological processes; (3) causing the release of naturally occurring substances that alter the function of cells and organs; and (4) interacting with the genetic material of cells, causing mutations.

■ Numerous factors influence the effects of a given chemical, including: (1) dose, (2) duration of exposure, (3) biological reactivity of the chemical in question, (4) route of exposure, (5) age, and (6) chemical interactions.

■ Some chemical substances produce an **additive response**—that is, an effect that is the sum of the individual responses.

■ Others may produce a **synergistic response**—one that is greater than the sum of the two individual ones.

■ Some chemicals with no toxic effect, when combined with another, make the latter even more toxic. This is called **potentiation**.

■ Certain chemicals can negate each other's effects, a process called **antagonism**.

■ Toxicity is also profoundly influenced by **bioaccumulation**, the ability of an organism to accumulate certain chemicals selectively within its body in certain tissues.

■ **Biological magnification**, the buildup of chemicals within food chains, also affects toxicity. Biological magnification occurs with chemicals that are concentrated in tissues (such as fat), poorly excreted, and are resistant to chemical breakdown.

■ **Mutations** are structural changes in the genetic material of cells caused by chemical, physical, or biological agents. If not repaired, mutations in **somatic cells** may lead to cancer.

■ Mutations in **germ cells** may be passed on to offspring, leading to birth defects, stillbirth, and cancer.

14.2 Mutations, Cancer, and Birth Defects

■ **Cancer** is an uncontrolled proliferation of cells. The formation of a cancerous tumor involves two separate steps: (1) **conversion** and (2) **development** and **progression**.

■ **Conversion** occurs when the DNA of a cell is mutated by a **mutagen**. These agents are known as **cocarcinogens**. When a cocarcinogen causes a mutation in a gene that controls cell division, a cell may be partially or completely released from normal growth control. Most cells are held in check by chemical substances released by neighboring cells.

■ Partially released cells often remain dormant for long periods. Eventually, they may be spurred on to divide uncontrollably by chemical substances called **promoters** or by errors in DNA replication that unleash cell division. This occurs during the development and progression phase.

■ **Birth defects** are structural and functional defects present in 7% to 12% of all newborn American children. The study of birth defects is known as **teratology**. Agents that cause them are called **teratogens**. These agents exert their effect during **organogenesis**.

■ In humans, few teratogens are known, although many substances are suspected.

■ **Reproductive toxicity** is the study of the toxic effects of chemical and physical agents during the reproductive cycle. Numerous chemical agents impair human reproduction.

14.3 Controlling Toxic Substances

■ The **Toxic Substances Control Act** was passed by Congress in 1976.

■ TSCA required premanufacture notification to the EPA of all new chemicals to be produced or imported, called for testing of existing chemicals suspected to be harmful, and established specific controls on several existing chemicals.

■ California passed a powerful new law in 1988 that required the state to establish safety levels for potentially toxic chemicals and required manufacturers to print toxin levels on food containers if they violate state standards, thus creating a market force to control toxins in consumer products.

■ Numerous other market-based policies can stimulate measures that reduce production and use of toxic chemicals, among them pollution charges and tradable permits.

14.4 Determining the Risks

■ **Risk assessment** is a science dedicated to understanding and estimating risk from various hazards.

■ Risk assessors first identify actual and potential hazards, then determine the probability, or likelihood, and severity of the hazard.

■ Once these are determined, a statement regarding risk can be made. It is then up to risk managers, usually public officials, to determine how best to deal with the risk.

■ **Risk acceptability** is determined by many factors; the most important is the **perceived benefit**, the benefit people think they will gain, and the **perceived harm**, the harm they expect to suffer.

■ Risk assessment is designed to manage risks in the most cost-effective manner. To do so, the **perceived risk**, the amount of risk people think is posed, must be equal or very close to the **actual risk**. The actual risk may be difficult to determine, especially in the case of new technologies with which society has had little experience.

■ All decision making regarding risk entails ethical considerations. Space and time are two important components of ethics. Most people tend to be concerned with the immediate future and self-interests. The interests of future generations, therefore, are often neglected.

■   Responsible sustainable decision making requires a broader ethic with more expansive space-time values.

Discussion and Critical Thinking Questions

1.   Define the terms *toxin, carcinogen, teratogen,* and *mutagen.*

2.   Describe the different effects toxins may have.

3.   What is cancer? Discuss how it forms and how it spreads.

4.   List some of the possible consequences of somatic and germ-cell mutations in humans.

5.   What is teratology? Do teratogenic chemicals always create birth defects when prescribed during pregnancy? Why or why not?

6.   List the factors that influence the toxicity of a chemical in a given individual.

7.   Define the terms *synergism, antagonism,* and *potentiation.* What is the difference between synergism and an additive effect?

8.   Define the terms *bioconcentration* and *biological magnification.* Based on your knowledge gained in this chapter, what factor(s) can be used to predict whether a chemical will be biologically magnified?

9.   What is meant by a threshold level? Explain the reasons why threshold levels exist for many toxins.

10.   Describe the major provisions of the Toxic Substances Control Act.

11.   What are the two major types of risk? Give examples.

12.   Describe the steps involved in determining the level of risk posed by technology.

13.   What factors determine whether a risk is acceptable to a population? What is the difference between voluntary and involuntary risks? What is the difference between actual and perceived risks?

14.   Many more people die in Montana and Wyoming from falls while hiking than are killed by grizzly bears. Why, then, are people so concerned about being killed by a bear when their chances of being killed in a fall are much greater?

15.   What are space-time values? In general, where does your concern lie in space and time?

16.   Some people think that because a substance is a cocarcinogen, they needn't really worry about it. Use your critical thinking skills to evaluate this position.

Suggested Readings

Amdur, M. O., Doull, J., and Klaassen, C. D., eds. (1991). *Casarett and Doull's Toxicology: The Basic Science of Poisons* (4th ed.). New York: Pergamon. Superb reference.

Chiras, D. (1982). Risk and Risk Assessment in Environmental Education. *Amer. Biol. Teacher* 44 (4): 460–65. A more technical presentation of risk and risk assessment.

———. (1992). *Lessons from Nature: Learning to Live Sustainably on the Earth.* Washington, D.C.: Island Press. Describes several market-based strategies to reduce pollution.

———. (1993). *Biology: The Web of Life.* St. Paul: West. Contains an important point/counterpoint on the health effects of magnetic fields.

Goldbaum, E. (1987). Can Cell Cultures Predict Toxicity? *Industrial Chemist* January: 34–37. Interesting look at an alternative way to test toxicity.

Kamarin, M. A. (1988). *Toxicology: A Primer on Toxicology Principles and Applications.* Chelsea, MI: Lewis Publishers. Good introduction to the subject.

Klaassen, C. D. and Eaton, D. L. (1991). Principles of Toxicology. In *Casarett and Doull's Toxicology: The Basic Science of Poisons.* Amdur, M. O., Doull, J., and Klaassen, C. D., eds. (4th ed.) New York: Pergamon. Detailed coverage of many important concepts.

Manson, J. M. and Weisburger, L. D. (1991). Teratogens. In *Casarett and Doull's Toxicology: The Basic Science of Poisons.* Amdur, M. O., Doull, J., and Klaassen, C. D., eds. (4th ed). New York: Pergamon. Detailed coverage of teratology.

Mausner, J. S. and Kramer, S. (1985). *Epidemiology: An Introductory Text.* Philadelphia: Saunders. Excellent reference.

Mossman et al. (1990). Asbestos: Scientific Developments and Implications for Public Policy. *Science* 247(January 19): 294–301. Excellent review.

Postel, S. (1988). Controlling Toxic Chemicals. In *State of the World 1989.* Starke, L., ed. New York: Norton. Excellent overview of toxic chemicals and their control.

Stavins, R. N. and Whitehead, B. W. (1992). Dealing with Pollution. Market-Based Incentives for Environmental Protection. *Environment* 34(7): 7–11, 29–42. Examines several different market-based policies.

Wilson, R. and Crouch, E. A. C. (1987). Risk Assessment and Comparisons: An Introduction. *Science* 236 (April 17): 267–70. Excellent introduction.

Global Lead Pollution

Lead is one of the most useful metals in modern industrial societies. Used by humankind for over 3000 years, lead is found in ceramic glazes, batteries, fishing sinkers, solder, and pipe. In gasoline, lead has been added to help reduce engine knocking.

Despite its value, lead is also a highly toxic poison. It enters the body primarily through inhalation and ingestion. Although it affects many organs, lead has a special affinity for bone and brain tissue. High-level exposure in certain factory workers has been linked to neurological symptoms: fatigue, headache, muscular tremor, clumsiness, and loss of memory. If exposure is discontinued, patients may slowly recover, but residual damage, such as epilepsy, idiocy, and hydrocephalus (fluid accumulation in the brain), often results. Continued high exposure may lead to convulsions, coma, and death. Interestingly, some scientists believe that lead drinking vessels and lead pipes in water systems may have caused a decline in birth rates and increased psychosis in ancient Rome's ruling class, contributing to the fall of the Roman Empire.

Today, because of better controls on lead in the workplace and in commercial products, acute lead poisoning is rare in most countries. Nonetheless, many people throughout the world are regularly exposed to low levels of lead with potentially serious consequences. Italian researcher Sergio Piomelli and his colleagues estimate that blood levels in humans before the advent of lead pollution were about 100 times lower than the normal range found today in people living in industrial societies, where the air is polluted by automobiles, power plants, and smelters and whose food is contaminated by lead solder and atmospheric fallout. However, even residents of Nepal have levels ten times higher than those estimated to be present before the widespread use of lead. Where does lead come from?

Sources of Lead

Lead is found in our food, water, air, and soils. Thus, no one is free from this potentially toxic metal. For most of us, food tops the list. Until quite recently, some of the lead in our food came from lead arsenate pesticides, which were banned in the United States in 1991. Lead in food also comes from emissions from automobiles, power plants, and smelters. Lead emitted by these various sources is frequently deposited in the soil, where it can be taken up by food crops. In recent years, the recycling of used motor oil has resulted in widespread emission of lead. According to the Natural Resources Defense Council, 85% of used motor oil is burned as fuel in unregulated boilers. This may represent the largest source of lead in our air. Until 1991, about half the lead in the human diet came from the solder in cans.

Although food is a major source of lead, most of the concern for lead exposure in humans has focused on automobile exhausts. In one study, the EPA estimated that 88% of the lead in the air we breathe comes from automobile exhaust, except around lead smelters and steel factories, which release large quantities of this harmful metal into the atmosphere. Regulations to eliminate lead from gasoline have markedly decreased concentrations in and around our cities.

Today, much of the focus on lead centers on drinking water. In 1986, a major study by the EPA revealed that lead levels in drinking water in many cities exceeded federal standards, potentially threatening the health of millions of Americans. Lead in drinking water comes from lead-based solder and from lead pipes often found in older homes.

Lead victims are primarily children and, among them, mostly children of poor black families, who tend to live in regions where pollution from automobile

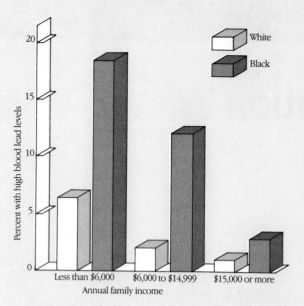

Figure S14.1 *High lead levels in U.S. children from six months to five years old, according to parents' income and race (1976–1980).*

exhaust is prevalent (Figure S14.1). Children may also eat dirt contaminated with lead from vehicles or may inhale lead in the atmosphere near highways. Even now that lead levels in automobile exhaust have fallen, children living in old, neglected buildings often ingest flakes of lead-based paint, which was applied before a ban was enacted in the 1940s.

Children are more susceptible to lead in part because they absorb more in the intestines. While only about 8% to 10% of the lead ingested by adults is absorbed by the intestines, the absorption rate in children may be as high as 40%. In addition, children are more sensitive to lead than adults. The developing brain seems to be the most sensitive organ. Furthermore, the toxicity of lead is increased in malnourished and iron-deficient children, who often come from poor urban families.

Effects of High-Level Exposure

As noted earlier, high-level lead exposure can cause numerous neurological disorders, including fatigue, headache, and muscular tremor. It may also reduce one's appetite, cause clumsiness, and result in a loss of memory. These symptoms are a result of damage caused by inorganic lead in the brain and spinal cord. Organic lead, alkyl lead in gasoline, for example, may also cause a number of psychological disorders, including hallucinations, delusions, and overexcitement, and may lead to death.

Lead exposure can also affect the nerves that arise from the brain and spinal cord. The most common symptom in individuals exposed to high levels of lead is weak-

ness of the extensor muscles, which cause the joints to open. On a cellular level, lead destroys the insulation (myelin sheath) of nerve cells. This may explain the reduction in nerve-impulse speed commonly seen in patients who have been exposed to high levels of lead.

Lead also damages the kidneys, causing a disturbance in the mechanisms that help us conserve valuable nutrients (such as glucose and amino acids) that might otherwise be lost in the urine. Prolonged, high-level exposure causes a progressive buildup of connective tissue in the kidney and degeneration of the filtering mechanisms that separate wastes from the bloodstream, the glomeruli.

Lead has a profound effect on reproduction, in laboratory animals and humans alike. Numerous reports show that the rate of spontaneous abortion is much higher in couples when one or both has been exposed to high levels of lead in the workplace. Recent studies show decreased fertility and damaged sperm in male workers with high to medium levels of lead in their blood. According to one study, exposure of a pregnant woman to high levels of lead in household drinking water nearly doubles the risk of her having a retarded child.

Effects of Low-Level Lead Exposure

The toxic effects of large doses of lead have long been known, but only recently have scientists begun to learn about the effects of low-level exposure, especially on the central nervous system.

U.S. researcher Herbert Needleman and his colleagues studied over 3000 children in the first and second grades in two towns near Boston with different levels of lead in their bodies. Children with high lead levels, but still below toxic levels, had significantly lower IQ scores than those with low levels. Attention span and classroom behavior were also significantly impaired. Several other studies have shown that lead levels in the blood of greater than 40 micrograms per 100 milliliters diminish intelligence and mental capacity in children under six years of age. A recent study in Great Britain showed that at an early age even marginally elevated levels of lead may have lasting adverse effects on intelligence and behavior.

In another important study, researchers found that exposure to small amounts of lead before birth, even at levels once considered safe, seriously affect mental development. Low-level exposure slows important aspects of mental development during the first two years of life and possibly beyond that. If additional studies support these results, researchers believe that the federal standard for acceptable blood levels should be lowered for fetuses.

In 1988, EPA scientist Joel Schwartz published results of a study that showed that elevated lead levels in the blood of men causes elevated blood pressure. An additional study by Schwartz the following year showed that relatively low levels of lead in a man's blood increased his risk of heart disease.

In 1990, Needleman reported that childhood exposure to lead levels once considered moderate or low can seriously and permanently alter the intelligence of adults. He and his colleagues measured levels of lead in baby teeth of 122 individuals, then divided the subjects into three groups according to lead levels. They found individuals from the group with high lead levels performed more poorly than those from the lower groups on tests of grammar, vocabulary, and reading.

Lead contaminates livestock and wildlife as well as humans. Studies in Illinois, for example, have shown that lead levels in urban songbirds were significantly higher than those in their rural counterparts, although concentrations did not approach toxic levels. A similar study of mice and voles showed that rodents living near major highways had significantly higher levels of lead than those living near less frequently used roads, but these levels were apparently not toxic. Possible long-term effects on reproduction were believed to be minimal.

Controls on Lead

Alarmed by the mounting evidence regarding the effects of lead in children, the EPA in 1973 began a progressive restriction of the lead content of gasoline (Figure S14.2). Between 1974 and 1980, lead consumed in gasoline dropped by 62%, and lead levels in the air have decreased by 54%. According to a study released in 1983, blood levels in over 27,000 Americans living in 64 areas have dropped from an average of 14.6 micro-

grams per 100 milliliters in February 1976 to 9.4 micrograms in February 1980.

Since the late 1960s, the use of lead additives in the United States has declined by 90%. Other countries have also taken an active role in reducing lead emissions. Claude F. Boutron and his colleagues, at Grenoble University in France, reported that ice samples from a remote area in Greenland showed that global lead levels have dropped more than 85% from 1967 to 1989. Earlier studies in Greenland showed that the lead levels in the ice rose dramatically during the 1950s and 1960s, largely as a result of the increasing use of lead additives in gasoline.

Studies that found a strong statistical link between lead levels in blood and high blood pressure spurred the EPA to impose a 90% reduction in the lead in gasoline by the end of 1985. Today, leaded gas is nearly gone from American gas stations. Because of the research showing the harmful effects of lead exposure to fetuses, the EPA recently announced a complete ban on leaded gasoline to be in effect by the mid-1990s. So far, though, the EPA has resisted controls to reduce lead emissions from furnaces that burn recycled motor oil.

While the United States has aggressively reduced lead in gasoline, most Western European nations have reacted more slowly. Eastern European nations are even more remiss. Fortunately, things are changing. By 1994 or 1995, the European Economic Community (EEC, a consortium of Western European nations) expects to ban leaded fuel altogether. Little progress is being made in the former Eastern bloc countries of Eastern Europe, who are focusing on other problems.

Cities in developing nations are even further behind. In these countries, the lead content of their gasoline is,

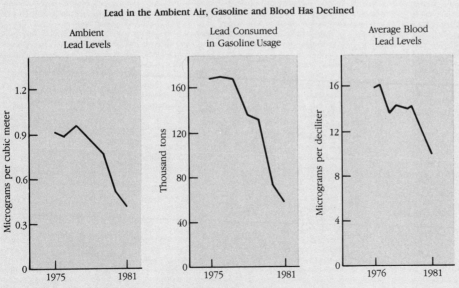

Lead in the Ambient Air, Gasoline and Blood Has Declined

Figure S14.2 *Reductions in lead in gasoline have resulted in marked decreases in lead in the air and blood of Americans.*

Source: *National Air Quality and Emissions Trends Report, 1982,* USEPA

on the average, twice that of the developed countries. Malnutrition and high levels of lead in the air will almost certainly have serious effects on their children.

As noted above, until quite recently, food remained the major source of lead in the United States. In fact, the lead concentration in the average American diet was 100 times that of our prehistoric ancestors. In 1979, the Food and Drug Administration issued an advance notice of allowable lead levels in food, aimed at reducing the intake of lead from lead-soldered cans by one-half over a five-year period. Since 1979, when 90% of all American-produced food cans contained lead solder, the number has dropped to about 4%. Food industry experts expect the number to drop to zero soon.

In 1986, Congress also banned the use of lead solder in pipes. Because drinking water accounts for about 20% of the lead exposure in Americans, the EPA recently adopted regulations that will reduce lead in public drinking water supplies, which could lower lead exposure in drinking water for about 138 million people. One regulation requires public water suppliers to treat their water with alkaline additives if lead is a problem or if the water is slightly acidic. This measure is expected to reduce lead leaching from pipes. The new EPA regulations also lower allowable levels of lead in the drinking water to about 5 parts per billion (ppb), down from 50 ppb. This measure is in all likelihood not going to result in a tenfold decrease because the current limit of 50 ppb is water measured at the tap, while the new standards measure water lead levels exiting treatment plants. Because most lead enters water after it leaves the treatment plant, the new ruling may not lower public exposure very much. Only time will tell.

Suggested Readings

Leyden, J. (1985). Nobody Wins with Lead. *National Wildlife* 23 (1): 46–48. Clearly written overview of the lead controversy.

Needleman, H. L. (1991). *Human Lead Exposure*. Chicago: CRC Press. Excellent summary of lead research.

Schwarz, J. (1988). The Relationship Between Blood Lead and Blood Pressure in the NHANES-2 Survey. *Env. Health Persp.* 78: 15–22. Excellent reference.

Singhal, R. and Thomas, J. A. (1980). *Lead Toxicity*. Baltimore: Urban and Schwarzenberg. Excellent technical review.

Chapter 15

Air Pollution: Protecting a Global Commons

Not life, but a good life, is to be chiefly valued.

—Socrates

Critical Thinking Exercise

A business magazine article notes: "On the issue of global warming, the scientific community is divided." In support of this assertion, it quotes two prominent scientists. One says he's "convinced the world is in a human-induced warming phase"; the other argues "there's simply not enough evidence to support such a conclusion." The author goes on to say that because of the uncertainty among the scientific community it makes no sense to launch a global effort to reduce carbon dioxide emissions, a position supported by the U.S. government under former President George Bush. Can you detect any problem in this reportage? What critical thinking rules were helpful to you in this examination?

Table 15.1 Natural Air Pollutants

Source	Pollutants
Volcanoes	Sulfur oxides, particulates
Forest fires	Carbon monoxide, carbon dioxide, nitrogen oxides, particulates
Wind storms	Dust
Plants (live)	Hydrocarbons, pollen
Plants (decaying)	Methane, hydrogen sulfide
Soil	Viruses, dust
Sea	Salt particulates

At the Ford automobile manufacturing plant in Cologne, Germany, plant managers modernized the paint-spray line. Their efforts not only cut pollution generated by their operation by 70% but also reduced the cost of painting a car by about $60, creating a sizable annual savings. Another German company that makes plastic films reduced emissions by 70% by instituting controls that capture 90% of solvents previously lost to the atmosphere. The savings will pay for the technological improvements, which are expected to save the company considerable sums of money in years to come.

These efforts, part of a growing movement to protect the air we breathe, generate considerable economic savings to companies.

This chapter examines air pollution, discussing principles needed to understand and solve local and global air pollution problems. It looks at traditional strategies and sustainable approaches that serve both the economy and the environment. Chapter Supplement 15.1 and Chapter 16 give details on some of the most important air pollution issues.

15.1 Air: The Endangered Global Commons

Air is a mixture of gases, including nitrogen (78%), oxygen (21%), carbon dioxide (0.04%), and several inert substances, among them argon (almost 1%), helium, xenon, neon, and krypton. Water vapor exists in varying amounts. Air also contains various amounts of pollutants.

The air we breathe is a renewable resource that is replenished by natural processes. Oxygen, for example, is replenished by plants. Air is also cleansed by natural processes, for example, rain, which rid the air of many but not all pollutants. Satellite pictures show that huge air masses sweep across the Earth's surface, picking up

moisture and pollutants in one region and depositing them sometimes hundreds of kilometers away.

Transparent, powerful, and nurturing, air is a global resource, akin to the commons described in Chapter 9. Owned by no one, it is used by all, as a source of oxygen vital to animals and carbon dioxide needed by plants. Industrial societies, however, have traditionally used the air as a waste dump for hundreds of harmful pollutants. Like other commons, no one has sole responsibility for protecting our air, and unfortunately, even those who do not pollute it suffer from the disregard of others.

Natural and Anthropogenic Sources

Pollutants in the atmosphere arise from two major sources: natural and anthropogenic. Globally, the largest sources are natural events: volcanoes, dust storms, forest fires, and the like (Table 15.1). In sheer quantity, natural pollutants often outweigh the products of human activities, the **anthropogenic pollutants**. Nevertheless, the anthropogenic pollutants generally create the most significant long-term threat to the biosphere. Why? Natural pollutants, except those from volcanoes, come from widely dispersed sources or infrequent events. Therefore, they generally do not substantially raise the ambient pollutant concentration, and, as a result, have little effect on biological systems. In contrast, power plants, automobiles, factories, and other human sources emit large quantities in restricted areas, making a significant contribution to local pollution levels. The higher the concentration, the greater the effect.

Anthropogenic Air Pollutants and Their Sources

Take a deep breath. If you live in a city, the chances are you just inhaled tiny amounts of dozens of different air

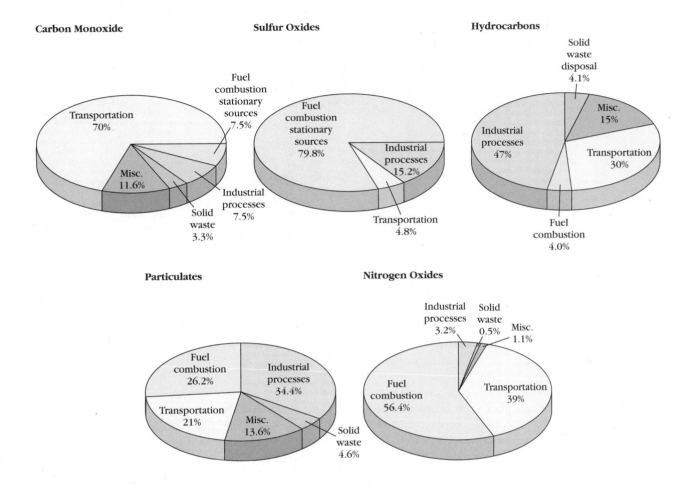

Figure 15.1 *Sources of the five regulated air pollutants in the United States.*

pollutants, most in concentrations too small (we think) to be harmful. This chapter concerns itself with six major pollutants: carbon monoxide, sulfur oxides, nitrogen oxides, particulates, hydrocarbons, and photochemical oxidants. Lead, an important air pollutant, was discussed in Chapter Supplement 14.1; radiation was examined in Chapter Supplement 11.1.

In 1980, the United States produced 160 million metric tons of the five primary pollutants: carbon monoxide, sulfur oxides, nitrogen oxides, particulates, and hydrocarbons. As a result of conservation, better pollution control, and a ban on lead, by 1988, U.S. production of the five major pollutants fell to 135 million metric tons. By 1991, levels fell to about 125 million metric tons.

Figure 15.1 shows that air pollutants come from three principal sources: transportation (autos, trucks, jets, and trains), fuel combustion at stationary sources (mostly power plants and factories), and various industrial processes. Air pollutants arise from vaporization (or evaporation), attrition (or friction), and combustion. Combustion is by far the major producer. Among combustion sources, the burning of fossil fuels (coal, oil, and natural gas) and products refined from them (gasoline, diesel, and jet fuel) pose the most significant threat

to the environment. In fact, our heavy reliance on fossil fuels is one of the root causes of the crisis of unsustainability.

Fossil fuels consist primarily of carbon and hydrogen atoms linked by chemical bonds. When ignited, a curious thing happens. The initial source of heat, say, a match, breaks some of the bonds in the organic compounds in a fuel. This releases energy in two forms: light and heat. Heat released in the process breaks other bonds, permitting the burning to continue until the fuel runs out. During combustion, oxygen reacts with carbon and hydrogen. Complete combustion, which rarely occurs, produces carbon dioxide (CO_2) and water (H_2O). Incomplete combustion produces carbon monoxide (CO) gas and unburned hydrocarbons (Figure 15.2).

Most fuels contain some mineral contaminants, such as mercury or lead. These unburnable contaminants may be carried off by hot combustion gases, escaping into the air as particulates. Other contaminants, such as sulfur, react with oxygen at high combustion temperatures, forming sulfur oxide gases, notably sulfur dioxide (SO_2) and sulfur trioxide (SO_3). In the absence of pollution control devices, these gases escape with the other smokestack gases.

Figure 15.2 *Products of fossil fuel combustion.*

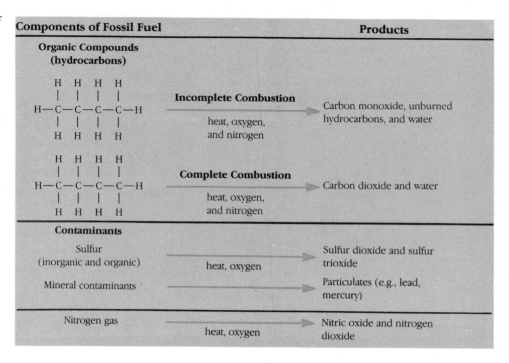

Combustion must take place in air because air provides a source of oxygen. But, air also contains nitrogen. During high temperature combustion, nitrogen (N_2) reacts with oxygen to form nitric oxide (NO). Nitric oxide is quickly converted to nitrogen dioxide (NO_2), a brownish orange gas seen in many modern cities. Nitrogen dioxide is a key reactant in the formation of photochemical smog, described shortly. The effects of these pollutants are listed in Table 15.2 on page 356.

Primary and Secondary Pollutants

The atmosphere is, in many ways, a chemist's nightmare; it contains hundreds of air pollutants from natural and anthropogenic sources. These pollutants, called **primary pollutants**, often react with one another or with water vapor. These reactions spawn a whole new set of pollutants, called **secondary pollutants**, chemical substances produced from the chemical reactions of primary pollutants from natural or anthropogenic pollutants, reactions often powered by energy from the sun. For example, sulfur dioxide gas is a primary pollutant released from a variety of sources, such as coal-fired power plants and automobiles. In the atmosphere, sulfur dioxide reacts with oxygen and water to produce sulfuric acid (H_2SO_4), a toxic and corrosive secondary pollutant with far-reaching effects (Chapter 16).

Toxic Air Pollutants

Health officials and environmental activists have long been concerned about the hundreds of potentially toxic pollutants released into the atmosphere each year in the United States from factories and other sources. Although generally emitted in much smaller quantities than the major pollutants discussed above, toxic air pollutants may be responsible for numerous cancer deaths. A recent study by the EPA suggested that 45 toxic air pollutants may cause as many as 1700 cases of cancer each year. By some estimates, 400 toxic air pollutants are released into the atmosphere in the United States.

15.2 Effects of Climate and Topography on Air Pollution

Gray-Air and Brown-Air Cities

If you are like most Americans, you live in or near a city. That city generally falls into one of two categories, based on the climate and the type of air pollution. Older, industrial cities, such as Nashville, New York, Philadelphia, St. Louis, Pittsburgh, and London, belong to a group of **gray-air cities** (Figure 15.3a); newer, relatively nonindustrialized cities, such as Denver, Los Angeles, Salt Lake City, and Albuquerque, belong to the group of **brown-air cities** (Figure 15.3b).

Gray-air cities are generally located in cold, moist climates. The major pollutants are sulfur oxides and particulates from factories. These pollutants combine with atmospheric moisture to form the grayish haze called **smog**, a term coined in 1905 to describe the mixture of smoke and fog that plagued industrial England. Gray-air cities primarily depend on coal and oil and are usually heavily industrialized. The air quality in these cities is especially poor during cold, wet winters, when

(a)

(b)

Figure 15.3 *Two types of air pollution:* (a) *Gray-air smog in Detroit;* (b) *Brown-air smog in Los Angeles.*

the demand for home heating oil and electricity as well as the atmospheric moisture content is high.

Brown-air cities are typically located in warm, dry, and sunny climates. The major sources of pollution are the automobile and the electric power plant; the major primary pollutants are carbon monoxide, hydrocarbons, and nitrogen oxides.

In brown-air cities, atmospheric hydrocarbons and nitrogen oxides from automobiles and power plants react in the presence of sunlight to form a witch's brew of harmful secondary pollutants. Among some of the worse are ozone, formaldehyde (the same chemical in embalming fluid), and peroxyacylnitrate (PAN). The reactions that form these and other potentially toxic chemicals are called **photochemical reactions** because they involve both sunlight and chemical pollutants. The resulting brownish orange shroud of air pollution is known as **photochemical smog**. Ozone (O_3) is one of the most prevalent chemicals in photochemical smog. This molecule is highly reactive. It erodes rubber, irritates the respiratory system, and damages trees.

In brown-air cities, early morning traffic provides the ingredients for photochemical smog, which usually reaches the highest levels in the early afternoon (Figure 15.4). Because the air laden with photochemical smog often drifts out of the city and because the reactions require sunlight and take time to occur, the suburbs and surrounding rural areas frequently have higher levels of photochemical smog than the cities themselves. Major pollution episodes in brown-air cities usually occur during the summer months, when the sun is most intense.

Today, the distinction between gray- and brown-air cities is disappearing. Most cities have gray air in the

winter, when pollution from wood stoves and oil burners and the moist, wet air conspire to darken the skies, and brown air in the summer, when sunlight and automobile pollutants are prevalent.

Although anthropogenic pollutants are the most significant sources of air pollution, researchers have found some instances where naturally occurring pollutants noticeably affect air quality. In Atlanta, for example, the trees emit a number of highly reactive hydrocarbons. These react with nitrogen dioxide, a gas emitted from automobiles and other combustion sources, producing ozone. New research shows that hydrocarbons from trees are 50 to 100 times more reactive than hydrocarbons from human sources.

Pollution control often requires systemswide analyses of both natural and anthropogenic sources. Such studies yield valuable information useful in controlling pollution. For instance, the EPA once thought that the city of Atlanta could meet federal air-quality standards for ozone by reducing the level of hydrocarbons from human sources, mostly automobile use, by 30%. New data, however, suggest that when the contribution from trees is added, the human-related sources need to be cut by 70% to 100% to significantly reduce ozone levels. Further studies suggest that efforts to cut nitrogen oxide emissions would be more effective and more feasible.

Factors Affecting Air Pollution Levels

Wind and Rain   Pollution is an enigma to many people. One day is clear. The next day, the skies fill with

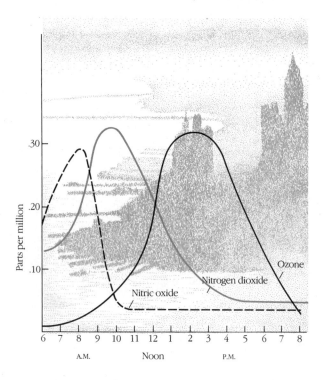

Figure 15.4 *Nitrogen oxides and hydrocarbons (not shown here) react to form ozone and other photochemical oxidants. Because sunlight and time are required for the reactions to occur, maximum ozone concentration occurs in the early afternoon. Hydrocarbon levels would follow the same pattern as nitric oxide levels.*

ugly crud. Numerous factors contribute to this puzzle. For example, wind sweeps dirty air out of cities. Rain washes pollutants from the sky. The residents in Seattle often brag about their clean air. The fact is Seattle produces as much pollution as any other city of similar size. However, it tends to blow away or be washed from the sky. Contrary to what many people think, the pollutants do not disappear. They are merely shifted elsewhere or deposited onto another medium, for example, surface waters. This transfer is typically referred to as **cross-media contamination.**

Airborne pollutants can travel hundreds, perhaps thousands, of kilometers to other cities or unpolluted wilderness, where they are deposited. The acidic pollutants responsible for the massive fish kills in Sweden's Tovdal River basin, for example, are blown into Scandinavia from industrial England and Europe.

Mountains and Hills   Salt Lake City stretches out below a giant mountain range. Granite peaks and jagged cliffs make this one of the most scenic American cities—that is, when the mountains are visible through the air pollution. Residents of Salt Lake City would tell you that mountain ranges can be an asset to a city, but they can also be a curse because they often block the flow of winds and trap pollutants for days on end. Mountains

also block the sun, which helps disperse pollutants, as explained below.

Temperature Inversions   Suppose you were riding in a hot-air balloon on a normal summer day. You would find that warm ground air rises and expands. As it expands, it cools. Thus, temperature decreases with altitude.

If you had proper equipment to measure pollution, you would also find that pollution rises with the warm ground air. Because of this, the atmosphere is gently stirred, and ground-level pollution is reduced (Figure 15.5a).

Atmospheric mixing is brought about by sunlight. Striking the Earth, sunlight heats the rocks and soil. This heat is transferred to the air immediately above the ground. The warm air then rises, mixing with cooler air.

Under certain atmospheric conditions, however, a different temperature profile is encountered. On some days, air temperature decreases with altitude but only to a certain point. After that point, the temperature begins to increase. This is called a **temperature inversion.** Temperature inversions create warm-air lids over cooler air (Figure 15.5b). Because the cool, dense ground air cannot mix vertically, pollutants become trapped in the air below it, often reaching dangerous levels.

Temperature inversions fit into two categories. A **subsidence inversion,** the first type, occurs when a high-pressure air mass stalls and forces a layer of warm air down over a region. These inversions may extend over many thousands of square kilometers (Figure 15.6). A **radiation inversion,** the second type, is typically local and usually short lived. It is a phenomenon generally witnessed on cold winter days. Radiation inversions begin to form a few hours before the sun sets. As the day ends, the air near the ground cools faster than the air above it. Thus, warm air lies over the cooler ground air, preventing it from rising and causing pollutants to accumulate. Making matters worse, radiation inversions also correspond with the evening commute when traffic and pollution emissions peak. Radiation inversions usually disperse in the morning when the sun strikes the Earth and vertical mixing begins. Radiation inversions are common in mountainous regions, especially in the winter when the sun is obstructed by the mountains and therefore unable to warm the ground enough to stimulate vertical mixing.

15.3 Effects of Air Pollution

In 1991, the EPA announced that 140 million Americans—over half of the U.S. population—were breathing unhealthy air, air that violated the ozone standard. Nearly 100 metropolitan areas failed to meet federal

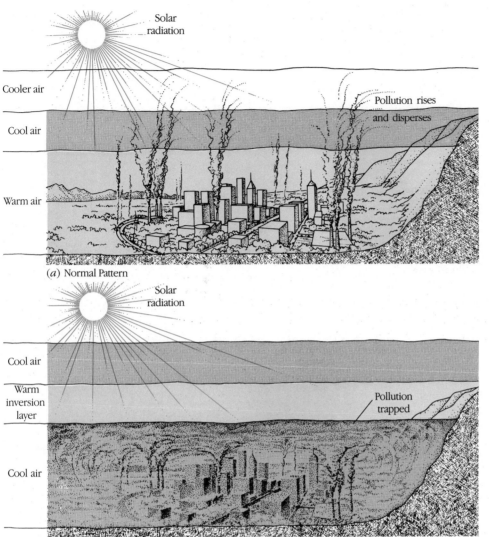

Figure 15.5 (a) *During normal conditions, air temperature decreases with altitude; thus, pollutants ascend and mix with atmospheric gases.* (b) *In a temperature inversion, however, warm air forms a "lid" over cooler air, thus trapping air pollution.*

air-quality standards, even though the standards had been eased and the deadline for meeting them had been extended repeatedly in the past 20 years. Ozone is the biggest problem. Particulates rank second, despite dramatic declines since 1970. All told, about 50 million Americans breathe air containing potentially harmful levels of particulates. According to a recent government report, by the year 2000, air pollution will be killing 57,000 Americans per year.

The American Lung Association conservatively estimates that air pollution costs Americans $50 billion per year in health costs, or about $200 per year for every man, woman, and child. Air pollution damages crops and buildings, adding billions each year to the price tag. Estimates in the late 1980s put ozone crop damage alone at $10 billion per year. Many other costs cannot be calculated: the loss of scenic view, the destruction of a favorite fishing spot, the erosion of a valuable statue.

Air pollution causes extraordinary environmental and economic damage worldwide, even in the developing countries that, with the exception of China, produce only a fraction of the world's air pollutants. In Europe, where three-fourths of the forests are affected by air pollution, the cost from damage is estimated to be about $30 billion per year, according to the International Institute of Applied Systems Analysis in Austria. Although no statistics are available, air pollution no doubt also costs developing countries huge sums in similar damage.

This section examines the impacts of pollution to human health, other organisms, and materials. These costs are typically referred to as **external costs**, as noted in previous chapters, because they are almost universally not factored into the costs of the activities, such as driving automobiles, that produce them. Such an omission, some say, leads to a false sense of prosperity.

Table 15.2   Major Air Pollutants—Their Sources and Health Effects

Pollutant	Major Anthropogenic Sources	Health Effects
Carbon monoxide	Transportation industry	Acute exposure: headache, dizziness, decreased physical performance, death
		Chronic exposure: stress on cardiovascular system, decreased tolerance to exercise, heart attack
Sulfur oxides	Stationary combustion sources, industry	Acute exposure: inflammation of respiratory tract, aggravation of asthma
		Chronic exposure: emphysema, bronchitis
Nitrogen oxides	Transportation, stationary combustion sources	Acute exposure: lung irritation
		Chronic exposure: bronchitis
Particulates	Stationary combustion sources, industry	Irritation of respiratory system, cancer
Hydrocarbons	Transportation	Unknown
Photochemical oxidants	Transportation, stationary combustion sources (indirectly through hydrocarbons and nitrogen oxides)	Acute exposure: respiratory irritation, eye irritation
		Chronic exposure: emphysema

Figure 15.6 *This large subsidence inversion occurred in August 1969, covered more than a dozen states, and, in some regions, persisted for 10 days, greatly increasing pollution levels. In most areas, particulate levels doubled and tripled.*

Health Effects

Immediate Health Effects   Toxins may cause many immediate effects on human health, ranging from shortness of breath to eye irritation to death (Table 15.2). The most noticeable health effects have been observed during **pollution episodes**, periods when atmospheric concentrations reached dangerous levels for short periods. Several notorious episodes have been documented. The very first occurred in the Meuse Valley in Belgium in 1930, the second in Donora, Pennsylvania, in 1948, and the third in London in 1952.

In each incident, pollution rose to dangerous levels as a result of temperature inversions that lasted three to four days. In addition, in each episode, pollution came primarily from burning coal in homes and factories. In Belgium, 65 people died; in Pennsylvania, 20 succumbed; and in London, 4000 people may have died as a result of toxic levels of particulates and sulfur dioxides. In all locations, many other people became ill. Those who died or became ill were usually old or suffering from cardiovascular disease and, therefore, were unable to cope with the added stress caused by the heavily polluted air.

Case Study 15.1

MPTP Pollution and Parkinson's Disease

In 1983, Dr. J. W. Langston made a bizarre discovery. Several young men and women were admitted to his hospital in a stupor. Completely immobile, the patients were unable to feed themselves, to move their limbs, or to talk.

Several months of intensive detective work revealed that the hopelessly "frozen" patients were drug addicts who had happened on a laboratory-synthesized heroin, one of the many new "designer drugs." This particular batch had been improperly made in the basement of the supplier's home and was contaminated with a chemical substance called MPTP, a pyridine compound.

Researchers have since discovered that MPTP attacks the substantia nigra, a portion of the midbrain involved in muscle movement. The substantia nigra is the region that gradually deteriorates in victims of the debilitating malady known as Parkinson's disease, which develops later in life, usually after age 45. Parkinson's victims suffer from tremor and partial or complete paralysis.

Individuals exposed to MPTP, even tiny amounts, develop symptoms of Parkinson's disease within a few days of exposure. Monkeys exposed to MPTP also develop Parkinson-like symptoms. Studies show that MPTP causes cell degeneration in the substantia nigra.

What is disturbing about this discovery is that MPTP is strikingly similar to a number of common industrial chemicals and widely used pesticides. It most closely resembles the herbicide paraquat. This has led some researchers to believe that Parkinson's disease may actually be caused by environmental pollution from industry and from agricultural pesticides.

Opponents of this view argue that, at the cellular level, Parkinson's disease is not that similar to paraquat toxicity. A number of studies have failed to show a link between industrialization and Parkinson's disease. Other studies have shown no increase in the disease in the United States since the 1940s, at which time pollution levels began climbing. Thus, at a 1985 meeting held at the National Institutes of Health, researchers largely agreed that the disease is not produced by environmental contamination.

No sooner had these researchers left their meeting than André Barbeau and his colleagues at the Research Institute of Montreal announced the startling results of a study on the incidence of Parkinson's disease in Quebec province. The researchers showed, quite to the surprise of the academic community, a remarkable correlation between the use of pesticides and the incidence of the disease. The statistical correlation was so high as to be irrefutable.

The researchers stressed, however, that herbicides are just one of many neurotoxins capable of destroying the nerve cells of the substantia nigra and causing this disease. They noted that a large number of industrial pyridines were also suspect. Many of these chemicals come from the chemical industry.

Further evidence supporting an environmental cause of Parkinson's disease, according to Barbeau and his colleagues, is that the malady was nonexistent before the Industrial Revolution. The incidence of the disease increased sharply through the 1800s and reached a plateau in the early decades of the 1900s, for unknown reasons. However, Barbeau warns that the rising use of paraquat and other chemically similar substances could cause an increase in Parkinson's disease.

Clearly, the jury is still out. At this time, Barbeau's work stands alone in showing a link between MPTP and Parkinson's disease. Further research is needed to determine if the link does indeed exist and what we can do about it.

Source: Adapted from Lewin, R. (1985). Parkinson's Disease: An Environmental Cause? *Science* (July 19): 257–58.

Another serious episode occurred in Bhopal, India, in December 1984, when approximately 40 metric tons of methyl isocyanate were accidentally released into the atmosphere from a pesticide facility. At least 2000 people died in the accident.

Less serious pollution episodes occur on a regular basis. In 1966, for example, an increase in sulfur dioxide levels in New York City resulted in a dramatic increase in the incidence of colds, coughs, rhinitis (nose irritation), and other symptoms, which increased fivefold almost overnight. When the air cleared, the symptoms disappeared. More recent studies have shown that pollution episodes occurring in the summer months in the northeastern United States and Canada, which are characterized by an increase in sulfate and ozone, increase the incidence of hospital admissions for respiratory diseases.

Many other less noticeable effects result from exposure to air pollution. Visitors to Los Angeles often complain of burning or itching eyes and irritated throats caused by photochemical smog. Commuters in heavy traffic are familiar with the headaches caused by carbon monoxide from automobile fumes. These effects, however, are generally ignored or simply seen as part of the price we pay for city living.

Chronic Health Effects   Long-term exposure to air pollution may result in a number of diseases, including chronic bronchitis, emphysema, asthma, and lung cancer.

One of every five American men between the ages of 40 and 60 has **chronic bronchitis**, a persistent inflammation of the bronchial tubes, which carry air into the lungs. Symptoms include a persistent cough, mucus buildup, and difficult breathing. Cigarette smoking is the primary cause of this disease, but urban air pollution is also a major contributing factor. Studies of children ages 10 to 12 in four U.S. cities showed a noticeable increase in the prevalence of chronic bronchitis as pollution levels rise. Sulfur dioxides, nitrogen dioxides, and ozone are believed to be the major causative agents.

Emphysema, another chronic effect, kills more people than lung cancer and tuberculosis combined. Emphysema is the fastest-growing cause of death in the United States. Over 2 million Americans suffer from this incurable disease, according to the American Lung Association. As we age, the small air sacs, or alveoli, in our lungs break down. This reduces the surface area for the exchange of oxygen with the blood. Breathing becomes more and more labored. When lung surface area is reduced by about 40%, victims suffer shortness of breath even when exercising lightly.

Emphysema is caused by cigarette smoking and may be caused by urban air pollution as well. One study, for instance, showed that the incidence of emphysema was higher in relatively polluted St. Louis than in relatively unpolluted Winnipeg (Figure 15.7). Studies in Great Britain showed that mail carriers who worked in polluted urban areas had a substantially higher death rate from emphysema than those who worked in unpolluted rural areas. Ozone, nitrogen dioxides, and sulfur oxides are the chemical agents believed responsible for this disease.

More recently, several studies have shown that bronchial asthma attacks may be associated with elevated levels of pollution. **Asthma** is a chronic disorder, marked by periodic episodes of wheezing and difficult breathing. Most cases of asthma are caused by allergic reactions to common stimulants, such as dust, pollen, and skin cells (dander) from pets. In some individuals, pollution may trigger asthma attacks. In an asthmatic attack, several passageways that carry air to the lungs, bronchi and bronchioles, fill with mucus, making breathing difficult. Irritants also stimulate the contraction of the smooth muscle cells in the walls of the bronchioles, making it even more difficult for an asthmatic to breathe. Periodic

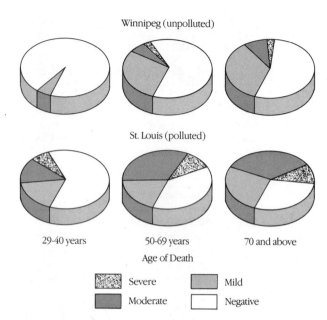

Figure 15.7   *Incidence of emphysema in St. Louis and Winnipeg. Note the increased incidence of emphysema in all three age groups in the more polluted urban environment of St. Louis.*

attacks can be quite disabling and may even lead to death. By one estimate, several thousand Americans die each year from severe asthma attacks. Victims are generally elderly individuals who are also suffering from other diseases.

A number of studies have shown that lung cancer rates are higher among urban residents than among rural residents even after the influence of cigarette smoking has been taken into account. Critical thinking recommends caution because health studies of human populations, or epidemiological studies, rely on statistical methods with some unavoidable shortcomings (Chapter Supplement 1.1). In lung cancer studies, for example, researchers usually compare death certificates of urbanites with death certificates of rural residents. If the urban population shows a higher incidence of lung cancer, it is tempting to conclude that the cause was urban air pollution. But, researchers must be careful to eliminate other causative agents, such as smoking and occupation. Most researchers eliminate smokers from studies but not all take into account urban occupations (factories), where men and women are often exposed to high levels of cancer-causing pollutants. Thus, the slightly higher incidence of lung cancer in some urban settings may result from occupational exposure or some other factor as well as pollution. Only time and further research will confirm the studies.

High-Risk Populations   Individuals are not equally affected by air pollution. Particularly susceptible are the old and infirm, especially those with preexisting lung and cardiovascular disorders. Carbon monoxide is dangerous to people with heart disease because it binds strongly to hemoglobin, the oxygen-carrying protein in red blood cells. This reduces the oxygen-carrying capacity of the blood. For sufficient oxygen to be delivered to the body's cells, the heart must pump more blood during a given period. This puts a strain on the heart and may trigger heart attacks in those whose hearts are already weakened.

As you may recall from Chapter 14, toxins often affect the young more severely than adults. Air pollution is no exception. In fact, some estimate that the health risk from air pollution is six times greater for children than for adults. Several reasons may account for this difference. First, children may be more susceptible because they are more active and therefore breathe more. As a result, they may be exposed to more pollution. Second, children typically suffer from colds and nasal congestion and thus tend to breathe more through their mouths. Because air bypasses the normal filtering mechanism of the nose, more pollutants enter the lungs.

Effects on Other Organisms   Air pollutants at levels humans are exposed to have a wide range of effects on experimental animals. Unfortunately, very little research has been performed on livestock or wild species. Scientists do know, however, that fluoride and arsenic released from smelters can seriously poison cattle grazing downwind. Acids produced from power plants, smelters, industrial boilers, and automobiles have been shown to be extremely harmful to wildlife, especially fish and forests (Chapter 16). In Southern California, millions of ponderosa pines have been damaged by air pollution, mostly ozone, from Los Angeles (Figure 15.8).

Ozone, sulfur dioxides, and sulfuric acid are the pollutants most hazardous to plants. Ozone, for instance, makes plants more brittle and likely to crack. Farms in Southern California and on the East Coast report significant damage to important vegetable crops. City gardeners also report damage to flowers and ornamental plants. Urban landscapers must plant pollution-resistant trees, notably locusts, along heavily traveled roads, but even they only last a fraction of their normal life span.

Sulfur dioxide damages plants directly, causing spotting of leaves. Recent studies show that air pollutants may actually enhance some plant species, making them more desirable to leaf-eating insects. According to botanists from Cornell University, air pollution and other stresses cause plants to produce the chemical called glutathione, which protects leaves from pollution but also attracts insects that normally have no interest in these plant species.

One group of organisms that is extremely sensitive to air pollution is lichens, which grow on rock, wood, or soil. A lichen is a composite organism, consisting of

Figure 15.8 *Air pollution damages trees like these pines in the mountains east of Los Angeles.*

a fungus with an internal alga. Because they are amazingly hardy and resilient, lichens can endure a wide variety of climates, often living where few other life forms exist. However, lichens obtain nourishment from the air, rain, and snow, which makes them extremely sensitive to air pollution.

In the 1800s, scientists began to notice that lichens were being killed by pollution in many European cities. In the 1860s, in fact, 32 species of lichens were collected in the Jardin de Luxembourg in Paris. By 1896, not a single species survived. In the town of Mendlesham, England, 129 species were identified between 1912 and 1921. In 1973, only 67 remained. In England, researchers found that lichens were absorbing heavy metals and pollution, and wherever pollution increased, the lichens died.

So sensitive are lichens to air pollution that air quality and the spread of pollution from an industrial source can be determined by mapping the presence or absence of lichens or by chemically analyzing the lichens in the area. In fact, an air-quality map of the British Isles was based on a lichen survey by 15,000 British schoolchildren in 1971.

Effects on Materials   Air pollutants severely damage metals, building materials (stone and concrete), paint, textiles, plastics, rubber, leather, paper, clothing, and ceramics (Table 15.3). The four most corrosive and harmful pollutants are sulfur dioxide, sulfuric acid, ozone, and nitric acid.

Table 15.3 Damage to Materials from Air Pollution

Material	Damage	Principal Pollutants
Metals	Corrosion or tarnishing of surfaces; loss of strength	Sulfur dioxide, hydrogen sulfide, particulates
Stone and concrete	Discoloration, erosion of surfaces, leaching	Sulfur dioxide, particulates
Paint	Discoloration, reduced gloss, pitting	Sulfur dioxide, hydrogen sulfide, particulates, ozone
Rubber	Weakening, cracking	Ozone, other photochemical oxidants
Leather	Weakening, deterioration of surface	Sulfur dioxide
Paper	Embrittlement	Sulfur dioxide
Textiles	Soiling, fading, deterioration of fabric	Sulfur dioxide, ozone, particulates, nitrogen dioxide
Ceramics	Altered surface appearance	Hydrogen fluoride, particulates

The Statue of Liberty, which was recently restored, had been pitted by sulfuric and nitric acids. The Taj Mahal in India, like many other buildings in the world, is being defaced by sulfur oxides air pollution from local power plants. Sulfuric and nitric acids cause cosmetic damage to metals and reduce their strength. In the Netherlands, bells that had been ringing true for three or four centuries have recently gone out of tune because of acid pollutants that have eaten away at them, lowering their pitch and rendering once familiar tunes indecipherable. Ozone cracks rubber windshield wipers, tires, and other rubber products, necessitating costly antioxidant additives. Particulates also cause damage. For example, particulates blown in the wind erode the surfaces of stone, causing significant damage.

Society pays a huge price tag for cleaning sooty buildings, repainting pitted houses and automobiles, and replacing damaged rubber products and clothing. The damage can also be tragic, for air pollution attacks irreplaceable works of art like the stone in the Parthenon in Athens, which has deteriorated more in the last 50 years than in the previous 2000 years because of air pollution.

15.4 Air Pollution Control: Toward a Sustainable Strategy

To date, most of the efforts to control pollution have relied on end-of-pipe solutions, mostly pollution control devices that capture pollutants or convert them into supposedly less harmful substances. We begin our study of pollution control with a look at these approaches, pointing out the pros and cons of this strategy. We then turn to the alternative—and potentially more sustainable—strategies.

Cleaner Air Through Better Laws

Society's laws are like the Earth's creatures: capable of evolving to fit present conditions. This is true of U.S. clean-air legislation, now entering its fourth decade. Early clean-air laws were fairly weak and ineffective, but they laid the foundation for some of the most progressive environmental legislation in the world.

Today, the federal **Clean Air Act** and its amendments seek to reduce pollution through various measures. The first major advances in protection came with a sweeping set of amendments to the act in 1970. They resulted in: (1) emissions standards for automobiles, (2) emissions standards for new industries, and (3) ambient air-quality standards for urban areas. The ambient air-quality standards established by the EPA covered six pollutants: carbon monoxide, sulfur oxides, nitrogen oxides, particulates, ozone, and hydrocarbons. These standards were designed to protect human health and the environment.

The 1970 amendments successfully reduced pollution from automobiles, factories, and power plants. In addition, they stimulated many states to pass their own air pollution laws, some with regulations even tighter than federal ones. Despite these gains, the amendments created some problems. For instance, in regions that exceeded ambient air-quality standards, the law prohibited the construction of new factories or the expansion of existing ones. The business community objected. In addition, some of the wording of the 1970 amendments was vague and required clarification. Of special interest were provisions dealing with the deterioration of air quality in areas that were already meeting federal standards.

Because of these and other problems, the Clean Air Act was amended again in 1977. To address the apparent limits on industrial growth in areas that were violating air-quality standards, or **nonattainment areas**, lawmakers devised a creative policy. This strategy allowed factories to expand and new ones to be built in nonattainment areas *only* if they met three provisions: (1) new sources achieved the lowest possible emission rates; (2) other sources of pollution under the same ownership in that state complied with emissions-control provisions; and (3) unavoidable emissions were offset by pollution reductions in other companies in the same region.

Because of the last provision, known as the **emissions offset policy**, newcomers were forced to request existing companies in the region to reduce their pollution emissions. In most cases, the newcomers actually paid the cost of pollution control devices. The emissions offset policy was used to produce an overall decrease in regional air pollution because the air pollution emissions permitted from both the new and existing facilities were set below preconstruction levels.

How to protect air quality in areas that were already meeting federal air-quality standards sparked considerable debate in Congress. Environmentalists felt that the ambient air-quality standards, in effect, gave industries a license to pollute up to permitted levels. The 1977 amendments, therefore, set forth rules for the **prevention of significant deterioration** (PSD) of air quality in clean-air areas. However, PSD requirements apply only to sulfur oxides and particulates, two pollutants viewed by Congress as the two most deserving of immediate action. PSD regulations essentially prevent significant increases in ambient air pollution in **attainment regions**, regions where air quality meets federal standards. Many air pollution experts think that the PSD requirements should be expanded to include other pollutants, such as ozone.

The 1977 amendments also strengthened the enforcement power of the EPA. In previous years when the EPA wanted to enforce compliance, it had to initiate a criminal lawsuit. Violators often engaged in protracted legal battles, since court costs were frequently lower than the cost of installing pollution control devices. Thanks to the 1977 amendments, the EPA can now initiate civil lawsuits, which do not require the heavy burden of proof needed for criminal convictions.

More important, the EPA was allowed to levy **noncompliance penalties** without going to court. These penalties are assessed on the grounds that violators have an unfair business advantage over competitors who comply with the law. Penalties equal to the estimated cost of pollution control devices eliminate the cost incentive of polluting.

The Clean Air Act and additional regulations helped reduce U.S. ambient air pollution (Figure 15.9). Between 1970 and 1991, particulate emissions in the United States dropped by 61%. During the same period, carbon monoxide emissions fell by 50%, sulfur oxide gases fell by 27%, and hydrocarbons declined 38%. The only major air pollutant to show little decline was nitrogen oxide, which dropped only 1%.

Additional gains in urban air quality may result from new regulations on wood-burning stoves. In 1990 and 1992, for instance, the EPA put into effect new regulations requiring all new fireplace inserts and freestanding wood stoves to produce far less pollution. In fact, EPA-certified wood stoves emit 70% to 90% fewer particulates than older models. Many cities and towns have adopted regulations that prohibit wood burning during high-pollution days.

In 1990, the Clean Air Act was amended once again to address inadequacies in the previous amendments. The new law, for example, set deadlines for establishing emission standards for 190 toxic chemicals. More important, it established a system of **pollution taxes**, (a market-based measure) on toxic chemical emissions that provide a powerful incentive for companies to reduce their release.

The 1990 Clean Air Act amendments also tightened emission standards for automobiles and raised the average mileage standards for new cars, a step that will improve automobile efficiency and attack the pollution problem at its roots.

In addition, the 1990 amendments established a market-based incentive program to reduce nitrogen and sulfur oxides, primary contributors to acid deposition (Chapter 16). According to the law, the EPA will issue tradable permits to companies throughout the United States. These permits stipulate allowable emissions for nitrogen and sulfur oxides. These rates are lower than present emission rates to improve air quality. Companies that find innovative and cost-effective means to reduce pollutants below their permitted allowance can sell their unused credits. Starting in 1994, these credits will be bought and sold on the commodities market. This system could encourage companies to develop cost-effective ways to prevent pollution. In fact, if a company can find inexpensive means to reduce its output, it could benefit economically from the sale of pollution credits.

·Finally, in a move that delighted environmentalists worldwide, the 1990 amendments called for a phase-out of ozone-depleting chemicals. (More on this in Chapter 16.)

Cleaner Air Through Technology

The Clean Air Act and numerous state laws and regulations passed in the last two and a half decades have prescribed many technological controls to cut down on air pollution. These devices either remove harmful sub-

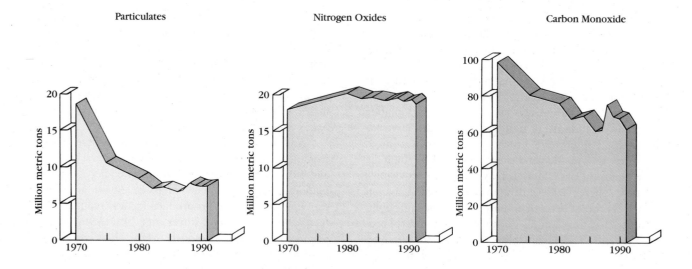

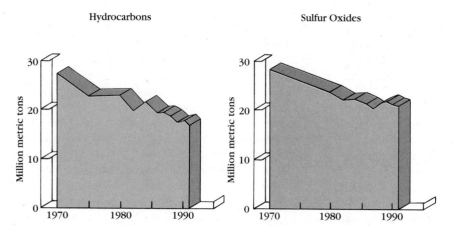

Figure 15.9 *Because of tighter pollution controls and increased efficiency, emission of many pollutants has decreased. Nitrogen oxide levels have climbed because of a lack of U.S. technology to eliminate the gas from combustion sources.*

stances from emissions gases or convert them into harmless substances. The first strategy is the most common for stationary combustion sources.

Stationary Sources   In some electric power plants, for example, **bag filters** separate particulate matter from stack gases (Figure 15.10a). In these devices, smoke passes through a series of cloth bags that simply filter out particulates. Filters often remove well over 99% of the particulates but do not remove gaseous pollutants, such as sulfur dioxide.

Cyclones are also used to remove particulates, generally in smaller operations (Figure 15.10b). In the cyclone, particulate-laden air is passed through a metal cylinder. The particulates strike the walls and fall to the bottom of the cyclone, where they can be removed. Cyclones remove 50% to 90% of the large particulates

but few of the small- and medium-sized ones. The small-sized particulates are the ones that cause the most harm because they can enter the lungs. Like filters, cyclones have no effect on gaseous pollutants.

Electrostatic precipitators, also used to remove particulates, are about 99% efficient and are installed in many U.S. coal-fired power plants (Figure 15.10c). In electrostatic precipitators, particulates first pass through an electric field, which charges the particles. The charged particles then attach to the wall of the device, which is oppositely charged. The current is periodically turned off, allowing the particulates to fall to the bottom.

The **scrubber**, unlike the other methods, removes particulates *and* gases, such as sulfur dioxide (Figure 15.10d). In scrubbers, pollutant-laden air is passed through a fine mist of water and lime, which traps over 99% of the particulates and 80% to 95% of the sulfur

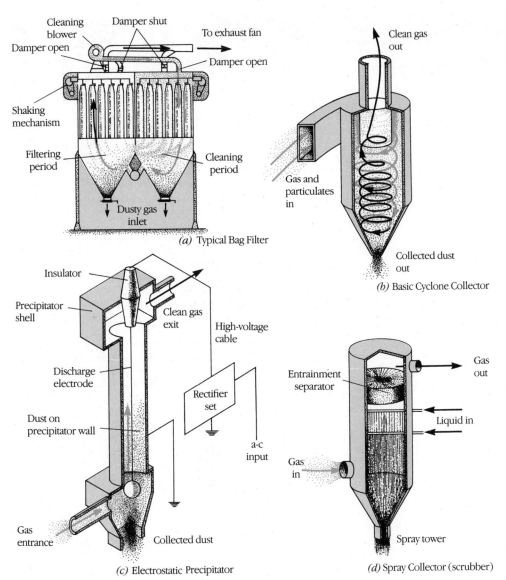

Figure 15.10 *Four pollution control devices used on stationary combustion sources.*

oxide gases. Nitrogen oxides and carbon dioxide are not removed by these devices.

Removing air pollution from stack gases helps clean up the air, but it creates a problem that many people fail to consider: hazardous wastes (Chapter 19). Particulates from power plants trapped by the various pollution control devices, for instance, contain harmful trace elements, such as mercury. Scrubbers produce a toxic sludge rich in sulfur compounds and toxic heavy metals. Improper disposal can create serious pollution problems elsewhere.

Mobile Sources Reducing pollution from mobile sources, such as cars, trucks, and planes, can also be achieved by changes in engine design. However, most new engine designs do not reduce emissions of carbon monoxide, nitrogen oxides, and hydrocarbons to acceptable levels. To reduce emissions further, it is necessary to pass the exhaust gases through **catalytic converters**, special devices that are attached to the exhaust system. Catalytic converters transform carbon monoxide and hydrocarbons into water and carbon dioxide. While this once seemed like a logical strategy, today many people realize that it eliminates two pollutants that contribute to urban air pollution, but does nothing to control carbon dioxide or nitrogen oxides.

Cars with catalytic converters can meet emissions standards if they are well tuned and if unleaded gasoline is burned in them. Leaded gasoline, which is being rapidly phased out in the United States, destroys the catalytic surface, resulting in emissions that greatly exceed standards. Unfortunately, recent statistics show that one of every ten American vehicles equipped with a catalytic converter is run on leaded gasoline.

Viewpoint 15.1

What's Sacrificed When We Arm?

Michael G. Renner
The author is a senior researcher at Worldwatch Institute and author of National Security: The Economic and Environmental Dimensions.

Our Common Future, the highly acclaimed report of the World Commission on Environment and Development, argues that Ethiopia could have reversed the steady advance of desertification threatening its food supply in the mid-seventies by spending no more than $50 million per year to plant trees and fight soil erosion. Instead, the government in Addis Ababa pumped $275 million per year into its military machine between 1975 and 1985 to fight secessionist movements in Eritrea and Tigre. When famine struck in 1985, more than a million Ethiopians died. Emergency relief measures alone carried a price tag of $500 million.

This is only one example of how governments preoccupied with military dangers ignore the equally or perhaps even more serious threats to their national security posed by environmental degradation. Facing internal or external foes armed with increasingly larger and more sophisticated arsenals, most leaders feel they have little alternative but to keep up. The result is a never-ending arms race and a waste of precious resources. A

seemingly limitless flow of weapons exports gives almost any nation or insurgency access to advanced conventional weaponry that is almost impossible to defend against—ironically, this makes all parties much less secure. For the nuclear-weapons states, the situation is even worse.

Compared to the imperative to guard against potential and real enemies at home and abroad, environmental protection is perceived as a luxury. National security is a meaningless concept, though, if it does not encompass the preservation of livable conditions within a country. Increasingly, states are finding their well-being undermined by environmental threats, such as soil erosion on their croplands, pollution of their air and water, or cataclysmic floods unleashed by denuded watersheds.

Particularly in the last two cases, these problems can be exacerbated by activities beyond their borders and, therefore, beyond their direct control. On a global scale, climate change and ozone depletion pose serious challenges to the safety and well-being of every nation.

Unfortunately, the Ethiopian famine is only one example of the fact that most leaders still see national security as primarily guaranteed by force of arms. The United States, for example, annually spends close to $300 billion on its military—even though the Cold War has ended. Yet, only one-third as much (almost two-thirds of which are private funds) are needed to deal with very real environmental pollution threats.

Because the pursuit of military power is such a costly endeavor, it drains resources needed to protect the environment and, thus, is beginning to lessen the security of many nations. This is made clear by a look at how portions of military budgets applied elsewhere could do wonders to combat serious environmental problems. For example, the estimated costs of $100 billion to meet U.S. clean-water goals by the year 2000 are

just slightly higher than the amount of money budgeted by the federal government to procure major weapons systems during fiscal years 1991 and 1992.

The Worldwatch Institute has estimated that a cumulative sum of about $774 billion would have to be expended worldwide during the final decade of this century to turn around adverse environmental trends in four priority areas: protecting topsoil on croplands from further erosion, reforesting the Earth, raising energy efficiency, and developing renewable sources of energy. This is at most 8% to 9% of current annual world military spending. Many solutions to environmental problems may lie in developing new technologies. But worldwide spending on military research and development is diverting funding in this critical area as well, growing from $13 billion per year in 1960 to $100 billion in 1986. That amount exceeds the combined governmental outlays on developing new energy technologies, improving human health, raising agricultural productivity, and controlling pollution.

National defense is a universally recognized and legitimized objective. Entrenched and powerful institutions guarantee a continuous flow of money in that direction. By contrast, the environment still does not have an adequate voice in most parliaments, cabinet rooms, or ministries.

For instance, national leaders usually regard building public transportation systems that save energy and reduce pollution as a local task. Dealing with hazardous wastes is often viewed as a regional rather than a national responsibility, while developing renewable energy sources or new production technologies is left to the private sector, which may not have much incentive to get involved.

Thus, the trade-offs can be relatively invisible to national leaders, particularly those facing insurgencies or aggressive neighbors. As the example of Ethiopia shows, they are, nevertheless, real.

Models of Global Sustainability 15.1

Germany's Environmentally Sustainable Approach Pays Huge Dividends

In Iphofen, Germany, the Knauf gypsum manufacturing plant produces drywall, which is used for building homes and offices. Although there's nothing unusual about this—drywall manufacturers are found throughout the world—the source of their raw material is worthy of note. Knauf makes drywall from scrubber sludge from a nearby coal-fired power plant. This drywall manufacturing plant has been so successful, in fact, that Knauf recently opened a second facility in England to supply British builders. The operation generates $48 million in revenue from scrubber sludge, a material that in the United States is dumped as useless waste.

This example is just one of many environmentally sustainable economic projects in Germany, indisputably the world's leader in environmental protection. Alan Miller of the Center for Global Change, a policy-analysis institute at the University of Maryland, contends, "There is no field of environmental protection where Germany does not stand out. It has the most rigorous controls of any nation, bar none. . . ."

In an article in *International Wildlife*, U.S. Senate committee on Environment and Public Works counsel Curtis Moore wrote: "The world's future begins in Germany. More than anywhere else on Earth, Germany is demonstrating that the greening of industry, far from being an impediment to commerce, is . . . a stimulus."

Germany is finding that efforts to use energy more efficiently (conserve), to recycle, and to tap into the Earth's generous supply of renewable resources, all vital components of the sustainable strategy, are creating enormous economic opportunities in the country. Consider a few examples.

Germany passed legislation that requires all car manufacturers to take back and recycle old cars. To facilitate this process, new cars rolling off the assembly line contain bar-coded parts that identify the material used in their manufacture. German auto manufacturers are also making cars so that they can be disassembled in 20 minutes.

By 1994, take-back requirements go into effect for virtually all products, helping to close the loop and reduce waste. By 1995, 72% of all glass and metals and 64% of all paperboard and plastic must be recycled. Even old refrigerators are hauled free of charge to recycling centers where chlorofluorocarbons are extracted for reuse.

In the village of Neunburg stands a $38 million plant owned by a consortium of governments and industries. This facility, which produces hydrogen from solar-generated electricity (Chapter 12), is really an experimental testing ground for new technologies. The Germans hope to be in the forefront when fossil fuel supplies run out or as nations look for alternatives to reduce their contribution to global warming and other air pollution problems (Chapter 16). If successful, Germany could power its entire economy by solar electricity and hydrogen.

Knowing that the future depends on clean, renewable energy, the German government also initiated the Thousand Roofs Program, which offers a 75% tax credit to homeowners who install photovoltaics on their roofs. The hope is that government funding will drive the costs downward and provide valuable hands-on experience in manufacturing that could position Germany well in ensuing years as photovoltaics come into wider use.

The Germans have found that even conventional pollution control strategies can pay off. For example, Germany adopted rules that require all power plants within its borders to cut sulfur oxide emissions by 90%, a goal achieved in six years. These rules stimulated considerable innovation on the part of Germany's industrial technologists. Today, technological innovators stand to make huge profits selling antipollution technology and know-how to Americans and others.

Source: Adapted from Moore, C. A. (1992). Down Germany's Road to a Clean Tomorrow. *International Wildlife* 22(5): 24–28.

In the United States, conventional catalytic converters do not remove nitrogen oxides, leaving this pollutant largely uncontrolled. American auto manufacturers once argued that an affordable converter cannot be developed. But Volvo, the Swedish automobile manufacturer, introduced a catalytic converter in 1977 that lowered nitrogen oxide emissions to well below current U.S. automobile standards. Researchers at Argonne National Laboratories in Illinois are now experimenting with a

process that will allow power plant operators to remove nitrogen oxides from smokestacks. A chemical added to scrubbers removes 70% of the nitrogen oxides and could cost much less than currently available technologies.

New Ways to Burn Coal and Natural Gas Another approach to pollution control is to develop new technologies that burn coal more efficiently. One of these is **magnetohydrodynamics (MHD)** (Figure 15.11). Coal is

Figure 15.11 *Magneto-hydrodynamics. Coal is mixed with an ion-producing "seed" substance, such as potassium, and burned. A hot ionized gas is emitted and shot through a magnetic field. The movement of the ionized gas through the magnetic field creates the electric current. Air or water is also heated and used to run an electric generator.*

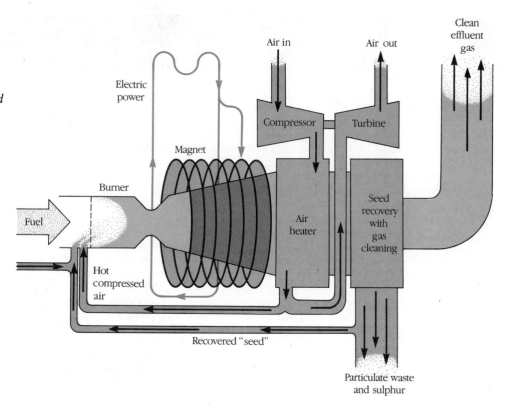

first crushed and mixed with potassium carbonate or cesium, substances that are easily stripped of electrons. The mixture, burned at extremely high temperatures, produces a hot ionized gas—a plasma—containing electrons. The plasma is passed through a nozzle into a magnetic field, generating an electric current. The heat of the gas creates steam, which powers a turbine.

MHD is about 60% efficient, compared with the 30% to 40% efficiency of a conventional coal-burning power plant. MHD systems remove 95% of the sulfur contaminants in coal, have lower nitrogen oxide emissions, and produce fewer particulates than conventional coal plants, but they release more fine particulates.

Coal may also be burned in **fluidized bed combustion (FBC)**, a technology that is also more efficient and cleaner than conventional coal-fired burners. In FBC, finely powdered coal is mixed with sand and limestone and then fed into the boiler. Hot air, fed from underneath, suspends the mixture while it burns, thus increasing the efficiency of combustion. The limestone reacts with sulfur, forming calcium sulfate and reducing sulfur oxide emissions. Lower combustion temperatures in FBC reduce nitrogen oxide formation.

Chapter 11 described the importance of shifting to natural gas and using more efficient combustion technologies, in particular natural gas turbines similar to jet engines. These efforts could dramatically reduce the emission of sulfur dioxide, particulates, and carbon dioxide from power plants and other facilities.

Cost of Air Pollution Control

Air pollution control strategies based on end-of-pipe solutions cost money. Smokestack scrubbers and electrostatic precipitators, for instance, cost millions to install and operate. Catalytic converters add $300 or more to the cost of a new car. Some controls may reap benefits far lower than their actual costs. One study, for instance, suggests that pollution controls on automobiles and other mobile sources cost about $15 billion per year but create benefits of under $1 billion per year. Although such analyses often grossly underestimate the economic and environmental savings of benefits, such as the cost of global warming, they do suggest the need for less expensive alternatives.

In some cases, pollution control can be very profitable. The Long Island Lighting Company, for example, began recovering vanadium from particulates collected at its power plants. In 1976, when the company began its program, it sold 362 metric tons of vanadium—about 9% of the total U.S. vanadium production—for $1.2 million. The company continues to extract vanadium today but only makes $10,000 to $20,000 per year because it is using a cleaner fuel. The Chemical Division of the Sherwin-Williams Company installed a pollution control system that captures solvents at a Chicago plant, saving that company $60,000 per year.

The EPA estimates that an annual expenditure of $5 billion in the United States would eliminate or sharply decrease most air pollution damage, saving $7 billion to $9 billion per year.

(The Viewpoint in this chapter suggests a possible source of money for increasing pollution control.) This expenditure would improve the quality of life, better people's health, improve visibility, reduce damage to buildings and statues, and protect wildlife. Such improvements would have an enormous impact on the habitability of the planet for present and future generations, a benefit with inestimable economic value. (Chapters 20 and 21 explain why society is often reluctant to invest in pollution control even though it could save billions of dollars.)

Toward a Sustainable Strategy

Air pollution control strategies are clearly improving the quality of the environment. But, as the previous discussion suggests, these end-of-pipe solutions treat the symptoms, not the underlying causes. In some cases, devices merely shift a pollutant from one medium (air), to another (groundwater), as in the case of toxic ash from coal-fired power plants. In others, such as the catalytic converter, they merely convert a pollutant that is toxic to humans (carbon monoxide) into a "climatic toxin" (carbon dioxide). Moreover, these pollution control strategies probably cost business far more than is necessary.

In sum, current pollution control strategies reduce the damage to the Earth's life-support system, postponing the day of reckoning. But, they do not eliminate the threat as is necessary for a sustainable future. These and other problems with the conventional air pollution control strategies have led many people worldwide to seek alternatives, that is, to develop a sustainable approach based on several of the key biological principles of sustainability: conservation, recycling, renewable resources, and population control.

As you learned in Chapter 12, conservation, recycling, and renewable energy sources can make substantial reductions in air pollution. Cutting energy demand by half through conservation measures, for instance, usually reduces air pollution output by half. Recycling aluminum rather than making it from raw ore cuts air pollution by 95%. Similar gains are made by recycling paper. Japan and Germany are leading the world in such efforts and could reap enormous economic benefits by tapping into the largely undeveloped potential of sustainable pollution control strategies. (Models of Global Sustainability 15.1 outlines innovative measures German companies are adopting to cut pollution.) Renewable energy represents a more effective means of reducing pollution.

Through individual, corporate, and government efforts, global demand for fossil fuel consumption and global pollution can be substantially lowered. These strategies are as important for businesses as they are for individuals. Each unnecessary product we buy, each bottle or can we toss out, and each gallon of gas we waste contributes to global pollution. This text promotes individual responsibility as an effective means of reducing resource demand and pollution. Individual actions, added together, are vital to the task of building a sustainable future.

The power of man has grown in every sphere, except over himself.

—Winston Churchill

❧   ❧   ❧

Critical Thinking Exercise Solution

Critical thinking rules encourage us to question sources of information and their conclusions. In this example, we find that the author of this article is grossly in error when he claims that the scientific community is divided on the issue. About 99% of the world's 700 atmospheric scientists believe that global warming is a reality; only a handful embrace the opposite view. The evidence the author introduces to support his assertion—a quote from each side of the issue—is terribly misleading.

This type of reporting is quite common in newspapers, television, and magazines. Although it is intended to give a balanced view of issues, in reality, it provides an extremely unbalanced view.

A more accurate conclusion would have been that not all scientists agree that global warming is occurring, but the vast majority do. It's important to note that even though the majority of the world's atmospheric scientists agree that global warming is happening, it doesn't mean they're right.

This exercise shows that digging a little deeper to find out more often throws conclusions into question.

Summary

15.1 Air: The Endangered Global Commons

■ Air pollutants come from a variety of natural and anthropogenic sources. As a rule, **anthropogenic pollutants**—the products of human activities—represent the most significant threat to the environment and its inhabitants.

■ The six major pollutants are carbon monoxide, sulfur oxides, nitrogen oxides, particulates, hydrocarbons, and photochemical oxidants. They come primarily from transportation, power plants, and industry.

15.2 Effects of Climate and Topography on Air Pollution

■ Cities fall into two categories. **Gray-air cities** are older, industrialized cities, often in colder climates. Their major pollutants are particulates and sulfates; they combine with moisture to form **smog**. **Brown-air cities** are new and relatively nonindustrialized and are located in sunny, dry climates. Their major pollutants are carbon monoxide, ozone, and nitrogen oxides.

■ In brown-air cities, hydrocarbons and nitrogen oxides react in the presence of sunlight to form **secondary pollutants**, major components of photochemical smog.

■ Numerous factors, such as wind, precipitation, topography, and temperature inversions, affect regional air pollution.

■ A **temperature inversion** occurs either when a high-pressure air mass stagnates over a region and forces a layer of warm air down (subsidence inversion) or when ground air cools faster than the air above it (radiation inversion). Both inversions result in the buildup of air pollution.

15.3 Effects of Air Pollution

■ Air pollution affects human health in many ways. Pollution episodes have numerous immediate effects, including discomfort, burning eyes and throats, colds, and coughs. Heart attacks and death may occur in patients with preexisting heart and lung disease.

■ Delayed effects also result from chronic exposure to air pollution, among them chronic bronchitis, emphysema, asthma, and possibly lung cancer.

■ Domestic animals are affected by many air pollutants, but the most noticeable impacts of air pollution are on wild animals and on materials, such as rubber, stone, and paint. Ozone, sulfur dioxide, and sulfuric acid are the most damaging pollutants.

■ Crops and forests are also damaged by air pollutants, particularly ozone, sulfuric acid, and sulfates, although the exact magnitude of the effect is unknown.

15.4 Air Pollution Control: Toward a Sustainable Strategy

■ To date, efforts to control pollution have relied primarily on end-of-pipe solutions, pollution control devices that capture pollutants or convert them into supposedly less harmful substances.

■ In the United States, the **Clean Air Act**, its amendments, and the regulations it spawned created emissions standards for automobiles, national ambient air-quality standards for major pollutants, standards for emissions from new stationary pollution sources, and a program to prevent significant deterioration of air.

■ The most recent amendments to the Clean Air Act, passed in 1990, set new deadlines for establishing emission standards for nearly 200 toxic chemicals and established a system of **pollution taxes** on toxic chemical emissions that could create an incentive for companies to cut back on their release. They also established a market-based incentive program, consisting of tradable permits, to reduce nitrogen and sulfur oxides. In addition, they called for a phaseout of ozone-depleting chemicals.

■ The recent amendments also tightened emission standards for automobiles and raised the average mileage standards for new cars.

■ The 1990 amendments are an important step forward because they call for several sustainable strategies.

■ Most U.S. clean-air legislation has promoted the use of pollution control devices, such as filters, cyclones, electrostatic precipitators, and scrubbers, which can eliminate air pollution but generally only shift the pollutant from one medium to another.

■ Pollution control strategies cost money but, in some cases, they can become an economic asset. The EPA estimates that an annual expenditure of $5 billion in the United States would eliminate or sharply decrease most air pollution damage, saving $7 billion to $9 billion per year.

■ This expenditure would improve the quality of life, better people's health, improve visibility, reduce damage to buildings and statues, and protect wildlife. Such improvements would have an enormous impact on the habitability of the planet for present and future generations.

■ Although air pollution control strategies are clearly improving the quality of the environment, they tend only to reduce the damage to the Earth's life-support system, not eliminate the threat. This and other problems with conventional air pollution control strategy have led many people to seek an alternative, a sustainable approach based on some of the biological principles of sustainability, notably, conservation, recycling, renewable resources, and population control.

Discussion and Critical Thinking Questions

1. "Natural pollutants are produced in greater quantities than anthropogenic air pollutants," says one expert. "So what are we worried about? Why should we spend billions to reduce air pollution?" Using your knowledge of air pollution and critical thinking skills, analyze this statement.

2. Describe the pollutants produced by burning fossil fuels. How is each product formed? Why is coal a "dirtier" fuel than gasoline?

3. Define the terms *primary* and *secondary pollutant.*

4. In what ways are gray-air and brown-air cities different? What are the major air pollutants in each type of city?

5. What is photochemical smog? How is it formed? Why are suburban levels of photochemical smog often higher than urban levels?

6. Nine eastern states recently agreed to adopt tough California standards for automobile emissions to control photochemical smog. These regulations will primarily cut down on hydrocarbon emissions. Given what you know about photochemical smog, critically analyze this effort. Do you think it is needed or would other steps be more effective, given the large number of trees along the eastern seaboard?

7. Describe some factors that affect air pollution levels in your community.

8. What is a temperature inversion?

9. What are the health effects of: (1) carbon monoxide, (2) sulfur oxides and particulates, and (3) photochemical oxidants?

10. List the major chronic health effects of air pollution. Which pollutants are thought to be the cause of these effects?

11. Of all the impacts of air pollution, which concern you the most? How do your education, home life, and religious beliefs affect your answer to this question?

12. Discuss the air pollution control legislation enacted by Congress, highlighting important features of the Clean Air Act and its amendments. How would you characterize the conventional approach to air pollution? How has this approach contributed to the prevalent notion that environmental regulation is bad for the economy?

13. List the ways you can help reduce air pollution.

Suggested Readings

Amdur, M. O. (1991). Air Pollutants. In *Casarett and Doull's Toxicology: The Basic Science of Poisons.* New York: Pergamon. Detailed analysis of the toxic effects of air pollutants.

Bruemmer, F. (1991). In Praise of the Lowly Lichen. *International Wildlife* 21(6): 30–33. Delightfully interesting account on lichens and their vulnerability to air pollution.

Cannon, J. (1990). *The Health Costs of Air Pollution: A Survey of Studies 1984 to 1989.* New York: American Lung Association. Excellent review.

Environmental Protection Agency. (1992). *National Air Quality and Emissions Trend Report 1991.* Research Triangle Park, North Carolina. Current statistics on air quality and emissions of major regulated pollutants.

Howe, C. W. (1991). An Evaluation of U.S. Air and Water Policies. *Environment* 33(7): 10–15, 34–36. Excellent analysis of U.S. policy.

Moore, C. (1992). Down Germany's Road to a Clean Tomorrow. *International Wildlife.* 22(5): 24–28. Extraordinary piece on the economic benefits of the sustainable strategy.

Renner, M. (1989). *National Security: The Economic and Environmental Dimensions.* Worldwatch Paper 89. Washington, D.C.: Worldwatch Institute. Detailed discussion of the relationship between environmental protection and national security.

White, A. L. (1991). Venezuela's Organic Law. *Environment* 33(7): 16–20, 37–42. Describes legislative actions to protect Venezuela's environment.

Indoor Air Pollution

Recent studies show that many Americans are exposed to high levels of potentially harmful substances in their homes and offices. These are known as **indoor air pollutants**. According to the EPA, the four most serious ones are cigarette smoke, radon (a radioactive substance), formaldehyde, and asbestos. As you learned in the last chapter, even low-frequency magnetic fields found in and around homes may pose a threat to human health and reproduction (Case Study 14.1).

Sometimes found at remarkably high levels, indoor air pollutants enter our bodies through the lungs or skin, sometimes both. These substances come from various sources, among them stoves, heating systems, and furniture. Even wood paneling, plywood, and carpet contribute to dangerous levels of pollution.

How serious a problem is indoor air pollution? In 1986, the EPA estimated that indoor air pollutants may cause several hundred U.S. cancer deaths per year. In 1988, the EPA revised that figure upward after finding that one pollutant alone, radon, may cause as many as 2000 to 20,000 cases of lung cancer each year. Indoor air pollution, in fact, may cause more lung cancer than ordinary air pollutants.

EPA officials believe that toxic substances in homes and offices are much more likely to cause cancer than ambient air pollutants for two main reasons. First, indoor levels are often much higher than outdoor levels, in some cases up to 100 times higher. Even in pristine rural areas, indoor air can be more polluted than outside air next to a chemical plant. Second, people, on average, spend about 90% of the time indoors. As you might expect, the people with the highest risk are the young, elderly, and infirm—that is, individuals who are sick or have heart and lung disease. Pregnant women are also at greater risk, as are office or factory workers who spend inordinate amounts of time indoors in polluted buildings.

The EPA estimates that one-fifth to one-third of all U.S. buildings, including its own headquarters in Washington, D.C., contain unhealthy levels of indoor air pollutants. These are often classified as **sick buildings**. The agency also estimates that 100,000 to 200,000 workers die prematurely each year as a result of exposure to indoor air pollution at work—either in offices or, more likely, in factories.

The EPA estimates that 10 million to 20 million Americans suffer from **sick building syndrome**, an assortment of debilitating health effects caused by polluted indoor air. These individuals display a variety of health problems, among them chronic respiratory problems, sinus infections, sore throats, and headaches. People may also complain of dizziness, rashes, eye irritation, and nausea. These symptoms are largely caused by exposure to formaldehyde in buildings, present in furniture, plywood, and other sources.

Although estimates of the effects of indoor air pollution are crude, they do suggest the need to take action. In order to understand various options, we will examine three of the major indoor air pollutants—products of combustion, formaldehyde, and radon.

Products of Combustion

Although most of us are unaware of it, a greal deal of combustion occurs in homes and other buildings.

Tobacco Smoke

Some combustion sources, such as cigarettes, pipes, and cigars, are relatively small but release large amounts of pollutants directly into the air. Rooms polluted by tobacco smoke contain high levels of particulates. In one study, researchers found that the particulate level in urban homes whose residents are nonsmokers averages about 40 micrograms per cubic meter. Smokers raised particulate levels in some cases to 700 micrograms, over ten times greater than the level allowed by ambient air-quality standards set by the Clean Air Act.

Combustion of tobacco also releases carbon monoxide (CO). As you may recall, CO binds to hemoglobin molecules in red blood cells and reduces their oxygen-carrying capacity. This effect is especially harmful to individuals with heart and lung diseases. Sulfur oxides and nitrogen oxides are also emitted by tobacco smoke. These lung irritants are believed to be responsible for emphysema and chronic bronchitis. Finally, cigarette smoke contains a number of carcinogens.

Vented Sources: Water Heaters, Furnaces, and Wood Stoves

Other indoor combustion sources, such as water heaters, furnaces, and wood stoves, burn large amounts of fuel but are usually vented to the outside. It is leaks in these devices that contribute to indoor air pollution.

Unvented Sources: Gas Stoves and Kerosene Heaters

Several combustion sources burn large amounts of fuel and release them directly into the air. Included in this group are gas stoves and kerosene heaters. Thus, as the fuel burns, pollutants enter the air.

Gas stoves are one of the most common unvented sources of indoor air pollution. They produce large amounts of carbon monoxide and nitrogen dioxide. As shown in Figure S15.1, levels in a kitchen with the windows closed can increase from a few parts per million (ppm) to over 40 ppm when four burners are in operation for half an hour or so. CO levels also increase appreciably in neighboring rooms. Nitrogen dioxide levels often follow the same pattern. (For a summary of the health effects, see Table 15.2.)

Concentrations of indoor air pollutants fall after combustion sources are turned off, of course, but it may take several hours before normal levels are reached in conventional, poorly sealed homes (Figure S15.2). In well-sealed, energy-efficient homes that haven't made allowances for ventilation, it can take much longer. If several lengthy meals are cooked during a day, exposure to these pollutants can be quite high.

Researchers have found that in certain homes the nitrogen oxide and carbon monoxide levels can exceed the limits set to protect human health. What the long-term implications are, no one knows.

Sulfur dioxide is not generally a problem in homes unless kerosene heaters are used. A popular source of heating in the 1970s and early 1980s, because of their relatively low cost, kerosene space heaters can release

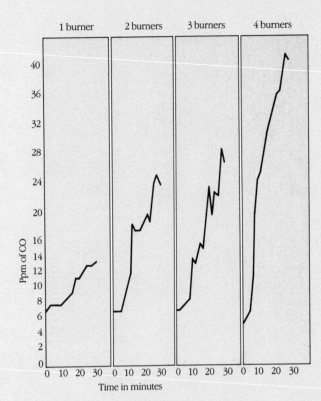

Figure S15.1 *Carbon monoxide levels in a kitchen with one to four gas burners turned on.*

significant amounts of carbon monoxide, nitrogen dioxide, and sulfur dioxide, even when functioning properly. When they malfunction, they release even more. For example, if the wick of a kerosene heater is damaged or covered with soot or if there is water in the fuel, pollution levels can become intolerable, creating headaches, coughing, and irritation of the throat. Because of these problems, kerosene heaters have been banned for indoor use in many states.

Reducing Pollution from Combustion Sources

The strategy required for each source of pollution varies. For tobacco smoke, the best policy is to prevent indoor smoking. To reduce pollution from leaky vented sources, the best strategy is to plug up the leaks or replace the appliance. For unvented combustion sources, like stoves, exhaust fans should be used when houses are closed up. On warm days, the windows should be cracked open. If ventilation is impossible, one might want to replace their gas stove (trade it or donate it) with an electric stove, bearing in mind that electric ranges are more inefficient and costly than natural gas or propane. Kerosene stoves should be avoided altogether.

Sustainable strategies, such as passive solar space and water heating, can help reduce indoor combustion by water heaters and furnaces and reduce ambient air

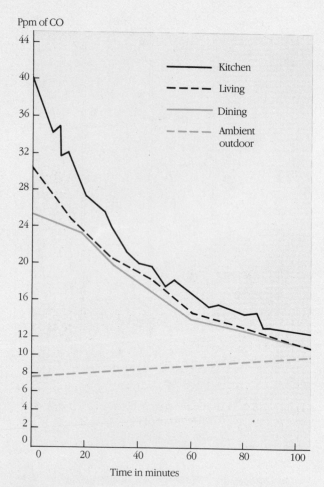

Figure S15.2 *Drop in carbon monoxide level in the kitchen, dining room, and living room after the stove is turned off.*

pollution as well. In tightly sealed, energy-efficient homes, air-exchange systems can be installed for use when the house is closed off. These systems periodically replace room air with outside air. Some systems merely bring in new air and vent old air. Others contain heat exchangers that transfer heat to the incoming air, reducing the amount of heat lost to the outside. Certain plants also help reduce indoor air pollution. For example, one spider plant per room is sufficient to prevent the buildup of nitrogen oxide.

Formaldehyde

Formaldehyde is a preservative for biological specimens. It is also a common indoor air pollutant. About 3.7 million metric tons of formaldehyde are used in the United States each year. Formaldehyde is found in the adhesive in plywood, particleboard, and paneling. Particleboard and other formaldehyde-containing wood products are used to build countertops, cabinets, subflooring, and about 90% of the furniture sold in the United States.

Drapes and upholstery also contain formaldehyde, as do the adhesives in wallpaper and carpeting.

One of the most notorious sources of formaldehyde exposure is the mobile home. According to EPA estimates, at least half of the mobile homes in the United States have levels sufficient enough to cause health problems. Two of every 1000 people living in a mobile home, according to the EPA, will develop lung cancer within ten years as a result of their exposure to formaldehyde.

Clearly, it is hard to avoid this chemical. Many new buildings, among them condos, homes, and townhouses, are built with wood containing large amounts of this substance. It is even found in paper products, toothpastes, shampoos, waxed paper, grocery bags, and some medicines, to which it is added to kill bacteria, fungi, and viruses.

One source, now banned, is urea-formaldehyde foam insulation. This product was once sprayed in the spaces between walls and was used heavily in mobile-home construction. The levels of formaldehyde in homes insulated with this material are four times higher than those in homes with other types of insulation. Furniture containing formaldehyde in wood and fabric has been shown to increase the levels in a previously unfurnished house by three times.

People with the greatest risk are those living in mobile homes or in newly built homes containing particleboard, plywood, or paneling (Table S15.1). In conventionally built homes with an abundance of cracks and poor insulation, air exchange between the inside and the outside occurs roughly once every hour. In airtight homes, the turnover is much slower, perhaps once every five hours. The tighter the home, the greater the concentration of formaldehyde unless special precautions are taken to avoid products containing this harmful substance or periodically to vent the indoor air.

Formaldehyde irritates the eyes, nose, and throat, but sensitivity varies among people. Some are sensitive to levels of 1.5 to 3 ppm, but others, who have been exposed to formaldehyde for long periods, become sensitized and respond to levels as low as 0.05 ppm.

At high levels, formaldehyde causes nasal cancer in rats and possibly mice. In monkeys who receive levels to which humans are typically exposed at work and at home, formaldehyde causes cellular changes believed to be the early stages of cancer within the linings of the respiratory tracts. It has also been shown to cause mutations in bacteria and many other organisms; many mutagens are also carcinogens. One epidemiological study showed a possible link between formaldehyde exposure and skin cancer in humans, but the evidence is sketchy.

Since 1987, risk assessment for cancer has been revised downward by a factor of 50 for indoor exposure from pressed wood products. However, the EPA believes that noncancer effects are now more important and that people are variably sensitive to levels from 0.1 ppm to 3 ppm.

Reducing Formaldehyde Exposure

The U.S. Consumer Product Safety Commission (CSPC) banned urea-formaldehyde insulation in 1981. Although the ban was overturned in court a few years later, almost no one is manufacturing or using it anymore. Like the CSPC, the EPA has taken little regulatory action; its only regulation is one that requires pesticides containing formaldehyde to disclose this fact on the label. In recent years, both the EPA and CSPC have worked with industry to promote voluntary reductions in formaldehyde in particleboard used for flooring and fiberboard used to make cabinets and furniture. Voluntary reductions in particleboard alone could reduce emissions by two-thirds by the end of the century.

The EPA's past refusal to regulate formaldehyde reflects a new attitude. In previous regulatory decisions, the fact that a chemical had caused cancer in any laboratory animal was enough to warrant controls. Under the Reagan administration (1981–1989), though, the EPA took a more conservative approach, maintaining that controls were unwarranted without conclusive proof from studies on humans exposed to formaldehyde. Animal studies were considered inadequate.

Many critics believe that this approach could weaken protection of human health. Norton Nelson, a highly regarded health scientist, contends: "Epidemiological studies must be regarded as a crude and insensitive tool. Only the most violent and intense carcinogens are likely to be detected by epidemiological techniques." Many critics agree and have pressed for an EPA ban on the use of formaldehyde in plywood, particleboard, paneling, and textiles. The U.S. Department of Housing and Urban Development now requires mobile-home builders to disclose any use of formaldehyde-containing products.

Homeowners can take steps by insisting that exterior-grade pressed wood be used for building or remodeling homes. They can also ventilate their homes when a new formaldehyde-containing product, a carpet or couch, for instance, is introduced. Air-exchange systems may be of assistance as well.

Radon

A more difficult indoor air pollutant to control is the naturally occurring radioactive gas **radon**. A daughter product of radium, found in rocks and soils, radon enters homes through cracks in the foundation or from the soil in homes without foundations. In some cases, stone, brick, and cement contain small quantities of radium that emit radon into homes and other buildings. Well water may also be contaminated with radon, which is emitted while showering, bathing, or washing dishes.

Inhaled, radon gas may emit radiation in the lungs that can cause mutations and lung cancer. Of greater concern, however, are the radioactive decay products of

Table S15.1   Formaldehyde in Mobile Homes and Houses Insulated with Urea-Formaldehyde Foam (UFF)

Type	Average Level of Formaldehyde (parts per million)
Homes without UFF	0.03
Homes with UFF	0.12
Mobile homes with UFF	0.38
Background[1]	0.01

[1]The background level is the normal atmospheric concentration of formaldehyde.

radon. When radon decays, or emits radiation, it is converted into radioactive lead, which can become lodged in the lung, providing long-term internal exposure to radiation. Lead and other decay products may also adhere to airborne particles that can be breathed into the lungs and also become lodged.

In 1988, a report by the U.S. assistant surgeon general noted that smoking raises a person's radon-related cancer risk 15 times higher than that of nonsmokers. A preliminary study, released in 1992, suggests that even nonsmokers face a higher risk from radon if they live or work with smokers.

Two health physicists performed a series of two- and three-day tests in which they found that room levels of radon's radioactive decay products increased dramatically in the presence of fine particulates from cigarette smoke. For example, within five hours of lighting a single cigarette in one nonsmoking family's basement, researchers found that the radioactive decay product levels in the room increased approximately 25%, although levels of radon gas remained the same. This effect lasted approximately nine hours before tailing off. A second cigarette lit 24 hours after the first increased the levels 40%.

How prevalent is radon? In the late 1980s, the EPA surveyed radon levels in houses in ten states and found that one of every five homes exceeded the *action level*, that is, the level at which efforts are recommended to lower radon exposure. The action level (4 picocuries per liter of air) poses the same lung cancer risk to individuals as smoking half a pack of cigarettes per day or receiving 200 to 300 chest X rays per year. In 1988, a new survey of 11,000 homes in Arizona, Indiana, Massachusetts, Minnesota, Missouri, North Dakota, Pennsylvania, and several midwestern states found that one of every three homes exceeded the action level. As a result, Lee Thomas, then administrator of the EPA, recommended that virtually everyone in the United States test their residence for radon.

Bernard Cohen, a health physicist from the University of Pittsburgh, believes that radon exposure in homes may cause more deaths than all other types of radiation exposure—natural and anthropogenic—combined. As with other forms of cancer, the link between cause and effect is difficult to prove. The risks associated with low-level radon exposure in American homes are generally extrapolated from studies of workers who have been exposed to very high levels, among them survivors of the atomic bomb blasts, recipients of high-dose X rays, and workers in uranium mines.

As noted in Chapter 14, some researchers believe that there is a threshold to radiation effects—that is, a level below which no hazard exists. Erring on the conservative side, EPA officials have assumed that even tiny exposures pose some risk. Controversial studies by Cohen suggest that there is a threshold level below which radon exposure is harmless. Similar studies in Scandinavia, Florida, South Carolina, New Jersey, and New York have turned up the same result, suggesting that the dangers of low levels of radon may be exaggerated. To date, though, the EPA has not altered its acceptable level.

Radon levels in existing and new buildings can be controlled in a variety of ways. In existing homes and offices, for example, owners can install air-to-heat exchangers that vent indoor air while saving heat. Owners can also seal cracks in the foundation to prevent radon from entering. In new construction, builders can install a system of porous pipes in a layer of gravel under the foundations of new homes. The pipes draw the radon away, preventing it from entering the house. (Figure S15.3.)

Although most of the radon that seeps into American homes comes from the underlying soil and rock, this is not the only source. In some instances, radon may be coming from groundwater and may be released into a room when the faucets are turned on. When the taps are turned on, water flows and radon gas escapes into the room. To control such emissions is difficult. Homeowners may have to connect to public water supplies if they are available or drill new wells.

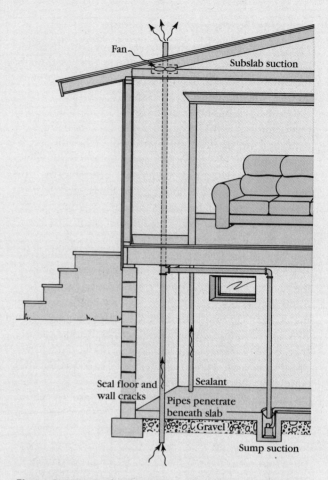

Figure S15.3 *Radon levels in existing and new buildings can be controlled in a variety of ways. Owners can seal cracks in the foundation to prevent radon from entering. In new construction, builders can install a system of porous pipes in a layer of gravel under the foundations. The pipes draw the radon away, preventing it from entering the house.*

Controlling Indoor Air Pollutants

As an environmental problem, indoor air pollution is a relatively new one. The novelty of this problem is partly responsible for the fact that no U.S. laws address it. Some legal experts argue that the Clean Air Act could be applicable to indoor air pollutants. For example, one provision of the Clean Air Act that could be applied to indoor air pollutants authorizes the EPA to draw up emissions standards for hazardous air pollutants. Some critics believe that the EPA, under authority loosely granted by these provisions, could develop formaldehyde emissions standards for plywood, particleboard, paneling, and other household products. Unfortunately, the overworked and underfunded EPA has its hands full with other responsibilities and has only taken steps to regulate one indoor air pollutant, asbestos (Chapter 14).

Some proponents call for a whole new set of amendments to the Clean Air Act to address indoor air pollutants through a series of indoor air pollution standards. As you can well imagine, the technical and legal problems in enforcing indoor standards would be enormous.

A second, more likely approach might be to use the Toxic Substances Control Act (Chapter 14). This law gives the EPA broad authority to control the production, distribution, and disposal of potentially hazardous chemicals. Bans on plywood, carpets, furniture, and other products containing formaldehyde could be implemented. Outright bans would seem less desirable than emissions controls. To date, though, the EPA Toxic Substance Office has not taken any steps in either direction.

A final potential weapon is the **Consumer Product Safety Act**. This law gives the **CSPC** the authority to regulate consumer products deemed hazardous to the public. Products that generate indoor air pollution certainly qualify. Under the law, the commission could develop safety standards for various products. Emission standards for stoves, for example, could indicate the permissible emissions of carbon monoxide, and standards for plywood, textiles, and furniture could set acceptable formaldehyde emissions. The commission could require manufacturers to warn the public of potential dangers associated with the use of their products.

Indoor air pollution is an emerging problem that has only recently been brought to the public's attention in the United States. It is also a growing problem elsewhere, especially in developing countries, where environmental standards in factories are much looser or nonexistent. In countries where wood, dung, and crop residues are the primary source of cooking and heating fuel, levels of toxic pollutants in homes make U.S. concerns seem trivial. More efficient stoves, vented to the outside, and the use of solar cookers, small solar-powered stoves that are inexpensive and easy to build, could greatly improve conditions by reducing exposure in the homes of the rural poor.

Suggested Readings

Environmental Protection Agency. (1992). *A Citizen's Guide to Radon* (2nd ed.) Washington, D.C.: Environmental Protection Agency. Overview of steps individuals can take to reduce radon exposure.

Lipske, M. (1987). How Safe Is the Air Inside Your Home? *National Wildlife* 25 (5): 34–39. Excellent reference.

Sheldon, L. S. et al. (1988). *Indoor Air Quality in Public Buildings*. Vol. I. Washington, D.C.: United States Environmental Protection Agency. Technical report.

Sheldon, L. S. et al. (1988). *Indoor Air Quality in Public Buildings*. Vol. II. Washington, D.C.: United States Environmental Protection Agency. Technical report.

Smith, K. R. (1988). Air Pollution: Assessing Total Exposure in the United States. *Environment* 30 (9): 10–15, 33–38. Outlines efforts to study indoor air pollution and personal exposure.

U.S. House Committee on Science and Technology. (1985). *Radon and Indoor Air Pollution. Hearing before the Subcommittee on Natural Resources, Agriculture Research, and Environment*. 99th Congress. October 10, 1985.

Vietmeyer, N. (1985). Plants That Eat Pollution. *National Wildlife* 25 (5): 10–11. A look at plants thought to lower pollution.

Chapter 16

Global Environmental Challenges: Ozone Depletion, Acid Deposition, and Global Warming

For 200 years we've been conquering nature. Now, we're beating it to death.

—*Tim McMillan*

Critical Thinking Exercise

In 1987, the industrial nations of the world met to negotiate a worldwide treaty to limit ozone-depleting chemical production and use. A few years later, faced with ever-worsening news on the depletion of the ozone layer, they met again and agreed to a complete ban on many ozone-depleting chemicals by the year 2000. On February 12, 1992, the *New York Times* announced that then President George Bush, who was concerned about the decline in the ozone layer, "directed American manufacturers to end virtually all production of chemicals that destroy ozone" by 1995, well ahead of schedule. For this move, the president received high praise from supporters.

However, there was a loophole. The president decided to permit limited production and use of these chemicals. Read the quote again and look for a key word or words that suggest the loophole. What critical thinking rule or rules does this exercise underscore?

In June 1992, 178 nations convened in Rio de Janeiro to hammer out the final language of Agenda 21, a massive blueprint for sustainable development. They also negotiated final language for various agreements and treaties on global warming, biodiversity, and forest protection, among others. (The outcome of the Earth Summit is described in Chapter 23.) This meeting was held in large part because of widespread recognition that human society is destroying key life-support systems of the planet and that without global cooperation we could greatly decrease the habitability of the planet for all life forms.

When I began studying environmental problems, over 20 years ago, most issues were local or regional in scale. Today, many problems have reached global proportions and require massive global cooperation to solve. This chapter looks at three global issues: ozone depletion, acid deposition, and global warming. Nuclear war is discussed in Chapter Supplement 16.1.

16.1 Stratospheric Ozone Depletion

Encircling the Earth is a thin, protective layer of **ozone gas** (O_3), which screens out 99% of the sun's harmful **ultraviolet (UV) radiation**. The ozone layer occupies the outer two-thirds of the stratosphere, 20 to 50 kilometers (12 to 30 miles) above the Earth's surface. The screening effect of the ozone layer protects all organisms from ultraviolet radiation, which is known to be both mutagenic and carcinogenic.

When ultraviolet radiation strikes ozone molecules, it causes them to split:

$$UV + O_3 \rightarrow O + O_2$$

The products, however, quickly reunite, reforming ozone and emitting heat:

$$O + O_2 \rightarrow O_3 + heat$$

Thus, the ozone layer is a renewable form of protection that converts harmful ultraviolet radiation into heat.

Life on Earth depends on the ozone layer. Without it, terrestrial life would vanish. Some animals would suffer serious burns and would develop cancer and lethal mutations. As described shortly, humans would be especially vulnerable. Plants would suffer as well. Unable to cope with the intense influx of ultraviolet radiation, most plants would perish and with them millions of species that depend on them for nutrition.

Activities That Deplete the Ozone Layer

Human civilization is systematically destroying the ozone layer through two principal activities: (1) the use of spray cans and refrigerants containing Freon gas and (2) jet travel through the stratosphere.

Freons In 1951, Freon spray-can propellants entered the U.S. market. **Freons**, described in Table 16.1, are also known as **chlorofluorocarbons (CFCs)**. Two CFCs were commonly used: (1) **Freon-11**, a propellant now banned in several countries, including the United States, and (2) **Freon-12**, still used in refrigerators, air conditioners, and freezers.

Until the early 1970s, chemists listed Freons as inert (unreactive) chemicals. Conventional wisdom held that Freon gases simply diffused into the upper layers of the atmosphere, where they were broken down by sunlight, without harm. Their release into the atmosphere was, therefore, of little concern.

In the early 1970s, however, two U.S. scientists reported that chlorine (Cl) atoms produced by the breakdown of Freons could react with stratospheric ozone (Figure 16.1). Shortly after their announcement, three research teams reported that highly reactive chlorine atoms produced when Freons are broken down, known as **chlorine free radicals**, could indeed react with and destroy many molecules of ozone (Figure 16.1b). As the chemical equation in Figure 16.1 illustrates, when chlorine reacts with ozone, it produces **chlorine oxide (ClO)**. This chemical can react with oxygen free radicals, freeing up chlorine to react with more ozone molecules (Figure 16.2a). Chlorine oxide can also react with ozone directly (Figure 16.2b). Because of these reactions, a single CFC molecule can destroy 100,000 molecules of ozone.

Table 16.1   Commonly Used Freons

Generic Name	Use	Chemical Name	Chemical Formula
Freon-11	Spray-can propellant	Trichloromonofluoromethane	$$\begin{array}{c} \text{Cl} \\ | \\ \text{Cl}-\text{C}-\text{Cl} \\ | \\ \text{F} \end{array}$$
Freon-12	Coolant in refrigerators, freezers, and air conditioners	Dichlorodifluoromethane	$$\begin{array}{c} \text{Cl} \\ | \\ \text{F}-\text{C}-\text{Cl} \\ | \\ \text{F} \end{array}$$

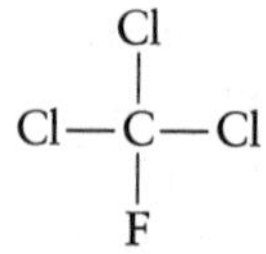

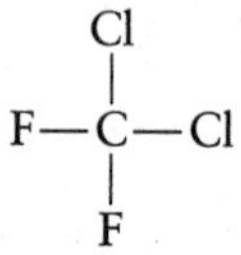

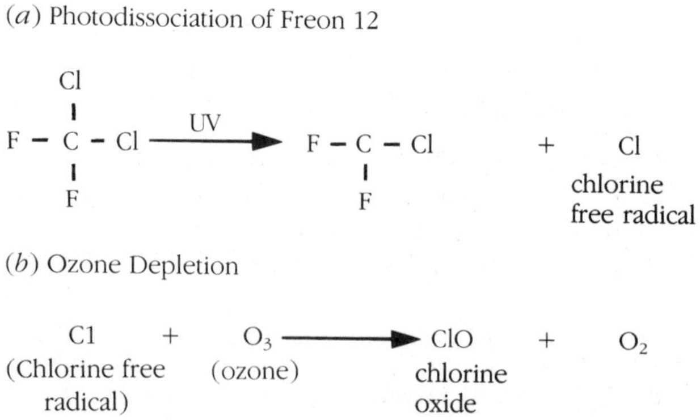

(*a*) Photodissociation of Freon 12

$$\begin{array}{c} \text{Cl} \\ | \\ \text{F}-\text{C}-\text{Cl} \\ | \\ \text{F} \end{array} \xrightarrow{\text{UV}} \begin{array}{c} \\ \text{F}-\text{C}-\text{Cl} \\ | \\ \text{F} \end{array} + \underset{\substack{\text{chlorine} \\ \text{free radical}}}{\text{Cl}}$$

(*b*) Ozone Depletion

$$\underset{\substack{\text{(Chlorine free} \\ \text{radical)}}}{\text{Cl}} + \underset{\text{(ozone)}}{\text{O}_3} \longrightarrow \underset{\substack{\text{chlorine} \\ \text{oxide}}}{\text{ClO}} + \text{O}_2$$

Figure 16.1   (a) *Freons, or chlorofluorocarbons (CFCs), are dissociated by ultraviolet radiation in the strato-sphere. This produces a highly reactive chlorine free radi-cal.* **(b)** *The free radical can react with ozone in the ozone layer, forming chlorine oxide. This reaction reduces the ozone concentration.*

(*a*) $\underset{\text{(Chlorine oxide)}}{\text{ClO}} + \underset{\substack{\text{(oxygen free} \\ \text{radical)}}}{\text{O}} \longrightarrow \underset{\substack{\text{(chlorine free} \\ \text{radical)}}}{\text{Cl}} + \underset{\substack{\text{(molecular} \\ \text{oxygen)}}}{\text{O}_2}$

(b) $\text{ClO} + \text{O}_3 \longrightarrow \text{ClO}_2 + \text{O}_2$

Figure 16.2   (a) *A single molecule of Freon gas can elim-inate many thousands of molecules of ozone, because chlorine oxide breaks down, re-forming the chlorine free radical.* **(b)** *Note, too, that chlorine oxide also reacts with ozone molecules, destroying them.*

$\underset{\substack{\text{Supersonic} \\ \text{Transport} \\ \text{Jet}}}{} \longrightarrow \underset{\substack{\text{(nitric} \\ \text{oxide)}}}{\text{NO}} + \underset{\text{(ozone)}}{\text{O}_3} \longrightarrow \underset{\substack{\text{(nitrogen} \\ \text{dioxide)}}}{\text{NO}_2} + \underset{\text{(oxygen)}}{\text{O}_2}$

Figure 16.3 *Supersonic and subsonic jets produce nitric oxide, which can react with ozone and reduce the ozone layer.*

High-Altitude Jets   The ozone layer may be vulner-able to other human activities as well. For instance, sci-entists believe that aircraft, such as the supersonic transport (SST) and military jets that fly in the strato-sphere, may destroy ozone through the release of nitric oxide (NO) produced by jet engines. Nitric oxide gas reacts with ozone to form nitrogen dioxide (NO_2) and oxygen (O_2) (Figure 16.3).

Because of this, in 1971, Congress killed plans to subsidize the construction of 300 to 400 SSTs; however, the British-French Concorde is in use today on a limited scale. Because the Concorde flies lower and burns less fuel than the proposed U.S. SST, it has less impact on the ozone layer. Ordinary commercial jets also produce nitric oxide and thus contribute to the destruction of the ozone layer but to a lesser degree.

Other Sources of Destruction

Other activities can also destroy the ozone layer. The detonation of nuclear weapons in the atmosphere, for example, produces ozone-destroying nitric oxide. In fact, an active period of atmospheric testing of nuclear weap-ons in the 1940s and 1950s caused a moderate but short-lived decrease in ozone. In the event of a nuclear war, ozone levels could fall dangerously low (Chapter Sup-plement 16.1).

Nitrogen fertilizer, which farmers apply to their fields, may also threaten the life-protecting ozone layer because nitrogen in fertilizers can be converted into nitric oxide gas in the soil. Nitric oxide gradually diffuses into the stratosphere where it reacts with ozone molecules.

Although the use of fertilizers has risen dramatically in the last decade, no one knows what harm they may cause in the long run. Other industrial pollutants, such as methyl chloride and carbon tetrachloride, also diffuse into the ozone layer where they eliminate ozone mole-cules. Finally, natural pollutants, such as nitrogen oxides from volcanoes and chloride ions from sea salt, also destroy the ozone layer, although these processes are presumably in balance with natural ozone replenishment.

Extent and Effects of Ozone Depletion

Concern over ozone depletion originated from the work of two chemists, F. Sherwood Rowland and Mario Molina, who in 1974 shocked the scientific community with their announcement that the previously considered safe CFCs could actually dismantle ozone molecules high in the stratosphere. Their projections indicated that CFCs could eventually destroy 20% to 30% of the ozone layer, imperiling life on Earth.

In the years following their report, many people grew skeptical of the early projections because other estimates indicated that the effect would be much smaller than Rowland and Molina had suggested. By 1984, most experts on the issue believed that a decline of only 2% to 4% sometime in the 21st century would result from the release of CFCs into the atmosphere. As a result, stratospheric ozone depletion became a "dead" issue for a while. In the mid-1980s, however, EPA scientists delivered another shocking report, which said that a 60% decline in stratospheric ozone levels by the year 2050 was likely if the production of CFCs continued to grow by 4.5% per year. Even a 2.5% increase, they contended, would deplete the ozone layer 26% by the year 2075.

These projections came in the wake of startling results from satellite studies of stratospheric ozone and CFC levels. One such study showed that the CFC concentration in the stratosphere had doubled in ten years. More alarming, though, were reports in the 1980s of a huge **ozone hole** about the size of the United States over Antarctica. Much to the surprise of scientists, this mysterious hole in the ozone layer began to appear every other spring over the southernmost tip of the globe and appeared to be growing worse with each episode.

Scientists found that CFCs and several natural climatic conditions could explain the now infamous ozone hole. One such weather phenomenon was a vortex of wind (a whirlpool in the atmosphere) that circles the pole during the winter months, blocking out warmer air. Studies showed that the vortex contributes to the formation of polar stratospheric clouds. Scientists found that ice crystals in these clouds facilitate the breakdown of ozone. Chemical reactions that destroy the ozone may occur on the surfaces of the ice crystals.

In December 1990 (Antarctica's summer), studies of ultraviolet radiation reaching the ground at Palmer Station, a U.S. base on Antarctica, showed that radiation reached a record high—twice the normal value. The researchers suggest that high levels of ultraviolet radiation result from the longer-than-normal persistence of the springtime ozone hole. In October 1991, studies

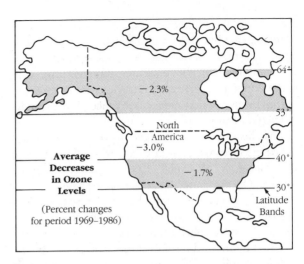

Figure 16.4 *Map of North America showing average annual ozone depletion at different latitudes.*

showed that ozone depletion over Antarctica was even greater, marking the third year in a row that a severe ozone hole had formed. This three-year trend may signal a dramatic shift from the pattern observed in the 1980s.

More recently, researchers found a similar hole in the ozone layer over the Arctic. Ozone levels there, however, are not as severely depleted because the vortex is not as strong as it is in the south. Therefore, invading winds can break through, keeping the Arctic air much warmer and reducing ozone destruction.

Although some critics believe that ozone depletion is a hoax, in 1987, several government agencies, including NASA, convened an international panel of more than 100 scientists to study the satellite data that had been gathered since 1962. In March 1988, the scientists confirmed that ozone had declined 1.7% to 3% over the Northern Hemisphere since 1969 (Figure 16.4). Over the heavily populated areas of North America and Europe, ozone levels had fallen 3%. The greatest declines, ranging from 5% to 10%, were recorded over Antarctica and the southern tip of Argentina. Do these declines result in an increase in ultraviolet radiation striking the Earth?

As the previous discussion showed, in Antarctica, the ozone hole resulted in a substantial increase in the amount of ultraviolet radiation striking the Earth. In most midlatitude regions, no increases have been recorded. In December 1991, Shaw C. Liu and his colleagues at the National Oceanic and Atmospheric Administration in Colorado concluded that atmospheric haze (fine particulates from human sources) in midlatitude nations currently filters out an amount of ultraviolet radiation equal to or greater than the excess now penetrating the stratosphere because of ozone depletion. The bad news, say the researchers, is that this "benefit" will probably prove transitory as industrial nations tighten controls on the pollutants responsible for haze. This could increase the amount of ultraviolet radiation striking the Earth. In reasonable amounts, ultraviolet radiation tans light skin and stimulates vitamin D production in the skin. However, excess ultraviolet exposure causes serious burns and may induce skin cancer. It is also lethal to bacteria and plants.

Medical researchers believe that a 1% depletion of the ozone layer would lead to a 2% to 4% increase in skin cancer. In November 1991, a panel of scientists convened by the United Nations released a report on the predicted effects of ozone depletion throughout the atmosphere. The scientists estimated that a 10% decrease in stratospheric ozone concentrations, which is expected by the end of this century, will cause 300,000 additional cases of skin cancer per year worldwide. Assuming a 4% mortality rate, about 12,000 additional deaths would occur annually in the United States alone. The scientists also predicted that a 10% decrease will cause 1.6 million additional cases of cataracts (a clouding of the lens of the eye) each year.

Studies of skin cancer show that light-skinned people are much more sensitive to ultraviolet radiation than more heavily pigmented individuals. In addition, some chemicals commonly found in drugs, soaps, cosmetics, and detergents may sensitize the skin to ultraviolet radiation. Thus, exposure to sunlight may increase the incidence of skin cancers among light-skinned people and users of many commercial products.

The 1991 report further noted that land- and water-dwelling plants could suffer from the depletion but emphasized that very little is known about effects of additional ultraviolet light on plants. Intense ultraviolet radiation is usually lethal to plants; smaller, nonfatal doses damage leaves, inhibit photosynthesis, cause mutations, or stunt growth.

In November 1991, Susan Weiler, head of the American Society of Limnology and Oceanography in Walla Walla, Washington, testified before Congress that studies in Antarctica by other scientists showed that phytoplankton (algae and other free-floating photosynthetic organisms) populations decrease about 6% to 12% when stratospheric ozone concentrations over the region drop by 40% (Chapter 2). Since phytoplankton form the base of the aquatic food chain, damage to them could cause widespread ecological problems. One scientist thinks that ozone depletion and subsequent effects on the food chain may be the reason why two species of penguin are declining in Antarctica.

Preventing Ozone Depletion

In the 1970s, fears of early projections of ozone depletion moved several nations, including the United States, Sweden, Finland, Norway, and Canada, to decrease CFC emissions. In 1978, for example, the United States

banned Freon used in spray cans. Freon-12, a refrigerant, coolant, and blowing agent used in the production of plastic foam, was not affected by the ban.

In light of the more recent findings on CFCs and ozone depletion, it became evident that worldwide cooperation was needed. In 1987, the United Nations sponsored negotiations aimed at reducing global CFC production. In September of that year, 24 nations signed a treaty, called the **Montreal Protocol**, which would cut production of five CFCs in half by 1999 and freeze production of halons, which are used in fire prevention systems, at 1986 levels. (In halons, bromine atoms replace some or all of the chlorine atoms.) Although halons are used in much smaller quantities worldwide, they are far more detrimental than CFCs.

This agreement paved the way for a gradual decline in CFC production in industrial nations. But, critics argued that it had too many loopholes. Like so many other pollution control strategies, it would only slow the rate of destruction, not stop it. EPA officials were disappointed because their computer projections showed that an 85% reduction in CFC emissions was needed to stabilize CFC levels in the atmosphere.

Before the Montreal Protocol went into effect something unusual happened. In March 1988, an international panel (described above) announced that ozone levels had fallen throughout the world. Two weeks later, DuPont, a major producer of CFCs, called for a total worldwide ban on CFC production when only two weeks earlier it had said that it would not support such a ban.

Continuing bad news about ozone depletion brought negotiators to the table once again, this time in London, where in June 1990, they reached a new agreement. This treaty was signed by 93 nations and called for the complete elimination of CFCs and halons by the year 2000 if substitutes were available by then. The signatories also agreed to phase out other ozone-depleting chemicals, among them carbon tetrachloride, methyl chloride, and even HCFCs, a class of chemicals once thought to be an excellent substitute for CFCs.

Unfortunately, the news about the ozone layer continues to worsen. In 1992, a team of 40 scientists announced record-high concentrations of chlorine oxide (ClO) in New England and Canada. Concentrations such as these had never been seen before, even in the Antarctic ozone hole. If chlorine levels continue to climb, chances are good that a severe Arctic ozone hole will begin to appear with great regularity, exposing Canada and parts of the United States, Europe, and Asia to dangerous levels of ultraviolet radiation.

Aircraft measurements in 1992 also showed rather disturbing findings about global ozone levels outside the Arctic. In flights as far south as the Caribbean, scientists detected chlorine oxide concentrations of up to five times the amount they had anticipated.

Because of continuing reports indicating faster-than-anticipated depletion of the ozone layer, representatives from 74 countries met in Copenhagen in November 1992 to hammer out another amendment to the Montreal Protocol. The new agreement accelerates the reduction in the production of chlorofluorocarbons and carbon tetrachloride by four years. It also phases out halons six years ahead of schedule and would end the production of methyl chloroform, a dry-cleaning agent, nine years faster than previously agreed. The plan initiates controls on hydrochlorofluorocarbons, or HCFCs, chemicals that were once viewed as a major substitute for CFCs. Although these substances are less damaging to the ozone layer, they still cause significant problems. The new treaty freezes HCFCs at 1991 levels and eliminates them by 2030. Despite progress, the agreement has a huge loophole that will allow companies to continue to use the banned chemical substances indefinitely for "essential uses" and for servicing existing equipment.

The Good News and Bad News About Ozone

The ozone story is an encouraging one. It illustrates how scientific knowledge can be used by society for the common good of humans and all other species. It also illustrates how disparate factions can work cooperatively toward a common future and undoubtedly served as a model for the international agreement on global warming signed at the Earth Summit in 1992.

A sobering word is in order, however. CFCs take about 15 years to migrate into the stratosphere; therefore, all of these chemicals produced in the past decade and a half have yet to reach the ozone layer. As they diffuse into this layer and begin breaking down, serious ozone depletion is likely to occur. Because many millions of metric tons of CFCs are already in the atmosphere and because they last about 100 years, it might take a century or more for the ozone layer to return to preindustrial concentrations.

16.2 Acid Deposition

In the 1960s, forest ranger Bill Marleau built the cabin of his boyhood dreams on Woods Lake in the western part of the Adirondack Mountains in New York State. Isolated in a dense forest of birch, hemlock, and maple, the lake offered Marleau excellent fishing. Ten years after Marleau finished his cabin, however, something bizarre happened: Woods Lake, once a murky green suspension of microscopic algae and zooplankton, teeming with trout, began to turn clear. As the lake went through this mysterious transformation, the trout stopped biting and soon disappeared altogether. Then, the lily pads turned brown and died; soon afterward, the bullfrogs, otters, and loons disappeared.

What happened to Woods Lake? What destroyed the web of life at this small, isolated lake, far from any sources of pollution? Scientists from the New York Department of Environmental Conservation say that Woods Lake is "critically acidified." As a result, virtually all forms of life in and around it have perished or moved elsewhere. The lake became acidified from acids and acid precursors deposited from the skies.

The deposition of acidic substances from the sky, generated from pollution largely human in origin—**acid deposition**—is commonplace today, as are lakes like Marleau's. In fact, Woods Lake is only 1 of about 375 lakes and ponds in the western Adirondacks that has turned acidic and hazardous to virtually all forms of life by acid deposition. In eastern Canada, 100 lakes have met a similar fate. In Scandinavia, the death count is 10,000. Across the globe, thousands of lakes are threatened unless something is done, and quickly.

Widely publicized as one of the most serious environmental threats facing us today, acid deposition is a phenomenon not only of environmental interest but also of grave economic importance. Studies show that acid deposition turns lakes acidic, kills fish and other aquatic organisms, damages crops, destroys forests, alters soil fertility, and destroys statues and buildings. Moreover, scientists are finding that acid precipitation is more widespread than once thought and is taking a larger toll on the environment and economics than originally imagined. Recent reports indicate that it poses a universal threat, affecting the developed countries as well as many developing nations. Earthscan, an international environmental group, reports that acid deposition is already damaging soils, crops, and buildings in much of the developing world. Rapidly growing urban centers with their poorly regulated industry and traffic congestion are largely the culprits. Ironically, tough pollution laws in developed countries have given multinational corporations incentives to set up operations in developing nations, whose pollution laws are, if existent, certainly much weaker.

What Is Acid Deposition?

To understand acid deposition requires a closer look at acids: how they form in the atmosphere and how they are deposited. Acidity is related to the amount of free hydrogen ions (H^+) in solution. The degree of acidity is measured on the potential hydrogen or pH scale, which ranges from 0 to 14 (Figure 16.5). Acidic substances, such as vinegar and lemon juice, have low pH values, that is, less than 7. Basic, or alkaline, substances, such as baking soda and lime, have high pH values on the scale, greater than 7. Neutral substances, such as pure water, have a pH of 7.

The pH scale, like the decibel scale, which measures loudness of a sound, is logarithmic. Thus, a change of

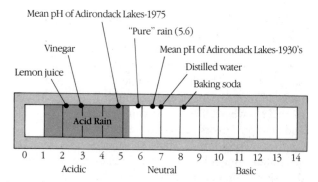

Figure 16.5 *A pH scale to indicate acid-base level.*

1 pH unit represents a tenfold change in the level of acidity. Therefore, rain with a pH of 4 is 10 times more acidic than rain with a pH of 5, 100 times more acidic than rain with a pH of 6, and 1000 times more acidic than rain with a pH of 7.

In an unpolluted environment, rainwater is slightly acidic, having a pH of about 5.7; the normal acidity of rainwater is created as atmospheric carbon dioxide is dissolved in water in clouds, mist, or fog and converted into a mild acid, carbonic acid. Acid precipitation is rain and snow with a pH below 5.7.

Wet Deposition   Acid deposition includes two broad categories: wet deposition and dry deposition. **Wet deposition** refers to acids deposited in rain and snow. These acids are formed when two gaseous air pollutants, sulfur and nitrogen oxides, combine with water in the atmosphere. Sulfur oxides form sulfuric acid; nitrogen oxides react with water to form nitric acid. Sulfuric and nitric acid are two of the three strongest acids known to science.

Produced in the atmosphere, these acids may accumulate in clouds and fall from the sky in rain and snow. Even coastal fogs may contain droplets of acid that, when deposited on buildings or plants, can cause noticeable damage. One study shows that moisture droplets in low-lying clouds (fog) tend to contain higher concentrations of acid than rain or snow that falls from them. Therefore, fog and clouds may bathe trees in highly acidic water. Making matters worse, recent studies suggest that the evaporation of recently deposited cloud water from forest canopies may result in acid concentrations on leaf surfaces much higher than those found in the cloud droplets themselves.

Dry Deposition   Sulfate and nitrate particulates are also present in the atmosphere. These pollutants may settle out of the atmosphere much like fine dust particles. This process is one form of **dry deposition**. Settling onto surfaces, these particulates can combine with water to form acids. Sulfur and nitrogen oxide gases may also be adsorbed onto the surfaces of plants or solid surfaces, where they, too, combine with water to form acids. This is another type of dry deposition.

Where Do Acids Come From?

Acid precursors, the chemical substances (primary pollutants) that give rise to acids, come from both natural and anthropogenic sources. The natural sources of sulfur oxides include volcanoes, forest fires, and bacterial decay. Anthropogenic sources are of major concern, however, because they are often concentrated in urban and industrialized regions, causing local levels to be quite high. About 70% of all anthropogenic sulfur dioxide comes from electric power plants, most of which burn coal. Like sulfur oxides, nitrogen oxides arise from a wide variety of sources. The most important anthropogenic sources are electric power plants and motor vehicles.

American factories, cars, and power plants currently produce approximately 21 million metric tons of sulfur dioxide and about 20 million metric tons of nitrogen oxides a year. The EPA estimates that because of regulations annual sulfur dioxide emissions will decrease to about 15 million metric tons by the year 2000 and that annual nitrogen oxide emissions will decrease to 16 million metric tons.

The Transport of Acid Precursors

Acids deposited in many regions of the world often originate from pollution sources many hundreds of kilometers away. Why? Acid precursors and acids can remain airborne for two to five days and may travel hundreds, perhaps even thousands, of kilometers before being deposited. Studies have shown that acids falling from the sky in southern Norway and Sweden usually originate in Great Britain and industrialized Europe. In the United States, acid precipitation falling in the Northeast usually comes from the industrialized Midwest, primarily the upper Mississippi and Ohio River valleys. Indiana and Ohio are the two major producers. Moving eastward, the mass of pollutants tends to converge on New York State and New England, where 50% of the lakes are in jeopardy because of low acid-neutralizing capacities.

Acid deposition is a global phenomenon. It is found in the United States, Canada, the Amazon Basin, Europe, and the Netherlands, to name a few. All of these areas share several features: they are downwind of heavily polluted areas and experience relatively high levels of acidic deposition. In Europe and Scandinavia, rain and snow samples frequently have pH values between 3 and 5. In the White Mountains of New Hampshire, the average annual pH of rainfall is about 4 to 4.21, nearly 100 times more acidic than normal precipitation. In the

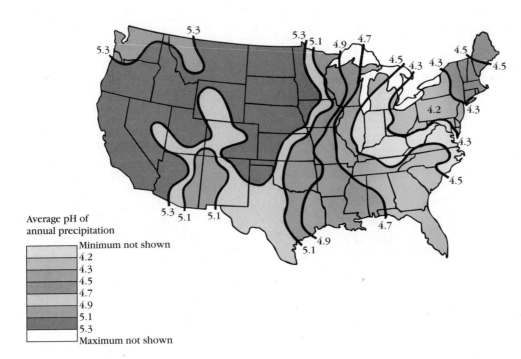

Figure 16.6 *Acid precipitation in the eastern United States, 1955 and 1990. Note the worsening of acid rain and the wider area experiencing it.*

Average pH of annual precipitation

Minimum not shown
4.2
4.3
4.5
4.7
4.9
5.1
5.3
Maximum not shown

1970s, rain samples collected in Pasadena, California, had an average pH of 3.9. One of the lowest pH measurements was made in Kane, Pennsylvania, where a rainfall sample with a pH of 2.7 was recorded—rain as acidic as vinegar. The grand prize for acidic rainfall, however, goes to Wheeling, West Virginia, where a rainfall sample had a pH of 2—stronger than lemon juice. More recent studies of acid fog downwind from the Los Angeles Basin, however, show fog water with pH levels as low as 1.7.

In southern Norway and Sweden and the northeastern United States, two ominous trends have been observed. First, acid precipitation is falling over a wider area than it was 10 to 20 years ago; second, areas over which the strongest acids are falling are expanding (Figure 16.6).

Impacts of Acid Deposition

Acidification of Lakes   Throughout the world, lakes and rivers and their fish are dying at an alarming rate. In the 1930s, for example, scientists surveyed lakes in the western part of the Adirondacks. Sampling the pH of 320 lakes, they found that most had pHs ranging from 6 to 7.5. In a 1975 survey of 216 lakes in the same area, a large number had pH values below 5, a level at which most aquatic life perishes (Figure 16.7). Of the acidified lakes, 82% were devoid of fish life. A new study of 1500 lakes in New York's Adirondack Park found that 25% of the lakes are so acidic that fish no longer live in them. Another 20% of lakes are acidic enough to be endangered.

In 1988, the National Wildlife Federation published a list of U.S. lakes that have become acidified. The study shows that eastern lakes have been particularly hard hit. One of every five lakes in Massachusetts, New Hampshire, New York, and Rhode Island is acidic enough to be harmful to aquatic life. As acid deposition continues, these lakes could become lethal to virtually all forms of life.

Recent studies have shown that acid deposition also occurs widely over the northern and central portions of Florida, with precipitation ten times more acidic than normal (Figure 16.8). One-third of the lakes in Florida are now acidic enough to be harmful to aquatic life.

In the mountains of southern Scandinavia, acidification of surface waters has occurred at a rapid rate for more than 40 years. In Sweden, approximately 20,000 lakes are without or soon to be without fish. Salmon runs in Norway have been eliminated because of the impact of acid precipitation on egg development, putting an end to inland commercial fishing in some areas.

In Canada, nine of Nova Scotia's famous salmon-fishing rivers have already lost their fish populations because of acidity. Eleven more are teetering on the brink of destruction. In southern Ontario and Quebec, acid precipitation has destroyed at least 100 lakes. By the year 2000, scientists predict nearly half of Quebec's 48,000 lakes will have been destroyed.

Effects on Parks and Wilderness Areas   Parks and wilderness areas in the United States are endangered by acid precipitation because many lie downwind from major industrial centers and have thin soils and waters

Figure 16.7 *The pH level in Adirondack lakes. In the 1930s, most lakes had a pH of 5.0 or higher, but in the 1970s, a large percentage had a pH of less than 5.*

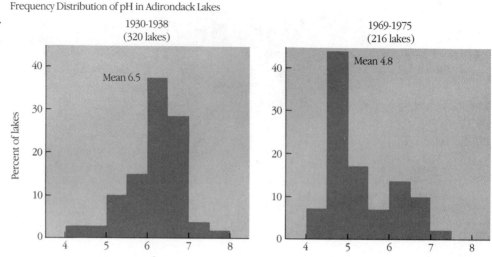

Frequency Distribution of pH in Adirondack Lakes

1930-1938 (320 lakes) — Mean 6.5

1969-1975 (216 lakes) — Mean 4.8

Figure 16.8 *Rainfall pH values in Florida. Lines show average pH within a region. Dots with pH values indicate specific readings at various sites in the state.*

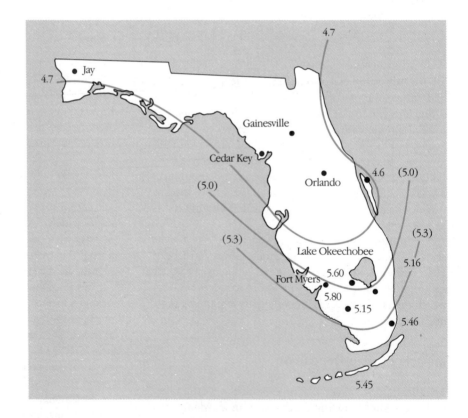

low in **buffers**, chemical substances that allow aquatic systems to resist changes in pH. When hydrogen ion levels increase, buffers combine with the free ions and eliminate them. When levels fall, they release them, thus maintaining a constant pH.

Preliminary data suggest that acid precipitation in the Great Smoky Mountains National Park has already stressed existing trout populations in the poorly buffered lakes. The average annual pH for rainfall is 4.3, over ten times more acidic than normal rain. Also of particular concern is the Quetico–Superior Lake country of Canada and northern Minnesota—home to three huge parks: Quetico Provincial Park in Canada and Voyageurs National Park and the Boundary Waters Canoe Area Wilderness in the United States. The EPA reports that one-fourth to one-third of the lakes in the two U.S. parks have so little buffering capacity that fish and other life forms are in danger.

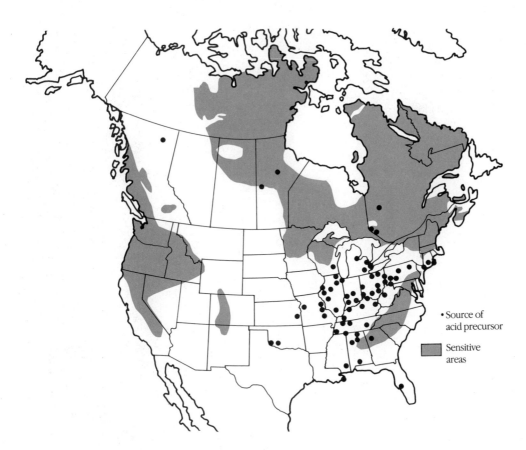

Figure 16.9 *Acid-sensitive areas in the United States and Canada (shaded areas) and major sources of acid precursors (dots).*

• Source of
acid precursor

Sensitive
areas

In the first comprehensive study of parks in the U.S. West, scientists found that Yosemite, Sequoia, Mount Rainier, North Cascades, and Rocky Mountain national parks were already affected by acid precipitation. A number of others were not affected but could be if western precipitation continues to increase.

Acidification of lakes and streams occurs in areas with a common geological denominator: thin soils with low acid-neutralizing capacities. Acid-sensitive areas in the United States, shown in Figure 16.9, include many mountainous regions. Ironically, the major producers, Indiana and Ohio, are the least vulnerable. Their thick topsoils contain abundant buffers.

Effects on Aquatic Ecosystems   Many species of fish die when the pH drops below 4.5 to 5 (Figure 16.10). Falling pH is only part of the reason fish die. Scientists have found that acidic rainwater or snowmelt dissolves toxic elements, such as aluminum, from soil and rocks. The acidic waters carry the metals to streams and lakes. Aluminum irritates the gills of brook trout, causing a buildup of mucus and, ultimately, death by asphyxiation (Figure 16.11).

Spring poses a special threat to fish and other aquatic organisms. When the snow begins to melt, the surface melts first. This water drains through the unmelted snowpack, leaching out most of the acids. In

fact, the first 30% of the meltwater contains virtually all of the acid and typically has a pH of 3 to 3.5, which is toxic to eggs, fry, and adult fish. Thus, when snow begins to melt, the acid concentration in nearby lakes and streams rises rapidly. This surge of acids coincides with the sensitive reproductive period for many species of fish.

Widening the Circle of Destruction   Many other species are affected by acid deposition. Songbirds living near acid-contaminated lakes in Scandinavia, for instance, lay eggs with softer shells than birds feeding near unaffected lakes. Scientists have found elevated levels of aluminum in the bones of affected birds and hypothesize that it comes from eating aquatic insects living in acidified waters. The aluminum interferes with normal calcium deposition, resulting in defective eggshells and fewer offspring.

Acidification of surface waters may also be partly responsible for a 60% decline in the population of black ducks on the East Coast in the last three decades. Although other factors have accounted for part of this decline, it appears that acids are killing the aquatic insects needed by female ducks and their offspring.

Spotted salamanders are also adversely affected by acidity. In the laboratory, exposure to water with a pH of 5 prevents normal embryonic development and results

Figure 16.10 *The sensitivity of fish and other aquatic organisms to acid levels varies. The figure indicates the lowest pH (highest acidity) at which the various organisms can survive. The yellow perch, for example, can withstand a pH of 4.5, but populations plummet if the pH falls any further. Smallmouth bass and mussels are more acid sensitive, perishing at levels below 5.0 and 5.5, respectively.*

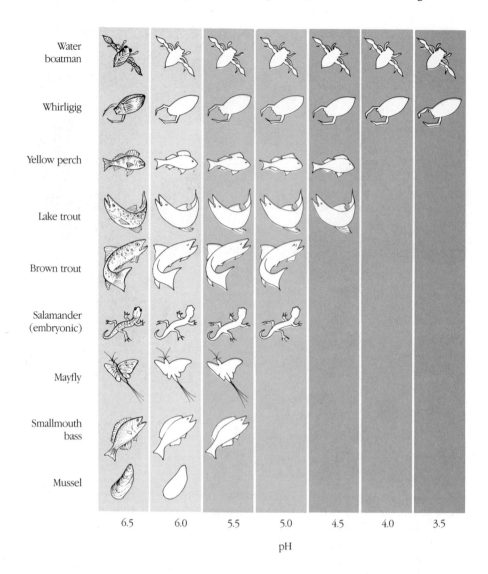

in gross deformities that are usually fatal. One study found that the mortality of fertilized eggs was 60% at pH 6 but only 1% at pH 7.

In the wild, spotted salamanders breed in "temporary ponds" created by melted snow. These ponds are likely to be highly acidic in regions where acid precipitation is prevalent; as a result, the fate of the spotted salamander is bleak.

The spotted salamander is as important as the birds and small mammals in the food chain. A drastic change in its population would very likely have serious repercussions in the entire ecosystem. Many other amphibian species have also disappeared or severely declined in their natural habitat, largely as a result of acid deposition (Chapter 8).

Damage to Recreational Resources In addition to damaging aquatic ecosystems, acids also damage recreation industries, forests, crops, buildings, and statues.

A report by the Ohio State government contends that if efforts to control acid precipitation are not undertaken soon, 2500 lakes per year will die in Ontario, Quebec, and New England throughout the remainder of this century. Many of the areas most susceptible to acid precipitation depend economically on recreation. In Ontario alone, approximately 2000 fishing lodges contribute $150 million per year to the economy. A spokesperson for the Ontario Ministry of the Environment says that even at the current rate of acid deposition, "there could be a $64 million loss and serious survival problems for 600 lodges over the next 20 years."

Forest Damage Numerous studies show that acid precipitation damages forests, causing significant decreases in productivity. Acid precipitation causes foliar damage to birch and pines, impairs seed germination of spruce seeds, erodes protective waxes from oak leaves, and leaches nutrients from plant leaves. In the former

Figure 16.11 *These fish were confined to a cage in a stream affected by acid rain. They died of asphyxiation caused by acid leaching of aluminum from the soil. Aluminum irritates the gills and causes mucus buildup, which blocks oxygen influx and kills the fish.*

Czechoslovakia, researchers estimate that 120,000 hectares (300,000 acres) of forest has been destroyed by pollution, mostly acid precipitation. In West Germany, 500,000 hectares (1.25 million acres) of forest is dead or dying. Even the famous Black Forest is now severely damaged by acidic pollutants from industry and automobiles. In Vermont's Green Mountains, half the red spruce, a high-elevation tree, have died from acid precipitation and acid fog. Lower-elevation sugar maples are also on the decline.

Swiss scientists believe that damage to trees may increase the likelihood of avalanches because trees help retain snow on steep mountainsides. In the next few years, 10% of the "barrier forests" may be lost, endangering the safety of mountain residents, skiers, and highway travelers.

Why are forests dying? In 1988, Robert Brock, a forest epidemiologist, reported findings from studies on Mt. Mitchell in North Carolina that help to explain the tragedy. He found that low-lying clouds that often bathe spruce and fir trees on the mountain are considerably more acidic than vinegar. Two days after a two-day cloudy period, Brock found that needle tips looked singed. These needles contained 7 to 11 times more sulfate than healthy ones.

In 1988, Professor Lee Klinger from the University of Colorado proposed another hypothesis to help explain why the world's forests are dying. One of the chief culprits, he says, may be an acid-loving moss that grows on the forest floor. Klinger has studied 100 regions in 30 states where forests are dying. In each one, he found a thick layer of moss carpeting the forest floor. These mosses act as sponges, holding so much water that the surface soils become saturated. In affected areas, the feeder roots of the trees and the trees themselves die for the same reason that a houseplant dies when it is overwatered: water eliminates air from the soil. Plants literally suffocate. Mosses may also kill mycorrhizal fungi that help trees absorb nutrients. Mosses acidify the

water passing through them. Acidic water dissolves toxic trace metals like aluminum found in the soil, which can also kill the root system. It is likely that direct foliar damage and root damage may combine forces.

Crop Damage Concern for agriculture has been raised by numerous researchers. Current estimates hold the economic costs of crop damage from acid precipitation to be around $5 billion per year in the United States. However, the results of many studies are inconclusive. Some researchers have reported that simulated acid precipitation decreases crop productivity, but others have found increases, and still others, no effect.

Scientists know that acid precipitation with a pH less than 3 damages leaves on bean plants. Laboratory studies of tobacco plants show that simulated acid precipitation leaches calcium from leaves.

Acid precipitation is particularly harmful to buds; therefore, acids falling on plants in the spring might impair growth. In addition, acid precipitation appears to inhibit photosynthesis, a process by which plants produce carbohydrates and other important chemicals.

Acids may damage plants by altering the soil. For example, acid rain may leach important elements from the soil, resulting in lower yield and reduced agricultural output. Acidification of soils may also impair soil bacteria and fungi that play an important role in nutrient cycling and nitrogen fixation, both essential to normal plant growth. Recent evidence shows that acids dissolve aluminum from the soil; aluminum damages cells in the water-transporting tubules of trees, closing off water transport and killing the trees. In some areas, sulfur and nitrogen from acid rain may enhance soil fertility. However, direct damage to growing plants and damage to the soil could easily offset the fertilizing effect.

Damage to Materials Acid precipitation also corrodes man-made structures and has taken its toll on some of special importance, such as the Statue of Liberty, the

Canadian Parliament building in Ottawa, Egypt's temples at Karnak, and the caryatids of the Acropolis—not just architectural works but works of art, priceless treasures.

Acid rain may also damage house paint and etch the surface of automobiles and trucks. One U.S. report claims that acid precipitation causes an estimated $5 billion in damage to buildings in 17 northeastern and midwestern states. The price tag includes the cost of repairing mortar, galvanized steel, and stone structures as well as the cost of repainting. It does not include damage to automobile paint, roofing materials, and concrete, potentially adding billions of dollars to the cost.

New Acid Rain Threat   The Environmental Defense Fund (EDF) recently found that airborne nitrates or nitric acid deposited in Chesapeake Bay stimulates the growth of algae and aquatic plants, which can impair navigation. In addition, surface plants reduce light penetration, blocking sunlight needed for photosynthesis in plants and algae in deeper layers. The plants help maintain oxygen levels; without them, sea life may perish. In addition, when aquatic vegetation dies in the fall, it decays. The bacteria that break down this organic matter also rob the water of oxygen, killing many aquatic organisms.

According to the EDF, acid deposition contributes at least one-fourth of the nitrogen entering Chesapeake Bay each year from human activities. Acid deposition, in fact, ranks second only to fertilizer runoff as a source of nitrogen. EDF scientists believe that acid precipitation may be a significant source of nitrogen pollution along the entire eastern seaboard and could negate efforts to reduce surface runoff.

Solving a Growing Problem

In 1984, the New York State legislature passed a bill that required utilities to reduce sulfur emissions by 30% by 1991. In that same year, nine European nations and Canada signed an agreement to make similar reductions but over a ten-year period. Minnesota also passed legislation to curb the growing problem.

The first significant U.S. legislation came with the passage of the 1990 amendments to the Clean Air Act (Chapter 15), which contained provisions for significant reductions in sulfur dioxide emissions and more modest cuts in nitrogen oxides by the year 2000.

International efforts to control sulfur oxide emissions generally rely on three strategies: (1) the installation of scrubbers on new and existing coal-fired power plants, (2) the combustion of low-sulfur coal or natural gas in utilities, and (3) the combustion of desulfurized coal—that is, coal from which the sulfur has been removed. Nitrogen dioxide levels are more difficult to

control because nitrogen oxide gases come from air that reacts with oxygen in high-temperature furnaces.

Another approach to the problem of acid deposition involves treating the lakes directly. In 1977, for example, the Swedish government embarked on an expensive program to neutralize acidic lakes by applying lime to thousands of lakes and rivers. These actions improved the water quality in many lakes, saving fish populations. In 1988 and 1989, Sweden spent over $18 million liming lakes, rivers, streams, and forests. In 1992 and 1993, the country budgeted $32 million.

Critics argue that liming is a short-term, stopgap solution, a little like administering CPR to a heart attack victim. In Canada, liming costs $120 per hectare ($50 per acre). The cost of treating a single lake, therefore, ranges between $4000 and $40,000. In five years, treated lakes often turn acidic again.

Pursuing another stopgap approach, Cornell's Carl Schofield is developing a strain of acid-resistant brook trout. Despite its immediate logic, this approach is doomed to fail. Questions remain: Even if a strain of trout that could survive at pH 4.8 were developed, what would happen when the pH dropped to 4.5? And what about the trout's food supply?

Toward a Sustainable Strategy

The approaches outlined above are symptomatic of an unsustainable society. Although important, stopgap measures treat the symptoms of the problem while ignoring the underlying root causes: our heavy dependence and terribly inefficient use of fossil fuels.

Recognizing that we must confront the root causes, some countries, such as Germany, Norway, and Sweden, have sought to reduce their dependence on fossil fuels. These countries and others have embarked on ambitious programs to promote energy efficiency and renewable energy sources, such as solar energy (Chapter 12). In Mexico, utilities, government, and industry are working together to make energy efficiency a cornerstone of the country's development. In India, private businesses have launched a program to install thousands of energy-efficient compact fluorescent light bulbs in homes and businesses. Such efforts provide a model for the rest of the world to follow.

16.3   Global Warming/Global Change

Scientists have long known that air pollution can affect local weather. For example, smoke from factories can substantially increase rainfall in areas downwind. In recent years, however, a growing body of evidence shows that air pollution can also affect global climate. In order to understand why, we must first look at the global energy balance, the basis for the planet's climate.

Global Energy Balance

Each day the Earth is bathed in sunlight. As you may recall from Chapter 2, approximately one-third of the sunlight striking the Earth and its atmosphere is reflected back into space. The rest is absorbed by the air, water, land, and plants. Absorbed sunlight is converted into heat, or **infrared radiation**, which is slowly radiated back into the atmosphere. Eventually all heat escapes the Earth's atmosphere and returns to space. As a result, energy input is balanced by energy output.

Scientists have discovered that this delicate balance may be altered by certain air pollutants, notably carbon dioxide (CO_2), nitrous oxide (N_2O), methane (CH_4), and CFCs. How do these gases affect the Earth's energy balance?

Consider carbon dioxide. Naturally occurring carbon dioxide allows sunlight to pass through the atmosphere and heat the Earth but absorbs infrared radiation escaping from the Earth's surface and radiates it back. This process helps maintain the Earth's temperature. The balance can be upset, though, if concentrations of carbon dioxide exceed normal levels. As they rise, more heat is reradiated to the Earth, causing the atmosphere to warm. Glass in a greenhouse behaves similarly and thus helps reduce heat escaping from the interior of the structure. Consequently, carbon dioxide and others that act similarly are known as **greenhouse gases**.

Upsetting the Balance: The Greenhouse Effect

Three of the four greenhouse gases—carbon dioxide, nitrous oxide, and methane—are emitted by natural sources. These gases, however, are also released in large quantity from anthropogenic sources. CFCs come solely from human sources. The production and release of anthropogenic greenhouse gases have risen rapidly in the past 40 years, creating concern among many scientists and environmentalists over a phenomenon called the **greenhouse effect** or **global warming**, a possible global rise in temperature caused by a buildup of gases that upset the Earth's heat balance.

The thermometer in the greenhouse that heats my office and bedroom in winter reads 38° C (100° F) on a sunny winter day, even when the outside temperature is well below freezing. Imagine what the global temperature would be if we were to encompass the Earth in a sphere of glass.

In many ways, we may be doing something similar. Although the average temperature would not be as high as it is in a greenhouse, it would be high enough to cause significant changes in average temperature, rainfall patterns, agricultural productivity, and natural vegetation, to name a few.

Table 16.2 lists the four greenhouse gases and several important facts about each, including their sources, the rate at which they are increasing in the atmosphere, and their life span. It also lists their relative greenhouse efficiency—that is, how they compare as a greenhouse gas to carbon dioxide. As shown, 1 molecule of CFC is equivalent to 15,000 molecules of carbon dioxide, partly because CFCs last so long.

Between 1870 and 1991, global concentrations of carbon dioxide increased about 23% (from 290 to 355 ppm). This rise is attributed to increasing global consumption of fossil fuels, deforestation, and burning of forests, especially in the tropics. Many scientists believe that global carbon dioxide levels could double between the years 2030 and 2050 if fossil fuel consumption continues to increase at the current pace.

To predict the effects of such a doubling, scientists use **global climate models**, computer programs that contain mathematical equations that attempt to simulate the various aspects of the climate and how they change. Using results of such models, the U.N.-sponsored Intergovernmental Panel on Climate Change (IPCC) predicts that if greenhouse gas emissions continue at the current rate, the Earth's temperature will increase approximately 0.5° F per decade, or 3° F to 8° F in the next century. At first glance, this increase seems insignificant. However, such a change could drastically alter global climate (Figure 16.12). According to scientific models, much of the United States and Canada would be drier than normal. If this happened, many midwestern agricultural states, now barely able to support rain-fed agriculture, would suffer crippling declines in productivity. The United States and Canada, now major food exporters, could become food-importing nations.

The Climatic Effects of Greenhouse Gases

Although the greenhouse effect may sound like science fiction, it is not. Scientists have known about it for over 100 years. Early atmospheric scientists, in fact, noted that normal levels of carbon dioxide found in the atmosphere make the Earth habitable. Without carbon dioxide, the Earth would be about 30° C (55° F) cooler, probably so cold that most life forms could not exist. Like so many things in the world around us, a little carbon dioxide is good; too much may be devastating.

Some scientists think that the global warming has already begun. They note that seven of the hottest years in the past 100 years occurred between 1980 and 1992 (Figure 16.13). In 1988, one particularly hot year, drought and heat cut grain production in the United States by one-third. In a normal year, U.S. farmers produce 300 million metric tons of grain. Americans normally consume 200 million metric tons of grain. The remaining 100 million metric tons are exported or stored. But, in 1988, severe drought reduced domestic

Table 16.2 Major Greenhouse Gases and Their Characteristics

Gas	Atmospheric Concentration (ppm)	Annual Increase (percent)	Life Span (years)	Relative Greenhouse Efficiency ($CO_2=1$)	Current Greenhouse Contribution (percent)	Principal Sources of Gas
Carbon dioxide (CO_2) (fossil fuels)	351.3	0.4	x^1	1	57 (44)	Coal, oil, natural gas, deforestation
(Biological)					(13)	
Chlorofluoro-carbons (CFCs)	0.000225	5	75–111	15,000	25	Foams, aerosols, refrigerants, solvents
Methane (CH_4)	1.675	1	11	25	12	Wetlands, rice, fossil fuels, livestock
Nitrous oxide (N_2O)	0.31	0.2	150	230	6	Fossil fuels, fertilizers, deforestation

[1]Carbon dioxide is a stable molecule with a 2–4 year average residence time in the atmosphere.
Source: World Watch Institute, U.S. EPA, and Journal of Geophysical Research.

production to 195 million metric tons, just below domestic consumption. Export obligations were met by emptying storage bins, and by the end of 1988, U.S. bins were all but empty for the first time in many years. Although subsequent years have been even hotter, grain production has not declined so drastically. However, should a repeat performance of the 1988 drought occur several years in a row, world food production could be crippled. Moreover, America might experience another devastating Dust Bowl that could last for decades.

The drought of 1988 also brought record forest fires and grassland fires, another potential impact of global warming. The most widely publicized fires were those that swept through Yellowstone National Park (Chapter 9). Wildfires raged out of control in the western United States in 1989 as well. Hundreds of thousands of hectares burned in that summer in California, Wyoming, Nebraska, Colorado, Idaho, and other states.

In June 1988, James Hansen, a cautious and highly respected scientist at NASA, told the press that he was nearly certain that global warming had begun. But, not all atmospheric scientists are convinced. The exceedingly hot years, they say, may be due to normal climatic variation. Although heat waves and droughts are not uncommon, some point out that the probability of seven hot years in a decade is quite small.

Global climate models predict that global warming could spawn bizarre and violent weather, notably hurricanes and severe floods, such as those witnessed in the U.S. and elsewhere in growing number in recent years. Warming seas impart more energy to the atmosphere, which generates such storms. An increase in devastating storms could have potentially serious economic impacts as well by destroying crops and making food more expensive. Property insurance would also rise.

As noted earlier, some experts predict that in the next 40 to 50 years rain-fed agriculture in the midwestern states will end. But don't expect irrigation to take up the slack. River flows and groundwater are likely to fall substantially because of global warming. The northern states may take up some of the slack as their climate becomes warmer, but the exchange will very

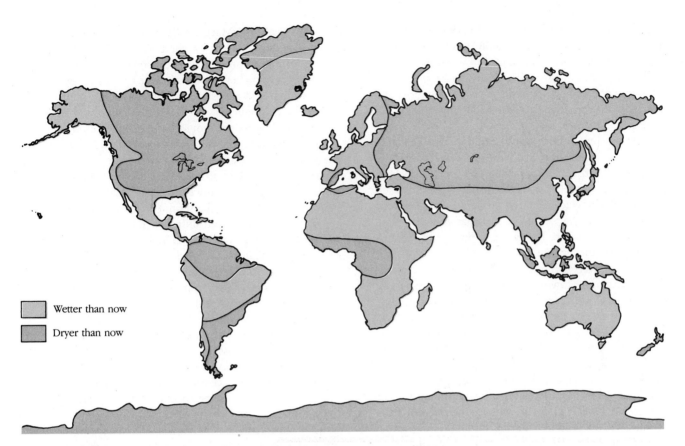

Figure 16.12 *Possible future changes in climate resulting from the greenhouse effect. This map is based on climatic conditions thought to have existed 4500 to 8000 years ago, when the average temperature was highest. Note that this is a generalized map, which may miss regional changes.*

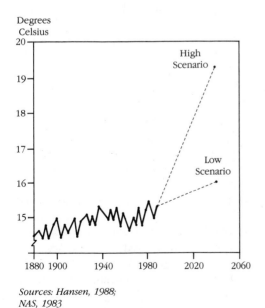

Sources: Hansen, 1988;
NAS, 1983

Figure 16.13 *Graph of average global temperature since 1880 with possible projections to the year 2040.*

likely be unequal for at least two reasons. First, northern soils are generally much poorer than those in Kansas, Iowa, Indiana, Ohio, and Illinois. Shorter growing seasons on the northern grasslands for many hundreds of years have retarded soil formation. Second, even though northern regions may be warmer, they might not receive adequate rainfall to support farming (Figure 16.12).

Desertlike conditions may spread. Agricultural regions neighboring deserts of New Mexico, Texas, Arizona, California, and Nevada could find their cropland and pastureland rendered useless. Snowfall in the Rockies may decline substantially, reducing the lucrative skiing and winter recreation industries. Global climate models suggest that rainfall may increase on both coasts, increasing flooding and damage. But, increased rainfall may not result in greater crop production if the timing is off. Heavy spring rains in western New York, for example, recently flooded fields and prevented farmers from planting their crops on time. By the Fourth of July, corn was only ankle high.

Global warming could make cities unbearable in the summer. Global climate models predict that by the year

2030 the number of days above 32° C (90° F) in Washington, D.C., will increase from 36 to 87. (That ought to get some action in the capital!) In Dallas, the number could increase from 17 to 78, greatly increasing utility costs for cooling, and water demand for irrigation, and generally making life uncomfortable for millions of people.

Ocean communities may also suffer. The IPCC predicts that by the year 2100 the sea level could rise between 30 cm (12 inches) and 110 cm (43 inches), perhaps even more. The rise in sea level would result from melting of glaciers and of the land-based Antarctic ice pack and an expansion of the seas resulting from warmer temperatures. Rising sea levels would threaten coastal cities throughout the world.

Today, over half the world's population live in coastal cities and towns. In fact, over 40 of the largest cities in the world are situated in coastal regions. Among cities most at risk are Miami, New Orleans, Bangkok, Hamburg, London, Leningrad, Shanghai, Sydney, Alexandria, and Dhaka. In the United States, more than half of the population live within 83 kilometers (50 miles) of the ocean. Even a modest increase in sea level would flood coastal wetlands, low-lying fields, and cities. The rise in sea level would also worsen the damage from storms. Waves produced during hurricanes and other storms would sweep further inland, damaging more homes and cities than they do today. Many people would have to relocate. Cities may be forced to build levees to restrain the seas or gradually move to higher ground as buildings are retired. The inland creep of the ocean is also bound to usurp farmland and wildlife habitat and create more crowding as people compete for a limited land base.

Developing nations would also suffer enormously as the oceans rise. In Asia, for instance, rice is produced in many low-lying regions that might be reclaimed by the sea. Storm surges could carry saltwater onto some of the remaining fields, killing crops and poisoning the soil. Bangladesh, a country with 118 million people in 1990, would be especially hard hit because much of its land is barely above sea level. In 1991, a devastating storm ravaged the coast of this poor country, killing an estimated 60,000 to 140,000 people, and flooding many rice fields, destroying crops. By some estimates, 17% of the land area of Bangladesh could be under water by the year 2030, worsening crowding inland.

A great many plants and animals could face difficult times as the planet warms. If the temperature change continues at the predicted rate, many species will become extinct. Others will suffer incredible declines in their populations. Only a limited number will be able to adapt or migrate to suitable habitat.

Interestingly, most species and habitats have dealt with changing conditions for eons, so global warming itself is less a concern than the rate at which it is likely to occur. Professor Margaret Davis at the University of Minnesota shows why in a computer simulation study she performed to predict the effects of a global temperature increase on several eastern tree species. If global carbon dioxide doubles by the year 2050 and temperatures rise as predicted, she found that hardwood trees east of the Mississippi would be hard hit. Her studies predict that beech trees would probably disappear from the southeastern United States, except in a few mountainous regions. Suitable beech tree habitat would shift north to New England and southeastern Canada, which is now the extreme northern limit of the tree's range.

Although species can shift their range, most will be unable to move as fast as required. At the end of the last Ice Age, for instance, beech trees "migrated" northward by dispersing seeds at a rate of about 10 kilometers per 50 years—far slower than the 500-kilometer migration needed to avoid destruction from the global warming predicted within the next 50 years. Many other trees in the United States face a similar fate. Over time, these trees will die out, possibly being replaced by less desirable heat-resistant species.

Practically every ecosystem on Earth will be affected by global warming. Some of the most important and most threatened are coastal ecosystems, notably mangrove swamps and coastal marshes. The future of these areas and the services they provide, among them protecting coastal regions from erosion and providing habitat for commercially important food species, is not bright. Two scientists at the University of California note that the predicted rates of sea-level rise over the next 100 years make it inevitable that most mangroves will be destroyed.

Some studies suggest that plants might thrive in a carbon dioxide-rich world. After all, carbon dioxide is essential to plant growth. A closer examination of the issue, however, suggests that the stated benefits of increasing carbon dioxide levels are overstated. In fact, studies show that while an elevated carbon dioxide level can benefit certain plants it can harm others, among them corn, sugarcane, and many grasses. In addition, elevated levels of carbon dioxide result in a reduction in nitrogen levels in virtually all plants for unknown reasons. Changes in the nutritional quality of plants could affect the entire food web. Studies show that insects compensate for the lower nutritional value of plants grown in elevated carbon dioxide by eating more. Possible declines in insect populations could have devastating effects on birds and other insect-eating species.

Animals, like plants, respond to warming trends by shifting to new habitats. But, not all animals are capable of moving as far as necessary. Stanford University researcher Dennis Murphy studied the Great Basin Mountains, lying between the Cascades and Sierra Nevadas on the west and the Rockies on the east. His studies indicated that 44% of the mammals, 23% of the but-

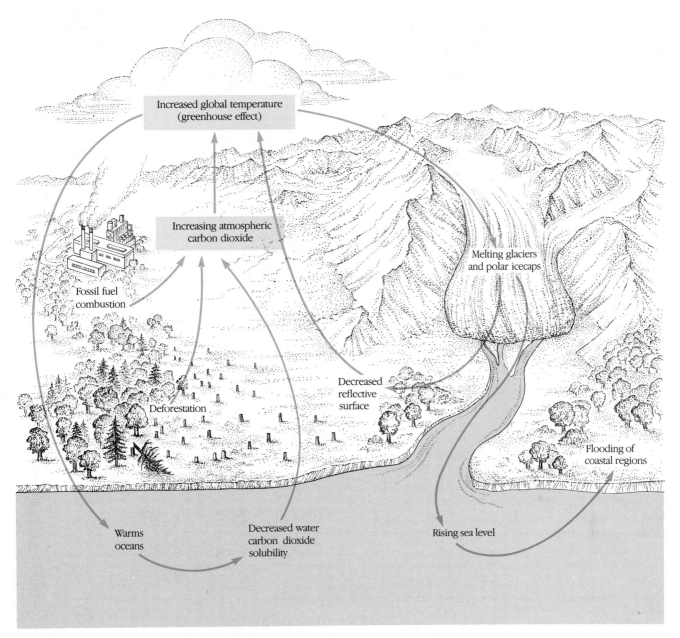

Figure 16.14   *Possible effects of global warming and potential positive feedback mechanisms.*

terflies, and a smaller percentage of birds would die as a result of a 3° C increase in global temperature.

Some species may actually thrive amid global warming, for example, organisms that are responsible for infectious diseases. Robert Shope at the Yale University School of Medicine predicts that some diseases now restricted to tropical areas could invade new territory as the planet warms. For example, one form of rabies now transmitted by vampire bats could spread northward from Mexico, causing damage of about $1 billion per year to the Texas cattle industry.

Global climate models suggest that conditions could deteriorate more quickly than anticipated because

changes now apparently in motion may stimulate dangerous positive feedbacks (Figure 16.14). A **positive feedback** occurs when one factor leads to an increase in another. In nature, positive feedback mechanisms are rare. Homeostatic mechanisms, discussed in Chapter 3, tend to operate through negative feedback mechanisms that maintain more or less constant conditions. In natural systems altered by humans, positive feedback mechanisms are dangerously common. These divisive forces could erode the life-support systems vital to sustainability. Consider some examples.

The oceans currently serve as a major reservoir for carbon dioxide. In fact, they store 60% more carbon dioxide than the atmosphere. Without the oceans, carbon dioxide levels in the air would be considerably higher than they are today. As the Earth's temperature rises, however, the ocean's ability to dissolve and hold carbon dioxide will likely decline (Figure 16.14). The oceans will then release much of the carbon dioxide they have absorbed. Additional atmospheric carbon dioxide could accelerate the warming trend, further reducing the amount of carbon dioxide held in ocean waters.

Several additional positive feedback mechanisms may be initiated. Melting glaciers, for instance, might decrease the reflective surface of the Earth, resulting in an increase in sunlight absorption. This would result in more heat generation, accelerating global warming. Rising temperatures might also require more energy to be spent in summer cooling, which adds more carbon dioxide to the atmosphere, once again accelerating the rise in temperature.

As noted earlier, the rise in atmospheric carbon dioxide concentrations is partly due to deforestation, especially in the tropics but also in parts of the United States, such as the Pacific Northwest. Trees absorb carbon dioxide, which they use to produce food molecules and structural tissues. Worldwide, forests are being cut much faster than they regrow. As a result, deforestation "contributes" approximately one-fourth of the annual global increase in carbon dioxide and, as noted on Table 16.2, about 13% of the annual rise in global temperature. Trees may also be destroyed by climatic change (rising temperature and drought) and from fires. This loss could also become part of a wildly accelerating positive feedback mechanism that causes a rapid increase in atmospheric carbon dioxide in the atmosphere and a more rapid shift in global temperature.

Several factors may counteract these forces, either negating the increase in temperature or lessening it. The oceans, for instance, tend to act as a temperature buffer, keeping temperatures from rising. In addition, rising temperatures may result in an increase in cloudiness, which could reduce sunlight penetration and reduce the rise in global temperature. Increased particulates from denuded, parched lands could also reflect incident solar energy.

In October 1991, a U.N. scientific panel announced a reduction in ozone concentrations in the lower stratosphere. Current evidence suggests that an ozone loss from this region will have a cooling effect on global climate. Temperature records over the last two decades indicate that the lower stratosphere has indeed cooled slightly during that period. By including this new information on ozone loss, predictions of global temperature increase may be lowered by about 20%.

Other new information suggests an adjustment in the opposite direction. This study showed that concentrations of certain chemicals in the atmosphere known as hydroxyl radicals (OH) may be lower than once thought. These molecules react with some greenhouse gases and remove them from the atmosphere. The lower-than-expected levels of hydroxyl radicals suggest that the atmosphere is not cleansing itself of harmful pollutants as rapidly as once thought, forcing an upward revision of global temperature.

In 1991, a study of 17 global climate models published in *Science* suggested that global warming may affect snowfall and average temperature in unanticipated ways. As a rule, climate modelers think that global warming will result in a net decrease in planetary snow cover. This, they predict, would increase absorption and global warming. However, five of the world's best climate models show that a reduction in snow cover would cool the atmosphere and counteract a small part of the greenhouse warming. The cooling effect may result from an increase in cloud cover or an increase in heat radiation into space.

In sum, many climate modelers acknowledge that current models don't do a very good job of simulating cloud behavior. With this much uncertainty, some people prefer to take a wait-and-see approach. They often oppose measures to cut greenhouse gas emissions. Critics argue that the stakes are too high and the economic and environmental benefits of cutting fossil fuel combustion are so valuable that the conversion should be made despite the uncertainty. In short, the cost of accepting the greenhouse hypothesis and acting accordingly is far lower than the potential cost of rejecting it.

Solving the Problem Sustainably

With this issue, as with others, numerous solutions are available. The most effective are those that strike at the roots of the problems, eliminating the harmful emissions of carbon dioxide, methane, CFCs, and nitrous oxide. This section discusses some of the most effective measures. Not surprisingly, the most effective approaches turn out to be applications of the biological principles of sustainability discussed in earlier chapters: conservation, recycling, renewable resource use, restoration, and population control.

Restoration   To reduce atmospheric carbon dioxide levels and stave off or stop the possible increase in global temperature will require a massive reforestation of the Earth. Australia recently announced that it was embarking on an ambitious program to plant 1 billion trees by the year 2000, partly to offset global warming. A few other countries are following suit, among them China, although efforts are well below that required to have any appreciable effect.

Norman Myers, an international expert on tropical forests, argues that replanting 2.6 million square kilometers (1 million square miles) of tropical rain forest would reduce annual emissions of carbon dioxide by 2.5

billion metric tons, or about 41%. Although this ambitious project would cost approximately $100 billion, Myers argues that it is a small price to pay, especially when one takes into account the potential economic damage caused by global warming. According to the U.S. EPA, protecting the East Coast from rising seawaters could cost the United States $75 to $110 billion. Global warming would necessitate costly modifications of irrigation systems and hydroelectric dams estimated at $100 billion. Replanting could also bring direct economic benefits to developing nations from sustained timber yields, reduced soil erosion, and sustainable harvest of forests. (See Models of Global Sustainability 16.1 for a discussion of one company's actions to offset global warming.) Individuals can help plant trees in clear-cuts, roadsides, abandoned fields, and backyards. If you really want to offset the carbon dioxide produced by your life-style you would have to plant 400 trees. A family of four would need to plant 2.5 hectares (6 acres) of fast-growing trees to offset its lifetime carbon dioxide production. Obviously, not everyone can replant trees, but individuals can help by supporting public and private reforestation projects.

Recycling and Energy Efficiency Individuals and businesses throughout the world can also help reduce global carbon dioxide levels by recycling and reusing all materials to the maximum extent possible. Table 16.3 compares energy use in cans and bottles. It shows that recycled glass and aluminum require far less energy, and consequently produce less carbon dioxide air pollution, than cans made from raw ore and used only once. Refillable bottles use much less, even if they are only used ten times. (For more on recycling and reuse, see Chapter 18.)

Energy efficiency is another means of reducing carbon dioxide and methane emissions. According to energy experts Amory Lovins of the Rocky Mountain Institute and Christopher Flavin of the Worldwatch Institute, global emissions could be cut by 3 billion metric tons per year within two decades by cost-effective and profitable technologies. No technical breakthroughs are needed, either. Combined with reforestation efforts outlined above, these efforts could nearly eliminate global carbon dioxide emissions.

Individuals can help reduce energy demand by walking, bicycling, or riding a bus to school or work; building smaller, energy-efficient homes; insulating existing homes; recycling; and using efficient appliances or doing some things by hand (for example, mixing by hand rather than using an electric mixer or drying clothes on a line rather than using a dryer). They can purchase the most energy-efficient appliances. For example, new energy-efficient refrigerators, the leading user of electricity in homes, can cut electric demand from 1200 kilowatt-hours per year to 750. Soon-to-be-released

Table 16.3 Energy Consumption per Use for 12-Ounce Beverage Containers

Container	Energy Use (Btus)
Aluminum can, used once	7050
Steel can, used once	5950
Recycled steel can	3880
Glass beer bottle, used once	3730
Recycled aluminum can	2550
Recycled glass beer bottle	2530
Refillable glass bottle, used 10 times	610

Source: From Brown, L. R., Flavin, C., and Postel, S. (1991). Saving the Planet: How to Shape an Environmentally Sustainable Economy. New York: Norton.

models will cut use to 240 kilowatt-hours per year. (See Chapter 12 for more ideas on energy efficiency.)

Energy efficiency could be stimulated in part by market-based incentives. For example, removal of subsidies from oil, coal, and natural gas or taxing of fossil fuels would motivate many companies to reduce their consumption of carbon dioxide-emitting fossil fuels. A study by the Congressional Budget Office (CBO) studied the potential effect of a carbon tax on fossil fuels starting at $11 per metric ton of carbon and rising to $110 per metric ton in the year 2000. This tax, it said, would cut U.S. carbon dioxide emissions by 37% and would make companies, on average, 23% more efficient. The project would reduce the nation's economic output by $45 billion per year, about 0.6% of the GNP, because companies would be spending more for fossil fuel. Phasing in the tax over a five- to ten-year period would ease the economic burden and might stimulate considerable voluntary action that could cut energy demand and start saving money, thus offsetting any losses. Advocates argue that carbon taxes could be balanced by reductions in income taxes and other forms of tax. This would eliminate or greatly lessen the economic impact of the carbon tax. To date, Sweden, Finland, and Norway have implemented carbon taxes. At this writing, the United States is considering a Btu tax on all energy.

Renewable Energy and Family Planning Efforts to slow down, perhaps halt, the growth of the human population and shift to clean, renewable energy sources could dramatically reduce carbon dioxide and methane emissions. As pointed out in Chapters 6 and 12, these steps are also vital to solving other environmental problems and creating a sustainable society.

Finally, the elimination of CFCs now underway could also help curb global warming, especially since

Models of Global Sustainability 16.1

Offsetting Global Warming: Planting a Seed

In 1988, Applied Energy Services (AES) of Arlington, Virginia, announced plans to help finance the planting of 50 million trees in Guatemala. AES is the first utility company ever known to take direct action to offset carbon dioxide emissions from a fossil fuel power plant it planned to build.

AES currently owns or has part ownership in six power plants in the United States and Great Britain. When it decided to build two new ones, its chief executive officer, Roger Sant, decided to do something about emissions. "Given the scientific consensus on the seriousness of the greenhouse problem, we decided it was time to stop talking and act," Sant remarked. AES also wanted the project to confer additional benefits, for example, improving the economic conditions of communities currently affected by deforestation and preserving endangered plant and animal species.

AES contacted the World Resources Institute (WRI) for suggestions. WRI found an ongoing project in Guatemala sponsored by CARE, an international relief and development agency. The CARE Guatemala Agroforestry Project was created to replant rain forests to reduce soil erosion and flooding. Over a ten-year period, the agency, now with financial support from AES, hopes to plant enough trees to remove 15 million metric tons of carbon dioxide from the atmosphere, approximately the same amount that AES's planned 180-megawatt coal-fired plant in Connecticut will emit over its 40-year life. The trees will also provide sustainable harvest of fruit, lumber, and fuelwood, and will help reduce soil erosion and flooding.

The price tag for this project is huge. AES will provide $2 million. CARE, the U.S. Agency for International Development, and the government of Guatemala each will donate $2 million. Peace Corps volunteers will plant the trees; their service and training are estimated to be valued at $7.5 million.

Some critics think that American companies would be more inclined to invest in domestic tree planting. One approach might be to plant trees on marginally productive, highly erodible farmland taken out of use in the United States in order to reduce soil erosion. Instead of the government subsidizing the planting of trees, power companies could, says Daniel Dudek of the Environmental Defense Fund. His proposal calls for utilities to lease the land currently being set aside by farmers, with federal assistance, and plant fast-growing trees on it. The cost would be about 70 cents per metric ton of carbon dioxide emissions removed. A utility wishing to offset emissions from a 1000-megawatt coal-fired power plant would need to plant a forest with a 48-kilometer (30-mile) diameter.

AES is a model of sustainable business. This innovative power company is "values driven" rather than solely profit or competition driven. Carrying its mission further, AES has also taken steps to offset pollution from a coal-fired power plant on Oahu, Hawaii, that came on-line in 1992. Instead of planting trees, though, AES has taken actions to protect a forest in Paraguay from destruction, halting the release of millions of metric tons of carbon sequestered in the trees. AES agreed to donate up to $2 million for the purchase and preservation of the 58,000-hectare (143,000-acre) Mbaracayu forest. Located in eastern Paraguay, the Mbaracayu is one of South America's few remaining tracts of dense, humid subtropical forest. The area is vital to the survival of many species. It includes 19 distinct plant communities, at least 300 bird species, and threatened and endangered animals including tapirs, jaguars, giant armadillos, peccaries, the rare bush dog, king vultures, and macaws. The forest is the traditional hunting and gathering area for the Ache tribe. AES is partners in this project with Nature Conservancy and a Paraguayan group that promotes sustainable development.

AES plans to continue its carbon-offset programs. At this writing, AES is conducting an evaluation of projects to offset carbon emissions from a coal-fired facility in eastern Oklahoma that began operation in 1991 and another being constructed in Florida.

Tree planting will not prevent global climate change. Current emissions from fossil fuel combustion are far too great for this strategy to work by itself. However, combined with measures to reduce deforestation and other approaches, such as conservation, recycling, and renewable fuels, replanting can play an important part in reversing the dangerous global climate change now underway.

---

these ozone-depleting chemicals are responsible for about one-fourth of the greenhouse effect.

International Cooperation to Halt Global Warming
This chapter opened with a discussion of the Earth Summit, the highly acclaimed global conference that addressed such pressing problems as forest protection and global warming, among others. One outcome of the meeting was a treaty calling on nations to reduce carbon dioxide emissions. The nations also signed an agreement to protect forests. (Details of these agreements are outlined in Chapter 23.)

Although both agreements are important, they fall seriously short of the job at hand. Many scientists hope that, as in the case of the first treaty on ozone protection (Montreal Protocol), nations will soon recognize the need to take more drastic steps to curb, even stop, global warming.

In closing, it is important to note that these three global environmental problems—ozone depletion, acid deposition, and global warming—are occurring in concert with one another and in synch with at least a dozen others. Urgency is advisable in part because so many problems seem to be converging on the present. Their combined effect could be synergistic, producing consequences far greater than expected.

Convergence and synergy suggest the need for root-level solutions, conservation, recycling, restoration, renewable resources, and the like. Interestingly, these solutions provide an elegant synergy of their own. Recycling an aluminum can, for instance, saves enough energy to run a conventional electric light bulb for 4 hours. Substituting a compact fluorescent for an incandescent bulb increases the energy savings from recycling an aluminum can to operate a bulb for 16 hours. Just as problems may synergize to accelerate the decline, solutions can work together to decrease our deleterious impacts and build a sustainable society.

You cannot escape the responsibility

of tomorrow by evading it today.

—Abraham Lincoln

❧   ❧   ❧

Critical Thinking Exercise Solution

The key phrase in the newspaper article that piqued my curiosity was "virtually all production." Upon reading a few more articles, I found that this phrase means that American companies can continue to produce ozone-depleting chemicals after the 1995 deadline if it is economically unfeasible to shift to safer alternatives. This loophole allows production of ozone-destroying chemicals at 15% of 1987 levels. In other words, the president's plan would not have eliminated the production of ozone-destroying chemicals by the end of 1995. This plan would have only accelerated the United States' complete phaseout called for by the London agreement by one year.

One of the critical thinking lessons you can learn from this exercise is to dig deeper. Be careful, define all terms. Politics can be a deceptive business, and our representatives are often experts at mild deceit—telling us one thing and taking credit for great changes when, in fact, they're not really doing much.

Read very carefully and seek information. Beware of television and newspaper reports as well because they often gloss over important details.

Summary

16.1 Stratospheric Ozone Depletion

■ The **ozone layer,** located in the upper stratosphere, contains a slightly elevated concentration of **ozone** molecules, which filter out **ultraviolet (UV) radiation** in sunlight.

■ Humans have caused a thinning of the ozone layer principally through two activities: (1) the use of spray cans and refrigerants containing **Freon** gas and (2) jet travel through the stratosphere.

■ Until the early 1970s, chemists thought that Freons, or **chlorofluorocarbons (CFCs),** were inert gases that would diffuse into the upper layers of the atmosphere and break down harmlessly, but research has shown that chlorine from CFCs broken down in the stratosphere destroys ozone molecules.

■ In the 1980s, a massive hole in the ozone layer was discovered over Antarctica. A smaller hole has been discovered over the Arctic. In March 1988, 100 scientists convened by NASA confirmed that ozone had declined 1.7% to 3% over the Northern Hemisphere since 1969.

■ In reasonable amounts, ultraviolet radiation tans light-colored skin and stimulates vitamin D production in the skin. Excess exposure can cause serious burns and skin cancer and can be lethal to bacteria and plants.

■ In 1991, a panel of scientists predicted that a 10% decrease in stratospheric ozone concentrations, expected by the end of this century, will cause 300,000 additional cases of skin cancer per year and an additional 1.6 million cases of cataracts each year worldwide.

■ Some studies in Antarctica have shown that phytoplankton populations decrease substantially when ozone concentrations fall. Since phytoplankton form the base of the aquatic food chain, damage to them could cause widespread ecological problems.

■ Efforts to slow down the erosion of the ozone layer were fairly modest in the 1970s. In light of more recent findings, the United Nations sponsored new negotiations in 1987 aimed at reducing CFC production worldwide. In September 1987, 24 nations signed a treaty, the **Montreal Protocol,** which would cut CFC production in half by 1999.

■ Continuing bad news about the ozone layer resulted in the signing of the **London Agreement** in 1990, which called for a complete ban on CFCs and halons by the year 2000 and for efforts to phase out other ozone-depleting chemicals.

■ In November 1992, 74 nations met in Copenhagen to hammer out a new agreement that accelerates the phase-out schedule and adds new chemicals to the list.

16.2 Acid Deposition

■ **Acid deposition** is the deposition of acidic substances, in large part produced from human pollution.

■ Acid deposition turns lakes acidic, kills fish and other aquatic organisms, and damages crops. It can also destroy forests and deface statues and buildings.

■ Acid deposition includes two broad categories: wet deposition and dry deposition. **Wet deposition** refers to acids deposited in rain and snow. These acids form when sulfur and nitrogen oxide gases combine with water in the atmosphere to form sulfuric acid and nitric acid, respectively.

■ **Dry deposition** occurs when sulfate and nitrate particulates are deposited from the atmosphere. It also refers to the adsorption of sulfur and nitrogen oxide gases on solid surfaces. When combined with water, these gases form acids.

■ **Acid precursors,** the chemical substances that give rise to acids, arise from both natural and anthropogenic sources. Anthropogenic sources of this pollutant are of greatest concern because they are often concentrated in urban and industrialized regions, causing local levels to be quite high.

■ About 70% of all anthropogenic sulfur dioxide comes from electric power plants, most of which burn coal. Like sulfur oxides, nitrogen oxides arise from a variety of sources, but the most important anthropogenic sources are generally electric power plants and motor vehicles.

■ Acid precursors and acids can remain airborne for two to five days and may travel hundreds, perhaps even thousands, of kilometers before being deposited.

■ Several U.S. states, nine European nations, and Canada have made strides to reduce acid deposition. The first significant federal action in the United States came with the passage of the 1990 Clean Air Act, which called for significant reductions in sulfur oxide emissions and modest cuts in nitrogen oxides by the year 2000.

■ Most international efforts to control sulfur oxide emissions rely on three strategies: (1) the installation of scrubbers on new and existing coal-fired power plants, (2) the combustion of low-sulfur coal or natural gas in utilities, and (3) the combustion of desulfurized coal.

■ Nitrogen dioxide emissions are more difficult to control because nitrogen oxide gases come from air that reacts with oxygen in high temperature furnaces.

■ Another approach to the problem involves treating lakes themselves or breeding acid-tolerant fish species.

However, these are stopgap measures. Although these and other stopgap measures may be important in the short term, they treat the symptoms of the problem while ignoring the underlying root causes: our heavy dependence on and inefficient use of fossil fuels.

■ Recognizing that we must confront the root causes, we must turn to population control, energy efficiency, recycling, and renewable energy use.

16.3 Global Warming

■ Carbon dioxide released from the combustion of fossil fuels reradiates heat escaping from the Earth's surface, warming the atmosphere. This is called the **greenhouse effect.**

■ Problems arise when carbon dioxide from anthropogenic sources increases in concentration, making the Earth's atmosphere warmer than normal.

■ A gradual warming may cause significant long-term global climate changes, melting of glaciers and Antarctic ice, an increase in sea level, a massive disruption of agriculture, and widespread species extinction.

■ At least three other pollutants contribute to greenhouse warming, including CFCs, methane, and nitrous oxide. Controlling these pollutants may be just as important as controlling carbon dioxide.

■ Signs of warming are already present. Seven of the hottest years in a century occurred between 1980 and 1992. Long-term drought has struck many parts of the world. Violent storms and unpredictable weather also appear with greater frequency.

■ Global warming could have potentially serious economic impacts as well, caused by increased coastal flooding, storm damage, forest fires, and decreasing crop productivity in major agricultural regions.

■ Global warming may cause a rise in sea level resulting from the melting of glaciers and of the land-based Antarctic ice pack and an expansion of the seas resulting from warmer temperatures.

■ Since most of the world's people live near the ocean, large numbers of people could be displaced at enormous cost. Alternatively, expensive levees might need to be constructed to hold back floodwaters. Developing nations would suffer enormously as the oceans rise.

■ If the change in temperature continues at the predicted rate, many species will become extinct. Others will suffer incredible declines in their populations. Only a limited number will be able to adapt or migrate to suitable habitat.

■ **Global climate models** suggest that conditions could deteriorate more quickly than anticipated because changes now apparently in motion may stimulate dangerous positive feedbacks.

■ With this issue, as with others, the most effective solutions will very likely be those that go to the root of the problems, eliminating the harmful emissions of carbon dioxide, CFCs, methane, and nitrous oxide.

■ The most efficacious approaches turn out to be applications of the biological principles of sustainability discussed in earlier chapters: conservation, recycling, renewable resource use, restoration, and population control.

Discussion and Critical Thinking Questions

1. What is the ozone layer? Why is it important to life on Earth?

2. Describe the major activities and pollutants that deplete the ozone layer.

3. What measures have been taken to reduce the destruction of the ozone layer? Will they reverse the decline quickly? Why or why not?

4. Describe the two major types of acid deposition.

5. What are the major types of acid found in acid deposition?

6. Using your critical thinking skills and your knowledge of biology, how would you assess efforts to date aimed at reducing acid deposition or counteracting its impact (e.g., smokestack scrubbers, liming lakes, and breeding acid-resistant fish)?

7. What is the greenhouse effect? What gases are responsible for it?

8. Describe some of the impacts of the continued rise of greenhouse gases.

9. Describe the factors that could accelerate greenhouse warming and those that may lessen it.

10. Compare the difference between heeding warnings about global warming and ignoring them. What action do you recommend? Why?

Suggested Readings

Ozone Depletion

Meadows, D. H., Meadows, D. L., and Randers, J. (1992). *Beyond the Limits: Confronting Global Collapse, Envisioning a Sustainable Future.* Post Mills, VT: Chelsea Green. Chapter 5 contains an excellent historical view of the ozone issue.

Shea, C. P. (1988). *Protecting Life on Earth: Steps to Save the Ozone Layer.* Worldwatch Paper 87. Washington, D.C.: Worldwatch Institute. Extraordinary coverage of the ozone controversy.

Acid Deposition

Flavin, C. (1990). Slowing Global Warming. In *State of the World 1990.* New York: Norton. Detailed account of ways to cut greenhouse gases.

Jacobson, J. L. (1990). Holding Back the Sea. In *State of the World 1990.* New York: Norton. Insightful look into the effects of rising sea level.

Johnson, A. H. (1986). Acid Deposition: Trends, Relationships, and Effects. *Environment* 28 (4): 6–11, 34–43. Comprehensive summary of the National Academy of Sciences report. Well worth reading.

Luoma, J. R. (1987). Black Duck Decline: An Acid Rain Link. *Audubon* 89 (3): 19–24. Extraordinary piece on the links between acid deposition and the dramatic decline in black ducks.

Mello, R. A. (1987). *Last Stand of the Red Spruce.* Washington, D.C.: Natural Resources Defense Council and Island Press. Highly readable account of forest damage and the underlying causes.

Ryan, J. C. (1992). When Nature Loses Its Cool. *WorldWatch* 5(5): 10–16. Interesting discussion of biological effects of global warming.

Global Warming

Bazzaz, F. A. and Fajer, E. D. (1992). Plant Life in a CO_2-rich World. *Scientific American* 264: 68–74. Describes some of the effects of rising carbon dioxide levels on plants, outside of temperature-induced changes.

Cohn, J. P. (1989). Gauging the Biological Impacts of the Greenhouse Effect. *Bioscience* 39 (3): 142–146. Interesting look at some of the projected biological impacts of global warming.

Hammond, A. L., Rodenburg, E., and Moomaw, W. R. (1991). Calculating National Accountability for Climate Change. *Environment* 33 (1): 11–15, 33–35. Shows the major contributors to global warming.

Jacobson, J. L. (1989). Swept Away. *Worldwatch* 2 (1): 20–26. Exceptional account of the problems arising from flooding as a result of global warming.

Myers, N. (1989). The Heat Is On: Global Warming Threatens the World. *Greenpeace* 14 (3): 8–13. Excellent overview of the threats of global warming.

———. (1991). Trees by the Billions: A Blueprint for Cooling. *International Wildlife* 21 (5): 12–15. Superb article showing the potential to replant tropical forests to reduce global carbon dioxide levels.

Schneider, S. H. (1989). *The Changing Climate. Scientific American* 261 (3): 70–79. Excellent overview.

Nuclear War: Environmental Nightmare

Ozone depletion, acid deposition, and global warming could result in major changes in the biosphere, but their effects tend to manifest themselves slowly. In contrast, a nuclear war would have grave immediate consequences on humans and a great many other species. A glimpse into the consequences of such an event provides a shocking view of the power we have to disrupt the biosphere.

The Nuclear Detonation

A nuclear explosion produces: (1) enormous amounts of heat and light, (2) an explosive blast and high winds, (3) direct nuclear radiation (radiation is described in Chapter Supplement 11.1), (4) a pulse of electromagnetic energy, and (5) radioactive dust called **fallout** (Figure S16.1).

Figure S16.1 *Mushroom cloud produced by a thermonuclear blast.*

Heat and Light

About one-third of the energy from a nuclear explosion is released as heat and light. The light flash may produce temporary blindness in victims looking at the explosion. Called **flash blindness**, it lasts only a few minutes in most people. However, individuals gazing directly at the fireball may never completely recover their eyesight.

The most common injury among survivors of atomic explosions is burns resulting from the intense heat flash. Even people standing 11 kilometers (7 miles) from a 1-megaton (1 million tons of TNT) explosion, for example, would suffer first-degree burns, which are similar to sunburn, on their exposed skin. At 9 kilometers (5.5 miles) their exposed skin would blister from the heat. At 7 kilometers (4.5 miles), their exposed skin would be charred by the heat from the blast.

In a major city, many burn victims would die from shock soon after an explosion. The heat flash would also ignite any combustible material on its path, turning the city into an inferno and killing many other people.

Explosive Blast and Winds

Several seconds after the heat and light flashes sweep through the area surrounding a bomb detonation site, air pressure rises. This increase, called **static overpressure**, is measured in pounds per square inch (psi, Table S16.1). Although the pressure increase is rather small, it is forceful enough to topple houses and office buildings and rupture eardrums. Tens of thousands of people would be crushed by collapsing buildings if a bomb exploded near a major city during working hours.

Accompanying the static overpressure are intense winds. A bomb with the explosive power of 1 megaton

Table S16.1   Blast Effects of Nuclear Explosions[1]

Distance from Explosion	1-Megaton Weapon	10-Megaton Weapon
1 mile	Overpressure: 43 psi Winds: 1700 mph Many humans killed	Above 200 psi Above 2000 mph Buried 20-centimeter-thick concrete arch destroyed
2 miles	Overpressure: 17 psi Winds: 400 mph Humans battered to death; lung hemorrhage; eardrums ruptured; heavy machinery dragged severely	50 psi 1800 mph Humans fatally crushed; severe damage to buried light corrugated steel arch
5 miles	Overpressure: 4 psi Winds: 130 mph Bones fractured; 90% of trees down; many buildings flattened	14 psi 330 mph Eardrums ruptured; lung hemorrhage; reinforced concrete building severely damaged
10 miles	Overpressure: 4.4 psi Winds: 150 mph Bones fractured; 90% of trees down; many buildings flattened	
20 miles	Overpressure: Below 1 psi Winds: Below 35 mph Many broken windows	1.5 psi 55 mph Cuts and blows from flying debris; many buildings moderately damaged

[1]The range of effects required dictates the optimum height at which the weapon would be detonated.
Source: Goodwin, P. (1981). *Nuclear War: The Facts on Our Survival.* New York: Rutledge Press, pp. 28–29.

produces winds of 290 kilometers per hour (180 miles per hour) 6.5 kilometers (4 miles) from the detonation site. This would fan fires and damage power lines, homes, and trees. Flying debris would injure any people and animals within range.

Direct Nuclear Radiation

A highly intense blast of radiation also spreads out in all directions from a nuclear explosion. For large weapons, the irradiated area is much smaller than the area affected by the heat or static overpressure; therefore, radiation would not be the principal cause of death. For smaller weapons, however, the range of intense radiation is greater than the range of heat and static overpressure; in such cases, radiation would be the major cause of death. (Some of the effects of radiation on human health are summarized in Table S16.2 and are discussed in Chapter Supplement 11.1.)

Electromagnetic Pulse

Nuclear detonations create a single momentary pulse of electromagnetic radiation that spreads in all directions. This energy is similar to the electric signal given off by lightning but much more powerful. The electromagnetic pulse would short-circuit radios, computers, and other electronic equipment, including telephone systems. The resulting blackout would occur at a time when communications would be badly needed to help coordinate rescue and health care. A single large bomb detonated in space 400 kilometers (250 miles) above Omaha, for example, could effectively cripple communications in the entire United States.

Fallout

The heat and force of a nuclear explosion thrusts thousands of tons of dust into the atmosphere. In the

Table S16.2 Effects of Radiation Exposure

Main Organ Involved	Distinguishing Signs	Convalescence Period	Incidence of Death	Death Occurs Within
	Above 50 rads: slight changes in blood cells			
Blood-forming systems: bone marrow, lymph glands	Moderate fall in white blood cell count	Several weeks		
Blood-forming systems: bone marrow, lymph glands	Severe loss of white blood cells; bleeding; infection; anemia (loss of red blood cells); loss of hair (above 300 rads)	1–12 months	0–90%	2–12 weeks
Blood-forming systems: bone marrow, lymph glands		Long	90–100%	2–12 weeks
Gastrointestinal tract	Diarrhea; fever; shock symptoms		100%	2–14 days
Central nervous system	Convulsions; tremor; involuntary movements; lethargy		100%	days

Source: Goodwin, P. (1981). *Nuclear War: The Facts on Our Survival.* New York: Rutledge Press, pp. 44–55.

process, much of the dirt becomes radioactive. Radioactive dust particles return to the Earth as fallout. Some of the dust particles may fall back immediately, landing near the site of the explosion and making the area intensely radioactive. Some fallout is carried higher into the atmosphere, only to fall back to Earth many days later. Some fallout may be transported so high that it circulates in the upper atmosphere for decades, settling from the sky very gradually over large areas. Radioactivity from fallout poses a threat to human health and the welfare of many other animal species.

Combined Injuries

One consequence of nuclear war that is sometimes overlooked is the effect of combined injuries. For example, victims of a nuclear exchange might be burned as well as exposed to nonlethal doses of radiation. Broken bones, cuts, and abrasions might add to their injuries. Studies on laboratory animals suggest that nonfatal injuries can add up, producing death.

Nuclear Winter

Imagine a heavy nuclear exchange in the Northern Hemisphere. Some people could survive the initial shock wave, the intense heat, the radiation, and the resulting fallout. But, awaiting them, according to some experts, would be a nightmarish world of cold and darkness called **nuclear winter**.

Multiple atomic blasts would carry millions of tons of dust and soot into the atmosphere. Rising columns of smoke from thousands of fires created by the explosions would create large, dark clouds. Dust from the explosions and smoke from the fires would block sunlight. Temperatures would plummet, with devastating effects on the biosphere.

The **nuclear-winter theory** was first published in 1982 by Paul Crutzen. Others, among them an American group including the renowned Carl Sagan, have arrived at similar conclusions.

No one knows for sure, but experts believe that about 40 million metric tons of soot would be required to produce a nuclear winter. Should 100 cities burn in a nuclear exchange, the world might be immersed in a dark, cold glacial age. The more smoke, the longer and more severe the winter would be.

Eventually, studies suggest, the clouds would coalesce and form a dense pall stretching from Florida to Alaska. The cloud could very easily spread south across the equator and plunge the Southern Hemisphere into a chilly twilight. Making matters worse, the smoke-laden air could absorb the sun's energy. This would cause the cloud to rise higher and thus lengthen its stay in the atmosphere.

Many scientists believe that the pall of smoke from an all-out nuclear war would plunge surface temperatures 20° to 30° C (36° to 54° F). Widespread extinction would result. Crop production throughout the world could halt. A drop in the average daily temperature of only 2° or 3° C during the growing season can cut wheat production in Canada and the Commonwealth of Independent States in half. Grains make up some 70% of the world's food energy; nuclear winter would, therefore, cause widespread global starvation. Climatic changes would devastate the developed countries, which are mostly in the temperate zone. According to a recent study by the International Council of Scientific Unions, however, the major environmental impacts would occur in the tropics and subtropics, where most of the developing nations are located. Reduced temperature and rainfall in these regions would wipe out thousands of species adapted to warm, constant climates.

Nuclear war by itself could disrupt world trade of fertilizers, pesticides, and the like, resulting in further declines in farm productivity worldwide. Nuclear explosions could also create a toxic nightmare as millions of metric tons of potentially toxic and carcinogenic substances spewed into the atmosphere from burning cities. Asbestos, dioxins, and PCBs, among others, would spread to surrounding areas, making them even more inhospitable to life. Finally, an all-out nuclear war would reduce the protective ozone layer, resulting in further damage. (See Chapter Supplement 15.1.)

Some scientists have criticized proponents of the nuclear-winter theory for underestimating the effects. Others, such as Edward Teller, known as the "father of the hydrogen bomb," argue that certain factors would "thaw" the nuclear winter, and they suggest that this new theory is exaggerated. Teller and his associates, for instance, contend that fire storms would probably not generate columns of smoke that rise into the upper atmosphere; thus, the duration of the nuclear winter would be lessened. Recently, scientists at the National Center for Atmospheric Research in Boulder, Colorado, included the effects of the oceans in the computer models for a nuclear winter. These new calculations indicate that the fall in temperatures worldwide would last days or, at most, weeks. In such a case, the effects of the cold would be considerably less drastic. Some scientists are therefore speaking of a **nuclear fall**. Undoubtedly, the debate will continue for years.

Predicting the Effects of Nuclear War

No one can accurately predict the effects of nuclear war because no one knows how many atomic weapons would be hurled between warring nations, where they would explode, what the weather conditions would be, or other imponderables. Clearly, a fraction of the total nuclear arsenals of the United States and the Common-

wealth of Independent States could devastate the major population centers of every country in the world.

Of this we can be certain: Any nuclear explosion over an urban area would have catastrophic effects. Widespread nuclear war would have environmental and health consequences beyond comprehension. Health care, transportation, and public sanitation would be crippled. Widespread climatic changes might grip the world in a winter- or fall-like cold that would destroy plants and wildlife and grind agriculture to a stop.

Paul Warnke, former U.S. coordinator for the Strategic Arms Limitation Talks (SALT II), noted: "In this the fourth decade of the nuclear age, it is tempting to assume that a nuclear exchange won't take place and that, in any event, there is nothing the average human being can do about it. But the fact is that a nuclear war could happen and very well may happen unless we, as citizens of a threatened world, decide that we will do something." Fortunately, something has happened.

In 1988, the United States and the then Soviet Union ratified a treaty to eliminate intermediate and shorter-range nuclear missiles in Europe. In 1991, the United States and the Commonwealth of Independent States signed the **Strategic Arms Reduction Treaty** to reduce the number of intercontinental ballistic missiles, submarine-launched ballistic missiles, and bombers. Further reductions were achieved in 1991 just before President Bush left office. By the year 2003, the total number of warheads in the United States and the Commonwealth of Independent States should be about 3000 to 3500 in each country, reduced from 12,600 and 11,000 in 1990, respectively. Like the ozone story, this potentially disastrous problem heads toward an optimistic solution.

Suggested Readings

Ehrlich, P. R. et al. (1984). *The Cold and the Dark: The World after Nuclear War*. New York: Norton. Thoughtful examination of nuclear winter.

Riordan, M., ed. (1982). *The Day After Midnight: The Effects of Nuclear War*. Palo Alto, CA: Cheshire Books. Detailed account worth reading.

Sagan, C. (1985). Nuclear Winter: A Report from the World Scientific Community. *Environment* 27 (8): 12–15, 38–39. Excellent overview of scientific reports on the nuclear-winter theory.

Turco, R. P. and Golitsyn, G. S. (1988). Global Effects of Nuclear War. *Environment* 30 (5): 8–16. Excellent overview of the possible effects of nuclear war.

Chapter 17

Water Pollution: Protecting Another Global Commons

It's a crime to catch a fish in some lakes, and a miracle in others.

—*Evan Esar*

Critical Thinking Exercise

A family living along Lake Superior argues that it doesn't need to conserve water. There's plenty of water in the lake, so why should they sacrifice? Do you agree with this assertion? What critical thinking rules did you use to analyze this issue?

"If there is magic in this planet, it is in water," wrote Loren Eiseley. Covering 70% of the Earth's surface and comprising two-thirds or more of the weight of most animals and up to 95% of the weight of plants, water is indispensable to life. Despite its crucial role in our lives, water is one of the most badly abused resources. Chapters 7 and 10, for example, described how unsustainable withdrawals of groundwater and surface water create regional shortages that are already seriously disrupting agriculture and society. Chapter Supplement 10.1 showed how pollution of estuaries was destroying food sources for humans and other species. This chapter covers water pollutants: where they come from; how they affect living organisms; and, finally, legal, technological, and personal measures to reduce them. As done in other chapters, it outlines the requirements of a sustainable response to this important problem.

17.1 Water and Water Pollution

Water pollution is any physical or chemical change in water that adversely affects organisms. Like many other problems, it is global in scope, but the types of pollution vary according to a country's level of development. In poorer, nonindustrialized nations, water pollution is predominantly caused by human and animal wastes, pathogenic organisms from this waste, pesticides, and sediment from unsound farming and timbering practices. Rich, industrial nations also suffer from these problems, but with their more extravagant life-styles and widespread industry, they create an additional assortment of potentially hazardous pollutants: heat, toxic metals, acids, pesticides, and organic chemicals. In between these two extremes are numerous countries with various levels of industrialization. They often have inadequate laws or none at all to combat water pollution. Or, if they do have good laws, they frequently lack adequate funding to enforce pollution laws. Their waters are, therefore, often badly polluted with an assortment of industrial and municipal wastes.

Like air pollutants, water pollutants come from numerous natural and anthropogenic sources. Because water respects no boundaries, pollutants produced in one country often end up in another's water supply. The thoughtless dumping of wastes in rivers, accidents (such as oil spills), and uncontrolled population growth can have dire consequences on lakes, streams, and oceans. For example, for many years, the Mediterranean Sea was treated as an unlimited dump for domestic and industrial wastes. Along Italy's northern coast, for example, the waters became a cesspool. Fortunately, internal actions and negotiations among the nations that border the Mediterranean Sea have spawned a regionwide plan that is beginning to reverse the years of decline.

Movement of pollutants from lakes and rivers to oceans is only half the problem. In recent years scientists have revealed cross-media contamination, the movement of a pollutant from one medium, for example, air, to another, such as water (Chapter 4). As an example, pesticides sprayed on crops may drift to nearby lakes and, from there, flow to the oceans. Toxic organics once dumped in evaporating ponds around factories dissipate, only to be washed out of the sky by rains. Hazardous wastes buried in the ground leak into aquifers, whose waters replenish streams. A joint committee of the U.S. National Research Council and the Royal Society of Canada announced in 1986 that levels of DDT and PCBs in the water of the Great Lakes had declined only slightly since 1978, despite tight controls on industrial releases into the lakes. Significant levels of these and other toxic chemicals persist in the lakes because of contaminated groundwater and their deposition from atmospheric pollution (Case Study 17.1). Unfortunately, very little has been done since 1986 to eliminate airborne toxins.

Point and Nonpoint Sources

When we ponder the sources of water pollution, we generally think of factories, power plants, and sewage treatment plants that pour tons of sometimes toxic chemicals into sewers, lakes, and rivers (Figure 17.1a). These **point sources**, so named because they are in discrete locations, are relatively easy to control. But, they are only half the problem. The other half includes sources we rarely think about, the **nonpoint sources**—less distinct sites, such as farms, forests, lawns, and urban streets (Figures 17.1b). For example, rainwater carries oil from driveways and streets, and pesticides and fertilizers from urban lawns, into storm sewers, then into streams (Table 17.1).

Figure 17.2 shows that agriculture is the predominant nonpoint pollution source in the United States, affecting nearly 60% of the nation's streams. In the United States, in fact, nonpoint sources release more than half the pollutants that end up in the waterways (Figure 17.3). Some of the substances include dust, sediment, pesticides, asbestos, fertilizers, heavy metals, salts, oil, grease, litter, and even air pollutants washed out of the sky by rain. Because the sources are many and spread out, control has proved difficult.

Figure 17.1 *Water pollution. (a) This factory dumping waste into a nearby river is an example of a point source. (b) Urban streets, lawns, and gardens are non-point sources.*

Some Features of Surface Waters

This chapter is concerned primarily with pollution of surface water. Therefore, it is important to examine some features of surface waters that affect pollution. (Information on groundwater can be found in Chapters 7 and 10.)

Freshwater ecosystems fall into two categories. **Flowing systems** include rivers and streams. Because water flows more quickly in them than in lakes, they can self-cleanse. However, this purging effect is useless if the supply of pollutants is constant or is spread evenly along a river's banks, as is common along rivers in many industrial countries, such as those in Europe.

Table 17.1 Major Nonpoint Pollution Sources in the United States

Activity	Explanation
Silviculture	Growing and harvesting trees for lumber and paper production can produce large quantities of sediment.
Agriculture	Disruption of natural vegetation leads to increased erosion; pesticide and fertilizer use, coupled with poor land management, can pollute neighboring surface water and groundwater.
Mining	Leaching from mine wastes and drainage from mines themselves can pollute surface and groundwater with metals and acids; disruption of natural vegetation accelerates sediment erosion.
Construction	Road and building construction disrupts vegetation and increases sediment erosion.
Salt use and groundwater overuse	Salt from roads and storage piles can pollute groundwater and surface water; saltwater intrusion from groundwater overdraft pollutes ground and surface water.
Drilling and waste disposal	Injection wells for waste disposal, septic tanks, hazardous waste dumps, and landfills for municipal garbage can contaminate groundwater.
Hydrological modification	Dam construction and diversion of water both can pollute surface waters.
Urban runoff	Pesticides, herbicides, and fertilizers applied to lawns and residues from roads can be washed into surface waters by rain.

Nonpoint Pollution Sources Affecting Streams

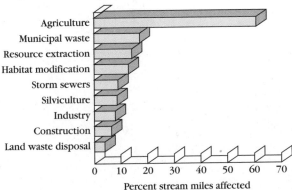

Source: U.S. Environmental Protection Agency, 1990.

Figure 17.2 *When pollution sources in the United States are listed by their impact, notably, the percentage of kilometers of streams they affect, agricultural activities and municipal wastes (sewage) turn out to be number one and two.*

Major Sources of U.S. Stream Pollution

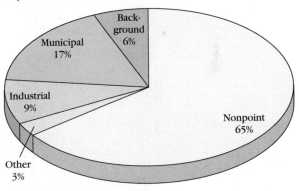

Source: U.S. Environmental Protection Agency.

Figure 17.3 *Contribution of various sources to U.S. water pollution. Note that nonpoint sources are by far the greatest polluter of U.S. waters.*

Standing systems, such as lakes and ponds, are usually more susceptible to pollution because water is replaced at a slower rate. A complete replacement, or turnover, of a lake's water may take 10 to 100 years or more; thus, pollutants can build up to hazardous levels. (For a discussion of the importance of turnover, see Models of Global Sustainability 17.1.)

As shown in Figure 17.4, lakes generally consist of three zones: (1) the **littoral zone,** the shallow waters along the shore where rooted vegetation, such as cattails and arrowheads, can grow; (2) the **limnetic zone,** the open water that sunlight penetrates and where phytoplankton, such as algae, live; and (3) the **profundal,** or deep, zone, into which sunlight does not penetrate.

Figure 17.4 (a) *The three ecological zones of a lake. Note the difference between (b) shallow eutrophic lakes, which tend to have high levels of plant nutrients, and (c) deep oligotrophic lakes, which support fewer fish because they lack nutrients that stimulate plant growth.*

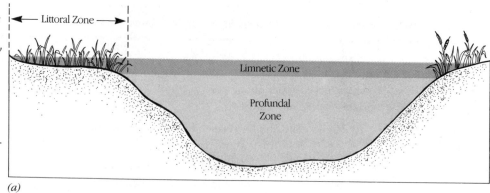

(a)

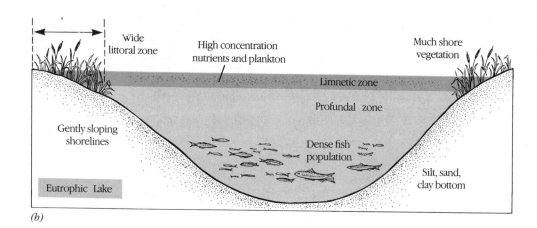

(b)

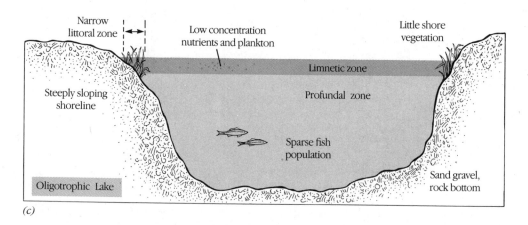

(c)

The littoral and limnetic zones are vital to a lake's health for they are home to many photosynthetic organisms that provide food and oxygen for aquatic life. Consequently, pollution in these zones can severely disrupt the ecological health of a lake and the welfare of those who depend on them for food. The profundal zone is also an ecologically important zone. In the mud and sediment along the bottom, billions of bacteria and fungi live. They decompose organic matter deposited from upper layers and thus help liberate nutrients needed for growth of plants in the littoral and limnetic zones. Pollutants that poison these organisms disrupt a vital ecological process.

In the warmer months, lakes can also be divided into three temperature zones (Figure 17.5). The upper, warm water is called the **epilimnion**; the deeper, cold water forms the **hypolimnion**; and the transition between the two is called the **thermocline**, or **metalimnion**.

In temperate regions, lakes go through an important mixing process twice a year, which allows for a complete mixing of oxygen, nutrients, and phytoplankton. In the fall, the air temperature begins to drop, and the surface water cools. When the surface water reaches 4° C (39° F), it becomes cooler and heavier than the water below. The denser surface water then begins to sink to the bottom. Winds also help churn the waters and create

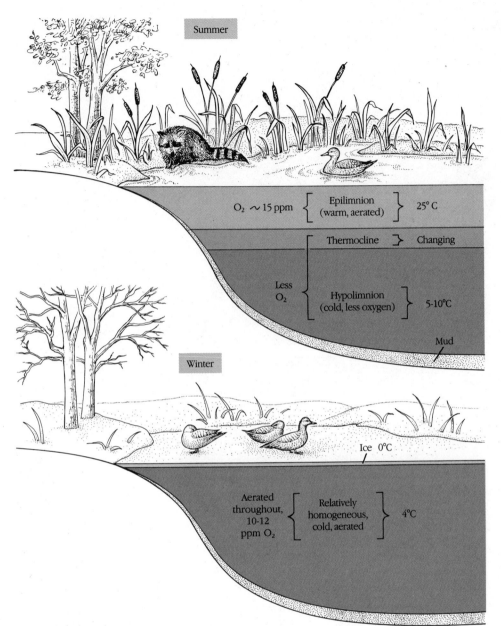

Figure 17.5 *The three thermal zones of a lake in the summer disappear in the winter. Oxygen-rich surface waters are mixed with oxygen-poor waters from deeper layers in the fall overturn, not shown here.*

a thorough mixing, known as the **fall overturn**. As a result of these forces, the summertime thermal stratification of a lake disappears.

During the winter, the water temperature is fairly constant from top to bottom, as are the oxygen and nutrient concentrations. Fish are fairly uniformly distributed throughout a lake. In the spring, though, as the ice starts to melt, the lake turns over again. Water expands when it freezes, which explains why ice floats on a lake. In the later winter and early spring, ice at 0° C begins to warm. When the meltwater reaches 4° C, all of the water is of uniform temperature. Winds mix the waters once again.

The seasonal turnover of lakes is important because it helps circulate oxygen from surface waters to deeper

waters in the fall, which allows organisms to survive in the profundal zone. In the spring, it carries important nutrients from the lower levels to the upper levels, where they can be used by plants and algae. With these basics in mind, let's look more closely at water pollution.

17.2 Types of Water Pollution

This section outlines the many forms of water pollution and their effects.

Nutrient Pollution and Eutrophication

Rivers, streams, and lakes contain many organic and inorganic nutrients needed by the plants and animals

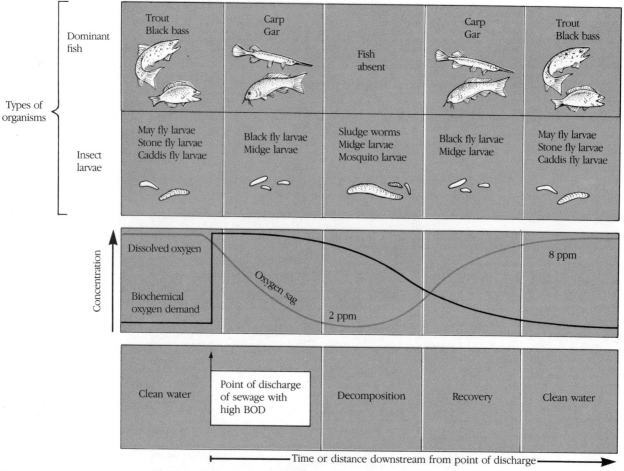

Figure 17.6 *The oxygen sag curve. Oxygen levels and biochemical oxygen demand are shown below a point source of organic nutrients. Note that oxygen levels drop immediately below the pollution source as a result of bacterial decay of organic matter. Oxygen levels recover downstream once the organic matter is removed from the water.*

that live in them. In higher-than-normal concentrations, however, these nutrients become pollutants.

Organic Nutrients   In developed countries, feedlots, sewage treatment plants, and some industries, such as paper mills and meat-packing plants, may release large quantities of organic chemicals into waterways. In developing countries, human sewage and animal wastes often enter streams and lakes from villages and farms. In surface waters, these organic pollutants are consumed by bacteria. As a result, bacterial populations tend to proliferate. This natural ecological mechanism is one of nature's many homeostatic systems, because as bacteria consume the organic matter, they help to purify the waters. Unfortunately, there's a catch.

During the degradation of organic pollutants, bacteria consume oxygen dissolved in water (Figure 17.6). As oxygen levels drop, fish and other aquatic organisms may perish. When oxygen levels become very low, anaerobic (nonoxygen-requiring) bacteria take over, breaking

down what's left. In the process, they produce the foul-smelling, toxic gases methane and hydrogen sulfide.

Oxygen depletion in rivers and streams occurs more readily in hot summer months because stream flow is generally lower and organic pollutant concentrations are higher. In addition, increased water temperatures speed up bacterial decay.

As the organic matter is depleted, oxygen levels may return to normal as a result of plants and phytoplankton or mixing of air caused by water turbulence. When numerous sources of organic pollutants are found along the course of a river or when organic input exceeds the ability of the aquatic system to restore conditions, recovery may be impossible. Lakes can also recover from organic pollutants, but usually much more slowly than rivers, for reasons explained earlier.

The organic nutrient concentration in streams is estimated by a test that determines the rate at which oxygen is depleted from a sample. In this test, polluted water is saturated with oxygen and held in a closed bottle

Case Study 17.1

The Great Lakes: Alive but Not Well

Carved by ancient glaciers, the five mammoth Great Lakes hold one-fifth of the world's standing fresh-water. Approximately 40 million people live within their drainage basin. Like so many inland waters, the lakes have suffered years of abuse from pollution and poor land management. Especially hard hit were Lake Ontario and Lake Erie.

Cultural Stress: The Death of a Lake?

The story of Lake Erie serves as a reminder of human civilization's immense impact and a lesson on ways to prevent future deterioration of the world's waters. Lake Erie was once surrounded by dense forest. Streams ran clean.

Today, approximately 30 million people live in the lake's watershed. The dense woodlands that once protected the soil were cut down to make room for farms, homes, industries, and roadways. As a result, large quantities of topsoil have been washed into the rivers and the lake, clogging navigable channels and destroying spawning areas essential to the lake's once-rich fish life. In the early 1900s, many of the wetlands along the lake's shore were drained, and hundreds of small dams were built to provide power to mills, blocking the upstream migration of fish, such as the walleye and sturgeon.

By the 1960s, the lake's waters were polluted with a wide assortment of organic and inorganic nutrients. Raw sewage floated on the water's surface, and algal blooms were commonplace. Dissolved oxygen levels frequently dropped to low levels, especially in the pro-fundal zone that occupies the large central basin of the lake. Blue green algae proliferated in the warm summer months in the shallow western end of the lake, creating a foul-smelling, murky green water. Lead, zinc, nickel, mercury, and other toxins from industry polluted harbors and built up in near-shore sediments. In 1970 and 1971, mercury levels in fish from Lake Erie often exceeded safe levels set by the FDA. Reductions in mercury discharges in 1975 resulted in a decline in mercury in fish, but violations of health standards continued.

In Colonial days, numerous fish species inhabited the lake and its tributaries. Largemouth and small-mouth bass, muskellunge, northern pike, and channel catfish were common in the lake's tributaries. Lake herring, blue pike, lake whitefish, lake sturgeon, and others lived in the open waters. By the 1940s blue pike and native lake trout had vanished. Sturgeon, lake herring, whitefish, and muskellunge managed to hold on but in reduced numbers.

Lake Erie suffered from severe cultural stress caused by overfishing, introduction of alien species, pollution, and destruction of shorelines and spawning grounds. Algal blooms, beach closings, thick deposits of sludge, oxygen depletion, taste and odor problems, and contaminated fish were the legacy of years of disregard and mismanagement.

By the late 1950s, large areas of Lake Erie's central basin were without oxygen, anoxic, for weeks on end during the summer. Until the late 1970s, anoxia spread cancerously. The lake was pronounced dead; many feared that the other lakes would follow suit.

The Joint Cleanup Program—Not Enough

Alarmed by the condition of Lake Erie and other lakes, in 1972, the United States and Canada agreed to restrict the discharge of pollution into the Great Lakes. The **Great Lakes Water Quality Agreement**, updated in 1978, called for controls of point and non-point pollution sources. It demanded that releases into the lakes of "any or all persistent toxic substances" be "virtually eliminated." With cooperation from industrial and municipal polluters, the lakes began to show signs of recovery. Lake Erie, the shallowest and fastest-flowing of the lakes, made a quick recovery. Today, it is teeming with fish. Gone are the raw sewage discharges that once discolored the waters and the massive algal blooms. The other lakes also show signs of improvement.

Despite these efforts, 43 areas still fail to meet the standards established in the U.S.-Canadian accord. In 1987, a Buffalo-based environmental group, Great Lakes United, published a report noting that while highly visible and odoriferous pollutants, such as sewage, had been significantly reduced, many toxic substances not visible to the naked eye continue to pour into the lakes. For instance, PCBs and pesticides still persist at unacceptable levels. Officials administering the 1978 agreement conceded that the U.S. and Canadian governments had not fully lived up to the terms of the agreement.

Toxins enter the lakes from factories and sewage treatment plants; nonpoint pollution, including farm-land and urban runoff; toxic fallout, for example, pollutants deposited from the atmosphere; and resuspension of substances contained in the bottom sediments caused in part when harbors and river mouths are dredged.

One of the most significant avenues is atmospheric deposition. Trace metals, pesticides, phosphorus, nitric acid, nitrates, sulfates, sulfuric acid, and organic compounds are all deposited from the air. Approximately 60% to 90% of all PCBs entering Lakes Superior and Michigan come from the atmosphere. An estimated 14,000 metric tons of aluminum is deposited in Lake

(continued)

Case Study 17.1 (continued)

The Great Lakes: Alive but Not Well

Superior from the atmosphere, and nearly 29,000 metric tons annually falls from the skies into Lake Michigan. The atmosphere is also a major source of phosphorus, providing an estimated 59% of the total input to Lake Superior. Without controls on atmospheric deposition, it will be impossible to end the discharge of toxins into the Great Lakes.

What does the continued pollution assault mean? First, commercial fishing, once an economic mainstay in the region, has virtually ceased in all of the lakes except Superior. Even there, the commercial fishermen operate under continual uncertainty, never knowing when their catch will exceed safe limits and be declared unsafe by the FDA. Second, introduced Pacific salmon and reintroduced lake trout survive in the lakes, but the salmon population must be restocked each year. Lake trout do not reproduce successfully either, except in Superior. Restocking costs millions of dollars per year. Third, continued pollution has forced some states to issue warnings advising women who are pregnant, lactating, or of childbearing age not to eat certain fish. Parents are also advised not to feed their children lake-caught fish. This warning came after a scientific study on chronic low-level exposure,

which showed that infants of women who had eaten PCB-contaminated fish two to three times per month were smaller, more sluggish, and had slower reflexes than infants of women who had not eaten contaminated fish. Children also accumulate the toxins, which may impact future fertility. Acceptable levels for most of the 800 pollutants found in the lakes simply have not been established, primarily because of a lack of information on health effects.

Reducing the problems posed by toxic pollutants is complicated. Eight states, two provinces, numerous tribal councils, and two nations share an interest in managing the waters of the Great Lakes. The result is often conflict that hinders the implementation of the steps needed to restore the Great Lakes back to a full, productive, and healthy life.

In recent talks, Canada and the United States agreed on ways to control airborne pollutants. They also agreed to remove contaminated sediments from the lakes and to better control groundwater pollution that contributes to the degradation of the Great Lakes. At least 15 to 30 years of rehabilitation are needed, but improvements are already apparent.

---

for five days; during this period in the water, bacteria degrade the organic matter and consume the oxygen that was added to it. The amount of oxygen remaining after five days gives an indication of the organic matter present; the more polluted a sample, the less oxygen left. This standard measurement is called the **biochemical oxygen demand (BOD)**.

Inorganic Plant Nutrients   Whereas organic nutrients nourish bacteria, certain inorganic nutrients stimulate the growth of aquatic plants. These plant foods include nitrogen, phosphorus, iron, sulfur, sodium, and potassium.

Nitrogen, in the form of ammonia and nitrates, and phosphorus, in the form of phosphates, are often limiting factors for populations of algae and other plants (Chapter 2). Consequently, if levels increase, plant growth goes wild, choking lakes and rivers with thick mats of algae or dense growths of aquatic plants. In freshwater lakes and reservoirs, phosphate is usually the limiting nutrient for plant growth; marine waters are usually nitrate limited.

Excessive plant growth impairs fishing, swimming, navigation, and recreational boating. In the fall, most of these plants die and are degraded by bacteria, which

may deplete dissolved oxygen, killing aquatic organisms. As oxygen levels drop, anaerobic bacteria resume the breakdown and produce noxious products, as noted earlier. Thus, inorganic nutrients ultimately create many of the same problems that organic nutrients do.

Inorganic fertilizers from croplands are the major anthropogenic source of plant nutrients in fresh water. When highly soluble fertilizers are used in excess, as much as 25% may be washed into streams and lakes by the rain. More careful use could greatly reduce this problem.

Laundry detergents are the second most important anthropogenic source of inorganic nutrient pollution in the United States. Many detergents contain synthetic phosphates, called tripolyphosphates (TPPs). These chemicals cling to dirt particles and grease, keeping them in suspension until the wash water is flushed out of the washing machine. Unfortunately, the phosphates stimulate the growth of aquatic algae, causing sudden spurts in growth called **algal blooms**.

Nearly 60% of the U.S. population lives in soft-water regions where soap-based cleansing agents work as well as detergents. In the hard-water regions, harmless substitutes for TPPs can be used. For example, lime soap-dispersing agents have been used in bar soaps for years and could easily be used for laundry detergents.

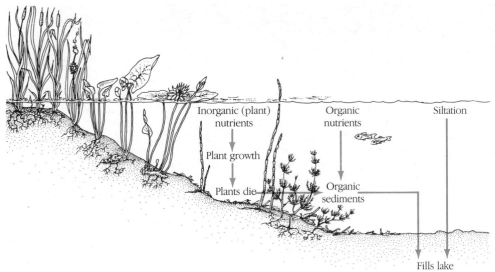

Figure 17.7 *Contributions of inorganic and organic nutrients and sediment to the succession of a lake into swampland. Inorganic nutrients contribute to eutrophication. Combined with sediment from natural or anthropogenic sources, they can cause a lake to fill in.*

Nitrates and nitric acid can also enter surface waters from the atmosphere (Chapter 16). According to the EDF, about 25% of the nitrogen polluting Chesapeake Bay comes from wet and dry deposition.

Eutrophication and Natural Succession   Lakes naturally pick up nutrients from surface runoff and rainfall. The amount becomes excessive and lakes begin to suffer when nutrient levels increase from poor farming practices, such as overfertilization, or from sewage treatment plant discharge.

The natural accumulation of nutrients in lakes is called **natural eutrophication**. Given sufficient time, natural eutrophication and natural soil erosion can transform shallow lakes into swampland and then into dry land, a process called **natural succession** (Chapter 3). In this process, inorganic nutrients stimulate plant growth. The plants eventually die and contribute organic sediment to lake bottoms (Figure 17.7). This sediment combines with silt from erosion, gradually filling in lakes.

Accelerated erosion, caused by human activities, and **cultural eutrophication**, resulting from inorganic nutrients released from farms, feedlots, and sewage treatment plants speed up the process. Good, productive lakes can become choked with vegetation that rots in the fall, depleting oxygen and emitting an offensive odor. Sediment and organic debris can fill the lake, destroying it.

The fate of lakes overfed with nutrients from sewage treatment plants and farms, however, is not as dim as scientists once believed; if nutrient inflow is greatly reduced or stopped, a lake may make a comeback. Lake Washington near Seattle, for example, became a foul-smelling, eutrophic eyesore after decades of abuse, during which millions of liters of sewage were dumped into

its waters. In 1968, local communities began to divert their wastes to Puget Sound, an arm of the sea with a greater capacity to assimilate the wastes. Lake Washington began a slow recovery.

Of course, the diversion of sewage to Puget Sound has had some negative effects. Although the sound cleanses itself more quickly than the lake, certain toxins in the waste and in surface runoff (nonpoint pollution) are having a harmful effect on marine life in the sound. The effects are especially noticeable in areas where industries have discharged their wastes for decades. Today, sediments in the sound contain high levels of toxic organic wastes and heavy metals, which may explain the high incidence of tumors in sole, a fish that lives on the bottom. Shellfish beds have been closed to prevent contamination. Fortunately, efforts are now underway to correct the problem.

Cultural eutrophication is the most widespread problem in U.S. lakes. In a 1990 report to Congress, 38 states reported that of 8000 lakes over half assessed were eutrophic or hypereutrophic, highly eutrophic. Nearly all receive wastes from industry and municipalities, but even if these sources were eliminated, many lakes would probably not improve significantly because of continued pollution from nonpoint sources.

Infectious Agents   Water may be polluted by pathogenic (disease-causing) bacteria, viruses, and protozoans. Waterborne infectious diseases are a problem of immense proportions in the less developed nations of Africa, Asia, and Latin America, as witnessed by the recent outbreak of cholera in many Central and South American nations. Infectious agents were once the major water pollutants of now-developed nations before sewage treatment plants and disinfection of drinking water became commonplace. Nonetheless, problems still

Models of Global Sustainability 17.1

The Helsinki Convention: International Cooperation to Protect the Baltic Sea

The Baltic Sea is surrounded by nine countries: Denmark, Finland, Norway, Germany, Poland, Sweden, Lithuania, Latvia, and Estonia. Covering 104,000 square kilometers (40,000 square miles), it receives runoff from a massive area covering about 163,000 square kilometers (630,000 square miles) and containing 71 million people. Approximately 15% of the world's industrial production occurs in the sea's basin.

Although the Baltic was a relatively clean body of water before the mid-1960s, today, it is one of the most polluted surface waters in the world. A wide assortment of pollutants come from factories, sewage treatment plants, agriculture, forestry, and shipping.

The Baltic Sea is extremely vulnerable to pollution because its waters stratify. Freshwater flowing into the sea from its many tributaries mixes incompletely with the saltwater, forming an upper layer with a slightly lower salt content. The saltier bottom layer is heavier. This stratification tends to reduce exchange between the layers, which results in extremely low oxygen levels in the lower layer. Any factors, such as pollution, that lower levels further can have a devastating effect on bottom-dwelling and deep-water sea life.

Because the Baltic's connection with the North Sea is a narrow channel, it takes about 50 years for the Baltic to exchange all of its water. The long turnover time and stratification result in a dangerous buildup of harmful substances.

The four issues of greatest concern are oil spills, eutrophication, the biomagnification of toxic pollution, and accidental discharges of hazardous or noxious substances. As you learned in this chapter, eutrophication occurs when inorganic nutrients, notably nitrogen and phosphorus, are present in excess amounts. In the Baltic Sea, more than 500,000 metric tons of nitrogen and about 50,000 metric tons of phosphorus enter yearly from fertilizer runoff, untreated sewage, and air pollution (cross-media contamination). This is four times the nitrogen and eight times the phosphorus that the sea received before 1900.

As noted in the chapter, excess inorganic nutrients result in an explosive growth of algae, which decay in the fall and early winter, depleting oxygen levels in large areas of the seafloor. This, in turn, has wiped out much of the bottom fauna (animal life). Scientists estimate that 25% of the bottom of the Baltic Sea is a marine desert.

The bioaccumulation of hazardous substances in fish, described in Chapter 14, has decreased reproduction in fish-eating birds, such as sea eagles, as well as sea mammals, including sea otters and seals. In 1900, the population of gray seals in the Baltic Sea was 1 million. Because of overhunting, the population decreased to 40,000 by 1940. In 1984, only 1000 to 1500 seals remained. This decline is largely attributed to pollutants in the ecosystem that reduce fertility,

cause tumors, and produce hormonal and metabolic changes.

In 1980, a monumental agreement known as the **Helsinki Convention** went into effect. Hailed as the first international agreement to reduce marine pollution in the Baltic, it was signed by all the nations bordering the sea. The agreement covers the sea itself, the seabed, as well as the sea life. It addresses all possible types of pollution, calling on signatories to take all appropriate measures to reduce and prevent pollution and to protect and enhance the marine environment.

Unfortunately, the Helsinki Convention has some loopholes that affect its performance. While the high seas (open waters) are covered by the convention, territorial waters remain the exclusive domain of the bordering states. Military aircraft and vessels are excluded, and atmospheric pollution is not addressed in a meaningful way. Curbing atmospheric emissions is a costly proposition, and Eastern Europe, which contributes a major portion of the atmospheric pollutants, is currently unable to afford pollution controls and prevention measures.

Although the Helsinki Convention has not been a complete success, it has instigated a protection process and, furthermore, serves as a model for other regional environmental pacts aimed at solving regional environmental problems created by neighboring countries.

In 1988, the environmental ministers of the signatory nations agreed to reduce by half the amounts of nutrient, heavy metal, and organic chlorine (PCBs and the like) discharged into the Baltic by 1995. And, in September 1990, the prime ministers and environmental ministers from the Baltic Sea nations met again to begin work on a program to restore the sea's ecological balance. In 1992, the Baltic countries agreed to pay $22.5 billion over 20 years to pay for environmental cleanup costs.

The success and significance of the Helsinki Convention will greatly depend on future cooperation and the role played by the Helsinki Commission (HELCOM). HELCOM has the authority to define objectives for pollution reduction, to observe implementation, and to promote additional protective measures. HELCOM is currently considering revisions that would, among other things, introduce more legally binding measures to reduce land-based pollution, including airborne emissions, and enlarge protective measures to cover the entire drainage basin of the Baltic Sea.

For years, cooperation between the Baltic States was hampered by politics. Today, the major obstacle is the immensity of the problem. Because of the slow turnover, at least 10 to 15 years will pass between substantial reductions in pollution and measurable decreases in pollution concentrations in the sea. Although a clean Baltic may eventually emerge, it will take billions of dollars and decades of effort.

remain. In a recent EPA report, 19% of the U.S. rivers studied had unacceptable levels of bacteria.

The major sources of infectious agents are: (1) untreated or improperly treated sewage; (2) animal wastes in fields and feedlots beside waterways; (3) meat-packing and tanning plants that release untreated animal wastes into water; and (4) some wildlife species, which transmit waterborne diseases. The major infectious diseases include viral hepatitis, polio (viral), typhoid fever (bacterial), amoebic dysentery (protozoan), cholera (bacterial), schistosomiasis (parasitic worm), and salmonellosis (bacterial). These diseases are especially harmful to the young, old, and infirm.

Measuring the level of each pathogenic organism would be costly and time consuming. By measuring levels of a naturally occurring intestinal bacterium, the **coliform bacterium**, water-quality personnel can determine how much fecal contamination has occurred. The higher the coliform count, the more likely the water is to contain some pathogenic agent from fecal contamination. About one-third of U.S. rivers now violates standards for coliform bacteria.

Toxic Organic Water Pollutants   About 10,000 synthetic organic compounds are in use today. Many of these find their way into our water. Concern over these pollutants are numerous: (1) Many toxic organic compounds, for instance, are nonbiodegradable or are degraded slowly, so they persist in the ecosystem. (2) Some are biomagnified in the food web (Chapter 14). (3) Some may cause cancer in humans and aquatic organisms; others are converted into carcinogens when they react with the chlorine used to disinfect water. (4) Some kill fish and other aquatic organisms. (5) Some are nuisances, giving water and fish an offensive taste or odor.

Unfortunately, knowledge of the effects of synthetic organics, which are often found in low concentrations, is rudimentary. Reports of diseases traceable to a single chemical are few, but many experts worry that cancer and genetic damage may result from long-term exposure.

Toxic Inorganic Water Pollutants   Inorganic water pollutants encompass a wide range of chemicals, including metals, acids, and salts. In most states, toxic metals, such as mercury and lead, are major water pollutants. Metals come from many sources, among them industrial discharge, urban runoff, sewage effluents, and mining. They are also derived from air pollution fallout, another example of cross-media contamination, and are released from natural sources. Recent surveys of U.S. drinking water show that some heavy metals, such as lead, come from pipes; others may come from groundwater supplies.

Mercury   One of the more common and potentially most harmful toxic metals is mercury. In the 1950s, mercury was thought to be an innocuous water pollutant, although it was known to have been hazardous to miners and to 19th-century hat makers, who frequently developed tremors, or "hatter's shakes," and lost hair and teeth.

In the 1950s, an outbreak of mercury poisonings in Japan raised awareness of the hazard. Residents who ate seafood from Minamata Bay, which was contaminated with methyl mercury, developed numbness of the limbs, lips, and tongue. They lost muscle control and suffered from other neurological defects, among them deafness, blurring of vision, clumsiness, apathy; mental derangement also occurred. Of 52 reported cases, 17 people died and 23 were permanently disabled.

Mercury is a by-product of the manufacture of the plastic vinyl chloride, of which beach balls, toys, and other products are made. It is also emitted in aqueous wastes of the chemical industry and from incinerators, power plants, research laboratories, and even hospitals. Worldwide, more than 10,000 metric tons of mercury are released into the air and water each year. Between 1977 and 1990, atmospheric concentrations of mercury increased about 1.5% per year in the Northern Hemisphere and 1.1% per year in the Southern Hemisphere.

In streams and lakes, inorganic mercury is converted by bacteria into two organic forms. One of these, dimethyl mercury, evaporates quickly from the water. But the other, **methyl mercury**, remains in the bottom sediments and is slowly released into the water, where it enters organisms in the food chain and is biologically magnified.

Nitrates and Nitrites   Nitrates and nitrites are common inorganic pollutants of water. Nitrates come from septic tanks, barnyards, heavily fertilized crops, and sewage treatment plants. Besides stimulating algal growth, nitrates are also converted to toxic **nitrites** in the intestines of humans. Nitrites combine with the hemoglobin in red blood corpuscles and form methemoglobin, which has a reduced oxygen-carrying capacity. Nitrites can be fatal to infants. Although rare today, nitrite poisonings usually occur in rural areas where drinking water is contaminated by septic tanks and farmyards.

Salts   Sodium chloride and calcium chloride are used on winter roads to melt snow. However, melting snow carries these salts into streams and groundwater. Salts kill sensitive plants, such as the sugar maple. In 1988, the Canadian Supreme Court ruled that the province of Ontario must compensate two farmers whose apple and peach orchards were damaged by salt applied on nearby highways.

In surface waters, salts may kill salt-intolerant organisms, allowing salt-tolerant species to thrive. However, fluctuations in the flow of salt lead to varying water concentrations, a condition that neither salt-tolerant species that thrive in high salt concentrations nor salt-intolerant organisms can survive.

Figure 17.8 *The cooling system of an electric power plant and its effect on surface waters and organisms. Note that water from the river is used to cool water that circulates through the reactor. Heated cooling water may be dumped back into the stream or lake, causing a dramatic increase in water temperature that is harmful to aquatic life.*

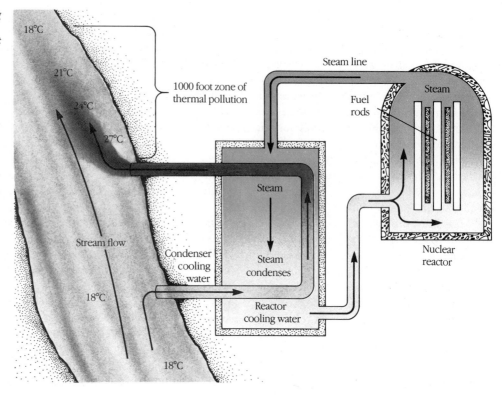

Chlorine

Chlorine is a highly reactive inorganic chemical commonly used: (1) to kill bacteria in drinking water, (2) to destroy potentially harmful organisms in treated wastewater released from sewage treatment plants into streams, and (3) to kill algae, bacteria, fungi, and other organisms that grow inside and clog the pipes of the cooling systems of power plants. Chlorine and some of the products it forms in water are highly toxic to fish and other organisms.

Chlorine reacts with organic compounds to form **chlorinated organics**. These chemicals may show up in drinking water downstream from sewage treatment plants and other sources. Many of them are known carcinogens and teratogens. However, medical studies indicate that the rates of certain cancers, such as liver or intestinal tract, are only slightly elevated in populations consuming water contaminated by these compounds.

Sediment

Sediment, the leading water pollutant in the United States in terms of volume, is a by-product of timber cutting (Chapter 9), agriculture (Chapter 7), mining (Chapters 11 and 13), and construction of roads and buildings. Agriculture increases erosion rates on average 4 to 8 times above normal, sometimes much more. Poor construction and mining may increase the rate of erosion by 10 to 200 times. Sediment is also a major problem in developing countries where poor farming practices, overgrazing, and careless logging practices occur.

Sediment destroys spawning and feeding grounds for fish, reduces fish and shellfish populations, smothers eggs and fry, fills in lakes and streams, and decreases light penetration, which destroys aquatic plants. The deposition of sediment in lakes speeds up natural succession. The filling in of streambeds, or **streambed aggradation**, results in a gradual widening of the channel. As streams fill, they become shallower. Water temperature may rise, lowering the amount of dissolved oxygen and making streams more vulnerable to organic pollutants that deplete oxygen. Lowered dissolved oxygen also wipes out species that require higher levels. Streambed aggradation makes streams more susceptible to flooding. In addition, sediment can fill shipping channels, which must then be dredged at considerable expense. As noted in Chapter 10, sediment fills in costly reservoirs throughout the world. Moreover, hydroelectric equipment associated with dams may be worn out by sediments. Finally, some pollutants, such as pesticides, nitrates, phosphates from agricultural fertilizers, and pathogenic organisms, bind to sediment. This extends their lifetime and impacts.

Sediment pollution can be checked, and even eliminated, by good land management and restoration (Chapters 7 and 9).

Thermal Pollution

Rapid or even gradual changes in water temperature can disrupt aquatic ecosystems. Industries frequently bring about such changes by dumping warm water into rivers and lakes, water that was used to cool various industrial processes. The U.S. electric power industry is a major contributor (Figure 17.8).

It uses about 86% of all cooling water in the United States, or about 730 billion liters (190 billion gallons) per day. Steel mills, oil refineries, and paper mills also use large amounts of water for cooling.

Small amounts of heat have no serious effect on the aquatic ecosystem, but large quantities can shift conditions beyond the range of tolerance, discussed in Chapter 3, thus killing heat-intolerant plants and animals outright and disrupting the web of life dependent on the aquatic food chain. Elimination of heat-intolerant species may allow heat-tolerant species to take over. These are usually less desirable species.

Thermal pollution lowers the dissolved oxygen content of water, at the same time increasing the metabolic rate of aquatic organisms. Since metabolism requires oxygen, some species may be eliminated entirely if the water temperature rises 10° C (18° F). At the Savannah River nuclear power plant, the number of rooted plant species and turtles was at least 75% lower in ponds receiving hot water than in ponds at normal temperature. The number of fish species was reduced by one-third.

Sudden changes in water temperature can cause **thermal shock**, the sudden death of fish and other organisms that cannot escape. Thermal shock is frequently experienced when power plants begin operation or when they temporarily shut down for repair. The latter can devastate heat-tolerant species that inhabit artificially warmed waters.

Fish spawn and migrate in response to changes in water temperature, and thermal pollution may interfere with these processes as well. Water temperature also influences the survival and early development of aquatic organisms. For instance, trout eggs may not hatch if water is too warm. Thermal pollution can also increase the susceptibility of aquatic organisms to parasites, certain toxins, and pathogens.

Thermal pollution can be controlled by constructing ponds for collecting and cooling water before its release into nearby lakes and streams. Cooling towers are an alternative way to dissipate heat (Figure 10.12).

17.3 Groundwater Pollution

Aquifers supply drinking water for about 120 million Americans. That water, scientists are now reporting, is increasingly threatened by pollution. In fact, many pollutants are present at much higher concentrations in groundwater than they are in most contaminated surface supplies. And, many contaminants are tasteless and odorless at concentrations thought to threaten human health.

About 4500 billion liters (1200 billion gallons) of contaminated water seeps into the ground in the United States every day from septic tanks, cesspools, oil wells, landfills, agriculture, and ponds holding hazardous wastes. Unfortunately, very little is known about the extent of groundwater contamination. Some experts believe that groundwater pollution is a minor problem. They estimate that 1% to 2% of U.S. groundwater is polluted. However, an EPA study completed in 1981 showed groundwater contamination in 28% of 954 cities with populations over 10,000. By more recent estimates, at least 8000 private, public, and industrial wells are contaminated. In 1989, the EPA launched a program to assess the extent of groundwater contamination.

Thousands of chemicals, many of them potentially harmful to health, turn up in water samples from polluted wells. The most common chemical pollutants are chlorides, nitrates, heavy metals, and various toxic organics, such as pesticides and degreasing agents. The low-molecular-weight organic compounds are particularly worrisome, since many of them are carcinogenic. Concern among medical experts is great because some fear that there is no threshold level for these compounds—that is, there is no level that's free from risk of cancer or other problems. Other experts fear that many chemicals may act synergistically, turning a potentially difficult problem into a health nightmare (Chapter 14). Beverly Paigen, a researcher in Oakland, California, renowned for her studies at Love Canal in Niagara Falls, New York, published a summary of health studies of Americans exposed to groundwater pollutants. The most common problems include miscarriage, low birth weight, birth defects, and premature infant death. Adults and children suffer skin rashes, eye irritation, and a whole host of neurological problems, including dizziness, headaches, seizures, and fainting spells. In a now widely publicized case in San Jose, California, pollutants from a leaky underground storage tank owned by the Fairchild Camera and Instrument Company are thought to have doubled the rate of miscarriage in pregnant women and tripled the rate of heart defects in newborns. In Woburn, Massachusetts, contaminated groundwater is blamed for a doubling in the childhood leukemia rate.

Many people think of groundwater as fast-flowing underground rivers. Nothing could be further from the truth. Groundwater typically moves from 5 centimeters (2 inches) to 64 centimeters (2 feet) per day. Since it moves so slowly, it may take years for water polluted in one location to appear in another. Additionally, once an aquifer is contaminated, it may take several hundred years for it to cleanse itself.

Detecting groundwater pollution is expensive and time consuming. Numerous test wells must be drilled to sample water and determine the rate and direction of flow. Despite intensive drilling, health officials can easily miss a tiny stream of pollutants that flows through one portion of a large aquifer. For example, liquids that do not readily mix with water may travel along the top or bottom of the aquifer in thin layers and are often difficult to detect.

Figure 17.9 *The anatomy of an oil spill from an improperly drilled hole. Note that the casing in this well did not extend to the oil deposit. As oil was pumped out, it began to leak out a fault, or fissure, and spill into the ocean.*

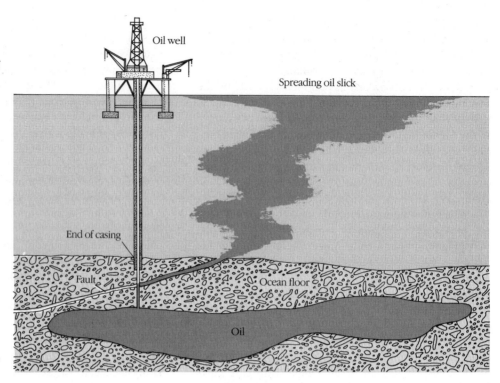

Groundwater supplies one-fourth of the annual water demand in the United States. Preventing groundwater pollution is generally the cheapest way to protect this vital resource (Chapter 18). Reducing the production of hazardous wastes would be an important first step. Improvements in waste disposal methods also help. Ways to achieve this goal are discussed in Chapter 18.

To reclaim polluted aquifers, it may be necessary to pump contaminated water to the surface, purify it, and then return it to the aquifer. New techniques are also being developed to use naturally occurring bacteria found in soil and groundwater to clean up some contamination. For instance, hydrocarbons, such as crude oil, gasoline, and creosote, that have leaked from storage tanks or are spilled from vehicles have polluted more groundwater used for drinking than any other class of chemicals in the United States. Microbiologists have known since the late 1970s that some bacteria can digest or break down hydrocarbons in the soil and groundwater, converting them into carbon dioxide and methane gases. Bacteria in the soil, however, as a rule can only degrade about 1% of the hydrocarbon pollution flowing past them because they lack key chemical nutrients needed for metabolism. By supplying these nutrients, researchers may be able to accelerate the bacterial decomposition of hydrocarbons.

17.4   Ocean Pollution

"When we go down to the low-tide line," Rachel Carson wrote, "we enter a world that is as old as the Earth itself—the primeval meeting place of the elements of Earth and water, a place of compromise and conflict and eternal change." Today, this compromise, conflict, and eternal change have taken on a new meaning as humankind forges out into the oceans in search of food, fuel, and minerals. Even the Kansas farmer and Minnesota factory have an impact on this vast body containing more than 1.3 billion cubic kilometers of water for many inland pollutants eventually make their way to the ocean.

The hazards of pollution in biologically rich coastal zones were discussed in Chapter 10. This section deals with two problems in the oceans: oil and plastic pollution.

Oil in the Seas

About 3.2 million metric tons of oil enters the world's seas every year. About half of the oil that contaminates the ocean comes from natural seepage from offshore deposits. One-fifth comes from well blowouts, breaks in pipelines, and tanker spills (Figure 17.9). The rest comes from oil disposed of inland and carried to the ocean via rivers.

Some oil entering the ocean comes from the transfer of oil from offshore platforms to shore and during normal operations. Contamination from this source has not captured the public attention, even though its effect on marine life and birds can be significant. What generally becomes headline news are the dramatic events, such as large spills—well blowouts or wrecked tankers that spill out tons of black, viscous oil. The number of oil spills in U.S. waters varies from year to year between 5000 and 10,000. In 1991, for instance, nearly 6000 oil spills in U.S. waters released 15 million kilograms of oil (about 115,000 barrels).

Biological Impacts of Oil   About 25% of the crude oil in a spill is volatile and evaporates within three months, becoming an air pollutant. Relatively nonvolatile compounds that are lighter than water, float on the surface, where they are broken down by bacteria over the next few months. Nearly 60% of the oil spill is destroyed in this way. The remaining 15% consists of heavier compounds that stick together and sink to the bottom in huge globs. In cold polar waters, oil decomposes very slowly. In some cases, it may become incorporated in sea ice and be released years afterward.

Before oil's chemical components can evaporate or be broken down, oil can be very harmful. Oil kills plants and animals in the estuarine zone. Especially hard hit are barnacles, mussels, crabs, and rock weed (a type of algae). Their recovery after a major spill may take two to ten years. Oil also washes up on beaches, killing organisms that live there. It settles to the ocean floor and kills benthic (bottom-dwelling) organisms, such as crabs. Those benthic organisms that survive may accumulate oil in their tissue, making them inedible. Oil poisons algae and may disrupt major food chains and decrease the yield of edible fish. It also coats birds, impairing flight or reducing the insulative property of feathers, thus making the birds highly vulnerable to hypothermia (Figure 17.10). Oil endangers fish hatcheries in coastal waters and can contaminate the flesh of commercially valuable fish, as it did in Prince William Sound in Alaska after the Valdez spill.

The amount of damage caused by oil pollution depends partly on the direction in which it is carried by the wind and ocean currents. If slicks reach land, they damage beaches and shorelines, recreational areas, and marine organisms. Oil may be driven over portions of the continental shelf, a highly productive marine zone, where it can poison clams, scallops, flounder, haddock, and other important food species. Oil driven out to sea has fewer environmental consequences because the open waters of the ocean support fewer species.

The damage resulting from an oil spill also depends on when it occurs. For example, the devastating spill off the coast of Alaska in 1989 occurred only two weeks before hundreds of thousands of ducks and other waterfowl migrated through the sound or came to nest. Had the spill occurred after the migration, the damage might have been much less.

Oil pollution of the oceans poses less of a threat to the overall marine environment than was once feared, a U.S. National Research Council committee concluded. Oil can have serious local effects, the group noted, and these can persist for decades. But overall, the marine

Figure 17.10 *This bird is beyond the help of volunteers, who attempted to save many oil-covered animals after a 1989 oil spill in Prince William Sound, Alaska.*

environment has not suffered irreversible damage from oil. The committee was quick to point out that scientists have only limited knowledge of the potential damage of oil to arctic and tropical regions, where much of the current oil development is occurring. Studies of the Alaska oil spill, for example, showed that because hydrocarbons in the oil are more stable and more persistent in cold waters than in warmer seas, their impact could be even greater.

In February 1993, the first results of government-funded research on the biological effects of the Alaskan oil spill were released. They showed substantial biological impacts. For instance, one report estimated that 500,000 birds died as a result of the spill, about ten times more than in any other spill in U.S. history.

The research showed that some species, such as the bald eagle, had recovered fully or were recovering well, while other species were recovering much slower than anticipated. Rockweed is a type of brown algae that once made up 90% of plant matter along the coastline. It is recovering at an extremely slow rate. Several bird species also seemed to be impaired. Reproduction in the population of common murres not killed by the oil spill, for instance, has nearly come to a halt. Several fish species suffered similar long-term reproductive effects.

Reducing the Number of Oil Spills   Thanks to public outcry and stricter controls, the number of oil spills began to fall after 1980 and has remained relatively stable in recent years. Tougher governmental standards for new oil tankers went into effect in 1979. New safety standards for older tankers were phased in between 1981 and 1985. Dual radar systems, backup steering controls, collision avoidance aids, and improved inspection and certification were instrumental in reducing spills. Under the new regulations, crude oil must be cleaned to eliminate sludge buildup in tanks. This sludge was once rinsed out at sea. New regulations also require tankers to have separate ballast tanks. These tanks are filled with saltwater to help ships keep their balance when returning after discharging their cargo. In older ships, empty oil tanks were once filled with water for ballast. When the ship arrived at port, the oil-contaminated water was dumped into the sea. The serious effects of oil washing and combined ballast-oil tanks must not be underestimated. Before regulations, about 1.3 million metric tons of oil were released each year during tank purging and ballast tank discharge, or over six times the amount released by tanker spills.

Although the number of oil spills has decreased worldwide since 1980, the quantity of oil released fluctuates wildly. In 1991, for example, large oil tanker accidents released approximately seven times as much oil as was released in 1990 from fewer spills. Much room for improvement remains.

In 1990, Congress passed the **Oil Pollution Act**. It establishes a $1 billion fund to be used to clean up oil spills and pay for damages. The money comes from a 3-cents-per-barrel tax on domestic and imported oil, thus passing the costs of cleanup and damage on to the consumer. Under the bill, oil companies would ultimately be responsible for cleanup costs but only to a point because the law sets strict limits on their financial liability. A company responsible for a spill, for instance, would pay only $10 million for cleanup unless it occurred near an onshore facility where damage is much greater. In such instances, companies would pay $350 million for each spill. The fund would be used to pay the remaining costs. Many critics are disappointed with the liability limits because the costs of cleaning oil are likely to be far greater. Cleanup in Prince William Sound to date has cost Exxon $2.5 billion and the federal government $154 million. In addition, in 1992, Exxon agreed to pay an additional $1.25 billion in criminal fees, restitution, and civil recovery to pay damages to salmon fishing companies and the like.

Although the law requires new tankers to be equipped with double steel hulls or effective double containment features to reduce the likelihood of spills, existing oil tankers are not required to be retrofitted with double hulls. However, existing tankers must be escorted by two towing vessels in high-risk areas and single-hull vessels will be phased out starting in 1995. By the year 2015, all tankers with single hulls will be banned.

In addition, the law establishes policies for regional oil-spill response teams that can be deployed immediately after a spill to coordinate cleanup efforts. It also allows states to set stricter guidelines. A state, for instance, could mandate double hulls for use in its harbors.

Plastic Pollution   A young seal swims playfully in the coastal waters of San Diego Bay. Floating in its watery domain is a piece of a plastic fishing net that has drifted with the currents for months. The seal swims around and around curiously, then plunges through an opening in the net only to be entrapped.

At first, the net is just a mild nuisance, but as the seal grows, the filament begins to tighten around its neck. Eventually, the net cuts into the seal's skin, leaving an open ring of raw flesh exposed to bacteria. Unless it is helped, the seal will perish. All told, about 400,000 sea mammals die each year as a result of the estimated 6.4 million metric tons of plastics, including nets, discarded into the ocean annually by commercial fishermen and sailors and military personnel. Tens of thousands of tons may also come from private boats, factories, and onshore facilities. No one knows the number of seabirds or fish that die each year from nylon fishing nets, plastic bags, six-pack yokes, plastic straps, and a myriad of other objects made from nonbiodegradable plastic.

Plastic nets entangle fish, birds, and sea mammals. They may strangle, starve to death, or drown their victims. Plastic bags, looking like jellyfish, are eaten by sea turtles. One scientist pulled enough plastic out of a leatherback turtle's stomach to make a ball reportedly several feet in diameter. In such cases, starvation is the usual result because the animal's stomach is packed with indigestible plastic that cannot pass through its digestive tract. When birds and fish gobble plastic beads resembling the tiny crustaceans that are a normal part of their diet, they may become poisoned and die. Discarded plastic eating utensils, when swallowed, may cut into an animal's stomach lining, causing it to bleed to death.

Growing awareness has created a groundswell of activity. The Oregon Fish and Wildlife Department organizes annual beach cleanups to remove plastic, netting many tons of plastic garbage. Italy recently placed a ban on all nonbiodegradable plastics, which went into effect in 1991. Oregon and Alaska have passed laws requiring that all six-pack yokes be biodegradable. In 1988, Congress passed the **Plastic Pollution Control Act**, which makes it unlawful for any U.S. vessel to discard plastic garbage in the ocean. This legislation also requires all manufacturers of six-pack yokes to use degradable plastic if technically feasible. The EPA is currently attempting to define *degradable*; a final ruling on this stipulation is not expected until 1994. In 1988, the Senate unanimously approved an international treaty banning the disposal of plastics in the ocean. The treaty also prohibits the dumping of other garbage within 20 to 40 kilometers (12 to 25 miles) of the world's coasts. The agreement, signed by 28 nations, went into effect in December 1988.

Table 17.2 Common Water Pollutants

Sources	Bacteria	Nutrients	Ammonia	Total Dissolved Solids	Acids	Toxics
Municipal sewage treatment plants	•	•	•			•
Industrial facilities						•
Combined sewer overflows	•	•	•	•		•
Nonpoint Sources						
Agricultural runoff	•	•		•		•
Urban runoff	•	•		•		•
Construction runoff		•				•
Mining runoff				•	•	•
Septic systems	•	•				•
Landfills spills						•
Forestry runoff		•				•

Source: U.S. Environmental Protection Agency.

Although the treaty may be difficult to enforce given the enormous size of the oceans, it could help reduce plastic pollution.

On another front, in what may prove to be a precedent-setting move, legislators in New Jersey introduced legislation to ban the release of nonbiodegradable balloons. Why? Many balloons released to celebrate national holidays and other occasions eventually end up in the ocean. Here, they are often mistaken for food by marine organisms. They accumulate in their digestive tracts and cause starvation.

Despite these steps, millions of fish, birds, and sea mammals will perish in coming years. Without stricter controls, biodegradable plastic, and widespread public cooperation, rising use of plastic will bring unnecessary death to innumerable sea creatures.

Medical Wastes and Sewage Sludge   In the summer of 1988, many Americans were shocked to learn that medical wastes were being illegally dumped into the ocean. Bloody bandages, sutures, vials of AIDS-infected blood, and used syringes washed up onto the eastern shores of the United States as well as the shores of Lake Erie.

Because there was no way to track the wastes to their source, Congress passed the **Medical Waste Tracking Act** in 1988. It created a two-year program that covered ten states, which required people generating, storing, treating, and disposing of medical wastes to keep records. That way if wastes washed up on shorelines,

they could be traced to their source so that responsible parties could be brought to justice. Unfortunately, this law expired in 1990; as of January 1993, it has not been renewed.

In 1988, Congress also passed the **Ocean Dumping Ban Act,** which prohibited the dumping of sewage sludge (organic material from sewage treatment plants, discussed later) in the ocean. Dumping officially ended on December 31, 1991.

According to the Natural Resources Defense Council (NRDC), 8.9 trillion liters (2.3 trillion gallons) of liquid waste generated from sewage treatment plants is also dumped directly into the ocean. Some of the waste receives little or no treatment before it is discharged. Much of it is industrial waste containing toxic organic chemicals and toxic metals.

17.5 Water Pollution Control

Table 17.2 offers a quick summary of the different sources of pollution and shows the major pollutants from each. This information is helpful in plotting control strategies. As in other areas, water pollution control has traditionally focused on end-of-pipe measures aimed at treating the symptoms, not the root causes, of the problem. This section looks at traditional approaches first, then examines the outlines of a more sustainable strategies.

Figure 17.11 *A vast sewage treatment plant in Des Moines, Iowa. The large circular devices are trickling filters for secondary treatment.*

Legal Controls

The cornerstone of U.S. water pollution policy is the **Clean Water Act.** Focused primarily on point sources, it has resulted in the construction and improvement of thousands of sewage treatment plants to handle municipal wastes (Figure 17.11) and installation of water pollution control equipment at factories to reduce the discharge of industrial waste into water bodies. Major provisions of the act are listed in Table 17.3.

Unfortunately, in many cases, nonpoint pollution from expanding cities offsets the gains from sewage treatment plants. Thus, water quality has not improved or has improved only slightly in many areas experiencing rapid population growth.

Controlling Nonpoint Pollution   In 1987, Congress amended the Clean Water Act to address nonpoint water pollution. Under the revised law, states were required to identify nonpoint sources and draft plans to control them. The amendments authorized the EPA to provide $400 million over four years to the states to set up their programs, an amount critics said would hardly put a dent in nonpoint pollution. As with numerous other environmental issues, laws passed to address a problem never bore fruit. As of 1992, only about one-third of the money earmarked for state programs was delivered; most of the programs exist only on paper. What happened? Congress failed to appropriate the money because of budgetary problems and because of a lack of support from the administration. It was also concerned that the EPA would not make sure that the states followed through.

Despite federal inaction, some state and local governments have taken actions on their own by passing zoning ordinances aimed at reducing agricultural and urban runoff. Maryland, for instance, limits the amount of pavement accompanying development along fragile regions of Chesapeake Bay. In Northern California, loggers are prohibited from clogging streams with silt. In addition, local soil conservation districts help identify trouble spots and work with farmers. Despite these gains, much work will be needed in future years.

New laws at all levels of government can help reduce nonpoint pollution. Since agricultural runoff is the largest unregulated source of nonpoint water pollution, affecting between 50% and 70% of all surface water and groundwater, laws that address pollution from farms are important. New regulations that require terracing of steep road banks, revegetation of denuded land, and use of mulches to hold soil in place while grasses root along newly constructed highways and housing sites are also needed. Laws that require sediment ponds to collect runoff before it can reach streams and porous pavements that soak up rainwater could also help reduce erosion and the influx of pollutants into nearby water bodies.

Preventing Groundwater Pollution   Several federal laws provide the EPA with authority to prevent and control sources of groundwater pollution and to clean up polluted groundwater. In 1984, the EPA adopted a groundwater protection strategy that, among many things, established an Office of Groundwater Protection. This strategy focuses on building state capacity—that is, state regulation and control of groundwater. Under the plan, the EPA provides technical assistance in analyzing problems and advice needed by states to establish their own groundwater-protection programs. Unfortu-

Table 17.3 Major Provisions of the U.S. Clean Water Act

Planning

The states receive planning grants from the EPA to review their water pollution problems and to determine ways to solve them by reducing or eliminating point and nonpoint water pollutants.

Standards development

The states adopt water quality standards for their streams. These standards define a use for each stream and prescribe the water quality needed to achieve that use.

Effluent standards

The EPA develops limits on how much pollution may be released by industries and municipalities. These limits are developed at the national level based on engineering and economic judgments. The EPA or the states are required to make the discharge limits for individual plants more stringent if necessary to meet the state water quality standards.

Grants and loans

The EPA provides financial assistance to state water programs for the construction of sewage treatment plants, permit applications, water quality monitoring, and enforcement.

Dredge and fill program

The EPA develops environmental guidelines to protect wetlands from dredge and fill activities. These guidelines are used to assess whether permits should be issued.

Permits and enforcement

All industries and municipal dischargers receive permits from either the EPA or the states. The EPA and the states regularly inspect these dischargers to determine whether they are in compliance with the permit and take appropriate enforcement actions if necessary.

Source: Environmental Protection Agency.

nately, say critics, states often lack necessary funds, experience, and expertise. Nonetheless, many states have adopted their own groundwater standards. Some have taken preventive measures by mapping out aquifers and banning industrial development over them. Today, the vast majority of the states have developed programs that at least minimally regulate discharges to groundwater. Clearly, much improvement is needed, especially in groundwater pollution prevention.

In 1986, Congress enacted laws that protect groundwater. Congress also recently passed legislation that requires the EPA to set standards for drinking water quality for 85 chemical substances, including pesticides and various industrial chemicals. The new law requires drinking water suppliers to test for these chemicals and maintain drinking water standards. It also requires them to monitor drinking water supplies for other substances that might pose a threat to human health. Even though the EPA is now monitoring these chemicals, hundreds more could be checked.

Control Technologies

The first sewage treatment plant in the United States was built in Memphis, Tennessee, in 1880; today, there are over 15,600. Sewage entering a treatment plant often contains pollutants from homes, hospitals, schools, and industries. It contains human wastes, paper, soap, detergent, dirt, cloth, food residues, microorganisms, and a variety of household chemicals. In some cases, water from storm drainage systems is mixed with municipal waste drainage systems to save the cost of building separate pipes for each, which can be exorbitant. Combined systems generally work well, but during storms, inflow may exceed plant capacity. Consequently, some untreated storm runoff and sewage passes directly into waterways, raising the coliform count in downstream waters and rendering them unfit for swimming. During 1989 and 1990, the Natural Resources Defense Council identified 2400 beach closings and pollution advisories in California and nine East Coast States, mostly New York, New Jersey, and Connecticut. The most consistent cause was overflow from outdated sewage treatment systems.

Primary Treatment   Sewage treatment can take place in three stages: primary, secondary, and tertiary. **Primary treatment** physically removes large objects by first passing the sewage through a series of grates and screens. Sand, dirt, and other solids settle out in grit chambers (Figure 17.12a). The solid organic matter, or sludge, settles out in a settling tank.

Secondary Treatment   The secondary stage destroys biodegradable organic matter through biological decay (Figure 17.12b). Sludge from primary treatment enters a large tank, where bacteria and other organisms decompose the waste. Another common way is to pass the liquid sludge through a **trickling filter** (Figure 17.13). Here, long pipes rotate slowly over a bed of stones and sometimes bark, dripping wastes on an artificial decomposer food web consisting of bacteria, protozoa, fungi, snails, worms, and insects. The bacteria and fungi consume the organics and, in turn, are consumed by protozoans. Snails and insects feed on the protozoans. Some

Case Study 17.2

The Case of the Dying Seals

In the spring of 1988, the harbor seals in the North Sea began to die in record number. Adult seals floated aimlessly in the water, too weak to eat. Pregnant females aborted their fetuses. The mysterious epidemic, which began off the coast of Denmark, quickly spread to seal colonies throughout the North and Baltic seas. By the middle of the summer, seals were dying along hundreds of miles of North Sea coastline. By September, the disease had spread to the Atlantic coast of Ireland.

Some people dubbed this incident the "black death of the sea," for it recalled the epidemics of bubonic plague, or black death, which devastated human population in Europe in the 1300s. In this tragic turn of events, though, the victims were helpless seals. In a few short months, the population of harbor seals, once numbering 18,000, fell to only 6000.

Studies showed that the massive die-off was caused directly by the phocine distemper virus (PDV), a new virus to science. Biologists believe, however, that the virus was only the most immediate cause of death. Pollution in the seas, they say, may have greatly weakened the immune systems of the seals, making them highly susceptible to infection.

Seals living in the waters off the coast of Germany and the Netherlands are heavily contaminated with toxic chemicals called polychlorinated biphenyls (PCBs), a substance once used as an insulator in electric devices. PCBs and other chemical contaminants are believed responsible for the reproductive problems and the suppression of the seals' immune systems. In fact, an emergency working group formed in London to deal with the epidemic. The group of biologists, veterinarians, and toxicologists concluded that ". . . persistent pollutants could not be excluded as an additional factor in the seal deaths." Though there is not a great deal of evidence to clarify the role of pollution, scientists have found that the highest seal mortality occurred in waters that were most polluted.

The North and Baltic seas have been polluted for years. The North Sea alone annually receives 60 billion liters (15 billion gallons) of wastewater from factories and waste treatment facilities in bordering industrial nations. The pollution problem in the North and Baltic seas is compounded by the fact that both seas are shallow and cleanse themselves very slowly. The North Sea, for instance, purges itself only twice every ten years. The Baltic Sea is even slower. (See Models of Global Sustainability 17.1.) Because they are cleansed so slowly, pollution can reach dangerous levels with serious impacts on fish and wildlife, and, possibly, people.

The seal plague may be the latest manifestation of a chronic pollution problem in the North and Baltic seas. Fortunately, many countries have decided to take action and have entered into cooperative agreements. Most of the countries bordering the North Sea, for instance, agreed to halve the amount of nutrient pollution (nitrates and phosphates) and toxic chemicals flowing into the sea by 1995. A similar agreement was reached to cut pollution entering the Baltic (Models of Global Sustainability 17.1).

Seal deaths off the coast of Europe are a symptom of a global problem. Similar tragedies are occurring elsewhere. Since June 1987, for example, as many as four out of ten dolphins off the Atlantic coast of the United States have perished. Studies of gulls in the Great Lakes have shown an alarming reproductive failure due to PCBs and other organic pollutants.

Despite an outpouring of laws to control pollution, the United States and other countries have hardly come to grips with the problem. Tens of thousands of hazardous waste sites litter the American landscape. Pollution control laws passed in the 1970s initially decreased water pollution nationwide, but since the early 1980s, pollution levels have remained more or less constant. Making matters worse, regulation and enforcement of hazardous waste laws have been lax.

Solutions to global problems require new laws and tighter controls. Critical thinking demands a search for additional systemic solutions. New technologies, for example, can help reduce waste. Individuals can also chip in. A personal commitment to conserve, to recycle, to use renewable resources (for example, paper rather than plastic), and to limit family size can go a long way in helping to solve the environmental problems. Individual actions, multiplied many times, must be a part of the solution.

inorganic nutrients are also removed. This step is followed by a secondary settling basin or clarifier to remove residual organic matter.

In most municipalities, the liquid remaining after secondary treatment is chlorinated to kill potentially pathogenic bacteria and protozoans, and is then released into receiving streams, lakes, or bays. The efficiency of primary and secondary treatment is shown in Table 17.4.

Tertiary Treatment Many methods exist for removing the chemicals that remain after secondary treatment. Most of these tertiary treatments are costly and, there-

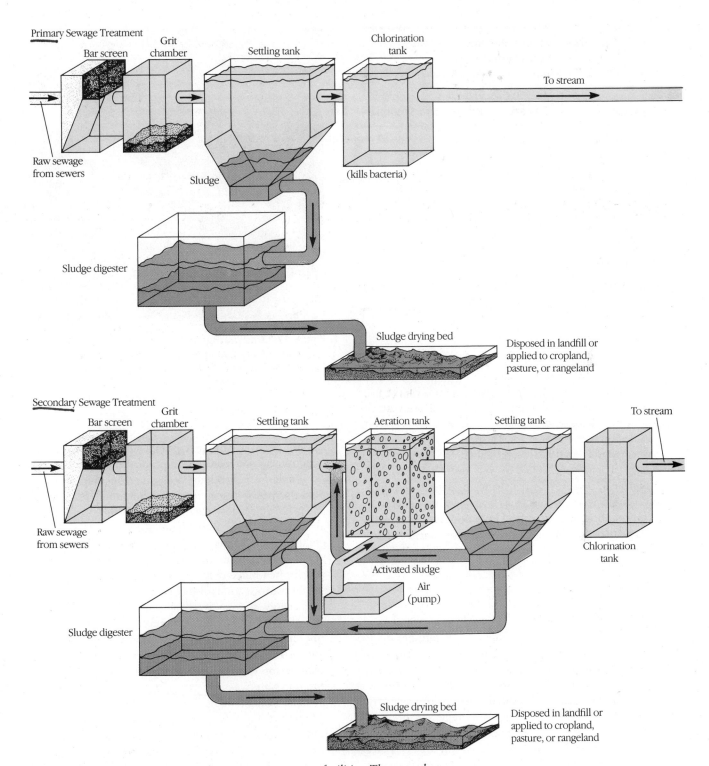

Figure 17.12 *Primary and secondary sewage treatment facilities. The secondary treatment facility contains all of the components of the primary system but has an additional aeration tank, or trickling filter, to further decompose organic matter.*

fore, are rarely used unless water is being released into bodies of water that require a high level of purity—for example, near the California resort towns surrounding Lake Tahoe. Fortunately, some cheaper options are gain-

ing recognition. For example, after secondary treatment, effluents can be transferred to holding ponds where algae and water hyacinths growing in the water consume the remaining nitrates and phosphates. Certain aquatic

Table 17.4 Removal of Pollutants by Sewage Treatment Plants

Substance	Percentage Removed by Treatment	
	Primary	Primary and Secondary
Solids	60	90
Organic wastes	30	90
Phosphorus	0	30
Nitrates	0	50
Salts	0	5
Radioisotopes	0	50
Pesticides	0	0

Figure 17.13 *Trickling filter. A decomposer food chain consisting of bacteria and other microorganisms in the rock or bark bed of the system consumes organic matter, nitrates, and phosphates in the liquid sewage.*

plants, such as duckweed, absorb dissolved organic materials directly from the water.

Aquatic plants grown in sewage ponds can be harvested and converted into food for humans or livestock. In Burma, Laos, and Thailand, duckweed has been consumed by farmers for years. The protein yield of a duckweed pond is six times greater than that of an equivalent field of soybeans. One of the problems with this approach, however, is that water hyacinths and duckweed also absorb toxic metals from the water. Therefore, consumption by humans and livestock must be carefully monitored. Another approach is the use of artificial wetlands, either indoor in greenhouses or outdoors. (These were discussed in Models of Global Sustainability 3.1.)

Sustainable Solutions

Most of the approaches outlined above are needed to protect our water. In making the transition to a sustainable society, considerable effort must be made to find solutions that strike at the root causes. These strategies do not merely capture pollutants bound for waterways and dispose of them in a slightly more acceptable manner, in the process contaminating another medium. Rather, they seek to prevent pollution in the first place.

Conservation, recycling, renewable resource use, and population control are four forms of pollution prevention. For example, by using energy more efficiently, individuals and companies can reduce their demand for electricity. This, in turn, decreases the amount of cooling water needed by utilities and cuts back on thermal pollution and chlorine pollution in streams and lakes. Farmers can devise ways to cut back on their use of fertilizers and pesticides (Chapters 7 and 18). They are already using new methods to control soil erosion (Chapter 7)

and are discovering that steps taken to reduce nonpoint pollution do not reduce food production.

Table 17.5 shows the dramatic reductions in water pollution possible from recycling aluminum, steel, paper, and glass. Since recycling these commodities also reduces air pollution and because air pollution can become water pollution, this strategy pays a double dividend. A third payoff comes from the reduction of mining wastes, which often make their way into streams.

In contrast to fossil fuel power plants, most renewable energy sources, such as photovoltaics and wind energy, require little, if any, water and produce little, if any, air and water pollution. Population control, of course, is the ultimate pollution prevention strategy. Each person less means that much less waste that goes into our waterways.

The sustainable strategy also seeks to weave human actions into nature's cycles. That is, it attempts to return nutrients to their site of origin or other useful purposes, for example, by using municipal sewage to fertilize crops. This approach is technically known as **land disposal**.

In ancient times, land disposal of human sewage was commonplace; it is still practiced in many developing nations, such as China and India. As the populations of many developing countries have grown and become more urbanized, this natural method of recycling wastes has gradually been phased out. Land disposal, discussed in Chapter 10 as a means of recharging groundwater, uses the surface vegetation, soil, and soil

Table 17.5 Environmental Benefits Derived from Substituting Secondary Materials for Virgin Resources

Environmental Benefit	Aluminum	Steel	Paper	Glass
	(percent)			
Reduction of				
Energy use	90–97	47–74	23–74	4–32
Air pollution	95	85	74	20
Water pollution	97	76	35	—
Mining wastes	—	97	—	80
Water use	—	40	58	50

Source: Lechter, R. C. and Sheil, M. T. 1986. Source Separation and Citizen Recycling. In *The Solid Waste Handbook*. Robinson, W. D., ed. New York: Wiley.

microorganisms as a natural filter for many potentially harmful chemicals. Sewage can be piped to pastures, fields, and forests (Figure 17.14). Organic matter in the effluent enriches the soil and improves its ability to retain water. Nitrates and phosphates serve as fertilizers. The water supports plant growth and helps recharge aquifers. Crops nourished by effluents from treated sewage show a remarkable increase in yield.

Land disposal of sludge from sewage treatment plants has some problems. First, treated sewage may contain harmful bacteria, protozoans, and viruses that could adhere to plants consumed by humans or livestock or become airborne after the effluent dries. To circumvent this problem, European countries, such as Switzerland and Germany, heat their sludge to destroy such organisms before applying it to pastures and cropland. Alternatively, sewage sludge can be decayed in compost piles before application. The heat given off during composting kills virtually all of the viruses, bacteria, and parasite eggs.

The second problem is that toxic metals found in some sewage may accumulate in soils and be taken up by plants and livestock. Metal-contaminated sewage usually comes from industries. By removing metals from their waste stream, an option called **pretreatment**, or preventing metals from entering the waste stream entirely (pollution prevention), industries can eliminate the problem.

The third problem is that transporting sludge to fields increases the cost of sewage treatment, limiting some of the incentive to use this method. Experts are quick to point out, however, that land disposal is ten times cheaper than building and operating a tertiary treatment plant.

Scientists at the University of Maryland recently developed an unusual way to put sludge to good use. By combining it with clay and slate, they formed odorless "biobricks," which look like ordinary bricks. Washington's Suburban Sanitary Commission recently tested the invention by building a 750-square-meter (8300-square-foot) maintenance building in Maryland with 20,000 biobricks. If successful, biobricks could help reduce land disturbance from mining materials for brick making and could help cut sewage disposal costs and environmental contamination.

Restoration efforts can also help clean up water supplies. Direct efforts are needed to clean up groundwater, as noted earlier in the chapter. In addition, restoration of wetlands can help reverse the decline of aquatic systems. In Florida, for example, drained swampland that was converted to farms is now pouring incredible amounts of plant nutrients into the Everglades. The native sawgrass that covers much of this immense wetland is adapted to a nutrient-poor environment. Unfortunately, in the presence of excess nutrients, cattails are taking over, outcompeting the slow-growing sawgrass. To date, about 20,000 acres of Everglades National Park has been destroyed by cattails, which are spreading at a rate of about 1.6 to 2.5 hectares (4 to 6 acres) per day. Debris from the plants rots in the water and creates anaerobic conditions that kill fish and other aquatic species. To reverse this trend, Florida officials have proposed taking 13,400 hectares (33,000 acres) of drained swampland, now being farmed, out of production. These lands would be flooded and set aside as a marshy area that would filter out nutrients from surrounding farmland, reducing the influx to the park by about 85%.

These are just a handful of steps needed to solve the water pollution dilemma sustainably. Of course, personal actions are also needed. Limiting family size and reducing the consumption of unnecessary goods are important actions individuals can take. By reducing the purchase of goods, individuals can greatly reduce hazardous waste production at factories. As a general rule, each metric ton of garbage generated by consumers results in 5 metric tons of manufacturing waste and 20 metric tons of waste at the source, the mine or forest where the raw materials were acquired. So every savings we make by buying less or buying more durable items can effect large upstream savings in waste and pollutants that may make their way into our waterways.

Another effective way to cut your personal contribution is to install a composting toilet. The composted wastes can be added safely to gardens, eliminating the need for synthetic fertilizers. If and when you become a homeowner, restrict your use of synthetic insecticides, herbicides, bleaches, detergents, disinfectants, and other chemicals, or find safe alternatives. Select low-phosphate

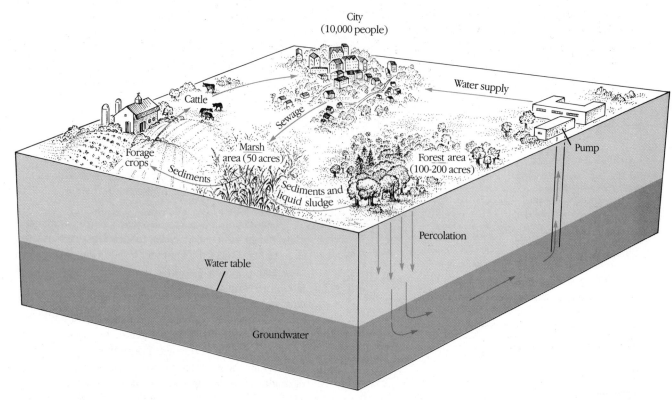

Figure 17.14 *Land disposal of sewage helps fertilize farmlands and forests, reduce surface water pollution, and replenish groundwater supplies.*

or no-phosphate detergents for washing clothes. You may also contact local and federal officials in support of further cleanup efforts.

It is astonishing with how little wisdom

mankind can be governed, when that

little wisdom is its own.

—W. R. Inge

Critical Thinking Exercise Solution

Water conservation in the home is not just about saving water. It's also about reducing energy demand, since it takes energy to pump water to treatment plants and it takes energy to heat water for domestic use. The less water you use, the less energy is needed in your home and at the water treatment plant. This, in turn, results in numerous environmental savings, from reduced habitat destruction from coal mining to less air pollution.

Water conservation in the home also means less water going down the drains to sewage treatment plants. That saves energy at the plant. It also reduces the demand for chlorine, which is used to disinfect the effluent of sewage treatment plants.

This exercise clearly shows how important it is to examine the big picture. It also illustrates one of the common myths of our society—that conservation means sacrifice. Water conservation measures in the home can be quite inexpensive and easy to install. Simple changes in behavior like not flushing the toilet every time or watering the lawn less frequently actually reduce work. Conservation measures can save you money as well, which can be used to enhance your life.

Summary

17.1 Water and Water Pollution

■ Water pollutants come from **point** and **nonpoint sources**. Their effects on aquatic systems largely depend on whether polluted waters are standing (lakes and ponds) or flowing (rivers). Standing systems are generally more susceptible because of slow turnover.

17.2 Types of Water Pollution

■ The major water pollutants are organic nutrients, inorganic nutrients, infectious agents, toxic organics, toxic inorganics, sediment, and heat.

■ **Organic nutrients** come from feedlots, municipal sewage treatment plants, and industry. They promote growth of natural populations of aquatic bacteria. Bacterial decomposition of organic materials results in declines in dissolved oxygen, with dire effects on other oxygen-requiring organisms.

■ Two **inorganic plant nutrients** of major concern are nitrogen and phosphorus. They come primarily from septic tanks, barnyards, heavily fertilized crops, and sewage treatment plants, and cause excessive plant growth that clogs navigable waterways. Bacterial decay of plants in the fall results in a drop in dissolved oxygen, which may suffocate fish and other organisms.

■ Water may contain pathogenic bacteria, viruses, protozoans, and parasites (**infectious agents**). Untreated or improperly treated sewage, animal wastes, meat-packing wastes, and some wild species are the major sources. Waterborne infectious diseases present a special problem in developing nations with poorly developed sewage treatment facilities.

■ **Toxic organic pollutants** include a large number of chemicals, such as pesticides and PCBs, many of which are nonbiodegradable or slowly degraded, biologically magnified, and carcinogenic.

■ **Toxic inorganic pollutants** include a wide range of chemicals, such as metals and salts, from a wide array of sources. **Mercury** is a particularly troublesome pollutant because it is converted into methyl and dimethyl mercury in aquatic ecosystems by aerobic bacteria. These forms are more toxic than inorganic mercury. Methyl mercury is biologically magnified in the food chain.

■ **Sediment,** the leading water pollutant in the United States, is a by-product of erosion resulting from poorly managed timber cutting, agriculture, ranching, mining, and construction.

■ Sediment destroys spawning and feeding grounds for fish, reduces fish and shellfish populations, destroys pools used for resting, smothers eggs and fry, fills in lakes and streams, and decreases light penetration, thus endangering aquatic plants.

■ **Thermal pollution** refers to the heating or cooling of water, both of which drastically alter biota in a body of water. Large quantities of heat can kill heat-sensitive organisms and harm organisms dependent on the aquatic ecosystem.

17.3 Groundwater Pollution

■ The concentration of many pollutants in groundwater is often higher than that in the most contaminated surface water supplies. Many of these chemicals are tasteless and odorless at concentrations believed to pose a threat to human health.

■ The major groundwater pollutants are chlorides, nitrates, heavy metals, and toxic organics.

■ Since groundwater usually moves slowly through an aquifer, it may take years for pollution to show up in areas adjacent to sources of contamination. And, once an aquifer is contaminated, the pollutants may remain for centuries.

17.4 Ocean Pollution

■ The oceans receive pollutants from many sources.

■ Oil pollution is one of the more serious problems. About half of the oil that contaminates the ocean comes from human sources: oil well blowouts, tanker spills, and inland disposal of oil.

■ Oil harms many organisms, especially if a spill occurs near an estuarine zone. It may take two to ten years for aquatic life to recover from a spill.

■ Thanks to public outcry and stricter controls, the number of oil spills has decreased substantially, although the problem is far from solved.

■ Plastic pollution has also become a major problem throughout the world. Plastic nets, plastic garbage, and plastic medical wastes are killing millions of marine mammals, turtles, and fish.

■ Animals may become tangled in the plastic debris or may eat it and die. Because of public outcry, many governments have banned the dumping of plastics in oceans.

17.5 Water Pollution Control

■ Most control efforts in the United States and abroad have sought to reduce point-source pollution.

■ Sewage treatment plants and limitations on factory discharges have been key strategies of the U.S. **Clean Water Act.**

■ Sewage treatment can take place in three stages. **Primary treatment** removes large objects by a series of grates and screens. Solid organic matter, or sludge, settles out in a primary settling tank.

■ **Secondary treatment** removes biodegradable organic material and some inorganic substances. Sludge may be decomposed by bacteria in large aerated tanks or may be passed through a **trickling filter**, where it is dripped over a bed of rocks or bark housing a detritus food chain.

■ In most municipalities, liquid remaining after secondary treatment is chlorinated to kill potentially pathogenic organisms and released into receiving streams.

■ Many methods exist for **tertiary treatment**, the final cleanup stage, but most are expensive and therefore rarely used. Fortunately, there are some cheaper options, such as algae and water hyacinth ponds and land disposal.

■ Progress in cleaning up waters in the United States and elsewhere has been slow. Continued population growth, industrial growth, and agricultural expansion have negated many gains. Nonpoint sources often offset gains at point sources, making it imperative that further controls be directed at them.

■ Preventive measures that strike at the roots of the problems, among them conservation, recycling, restoration, and population control, can go a long way toward solving these problems and building a sustainable society.

Discussion and Critical Thinking Questions

1. List the major types of water pollutants found in developing and developed nations.

2. Define the terms *point source* and *nonpoint source*. Give some examples of each, and explain why nonpoint sources of water pollution are often more difficult to control than point sources.

3. What are the three major ecological zones of a lake? Describe each one. What are the unique ecological problems of each?

4. Explain why a lake "turns over" in the spring and fall, and how this natural turnover benefits aquatic organisms.

5. Describe where organic nutrients come from, what effects they have on aquatic ecosystems, and how they can be controlled.

6. What are inorganic plant nutrients, and how do they affect the aquatic environment?

7. Define the term *eutrophication*. Describe how inorganic and organic nutrients accelerate natural succession of a lake.

8. What are the major sources of infectious agents in polluted water? How can they be controlled?

9. What are some of the major inorganic water pollutants? How do they affect the aquatic environment and human populations?

10. Why is chlorine used in the treatment of human sewage and drinking water? What dangers does its use pose? How would you determine if the risks of chlorine use outweigh the benefits?

11. What are the major sources of sediment, and how can they be controlled? What are the costs and benefits of sediment control?

12. A factory that has been polluting a nearby stream with toxic organic chemicals has agreed to stop. It is proposing the use of evaporation ponds, where the wastes will sit until they evaporate. Using your critical thinking skills, your systems thinking abilities, and your knowledge of water pollution, critically analyze this proposal.

13. What are the major sources of groundwater pollution? How can groundwater pollution be reduced or eliminated? What can you do?

14. Discuss the sources of oil in the ocean and ways to reduce oil contamination.

15. Describe primary and secondary wastewater treatment. What happens at each stage, and what pollutants are removed?

16. A housing development is being built in your town. The project's developers say that pollution in the river that flows through the town will not increase because they will pay to expand the town's sewage treatment plant. Do you agree or disagree? Why?

Suggested Readings

Borelli, P. (1989). Troubled Waters. Alaska's Rude Awakening to the Price of Oil Development. *The Amicus Journal* 11 (3): 10–20. Excellent overview of the Alaskan oil spill.

Dolin, E. J. (1992). Boston Harbor's Murky Political Waters. *Environment* 34 (6): 6–11, 26–33. Detailed case study of a pollution disaster.

Eder, T. and Jackson, J. (1988). *A Citizen's Guide to the Great Lakes Water Quality Agreement.* Buffalo: State University College. Excellent overview of the agreement between the United States and Canada to clean up the Great Lakes.

EPA. (1992). *The Quality of Our Nation's Water: 1990.* Washington, D.C.: EPA. Good survey of problems and solutions.

Maurits la Rivière, J. W. (1989). Threats to the World's Water. *Scientific American* 261 (3): 80–94. Up-to-date survey of global pollution problems.

Miller, S. K. (1992). When Pollution Runs Wild. *National Wildlife* 30 (1): 26–28. Interesting look at nonpoint pollution.

O'Hara, K. J., Iudicello, S., and Bierce, R. (1988). *A Citizen's Guide to Plastics in the Ocean.* Washington, D.C.: Center for Marine Conservation. Excellent survey.

Okun, D. A. (1991). A Water and Sanitation Strategy for the Developing World. *Environment* 33 (8): 16–20, 38–43. Explains that demand for water and sewage treatment in the developing world is not being met.

White, A. L. (1991). Venezuela's Organic Law: Regulating Pollution in an Industrializing Country. *Environment* 33 (7): 16–20, 37–44. Interesting look at the challenge of controlling pollution in a country undergoing industrialization.

WRI and International Institute for Environment and Development. (1988). *World Resources 1988–89.* New York: Basic Books. Chapter 9 contains an excellent survey of problems facing the oceans.

Chapter 18

Pesticides: Learning to Control Pests Sustainably

What we do for ourselves dies with us. What we do for others and the world remains and is immortal.

—Albert Pine

Critical Thinking Exercise

A newspaper recently reported that a pesticide commonly found in small amounts on fruits and vegetables caused cancer in mice. Food safety advocates quoted in the article said that the pesticide should be banned. EPA officials said it was safe because their calculations showed that it would only cause one additional case of cancer per million people in the population each year. How would you analyze this issue? What critical thinking rules are necessary?

Ron Rosmann is not a revolutionary, he's an Iowa farmer. But, like many of his cohorts, he is sowing the seeds of a revolution in farming. For years, Rosmann, like other corn and soybean farmers, believed that without chemical pesticides on his farm he'd be out of business. **Pesticides** are chemical substances that kill insects, weeds, and a whole assortment of organisms that reduce crop output.

In the 1980s, Rosmann visited several successful farms that grew crops without pesticides. Much to his surprise, their fields looked great. There were few weeds and insects. Encouraged by what seemed an impossibility, growing crops without his arsenal of sprays, Rosmann began to experiment with alternatives to pesticides the next year. On his farm, weeds were his biggest problem and **herbicides**, chemicals used to kill them, were a big expense. By using techniques he had seen the previous year, Rosmann found that he could virtually eliminate pesticides. In the nine years that followed, he has used herbicides only once, and then in small amounts because he had mistakenly let weeds get out of control. Today, Rosmann saves $4000 to $5000 per year on pesticides. Much to his surprise, his crop yields have increased. In other words, it costs him less to farm, and he's growing more food per hectare!

Unlike Rosmann, the majority of farmers in the industrial countries collectively spend billions of dollars on chemical pesticides to wipe out pests. To them, weeds and other pests, which reduce production and profits, are an impediment to efficient farming.

Each year, weeds, insects, bacteria, fungi, viruses, birds, rodents, mammals, and other organisms—often referred to as **pests**—consume or destroy an estimated 48% of the world's food production. This estimate includes both pre- and postharvest losses. The highest rate of destruction occurs in the tropics and subtropics, where as many as three crops are grown each year on the same field and where conditions for insect survival are optimal.

Crop destruction from pests is high even in the developed nations despite elaborate and costly control strategies. In the United States, for instance, preharvest losses are estimated to be about 37%, and postharvest losses amount to 9% of what remains. Together, about 45% of annual U.S. production is lost to various pests, or about $64 billion worth of food for humans and livestock. Some of the most common insect pests in the United States are shown in Figure 18.1.

Similar losses are reported in other developed countries. For example, wheat production in Saskatchewan has been reduced in some years by three-fourths as a result of wireworm infestation. In a world where about 20% of the population is hungry, such losses are tragic. Reducing losses through pest control can help increase world food supplies. This is, however, a task that must be embarked on judiciously. Ecological costs and benefits of various strategies must be scrutinized.

This chapter describes conventional pest control strategies and the many environmental problems that result from them. It also illustrates a potentially more sustainable approach that seeks cooperation with nature.

18.1 Discovering the Impacts of Pesticides: A Brief History

Pest control measures have been used throughout the centuries. In China over 3000 years ago, for example, farmers controlled locusts by burning infested fields. In the ancient Middle East, open ditches were used to trap immature locusts. In Greece, Pliny the Elder (A.D. 23–79) compiled a list of common compounds, including arsenic, sulfur, caustic soda, and olive oil, that were used to control crop pests. In 1182, Chinese citizens were required to collect and kill locusts in an effort to control an outbreak. Techniques similar to these have been in use since the beginning of agriculture in many other parts of the world.

Development of Chemical Pesticides

In recent years, **chemical pesticides** (biocides) have come to form the cornerstone of pest management. Early pesticides, known as **first-generation pesticides**, were simple preparations made of ash, sulfur, arsenic compounds, ground tobacco, or hydrogen cyanide. Lead, zinc, and mercury compounds were also used. Today, few of these remain in use because they are toxic, relatively ineffective, and environmentally persistent. In fact, many of the compounds remain in the soil for over 50 years.

In 1939, the Swiss chemist Paul Müller discovered the insecticidal properties of a synthetic organic compound called DDT (dichlorodiphenyltrichloroethane). DDT ushered in a new era of chemical control and was the first in a long line of **second-generation pesticides**, synthetic organic pesticides. For 25 years following its development, DDT was viewed by many as the savior

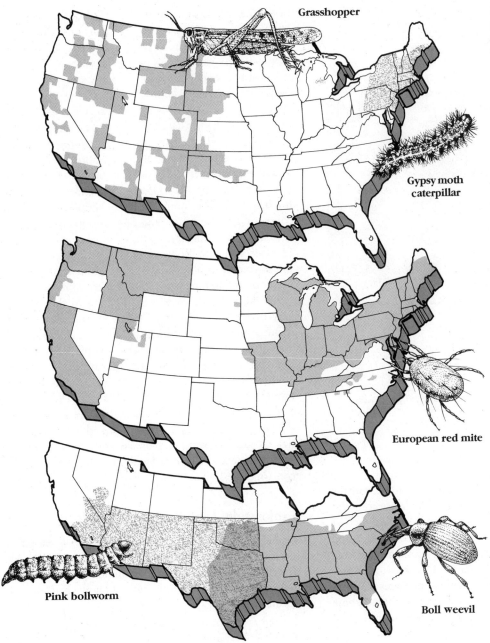

Figure 18.1 *Prominent pests in the United States and the areas where they are found.*

of humankind for it was quite lethal to insect pests and thus increased crop yield. Because it was relatively inexpensive to produce, DDT use became widespread. In 1944, in fact, Müller was awarded a Nobel Prize.

Over the years, thousands of new chemical pesticides have been synthesized and tested. Today, 1500 different substances are used in over 33,000 commercial formulations of herbicides, insecticides, fungicides, miticides, and rodenticides with one goal in mind: to reduce pests to tolerable levels. Some pesticides, like DDT, attack a wide variety of organisms and are called **broad-spectrum pesticides.** Others are **narrow-spectrum pes-**

ticides, which are used in controlling a few pests. Figure 18.2 shows the growth in pesticide production in the United States since World War II.

Most chemical pesticides fall into three chemical families: (1) chlorinated hydrocarbons (organochlorines), (2) organic phosphates (organophosphates), and (3) carbamates. The **chlorinated hydrocarbons** are a high-risk group, including DDT, aldrin, kepone, dieldrin, chlordane, heptachlor, endrin, mirex, toxaphene, and lindane. All of these have been banned, drastically restricted, or are being considered for such actions because of their ability to cause cancer, birth defects,

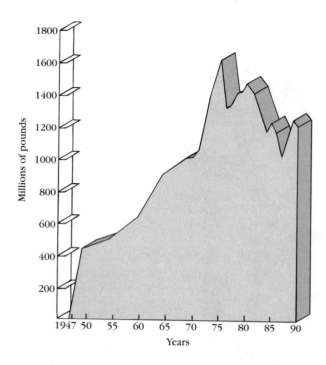

Figure 18.2 *Increasing pesticide production in the United States. Note that production began to decline in the 1970s and leveled off in the 1980s, in part because of a shift away from environmentally harmful pesticides.*

neurological disorders, and damage to wildlife and the environment. Generally, chlorinated hydrocarbons are extremely resistant to breakdown and therefore persist in the environment. They are also passed up the food chain and biomagnified and, therefore, remain for long periods in body fat.

The second group, the **organic phosphates**, consists of chemicals such as malathion and parathion. These toxic chemicals break down much more rapidly than chlorinated hydrocarbons. For example, the chlorinated hydrocarbon DDT has a half-life of two years; parathion has a half-life of two days. Organic phosphates are water soluble and are excreted in the urine. Because they are water soluble, they are less likely to bioaccumulate (Chapter 14). Despite their benefits, organic phosphates are still of great concern, for humans exposed to even low levels may suffer from drowsiness, confusion, cramps, diarrhea, vomiting, headaches, and difficulty breathing. Higher levels can cause severe convulsions, paralysis, tremors, coma, and death.

The third group, the **carbamates**, are widely used today as insecticides, herbicides, and fungicides. One of the most common is carbaryl, commonly known as Sevin. As a group, carbamates are, like organic phosphates, less persistent than chlorinated hydrocarbons, remaining only a few days to two weeks after applica-

tion. They are also water soluble and do not bioaccumulate. But, like chlorinated hydrocarbons and organic phosphates, they are nerve poisons that have been shown to cause birth defects and genetic damage.

Approximately 2.5 million metric tons of chemical pesticides are used annually throughout the world, approximately 22% in North America and about 57% in Europe and other developed countries. The remaining 21% is used primarily in developing countries.

The United States is a leading agricultural nation and a leader in pesticide use. U.S. farmers use about 3 million pounds per day. Of these pesticides, insecticides for insect control comprise 25%; herbicides for weed control comprise 60%. The rest are mostly fungicides used to reduce fungal growth.

According to Robert Metcalf, an entomologist at the University of Illinois and author of a standard college textbook on pest management, farmers apply more than twice the pesticide they need. Adding to the unnecessary environmental contamination are a cadre of misinformed homeowners who apply about 10% of all U.S. chemical pesticides on gardens, lawns, and trees in higher amounts per hectare than farmers do.

Pesticides constitute only about 3% of the commonly used commercial chemicals in the United States each year. Nonetheless, because they are released into the environment in large quantities and have the potential to alter ecosystem balance and threaten human health, their use has created widespread and often heated controversy.

Exploration, Exploitation, and Reflection

Pesticide use has progressed through three developmental stages: (1) exploratory, (2) exploitive, and (3) reflective. These periods illustrate a common progression that seems to follow the introduction of many chemicals and technologies.

Exploration   During the exploratory stage, starting in the 1940s, DDT and other new pesticides were applied to a variety of crops, with astonishing results. DDT was even used to delouse soldiers and civilians during World War II.

DDT and other chemical pesticides proved to be fast and efficient in controlling insects, weeds, and other pests. In India, for example, before the use of DDT in the 1950s to control malaria-carrying mosquitoes, there were over 100 million cases of malaria each year. By 1961, the annual incidence had been reduced to 50,000. DDT no doubt saved millions of lives.

Pesticides also allowed farmers to respond quickly to pest outbreaks, thus avoiding economic disaster. In general, the second-generation pesticides were cheap and relatively easy to apply, and their use resulted in substantial financial gains as yields increased. Some insec-

Figure 18.3 *Pesticide resistance. A tobacco budworm crawls through deadly DDT unaffected.*

ticides, such as DDT and dieldrin, persisted long after application, giving extended protection. The persistence of these pesticides, in fact, was thought to be one of their major advantages, for one application could have lasting effects.

Exploitation   The successes of the explorative phase led to the exploitive phase, during which pesticide production and use expanded considerably. During this period, researchers developed many new pesticides. In the rush to expand pesticide use, however, many problems arose.

One of the most troubling problems was that pesticides, especially broad-spectrum chemicals, often destroyed ecological pest controls—that is, insect predators and parasitic insects that naturally help control pests and potential pests. The loss of these beneficial insects resulted in a proliferation of pest species. In fact, some insect species that had previously been benign suddenly increased in number and began creating a need for additional pesticides. Agronomists call this population explosion of new pests an **upset.**

A classic example of such an upset took place in California. Spider mites, once only a minor crop pest, have become a major pest because of the use of pesticides that killed off many of their natural enemies, which were more sensitive to the sprays. Today, mites cause twice as much damage as any other insect pest in California and cost farmers in damage and control five times what they cost 25 years ago. Two of modern farming's most costly pests, the cotton bollworm and corn-root worm, were minor problems 50 years ago before widespread pesticide use reduced their natural predators. In the United States, one-third of the nation's 300 most de-

structive insect pests are secondary pests, that is, insects that previously caused little or no trouble at all.

Pesticides also destroy other beneficial insects, such as honeybees, which play an important role in pollination. Honeybees pollinate crops annually worth about $20 billion. Apple orchards are particularly hard hit. Over 400,000 bee colonies are destroyed or severely damaged in the United States each year by pesticides.

Another unanticipated effect of pesticides was the dramatic increase in **genetically resistant insects.** Because of genetic diversity, a small portion of any insect population, roughly 5%, is genetically resistant to pesticides and is not killed by a normal application (Figure 18.3). The insects may contain enzymes, for example, that destroy or detoxify the poison. Therefore, an initial application of pesticides will kill all but the genetically resistant members of a pest population. Although such pests do little damage initially, over time they reproduce and form a sizable population that can cause significant crop damage.

As farmers encountered genetic resistance, they began to try new approaches. The first strategy was to increase the amount of pesticide. However, because of genetic resistance, this solution is effective only in the short term. Even though the higher dose wipes out the majority of the insect pests, it invariably leaves behind a small subpopulation that is genetically resistant to the increased dose. The pests soon proliferate and cause farmers to apply even more, creating a vicious addictive cycle that many observers call the **pesticide treadmill.** When DDT and other insecticides were first introduced in Central America, cotton fields were sprayed eight times each growing season; today, because of genetic resistance, 30 to 40 applications are typical.

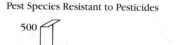

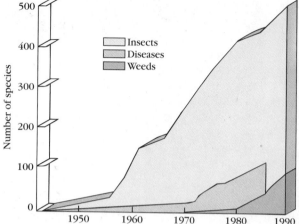

Source: George P. Georgton, University of California, Riverside, 1990.

Figure 18.4 *This graph shows the number of species of insects, disease organisms, and weeds that are resistant to at least one pesticide.*

The second strategy was the development of new pesticides. However, scientists found that it was expensive to create pesticides and that insects developed genetic resistance to these new chemicals almost as quickly as they were applied.

Genetic resistance to DDT was first reported in 1947 by Italian researchers. Today, over 500 insect species are resistant to at least one form of pesticide (Figure 18.4). More than 20 of the world's worst pests are now resistant to most types of insecticides. Moreover, farmers are finding that crop diseases and weeds are also developing resistance to the chemicals that are used to "control" them. Worldwide, 150 plant pathogens (microscopic organisms that cause disease) and 113 weeds have developed resistance to at least one pesticide.

Since the introduction of DDT, insects have never met a chemical pesticide they couldn't defeat. Despite the increased application of chemical pesticides, annual losses due to pests have continued to climb. Although insecticide use has increased tenfold since World War II, crop damage has doubled. Today, not only do insects take a larger percentage of the U.S. harvest than they did before the introduction of DDT but damage from fungi and weeds is climbing as well.

Another problem that emerged during the era of exploitation was that many of the second-generation pesticides proved harmful to nontarget species, birds, fish, and other animals. Trout in upstate New York, for example, showed elevated levels of DDT and DDE (dichlorodiphenyldichloroethylene), its chief breakdown product, because of spraying in nearby forests. These toxins did not affect the adults but reduced survival among newly hatched fry. Insect- and worm-eating birds also perished in areas where aerial spraying of insecticide had occurred. As a result of widespread pesticide use, populations of many birds plummeted.

Pesticide manufacturers argued that the chemicals were found in only minute concentrations in the environment and could not be the cause of declining populations of fish and wildlife. But, numerous experimental studies showed that certain persistent pesticides—even when present in small amounts in the environment—could drastically affect the reproduction and survival rate of birds and other animals through biomagnification, described in Chapter 14.

Early studies showed that although DDT and DDE levels in aquatic ecosystems were quite low, concentrations were higher in producers and still higher in consumers because of biomagnification. Fish-eating birds, the consumers at the top trophic level, had the highest concentrations of DDT and DDE. Although these levels were not lethal to adults, they impaired reproduction. In birds that feed on fish and other birds, such as peregrine falcons, brown pelicans, cormorants, bald eagles, gulls, and ospreys, DDE and DDT reduced calcium deposition in eggshells. Of the two compounds, biologists discovered that DDE posed the greater physiological threat to birds, in part because it persisted longer. Reduced eggshell calcium levels create a thinner, more fragile shell that cracks easily during incubation. As a result of widespread DDT contamination, many predatory populations were nearly wiped out. (For a more detailed discussion, see Chapter 8.) In one study of bald eagle reproduction, James Grier, a zoologist at North Dakota State University, showed that the number of young per nest in northwestern Ontario declined by about 70% between 1966 and 1974.

Other studies showed the presence of DDT in fish, beef, and other foods. DDT also appeared in the fatty tissues of seals and Eskimos in the Arctic, far from its point of use, indicating that it was traveling in the atmosphere to remote parts of the globe, being washed from the sky by rain, and passing through the food chain. DDT has also been detected in human breast milk, a discovery that caused considerable alarm, although the long-term effects of low levels on humans remain unknown.

Numerous farm workers suffered from direct exposure to pesticides on the job as well, indicating widespread misuse. Workers pick up pesticide on their clothing and skin through accidents and negligence or when they prematurely enter sprayed fields. In most cases, workers are poorly protected. They receive few instructions and no protective gear to minimize exposure.

Health problems from pesticide exposure are pronounced in the developing countries, where illiteracy and poor worker safety standards are commonplace. In most countries, workers are poorly protected, if at all. Workers often complain of being sprayed with insecticides from helicopters or planes while they are working in fields and on plantations. A recent study of the West Bank and the Gaza Strip uncovered another serious problem: the instructions on pesticide containers were printed in Hebrew, despite the fact that most of the farmers and farm workers read Arabic.

In developing countries, where farm labor is abundant and worker safety provisions minimal, workers are often viewed as an expendable commodity. Christopher Brady, who is active in development work in Latin America and Africa, writes, "As a result of increasing health risks to the sprayers, Tela (a Honduran subsidiary of Chiquita Brands, Inc., a U.S. company) has changed its hiring policy. Sprayers are now only hired on a six-month contract." They are "let go and new ones hired before any serious health problems may be detected." Workers spraying herbicides on banana plantations receive a thin cotton mask as their only protective gear. Moreover, empty barrels once containing concentrated pesticide are often used by peasants to hold water.

Insomnia, nausea, and loss of sex drive are only several of the many symptoms of poisoning. Some people exposed to pesticides complain of reduced powers of concentration, irritability, and nervous disorders. Spills on skin may cause rashes and a burning sensation. In severe cases, death can occur.

In the United States, at least 45,000 workers are poisoned each year, but many experts believe that this figure grossly underestimates the number of serious poisonings. Surveys in California, for instance, show that three-fourths of all serious poisonings go unreported. Each year, 200 to 1000 people die from pesticide poisoning in the United States. According to various estimates, worldwide, at least 500,000, and perhaps as many as 2 million, pesticide poisonings occur annually. These result in somewhere between 4000 and 19,000 deaths each year and numerous chronic and fatal illnesses, according to the World Health Organization (WHO). What is more, WHO expects pesticide poisonings to increase in the near future if pesticide use intensifies in developing countries.

In the former Soviet Union, heavy pesticide use on cotton crops in farmland near the Aral Sea resulted in exposures up to 25 times greater than the national norm. This, in turn, led to an assortment of medical problems, among them liver and kidney disease, rising cancer rates, and birth defects. Infant mortality in the region was reportedly 4 times the Soviet average.

Although farm and chemical workers are the groups most heavily exposed to pesticides, residents of rural and even suburban areas are often exposed to potentially dangerous levels if they live near agricultural lands. Families living near fields sprayed with herbicides and pesticides outside Scottsdale, Arizona, for example, suffered from persistent headaches, cramps, skin rashes, dizzi-

ness, high blood pressure, chest pains, persistent coughs, internal bleeding, and leukemia. Health officials are most concerned about possible long-term problems from exposure.

People living near pesticide-treated fields are heavily exposed because one-half to three-fourths of the sprayed material never reaches the ground but is carried away by light winds (Figure 18.5). In most cases, only 1% of the pesticide actually reaches the target species. Recent studies have also shown that pesticide applied to one crop may be able to vaporize under sunlight and drift to neighboring crops.

In their zeal to control mosquitoes, many cities in the southern United States routinely spray insecticides in neighborhoods and nearby breeding areas. In Florida, for instance, various mosquito control agencies argue that controlling the insect is vital to real estate interests and tourism because it minimizes the risk of encephalitis, a potentially deadly brain infection that is caused by organisms carried by mosquitoes. In 1990 to 1991, an encephalitis epidemic in Florida resulted in over 130 cases and more than 10 deaths. To protect against this disease, trucks and aircraft spew out tons of pesticides while residents sleep. What long-term health impact this has, if any, is unknown. Florida wildlife officials, however, think that pesticide use is largely responsible for a 70% decline in the population of snook, a popular sport fish. Adding to the problem, city officials use a variety of pesticides in city parks, and lawn-care companies and individuals douse lawns and trees with a variety of toxic substances, often incorrectly and without warning neighbors.

Consumers may also become the victims of pesticide poisonings. In the summer of 1985, for instance, 1400 people on the West Coast were stricken with nausea, diarrhea, vomiting, and blurred vision after eating watermelons contaminated with the pesticide aldicarb illegally used by farmers. The EPA permits aldicarb for use on cotton and vegetables that are cooked before consumption, such as beans and potatoes, but not on produce eaten without cooking, like watermelon. In 1992, 29 people were poisoned and 3 hospitalized in Ireland after eating cucumbers sprayed with aldicarb. Pesticide application controls worldwide are largely voluntary—that is, restrictions are printed on the label, with compliance left to the discretion of the farmers. The aldicarb incidents point out a major flaw in pesticide regulation.

Reflection As a result of growing concern over the biological and ecological effects of pesticides and growing skepticism regarding their effectiveness, industrial societies have entered a reflective stage, a period of tempered optimism. Although many farmers remain convinced of the benefits of pesticides, many realize that the benefits do not come without a cost. Others, like Ron Rosmann, realize that farmers can produce adequate

Rain 0.1-0.3 ppb

Tradewinds 0.1-0.3 ppb

Rivers and lakes
0.001-0.2 ppb

Groundwater
0.001-0.2 ppb

Fat of man
6-12 ppm

Fat of cows 0.5 ppm

Figure 18.5   *Pesticide sprayed from planes contaminates the ecosystem because much of the pesticide drifts away. Various avenues for the dispersal of pesticides are shown. Average values for DDT concentrations are indicated in parts per million (ppm) and billion (ppb).*

amounts of high-quality food without the costly and potentially dangerous arsenal of chemical pesticides.

Limited caution regarding pesticides began in 1962 in the United States with the publication of Rachel Carson's book *Silent Spring*, which pointed out many of the real and potential impacts of pesticide use. But public attention was not sufficiently aroused until a decade later with the near extinction of peregrine falcons, brown pelicans, cormorants, bald eagles, and other bird species.

Our understanding of pesticide impacts and clean, environmentally sustainable alternatives continues to grow, casting further doubt on conventional pest control practices. Recently, researchers found that herbicides penetrate the soil up to ten times deeper than earlier laboratory tests had indicated. Such findings explain why groundwater is often polluted with these chemicals.

Attention has also recently focused on the use of pesticides on golf courses, lawns, and gardens. Americans and many other people, among them the Japanese, love vast, green expanses of lawn and will go to great lengths to create the perfect lawn or golf course. Unfortunately, numerous studies have shown that birds are often the victim of this obsession. In the United States,

for example, more than 30 kilograms (67 million pounds) of chemicals are used each year to control fungus, insects, and weeds on lawns and golf courses.

One pesticide, diazinon, was once commonly used on sod farms and golf courses throughout the United States to control insects. In 1984, three fairways were treated with diazinon in Hempstead, New York. Two days later, 700 Atlantic brant geese, 7% of the state's entire population, died of acute diazinon poisoning.

Diazinon was also used to treat nine fairways at a golf course in Bellingham, Washington. After application, the area was irrigated to decrease pesticide concentrate on the surface. Nevertheless, 85 American wigeon ducks perished after eating grass on one of the treated fairways on the day of application.

In 1986, the EPA banned the use of diazinon on sod farms and golf courses because of similar incidents. However, because the bird kills appeared to be associated with large areas of diazinon use, the pesticide is still available for use on home lawns.

Birds have also been dramatically affected by the use of granular carbofuran, which was developed in 1970 by the FMC Corporation of Philadelphia. Farmers throughout the United States use carbofuran to eradicate nematodes and insects from corn, rice, and other crops. Although carbofuran apparently poses no threat to humans when applied to crops, it is lethal to songbirds. In fact, a songbird can die after ingesting a single granule.

In 1989, EPA records showed that about 2 million birds were dying of carbofuran poisoning each year. In 1990, more than 200 songbirds were poisoned in eastern Virginia near the Rappahannock River. FMC acknowledged a problem but faulted the farmers for misusing carbofuran or for mishandling it.

In 1991, FMC issued new instructions for the use of carbofuran. Still suspicious, citizens in Virginia decided to monitor the results of that year's pesticide application on 3600 hectares (900 acres) of treated farmland. They found 62 bird carcasses, 10 sick birds, and 47 "feather spots," indicating a bird had died but been eaten by a scavenger, on the 36-hectare (900-acre) field, probably only half of what actually died. Considering that hundreds of thousands of hectares of farmland in the state had been treated with carbofuran that spring, it was estimated that tens of thousands of birds probably died from the pesticide.

In 1991, Virginia banned the use of carbofuran and since that time, carbofuran has been banned nationwide by the EPA. By September 1994, only 1140 kilograms (2500 pounds) can be sold in the United States, a drastic reduction from annual sales of 4.5 million kilograms (10 million pounds) in previous years. Though the ban will likely have a positive impact on birds, it is not faultless. Export of carbofuran will continue and will probably increase to compensate for shrinking domestic sales.

U.S. researchers have found that bans on pesticides have greatly benefited U.S. wildlife. The endangered bald eagle appears to be on the upswing. Recent studies show that DDE and DDT levels have dropped in wild populations and that normal reproductive rates have returned. Researchers caution, however, that domestic pesticide bans are only part of the answer. Continued use of pesticides outside of the United States, as in the case of carbofuran, poisons migratory species, such as songbirds that spend fall and winter in South and Central America. In addition, much of the produce imported into the United States has recently been found to be contaminated with pesticides, many of which have been banned in this country. This phenomenon is sometimes called the *circle of poison*. Since one-fourth of the fruit and vegetables sold in the United States comes from foreign soil, global bans of harmful pesticides are needed to protect not only humans but also the many species that share this planet with us.

Pesticide use is also proving to be quite costly. In the 1991 growing season, for instance, a fungicide called benomyl, produced by DuPont and sprayed on a variety of crops, allegedly destroyed millions of dollars worth of crops in the United States and several other countries.

Research shows that in hot, humid climates benomyl is converted into several toxic compounds that may kill or severely stunt plants (Figure 18.6). Florida was one of the worst-hit areas, with damages estimated at $1 billion. To date, more than 1600 U.S. farmers in 40 states have filed claims against the company; three-fourths has been settled out of court at a cost of over $500 million. Similar problems have been reported in Costa Rica, Puerto Rico, and Jamaica.

Pesticides may also end up in groundwater in rural communities, causing a potential health risk and necessitating costly cleanup. The Monsanto Company, which produces herbicides, estimated that of 6 million wells in a surveyed area in the United States, 770,000 wells, or 13%, were contaminated with one or more of five herbicides; 6600 wells contained levels exceeding the EPA's maximum contamination level. Although the company insists that none of the herbicides poses a threat to health, it is offering well owners whose water is contaminated above specified concentrations up to $2000 per well to make corrections.

Pesticides will probably remain in use despite their many drawbacks. However, many experts believe that pesticides will play a much smaller role in future agriculture. For example, the red spider mite is kept under control in apple orchards by applying insecticides early in the season, well before the mite's natural predators emerge. Throughout the rest of the season, farmers refrain from pesticide use, letting the natural predators do their work. Using a similar approach on cotton, researchers at Texas A&M have cut pesticide use by 70% while maintaining normal cotton crop production.

18.2 Regulating Pesticides: Treating the Symptoms

In the United States, pesticides are currently regulated by two laws: the **Federal Insecticide, Fungicide, and Rodenticide Act (FIFRA)** and the **Federal Food, Drug, and Cosmetic Act (FFDCA)**.

FIFRA is currently administered by the EPA. When passed in 1947, long before the EPA came into existence, this law had a rather narrow purview. Its chief goal was to protect farmers and farm workers from dangerous and also ineffective chemicals. FIFRA simply required manufacturers to register pesticides transported across state boundaries. Those produced and used within state boundaries were free from federal scrutiny.

In 1972, the scope of FIFRA was broadened to protect public health and the environment from *new* chemical pesticides. From that point on, manufacturers were required to test new pesticides for health effects before they could be registered for use. Previous chemical pesticides were unaffected.

Figure 18.6 *Researchers used the fungicide Benlate in 1974 to treat young tree seedlings. In this photo, the two alder buckthorn on the left were grown in Benlate-treated soil.*

Registration is a kind of permitting process managed by the EPA. When a company develops a new pesticide that it wants to market, it must perform routine tests on plants and animals to determine the toxicity of the chemical. Test plots are also sprayed with the pesticide to determine residues—that is, how long the chemical remains. These data are then submitted to EPA scientists for review. Taking into account the average American diet, residue levels, and toxic effects, the EPA then determines on which crops, if any, the pesticide can be used.

Critical thinking skills you have learned may suggest a problem right away: is the average American diet accurately determined? The truth be known, the average American diet includes a great deal of red meat, chicken, and other meat products. But, that automatically excludes vegetarians, who tend to consume large quantities of fruits and vegetables. Ironically, many vegetarians who select a special diet for health reasons are inadvertently exposed to the highest pesticide levels.

Pesticides registered by the EPA are approved for general or restricted use. General use means that anyone can purchase and use them. Technically, restricted-use pesticides are to be used by licensed applicators, farmers, and lawn-care companies.

The problem with this system is that pesticide use is largely an honor system. Labels on restricted-use products describe their legal uses, and only licensed applicators can buy them, but other than these limitations, very little, if anything, prevents people from using pesticides any way they please. The aldicarb poisonings cited earlier illustrate this common problem.

To help protect public health, FIFRA also authorizes the EPA to set tolerance levels for pesticides on foods.

Tolerance levels are concentrations in or on foods that are believed to pose an acceptable health risk. For cancer, the EPA sets the concentration at a level it thinks will cause no more than one additional cancer death in 1 million people. This determination is obviously fraught with difficulty (Case Study 18.1 and Chapter 14).

In 1988, after several years of struggle, FIFRA was amended to correct some of its most glaring weaknesses. One of the most significant gains was a plan to register many pesticides introduced before the EPA took over the process. Pesticide registrations in the 1950s and 1960s were made with very little, if any, sound toxicological data, say critics. Reregistering the over 300 chemical pesticides should be completed by 1997 and will be funded by fees paid by the chemical companies that produced them.

Some experts see this as a weeding-out process and hope that many pesticides will be canceled. Companies afraid that their products won't be approved for health and environmental reasons won't invest the money needed to reregister them, especially if the products are marginally profitable.

Despite this improvement, critics say additional reform is necessary. For example, pesticide registration does not currently require manufacturers to test for neurotoxicity—toxic effects on the brain, spinal cord, and nerves. To many critics, this is a glaring omission since about half of the pesticides, especially the organic phosphates, are insect neurotoxins that also affect the human nervous system.

New research also shows that some pesticides damage the immune system, the body's defense against bacteria, viruses, and even cancer. Pesticide registration, however, requires no test of immune system effects, another glaring omission, say critics. In addition, pesticide registration does not account for possible synergism, the superadditive effect described in Chapter 14.

Perhaps the most glaring problem, though, is that FIFRA provides virtually no monitoring of end use. End users can apply as much pesticide as they want and can apply it wherever they want because no one oversees them. Moreover, there is no one to see that solutions are mixed correctly or that equipment is working properly. As a result, farmers frequently apply much more than is needed.

Licensing and training are two avenues available to address this problem. To be licensed, most states require farmers to take a test. In Colorado, for example, farmers must study a booklet provided by the EPA, then take an open-book test. If they pass, they're licensed to spray restricted pesticides on their fields. Critics argue that more rigorous testing and education are needed.

Governments, pesticide companies, farmer's groups, and universities could provide additional hands-on training and monitoring to promote safe pesticide use. As mentioned earlier, programs that teach farmers to become better at identifying pests and monitoring pest populations in the field could decrease insecticide use and costs. Special crop scouts, such as those now in use in Indonesia, could be trained to monitor pest populations and determine when they have reached the threshold level (Models of Global Sustainability 18.1). They could also assist farmers in finding alternative methods to control pests.

As noted earlier, under FIFRA, the EPA sets tolerance levels for pesticide residues on fruit and vegetables and other foods. But it is up to the Food and Drug Administration (FDA) and state agencies to monitor the nation's food supply and to enforce tolerance levels. The FDA can, for example, seize and condemn foods containing residues that exceed EPA levels or that contain illegal pesticides, but only when they are shipped from one state to another. Within the states, the responsibility for monitoring food lies with state agencies.

Both the FDA and state agencies suffer from common problems: a chronic lack of funds and a shortage of inspectors. Given these limits and the massive amount of food consumed by the American public, it can be no surprise that only a small portion of food is actually tested. Furthermore, examiners only test for the presence of a handful of the pesticides that could be on our food. Unless you grow your own food, you are probably eating more pesticide than you would like. In a recent study, more than 50% of all food in supermarkets contained detectable levels of pesticides.

Similar problems occur in other countries as well. In Ireland, for instance, the Pesticide Control Service has only 11 scientists to test food. Consequently, few farmers are inspected, and only 2000 food samples are tested annually.

In 1987, the EPA reported that at least 55 pesticides that leave residues on food are thought to be carcinogenic. In 1987, the National Academy of Sciences issued a report concluding that 1 million Americans alive today will develop cancer as a result of pesticide contamination of their food—that's 1 of every 250 Americans. Add to that possible birth defects, miscarriages, mutations, neurological effects, and other milder symptoms, and it is little wonder that the EPA ranks pesticides in food as one of the nation's most serious health problems.

Despite their recognition of potential problems, the EPA has a long way to go. Most critics think that the tolerance levels set for pesticide residues are inadequate because they fail to take into account the special diets of vegetarians and perhaps even children. The EPA rarely revises tolerance levels when new scientific data about risks become available, say its critics. And, it rarely bans a pesticide if it is harmful to wildlife but not to people.

Pesticide residues may or may not cause cancer. The evidence is controversial. For some people, the mere presence of a potential carcinogen on their food is reason enough for concern, indeed, outrage. Some scientists, manufacturers, and regulators argue that low levels of pesticides are not worth worrying about. A slightly elevated risk of cancer, a poisoned bird or fish are the price we pay for progress. In short, pesticides are one of those necessary evils.

The next section shows that pesticide use can be reduced and even eliminated through integrated pest management. In the interim, individuals can avoid pesticides by growing some of their own fruits and vegetables or by purchasing organically grown produce. Although washing fruits and vegetables can help, it won't eliminate all pesticide residues. Beyond that, individuals can exercise their democratic prerogatives by writing to local, state, and federal officials and asking for tougher laws to regulate pesticide registration and use. Regulations requiring store owners to label produce indicating the pesticides that have been used could inspire farmers to find alternatives to these potentially harmful substances.

18.3 Integrated Pest Management: A Sustainable Solution?

Integrated pest management (IPM) is a new strategy of pest control gaining popularity throughout the world. It depends on four basic means of pest control: environmental, genetic, chemical, and cultural. Two or more of these approaches may be used simultaneously to control pests with minimum damage to the environment.

Environmental Controls

Environmental control methods are designed to alter the biotic and abiotic environment, making it inhospitable

Figure 18.7 *In this photo, alfalfa and soy beans are grown side by side in the same field to reduce soil erosion. This technique, called strip cropping, also increases diversity and reduces pest and disease outbreak.*

to pests. Because they generally rely on knowledge more than costly technology, these practices are well suited for developing countries. Still, they can be equally effective if used properly in modern agricultural societies.

Increasing Crop Diversity   In Chapter 7, we saw that monocultures generally promote pest and disease outbreaks. Crop diversity, on the other hand, reduces the amount of food available to any one pest and helps prevent such rapid population growth. Several techniques can increase crop diversity, among them heteroculture and crop rotation.

A farmer who plants several crops side by side, rather than huge expanses of one massive crop covering all of his or her land, is practicing **heteroculture** (Figure 18.7). This simple but effective measure works because it provides environmental resistance to pests. That is, pest populations are often much smaller in heterocultures than in monocultures because there is less to eat. In addition, some crops harbor predatory insects that feed on pests in nearby crops. Corn and peanuts grown in adjacent fields, for instance, can reduce corn borers by as much as 80%. Part of the reason for this success may be that predatory insects that feed on the corn borer live in peanut crops.

One of the most recent and innovative techniques of intermixing crops is to alternate strips of corn and soybeans, each with a dozen or more rows, side by side in the same field (Chapter 7). This not only decreases

insect pests but also increases yield because the corn protects the soybeans from wind and the openness of the field provides more sunlight to the corn.

Heteroculture not only decreases pesticide use; it provides a means of diversifying farm production. In some ways, then, it is a form of insurance. Ron Rosmann, for example, plants 100 hectares (250 acres) of his 200-hectare (500-acre) farm in corn and soybeans, and devotes the rest to hay, pasture, cattle, hogs, and chickens, and a tree nursery that he hopes will help pay for his three sons' college educations. A bad year for corn will not wipe him out.

Heteroculture can also be practiced by home gardeners with great success. For example, I have been growing vegetables in a pesticide-free garden for the better part of 12 years and have had virtually no trouble with insects in large part because I intermix species. Small patches of carrots are planted next to peas, which are next to spinach, and so on. I also plant onions, marigolds, and other species that repel pests.

Crop rotation, discussed in Chapter Supplement 7.1 as a means of reducing soil erosion and increasing soil fertility, also helps control pests for at least two reasons. First, the healthier the soil, the healthier and more resistant the plants are to insects and disease. Second, it helps suppress pest populations. For instance, wireworms feed on potatoes but not alfalfa. Therefore, if potatoes and alfalfa are alternated from year to year in the same field, wireworm offspring that hatch in the alfalfa patch will have little to feed on. Their numbers will severely decline. When potatoes are planted the next year, few wireworms will be around. Although the population may increase during the growing season, it will generally not reach a harmful level. The next year, alfalfa is planted and wireworm offspring once again perish. Gardeners can also practice crop rotation on a small scale to hold insects in check.

Altering the Time of Planting   Some plants naturally escape insect pests by sprouting early or late in the growing season. A good example of this adaptation is the wild radish, which sprouts early in the season before the emergence of the troublesome cabbage maggot fly.

Agriculturalists can use their knowledge of an insect's life cycle to coordinate plantings with the expected date of hatching. A slight delay in planting of wheat, for example, helps protect this crop against the destructive Hessian fly. In general, if a pest emerges early in the spring, planting can be delayed to avoid that pest within the limits of the growing season. Without food, the pest will perish. If the pest emerges late in the growing season, a slightly earlier planting may prove effective.

Home gardeners can also foil pests by planting seedlings, instead of seeds. Certain crops, such as lettuce, can be mowed down by hungry slugs and other insects when they first emerge from the ground. But, if larger

Case Study 18.1

The Alar Controversy: Apples, Alar, and Alarmists?

In the spring of 1989, the Natural Resources Defense Council (NRDC) announced the results of a two-year health study on children and pesticides. It concluded that U.S. children are exposed to dangerous levels of pesticides in fruits and vegetables and that 5500 to 6200 children alive at the time of the study will develop cancer in their lifetimes from just eight of the many pesticides to which they are exposed in their preschool years.

The NRDC also charged the EPA with routinely neglecting children in setting their standards. Children face a higher risk from pesticides in part because they eat more fruit than adults. Although the EPA began taking into account the higher level of fruit consumption in children several years ago, the agency still relies on 1977 consumption data that are grossly out of date and that lead to erroneous estimates of risk. A 1985 survey by the U.S. Department of Agriculture (USDA), for example, shows that fruit consumption in preschoolers has increased 30% since 1977. Today, approximately 30% of preschoolers' diets and 20% of their mothers' diets consist of fruit. Safety standards should reflect these facts.

Several researchers who reviewed the NRDC study thought that the group underestimated the cancer risk from pesticide residues for several reasons. First, there are at least 500 pesticides in common use that leave residues on fruit and vegetables. The NRDC only looked at 27 pesticides and calculated the risk for 8. Second, the NRDC research team omitted several foods, such as milk, which are an important component of children's diets. Third, the NRDC study only examined the effects of childhood exposure, ignoring the continued exposure to pesticide residues in the adult diet. In addition, some research suggests that children are more sensitive to chemical exposure than adults. Rapidly dividing cells are prime targets for carcinogens (Chapter 14). Furthermore, enzymes needed to detoxify chemicals may not have fully developed in children. Early exposure also carries a greater cancer risk than exposure later in life because it results in a longer lead time to develop cancer.

The main culprit in pesticide-caused cancer, the NRDC charged, is a chemical called Alar, a growth regulator that delays ripening so apples do not prematurely fall off the tree. Alar also promotes the reddening of apples and delays overripening so that apples stay fresher in storage. Alar penetrates the flesh of apples and cannot be washed off.

Based on what it considered the best available information at the time, the NRDC estimated that 86% to 96% of the total pesticide cancer risk arises from one chemical, Alar, and its breakdown product, UDMH. This conclusion was based on risk assessment derived from toxicology studies performed in the 1970s.

The EPA's safety standards seek to limit cancer risk to 1 in 1 million. The estimates used by the NRDC in its Alar campaign suggest that UDMH poses a cancer risk of 1 in 4200—240 times greater than the routinely accepted standard.

The NRDC revelations spurred a great deal of public interest. Actress Meryl Streep began speaking publicly against Alar and CBS's Ed Bradley delivered an exposé on it. Not everyone agreed. Dr. Bruce Ames of the University of California at Berkeley argued that Alar posed a much lower threat than some naturally occurring chemicals found in some foods. He argued that the ban on Alar, which followed the NRDC report, would require orchard owners to increase pesticide use. An insect called the leafminer, for instance, causes apples to fall prematurely. To control them, apple growers may turn to more pesticides, increasing human exposure to yet another potentially carcinogenic substance. Molds may increase in apples on trees and in storage, said Ames, because apples are less firm and more susceptible. Naturally occurring mold toxins could increase, exposing people to a greater danger than Alar itself. Finally, Ames pointed out that regulatory agencies choose safety limits for low-level human exposure based on high-level animal exposure, making the former quite speculative.

The Alar controversy has taken some interesting turns since the chemical was pulled off the market by its manufacturer. Since that time, the EPA has twice lowered its estimate of cancer risk because of new information. The first reduction resulted in a cancer risk value ten times lower than the one NRDC used. Still maintaining that Alar is carcinogenic, the EPA halved its estimate again in 1991. Although this won't resurrect Alar, it has invigorated the apple growers of Washington State, who were hit hard by the Alar ban. In 1991, they filed a $200 million lawsuit against CBS, NRDC, and its media advisers. The suit claims that these parties knowingly hyped the risk of the chemical to get attention, an accusation the defendants deny.

Another offshoot of the controversy is an attempt to overthrow the infamous Delaney Amendment of the Food and Drug Act, which forbids the marketing of any product judged to be carcinogenic. The revised legislation would permit products to be kept in use if the health risks are deemed negligible.

Part of the blame for this debacle, say some, is the EPA's, for issuing cancer risk estimates in 1987 based on 1970 data. In 1985, the EPA's own Scientific Advisory Board had ruled that the technical basis for labeling UDMH a carcinogen was flimsy. If the agency had waited for studies they ordered in 1985, the whole Alar scare might not have materialized. That's not a comforting thought to the apple industry and many

(*continued*)

Case Study 18.1 (continued)

The Alar Controversy: Apples, Alar, and Alarmists?

private growers, who lost an estimated $100 to $150 million as a result of the sharp decrease in apple consumption in 1989.

One of the chief lessons that can be learned from this debate is that the study of human cancers based on animal studies is an imperfect science. Risk management is a political, not a scientific, process. Science only provides the data. Most people would rather err on the conservative side, especially when it comes to children's health issues. The appearance of pesticides and other additives in our foods without our choice

worries many people. Most people do not know whether Alar causes cancer, but they know that a few experts think that it may. That's enough for them.

Edward Groth III of the Consumer Union summed it up best: "We must teach people to see risks in perspective. At the same time, we (scientists and public policy makers) must listen to what people say about risks. It is not the size of the risk but its moral offensiveness that makes the public respond so strongly."

seedlings are planted, the plants have a better chance of surviving.

Altering Plant and Soil Nutrients The levels of certain nutrients in soil and plants can also affect pest population size. Thus, by regulating soil nutrients, a farmer may be able to control pests.

Nitrogen is one of the important nutrients that insects and parasites derive from plants. Too much or too little of this key element can alter the population size of various pests. For example, grain aphids reproduce more successfully on grain high in nitrogen. Other insects, such as the greenhouse thrip and mites, do poorly on high-nitrogen spinach and tomatoes, respectively. Therefore, knowledge of pest nutrient requirements, soil nutrient levels, and plant nutrient levels can be helpful in controlling pests. Plants rich or poor in nitrogen can be selected to control pests as long as the nitrogen level is adequate for human consumption.

Controlling Adjacent Crops and Weeds In some cases, adjacent crops and patches of weeds may provide food and habitat for harmful insects and other pests. For example, certain plants adjacent to valuable food crops may harbor viruses that can infect pest species and therefore be transmitted to crops. Eliminating such crops and weeds can prove helpful in the control of insects and other pests.

In other instances, adjacent low-value crops attract pests away from more valuable crops. These are called **trap crops**. Alfalfa is a good example. When planted adjacent to cotton, it lures the harmful lygus bug away from the cotton plants and thus prevents serious damage to the cotton. Some farmers may even spray the alfalfa with pesticide to get rid of the bug, using far less chemical

than would be necessary if the entire cotton crop had to be sprayed.

Introducing Predators, Parasites, and Disease Organisms In nature, thousands of potential insect pests never become real pests because of natural controls exerted by predators, diseases, and parasites, that is, biotic components of environmental resistance. Farmers can capitalize on this knowledge to manage weeds, insects, rodents, and other pests.

To date, scientists have documented over 300 examples of partial or complete control of crop pests through natural predators and parasites. This technique is generally referred to as **biological control**. One classic example of an effective biological control is the case of the prickly pear cactus in Australia. The prickly pear cactus was introduced in Australia from its native Mexico. By 1925, over 24 million hectares (60 million acres) of land had been badly infested; half of this land was abandoned because of the thick carpet of cactus. Farmers introduced a cactus-eating insect to Australia to eradicate the pest; seven years later much of the land had been cleared and was available for cultivation once again.

The predatory lady beetle was introduced into California from Australia in the 1880s to control an insect that destroyed citrus trees. Parasitic insects from Iran, Iraq, and Pakistan have been introduced to control the olive scale, an insect that once threatened the state's olive trees. Both lady beetles and the predatory insects now exert complete control on their prey, keeping their populations at manageable levels without the use of pesticides.

Entomologists in the United States are currently experimenting with a new method of controlling mosquitoes using *Toxorhynchites rutilis*, called "Big Tox" for short, a large, nonbiting mosquito whose larvae feed on the larvae of other mosquitoes. Bred in captivity, this predatory mosquito will be released in infested regions to control biting mosquitoes.

A few insect pests can also be controlled by birds, a natural control organism whose potential has been overlooked. Brown thrashers can eat over 6000 insects in one day. A swallow consumes 1000 leafhoppers in 12 hours, and a pair of flickers can snack on 500 ants and go away hungry. In China, thousands of ducklings are driven through rice fields; in some places, they decrease the insect populations by 60% to 75%, allowing farmers to reduce insecticide use considerably. Their droppings also provide fertilizer for the crops.

Bacteria and other microorganisms can be brought to bear on pests. One common example is the bacterium *Bacillus thuringiensis* (BT), used to control many leaf-eating caterpillars. Cultivated in the lab and sold commercially, it is available as a powder that is dusted on plants or mixed with water, then sprayed on plants. Caterpillars that eat the bacteria die because BT produces a toxic protein that paralyzes their digestive system. Humans and other organisms are usually unaffected.

BT is used by organic gardeners with considerable success. It has been sprayed in China to control pine caterpillars and cabbage army worms. In California, it has been used for more than 20 years to control various crop-eating caterpillars, and it is currently applied in the northeastern United States to help control gypsy moths, which devastate forests. Another strain of BT has also been employed in the battle against mosquitoes in Colorado and other states. The use of BT and other microorganisms has resulted in a measurable reduction in insecticide use in certain crops, especially almonds and tomatoes.

Researchers have also successfully inoculated corn plants with genetically altered bacteria containing the BT gene. The bacteria multiply in the corn plant as they grow and kill European corn borers, which feed on the stalks. Studies show that the bacteria do not migrate into the kernels of the corn plant.

Viruses and fungi may be used similarly. In Australia, after years of fruitless efforts to control rabbits, scientists introduced a pathogenic myxoma virus, which eliminated almost all of the rabbits within one year. Unfortunately, the virus has evolved to an avirulent form and the rabbit has evolved resistance. Control is no longer as effective as it once was. Cabbage loopers can be controlled with 0.5 gram of an experimentally produced virus applied to a hectare of cropland. Other viruses are being used to control pests, such as the pink bollworm, which damages cotton, and the gypsy moth, mentioned above.

U.S. scientists are also testing a fungus, *Tolypocladium cylindrosporum*, to manage mosquitoes. Special traps are set out to attract adult females. Females enter the traps and are contaminated with spores, but they are

allowed to escape, thus carrying the fungal spores back into the environment, where they infect eggs and larvae.

A word of caution. Biological control agents must be carefully developed to ensure that they do not pose a threat to humans, livestock, and natural ecosystems. (Chapter 8 describes the problem of alien species.) In the early 1980s in sub-Saharan Africa, for example, an insect known as the mealybug became a major pest, attacking cassava plants. This plant produces an edible root that is a staple for about 200 million people. To find a control for this troublesome pest, researchers scoured the mealybug's South American homeland for natural enemies. They eventually located a small parasitic wasp that injects its eggs into the larvae of the mealybug. When the eggs hatch, they devour the larvae. Before the researchers released the wasp, they performed extensive tests to see if the wasp would survive in its new home and if it would become a pest itself. They were convinced that it wouldn't, and the bug was released in 1986 and today is providing protection in 24 African countries. So far, the wasp seems to be working fine.

Another problem with biological control is that target organisms can develop genetic resistance to biological controls, as did the Australian rabbits mentioned earlier. Researchers in Kansas recently found that larvae of the Indian meal moth that feed on grain stored in sheds and bins develop genetic resistance to BT. In such cases, new controls could be introduced or alternated with BT. In some instances, biological control agents themselves may undergo genetic changes that offset the newly acquired resistance of the pest, a process called **coevolution**. As yet, there is no record of such changes in biological control agents, but some scientists think that coevolution is inevitable.

Genetic Controls

IPM includes two major genetic control strategies, the sterile male technique and breeding genetically resistant crops and animals. Both are important components and can be used in conjunction with other methods.

Sterile Male Technique   The **sterile male technique** has been used effectively against several species of insect pests, including the screwworm fly in Mexico and the United States, the Mediterranean fruit fly in Capri, the melon fly on the island of Rota (near Guam), and the Oriental fruit fly in Guam.

In this technique, males of the pest species are raised in captivity and sterilized by irradiation or by exposure to certain chemicals. The sterilized males are then released in large numbers in infested areas, where they mate with wild females. Since many insect species mate only once, eggs produced by such a union are infertile. If the population of sterilized males greatly exceeds that of the wild males, most of the matings will be with sterile

Figure 18.8   *The USDA controls screwworm populations by releasing sterile male flies in Texas.*

males. Consequently, insect populations can be brought under control swiftly, that is, reduced in number to prevent significant economic damage.

In the United States, screwworms have been controlled by this method, saving millions of dollars each year (Figure 18.8). The screwworm fly lays eggs in open wounds of cattle and other warm-blooded animals. The eggs hatch within a few hours, and the larvae feed off blood and tissue fluid. This keeps the wound open, allowing a bacterial infection to set in, which may kill the host.

In 1976, over 29,000 cases of screwworm infestation were reported. Because of inclement weather and an extensive release of sterile males, costing $6 million, the following year only 457 cases were reported. The screwworm has practically been eliminated from the southeastern United States, but it still remains in the Southwest, where new flies migrate from Mexico. With continued cooperation between Mexico and the United States, the future of the screwworm as a major pest may be a short one.

Controlling the screwworm has resulted in substantial economic benefits at a relatively low cost. In fact, the program saves the cattle industry about $120 million annually.

Sterile males have been introduced in other instances with much less success. For example, California imported sterilized Mediterranean fruit flies (medflies) from Hawaii in efforts to control this pest in 1980 and 1981. The medfly lays its eggs in 235 different fruits, nuts, and vegetables; its larvae develop in the ripening fruits and eventually destroy them. Agricultural interests argued that if the medfly proliferated, it would cause $2.6 billion in damage per year in California alone.

At first, the state tried a combined approach, using sterile males and baited traps laced with malathion to avoid widespread aerial spraying, which, scientists had argued, would cause undue health risk. Combined programs worked well in two areas in the state in 1975 and 1980. In the 1980–1981 incident, however, poorly funded control efforts were inadequate and applied too late. Fertile medflies kept appearing both inside and outside of areas believed to be infested. Farmers became nervous, and the U.S. Department of Agriculture threatened to quarantine all California produce, forcing the governor to order a massive aerial spraying program that eradicated the medfly. Follow-up studies showed that the spraying, although successful in controlling the medfly, had also reduced many predatory and parasitic insects that control populations of other insect pests. Researchers found, for instance, that aphids and whitefly populations in home gardens had increased dramatically because of the loss of natural controls. Insects that destroy olive trees increased substantially in sprayed agricultural regions for similar reasons.

The sterile male technique also proved unsuccessful in mosquito control. Scientists believe that the chief reason for these failures is the lower sexual activity of sterilized males compared with wild males. Other reasons may include an inadequate number of sterile males, ignorance of the insect pest's breeding cycle, and the inmigration of additional pests. Some researchers also suggest that through natural selection a new race of insects may evolve that recognizes and avoids sterile males.

Despite these problems, the sterile male technique is an important tool in integrated pest management. It is species-specific, can be used with environmental controls, and can be effective in eliminating pests in low-density infestations.

Developing Resistant Crops and Animals   Genetically resistant crops and animals can be developed through genetic engineering and artificial selection.

In a recent example of artificial selection, scientists found that certain oils in the skins of oranges, grapefruits, and lemons are highly toxic to the eggs and larvae of the Caribbean fruit fly, which lays its eggs in the skins of these fruits. The flies' larvae destroy the fruit. Scientists may now be able to selectively breed citrus fruits to increase the amount of toxic oils in their peels.

Scientists at Cornell University are developing a new type of potato plant whose leaves, stems, and sprouts are covered with tiny, sticky hairs that trap insects and immobilize their legs and mouth parts. Field tests show that this plant can reduce green peach aphids, which also attack potatoes, by half. The new variety was developed by crossing cultivated potatoes with a wild species with sticky hairs, which grows as a weed in Bolivia.

Other genetic research has led to Hessian fly-resistant wheat and leafhopper-resistant soybeans, alfalfa, cotton, and potatoes. Work on chemical factors that attract insects to plants may help scientists selectively remove them to make plants unappealing.

The Monsanto Company recently announced another promising weapon in the fight against pests. Robert Kaufman and his colleagues isolated the gene that gives BT its pesticidal action. The scientists transferred that gene to another bacterium, *Pseudomonas fluorescens*, that lives on the roots of corn and several other plants. The transplanted gene renders the host bacterium lethal to insects and other organisms, such as the black cutworm, that feed on the roots of important commercial plants. Simply by planting seeds that have been pretreated with *P. fluorescens* bearing the toxic gene, farmers may be able to provide long-term protection without many of the dangers of pesticides. However, widespread use of BT in corn and other crops may result in the rapid development of resistance and loss of the BT control farmers now enjoy.

Monsanto hopes that more insecticidal genes can be added to *P. fluorescens* in the future, giving corn a wider range of protection and reducing chemical pesticide use, thus protecting wildlife from the harmful toxic pesticides that have been the mainstay of agriculture for decades.

Root-zone protection is not the only strategy that geneticists are developing. Numerous bacteria colonize above-ground plant parts; fitted with insecticidal genes from BT and other naturally occurring biological agents, these bacteria could create a protective barrier to ward off dozens of insect pests.

Genetic resistance is a necessary element of effective pest management. The major problem is the time, money, and labor involved in producing resistant varieties. Furthermore, genetic resistance can be overcome when pests adapt. In this case, scientists must be ready with new varieties.

Chemical Controls

Chemical controls may also be a part of IPM, but as you shall soon see, a whole new arsenal of natural (presumably nontoxic) chemicals may come to play a significant role in controlling pests. First, though, we look at conventional pesticides.

Second-Generation Pesticides   Even with wider use of biological control agents and other strategies of IPM, second-generation pesticides will likely remain a part of our pest control strategy for many years. However, several principles should guide their use: (1) they should be applied sparingly; (2) they should be applied at the most effective time to reduce the number of applications; (3) they should destroy as few natural predators, nonpest species, and biological control agents as possible; (4) they should not be applied near drinking water sup-

plies; (5) they should be carefully tested for toxic effects; (6) they should be avoided if they are persistent and tend to bioaccumulate; (7) they should be used in ways that reduce exposure to workers and nearby families; and (8) they should be used to reduce populations to low levels initially and then environmental, genetic, and cultural control measures should be used to keep populations reduced.

One way to minimize the use of insecticides is to spray only affected areas. Another technique useful for herbicides is the use of special wick applicators, rather than sprayers, which deliver a small dose directly to the target species with minimum environmental contamination.

Developed nations have an important role to play by discouraging companies from exporting banned pesticides to developing countries, from which they often return on imported produce. The rich can also help the poor develop a sustainable pest management program through technical and financial assistance.

Third-Generation Pesticides · Today, several new chemical agents have been developed, such as pheromones, insect hormones, and natural chemical repellents, that could further displace potentially harmful second-generation pesticides. These are **third-generation pesticides**.

Insects and other animals release chemicals called **pheromones**, which provide a chemical means of communication. One well-known group of pheromones is the **sex attractants**, which are emitted by female insects to attract males at the time of breeding. Effective in extraordinarily small concentrations, pheromones draw males to females, an evolutionary adaptation that ensures a high rate of reproductive success.

Some sex attractants are produced commercially and are available for pest control. They are used in **pheromone traps**, which lure males. These traps may contain a pesticide-laden bait or a sticky substance that immobilizes insects (Figure 18.9). Pheromones can also be sprayed widely at breeding time. This is known as the **confusion technique**, as the males are drawn by the pheromone from all directions and may never find a partner. One modification of this technique involves the release of wood chips treated with sex attractants. Males are attracted to the wood chips and may attempt to breed with them. Pheromone traps can also be used to pinpoint the time when insect eggs hatch. Males emerging at this time are attracted to the traps. By knowing precisely when insects appear, farmers can time their pesticide applications for maximum effectiveness. This technique helps reduce the amounts of pesticide applied. Finally, pheromones can be used to lure beneficial insects from fields so that pesticide sprays can be applied.

Pheromone traps of various sorts have been used to control at least 25 insect species and can be practiced

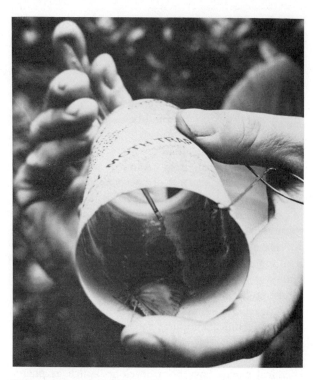

Figure 18.9 *Pheromone trap, containing a sticky substance to immobilize male gypsy moths in search of mates.*

with other IPM methods. As you can see from the previous discussion, the use of pheromones offers several advantages over second-generation pesticides. First, they are nontoxic and biodegradable. As a result, they are not expected to have any significant environmental impacts. Second, they can be used at low concentrations. Third, they are highly species-specific. The major disadvantage is the high cost of developing new pheromones.

The life cycle of many insects is shown in Figure 18.10. As illustrated, adult insects lay eggs, which develop into larvae, the caterpillar stage. Larvae are voracious eaters, and are often the most troublesome form of pest species. Eventually, larvae spin a cocoon in which they undergo incredible change, transforming from a caterpillarlike creature into a flying form, such as a moth or butterfly, the adult form.

The entire life cycle of insects is regulated by two hormones, **juvenile hormone** and **molting hormone**. Hormones are chemical substances that are produced by specific cells in the body and travel through the bloodstream to distant sites, where they exert some effect. Altering the levels of juvenile and molting hormones disrupts an insect's life cycle, sometimes resulting in death. For example, larvae treated with juvenile hormone are prevented from maturing and eventually die. If given molting hormone, they will enter the pupal stage too early and die. Interestingly, some plants have evolved

chemicals structurally similar to juvenile hormone. When ingested by hungry larvae, these chemicals prevent the larvae from pupating. This, in turn, prevents the formation of the adult form that produces eggs and additional generations.

Insect hormones applied to crops offer many of the same advantages that pheromones offer, including biodegradability, lack of toxicity, and low persistence in the environment. Like pheromones, however, they have a high cost and long production time. In addition, insect hormones act rather slowly, sometimes taking a week or two to eliminate a pest, by which time extensive damage may occur. In addition, insect hormones are not as species-specific as pheromones and therefore may affect natural predators and other nonpest species. The timing of application is also critical because hormones are effective only at certain times in an insect's life cycle.

Researchers recently discovered a plant from Malaysia that produces juvenile hormone. They hope that the genes responsible for the production of this hormone can be transferred to commercially important crops, offering another avenue of protection.

Natives of the South American tropics have used the seeds and leaves of the neem tree for many years to control pests. Researchers found that this tree produces chemicals that kill or repel a variety of insects. This extract may become useful in the control of larvae that feed on vegetables and ornamental crops.

On another front, Egyptian researchers discovered that flies ignored a species of brown algae left out on a counter to dry. Curious, researchers extracted a mixture of chemicals from the algae and found that these chemicals also repel a variety of insects that attack cotton and rice.

Natural chemicals like these may prove beneficial. Like other third-generation pesticides, they are biodegradable and nonpersistent.

Researchers also found that some plants produce chemicals that alter insect metabolism. Petunias, for example, synthesize a chemical that dramatically stunts the growth of corn earworms. Scientists hope that they can transfer the genes to crop plants either through genetic engineering or more conventional means to provide a natural protection against pests, thus providing an on-site means of control. Nicotine, caffeine, and citrus oil are all natural insecticides under investigation today.

Cultural Controls

The final component of IPM is the **cultural controls**, any one of a dozen techniques to control pest populations that do not fall under previous categories. These methods include cultivation to control weeds, noise-makers to frighten birds, and manual removal of insects from crops, which is especially suitable for smaller gardens. Also included in this group are such measures as

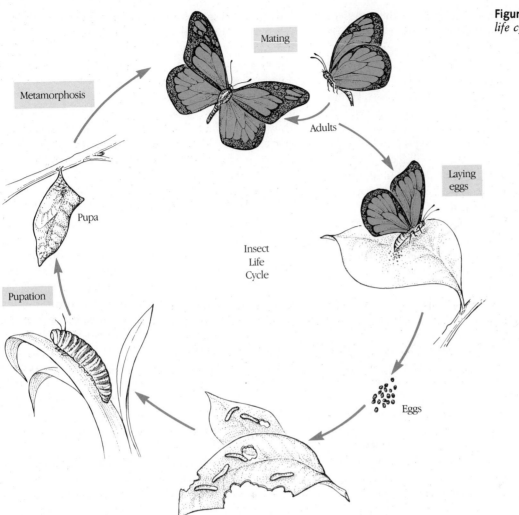

Figure 18.10   *The insect life cycle.*

the destruction of insect breeding grounds, improved forecasting of insect emergence, quarantines on imported foods to prevent the spread of pests, and water and fertilizer management to ensure optimum crop health and resistance to pests. One of the most badly needed cultural controls is monitoring.

Monitoring   A farmer spots infestation of pest insects in his field. The next day, he hooks up the sprayer, fills it with insecticide, and heads to the field to spray the entire crop. The farmer's neighbor consults the schedule provided to him by the pesticide salesman. According to the calendar, today is the day to spray his crops. He, too, fills up the sprayer and heads out for an afternoon of work.

These are typical scenarios. The first is a knee-jerk reaction to an infestation that may be affecting only a tiny portion of a farmer's entire crop. If the farmer were to check more carefully, he might find that only a portion of his crop needed spraying, a discovery that could result in considerable savings of time and money.

The second is also a knee-jerk response, but this time to schedules that may not reflect the needs of a farmer's crops. Were a farmer to check, he might find that the pests are under control or nonexistent. He, too, might save lots of time and money were he to spend a little more time determining the status of his crops and the insect pests—in short, monitoring for pest outbreaks.

Successful IPM requires a better understanding of insect biology as well as better skills in recognizing and counting insects. By knowing what insects are present, and where, farmers can institute better controls. Farmers especially need training in methods to monitor the size of insect populations so they can determine when they really have a problem. This way, they won't overreact to a stray pest or two.

Educating the World on Alternative Strategies

According to Worldwatch Institute's Peter Weber, "Farmers seem to see in pesticides . . . the illusion of a guarantee, which they can never get from the weather, markets, or politicians." In fact, some banks even require farmers to use pesticides to qualify for crop loans.

With a little imagination and thought, farmers are finding that they can reduce pesticide use by 50% without lowering harvests or significantly raising costs. Farmers like Ron Rosmann can curtail pest damage without chemicals.

Training the world's farmers in IPM is vital to efforts to build a sustainable society. But, it will require a massive global educational effort. Schools and universities, extension services, and even agricultural magazines can all play a role in retraining farmers.

In virtually all countries, farmers receive most of their advice on pest control from the sales representatives of the chemical manufacturers that produce pesticides, who have an obvious conflict of interest. In some universities, much of the research on pest control is sponsored by pesticide manufacturers, which obviously biases the system. In the United States, for example, the U.S. Department of Agriculture spends a paltry 1% of its $1.6 billion annual budget on IPM and sustainable agriculture. In contrast, pesticide companies shell out $1.7 billion per year for research and development of chemical controls.

Fortunately, several U.S. universities have recently developed sustainable agriculture programs that teach their students about IPM, among other subjects. Students at Iowa State University and the University of California at Davis, for example, can learn cost-effective ways of controlling pests without chemicals or with minimal use. State university extension programs also provide information to farmers and gardeners on alternatives to chemical pesticides.

Universities in developing countries are also beginning to train farmers and students. Birzeit University on the West Bank, for instance, recently embarked on a six-year program to help reduce the use of pesticides in Palestine and introduce farmers to IPM through an ambitious educational campaign.

In developing countries, agricultural departments have trained farmers in IPM and encouraged them to educate fellow farmers. In Indonesia, the government recently trained and hired 2000 crop scouts to work with farmers to monitor insect populations and to teach them the techniques of IPM. (See the Models of Global Sustainability 18.1 for more on Indonesia's program.)

Farmers worldwide can also work together without government support. Ron Rosmann, for example, is a member of Practical Farmers of Iowa (PFI), a group of some 400 farmers interested in not only cutting costs but also preventing soil erosion and using fewer chemicals—ultimately, farming more sustainably. Groups like PFI have been started in several other states as well. Farmers who belong to such groups are pioneering new ways to produce crops that could result in dramatic

decreases in pesticide use while maintaining or increasing yields. By sharing their ideas with nonmembers, they can spread the word to a broad range of farmers.

Farmer groups are also beginning to form in developing countries. Bolivian farmers, for instance, recently formed a group called the Association of Ecological Producers. A similar group was formed in Mexico and draws off a long tradition of pesticide-free agriculture. During 1992, Mexico's 13,000 organic farmers will export an estimated $20 million worth of food to the United States and Europe.

Nongovernmental organizations (NGOs)—among them, consumer and environmental groups—can play a major role in educating farmers and government officials worldwide. A recent ban on several widely used and highly toxic pesticides in the Philippines, for instance, is the result of the efforts of numerous NGOs. Concerted efforts on their part resulted in a dramatic shift in official and public opinion away from chemical pesticide toward safer alternatives. The success of these groups can be attributed in large part to the efforts of the NGOs to demonstrate alternatives that are available to farmers. NGOs have also publicized research showing that crop yields don't have to fall when pesticides are abandoned. Through their work, they have successfully promoted the notion of IPM, which has gained wider acceptance than the chemical pesticide dogma promoted by chemical companies.

In summary, education is a prerequisite to sustainable agriculture relying in part on IPM. Without it, the heavy dependence on chemical pesticides is bound to continue, as are the dangerous effects of their misuse.

Rethinking Chemical-Intensive Agriculture: Some Final Thoughts

The central goal of pest control is to reduce pest populations to levels that do not cause economic damage. The goal is not to eliminate pests entirely, which is, of course, an impossible task.

As the previous material has shown, pest control by pesticides incurs economic, environmental, and health costs. Ideally, these costs—or adverse impacts—should be much less than the economic benefits realized by increased yields (Chapter 14). Many would argue that in today's world, the benefits of pesticide use do not justify the many harmful effects borne by society and the environment.

The standard line, however, is that pesticides increase crop yield and provide food for millions of people. The fish, birds, and other animals and even the thousands of farm workers who die from pesticide poisoning are justified because millions are saved.

Proponents of pesticide-free agriculture argue that this is faulty reasoning. They point out, as noted earlier,

Models of Global Sustainability 18.1

Indonesia Turns to Biological Pest Control

Indonesia is a country of islands—nearly 14,000 of them—in Southeast Asia. In 1983, this rural nation, once the world's leading importer of rice, succeeded in growing enough rice to feed its own people. New strains of rice, fertilizers, and an intricate irrigation system deserve much of the credit for the success. In 1985, however, the notorious brown planthopper threatened the progress of the previous years. This insect causes rice to dry out, rot, and fall in the field. Infestations of the insect can cause enormous damage. From 1975 to 1979, in fact, the brown planthopper destroyed 4 million hectares (10 million acres) of rice crop in Indonesia.

To combat the planthopper, in 1985 the government decided to try IPM. It was advised to do so by an Indonesian entomologist, Dr. Ida Oka, who had received his Ph.D. from Cornell University under David Pimentel, a leading expert on natural insect control. Oka had been in charge of pest management in the late 1970s and early 1980s and had implemented sound IPM for rice and other crops. During that time, pesticide use had been greatly reduced and rice yields soared. But Oka had resigned when a new minister of agriculture was appointed. This official's pro-pesticide policies in which government paid about 85% of the cost to farmers resulted in widespread use of chemical pesticides and clearly placed the country on the pesticide treadmill. This policy was largely responsible for the outbreak of the brown planthopper in 1985.

Indonesian scientists found that pesticide use to control the planthopper and other insects killed many beneficial insects that preyed on the pest. One of the beneficial insects destroyed in the spraying is the wolf spider, which can devour 5 to 20 brown planthoppers a day. If left alone, the beneficial insects can often control the harmful ones. Researchers also found that farmers sprayed fields regularly, whether they needed it or not. The overuse of pesticides actually increased the severity of infestations, so much so that by 1986 the country was in danger of becoming a rice importer once again.

Convinced of the danger of the continuing use of pesticides, the Indonesian government asked the United Nations Food and Agricultural Organization (FAO) to help them promote an IPM program. In

1986, with the help of the FAO, the government embarked on an ambitious crash program to educate farmers on IPM and the dangers of pesticides. Experts ventured into the rice paddies where they showed farmers how to diagnose problems, calculate the ratio of good bugs to bad ones, and decide how much damage the crop could stand without a decrease in yield. They then taught farmers ways to reduce spraying and methods to protect beneficial predatory insects.

Early results showed that IPM worked. IPM reduced pesticide use substantially. Trained farmers, for example, apply one-ninth as much pesticide as they did before training, with no decrease—sometimes even an increase—in crop yield. Farmers have also learned to discern insect damage from fungal damage more carefully, and this helps to reduce pesticide use.

The average yield on farms using pesticides was 2.47 tons per acre compared with 2.55 tons per acre on fields controlled by IPM. Despite high subsidies for insecticides, the farms using IPM proved more profitable than those sprayed more frequently. The government saves an estimated $120 million per year on pesticide subsidies. Indonesia's streams and wildlife are also showing signs of recovery.

The success of the pilot program in Indonesia convinced the government to adopt IPM as a national pest control strategy. The government, in fact, banned 56 of the 57 pesticides previously approved for farming in Indonesia to help protect predatory insects. The government then launched a massive campaign to educate Indonesia's farmers in IPM. In 1992, the country hired 2000 crop scouts to educate farmers on IPM and hopes that by 1994 all farmers will be using this method.

IPM has spread to Thailand, Bangladesh, Sri Lanka, Malaysia, India, and China. Farmers in Indonesia who have not yet been introduced to the technique are calling for action. They want to be included in this experiment, which, if successful, could help the world move to a more sustainable, environmentally safe form of farming. Agricultural experts believe that IPM could be used on fields that provide 45% of the rice for people living in southern and Southeast Asia and could save millions of dollars, preserve wildlife, and protect human health without endangering high crop yields.

that pest damage is increasing despite the increasing use of pesticides, making less food available per hectare. In addition, crops can be grown without pesticides to feed the millions of hungry people. Numerous examples cited in this chapter show that pesticide-free farming can, if

properly executed, be more profitable than chemical-intensive farming. Ultimately, the final and unavoidable measure of its success has to be its sustainability. Chemical-intensive agriculture is very likely not sustainable in the long run.

The shift to pesticide-free farming will require attitude changes on the part of consumers. All of us will have to rethink and change our buying habits. We will have to learn to accept slightly blemished fruits and vegetables rather than demand picture-perfect produce made possible only by the heavy application of chemical pesticides. A few more spots on our oranges could mean many more birds overhead, cleaner waterways, improved health for workers and the general public, and cheaper oranges. The blemishes don't change the nutritional value of the produce one bit.

Governments can help reduce exposure to pesticides by passing ordinances that require pesticide applicators who spray trees, shrubs, and lawns to post notices or notify chemically sensitive individuals in advance. Governments can also help by providing low-cost crop insurance to farmers who are making the transition to IPM. As farmers shift from chemical-intensive use to IPM, losses can be severe because the ecological balance of their farms may be severely out of kilter. With the safeguard of insurance, farmers can wean themselves from pesticides and feel confident that they won't go bankrupt in the process.

Governments can also help by promoting **organic certification programs**. Developed nationally and in several states, such as Colorado and California, these programs set standards farmers must meet to permit them to label their produce "organically grown." Certification will avoid dishonesty and help consumers determine which produce is truly pesticide free. It could also stimulate other farmers to think about switching to chemical-free methods of farming. To the delight of many people, numerous U.S. farmers have already switched. In 1987, the U.S. Department of Agriculture announced that 30,000 American farmers had forsaken pesticides and artificial fertilizers altogether. At least 100,000 more have made substantial reductions in chemical use. In recent years, the shift has continued, although statistics are not available. That's good news to some, but it is only a small fraction of the 2 million U.S. farmers.

Our doubts are traitors and make us lose the

good we oft might win by fearing to attempt.

—*Shakespeare*

≈❧   ≈❧   ≈❧

Critical Thinking Exercise Solution

As with many other issues, it is important to dig deeper, consider the big picture, and uncover bias that may enter into the arguments of some proponents.

Before you go too far, though, it is important to scrutinize the experiment itself. Was it performed correctly and are the results applicable to humans?

Suppose you find that the results are valid. You would then want to look at the next most important question: should the pesticide be banned? In the public policy arena, this question pivots on another relatively simple question: are the risks worth the benefits?

In order to answer this, it is necessary to seek more information and viewpoints. If you do, you might find that farmers and pesticide manufacturers would argue that this insecticide helps them prevent crop damage, which saves farmers millions of dollars a year. They also may say that pesticides make food cheaper for consumers, and the pesticide manufacturing industry provides thousands of jobs.

As for the costs (the risks), if EPA estimates are correct, 10 to 20 additional U.S. citizens will contract cancer each year and many of them will die. The pesticide may also contaminate groundwater, kill fish and birds, and have other adverse environmental effects.

When you dig deeper, you will find that alternative methods of controlling insects are available. These methods may have the added benefit of reducing groundwater contamination and mortality in birds, fish, and other wildlife. In some cases, alternative methods lower crop yield, but because of lower input costs, farmers can make as much if not more money.

Can you think of any other costs and benefits? After analyzing both sides of the argument, what is your opinion? Should the pesticide be banned? Would your opinion change if you were one of the cancer victims?

Summary

18.1   Discovering the Impacts of Pesticides: A Brief History

■   Each year, insects, weeds, bacteria, fungi, viruses, parasites, birds, rodents, and mammals consume or destroy an estimated 48% of the world's food crop. Huge gains in food supply could be made by reducing losses through pest control.

■   In recent years, **chemical pesticides** have come to form the cornerstone of pest management. Pesticides are chemicals used to kill troublesome pests.

■   **First-generation pesticides** were simple preparations made of ashes, sulfur, and other chemicals.

■   **Second-generation pesticides** are synthetic organic compounds like DDT. Some are **broad-spectrum chemicals**, able to kill a wide variety of organisms. Others are **narrow-spectrum pesticides**, used to control a few pests.

■ Second-generation pesticides can be categorized into three groups: (1) **chlorinated hydrocarbons**, a high-risk group that is largely banned today; (2) **organic phosphates**, less risky but still quite toxic; and (3) **carbamates**, widely used today because they tend to break down quickly and do not bioconcentrate as chlorinated hydrocarbons do.

■ Pesticide use may help control pests but not without problems. Some broad-spectrum pesticides, for instance, kill predatory and parasitic insects that naturally control pests. Pesticides also destroy beneficial insects, such as honeybees.

■ Perhaps the most striking problem is **genetic resistance** among pest populations, which has been countered with increasing doses and new chemical preparations, generally referred to as the **pesticide treadmill**. The result is considerable environmental pollution.

■ DDT proved especially harmful to predatory bird populations because it impaired eggshell deposition. Growing evidence shows that many pesticides are harmful to people, especially farm workers.

18.2 Regulating Pesticides: Treating the Symptoms

■ Pesticides in U.S. foods are currently regulated by the EPA and FDA through the **Federal Insecticide, Fungicide, and Rodenticide Act (FIRA)** and the **Federal Food, Drug, and Cosmetic Act**, respectively.

■ FIFRA creates a system of registration intended to protect humans and the environment from adverse effects.

■ Under powers granted by FIFRA, the EPA sets **tolerance levels** for pesticides in food, but many critics think that they do not accurately reflect health risks and need to be lowered. Better regulation of pesticide end use is also needed.

■ The FDA monitors the food supply for pesticide residues and has the power to seize shipments containing levels that exceed those permitted by the EPA. Unfortunately, the FDA does not test for all pesticides and tests only a fraction of our food.

18.3 Integrated Pest Management: A Sustainable Solution?

■ The problems with pesticide use have led many to suggest the need for an **integrated pest management** strategy that relies on four basic means of control: environmental, genetic, chemical, and cultural.

■ **Environmental controls** are designed to alter a pest's immediate environment, making it lethal or inhospitable. Some options include increasing crop diversity, varying the time of planting, altering plant and soil nutrients, controlling adjacent crops and weeds, and introducing predators, parasites, and disease organisms.

■ **Genetic controls** are alterations in the genetic composition of pests, crops, and livestock. In the **sterile male technique**, for example, captive-raised males of the pest species are sterilized and then released into fields, where they greatly outnumber wild fertile males. This technique ensures a high percentage of infertile matings.

■ **Genetic resistance** can be bred into crops and livestock or introduced through genetic engineering.

■ **Chemical controls** rely on sparing use of chemical pesticides, applied only when and where they are needed. New applicators can also minimize use. Naturally occurring chemicals may also play a role in pest control in the future.

■ **Sex-attractant pheromones**, chemicals released by females to attract males, have been synthetically produced and marketed. They can be used to draw males into traps laced with poison or sticky materials.

■ **Insect hormones** that control vital life processes can also be applied to crops, disrupting the normal developmental cycle and killing the pests.

■ The central goal of pest control is to reduce pest populations to levels that do not cause economic damage. Logic dictates that the cost of pest control should not exceed the economic benefits. External costs must be factored into the cost equation.

Discussion and Critical Thinking Questions

1. List and discuss reasons why pest damage is high in developed nations despite extensive use of chemical pesticides.

2. Critically analyze the statement: "Without pesticides, crop damage in the United States would be much higher."

3. Describe some of the environmental and health problems caused by the use of pesticides. How can they be reduced or avoided?

4. How are beneficial species affected by insecticide use? Give some examples.

5. What is the relationship between pesticide treadmill and genetic resistance?

6. Why do DDT and other chlorinated hydrocarbons persist in the environment? Why do they cause problems even though they are found in low concentration in water?

7. List the major components of IPM. What advantages does this strategy offer over current management techniques?

8. Explain why crop rotation and increasing crop diversity reduce pest populations.

9. Describe some of the biological control methods. Give examples. Using your knowledge of biology

and evolution, would you expect pest species to develop resistance to biological controls?

10. Describe why the sterile male technique works.

11. Discuss some ways in which genetic engineering may be used to help cut down on pest damage.

12. You are appointed director of the state agricultural department. Outline a plan to encourage farmers to minimize and eliminate pesticide use.

Suggested Readings

Ames, B. N. and Gold, L. S. (1989). Pesticides, Risk, and Applesauce. *Science* 244 (4906): 755–67. Looks at natural carcinogens and important considerations for controlling human-produced pesticides.

Bosch, R. van der. (1978). *The Pesticide Conspiracy.* New York: Doubleday. A classic study of the influence of pesticide manufacturers on IPM.

Brady, C. (1992). The Perfect Poison for the Perfect Banana. *Global Pesticide Campaigner* 2 (3): 8–9. Alarming interview with a Honduran farm worker who applies herbicides.

Carson, R. (1962). *Silent Spring.* Boston: Houghton Mifflin. The book that raised worldwide alarm over the use of pesticides.

Curtis, J., Mott, L., and Kuhnle, T. (1991). *Harvest of Hope.* New York: Natural Resources Defense Council. Documents potential for reductions in pesticide use on nine major U.S. crops.

Dreistadt, S. H. and Dahlsten, D. L. (1986). California's Medfly Campaign: Lessons from the Field. *Environment* 28 (6): 18–20, 40–44. Sober look at pest control.

Gup, T. (1991). Getting at the Roots of a National Obsession. *National Wildlife* 29 (4): 18–20. Excellent coverage of the use of pesticides on lawns.

Lipske, M. (1990). Natural Farming Harvests New Support. *National Wildlife* 28 (3): 19–23. Highlights some alternative farming methods.

Marshall, E. (1991). A is for Apple, Alar, and . . . Alarmist? *Science* 254: 20–22. Good historical perspective on the alar controversy.

Sewell, B. H. et al. (1989). *Intolerable Risk: Pesticides in Our Children's Food.* Washington, D.C.: Natural Resources Defense Council. Controversial paper worth reading.

Siedenburg, K. (1992). Philippine Pesticide Bans Show NGO Strength. *Global Pesticide Campaigner* 2 (3): 1, 6–7. Excellent case study showing how nongovernmental organizations can impact pesticide use.

Weber, P. (1992). A Place for Pesticides? *World-Watch* 5 (3): 18–25. Superb overview of the pesticide issue with a frank discussion of options.

Have You Received Your *Environmental Action Guide?*

If you've purchased a new book, your *Environmental Action Guide* should appear between pages 454 and 455. We've bound the *Environmental Action Guide* into the text and have perforated it for easy removal. This avoids using plastic shrink-wrap and helps reduce waste. We hope you will find the *Environmental Action Guide* useful and informative.

Chapter 19

Hazardous and Solid Wastes: Dealing with the Problems Sustainably

There is nothing more frightful than ignorance in action.

—Goethe

Critical Thinking Exercise

As a hazardous waste manager of a chemical company, you are faced with a dilemma. Your company produces over 4000 metric tons of highly toxic waste. The cost of disposal is several million dollars. An official from a new waste disposal firm contacts you and says that he can dispose of the waste at half the cost. When you ask him where it is going, he says that it will be shipped to Niger, an African nation, where it will be incinerated. He tells you that pollution controls on the incinerator where the waste will be burned are not as sophisticated as the one owned by the U.S. disposal company you've always used, but it doesn't really matter because there's so little toxic waste burned in Niger anyway. What choice do you make? What factors will affect your choice?

The 3M Corporation, based in Minnesota, is a model of corporate ingenuity and foresight. In 1975, the company started a Pollution Prevention Pays Program, designed to reduce its solid and hazardous waste. By substituting less toxic or nontoxic chemicals for solvents, by recycling everything possible, and by modifying manufacturing processes, 3M has cut the pollution and waste generated per unit of production by 50% in the past 15 years.

The company's program annually eliminates nearly 90,000 metric tons of air pollutants. It reduces the output of solid waste by 243,000 metric tons per year and the production of wastewater by 5.8 billion liters (1.5 billion gallons) per year. What is more, these and other measures implemented by the company have saved approximately $500 million in 15 years. Although interest in reducing waste by recycling and using resources more efficiently is increasing, businesses are a long way from tapping into the full potential of this strategy. Many continue to discard millions of tons of perfectly usable material, including cardboard, wood, office paper, and metals. Moreover, some companies continue to illegally and irresponsibly dump hazardous waste into the air, water, and soil (Figure 19.1).

(a)

(b)

Figure 19.1 (a) *In May 1984, workers from the Neville Chemical plant in Santa Fe Springs, California, began covering soil believed to be contaminated with highly toxic chemical wastes. The covering prevented the toxic soil from blowing off the site. Company officials were recently found guilty of illegally disposing of hazardous wastes.* (b) *Pollution Abatement Services chemical dump at Oswego, New York. The dump is now inactive and awaiting cleanup.*

This chapter discusses solid and hazardous wastes. It illustrates how individuals, businesses, and governments have traditionally addressed the problem of waste, and the limitations of these strategies. It also presents more sustainable approaches that could help eliminate the waste of the Earth's resources and drastically reduce pollution of the environment.

19.1 Hazardous Wastes: Coming to Terms with the Problem

Love Canal: The Awakening

Hazardous wastes are waste products of businesses, homes, and other facilities that if not handled or disposed of properly or destroyed pose a threat to people and environment. For many years, pollution was seen as a sign of progress. Today, many individuals view pollution in general and hazardous waste in particular as signs of failed technologies or failed industrial systems.

The severity of the U.S. hazardous waste problem caught the attention of the American public in the 1970s when toxic chemicals began to ooze out of a dump known as Love Canal, in New York near Niagara Falls. The story of Love Canal began in the 1880s when William T. Love began digging a canal that would run from the Niagara River just above Niagara Falls to a point on the river below the falls. Built to divert water to an electric power plant to supply future industrial facilities along its banks, the canal was never completed. Only a small remnant of the canal remained by the early 1900s. In 1942, the Hooker Chemical Company signed an agreement with the canal's new owner, the Niagara Power and Development Corporation, to dump hazardous wastes in the abandoned canal. In 1946, Hooker bought the site, and from 1947 to 1952, it disposed of over 20,000 metric tons of highly toxic and carcinogenic wastes, including dioxin.

In 1952, the story took an ironic twist. In that year, the city of Niagara Falls began condemnation proceedings on the property that would allow it to use the land for an elementary school and residential community. With no other choice, Hooker sold the land for $1 in exchange for a release from any future liability. Hooker insists that it warned against construction on the dump site itself, but it allegedly never disclosed the real danger of building on it. Before turning the land over to the city, Hooker sealed the pit with a clay cap and topsoil, once thought sufficient to protect hazardous waste dumps.

Troubles began in January 1954, however, when workers removed the clay cap during the construction of the school. In the late 1950s, rusting and leaking barrels of toxic waste began to surface. Children playing near them suffered chemical burns; some became ill and died. Hooker said that it warned the school board not to let children play in contaminated areas but apparently made no effort to warn local residents of the potential problems.

The problems continued for years. Chemical fumes took the bark off trees and killed grass and plants in vegetable gardens. Smelly pools of toxins welled up on the surface. In the early 1970s, after a period of heavy rainfall, basements in homes near the dump began to flood with a thick, black sludge of toxic chemicals. The chemical smells in homes around the dump site became intolerable.

Tests in 1978 on water, air, and soil in the area detected 82 different chemical contaminants, a dozen of which were known or suspected carcinogens. In that same year, the N.Y. State Health Department found that nearly 1 of every 3 pregnant women in the area had miscarried, a rate much higher than expected. Birth defects were observed in 5 of 24 children. Another study, released in 1979 by Dr. Beverly Paigen of the Roswell Cancer Institute, showed that over half of the children born between 1974 and 1978 to families living in areas where groundwater was leaching toxic chemicals from the dump had birth defects. In this study, the overall incidence of birth defects in the Love Canal area was 1 in 5, compared with a normal rate of less than 1 in 10 (Chapter 14). The miscarriage rate was 25 in 100, compared with 8 in 100 women moving into the area. Asthma was four times as prevalent in wet areas as dry areas in the region; the incidence of urinary and convulsive disorders was almost three times higher than expected. The incidence of nasal and sinus infections, respiratory diseases, rashes, and headaches was also elevated.

As a result of public outcry, the school was soon closed. The state fenced off the canal and evacuated several hundred families (Figure 19.2). Then President Jimmy Carter declared the site a disaster area. In May 1980, a new study revealed high levels of genetic damage among residents living near the canal. An additional 780 families were evacuated from outlying areas.

As of 1991, Love Canal had cost the state of New York and the federal government approximately $150 million for cleanup, research, and relocation of residents. In 1987, the EPA announced plans to clean the sewers and dredge two creeks in the Love Canal area to remove sediments contaminated with toxins. The sewers were cleansed with power washers. In 1988 and 1989, the creeks were diverted so they would dry up. Bulldozers then removed the top 46 centimeters (18 inches) of mud, which at this writing (January 1993) are being stored by Occidental Petroleum (formerly Hooker) to be burned by the company in a special incinerator built especially for this project. All told, about 35,000 cubic meters of sediment will be burned, making this the largest single application of thermal destruction in modern

Figure 19.2 *House being bulldozed in the Love Canal area of Niagara Falls, New York.*

history. The EPA estimates that the dredging and incineration will cost $26 to $31 million.

A 1980 study by the EPA showed that chemical contamination was pretty much limited to the canal area (the actual dump), an area immediately south of it, and two rows of houses on either side of the canal (Figure 19.3). The last group of residents to be evacuated, the report said, were probably moved out unnecessarily. The EPA study also showed that the dump had contaminated shallow groundwater but not deeper aquifers. The EPA concluded that further migration of toxic chemicals was highly unlikely. Based on this study and other work, the EPA and the state of New York declared two-thirds of the evacuated Love Canal site "habitable" and proposed to sell the houses. About 40 to 50 houses have been sold as of January 1993. Lois Gibbs, the Love Canal resident largely responsible for drawing public attention to the disaster and getting the state and federal governments to take action, argues that the decision to resettle the area has been improperly made. In fact, she claims that in assessing the habitability of the Love Canal site the N.Y. State Health Department compared it to two other sites, both badly contaminated by industrial wastes, and deemed it suitable for resettlement. Gibbs warns that resettling Love Canal will put more people at risk.

The Dimensions of a Toxic Nightmare

Love Canal began the frenetic search for hazardous waste dumps and illegal waste disposal practices that continues today in the United States and abroad. Many people, previously unable to explain bizarre diseases in their family, have found the answer in nearby waste dumps or contaminated factories that leaked hazardous wastes into groundwater, nearby streams, or the air.

In the years following the Love Canal incident, the American public has been barraged by a list of startling statistics showing that what appeared to be an isolated incident was, in fact, just the tip of the iceberg. The EPA, for instance, estimated that there were 14 other sites in Niagara Falls alone that it considered an "imminent hazard." Nationwide, the EPA announced, Love Canal was 1 of 10,000 sites in need of a cleanup. In 1989, it estimated that the actual number of hazardous waste sites was more like 32,000. The U.S. General Accounting Office (GAO) estimates that the number of hazardous waste sites could be much higher, perhaps 100,000 to 400,000. Yet, these estimates do not include the 17,000 hazardous waste "hot spots" on U.S. military bases. Tens of thousands, perhaps hundreds of thousands, of badly contaminated sites may exist in Europe, especially the former Eastern bloc nations, and the Commonwealth of Independent States.

Making matters worse, each year U.S. factories create an estimated 200 to 250 million metric tons of hazardous waste, or about 1 metric ton for every man, woman, and child. And, the United States is not alone. European countries and many developing nations also produce tens of millions of tons of hazardous waste each year.

In 1985, 90% of the hazardous wastes in the United States was improperly disposed of, ending up in abandoned warehouses; in rivers, streams, and lakes; in leaky landfills that contaminate groundwater; in fields and forests; and along highways. Because of improvements in hazardous waste management, smaller amounts of toxic waste are being improperly discharged, although no estimates were available. Even more encouraging are efforts to eliminate wastes in the first place through pollution prevention.

Don't be misled by improvements; the United States and other nations must do a great deal more to curb this problem. As an indication, in 1987, in California alone, businesses produced 2.5 million metric tons of pollution. Here's where it ended up:

1. 65% was discharged into the ocean, lakes, and rivers.

2. 27% was injected into deep wells.

3. 4.3% went to public sewers and sewage treatment plants.

4. 1.8% went to treatment, storage, and disposal facilities.

5. 1.3% escaped into the air.

Improper waste disposal has left a legacy of costly effects: (1) groundwater contamination, (2) well closures, (3) habitat destruction, (4) human disease, (5) soil

Figure 19.3 *Love Canal. Area within the (green) rectangle was closed off, and citizens were evacuated after studies showed elevated incidence of birth defects, stillbirths, and a variety of symptoms most likely attributable to toxic wastes. Citizens were also evacuated from the declaration area (gray), but tests have shown that hazardous wastes have not migrated into this area and these citizens may have been unnecessarily evacuated.*

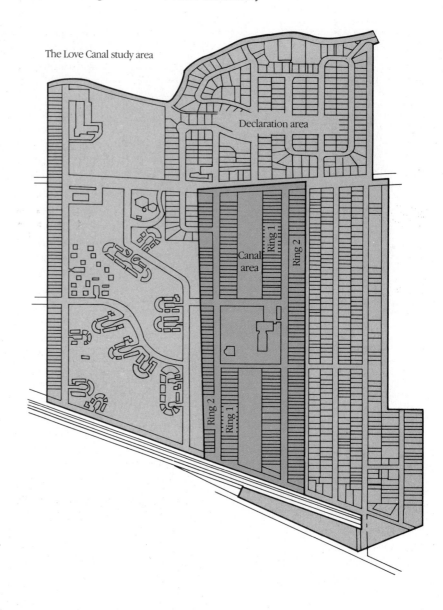

The Love Canal study area

contamination, (6) fish kills, (7) livestock disease, (8) sewage treatment plant damage, (9) town closures, and (10) difficult or impossible cleanups. Irresponsible and ill-conceived waste disposal continuing today will create a legacy of polluted groundwater and contaminated land that could persist for decades, perhaps centuries.

A decade since the United States first awakened to the hazardous waste issue, most experts now expect a longer, more difficult cleanup battle than originally anticipated. One reason is the original estimate of both number of sites and severity of damage was too low. The price tag could also be much higher than anticipated. The U.S. Office of Technology Assessment (OTA) estimates that it will cost $100 billion to clean up the 10,000 sites in the United States that pose a serious threat to health (Figure 19.4). A recent estimate by researchers at the University of Tennessee puts the cleanup cost of all

hazardous waste sites in the United States by state and federal programs at $750 billion.

LUST—It's Not What You Think

You feel dizzy. Your head spins. Your insides ache. You haven't been yourself for weeks. What may be ailing you is LUST—but not the usual kind. Your symptoms may be caused by the latest in a long list of hazardous chemical problems, groundwater pollution from a leaking underground storage tank, which EPA's top acronymists dubbed LUST. Some time later, they dropped the L, leaving UST for underground storage tanks.

Moisture and soil acidity gradually corrode older tanks, causing them to leak petroleum by-products, toxic chemicals, and hazardous wastes. The main concern is the potential effect on groundwater and human health. Even a small leak can contaminate large quantities

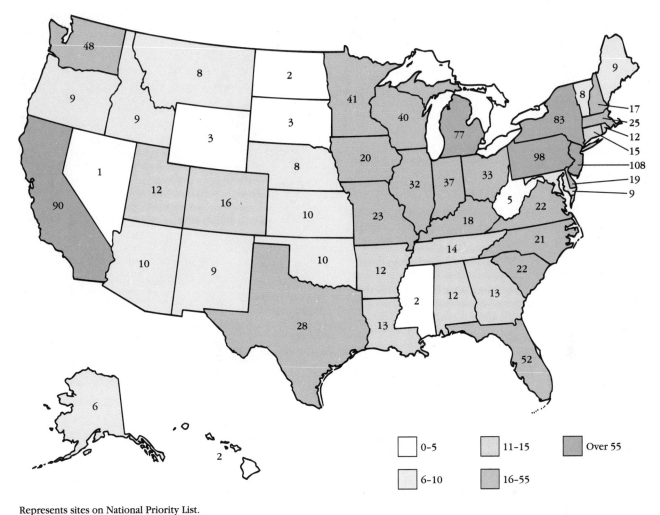

0–5	11–15	Over 55
6–10	16–55	

Represents sites on National Priority List.
[1]Includes nine in Puerto Rico; and two in Guam.

Figure 19.4 *Hazardous waste sites on the EPA's 1992 priority list for cleanup.*

of groundwater. For example, 1.5 cups of hazardous liquid leaking out of a tank per hour can contaminate nearly 4 million liters (1 million gallons of groundwater) in a day. Contaminated groundwater is very difficult—sometimes impossible—and expensive to clean up (Chapter 17).

Many people in affected areas have switched to bottled water, a temporary solution at best. Contaminated water used for baths and showers can also be dangerous. Benzene, a component of gasoline that can cause cancer, is absorbed through the skin when bathing. Showering generates dangerous vapors that can cause skin and eye irritation.

A report by the New York Department of Environmental Conservation suggests that at least half of the state's underground steel tanks containing petroleum products over 15 years old may be leaking. Nationwide, 3 to 5 million underground storage tanks containing

hazardous materials dot the United States. The EPA estimates that 200,000 to 400,000 of these tanks are leaking.

Major oil companies have already spent millions to clean up polluted groundwater and soil and install new tanks. The cost of such actions can be exorbitant. Chevron alone estimates its replacement costs at about $100 million. Unfortunately, half of U.S. service stations are owned by independent dealers, who generally are financially unable to replace the leaking tanks. Many tanks that could be leaking are under schools, police stations, and homes.

About 90% of the cleanup and replacement of leaking underground storage tanks is being financed and performed by private industry. States must assume the remaining 10%. The EPA sets guidelines for cleanup and replacement and also provides financial assistance to help states. Today, nearly 30 states have their own funds to pay for part of the cleanup cost and 7 more could have programs in the near future. State and federal funds are derived from taxes on gasoline.

19.2 Attacking Hazardous Wastes on Two Fronts

Two hazardous waste problems face virtually all industrial nations and, to a lesser extent, developing countries. First, how do they clean up existing hazardous waste sites and leaking storage tanks? Second, how do they deal with the enormous amounts of hazardous waste produced each year to avoid creating new contaminated sites?

The first problem requires immediate action. It's one place where the Band-Aid approach is appropriate. That said, all solutions that fall into this category must be sustainable ones. That is, they must not merely shift the problem from one location (a contaminated factory site) to another (a landfill where the hazardous wastes are dumped). The second problem calls for long-term, preventive measures that eliminate the production of wastes.

Sustainably Cleaning Up Past Mistakes

In June 1983, the 2400 residents of Times Beach, Missouri, agreed to sell 800 homes and 30 businesses to the federal government for $35 million. Why? The roads in Times Beach had been sprayed with oil containing hazardous wastes, including dioxin. How did the oil get contaminated?

As in other locations, unscrupulous hazardous waste disposal companies had mixed toxic chemical wastes with waste oil, then spread it on dirt roads to control dust. In Times Beach, the dioxin levels in the soil were 100 to 300 times higher than levels considered harmful during long-term exposure (Figure 19.5). The town had to go and the federal government bought it. Today, Times Beach is a ghost town bordered by a tall chain-link fence. Its only occupants are occasional EPA officials or scientists from companies that are looking for ways to detoxify the soil.

The $35 million purchase price for this contaminated piece of real estate came from a special fund known as the **Superfund**. It was created in 1980 by the **Comprehensive Environmental Response, Compensation, and Liability Act (CERCLA)**. Commonly called the Superfund Act, it and its two amendments in 1986 and 1990 established a $16.3 billion fund financed by state and federal governments and taxes on chemical and oil companies. The money is earmarked to clean up leaking underground storage tanks that are deemed a threat to human health and abandoned, inactive hazardous waste sites, including hazardous and municipal dumps and contaminated factories or mines and mills.

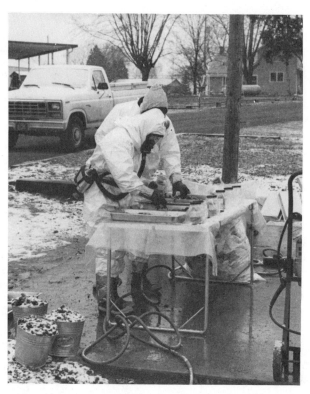

Figure 19.5 *Times Beach, Missouri. Two EPA workers testing the soil for dioxin.*

Under CERCLA, the EPA is authorized to collect fines from parties responsible for the contamination—totaling up to three times the cleanup cost. Thus, this law makes owners and operators of hazardous waste dump sites and contaminated areas, as well as their customers, responsible for cleanup costs and property damage. Under the law, all businesses, hospitals, schools, cities, and other parties that deposited hazardous waste at a site are liable for a portion of the cleanup cost based on the type and the volume of waste they deposited. Although it is a bone of contention to some, CERCLA requires a sharing of costs even at licensed hazardous waste disposal facilities where hazardous wastes were legally disposed of in previous years. As one industry representative put it, "You are liable for your waste forever." If a company or party is no longer in business, the remaining parties must pay the cost.

Fifteen years since its inception, Superfund has clearly had an impact. By the end of 1991, for instance, the EPA had surveyed more than 30,000 potential Superfund sites and completed more than 2700 emergency actions—steps to reduce immediate threats, for example, removing barrels of waste stored in abandoned warehouses. It had placed 1211 sites on a National Priority List (NPL) because of their potential health threat. Remedial action had been taken on 1161 NPL sites, but cleanup has proven costly and slow. By March 1992, only 71 sites had been completely cleaned up.

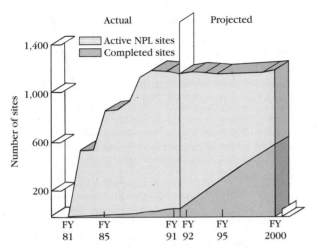

Figure 19.6 *Projected cleanup schedule for U.S. toxic hot spots. Some critics think that the accelerated rate of cleanup shown here may not materialize.*

Due to public pressure and its own realization that cleanup was progressing much too slowly, the EPA identified ways to cut the time for a typical cleanup by two years. The current average is seven to ten years. By the year 2000, the EPA hopes to have 650 sites restored (Figure 19.6).

Despite these successes, the Superfund Program has been riddled with problems. One of the most significant is cost. Stabilizing a leaking pond designed to hold hazardous wastes costs the EPA $500,000. A study to determine what chemicals are leaking from another site can cost the agency $800,000.

A second problem is that it has created a legal nightmare, which an employee of the insurance industry called a "massive web of litigation between the EPA, waste depositors, and insurance companies." Nearly 60% of the funds paid to date has been for legal fees. This money has not been spent on cleanup but on identifying liable parties and getting them to pay their fair share.

Initial mismanagement by top EPA officials also delayed serious action by the agency in the early years. Officials negotiated with owners of hazardous waste dumps to begin private cleanups, but, say critics, let some companies off too easily and only required superficial cleanups. Officials waived future liability in some cases. Thus, if problems develop in the future, companies will bear no responsibility. Investigations conducted in 1983 led the EPA's top leadership to resign or be fired because of the issue. One EPA official went to jail for perjury.

The Superfund Act has also been criticized for providing money for cleanup and financial compensation for property damage, but failing to provide avenues by which victims of illegal dumping of hazardous wastes can be compensated for personal injury or death. According to Senator George Mitchell of Maine, "Under the legislation, it's all right to hurt people but not trees." Many people believe that a fund similar to worker's compensation is needed to compensate injured parties.

According to the OTA, the EPA has often opted for quick-fix solutions to clean up areas. Three-fourths of the cleanups, it says, are inadequate in the long term. In Love Canal, for instance, the EPA simply put a clay cap over the dump site and dug a ditch around it with monitoring wells to determine if hazardous wastes were leaking out. In other sites, contaminated soil was excavated and hauled off to another landfill. Incineration and biological destruction of the wastes might have been more lasting solutions.

Alternative Funding Options   Some critics of CERCLA argue that the law has created an adversarial relationship among many parties that is counterproductive and inefficient. Too much money is spent on litigation and too little on cleanup. Although critical of CERCLA, these individuals do not suggest we should abandon funding of cleanup efforts, only that we find options that put our money to better use.

One proposal calls for a federal hazardous waste tax levied on each ton of hazardous waste disposed of, incinerated, or treated. This money would generate revenue for cleanup of the most heavily contaminated sites. Responsible parties would be off the hook, and litigation would be eliminated.

The American Association of Property and Casualty Insurers has proposed funding cleanup via a small fee on each new commercial insurance policy written in the United States. Their calculations suggest that this would generate more than $4 billion per year that could be used for cleanup. Both these alternatives might allow the EPA to redirect its efforts to cleanup.

What to Do with Today's Waste: Preventing Future Disasters

The high cost of cleanup strongly suggests the need for active preventive measures to avoid further contamination. Several possibilities exist.

Preventing Improper Disposal   In 1976, Congress passed the **Resource Conservation and Recovery Act (RCRA)**, which is designed to eliminate illegal and improper waste disposal. Under the law, the EPA was designated the nation's hazardous waste watchdog. The EPA's first role was to determine which wastes were hazardous. RCRA also called on the agency to establish a nationwide reporting system for all companies handling hazardous chemicals. This requirement created a trail of paperwork that follows hazardous wastes from the moment they are generated to the moment they are disposed of—a so-called cradle-to-grave tracking. Congress believed that this requirement would make it difficult for waste generators to dump hazardous wastes improperly. RCRA also directed the EPA to set industrywide standards for packaging, shipping, and disposal of wastes. Only licensed facilities could receive wastes.

Unfortunately, RCRA's implementation has been slow. It was not until four years after Congress adopted the law that the EPA established its first hazardous waste regulations. To the dismay of many, the regulations were full of loopholes, and about 40 million metric tons of pollutants annually escaped control.

Because of public pressure, in 1984, Congress passed a set of amendments to eliminate RCRA's loopholes and ensure proper waste disposal. For example, under the original law, if a company produced less than 1000 kilograms (2200 pounds) of hazardous waste per month, it could dump them in a local landfill. The amendments changed the rules so that any individual or company that generated more than 100 kilograms (220 pounds) of hazardous waste per year must follow the same guidelines imposed on large waste producers.

The 1984 amendments also declared a national policy to reduce or eliminate land disposal of hazardous waste. Congress made it clear that land disposal technologies must be a last resort. The 1984 amendments gave preference to reuse, recycling, detoxification (e.g., incineration), and other measures discussed below. These approaches could bring the United States closer to a sustainable waste management system.

Preventing Leaking Underground Storage Tanks

The 1984 amendments to RCRA also addressed leaking underground storage tanks. After May 1985, for example, the amendments required that all newly installed underground tanks must be protected from corrosion for the life of the tank. The lining of the tank must consist of materials compatible with stored substances. Furthermore, owners and operators must have methods for detecting leaks, must take corrective action when leaks occur, and must report all actions.

Weaknesses in RCRA

Despite these changes, RCRA still has many loopholes. For one, its definition of hazardous wastes remains too narrow. Michael Picker of the National Toxics Campaign, for example, thinks that municipal waste should be classified as hazardous waste because it contains toxic chemicals, such as pesticides. Leachates from some municipal landfills are as toxic as those found at regulated hazardous waste facilities. According to John Young of the Worldwatch Institute, more than one of every five hazardous waste sites on the Superfund cleanup list is a municipal landfill.

Picker also thinks that sewage and untreated wastewater handled by publicly owned sewage treatment plants should be considered a hazardous waste. Toxic chemicals in the sewer system, released by factories and homeowners, can escape into the air and into waterways. Agricultural wastes, mostly pesticides, are also not regulated. In California, for example, rules require that leftover pesticides be diluted and sprayed into the environment. Mill and mine tailings are also excluded from most regulatory control. By expanding the definition of what is toxic and instituting better controls, the government could greatly reduce the influx of hazardous materials into the increasingly poisoned environment.

One lesson learned in the past 20 years is that passing a law is not a guarantee of protection. One reason is that agencies responsible for administering and enforcing new laws don't always perform as they are instructed. Some drag their feet because they don't approve of the law. A second reason is that agencies may be so underfunded and so overworked that they can't take on new responsibilities, or if they do, they do an inadequate job. The EPA is a case in point. Understaffed and underfunded, the EPA today struggles to implement RCRA and the handful of other laws aimed at protecting public health and the environment. Since the agency was formed, its work load has more than doubled, but funding has, until quite recently, remained at more or less the same level (when one adjusts for inflation). Even now, increases in funding only partly offset the remarkable increase in the work load.

Exporting Toxic Troubles

Another lesson becoming painfully clear is that tough environmental legislation often has unanticipated results. For example, regulations that increase the cost of hazardous waste disposal have caused some unscrupulous companies to illegally dump their wastes. Higher costs have also led some companies and cities to export toxic wastes, including incinerator ash, abroad. In the 1980s, these wastes often ended up in cash-hungry developing countries or Eastern bloc nations, none of which had adequate laws requiring proper hazardous waste disposal.

European and U.S. companies are believed to be the most heavily involved in exporting toxic wastes. Because the trade is presumed to take place illicitly, no reliable records exist regarding the quantities of materials being exported. The international environmental group Greenpeace, however, estimates that about 4.5 million metric tons of hazardous chemicals were imported annually by former East Germany under Erich Honecker's regime, supplying an estimated 10% to 20% of the country's annual income. Researchers studying West Africa estimate that in 1988 more than 22 million metric tons of hazardous waste were imported. Today, in the United States, the export of hazardous wastes is on the rise, according to Hilary French of the Worldwatch Institute.

The problem with exporting waste is that many of the countries that receive waste don't know what is in it, don't know how toxic the materials really are, and don't have facilities to store it or dispose of it properly.

In March 1988, a Norwegian freighter arrived on Africa's west coast to deliver a cargo listed as "raw materials for bricks." A Guinea concrete manufacturer

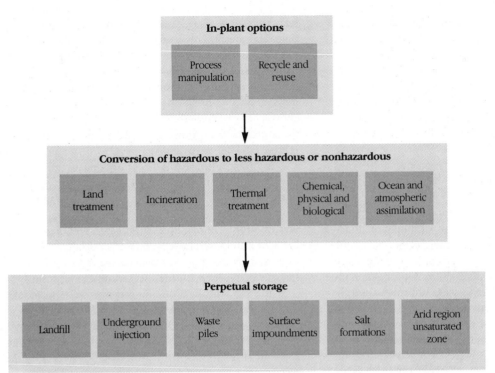

In-plant options

Process manipulation | Recycle and reuse

Conversion of hazardous to less hazardous or nonhazardous

Land treatment | Incineration | Thermal treatment | Chemical, physical and biological | Ocean and atmospheric assimilation

Perpetual storage

Landfill | Underground injection | Waste piles | Surface impoundments | Salt formations | Arid region unsaturated zone

Figure 19.7 *A three-tier hierarchy of options for handling hazardous wastes. The top tier includes in-plant options that reduce the production of hazardous waste in the first place. It contains the most desirable options. The middle tier converts hazardous materials into nonhazardous or less hazardous substances. Perpetual storage, the lowest tier, is the least desirable, but often the cheapest, alternative.*

had purchased the material to build roads and cinder blocks. But, bricks made from it crumbled in the hands of laborers and trees near piles where the material was stored died. Why? The brick material turned out to be ash from a Philadelphia trash incinerator containing heavy metals and dioxins. The ash had originally been destined for Panama, where it was to be used to build a road through a wetlands, but Greenpeace warned the government that the material was hazardous. The government halted the shipment, and the seller found a new buyer in Africa.

In Tecate, Mexico, Mexican officials found that 400,000 liters (100,000 gallons) of hazardous materials had been dumped on the ground. It came from the United States and had been shipped to a bogus, unlicensed Mexican recycling company and dumped near Tecate.

These incidents have stirred debate on what the United States can do to regulate the export of hazardous materials. In 1986, Congress again amended RCRA, establishing procedures to notify importing countries and obtain prior written consent. These regulations, however, may be insufficient because EPA officials think that hundreds of tons of hazardous wastes are still being exported illegally.

Exporting hazardous waste to a nation without its full consent violates principles of international law. Accordingly, 81 African nations have passed laws banning the import of hazardous wastes. In some countries, importing hazardous wastes is punishable by stiff jail terms and multimillion dollar fines. In Nigeria, an importer can be sentenced to death. Through the Organization of Eastern Caribbean States, 22 Latin American countries have also joined forces to stop the dumping of hazardous wastes on their soils.

In 1990, the European Economic Community (EEC) agreed to ban exports of toxic and radioactive waste to 68 former European colonies. Many developing nations who were part of the agreement agreed not to import hazardous wastes from non-EEC members. Although these are important steps forward, many developing nations are still open to exports, representing a huge repository for hazardous wastes from industrial nations.

In the United Nations, talks are underway for the development of global standards to regulate transboundary movement of hazardous waste. This action may be a step in the wrong direction, for it could end up promoting export and not encouraging waste reduction. Furthermore, regulating and enforcing such a program could prove to be a nightmare. Thus, some critics believe that a complete ban on the international movement of waste is the best answer. Such a ban would help protect the environment from inadequate disposal methods and would help nations develop long-term solutions that reduce hazardous waste production.

Dealing with Today's Wastes: Getting to the Roots of the Problem

In 1983, the National Academy of Sciences issued a report outlining options for handling hazardous wastes (Figure 19.7). At the top of the list are **in-plant options,** generally relatively simple and cost-effective changes that can reduce hazardous waste production.

Process Manipulation, Reuse, and Recycling In-plant options include three general actions. The first, **process manipulation**, involves alterations of manufacturing processes that cut waste production. One alteration is **substitution**, the use of a nontoxic or less toxic substitute in manufacturing. Cleo Wrap, the world's largest producer of gift wrap, for example, switched inks and cut its annual production of hazardous waste by 140,000 kilograms (300,000 pounds). Industries can also change the chemical composition of their products, eliminating ones that are harmful or might produce harmful by-products during manufacturing. Nontoxic household cleaners are a good example. Another type of process manipulation involves the monitoring of manufacturing processes to locate leaks that are emitting toxic chemicals into the environment and then fixing them. Exxon Chemical, for example, installed floating lids over some vats that contained volatile organic compounds, greatly reducing losses from evaporation. According to the OTA, U.S. industries could reduce or prevent more than 50% of their hazardous waste generation through process manipulation.

The second and third in-plant options are the **reuse** and **recycling strategies**. In some instances, companies can capture toxic wastes and, with little or no purification, reuse them to manufacture other products or sell them to other companies for reuse. In metal finishing, nickel that's left behind in rinse water can be recovered and reused or sold. For plants that cannot afford on-site technologies, shared facilities or third-party recyclers can provide an economical alternative. Regional waste exchanges—both private and government operated—can assist companies in buying and selling wastes.

Reuse and recycling strategies help reduce waste output by putting perfectly good materials to use. These strategies may save companies millions of dollars per year in material costs. In addition, they eliminate the cost of waste disposal, cut down on potential environmental and health damage, and eliminate costly cleanups.

Conversion to Less Hazardous or Nonhazardous Substances Waste reduction, reuse, and recycling can make a significant dent in waste production. Unfortunately, not all waste can be eliminated, reused, and recycled. Some waste will always be produced. The NAS recommends that, where appropriate, remaining wastes be destroyed or detoxified—that is, converted to less hazardous or nonhazardous materials.

Detoxification can be accomplished for certain types of waste by **land disposal**, applying them to land. When mixed with the top layer of soil, some waste materials can be broken down by chemical reactions, oxidation by sunlight, or by bacteria and fungi in the soil. Some

nondegradable wastes may be absorbed onto soil particles and presumably held there indefinitely. Others may migrate into deeper layers. A word of caution: land treatment is an expensive option, which requires care to avoid polluting ecosystems, poisoning cattle and other animals, and contaminating groundwater. As recent studies show, changes in conditions that bind toxins to soil particles may change, causing a sudden and unanticipated release into the environment.

Another option available for organic wastes is **incineration**. High-temperature furnaces at stationary waste disposal sites, on ships that burn wastes at sea, and on mobile trailers can be used to burn toxic organic wastes—for example, PCBs, pesticides, and dioxin (Figure 19.8). In these facilities, oil and natural gas are used as a fuel. Hazardous substances are injected into the furnace or mixed with the fuel before combustion.

In 1985, the EPA announced that its new mobile incinerator destroyed 99.9999% of the dioxin wastes in soils and liquid wastes. Officials at the EPA are optimistic that the incinerator will be useful in cleaning up many contaminated sites. Critics point out, however, these results come from tests run under ideal conditions. In real life, incinerators usually handle a variety of wastes that may not burn so efficiently.

Hazardous waste incinerators can also provide energy for communities and factories. However, many communities object to hazardous waste incinerators, fearing the release of toxins from spills during transport or leaks at the plants. Incinerators may not always perform adequately and operating personnel may bypass regulations. Low-level releases from smokestacks may also result in long-term exposure to hazardous chemicals.

Low-temperature decomposition (pyrolysis) of some wastes including cyanide and toxic organics, such as pesticides, offers some promise. In this technique, wastes are mixed with air and maintained under high pressure while being heated to 450° C to 600° C (840° F to 1100° F). During this process, organic compounds are broken into smaller biodegradable molecules. Valuable materials can be extracted and recycled. One advantage of this process is that it uses less energy than incineration.

Chemical, physical, and biological agents can also be used to detoxify or neutralize hazardous wastes. For example, lime can neutralize sulfuric acid. Ozone can be used to break up small organic molecules, nitrogen compounds, and cyanides. Toxic wastes can be encapsulated in waterproof plastic and disposed of in landfills. Many bacteria can degrade or detoxify organic wastes and may prove helpful in the future. New strains capable of destroying a wide variety of organic wastes may be developed through genetic engineering.

Perpetual Storage In-plant modifications and conversion technologies that destroy or detoxify wastes can-

Case Study 19.1

Redefining National Security: Waste from the Nuclear Weapons Industry

It is a strange irony. For many years, the federal government spent billions of dollars to build atomic weapons at federal sites nationwide. Part of a nearly $300 billion national security system, the weapons presumably protect the country from foreign aggression. However, the weapons plants themselves rank among some of the most hazardous facilities in the country, potentially poisoning people while protecting the country.

The Department of Energy builds nuclear weapons at 45 sites, covering a land area equal to Delaware and Rhode Island combined. Recent disclosures of ineptitude and careless waste management at the nuclear facilities have created a storm of controversy. In July 1988, the General Accounting Office estimated that cleaning up 17 of the DOE's defense facilities would cost taxpayers about $20 billion. More recent estimates put the cost of cleanup at $100 billion. Most cleanup funds would be designated for the three most heavily contaminated sites: the Rocky Flats plant near Golden, Colorado; the Hanford plant in Richland, Washington; and the Fernald Feed Materials Production Center in Fernald, Ohio. Cleanup could take 20 to 30 years to complete.

For years, the nation's nuclear weapons industry has been supervised by lax government officials and, some say, even ordered by government officials to stifle concerns over health and safety. Private contractors run the weapons facilities under federal supervision.

Studies show that some of the facilities have released large quantities of radioactive material into the air and have dumped tons of potentially carcinogenic waste into nearby water bodies and leaking pits. This has contaminated groundwater supplies. One of the worst problems occurred at the Hanford facility in Washington. Documents secured by a Spokane environmental group under the Freedom of Information Act showed that between 1944 and 1956, huge amounts of radioactive iodine were released into the air by the Hanford facility. Radioactive iodine was deposited on the land around the facility for as far as 25 kilometers (15 miles).

Some local residents have noted an unusually high number of cancer deaths and thyroid problems in that area and wonder if there is a connection. They also note that residents were never informed of the releases.

Tom Bailie was born near Hanford in 1947. His father had surgery for colon cancer at age 39. His mother had skin cancer. His two sisters have had their lower colons removed. Bailie himself is sterile and has only 90% of his lung capacity. He lays the blame on the Hanford facility.

The Centers for Disease Control in Atlanta estimates that 20,000 children in eastern Washington may have been exposed to potentially harmful levels of radioactive iodine in milk produced by cows that graze on contaminated grasslands.

The Feed Materials Production Center in Ohio is a uranium processing plant. Its name and the red and white checkerboard design on a nearby water tower led some local residents to think that the facility produced cattle or pet food. To their dismay, they learned that the company produced uranium fuel rods for nuclear reactors and components for nuclear warheads. The plant also released substantial amounts of radiation.

Richard Shank, the director of Ohio's EPA, estimates that the Fernald plant has released nearly 140,000 kilograms (300,000 pounds) of uranium into the air since it began production. The plant has also deliberately discharged nearly 77,000 kilograms (170,000 pounds) of waste into a nearby river over the past 37 years and dumped 6 million kilograms (12.7 million pounds) of waste into open pits.

The Fernald plant began operation in 1953. The contractor, National Lead of Ohio, was told by the Atomic Energy Commission to dump radioactive waste into pits in the ground, a standard practice at the time. But, rainwater caused the pits to overflow; National Lead suggested ways to fix the problem, but their efforts were reportedly thwarted by the Atomic Energy Commission.

In 1958, the company warned the Atomic Energy Commission that concrete storage tanks that held radioactive waste were also leaking. The Atomic Energy Commission reportedly suggested that the contractor not fill the tanks above the leaks. The flawed tanks are still in use today.

Problems also abound at the DOE's Savannah River operation. Studies show that many shallow aquifers in the region are contaminated with radioactive wastes. A deeper aquifer is contaminated with chemical toxins.

The facility's manager, DuPont, has been accused of numerous wrongdoings over the past 30 years. On May 10, 1965, for instance, operators ignored an alarm indicating that radioactive water was spilling out of a reactor onto the floor for 15 minutes. About 8000 liters (2000 gallons) of fluid leaked out of the reactor.

In 1982, a technician at the Savannah River plant left a valve open for 12 hours. Radioactive liquid flooded a plutonium processing room, filling it .6 meter (two feet) deep. Radioactive liquid waste has also been stored in tanks with corroded bottoms, which leak into groundwater.

A third trouble spot is the Rocky Flats facility, which made nuclear triggers for hydrogen bombs.
(continued)

Located upwind from Denver, Colorado, the plant has a long record of problems caused in large part by extremely bad management. For many years, barrels containing oil contaminated with plutonium were buried on site, where they leaked into the soil. Cleanup could cost as much as $750 million. The Rocky Flats plant also releases small but significant amounts of plutonium into the air every day. Several fires at the plant have released huge amounts of plutonium as well. Damage from a fire in May 1969 cost $21 million. Toxic wastes and radioactivity have been detected in ponds on the property, worrying health officials about nearby water supplies. Studies of nearby residents show a higher-than-average rate of leukemia and other cancers.

The DOE assumed responsibility of the nuclear weapons network in 1977 from the now-defunct Atomic Energy Commission. At least on the surface, the DOE seems intent on reforming the nuclear weapons industry. But, in the meantime, it has a huge task to perform: cleaning up thousands of sites contaminated not only by radioactive waste but by hazardous wastes as well. By various estimates, defense facilities, including military bases, collectively house about 17,000 hazardous waste hot spots.

not rid us of all of our waste. By various estimates, 25% to 40% of the waste stream will remain even after the best efforts to reduce, reuse, recycle, and destroy it.

Residual hazardous waste could be stored by one of a half-dozen options (Figure 19.7). For example, residual waste could be dumped in **secured landfills**, excavated pits lined by impermeable synthetic liners and thick, impermeable layers of clay. To lower the risk of leakage, landfills should be placed in arid regions—neither over aquifers nor near major water supplies. Special drains must be installed to catch any liquids that leak out of the site. Groundwater and air should be monitored regularly to detect leaks.

Growing public opposition to this strategy makes it more difficult for companies to find dump sites. Some observers have labeled this the **NIMBY syndrome**: get rid of the stuff but *not in my backyard*. Ironically, it seems that most people want the products available in an industrial economy that inadvertently generate waste, but few of us want the wastes dumped or even burned nearby.

Even though the EPA has issued tough regulations for hazardous waste landfills, critics argue that landfills are only a temporary solution. No matter how well constructed, they will eventually leak. In an attempt to reduce problems for future generations, the EPA has drawn up a list of chemicals that cannot be disposed of in landfills.

Because landfills are one of the cheapest waste disposal practices in use today, they are often favored by industry. But, the savings they offer today are very likely to be charged to future generations.

Other methods of perpetual storage include: (1) use of surface impoundments and specially built warehouses that hold wastes in ideal conditions and prevent any material from leaking into the environment, (2) deposition in geologically stable salt formations, and (3) deposition deep in the ground in arid regions where groundwater is absent.

Barriers to Waste Reduction   Obviously, the most effective and sustainable solution to the waste problem is to find ways of reducing production in the first place. However, when the EPA began its hazardous waste control program in the United States, it spawned a large and lucrative treatment and disposal industry. Some observers believe that this politically powerful industry may be thwarting waste minimization policies. In fact, the EPA today spends less than 5% of its hazardous waste budget on preventing pollution.

Decreasing hazardous waste production would harm the hazardous waste management industry and discourage investment in the industry. "Hazardous waste regulators tied to the needs of the industry they foster have a powerful incentive to encourage waste production," says the National Toxics Campaign's Michael Picker.

Disposing of High-Level Radioactive Wastes

Although high-level radioactive wastes are some of the most hazardous wastes, they have long been ignored by many countries. High-level radioactive waste is generated by commercial nuclear power plants and weapons production facilities. Lower-level wastes come from research laboratories and hospitals. Many radioactive wastes have a long lifetime. Others can concentrate in animal tissues. Virtually all pose a serious threat to human health (Chapter 12).

Figure 19.8 *A mobile hazardous waste incinerator owned and operated by the EPA avoids the problem of transporting exceptionally dangerous materials.*

High-level radioactive waste is not a problem that will disappear, even though the U.S. nuclear power industry is on the decline (Chapter 12). By the year 2000, experts predict, 60,000 metric tons of radioactive wastes will have been generated by commercial plants. The continued operation of existing power plants and nuclear weapons facilities—and possible expansions—necessitate long-term, low-risk storage of nuclear wastes.

Recognizing the need to do something with nuclear wastes rather than continuing to stockpile them at power plants or weapons facilities, Congress passed the **Nuclear Waste Policy Act** in 1982. This law established a timetable for the U.S. Department of Energy (DOE) to select a deep underground disposal site for high-level radioactive wastes to be in operation by 1998. The DOE initially identified potential sites in Nevada, Washington, and Texas, but the latter two were dropped by Congress in 1986 and 1987, largely as a result of public protest and budget concerns. (Studying a single site could cost over $1 billion!)

The DOE focused its attention on the Yucca Mountain site in Nevada. Located on federal land about 160 kilometers (100 miles) northwest of Las Vegas, Nevada, Yucca Mountain could someday be home to a huge underground storage site excavated in the volcanic rock, costing $10 billion.

The DOE had hoped to begin construction on the site by 1998, after adequate testing, and open the site by 2003. In order for federal decision makers first to approve the site, though, they must be relatively sure that earthquakes, volcanic eruptions, and climate change will not threaten the stability of the repository.

If the Nevada site passes muster and the state doesn't oppose its construction, workers will dig a repository 600 meters (2000 feet) below the surface. High-level nuclear wastes from power plants and defense facilities all over the United States, and possibly from other countries, will be shipped in casks to their permanent home. However, if the Nevada site proves unsatisfactory, the process would have to begin again, further delaying construction.

In 1990, the DOE announced that it was postponing the opening of the Nevada site until 2010. Some observers believe that it may never open, because scientists discovered a fairly active volcano only 11 kilometers (7 miles) away. DOE scientists also identified numerous active earthquake fault lines at the site. The presence of faults and underground fractured rock has some scientists concerned about the intrusion of groundwater into the waste repository.

Given the uncertainties plaguing the Yucca Mountain site, the DOE has asked Congress to authorize it to build an above-ground intermediary storage facility for high-level radioactive wastes. Some people oppose this idea, fearing that it would thwart efforts to find a permanent facility and that intermediate storage would become permanent.

The DOE also recently proposed that the entire matter of high-level nuclear waste disposal be turned over to private industry. This could reduce costs and government involvement. However, critics point out that it would also reduce public input and relegate an important function to an industry that might cut corners to save money.

One space-age waste disposal proposal is to shoot high-level wastes into the sun where it will be destroyed. A critical analysis of this idea shows that the cost, energy requirements, and material requirements would be astronomical. Disposal of radioactive wastes from a single 1000-megawatt nuclear plant would cost over $1 million per year. Furthermore, radioactive capsules shot into space might someday return to Earth or, in a replay of the Challenger tragedy, never make it out of the atmosphere.

Others have suggested dumping radioactive wastes on uninhabited lands in the Arctic and Antarctica. Too little is known about the effects of this disposal technique for experts to assess its costs and benefits.

Radioactive waste can be bombarded with neutrons in special reactors to transmute, or convert, some of it into less harmful substances. However, existing reactors do a poor job of altering cesium-137 and strontium-90, two of the more dangerous by-products of nuclear fission.

Seabed disposal was once used by the United States and European countries but is now forbidden. Still, some scientists suggest that the seabed may provide a site for radioactive wastes; the effects are difficult to predict.

The problems of disposal suggest to some that nuclear power should be phased out. Cleaner, less costly measures to produce energy should be developed. As noted earlier, cost-effective and low-risk methods are needed to dispose of the vast amounts of waste that have already been generated.

Disposing of Low- and Medium-Level Radioactive Wastes

Low-level waste from hospitals and research laboratories is packaged and shipped to three disposal sites—in Nevada, Washington State, and South Carolina—where it is buried in the ground. In 1992, Congress defeated efforts by the Nuclear Regulatory Commission, the government body charged with regulating the nuclear industry and waste disposal, to reclassify low-level wastes so they could be discarded in municipal landfills with municipal garbage.

Medium-level waste from nuclear power plants and weapons facilities is another matter altogether. In the 1980s, the DOE began construction on a medium-level radioactive waste depository in New Mexico, called WIPP (Waste Isolation Pilot Plant). WIPP is an experimental project, which, if successful, could be expanded to full operation. In August 1988, however, two months before the site was to begin receiving wastes, the DOE announced that it was postponing the opening because researchers were concerned about the safety of the $700 million facility that had been carved out of salt deposits 630 meters (2100 feet) below the ground near Carlsbad, New Mexico. Early in 1988, groundwater began to leak

from the facility where it might corrode the steel canisters that stored radioactive waste, allowing radiation to leak. Later in 1993, the DOE expects to start receiving wastes for a 5- to 7-year test period. If the tests prove successful, the facility will eventually hold 6.2 million cubic feet of radioactive waste.

Unlike many other hazardous wastes, most nuclear wastes cannot be reused or recycled. Process modification can help reduce waste output, but, for the most part, the problem has to be addressed at the end of the pipe. The end of the Cold War and decline in the nuclear arsenal of the United States and former Soviet Union could greatly reduce production at nuclear weapons facilities. Nuclear waste from power plants could also decline as plants are phased out in the United States and elsewhere (Chapter 11). But, the huge volumes of waste that already exist must be dealt with somehow.

Some Obstacles to Sustainable Hazardous Waste Management

Hazardous waste production has dropped steadily in the United States since 1980. This encouraging trend has resulted partly because of a downturn in the chemical manufacturing industry and also because of efforts to reduce hazardous waste production by process modification, reuse, and recycling. The Chemical Manufacturers Association's (CMA) sixth annual hazardous waste survey of 221 plants studied from 1981 to 1986 showed that manufacturers had reduced solid hazardous waste production by 56%. The plants had also reduced hazardous waste released into water by nearly 10%. (More recent statistics are not available.) The survey suggests a trend away from the disposal of wastes in landfills in favor of more permanently effective techniques, such as process redesign and recycling. Critical thinking, however, suggests a closer look. Some think that industry surveys may overstate progress in reducing hazardous wastes. As noted above, a downturn in chemical production during this period may be partly responsible for the apparent reduction in solid hazardous wastes.

A study by the National Wildlife Federation (NWF) suggests that most of the reductions in toxic emissions claimed by the nation's top industrial polluters are the results of creative accounting, not improved pollution control. NWF studied 29 facilities and concluded that 39% of the reduction in emissions from 1987 to 1988 was almost entirely the result of changes in reporting requirements or in the methods of analysis. There were, however, some genuine success stories. Eastman Kodak, for example, made alterations in its plant in Kingsport, Tennessee, that resulted in a 0.5-million-kilogram (1.2-million-pound) reduction in toxic emissions.

Another problem is that much of the hazardous waste released by various industrial processes in the United States (85%) is highly diluted in water. This dilute waste stream is typically pumped into deep wells, from which hazardous substances may leak into groundwater. Besides contaminating drinking water, industrial wastes pumped into deep wells can increase the incidence of earthquakes. Two Ohio geologists believe that industrial waste injected into the ground by a chemical company in Ohio may have created underground pressure, triggering the earthquake that shook northeastern Ohio and a nearby nuclear power plant in 1986. It measured 4.9 on the Richter scale. The geologists contend that pumping waste into the sandstone increased fluid pressure in the pores of a bedrock located directly above the crystalline bedrock, causing the earthquake along a fault in the latter.

The sheer volume of polluted water makes it difficult to regulate. Because removing hazardous substances from the water is extremely costly, most plant owners are unwilling to make the investment in wastewater treatment. To cut down on deep-well discharges, inexpensive techniques must be developed to separate the hazardous wastes.

The push to reduce hazardous waste disposal on land and in water has stirred considerable interest in incinerators. On the surface, incineration seems like an attractive option. Incinerators are equipped with pollution control devices and can remove many of the chemical pollutants. They can also be used to generate energy and reduce landfilling of hazardous materials. In rural areas, they provide badly needed jobs. Companies that want to locate them in rural areas often pay huge sums of cash to support schools, roads, water supply systems, and other needed projects as incentives.

One person's solution, however, often becomes another person's problem. For example, in Ritzville, Washington, a farming town in the eastern part of the state, a hazardous waste incinerator was proposed by an incinerator company in the late 1980s. It promised jobs and more tax revenues for the economically depressed region, but not all residents thought it was a good idea. One problem was cross-media contamination. The incinerator would have required a landfill to dispose of the highly toxic residue it produced. Some residents were wary of leaks that might have polluted their groundwater, which is used for irrigation and drinking. Residents were also concerned about toxic emissions. Although the plant's design incorporated state-of-the-art pollution control equipment, small amounts of toxins would have been emitted and could have accumulated in the soil and crops downwind. Finally, residents were concerned about hazardous waste shipments, which would increase dramatically if the incinerator had been approved. Fortunately for opponents, the company lost interest in the project.

The story in Ritzville is being repeated all over the United States. Because urban residents don't want incinerators, companies are scouring rural regions, where there is less political opposition and where the promises of economic benefits may be better received.

Proponents of incineration think that the rural siting strategy may be shrewd business, but it ultimately diverts attention from finding permanent solutions, like process modification, reduction, reuse, and recycling, which are environmentally more acceptable and, in the long run, are more sustainable strategies for dealing with hazardous wastes.

Individual Actions Count

Individuals can participate in many ways. For example, each of us can contribute by recycling oil, not dumping it in sewers or in vacant lots. We can properly dispose of paints, paint thinners, and other potentially toxic wastes. Many counties sponsor periodic toxic roundups during which they will collect toxic household products for disposal free of charge. One of the best strategies is to avoid toxic chemicals, such as pesticides and cleaning agents, in the first place. You can cut back on potentially hazardous materials by purchasing environmentally safe cleaning products and insecticides. Reducing your consumption of nonessential goods, whose production invariably creates toxic wastes that poisons land and water, can also help.

You can also learn about waste sites in your area and become active in grass-roots organizations working on reduction, recycling, and safer disposal methods. Together, millions of Americans using resources wisely can make significant inroads into the hazardous waste problem.

19.3 Solid Wastes: Understanding the Problem

Each year, human society produces mountains of **municipal solid wastes**: paper, metals, leftover food, and other items discarded by businesses, hospitals, airports, schools, stores, and homes (Figure 19.9). In 1990, 175 million metric tons of municipal solid waste were generated in the United States—enough garbage to fill the Superdome in New Orleans nearly 2.5 times every day. Each year, Americans produce approximately 1140 kilograms (2500 pounds) per person—about 1.8 kilograms (4 pounds) per person per day. A city of 1 million people could fill the Superdome once every year.

Up sharply from 1980, municipal solid wastes comprise only about 4% of the total solid waste discarded in the United States each year. However, municipal waste is growing at a rate of 2% to 4% per year. At a 2% growth rate, the output of garbage will double in 35 years. At 4%, the output will double in 17.5 years. The continued growth and the sheer volume of waste create major problems in many cities, where land for

Models of Global Sustainability 19.1

One Man Changes a Nation

What the Earth needs most is a variety of useful models, says Karl-Henrik Robert, a leading cancer researcher in Sweden. By that, he means model homes, buildings, companies, communities, and, indeed, countries that help demonstrate how to live and work sustainably on the planet. In a similar vein, Worldwatch Institute's Michael Renner notes that on an international scale one way of inspiring global cooperation is for countries to take action on their own—that is, to become models for the rest of the world.

One stellar example is Sweden. Already divesting itself of nuclear power and seeking to make up the losses by using energy more efficiently, Sweden is a world leader in sustainability. In an effort to help transform his country to an even better role model for the world, Robert recently embarked on an ambitious project to change the thinking of the entire nation.

Convinced that the root cause of the environmental crisis is modern society's dependence on linear systems and linear ways of thinking, Robert decided to build a national consensus on the need for recycling of all wastes—not only municipal solid wastes but hazardous industrial wastes as well. To do this, he met with influential scientists, musicians, artists, politicians, educators, and even the king of Sweden to share his ideas.

The success of his approach lay in seeking common ground where all kinds of people could agree—that is, working on root causes and refusing to get mired in the scientific debate over details of the symptoms, which is, unfortunately, where much of the discussions worldwide remain. After reaching agreement with a wide number of influential people, Robert and his new-found colleagues began issuing "consensus reports" on the condition of the environment and on root-level strategies needed to reverse the deterioration. Through booklets and audio cassettes sent to all of Sweden's schools and households, seminars for members of Parliament, television programs, and journal articles, they spread the word deep and wide in an unprecedented effort to reshape the thinking of a people.

Other nations may follow suit. And fortunately, says Robert, "it appears . . . that there is now a growing core of thoughtful decision-makers who understand that the time to act is now. Whether we want to help others or ourselves, to conduct our affairs ethically or compete in tomorrow's markets, the possibility of success rides on the shoulders of well-informed business and political leaders who are supported in their efforts to base the foundations of society on natural laws."

Robert also points out that despite all the quibbling over peripheral issues, enough scientific consensus already exists to get on with the necessary work. In most cases, more research is not needed. Furthermore, since environmentally sound technology is already available, the pace of transition to cyclic processes (industrial ecosystems) like those discussed in Section 19.3 is limited only by our willingness to act. The longer we delay, the more painful the sacrifices we will have to make down the road.

The Netherlands and Norway are also becoming models of thoughtful, environmentally responsible living. For example, both have unilaterally adopted policies to either freeze or cut carbon dioxide emissions, helping to curb global warming. Although these policies will probably have very little effect on climate, given the massive production of pollutants by the United States, Europe, and the Commonwealth of Independent States, they may motivate other countries to emulate them.

Model countries provide inspiration; they also provide a source of useful experience that could help steepen the learning curve in countries that choose to follow suit. In addition, model countries position themselves to become innovators and exporters of technologies and technological know-how for which the rest of the world may soon be clamoring. Those countries that insist on quibbling over global warming, say some critics, may therefore be losing an opportunity to become leaders in resource-efficient technologies and other measures vital to building a sustainable world.

Source: Adapted from Chiras, D. D. (1992). *Lessons from Nature: Learning to Live Sustainably on the Earth.* Washington, D.C.: Island Press.

disposal is growing more scarce and more costly by the day.

Garbage disposal is also of concern to those interested in building a sustainable future because it squanders Earth's resources. The more that is thrown away, the more minerals must be mined. The more paper we throw away, the more trees must be cut. The more plastic we discard, the more oil must be drilled. Each of these activities produces incredible waste itself and equally impressive amounts of environmental damage.

Science fiction writer Arthur C. Clarke noted that "solid wastes are only raw materials we're too stupid to

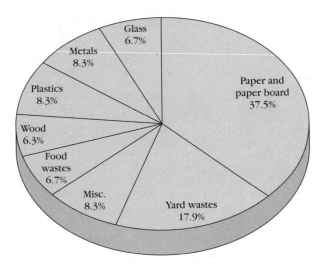

Figure 19.9 *Composition of U.S. municipal solid waste by weight after recovery of some recyclable and compostable materials (1990).*

use." In the United States, only about 25 million metric tons of American trash is currently recycled and composted, or about 14% of the total. Another 16% is burned in incinerators. The remaining 70%—millions of metric tons of perfectly recyclable paper, glass, metals, tires, and plastics—is dumped into the ground and buried.

Landfill disposal not only wastes valuable resources, it costs communities millions of dollars each year. Landfilling the disposable diapers used in the United States, for example, costs the nation an estimated $300 million per year. For most local governments, the cost of trash disposal is usually exceeded only by the cost of education and highway construction and maintenance.

Like so many other problems, municipal solid waste is the product of many interacting factors: (1) large populations, (2) high per capita consumption, (3) low product durability, (4) a rash of disposable products, (5) low reuse and recycling rates, (6) a lack of personal and governmental commitment to reduce waste, (7) widely dispersed populations where producers of recyclable and reusable items are separated from those willing to purchase these materials, (8) cheap energy and abundant land for disposal, and (9) powerful business interests that make large profits from waste disposal and disposable products.

19.4 Solving a Growing Problem Sustainably

Actions to reduce the output of solid waste generally fall into three broad categories (Figure 19.10). The traditional response to solid waste is known as the **output**

approach. It consists of ways to deal with trash flowing out of cities and towns. Most often, this means incinerating trash or dumping it in landfills. A more sustainable strategy is known as the **input approach**. It consists of activities that reduce the amount of materials entering the production-consumption cycle, for example, efforts to reduce consumption and waste by increasing product durability. The third approach, also essential to building a sustainable society, is the **throughput approach**. It consists of ways to direct materials back into the production-consumption system, creating a closed-loop system akin to the cyclic systems found in nature. Reuse and recycling fall under this category.

The Traditional Strategy: Output Approach

The most widely used strategy throughout the world is the output approach, landfilling, ocean dumping, and incineration. In Ireland, for instance, 100% of all garbage is landfilled. Australia and Canada trail only slightly behind Ireland with rates of 98% and 93%, respectively. Great Britain landfills 90% and France 54%. Landfilling and incineration are typical end-of-pipe solutions. Like many others discussed in this text, they treat the symptoms of the problem while ignoring the root causes.

Dumps and Landfills The garbage dump. By any other name it would still smell as bad. Until the 1960s, garbage dumps were prevalent features of the American landscape. But public objection to wafting odors, rat- and insect-infested midden heaps, and dark plumes of smoke that billowed out of burning dumps forced cities to look for alternative ways to deal with their growing trash problem. The federal government contributed to the demise of the dump by passing RCRA. It required all open dumps to be closed or upgraded by 1983.

The open dump has been replaced by a second cousin from a better part of town, the sanitary landfill. A **sanitary landfill** is a natural or man-made depression into which solid wastes are dumped, compressed, and daily covered with a layer of dirt. Because solid wastes are no longer burned, as they were in many open dumps, air pollution is greatly reduced. Because trash is covered each day with a layer of dirt, odors, flies, insects, rodents, and potential health problems are eliminated or sharply reduced.

Despite their immediate benefits, landfills have some notable problems. First, and most important, landfills require land. The trash from 10,000 people in a year will cover 1 hectare 1.2 meters deep (1 acre 10 feet deep). Around many cities usable land is in short supply or is expensive. Second, landfills, like dumps, require a lot of energy for excavation, filling, and hauling trash. Third, they can pollute groundwater. Toxic household wastes (paint thinner, pesticides, and other poisons) and feces

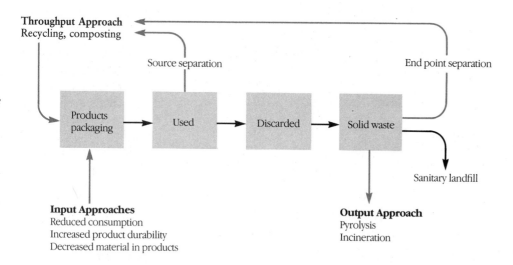

Figure 19.10 *Three strategies for reducing solid waste. A combination of all three must be applied to alleviate the solid waste problem, but efforts should concentrate on the input and throughput solutions.*

(from disposable diapers, kitty litter, and backyard cleanup of Rover's messes) are discarded in municipal landfills where they can leak into groundwater. Fourth, they produce methane gas from the decomposition of organic materials. Methane can seep through the ground into buildings built over or near reclaimed landfills. Methane is explosive at relatively low concentrations. Fifth, they sink or subside as the organic trash decays, requiring additional regrading and filling. Buildings constructed on top of reclaimed landfills may suffer serious structural damage. Sixth, they have low social acceptability. Quite understandably, most people don't want the noise, traffic, and blowing debris that accompanies even the best-managed landfills.

Although landfilling is ultimately an unsustainable strategy, there are many ways to make it more environmentally acceptable. Energy requirements, for example, can be cut by new methods of waste collection. Packer trucks now reduce waste volume by 60%, meaning that fewer trucks are needed to haul garbage to landfills. Vacuum collection systems can also be used to save energy. In these systems, solid waste is dumped into pipes that carry it to a central collection point.

Vacuum collection systems are feasible in urban areas where population density is high. Apartment complexes are extraordinarily well suited to this method of trash disposal. One such system is in operation in Sundbyberg, Sweden. Garbage is whisked away from wall chutes to a central collection facility, where the glass and metals are removed by an automated process. The burnables are incinerated, providing heat for the 1100 apartments using the system. A similar system handles 45 metric tons of waste per day at Disney World, Florida. Today, over 400 such systems are in operation in Europe in hospitals, apartment buildings, and housing tracts.

Water pollution problems can be reduced by locating landfills away from streams, lakes, and aquifers. Test wells around the site can be used to monitor the move-

ment of pollutants, if any, away from the site. Special drainage systems and careful landscaping can decrease the flow of water over the surface of a landfill, thus reducing the amount of water penetrating it. Impermeable clay caps and liners can reduce water infiltration and the escape of pollutants. In addition, pollutants leaking from the site can be collected by specially built drainage systems and then detoxified. The toxic seepage can then be shipped to hazardous waste facilities.

Methane gas produced in landfills can be drawn off by special pipes and sold as fuel, supplementing natural gas. Subsidence damage to buildings built on reclaimed sites can be reduced by removing organic wastes before disposal and by allowing organic decay to proceed for a number of years before construction.

Ocean Dumping   When most of us think about the ocean, we conjure up images of white beaches, gentle waves, and the smell of coconut oil. For years, however, many city officials have held a different view of the ocean. They looked at the world's vast oceans as a huge garbage dump for a variety of wastes, including municipal garbage and human sewage. Until a few years ago, U.S. municipalities relied on 126 offshore dump sites to dispose of their wastes. Although most of the solid waste dumped at sea was mud and sediment from dredging of harbors, estuaries, and rivers, considerable amounts of industrial hazardous wastes, municipal wastes, and sludge from sewage treatment plants were also dumped at sea. Even today, city planners are trying to find ways to justify dumping some of their trash offshore. Some propose building offshore "islands of trash" from discarded automobiles, demolished buildings, and other solid wastes to support hotels and airports. Others talk of (and build) artificial reefs of discarded stone, toilets, cement slabs, and automobiles.

Ocean dumping of many wastes off U.S. waters began a steady decline after passage of the **Marine Pro-**

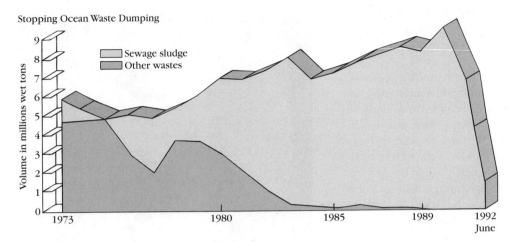

Stopping Ocean Waste Dumping

Figure 19.11 *Ocean dumping of sewage sludge only recently decreased. Other oceanic waste dumping was largely brought to a halt in the early 1980s.*

tection, Research, and Sanctuaries Act in 1972 (Figure 19.11). The long-term goal of this act was to phase out all ocean dumping, especially sewage and solid wastes. In 1977, the EPA issued regulations calling for an end by 1981 to all dumping of industrial wastes and sludge in the Atlantic Ocean, where 90% of the waste disposal was occurring. While hazardous waste and solid waste dumping declined sharply, sewage sludge disposal remained a significant problem because New York City and others continued to dump sludge in the ocean for many years.

Appalled by the illegal dumping of medical wastes and other garbage, Congress passed two laws in 1988 to help clean up the oceans. The first was the **Medical Waste Tracking Act** (1988) (Chapter 17). The second was the **Ocean Dumping Ban Act** (1988), which called for an end to sewage sludge disposal at sea by December 21, 1991. The EPA negotiated agreements with local jurisdictions that phased out ocean dumping of sewage sludge by June 1992 (Figure 19.11).

Incineration   Incineration is another output control currently widely used in some countries, such as Denmark, which burns over 60% of its solid wastes. The Netherlands and Sweden each burn about one-third of theirs. The United States, on the other hand, burns about 16% of its solid waste, up considerably in the past few years.

Burning trash can reduce the waste volume by two-thirds and save landfill space. It can also be used to produce heat and electricity, thus turning waste into energy. For that reason, incinerators are often called **waste-to-energy** (WTE) plants. One ton of garbage is equivalent to about one barrel of oil. But, even with the energy gain, WTE plants are often more costly to build and operate than landfills. As a result, they can increase the cost of waste disposal. WTE plants are also much more expensive than recycling programs. The Seattle curbside recycling program, for example, costs one-tenth

as much as a proposed incinerator. Other recycling projects cost about one-third as much as incinerators.

Each incinerator must be individually designed to accommodate the local mixture of burnable and non-burnable refuse. Operating these incinerators is made more difficult because the mixture varies from season to season. In the spring and fall, for example, yard and garden waste increases dramatically. Wet leaves, however, do not burn well. To reduce this problem, municipalities may require homeowners to separate combustible material from wet organic matter. Incinerators also emit toxic pollutants, especially when plastics are burned.

Another major problem with garbage incinerators is that the ash they produce may be hazardous to human health. Test data compiled by the Environmental Defense Fund (EDF) show that incinerator ash contains toxic metals, such as lead and cadmium, and dioxin, in concentrations considered hazardous. The EDF and others, therefore, argue that ash from garbage incinerators should be reclassified as a hazardous substance and disposed of in hazardous waste facilities, which is much more costly than dumping in ordinary landfills.

Should the EPA reclassify incinerator ash, many municipalities currently operating waste-to-energy plants will see the cost of running these facilities increase substantially. Many may come to regret their decision to adopt WTEs. Many may find that ash disposed of in ordinary landfills over the years may begin to leak into groundwater, polluting public and private drinking water supplies. Cleaning up these sites could cost many millions of dollars.

Many municipalities are opting for refuse-derived fuel plants in which nonburnable materials are first removed before combustion, rather than mass burn facilities, where the entire waste stream is burned without separation. Separation of noncombustible wastes increases the combustion efficiency, is much cleaner, and is more efficient than mass burn, but it still produces hazardous materials. Refuse-derived fuel plants are a step in the right direction but are not as cost-effective as programs that rely on waste reduction, composting, and recycling.

Incinerators are also riddled with problems of social acceptability. Residents of Lowell, Massachusetts, defeated an incinerator that its city council was planning to build because they found, among other things, that Lowell only produced 225 metric tons of waste per day while the plant would require 1350 metric tons to operate. That meant that the city council would have had to enter into agreements with neighboring towns to accept their trash to meet the plant's needs. Lowell would have become a "waste magnet." Citizens in Spokane, Washington, were not so lucky. Their city council approved a massive incinerator that will require a considerable amount of imported trash to keep it running.

Incinerators may seem like a good way to solve the growing trash problem, but critical thinking suggests the need for a long-term view of the problem at hand. The most important question today is not just how to reduce landfilling. It is how to cut waste; save valuable resources; protect the air, water, and land; and ensure vital wildlife habitat. In short, how is it possible to create a sustainable system of waste management? Clearly, incinerators reduce trash, but they don't contribute significantly to the goals of sustainability.

Sustainable Options: The Input Approach

Sustainable solutions require efforts to approach the root causes of environmental problems. In solid waste, as in hazardous waste, some of the most important strategies are those that seek to reduce the amount of materials entering the production-consumption system. Three initial ways of minimizing waste are: (1) increasing product durability, (2) reducing the amount of material in goods and their packaging, and (3) reducing consumption.

Increasing Product Durability   More durable toys, garden tools, cars, and clothing require less frequent replacement and thus decrease resource use. The trouble is, as John Ruskin noted, "There is hardly anything in the world that some man cannot make a little worse and sell a little cheaper." In the long run, however, cheaply made goods end up costing consumers more than slightly more expensive, well-made goods. The rapid turnover may be profitable to businesses but unsustainable from an ecological standpoint. Planned obsolescence of products destroys the air, water, and land.

Reducing the Amount of Material in Products and Packaging   In the United States, packaging (bottles) accounts for 90% of the glass consumed in the United States, 50% of the paper, 11% of the aluminum, and

Figure 19.12 *Aseptic containers like this one are convenient and energy efficient but unrecyclable.*

8% of the steel, according to Concern, Inc., a citizen's group based in Washington, D.C. All told, about 30% of the municipal solid waste (by weight) is discarded packaging.

Of course, packaging is necessary, but much of it is superfluous and wasteful. The Campbell Soup Company realized this and redesigned its soup cans; today, cans use 30% less material than they did in the 1970s. Some beverage companies now package drinks in **aseptic containers**, boxes constructed of several thin layers of polyethylene, foil, and paper (Figure 19.12). The containers, which hold milk, juices, and wine, keep the contents fresh for several months without refrigeration. Beverages in aseptic containers cost the consumer about 10% less than those in cans. The main reason for this discrepancy is that canned drinks must be pasteurized for 45 minutes, whereas the contents of aseptic packages are sterilized out of the package for only 1 minute. This greatly reduces energy demand and preserves flavor. Furthermore, beverages in aseptic containers do not require refrigeration during transportation and storage, which also lowers the energy demand. Lighter-weight containers also help cut down on transportation costs. The only drawback—and it is a big one—is that aseptic containers are completely unrecyclable.

Some manufacturers are experimenting with biodegradable packages. Most notable are plastics and food containers that dissolve in water when cooked, thus reducing solid waste and also adding nutrients to the food. Biodegradable plastic shopping bags began to appear in grocery stores throughout the United States in 1989. Most of these bags are a blend of corn starch, which is biodegradable, and polyethylene, a plastic that is not. Bacteria in landfills can break down the starch, leaving behind tiny particles of plastic. Initial research shows that the breakdown occurs very slowly in the oxygen-free environment of the landfill. Thus, bags that may "decompose" in a few weeks in lab tests may take more than 100 years to decay in a landfill. Environmentalists are concerned that the plastic particles remaining after biodegradation may end up polluting groundwater. A far better approach might be reusable shopping bags—sturdy nylon or canvas bags that individuals can use over and over again.

Virtually all products can be redesigned to reduce waste. Many large newspapers, for example, have been redesigned to a smaller, more economical format that has cut the use of newsprint by 5%. Smaller cars and trucks have emerged. Smaller computers and calculators have also helped save valuable materials. Smaller houses could help as well.

Reducing Consumption   G. B. Shaw wrote, "Our necessities are few, but our wants are endless." Ceaseless efforts to satisfy endless wants are a big part of the solid waste dilemma facing the United States, Canada, and other countries. By cutting back on consumption, individuals can help reduce solid wastes and many other environmental problems.

Reductions in consumption require shifts in our attitude. The notion that "new is better" leads many consumers to purchase new goods when old ones still work. The fashion industry thrives on its ability to convince the public that new fashions are "in"—and that anyone wearing the old is "out." Advertisers capitalize on this strategy as well, and many consumers fall into the trap. (More on this subject in Chapter 21.)

Another way to reduce consumption is through outright bans. Some cities, for example, have taken steps to reduce waste by banning objectionable materials—notably plastics. In 1988, Suffolk County, New York, passed an ordinance banning plastic grocery bags and many plastic food containers in the county that went into effect in 1989. Fast-food restaurants had to switch to biodegradable products. Minneapolis-St. Paul, Minnesota; Berkeley and Palo Alto, California; Newark, New Jersey; and other cities have passed similar ordinances, prompting the plastics industry to take measures to develop recycling facilities.

Individuals can make a personal effort to reduce consumption with little noticeable change in life-style. The possibilities are limitless. (See the *Environmental Action Guide* for suggestions.)

Table 19.1   Reuse and Recycling of Solid Wastes

Material	Reuse and Recycling
Paper	Repulped and made into cardboard, paper, and a number of paper products. Incinerated to generate heat. Shredded and used as mulch or insulation.
Organic matter	Composted and added to gardens and farms to enrich the soil. Incinerated to generate heat.
Clothing and textiles	Shredded and reused for new fiber products, or burned to generate energy. Donated to charities or sold at garage sales.
Glass	Returned and refilled. Crushed and used to make new glass. Crushed and mixed with asphalt. Crushed and added to bricks and cinderblocks.
Metals	Remelted and used to manufacture new metal for containers, building, and other uses.

Source: Modified from Nebel, B. J. (1981). *Environmental Science.* Englewood Cliffs, NJ: Prentice-Hall, p. 297.

Another Sustainable Strategy: The Throughput Approach

Throughput approaches—reuse and recycling—remove useful materials before they enter the waste stream and channel them to manufacturers (Table 19.1).

The Reuse Option   Reuse is the return of operable or repairable goods into the market system for someone else to use. In most cities, organizations such as Goodwill and the Disabled American Veterans pick up usable discards or slightly damaged products, including clothes, furniture, books, and appliances. Many of them provide drop-off stations as well. These products are cleaned, repaired, then resold to the needy and the frugal. Secondhand stores also provide consumers with options for buying children's clothing and toys, furniture, appliances, and a host of other products. So before you throw out a still-usable product, why not give one of these organizations a call? You will be helping increase the product's useful life span and helping to reduce resource demand.

Packaging materials, such as cardboard boxes, bottles, and grocery bags, can also be reused, saving both energy and materials. Shopping bags can be reused by

Table 19.2   Energy Consumption Per Use for 12-Ounce Beverage Containers

Container	Energy Use (Btus)
Aluminum can, used once	7050
Steel can, used once	5950
Recycled steel can	3880
Glass beer bottle, used once	3730
Recycled aluminum can	2550
Recycled glass beer bottle	2530
Refillable glass bottle, used 10 times	610

Source: Brown, L. R., Flavin, C., and Postel, S. (1991). *Saving the Planet: How to Shape an Environmentally Sustainable Economy.* New York: Norton, Table 4-1, p. 71.

consumers for their own groceries. Reusable beverage containers can be sterilized, refilled, and returned to the shelf, sometimes completing the cycle as many as 50 times. Unfortunately, disposable and recyclable bottles and cans have nearly eliminated the reusable container from the market.

Table 19.2 compares the energy demand of refillable glass bottles, for example, used ten times, with various other packaging options. As illustrated, when it comes to energy demand, recycling is generally more efficient than one-time use. Reusing glass bottles only ten times, though, is four times more efficient than recycled glass.

Besides saving energy, the reuse option: (1) reduces the land area needed for solid waste disposal, (2) creates jobs, (3) provides inexpensive products for the poor and the thrifty, (4) reduces litter, (5) decreases the amount of materials consumed by society, and (6) helps reduce pollution and environmental degradation.

The Recycling Option   In human societies, recycling refers to the return of materials to manufacturers, where they can be melted down and reincorporated into products. Recycling alleviates future resource shortages, reduces energy demand, cuts pollution, saves water, and decreases solid waste disposal and incineration.

Consider some examples of the benefits: Each 1.2-meter (4-foot) stack of newspaper you recycle saves a 12-meter (40-foot) Douglas fir tree. Recycling a ton of newspaper saves 17 trees. Paper recycling uses one-third to one-half as much energy as the conventional process of making paper from wood pulp. Small savings can add up to make incredible changes. If the United States, for

example, increased paper recycling by 30%, it would save an estimated 350 million trees each year. Paper recycling has an added benefit of reducing air pollution by 95%.

Aluminum recycling offers great benefits as well. As noted in two previous chapters, aluminum recycling requires 95% less energy than making aluminum from raw ore (bauxite). Thus, a manufacturer can make 20 aluminum cans from recycled metal with the same energy it takes to make 1 can from bauxite ore. Aluminum recycling produces 95% less air pollution as well. Similar environmental benefits are available from recycling other metals and plastics.

Japan is the world's leader in recycling. Currently, about 50% of its waste is composted and recycled. In the United States and many other countries, however, recycling efforts fall far short of their full potential, although there are exceptions. One exception is the automobile. In the United States, approximately 90% of all American cars are recycled. Lead in car batteries is another exception with about 95% of all batteries making their way back into the manufacturing process. Despite these impressive statistics, in 1990, nationwide only about 2% of the plastic, 20% of the glass, 29% of the paper and cardboard, and 38% of the aluminum was recycled. Recycling rates for many products could easily double or triple. Plastic recycling could increase 10 to 20 times.

Obstacles to Recycling   From an energy and resource standpoint, recycling is generally not as good an option as reuse, but it is far better than burning materials and infinitely better than throwing them away. If recycling is such a good idea, why don't Americans, Canadians, and other peoples do more of it? The reasons are complex.

First, many industrial societies grew up with abundant resources. They saw little need for recycling, except perhaps in times of war, when recycling became commonplace. Given the seeming inexhaustible supply of materials, factories were primarily set up to handle virgin material. The entire production-consumption system was built without recycling in mind. Corporate empires were built on the profits of extractive industries and those companies today wield enormous political power. Changing this ingrained and wasteful system will not be easy.

Second, the nation's tax laws today support extraction. As pointed out in Chapter 13, U.S. laws work against recycling. Even today, for example, mining companies receive generous tax breaks, called depletion allowances, that give them an unfair advantage over recyclers. These tax breaks often make virgin materials cheaper than recycled ones. Logging companies that supply wood for paper mills and other uses are also heavily subsidized by the federal government—and thus by the taxpayer. Logging roads in national forests, for instance, are built with public money, and timber sales on public land are frequently made below cost (Chapter 9), further benefiting a virgin paper industry over recycling. Federal subsidies create unfair economics. Hard-rock mining companies can also purchase federal land at $2.50 an acre, as a result of the Mining Act of 1872. They are also not required to pay any royalty to the government for extracted minerals, which artificially reduces the price of minerals and metals.

The traditional extractive industries receive another hidden subsidy that's far more difficult to quantify. It is called an economic externality—a cost that is passed on to the public from pollution and other harmful effects of these activities. Because the traditional ways of making paper, tin cans, and other products produce more pollution, they have a bigger impact on our health and the environment than recycling, which uses less energy and produces far less pollution. But, the higher costs of producing products from virgin materials is not reflected in their prices. They are, instead, paid in federal taxes that go to clean up the air and water. They are paid in higher health bills and dozens of other ways of which few of us are aware.

Another difficulty is the built-in transportation price differential, mandated by law, for scrap metal and ore. For example, ore travels more cheaply than metal bound for a recycling mill.

In some locations, such as the U.S. West, cheap landfill costs are proving to be a difficult barrier to overcome. In Colorado, for example, landfill tipping fees—the cost to dump a ton of trash in a landfill—range from $4 to $7 per ton. Recycling may cost $30 to $50 per ton. On the East and West Coasts, landfill costs may be $40 to $100, making recycling far more profitable. Cheap landfilling is bound to end in the next decade as landfills are closed and new, more expensive sites are developed.

Another problem in interior states like Colorado is that they are located a great distance from the nation's paper recycling mills, making it more expensive and less profitable to recycle goods, such as paper, where local manufacturers are not available.

Recycling suffers from an image problem as well. In the 1970s, many recycled paper products were of inferior quality. Many people who used them were dissatisfied and soon returned to virgin materials. Since that time, however, recycled paper products have improved immensely. Recycled office paper, stationery, and computer paper are indistinguishable from virgin stock. Still, the notion that recycled paper is inferior persists.

Plastics pose a special problem for recycling. Most plastic is perfectly recyclable. The problem is that there are more than 45 different types of plastic commonly used for packaging. Making matters worse, some packages contain two or more types. A plastic ketchup bottle, for example, has five layers of plastic, each one different,

Figure 19.13 *The plastic codes on the bottom of virtually all plastic containers tell the type of plastic the product contains, making it easier for recyclers to distinguish one from another.*

each one providing a special feature needed to make a perfectly squeezable bottle.

One way plastic manufacturers have helped promote plastic recycling is by placing codes on plastic packaging. The codes tell individuals and recyclers exactly what type of plastic has been used (Figure 19.13).

Overcoming the Obstacles Despite these barriers, U.S. recycling efforts are on the rise and are bound to increase substantially in the 1990s. In 1988, the EPA announced a nationwide goal of reducing the solid waste stream by 25% through waste reduction and recycling by 1994. But, government goals are only part of the equation. Rising energy prices, decreasing landfill space, and depletion of high-grade ores will surely increase recycling. Education through schools, environmental groups, and the media can increase public awareness. Policy changes are also needed to do away with preferential freight rates, eliminate subsidies for timber harvesting and mining, and increase demand for products made from recycled materials.

Many major U.S. cities have developed recycling programs to reduce landfilling. A study in 1992, for example, estimates that nearly 4000 cities and towns collect recyclables; three-fourths of the programs came on-line in the last three years. In 1985, for instance, Philadelphia topped out its last landfill. The next closest one is a 335-kilometer (210-mile) round-trip. Because

Point/Counterpoint 19.1

Are We Facing an Epidemic of Cancer?

Lewis G. Regenstein
Lewis Regenstein, an Atlanta writer and conservationist, is author of How to Survive in America the Poisoned *and is president of the Interfaith Council for the Protection of Animals and Nature.*

America's Epidemic of Chemicals and Cancer

America is in the throes of an unprecedented cancer epidemic, caused largely by the pervasive presence in our environment and food chain of carcinogenic pesticides and industrial chemicals.

Today, significant levels of hundreds of toxic chemicals known to cause cancer, miscarriages, birth defects, and other health effects, are found regularly in our food, our air, our water—and our own bodies. Accompanying this widespread pollution has been a dramatic and alarming rise in the cancer rate in recent decades.

Each year, over a million Americans (about 1,130,000 in 1992, not including 450,000 skin cancers) are diagnosed as having cancer—over 3000 people a day! The disease now strikes almost 1 American in 3, and kills over a thousand of us every day! This means that of the Americans now alive, some 70 to 80 million people can expect to contract cancer in their lifetimes. More Americans die of cancer every year (an estimated 520,000 in

1992) than were killed in combat in World War II, Korea, and Vietnam combined. And cancer has now become a common disease of the young as well as the old, with incidence of childhood cancers, especially leukemia and brain tumors, mounting sharply in recent years.

Man-made chemicals are also depleting the Earth's protective ozone layer, which shields us from most of the sun's ultraviolet rays. The EPA has projected that the increase in radiation hitting the Earth will cause Americans to suffer 40 million cases of skin cancer, 800,000 deaths in the next 88 years, and 12 million incidences of eye cataracts.

In 1978, the president's Council on Environmental Quality (CEQ) reported unequivocally that "most researchers agree that 70% to 90% of all cancers are caused by environmental influences and are hence theoretically preventable."

Evidence continues to mount demonstrating that toxic chemicals are heavily contributing to the cancer epidemic. In general, the most polluted areas of the country have the highest cancer rates. Heavily industrialized New Jersey has the greatest concentration of chemical and petroleum facilities. In July 1982, the University of Medicine and Dentistry of New Jersey released a study showing a correlation between the presence of toxic waste dumps and elevated cancer death rates (up to 50% above average) in areas of the state.

In February 1984, a report by researchers at the Harvard School of Public Health demonstrated a link between the consumption of chemically contaminated well water near Woburn, Massachusetts, and the extraordinary incidence of childhood leukemia, stillbirths, birth defects, and disorders of the kidneys, lungs, and skin among local residents.

Dr. Samuel Epstein of the University of Illinois Medical Center, perhaps the foremost authority on the subject, points out that apart from AIDS, "cancer is the only

major killing disease which is on the increase," with incidence rising by at least 2% a year, and death rates at 1% annually over the last decade. He concludes that "the facts show very clearly that we are in a cancer epidemic now," in large part because of "the carcinogenizing of our environment, the increasing contamination of our air and our water and our food and the workplace."

Today, every American is regularly and unavoidably exposed to a variety of dangerous chemicals. Dozens of pesticides used on our food are known or thought to cause cancer and birth defects in animals. By the time restrictions were placed on some of the deadliest chemicals, such as DDT, dieldrin, BHC, and PCBs, these carcinogenic poisons were being found in the flesh tissues of literally 99% of all Americans tested, as well as in the food chain and even mother's milk. In fact, breast milk is heavily contaminated with high levels of banned, cancer-causing chemicals. And, virtually all Americans carry in their bodies traces of dioxin (TCDD), the most deadly man-made chemical known.

The response of the U.S. government has been largely weak or nonexistent enforcement of the nation's health and environmental protection laws. For example, with few exceptions, the EPA has refused to carry out its legal duty to ban or restrict pesticides known to cause cancer. Nor has the government adequately implemented or enforced the laws regulating hazardous waste, which is being generated at a rate of up to 275 million metric tons a year—over a ton for every man, woman, and child in the nation. Much of this is disposed of in a manner that will ultimately threaten the health of nearby residents.

Thus, we are even now sowing the seeds for cancer epidemics of the future. Only time will tell what will be the effect on this generation, and future ones, of Americans—the chemical industry's ultimate guinea pigs.

David L. Eaton

David L. Eaton is professor of Environmental Health and Environmental Studies, and director of Toxicology at the University of Washington. He has an active research program on the mechanisms by which chemicals cause cancer.

"America's Epidemic of Chemicals and Cancer"— Myth or Fact?

There is no debate that cancer is a devastating and deadly disease. One in three people living in the United States today will contract some form of cancer in his or her lifetime, and one in four will die from it if current rates continue.

Are cancer rates increasing in epidemic proportions? The total number of people and the fraction of all deaths attributable to cancer have increased dramatically in the past 50 years. However, cancer is largely a disease of old age, and thus it is necessary to adjust such statistics for changes in the age distribution of our population. A 1988 report from the National Cancer Institute states that "the age adjusted mortality rates for all [types of] cancers combined, except lung cancer, have been declining since 1950 for all individual age groups except 85 and above." Statistics from the American Cancer Society yield the same conclusion.

The incidence of some childhood leukemias and brain tumors has indeed increased significantly in the past decade. How much of this increase is a result of better diagnosis and reporting, rather than a true increase, remains controversial. We are in an "epidemic" of lung cancer. For most of the first half of this century, lung cancer mortality was not even in the "top five" types of cancer-related deaths. Lung cancer is now the leading cause of cancer-related deaths in both men and women. About 85% to 90% of all lung cancers in men, and perhaps 70% in women, is directly attributable to smoking. Per capita consumption of cigarettes increased five-fold in men from 1900 to 1960 and with it, a concomitant increase in lung cancer. This pattern was repeated 20 years later in women. The risks of several other types of common cancers are also increased by smoking (e.g., cancers of the bladder and esophagus). Approximately one-third of all cancer deaths could be eliminated by eliminating smoking from our society.

Of the variety of environmental factors other than smoking, dietary factors are now generally thought to represent the largest source of cancer risk, perhaps related to 30% to 40% of all cancers. Although synthetic chemicals, such as industrial pollutants and pesticides present in trace amounts in our food supply, may contribute to dietary risk, recent studies have suggested that this contribution is trivial relative to other "nonpollutant" factors. For example, the risk of breast cancer in women (second only to lung cancer in incidence and mortality) is significantly increased by high fat diets, and the amount of fiber in the diet substantially influences the risk of colon cancer, a major site of cancer in both men and women.

The largest source of exposure to cancer-causing chemicals may not be industrial pollution but chemicals that occur naturally in our diet. All plants produce toxic chemicals as a means of protection against insects, fungi, and animal predators. It has been estimated that we ingest in our diet about 10,000 times more of "nature's pesticides" than man-made chemical residues. Many of these chemicals are potent mutagens and carcinogens, and are frequently present at levels thousands of times higher than the trace levels of synthetic pesticide residues and industrial chemicals sometimes found in food crops. Taken together, the dietary risk factors from natural sources, often present in relatively high amounts, are far more important than the pesticide residues and industrial chemicals that can often be detected at exceedingly small concentrations in our diets. Unfortunately, because of the relatively high exposure to carcinogens from natural sources, the complete elimination of synthetic industrial chemicals from our diet, if it were possible, would not likely have any significant beneficial effect on cancer incidence and mortality.

Finally, recent advances in the understanding of the biology of cancer suggest that "spontaneous" or "background" alterations in DNA may explain much of the cause of cancer. The use of modern techniques in molecular biology has revealed that DNA is inherently unstable and can be altered by normal errors in DNA replication. DNA is subject to extensive damage from processes associated with normal cellular metabolism. Within our life span, our cells undergo about 10 million billion cell divisions. Spontaneous errors in this process, which lead to mutations and cancer, accumulate with age. It is not surprising then that cancer seems to be a frequent outcome of old age.

The views that we are in an overall cancer epidemic and that these cancers are largely a result of industrial chemicals are not supported by the vast majority of cancer researchers throughout the world. Unfortunately, it will take some time for the political arena, influenced greatly by public fears, *(continued)*

Point/Counterpoint 19.1 (*continued*)

Are We Facing an Epidemic of Cancer?

to come to grips with the fact that further reduction in public exposure (i.e., nonoccupational) to synthetic chemicals will not have much impact on cancer incidence and that such results will come only at great social and economic expense. The United States is currently spending about $80 billion per year on pollution reduction, about nine times the total budget for all basic scientific research. I believe that much of this is justified to enhance the quality of our environment, and ensure the habitability of our planet for our children, their children's children, and other species. However, there is also little question that huge sums of money are spent each year to re-

duce what is in all likelihood a trivial cancer risk, with few other environmental benefits. If our society is truly concerned about reducing the human tragedy from cancer, more efforts should be focused on eliminating smoking and alcohol abuse, better research and education on dietary risk factors, more research into the biochemical and molecular events that lead to cancer (which in turn will lead to more effective preventive and curative measures), and continued identification and reduction in occupational exposures to those chemicals that pose a significant cancer risk.

Critical Thinking

1. Summarize each author's major points and supporting data. In your view, how well has each of the authors supported his contentions? Point out potential weaknesses in arguments.
2. Review the critical thinking skills presented in Chapter Supplement 1.1. How can they help you in analyzing the facts and fallacies in the arguments presented here?
3. Do you agree with Eaton that to determine whether we are in a cancer epidemic caused by synthetic chemicals, such as pesticides, we should eliminate lung cancer from the calculation? Why or why not?

transportation costs would be prohibitive, the city began an active recycling program. Faced with similar problems, Seattle instituted a curbside recycling program in 1988. In Seattle, recyclables are picked up once a month and sent to a recycling facility. In 1992, Seattle residents recycled 40% of their garbage, and by 1998, the city hopes to recycle 60%.

Most of the northeastern states are planning to recycle large amounts of their waste streams, between 20% and 50% in the not-too-distant future. Through a variety of measures, New York City plans to recycle half of its solid waste by 1997.

The trash crisis has struck many American cities, including New York City, Chicago, Miami, Los Angeles, Minneapolis, and Berkeley. Shrinking land and public opposition to landfills in their neighborhoods have forced city officials to look for alternatives; not surprisingly, recycling often comes out on top.

Cities and towns are trying several options. Some are using drop-off sites, where residents can deposit their recyclables on the way to work or to the grocery store. San Francisco now has six redemption centers in operation. In California, hundreds of igloo-shaped glass banks are situated near busy intersections. Dozens of paper redemption centers are also scattered throughout Los Angeles and San Diego. Drop-off centers can be successful but only if containers are conveniently placed,

for example, at train stations, near parks, or near heavy-use intersections. Compared to curbside recycling, however, drop-off programs return only a small fraction of potentially recyclable materials.

Curbside recycling is by far the most successful type of recycling program. Participation rates as high as 60% to 80% can be expected if recyclables and trash are picked up on the same day. One study in Canada showed that curbside recycling requires about 10% less energy than a drop-off program.

Because of declining landfill space and rising costs, recycling is catching on like wildfire in the United States. According to the EPA, 23 states currently have comprehensive recycling programs. Several others are expected soon. A growing number of recycling programs are being run by private haulers. Toronto, Canada, for example, is home to a program run by Waste Management, Inc. In some areas, all trash haulers offer curbside recycling.

Recycling is but one of the key strategies used in building a sustainable society. It must be convenient and cost-effective. It will, however, demand individual action. Separating trash before collection requires a little effort, but not much more than the effort expended to separate white clothes from colored clothes when doing laundry. Source separation is a low rent to pay for the riches the Earth gives us. It is a small sacrifice in return for a better world and a sustainable future.

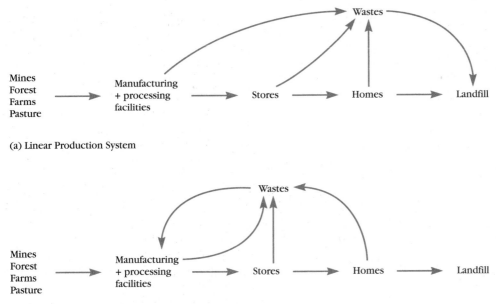

(a) Linear Production System

(b) Cyclic Production System: The Industrial Ecosystem

Figure 19.14 *By reducing waste, reusing, composting, and recycling, individuals and businesses can convert the linear production system (a) into a cyclic one (b). Jobs will shift away from the ends of the linear system to the center, where recycling and composting occur. Jobs at all levels of income will be created in recycling and composting industries.*

Raw trash can also be collected and shipped to resource recovery centers, where it is separated by machines or people. This technique, called end-point separation, is generally more costly than source separation. Moreover, complete separation may not be possible at these facilities, thus lowering the value of recyclable materials.

Procuring Recycled Materials Japan currently recycles about half of its garbage. Although impressive, the country's recycling rate could climb as high 80% to 90%. One suburb of Tokyo, for instance, currently recycles and composts 90% of its garbage. Similar gains are possible elsewhere. In a pilot project involving 100 families in East Hampton, New York, the recycling rate was 84%!

Although impressive recycling rates are possible, they will not occur without changes in attitude and policy. One change in attitude is a shift from the exploitive one (frontierism), so prevalent throughout the world today, to a nurturing one in which people are willing to act sustainably for the good of the biosphere, future generations, and other species.

A second important change is a shift in the production consumption system. As pointed out in Chapter 1, human economies are largely linear (Figure 19.14). A sustainable economy will most likely be cyclic. Instead of traveling in a straight line from mine to landfill or incinerator, materials will flow back through the system as much as possible. Although some raw materials will need to be extracted and some wastes will invariably be produced, the goal of a sustainable economy is to max-

imize the cycling within the economy. As you shall soon see, this will mean a shift away from employment at either end of the system toward the middle, where the recycling occurs both in pickup and remanufacturing.

Building an industrial ecosystem, that is, a system of commerce that is an analog of natural ecosystems, means a dramatic shift to recycling. But "recycling," as currently practiced, is not enough. Companies must be willing to use the materials picked up in curbside programs or drop-off sites. Financial incentives and disincentives can be used to promote remanufacturing. For instance, Florida levied a 10% tax on newspaper and book publishers using virgin paper to discourage its use. Colorado offered a small tax credit to companies that purchase equipment that allows them to incorporate recycled materials into the manufacturing process. In addition, individuals must be willing to purchase recycled paper and other materials to help support manufacturers. Recyclers must have a market for their materials.

In November 1990, Germany's minister for the environment issued a decree calling for an 80% reduction in the amount of packaging entering the waste stream. To meet these goals, retailers are required to collect packaging materials and recycle them. This puts pressure on manufacturers to reduce excess packaging wherever possible and perhaps helps develop markets to use recyclables. To promote public participation, the government is also levying a small deposit, equivalent to about 30 cents in U.S. currency, on most packages. Retailers can avoid the deposit and take-back requirements if industry establishes alternative collection and recycling systems that meet the established goals.

Laws requiring deposits on beverage containers, commonly known as bottle bills, can also dramatically increase recycling rates. At this writing, nine U.S. states have bottle bills. In such states, over 90% of the glass and aluminum containers are returned for recycling. Container-deposit legislation is also on the books in the Netherlands, Scandinavia, the Commonwealth of Independent States, Canada, Japan, parts of Australia, and a handful of developing nations.

Individuals can help promote recycling by buying recycled products. Governments can also help. When added together, local, state, and federal governments account for about 20% of the U.S. **gross national product (GNP)**—that is, the nation's total annual output of goods and services. In other words, government is the single most important purchaser in the economy. Think of the impact that local, state, and federal governments could have if they all started buying recycled paper. The demand created would inspire many business owners to switch from conventional practices and would greatly increase markets and create supplies for others.

In 1986, 13 states had procurement programs of some sort, which favored recycled products. In 1992, all 50 states and the District of Columbia had programs, purchasing millions of reams of recycled office paper and paper products. New York, for example, allows state agencies to purchase recycled paper if it comes within 5% of the cost of virgin stock; California allows purchase if it comes within 10%. These are called **price preference policies**. In these two states, about one-fourth of all the office paper, tissue, paper towels, and cardboard purchased by state agencies is made from recycled stock. On average, the states pay only about 2% more for office paper. Who knows what they save in reduced pollution and health bills?

In October 1991, then President George Bush signed an executive order that required all federal departments and agencies to purchase products made with recycled materials wherever possible. This order also requires federal agencies to appoint recycling coordinators to increase recycling of discarded items by the nation's 3 million federal employees.

Unknown to many, the Resource Conservation and Recovery Act called on federal agencies to purchase recycled materials in 1976, but only if the products were reasonably priced. Unfortunately, two problems thwarted progress. First, the law called on the EPA to publish a list of guidelines for a dozen or so recycled products. However, the EPA did not begin drawing up the guidelines for over a decade—and only then after it was sued by environmentalists. The job was not completed until 1989.

The second problem was interpretation of the "reasonable cost" provisions to mean the "lowest cost." Given federal transportation policies, smaller markets, and the failure of current pricing systems to reflect exter-

nal costs, among other problems, recycled products are often slightly more expensive. Colorado officials, however, have found that some items are cost competitive, even cheaper. Money saved on one product, for instance, can be used to pay the slightly higher price of another, thus keeping state spending constant.

Recycling programs in the United States are currently hindered by small markets. For example, in the past decade, municipalities starting many new recycling programs discovered that some materials have no markets or that prices they will receive are inadequate to support their program. When New York State began recycling plastic soft-drink bottles, it couldn't find a buyer; the recycling program became an expensive trip to the landfill. But, over time markets developed, and the soft-drink bottles are now ground up and used to make filler for pillows and jackets, among other products. Many cities found that newspaper recycling efforts were crippled by a fall in newspaper prices. Researchers at EDF note, however, that "gluts" of recyclable materials are only temporary because markets develop in response to rising secondary material supplies. At this writing nearly 70 mills in Canada and the United States are being built or modified and will soon come on-line to handle a wide range of recycled paper products, greatly increasing market demand for recycled newsprint, cardboard, and other papers.

Recycled newsprint can be used to make a variety of useful products in addition to new newsprint. For instance, it can be used to make egg cartons, cereal boxes, map tubes, drywall, ceiling tiles, animal bedding, and insulation. Cities and towns can encourage companies to manufacture these products from waste, thus putting locally available wastes to good use and helping build stable self-reliant regional or state economies.

To promote recycling among citizens, cities and towns can create several economic incentives. In Seattle, residents who recycle pay $5 per month less if they fill only one trash can each week and superrecyclers get an even lower rate. The city of High Bridge, New Jersey, at one time charged residents an annual flat fee of $280 for garbage collection, for which they received virtually unlimited trash pickup. Today, though, residents have to purchase stickers, which they attach to every trash bag left by the curb. Fifty-two stickers cost $140. Additional stickers cost $1.25 each. Since this program began waste output has dropped by 25%, in large part because of increased efforts on the part of citizens to compost, recycle, and compact their trash.

As recycling becomes more in vogue, more and more manufacturers are bound to shift to recycled materials and more recycling facilities will open, creating jobs and economic opportunities. By the time most of you graduate and start your family, recycling could well be the chief source of materials.

It may be surprising to many, but recycling is much

more prevalent in the developing countries than in the industrialized nations. The poor raid the dumps for food, clothing, and materials for shelter and also seek out discarded metals and other goods they can sell. The developed nations can help promote recycling as they assist developing nations in their efforts to raise their economic well-being. Information on the energy and material savings as well as the technologies for recycling could be built into global industrialization. To be credible, however, the developed nations will need to increase their own recycling efforts.

Composting   Another valuable throughput strategy that reduces waste and recycles materials is **composting**, the process in which organic wastes, such as leaves, grass clippings, cardboard, and paper, are stockpiled, then allowed to decay. The resulting product, **compost**, is a nutrient-rich organic material much like humus that can be used to build soil fertility. Composting may occur in backyards, in neighborhood facilities, or in large municipalities.

The United States has several successful municipal composting operations. In Seattle, for example, zoo officials compost all of the 900 cubic yards of manure produced by the zoo's many animals. When decomposed, the manure forms a rich organic soil supplement they call "zoodo," which is sold to gardeners and homeowners. The program, which costs about $30,000 per year, saves $35,000 in landfilling costs. Moreover, zoodo is sold at $12 or more per cubic yard. Palo Alto, California,

has had a successful composting program for yard wastes since 1979. Portland, Oregon, and Berkeley, California, recently started yard-waste composting programs. Programs exist in 19 other cities in eight states: Arizona, Delaware, Kansas, Florida, Minnesota, Texas, Washington, and Wisconsin. Most of the large-scale composting programs are in Europe—in the Netherlands, Belgium, England, and Italy—and in Israel (Figure 19.15).

Widespread composting, combined with recycling, can result in substantial reductions in municipal waste. Nationwide, organic wastes constitute about 20% of the garbage dumped into landfills or burned in incinerators. During the fall, compostable waste can make up 75% of the waste stream. Composting these wastes not only reduces landfilling but saves considerable amounts of energy needed to transport wastes to landfills. Composting is often cheaper than landfilling and can help recycle nutrients to farmland, closing the loops of an extremely important nutrient cycle. In New Jersey, composting in Morris County costs municipalities $16 to $32 per ton, much less than landfilling, which costs over $110 per ton. Collecting and composting leaves and yard waste in Minneapolis-St. Paul costs about $65 per ton compared to landfilling at $90 per ton.

Despite their obvious benefits, large-scale composting operations have some drawbacks. First, they require large tracts of land because they can produce odor and create breeding sites for pests. Because of this, composting facilities make poor neighbors and are usually sited long distances from homes. This adds significantly to transport costs and energy consumption. Large-scale composting facilities are often expensive undertakings, in part because of the need to sort out the noncompostable materials, such as plastics, metals, and glass. Their expense is also attributable to the large investment in machinery needed to turn the compost regularly to accelerate decay.

To make municipal composting more cost-effective, cities and counties could rely on labor provided by convicts, welfare recipients, or the unemployed. Citizens could be required to sort out recyclable metals, plastics, and glass, thus eliminating the cost of separating the wastes later.

Composting can be practiced very successfully at home. Gardeners can make their own compost piles of leaves, grass clippings, and vegetable wastes from the kitchen. By mixing these materials with a little soil, which contains the bacteria that do the breakdown, and watering the pile from time to time to keep it moist, homeowners can produce a nutrient-rich soil supplement for lawns and gardens, eliminating the need for artificial fertilizers. A commercially available container or a simple wooden enclosure helps to deter curious dogs, raccoons, and skunks. Home composting also eliminates the need to haul wastes to central facilities.

Another municipal waste product of cities and towns is sewage sludge. As noted in Chapter 17, sludge is typically dumped in landfills. But, sludge can be combined with municipal solid waste, such as leaves, paper, grass clippings, etc., and composted. This process, called **co-composting**, destroys viruses and bacteria in the sludge, so the product can be sold for use on farms or on gardens and lawns. Co-composting is expected to become increasingly popular in the United States and other countries because it costs less and is more ecologically sound than landfilling.

John Young of the Worldwatch Institute wrote: "More efficient use of materials could virtually eliminate incineration and dramatically reduce dependence on landfills. It could also substantially lower energy needs. . . . Taken together, source reduction, reuse, and recycling can not only cut waste but also foster more flexible and self-reliant economies. Decentralized collection and processing of secondary materials can create new industries and jobs." In fact, numerous studies show that recycling rather than landfilling creates far more jobs than the traditional strategies, landfilling and incineration. A study in New York showed that recycling 10,000 tons of trash through curbside recycling programs produced, on average, 36 new jobs. Landfilling or incinerating that trash produced only about 1 job each. While some jobs will be lost in the front end of the production-consumption system—that is, at the mines and in the forests where raw materials are extracted—many more jobs will open up in the middle,

the recycling portion. As Young points out, while the economic health of nations is often measured by the amount of materials they consume, prosperity doesn't have to be so tightly linked to consumption. We can live well and live sustainably.

We are poisoning ourselves and our posterity.

—*Barry Commoner*

Critical Thinking Exercise Solution

This is a clear-cut case of dollars versus environmental protection that hinges on values. You are fairly certain that the incinerator in Niger will produce toxic emissions and that spills may occur en route. Furthermore, waste from the incinerator will very likely be dumped in an open pit, where it could leak into the groundwater. In essence, your economic savings will be passed on to the unsuspecting people of Niger.

A savings of $1 million is fairly substantial, however. Your boss will no doubt take notice, and may even give you a sizable raise for saving the company so much money. How do you decide? What's more important—protecting the environment of another country or saving your company money and maybe setting yourself up for a raise? Do you feel a sense of intragenerational equity?

Summary

19.1   Hazardous Wastes: Coming to Terms with the Problem

■   The problem of **hazardous waste** caught the attention of the American public in the 1970s when toxic chemicals began to ooze out of a dump known as Love Canal, near Niagara Falls.

■   Shortly after Love Canal, the true proportions of the hazardous waste problem became evident. First, estimates suggested that 10,000 toxic hot spots needed cleaning up. However, newer studies put the number much higher.

■   Ill-conceived and irresponsible waste disposal has left a legacy of polluted groundwater and contaminated land. Cleanup could take 50 years and cost $100 billion or more.

■   The latest in a long list of hazardous chemical problems stems from **leaking underground storage tanks**.

Resources Misuse

Throughout much of history, humans have been able to ignore the impact of their activities on natural resources. However, in recent years, we have been forced to recognize the error of our ways; air and water pollution, unsafe disposal of toxic wastes, the near catastrophe at Three Mile Island, depletion of precious minerals, extinction of endangered species, and more. These misuses threaten the resources needed to sustain not only our civilization, but our planet as well. Today, the earth's future lies in a combined and difficult task; not only must we correct our past mistakes, we must *understand* how and why they happened, and take steps to ensure that they don't happen again.

1 There goes the neighborhood... A wet cooling tower stands in bold contrast to a farmhouse in rural northern Ohio.

2 A denuded hill in Alaska has been clear-cut to remove timber for domestic and foreign markets.

3 Controlled fires burn the grassy vegetation of cleared rain forest. Some experts feel that removal of tropical rain forests will have a direct effect on climatic conditions. The release of great amounts of carbon dioxide into the atmosphere is said to contribute to the greenhouse effect, a phenomenon which may produce an unfavorable warming trend in the earth's atmosphere.

4 Land in a tropical rain forest is cleared for farming or grazing. After removal of vegetation, the nutrient-poor soil is often washed away or baked into a brick-like consistency, rendering it useless.

5 Overgrazing turned a once-rich grassland into a lifeless desert. Without the plant-root network that once bound it, the soil is unable to retain moisture and is easily blown away.

6 An infrared aerial view of Love Canal, Niagara Falls, New York. The dump, home to dozens of chemicals including PCBs and dioxin, was sold to the city in the 1950s as a school and playground site. Healthy vegetation appears red; brown vegetation indicates contamination. The toxic wastes were spread throughout the neighborhood via underground streams; 237 families were evacuated from the area.

7 Toxic wastes are stored in steel drums, then buried or stored in toxic waste dumps like this one. The steel drums, which are supposed to separate the dangerous wastes from you and me, rust due to their corrosive contents.

8 Even in the wilderness of Antarctica massive amounts of solid waste, unhidden by the snow, are a constant ecological threat.

9 Industrial society produces massive quantities of refuse, some of which gets buried, burned, or recycled. The rest ends up along highways, in empty lots, or, as in this case, in someone's backyard.

10 Boats at dock on the Amazon sit immersed in a sea of floating garbage.

2

3

4

5

6

7

8

9

10

11 Waste from an iron ore processing plant colors the water of this artificial pond a bright orange. Such wastes seep into the earth, contaminating groundwater supplies.

12 Nitrogen fertilizer added to irrigation water to boost crop production in the Oro Valley, Arizona, may contaminate groundwater in this area.

13 New York City lies enshrouded in a layer of filthy air.

14 At McMurdo Station (U.S.A.) solid refuse covers the embankment, while raw sewage flows into the Antarctic Ocean.

13

11

12

14

Thousands of tanks holding gasoline, other petroleum by-products, and hazardous wastes of all kinds may be leaking. Health officials' main concern is the potential effect on groundwater and human health.

19.2 Attacking Hazardous Wastes on Two Fronts

■ Industrial societies must find ways to deal with: (1) leaking and contaminated waste disposal sites and (2) enormous amounts of waste produced each year.

■ To deal with leaking dumps and contaminated areas, Congress passed the **Comprehensive Environmental Response, Compensation, and Liability Act (CERCLA** or **Superfund Act)** in 1980.

■ The act holds owners and operators of hazardous waste disposal sites, as well as their customers, liable for cleanup and property damage. This legislation established the Superfund, which is used to clean up imminent hazards.

■ Superfund has some major drawbacks. One of the most critical weaknesses in the Superfund legislation is that it fails to address human health impacts.

■ In 1976, Congress passed the **Resource Conservation and Recovery Act (RCRA)**, aimed at cutting back on improper waste disposal.

■ RCRA called on the EPA to determine which wastes were hazardous and how they should be handled and disposed of. A reporting network was established to trace hazardous wastes from cradle to grave. The EPA also began issuing permits for waste disposal. Unfortunately, too many loopholes existed in the legislation, allowing much hazardous waste to escape control. New amendments in 1984 closed many of the loopholes.

■ Many technological options exist for controlling hazardous wastes. The first line of attack is to reduce hazardous waste production through **process manipulation, reuse,** and **recycling** of waste products.

■ **Detoxification** and **stabilization** are the second line of attack. Detoxification includes **land disposal, incineration,** and **low-temperature decomposition.**

■ Hazardous wastes that cannot be broken down can be **stored permanently** in **secured landfills** lined by synthetic liners and thick, theoretically impermeable layers of clay.

■ Although landfills are the cheapest option, they have many disadvantages, the most noteworthy being inevitable leaks.

■ Disposal of radioactive wastes creates a special problem for society. Many experts believe that deep geological burial is the best technical option for disposal of most dangerous radioactive wastes.

■ In 1982, the **Nuclear Waste Policy Act** established a strict timetable for the DOE to choose an appropriate site for disposal of high-level radioactive wastes.

■ The U.S. government concentrated its efforts on a site in Nevada, but new evidence of an active volcano and numerous fault lines suggests that this site may never be developed.

■ Medium- and low-level radioactive wastes will soon be shipped to a new facility in New Mexico.

■ Solving the hazardous waste problems is an enormous task that could be made easier by individual contribution. Reduced consumption, especially of unnecessary items, could cut back on currently generated waste.

19.3 Solid Wastes: Understanding the Problem

■ **Municipal solid waste** consists of paper, metals, leftover food, and other items discarded by businesses, hospitals, airports, schools, stores, and homes.

■ Municipal solid wastes make up about 4% of the total solid waste discarded in the United States each year, but the output is growing at a rate of 2% to 4% per year.

■ Landfilling, the prevalent means of disposal, squanders the Earth's resources and augments environmental damage. Landfill disposal also costs communities millions of dollars each year.

19.4 Solving a Growing Problem Sustainably

■ Actions to reduce solid waste generally fall into three broad categories: input, output, and throughput.

■ The traditional response is the **output approach.** It consists of ways to deal with trash flowing out of cities and towns. Most often, this means incineration or landfilling.

■ A more sustainable strategy is known as **input approach.** It consists of activities that reduce the amount of materials entering the production-consumption cycle, for example, efforts to reduce consumption and waste by increasing product durability.

■ The third approach is the **throughput approach.** Also vital to sustainability, it consists of ways to direct materials back into the production-consumption system, creating a closed-loop system akin to the cyclic kind found in nature. Reuse and recycling fall under this category.

■ From the standpoint of sustainability, the input and throughput approaches are the most beneficial.

■ Decentralized collection and processing of secondary materials can create new industries and jobs, as supported by numerous studies.

■ While the economic health of nations is often measured by the amount of materials they consume, prosperity doesn't have to be tightly linked to consumption. We can live well and live sustainably.

Discussion and Critical Thinking Questions

1. Summarize the major events occurring at Love Canal. Who was to blame for this problem? How could it have been avoided?

2. You are appointed to head a state agency on hazardous waste disposal. You and your staff are to make recommendations for a statewide plan to handle hazardous wastes. Draw up a plan to eliminate dumping. Which techniques would have the highest priority? How would you implement your plan?

3. Discuss the major provisions of the Resource Conservation and Recovery Act (1976) and the Comprehensive Environmental Response, Compensation and Liability Act (1980), the so-called Superfund Act. What are the weaknesses of each?

4. Describe the pros and cons of the major technological controls on hazardous wastes, including process modification, reuse and recycling, conversion to nonhazardous or less hazardous materials, and perpetual disposal.

5. Debate this statement: "All hazardous wastes should be recycled and reused to eliminate disposal."

6. A hazardous waste site is proposed within your community. What information would you want to know about the site? How would you go about getting the information you need? Would you oppose it? Why or why not?

7. List personal ways in which individuals contribute to lessening the hazardous waste problem.

8. Discuss some of the options for disposing of radioactive wastes. Which ones seem the most intelligent to you? Why?

9. Debate this statement: "Victims of improper hazardous waste disposal practices should be compensated by a victim compensation fund developed by taxing the producers of toxic waste."

10. Describe the three basic approaches to solving the solid waste problem. Give examples of each one. Which is (are) the most sustainable? Why?

11. Describe the pros and cons of landfilling, incineration, source reduction, composting, reuse, and recycling.

12. Changing the linear production-consumption system to a cyclic one means shifting to what activities? How will this affect jobs? In your opinion, are the shifts required necessary? How can the changes be accomplished more easily?

Suggested Readings

Hazardous Waste

Burns, M. E., ed. (1987). *Low-Level Radioactive Waste Regulation: Science, Politics and Fear*. New York: Lewis Publications. A comprehensive analysis of radiation.

Carlin, A., Scodari, P. F., and Garner, D. H. (1992). Environmental Investments: The Cost of Cleaning Up. *Environment* 34 (2): 12–20, 38–45. Summary of EPA report to Congress.

Center for Neighborhood Technology (1990). *Sustainable Manufacturing*. Chicago: Center for Neighborhood Technology. Excellent discussion of efforts needed to reduce hazardous waste.

Fischhoff, B. (1991). Report from Poland: Science and Politics in the Midst of Environmental Disaster. *Environment* 33 (2): 12–17, 37. Describes the dimensions of the hazardous waste problem in Poland.

French, H. (1990). A Most Deadly Trade. *World-Watch* 3 (4): 11–17. Documents the movement of hazardous materials to the developing countries and Eastern Europe.

Friedlander, S. K. (1989). Pollution Prevention: Implications for Engineering Design, Research, and Education. *Environment* 31 (4): 10–15, 36–38. Important source of information on the pollution prevention strategy.

Frosch, R. A. and Gallopoulos, N. F. (1989). Strategies for Manufacturing. *Scientific American* 261 (3): 144–52. Outlines the concept of the industrial ecosystem.

Russell, M., Colglazier, E. W., and Tonn, B. E. (1992). The U.S. Hazardous Waste Legacy. *Environment* 34 (6): 12–15, 34–39. Discusses the cost of cleaning up America's hazardous wastes.

Steinhart, P. (1990). Innocent Victims of a Toxic World. *National Wildlife* 28 (2): 20–27. Discusses effects of toxins on wildlife.

Stigliani, W. M. et al. (1991). Chemical Time Bombs. Predicting the Unpredictable. *Environment* 33 (4): 4–9, 26–30. Shows how chemical contamination builds, then surpasses critical threshold levels, suddenly creating poisonous conditions.

Solid Waste

Brown, L. R., Flavin, C., and Postel, S. (1991). *Saving the Planet: How to Shape an Environmentally Sustainable Economy*. New York: Norton. See Chapter 4 for an excellent discussion of recycling and reusing waste.

Carless, J. (1992). *Taking Out the Trash: A No-Nonsense Guide to Recycling*. Washington, D.C.:

Island Press. Practical guide showing how individuals, businesses, and communities can help alleviate the solid waste crisis.

Chiras, D. D. (1992). *Lessons from Nature: Learning to Live Sustainably on the Earth*. Washington, D.C.: Island Press. See Chapter 10 for a discussion of ways to build a sustainable waste management system.

Concern, Inc. (1988). *Waste: Choice for Communities*. Washington, D.C.: Concern, Inc. Excellent booklet on waste management, including recycling.

Durning, A. (1990). How Much Is Enough? *World-Watch* 3 (6): 12–19. Hard-hitting article on the impacts of and reasons for consumption.

Environmental Defense Fund. (1988). *Coming Full Circle: Successful Recycling Today*. Washington, D.C.: Environmental Defense Fund. Superb study of successful recycling programs.

Renner, M. (1992). Creating Sustainable Jobs in Industrial Countries. In *State of the World 1992*. New York: Norton. Discusses the job potential of a sustainable economy, based in part on recycling.

Young, J. (1991). Reducing Waste, Saving Materials. In *State of the World 1991*. New York: Norton. Describes several strategies to reduce material waste.

———. (1991). Tossing the Throwaway Habit. *World-Watch* 4 (3): 26–33. Describes how states and nations are trying to cut waste and promote recycling.

Part 5

Environment and Society

Environmental Ethics: The Foundation of a Sustainable Society

Modern man is the victim of the very instruments he values most. Every gain in power, every mastery of natural forces, every scientific addition to knowledge, has proved potentially dangerous, because it has not been accompanied by equal gains in self-understanding and self-discipline.

—Lewis Mumford

Critical Thinking Exercise

Many people assert that the only way to get others to act reasonably to protect the environment is through economic incentives or disincentives. "People just won't act unless they see that it will help them financially," say some. "Either penalize them or give them some economic incentive if you want action." What is your opinion of these statements? Do they stand up to critical thinking?

This text began with an outline of today's environmental crisis. Each part explored a particular facet of it and explained the role of economics, ethics, and government in creating problems and solving them. Part 2, for instance, covered the population question, Part 3 surveyed resource problems, and Part 4 described the many faces of pollution. Each chapter outlined traditional strategies, which often fail to address the root causes of the deepening crisis, and a set of sustainable remedies, that is, solutions designed to confront the root causes and create lasting solutions. Included in the sustainable approach are ways to promote conservation, recycling, renewable resource use, restoration of renewable resources, population control, and adaptability.

In order to effect change, we also need a set of values that serves people and the planet equally well. Chapter 1 described a set of values called the sustainable ethic. These values are based on the realization that current patterns of economic activity are not sustainable. This chapter reexamines the subject, highlighting key points already made and expanding the discussion to deepen your appreciation of ethical options. The chapter begins with a discussion of the frontier mentality, what it is, how it affects us, and how it evolved. Next, it explores in greater depth the sustainable ethic and discusses some methods to develop and implement a global sustainable ethic.

20.1   The Frontier Mentality Revisited

As pointed out in Chapter 1, most industrial countries operate under a frontier mentality (Table 20.1). The frontier mentality is characterized by three tenets: First, the Earth is an unlimited supply of resources for exclusive human use; in other words, "There is always more, and it's all for us." Embodied in this notion is the belief the Earth has an unlimited capacity to assimilate pollution from human activities. Second, humans are apart from nature rather than a part of it. In other words, many seem to think that humans are immune to the natural forces and ecological laws that affect all other organisms. Third, human success is best achieved through the domination and control of nature.

The frontier mentality has been a part of human thinking for many tens of thousands of years. It may have begun to emerge in the hunting and gathering societies and was clearly present in agricultural societies. It lives on today in the industrial world. The damage created by this outmoded way of thinking has reached enormous proportions in large part because human numbers have increased dramatically and many technologies now produce enormous, life-threatening changes in the environment.

A prime example of the frontier mentality can be illustrated in the settling of North America. Early frontier people cut down forests to make room for farms and grew crops on the soil until the nutrients had been depleted or the soil eroded away. They then moved on, forging into new territory to repeat the cycle (Chapter 9). Modern societies display the same brand of careless frontierism in their quest for wealth. By cutting down forests without replanting them, depleting ocean fisheries with little concern for long-term population stability, mining minerals as if there will always be more, and polluting the air, water, and soil, modern societies have created the potential for widespread ecological collapse.

The frontier ethic is not unique to the Western world. Similar views are evident in Latin America, Africa, and Asia. In Panama, for example, past military leaders in promoting economic development have called for "the conquest of the forest." In 1940, Brazil's president wrote of the need to "conquer the land, tame the waters, and subjugate the jungle." In Asia, Japan has become a world leader in exploiting the global environment. Thailand and Indonesia have suffered mightily under resource development. In the 12th century, the king of Sri Lanka wrote, "Let not a single drop of water that falls on the land go into the sea without serving the people."

Perhaps the most troubling aspect of the frontier ethic—the notion of unlimited resources, feelings of separation from nature, and an insistence on conquering nature—is the disregard for the Earth it fosters. Why worry about soil erosion or water pollution? Why worry about overfishing? Why worry about tropical rain forests? There is always more and it's ours for the taking.

Another troubling aspect of the frontier notion that humans succeed best through the subjugation of nature is that this strategy often backfires, creating severe ecological backlashes, such as pest resistance, increased flooding, wildlife extinction, and severe soil erosion. Ecological backlashes affect our economic well-being and our own survival, underscoring an important but often overlooked fact: The future of people and nature are closely intertwined.

Table 20.1 Frontier and Sustainable Ethics Compared

Frontier Ethic	Sustainable Ethic
*The earth is an unlimited bank of resources for exclusive human use.	*The earth has a limited supply of resources used by all species.
When the supply runs out, move elsewhere.	Recycling and the use of renewable resources will prevent depletion.
Life will be made better if we just continue to add to our material wealth.	Life's value is not simply the sum total of our banking accounts.
The cost of any project is determined by the cost of materials, energy, and labor. Economics is all that matters.	The cost is more than the sum of the energy, labor, and materials. External costs such as damage to health and the environment must be calculated.
*The key to success is through domination and control; nature is to be overcome.	*We must understand and cooperate with nature.
New laws and technologies will solve our environmental problems.	Individual efforts to solve the pressing problems must be combined with tough laws and new technologies.
*Humans are apart from nature; we are above nature, somehow separated from it and superior to it.	*We are a part of nature, ruled by its rules and respectful of its components. We are not superior to nature.
Waste is to be expected in all human endeavor.	Waste is intolerable; every wasted object should have a use.

*Indicate key ethical principles of each set of ethics.

Today, people are learning to cooperate with nature—to weave human activities into the economy of nature. Biological pest control, watershed protection to control flooding, solar energy, growth management to protect valuable farmland, and a host of other measures are good examples of ways we can meet human needs without bankrupting the biosphere and foreclosing on future generations. Refraining from building homes on barrier islands or in the floodplains of rivers, grazing cattle at the carrying capacity of grasslands, and letting natural forest fires burn are additional means of living and thriving within limits. They all pay huge dividends in the long term and represent a kind of insurance policy for the future. The Case Studies and Models of Global Sustainability included in many chapters in this text illustrate how people are learning to fit into nature's grand scheme, abiding by immutable ecological laws.

Despite these gains, the frontier mentality is still the dominant social paradigm of modern society. (Paradigms are discussed in Chapter Supplement 1.1.) If you listen to politicians or newscasters, you will hear the incessant chatter of unlimited possibilities—for continued growth, bold new frontiers, and vast, virtually limitless resources. So deeply imbedded is the frontier mentality that it affects how most of us view our problems and how we attempt to solve them. What is more, nearly all social

and political institutions function to maintain the chief goal of frontierism—continued growth. This deep entrenchment makes the frontier mentality difficult to dislodge.

20.2 Roots of the Frontier Mentality

Where did the frontier ethic come from and why does it persist? Philosopher Bertrand Russell wrote: "Every living thing is a sort of imperialist, seeking to transform as much as possible of the environment into itself and its seed." Although organisms do not intentionally seek to convert the resources of the environment into their own kind, many populations naturally expand in proportion to available resources—water, soil, sunlight, and others. I call this tendency to expand **biological imperialism.**

The human population, like that of any other species, expands in response to favorable environmental conditions. Like the water hyacinth (Chapter 8), *Homo sapiens* has been a remarkably successful colonizer. Too successful, in some cases.

From our meager origins in Africa, we have expanded to capitalize on the Earth's resources. Today, human civilization in one form or another inhabits virtually every biome on Earth and makes use of many aquatic life zones. Hardly a square meter of the planet has been spared the human imprint. In large part, major environmental transgressions wrought by human society owe their origin to biological imperialism.

Early environmental disasters, such as the Pleistocene extinction of many species of North American mammals, can be attributed in large part to the steady outward expansion of humanity aided by the use of primitive technology, such as fire and early weapons. With the advancement of technology in the Agricultural and Industrial revolutions, the pace of environmental destruction quickened. In the past 40 years, we have seen a further hastening of the Earth's destruction as technology and per capita consumption, particularly in the Western world, have surged forward.

Over the years, technology has served to mitigate natural checks on human population expansion. And, it has helped humans expand the carrying capacity of the environment to support humans. In effect, technology unleashes biological imperialism, permitting us to expand our numbers and dramatically reshape the planet to our liking. Agricultural technologies, for instance, dramatically increase food production, and medical technology has eradicated many infectious diseases, such as the plague, which once took a massive toll on human populations. Together, these and other advances have led to an explosion in human numbers previously unwitnessed on the planet.

Biological imperialism and technological liberation combine to make our species the most dangerous life form on the planet. In short, technological liberation has translated a natural biological urge into a global environmental disaster.

While humans may be biologically predisposed to imperialistic tendencies, somewhere in human evolution, a frontier ethic began to be articulated. The frontier ethic may have simply been a social justification for a deep-seated biological urge.

Most historians exploring the roots of the frontier mentality have focused on the origin of the idea that humans are "apart from and above nature" and "succeed by domination and control." But, equally important are the roots of the attitude that "there is always more." Let's first examine the origin of the attitudes of separation, domination, and control.

University of California historian Lynn White argues that Christian teachings are at the root of Western attitudes toward nature. In support of his case, he cites Genesis 1:28, which instructs us to "Be fruitful and multiply, and replenish the Earth, and subdue it; and have dominion over the fish of the sea, and over the fowl of the air, and over every living thing that moves upon the earth." This quotation, says White, proclaims human supremacy over nature. In a classic paper, "The Historical Roots of Our Ecologic Crisis," White argues that

the Judeo-Christian ethic maintains that nature has only one purpose: to serve humans. Furthermore, he argues, Christians believe it to be God's will that we exploit nature for our own purposes. White also argues that Christian teachings fostered a dualism of humans and nature. This sense of separateness allowed people to exploit nature with indifference.

White's main argument is that Christian teachings (dualism, separation, and dominion) greatly influenced the development of science and technology. The emergence of science and technology, directed by Christian teachings, spawned an era of environmental destruction that continues today, says White.

Critics of White's thesis point out that the Hebrew word translated as "dominion" implies a supremacy that is tempered with love, concern, and responsibility. Moreover, religious scholars point out that passages in the Bible call on humankind to protect the Earth. Revelations, for example, says "Hurt not the earth, neither the sea, nor the trees" (7:3). In Genesis, Adam and Eve are placed in the garden of Eden and told to "till it and keep [*shomer*, meaning "guard" in Hebrew] it" (2:15). They are instructed not to destroy but to supervise and maintain it—to protect the environment. White, say some scholars, has taken one verse from the Bible and built an argument on it, ignoring others.

Biologist René Dubos and others pointed out that exploitation of the natural world has occurred since antiquity, long before the advent of Christianity. Soil erosion, species extinction, excessive exploitation of natural resources, and other environmental disasters are not unique to Judeo-Christian peoples. Two examples of such damage were the widespread extinction of many large animals, such as the wooly mammoth and saber-toothed tiger, caused by early North American hunters and gatherers about 10,000 years ago, and overgrazing and general mismanagement of the once-fertile Tigris-Euphrates region, which led to widespread desertification. Such changes are occurring today even in countries with religious beliefs that purportedly foster reverence and protection of the natural world (Figure 20.1).

White's assertions about the roots of our attitudes toward nature illustrate a problem common to many debates—that is, they tend to simplify matters, searching for single causes, when, in fact, several factors may be involved. Critical thinking instructs us to dig deeper.

Psychologists shed additional light on the roots of our attitudes toward nature. One particularly important insight stems from an understanding of the mental models of reality that humans create. Psychologist Alan Watts noted that for most of us what is inside our skin we call "I." That which is outside is "not I." Today, and in years past, most people operate with the view of nature as "not I." This dualistic concept of humans and nature is a natural psychological development that results in an ethic of separation. Thus, what we do to the Earth is

Figure 20.1 *Eastern religions often proclaim a deep and abiding respect for nature, yet natural systems suffer enormously as shown here in this deforested region in Thailand.*

viewed as little consequence to us. Knowing how dependent humans are on the natural world, it should be painfully evident that this dualism misrepresents reality and misdirects our actions.

Our attitudes toward nature, especially separation and domination, also stem from ignorance, which not only shapes views but keeps them from changing. Two areas of ignorance are particularly important. First is the widespread lack of understanding of the supporting role that natural systems play in our lives. Even today, despite massive growth in information, only a small number of individuals realize the importance of natural systems to our economic and personal well-being. Without such an understanding, the frontier notion of separation persists. Second is ignorance of the long-term environmental consequences of human action. Ignorance extends back to the time when the first humans evolved. From an environmental standpoint, such ignorance was fairly inconsequential until human numbers reached a threshold of population size and technological development.

Having viewed some roots of separation and domination, we now turn our attention to the dangerous notion that "there is always more." Tracing this viewpoint is fairly easy. For most of human history, population size was small in comparison with the Earth's resources and its capacity to assimilate wastes (Figure 20.2). For hundreds of thousands of years, humans usually found what they wanted. Even when local supplies were exhausted, humans were able to find what was needed elsewhere. In short, there always seemed to be more.

Today, however, the human population has climbed to 5.5 billion, species extinction is accelerating, many

natural resources are becoming depleted, and global pollution threatens the long-term health of the biosphere. In sum, human society has reached critical limits and, in some cases, exceeded them. Nevertheless, continued ignorance of limits runs rampant. Combined with denial, an abiding faith in technological fixes, difficulties determining exact supplies (Chapter 13), and other factors, many people hold steadfast to the idea of unlimited resources.

In summary, the frontier mentality may be a conscious expression of biological imperialism. It may have roots in religious teachings and psychological development, especially the dualistic model of humans and nature. All the while, ignorance and denial prevent us from correcting fundamental misconceptions. Our challenge today is to overcome the ignorance through education and go beyond denial to change the way we see ourselves and our relation to the planet.

20.3 The Sustainable Ethic: Making the Transition

As entrenched as our belief system may seem, it can be changed. Colonialism and slavery and the beliefs that supported them, for instance, have fallen by the wayside. Communism fell dramatically and almost without warning, despite 60 years of rule. New paradigms replaced these outmoded systems when it became clear that they did not serve humanity well. Although their demise was not without turmoil, they did fall (Figure 20.3). The same may hold true for the frontier ethic, which a growing number of people are recognizing has outlived its

Figure 20.2 *For a long time in human evolution, the Earth consisted of huge, unpopulated expanses. Resources must have seemed inexhaustible.*

welcome. As the futurist John Naisbitt wrote, "Change occurs where there is a confluence of both changing values and economic necessity." What will replace the frontier ethic?

The Environmental Ethic

Some people think all that's needed is an environmental ethic that views the environment as important to our lives and worthy of protection. One of the most influential proponents of environmental ethics was the late Aldo Leopold, a wildlife ecologist, best known for his book *A Sand County Almanac*. Leopold carried on the battle of John Muir, the founder of Sierra Club and a long-time crusader for wilderness. Leopold described the need to include nature in our ethical concerns, more specifically to extend our concerns beyond people. He called his ethic a **land ethic**. It proposed that humans were a part of a larger community that included soil, water, plants, animals—in short, the land. Leopold suggested caution and deferred rewards in our use of natural resources.

Leopold first suggested the need for a land ethic in 1933. His book, which was published in 1949, took the message further. Charles E. Little, author and founder of the American Land Resource Association, calls the land ethic "one of the most important ideas of the century."

Figure 20.3 *Communism fell dramatically, despite 60 years of domination. Change may occur when economic and ecological values change.*

Leopold's view was considerably more encompassing than the view of Theodore Roosevelt-style conservationists, who protected resources principally because they were of value to humans. The land ethic has helped to change the thinking of many people worldwide.

The land ethic teaches us to respect the land and ecosystems. It instructs us, in Leopold's own words, to enlarge "the boundaries of the community to include soils, water, plants and animals, or collectively: the land." In so doing, Leopold thought that the role of *Homo sapiens* would shift from "conqueror . . . to plain member and citizen of it."

For Leopold, conservation required equal amounts of reflection and action. He called for individual responsibility in maintaining the health of the land, but his guidelines revolved mostly around wildlife management. As such, his land ethic falls short of the complex challenges facing the world today.

A New Ethic to Meet Today's Challenges

Enter the sustainable ethic. The main tenet of the sustainable ethic is that "there is *not* always more." As previous chapters have shown, the Earth has a limited supply of nonrenewable resources, such as metals and oil. Even renewable resources can be depleted if not properly harvested—that is, harvested at a rate that does not exceed the replacement rate. Furthermore, the Earth has a limited capacity to absorb our waste.

To develop a sustainable society, we must recognize that infinite growth of material consumption in a finite world is impossible. We must also accept the fact that ever-increasing production and consumption in a world of limits is very likely a prescription for disaster. It could destroy the life-support systems of the planet upon which we depend.

The sustainable ethic also holds that humans are not *apart from* but rather *a part of* nature. We are subject to the laws that govern all life on Earth, not immune to them. This notion is articulated in a quote attributed to the Indian chief Seattle, who watched in despair as white people usurped the Indian lands of the Pacific Northwest:

> You must teach your children that the ground beneath their feet is the ashes of our grandfathers. So they will respect the land, tell your children that the Earth is rich with the lives of our kin. Teach your children what we have taught our children—that the Earth is our mother. Whatever befalls the Earth, befalls the sons of the Earth. If men spit upon the ground, they spit upon themselves.
>
> This we know. The Earth does not belong to man; man belongs to the Earth. This we know. All things are connected like the blood which unites one family. All things are connected. What befalls the Earth befalls the sons of the Earth. Man did not weave the web of life; he is merely a strand in it. Whatever he does to the web, he does to himself.

The sustainable ethic also maintains that the key to success is through cooperation rather than domination

Viewpoint 20.1

Why We Should Feel Responsible for Future Generations

Robert Mellert
The author teaches philosophy and future studies at Brookdale Community College in Lincroft, New Jersey. He has published numerous articles on process philosophy, ethics, and future studies.

"Why should I feel obligated to future generations? We're inevitably separated by time and space. My presence here on Earth now will have no influence on someone living 200 years from now."

You may have heard this opinion expressed by your friends—perhaps you even hold it yourself. But, if you have ever explored a wilderness preserve, used a library, or visited a historical monument, you already have some reasons for being responsible. Much of what we value in our family, our society, and our world has been provided by our predecessors, sometimes at considerable cost and effort on their part.

In today's world, we face a number of issues that will affect future generations even more profoundly than they affect us now. Exploding world populations, shrinking nonrenewable resources, and plant and animal organisms threatened with extinction all add up to one thing—an ailing environment.

These are not isolated problems. Each stems from a common perception of our relationship to the world and our future. This perception can be characterized by a description of people and things as unique, immediate, individual, and separate from everything else. Any solutions that we might use to resolve our problems would have to start by challenging this perception.

Let me propose four basic considerations that may suggest a new paradigm for understanding our relationship to the world and our future. I believe our moral responsibility to coming generations will follow directly from these.

1. Future generations will be essentially the same as we are. They may have different wants and priorities, but they will manifest the same basic needs for food, water, air, and space. In addition, they will have the same basic physical and mental capacities with which to interact with their environment. Once born, they, too, will claim a right to life and protection from life-threatening conditions, such as extreme temperatures, toxins, famine, and disease. To give them life without also providing the basic means to sustain and enhance life would be cruel. If we expect the species to continue, we are obliged to leave a hospitable environment for those still to come.

2. One is born into a given generation by historical accident. None of us chose when to be born, or to whom. Because we have no special claim to the time and place of our birth, justice would require that we have no more rights over the world and its resources than anyone else.

3. Our survival as a species is more important than our individual survival. This is confirmed in nature every day; parents, whether they be rabbits, wolves, whales, or humans, spend their energies to reproduce and care for their young before they themselves die. Many will even risk their own lives for their offspring. This is because life is not ours to keep, but to share with others.

4. Even after we die, the effects of our life continue. We will be present in the memories of others and in the habits and traditions we shared with them. Our ideas will continue to enlarge the range of options for others; and even when these memories and ideas are no longer consciously a part of the future, they will ripple onward, actively influencing the course of future events and people. What has been can never die. We are what we have been given and what we have chosen to make of these "gifts." In short, we are the product of our ancestors—all that they died for and believed in—and we are the product of our decisions. Future generations will be the result of what we are now and how they use what we leave them.

If we accept these four simple ideas, it is easy to see why we have an obligation to future generations. Our obligation is based on the truth that we are more than unique and separate individuals, living only the immediacy of the now. We are, rather, parts of a much larger whole, one that transcends space and time. As John Locke, the great English philosopher, once said, we owe the future "enough and as good" as we received from the past.

and control. In short, we must learn to fit the human economy within nature's economy. Put another way, all human action must be ecologically sound. Just as all flying machines must obey certain laws of aerodynamics, human action must obey the laws of nature, respecting limits and protecting the life-support systems upon which our economy and our lives depend.

The core value of the sustainable ethic is respect and care for the community of life. This value reflects a duty to one another and to other life forms, now and in the future. The core value implies that human activity should not occur at the expense of other species or other people. As this text has pointed out many times, managing our activities so that they do not threaten the survival of other species or eliminate their habitats is as much a matter of ethics as it is a practicality.

One outgrowth of a change to a sustainable ethic is restraint. In regard to technology and development, the ability to say "I can" would not inevitably be followed by "I will." Instead, new questions would be asked: "Should we build this dam? Should we introduce this product? Should we build more nuclear weapons? Should we have another child?" Instead, "I can" would be followed by two important questions: "What are the environmental consequences?" and "Should I?"

The sustainable ethic entails turning away from self-centered thinking and favoring what is good for the whole of society and the Earth. Restraint is exercised because it benefits all people alive today, future generations, and the many species that share the planet with us. Restraint could help create a high-synergy society in which the individual parts function for the good of the whole.

As noted in the first chapter, the sustainable ethic outlines six principles by which society can operate, thus putting ethical guidelines into action. They are conservation, recycling, renewable resource use, restoration, population control, and adaptability. These operating principles could help us reshape all human activities and systems, from the way we acquire food to the way we transport ourselves from home to work.

The sustainable ethic and its practices are growing worldwide, as witnessed in 1992 by the Rio Conference, which brought leaders from virtually every nation of the world together toward a common goal.

The sustainable ethical system is a new paradigm that lays the foundation for a sustainable society. However different a sustainable society may be, it does not necessarily renounce all technology, all growth, or all material goods. Instead, it advocates a thoughtful look at the long-term health of the planet and an evaluation of the consequences of technology, population growth, and materialism. All decisions are passed through the sustainability filter to determine if they will enhance our present and our future.

Toward a Humane, Sustainable Future

In 1992, Donella Meadows, Dennis Meadows, and Jorgen Randers, three of the four authors of *Limits to Growth*, which was discussed in Chapter 6, published a sequel entitled *Beyond the Limits*. In it, the authors wrote, "The human world is beyond its limits. The present way of doing things is unsustainable. The future, to be viable at all, must be one of drawing back, easing down, healing." The sustainable ethic can provide an ethical framework for these changes.

To live sustainably, though, it is not enough to change the way we treat the Earth. It is not enough to become better stewards of the Earth's resources and ecosystems. We must also dramatically improve the way we treat one another. In other words, it is essential to create a humane human existence, reshaping social structures so they foster human dignity and welfare. As a rule, those countries where people are the most oppressed—where freedoms and justice are oppressed by military regimes, corrupt governments, or powerful economic forces—have some of the worst environmental conditions.

Although individuals differ in the goals they set for economic and social development, several common goals are universal, say the authors of *Caring for the Earth*. These include a long and healthy life, education, and access to resources needed for a decent standard of living. They also include political freedom, a guarantee of human rights, and freedom from violence. Thus, many observers believe that sustainability requires an ethic calling for universal justice, freedom, and compassion for one another.

Because ethics are the foundation of human behavior, changes in the way we think could begin to change the way we act. If actions are based on a set of principles conducive to a humane, sustainable existence, our future could follow accordingly. But how does one go about changing the way the world thinks?

20.4 Developing and Implementing the Sustainable Ethic

The daunting task of changing the way people think and act worldwide elicits great pessimism. Nonetheless, examples of the shift in ethical values abound, testing the resolve of even the most ardent pessimists. This section describes several ways by which we can literally change the way the world thinks.

Promoting Models of Sustainability

The many Models of Global Sustainability in this text highlight successes of people, businesses, and governments that have taken significant actions based on their commitment to the Earth and one another to build a sustainable world. Models such as these are extremely important catalysts in the sustainable transition. Similar examples are often highlighted in other books, television news reports, newspaper articles, documentaries, magazine articles, government reports, and other avenues. Not only do they inspire other people to take action, they also offer practical guidance on the ways to achieve similar goals and to prosper within limits. The successes of models are also broadcast by word-of-mouth communications and via computer networks.

Education

Educators can also play an important role in changing the thinking of the world's people. Recognizing the importance of environmental education, some states, such as Wisconsin, have made environmental education mandatory. Colleges, such as Prescott College in Arizona, include environmental education as part of their mission statement.

Steve Van Matre, head of the Institute for Earth Education, has proposed an alternative to issue-based environmental education for school children. He argues that environmental education for school-age children is often disorganized and haphazard. It fails to develop a true understanding of the importance of the environment in supporting our lives. It also tends to overlook how we individually impact those systems and what we can personally accomplish by changing our habits. By focusing on these three areas and developing strong connections with the environment, Van Matre believes we can make important gains in environmental education.

Churches

Churches can also become an important force in creating global change. Religious leaders who are not already involved, for example, may want to explore a broader sustainable ethic with their parishioners and suggest ways to change. Many churches sponsor adult education classes that deal with social, political, and environmental issues. These, too, can be an excellent route for raising awareness and promoting a new Earth ethic.

Declarations of Sustainable Ethics and Policy

Nations can help redirect human values and actions on a broad scale by adopting declarations of sustainable development and ratifying constitutional amendments that make the environment a leading consideration (Chapter 21).

A World Organization Dedicated to Sustainable Development

In 1991, the authors of *Caring for the Future* proposed a novel idea for educating the world's people and bringing about the sustainable transition. They recommended the establishment of a world organization dedicated to sustainable development, akin to Amnesty International, which defends and ensures human rights in all countries. This organization would seek ways to secure the observance of sustainable ethics and practices in all countries.

As farfetched as this may seem, the 1992 Earth Summit in Rio actually produced an organization with similar goals, known as the **U.N. Commission on Sustainable Development (UNCSD)**. The commission could become an international watchdog, overseeing promises made at the Rio Conference, "shining the spotlight on countries that renege and pushing through stronger commitments as necessary," says Worldwatch Institute's Hilary French.

UNCSD will consist of high-level government representatives, very likely heads of environmental protection branches of governments, with the power to make decisions. The commission will use its powers of publicity and peer pressure to encourage the implementation of **Agenda 21**, a massive blueprint for sustainable action. It will also promote the **Rio Declaration**, a statement of 27 principles regarding environment and development, and an agreement on forest protection.

Equally important, the commission will help coordinate the many programs and agencies of the United Nations that deal with the environment, among them the Development Program, the World Bank, the Food and Agriculture Organization, and the International Monetary Fund. All U.N. entities will report their activities to UNCSD to integrate the sustainable policies adopted at the Earth Summit.

Although it is too early to judge the success, this effort could be very instrumental in instilling sustainable beliefs and practices among important players in the world community.

A Role for Everyone

Political leaders can play a major role in reeducating people through speeches and television advertisements. Parents can help by teaching their children. Children can help by teaching their parents. The opportunities are limitless; the time to make the changes is not.

20.5 Overcoming Obstacles to Sustainability

Promoting a global sustainable ethic will be slow, uphill work. Many people do not see the need for change, and others resist change because it threatens their personal interests.

Clearly, many obstacles lie on the road to sustainability, among them our faith in technological fixes, apathy, feelings of insignificance, self-centeredness, ego gratification, and economic self-interest and outmoded governmental policies. This section explores some ways to overcome these obstacles.

Faith in Technological Fixes

Many people express extreme optimism in the power of technology to solve our problems, a phenomenon labeled **technological optimism**. Economist Julian Simon, for instance, says there are no limits (Point/Counterpoint 5.1). As long as we exercise our ingenuity, we can develop new technologies to expand our reserves or find substitutes. So why change our views? The world truly is unlimited!

In their zeal to find high-tech solutions, logical, inexpensive, and Earth-friendly low-tech solutions, such as conservation, are often overlooked. As you have seen in previous chapters, conservation is a far more effective way of reducing global carbon dioxide and is far cheaper than the proposed nuclear solution. A dollar invested in conservation, in fact, reduces carbon dioxide emissions seven times more than a dollar invested in nuclear power. Recycling is a far cheaper solution than high-tech trash incinerators (Chapter 18). But, many municipalities are building incinerators.

Unfortunately, our unqualified optimism in technology blinds us to the obvious solutions. Spurred by thoughts of limitless resources, many applied scientists, technologists, politicians, and business people promise an unlimited future. Sure that these experts can't be wrong, many people place their faith in their proposed solutions and resist changing their views.

Today, more and more people are beginning to see the fallacy of technological optimism and are calling for simpler, more effective solutions that strike at the root causes of the crisis of unsustainability.

Apathy, Powerlessness, and Despair

Many people, while understanding that the Earth is finite, remain apathetic about the course of modern society. Involved in their own lives, they see resource limitations and pollution as problems for which someone else must take responsibility. Apathy is effortless, noncontroversial, and cheap.

Where does apathy come from? In part, many of us are taught or conditioned to be apathetic—not to rock the boat or make waves. Someone bulldozes a favorite forest to put up a shopping mall and someone calms our indignation by saying, "That's the price of progress."

Government and history professors often fail to teach students about participating in government. For many people, democracy means freedom. But, democracy also entails certain responsibilities. First and foremost, we have a responsibility to participate: to vote, to write congressional representatives, to sit on citizen advisory committees. But, many people don't even vote. Why bother?

Another cause of apathy is powerlessness. In the United States and other countries, people are often paralyzed by a sense of powerlessness—feelings of insignificance. If you can't do anything, why worry?

This pervasive feeling has a powerful effect on our lives and the environment. Feelings of insignificance create many of our problems and keep us from solving them. I call this the *paradox of inconsequence*. Being just one of billions of people on Earth is an excuse many of us use for continuing to do what we have always done, living our lives somewhat wastefully. What difference, we ask, does it make if I drive 10 miles per hour over the speed limit and waste a little gasoline? I'm just 1 of 255 million Americans. I'm just a small fraction of the problem.

The trouble with this thinking is that millions of people think the same thing. Together, their actions add up to a lot of waste and pollution.

Ironically, feelings of insignificance also keep us from solving many problems. If I drive the speed limit, recycle aluminum and paper, and keep my thermostat at a reasonable temperature, what difference will it make? Because I'm only 1 of 255 million, my contribution to the resource depletion and pollution is insignificant. So why do it?

In a nutshell, feelings of insignificance create many of our problems and keep us from solving them. But, all of this can be turned around. Millions of people acting responsibly can have huge impacts. Ascend a marble stairway that tens of thousands have ascended before you. Feel the grooves worn by their feet. Although each one would say he or she had no effect, together they had an enormous impact. If we can have a negative impact, we can also have a positive impact by acting responsibly. Education can greatly help to convince individuals of the mathematics of responsible living.

Figure 20.4 shows a hypothetical curve that plots apathy and its close cousin, despair, against empowerment—taking control of your life and taking action. It suggests, and with good reason, that as people take control of their lives and work toward change, apathy and despair tend to fall.

The Self-Centered View

Some people have a self-centered view. Their economic and noneconomic welfare governs their actions: what kinds of homes they buy, what size cars they buy, how

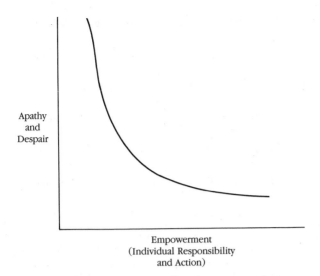

Figure 20.4 *Hypothetical curve showing that apathy and despair are at their peak when people are powerless or feel powerless to make changes. When individuals take action, apathy and despair often fall, opening the doors to positive change.*

much luxury they surround themselves with, and so on. Author and social critic Tom Wolfe coined the phrase the *me-generation* to describe the U.S. citizens of the 1980s who seemed intent on self-indulgence. Many people of the me-generation buy what they can afford, giving little thought to the effects of consumption. Replacing a self-centered approach with a global environmental perspective is the main thrust of the sustainable ethic.

Ironically, many environmental groups have failed to promote individual responsibility and action among their members and the general public. Much of their educational outreach is through direct mail, appeals that generally outline the problems we face and proposals to solve them through legal and legislative action. Individuals are asked to help, but individual involvement is generally limited to writing a check. Wendell Berry, philosopher and author of *The Unsettling of America*, writes: "The giving of money has . . . become our characteristic virtue. But to give is not to do. The money is given in lieu of action, thought, care, time." I call it the *cash conscience*. Environmental groups also spend much of their time pointing accusatory fingers at lawmakers, inept government regulators, and avaricious business executives.

This lopsided approach nurtures the idea that the blame for the environmental crisis lies almost entirely in someone else's hands and can only be solved by regulating someone else's activities or by applying new and more efficient technologies to control someone else's pollution. It nurtures the idea that money alone can solve our problems. The legislative and legal work of environmental organizations, such as the Environmental

Defense Fund, the Natural Resources Defense Council, Greenpeace, the Sierra Club, and others is vital to our efforts to improve the environment, but individual action is equally important. The me-generation will never get the message unless someone starts saying it.

Ego Gratification

According to psychologists, as we pass from infancy to adulthood, we develop a sense of the self, our own separate identity. Psychologists call this the *derived self.* Once the derived self becomes established, it needs reaffirmation. We can get that from our parents, friends, teachers, and ourselves. If not, we may seek it in new clothes, fast cars, and extravagant homes, self-rewarding behavior that builds our egos or reaffirms that we are important. Reaffirmation may be at the roots of materialism and overconsumption, which depletes essential resources, causes pollution, and reduces wildlife habitat. What can a society do to build feelings of self-worth?

Adult education programs and better training in health classes can help us break patterns of child-rearing and create new ones. Religious leaders can help as well.

Economic Self-Interest and Outmoded Governmental Policies

Most of us live in countries whose socioeconomic systems are out of line with the natural world—that is, whose human systems are operating unsustainably. Greed, ignorance, and a host of other factors drive us farther down the road to ecological ruin. Our governments and the economic systems they often serve are a barrier to sustainability. Previous chapters have pointed out some of the flaws in both government policy and economics. Chapters 21 and 22 go into more depth, showing where reform is needed to place these human activities back onto a sustainable course.

Value Judgments and Decision Making

Each day people are faced with dozens of decisions, many of which influence the environment. For example, should you ride your bicycle or drive your car to school or work? Should you turn up the heat or throw out an old pair of pants? The sum of the decisions made by many millions of individuals like yourself add up. Corporations and governments face many decisions as well. Single decisions at this level have impacts as profound as many millions of individual decisions. For example, should the government develop nuclear power or push for energy conservation? Should the government use recycled paper or continue buying virgin stock? Clearly, understanding how decisions are made and discovering alternative ways of making decisions can help us, individually and collectively, learn to be more responsible.

Ultimately, most of our decisions are influenced by our values—what we view as right or wrong, desirable or undesirable. Values are learned from our parents, relatives, peers, teachers, religious leaders, politicians, writers, and even news commentators and reporters. Values may shift over the years, sometimes drastically.

Many people today base their judgments on a doctrine by which the worth of anything is determined by its usefulness, **utilitarianism**. This view tends to put human needs above most, if not all, others. Economics, discussed in the next chapter, is the yardstick of many utilitarian decisions. Utilitarian resource management, for example, finds the fastest, cheapest ways to acquire resources. It puts human needs above all else. Thus, forests may be reseeded, not so much to provide habitat for animals as to provide wood for future generations.

Interestingly, at first glance, the frontier ethic appears to be heavily utilitarian, serving immediate human needs. Upon closer examination, though, it is clear that its tenets are anything but utilitarian, for they lead to destructive behavior that forecloses on our future.

At the opposite end of the spectrum is a new and controversial view of **natural rights** (described earlier in this chapter and also in Chapter 8). It says that all living things have rights irrespective of their value to human society. Wilderness should be set aside, not just so people can use it, but to protect the species that have lived on the land for thousands of years. Animals, plants, and insects should be preserved, not because they are of use to humans, but because they have a right to live.

Another value system is based on the word of God as written in the Bible, **divine law**. For the most part, divine law dictates personal behavior—interactions between people. The Ten Commandments instruct us how to treat one another: Thou shalt not kill. Thou shalt not commit adultery. One religious group has recently added an 11th commandment: Protect the Earth.

These are some of the principal value systems that affect our decisions. As you think about them, you may realize that no one system is at work at all times. In some cases, you may act out of utility. In others, you may make a sacrifice for the good of the whole.

The sustainable ethic presents an alternative system of values. It calls on us to consider a new set of parameters in making decisions. It reminds us of our place in the world and implores us to act in cooperation and consideration rather than in isolation and strict self-interest. At first, this may seem anything but utilitarian. It is not. The sustainable ethic is a means of serving human needs and the needs of nature, which are the same.

A human being is part of the whole, called by us "Universe." . . . He experiences himself, his thoughts and feelings as something separated from the rest—a kind of optical delusion of his consciousness. This delusion is a kind of prison for us, restricting us to our personal desires and to affection for a few persons nearest to us. Our task must be to free ourselves from this prison by widening our circle of compassion to embrace all living creatures and the whole of nature in its beauty.

—Albert Einstein

Critical Thinking Exercise Solution

You, your friends, and your parents probably do many things to help protect the environment that don't benefit you economically. For example, you may recycle. Although you probably don't earn enough from the bottles and cans you return to make it worth your while, you do it anyway. So do millions of other people in the United States and other countries. Therefore, you and your family are living proof that some people do act for unselfish reasons.

The cynic who asserts that people respond only when they're financially benefited is making a broad generalization that fails to hold up to scrutiny. A more accurate statement might be that many people will take actions on their own; others seem to respond best when they're financially motivated. Those who make statements such as the ones that launched this exercise may be motivated by money, and consequently may think everyone else is. Their bias clearly affects their view of the world. Beware of powerful generalizations; they can become crippling thought stoppers.

Summary

20.1 The Frontier Mentality Revisited

■ Many people operate with a **frontier ethic**, which is based on three tenets: (1) The world has an unlimited supply of resources for exclusive human use, (2) humans are apart from nature, immune to its laws, and (3) success stems from the control and domination of nature.

■ The frontier ethic is one of the root causes of the environmental crisis. It has been a part of human thinking for thousands of years.

20.2 Roots of the Frontier Mentality

■ Humans are biological imperialists. We naturally expand to make use of available resources. Biological imperialism has been aided by technological advances that unleash us from limits in ways that could result in global ecological collapse.

■ The frontier ethic may be a justification for a deep-seated biological impulse.

■ Some claim that the frontier ethic is embodied in Judeo-Christian teachings, which teach dualism (separation) and domination and control. But opponents argue that Biblical teachings also call for respect and wise stewardship.

■ Human psychological development tends to promote a view of humans as apart from nature.

■ The roots of the attitude that there is always more are easily seen. For most of human history, population size has been small compared with the Earth's resource supplies. There has always been more, until recently.

20.3 The Sustainable Ethic: Making the Transition

■ The **sustainable ethic** holds that the Earth has a limited supply of resources and that humans are a part of nature. It also maintains that success comes from efforts to fit our activities into the economy of nature, not dominate and control natural systems.

■ Putting these lofty ideals into action requires: conservation, recycling, renewable resource use, restoration, and population control. It will also require cultural adaptability.

20.4 Developing and Implementing the Sustainable Ethic

■ The task of changing the way people think and act worldwide elicits great pessimism. But, examples of the shift in ethical values abound, testing the resolve of even the most ardent pessimists.

■ Models of sustainable actions are extremely important catalysts in spurring the sustainable revolution. They can be promoted through books, classroom lessons, sermons, political speeches, and television specials.

■ Nations can help by adopting declarations and constitutional amendments based on the sustainable ethic.

■ At the Earth Summit in Brazil in 1992, the United Nations established the international organization known as the **U.N. Commission on Sustainable Development**, which could play a major role in promoting sustainable activities throughout the world.

20.5 Overcoming Obstacles to Sustainability

■ Many barriers lie in the way of changing attitudes and building a sustainable society.

■ One barrier is apathy. Another is the fact that many people hold self-centered views: What they can afford determines what they buy. Little thought is given to what the Earth can afford or one's obligations to future generations.

■ Feelings of insignificance also contribute to the problems we suffer and can prevent us from solving them.

■ Human decisions about the environment are affected by our values. Many people base their decisions on the value of things by considering only their usefulness. This is called **utilitarianism**.

■ **Natural rights** hold that all things have value and rights, irrespective of their usefulness to humans.

■ **Divine law** is contained in the Bible. It dictates how some people act.

■ The sustainable ethic is an alternative system with great importance. It is quite utilitarian, although this isn't obvious at first, for it seeks to protect natural systems that are the life-support systems of the planet and the source of all economic wealth.

Discussion and Critical Thinking Questions

1. Describe the three major tenets of the frontier ethic. For one week, when you listen to the news, watch television, read articles, and listen to people talk, note examples of the frontier mentality and make a list of them.

2. Is your personal ethic closer to a frontier or a sustainable ethic?

3. Debate this statement: "The frontier ethic comes from Judeo-Christian teachings, which give humans dominion over the world."

4. Describe the terms *biological imperialism* and *derived self*. How are these related to environmental damage?

5. Discuss the tenets of the sustainable ethic. Indicate which of these tenets coincide with your personal beliefs. Which ones don't? Why?

6. Critically analyze the tenets of the sustainable ethic. Do you see any flaws in them?

7. Describe modes by which the sustainable ethic can be promoted worldwide. Would you approve of such an activity?

8. Describe the role apathy, self-centeredness, feelings of insignificance, and technological optimism play in creating environmental problems and blocking change. How can each one be changed?

9. "Animals and plants have rights," says a leading philosopher, "irrespective of their value to humans." Do you agree? Why or why not?

Suggested Readings

Berry, W. (1987). *Home Economics*. Berkeley: North Point Press. Thoughtful collection of essays on wise stewardship.

Chiras, D. D. (1990). *Beyond the Fray: Reshaping the American Environmental Response*. Boulder, CO: Johnson Books. Describes important changes needed in the environmental movement.

———. (1992). *Lessons from Nature: Learning to Live Sustainably on the Earth*. Washington, D.C.: Island Press. See Chapters 2 and 3 for a detailed discussion of sustainable ethics.

———. (1992). An Inquiry into the Root Causes of the Environmental Crisis. *Env. Carcinogenesis and Ecotoxicology Reviews* C10 (1): 73–119. Reviews the thinking on the root causes of the environmental crisis, including human attitudes.

DeVall, B. (1988). *Simple in Means, Rich in Ends. Practicing Deep Ecology*. Salt Lake City: Peregrine Smith. Important discussion of deep ecology and ways to put reverence for nature into action.

Hardin, G. (1985). *Filters against Folly*. New York: Viking. Thoughtful treatise on ethics. Important reading.

LaMay, C. L. and Dennis, E. E. (1991). *Media and the Environment*. Washington, D.C.: Island Press. Insightful and useful critique of environmental reporting by the media.

Leopold, A. (1966). *A Sand County Almanac*. New York: Ballantine. Collection of essays on nature and conservation.

Milbrath, L. W. (1989). *Envisioning a Sustainable Society. Learning Our Way Out*. Albany: State University of New York Press. Early chapters discuss values of the dominant paradigm (frontierism) and need for a new value system.

Nash, R. F. (1989). *The Rights of Nature. A History of Environmental Ethics*. Madison: University of Wisconsin Press. Important reading.

Regenstein, L. (1991). *Replenish the Earth*. New York: Crossroad. Discusses organized religion's treatment of animals and nature.

Rolston, H. (1987). *Environmental Ethics. Duties to and Values in the Natural World*. Philadelphia: Temple University Press. Explains the rights of other creatures.

Schumacher, E. F. (1973). *Small Is Beautiful: Economics as if People Mattered*. New York: Harper and Row. One of the best books ever written on the subject of a sustainable society and new ethical systems.

Van Matre, S. and Weiler, B. (1983). *The Earth Speaks*. Warrenville, IL: The Institute for Earth Education. Superb collection of writings on nature.

Toward a Sustainable Human Economy: Challenges of the Industrial World

A penny will hide the biggest star in the Universe if you hold it close enough to your eye.

—Samuel Grafton

Critical Thinking Exercise

"In order to improve our economy, we need to grow—that is, manufacture and sell more goods," say many business economists. "If we can't do it here, we need to tap into global markets. Without economic growth, jobs won't be available for those who need them. We have to keep growing, even if it means lowering environmental standards."

This economic growth logic is widespread in the United States and other countries. Critics point out, however, that it is flawed for several reasons. Can you think of any reasons that suggest that economic growth is not all it is cracked up to be? Can you think of any other alternatives to growth? What critical thinking rules come into play?

In southern India, people once made traps for monkeys by drilling small holes in coconuts, filling the shells with rice, and chaining them to trees. The success of this trap was based on a simple principle: The hole was large enough for a monkey to insert its empty hand but too small for it to pull out a handful of rice. As monkeys clung tenaciously to their rice, villagers threw nets over them. In essence, then, monkeys were trapped by their own refusal to let go.

Many observers believe that humankind is caught in a similar dilemma. Clinging tenaciously to an unsustainable way of living and conducting business, many ignore the critical environmental trends. As a result, they often resist changes needed to build a sustainable human economy. This chapter begins with an overview of some key economic principles, describes the economics of pollution control and resource management, and then discusses major weaknesses in economic systems from an environmental standpoint. It concludes with a discussion of ways to build sustainable economic systems in industrial nations. (Chapter 22 tackles developing nations.)

21.1 Economics and the Environment

Manufacturing and trade began with the advent of towns and villages (Chapter 4). These activities owe their origin to the agricultural surpluses that allowed people to take up various crafts and trades to earn a living. But, the science of economics began in earnest only two centuries ago, with the publication of Adam Smith's book *The Wealth of Nations.*

Economics is the study of the production, distribution, and consumption of goods and services. It concerns itself primarily with two factors: inputs and outputs. **Inputs** include labor, land, and commodities, such as energy or minerals, that companies require to produce their products. **Outputs** are the products—the goods and services that companies produce for consumption or for further production.

Economics, like environmental science, is concerned with relationships. It employs scientific tools to discover the laws that regulate economies. The description of economic facts and relationships falls within the purview of **descriptive economics**, so named because it is supposed to be free of judgment. Descriptive economics is, relatively speaking, a pure science. Its questions can be answered only by facts. Economics melds with political science and sociology when it attempts to answer value-laden questions. For example, should companies pay for pollution controls? How rapidly should the economy grow? Such questions cannot be answered by empirical facts and figures. There are no right or wrong answers to them, for they are value judgments and are left to the political process. This realm of economics is called **normative economics.**

More and more, people are demanding a new kind of economics, an **ecological economics**, one that concerns itself with supplying the legitimate needs of people while protecting the environment. One of the principal goals of such an economy is to ensure that economic development truly serves people, and serves people equitably. In this chapter, this new discipline is referred to as **sustainable economics**. Sustainable economics represents a new partnership between humans and nature. It is a system of supplying goods and services that is in synch with natural systems, not in opposition. Before we examine what a sustainable economic system might look like and study ways to make this vision a reality, we must focus on conventional economics, looking first at economic systems, then studying some basic principles of economics.

Economic Systems

Economics is a tool that helps societies answer three fundamental questions: (1) What commodities should it produce and in what quantity? (2) How should it produce its goods? (3) For whom should it produce them? Of course, a society can solve these three basic problems in many ways. In **command economies**, such as those found in Cuba and in the former Soviet Union and Eastern bloc nations, governments dictate production and distribution goals. Bureaucrats make all the decisions, even those concerning the amount of fertilizer to be applied to farmland in a region.

In **market economies**, governments (theoretically) take a backseat to the marketplace. That is, companies generally produce the goods and services that yield the highest profit, thus answering the first question: What goods and services should be produced and in what quantity? Profit dictates the answer to the second question: How are goods produced? Generally, the least costly method of production yields the greatest profit. In a market economy, the answer to the third question—for whom?—is determined by money. In general, whoever can afford a good or service will get it.

One of the key principles of economics is the **law of scarcity**. It states that most things that people want are limited. As a result their sale is rationed. In a market economy, price is the primary rationing mechanism. For instance, few people drive Porsches because the price greatly exceeds their ability to pay for them. In command economies, governments ration most of the output, although prices do play a role.

In truth, most economies are mixed—that is, they contain elements of command and market economies. In the market economies of Great Britain, Canada, and the United States, for example, free enterprise is the rule. But, in these and other countries, governments also influence economic behavior.

As you saw in Part 4, one way governments regulate the economy is through laws and regulations that require companies to control pollution. Laws that stipulate how much pollution a company can emit affect the price of goods and services. Outright government bans on dangerous products limit product availability and, therefore, dictate production and consumption, interfering in a free economy in order to protect health or the environment.

Another way of influencing the market is through federally mandated freight rates. As noted in Chapter 13, raw ore travels cheaper than scrap for recycling, which benefits the mining industry and deters recycling.

As you may recall from Parts 3 and 4, state and federal governments also subsidize various activities by special tax breaks or by sponsoring or funding research. These and other subsidies create an uneven playing field that benefits some and harms others. Last, but far from least, governments impose taxes on imports, tariffs, to regulate the flow of goods into a country, thus stifling free international competition.

In sum, regulations, bans, subsidies, and other policy instruments are levers through which governments influence market economies to protect natural resources, people, and special economic interests, such as the mining industry. Those who argue for a pure market economy, without government interference whatsoever, believe that markets can solve all problems.

Although they're nearly extinct, command economies also generally consist of a mixture of market and command practices. China, for example, allows some free market enterprise within an economic system generally controlled by the central government.

Before turning to the principles that govern market economies, it is important to note that both types of economic systems have enormous impact on the environment. The command economies of Eastern Europe and the former Soviet Union, for example, have produced incredible amounts of air and water pollution leaving behind a trail of toxic hot spots not unlike those found in the Western world, where free enterprise has been the rule.

The Law of Supply and Demand

In market economies, the three essential questions posed above are generally solved by price. In turn, the price of a good or service is determined by two other factors: supply and demand. **Supply** refers to the amount of a resource, product, or service that's available. **Demand** refers to the amount people want. The predictable interplay of price, supply, and demand constitutes the **law of supply and demand**.

To understand the relationship between these three factors, take a moment to study Figure 21.1a, which shows a demand curve for rice. On the vertical axis of the graph is the price (P) of rice in dollars per bushel. On the horizontal axis is the quantity (Q) that people will buy at each price. This graph shows that as the price rises, the demand decreases. Conversely, a lowering of price generally increases demand. The relationship between P and Q is inverse—that is, as one increases, the other decreases.

Most of us are familiar with the interplay of price and demand. For example, in the weeks before a sale on blue jeans goes into effect at a local store, sales might be sluggish. The day the price drops, though, business picks up as people flock to avail themselves of the bargains.

The supply curve is of interest chiefly to producers (Figure 21.1b). It shows the relationship between price and the quantity that suppliers will produce. It illustrates an intuitively simple concept: The higher the price, the more producers are generally able to produce. At lower prices, producers often have to reduce production. The fall in oil prices in the early 1980s, for example, put many oil drillers out of business and caused incredible economic hardship throughout the world. What happened? As oil prices fell as a result of conservation, which reduced demand, and cheaper foreign oil, many companies went bankrupt or shut down wells because they couldn't produce oil profitably at the lower prices.

The economy is a balancing act (offset by some government tampering as mentioned earlier) between two principal players, supply and demand. Supply and demand interact to determine the price. Graphically, this is represented by the intersection of the supply and demand curves and is known as the **market price equilibrium** (Figure 21.1c). The market price equilibrium represents the price at which consumers can afford to buy a product and the price at which producers can afford to produce it.

Imagine an economy in which literally millions of prices are set by this kind of interaction without interference by governments. That system would be a **free**

Figure 21.1 *Supply and demand curves for rice.* (a) *The demand curve shows the relationship between the price (P) on the vertical axis and demand (Q) on the horizontal. This graph shows that a rise in prices reduces the demand. Falling prices increase demand.* (b) *The supply curve shows the relationship between the price and supply, or amount produced. The higher the price, the more farmers will produce.* (c) *The market equilibrium point is the intersection of the supply and demand curves. It's the price people will pay for rice and the amount farmers will produce.*

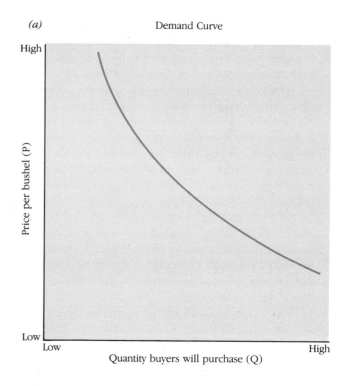

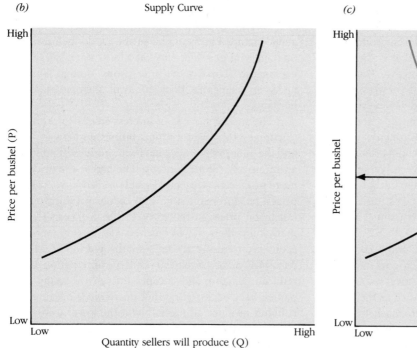

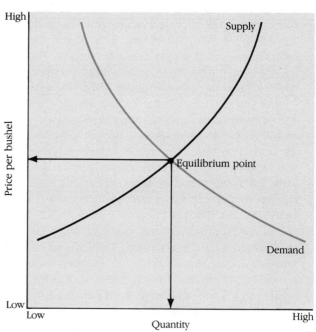

market system. But, as noted above, in most countries, governments do meddle in economic affairs, sometimes benefiting through subsidies and sometimes thwarting free enterprise through regulations.

The free market system may also be tampered with by business itself. Some businesses, for instance, push for protectionist trade policies that limit or eliminate imports from foreign producers in an effort to keep markets to themselves. Others buy up competitors, creating monopolies. Monopolies eliminate competition and can result in prices that are set at any level the monopoly holder wants, provided the public will pay. Antitrust laws in the United States are aimed at protecting individuals from monopolies. Companies may try to drive others out of business. For example, Browning Ferris Industries (BFI), a major U.S. trash hauler, was found guilty of illegally trying to drive a smaller Vermont trash hauler out of business. BFI was fined $51,000 in damages and $6 million in punitive damages.

Environmental Implications of Supply and Demand

The law of supply and demand has some very real implications for sustainability. First, consider the impact of declining supplies by looking at the ivory trade. For years, elephants were slaughtered by the thousands in Africa to support the profitable ivory trade. When elephant populations began to decline, ivory prices rose. Even after African countries made it illegal to shoot elephants, poachers continued. Why? Bans on ivory export and the dramatic decline in wild elephant populations caused by overhunting increased the price of ivory. That, in turn, gave poachers a considerable economic incentive to continue illegal slaughter, even at the risk of being killed by game wardens. The supply graph predicts such activity, showing that the higher the price, the more willing someone is to produce a given product. People become rich, and the elephant is pushed toward extinction. It's a simple line of cause and effect with devastating consequences. Fortunately, worldwide government regulations have recently put a stop to the sale of ivory products. This has caused the market for raw ivory to evaporate, making poaching unprofitable and brightening the future of the elephant.

Supply and demand economics also have considerable impact on conservation efforts. For instance, when the price of oil climbed in the late 1970s because of the artificial shortages created by embargoes on foreign oil, many industrial nations found ways to improve efficiency of factories, automobiles, and homes (Chapter 12). But, supply and demand can also spawn wasteful behavior. The fall in oil prices in the early 1980s, created by conservation and increases in production by Great Britain and other countries, eroded many people's resolve to save energy. The U.S. government, in fact, lost interest in energy efficiency. As a general rule, abundant supplies, at least in the near term, lead to low prices, which often foster wasteful practices. "We've got plenty, so why conserve?" seems to be the attitude of many.

Measuring Economic Success: The GNP

Economists need to measure economic activity. The most widely used measure of a nation's economy is its gross national product (GNP). The GNP is the market value of the nation's output, in other words, the value of all goods and services that a nation produces and sells, including government purchases, in a given year. Real GNP is the GNP adjusted for inflation. Per capita GNP is defined as follows:

Per capita GNP = Real GNP/Total population

Widely used to track economies, the GNP gives a general picture of the wealth of nations and the living standards of their people. As you will see in Section 21.4, the GNP, like the law of supply and demand, is somewhat flawed and in need of corrections to make it sensitive to environmental concerns and sustainability.

21.2 Economics of Pollution Control

The free market economic system in the United States and other countries, as well as the command economies of Socialist and Communist nations, treated pollution with almost uniform disregard until the late 1960s and early 1970s. Some continued even through the 1980s. Pollution symbolized progress. Where it landed, no one seemed to know or care. Under such a system, pollution and its impacts are considered a cost to society not paid by the manufacturer or its customers, an **economic externality**. Economic externalities include damage to human health, fish and wildlife populations, vegetation, climate, and others. Many examples have been pointed out in the text.

In many instances, businesses were simply unaware of the external costs of their activities. Gradually, though, citizens throughout the world began suing polluters if their system of government allowed it. Governments established pollution standards to protect people and the environment, and businesses began to curb pollution, usually by installing pollution control devices. Some of the economic externalities came home to roost as costs were internalized. Reducing external costs became part of the cost of doing business, and consumers paid a slightly higher price for electricity, cars, and a host of other products.

Chapters 14 through 18 described U.S. pollution laws that have forced significant cost internalization, among them the Clean Air Act, the Clean Water Act, the Resource Conservation and Recovery Act, and the Surface Mine Control and Reclamation Act. Although these laws are important, they have not resulted in a full internalization of costs. As pointed out in previous chapters, environmental damage still occurs at unsustainable levels.

The reason environmental destruction continues is largely one of policy. Most government policy designed to control pollution relies on end-of-pipe controls (Chapter 1). In most cases, such controls only reduce the output and lessen problems or slow their rate of development. To a large extent, economics determine the level of control—that is, just how much money people are willing to spend on controlling pollution.

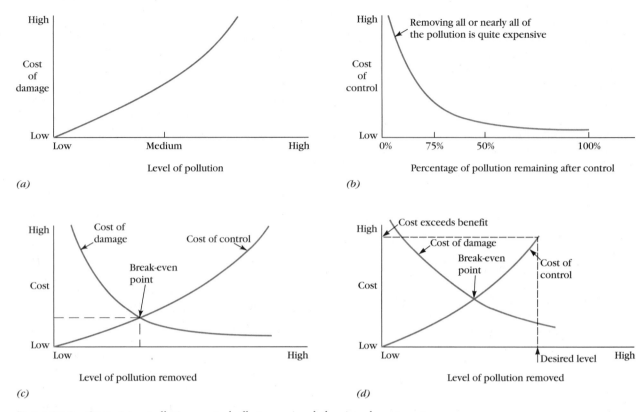

Figure 21.2 *Optimizing pollution control efforts requires balancing cleanup costs against economic benefits of control.* (**a**) *This graph shows that as pollution levels rise, the amount of damage increases, as reflected in rising costs.* (**b**) *Removing pollution requires an economic expenditure when pollution control devices are used. Removing 100% of the pollution is quite expensive. Removing 75% costs less per unit removed. This graph shows the relationship between cost of control and percentage of pollution remaining.* (**c**) *This graph shows the point at which pollution control costs equal economic damage. This is the break-even point. Unfortunately, determining the full costs of pollution is not easy, and estimates may be in serious error.* (**d**) *Society may wish to lower pollution levels past the break-even point. The desired level of pollution may technically cost society more than it benefits in reducing damage.*

Cost-Benefit Analysis and Pollution Control

As you may recall from discussions of risk assessment, the chief goal of pollution control is to reduce pollution in ways that yield the maximum benefit at the lowest cost (Chapter 14). This goal is made possible by cost-benefit analysis.

The cost of pollution control and the cost of harmful effects of pollution can be plotted on graphs (Figure 21.2). Figure 21.2a illustrates an intuitive concept: the higher the level of pollution, the higher the cost. Thus, the more acid deposition, the more damage one can expect.

By the same token, removing pollutants decreases the external costs borne by society. Figure 21.2b shows that a little bit of effort can remove large quantities of pollutants cheaply. As more and more pollutants are removed, however, the cost to remove each additional increment rises. In the initial phase, for example, remov-

ing 100 units of pollution may cost only $1, or one cent per unit. Later on, the cost of removing pollution increases rapidly, for example, to $1 per unit of pollution. This phenomenon is called the **law of diminishing returns**. It states that for each dollar invested, a smaller amount is returned.

If both graphs are placed together, the point where the two lines intersect is the level at which costs and benefits are equal, the break-even point (Figure 21.2c). If a society wants even lower pollution levels, it has to pay more (Figure 21.2d).

Although this sounds very easy, it isn't. Determining the actual cost of pollution is very difficult (Chapter 14). One of the first problems in this exercise is determining the amount of damage. For example, how many people will die from cancer induced by air pollution? How many fish will be poisoned? The second problem comes in assigning a monetary value to lost lives, lost wilderness, polluted air, extinct species, or obstructed views. How much is a person's life worth? How much are the many free services, oxygen production and carbon dioxide trapping, for example, provided by nature really worth?

One way to circumvent the dilemma of the unpriceable good is to calculate the cost of offsetting damage, the mitigation costs. How much would it cost to restore an eroded statue? How much would it cost to move an endangered species to a new habitat? Another means is to calculate maintenance costs—that is, how much would it cost to maintain and protect natural services?

Reducing all life forms to dollars and cents bothers many people. It strikes many as morally wrong to consider sacrificing people's lives so that society can have an endless supply of disposable pens, diapers, and razors. Others consider it a cruel form of business logic. How can people or species like the spotted owl be sacrificed for profit? Clearly, it is not an easy issue.

Because of the difficulties in assessing damage and the elusive nature of determining value, modern society is bound to find itself torn by two factions. On the one side are those who would allow some acceptable damage as a trade-off for the benefits of modern technology. On the other are those who argue for eliminating harmful pollutants altogether. Torn between polar opposites, we muddle along, using the best scientific information available on environmental and health damage and battling one another in courts and legislatures over the proper levels of control. Interestingly, all this conflict could be greatly reduced by applying sustainable principles of conservation, pollution prevention, recycling, and renewable resource use. These alternative solutions greatly eliminate many adverse impacts because they eliminate the source of the problem.

Who Should Pay for Pollution Control?

In today's world, pollution control devices remain the dominant means of dealing with pollution. One of the most frequently asked economic questions is: "Who should pay for pollution control?" Should corporations be required to pay for scrubbers and other devices to clean up the environment? In such cases, the costs will be passed on to consumers. This strategy is known as the **consumer-pays option.** Another option, the **taxpayer-pays option,** places the financial responsibility on taxpayers. Government payments to coal miners who suffer from black lung disease represent an example of this option. Government programs to lime lakes to neutralize acids is another.

Individuals who favor the consumer-pays option argue that the people who use the products that create pollution should bear the cost. The more one buys and thus pollutes and depletes, the more one should pay. Frugal individuals should not have to subsidize the cost of cleanup through taxes. Passing the cost directly to

the consumer, they add, could create more disciplined buying habits.

Those who support the taxpayer-pays option argue that taxpayers have allowed industry to pollute with impunity for years. Today, new standards imposed on industry place costly burdens on companies that have been operating under the law for long periods. A good example of this is the Superfund Act described in Chapter 19. As pointed out in that chapter, some think that when society changes its rules, it ought to be required to pay for controls. Advocates of the taxpayer-pays option also argue that society has elected and continues to elect officials who make deals with polluters to entice them and their polluting businesses into the community. In other cases, elected officials have overlooked flagrant violations of environmental laws. If society is responsible for elected officials who permit pollution and other forms of environmental destruction, then society must bear at least some of the cost.

As in most controversies, both arguments have validity. In the case of older industries suddenly confronted with new laws, taxpayers might bear the economic burden of new controls. However, in new industries, pollution control costs should probably be borne by the corporation and consumers.

This discussion may leave the impression that pollution prevention always costs inordinate sums of money. Far from it. In some cases, redesigning chemical and industrial processes can sharply reduce energy use and waste. Pollution control devices in many industries can be used to capture useful products that otherwise might be dispersed into the air. Some projects can generate profits by the sale or reuse of these materials. Savings in raw materials or revenues from such sales can pay for the cost of installing and maintaining pollution control equipment and may also generate a profit. Preventing pollution, in many cases, pays.

Nowhere is this case for pollution prevention more obvious than in the 1989 oil spill in Alaska. Such accidents have three costs: direct costs, indirect costs, and repercussion costs.

Direct costs from an oil spill are those that oil companies incur in the weeks and months following an accident. These include the costs of lost oil, cleanup, waste disposal, and ship repairs. The oil lost from the *Exxon Valdez* was estimated at $4.8 million. Cleanup costs were $2.5 billion the first year. The company also agreed to pay an additional $1.25 billion in fines and penalties. Clearly, some preventive measures would have saved the company vast sums of money.

Indirect costs are those costs picked up by state and federal agencies. Oil companies usually are required to reimburse the government for some or all of these costs. Another indirect cost results from damage to the local economy, such as reductions in tourism and fishing revenues, and damage to wildlife. The Clean Water Act

Case Study 21.1

Washington's Historic Timber, Fish, and Wildlife Agreement

In the Pacific Northwest, conflict and litigation have frequently surrounded the management of public forests. Since the early 1970s, a succession of battles has been waged between timber companies, environmentalists, anglers, Indian tribes, wildlife managers, and white-water enthusiasts. The battles are complex and often bitter, with rivals arguing over ways to harvest timber while protecting recreation, wilderness, fisheries, and other important environmental values.

In 1986, the warring factions began to explore another way to manage forests. The process began when representatives from the Indian tribes and the timber industry decided to put their differences aside and to find creative ways to solve several thorny resource management problems.

Two of the thorniest issues were timber cutting bordering lakes and streams—that is, riparian zones—and construction of logging roads, both practices that can lead to excessive erosion. The Forest Practices Board of the state of Washington's Department of Natural Resources had been struggling for four years to draft rules for timber cutting in riparian zones. They had written technical reports, sponsored workshops, and were beginning to develop some management recommendations when they were approached by a few individuals who were exploring a new way to negotiate solutions to the same problems under consideration by the board. Commissioner of Public Lands Brian Boyle liked what he heard so much that he temporarily suspended the Forest Practices Board's activities to give the new proponents some time to work out an agreement.

The first step was for the technical people—the researchers—from the state of Washington, the Indian tribes, environmental groups, and timber companies to sit down and come up with some resolutions. This process clarified individual interests.

To help the process, certain ground rules were established. First, no votes would be cast. All decisions would be achieved by consensus. Second, each group would have all the time it needed to present its side. Spokespersons for fish, timber, water, wildlife, archeological, and cultural resources were allowed time to outline their interests. The combined group focused on new ways to achieve win/win answers and solutions to a broad range of issues. Posturing and advocacy were set aside, and old enemies found that they could actually agree.

Within six months, the Indian and timber representatives, working with environmentalists, government agencies, and other individuals, proposed the Timber, Fish, and Wildlife (TFW) agreement. The TFW agreement established a new kind of management process called **adaptive management**, an experimental approach that allows officials to monitor and evaluate forestry practices. Using the information from experience with different management techniques, researchers are able to determine what works and what doesn't. This information is then reviewed by a TFW policy group. If the forestry practices are found wanting, the policy group can suggest changes in the regulations, which, if adopted, could improve forest management and better protect the environment.

After public hearings, the Forest Practices Board adopted the regulations recommended by the TFW participants, who then convinced the state legislature to appropriate $4.5 million, an unprecedented amount of money, to implement the agreement. The rules adopted by the state became effective January 1, 1988, and seem to be working well. When the program started, only 600,000 hectares (1.5 million acres) of commercial forestland were placed under TFW management. Today, however, approximately 1.6 million hectares (4 million acres) are under the TFW program.

Adaptive management improves water quality in streams and lakes, boosts fish populations, and improves wildlife habitat with substantial economic impacts on the state. It also gives the environmental community unprecedented involvement in timber harvesting on both state and private lands. Not to be forgotten, it also decreases litigation costs.

The TFW agreement was developed by the affected parties themselves, not by a government regulator. During the process, new working relationships were formed, which united traditional foes for the common good. Those who participated in the project think that it worked because of a willingness on the part of the participants to go beyond confrontation to cooperation, movement vitally needed to build a sustainable society.

The TFW agreement may provide a model for other conflicting groups throughout the United States to resolve issues and reach creative solutions that allow for change and wide participation.

holds companies responsible for oil spills that damage natural resources. The law permits federal agencies to collect money for lost sea otters, waterfowl, and eagles. All told, indirect costs from an accident could total several hundred million dollars.

Repercussion costs, the image problems arising from a spill, may cause people to boycott the company or reduce their patronage. Adverse publicity may also result in more costly restrictions on oil tankers and has certainly hampered oil companies in their plans to explore for oil in the Arctic National Wildlife Refuge, at least for the time being (Case Study 11.1).

Companies invariably balance the costs of prevention against the possible costs of an accident. But, if they don't incorporate the full costs, cost-benefit analyses are likely to be flawed and may end up costing them hundreds of millions of dollars. Clearly, pollution prevention pays—not just in dollars and cents but in a cleaner, healthier environment for all living things.

21.3 Economics of Resource Management

The previous section looked at the economics of pollution control, as it is typically practiced. This section examines the economics of resource management, offering additional insights into the problems of modern society as well as some new solutions.

Many decisions about natural resources are influenced by basic economic considerations, among them a factor called time preference.

Time Preference

Time preference is a measure of one's willingness to postpone some current income for greater returns in the future. For example, suppose a friend offers you $100 today or $108 a year from now. If you are short on cash and need to pay for schoolbooks, you may take the money now. Your decision to accept the money now is based on your current needs, which, in this case, outweigh the benefits of waiting the year, even though you would be $8 ahead. Economists would say that your need for current income outweighs greater returns in the future.

Time preference is also influenced by uncertainty. In the previous example, how certain are you that your friend will be around in a year to give you the promised $108? A third factor affecting time preference is the rate of return. The higher the rate of return, the more likely you will wait for the income. For instance, if your friend offered you $150 a year from now, you'd probably wait. You could borrow $100 from your parents at 10% interest, repay them at the end of the year with the $150 your friend gave you, and still have money left over.

Inflation also affects time preference. In inflationary times, people are apt to invest now to avoid higher costs later. But, inflation can also drive interest rates up, making savings more appealing.

Time preference applies equally well to the ways in which we manage many natural resources, such as water, farmland, and forests. Take agriculture as an example. Farmers have two basic choices when it comes to land management. They can choose a depletion strategy to acquire an immediate high rate of return for a short period. This might involve the use of multiple cropping, artificial fertilizers, herbicides, and pesticides to maximize production (Chapter 7). Soil erosion control and other techniques might be ignored. Alternatively, farmers may choose a conservation strategy, techniques to conserve topsoil and maintain soil fertility. These actions require immediate monetary investments. They may cost the farmers a little more in the short run and cut into immediate profits. When one takes into account the possible loss of future income from soil erosion, the conservation strategy may make the most sense.

The choice of strategy in this example depends on the time preference. Will farmers choose the cheapest route of production, which gives the highest profit in the short term? Or, will they choose a slightly more expensive route, forgoing immediate high profits in favor of sustainable profits in the future? The economic needs of farmers determine, to a large extent, their time preference. For example, a young farmer looking forward to a productive career may opt for the conservation strategy. His immediate needs may be small. He may have no family and few debts. He can sacrifice income now for larger returns in the long run. However, an established farmer may have a family to support and excessive debt. He may, therefore, maximize his profits through a depletion strategy.

Farmers' willingness to give up potentially higher income from conservation may also result from uncertainty about future prices, long-range prospects for farming, and interest rates. If the price of corn is high this year but expected to drop significantly in coming years, farmers may choose to make their money now. If the bottom falls out of the market in the next few years, they will have made the most of this short-term opportunity. If interest rates are likely to rise, short-term profit making may be the preferable choice. High interest rates on land and machinery that farmers purchase tend to encourage the depletion strategy.

Opportunity Cost

Another factor that greatly influences economic resource management decisions is the opportunity cost. **Opportunity cost** is the cost of lost opportunities. For instance, the conservation strategy requires a monetary investment. The money put into conservation could have been invested in the stock market or a new business venture, possibly yielding more profit with less work than the conservation strategy. As a result, when opportunity costs are high, farmers are likely to choose options other than conservation.

Opportunity costs are also incurred when resources are wantonly destroyed. Chapter 8 on wildlife extinction, for example, described the economic benefits of medicines derived from plants growing in tropical rain forests. Losing them creates a significant opportunity cost—both a loss of profit and potentially life-saving drugs. As another example, many of the world's ocean fisheries have been badly depleted. Salmon runs have also been ruined by dams and water diversion projects, pollution, and outright habitat destruction. The economic loss to commercial fishing interests and lost recreational opportunities are enormous.

These losses suggest that a broader view of opportunities be considered during resource management decisions and when making new laws and regulations.

Discount Rates

Economists rely on a decision-making tool that reflects time preference and opportunity costs. Known as the **discount rate**, it allows investors and economists to determine the present economic value of different profit-making options. Under this complicated practice, immediate profit is the main concern. Thus, it is perfectly rational to liquidate a forest or a fishery to achieve the maximum profit. Colin Clark, a professor of applied mathematics at the University of British Columbia, notes that if the interest rate of one's money in a bank is growing faster than a natural resource, which is almost always the case, it makes perfect sense from an economic standpoint to liquidate the resource as quickly as possible and put the money in the bank. If the natural resource is a forest, then, this exercise would lead one to cut down the trees at once and invest the money in avenues that yield higher returns.

The World Bank, which lends about $20 billion per year for projects in developing countries, uses a discount rate of 10% as its measure of investment wisdom. That is, if a project yields a 10% return on investment, it is deemed suitable. Clearly, a forest growing at a rate of 2% to 3% per year hasn't much of a chance. It will be cut and the money put into other money-making ventures.

Ethics

For many people, money is a driving force. They can't be convinced that other people will act out of a sense of duty to future generations or other species. Others point out that many noneconomic factors influence economic decisions. One of the key factors is ethics. In some cases, ethics can be as powerful as—or even more powerful than—profit and other economic factors.

In building a sustainable society, ethics writers, educators, business leaders, and government leaders can play an important role in creating a long-term view, one that seeks to ensure the survival and well-being of all life. This view, if widely held, could help foster wiser management of the Earth's resources. Most important, it could help shift time preference and help people reconsider opportunity costs and discount rates.

21.4 What's Wrong with Economics: An Ecological Perspective

Herman Daly, a leader in the fight to create an Earth-friendly system of economics, studied three leading economics textbooks and found that not one of them mentioned pollution, environment, or natural resources. Some economists see this almost complete disregard for the environment as a fundamental flaw in their discipline. Worldwatch Institute's Sandra Postel writes, "While the environment and the economy are tightly interwoven in reality, they are almost completely divorced from one another in economic structures and institutions."

In my book *Lessons from Nature*, I outline four major "flaws" in economic thinking when viewed through the lens of sustainability: It is shortsighted; it is obsessed with growth; it promotes dependency; and it tends to exploit people and the environment. This criticism is not meant to be a denouncement of capitalism or a condemnation of those who are part of the economic system, which includes all of us. It is intended to show how we need to realign the human economy with the economy of nature.

Lack of Vision

Earlier in the chapter, you learned that the law of supply and demand governs modern economic transactions and profoundly influences our thinking. Proponents of sustainable economics, however, point out that this crucial law of economics is shortsighted, for it fails to take into account the finite supplies of many natural resources: oil, natural gas, and minerals. As a rule, supply and demand economics focuses on immediate supplies and is blind to long-term stocks. Why worry about limits, ask supply and demand proponents, because as a resource is depleted, rising prices will stimulate exploration and more discovery, thus creating new supplies. Or, advocates say, falling supplies will force us to find substitutes, permitting society to continue on the endless treadmill of production and consumption.

At some point, nonrenewable resources become economically depleted—that is, they fall into such short supply that they are no longer affordable (Chapter 13). If substitutes are not available, and there appear to be a number of important minerals for which there are no substitutes, hard times are likely. Long before that point, though, the rising prices of declining resources could cripple the global economy.

Soil, the ozone layer, and the current climate are vital Earth assets. Exploited and abused, these resources have no substitutes. Their exploitation is clearly part of a short-term economic thinking that looks upon them as vast and inexpensive. Only now, we're finding that their supplies were limited and that previous signals of their abundance were misleading.

As practiced today, supply and demand economics is a serious impediment to sustainability. To overcome it, supply and demand economics must be adjusted to reflect ecological realities. Key economic players, among them business and economics professors, can assist by pointing out the limitations of the supply and demand theory. Such activities could temper our obsession for growth and stimulate efforts to recycle and use resources more efficiently.

Changes are also needed in public policy to help adjust economic activity to honor limits. The basic goal of these policy measures is to adjust current prices to reflect long-term supplies. Several market tools are available, including a range of incentives and disincentives. One of the most important adjustment tools is the user fee, or green tax. **User fees**, or **green taxes**, are taxes on raw materials paid by producers and ultimately passed on to consumers. One example is the severance tax charged to coal companies on each ton of coal they mine. The carbon tax currently used in several European countries is another (Chapter 16).

User fees artificially increase the cost of raw materials and finished products. This, in turn, helps promote conservation and raises revenues that can be used to develop alternative supplies. For instance, money raised by a tax on coal could be used to promote energy conservation or renewable energy, such as solar. Thus, a tax can help ensure future generations that they, too, will have access to resources needed for a healthy, productive life.

Another means of instilling vision in the economic system is to revamp cost-benefit analysis. Efforts are needed to identify all environmental, social, and human health impacts, present and future, and to quantify them. Companies should try to ensure that all costs are incorporated into the price of the product or service. This is known as **full-cost pricing**. Incorporating the external costs into the cost of a good or service can narrow the gap between the market price and the real cost. (Models of Global Sustainability 16.1 showed how Applied Energy Services offsets carbon dioxide pollution from its coal-fired power plants, and is an example of full-cost pricing in action.)

Full-cost pricing could stimulate economic change. For instance, it might compel companies to find ways to prevent problems at the outset, through sustainable practices, such as pollution prevention, energy efficiency, recycling, and renewable energy use. Government agencies, such as the EPA and DOE, could widen their research efforts to determine the full costs of various economic activities, thus helping businesses adjust their cost-benefit analyses. Business and economics professors can train future business leaders in the practice of full-cost pricing.

Governments can also play a role by requiring least-cost policies. Numerous states now require utilities to choose the least costly means of producing new energy (Chapter 12). To reflect the cost of various strategies more accurately, some states require utilities to add 15% to the cost of conventional strategies, among them coal, oil, and nuclear. This helps account for economic externalities and usually makes environmentally sustainable strategies more competitive.

Obsession with Growth

In the 1967 edition of his popular economics textbook, Paul Samuelson, a Nobel Prize-winning MIT economist, defines economics as "the science of growth." Subsequent editions have dropped the wording, but the bias remains. In the 1984 edition, for example, Samuelson wrote, "Today, the ultimate measure of economic success is a country's ability to generate a high level of and rapid growth in the output of economic goods and services. Greater output of food and clothing, cars and education, radios and concerts—what else is an economy for if not to produce an appropriate mix of these in high quantity and fidelity?" Although some economists do not subscribe to this view, many do, especially those in the business world and in government. To them, economic growth is synonymous with progress.

Critics label this undying dedication to economic growth "growthmania." Its roots are firmly embedded in the frontier notion that "there is always more."

Economic growth is based on an increasing consumption of goods and services. Such an increase generally arises from an increase in population size and an increase in the amount each individual buys, or per capita consumption. Because it means greater production and, presumably, greater economic wealth, population growth has traditionally been viewed as an asset to society with each new baby viewed as a potential new consumer.

The dangerous aspect of growthmania is, as mentioned above, that it tends to equate economic growth with human progress. The faster an economy is growing, many assume, the better off people are. Thus, a rising GNP is interpreted to mean that a country's health and its citizens' lives are improving. Over the years, so many people have bought into this logic that economic growth has become a way of life. In fact, it is no exaggeration to say that economic growth has become the abiding principle of economics and business and the central focus of political campaigns and government policy. With the growth-is-essential philosophy so deeply imbedded in our society, many people are blind to the outcome of continual growth. Few people today recognize that progress based solely on continuous economic growth is ultimately incompatible with the economy of nature.

What's Wrong with Relying on the GNP?   MIT's Samuelson and Yale's William Nordhaus note that the "GNP is a flawed index of a nation's true economic welfare." Why? The GNP includes many goods and services that make no contribution to the welfare of the people. That is, it fails to differentiate between "good" and "bad" expenditures. For example, the GNP includes all expenditures on homes, books, concerts, and food—deemed beneficial because they improve the standard of living. It also includes expenditures on oil spill cleanup, cancer treatment, air pollution damage, and water pollution projects—all necessary evils created by pollution. Therefore, a country with filthy air and polluted water faced with rising cancer rates might register a high GNP. Politicians looking on from the sidelines might mistake the high GNP for an enviable condition of living, while residents would decry the conditions in which they are living.

By carefully defining terms, as required by critical thinking, we can see the fallacy of overdependence on GNP as a measure of success. A telling example is the state of Alaska, whose economic output increased by $1 billion in 1989, the year of the *Valdez* oil spill.

Nordhaus and another Yale economist, James Tobin, devised a measure that adjusts the GNP to make it a more accurate representation of the good that people receive from their nation's economic growth. This measure, called the **net economic welfare** (**NEW**), subtracts the "disamenities" of an economy—the cost of pollution, the cost of medical care for victims of lung disease caused by urban air pollution, and other costs resulting from economic growth—from the GNP. NEW also adds the cost of certain activities, such as household services provided by men and women, that are not part of the traditional GNP calculations but that improve well-being.

Figure 21.3 shows the relationship between NEW and GNP. As you might expect, NEW is lower than its real GNP. That is to be expected. Moreover, NEW is growing slower than GNP. In other words, as the nation's output grows, the economic benefits of growth fall behind, largely as a result of rising pollution and environmental destruction.

Economist Daly proposed an alternative measure, the **Index of Sustainable Economic Welfare** (**ISEW**). It

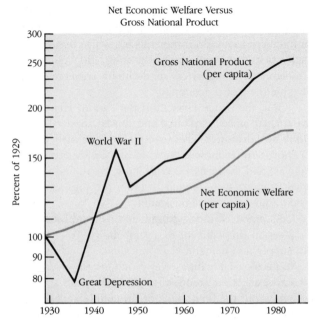

Figure 21.3   *The GNP is the market value of all goods and services produced by a country. U.S. per capita GNP has risen continuously for many years. The net economic welfare (NEW) is a measure of beneficial goods and services. It is derived by subtracting the negative aspects of the GNP that do nothing to improve the quality of life, such as damage from air pollution. As shown here, the per capita NEW is lower than the GNP and rises at a slower rate, suggesting diminishing returns from economic growth.*

includes factors such as the cost of air and water pollution, cropland and wetland losses, and other forms of environmental decay. It also entails the costs of car accidents and a host of other factors that affect human welfare. As in the case of the GNP and NEW, ISEW and GNP do not compare favorably over the past 20 years. In fact, in the 1970s, the per capita GNP increased about 2% per year and about 1.8% per year in the 1980s (Figure 21.4). In contrast, the per capita ISEW increased only about 0.7% in the 1970s and declined 0.8% per year in the 1980s, a decrease largely attributed to the rapid deterioration of the environment.

Another problem with the GNP is that it fails to account for the destruction of natural assets. For example, suppose that lumber companies cut down a nation's forests to increase exports but fail to replant them. The GNP of the country during the cutting phase would reflect a prospering economy. But, if the destruction of the forests was not taken into account, the GNP would be a "false beacon" that can "draw those who steer by it onto the rocks," according to economist Robert Repetto.

Nigeria is a country that fell into this trap. Once one of the world's largest exporters of logs from tropical rain forests, today its lumber shipments have fallen to a

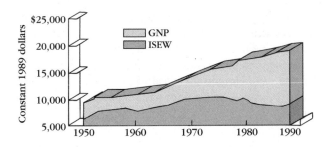

Source: Herman E. Daly and John B. Cobb, Jr., *For the Common Good: Redirecting the Economy Toward Community, the Environment, and a Sustainable Future* (Boston: Beacon Press, 1990).

Figure 21.4 *Herman Daly's Index of Sustainable Economic Welfare (ISEW) shows that economic progress of the past decade is not all that it is touted.*

mere trickle. In 1988, the last year for which data was available, Nigeria exported $6 million worth of forestry products and imported $100 million.

Unfortunately, many other countries are failing to take into account the loss of natural assets, including Bolivia, Colombia, Ethiopia, Ghana, Indonesia, and Kenya. Each of them depends on primary resources, such as minerals, timber, and crops, for 75% or more of their export. Throughout the world, many countries are treating the Earth as if it were a corporation in liquidation. But, we're not selling off assets, we're destroying them. While present income, measured by the GNP, may look good, soon there will be no assets and no source of income.

Yet another largely unrecognized problem with the GNP is that it is blind to accumulated wealth. Modern economies today compare one another on the basis of their annual growth in the GNP. This criterion, however, fails to take into account the accumulated wealth of a country—how much people already have. A newly developed nation may have a 5% annual growth in its GNP, compared with a 2% growth rate in a developed country, such as the United States. Does this mean the newly developed nation is doing better? Not necessarily. It only means that its economy is expanding more rapidly. Most likely, it has more room to expand, but its people are far less wealthy than those of the United States. Keeping in mind their accumulated wealth may help nations temper their insatiable lust for more growth.

Finally, the GNP pays no attention to the distribution of wealth in a society. Many projects that improve a nation's GNP fail to improve the lives of the people they are designed to help. A select few benefit, and, in some cases, the masses may actually be harmed. (Some people call this the *trickle-up theory*.) In developed countries, the GNP hides frightening inequities. In the United States, for example, a country deemed rich by GNP stan-

dards, one-fifth of all children live in poverty and nearly 40 million Americans have no health insurance.

Author Wendell Berry calls the GNP the "fever chart of our consumption." It is much more than that. It is an indiscriminate measure of our waste and our disregard for environmental conditions. It has become a measure of our unsustainability. The higher the GNP climbs, the closer we are to collapse. Clearly, something else is needed to measure progress.

Ending Our Obsession with Growth   New measures of economic success, such as NEW and the ISEW, are needed to help end our obsession with growth. Alternative measures that include social, economic, and environmental indicators can provide a more accurate picture of progress. By continuing to ignore costs, we perpetuate an illusion of progress that allows political leaders and the rest of society to avoid hard decisions needed to put the economy onto a sustainable path.

Fortunately, some efforts are underway to recalculate national GNPs in ways that consider the depletion and deterioration of natural resources. According to the Worldwatch Institute, Australia, Canada, France, the Netherlands, and Norway have begun to inventory their natural resources, a small but important step needed to determine the loss of natural capital. Germany and the United States are planning to calculate alternative GNPs that take into account environmental damage, although the alternative measures won't be available until the mid-1990s.

The U.N. Statistical Commission recently agreed to develop guidelines for nations that want to calculate alternative GNPs. However, they still view the existing GNP as an acceptable measure. Because this group will not revisit the issue for 20 years, critics think a stronger stand is needed.

In closing, singular measures of economic, social, and environmental health of a nation may be helpful but could hide important distinctions within countries. What's needed are new national report cards based on factors such as health, literacy, environmental conditions, resource supplies, and income distribution. Pertinent measures of sustainability could also be included, among them efficiency of resource use, reliance on renewable energy, level of recycling, and expenditures on restoration. Population control measures and public policy supporting sustainable goals could also be included.

Making the Economic System Work   Some economists believe that the continued divergence of the GNP and measures of economic welfare and sustainability discussed previously may be inevitable as the world becomes more congested and more dependent on fossil fuels, nuclear power, large-scale technologies, synthetic chemicals, and disposable goods.

Advocates of a sustainable future argue that nations should strive to reduce the difference between the GNP and NEW or ISEW. That is, economic systems should work for our benefit, not to our detriment. For every gain in economic output, we should receive a similar social benefit, not a shot of pollution requiring future cleanup.

One way of maximizing the social benefit of economic activity is to reduce the pollution, waste, and environmental destruction per dollar of GNP. This can be accomplished by applying the biological principles of sustainability. By becoming more efficient in our use of resources, by recycling and reusing all materials, by converting to clean, renewable energy supplies, we can reshape our economic system so that economic activity translates into measurable improvements in our lives and the ecosystems upon which we depend.

Corporations, small businesses, and even individuals can help promote these activities by adopting new ways. As many previous examples in this text have shown, such changes can result in substantial economic savings. Governments can also assist by providing a range of incentives and disincentives.

Development Versus Growth   Critics of growth-mania point out that our economy is like a cancer. And, they remind us, all cancers kill their hosts if they are not brought under control or removed. When a nation continues growing after maturity, it also becomes something of a cancer. Environmental degradation, resource depletion, and loss of a sense of community are three symptoms of cancerous economic growth.

Critics of economic growth argue that the future of industrial nations depends more on development than on continued growth. Continued economic growth is dangerous, perhaps even suicidal, for it will very likely lead to the ruin of the Earth's many life-support systems. Many observers assert that the continued extraction of resources and ever-increasing production and consumption cannot be sustained.

Sustainable economies need not be stagnant. Rather, a sustainable economic system affords opportunity for economic and social development. As Michael Kinsley of the Rocky Mountain Institute notes, "... development is very different from growth. After reaching physical maturity, we humans can continue to develop in many beneficial and interesting new ways—learning new skills, gaining deeper wisdom and much more." Likewise, a nation can develop without increasing its material consumption. It can still create jobs and expand cultural and educational opportunities without expanding its demand for resources.

The developed world, then, must strive to create a higher quality of life without depleting resources or causing environmental impacts that undercut future generations and other species. Many of the strategies outlined in this chapter will help us reach this goal.

In developing nations, development and limited growth within ecological limits are probably required. Limited economic growth is permissible because it is essential to raise the standard of living of the poor. Such efforts will help end the cycle of poverty and environmental decay. (Strategies for sustainable development in developing nations are detailed in Chapter 22.)

The Economic System and Dependency

The economic system of much of the world also tends to foster dependency between individuals, regions, and nations. Economic interdependence, while desirable in some respects, creates very real problems. First, it tends to separate producers of goods and services from consumers. As a result, consumers are often blind to the source of human wealth, the Earth, and oblivious to the environmental costs of satisfying seemingly endless desires. Nations are also highly dependent on one another for energy, minerals, food, and many other resources. This leads to the second problem: Economic interdependency often allows human populations to flourish beyond the regional carrying capacity of communities. Water imported from the Colorado River to the cities and towns of southern Arizona, for example, has resulted in explosive population growth. Food imports allow populations in regions with poor agricultural potential to flourish. Without infusions of fuel, food, and water, many regions could not support large human settlements.

Economic dependency has created unsustainable enclaves of humanity. If resource supplies dwindle, many of these regions could suffer extreme economic and social hardship. Studies of global warming suggest that water flow in the Colorado River, which currently serves an estimated 20 million people, could fall by 30% with devastating effects on water-dependent cities. This could cripple agriculture in arid Southern California.

Despite these problems, business economists usually refer to the global marketplace, and interdependence in general, as a desirable goal. From an ecological view, however, the global marketplace and rising interdependence that accompanies its development could be a dangerous bargain.

Some leading thinkers on the subject argue that from a sustainable vantage point, the future lies not in globalization, but rather in local or regional self-reliance—that is, human communities living within the means of a region. Food, water, energy, and other resources would come from local sources, not be imported from sources hundreds, even thousands, of kilometers away. Regional and local self-reliance essentially means living within the limits of one's region. This approach does not eliminate all imports and exports, but rather it seeks self-sufficiency and reliance on resources that are immediately available. Such an approach could reduce damage and spur interest in sustainable resource management.

The transition to a sustainable society may ultimately require increasing regional self-reliance. For the United States, regional self-reliance means taking steps to end the nation's dependency on foreign oil. For individual states, it means developing diverse economies that produce many of the goods and services needed by people on a day-to-day basis. In fact, many products could be manufactured locally from recycled materials gleaned from wastes. Besides promoting self-reliance, diversification of regional economies could help make communities and states more recession proof.

Local self-reliance, while in opposition to the dominant view of economic success, is beginning to emerge throughout the world. In Brazil, for instance, a program to produce ethanol for cars and trucks was introduced to reduce the nation's dependence on foreign oil (Chapter 12). California's public utilities have developed an energy strategy that relies principally on energy efficiency and local supplies of renewable energy, such as wind and geothermal. Efforts to promote domestic-produced goods are another sign of emerging self-reliance.

Exploitation of People and Nature

The human economy tends to exploit people. Many economic practices, in fact, widen the gap between the rich and the poor, the powerful and the weak. Income statistics support this assertion. In 1980, for instance, the average salaries of the chief executive officers (CEOs) of the 300 largest companies in the United States were 20 times greater than the average salaries of manufacturing employees; by 1990, the CEOs were earning 93 times more than the average manufacturing employees. As further proof of the widening gap, from 1980 to 1990, the income of the top 1% of the U.S. population, those earning over $250,000 per year, doubled, while middle income increased only slightly. Meanwhile, the poor got poorer.

Economic exploitation is part of the reason why developing nations remain so poor. As in colonial times, many wealthy nations continue to reap the benefits of natural resources in developing countries, often paying only a fraction of their real value. Gold mined in several African countries, for example, provides little economic benefit to poor miners and their local economies, while the middlemen who purchase and sell it to others reap huge profits. Poverty caused in part by this form of exploitation is a key element in the complex equation of rapid population growth and environmental decay in many developing countries.

The human economy also exploits the environment. Chapter 9, for instance, noted that in the 1800s the timber industry cut down most of the white pine trees from New England to Minnesota. When the forests were depleted, settlers moved to the Southeast and Northwest to continue the cycle. Widespread overharvesting of many whale species is another example of economic exploitation, as is commercial fishing, which has already depleted at least two dozen ocean fisheries in the North Atlantic in the past 40 years (Chapter 7).

Making economies less exploitive of people and nature is an enormous challenge. Virtually all of the practices outlined in this text that promote sustainability and those that help correct the flaws in the economy could greatly reduce our exploitation of nature. Some measures could reduce exploitation of people as well. For instance, alternative measures of progress that indicate income distribution or education and illness among different socioeconomic strata could help adjust economic activities so that they serve all people, not just a select few. Conservation can also help. Protecting forests, for instance, by reducing demand for materials that are extracted from them (timber) or under them (minerals) protects the homes of a large segment of humanity that make the forest their home. When a rain forest is converted to a coffee plantation, the owners often reap a sizable profit, but poor rural families that depend on the forest for food and fuel suffer.

Other changes are also needed to make the economic system fairer. In mineral-exporting countries, for instance, eliminating the middlemen can bring more income to the people who extract the ore. (For an example, see Models of Global Sustainability 21.1.) Furthermore, developing factories in which people can convert raw materials (ore) into intermediate (metal) or finished products (hubcaps) can shift the wealth to those who supply the raw materials. (Further ideas on reducing the exploitation of people in developing countries are presented in Chapter 22.)

21.5 Creating a Sustainable Economic System: Challenges in the Industrial World

Economist Kenneth Boulding was one of the first to write about changes needed in the U.S. economy. In 1966, he coined the phrase *cowboy economy* to describe the present economic system, characterized by maximum production, consumption, resource use, and profit. Boulding suggested that the cowboy, or frontier, economy be replaced by a *spaceship economy*—an economic system that recognizes that the Earth, much like a spaceship, is a closed system wholly dependent on a fragile life-support system.

In this text, an economic system that seeks to meet human needs while protecting the life-support systems of the biosphere is referred to as a **sustainable economy** (Table 21.1). The notion of a sustainable economy was not widely accepted when Boulding first proposed it. Today, however, as more and more nations face limits, experience environmental deterioration firsthand, and ponder threats such as global warming, the need to reshape economic systems is becoming clear. The Earth

---

Table 21.1   Characteristics of a Sustainable Economy

Recognizes the importance of natural systems to human well-being

Ensures economic activities improve the quality of life for all, not just a select few

Seeks ways to ensure that economic activities maintain or improve natural systems

Uses all resources efficiently

Promotes maximum recycling and reuse

Relies heavily on clean, renewable technologies

Restores damaged ecosystems

Promotes regional self-reliance

Relies on appropriate technology

Summit in Rio de Janeiro is a symbol of the global realizations. How do we do it?

Section 21.4 outlines a handful of ways to make the economy more sustainable, among them user taxes, full-cost pricing, and better cost-benefit analysis. It also calls for new measures of progress and efforts to increase regional self-reliance. New laws and private initiatives that promote conservation (efficiency), restoration, and recycling are required, as are efforts that encourage the use of renewable resources and promote population control. These steps could help revamp major sectors of society, such as agriculture, transportation, housing, and energy. No one of these changes will work alone. All are needed; each one is vital to the effort.

Because many private and government initiatives to promote these activities have been outlined in previous chapters, this section examines four major ideas: harnessing market forces, corporate reform, green products and green seals of approval, and appropriate technology. Each of these is an important element in the complex equation of sustainability.

Harnessing Market Forces to Protect the Environment

Companies often argue that left to their own devices, they could find ways to reduce pollution at a much lower cost than governmentally mandated controls. Because of this complaint and the growing cost of regulation, former Democratic Senator Tim Wirth (Colorado) and a Republican colleague, the late John Heinz (Pennsylvania), assembled a multidisciplinary bipartisan team from businesses, colleges and universities, the environmental community, and government. Their goal was to propose ways that government could tap into economic forces

to increase environmental protection and sustainable resource management. Such measures could supplement traditional laws aimed at regulating pollution and resource use.

The report, issued in 1988, outlined a number of "marketplace solutions," including: (1) economic disincentives, (2) economic incentives, (3) tradable and marketable permits, (4) laws that eliminate market barriers to efficient resource use, and (5) laws that remove unwarranted subsidies for environmentally destructive activities.

Economic Disincentives   Governments have many legal means to force companies to control pollution, reclaim damaged lands, or reduce resource consumption. In the United States and many other industrial countries, the mainstay of environmental control is a complex and costly set of rules and regulations. U.S. environmental regulations generally establish standards of conduct for a number of activities ranging from the release of pollutants from smokestacks to the restoration of surface-mined land. Some regulations stipulate how companies will meet standards, that is, what technologies they will use.

Regulations are generally backed by legal recourse, among them fines and prison sentences. Fines are a type of economic disincentive to discourage environmentally destructive activities.

The 1988 report proposed a different set of economic disincentives, among them user fees and pollution taxes. **User fees** and **pollution taxes** impose a fee on products that waste resources, pollute the environment, or damage ecosystems. A good example is a tax on cars that get poor gas mileage, the **gas-guzzler tax**. Carbon taxes are another good example of pollution taxes (Chapter 16).

Economic disincentives can work. The 1990 Clean Air Act, for instance, imposed a steep tax on the use of ozone-depleting CFCs. It provided considerable incentives for companies to find alternatives.

Hazel Henderson, a leader in rethinking economics, calls user fees and pollution charges green taxes and argues that they should be applied to a wide variety of products and activities, among them disposable goods, airplane travel, international tourism, and oil consumption. Interestingly, some European nations that embrace the idea that present generations have an obligation to future generations have instituted a number of green taxes. Henderson points out that they have found that green taxes can generate a substantial source of revenue. She also notes that many business executives find green fees more acceptable than laws and regulations. Surprisingly, most economists approve of the idea.

Economic Incentives   In recent years, government officials have also explored a variety of economic incentives that induce companies to comply with pollution

Models of Global Sustainability 21.1

Cultural Survival: Serving People and Nature Sustainably

Throughout the tropics, efforts are underway to improve the lives of indigenous people without destroying the rich diversity of the rain forest. One way to protect resources is to set them off limits to people. Such reserves, however, stand little chance of remaining intact if local people are robbed of the resources they need to survive. Recognizing that in some instances land must be used to be preserved, some countries have established extractive reserves (Chapter 9).

Extractive reserves are protected regions where local residents harvest the natural products of the forest, among them fruits and nuts, rubber, oils, fibers, and medicines. Harvested in ways that do not harm the forest, production can be maintained far into the future. In fact, indigenous peoples have depended on a sustainable harvest of forest resources throughout history.

Although numerous countries have set aside huge tracts of land, the success of extractive reserves depends on efforts to get a better deal for local residents. Jason Clay, an anthropologist with the U.S.-based group known as Cultural Survival, is one of many volunteers helping to make this happen. Clay manages a rain forest marketing project. He, like other members of the group, encourages indigenous people to maintain their culture and their independence as they come into contact with the outside world. He also helps rain forest inhabitants process and sell the nuts they harvest for profit. Instead of receiving 4 cents per pound, the typical price for unprocessed nuts, processing and marketing raise the price to $1.00 per pound. Cultural Survival charges its commercial clients an additional 5%, which it uses to support the group's activities throughout the world.

One of Cultural Survival's biggest customers is a company called Rainforest Products, which currently markets two cereals in the United States, Rainforest Crisp and Rainforest Granola. Both cereals contain Brazil nuts and cashews harvested by indigenous Amazonians. Ben and Jerry's Ice Cream of Vermont also markets the popular ice cream called Rainforest Crunch, made with hand-picked Brazil nuts and cashews from the Amazon.

"For the forests, it is a question of use it or lose it," says Cultural Survival's Clay. "The value of the rain forest will have to be tested in the marketplace. But the point is to change the market, not the forest."

By that, Clay means the rain forest needs to be viewed differently—as a source of sustainably harvested fruits, nuts, and other products rather than a source of wood and pastureland. But, do such non-wood forest products have adequate economic value?

In 1982, the World Bank stated that "the extractive reserves are *the* most promising alternative to

land clearing and colonization schemes" [italics mine]. What is the value of a standing forest? In 1989, Charles M. Peters of the N.Y. Botanical Gardens' Institute of Economic Botany and his associates evaluated a 1-hectare plot of Amazonian forest in Peru to determine the economic value of a variety of options. An inventory of plant life turned up 842 trees with diameters greater than or equal to 10 centimeters (4 inches). The trees include specimens from 275 species. Seventy-two of the species yield products that are currently marketed by natives. Sixty species produce commercial timber, one species produces rubber, and 11 species produce edible fruits. Peters and his associates found that the 1-hectare survey area produced fruit whose annual net worth was about $400 and rubber worth more than $22.

The 1-hectare plot also contains 93.8 cubic meters of marketable timber, which if clear-cut would be worth about $1000. However, clear-cutting would destroy the fruit and rubber trees, and would put an end to future timber production. The $1000 net income would be a one-time profit.

The researchers also examined selective cutting of marketable trees (Chapter 9), which would be compatible with fruit and rubber harvesting. They found that this option by itself yielded only about $310 per year.

As pointed out in the chapter, economists often rely on the discount rate to determine the present economic value of different profit-making strategies. Net present value is equal to the net revenue produced each year divided by a discount rate. The equation is $NPV = V/r$, where V is the annual revenue and r is the discount rate. Peters and his colleagues used a 5% inflation-free discount rate to determine the NPV of the fruit and latex resources of all future harvests. They found that the NPV of sustainable fruit and latex harvests was $6330 per hectare, assuming that 25% of the fruit crop was left in the forest for regeneration.

The researchers used another equation to determine the NPV of a perpetual series of sustainable timber harvests. Using that equation, they found that the NPV of sustainable harvest by selective cutting was $490. The fruit, rubber, and sustainable timber harvests of the 1-hectare plot had a combined net worth of $6820. The value would increase even more if the revenues from medicinal plants, small palms, and other plants were included.

Clear-cutting had a much lower NPV because, as noted above, the practice is not sustainable. Interestingly, even conversion of the forest to tree farms was found to have a lower NPV than a sustainable harvest. Timber and pulpwood produced on a 1-hectare

(*Continued*)

Models of Global Sustainability 21.1 (*continued*)

Cultural Survival: Serving People and Nature Sustainably

plantation of a commercially valuable tree called *Gmelina arborea* in Brazil was estimated at $3184, less than half that of the natural forest. Calculations also showed that a fully stocked cattle ranch in Brazil has a present value of only $2960 per hectare, even if the costs of weeding, fencing, and animal care are excluded.

This economic assessment clearly shows that the value of a standing forest is much higher than the most common alternatives. Except for a few cases, though, decisions about the forests seem to favor eco-

nomically less productive and environmentally unsustainable options.

If the goal of economics is to serve people and protect, even enhance, the environment, extractive reserves are essential. But, to be successful, people must have more control of the market. The middlemen must be eliminated to generate more income for collectors. If extractive reserves are successful, not only will life-styles improve but large tracts of rain forest and the species that live in them could be saved.

laws. Billions of dollars of grants, for example, were provided by the Clean Water Act to assist cities and towns to improve their municipal sewage treatment plants (Chapter 17). Tax credits can also be effective incentives. A tax credit works this way: A government gives a company a tax credit—say 10%—for investing in recycling equipment or buying recycled material. This lowers the cost of business and encourages "responsible" practices. The 10% credit for a $100,000 purchase would amount to a $10,000 savings on taxes. Tax credits can be given to individuals who invest in environmentally responsible products, for example, solar energy, wind energy, and conservation. Wisconsin and Colorado offer a 5% tax credit for companies that invest in recycling equipment.

Faced with budget shortfalls, governments are often wary of tax credits because they can lower tax revenues. Therefore, careful analysis is a prerequisite when considering offering tax credits to avoid losing revenues, investing in businesses that could prosper without support, and investing in activities that could not survive even with tax credits. Creative leaders can find ways to help environmentally responsible businesses get started that save the city money. For example, a city that is responsible for its trash pickup might find it advantageous to subsidize private recycling businesses. In the process, the city can save enormous amounts of money by reducing the amount of garbage it needs to landfill. This cuts landfill tipping fees, labor costs, wear and tear on trucks, and fuel consumption in vehicles that transport garbage to landfills.

The Permit System Another incentive is the marketable permit. A marketable permit works this way: The EPA grants a permit to Company A, allowing it to emit 10,000 metric tons of sulfur dioxide each year, but

the company finds a cheap way to cut emissions to 5000 metric tons. It can then sell its permit for 5000 metric tons to Company B. For Company B, purchasing the permit may be cheaper than installing pollution control devices. Under this system, the total emission of pollution in the region remains the same. However, permit systems can also be designed to reduce pollution levels. In this example, the EPA would simply lower permitted levels of emissions.

In Colorado, a tradable permit system was used to control water pollution entering the Dillon reservoir. Nitrogen and phosphorus from farms, sewage treatment plants, and other sources had begun to make the reservoir eutrophic. A study of various options to reduce phosphorus pollution showed that additional controls in treatment plants would cost the towns about $1800 per kilogram of phosphorus removed. Controlling nonpoint pollution would cost only about $150 per kilogram—over ten times less. The legislature and the EPA approved a tradable permit plan, allowing the publicly owned sewage treatment plant to pay for nonpoint pollution controls on farmland, saving an estimated $1 million per year.

Removing Market Barriers and Subsidies Numerous laws and regulations create market barriers and provide subsidies for environmentally destructive activities, such as mining, forestry, and solid waste. Depletion allowances, for example, provide tax breaks for fuel and mineral companies (Chapter 13). As these companies deplete their reserves, they are allowed a tax credit. The money is supposed to be used to invest in more exploration, but many companies use it to diversify, that is, to buy other companies unrelated to fuel and mineral production. Another barrier, preferential freight rates mandated by federal regulations, make it cheaper to haul virgin materials than scrap bound for recycling (Chapter 18). Subsidies to oil, gas, coal, and nuclear industries give them a considerable advantage over renewable fuels (Chapter 11).

By removing the subsidies and shipping regulations, economically inefficient and environmentally unsustainable practices can be eliminated. Removing these barriers can help protect the environment, ensure sustainable practices, and reduce government spending.

Corporate Reform: Greening the Corporation

A sustainable economy depends on the emergence of companies that operate sustainably. Marketplace solutions and the traditional command and control legislation can promote such operations. A sense of corporate environmental responsibility may also inspire corporate change.

In 1989, a number of environmental groups and several managers of major pension funds in the United States joined to form the Coalition for Environmentally Responsible Economies (CERES). CERES proposed a set of guidelines for responsible corporate conduct. Called the CERES Principles, they call on companies to:

1. protect the biosphere, by reducing and eliminating pollution,

2. promote the sustainable use of natural resources by ensuring sustainable management of land, water, and forests,

3. reduce the production and disposal of wastes by recycling, waste minimization, and other measures, and use safe disposal methods for wastes that cannot be handled otherwise,

4. employ safe and sustainable energy sources and use energy efficiently,

5. reduce risk to the environment and to workers,

6. market safe products and services—those that have minimal environmental impacts—and inform consumers of the impacts of the products and services they offer,

7. restore previous environmental damage and provide compensation to persons who have been adversely affected by company actions,

8. disclose accidents and hazards and protect employees who report them,

9. employ environmental directors with at least one member of the board of directors qualified to represent environmental interests and employ environmental managers with a senior executive responsible for environmental affairs,

10. assess and audit progress in implementing the CERES Principles.

Unfortunately, only a handful of companies have adopted the CERES Principles in their entirety.

Take a moment to study them, and you will see almost all of the principles of sustainability and many other ideas discussed in this chapter embodied in them. Number 9 is particularly important. It instructs companies to appoint high-ranking executives to the task of environmental management. IBM, for example, appointed a vice president of environmental health and safety with a staff of about 30 people to ensure that corporate environmental policy is carried out. IBM has also upgraded the status of its environmental staff, creating a more powerful and autonomous group that ensures production goals are met while minimizing pollution and reducing environmental damage. Environmental management staff are responsible for periodic checks to ensure compliance with federal laws and enactment of sustainable practices, **environmental auditing**. Most agree that environmental auditors must have the autonomy of corporate financial auditors. Ciba-Geigy, a manufacturer of drugs, pesticides, and other products, recently placed all environmental auditors in an independent group that reports directly to the CEO.

The success of corporate programs depends in large part on leadership from CEO. Officers or presidents of corporations who look favorably upon corporate environmental policy can make or break a company's program. In May 1989, when the chairman of DuPont, Edgar S. Woolard, Jr., proclaimed a new policy of "corporate environmentalism," the vice president of safety, health, and environmental affairs, who had faced stiff resistance from plant managers, noticed a sudden change in interest from employees.

Individuals can help influence the direction of business by investing in socially and environmentally responsible companies and mutual funds. The New Alternatives Fund, for example, is a mutual fund that invests in companies partaking in many environmentally sustainable activities, among them cogeneration, insulation, efficient light bulbs, and other forms of energy conservation. Offering a respectable rate of return, it and other environmentally responsible investments illustrate that good business and environmental protection can go hand in hand.

Green Products and Green Seals of Approval

Individuals can also influence corporate policy and make an enormous impact on the environment by avoiding environmentally unfriendly products, among them disposable diapers and pens, and purchasing environmentally friendly goods and services or **green products**. Green products may be goods made from recycled materials, for example, books, toilet paper, and paper towels made from recycled paper. They also include nontoxic cleaners, high-efficiency showerheads or light bulbs, and

Figure 21.5 *Eco-labels like this one help consumers make wise purchasing decisions.*

a wide range of reusable products, such as shopping bags.

Individuals can also purchase products from companies with good environmental records. The Council on Economic Priorities, a nonprofit consumer/environmental group, published a handbook in 1988 called *Shopping for a Better World*. It rates hundreds of products that people routinely buy. Companies are scored in ten categories, including environmental policy, charitable contributions, minority advancement, and women's advancement.

In the early 1980s, the government of West Germany instituted a product-labeling program known as the Blue Angel Program. A Blue Angel label on a product ensures the customer that the product is environmentally acceptable and was produced in an environmentally acceptable manner. Since its beginning, over 3500 products in 50 different categories have been scrutinized by the program.

The Blue Angel Program and a German magazine that offers even more complete analysis of the environmental acceptability of products have had a profound influence on German industries. In some cases, companies whose products received poor ratings altered their manufacturing processes or made drastic changes in the products themselves to receive a seal of approval.

Eco-labeling programs have emerged in other European countries, Canada, Japan, and the United States. The Green Seal Program in the United States, headed by Earth Day cofounder Denis Hayes, for instance, passes judgment on products using criteria similar to those used in Canada (Figure 21.5). A second group, Green Cross Certification Company, however, relies on a more rigorous set of criteria.

Table 21.2 Characteristics of Appropriate Technology

Machines are small to medium-sized.

Human labor is favored over automation.

Machines are easy to understand and repair.

Production is decentralized.

Factories use local resources.

Factories use renewable resources whenever possible.

Equipment uses energy and materials efficiently.

Production facilities are relatively free of pollution.

Production is less capital intensive than conventional technology.

Management stresses meaningful work, allowing workers to perform a variety of tasks.

Products are generally for local consumption.

Products are durable.

The means of production are compatible with local culture.

A truly sustainable economy would require that all products be produced in efficient, environmentally sustainable ways. In the meantime, individuals can send a strong signal to the marketplace by purchasing products that are environmentally acceptable.

Although green products and green seals of approval are an important step along the road to sustainability, some critics point out that green consuming is still consuming. To build a sustainable world, we need to reduce consumption, that is, learn to do less with less.

Appropriate Technology and Sustainable Economic Development

In his classic book, *Small Is Beautiful*, the late E. F. Schumacher popularized the term **appropriate technology**. Summarized in Table 21.2, appropriate technology puts people to work in meaningful ways. Compared with many modern forms of technology, it is efficient on a small scale. It uses locally available resources, uses less energy to operate, and produces minimal amounts of waste. Some examples of appropriate technologies geared to a less exploitive life-style include passive solar heating for homes and business, wind generators and photovoltaic panels for making electricity, bicycles for commuting to work when the weather permits, and compost piles to decompose yard wastes.

Appropriate technology could help developed nations reduce their resource demand and impact on the environment. It could also free up resources for other people, making a better life possible for a larger number of people.

In developed countries, the shift to appropriate technology will eliminate some jobs that have been a part of the economy for decades. Auto- and steelworkers and miners, for example, may be phased out. But, many new jobs will emerge as our priorities shift.

Appropriate technology, like other suggestions in this chapter, is not a panacea. It is one part of the solution. (See Table 21.3 for a summary of the ideas presented here.)

A Hopeful Future

"Throughout most of its tenure on Earth," economist Herman Daly points out, "humanity has existed in near steady-state conditions. Only in the past two centuries has growth become the norm." If we are to create an enduring human presence, our economic system must conform to the design principles of the ecosystem. That does not mean that the economic future is bleak. A sustainable economy need not translate into dull living and an absence of growth. What it implies is that growth will occur in some sectors, such as solar energy and energy efficiency, while others, such as oil and steel production, will be phased out. A sustainable economy, based largely on renewable resources, does not mean a retreat into the Dark Ages. Many nonrenewable resources, metals, for instance, would remain in use, recycling through the system many times. Some mining would be necessary to replace what is lost or locked up permanently in structures. Renewable fuels such as ethanol and perhaps even hydrogen could power the transportation system. Solar energy could heat well-insulated homes. Photovoltaics, along with wind energy, might provide electricity.

Clearly, today's economy and that proposed by a growing number of people are worlds apart. One of the outstanding differences is the continued high level of consumption and waste in the former, a characteristic that is bound to deplete nonrenewable (oil and aluminum) and renewable (wood from the tropical rain forest) resources and possibly create a population crash. In contrast, a truly sustainable economy achieves a level of production and consumption that can be sustained forever.

Beyond the Limits   Computer studies by Donella Meadows and her colleagues, discussed in their book *Beyond the Limits*, show that a sustainable world economy is possible. Their studies show that everyone in the

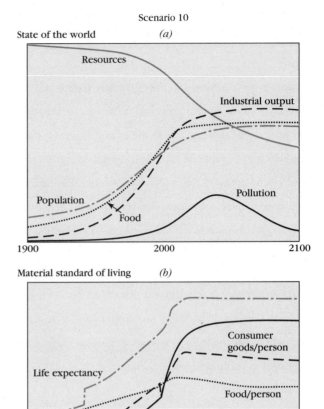

Figure 21.6 *Two graphs depicting trends in (a) industrial output, population, food production and (b) per capita food availability, etc., resulting from a concerted global effort to steer onto a sustainable course.*

world could live at more frugal consumption levels. But, we can't get there just by dramatic improvements in pollution control, increases in the amount of food produced per hectare, reductions in soil erosion, and increases in the use of resource efficient technologies. While all of these steps are essential, they are not enough. Meadows and her colleagues argue that markets and new technologies can't keep us from collapsing.

A sustainable future is possible only if all couples have access to effective birth control and decide to limit their family size to two children. Sustainability also hinges on maintaining a standard of living equivalent to that of the Europeans. If we could do all this by 1995, the resulting society could sustain a population of 7.7 billion at a comfortable standard of living with high life expectancy at least until the year 2100 (Figure 21.6). If we set our consumption goals higher, though, the computer runs show that we won't make it.

Table 21.3 Steps Essential to Building a Sustainable Economy

Goals

Reduce population growth, then gradually reduce population size through attrition

Reduce resource consumption and waste by reducing demand, increasing product durability, and increasing efficiency

Recycle, reuse, and compost to the maximum extent possible

Develop a wide range of renewable energy resources

Protect and conserve renewable resources, such as farmland, fisheries, forests, grasslands, air, and water

Improve renewable resource management to ensure sustainability

Repair past damage to natural resources by replanting forests and grasslands, reseeding roadsides, restoring streams, cleaning groundwater, and reducing overgrazing

Increase national and regional self-sufficiency by using renewable resources whenever possible

Support sustainable development projects in developing nations

Develop sustainable ethic and promote individual and corporate responsibility and action

Improve social conditions by promoting democracy, justice, and a more equitable distribution of wealth

Work for global peace and cooperation

Some Policy Tools

Environmental education programs

Green taxes and full-cost pricing requirements for business

Alternative measures of progress, such as NEW and the ISEW, that look at a wide range of economic, social, and environmental conditions

Laws that harness market forces including: (1) economic disincentives, (2) economic incentives, (3) tradable and marketable permits, (4) laws that eliminate market barriers that promote inefficient resource use, and (5) laws that remove subsidies of environmentally destructive activities.

Computer modeling studies show trends, not absolute truths. That is, they show what might happen within an approximate time frame. All in all, they're a useful policy tool, but, like any projection, one should study them critically. In the Beyond the Limits computer run that leads to a sustainable world, many improvements must come on-line almost immediately. Unfortunately, they will not. Progress toward sustainability, while encouraging, is occurring at a much slower rate. Furthermore, the computer model does not account for wars, which drain capital needed for environmentally sustainable improvements. Wars also waste resources and damage the environment directly. Nor does the model take into account civil strife, corruption, or natural events, such as earthquakes, which also drain capital and have considerable environmental impact. Because of this, the authors' conclusion that we can achieve a sustainable state may be wildly optimistic. Some note that we may have already pushed too far. Gazing into

the crystal ball, they see considerable turmoil and collapse of ecosystems and the populations that depend on them. However, this should not deter us from trying. Those countries that take steps to become sustainable may have a fighting chance of surviving and flourishing. No one knows for sure.

21.6 Environmental Protection Versus Jobs: Problem or Opportunity?

During the 1992 presidential election campaign, former President George Bush repeated a theme many times: We need environmental protection but not at the cost of jobs. He pointed out that we can't have higher mileage cars because it would put thousands of American autoworkers out of work. We can't save the spotted owl because it will cost jobs! Ignoring the loss of employment from automation and other factors, which cost the timber industry 12,000 jobs from 1977 to 1987, Bush and others hammered the point over and over while Bill Clinton and Al Gore pointed out that the choice between jobs and environmental protection is a false choice. (Remember the critical thinking rule that warns to avoid dualistic thinking!) Who's right in this debate? Do environmental regulations cost jobs? Do they cripple businesses?

Environmental regulations and permits can delay projects, such as dams and power plants. However, several studies show that delays are often the result of poor planning on the part of the companies and government agencies overseeing the work. Delays could be greatly reduced if corporations and governmental agencies were willing to invite the public or the government to participate in the early planning stages of projects.

Business leaders argue that environmental regulations also cut productivity and reduce income, resulting in loss of jobs. **Productivity** is the dollar value of goods per hour of paid employment. This figure is often used to determine how healthy an economy is. Businesspeople argue that environmental regulations divert workers from productive jobs, such as miners, to nonproductive ones, such as mine safety inspectors, raising the cost of doing business or lowering productivity. Careful analysis shows that environmental regulations do indeed diminish the output of industry. By various estimates, environmental regulations decrease productivity by an estimated 5% to 15%. However, they decrease output much less than other factors, such as high energy prices and the general shift to a service economy.

In an article in *International Wildlife*, Curtis Moore described the efforts of Germany to protect the environment. He wrote: "More than anywhere else on Earth, Germany is demonstrating that the greening of industry, far from being an impediment to commerce, is in fact a stimulus." He went on to say, "Precisely because it is so environmentally advanced, Germany is lean, competitive, and poised to dominate the global marketplace." Through end-of-pipe controls and numerous sustainable strategies that address the root causes of the problem, Germans have shown that cleaning up industry can stimulate business. Tough standards stimulate innovation, and nations with the most rigorous requirements for pollution control often become leading exporters in control technologies. In addition, environmentally responsible businesses can result in better employment opportunities.

Several studies show that far more jobs are created by environmental controls than are lost. The Clean Air Act and Clean Water Act have created at least 300,000 jobs in pollution control. Studies by the EPA indicate that very few jobs are lost as a result of environmental regulation. Only companies that are in great financial jeopardy are affected. Even the AFL-CIO, one of the nation's leading labor unions, admits that not one

plant shutdown can be attributed to environmental regulations.

This is not to say that tighter controls on pollution, reductions in timber cutting, or other environmental regulations are benign. They do affect communities, sometimes profoundly. Environmental protection measures, especially the kinds of changes needed to develop a sustainable system, will result in a massive shift in employment. Employment will very likely shift from the ends of the present linear system of production and consumption (mines, forests, and waste dumps) to the center of an efficient cyclic system.

Numerous studies suggest that this shift will result in a net increase in the number of jobs. That is, restructuring the economy and our way of life to rely more on energy efficiency, recycling, mass transit, and alternative fuels will create far more jobs than are lost. As Chapter 12 notes, a large nuclear power plant employs 100 workers. A coal-fired power plant that generates the same amount of electricity employs 116 workers, while a solar thermal facility employs twice as many workers, about 250. A wind farm employs twice as many people again, 540 in all. Furthermore, wind and solar thermal energy are cheaper and more environmentally benign than nuclear power.

Studies in the United States show that energy efficiency also creates jobs. Michael Renner of the Worldwatch Institute estimates that weatherizing all U.S. households could create 6 to 7 million job-years—that's 300,000 jobs lasting 20 years each!

A European study in 1985 that examined the employment potential of six energy conservation and renewable energy technologies in 4 countries—Great Britain, Denmark, France, and former West Germany—found that these countries could create 142,000 job-years—7000 jobs lasting 20 years. A full-fledged program for all 12 European nations could create 530,000 job-years.

Renner also notes that "although a shift from fossil fuels to solar energy entails job losses in the oil, coal, and gas industries, there are overlaps among the kinds of supplies and skills required by the solar industry that will minimize overall job loss." Companies that produce the materials used to manufacture solar panels are already in existence. Many of the skills needed to tap renewable energy supplies are similar to those required for conventional construction and heating system installation. Work opportunities could spring up in a variety of occupations, among them carpentry, plumbing, and construction.

Shifting to mass transit may also create more jobs than are created by the current system of automobile transit. A German study showed that spending $1 billion on highway construction yields only 24,000 to 33,000 direct and indirect jobs. Spending the same amount on railway and light rail construction creates 38,000 to

40,000 jobs. Furthermore, the shift from cars and trucks to railroads, subways, light rail lines, and buses offers alternative job opportunities for much of the work force.

Recycling offers similar employment opportunities. ALCOA estimates that at least 30,000 people in the United States are employed in aluminum recycling, nearly twice as many as in aluminum production. Vermont's recycling facilities employ 550 to 3000 people for each 1 million tons of materials they handle. Incinerators, on the other hand, equipped to handle the same amount, employ only 150 to 1100 people, and landfills employ only 50 to 360.

One final point often overlooked in the debate over economics and environmental protection: Environmental laws can help save companies money. For example, pollution prevention measures reduce hazardous waste production and waste disposal costs. They also eliminate future liability for cleanup and may reduce or eliminate employee health effects and future claims for health impacts. Pollution prevention can reduce lawsuits and greatly slash the cost of compliance—installation and operation of pollution control equipment. Many examples cited in this text highlight companies that have saved considerable amounts of money by preventing pollution, using energy more efficiently, and recycling and reusing wastes.

In closing, Worldwatch Institute's Cynthia Pollock Shea notes: "Businesses that protect the environment can make a healthy profit." Germany and Japanese companies are showing us that environmental protection is a precondition for success.

Civilization is a slow process of adopting

the ideas of minorities.

—Anonymous

        ↂ   ↂ   ↂ

Critical Thinking Exercise Solution

Critical thinking rules advise us to examine our biases very carefully and to consider the big picture. Many people in positions of power are wed to the notion that all growth is good. As you learned in this chapter, however, much of the economic growth occurring in the United States has not improved society. In other words, much of the growth in the U.S. GNP has resulted from bad economics—pollution cleanup, repair of damage from pollution damage, and the like. Statistics also show that economic growth often bene-

fits the wealthiest members of society. It is not trickling down as intended.

In this chapter, you also learned that to improve an economy it is not always necessary to manufacture more goods. One of the fastest and most profitable ways of improving an economy is to use resources more efficiently. By using energy, water, and materials more efficiently, for instance, companies can produce goods cheaper, make a higher profit, and compete more effectively. Pollution prevention is another strategy with extraordinary economic benefits. These strategies permit companies to produce goods while improving their bottom line and the environment.

Summary

21.1 Economics and the Environment

■ **Economics** is a science that concerns itself with the production, distribution, and consumption of goods and services.

■ Economics helps society answer three fundamental questions: (1) What commodities should it produce and in what quantity? (2) How should it produce its goods? (3) For whom should it produce them?

■ In **command economies**, governments dictate production and distribution goals. In **market economies**, prices, supply, and demand dictate economic activity. Most economies have features of both.

■ One of the key principles of economics is the **law of scarcity**, which states that most things people want are limited. As a result, items must be rationed, usually by price determined by the interaction of supply and demand.

■ The most widely used measure of social progress is the gross national product (GNP), the value of a nation's total output of goods and services.

21.2 Economics of Pollution Control

■ Pollution control seeks to achieve the maximum reduction of pollution that yields the maximum benefit at the lowest cost.

■ Cost and benefits are determined through **cost-benefit analysis**, but calculating the true cost of pollution, including externalities, is very difficult.

■ Who pays the cost of pollution control is an issue of much debate. Some argue that consumers should bear the burden; others argue that taxpayers should pay. In reality, there are times when consumers should pay. For example, when new plants are built, the cost of pollution control should be incorporated into prices. In older industries now under new laws and regulations, the taxpayer should at least help defray costs.

21.3 Economics of Resource Management

■ Resource management is influenced by economics in many ways. Two of the most important factors are **time preference**, a measure of one's eagerness to receive a return on investment, and **opportunity cost**, the cost of lost money-making opportunities.

■ Economists use the **discount rate** to assess various business strategies. It reflects time preference and opportunity costs. Unfortunately, decisions based on discount rates often result in excess exploitation to maximize immediate return.

21.4 What's Wrong with Economics: An Ecological Perspective

■ A fundamental problem with economics is an almost complete disregard for the environment. Economic thinking tends to be shortsighted, obsessed with growth, and exploitive, and promotes dependency.

■ Proponents of sustainable economics point out that reliance on supply and demand is a serious impediment because it focuses on immediate supplies and is blind to long-term stocks.

■ Supply and demand economics can be adjusted to reflect reality by several market incentives, including **user fees** and **pollution taxes**. Both add to the cost of raw materials and finished products, thus helping to promote conservation and raise money for developing alternative supplies.

■ Another way to revamp the economic system is **full-cost pricing**, incorporating the full cost of a product into its price.

■ Economic systems are almost universally obsessed with growth as measured by the GNP. Critics point out that growth of the GNP is not an unqualified good because this measure fails to differentiate between "good" and "bad" expenditures.

■ Two alternative measures of economic progress are the **net economic welfare** (NEW) and the **Index of Sustainable Economic Welfare** (ISEW). Both include factors such as the cost of air and water pollution, cropland and wetland losses, and other forms of environmental decay. Both indicate that much of the economic growth over the past 20 years has been offset by environmental deterioration.

■ Another problem with the GNP is that in tracking the income of a nation it fails to account for the destruction of natural assets. The GNP is also blind to accumulated wealth and pays little attention to the distribution of wealth in a society.

■ New measures of economic success such as the NEW and the ISEW, which include social, economic, and environmental indicators, provide a more accurate picture of human progress.

■ The economic system of much of the world also tends to foster economic interdependence. While beneficial, it often allows human populations to flourish beyond the regional carrying capacity of communities.

■ From a sustainable vantage point, the future of the human society may lie in a movement toward greater self-reliance.

21.5 Creating a Sustainable Economic System: Challenges in the Industrial World

■ A sustainable economy, based on conservation, recycling, renewable resource use, restoration, and population control, seeks to meet human needs while protecting the life-support systems of the biosphere.

■ Numerous steps are needed to make economies more sustainable. Efforts are also needed to harness market forces, reform corporations, increase the number of green products and implement green seals of approval programs, and employ appropriate technology.

■ Marketplace solutions include: (1) disincentives, (2) incentives, (3) tradable and marketable permits, (4) laws that eliminate market barriers that promote inefficient resource use, and (5) laws that remove unwarranted subsidies for environmentally destructive activities.

■ A sustainable economy depends on the emergence of companies that put into practice the principles of sustainability. Individuals can influence corporate policy by investing in socially and environmentally responsible companies, purchasing green products, and avoiding environmentally unfriendly ones.

■ Appropriate technology is also needed to build a sustainable world. Appropriate technology is efficient on a small scale, uses locally available resources, and produces minimal waste.

■ Appropriate technology can help developed nations reduce their demand for resources, reduce their impact on the environment, and could also free up resources for other people, making a better life possible for a larger number of people.

21.6 Environmental Protection Versus Jobs: Problem or Opportunity?

■ Environmental regulations can delay projects and add to their cost. Studies show that many delays could be avoided if businesses and government sought community input from the start.

■ Studies also show that environmental regulations diminish the output of industry by 5% to 15%. However, some countries, such as Germany, have shown that strong commitment to environmental controls can stimulate innovation and that nations with the most rigorous requirements for pollution control often become leading exporters of control technologies.

■   Environmental controls can also create many employment opportunities. Several studies show that far more jobs are created by environmental protection than have been lost.

■   Environmental laws can help save companies money, increasing their profitability.

Discussion and Critical Thinking Questions

1. Define the following terms: *command economy, market economy,* and *law of scarcity.*

2. Describe the law of supply and demand. What is the market price equilibrium? What are the major weaknesses of the law of supply and demand? How can they be corrected?

3. Define your own economic goals. Would you classify them as consistent with a frontier economy or a sustainable economy?

4. Using your critical thinking skills, analyze this statement: "Population growth is economically beneficial."

5. In your view, is continued economic growth in developed nations possible? If so, why and for how long?

6. Describe the gross national product (GNP) and its strengths and weaknesses.

7. Define the term *net economic welfare.* How does it differ from the GNP? Is it greater than the GNP or less? Does it grow as quickly or more slowly than the GNP? Is this good or bad?

8. Describe the Index of Sustainable Economic Welfare. How does it differ from the GNP?

9. Discuss how time preference, opportunity costs, and discount rates are related. How do they affect the ways in which people manage natural resources?

10. Define the term *economic externality* and describe how externalities can be internalized. What are the benefits of this action?

11. What is the economically optimal level of pollution control? Why is it impractical to consider reducing pollution from factories and other sources to zero via pollution control devices?

12. Describe the law of diminishing returns. How does it apply to pollution control? Can you think of any other examples where the law applies?

13. The analysis of the economic system and economic thinking suggests a number of weaknesses when viewed through the lens of sustainability. What are they? Describe each one and how each can be corrected.

14. Describe a sustainable economy. What are its main goals? In your view, is it a practical alternative to the current economic system? What are its strengths and weaknesses?

15. Using your critical thinking skills, analyze the following statement: "Cleaning up the environment will put thousands of people out of work. We simply can't afford to do it."

Suggested Readings

Brown, L. R., Flavin, C., and Postel, S. (1991). *Saving the Planet: How to Shape an Environmentally Sustainable Global Economy.* New York: Norton. Basic account of ways to reshape the human economy.

Chiras, D. D. (1992). *Lessons from Nature: Learning to Live Sustainably on the Earth.* Washington, D.C.: Island Press. See Chapters 4 and 5 for a discussion of sustainable economics.

Daly, H. E. and Cobb, J. B. (1989). *For the Common Good: Redirecting the Economy toward Community, the Environment, and a Sustainable Future.* Boston: Beacon Press. A challenging but worthwhile book for those interested in probing deeper.

Johnston, C. M. (1992). The Wisdom of Limits. *Context* 32: 48–51. Important reading on reaching cultural maturity by recognizing limits and making appropriate adjustments.

Meadows, D. H., Meadows, D. L., and Randers, J. (1992). *Beyond the Limits: Confronting Global Collapse, Envisioning a Sustainable Future.* Post Mills, VT: Chelsea Green. Contains important information on economics.

Meeker-Lowry, S. (1988). *Economics as if the Earth Really Mattered.* Philadelphia: New Society Publishers. Contains ideas on ways individuals can contribute to creating a sustainable economic system.

Postel, S. and Flavin, C. (1991). Reshaping the Global Economy. In *State of the World 1991.* New York: Norton. Important reading that outlines ways to help convert to a more sustainable economy.

Renner, M. G. (1992). Saving the Earth, Creating Jobs. *World-Watch* 5 (1): 10–17. Essential reading on the job potential of the sustainable strategy.

Repetto, R. (1992). Earth in the Balance Sheet: Incorporating Natural Resources in National Income Accounts. *Environment* 34 (7): 12–20, 43–45. Very important reading.

Stavins, R. N. and Whitehead, B. W. (1992). Dealing with Pollution: Market-Based Incentives for Environmental Protection. *Environment* 34 (7): 6–11, 29–42. Detailed analysis of various market-based incentives with an analysis of their effectiveness.

Chapter 22

Sustainable Economic Development: Challenges Facing Developing Nations

To change and change for the better are two different things.

—German proverb

Critical Thinking Exercise

Many experts believe that sustainable economic development will require massive improvements in the lives of people in the developing countries. They see industrialization and attainment of life-styles akin to those found in the West as the key to success. Critically analyze these ideas.

A huge dam constructed with international financial assistance now spans India's Tawa River. The project, like many others in the developing world, has proved to be a mixed blessing. Although it did provide electricity as promised, the irrigation canals built to transport water from the reservoir to nearby farms were constructed in porous soils. As water flowed through the canals, much of it drained into nearby farm fields, making it difficult for farmers to work the land. In some instances, water filled the soil pores and suffocated the plants, killing crops. Tragically, the farmers who were supposed to have benefited from the project have experienced a marked decrease in their food production.

Near the village of Sukhomajri, India, livestock overgrazed fields, reducing vegetative cover by 95%. This, in turn, led to a marked increase in soil erosion. To combat the problem, local villagers built small earthen dams to capture water from rain. The water was used to irrigate the fields and coax the land back to life. A community-wide agreement to reduce overgrazing is also giving the land a chance to recover.

The first example is typical of most efforts to improve conditions in developing countries. Unfortunately, many such projects do more harm than good. The second example represents a new form of development, **sustainable development**, aimed at meeting the needs of people while protecting the environment upon which humans depend. It is part of a growing list of ventures that could revolutionize the way developing countries progress. Like other projects, it is aimed at enabling people to enjoy long, healthy, and fulfilling lives. In short, sustainable development is both people centered and conservation based.

This chapter looks briefly at traditional economic development strategies in developing nations and their many impacts. It also offers guidelines on sustainable economic development strategies and concludes with a discussion of some of the barriers to sustainable development.

22.1   Impacts of Conventional Economic Development Strategies

Many developing countries are steeped in economic and environmental problems. The sources of their dilemma are many. For example, a large number of countries strain under the burden of rapid population growth. Many suffer from political corruption. Others are enmeshed in civil wars or on-going military conflicts with neighboring countries, two activities that waste precious, limited resources and destroy valuable land (Figure 22.1). Economic and environmental problems also

Figure 22.1   *Civil wars and conflicts with neighboring countries threaten national security, but so does environmental deterioration. Resolving conflicts peacefully is essential to living sustainably.*

stem from exploitation by international corporations and foreign governments (Chapters 7 and 9). Finally, many well-meaning forms of assistance offered by developed countries and international lending agencies add to the environmental and economic nightmare. This chapter looks at this last aspect.

During the past 50 years, developing countries have received considerable economic and technical assistance from the developed world. This aid has primarily been aimed at improving health care, increasing food production, providing energy (particularly electricity), and building an industrial base. Unfortunately, many ill-conceived development projects funded by the West have spawned enormous social, economic, and environmental damage in recipient nations. The environmental impacts include massive deforestation, soil erosion, and desertification. Some projects have resulted in widespread pesticide poisoning of the population and the environment. Valuable wilderness and farmland have been flooded by dams or destroyed by mines. Given the importance of the environment to the well-being of people, these problems have contributed to massive social and economic ills.

What is the problem with Western development assistance? In general, most development strategies attempt to transpose Western ways on developing nations. Large-scale agricultural projects and dams, such as the one on the Tawa River, are examples. These and other Western-style schemes seek to remold nature to benefit humans. But, as you have seen, many efforts to reshape nature have severe ecological backlashes with tremendous economic repercussions.

As a rule, Western-style development projects tend to centralize wealth in cities and in the hands of a select few. In many cases, such projects benefit a small segment of the population while ignoring or, even worse, harming the masses they were meant to serve. Huge dams built for hydropower, for example, often flood productive farmland and displace farmers, forcing them to move to marginal land that frequently cannot withstand heavy use. Dams also inundate forests, destroying the homes and food source of native peoples (Chapter 10). In arid regions, irrigation water from reservoirs increases salinization and waterlogging of agricultural soils, both of which decrease the productivity of the land and can render it useless (Chapter 7). Salinization and waterlogging translate into economic disaster for small farmers. Today, over half of the irrigated cropland in developing countries suffers from salinization.

Some development projects negate gains achieved by others. For example, deforestation projects in Guatemala to increase agricultural output have resulted in massive soil erosion. This tends to increase flooding and sedimentation in reservoirs, greatly reducing their useful life span.

Another common form of development involves the introduction of high-yield crops to increase food production. While potentially valuable, these projects tend to create a costly cycle of dependency. In order for many of these crops to survive, they must be doused with pesticides and fertilized, then irrigated. Pesticides and fertilizers are usually imported from developed nations. Their costs are often prohibitive, and their impact on the environment and health of the workers profound. In Central America, chronic and acute pesticide poisoning are among the region's most serious problems (Chapter 18). In order to promote pesticide use, the governments of some developing nations subsidize them, thereby creating a costly economic dependency that drains money needed for education and environmental protection. In Nigeria, a factory was built to produce fertilizer for farmers, but developers discovered that the nation's transportation system was inadequate to transport the fertilizer to rural farmers, who didn't have enough money to pay for fertilizer anyway. The factory is now abandoned. Such projects may have been the stimulus for Walter Reid's comment: "Developing countries are littered with the rusting good intentions of projects that did not achieve social or economic success" (Figure 22.2).

Western-style development with its disregard for limits can decimate potentially sustainable sources of food, timber, and fiber. For instance, before 1960, most of Thailand's commercial fishing fleet consisted of small-scale family operators. Today, modern fishing vessels dragging trawl nets (a Western technology) cruise the shorelines, catching virtually everything that swims and leaving nothing to sustain fisheries. This development could bring an end to a once-lucrative commercial fishing industry.

International development assistance has also helped finance road building to colonize remote regions. In western Brazil, for instance, tropical rain forests were cut to make way for farmers and ranchers, but 80% of the people soon left because the soils quickly washed away or lost their fertility (Chapter 9).

Finally, Western development often undermines sustainable practices and sustainable cultures. Some refer to this as *cultural erosion*. In a remote area of northern India, Helena Norberg-Hodge, who has worked in the area for many years, notes that people once lived sustainably and well. Their community was close knit, their food locally produced, and their clothing made from local materials. The culture was shaped by a close, intimate relationship with the environment.

Because of exposure to Western ideology and Western ways of doing things, the people now often feel inadequate, even inferior. Many have rejected their own culture. Today, they are struggling with the choice of preserving an economy that was once self-reliant and

Figure 22.2 *This tractor in Niger, West Africa lies rusting in a field. When it broke down, knowledge, skills, and the parts to repair it were not available.*

sustainable or turning to the alluring but unsustainable ways of the West.

Who's Financing International Development?

Financial backing for international development, which comes from thousands of sources, falls into one of four categories: (1) multilateral development banks (MDBs), (2) private commercial banks, (3) development agencies of industrial nations, or (4) private foundations.

Four MDBs provide the bulk of the money: the World Bank, the Inter-American Development Bank, the Asian Development Bank, and the African Development Bank. All receive funding from the developed nations. The World Bank, for example, is headquartered in New York City and is supported by the United States, France, Germany, the United Kingdom, and others. The World Bank lends about $200 billion per year to support a wide range of projects from road construction to farming to dam building and very recently family planning.

Additional money comes from private commercial banks who make loans directly to developing countries. International development agencies also play a key role in development. The U.S. Agency for International Development, for example, provides outright grants for development projects. Although private foundations also help support international development, their contribution is small.

As pointed out in the previous material, many projects sponsored by the various organizations are unsustainable. Thus, they not only fail to improve the lives of those they are designed to help; they worsen them. In fact, volumes of horror stories could be written about well-intended projects gone awry. Fortunately, many funding agencies are beginning to see the need for different practices—for sustainable development. Sustainable development does not require new technologies and new knowledge; they already exist. But, it does require steps to put new ideas, as well as some old ones, and new technologies into practice.

22.2 Sustainable Economic Development Strategies

To create a sustainable future, developing nations must first slow the population growth rate. Dramatic improvements are needed in education and health care. Freedom from repression and gains in women's rights are also essential. Ultimately, developing nations must build sustainable systems of agriculture and commerce. Demand for energy, food, water, waste, housing, and transportation must all be met in ways that protect and enhance natural systems.

Some progressive thinkers believe that the developing nations, like the industrial nations, must become more self-reliant (Chapter 21). In fact, many proponents of sustainable development believe that self-sufficiency is one of the cornerstones of success. Increasing self-reliance clearly opposes current efforts to create a global economy. Nonetheless, in order to create a sustainable future where people live within the limits of the environment, it may be necessary and prudent to find ways in which developing nations can become less dependent

Figure 22.3 *Young agricultural students in Burkina Faso (West Africa) learn to use and maintain plows, an appropriate technology in many developing countries.*

on the outside world. To achieve this wherever possible, development projects should tap into local resources to meet local needs.

Sustainable economic development also requires appropriate policies—that is, laws, regulations, and subsidies that promote sustainable practices. Inappropriate policies, among them subsidies for pesticide use or tax breaks for companies that destroy natural systems, need to be eliminated.

Sustainable development can also be fostered by routine analyses of social, economic, and environmental impacts of proposed projects. Until very recently, concern over the long-term environmental sustainability of development projects received little, if any, attention by developers and lending institutions. Social costs and benefits received surprisingly little thought. Impact analyses could help countries avoid projects that are likely to fail or are destined to create irreparable environmental damage.

Employing Appropriate Technology

One means of promoting economic development while protecting the environment is appropriate technology. Appropriate technology refers to an assortment of environmentally compatible technologies (devices, machines,

and factories), among them hand-operated tools, solar cells, and methane digesters (Chapter 21). Appropriate technologies are generally small-scale solutions to local problems. They rely on locally available resources and local knowledge for repair and service. As a rule, appropriate technologies depend on people, human labor, to produce goods and services. In contrast, highly automated factories and machines require large amounts of fossil fuel and capital. Relatively speaking, they put fewer people to work than appropriate technologies.

Appropriate technologies are suitable for the developing nations because most lack the financial and fossil fuel resources required to support modern Western-style industries equipped with an assortment of labor-saving energy-intensive devices. Because appropriate technologies rely more on human labor, they are quite suitable to many highly populated developing nations, such as India and China.

Appropriate technology, such as oxen-drawn steel plows and solar-powered water pumps, are badly needed on farms throughout the developing world (Figure 22.3). Western alternatives, such as tractors and diesel-powered water pumps, are inappropriate technologies for developing nations. Although tractors increase food output, they displace farm workers. In many cases, displaced farm workers move to overpopulated cities in search of work, creating serious problems in burgeoning cities (Chapter 5). In addition, the costs of tractors and diesel-powered pumps, repair, and energy to run them are prohibitive in poor, rural regions of developing countries.

One inexpensive appropriate technology is the solar box cooker, which uses sunlight to cook meals (Figure 22.4). Constructed from aluminum foil, glass, and cardboard boxes, solar box cookers are suitable in regions where firewood is scarce and sunlight abundant, among them northern Pakistan, Nepal, and parts of Africa. This relatively inexpensive device can reduce deforestation (Chapter 9). Since many families cook over wood in unventilated or poorly ventilated spaces, solar box cookers can reduce health problems from smoke inhalation.

Another appropriate technology is the photovoltaic cell (Chapter 12). Photovoltaics are far cheaper than centralized coal-fired power plants and transmission lines in rural regions in developing countries. Installed in villages, they can provide electricity for refrigerators and freezers to store medicines. They can also power water pumps.

Gandhi summed up the challenge facing developing nations when he said that the poor of the world cannot be helped by mass production, only by production of the masses.

Many experts agree that the world cannot support economic development in the developing nations on a scale similar to that in the developed nations. By one estimate, if the rest of the world lived like people in the

Figure 22.4 *This solar box cooker is inexpensive and easy to make. It helps reduce the need for firewood in sunny, arid regions and eliminates pollution generated by cooking with wood.*

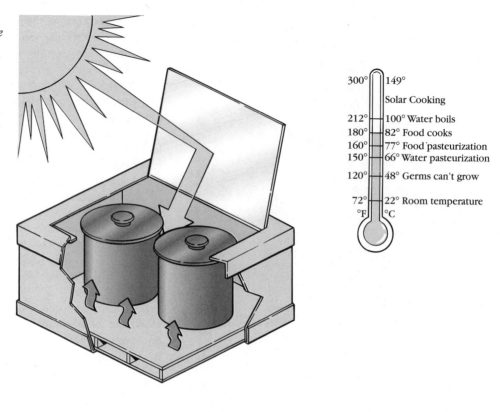

°F	°C	
300°	149°	
		Solar Cooking
212°	100°	Water boils
180°	82°	Food cooks
160°	77°	Food pasteurization
150°	66°	Water pasteurization
120°	48°	Germs can't grow
72°	22°	Room temperature

United States, the world would run out of resources in about two to three months. (It is doubtful that the developed nations can support this level of economic activity in the long term either.) But, without some improvement in life-styles, hunger, disease, environmental destruction, and death are bound to continue.

Creating Environmentally Compatible Systems of Production

Over the years, the frontier ethic has driven Western-style development. As you may recall from Chapters 4 and 20, the frontier ethic insists that the key to human success lies in the domination and control of nature. Western-style agricultural projects in the rain forest, for instance, remove trees from huge areas and introduce equipment and farming practices that, although suitable in Iowa, fail miserably in the tropics.

The key to success lies in modifying production systems to fit the environment, not in trying to modify the environment to fit the production system. That is, developers must design with nature rather than attempt to redesign nature to fit human needs. Growing crops in small clearings in tropical rain forests or harvesting food and fiber from relatively undisturbed forests, for example, are both sustainable activities that fit within the workings of nature, whereas leveling huge tracts of forest to grow row crops does not.

Sustainable development seeks to work within the constraints of natural systems. It may require many regional small-scale projects that honor limits and permit people to benefit from nature's opportunities. Such projects require little, if any, financial backing and can pay for themselves many times over, in contrast to large Western-financed projects that often cause tremendous economic and environmental damage and require an enormous burden of debt.

Tapping Local Expertise and Encouraging Participation

Convinced of the superiority of Western ways, many international development experts overlook the abundance of knowledge held by indigenous peoples. For instance, the Lancandons of Mexico, believed to have descended from the Mayans, have a profound knowledge of tropical rain forest agriculture. They once grew as many as 80 different crops in the rain forest until the Mexican government forced them off their land.

Although many native people have abandoned sustainable farming practices for one reason or another, the knowledge still exists. Tapping into this vast reservoir could help countries find sustainable solutions to local problems. A good example of a sustainable agricultural practice is agroforestry, growing crops and trees together, as discussed in Models of Global Sustainability 4.1.

Besides overlooking local knowledge, many development projects fail because they address issues not viewed as problems by local people. By discerning what

Models of Global Sustainability 22.1

Bhutan: Pursuing a Sustainable Path

The country of Bhutan, nestled in the majestic Himalayas, is a country in transition. Its air is clean, its forests are largely intact, and its wildlife are abundant and diverse. About the size of Switzerland, Bhutan has a population of only 700,000.

In Bhutan, land is distributed relatively equitably so that landlessness and poverty are rare. Today, 90% of the people live in rural areas, mostly on small farms where they grow corn and rice to feed their families. They use draft animals to pull their plows. Although most basic needs are met, infant and maternal mortality rates are among the highest in the world, largely due to a lack of medical facilities and clean water. Additionally, 70% of the population is illiterate.

Bhutan's 37-year-old monarch, King Wanchuk, wanted to improve the condition of his people. To do this, he realized, Bhutan could have followed the economic development strategy of nearby countries, which have paid little attention to the environment. In neighboring India and Nepal, the results of this approach are painfully evident: The forests are severely overcut, much of the soil is eroded, and more lives have been impoverished than improved. These examples and factors discussed below convinced Bhutan to follow a sustainable path, one that would sustain human progress not just in a few places for a few years but for the entire country into the distant future.

In May 1991, the king said, "We would like to develop rapidly, but we would also like to ensure that there is a certain amount of harmony between rapid development and our culture and environment." Fortunately, culture and environment are closely linked in Bhutan. The religion of Buddhism, practiced by many of the nation's people, dictates respect for the environment. In fact, some think that the practice of Buddhism in Bhutan is one reason the natural environment is still largely intact.

Today, 64% of Bhutan's land area is covered by forests. When it came to the government's attention that forest resources were being overcut in some areas, officials responded by nationalizing most logging operations. In 1985, the government enacted a National Forest Policy, which made the conservation of forests a top priority and economic considerations secondary. This policy, designed to ensure that 60% of the country remains forested, requires that the rate of tree harvesting be equal to the rate of replanting. To preserve part of the forests, the government of Bhutan has set aside more than 20% of its land in ten reserves. The Royal Manas National Park covers 45,000 hectares (165 square miles) and provides sanctuary for many of the endangered species of south Asia. Elephants, golden langur monkeys, tigers, wild buffalo, and more than 500 species of birds can be found in the park.

In 1990, Bhutan's highest officials passed the Paro Resolution, a call for a national sustainable development strategy. It recognizes that the key to sustainable development is to find a path that will allow the country to meet the needs of the people without undermining the natural resource base of the nation. According to the agreement, new industries, new agricultural markets, and new forestry products need to be carefully developed with respect to the environment.

Development in Bhutan is slated to proceed slowly and to be based primarily on renewable resources, among them hydropower and solar energy. Officials see hydropower as the country's largest source of potential earnings. Today, most rivers still run free in Bhutan. The largest dam in the country produces $25 million worth of electricity, which is sold to India. It also supplies power to approximately 23,000 families in and around Bhutan's two largest towns. Bhutan hopes to build more dams in the future with financing from the World Bank. Aimed at increasing export income, these projects will be built only when the king and his advisers are sure that such prospects will cause minimal environmental damage.

If the goal of development is to improve the lives of people, a better alternative might be photovoltaics, which produce electricity from sunlight. Used to power thousands of isolated farms and villages, photovoltaics could provide a sustainable supply of energy for local needs without having to dam the nation's rivers. Unfortunately, foreign funding sources for this approach are scarce, so photovoltaics are not being aggressively pursued by the government at this time.

Bhutan has also begun to develop extractive industries. In the north, the conifer forests are selectively harvested by the government-owned Bhutan Logging Corporation. In the south, a joint government-private company operates a large plywood plant. The company is allowed to clear-cut hardwood forests in the area as long as it replants.

Bhutan has also opened its borders to tourists but limits the number to 1500 to 2000 per year. To limit the impact of tourism, those allowed into the country must follow a strict, government-set itinerary.

To deal with its rapidly growing population and fears that it will have to increase grain imports more than the current 10% to feed its people, the nation hopes to limit the growth rate of its population to 2% per year in part by limiting families to two children. Contraceptives and family planning advice are available in 70 clinics across the country.

If industrialization is allowed to progress too far or too rapidly, the Bhutanese government knows that it is likely that natural resources will be irrevocably destroyed and people impoverished. Rapid economic
(continued)

Models of Global Sustainability 22.1 (*continued*)

Bhutan: Pursuing a Sustainable Path

growth may be accompanied by growing inequality, which could tear the social fabric of the country. Increasing consumerism may lead to the pursuit of narrow economic goals.

The country of Bhutan has the option of creating a prosperous and sustainable society from scratch. Few countries have such a choice. A fundamentally sustainable course is already in place. Environmentally damaging infrastructure and polluting industries have not been developed. The country is not hindered by debt as many other nations are. Its cultural values remain strong and intact. Income is evenly distributed. There are few economic or ethnic tensions. Moreover,

the government has strong convictions about sustainability and has not been corrupted by business interests.

Even with such strong convictions, the road to sustainability will require constant vigilance. Most think that the effort will be well worth it. As Worldwatch Institute's Christopher Flavin notes: "It will involve a process of continual adaptation and change, preserving what is best in Bhutanese society while improving living standards and protecting the natural resource base on which the country depends. Such an endeavor would truly make Bhutan a model for the world."

local people want, then involving them in the project design, developers are more likely to succeed. Development is unlikely to be sustained unless the needs of people are identified and local residents support the project.

Finally, development projects must be culturally sensitive. In many countries, rural people value a sense of community and cooperation. Unlike the West, where competition reigns supreme, people in many communities in the developing world tend to work together for the good of the entire citizenry. Western-style projects that promote individual gain may be viewed as a hindrance to progress. In such instances, developers might be advised to find community-wide development projects that promote the good of the whole rather than private enterprise that promotes competition.

For years, local cultures have been viewed as an obstacle to success. Today, more and more people in the development community realize that for development efforts to be successful, they must preserve cultures. Projects must be woven into the social fabric of the people they are intended to help, not forced upon them in a culturally disruptive manner.

Promoting Flexible Strategies

Another problem with traditional development strategies is that they tend to be organized from the top down. That is, most decisions are made by a select few—those with the training and those with the money. Relying on the top-down approach, large bureaucracies frequently attempt to manage projects many miles from their headquarters.

This organizational structure tends to be unwieldy and inflexible. Soon after they begin, most projects face obstacles. Without flexible goals and some flexibility in achieving them, bureaucracies tend to resist necessary adjustments. The result is failure.

To be successful, development programs could be viewed as experiments. Small pilot projects could be run first to determine problems and solutions. At the very least, development projects require flexibility. If troubles arise, managers need to be able to make adjustments to address them.

One way of increasing flexibility is to remove the actual management of projects from the hands of international development agencies and banks. Who would take over? One candidate for management is private groups involved in sustainable development, **nongovernmental organizations** (NGOs). With funding from development agencies, NGOs could provide flexibility in managing projects to ensure adequate safeguards to protect the environment while improving the lives of the people. Fortunately, more and more MDBs are involving NGOs in project management.

Improving the Status and Expanding the Role of Women

In most developing countries, women are important managers of natural resources. They collect firewood and maintain gardens to feed their families. And, they bear the children. Women can play a huge role in restoring and sustaining the environment (Figure 22.5). Yet, in most countries, women have little access to education, credit, and land. They are poorly trained, and, despite

Figure 22.5 *Women play a key role in environmental decision making and restoration. These women in Maathai, Wangari Africa are part of a movement to replant the nation's forests.*

the fact that they perform most of the work, their labors account for only a small portion of the total income.

Sustainable development clearly requires steps to improve the status of women. Many different reforms are needed, among them measures to eliminate discrimination and cultural traditions that cast women in an inferior role. Proponents of this movement call for legislation to ensure equal pay for equal work, job training, maternity leave, and other benefits. Efforts are needed to ensure that women have a full voice in decision making. Women also need access to credit at a reasonable rate of interest. In India, for instance, uneducated rural women are often forced to borrow money from loan sharks at 10% per day. Some use the money to buy fish, which they sell. Today, thanks to the Working Women's Forum, a national Indian organization, rural women can receive credit at a reasonable rate to start small businesses to generate additional income for their families. Many have proved to be highly successful entrepreneurs. The Working Women's Forum boasts a 95% to 100% payback on loans.

Women also need access to better education, health care, and family planning. Once established in a region, the Working Women's Forum helps women acquire government services for prenatal and child care, immunization, and family planning. As the income and health of the family improves, the number of children born drops drastically.

Programs similar to the Working Women's Forum are forming all over the world. To many observers, this movement represents another ray of hope that population growth and environmental decline may be brought under control.

Preserving Natural Systems and Their Services

Sustainable development also requires efforts to protect natural systems that serve as a source of goods and services. Appropriate technology and many efforts to design with nature help protect natural systems. Developers and local people can also protect the environment by establishing guidelines that minimize the destruction of wildlands.

The World Bank recently adopted a set of guidelines that could serve as a model for all international development activities. The bank's policy strictly prohibits supporting projects that will destroy wildlands of special concern, including wetlands and virgin rain forest. Instead, the bank favors projects on land that has already been disturbed, for example, forestland that has already been cut or grassland that has been farmed. This stipulation minimizes the destruction of undisturbed land. When the bank believes that undisturbed land is justifiably developed, it favors projects on lands that are less valuable, for instance, forests with lower species diversity or land not essential to protect water supplies. But, even in such instances, the bank calls on developers to offset losses by improving management on similar lands. Finally, the bank calls on developers to protect lands surrounding development. It might, for example, provide financial support for a dam project, but only if efforts are made to protect its watershed from deforestation.

This set of guidelines reflects the importance of wildlands and the "free" services they provide. Replacing free services, such as watershed protection, can be costly. Corrective measures to regain nature's gifts may require massive outlays of money for reforestation, flood control, water purification, and others.

Improving the Productivity of Existing Lands

The efforts outlined above protect land from unnecessary development. Better management of existing lands can also save land from the ravages of development (Chapter 9). Erosion control and grazing management, for instance, can protect farmland and grassland, reducing the stress on undisturbed ecosystems. Efforts are needed to improve the fertility of existing cropland to increase crop output. This reduces the demand for new farmland to feed the swelling human population (Chapter 7). International lending agencies and development agencies could support projects that would improve the condition of lands already cultivated or lands already being grazed.

In addition, they could promote better forest management and efforts to replant tropical rain forests to reduce the pressure on undisturbed lands (Chapter 9).

22.3 Overcoming Attitudinal and Economic Barriers

The previous section outlined many barriers to sustainable development, among them the unflagging efforts of developed countries to transpose Western ways into foreign cultures and the failure of developers to involve local people and tap local knowledge. Several substantial attitudinal and economic barriers also lie in the way of sustainable development.

Attitudinal Barriers

The first barrier is the predominant notion that environmental protection is a luxury—that is, a matter governments can address *after* people have achieved prosperity. The prevalent view is that poor people, and their leaders, can't be concerned about protecting the environment if people are starving and living without basic necessities.

Fortunately, many people recognize that environmental protection is essential to the survival and prosperity of all people. Moreover, increasing evidence shows that poverty and environmental deterioration form a vicious cycle. Environmental decay drives many people into poverty. As desperate people consume and destroy the resource base upon which they depend, poverty worsens and increases environmental problems, creating a vicious downward spiral that undermines their long-term prospects. Lester Brown, Christopher Flavin, and Sandra Postel note in their book *Saving the Planet*, "Rather than a choice between the alleviation of poverty and the reversal of environmental decline, world leaders now face the reality that neither goal is achievable unless the other is pursued as well."

Environmental protection, therefore, is not an indulgence afforded only by the wealthy. It is a prerequisite for survival and a decent standard of living (Figure 22.6). The goal of sustainable development is to meet human needs in ways that satisfy the need to protect and enhance natural systems.

A second, similar barrier is the belief that efficiency is a luxury among developing nations. Today, developing countries spend nearly two-thirds of their export earnings on oil imports. Making more efficient use of oil and other fossil fuels is vital to the future of these countries and could reduce sizable trade deficits that cripple their economies.

Efficiency measures cost far less than new power plants and can improve the lives of many people. For example, if India replaced 20% of its incandescent light bulbs with high-efficiency compact fluorescent bulbs over the next 20 years, the country could avoid building eight huge power plants at a total cost of $4 to $8 billion. This project would cost less than $1 billion and would save substantial sums of capital that could be put to other uses.

Considerable economic and environmental gains can be made in the developing nations by improving energy efficiency. High-efficiency wood cook stoves, for instance, could cut demand for wood and help reduce deforestation. Energy efficiency, combined with projects to replant trees and better manage forests, could be instrumental in helping people meet their needs sustainably. Today, however, energy-efficiency projects constitute less than 1% of all international aid.

Economic Barriers

Developing nations will find it to their advantage to concentrate financial resources on many underfunded areas, including housing, education, sanitation, clean water, sustainable agriculture, and family planning (Chapter 7). By reducing their import of fossil fuels, primarily through conservation, they can generate money to fund programs in these areas.

Unfortunately, the poorest countries have the greatest difficulty securing money to meet the needs of people sustainably. Many of them spend surprisingly large amounts of their national budgets on military protection. As the Commonwealth of Independent States and the United States recently learned, reducing expenditures on weapons and armies is important to sustainable economic development. In fact, reduced military expenditures are widely regarded as a source of money that could be deployed for civil uses, among them family planning and sustainable agriculture. (Viewpoint 15.1 described some cuts and how the money could be used otherwise.)

Successfully reducing military expenditures depends on satisfying governments that their borders are secure from invasion and that internal strife can be managed with less investment. It also rests on efforts to convince countries that their economic and environmental well-being could improve by such steps. In addition, nations whose economies depend on the export of arms need to be persuaded that their interests could be better served by converting to other economic activities. Although these efforts will not be easy, they are in the interest of the entire world's people.

Two international environmental groups and the U.N. Environment Programme outlined measures needed to demilitarize the world, among them, an acceleration of agreements to limit weapon production. International agreements to regulate the arms trade and treaties to reduce financial assistance for military expenditures could also help. International conventions that limit certain types of action, like the destruction of

Figure 22.6 *Environmental protection is not a luxury. It is a condition for existence.*

Kuwait's oil fields in the 1991 Iraqi invasion, are also needed. They also call on the world community to work together to redeploy military personnel from wartime activities to peacetime activities, among them disaster relief and conservation.

Wealthy nations can help improve the economic conditions of developing nations by ending their exploitation, which tends to rob the poorer nations of their natural resources. (For a few ideas, see Models of Global Sustainability 21.1.) As pointed out in Chapter 13, multinational corporations, wealthy landowners, and industrialists in developing nations often become rich at the expense of the poor and their natural resource base.

Besides reducing economic exploitation, wealthy nations can help by relieving the burden of international debt—money lent to support many unsustainable development projects. Many of the poorest developing nations are also the most debt ridden. Servicing the debt has forced many developing nations to convert food crops for domestic production to cash crops, tea and citrus fruit, for instance, for export (Chapter 7). Debt-for-nature swaps discussed in Chapter 8 are one means of reducing debt.

Many individuals maintain that a more equitable distribution of world resources could help poorer nations feed, clothe, and house their people. How that would be done and how successful it would be is debatable. Food handouts, while helping to alleviate suffering, don't get at the root causes of the problems: rapid population growth, poverty, ill-conceived agricultural systems, and other unsustainable activities, among them

forestry and fishing. More effective might be assistance aimed at helping developing nations become more self-reliant and sustainable.

The changes and policies outlined in this chapter could help the developing nations progress along a sustainable path, operating within the limits posed by the Earth and availing themselves of many opportunities. As many astute observers point out, they could also serve as a set of guidelines for development in industrial nations, where widespread use of appropriate technology and design with nature are especially needed. The World Bank ecosystem policy that seeks to protect untouched wildlands by diverting development to already disturbed lands is also essential, as are efforts to sustainably improve the productivity of existing lands.

The West can learn a great deal from the cultures of people in the developing countries. In fact, many modern writers champion a new world view based on cooperation and holistic (systems) thinking (Chapter 1). The new world view is anticipatory in that it seeks to understand long-term, systems-wide impacts. It is also participatory in that it seeks to involve many people in decision making. These values are common in what people in the West disparagingly call "primitive cultures." On this matter, anthropologists Margaret and William Ellis wrote, "There are new future concepts already conceived and still practiced by various people around the world that we need to understand, adapt, and adopt." Our future depends in part on learning important lessons from the people we have inadvertently victimized in the past with our generous development assistance.

He helps others most, who shows

them how to help themselves.

—A. P. Gouthey

ॐ   ॐ   ॐ

Critical Thinking Exercise Solution

Raising the standard of living in developing countries is probably essential to creating a sustainable future, for poverty and environmental decay tend to go hand in hand. Numerous examples show that people who are poor and hungry do not treat the environment well. Furthermore, raising the standard of living, and along with it education levels, health care, and job opportunities, will have a positive effect on controlling population growth. However, raising the standard of living puts additional strain on the environment's source and sink functions. In fact, developed countries with their high standard of living are responsible for enormous environmental damage.

Ultimately, the Earth and global ecological systems cannot support a world population with a standard of living equivalent to that of the United States and Canada. Although many experts believe that some economic development is necessary, it will probably have to be limited. Does this seem fair to you? Can you think of ways to allow greater economic development? One proposal calls on the developed nations to make massive improvements in efficiency, recycling, renewable resource use, restoration, and population control to make room for the developing nations. By greatly reducing stress on the environment, we in the developed world free up resources needed by the developing nations, and we permit the developing nations access to global sink functions—pollution assimilation. The bottom line, though, is whatever level of development does occur, human society must fit within global ecological constraints. In the final analysis, that's the challenge: to redirect all of society to create a humane, sustainable future.

Summary

22.1 Impacts of Conventional Economic Development Strategies

■ Many economic and environmental problems in developing countries stem from development assistance provided by the industrial nations.

■ Development assistance has often resulted in massive deforestation, soil erosion, desertification, pesticide contamination and poisoning, farmland destruction, and loss of natural lands and the services they provide. Environmental ills tend to deepen social and economic troubles.

■ One failing of economic development is its attempt to impose Western ways on developing nations. Large-scale, capital-intensive projects tend to remold natural systems to benefit humans, often with serious ecological backlashes.

■ Western-style development projects tend to benefit a small segment of the population while ignoring, sometimes even harming, the people they were designed to serve.

■ Western-style development can result in the decimation of potentially sustainable resources and sustainable cultures.

22.2 Sustainable Economic Development Strategies

■ In order to create a sustainable future, the developing nations must slow the population growth rate, improve education and health care, and promote democracy.

■ Developing nations must build sustainable systems that provide energy, food, water, transportation, waste, housing, and other needs without damaging the environment.

■ **Sustainable development** may require that the developing nations become more self-sufficient.

■ Sustainable economic development requires laws, regulations, and subsidies that promote an assortment of sustainable practices. Policies that promote unsustainable practices must be abolished.

■ Sustainable development can also be promoted by routine analyses that assess the social, economic, and environmental impacts of proposed projects.

■ One means of promoting economic development while protecting the environment is appropriate technology, most of which represents small-scale solutions to local problems that rely on locally available resources and local labor and knowledge.

■ Appropriate technologies are suitable for the developing nations, which lack the financial and fossil fuel resources needed to support Western-style industries.

■ Appropriate technologies rely more on human labor, which is found in abundance in developing nations.

■ Sustainability also depends on designing human systems, like food production, to fit within the constraints of natural systems rather than trying to redesign nature to our liking.

■ Tapping into local knowledge and skills found in local people can help countries find sustainable solutions to local problems.

■ Successful development projects need to address problems viewed as critical by local people. Development is unlikely to be sustained unless the needs of people are identified and local residents support the project.

■ Development projects must be culturally sensitive. Developers are advised to find development projects that support rather than destroy local cultures.

■ To increase the success rate of development projects, pilot projects could be run to determine problems and solutions. Development projects are more successful if they are managed flexibly, so that if troubles arise, managers can make adjustments to address them.

■ Sustainable development requires steps to improve the status of women. Efforts are also needed to ensure that women have a full voice in all decision making. Women need access to credit at a reasonable rate of interest. Women also need access to better education, health care, and family planning.

■ Developers and local people can protect the environment by establishing guidelines that minimize the destruction of wildlands.

■ Development should be prohibited from destroying wildlands of special concern. Projects should be carried out on land that has already been altered.

■ When undisturbed land is to be justifiably developed, projects should be placed on lands that are less valuable. In such instances, developers can offset losses by improving management on similar lands.

■ Efforts are needed to protect lands surrounding development.

■ Instead of focusing entirely on new development projects, international lending agencies and development agencies should support projects that seek to improve the condition of lands already under the plow or lands already being grazed.

22.3 Overcoming Attitudinal and Economic Barriers

■ Many obstacles lie in the way of sustainable development. One barrier of note is the prevalent notion that environmental protection is a luxury, a matter governments can address *after* people have achieved prosperity. Fortunately, many recognize that environmental protection is essential to the survival and prosperity of a people.

■ Another similar belief is that efficiency is a luxury among developing nations. Making more efficient use of resources, especially oil and other fossil fuels, is not a luxury; it is vital to the future of these countries. It could reduce trade deficits and costly expenditures for new power plants that cripple the economies of many developing nations. Such efforts can free up money for other pressing needs.

■ Money to support these programs can also come from reductions in military expenditures.

■ Reducing or eliminating economic exploitation by the industrial world could improve the economic condition of developing nations, as will efforts to forgive foreign debt.

■ The changes and policies outlined in this chapter could help the developing nations progress along a sustainable path, operating within the limits posed by the Earth and availing themselves of many opportunities. They could also serve as a set of guidelines for development in the industrial nations as well.

Discussion and Critical Thinking Questions

1. Describe the goals of sustainable economic development.

2. Traditional development schemes in poor, nonindustrialized nations often result in noticeable increases in the GNPs of the target nations. Why is an increase in GNP not necessarily a sign of progress? Give some examples from this chapter.

3. You are appointed head of development in a poor African nation. Your population of 35 million is growing at a rate of 3% per year. Many of your rural people are poor. Outline a plan to improve the lives of the people.

4. List several ways to make economic development strategies more sustainable. Give examples of each one.

5. List the barriers currently hampering sustainable economic development and ways to address them.

6. Using your critical thinking skills, debate the following statement: "In order to help the people of developing nations, we must help them build factories to make tractors, bulldozers, and other equipment that have been the key to success in the United States and other developed countries."

Suggested Readings

Brown, L. R., Flavin, C., and Postel, S. (1991). *Saving the Planet: How to Shape an Environmentally Sustainable Global Economy*. New York: Norton. Contains many important ideas for sustainable economic development.

Chiras, D. D. (1992). *Lessons from Nature: Learning to Live Sustainably on the Earth*. Washington, D.C.: Island Press. See Chapters 6 and 7.

Dobb, E. (1992). Solar Cooker. *Audubon* 94 (6): 100–105. Interesting article that tells about an appropriate technology.

Holmberg, J. (1992). *Making Development Sustainable: Redefining Institutions, Policy, and Economics.* Washington, D.C.: Island Press. Important collection of essays on sustainable development.

IUCN, UNEP, and WWF. (1991). *Caring for the Earth: A Strategy for Sustainable Living.* Gland, Switzerland: Earthscan. Contains many ideas for sustainable economic development.

Panayotou, T. and Ashton, P. S. (1992). *Not by Timber Alone: Economics and Ecology for Sustaining Tropical Forests.* Washington, D.C.: Island Press. Examines tropical rain forests as a source of many products that could be harvested sustainably.

Plotkin, M. and Famolare, L. (1992). *Sustainable Harvest and Marketing of Rain Forest Products.* Washington, D.C.: Island Press. Describes products that can be extracted with little if any damage to the forest ecosystem.

Reid, W. V. C., Barnes, J. N., and Blackwelder, B. (1988). *Bankrolling Success: A Portfolio of Sustainable Development Projects.* Washington, D.C.: Environmental Policy Institute and National Wildlife Federation. Study of successful development projects.

World Commission on Environment and Development (1987). *Our Common Future.* Oxford: Oxford University Press. See Chapter 2 on sustainable development.

Chapter 23

Government in a Sustainable World

The loftier the building, the deeper must the foundation be laid.

—*Thomas à Kempis*

Critical Thinking Exercise

In the U.S. *National Report*, a publication prepared for the U.N. Conference on Environment and Development, the authors frequently bragged about the billions of dollars the nation was spending on pollution control. They cited case studies that showed specific actions on the part of businesses and government agencies. Overall, the *National Report* seemed to imply that because the United States is spending lots of money on environmental problems and putting a lot of effort into solving them, the nation is acting responsibly and with sufficient vigor. Is there a fallacy in this kind of reasoning?

We need a future we can believe in, one that is neither so optimistic as to be unrealistic nor so grim as to invite apathy or despair. In short, we need a future that is not only hopeful but also attainable. This text has outlined one of many possible futures, a sustainable one in which human society lives within the limits of nature. Creating such a system requires an understanding of basic principles of ecology and an appreciation of how vital natural systems are to our lives. It also depends on a decrease in population growth (Chapters 5 and 6) and dramatic changes in the resources we use and the way we manage them (Chapters 7 through 13). To create a sustainable human presence, we must also alter the way we deal with pollution (Chapters 14 through 19). The changes needed depend on a shift in people's attitudes (Chapter 20) and dramatic changes in economic systems (Chapters 21 and 22). Such modifications may come about in many ways. One of the most effective means of social change is through government.

This chapter examines government: the roles of government, the participants in government policy-making, and how decisions are made. It discusses some key obstacles to a sustainable future and ways to overcome them. Finally, it outlines important changes in international governance needed to make a sustainable future a reality.

23.1   The Role of Government in Environmental Protection

Chapter 21 described two basic economic systems: market economies, in which the marketplace determines the availability and price of goods and services, and command economies, in which the government plays a key role in these functions. Most countries have features of both economies but tend to lean in one direction or the other.

Forms of Government

For the most part, governments and economic systems go hand in hand. Countries with market economies are generally **democratic nations** with representative governments—that is, governments largely staffed by officials that are elected by the people. Representative governments operate by rules agreed on by the majority. In democratic nations, the means of production and distribution of goods and services are, for the most part, privately held and privately managed. Profit is a major driving force of this system. However, democratic nations have found that varying degrees of governmental control are needed to reduce an assortment of unlawful activities.

Countries with command economies are, as a rule, **Communist** or **Socialist nations** in which society technically owns and operates the production and distribution network. Such nations emphasize the requirements of the state rather than individual liberties. Profit is not a motive in such systems; the nations believe that goods should be distributed equally, although in practice this rarely occurs.

Just as economies are mixed, so are governments. New Zealand, Germany, and Sweden are all democratically governed, but in all three nations, the government provides health care and other services that are paid for through taxes. In China, one of the few remaining Communist nations, the private sale of products for profit is currently permissible, and growing.

How Government Policies Influence the Quality of the Environment

Democratic governments regulate activities, such as environmental protection, via three measures: (1) tax policies, (2) direct financial support, and (3) laws and regulations.

Taxes are levied on personal property, commodities, and income, both personal and corporate. Taxes are a means of generating funds for public services, among them highways, road maintenance, police protection, pollution control, wildlife programs, and the like. Taxes can be used to promote or discourage activities (Chapter 21). For instance, a sizable gasoline tax in European nations promotes the purchase of energy-efficient automobiles and discourages driving. It's not surprising that some of the best mass transit systems are found in Europe.

Tax policies of most nations also allow for tax breaks, lower tax rates for the purchase of certain products. Tax breaks are designed to promote certain activities and can be used to move society onto a sustainable path. In some states, for example, recycled paper is taxed at a lower rate than paper made directly from wood pulp.

Two more common tools in a government's bag of tricks are tax credits and income tax deductions. A **tax credit** is an amount of the purchase price that one can deduct from income or corporate tax. In the United States in the early 1980s, for example, the federal government offered individuals and businesses sizable tax credits if they invested in energy conservation, such as insulation or weatherstripping, and renewable energy, such as solar panels and wind energy. Many states offered similar tax credits. In Colorado, for instance, the combined federal and state tax credit for solar energy came to 70%. Thus, 70% of the cost of a solar system could be deducted from one's tax bills. This form of tax policy passes the cost on to the taxpayer.

Income tax deductions work similarly. When a government offers a deduction for an activity, the amount of the purchase or some percentage of the purchase price is deducted from one's total income. This lowers income and therefore lowers the amount of tax one pays. The U.S. government currently offers income tax deductions for children. Couples with children receive a small tax break for each child. Interest on loans for second homes was, until recently, also tax deductible. Tax policies that favor undesirable practices such as this can also have significant impacts on the future.

Direct government expenditures can also have a profound effect on the environment. For instance, in the 1980s, government grants to communities for sewage treatment plant construction created thousands of jobs and helped reduce water pollution (Chapter 17). Additionally, government-funded research programs in energy conservation and solar energy yielded knowledge and technologies essential to a sustainable future. Like investments in construction and research and development, government procurement programs for environmentally beneficial goods, such as photovoltaics and recycled products, can help create larger markets (Chapters 12 and 19). This, in turn, could promote mass production, which drives prices down and makes these products more affordable.

Government expenditures on environmentally destructive technologies, nuclear energy research, for instance, and activities, such as war and oil development, can have the opposite effect. The most obvious reason is that they direct society down the wrong path of development. The further society travels down that path, the harder it will be to shift onto a more sustainable one. Such expenditures also divert money from other important tasks, such as energy efficiency and renewable energy development.

Cutting government expenditures on environmentally destructive activities could free up considerable sums of money, creating funds that could be invested in promoting sustainability (Chapters 11 and 22). Even modest reductions in global military spending could finance a massive campaign to restore the Earth and build a sustainable world community living in harmony with nature (Viewpoint 15.1).

The final and probably the most widely used governmental tools are laws and regulations. **Laws** emanate from legislative bodies: congresses, parliaments, town councils, and the like. They are generally designed to ban or curb activities deemed undesirable by the people and their representatives. For example, the **Corporate Average Fuel Economy Act** was passed in 1975. This law established a set of goals for gas mileage standards for automobile manufacturers.

However impressive, a nation's laws are only as good as their enforcement. Enforcement depends on having adequate personnel and funds. If an agency lacks funds, human power, or the will to enforce a new law, little will come out of it. As noted in Chapter 19, the EPA took 12 years to make its first recommendations on the acceptable recycled content of a dozen or so products that could be used by the federal government, an act mandated by RCRA. It did so only under pressure from environmental groups.

The Corporate Average Fuel Economy Act is an example of a law whose utility was greatly reduced by lack of interest and heavy pressure from private industry. The U.S. Department of Transportation was given the responsibility for implementing and enforcing mileage standards, but it bowed to heavy pressure from Ford and General Motors, rolling back the mileage standards twice during the Reagan administration. In 1985, the average fleet mileage was far below Congress's goal of 27.5 mpg. When President George Bush took office in 1989, however, he pledged to raise standards, and by 1990, the average new car got 27.5 mpg. Further improvements called for in the ensuing years, however, have been slow and efforts to raise the mileage standard in the United States to 40 mpg by the year 2000 were defeated in 1991.

The president of a nation can have a powerful impact on environmental policy. Unsympathetic leaders can deter progress. President Ronald Reagan, for instance, who was in office from 1981 to 1989, openly opposed environmental protection goals of previous administrations. Reagan actively sought to undermine protection by weakening existing laws and agencies, such as the EPA and the Council on Environmental Quality. President Bush, who laid claim to the title of environmental president, improved matters somewhat, but behind the scenes, the Council on Competitiveness (chaired by Vice President Dan Quayle) systematically worked to undermine the nation's environmental laws and regulations. The council, for instance, proposed dramatic changes in wetland designation that would have eliminated about half of the nation's wetlands from protection. After passage of the 1990 Clean Air Act Amendments, the council unsuccessfully proposed a strategy for companies to duck compliance. If a company felt it couldn't meet the law's requirements, all it had to do was write the governor of the state. If the governor didn't reply in a week, the company was exempt from the law.

The United States is not the only country in which environmental laws fail to accomplish their goals because of lack of funds or political shenanigans. Before it disbanded, the former Soviet Union had some of the most stringent environmental standards in the world, but enforcement was virtually nonexistent. The country lacked the technical expertise and the financial resources to control pollution. Furthermore, the economic and social problems were so enormous that they were ignored by the government.

Environmental lawmakers usually set broad goals in their legislation and turn the job of writing regulations over to various agencies that have the expertise to draft "regs" to achieve congressionally mandated goals. In the United States, the EPA is the principal rule maker. A host of other agencies also participate in this process. These agencies draft regulations, then open them to public comment. After receiving input, they may modify the regs or put them into effect.

The EPA is currently empowered by nine federal laws, among them the Clean Water Act, the Clean Air Act, the Superfund Act, and the Resource Conservation and Recovery Act. Each law gives the agency authority to enact regulations governing activities ranging from hazardous waste to emissions standards for new factories.

Through the mechanisms noted above, the U.S. government has made tremendous strides in protecting the environment, managing resources, and controlling population. Many chapters in this text have pointed out the most significant advances. (For a quick review of U.S. progress, see Table 14.2, which lists nearly two dozen environmental laws and amendments enacted since 1958.) Similar progress has been made in most other developed nations. While environmental conditions are better than they would have been in the absence of these measures, most governments have dealt with environment problems after the fact. That is, governments have typically relied on end-of-pipe solutions or regulations that merely slow down the destruction of global ecosystems. This chapter describes policies needed to confront the roots of the problems to create a sustainable way of life.

But what about the developing world, where over 4 billion people live? Many developing nations have adopted policies to control population growth, save wildlife, and reduce pollution. However, progress has been hindered in many cases by a lack of money, rapid population growth, corruption, civil war, war with neighboring nations, and governmental apathy and ineptitude. Interference by foreign interests and misguided development projects have also worsened environmental quality. In many cases, hunger and poverty are so severe that environmental matters are relegated to the backseat. Without population stabilization and programs to raise the standard of living and the availability of food, little progress can be expected.

23.2 Political Decision Making: The Process and the Players

Now that we've seen the basic tools by which governments influence the environment and shape the future of a country, indeed, the entire world, this section takes a closer look at the process and the players.

Government Officials

In Communist governments, policy is largely determined by a ruling elite. Ironically, in such systems, the people who government is meant to serve generally have little or no input. In fact, they often live under severe oppression. In contrast, in democratic nations, a wide assortment of government officials make important decisions that affect the environment. Some government officials, senators, for instance, are elected by the people; others, such as agency heads, are appointed by the president, governors, or mayors.

The 1989 nuclear weapons scandal, described in Case Study 19.1, showed that even in democratic nations government officials can work against the good of the whole. As you may recall, over the past 50 years, dozens of nuclear weapons facilities have released large amounts of radiation into the air, water, and soil. Government regulators either turned a deaf ear to reports of misdoings or openly endorsed illegal acts. A case in point: In his first term, President Reagan signed an executive order that rendered government weapons facilities immune from U.S. environmental laws. Officials in the EPA and state health departments have also been known to overlook violations of environmental and health laws at these facilities. However, it is important to note that dedicated, hard-working men and women at all levels of government have also made important contributions to bettering the environment.

The Public

In a democratic society, voters generally have a say in policy-making through their elected representatives (Figure 23.1). Letters, phone calls, attendance at town meetings, and responses to surveys are four lines of communication that permit citizens to voice their opinions on a variety of issues.

Although it is sometimes forgotten, elected representatives are a society's employees. Their job is to

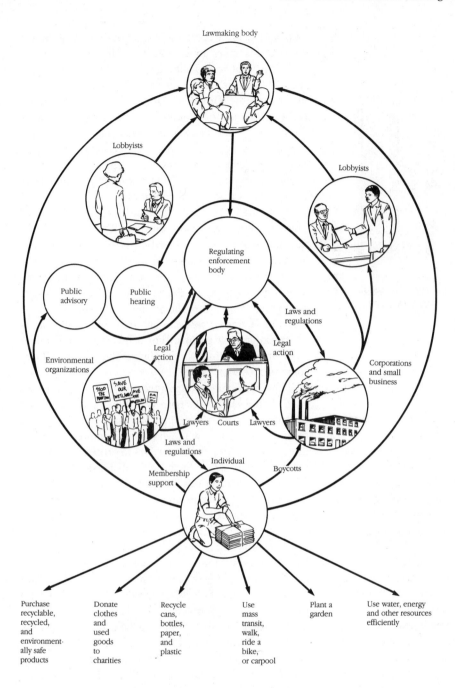

Lawmaking body

Lobbyists

Lobbyists

Regulating enforcement body

Public advisory

Public hearing

Laws and regulations

Legal action

Legal action

Environmental organizations

Corporations and small business

Lawyers   Courts   Lawyers

Laws and regulations

Membership support

Individual

Boycotts

| Purchase recyclable, recycled, and environmentally safe products | Donate clothes and used goods to charities | Recycle cans, bottles, paper, and plastic | Use mass transit, walk, ride a bike, or carpool | Plant a garden | Use water, energy and other resources efficiently |

Figure 23.1 *Diagram showing the many lines of connection between government, individuals, and corporations. Laws are passed by lawmakers. These, in turn, affect individual behavior and business behaviors. Individuals, environmental groups, and businesses also influence the laws through lobbyists and the court system.*

interpret public preferences and find ways of enacting policies that satisfy the needs of the electorate. In many cases, however, the populace is divided on issues. Elected representatives may even choose to vote against their constituency's wishes. Public officials are also frequently influenced by special interests. Because of this and other reasons, many people have grown cynical about their influence on the government and the future. Many people feel victimized and disempowered by the political process, which leads to apathy and poor voter turnout at elections.

Apathy is counterproductive to the political process. Citizens can make a difference. One letter to a congres-

sional representative may not change a vote, but many letters might. As a rule, for each letter they receive, legislators generally estimate that there are 5000 to 10,000 people who feel the same way. Thus, 1000 letters mean 5 million people think similarly, which is pretty compelling to an elected official.

A letter to a state legislator can have special impact because state legislators rarely hear from the public. Twenty or 30 letters are cause for concern. In their minds, 40 or 50 letters indicate a disaster requiring action. So, write and vote. Let your voice be heard. (The *Environmental Action Guide* has more information on becoming politically active.)

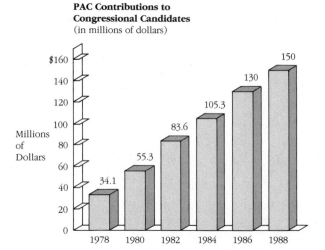

PAC Contributions to Congressional Candidates
(in millions of dollars)

Figure 23.2 *Political action committee (PAC) contributions to congressional candidates from 1978 to 1988.*

Special Interest Groups

The **theory of public choice** states that politicians act in ways that maximize their chances of reelection. To do so, they must appease voters and special interest groups (Figure 23.1). Special interest groups, including environmental organizations, often work through **lobbyists**, individuals who convince legislators and their aides of the merits of their views. In some cases, lobbyists even draft legislation. When Harry Truman was president (1945 to 1953), about 450 lobbyists worked the halls of Congress; today, an estimated 23,000 lobbyists swarm the nation's Capitol.

Besides their physical presence, special interest groups often reward politicians monetarily, providing funds needed to run costly political campaigns. To make an even larger impact, many special interests band together to form **political action committees (PACs)**. A PAC is a consortium of individuals or organizations that pools financial resources to create a single large donation. In the 1988 elections, PACs donated more than $150 million to congressional candidates, up from $34 million a decade earlier (Figure 23.2). In the 1992 election, they contributed an estimated $254 million.

According to the nonprofit citizen's lobbying group Common Cause, PAC contributions pay off "in billions of dollars worth of government favors for the corporations and other special interest groups that make them." PAC contributions from business, for example, may help defeat or dilute important environmental legislation. In the 1980s, PAC contributions and coal and utility industry lobbyists allegedly stalled the enactment of legislation to reduce acid deposition, toxic air pollution, and other important problems. In 1984, auto manufacturers successfully lobbied Congress to pass import quotas on energy-efficient Japanese automobiles,

which yielded the companies $300 million in profit and cost the consumer an estimated $2 billion or more.

One of the most insidious political games played by special interests, and especially business interests, is known as the **Double-C/Double-P game.** Double-C/Double-P stands for "commonize the costs and privatize the profits." Here's how it is played: Business lobbyists support and sometimes propose legislation that promotes government programs that will generate huge profits for the private sector. Thus, taxpayers end up paying for programs that support businesses.

Although democracy can be distorted by special interest groups, especially business PACs, money does not always reign supreme. In the United States, powerful antipollution laws, auto safety standards, hazardous waste laws, and other important environmental laws have been enacted in the last three decades, despite strong opposition from the business community. A good measure of this success can be attributed to the efforts of environmental groups. (For a listing of some major environmental groups in the United States, see the *Environmental Action Guide*.)

Environmental groups affect public policy in several ways. Some, such as Greenpeace, take an active role on the front lines, protesting environmental injustices, often meeting face-to-face with whalers, seal hunters, and polluters, endangering their own lives to protest actions they oppose (Figure 23.3). Such public displays have proved highly successful in raising public awareness. In 1980, a radical group of environmentalists formed. They called themselves **Earth First!** Under the leadership of Dave Foreman, they took more active, sometimes illegal steps to protect the environment. They put their bodies on the line to stop bulldozers. They embedded spikes in old-growth trees, which thwarted chain saws. They put dirt in the fuel tanks of heavy equipment. Their tactics were called **ecotage**—sabotage in the name of the environment. (For a debate on the pros and cons of ecotage, see Point/Counterpoint 23.1.)

Some years later, Paul Watson, one of the original founders of Greenpeace, started the oceanic equivalent of Earth First! Called the Seashepherd Conservation Society, members roam the high seas, ramming ships that are illegally whaling, flying miniature airplanes to disrupt sonar, and generally disrupting illegal activities. They have never been arrested or charged with a crime because they stop people engaged in illegal activities who don't want to attract public attention.

Most environmental groups operate within legal means. Some groups, for instance, are involved in education. They may prepare educational materials for schools or may deliver talks to public gatherings. Some groups work to preserve endangered lands. The Nature Conservancy, for instance, purchases land that is then set aside to protect endangered wildlife and plants, and to provide future generations with opportunities to enjoy

Figure 23.3 *Greenpeace activists position themselves between Soviet whaling ships and whales to thwart the slaughter of whales.*

the out-of-doors. Other groups are involved in lobbying efforts, writing new laws, getting laws passed, or strengthening existing laws. Still others serve as watchdogs, making sure that polluters obey the rules and that government agencies do their jobs.

Many environmental groups have become involved in proactive planning, devising sustainable alternatives to environmentally harmful projects. The Environmental Defense Fund (EDF) has been a leader in this arena. Its staff of economists, public policy experts, and scientists research issues and offer economically and environmentally sound alternatives. The EDF worked with the McDonald's Corporation to institute the company's recycling and waste reduction program.

Some environmental groups spend much of their time in court, suing governmental agencies when they are not doing their job or are doing it poorly, or suing businesses that violate environmental laws.

Many environmental groups operate on many levels. The EDF, for instance, began as a litigator, suing government agencies and businesses. Today, it works closely with business and government to find alternative practices that serve people and the environment. The EDF also publishes books and newsletters that influence public opinion and public policy.

Environmental groups can offset the disproportionate influence of business, but they are usually at a disadvantage. Corporations like Exxon, which boasts a $5 billion annual profit, usually wield more influence in the political process.

Unfortunately, a strong antienvironmental movement is cropping up in the United States. Called the **Wise Use Movement**, it is heavily funded by timber compa-

nies, mining companies, oil and coal companies, cattlemen, and the like. Representatives spread half-truths about the environment and actively try to dismantle environmental protection. (For a discussion of the Wise Use Movement, see Viewpoint 23.1.)

23.3   Creating Governments That Foster Sustainability

From an environmental standpoint, democratic governments display the same weaknesses as the economy. That is, they are generally shortsighted, obsessed with growth, and exploitive of people and nature. They also tend to promote dependency or undermine self-reliance.

Why do governments and economies have the same flaws? In a nutshell, the frontier ethic of many Americans is largely responsible for creating mutually supporting systems of government and economics.

To build a humane, sustainable society, governments and the economies they serve must be made responsive to the needs of future generations and other species that share this planet with them. First and foremost, governments must adopt the goal of sustainability as a central organizing principle. This section offers additional ideas on ways to correct the "flaws" in governments and economies. The following section looks at international governance.

Creating Government with Vision

Responding to problems requires a consensus for action, which may be translated into appropriate governmental

Randal O'Toole

As an economist for Cascade Holistic Economic Consultants, of Portland, Oregon, Randal O'Toole has examined management practices on over 60 national forests. He is the author of Reforming the Forest Service, *a comprehensive package of reforms based on his research on the national forests.*

Direct Action Wastes Limited Resources of the Environmental Community

Civil disobedience, ecotage, and other direct actions aimed at stopping ecological destruction have scored a few spectacular victories. In Australia, civil disobedience successfully halted construction of the economically and environmentally costly Franklin Dam. In the Pacific Northwest, direct action brought national attention to the spotted owl and other old-growth forest dependent wildlife, which may yet lead to their permanent protection.

For the most part, however, direct action has proven largely ineffective, even counterproductive, at protecting the environment. Moreover, it raises important ethical questions that its advocates often ignore.

Of the two basic types of direct action—civil disobedience and ecotage—ecotage is the least effective because all it does is increase the cost to resource developers. If millions of dollars can be made from cutting old-growth forests or drilling for oil, adding a few thousand dollars to the costs will not stop the cutting or drilling. Ecotage may make the ecosaboteurs feel good, but it does little to protect the environment.

The most successful direct actions have all used civil disobedience—a human blockade in the case of Franklin Dam, tree-sitting in the case of old-growth forests. The goal of such actions is to gain media exposure so that people learn about environmental problems. Yet, civil disobedience suffers from two major weaknesses. First, it only works a few times. The first person who sits in a redwood tree is news. The 100th person is boring.

More importantly, civil disobedience works because it polarizes people into action—letter writing, testimony, and other lobbying. This seems good at first because it encourages Congress or state legislatures to address environmental problems. In the long run, however, direct actions polarize people on both sides of an issue. They help opponents of environmental protection as much as they help supporters.

This is apparent in the West, which has recently seen the rapid growth of a counterenvironmental crusade known as the Wise Use Movement [Viewpoint 23.1]. Their followers are people upset by environmental tactics, government bureaucracy, taxes, and other problems. Direct action drives people into this camp as much as it gains environmental supporters.

Beyond the effectiveness of the direct action are questions of ethics. A few environmental problems may be life-and-death issues for the human race. But, to be honest, most are just questions of values: Is a fish worth more than a dam? Is an owl worth more than the houses that can be built from wood?

Environmentalists sometimes turn to direct action out of frustration that the current system does not allow them to express their values. But too often, direct action attempts to forcibly or legislatively impose their values on others, which threatens the stability of our society.

Instead of trying to impose values on others, the environment is better protected by finding ways to express environmental values. Many resources, such as air, wildlife, and outdoor recreation on the public lands, are free or nearly free. As a result, they are overused and abused. When a forest manager compares timber values, which directly add to the manager's income, with recreation values, which may be real but contribute nothing to the manager's income, timber values nearly always win out.

The solution is to find institutions that allow us to express our appreciation for the environment. Some of these institutions already exist. Groups like Nature Conservancy buy critical wildlife habitat. The Clean Air Act allows polluters to sell their pollution permits, thus giving them an incentive to reduce their pollution. Other institutions must be created. Recreation fees would give forest managers a reason to protect scenic beauty and other resources that recreationists value. Tradable water rights would allow fishing groups like Trout Unlimited to buy irrigation water and leave it in the stream.

Like anyone, environmentalists have limited resources to spend on actions to protect the environment. Those resources must be divided between lobbying, litigation, research, and other activities. To be as effective—and ethical—as possible, environmentalists should concentrate on promoting institutions that protect our environment as well as the stability of our society.

Howie Wolke

Howie Wolke is a wilderness backpacking guide based in western Montana. A longtime wildland activist, he is the author of Wilderness on the Rocks, *a critique of wildland conservation in America, and coauthor of* The Big Outside *(with Dave Foreman), a descriptive inventory of the remaining big roadless areas in the United States.*

Direct Action Fills an Empty Niche in Wildland Conservation

Civil disobedience, ecotage ("monkey-wrenching"), and other direct actions catapulted old-growth forest destruction into the national consciousness. Direct action helped stop Australia's Franklin Dam and the infamous Vegas-Barstow dirt bike race in the fragile California desert. The U.S. Forest Service has quietly withdrawn timber sales after actual or threatened monkeywrenching or civil disobedience. Guerrilla theater at various dams has forced many to consider the consequences of unrestrained dam building.

Traditional tactics (letters, lobbying, litigation, articles, slide shows, rallies, and boycotts) have saved many more wild places. The Grand Canyon remains un-

dammed and the National Wilderness System is expanding, proving that traditional tactics work. Or do they?

Despite impressive victories, habitat destruction is rampant. Consequently, species, subspecies, and populations become locally, regionally, or entirely extinct as native wildland ecosystems collapse. Dozens of species and subspecies in the U.S. have become extinct since the Endangered Species Act was enacted in 1973. The U.S. Forest Service annually destroys—mostly via logging and road building—over a million acres of unprotected wilderness, despite the growing National Wilderness System (mostly growing with rocky high-altitude additions).

Wildland ecosystems are under assault everywhere, by loggers, miners, cattlemen, power companies, dam builders, and subdividers. Urban sprawl abuts wilderness. Air pollution kills forests. Fences halt wildlife migrations. Poachers exterminate black bear and black rhino for body parts used for mythical aphrodisiacs. Idiots on all-terrain vehicles crush desert tortoises, bird nests, insects, and small mammals.

Today's global ecological crisis is complex and unprecedented. The late Cretaceous extinction 65 million years ago pales when compared with today's galloping loss of biodiversity. Globally, up to one species per hour is wiped out. The obvious concern, of course, is survival. How much biodiversity can we lose before the life-support web crumbles or before humans can no longer live comfortably? Or at all? In truth, we don't know and perhaps can never know in time to avert monumental disaster.

But, the question of survival obscures another profound consideration. Life has intrinsic value. Some economists reduce all conflicts to opposing values: cost versus benefits. "Is an owl 'worth' more than the houses derived from wood?" they ask. Such reductionism oppos-

es nearly all enlightened historic human thought. Do we value only human life? Are we appalled only when torture or genocide victims are human? I think not. The value of thriving diverse native life transcends the almighty buck. Those who resort to nonviolent eco-defense are imposing their values upon others no more than corporations that donate money to political campaigns or, for that matter, than those who threw the Boston Tea Party.

Three questions should always be confronted when considering direct action: (1) Is the action nonviolent toward life? (2) Will the action increase public awareness of a problem? (3) Will it slow or delay destruction so that long-term protection remains plausible?

The success of direct action is now apparent. Besides some obvious victories, direct action brings new activists into the wildland battles. It permits various mainstream environmental groups to take stronger positions without appearing "radical." And, despite contrary claims, civil disobedience still draws attention. Yes, the first tree-sitter is news; the 100th isn't. But, civil disobedience is effective when applied strategically: Save it for the worst atrocities.

The bottom line is this: *Nothing* yet has stopped today's spiraling loss of habitat and biodiversity. We need better laws, better incentives, better news media, better representatives, and better institutions. But, we can't afford to wait. We must seize any opportunity to save habitat. Life is falling through the cracks of the well intentioned. In these complex times, simple solutions won't work. Wildland conservation must be eclectic.

Just as lobbying isn't for everyone, neither is civil disobedience. To expand the realm of effective conservation activism, few avenues of ecological defense should be closed. It's okay to specialize. The Nature Conservancy buys land.

(continued)

Point/Counterpoint 23.1 (*continued*)

Ecotage: Does It Help or Hinder the Environmental Movement?

The Alliance for the Wild Rockies lobbies and educates. The Sierra Club Legal Defense Fund litigates. Earth First! specializes in direct action. Ultimately, no human or non-human society can sustain itself in a wasteland. The challenge is to protect habitat. Sustainability and the joy of unharnessed life will follow.

Critical Thinking

1. Summarize the major points of each author.
2. Do you detect any violations of critical thinking rules?

---

policies. In the United States, one of the key barriers to building a sustainable future is a lack of consensus about the severity of the world's problems. As Herbert Prochnow noted, "How can a government know what the people want when the people don't know?" In other words, how can we agree on a course for the future when few agree on what is best?

Increasing Public Awareness Through Research and Education   The lack of consensus on environmental issues results in part from widespread ignorance about issues among the general public. Many people who speak out against environmental concerns have very little knowledge of the issues. Many are compelled by a sense of optimism to dismiss claims of impending environmental disaster.

One solution to the problem of public ignorance is government support for scientific study of environmental problems. Research is also needed to identify the levels of consumption that are sustainable. Researchers can explore creative, sustainable solutions to help society reach these goals. Research is also needed to determine how values can be changed to achieve a sustainable society. Support for this work could come from governments and private foundations.

Information is important, but in order to make a difference, it must be communicated to the citizenry. Concerted efforts are needed to inform people worldwide of the problems facing the global ecosystem. Well-informed teachers and public officials can help disseminate this knowledge, as can the media (Figure 23.4). Environmental groups and religious leaders can also play a role. Education can assist in making solutions one of the immediate concerns of everyday citizens, elected officials, and business people (Chapter 20).

Getting Beyond Crisis Management   Public ignorance of issues also results from a tendency for public policymakers to practice **crisis politics**, that is, to deal primarily with crises while ignoring long-range problems. Henry Kissinger put it best when he said in gov-

ernment the "urgent often displaces the important." In crisis management, immediate problems, such as strikes, disaster relief, and immediate economic problems, tend to get all of the attention. Therefore, policymakers spend little time thinking about and addressing long-range problems, especially environmental ones. In addition, when new problems emerge, governments often ignore them until a small—and probably solvable—matter becomes a monstrous problem that defies solution. Governments are generally unwilling to devote money and resources to "small" problems because they're spending so much treating the latest list of crises. This results in a barrage of very costly problems that could have been solved initially at a fraction of the cost. A good example is the savings and loan scandal. Government representatives were warned about the problem in 1984 but ignored it until six years later when, by then, a minor problem had grown to a $200 billion headache.

When immediate issues take precedence over important long-term problems, governments lumber from crisis to crisis. It's not an efficient way to run a government or a society. Making matters worse, in most cases, legislators apply marginal corrections to present-day crises. Thus, they often end up treating the symptoms, not the causes.

In 1985, then Senator Al Gore introduced a bill to Congress called the **Critical Trends Assessment Act**. Had it passed, it would have required the federal government to establish an **Office of Critical Trends Analysis (OCTA)** to examine long-term trends, especially economic and environmental ones. OCTA would collect and analyze vast amounts of data from a variety of government agencies. Much of this information currently goes unused. Using knowledge from many sources, the OCTA would then advise the president on the effects of current policy on the nation's future. It would also issue regular reports that would make recommendations for changes in policies to avert problems and bring government and society in line with economic and ecological realities. State governments could also establish OCTAs whose central purpose would be to forge a sustainable economic and environmental future.

Figure 23.4 *Outdoor classroom discussion on wildlife preservation. Education of children of all ages is a key to building a sustainable society.*

OCTAs could help convince influential government leaders as well as business interests of the importance of many environmental issues, reducing ignorance and creating a consensus for change. Ultimately, with other changes outlined in this text, they could help shape a sustainable future.

Alternatively, special commissions on the future could be established. In 1977, then President Jimmy Carter directed several key agencies in the federal government to study population, resources, and pollution trends, and issue a report on their findings, which would serve as a foundation for longer-term planning. The resulting study, *The Global 2000 Report to the President*, is a gold mine of information about the future on which many decisions can be based. This report could be updated and revised.

Another way of creating vision in government is to establish special study sections in various governmental agencies, such as the U.S. Departments of Agriculture, Energy, and Interior. Reports from these agencies could be distributed to political leaders, teachers, and the public.

Internationally, long-range vision can be facilitated by the United Nations through the Environmental Programme (UNEP) or specially established commissions. In the mid-1980s, for instance, the United Nations established the **World Commission on Environment and Development**. Members of the commission were asked to produce a global agenda for sustainable economic development. Their report, *Our Common Future*, called on the U.N. General Assembly to find ways to achieve sustainable development and was the main stimulus for the U.N. Conference on Economic Development in Rio in 1992. (The U.N.'s role is described more fully in Section 23.4.)

Getting Beyond Limited Planning Horizons   Another barrier to sustainability is the limited planning horizon of many elected representatives. Limited planning horizons result from political shortsightedness. Critics of the political process often complain that concerns for reelection restrict the planning horizons of many politicians. Frequent reelection campaigns tend to force politicians to choose short-term solutions rather than long-term answers. To get reelected, politicians need results, even if they treat only the symptoms of a crisis.

Short terms of office are part of the problem. Members of the House of Representatives, for instance, serve two-year terms. Although this may have been appropriate 200 years ago when the nation's problems were few, today it is inadequate. Why? Some House members spend inordinate amounts of their time collecting money, plotting reelection strategies, and campaigning, time that they should devote to their job.

One way to address this problem is to institute longer terms of office for members of the House of Representatives, perhaps even for the president. Term limitations could also help. Fourteen states now have some form of term limitation. A six-year nonrenewable term for the president, for example, might reduce the reelection pressures that lead to shortsighted political decisions and take valuable time away from the office. It would also reduce the amount of time spent on the campaign trail while in office and increase the amount of work time. Longer terms of office, of course, could have damaging effects as well.

Shorter campaigns could also help democracies to work more efficiently. In Great Britain, campaigning begins six weeks before general elections. Incumbents spend less time on the campaign trail than do U.S. politicians. In addition, they spend considerably less money. In the United States, campaigns often run a full year and cost tens of millions of dollars. Presidential candidates spend $60 to $100 million in their bid for office. In order to finance such costly campaigns, they frequently take money from PACs or businesses, which, as pointed out earlier in the chapter, distorts the political process, making it bow to economic interests of business that may be shortsighted.

Overcoming the barrier of limited planning horizons will also require more farsighted leaders who are willing to implement policies that have implications well past the next election. Such policies may not favor the immediate electorate as much as the future electorate. But, citizens must be willing to elect men and women with a vision of the future and the skills to articulate and popularize that vision. If we expect visionary leaders, though, we must become more visionary ourselves.

Environmental groups at the national and state level are helping support responsible leadership. Individuals can help by supporting candidates who promote the efficient use of resources, recycling, renewable resource use, restoration and sustainable management of resources, and population control. Citizens can write or visit their elected officials. (For some tips on writing congressional representatives, see the *Environmental Action Guide*.) One effective way for citizens to make their voice heard is to join advocacy groups.

Becoming Proactive   A government that "lives and acts for today" is by definition a **reactive government**. Its laws and regulations are often ineffective, and in the long run, they can complicate matters, making a truly effective solution more difficult to reach and foreclosing options for future generations. Many of the laws passed by reactive governments are **retrospective**, that is, they attempt to regulate something that has gotten out of hand. For example, the Superfund Act provides money to clean up toxic hot spots in the United States. Retrospective laws are part of crisis management. They're necessary but less desirable in the long run.

A long-term outlook necessary for sustainability requires **proactive laws**, ones that attempt to prevent problems in the first place or ones that confront problems while they're still small and easily solvable. A good example is the Toxic Substances Control Act, with its provisions for screening new chemicals before they are introduced into the marketplace (Chapter 14). Laws that promote solar energy, conservation, recycling, and population control are additional examples. Ultimately, such laws could facilitate the transition from a wasteful frontier economy to an efficient sustainable society.

Many examples of future-oriented acts by Congress exist:

1. Establishment in 1946 of the National Science Foundation to promote research.

2. Passage in 1969 of the **National Environmental Policy Act**, which requires environmental impact statements (EISs) for all projects sponsored or supported by the federal government.

3. Creation in 1972 of the **Office of Technology Assessment (OTA)**, to examine the costs and benefits of new technologies.

4. Authorization of the Congressional Research Service to create a Futures Research Group.

5. Passage in 1987 of the **National Appliance Energy Conservation Act**, which establishes efficiency standards for appliances.

Western democracies today are generally a mix of reactive and proactive laws, with a strong leaning toward a reactive approach.

As we move into the last years of the 20th century and as the world population continues to grow exponentially, proactive government becomes more necessary than ever before.

Ending Our Obsession with Growth

Some of the steps outlined above could also help nations reduce their obsession with growth. Offices of critical trends analysis, for instance, could help show the potential outcome of current policies that promote continual growth in production and consumption. Education and research could have similar effects.

New measures of progress could also temper the obsession with growth (Chapter 21). Governments can help by promoting their use and publicizing real progress rather than the illusion of progress, as suggested by the rising GNP.

In 1949, Congress passed the **Full Employment Act**, which calls on the government and its various agencies to promote maximum production and consumption to stimulate full employment. Although full employment is a desirable goal, the notion that it can only be achieved through maximum production and consumption is erroneous and ecologically ill advised. Repealing this law, and replacing it with a **Sustainable Futures Act**, may be a symbolic act, but it could be an important step in building a sustainable future.

A Sustainable Futures Act would call on Congress to actively promote sustainability through conservation (efficiency and pollution prevention), recycling, renewable resource use, restoration, and population control. Moreover, it would require that all new policies be judged through the lens of sustainability. The act might also set up a special legislative body to review all existing policies and suggest amendments or new laws that would align human activities from international trade to energy to manufacturing with the dictates of sustainability.

Reducing Exploitation and Promoting Self-Reliance

Reducing the exploitation of people and natural resources is another goal of sustainable government. Many of the changes mentioned above would contribute to this goal. Steps that reduce our obsession with growth and create vision, for instance, tend to make societies less resource intensive and thus less exploitive. Additional steps to reduce our exploitive ways follow. (Others have been discussed in Chapters 21 and 22.)

Viewpoint 23.1

The Wise Use Movement: Dismantling the Public Lands

Kathy Kilmer

Kathy Kilmer is media coordinator for the Wilderness Society, a nonprofit organization working since 1935 to preserve America's public lands and to protect wildlife and biodiversity, clean water, public health, and quality of life.

They come wrapped in patriotic buzzwords: Center for the Defense of Free Enterprise; Citizens for the Environment; Alliance for America. Their letterheads read like the roster at a conservationists' convention: Our Land Society; National Wetlands Coalition; American Forest Resource Alliance.

Yet, these groups have a common goal that is neither patriotic nor earth-friendly: to continue to log, graze, mine, and develop America's public lands regardless of the environmental consequences.

Worse yet, American taxpayers help foot the bill for these activities in the form of subsidies amounting to some $800 million each year. This includes uncollected royalties on hard-rock mining, estimated to be $500 million annually; timber sales below cost on national forests, an average of $200 million per year; grazing fee program costs, about $100 million annually. The cost and damage to the public lands, in denuded hillsides, ruined watersheds, diminished biodiversity, and disappearing wetlands, is difficult to calculate in dollars and cents.

Industries that have profited at the expense of public lands are determined to block any reforms that will cut into their profits. They are equally committed to weakening the Endangered Species Act, opposing wetlands and wilderness protection, and blocking reform of federal grazing programs and the General Mining Law of 1872. These industries regularly spend millions of dollars on lobbyists and advertising. Now they're funding grass-roots organizing campaigns

designed to convince Congress that most westerners oppose environmental reforms. Misinformation, fear and intimidation, and slick public relations are the tactics most often employed by the [Wise] Use Movement, which targets rural communities hit hard by recession. Equating environmental regulations and endangered species protection with unemployment, they manage to deflect attention from real causes, such as a depressed economy, boom-and-bust industrial cycles, industries retooling for greater efficiency, or a complex mix of all of these things.

People for the West! is a casebook example of the tactics employed by the [Wise] Use Movement. The Colorado-based organization gets most of its funding from mining companies and suppliers. Its chief objective is to block mining law reform as well as environmental regulations that interfere with mining, and oil and gas leasing. Hard-rock mining currently is governed by the General Mining Law, which does not require reclamation. Nor do mining companies pay royalties for mining the public lands. Mining companies that support PFW! receive an added bonus: a built-in lobbying force in the form of employees who are asked—on the job—to sign petitions, attend rallies, and write to Congress.

Ron Arnold, of the Center for the Defense of Free Enterprise (CDFE), in Bellevue, Washington, is one of the most vocal leaders of the [Wise] Use Movement. It was he who first coined the term *Wise Use*, purloining it from the writings of Gifford Pinchot, the first Chief of the U.S. Forest Service. In the December issue of *Outside* magazine, Arnold was quoted as saying "The environmental movement is a rich, powerful menace to society and we intend to destroy it."

Arnold's Wise Use Agenda is an antienvironmentalist's wish list of

reforms: allow oil development in national parks and wilderness areas; weaken the Endangered Species Act; make grazing privileges, mining claims, and timber contracts "property rights"; and more. Arnold has been linked to the Unification Church of Rev. Sun Myung Moon, a Korean who believes himself destined to lead an ultraconservative, worldwide church-state.

The [Wise] Use Movement found ready access to the Bush administration. Leading federal officials openly supported their efforts, attended their conventions, and endorsed their aims. Powerful western lawmakers became staunch allies. As a result, the [Wise] Use Movement was able to weaken the definition of wetlands; block legislation to reform the federal grazing program; and stall federal action to protect the spotted owl and ancient forests. Under the Clinton administration, it is unlikely movement leaders will find such ready access. As a result, the [Wise] Use Movement will intensify efforts at the state and local level, trying to achieve their aims through local law and ordinance.

Leaders of the [Wise] Use Movement claim they merely are trying to wrest back some of the ground gained by environmentalists during the past two decades. But, it is not a question of taking back because development interests have dominated the land-use debate in the West for more than a hundred years. Today, they fight to ensure that the public has no voice at all in the way public lands are managed. At risk is a legacy of public trust and the most fundamental human right Americans hold dear: a clean and healthy environment—for themselves, their children, and the Earth itself.

Creating Offices of Sustainable Economic Development
One idea I present in *Lessons from Nature* is the establishment of **offices of sustainable economic development** (**OSEDs**) in state governments or, perhaps, at the national level. All state governments currently have offices of business development or offices of economic development that promote business, generally by recruiting new businesses and helping existing business become more profitable. As a rule, little thought is given to matters of sustainability.

OSEDs could serve as a catalyst for change. First, they could assist existing businesses in developing more sustainable practices, for example, using energy and other resources more efficiently or recycling and purchasing recycled goods. Besides promoting the use of locally available waste resources and helping build industries that divert wastes to useful purposes, OSEDs could promote pollution prevention (Chapter 19) and renewable energy resources (Chapter 12). OSEDs could identify economic opportunities from local restoration projects. Second, they could assist in promoting sustainable business opportunities, for example, helping identify locally available resources, such as wastepaper or aluminum scrap, around which new businesses could be developed. By focusing on sustainable economic progress, OSEDs could help build strong, stable, and regionally self-reliant economies.

Working with state offices of critical trends analysis and elected officials, OSEDs could help create a comprehensive strategy for sustainability. They could also help promote better land-use planning, which is vital to sustainable economic development throughout the world.

Sustainability Through Land-Use Planning   Chapter 22 pointed out that many new dams built in developing countries fail to provide water for irrigation for very long. Most often, upstream from the costly dams, hillsides have been stripped of trees or are so heavily overgrazed that when rains come, they wash tons of sediment into the reservoirs, rendering them useless in short order.

Countless examples of shortsighted land-use planning exist in developed countries as well. Building suburbs, cities, and airports on prime agricultural land, constructing homes in floodplains, and farming steep hillsides are three examples. Building homes in filled wetlands or factories upwind from cities and towns are additional ones.

Land-use planning allows government to put the land to its best use. Land-use regulations determine where people can live and do business, where water pipes, electric lines, roads, and shopping malls can go with the least amount of damage. Land-use planning can also be used to preserve farmland, recreational areas, wetlands, scenic views, watersheds, aquifer recharge zones, and wildlife habitat.

When applied to specific sites, land-use plans take into account the slope of the land, soil quality, water drainage, location of wildlife habitat, and many other features. This permits planners and developers to design with nature rather than redesigning nature. (Case Study 10.1 illustrates an extraordinary land-use planning success in Woodlands, Texas.)

When properly executed, land-use planning can help to achieve a more sustainable relationship with the Earth. Land-use plans can protect the foundation of tomorrow's civilization, renewable resources such as farmland, pastures, forests, fisheries, and wild species. Sustainable land-use planning, therefore, decreases unnecessary exploitation of natural systems and fosters local self-reliance.

The Japanese provide a model that many countries could adopt to protect their land. In 1968, the entire country was placed under a nationwide land-use planning program. Lands were divided into urban, agricultural, and "other" classes. Several years later, the zoning classifications were expanded to include forests, natural parks, and nature reserves. The Japanese plan's success lies in protecting land from the market system, which, left on its own, appropriates land irrespective of its long-term or intrinsic value.

Many European nations have adopted similar programs. In Belgium, France, the Netherlands, and former West Germany, national guidelines for land-use planning were established in the 1960s. Administered by local governments, they protect farmland, prevent urban sprawl, and help to establish undeveloped areas in or around cities and towns, **greenbelts**. The Netherlands has one of the best programs of all, for it governs water and energy use as well as land use.

Land-use planning at the federal level in the United States is rudimentary. Except for establishing national parks, wilderness areas, national forests, and wildlife preserves, the federal government has done little to systematically protect its land. Most zoning occurs on the community level, and much of that is inadequate. On the local level, planners primarily concern themselves with restrictions on land use for commercial purposes—housing developments and industrial development. Because of the United States' reliance on community-level planning, states are a patchwork quilt of conflicting rules and regulations. Some people believe that statewide land-use planning is needed.

However, statewide land-use planning is an idea that is slow in coming. Only a handful of states have comprehensive programs. Oregon passed such a program in the early 1970s (Models of Global Sustainability 6.1). Noteworthy programs now exist in Maine, New Jersey, Florida, Vermont, Washington, and Hawaii.

For years, the main tool of land-use planning has been **zoning regulations** that classify land according to use. In a city, zoning helps separate potentially noisy, smelly, or hazardous activities from residential areas. When used properly, zoning can also protect farmland and other lands from urban development. In rural Black Hawk County in Iowa, for instance, zoning laws prohibit housing developments on prime farmland, while permitting them on lands with lower productivity.

New approaches are also being adopted to protect valuable land, especially farmland. One public policy tool is the **differential tax rate**, which allows city officials to tax different lands at different rates. For example, farmland is taxed at a lower rate than housing developments.

Another technique that helps farmers keep their land instead of selling it to developers is the purchase of development rights. A **development right** is a fee paid to a farmer to prevent the land from being "developed," a euphemism for bulldozed, paved, and built on. To determine the cost of a development right, two assessments of the land are made, one of its value as farmland and one of its value for development. The difference between the two is the development right. A state or local government may buy the development right from the farmer and hold it in perpetuity. From then on, the land must be used for farming, no matter how many times it changes hands.

Still another way of reducing the spread of human populations onto valuable land is the **growth-pays-its-own-way-concept**. This idea, partially practiced in Boulder, Colorado, calls on developers and new businesses to pay the cost of schools, highways, water lines, sewer lines, and other forms of infrastructure that will be needed as the community expands. The rationale is that the cost of new development should be passed on to those who profit from it, not existing residents. A new home comes with a small development fee attached, which the new homeowner pays.

This keeps local taxes from rising to subsidize new development. It also encourages builders to locate new housing projects closer to existing schools, water lines, and highways, thus reducing sprawl and the destruction of farmland and other ecologically important sites. Development fees might also encourage developers to install water-efficient fixtures and pay for water-efficiency measures in existing homes and businesses, thus preventing an increase in overall demand.

Land-use planning is essential in developing nations as well. Urbanization in these countries is a major problem, and millions of hectares of farmland are destroyed each year. In some areas, land reform is badly needed. Wealthy landowners in many Latin American countries, for example, graze their cattle in rich valleys while peasants scratch out a living on the erodible hillsides. Consequently, hilly terrain that should be protected from erosion is being torn up by plows and washed away by rainfall. Some argue that sensible land use hinges on reform of these outdated land-holding systems.

These are just a handful of ideas that could help reshape government policy to foster sustainability. When combined with many other ideas given in previous chapters that promote sustainable ethics, revamp economics, and reshape unsustainable systems, they could form a national framework for making a dramatic realignment of human systems necessary to steer us onto a sustainable course.

23.4 Global Government: Toward a Sustainable World Community

For many years, environmental problems were largely local, regional, or national issues. Solutions required actions on a similar scale. Today, though, many environmental ills affect the entire globe and require international efforts to solve.

This realization and the realization that ozone depletion, deforestation, global warming, and other problems have serious impacts on the future of a country have led many nations to rethink outdated notions of **national security**. Once equated with military strength needed to protect a nation from outside aggression, national security today is beginning to be seen more broadly. Today, many countries recognize that although national security requires adequate military power, it also entails efforts to protect a nation's environment from internal threats, such as deforestation and soil erosion, as well as external threats, such as global warming. Achieving true national security, therefore, necessitates many of the changes described in this test, but also massive global efforts that address the roots of the current crisis of unsustainability. Consider national efforts first.

Models of Sustainable Development

In any transition, people and nations often become models of desirable action. In the sustainable transition, several European countries have recently emerged as models. One of the best examples is Germany (Models of Global Sustainability 15.1).

Germany probably has the highest environmental standards in the world. Part of the reason for this lies in the **Green Party**, a political party that advocates measures to protect the environment. The Greens support reductions in pollution and hazardous wastes, putting an end to nuclear power and nuclear weapons, efficiency measures, and the use of renewable resources. Rallying under the banner "We are neither right nor left, but in the front," the Greens have been instrumental in transforming German society (Figure 23.5). If the Green Party gets its way, Germany could become a nation of nonpolluting industries that concentrates mainly on socially necessary products.

Figure 23.5 *One member of Germany's Green Party shows up at parliament in a solar-powered vehicle to demonstrate this alternative, relatively clean form of transportation.*

The Greens represent a major shift in political thinking. They take a long-term view of the future, calling for redirection of policy consistent with the sustainable ethic.

In Western Europe, a dozen Green parties now exist, although in recent years, they have lost some influence. The Green Party has also spread to the United States. Colorado, Connecticut, Washington, and dozens of other states now have Green parties that promote social justice and a clean and healthy environment. Green Party members write letters and phone political leaders, stage peaceful demonstrations, and work to get their members elected to state legislatures and other offices.

Conventional political wisdom holds that in the United States popular ideas are absorbed by major political parties. They enter the mainstream of political thought, where they become integrated into public policy. This has also happened in Germany. Thus, no matter whether the U.S. Green Party survives or is simply absorbed into the mainstream, it may have begun a movement that could reshape American politics.

Like Germany, Sweden, Norway, and the Netherlands are also models whose policies and actions could inspire other countries to follow suit, especially when skeptical leaders see the economic benefits of pursuing a sustainable path. Besides providing inspiration, model countries can become a source of new technologies and

enjoy a competitive advantage over those countries that are resisting change.

Regional and Global Alliances

Because the task is global in nature, efforts are needed to create regional and global alliances to solve problems that afflict many nations. Several examples of environmental alliances have been discussed in this text. Chapter 8 described the International Whaling Commission (IWC). Composed of members from all of the whaling nations, the IWC sets quotas on whale kills and enacts outright bans. The IWC, however, has no enforcement power; it relies principally on cooperation. As noted in Chapter 8, not all nations comply with the regulations. To help give the IWC a little muscle, the United States and other nations prohibit trade with countries that violate the agreements.

Case Study 17.1 discussed two U.S. alliances involving a number of eastern states. One alliance is working to clean up and protect Chesapeake Bay; the other, which includes Canada as well, is seeking ways to restore and protect the Great Lakes. Case Study 17.2 discussed an agreement reached by the European nations that border the Baltic Sea. Chapter 16 described the Montreal and London agreements, international treaties aimed at protecting the ozone layer.

Some of the most recent alliances emerged from the Earth Summit in Rio in June 1992. This section examines the various accords and discusses further measures needed to improve them.

The Climate Convention   Signed by 154 nations, this agreement calls on the world community to hold greenhouse gas emissions at 1990 levels by the year 2000. Unfortunately, this agreement does not commit any nation to hold greenhouse gases to certain amounts by certain times. Although the European nations wanted limitations, the Bush administration opposed them and forced a compromise position.

The good news is that the agreement requires nations who signed to prepare progress reports that detail their actions to cut greenhouse gases. These reports will be reviewed by a special committee that could, at a later date, modify the treaty to establish specific emissions levels and target dates, an eventuality the European nations are already promoting.

Although the climate convention is an important first step, it will only slow the rate of climate change because greenhouse gas emissions already exceed the Earth's carrying capacity. Further cuts are needed.

The Biodiversity Convention   This agreement was signed by all participants except the United States, which refused because of language the Bush administration deemed unacceptable. This agreement calls on nations to prepare inventories of species to be preserved and to develop strategies to conserve and use biological resources sustainably. Like the climate treaty, no deadlines are set. In addition, the language of the proposal is quite weak.

Agenda 21   By far the most comprehensive document to emerge from the Earth Summit, Agenda 21 is a massive work plan for national action and international cooperation aimed at achieving sustainable development. Its 39 chapters of nonbinding recommendations are a blueprint for a sustainable future. For instance, Agenda 21 urges nations to adopt national strategies for sustainable development. It calls for community-based sustainable development. It also urges countries to adopt national strategies that eliminate obstacles to the full participation of women in sustainable development. It recommends that countries eliminate subsidies for environmentally destructive activities and encourages them to ensure public input.

Despite these and other important recommendations, Agenda 21 has some serious weaknesses. Language that would commit developed nations to donate a certain percentage of their annual GNP to sustainable development overseas was eliminated. In addition, Agenda 21 does not propose new ways of eliminating debt. Nor does it offer new ways to accelerate the transition to sustainable energy systems, because of pressure from the United States, Saudi Arabia, and Kuwait. All references to full-cost pricing were eliminated. Moreover, the chapter on forestry does not recommend policies of sustainable forest management. The chapter on population fails to underscore the importance of population control to sustainable development and avoids the term *family planning* altogether, reportedly because of pressure from the Vatican.

Forest Principles   Another nonbinding agreement adopted by attendees of the Earth Summit was the **Authoritative Statement of Principles on the World's Forests.** Although, in this instance, the U.S. government pushed for stronger efforts to preserve native forests, 77 developing nations, led by India and Malaysia, watered down the agreement. In fact, they eliminated language pointing out the value of forests to the global environment. In addition, Malaysia eliminated wording that encourages efforts to promote international trade of products from sustainably managed forests. According to Gareth Porter of the Energy and Environment Institute, the result is a set of principles that underscores sovereign rights of nations to exploit their forests, legitimizing existing policies in those countries that are currently endangering the world's forests.

Rio Declaration   The final outcome of the Earth Summit was the Rio Declaration. It includes recommendations for a number of legal principles vital to achieving sustainable development, among them public access to government information on the environment. It also calls on nations to use environmental impact statements and to exercise caution in their development plans. In addition, it encourages timely notification of activities that could exert effects across a nation's boundaries. The Rio Declaration further recognizes the special responsibility of the developed countries in achieving global environmental restoration because of their technological and financial capabilities and their consumptive patterns and pollution production.

Although the outcome of the Earth Summit is less than many hoped, the progress made there is a good start, especially considering the complexity of negotiations involving nearly 180 nations with very different goals and priorities. Because many environmental, economic, and social problems will inevitably worsen in the coming years, it is likely that considerable progress will be made on areas that were ignored or deleted because of their political sensitivity.

Strengthening International Government

To solve global environmental problems, national sustainable development plans and international agreements are badly needed. Also needed are efforts to strengthen existing international institutions, notably the United Nations (Figure 23.6). Some observers think that we may need some form of world government to address the problems facing humankind.

Strengthening the United Nation's Role in Sustainable Development   The United Nations has several key programs, such as the Fund for Population Activities (UNFPA), the Food and Agricultural Organization (FAO), and the U.N. Environment Programme (UNEP). Over the years, these programs have helped scores of developing nations in many ways. Conferences on population and its impact, sponsored by UNEP, for instance, have helped convince most developing nations that actions must be taken to control rampant growth. In 1987, UNEP facilitated negotiations that led to the Montreal Accord, and in 1990, it sponsored negotiations that produced the ground-breaking London agreement (Chapter 16). In 1988, the U.N. Commission for Europe negotiated a treaty that calls for a freeze on nitrogen oxide emissions at their 1987 levels by using the best-available control technology on new vehicles and power plants in the United States and Europe. The Rio Conference, which was sponsored by UNEP, is another area of notable success.

Despite its successes and its importance to the future of the world, UNEP operates on a meager $40-million-a-year budget. Its staff of 350 is, according to most

Figure 23.6 *The United Nations has helped increase global cooperation in pollution control and other important issues. It could provide a forum for more progress on building a sustainable society.*

calls for a world government consisting of three bodies: the house of peoples, which would consist of 1000 representatives elected by the people; a house of counselors, consisting of representatives appointed by universities and colleges; and a house of nations with representatives appointed or elected by national governments.

Although such a government could prove unwieldy at times and could threaten sovereign rights of nations, proponents argue that the world's problems are so serious and time is so short that the step is well worth the trouble. Others argue against the notion, convinced that existing approaches are sufficient.

In closing, a global sustainable society and government may seem like utopia. As writer Rolf Edberg reminds us, "The utopia of one generation may be recognized as a practical necessity by the next." It is hoped the present generation is beginning to recognize that sustainability is a practical necessity.

Edberg also notes that achieving such goals depends on society's ability to free itself from "ideas and emotions that once had a function in our battle for survival but have since become useless." Starving masses and the universal fear of environmental destruction may be the psychological forces that set the stage for a new world society governed by sustainable principles.

The great thing in this world is not so much where we stand as in what direction we are moving.

—Oliver Wendell Holmes

Critical Thinking Exercise Solution

One of the myths we propagate in our society is that results can be measured by the amount of money we spend on a problem. That is, if we are spending a lot of money, we must be doing a good job. Unfortunately, results don't always parallel expenditures. For example, billions of dollars have been spent on hazardous waste cleanup in the United States, but much of this money has ended up in the pockets of attorneys, not in actual cleanup efforts. Spending on water pollution control at sewage treatment plants keeps sewage out of streams, but sludge ends up in landfills.

The *National Report* also fails to address the most important question of all: Is the money we are spending enough to create a sustainable way of life? In other words, is it money spent on stopgap mea-

observers, much too small for the task at hand. Increasing the budget and the staff of UNEP could greatly improve its effectiveness. It would allow UNEP to gather more data and disseminate information more widely. In addition, it would allow UNEP to expand efforts to promote alliances around key issues, such as population growth, desertification, and deforestation.

Former foreign minister of the former Soviet Union, Edvard Shevardnadze called for the creation of a U.N. Ecological Council. Perhaps better called the U.N. Environmental Security Council (UNESC), this governing body could be invested with powers equivalent to those of the U.N. General Assembly or the U.N. Security Council. The UNESC might be given the power to pass resolutions condemning environmentally unsustainable activities, such as deforestation and overfishing. It could also pass resolutions that compel nations to adopt sustainable practices.

Creating a World Government   Some observers believe that international economic, social, and environmental problems suggest the need for a world government, a democratic parliament of nations. One proposal

Table 23.1 Summary of Recommendations for Sustainable Government

Adopt national sustainable development strategy that promotes conservation, recycling, renewable resource use, restoration, and population control

Repeal or modify existing laws that hinder sustainability

Increase support of research on environmental problems and sustainable solutions to them

Increase support of education on environmental issues and sustainable solutions

Pass a national Sustainable Futures Act

Establish offices of critical trends analysis at state and national levels

Establish offices of sustainable economic development in state governments

Evaluate all new and existing policies on the basis of sustainability

Adopt sustainable land-use planning at state and national level

Implement policies and actions outlined in Agenda 21 and other international agreements

Support sustainable development in nonindustrialized nations

---

sures, or does it go toward systemic solutions that foster a sustainable human presence?

The important lesson here is that when you dig a little deeper you find a simple answer, but appealing logic falls apart. The report fails to answer the most important question of all: Are solutions advancing the transition to a sustainable society or merely marking time?

Summary

23.1 The Role of Government in Environmental Protection

■ Democratic governments employ three major tools to regulate the economy and other activities, such as environmental protection: (1) tax policies, (2) direct expenditures, and (3) laws and regulations.

23.2 Political Decision Making: The Process and the Players

■ In democracies, political decisions are influenced by many people—government officials, the general public, and special interest groups (including business, labor, and environmentalists).

■ Because of sometimes vast financial resources, businesses often wield considerable power. But, environmental groups are also effective agents of social change.

■ Environmental groups operate in many ways: by purchasing land for protection, educating the public, pushing for new and tighter laws and regulations, acting as watchdogs over corporate and government organizations to ensure compliance, suing recalcitrant agencies and corporations that violate the laws, and staging public protests to draw attention to important issues.

23.3 Creating Governments That Foster Sustainability

■ From an environmental viewpoint, governments are often shortsighted, obsessed with growth, exploitive of people and nature, and tend to promote dependency.

■ To build a humane, sustainable society, governments and the economies they serve must be accountable to the needs of future generations and other species by adopting the goal of sustainability as a central organizing principle.

■ Shortsightedness results in part from widespread ignorance about issues. Government support for scientific study of environmental problems and solutions and dissemination of this information to the citizenry can help combat the problem.

■ Another reason for shortsightedness is the tendency for public policymakers to practice **crisis politics**, a management policy in which immediate problems get all of the attention and long-term problems are ignored until they become immediate. Consequently, small—and solvable—matters become monstrous problems that defy solution.

■ One way to address this problem is to create institutions or agencies that monitor long-term trends, such as the proposed **Office of Critical Trends Analysis**, special study sections in various agencies, or special commissions on the future through national governments and the United Nations.

■ Another expression of shortsightedness is the limited planning horizon of many elected representatives, which results in part from concerns about reelection and short terms of office. Longer terms for some and term limitations could help. Also needed are farsighted leaders who are willing to implement long-term policies and citizens willing to elect such representatives.

■ A government that practices crisis management is a **reactive government**. Its laws and regulations are often ineffective; they make lasting solutions harder to achieve.

■ Many of the laws passed by reactive governments are **retrospective**. These laws, which attempt to regulate problems that have gotten out of hand, are necessary, but less desirable in the long run.

■ A long-term outlook necessary for sustainability requires **proactive laws**, ones that prevent problems or confront them when they're small and solvable.

■ Western democracies are generally a mix of reactive and proactive laws, but as societies strive for sustainability, proactive governments become more necessary than ever before.

■ Some of the steps outlined above could help nations reduce their obsession with growth. Also needed are new measures of progress and passage of a **Sustainable Futures Act**, which would call on Congress to promote the principles of sustainability and require that all new policies be judged through the lens of sustainability.

■ Reducing the exploitation of people and natural resources in the name of economic progress is another goal of sustainable government. Many of the changes mentioned in this chapter would contribute to this goal.

■ **Offices of sustainable economic development** (OSEDs) might help. Working with elected officials, OSEDs could help formulate comprehensive strategies for sustainable economic progress, especially land-use planning.

23.4 Global Government: Toward a Sustainable World Community

■ Today, many environmental problems affect the entire globe and will undoubtedly require global action. This has led many nations to redefine national security to include the protection of a nation's environment from internal and external environmental threats.

■ Some nations are serving as models of sustainable development. Germany, Sweden, Norway, and the Netherlands are good examples. Besides providing inspiration, model countries are a source of new technologies that could promote sustainability.

■ Efforts are also needed to create regional and global alliances to solve problems that afflict many nations. Numerous examples are available.

■ Some think that solving global environmental problems requires efforts to strengthen existing institutions, like UNEP, or perhaps even creating a U.N. Environmental Security Council. Others think that some form of world government is required.

Discussion and Critical Thinking Questions

1. What are the major tools by which governments control economic activity and environmental protection? List and describe the pros and cons of each one.

2. In what ways are democracies unrepresentative? Can you give some examples?

3. Look around your community. What aspects of it could have been planned better? Why?

4. This chapter lists a number of weaknesses in the current system of government. What are they? What proposals are given to overcome them? Critically analyze each of the proposals. List the benefits of each one and point out problems and ways to solve them.

5. Critically analyze the following statement: "Nations should be left on their own. Others should not meddle in their internal affairs, even in the name of global environmental protection."

6. List the pros and cons of increasing the power of the United Nations in matters of environmental protection.

7. Debate this statement: "We need an international democratic government to address social, economic, and environmental problems."

Suggested Readings

Brown, L. R., Flavin, C., and Postel, S. (1991). *Saving the Planet: How to Shape an Environmentally Sustainable Global Economy*. New York: Norton. Highly readable account of ways to reshape the economy.

Brown, L. R. (1992). Launching the Environmental Revolution. In *State of the World 1992*. Norton: New York. Describes the pivotal role of governments in promoting a sustainable future.

Chiras, D. D. (1990). *Beyond the Fray: Reshaping America's Environmental Response*. Boulder, CO: Johnson Books. Describes alternative governmental solutions to achieve sustainability.

———(1992). *Lessons from Nature: Learning to Live Sustainably on the Earth*. Washington, D.C.: Island Press. See Chapters 8 and 9 for a discussion of government.

French, H. F. (1992). Strengthening Global Environmental Governance. In *State of the World 1992*. Norton: New York. Discusses ways to strengthen international cooperation.

Holmberg, J. (1992). *Making Development Sustainable: Redefining Institutions, Policy, and Economics*. Washington, D.C.: Island Press. Important collection of essays on sustainable development.

IUCN, UNEP, and WWF. (1991). *Caring for the Earth: A Strategy for Sustainable Living*. Gland, Switzerland: Earthscan. Contains many ideas for sustainable economic development.

Porter, G. and Islam, I. (1992). *The Road from Rio: An Agenda for U.S. Follow-up to the Earth Summit*. Washington, D.C.: Environmental and Energy Study Institute. Excellent survey of the agreements reached at the Rio Conference.

A Primer on Environmental Law

Our environmental laws are not ordinary

laws, they are laws of survival.

—Edmund Muskie

ঽ৯   ঽ৯   ঽ৯

An ancient Roman legal axiom proclaims, "The people's safety is the highest law." Many environmental laws discussed in this text recognize that the health and welfare of a people, their safety, is partly dependent on a clean, healthy environment. This realization, though, grew slowly, for it was not until the late 1960s and early 1970s that lawmakers at a national level realized that what we do to our environment we do to ourselves and that protecting our environment is one important form of preventive medicine.

During this time, environmental protection became increasingly important to U.S. citizens and their congressional representatives. Today, the United States has one of the toughest and most comprehensive systems of environmental laws and regulations in the world. The United States has, in fact, served as a model to many countries, which have patterned their environmental protection laws after U.S. statutes, benefiting from the country's experience as well as its errors. Unfortunately, in the 1980s, the U.S. leadership role dwindled. At the 1992 Rio Conference, in fact, the United States was a major impediment to global progress. Although at this writing it is too early to tell, the Clinton administration may change all that.

Evolution of U.S. Environmental Law

The progression of U.S. environmental laws and regulations followed a logical evolutionary root that began with local ordinances imposed by local governments. Interested in protecting health and environmental quality, officials of cities and towns passed local laws to limit activities of private citizens for the good of the whole. For example, municipal ordinances regulated burning of trash within city limits to reduce air pollution. These efforts began in the 1800s.

By the end of the 1800s, however, it became clear that efforts to control problems, such as water pollution, through local ordinances were often inadequate, especially in densely populated regions. Disputes arose between neighboring municipalities with different laws. Regions with strict laws were hampered in cleaning up their rivers by upstream cities with lax pollution laws. Thus, states began drafting legislation to regulate water pollution.

Soon, state laws also proved inadequate because air and rivers flow freely across state borders. Thus, interstate conflicts over pollution replaced the conflicts between neighboring municipalities. In addition, state programs were often inadequately funded and lacked the technical expertise needed to set pollution standards. State agencies also found themselves powerless against large corporate polluters with political influence in the courts and legislatures. Because of these problems and the growing effectiveness of special interest groups, environmental controls shifted to the federal government, gradually in the 1940s and 1950s and then more rapidly.

Initially, the federal government restricted itself to research on health and pollution control technologies. This approach met little opposition from state and local governments. Next, the federal government began offering grants to fund pollution control projects and the formation of state pollution enforcement agencies.

With increasing pressure from environmental groups and citizens, the federal government began to assume a larger role in enforcement. Today, in fact, much to the dismay of some, the federal government sets ambient pollution standards and standards for emissions from factories, automobiles, and other sources, and can take strict enforcement actions if needed. Stiff fines and penalties are frequently employed to coerce business into cooperating. Corporate executives have even been jailed.

The shift to federal control is based on at least two important principles of American federalism: (1) When it is important to maintain uniform standards, the federal government provides the best route. Uniform policies help minimize interstate conflicts and help create a fairer economic system for businesses. (2) The power of the federal government to tax is much stronger than a state's. To control pollution effectively requires expensive research, which the federal government can more easily afford. Furthermore, it would be costly, time consuming, and redundant for each state to carry out this extensive research.

The shift to the federal level has some disadvantages, however. First, the federal government may not always understand the problems of the regions it regulates as well as local officials. Another criticism is that states should have the right to do as they please with their own resources; in other words, federal control diminishes self-determination and self-governance. But without central controls, states impinge on one another's quality of life. For example, poor watershed management in the Rocky Mountain states leading to erosion could have long-term adverse impacts on downstream users to the east and west.

One way of addressing these problems is to develop federal standards but allow the states to manage their own programs. Thus, the Clean Air Act, the Surface Mining Control and Reclamation Act, and the Resource Conservation and Recovery Act all permit the states to run their own programs as long as they are at least as stringent as the one set up by federal law. These acts also provide money to assist the states in setting up their own programs.

National Environmental Policy Act

One of the most significant advances in environmental protection of the past three decades was passage of the **National Environmental Policy Act** (NEPA) in 1969. NEPA is a brief, rather general statute with several key goals. First, it declares a national policy to "use all practicable means" to minimize environmental impact of federal actions. More specifically, NEPA requires decisions regarding federally controlled or subsidized projects, such as dams, highways, and airports, to outline possible adverse impacts in an environmental impact statement (EIS). Among other things, an EIS must describe: (1) what the project is; (2) the need for it; (3) its environmental impact, both in the short term and the long term; and (4) proposals to minimize or, in the language of bureaucrats, to mitigate the impact, including alternatives to the project. Drafts of an EIS are available for public comment and review by federal agencies. Written comments from the public must be addressed in a final EIS, which is issued at least 30 days before undertaking the proposed action.

The EIS has been an effective way of getting businesses and governmental agencies to focus on the environmental impacts of their projects. That is, it has helped promote vision in economic development strategies. The underlying goal of the EIS is that individuals who become aware of their potential impact will act responsibly to avert it as much as possible. In this sense, the EIS is a political carrot (a gentle inducement) rather than a political stick (a punishment). Available early in the planning stage, it can help decision makers determine whether they are directing their policies, programs, and plans in compliance with the national environmental goals expressed in NEPA and other legislation. Between 1970 and 1983, approximately 24,000 EISs were prepared. As discussed in Chapter 3, the EIS could with certain modifications be a superb tool to promote sustainability.

NEPA also established the **Council on Environmental Quality** in the executive branch. The council publishes an annual report, *Environmental Quality*, on the environment and on environmental protection efforts of the federal government. It also develops and recommends new environmental policies to the president. In 1993, the Clinton administration abolished the CEQ, creating a one-person Office of Environmental Policy, a move that perplexed many environmental leaders.

NEPA has significantly affected federal decision making. It has led to hundreds of lawsuits filed by environmental groups against the government, perhaps more than any other environmental statute. In addition, several states and nations throughout the world have passed laws or issued executive or administrative orders patterned after NEPA. France, Canada, Australia, New Zealand, and Sweden all require EISs. California passed an **Environmental Quality Act** in 1970 that requires EISs for all projects—private and public—that will affect the environment in a significant way.

The success of NEPA has been great, but the law is not without flaws. One of the most frequent criticisms is that EISs are too lengthy and deal with too many peripheral subjects. Reports are often ignored; they may show serious adverse impacts, but the project will be approved and carried out anyway, often without ameliorative actions. A common complaint from environmentalists is that reports are often based on inadequate information. Projections of environmental impact are difficult to make and often too subjective. Practically no work has been done to see if projected impacts actually materialize; thus Americans continue to be unprepared to make sound projections about impact. EISs may be "doctored" by agencies or private consulting firms that write them for federal agencies to hide the real impacts. Some agencies can avoid writing EISs by simply stating that there will be no adverse environmental impact; it is then up to others to prove the need in court. Other critics of EISs contend that they are too costly and often lead to delays in important projects. The paperwork and time involved seem excessive.

To answer some of these complaints, the Council on Environmental Quality issued streamlined procedures for preparing EISs in 1979. The guidelines call for: (1) a maximum length of 150 pages, except for more involved projects; (2) a summary of no more than 15 pages, which describes major findings and conclusions; (3) documentation and referencing of projected impacts; and (4) the use of clear, concise, and plain language.

Some argue that EISs are after the fact, that is, they are designed to assess impacts of projects that have already been planned. A more appropriate method would be to require analyses of need and then seek sustainable ways to meet the need. Some environmentalists believe that NEPA should require agencies to select the most environmentally benign and cost-effective approach, both in the short term and in the long run.

Another unanswered problem is that environmental groups can sue an agency that they believe should have filed an EIS or one that has filed an inadequate EIS, but they cannot recover attorneys' fees from such suits. Recovery of attorneys' fees would relieve the costly burden of forcing governmental agencies to heed the law, although critics point out that it might also open the door to numerous costly lawsuits, which would be a burden on the taxpayer.

Although it has its critics and still stands in need of improvement, NEPA is the cornerstone of U.S. governmental policy. Numerous federal agencies have reported important environmental benefits from it and economic savings from recently revised rules.

Environmental Protection Agency

Another major environmental accomplishment is the establishment of the **Environmental Protection Agency (EPA)** in 1970. The EPA was founded by a presidential executive order calling for a major reorganization of 15 existing federal agencies working on important environmental issues.

The EPA was directed to carry out the Federal Water Pollution Control Act and the Clean Air Act. Having grown in size, it now manages many of the environmental protection laws that issue from Congress. Current responsibilities of the EPA include research on the health and environmental impacts of a wide range of pollutants as well as the development and enforcement of health and environmental standards for pollutants outside the workplace. The EPA is concerned with a variety of areas, including pesticides, hazardous wastes, toxic substances, water pollution, air pollution, radiation, and noise pollution.

The EPA can provide incentives to state and local communities through grants that help pay for water pollution control projects. Grants to universities have helped expand the research capability of the agency. The EPA carries on much of its own research at four National Environmental Research Centers, located at Cincinnati, Ohio; Research Triangle Park, North Carolina; Las Vegas, Nevada; and Corvallis, Oregon.

The EPA is often caught in cross fire between opposing groups, for example, between environmentalists who seek tighter controls and the businesses the EPA regulates, which commonly complain that regulations are too stringent and costly. In the late 1970s, for example, widespread interest in the environment was overridden by economic hardship brought on by the oil embargoes and inflation. A powerful political movement arose to dismantle or weaken the EPA. The most common argument from the business sector was that the cost of protection was too excessive and that stiff environmental protection laws were preventing a healthy economy. But, as you have seen, some countries, such as Japan and Germany, have tough laws, continue to fight pollution, and still do extremely well economically (Chapter 21). As Russell Peterson, former administrator of the EPA and president of the National Audubon Society, noted, "We cannot have a thriving economy without a thriving ecosystem."

The American people continue to express a strong concern for a healthy environment; they support maintaining existing laws or even strengthening them. New problems, such as acid precipitation and hazardous wastes, constantly crop up, demanding the attention of the EPA and other federal bureaucracies. A growing population and an expanding economy create an ever-increasing burden on the environment and on the agencies that regulate environmental issues. Thus, many argue that the EPA and other agencies involved in environmental problems should be expanded, not merely maintained or reduced.

Principles of Environmental Law

In the United States, government's authority to protect the environment is conferred by the federal and state constitutions. Environmental protection is conferred by common law, federal and state statutes and local ordinances, and regulations promulgated by state and federal agencies. Statutory law and common law are the mainstays of environmental protection.

Statutory Law

This text has presented many examples of state and federal laws for environmental protection and resource management. These are laws approved by legislators, which are known as **statutory laws.** Statutory laws generally establish broad goals, among them the protection of health and environment by reducing air pollution or the judicious use of natural resources. However, as pointed out in this chapter, Congress and the state legislatures lack the time and expertise to determine specifically how these goals can be met. Thus, Congress assigns the technical details—the setting of standards, pollution control requirements, and resource management programs—to federal agencies, such as the EPA.

Besides setting standards of acceptable behavior and calling on agencies to determine regulations, statutory laws also provide authority for various enforcing agencies to take legal action—to fine polluters or to take polluters to court to face criminal charges and possible jail sentences.

Common Law

Many environmental cases are tried on the basis of **common law**, a body of unwritten rules and principles derived from thousands of years of legal decisions. Common law is based on proper or reasonable behavior.

Common law is a rather flexible form of law that attempts to balance competing societal interests. As an example, a company that generates noise may be brought to court by a nearby landowner who argues that the factory is a nuisance. The landowner may sue to have the action stopped through an injunction. In deciding the case, the court relies on common-law principles. It weighs the legitimate interests of the company in doing business (and thus making noise) and the interests of society, which wants its citizens employed and wants to collect taxes from the company, against the interests of the landowner, who is trying to protect the family's rest, health, and enjoyment of property.

The court may favor the one who files the lawsuit, the **plaintiff**, if the damage, such as loss of sleep, health effects, and inconvenience, is greater than the cost of preventing the risk, such as costs of noise abatement, loss of jobs, and loss of tax revenues. But, the court may not issue an order to shut down the factory, an **injunction**; instead, it may simply require the one defending the case or whose actions are being contested, the **defendant**, to reduce noise levels within a certain period, striking a balance between competing interests. Cases such as this one illustrate the **balance principle**.

Many cases involving common law are settled on two legal principles: nuisance and negligence.

Nuisance   The most common ground for action in the field of environmental common law is nuisance. A **nuisance** is a class of wrongs that arises from the unreasonable, unwarranted, or unlawful use of a person's own property that obstructs or injures the right of the public or another individual. It may result in annoyance, inconvenience, discomfort, or hurt. What this means is that one can use one's personal property or land in any way one sees fit but only in a reasonable manner and as long

as that use of the property does not cause material injury or annoyance to the public or another individual.

Generally, two types of remedy are available in a nuisance suit: compensation and injunction. **Compensation** is a monetary award for damage caused. An **injunction** is a court order requiring the nuisance to be stopped.

Nuisances are often characterized as either public or private. Until recently, the two were distinctly different concepts. A **public nuisance** is an activity that harms or interferes with the rights of the general public. Typically, public nuisance suits are brought to court by public officials. A **private nuisance** is one that affects only a few people. For example, the pollution of a well affecting only one or two families is considered a private nuisance. A public nuisance would be pollution that affects hundreds, perhaps thousands, of landowners along a river's shores. The most common environmental nuisance is noise. Water pollutants and air pollutants, such as smoke, dust, odors, and other chemicals, are other major nuisances.

Historically, the distinction between public and private nuisance has hampered pollution abatement because the courts have traditionally held that an individual could bring a public nuisance suit only if the individual had suffered a unique injury, that is, one different from that suffered by others. For example, an individual would not be permitted to sue a company for polluting a river shared by many others. In this case, the only legal recourse would be a public nuisance suit brought by an official, the local health department, for instance. In such instances, though, public officials are often unwilling to file suit against local businesses that provide important tax dollars for the community and campaign support as well.

Increasingly, the distinction between public and private nuisance is fading; private persons can bring suit to stop a public nuisance. As a result, the private individual is gradually acquiring more power to stop polluters.

Several common defenses are used to fight nuisance suits. Since most nuisance suits are decided by balancing the rights and interests of the opposing parties, **good-faith efforts** of the polluter may influence the decision. For example, if a small company had installed pollution control devices and had attempted to keep them operating properly but was still creating a nuisance, the court might hold it liable but would probably be more lenient in damages or conditions of abatement. If, on the other hand, the company had made no attempt to eliminate the pollution, the court would generally be more severe.

The availability of pollution control must also be considered. If a company is using state-of-the-art pollution control and still creates a nuisance, the court may not impose damages or an injunction. In contrast, if the company has failed to keep pace with pollution control equipment, the court may order such controls installed.

In states that still distinguish between public and private nuisance, a class-action suit can be filed. A **class-action suit** is brought on behalf of many people and seeks remedy for damage caused to the entire group by a nuisance. In order for a federal class-action suit for compensation to be permitted, however, each person named in the suit must have suffered at least $10,000 in damage. If not, the suit can be dismissed.

One defense against nuisance suits is that the plaintiff has "come to the nuisance." **Coming to the nuisance** occurs when an individual moves into an area where a nuisance, such as an airport, animal feedlot, or factory, already exists and then begins to complain. An old common-law principle holds that if you voluntarily place yourself in a situation in which you suffer injury, you have no legal right to sue for damages or an injunction. In most courts, however, even though you purchase property and know of the existence of a nuisance, you still have the right to file suit to abate it or recover damages. This is based on another common-law principle: Clean air and the enjoyment of property are rights that go along with owning the property. Thus, if population expands toward a nuisance, it may be the responsibility of the party creating the nuisance to end it.

According to environmental attorney Thomas Sullivan, "The courts are moving to strict liability for environmental nuisances, so that practically speaking, there are no good defenses. The solution is: do not create nuisances."

Negligence A second major principle of common law is **negligence**. From a legal viewpoint, a person is negligent if he or she acts in an unreasonable manner and if these actions cause personal or property damage.

Negligence provides a basis for liability, just as nuisance does, but negligence is generally more difficult to prove. What is reasonable action in one instance may not be reasonable in another. Statutory laws and regulations help the courts determine whether behavior is reasonable. For example, regulations drawn up by the EPA specify how certain hazardous wastes should be treated. Failure to comply with those standards may be evidence of negligence. Negligence may be shown in instances where a company fails to use common practices in the industry. For example, a company may be found negligent if it fails to transport hazardous wastes in containers like those used by other companies.

In a much broader sense, the courts may decide that a company is negligent simply if it fails to do something that a reasonable person would have done. For example, negligence might be demonstrated if a company failed to test its wastes for the presence of harmful chemicals when a reasonable person would have done so. Likewise, negligence may stem from action that a reasonable person would not have taken. In summary, then, negligence can result from either inaction or action that may be deemed unreasonable considering the circumstances.

The **concept of knowing** is fundamentally important to negligence suits. Briefly, negligence can be determined on the basis of what a defendant knew or should have known about a particular risk. The standard of comparison is what a reasonable person should have known under similar circumstances. For example, a man on trial may argue that he is not negligent because he did not know that a harmful chemical was in the materials that he dumped into a municipal waste dump, which subsequently polluted nearby groundwater. His argument will be valid if a reasonable person in his position could not have known about the wastes.

Interestingly, in some cases, determinations of liability for damage or harm need not be based on proof of negligence. For instance, in cases where the risk is extraordinarily large, one need prove only that injury or damage occurred, not that the operator was negligent or acted unreasonably. Business interests often try to lessen their liability by getting Congress to pass laws that put ceilings on liability. The Price-Anderson Act, for example, frees utility companies from financial liability incurred by a nuclear power plant accident. The act requires the government to pay for all damages outside the plant but only up to $560 million. The airlines have a similar law; at this writing, the oil companies are pressing for similar limits on liability resulting from tanker spills.

Problems with Environmental Lawsuits

Legal actions to stop a nuisance or collect damages from a nuisance or act of negligence carry with them a **burden of proof**. That is, in order to win, plaintiffs must prove that they have been harmed in some significant way and that the defendant is responsible for that harm. This is not always easy for several reasons. First, the cause-and-effect connection between a pollutant and disease may not have been definitely established by the medical community. If doubt exists, the case is weakened. Second, diseases such as cancer may occur decades after the exposure (Chapter 14). This makes it extremely difficult to prove causation. As a rule, it is generally easier to link cause and effect with acute diseases. Third, it is often difficult or impossible to identify the party responsible for damage, especially in areas where there are many industries or where illegal acts, such as midnight dumping of hazardous wastes, have occurred. Any reasonable doubt about the party responsible for personal or property damage may weaken a lawsuit.

The **statute of limitations** limits the length of time within which a person can sue after a particular event. This provision also creates problems in cases of delayed diseases. Statutes of limitations are designed to reduce lawsuits in which evidence is unavailable or memories of potential witnesses may have faded. In latent-disease cases, though, they create a major obstacle to individuals seeking compensation for damage in states that apply the time limitations to the onset of exposure. This essentially makes it impossible for cancer victims to file suit. Other states start the judicial clocks from the time the victim learns of the disease. This makes it a little easier to collect compensation for diseases such as cancer, black lung, and emphysema.

Out-of-court settlements present another legal problem. Such settlements have hindered environmental law. Eager to avoid a costly settlement, a company may pay victims if they agree to dismiss the company from further liability. Out-of-court settlements may also benefit plaintiffs, saving them the time, headaches, and costs of environmental litigation.

While advantageous to both parties, these settlements provide no precedents for environmental laws. In short, the fewer cases that make it to court, the fewer guidelines courts have to settle future cases. This lack of precedents may discourage attorneys and citizens from filing court cases. Without clear examples from the past, they may simply be unwilling to face costly, time-consuming legal battles.

Resolving Environmental Disputes Out of Court

An increasing number of disputes between environmentalists and businesses are being settled out of court by dispute resolution, or **mediation**. This innovative approach often employs a neutral party who mediates the discussion between opposing parties. The mediator keeps the proceedings on track, encourages rivals to work together, and tries to resolve the dispute in a way that is satisfactory to both groups.

The benefits of mediation over litigation are many. Mediation is much less costly and time consuming. Mediation also tends to create better feelings among disputants, whereas court settlements create winners and losers and often leave bitter feelings. In addition, and perhaps most important of all, mediation may bring about a more satisfactory resolution. For instance, environmental lawsuits often hinge on specific points of law rather than substantive issues and thus may have little to do with what the plaintiff really wants. An environmental group might bring a suit over the adequacy of an EIS but, in reality, might want the government to ensure protection of a valuable species that would be affected by the project. In mediation, this will be the central issue. Mediation may therefore result in more appropriate solutions.

Mediation also tends to promote a more accurate view of problems. For instance, in lawsuits, each party tends to bring up evidence that favors its goal and ignore or dismiss unfavorable information or ambiguous information that might weaken its stand. In mediation, both parties are encouraged to openly discuss the uncertainties of their positions, discovering many points of agreement and building a better understanding of opposing positions.

Mediation also has its drawbacks. First, funding is inadequate. In the past, mediation has been financed largely by foundations. Federal, state, and local governments need to develop programs to fund mediation. Second, some groups fear they will lose their constituency if they enter into negotiations on certain issues because they will be giving the impression that they are failing at their stated goals and compromising with the "enemy." A third drawback is a lack of faith in the outcome of mediation. Unlike court orders, resolutions drawn up in mediation are not legally enforceable. Thus, months of discussion may produce nothing but a piece of paper that polluters will ignore.

Effective mediation requires the following: (1) A truly neutral party must serve as mediator. (2) A formally agreed-on agenda for discussion and a point of focus are also important. Resource and pollution issues should be the focus of discussion; disputes over values should not dominate the proceedings because they cannot be solved by mediation. Although values will surely emerge during the debate, they should not be the central point. (3) Both sides must be willing to explore new ideas and possibilities. (4) Disputants must deal honestly with each other. (5) An adequate representation of all interested parties must also be achieved. If someone is not represented, an unsatisfactory solution may result. This could lead to a lawsuit. (6) Strict rules should be imposed regarding news releases. The media should not be employed as a lever by either group.

Dispute resolution is growing in the United States, but it will not replace litigation. Still, it can play a valuable role in the future. Eight states have organizations that offer mediation services, and a growing number of private organizations have been formed to provide professional mediators with experience and knowledge in environmental issues. Thus, more and more disputes may be settled by this noncombative approach.

Suggested Readings

Arbuckle, J. G. et al. (1985). *Environmental Law Handbook* (8th ed.). Washington, D.C.: Government Institutes. Superb overview of environmental law.

Epstein, S. S., Brown, L. O., and Pope, C. (1982). *Hazardous Waste in America*. San Francisco: Sierra Club Books. See Chapter 10 for a good overview of principles of environmental law.

Findley, R. W. and Farber, D. A. (1981). *Environmental Law: Cases and Materials*. St. Paul, MN: West Publishing. In-depth presentation of important cases in environmental law.

Wenner, L. M. (1976). *One Environment Under Law: A Public-Policy Dilemma*. Pacific Palisades, CA: Goodyear. Superb reading.

Glossary

Abiotic factors   Nonliving components of the ecosystem, including chemical and physical factors such as availability of nitrogen, temperature, and rainfall.

Accelerated erosion   Loss of soil due to wind or water in land disturbed by human activities.

Accelerated extinction   Elimination of species due to human activities such as habitat destruction, commercial hunting, sport hunting, and pollution.

Acid deposition   Rain or snow that has a lower pH than precipitation from unpolluted skies, also includes dry forms of deposition such as nitrate and sulfate particles.

Acid mine drainage   Sulfuric acid that drains from mines, especially abandoned underground coal mines in the East (Appalachia). Created by the chemical reaction between oxygen, water, and iron sulfides found in coal and surrounding rocks.

Acid precursor   Refers two sulfur oxides and nitrogen oxides, chemical air pollutants that can be converted into an sulfuric and nitric acids, respectively, in the atmosphere.

Active solar   Capturing and storage of the sun's energy through special collection devices (solar panels) that absorb heat and transfer it to air, water, or some other medium, which is then pumped to a storage site (usually a water tank) for later use. Contrast with passive solar. Actual risk An accurate measure of the hazard posed by a certain technology or action.

Acute effects   In general, effects that occur shortly after exposure to toxic agents. Contrast with chronic effects.

Acute exposure   Short-term exposure usually to high levels of one or more agents. Symptoms generally appear soon after exposure.

Adaptation   A genetically determined structural or functional characteristic of an organism that enhances its chances of reproducing and passing on its genes.

Adaptive management   Experimental approach to resource management that allows officials to monitor and evaluate practices and change them as needed.

Adaptive radiation   Evolution of several life forms from a common ancestor.

Advanced industrial society   Post-World War II industrial society characterized by great rises in production and consumption, increased energy demand, and a shift toward synthetics and nonrenewable resources.

Age-specific fertility rate   Number of live births per 1000 women of a specific age group.

Agricultural land conversion   Transformation of farmland to other purposes, primarily cities, highways, airports, and the like.

Agricultural society   A group of people living in villages or towns and relying on domestic animals and crops grown in nearby fields. Characterized by specialization of work roles.

Algal bloom   Rapid growth of algae in surface waters due to increase in inorganic nutrients, usually either nitrogen or phosphorus.

Alien species   (or foreign species) Any species introduced into or living in a new habitat. Also known as an exotic.

Alpha particles   Positively charged particles consisting of two protons and two neutrons, emitted from radioactive nuclei.

Alveoli   Small sacs in the lungs where exchange of oxygen and carbon dioxide between air and blood occurs.

Ambient air quality standard   Maximum permissible concentration of a pollutant in the air around us. Contrast with emissions standard.

Annuals   Plants that grow from seeds, for example, domestic corn and radishes.

Antagonism   In toxicology, when two chemical or physical agents (often toxins) counteract each other to produce a lesser response than would be expected if individual effects were added together.

Anthropogenic hazard   A danger created by humans.

Appropriate technology   A term coined by the late E. F. Schumacher to refer to technology that is "appropriate" for the economy, resources, and culture of a region. It is characterized by small- to medium-sized machines, maximum human labor, ease of understanding, meaningful employment, use of local resources, decentralized production, production of durable products, emphasis on renewable resources, especially energy, and compatibility with the environment and culture.

Aquaculture   Cultivation of fish and other aquatic organisms in freshwater ponds, lakes, irrigation ditches, and other bodies of water.

Aquatic life zone   Distinct regions of the ocean akin to terrestrial biomes.

Aquifer   Underground stratum of porous material (sandstone) containing water (groundwater), which may be withdrawn from wells for human use.

Aquifer recharge zone   Region in which water from rain or snow percolates into an aquifer, replenishing the supply of groundwater.

Artificial selection   Selective breeding to create new plant and animal breeds to bring out desirable characteristics.

Asbestos   One of several naturally occurring silicate fibers. Useful in society as an insulator but deadly to breathe even in small amounts. Causes mesothelioma, asbestosis, and lung cancer.

Asbestosis   Lung disease characterized by buildup of scar tissue in the lungs. Caused by inhalation of asbestos.

Asthma   Lung disorder characterized by constriction and excessive mucus production in the bronchioles, resulting in periodic difficulty in breathing, shortness of breath, coughing. Usually caused by allergy and often aggravated by air pollution.

Atmosphere   Layer of air surrounding the earth.

Atom   A basic unit of matter consisting of a nucleus of positively charged protons and uncharged neutrons, and an outer cloud of electrons orbiting the nucleus.

Autotroph   Organisms, such as plants, that produce their own food generally via photosythesis. See producer.

Auxins   Plant hormones responsible for stimulating growth.

Bacteria   A group of single-celled organisms, each surrounded by a cell wall and containing circular DNA. Responsible for some diseases and many beneficial functions, such as the decay of organic materials and nutrient recycling.

Barrier islands   Small, sandy islands off a coast, separated from the mainland by lagoons or bays.

Beach drift   Wave-caused movement of sand along a beach.

Beta particles   Negatively charged particles emitted from nuclei of radioactive elements when a neutron is converted to a proton.

Big bang   Theory of the universe's formation. States that all matter in the universe was infinitely compressed 15 to 20 billion years ago and then exploded, sending energy and matter out into space. The matter was in the form of subatomic particles which formed atoms as the universe cooled over millions of years.

Bioaccumulation   Refers to the buildup of chemicals within tissues and organs of the body and results factors a resistance to breakdown and slow excretion rates.

Biochemical oxygen demand (BOD)   Measure of oxygen depletion of water (largely from bacterial decay) due to presence of biodegradable organic pollutants. Gives scientists an indication of how much organic matter is in water.

Biodegradable   Material that can be broken down by naturally occurring organisms such as bacteria in air, water, and soil.

Biogas   A gas containing methane and carbon dioxide. Produced by anaerobic decay of organic matter, especially manure and crop residues.

Biogeochemical cycle   Complex cyclical transfer of nutrients from the environment to organisms and back to the environment. Examples include the carbon, nitrogen, and phosphorus cycles.

Biological control   Use of naturally occurring predators, parasites, bacteria, and viruses to control pests. Also called food chain control.

Biological extinction   Disappearance of a species from part or all of its range.

Biological magnification   Buildup of chemical elements or substances in organisms in successively higher trophic levels. Also called biomagnification.

Biomass   As measured by ecologists, the dried weight of all organic matter in the ecosystem. In the energy field, any form of organic material (from both plants and animals) from which energy can be derived by burning or bioconversion such as fermentation. Includes wood, cow dung, agricultural crop residues, forestry residues, scrap paper.

Biomass pyramid   See pyramid of biomass.

Biome   One of several immense terrestrial regions, each characterized throughout its extent by similar plants, animals, climate, and soil type.

Biosphere   All the life-supporting regions (ecosystems) of the earth and all the interactions that occur between organisms and between organisms and the environment.

Biotic factor   The biological component of the ecosystem, consisting of populations of plants, animals, and microorganisms in complex communities.

Biotic (reproductive) potential   Maximum reproductive potential of a species.

Birth control   Any measure designed to reduce births, including contraception and abortion.

Birth defect   An anatomical (structural) or physiological (functional) defect in a newborn.

Bloom   See algal bloom.

Breeder reactor   Fission reactor that produces electricity and also converts abundant but nonfissile uranium-238 into fissile plutonium-239, which can be used in other fission reactors.

Broad-spectrum pesticide (or biocide)   Chemical agent effective in controlling a large number of pests.

Bronchitis   Persistent inflammation of the bronchi caused by smoking and air pollutants. Symptoms include mucus buildup, chronic cough, and throat irritation.

Brown-air cities   Newer, relatively nonindustrialized cities whose polluted skies contain photochemical oxidants (especially ozone) and nitrogen oxides, largely from automobiles and power plants. Tend to have dry, sunny climates. Contrast with gray-air cities.

Buffer zone   Region around a proected area in which limited human activity is permitted.

Calorie   Amount of heat needed to raise 1 gram of water 1 degree Celsius or 1.8 degrees Fahrenheit. One thousand calories is equal to one kilocalorie, commonly written as Calorie.

Cancer   Uncontrolled proliferation of cells in humans and other living organisms. In humans, includes more than 100 different types afflicting individuals of all races and ages.

Carbon cycle   The cycling of carbon between organisms and the environment.

Carcinogen   A chemical or physical agent that causes cancer to develop, often decades after the original exposure.

Carrying capacity   Maximum population size that a given ecosystem can support for an indefinite period or on a sustainable basis. Catalyst Substance that accelerates chemical reactions but is not used up in the process. Enzymes are biological catalysts. Also see catalytic converter.

Catalytic converter   Device attached to the exhaust system of automobiles and trucks to rid the exhaust gases of harmful pollutants.

Cation   Any one of many kinds of positively charged ions.

Cellular respiration   Process by which a cell breaks down glucose and other organic molecules to acquire energy. Also called oxidative metabolism.

Chlorofluorocarbons   Organic molecules consisting of chlorine and fluorine covalently bonded to carbon. CCl_3F (Freon-11) and CCl_2F_2 (Freon-12). Used as spray-can propellants and coolants. Previously thought to be inert, but now known to destroy the stratospheric ozone layer. Also called chlorofluoromethanes and freon gases.

Chlorophyll   Pigment of plant cells that absorbs sunlight, thus allowing plants to capture solar energy.

Chromosomes   Genetic material of organisms, containing DNA and protein. Carries the genetic information that controls all cellular activity.

Chronic bronchitis   Persistent inflammation of the bronchi due to pollutants in ambient air and tobacco smoke. Characterized by persistent cough.

Chronic effects   In general, the delayed health results of toxic agents, for example, emphysema, bronchitis, and cancer. Contrast with acute effects.

Chronic exposure   Exposure to one or more toxic agents generally over a long period. Symptoms (e.g., cancer) usu-ally appear long after exposure.

Chronic obstructive lung disease   Any one of several lung diseases characterized by obstruction of breathing. Includes emphysema, bronchitis, and diseases with symptoms of both of these.

Clear-cutting   Removal of all trees from a forested area.Climate The average weather conditions: temperature, solar radiation, precipitation, and humidity.

Climate   Average weather conditions, including temperature, rainfall, and humidity.

Climax community or ecosystem   See mature community.

Closed system   A system that can exchange energy, but does not exchange matter, with the surrounding environment. Example: the Earth. Contrast with open system.

Coal gasification   Production of combustible organic gases (mostly methane) by applying heat and steam to coal in an oxygen-enriched environment. Carried out in surface vessels or in situ.

Coal liquefaction   Production of synthetic oil from coal.

Coastal wetlands   Wet or flooded regions along coastlines, including mangrove swamps, salt marshes, bays, and lagoons. Contrast with inland wetlands.

Coevolution   Process whereby two species evolve adaptations as a result of extensive interactions with each other.

Cogeneration   Production of two or more forms of useful energy from one process. For example, production of electricity and steam heat from combustion of coal. Increases energy efficiency.

Coliform bacterium   Common bacterium found in the intestinal tracts of humans and other species. Used in water quality analysis to determine the extent of fecal contamination.

Commensalism   Relationship between two organisms that is beneficial to one and neither harmful nor helpful to the other.

Common law   Body of rules and principles based on judicial precedent rather than legislative enactments. Founded on an innate sense of justice, good conscience, and reason. Flexible and adaptive. Contrast with statutory law.

Commons   Any resource used in common by many people, such as air, water, and grazing land.

Community   Also called a biological community. The populations of plants, animals, and microorganisms living and interacting in a given locality.

Competition   Vying for resources between members of the same or different species.

Composting   Aerobic decay of organic matter to generate a humus-like substance used to supplement soil.

Confusion technique (of pest control)   Release of insect sex-attractant pheromones identical to pheromones released by normal breeding females to attract males for mating. Release in large quantities confuses males as to the location of the females, thus minimizing the chances of males finding females and helping to control pest populations.

Conservation A strategy to reduce the use of resources, especially through increased efficiency, reuse, recycling, and decreased demand.

Conservation biology See restoration ecology.

Consumer (or consumer organism) An organism in the ecosystem that feeds on autotrophs and/or heterotrophs. Synonym: heterotroph.

Continental drift Movement of the earth's tectonic plates on a semiliquid layer of mantle, forcing continents to shift position over hundreds of thousands of years.

Contour farming Soil erosion control technique in which row crops (corn) are planted along the contour lines in sloping or hilly fields rather than up and down the hills.

Contraceptive Any device or chemical substance used to prevent conception.

Control group In scientific experimentation, a group that is untreated and compared with a treated, or experimental, group.

Control rods Special rods containing neutron-absorbing materials. Inserted into a reactor core to control the rate of fission or to shut down fission reactions.

Convergent evolution The independent evolution of similar traits among unrelated organisms resulting from similar selective pressures.

Cosmic radiation High-energy electromagnetic radiation similar to cosmic rays but originating from periodic solar flare-ups. Possesses extraordinary ability to penetrate materials, including cement walls.

Cost-benefit analysis Way of determining the economic, social, and environmental costs and benefits of a proposed action such as construction of a dam or highway. Still a crude analytical tool because of the difficulty of measuring environmental costs.

Critical population size Population level below which a species cannot successfully reproduce.

Crop rotation Alternating crops in fields to help restore soil fertility and also control pests.

Crossing over Transfer of genetic material from one chromosome to another during the formation of gametes.Cross-media contamination The movement of pollution from one medium, such as air, to another, such as water.

Crude birth rate Number of births per 1000 people in a population at the midpoint of the year.

Crude death rate Number of deaths per 1000 people in a population at midyear.

Cultural control (of pests) Techniques to control pest populations not involving chemical pesticides, environmental controls, or genetic controls. Examples: cultivation to control weeds and manual removal of insects from crops.

Cultural eutrophication Eutrophication (see definition) due largely to human activities.

Daughter nuclei Atomic nuclei that are produced during fission of uranium.

DDT Dichlorodiphenyltrichloroethane. An organochlorine insecticide used first to control malaria-carrying mos-

quitoes and lice and later to control a variety of insect pests, but now banned in the United States because of its persistence in the environment and its ability to bioaccumulate.

Debt-for-nature swap Arrangement made between debtor nation, intermediary, and a lending institution in which intermediary buys up a debt of a debtor nation at a discount in exchange for funds or programs to protect vital habitats in the debtor nation.

Decibel (dB) A unit to measure the loudness of sound.
Decomposer Also microconsumer. An organism that breaks down nonliving organic material. Examples: bacteria and fungi.

Decomposer food chain A specific nutrient and energy pathway in an ecosystem in which decomposer organisms (bacteria and fungi) consume dead plants and animals as well as animal wastes. Essential for the return of nutrients to soil and carbon dioxide to the atmosphere. Also called detritus food chain.

Deforestation Destruction of forests by clear-cutting.

Demographic transition A phenomenon witnessed in populations of industrializing nations. As industrialization proceeds and wealth accumulates, crude birth rate and crude death rate decline, resulting in zero or low population growth. Decline in death rate usually precedes the decline in birth rate, producing a period of rapid growth before stabilization.

Demography The science of population.

Depletion allowance Tax relief given to extractive industries as they deplete reserves. Intended to allow the companies to invest more in exploration. Gives extraction industries unfair advantage over recycling companies.

Desert Biome located throughout the world. Often found on the downwind side of mountain ranges. Characterized by low humidity, high summertime temperatures, and plants and animals especially adapted to lack of water.

Desertification The formation of desert in arid and semiarid regions from overgrazing, deforestation, poor agricultural practices, and climate change. Found today in Africa, the Middle East, and the southwestern United States.

Detoxification Rendering a substance harmless by reacting it with another chemical, chemically modifying it, or destroying the molecule through combustion or thermal decomposition.

Detritus Any organic waste from plants and animals.

Detritus feeders Organisms in the decomposer food chain that feed primarily on organic waste (detritus), such as fallen leaves.

Detritus food chain See decomposer food chain.

Deuterium An isotope of hydrogen whose nucleus contains one proton and one neutron (a hydrogen atom has only one proton).

Developed country A convenient term that describes industrialized nations, generally characterized by high standard of living, low population growth rate, low infant mortality, excessive material consumption, high per capita energy consumption, high per capita income, urban population, and low illiteracy.

Developing country Term describing the nonindustrialized nations, generally characterized by low standard of living, high population growth rate, high infant mortality, low material consumption, low per capita energy consumption, low per capita income, rural population, and high illiteracy.

Dioxin   A large group of highly toxic, carcinogenic compounds containing some herbicides (2,4-D and 2,4,5-T) and Agent Orange. Once disposed of by mixing with waste crankcase oil that was spread on dirt roads to control dust.

Diversity   A measure of the number of different species in an ecosystem.

DNA (deoxyribonucleic acid)   A long-chained organic molecule that is found in chromosomes and carries the genetic information that controls cellular function and is the basis of heredity.

Dose-response curve   Graphical representation of the effects of varying doses of chemical or physical agents.

Doubling time   The length of time it takes some measured entity (population) to double in size at a given growth rate.

Ecological backlashes   Ecological effects of seemingly harmless activities, for example, the greenhouse effect.

Ecological equivalents   Organisms that occupy similar ecological niches in different regions of the world.

Ecological niche   See niche.

Ecological system   See ecosystem.

Ecology   Study of living organisms and their relationships to one another and the environment.

Economic depletion   Reduction in the supply of a resource to the point at which it is no longer economically feasible to continue mining, extracting, or harvesting it.

Economic externality   A cost (environmental damage, •illness) of manufacturing, road building, or other action that is not taken into account when determining the total cost of production or construction. A cost generally passed on to the general public and taxpayers; external cost.

Ecosphere   See biosphere.

Ecosystem   Short for ecological system. A community of organisms occupying a given region within a biome. Also, the physical and chemical environment of that community and all the interactions between organisms and between organisms and their environment.

Ecotone   Transition zone between adjacent ecosystems.

Element   A substance, such as oxygen, gold, or carbon, that is distinguished from all other elements by the number of protons in its atomic nucleus. The atoms of an element cannot be decomposed by chemical means.

Emigration   Movement of people out of a country to establish residence elsewhere.

Emissions offset policy   Strategy to control air pollution in areas meeting federal ambient air quality standards, whereby new factories must secure emissions reductions from existing factories to begin operation; thus the overall pollution level does not increase.

Emissions standard   The maximum amount of a pol-

lutant permitted to be released from a point source (see definition).

Emphysema   A progressive, debilitating lung disease caused by smoking and pollution at work and in the environment. Characterized by gradual breakdown of the alveoli (see definition) and difficulty in catching one's breath.

Endangered species   A plant, animal, or microorganism that is in immediate danger of biological extinction. See threatened and rare species.

Energy   The capacity to do work. Found in many forms, including heat, light, sound, electricity, coal, oil, and gasoline.

Energy pyramid   See pyramid of energy.

Energy quality   The amount of useful work acquired from a given form of energy. High-quality energy forms are concentrated (e.g., oil and coal); low-quality energy forms are less concentrated (e.g., solar heat).

Energy system   The complete production-consumption process for energy resources, including exploration, mining, refining, transportation, and waste disposal.

Entropy   A measure of disorder. The second law of thermodynamics applied to matter says that all systems proceed to maximum disorder (maximum entropy).

Environment   All the biological and nonbiological factors that affect an organism's life.

Environmental control (of pests)   Methods designed to alter the abiotic and biotic environments of pests, making them inhospitable or intolerable. Examples include increasing crop diversity, altering time of planting, and altering soil nutrient levels.

Environmental impact statement (EIS or ES)   Document prepared primarily to outline potential impacts of projects supported in part or in their entirety by federal funds.

Environmental phase (of the nutrient cycle)   Part of the nutrient or biogeochemical cycle in which the nutrient is deposited or cycles through the environment (air, water, and soil).

Environmental resistance   Abiotic and biotic factors that can potentially reduce population size. Environmental science The interdisciplinary study of the complex and interconnected issues of population, resources, and pollution.

Epidemiology   Study of disease and death in human populations, which attempts to find links between causes and effects through statistical methods.

Epilimnion   Upper, warm waters of a lake. Contrast with hypolimnion.

Estuarine zones   Coastal wetlands and estuaries.

Estuary   Coastal regions such as inlets or mouths of rivers where salt and fresh water mix.

Ethanol   Grain alcohol, or ethyl alcohol, produced by fermentation of organic matter. Can be used as a fuel for a variety of vehicles and as a chemical feedstock.

Eukaryotes   The first aerobic cells complete with nuclei and energy-releasing organelles.

Eutrophication   Accumulation of nutrients in a lake or pond due to human intervention (cultural eutrophication) or natural causes (natural eutrophication). Contributes to process of succession (see definition).

Evapotranspiration   Evaporation of water from soil and transpiration of water from plants.

Evolution   A long-term process of change in organisms caused by random genetic changes that favor the survival and reproduction of the organism possessing the genetic change. Through evolution, organisms become better adapted to their environment.

Exclusion principle   Ecological law holding that no two species can occupy the exact same niche.

Experimental group   In scientific experimentation, a group that is treated and compared with an untreated, or control, group.

Exponential curve   See J curve.

Exponential growth   Increase in any measurable thing by a fixed percentage. When plotted on graph paper, it forms a J-shaped curve.

Externality   A spillover effect that benefits or harms others. The source of the effect (say, pollution) does not pay for the effect.

Extinction   See biological extinction.

Extractive reserve   A forest protected for the sustainable harvest of fruit, nuts, and other products generally by local people.

Fallout   Radioactive materials produced during an atomic detonation and later deposited from the air.

Fall overturn   Annual cycle in deep lakes in temperate climates, in which the warm surface water and cool subsurface water mix.

Family planning   Process by which couples determine the number and spacing of children.

Feedlot   Fenced area where cattle are raised in close confinement to minimize energy loss and maximize weight gain.

First-generation pesticides   Earliest known chemical pesticides such as ashes, sulfur, ground tobacco, and hydrogen cyanide. Contrast with second- and third-generation pesticides.

First-law efficiency   A measure of the efficiency of energy use. Total amount of useful work derived from a system divided by the total amount of energy put into a system.

First law of thermodynamics   Also called the law of conservation of energy. States that energy is neither created nor destroyed; it can only be transformed from one form to another.

Fission   Splitting of atomic nuclei when they are struck by neutrons or other subatomic particles.

Fission fragments   See daughter nuclei.

Floodplain   Low-lying region along river or stream, periodically subject to natural flooding. Common site for human habitation and farming.

Fly ash   Mineral matter escaping with smokestack gases from combustion of coal.

Food chain   A specific nutrient and energy pathway in ecosystems proceeding from producer to consumer. Part of a bigger network called the food web. See decomposer food chain and grazer food chain.

Food web   Complex intermeshing of individual food chains in an ecosystem.

Foreign species   See alien species.

Fossil fuel   Any one of the organic fuels (coal, natural gas, oil, tar sands, and oil shale) derived from once-living plants or animals. Freons See chlorofluorocarbons.

Frontier mentality   A mind-set that views humans as "above" all other forms of life rather than as an integral part of nature and sees the world as an unlimited supply of resources for human use regardless of the impacts on other species. Implicit in this view are the notions that bigger is better, continued material wealth will improve life, and nature must be subdued.

Fuel rods   Rods packed with small pellets of radioactive fuel (usually a mixture of fissionable uranium-235 and uranium-238) for use in fission reactors.

Full-cost pricing   Technique to include all costs, including the cost of ecological damage, into the price of a product or service.

Gaia hypothesis   Term coined by James Lovelock to describe the earth's capacity to maintain the physical and chemical conditions necessary for life.

Galaxy   Grouping of billions of stars, gas, and dust, such as the Milky Way galaxy.

Gamma rays   A high-energy form of radiation given off by certain radionuclides. Can easily penetrate the skin and damage cells.

Gasohol   Liquid fuel for vehicles, containing nine parts gasoline and one part ethanol.

Gene   Segment of the DNA that either codes for proteins produced by the cell (structural gene) or regulates structural genes.

Gene pool   Sum total of all the genes and their alternate forms in a population.

Generalists   Organisms that have a broad niche, usually feeding on a variety of food materials and sometimes adapted to a large number of habitats.

Genetic control (of pests)   Development of plants and animals genetically resistant to pests through breeding programs and genetic engineering. Also, introduction of sterilized males of pest species (see sterile-male technique).

Genetic engineering   Isolation and production of genes that are then inserted in bacteria or other organisms. Can be used to produce insulin and other hormones. May someday also be used to treat genetic diseases.

Geopressurized zone   Aquifer containing superheated, pressurized water and steam trapped by impermeable rock strata and heated by underlying magma.

Geothermal energy   Energy derived from the earth's heat that comes from decay of naturally occurring radioactive materials in the earth's crust, magma, and friction caused by movement of tectonic plates.

GNP See gross national product.

Gradualism Theory of evolution holding that species evolve over long periods. Contrast with punctuated equilibrium. Grasslands Biome found in both temperate and tropical regions and characterized by periodic drought, flat or slightly rolling terrain, and large grazers that feed off the lush grasses.

Gray-air cities Older industrial cities characterized by predominantly sulfur dioxide and particulate pollution. Contrast with brown-air cities.

Grazer food chain A specific nutrient and energy pathway starting with plants that are consumed by grazers (herbivores).

Greenhouse effect Mechanism that explains atmospheric heating caused by increasing carbon dioxide. Carbon dioxide is believed to act like the glass in a greenhouse, permitting visible light to penetrate but impeding the escape of infrared radiation, or heat.

Green product General term referring to environmentally friendly products. They may be made from recycled materials or may be fully recyclable. They may be reusable or nontoxic or may help promote efficient use of resources.

Green Revolution Developments in plant genetics in the late '50s and early '60s resulting in high-yield varieties producing three to five times more grain than previous plants but requiring intensive irrigation and fertilizer use.

Green tax General term applying to user fees, severance taxes, and pollution taxes.

Gross national product (GNP) Total national output of goods and services valued at market prices, including net exports and private investment.

Gross primary productivity The total amount of sunlight converted into chemical-bond energy by a plant. This measure does not take into account how much energy a plant uses for normal cellular functions. See net primary productivity.

Groundwater Water below the earth's surface in the saturated zone.

Growth factor Any one of many biotic and abiotic factors that stimulate growth of populations. Contrast with reduction factors.

Habitat The specific region in which an organism lives.

Half-life Time required for one-half of a given amount of radioactive material to decay, producing one-half the original mass. Can also be used to describe the length of residence of chemicals in tissues. Biological half-life refers to the time it takes for one-half of a given amount of a substance to be excreted or catabolized.

Hard path A term coined by Amory Lovins to describe large, centralized energy systems such as coal, oil, or nuclear power, characterized by extensive power distribution, central control, and lack of renewability.

Hazardous waste Any potentially harmful solid, liquid, or gaseous waste product of manufacturing or other human activities.

Herbicide Chemical agent used to control weeds.

Herbivore Heterotrophic organism that feeds exclusively on plants.

Heteroculture Agriculture in which several plant species are grown simultaneously to reduce insect infestation and disease.

Heterotroph An organism that feeds on other organisms such as plants and animals. It cannot make its own foodstuffs.

Homeostasis State of relative constancy in organisms and ecosystems. A kind of dynamic equilibrium.

Hot-rock zones Most widespread geothermal resource. Regions where bedrock is heated by underlying magma.

Humus Mixture of decaying organic matter and inorganic matter that increases soil fertility, aeration, and water retention.

Hunting and gathering society People who lived as nomads or in semipermanent sites from the beginning of human evolution until approximately 5000 bc. Some remnant populations still survive. They gathered seeds, fruits, roots, and other plant products and hunted indigenous species for food.

Hybrid Offspring produced by cross-mating of two different strains or varieties of plants or animals.

Hydrocarbons Organic molecules containing hydrogen and carbon. Released during the incomplete combustion of organic fuels. React with nitrogen oxides and sunlight to form photochemical oxidants in photochemical smog.

Hydroelectric power Electricity produced in turbines powered by running water.

Hydrological cycle The movement of water through the environment from atmosphere to earth and back again. Major events include evaporation and precipitation. Also called the water cycle.

Hydrosphere The watery portion of the planet. Contrast with atmosphere and lithosphere.

Hydrothermal convection zone Rock strata containing large amounts of water heated by underlying magma and driven to the surface through cracks and fissures in overlying rock layers. Forms hot springs and geysers.

Hypolimnion Deep, cold waters of a lake. Contrast with epilimnion.

Hypothesis Tentative explanation for a natural phenomenon.

Immature ecosystem An early successional community characterized by low species diversity and low stability. Contrast with mature ecosystem.

Immigration Movement of people into a country to set up residence there.

Indoor air pollution Generally refers to air pollutants in homes from internal sources such as smokers, fireplaces, wood stoves, carpets, paneling, furniture, foam insulation, and cooking stoves.

Induced abortion Surgical procedure to interrupt pregnancy by removing the embryo or fetus from the uterus. In the first trimester, generally carried out by vacuum aspiration. Contrast with spontaneous abortion.

Industrial revolution   Period in history marked by a shift in manufacturing from small-scale production by hand to large-scale production by machine. Occurred in England in the 1700s and in the United States in the 1800s.

Industrial smog   Air pollution from industrial cities (gray-air cities), consisting mostly of particulates and sulfur oxides. Contrast with photochemical smog.

Industrial society   Group of people living in urban or rural environments that are characterized by mechanization of industrial production and agriculture. Widespread machine labor causes high energy demands and pollution. Increasing control over natural processes leads to feelings that humans are apart from nature and superior to it. Infant mortality rate Number of infants under 1 year of age dying per 1000 births in any given year.

Infectious disease   Generally, a disease caused by a virus, bacterium, or parasite that can be transmitted from one organism to another (example: viral hepatitis).

Infrared radiation   Heat, an electromagnetic radiation of wavelength outside the red end of the visible spectrum.

Inland wetlands   Wet or flooded regions along inland surface waters. Includes marshes, bogs, and river outflow lands. Contrast with coastal wetlands.

In-migration   Movement of people into a state or region within a country to set up residence.

Inorganic fertilizer   Synthetic plant nutrient added to the soil to replace lost nutrients. Major components include nitrogen, phosphorus, and potassium. Also called artificial fertilizer or synthetic fertilizer.

Input approach   A method of solving an environmental problem by reducing the inputs. For example, reducing consumption and increasing product durability can cut production of solid wastes, pollution, or hazardous wastes.

Insecticide   One form of pesticide used specifically to control insect populations.

Integrated pest management   Pest control with minimum risk to humans and the environment through use of a variety of control techniques (including pesticides and biological controls).

Integrated wildlife or species management   Control of populations through the use of many techniques, including the reintroduction of natural predators, habitat improvement, reduction in habitat destruction, establishment of preserves, reduced pollution, and captive breeding.

Interspecific competition   Competition between members of different species.

Ion   A particle formed when an atom loses or gains an electron.

Ionizing radiation   Electromagnetic radiation with the capacity to form ions in body tissues and other substances.

Isotopes   Atoms of the same element that differ in their atomic weight because of variations in the number of neutrons in their nuclei.

J curve   A graphical representation of exponential growth. Juvenile hormone Chemical substance in insects that stimulates growth through early life stages. Used with some success as an insecticide. When applied to infested fields, JH alters normal growth and development of insect pests, resulting in their death.

Kerogen   Solid, insoluble organic material found in oil shale.

Keystone species   Critical species in an ecosystem whose loss profoundly affects several or many others.

Kilocalorie   One thousand calories.

Kilowatt   One thousand watts. See watt.

Kinetic energy   The energy of objects in motion.

Kwashiorkor   Dietary deficiency caused by insufficient protein intake; common in children one to three years of age in less developed countries. Characterized by growth retardation, wasting of muscles in limbs, and accumulation of fluids in the body, especially in feet, legs, hands, and face.

Lag effect   The tendency for a population to continue growing even after it has reached replacement-level fertility. Caused by an expanding number of women reaching reproductive age.

Landfill   Depression in the ground or excavated site in which waste is deposited. Sanitary landfills are so named because trash is covered daily with soil to prevent odors and proliferation of rodents and flies. Can be used for hazardous wastes, but preferably as last resort. Must be lined with clay and impermeable linears to accept hazardous materials.

Land ethic   View that extends ethical concerns beyond humans to the ecosystem.

Land-use planning   Process whereby land uses are matched with the needs of the community and environmental considerations, for example, need for open space and agricultural land and for control of water and air pollution.

Laterite   Soil found in some tropical rain forests. Rich in iron and aluminum but generally of poor fertility. Turns bricklike if exposed to sunlight.

Least-cost planning   Process in which demand for a resource (e.g., electricity) is met in the least costly way. Generally includes only economic costs.

Life expectancy   Term referring to number of years people in a society live on average.

Light water reactor   Most common fission reactor for generating electricity. Water bathes the core of the reactor and is used to generate steam, which turns the turbines that generate electricity. Contrast with liquid metal fast breeder.

Light year   Astronomical unit that measures the distance that light can travel in a year.

Limiting factor   A chemical or physical factor that determines whether an organism can survive in a given ecosystem. In most ecosystems, rainfall is the limiting factor.

Limnetic zone   Open water zone of lakes through which sunlight penetrates; contains algae and other microscopic organisms that feed on dissolved nutrients.

Linearity   Refers to linear thinking and linear systems design.

Linear thinking   A mode of thinking that oversimplifies complex issues. Often ignores complex networks of cause and effect and systems impacts.

Liquefaction   Production of liquid fuel from coal.

Liquid metal fast breeder   Fission reactor that uses liquid metals such as sodium as a coolant.

Lithosphere   Crust of the earth. Contrast with hydrosphere and atmosphere.

Littoral drift   Movement of beach sand parallel to the shoreline. Caused by waves and longshore currents parallel to the beach.

Littoral zone   Shallow waters along a lakeshore where rooted vegetation often grows.

Macronutrient   A chemical substance needed by living organisms in large quantities (for example, carbon, oxygen, hydrogen, and nitrogen). Contrast with micronutrient.

Magma   Molten rock beneath the earth's crust.

Malnourishment   A dietary deficiency caused by lack of vital nutrients and vitamins.

Manganese nodules   Nodular accumulations of manganese and other minerals such as iron and copper found on the ocean floor at depths of 300 to 6000 meters. Particularly abundant in the Pacific Ocean.

Marasmus   A dietary deficiency caused by insufficient intake of protein and calories and occurring primarily in infants under the age of 1, usually as the result of discontinuation of breast feeding.

Mariculture   Cultivation of fish and other aquatic organisms in salt water (estuaries and bays).

Mature community   A community that remains more or less the same over a long period of time. Climax stage of succession. Also called a climax community.

Mature ecosystem   An ecosystem in the climax stage of succession, characterized by high species diversity and high stability. Contrast with immature ecosystem.

Measure of economic welfare   Proposed standard that takes into account the accumulated wealth of a nation.

Megawatt   Measure of electrical power equal to a million watts, or 1000 kilowatts. See watt.

Mesothelioma   A tumor of the lining of the lung (pleura). Caused by asbestos.

Metalimnion   See thermocline.

Metastasis   Movement of cancer cells to another location where new tumors are formed.

Methyl mercury   Water-soluble organic form of mercury formed by bacteria in aquatic ecosystems from inorganic (insoluble) mercury pollution. Able to undergo biological magnification.

Microconsumers   Bacteria and single-celled fungi that are part of the decomposer food chain.

Micronutrient   An element needed by organisms, but only in small quantities, such as copper, iron, and zinc. Contrast with macronutrient.

Migration   Movement of people across state and national boundaries to set up new residence. See immigration, emigration, in-migration, and out-migration.

Mill tailings   Residue from uranium processing plants. Spent uranium ore that is contaminated with radioactivity.

Mineral   A chemical element (e.g., gold) or inorganic compound (e.g., iron ore) existing naturally.

Minimum tillage   Reduced plowing and cultivating of cropland between and during growing seasons to help reduce soil erosion and save energy. Also called conservation tillage.

Molecule   Particle consisting of two or more atoms bonded together. The atoms in a molecule can be of the same element but are usually of different elements.

Monoculture   Cultivation of one plant species (such as wheat) over a large area. Highly susceptible to disease and insects.

Municipal solid waste   Refers to garbage from homes and businesses in cities and towns. Typically contains high percentage of recyclable and compostable materials.

Mutagen   A chemical or physical agent capable of damaging the genetic material (DNA and chromosomes) of living organisms in both germ cells and somatic cells.

Mutation   In general, any damage to the DNA and chromosomes.

Mutualism   Relationship between two organisms that is beneficial to both.

Narrow-spectrum pesticide   A chemical agent effective in controlling a small number of pests.

Natural erosion   Loss of soil occurring at a slow rate but not caused by human activities. A natural event in all terrestrial ecosystems.

Natural eutrophication   See eutrophication.

Natural gas   Gaseous fuel containing 50%-90% methane and lesser amounts of other burnable organic gases such as propane and butane.

Natural hazards   Dangers that result from normal meteorologic, atmospheric, oceanic, biological, and geological phenomena.

Natural resource   See resource.

Natural selection   Process in which slight variations in organisms (adaptations) are preserved if they are useful and help the organism to better respond to its environment.

Negative feedback   Control mechanism present in the ecosystem and in all organisms. Information in the form of chemical, physical, and biological agents influences processes, causing them to shut down or reduce their activity.

Net energy   See net useful energy production.

Net migration   Number of immigrants minus the number of emigrants. Can be expressed as a rate by determining immigration and emigration rates.

Net primary productivity Gross primary productivity (the total amount of energy that plants produce) minus the energy plants use during cellular respiration.

Net useful energy production Amount of useful energy extracted from an energy system.

Neutralism A relationship without ties.

Niche Also called an ecological niche. An organism's place in the ecosystem: where it lives, what it consumes, what consumes it, and how it interacts with all biotic and abiotic factors.

Nitrate (NO_3^-) Inorganic anion containing three oxygen atoms and one nitrogen atom linked by covalent bonds.

Nitrite (NO_2^-) Inorganic anion containing two oxygen atoms and one nitrogen atom. Combines with hemoglobin and may cause serious health impairment and death in children.

Nitrogen cycle The cycling of nitrogen between organisms and the environment.

Nitrogen fixation Conversion of atmospheric nitrogen (a gas) into nitrate and ammonium ions (inorganic form), which can be used by plants.

Nitrogen oxides Nitric oxide (NO) and nitrogen dioxide (NO2), produced during combustion when atmospheric nitrogen (N2) combines with oxygen. Can be converted into nitric acid (HNO3). All are harmful to humans and other organisms.

Noise An unwanted or unpleasant sound.

Nonattainment area Region that violates EPA air pollution standards.

Nonpoint source (of pollution) Diffuse source of pollution such as an eroding field, urban and suburban lands, and forests. Contrast with point source.

Nonrenewable resource Resource that is not replaced or regenerated naturally within a reasonable period (fossil fuel, mineral).

Nuclear fall Hypothesis suggesting that the effects on the Earth's climate of dust and smoke released in nuclear explosions would be more temporary and less severe than predicted by the nuclear winter hypothesis. Contrast with nuclear winter.

Nuclear fission Splitting of an atomic nucleus when neutrons strike the nucleus. Products are two or more smaller nuclei, neutrons (which can cause further fission reactions), and an enormous amount of heat and radiation energy.

Nuclear fusion Joining of two small atomic nuclei (such as hydrogen and deuterium) to form a new and larger nucleus (such as helium) accompanied by an enormous release of energy. Source of light and heat from the sun.

Nuclear power (or energy) Energy from the fission or fusion of atomic nuclei.

Nuclear winter Hypothesis suggesting that dust from nuclear explosions and smoke from burning cities would reduce solar radiation, resulting in a dramatic decrease in global temperature. Contrast with nuclear fall.

Nutrient cycle Same as biogeochemical cycle.

Off-road vehicle (ORV) Any vehicle used cross-country, especially in a recreational capacity (four-wheel-drive vehicles, dune buggies, all-terrain vehicles, snowmobiles, and trail bikes).

Oil See petroleum.

Oil shale A fine-grained sedimentary rock called marlstone and containing an organic substance known as kerogen. When heated, it gives off shale oil, which is much like crude oil.

Old growth forest Ancient forests with trees often 150 to 1000 or more years old. Also called primary forest.

Omnivore An organism that eats both plants and animals.

Open system A system that freely exchanges energy and matter with the environment. Example: any living organism. Contrast with closed system.

Opportunity costs Costs of lost money-making opportunities (and potentially higher income) incurred when we make a decision to invest our money in a particular way.

Ore Rock bearing important minerals, for example, uranium ore.

Ore deposit A valuable mineral located in high concentration in a given region.

Organic farming Agricultural system in which natural fertilizers (manure and crop residues), crop rotation, contour planting, biological insect control measures, and other techniques are used to ensure soil fertility, erosion control, and pest control.

Organic fertilizer Material such as plant and animal wastes added to cropland and pastures to improve soil. Provides valuable soil nutrients and increases the organic content of soil (thus increasing moisture content).

Organismic phase The part of the nutrient cycle in which nutrients are located in organisms: plants, animals, bacteria, fungi, or others.

Out-migration Movement of people out of a state or region within a country to set up residence elsewhere in that country.

Output approach A method of solving an environmental problem by controlling the outputs. For example, composting or burning trash reduces the land requirements for solid waste disposal. Control devices reduce air and water pollution.

Overgrazing Excessive consumption of producer organisms (plants) by grazers such as deer, rabbits, and domestic livestock. Indication that the ecosystem is out of balance.

Overpopulation A condition resulting when the number of organisms in an ecosystem exceeds its ability to assimilate wastes and provide resources. Creates physical and mental stress on a species as a result of competition for limited resources and deterioration of the environment.

Overshoot Refers to the phenomenon occurring when a population of organisms exceeds the carrying capacity of its environment.

Oxidants Oxidizing chemicals (for example, ozone) found in the atmosphere.

Oxygen-demanding wastes   Organic wastes that are broken down in water by aerobic bacteria. Aerobic breakdown causes the oxygen levels to drop.

Ozone (O3)   Inorganic molecule found in the atmosphere, where it is a pollutant because of its harmful effects on living tissue and rubber. Also found in the stratosphere, where it helps screen ultraviolet light. Used in some advanced sewage treatment plants.

Ozone layer   Thin layer of ozone molecules in the stratosphere. Absorbs ultraviolet light and converts it to infrared radiation. Effectively screens out 99% of the ultraviolet light.

Paradigm   A major theoretical construct that is central to a field of study. For example, the theory of evolution and the structure of DNA are two paradigms that are central to biological science.

Parasitism   Relationship in which one species lives in or on another, its host.

Particulates   Solid particles (dust, pollen, soot) or water droplets in the atmosphere.

Passive solar   Capture and retention of the sun's energy within a building through south-facing windows and some form of heat storage in the building (brick or cement floors and walls). Contrast with active solar.

PCBs   See polychlorinated biphenyls.

Perennial   A plant that grows from the same root structure year after year (for example, rose bushes).

Permafrost   Permanently frozen ground found in the tundra.

Permanent threshold shift   Loss of hearing after continued exposure to noise. Contrast with temporary threshold shift.

Pesticide   A general term referring to a chemical, physical, or biological agent that kills organisms we classify as pests, such as insects and rodents. Also called biocide.

Petroleum   A viscous liquid containing numerous burnable hydrocarbons. Distilled into a variety of useful fuels (fuel oil, gasoline, and diesel) and petrochemicals (chemicals that can be used as a chemical feedstock for the production of drugs, plastics, and other substances). pH Measure of acidity on a scale from 0 to 14, with pH 7 being neutral, numbers greater than 7 being basic, and numbers less than 7 being acidic.

Pheromone   Chemical substance given off by insects and other species. Sex-attractant pheromones released into the atmosphere in small quantity by female insects attract males at breeding time. Can be used in pest control. See pheromone traps and confusion technique.

Pheromone traps   Traps containing pheromones to attract insect pests and pesticide to kill pests or a sticky substance to immobilize them. These traps may be used to pinpoint the emergence of insects, allowing conventional pesticides to be used in moderation.

Photochemical oxidants   Ozone and a variety of oxygenated organic compounds produced when sunlight, hydrocarbons, and nitrogen oxides react in the atmosphere.

Photochemical reaction   A chemical reaction that occurs in the atmosphere involving sunlight or heat, pollutants, and sometimes natural atmospheric chemicals.

Photochemical smog   A complex mixture of photochemical oxidants and nitrogen oxides. Usually has a brownish-orange color.

Photosynthesis   A two-part process in plants and algae involving (1) the capture of sunlight and its conversion into cellular energy and (2) the production of organic molecules such as glucose and amino acids from carbon dioxide, water, and energy from the sun.

Photovoltaic cell   Thin wafer of silicon or other material that emits electrons when struck by sunlight, thus generating an electrical current. Also solar cell.

Phytoplankton   Includes single-celled algae and other free-floating photosynthetic organisms.

Pioneer community   The first community to become established in a once-lifeless environment during primary succession.

Pitch (or frequency)   Measure of the frequency of a sound in cycles per second (cps) (hertz, Hz)—compressional sound waves passing a given point per second. The higher the cps, the higher the pitch.

Pneumoconiosis (black lung)   A debilitating lung disease caused by prolonged inhalation of coal and other mineral dusts. Results in a decreased elasticity and gradual breakdown of alveoli in the lungs. Eventually leads to death.

Point source (of pollution)   Easily discernible source of pollution such as a factory. Contrast with nonpoint source.

Pollution   Any physical, chemical, or biological alteration of air, water, or land that is harmful to living organisms.

Pollution prevention   Any one of several methods to reduce pollution production, such as process modification and substitution.

Polychlorinated biphenyls (PCBs)   Group of at least 50 organic compounds, used for many years as insulation in electrical equipment. Capable of biological magnification. Disrupts reproduction in gulls and possibly other organisms high on the food chain.

Population   A group of organisms of the same species living within a specified region.

Population control   In human populations, all methods of reducing birth rate, primarily through pregnancy prevention and abortion. In an ecological sense, regulation of population size by a myriad of abiotic and biotic factors.

Population crash (dieback)   Sudden decrease in population that results when an organism exceeds the carrying capacity of its environment.

Population growth rate   Rate at which a population increases on a yearly basis, expressed as a percentage. For world population: GR = (crude birth rate - crude death rate) x 100. For a given country, population growth rate must also take into account the net migration rate.

Population histogram   Graphical representation of population by age and sex.

Positive feedback   Control mechanism in ecosystems and organisms in which information influences some process, causing it to increase.

Potential energy   Stored energy.

Predator   An organism that actively hunts its prey.

Presbycusis   Loss of hearing with age through natural deterioration of the organ of Corti, the sound receptor in the ear. Contrast with sociocusis.

Prey   Organism (e.g., deer) attacked and killed by predator.

Primary air pollutant   A pollutant that has not undergone any chemical transformation; emitted by either a natural or an anthropogenic source.

Primary consumer   First consuming organism in a given food chain. A grazer in grazer food chains or a decomposer organism or insect in decomposer food chains. Belongs to the second trophic level.

Primary forest   See old-growth forest.

Primary pollutant   Pollutant produced by combustion or other sources. Can be chemically modified after release, creating a secondary pollutant.

Primary succession   The sequential development of biotic communities where none previously existed.

Primary treatment (of sewage)   First step in sewage treatment to remove large solid objects by screens (filters) and sediment and organic matter in settling chambers. See secondary and tertiary treatment.

Proactive government   One that is concerned with long-range problems and lasting solutions. Contrast with reactive government.

Producer (autotroph or producer organism)   One of the organisms that produces the organic matter cycling through the ecosystem. Producers include plants and photosynthetic algae.

Productivity   The rate of conversion of sunlight by plants into chemical-bond energy (covalent bonds in organic molecules). See gross and net primary productivity.

Profundal zone   Deeper lake water, into which sunlight does not penetrate. Below the limnetic zone.

Prospective law   One designed to address future problems and generate long-lasting solutions. Contrast with retrospective law.

Punctuated equilibrium   A theory of evolution stating that species are fairly stable for long periods and that new species evolve rapidly over short periods of thousands of years that punctuate the equilibrium. Contrast with gradualism.

Pyramid of biomass   Graphical representation of the amount of biomass (organic matter) at each trophic level in an ecosystem.

Pyramid of numbers   Graphical representation of the number of organisms of different species at each trophic level in an ecosystem.

Quad   One quadrillion (1015) BTUs of heat.

Rad (radiation absorbed dose)   Measure of the amount of energy deposited in a tissue or some other medium struck by radiation. One rad = 100 ergs of energy deposited in one gram of tissue.

Radioactive waste   Any solid or liquid waste material containing radioactivity. Produced by research labs, hospitals, nuclear weapons factories, and fission reactors.

Radioactivity   Radiation released from unstable nuclei. See alpha and beta particles and gamma rays.

Radionuclides   Radioactive forms (isotopes) of elements.

Rain shadow   Arid downwind (leeward) side of mountain range.

Rangeland   Grazing land for cattle, sheep, and other domestic livestock.

Range of tolerance   Range of physical and chemical factors in which an organism can survive. When the upper or lower limits of this range are exceeded, growth, reproduction, and survival are threatened.

Reactive government   A government that lives and acts for today, addressing present-day problems as they arise. Shows little or no concern for long-term issues and solutions. Contrast with proactive government.

Reactor core   Assemblage of fuel rods and control rods inside a reactor vessel. Bathed by water to help control the rate of fission and absorb the heat.

Real price (or cost)   The price of a commodity or service in fixed dollars, that is, the value of a dollar at an earlier time. Helpful way to determine whether a resource has experienced a real increase in cost or whether higher costs are simply due to inflation.

Reclamation   As used here, the process of returning land to its prior use. Common usage: to convert deserts and other areas into habitable, productive land.

Recycling   A strategy to reduce resource use by returning used or waste materials from the consumption phase to the production phase of the economy.

Reduction factors   Abiotic and biotic factors that tend to decrease population growth and help balance populations and ecosystems, offsetting growth factors.

Relative humidity   The amount of moisture in a given quantity of air divided by the amount the air could hold at that temperature. Expressed as a percentage.

REM (roentgen equivalent man)   Measure that accounts for the damage done by a given type of radiation. One rad = one rem for X rays, gamma rays, and beta particles, but one rad = 10 to 20 rems for alpha particles, because they do more damage.

Renewable resource   A resource replaced by natural ecological cycles (water, plants, animals) or natural chemical or physical processes (sunlight, wind).

Replacement-level fertility   Number of children a couple must have to replace themselves in the population.

Reproductive age   Age during which most women bear their offspring (ages 14–44). Reproductive isolation Any of many mechanisms that prevent species from interbreeding or producing viable offspring.

Reserve Deposit of energy or minerals that is economically and geologically feasible to remove with current and foreseeable technology.

Residence time Length of time a chemical spends in the environment.

Resilience Ability of an ecosystem to return to normal after a disturbance.

Resource (in general) Anything used by organisms to meet their needs, including air, water, minerals, plants, fuels, and animals.

Resource (as a measurement of a mineral or fuel) Total amount of a mineral or fuel on earth. Generally, only a small fraction can be recovered. Compare with reserve.

Restoration ecology Study of restoring ecosystems to their natural state after human interference. Also called conservation biology.

Retorting Process of removing kerogen from oil shale, usually by burning or heating the shale. Can be carried out in surface vessels (surface retorting) or underground in fractured shale (in situ retorting).

Retrospective law One that attempts to solve a problem without giving much attention to potential future problems. Contrast with prospective law.

Reverse osmosis Means of purifying water for pollution control and desalination. Water is forced through porous membranes; pores allow passage of water molecules but not impurities.

Risk acceptability A measure of how acceptable a hazard is to a population.

Risk assessment The science of determining what hazards a society is exposed to from natural and human causes and the probability and severity of those risks.

Risk probability The likelihood a hazardous event will occur.

Risk severity A measure of the total damage a hazardous event would cause.

Salinization Deposition of salts in irrigated soils, making soil unfit for most crops. Caused by rising water table due to inadequate drainage of irrigated soils.

Saltwater intrusion Movement of saltwater from oceans or saltwater aquifers into freshwater aquifers, caused by depletion of the freshwater aquifers or low precipitation or both.

Sanitary landfill Solid waste disposal site where garbage is dumped and covered daily with a layer of dirt to reduce odors, insects, and rats.

Scrubber Pollution control device that removes particulates and sulfur oxides from smokestacks by passing exhaust gases through a fine spray of water containing lime.

Secondary consumer Second consuming organism in food chain. Belongs to the third trophic level.

Secondary pollutant A chemical pollutant from a natural or anthropogenic source that undergoes chemical change as a result of reacting with another pollutant, sunlight, atmospheric moisture, or some other environmental agent.

Secondary succession The sequential development of biotic communities occurring after the complete or partial destruction of an existing community by natural or anthropogenic forces.

Secondary treatment (of sewage) After primary treatment, removal of biodegradable organic matter from sewage using bacteria and other microconsumers in activated sludge or trickle filters. Also removes some of the phosphorus (30%) and nitrate (50%). See also tertiary treatment.

Second-generation pesticides Synthetic organic chemicals such as DDT that replaced older pesticides such as sulfur, ground tobacco, and ashes. Generally resistant to bacterial breakdown.

Second-law efficiency Measure of the efficiency of energy use, taking into account the unavoidable loss (described by the second law of thermodynamics) of energy during energy conversions. Calculated by dividing the minimum amount of energy required to perform a task by the actual amount used.

Second law of thermodynamics States that when energy is converted from one form to another, it is degraded; that is, it is converted from a concentrated to a less concentrated form. The amount of useful energy decreases during such conversions.

Secured landfill One lined by clay and synthetic liners in an effort to prevent leakage.

Sediment Soil particles, sand, and other mineral matter eroded from land and carried in surface waters.

Selective advantage An advantage one member of a species has over others by virtue of some adaptation it has acquired.

Selective cutting Restricted removal of trees. Especially useful for mixed hardwood stands. Contrast with clearcutting and shelter-wood cutting.

Sewage treatment plant Facility where human solid and liquid wastes from homes, hospitals, and industries are treated, primarily to remove organic matter, nitrates, and phosphates.

Shale oil Thick, heavy oil formed when shale is heated (retorted). Can be refined to produce fuel oil, kerosene, diesel fuel, and other petroleum products and petrochemicals.

Shelterbelts Rows of trees and shrubs planted alongside fields to reduce wind erosion and retain snow to increase soil moisture. May also be used to reduce heat loss from wind and thus conserve energy around homes and farms.

Shelter-wood cutting Three-step process spread out over years: (1) removal of poor-quality trees to improve growth of commercially valuable trees and allow new seedlings to become established, (2) removal of commercially valuable trees once seedlings are established, and (3) cutting remaining mature trees grown from seedlings.

Sigmoidal curve An S-shaped curve.

Simplified ecosystem One with lowered species' diversity, usually as a result of human intervention.

Sinkhole Hole created by sudden collapse of the earth's

surface due to groundwater overdraft. A form of subsidence.

Slash-and-burn agriculture Farming practice in which small plots are cleared of vegetation by cutting and burning. Crops are grown until the soil is depleted; then the land is abandoned. This allows the natural vegetation and soil to recover. Common practice of early agricultural societies living in the tropics.

Sludge Solid organic material produced during sewage treatment.

Smelter A factory where ores are melted to separate impurities from the valuable minerals.

Smog Originally referred to a grayish haze (combination of smoke and fog) found in industrial cities. Also pertains to pollution called photochemical smog, found in newer cities. See industrial smog. Social Darwinism The application (or misapplication) of the theory of evolution to social behavior.

Sociocusis Hearing loss from human activities. Contrast with presbycusis.

Soft path A term coined by Amory Lovins to describe such practices as conservation, efficient use of energy, and renewable energy systems such as solar and wind. Characterized by high labor intensity, decentralized energy production, and small-scale technology. Contrast with hard path.

Soil horizons Layers found in most soils.

Solar collector Device to absorb sunlight and convert it into heat.

Solar energy Energy derived from the sun (heat) and natural phenomena driven by the sun (wind, biomass, running water).

Solar system Group of planets revolving around a star.

Sonic boom A high-energy wake creating an explosive boom that trails after jets traveling faster than the speed of sound.

Spaceship earth Metaphor introduced in the 1960s to foster a greater appreciation of the finite nature of earth's resources and the ecological cycles that replenish oxygen and other important nutrients.

Specialist Organism that has a narrow niche, usually feeding on one or a few food materials and adapted to a particular habitat.

Speciation Formation of new species.

Species A group of plants, animals, or microorganisms that have a high degree of similarity and generally can interbreed only among themselves.

Species diversity Measure of the number of different species in a biological community.

Spontaneous abortion Loss of an embryo or fetus from the uterus not caused by surgery. Generally the result of chromosomal abnormalities. Contrast with induced abortion.

Spring overturn Annual cycle in deep lakes in temperate climates in which surface and subsurface waters mix.

SST (supersonic transport) Jet that travels faster than the speed of sound.

Stable runoff Amount of surface runoff that can be counted on from year to year.

Star Spherical cloud of hot gas, such as the sun, fueled by nuclear fusion reactions in its core.

Statutory law Law enacted by Congress or a state legislature. Contrast with common law.

Sterile-male technique Pest control strategy whereby males of the pest species are grown in captivity, sterilized, then released en masse in infested areas at breeding time. Sterile males far exceed normal wild males and mate with normal females, resulting in infertile matings and control of the pest.

Sterilization A highly successful procedure in males and females to prevent pregnancy. In males the ducts (vas deferens) that carry sperm from the testicles are cut and tied (vasectomy); in females the Fallopian tubes, or oviducts, which transport ova from the ovary to the uterus, are cut and tied (tubal ligation). Sterilization is not to be confused with castration in males (complete removal of the gonads).

Stratosphere Outer region of the earth's atmosphere, found outside the troposphere, extending 7 to 25 miles above the earth's surface. Outermost layer of the stratosphere contains the ozone layer.

Streambed aggradation Deposition of sediment in streams or rivers, thereby reducing their water-carrying capacity.

Streambed channelization An ecologically unsound way of reducing flooding by deepening and straightening of streams, accompanied by removal of trees and other vegetation along the banks.

Strip cropping Soil conservation technique in which alternating crop varieties are planted in huge strips across fields to reduce wind and water erosion of soil.

Subsidence Sinking of land caused by collapse of underground mines or depletion of groundwater.

Succession The natural replacement of one biotic community by another. See primary and secondary succession.

Sulfur dioxide Colorless gas produced during combustion of fossil fuels contaminated with organic and inorganic sulfur compounds. Can be converted into sulfuric acid in the atmosphere.

Sulfur oxides Sulfur dioxide and sulfur trioxide, common air pollutants arising from combustion of coal, oil, gasoline, and diesel fuel. Also produced by natural sources such as bacterial decay and hot springs. Sulfur dioxide reacts with oxygen to form sulfur trioxide, which may react with water to form sulfuric acid.

Supply and demand theory Also known as the law of supply and demand. Economic theory explaining the price of goods and services. The supply of and demand for goods and services are primary price determinants. High demand diminishes supply, creating competition for existing goods and services, thus driving up prices.

Surface mining Any of several mining techniques in

which all the dirt and rock overlying a desirable mineral (coal, for example) are first removed, exposing the mineral.

Surface runoff   Water flowing in streams and over the ground during rainstorm or snowmelt.

Sustainable development   Economic development that meets current needs without compromising ability of future generations to meet their needs. Relies on appropriate technology, efficient use of resources, recycling, renewable resource use, restoration, growth management, and other measures.

Sustainable economics   Economic system that seeks to meet human needs while protecting the life-support systems of the biosphere.

Sustainable ethics (mentality)   A mind-set that views humans as a part of nature and earth as a limited supply of resources, which must be carefully managed to prevent irreparable damage. Obligations to future generations require us to exercise restraint to ensure adequate resources and a clean and healthy environment.

Sustainable society   One based on sustainable ethics. Lives within the limits imposed by nature. Based on maximum use of renewable resources, recycling, conservation, and population control.

Sustained yield concept   Use of renewable natural resources, such as forests and grassland, that will not cause their destruction and will ensure continued use.

Sympatric speciation   Formation of new species without geographical isolation. Common in plants.

Synergism   Phenomenon occurring when two or more agents (often toxins) together to produce an effect larger than expected based on knowledge of the effect of each alone. Sometimes simply refers to two or more components acting together.

Synfuel   See synthetic fuel.

Synthetic fertilizer   Same as inorganic fertilizer.

Synthetic fuel   Gaseous or liquid organic fuel derived from coal, oil shale, or tar sands.

Systems thinking   Mode of thinking that takes into account complex webs of cause and effect. Contrast with linear thinking.

Taiga   Biome found south of the tundra across North America, Europe, and Asia, characterized by coniferous forests, soil that thaws during the summer months, abundant precipitation, and high species diversity.

Tar sands   Also known as oil sands or bituminous sands. Sand impregnated with a viscous, petroleumlike substance, bitumen, which can be driven off by heat, producing a synthetic oil.

Technological fix   A purely technological answer to a problem. Also called a technical fix.

Technological optimism   Undying faith in technological fixes.

Tectonic plates   Huge segments of the earth's crust that often contain entire continents or parts of them and that float on an underlying semiliquid layer.

Temperate deciduous forest   Biome located in the eastern United States, Europe, and northeastern China below the taiga. Characterized by deciduous and nondeciduous trees, warm growing season, abundant rainfall, and a rich species diversity.

Temperature inversion   Alteration in the normal atmospheric temperature profile so that air temperature increases with altitude rather than decreases.

Temporary threshold shift   Momentary dulling of the sense of hearing after exposure to loud sounds. Can lead to permanent threshold shift.

Teratogen   A chemical or physical agent capable of causing birth defects.

Terracing   Construction of small earthen embankments on hilly or mountainous terrain to reduce the velocity of water flowing across the soil and thus reduce soil erosion.

Tertiary treatment (of sewage)   Removal of nitrates, phosphates, chlorinated compounds, salts, acids, metals, and toxic organics after secondary treatment.

Thermal pollution   Heat added to air or water that adversely affects living organisms and may alter climate.

Thermocline   Sharp transition between upper, warm waters (epilimnion) and deeper, cold waters (hypolimnion) of a lake. Also called metalimnion.

Thermodynamics   The study of energy conversions. See the first and second laws of thermodynamics.

Third-generation pesticides   Newer chemical agents to control pests, such as pheromones and insect hormones.

Threatened species   A species whose population is declining and could become extinct.

Threshold level   A level of exposure below which no effect is observed or measured.

Throughput approach   A method of solving an environmental problem by recycling and reuse. For example, recycling or reusing hazardous wastes reduces their output.

Time preference   A measure of the value of an immediate gain in comparison with a long-term gain.

Tolerance level (for pesticides)   Level of residue on fruits and vegetables permitted by EPA because it is considered "safe."

Total fertility rate   Average number of children that would be born alive to a woman if she were to pass through all her childbearing years conforming to the age-specific fertility rates.

Toxin   A chemical, physical, or biological agent that causes disease or some alteration of the normal structure and function of an organism. Impairments may be slight or severe. Onset of effects may be immediate or delayed.

Tradable permit   Permit issued by government agency that allows companies to release certain amounts of pollution. Companies that find ways to reduce pollution can sell permits to other companies.

Transpiration   Escape of water from plants through pores (stomata) in the leaves.

Tree farms Private forests devoted to maximum timber growth and relying heavily on herbicides, insecticides, and fertilizers.

Tritium (hydrogen-3) Radioactive isotope of hydrogen whose nucleus contains two neutrons and one proton. Can be used in fusion reactors.

Trophic level Describes the position of the organism in the food chain.

Tropical rain forest Lush forests near the equator with high annual rainfall, high average temperature, and notoriously nutrient-poor soil. Possibly the richest ecosystem on earth.

Tundra (alpine) Life zone found on mountaintops. Closely resembles the arctic tundra in terms of precipitation, temperature, growing season, plants, and animals. Extraordinarily fragile.

Tundra (arctic) First major life zone or biome south of the North Pole. Vast region on far northern borders of North America, Europe, and Asia. Characterized by lack of trees, low precipitation, and low temperatures.

Ultimate production Total amount of a nonrenewable resource that could ultimately be extracted at a reasonable price.

Ultraviolet (UV) light or radiation Electromagnetic radiation from sun and special lamps. Causes sunburn and mutations in bacteria and other living cells.

Undernourishment A lack of calories in the diet. Contrast with malnourishment.

Variation Genetically based differences in behavior, structure, or function in a population.

Waste-to-energy plant Incinerator for rubbish that produces small amounts of electricity from heat given off by combustion.

Water cycle See hydrological cycle.

Waterlogging High water table causing saturation of soils due to poor soil drainage and irrigation. Decreases soil oxygen and kills plants.

Watershed Land area drained by a given stream or river.

Water table Top of the zone of saturation.

Watt Unit of power indicating rate at which electrical work is being performed.

Wave power Energy derived from sea waves.

Wet cooling tower Device used for cooling water from power plants. Hot water flows through rising air, which draws off heat. Cool water is then returned to the system.

Wetlands Land areas along fresh water (inland wetlands) and salt water (coastal wetlands) that are flooded all or part of the time.

Wilderness An area where the biological community is relatively undisturbed by humans. Seen by developers as an untapped supply of resources such as timber and minerals, seen by environmentalists as a haven from hectic urban life, an area for reflection and solitude.

Wilderness area An area established by the US Congress under the Wilderness Act (1964) where timber cutting and use of motorized vehicles are prohibited. Most are located in national forests.

Wildlife corridor Protected piece of land that connects protected habitat allowing animals to migrate.

Wind energy Energy captured from the wind to generate electricity or pump water. An indirect form of solar energy.

Wind generators Windmills that produce electrical energy.

Zero population growth A condition in which population is not increasing; the population growth rate is zero.

Zone of intolerance Range of environmental conditions that an organism cannot survive in.

Zone of physiological stress Upper and lower limits of range of tolerance where organisms have difficulty surviving.

Zooplankton Nonphotosynthetic, single-celled aquatic organisms.

Text, Photo, and Illustration Credits

Front Matter

Text, pp. x-xii: Reprinted with permission from Human Biology: Health, Homeostasis, and the Environment by Daniel D. Chiras; Copyright © 1991 by West Publishing Company. All rights reserved.

Chapter 1 Figures

1.1: Courtesy of Martin-Marieta, Inc.

1.2: Reprinted with permission of Macmillan Publishing Company from Natural Resource Conservation: An Ecological Approach, Fifth Edition by Oliver S. Owen and Daniel D. Chiras. Copyright © 1990 by Macmillan Publishing Co.

1.3: Courtesy of Native Forest Council

1.4: © Lois and George Cox/Bruce Coleman, Inc.

1.6: Adapted with permission from Daniel D. Chiras, Lessons in Nature: Learning to Live Sustainably on the Earth, fig. 1, p. 14, © 1992 by Island Press. (Washington, D.C. & Covelo, CA: Island Press, 1992.)

Chapter Supplement Table

S1.2: Adapted from Donella Meadows, The Global Citizen, Island Press, 1991 and Daniel D. Chiras, Beyond the Fray: Reshaping America's Environmental Response, © 1990 Johnson Books.

Chapter 2 Figures

2.1: (CS) © Valerie Hodgson/Visuals Unlimited

2.2: (S) © Jerry J. Hout/Bruce Coleman, Inc.

2.3: (S) © Len Rue Jr./VU

2.5a: National Parks and Wildlife Service, Sydney, Australia

2.5b: W. Perry Conway

2.9: Modified and redrawn from R.L. Smith, Ecology and Field Biology, Second Edition, Harper & Row, New York, 1974.

2.11: Modified and redrawn from B.J. Nebel, Environmental Science, Prentice-Hall, Englewood Cliffs, NJ, 1981.

2.14: From Fundamentals of Ecology, Third Edition by Eugene P. Odum, © 1971 by W.B. Saunders Publishing, a division of Holt Rinehart Winston. Reprinted by permission of the publisher.

Table

2.1: From Campbell, Biology, 2nd Edition, p. 1120, Table 49.1, © 1990 Benjamin/Cummings Publishing Company. Data: Whittaker, Communities and Ecosystems, 2e, Macmillan, 1975.

Chapter 3 Figures

3.1: © Hans Reinhard/Bruce Coleman, Inc.

3.5: From Simpson, G.G., "Species Density of North American Recent Mammals." Systematic Zoology vol. 13, pp. 57–73.

3.10: Johnson, Rayle, Wedberg, Biology: An Introduction (Redwood City, CA: Benjamin/Cummings Publishing Company, 1984). Copyright © 1984 by Benjamin/Cummings Publishing Co.

3.11: Johnson, Rayle, Wedberg, Biology: An Introduction (Redwood City, CA: Benjamin/Cummings Publishing Company, 1984). Copyright © 1984 by Benjamin/Cummings Publishing Co.

3.12: Adapted from print of Darwin's finches from American Museum of Natural History.

3.14: Ron Willcocks/Aerie Nature Series

3.15: W. Perry Conway

Table

3.1: From Odum, E., "The Strategy of Ecosystem Development," Table 1, SCIENCE 164: 262–270, Copyright © 1969 by the AAAS.

Chapter 4 Figures

4.1: Milton Tierney Jr./VU

4.2a: © Alan G. Nelson/Animals Animals

4.2b: © Gary Milburn/Tom Stack & Associates

4.2c: © M. Austerman/Animals Animals

4.3: Reprinted with permission from Human Biology: Health, Homeostasis, and the Environment by Daniel D. Chiras; Copyright © 1991 by West Publishing Company. All rights reserved.

4.4a: © Marjorie Shostak/Anthro Photo

4.4b: © Marjorie Shostak/Anthro Photo

4.6: Jean Whitney/Atoz Images

4.7: Carnegie Photographic Library of Pittsburgh

4.8: USDA/Soil Conservation Service

Chapter 5 Figures

5.1: © Frank D. Smith/Jeroboam, Inc.

5.2: UNICEF/Maggie Murray-Lee

5.8: © Ted Soqui

5.9: Modified and redrawn from Jeanne C. Bigarre, 1979, "The Sunning of America: Migration to the Sunbelt," Population Bulletin v. 34(1).

5.11: Kent Reno/Jeroboam, Inc.

Tables

5.2: Population Reference Bureau. World Population Data Sheet, 1992.

5.3: Population Reference Bureau. World Population Data Sheet, 1989.

5.4: Population Reference Bureau. World Population Data Sheet.

Chapter 6 Figures

6.1: From The Limits to Growth: A Report for the Club of Rome's Project on the Predicament of Mankind, Donella H. Meadows, Dennis L. Meadows, Jorgen Randers, William W. Behrens III, a Potomac Associates book published by Universe Books, 1972. Graphics by Potomac Associates.

6.2: From The Limits to Growth: A Report for the Club of Rome's Project on the Predicament of Mankind, Donella H. Meadows, Dennis L. Meadows, Jorgen Randers, William W. Behrens III, a Potomac Associates book published by Universe Books, 1972. Graphics by Potomac Associates.

6.3: Modified from A. Haupt and T. Kane (1978). Population Handbook. Washington, D.C.: Population Reference Bureau.

6.4: AP/World Wide Photos

6.5: UNICEF/Abigail Heyman

6.7: Graphic: Peter Donaldson & Amy Ong Tsui, "International Family Planning," Population Bulletin 45, no. 3 (Washington, D.C.: Population Reference Bureau, Inc., 1990). Data: UN, Levels & Trends of Contraceptive Use as Assessed in 1988 (New York: United Nations, 1989), p. 33.

Chapter 7 Figures

7.1: FAO/VU

7.2: UNICEF/Horst Max Cerni

7.3: (S) USDA/Soil Conservation Service

7.6: © Wayne Miller/Magnum Photos

7.7: Modified and redrawn with permission from L.R.

Brown (1982). "Fuel Farms: Croplands of the Future?" The Futurist 14(3):16–28. Published by the World Future Society, 4916 St. Elmo Avenue, Bethesda, MD 20814–5089.

7.8: USDA

7.9a: United Nations/Bill Graham

7.9b: USDA/Soil Conservation Service

7.10a: USDA/Soil Conservation Service

7.10b: F. Mattioli/FAO Photo

Models of Global Sustainability

7.1: Adapted from Ming, L., "Fighting China's Sea of Sand," International Wildlife 18(6); 38–45.

Table

7.1: Reichart, W., "Agriculture's Diminishing Diversity," Environment 24(9);6–11, 39–44. Reprinted with permission of the Helen Dwight Reid Educational Foundation. Published by Heldref Publications, 1319 Eighteenth St., N.W. Washington, D.C. 20036–1802. Copyright © 1982.

Chapter 8 Figures

8.1: © Karl Koford/Photo Researchers, Inc.

8.4: © John Lanois/Black Star

8.5: © M.N. Boulton/Bruce Coleman, Inc.

8.7: USFWS/U.C. Davis

8.8: Courtesy of the Tennessee Wildlife Resources Agency

8.9: USDA/Agricultural Resource Agency

8.10: Courtesy of Multimedia Publications, Ltd.

8.11: © David Houston/Bruce Coleman, Inc.

Chapter 9 Figures

9.1: © Eugenio Turrie/Bruce Coleman, Inc.

9.3: USDA/Soil Conservation Service

9.5a: U.S. Forest Service

9.5b: U.S. Forest Service

9.6: © Wayne Lankinen/Bruce Coleman, Inc.

9.8: From Conservation of Natural Resources, Guy Harold Smith. Copyright © 1965 by John Wiley & Sons. Reprinted with permission.

9.9: © Visuals Unlimited

9.9a: U.S. Forest Service

9.9b: U.S. Forest Service

9.9c: Steve Botti, National Park Service

Chapter 10 Figures

10.1: (CS) Florida State Archives

10.5: Modified and redrawn from A.N. Strahler and A.H. Strahler (1973), Environmental Science, Hamilton Publishing Co., Santa Barbara, CA © 1973 by John Wiley & Sons.

10.6: U.S. Geological Survey

10.7: USDA/Soil Conservation Service

10.8: © Novosti/Sovfoto

10.10: Science VU/Visuals Unlimited

Chapter 11 Figures

11.1: (CS) Steve McCutcheon/Alaska Pictorial Service

11.3: (S) © Gilles Peress/Magnum Photos, Inc.

11.4: Adapted from the BP Statistical Review of World Energy.

11.6: From Man's Domain, A Thematic Atlas of the World, Third Edition, Norman J.W. Thrower, ed. © 1968, 1970, 1975 by General Drafting Co., Inc. Published by McGraw-Hill Book Co.

11.7: From Man's Domain, A Thematic Atlas of the World, Third Edition, Norman J.W. Thrower, ed. © 1968, 1970, 1975 by General Drafting Co., Inc. Published by McGraw-Hill Book Co.

11.10b: U.S. Geological Survey

11.15: Tass from Sovfoto

11.16: Energy Strategies: Toward a Solar Future; © 1980 Union of Concerned Scientists. Reprinted with permission from Ballinger Publishing Co.

11.17: Washington Public Power Supply System

Chapter 12 Figures

12.1: (S) Courtesy of San Jose Dept. of Transportation

12.1a: © Pamela Freund/Solar Survival Architecture

12.1b: © Pamela Freund/Solar Survival Architecture

12.2: Courtesy of Osram Corp.

12.7: Courtesy of Solar Station

12.9: © Emilio A. Mercado/Jeroboam, Inc.

12.10: U.S. Forest Service

12.12: From Kendall, H.W. and Nadis, S.J. Energy Strategies: Toward A Solar Future. © 1980 Union of Concerned Scientists. Reprinted with permission from Ballinger Publishing Company.

12.13: Data: Worldwatch Institute.

Tables

12.1: Industrial country data: BP Statistical Review of World Energy, London, 1991 and Organisation of Economic Cooperation and Development, In Figures, Paris, 1991. Developing country data: United Nations, Yearbook of Energy Statistics, New York, 1991 and Central Intelligence Agency, Handbook of Economic Statistics, Washington, D.C., 1990.

12.3: Lester Brown et al., Saving the Planet: How to Shape an Environmentally Sustainable Global Economy, 1991, W.W. Norton, T3.2.

12.4: Public Citizen Critical Mass Energy Product and Worldwatch Institute.

12.5: From Kendall, H.W. and Nadis, S.J. Energy Strategies: Toward a Solar Future. © 1980 Union of Concerned Scientists. Reprinted with permission from Ballinger Publishing Company.

12.6: From Kendall, H.W. and Nadis, S.J. Energy Strategies: Toward a Solar Future. © 1980 Union of Concerned Scientists. Reprinted with permission from Ballinger Publishing Company.

Chapter Supplement Table

S12.1: Worldwatch Institute, Paper 63, Jan. 1985, p. 27

Chapter 13 Figures

13.2: From Man's Domain, A Thematic Atlas of the World, Third Edition, Norman J.W. Thrower, ed. © 1968, 1970, 1975 by General Drafting Co., Inc. Published by McGraw-Hill Book Co.

13.8: NOAA

Table

13.1: Data Sources: Manganese, Thomas S. Jones, U.S. Bureau of Mines, Oct. 15, 1991; Chromium, John F. Papp, USBM, Oct. 15, 1991; USBM Mineral Commodity Summaries 1991 Nonmetals, Worldwatch Institute estimates based on USBM Mineral Commodity Summaries.

Chapter 14 Figures

14.1: From D.L. Davis, B.H. Magee, "Cancer and Industrial Chemical Production," fig. 14.1, SCIENCE 206:1356–1358, Copyright © 1979 by the AAAS.

14.2: Adapted from Casarett and Doull, Toxicology: The Basic Science of Poisons, 2e, ed. J. Doull, M.D., C.D. Klaassen, Ph.D., M.O. Amdur, Ph.D. Macmillan Publishing Co., 1980.

14.4: Reprinted with permission from p. 520 of Human Biology: Health, Homeostasis, and the Environment by Daniel D. Chiras; Copyright © 1991 by West Publishing Co. All rights reserved.

14.7: Adapted from Casarett and Doull, Toxicology: The Basic Science of Poisons, 2e, ed. J. Doull, M.D., C.D. Klaassen, Ph.D., M.O. Amdur, Ph.D. Macmillan Publishing Co., 1980.

14.11: Bruce Larson/Tacoma News Tribune

Chapter Supplement Figure

S14.2: National Air Quality and Emissions Trends Report, 1982, US Environmental Protection Agency.

Table

14.2: Council on Environmental Quality.

Chapter 15 Figures

15.3a: David C. McElroy/FPG

15.3b: Dave Baird/Tom Stack & Assoc.

15.8: Tom McHugh/Photo Researchers, Inc.

Chapter 15 Supplement Figures

S15.1: From T.D. Sterling and E. Sterling (1979) "Carbon Monoxide Levels in Kitchens and Homes with Gas Cookers," Journal of Air Pollution Control Association, 29(3):238–241. Reprinted with permission.

S15.2: From T.D. Sterling and E. Sterling (1979) "Carbon Monoxide Levels in Kitchens and Homes with Gas Cookers," Journal of Air Pollution Control Association, 29(3):238–241. Reprinted with permission.

S15.3: From A Citizen's Guide to Radon, Second Edition, U.S. Environmental Protection Agency, May 1992.

Case Study

15.1: From Lewin, R., "Parkinson's Disease: An Environmental Cause?" SCIENCE 229:257–58, Copyright © 1985 by the AAAS.

Models of Global Sustainability

15.1: Adapted from Curtis A. Moore, "Down Germany's Road to a Clean Tomorrow," International Wildlife 22(5):24–28.

Chapter 16 Figures

16.1: (S) © AP Wide World Photos

16.8: Brezonik et al., "Acid Precipitation and Sulfate Deposition in Florida," SCIENCE 203, May 30, 1980. Copyright © 1980 by the AAAS.

16.10: © Ted Spiegel/Black Star

16.13: Data Sources: J. Hansen and S. Lebedeff, "Global Surface Air Temperatures: Update through 1987," Geophysical Research Letters, Vol. 15, No. 4, 1988; Helene Wilson, NASA Goddard Institute for Space Studies, New York, March 23, 1992.

Tables

16.2: Worldwatch Institute, based on sources including U.S. Environmental Protection Agency, Policy Options for Stabilizing Global Climate Change, 1989 draft; V. Ramanathan et al., "Trace Gas Trends and Their Potential Role in Climate Change," Journal of Geophysical Research, June 20, 1985; James Hansen et al., "Greenhouse Effect of Chlorofluorocarbons and Other Trace Gases," Journal of Geophysical Research, in press.

16.3: Lester Brown et al., Saving the Planet: How to Shape an Environmentally Sustainable Global Economy, 1991, Norton, table 4.1.

Chapter Supplement Tables

S16.1: From P. Goodwin, (1981), Nuclear War: The Facts On Our Survival. New York: Rutledge Press. pp. 28–29. Reprinted with permission.

S16.2: From P. Goodwin, (1981), Nuclear War: The Facts On Our Survival. New York: Rutledge Press. pp. 44–45. Reprinted with permission.

Chapter 17 Figures

17.1a: © John Hendry Jr./Photo Researchers, Inc.

17.1b: © Image Works

17.2: U.S. Environmental Protection Agency, 1990.

17.3: U.S. Environmental Protection Agency.

17.10: © David Rinehart/Greenpeace

17.11: The Des Moines Register

17.13: Bureau of Reclamation

Tables

17.2: U.S. Environmental Protection Agency

17.3: U.S. Environmental Protection Agency

17.5: Data source: Robert Cowles Letcher and Mary T. Sheil, "Source Separation and Citizen Recycling," Wm. D. Robinson, ed., The Solid Waste Handbook, John Wiley & Sons, 1986.

Chapter 18 Figures

18.1: Modified and redrawn with permission from A.A. Boraiko and F. Ward (1980) "The Pesticide Dilemma," National Geographic 157(2), p. 149.

18.3: © Fred Ward/Black Star

18.4: Based on Weber, "A Place for Pesticides," Worldwatch Institute, vol. 5, no. 3(graph); data from George P. Georghidu, University of California at Riverside, 1990.

18.6: Courtesy of W.K. Hock

18.7: © Grant Heilman Photography

18.8: USDA

18.9: Larry Lana/USDA

Chapter 19 Figures

19.1: Michael A. Leonard

19.1b: © Dennie Cody/FPG

19.2: © Joe Traver/Gamma-Liaison

19.5: Courtesy of the U.S. EPA, MSB-MERI, Edison, NJ

19.6: U.S. Environmental Protection Agency, Securing Our Legacy, p. 16. U.S. Government Printing Office.

19.8: Courtesy of the U.S. EPA, MSB-MERI, Edison, NJ

19.11: U.S. Environmental Protection Agency, Securing Our Legacy, p. 29. U.S. Government Printing Office.

19.13: © Dion Ogust/Image Works

19.15: © Louis Psihoyos/Contact Press

Models of Global Sustainability

19.1: Adapted with permission from Daniel D. Chiras, Lessons from Nature: Learning to Live Sustainably on the Earth, © 1992 Island Press. (Washington, D.C. & Covelo, CA: Island Press, 1992.)

Tables

19.1: Modified from Nebel, B.J. (1981) Environmental Science. Englewood Cliffs, NJ: Prentice-Hall, p. 297.

19.2: Lester Brown et al., Saving the Planet: How to Shape an Environmentally Sustainable Global Economy, 1991, W.W. Norton, table 4.1.

Chapter 20 Figures

20.1: Rainforest Action Network

20.2: Reprinted with permission from p. 3 of Human Biology: Health, Homeostasis, and the Environment by Daniel D. Chiras; Copyright © 1991 by West Publishing Co. All rights reserved.

20.3: © Gamma Liaison Network

Chapter 21 Figures

21.4: From FOR THE COMMON GOOD by Herman E. Daly and John B. Cobb, Jr. Copyright © 1989 by Herman E. Daly & John B. Cobb. Reprinted by permission of Beacon Press.

21.5: Courtesy of Green Seal

21.6: Scenario 10: Reprinted from Beyond the Limits, © 1992 by Meadows, Meadows, and Randers. With permission from Chelsea Green Publishing Co., Post Mills, VT. McClelland & Stewart Inc., Ontario, Canada; Earthscan Publications Ltd., London, U.K.

Chapter 22 Figures

22.1: © J. Hamilton-Spooner/Gamma-Liaison

22.2: ©P. Gontier/Image Works

22.3: © J. Van Acker/FAO Photo

22.4: From Edwin Dobb, "Solar Cooker," Audubon, Nov/Dec 1992, p. 102.

22.5: © Wendy Stone/Gamma Liaison Network

22.6: © Lindsay Hebberd/Woodfin Camp & Assoc.

Chapter 23 Figures

23.3: © Rex Weyler/Greenpeace

23.4: Scott Reuman/Keystone Science School

23.5: Reuters/Bettmann Newsphotos

23.6: © Forrest Anderson/Gamma-Liaison

Index

SOUTH

MODULE TO ACCOMPANY
ENVIRONMENTAL SCIENCE

FOURTH EDITION

DANIEL D. CHIRAS, EDITOR

Bruce C. Wyman, Consultant
McNEESE STATE UNIVERSITY

The Benjamin/Cummings Publishing Company, Inc.

Redwood City, California • Menlo Park, California • Reading, Massachusetts
New York • Don Mills, Ontario • Workingham, U.K. • Amsterdam
Bonn • Sydney • Singapore • Tokyo • Madrid • San Juan

Science Executive Editor: Robin Heyden
Project Editors: Valerie Kuletz and Kim Viano
Editorial Assistants: Thor Ekstrom and Roseann Viano
Marketing Manager: John Minnick
Art and Design Manager: Michele Carter
Manufacturing Supervisor: Jenny Rossi
Production Editor: Donna Linden
Photo Editor: Kelli d'Angona-West
Photo Researcher: Alisa Guttman
Permissions Editor: Marty Granahan
Research Assistants: Sally Almeria, Kathleen Chiras, and Dale Whitney-Boyd
Copyeditor: Elizabeth Gehman
Text Design, Illustrations, and Composition: Seventeenth Street Studios
Cover Design: Rob Hugel and Yvo Riezebos
Printing and Binding: Webcrafters, Inc.

Text and photo credits appear on page 90.

This book is printed on recycled, acid-free paper.

Copyright © 1994 by the Benjamin/Cummings Publishing Company, Inc.

All rights reserved. No part of this publication may be reproduced, stored in a database or retrieval system, distributed, or transmitted, in any form or by any means, electronic, mechanical, photocopying, recording, or otherwise, without the prior written permission of the publisher. Printed in the United States of America. Published simultaneously in Canada.

ISBN 0-8053-4230–3

1 2 3 4 5 6 7 8 9 10–WC–97 96 95 94 93

The Benjamin/Cummings Publishing Company, Inc.
390 Bridge Parkway
Redwood City, California 94065

Contents

Hazardous Waste

Solid Waste

Urbanization

Introduction

As the U.S. population, the economy, and per capita consumption have expanded, many environmental problems have progressed from local issues, affecting relatively small areas, to regional issues, whose impacts stretch across many jurisdictions. In recent years, many regional issues have expanded to become national in scope. In turn, some national issues have become global in nature. Water pollution is a good example. At one time, water pollution affected only small portions of streams. Then, as factories, farms, and cities sprung up, entire rivers turned sour, poisoned by pollution. Then, as the condition of rivers deteriorated, the oceans began showing signs of stress.

Despite the fact that numerous environmental problems have become global in nature, solutions still often require local actions. Buckminster Fuller coined the phrase "think globally, act locally" as a reminder of this important, but often overlooked message. To illustrate this point, consider the efforts needed to clean up the nation's rivers. Experience has shown that controls on a few major polluters are generally insufficient. What is needed are comprehensive watershed management programs that establish measures to control the numerous pollution sources within the confines of a river's drainage basin. This approach, which simultaneously reduces local, regional, and global water pollution, obviously requires action by dozens, sometimes thousands, of participants.

In the spring of 1992, my editors and I decided to publish four regional versions with Environmental Science. Our decision was based on the results of a survey completed by hundreds of environmental science professors throughout the United States. The survey indicated a strong interest in covering local and regional issues in more depth. And, it is our belief that environmental science courses could be enhanced by coverage of local and regional environmental issues. Why? Many local and regional issues are of intense interest to students and many seem more manageable than global issues, whose fate is tied up in international politics and corporate decision making.

These region-specific versions provide information on a variety of important environmental issues. They offer not only descriptions of the problems facing many areas but also offer many innovative local solutions devised by individuals, businesses, and government.

Rather than writing the material from scratch, we decided to use original articles from current literature. My research assistants, consultants, and I searched through many different local and national magazines and newspapers. We made numerous phone calls to experts in various issues, and called on government agencies for contributions as well.

Wherever possible, we sought articles that offered sustainable solutions—that is, responses that strike at the root causes in an effort to truly solve environmental problems, not merely patch them up and pass them along to the next generation. You will find that articles occasionally have examples from two or more regions, and sometimes articles are of a general nature. These general articles also illustrate important points that merit inclusion.

I selected the best articles we could find, which have been reprinted here with permission. Each issue is presented in the regional versions in the same order as the core text. In the South, the issues include: vanishing habitat and wetlands, air pollution, water pollution, hazardous waste, solid waste and urbanization. Each issue area begins with a brief introduction that offers an overview of the region and its problems.

Each issue is preceded by a brief list of background readings in the text. These readings introduce important concepts and terms that are often necessary to understand the articles. Ultimately, you will get a lot more out of the articles by first reading the cross-referenced material. Each article is followed by one or more questions meant to help you assimilate and think critically about the new information.

Unfortunately, space limitations made it difficult to cover every state and every issue in detail. If we had, this supplement could easily have been 1000 to 2000 pages long. What I attempted to do was to find articles that dealt with some of the most pressing problems and the most interesting solutions.

It is my hope that these regional versions will help make environmental science classes more relevant and that they might inspire students to become part of the solution through individual actions, such as using energy more efficiently, joining an environmental or citizens' group or taking an active role in the environmental-political process. I also hope that it will encourage some students to choose a career in environmental protection.

As a final note, I welcome articles from students and professors for future editions and ask that you send them to me in care of the publisher: Dan Chiras, c/o Benjamin/Cummings, 390 Bridge Parkway, Redwood City, CA 94065.

Vanishing Habitat and Wetlands

The nation's wetlands are among our most valuable and endangered natural resources. These inland and coastal areas serve as fish and wildlife habitats, stop-off points for migratory waterfowl, natural water purification zones, erosion prevention buffers, floodwater storage, and groundwater recharge zones. Wetlands are coveted recreational resources, providing a place for birdwatchers, anglers, and wildlife advocates to visit. And, lest we forget, fish and shellfish that live in wetlands, during at least part of their life cycle, are an important source of food for humans.

The South is well endowed with inland and coastal wetlands. In fact, 8 of the states in the southern region fall within the top 12 states in the lower 48 in total wetland acreage. Florida has the largest total wetland area (around 4.5 million hectares, or 11 million acres) and Louisiana is number two with approximately 3.6 million hectares (9 million acres).

Southern states have witnessed dramatic declines in wetland acreage in the past 200 years. For example, in a recent study that documented wetland loss in the United States from the late 1700s to the present, researchers estimated that the states in the southern region lost about one-half of their wetlands, a decline of about 22 million hectares (55 million acres). Inland wetlands have been especially hard hit. Three states experienced remarkable declines in inland wetlands: Arkansas (72%), Kentucky (81%), and Oklahoma (67%).

This loss continues today, despite wider recognition of the ecological and economic value of wetlands. Other habitats are also shrinking as the human population expands. With this expansion, hundreds of species have lost their homes.

Background Reading: Chapters 2, 3, 8, and 9.

Wetlands, Wonderlands

Wetlands Are Precious and Vanishing. What's Being Done to Save Them?

Barbara Sleeper

■

SOURCE
Animals, January/February 1991, Vol. 24, No. 1

Seattle-based writer and wildlife biologist *Barbara Sleeper* kayaked through Washington State wetlands last spring.

UNTIL RECENTLY, most people viewed wetlands as wastelands—shoe-sucking damp places home to mosquitoes, ooze, and pestilence. These muck lands, along with their bugs, strange smells, and mythical swamp monsters were best avoided—better yet, eliminated. As a result, over half of America's original wetlands have been destroyed—diked, drained, and filled in for housing developments and industrial complexes, converted to farmland, and used as receptacles for household and hazardous waste.

In reality, wetlands are sensitive transitional areas located between open water and dry land that are saturated with water at least part of the growing season—ideal environments for plants adapted to water-logged, oxygen-depleted wetland soils. Even when seemingly dry for long periods of time, wetland soils are usually saturated at or just below the surface. Due to a combination of factors from vegetation, water chemistry, and soil type to climate, these areas vary greatly.

Wetlands include everything from soggy bogs, fens, peatlands, marshes, and swamps to vernal pools, sloughs, and prairie potholes. From the borders of inland lakes, rivers, and ponds, to the Alaskan tundra (frozen wetlands), and the bottomlands of the Mississippi Delta, wetlands are a pervasive, previously low-profile ecosystem as critical to our planet's well-being as the much publicized tropical rainforests—which, in many cases, are also wetlands. In general, these habitats are broadly classified as either coastal or inland wetlands. The one element common to all is *water*.

"The intrinsic value of a wetland area is subtle," explains Daryl Scheibel, program coordinator for the Seattle-based Wetlands Watch, an innovative public-education project implemented by the Washington Environmental Council and the Puget Sound Alliance to increase public awareness of wetlands issues. "The positive attributes affect everyone's quality of life—and yet, to most people, a new shopping mall appears to have more immediate, tangible value than the wetlands it's being built upon," says Scheibel. "These are the attitudes we are trying to change."

Like giant sponges, wetlands store water in the wettest seasons of the year and release it at a constant rate to regulate stream flow. They prevent and control inland flooding by storing and slowing flood waters. They replenish water supplies by removing pollutants and recharging underground aquifers. They protect shorelines from erosion by ocean waves and storms, and provide critical habitat for waterfowl, shore birds, wading birds, and many other wildlife species. Nowhere else can one find a wood duck, a muskrat—or even a swamp rose.

Wetlands support many of the nation's coastal fisheries, too—to the tune of $10 billion annually. More than 70 percent of marine commercial species depend on the prime spawning and nursery areas provided by coastal wetlands. More than 50 species of fish spawn or feed in the lower Mississippi swamps alone. And, wetlands, produce other valuable commercial products such as shellfish, timber, and cranberries.

From birdwatching in the Florida Everglades to canoeing in Georgia's Okefenokee Swamp, wetlands offer numerous recreational opportunities enjoyed by an estimated 50 million people each year. They also help maintain the earth's atmospheric balance by producing and absorbing methane, nitrogen, and carbon dioxide gases. Each wetland works in combination with other wetlands as part of a complex, integrated system that delivers these benefits and others to society. Should these natural filtration systems be destroyed, not only is marine productivity reduced in the swamps—but downstream as well. The cumulative effects of wetlands degradation can become quite severe.

In December 1988, for example, more than 7,500 birds died in Texas during an unusual outbreak of avian cholera. Droughtlike conditions in the summer and fall reduced the available wetlands so that migrating birds from the north had to crowd into small areas, encouraging the spread of the epidemic. For a week, state agencies and volunteer groups pumped water into rice fields in southeast Texas—creating 1,500 acres of clean wetlands to which the geese and ducks could disperse.

In a report just released to Congress by the Department of the Interior's U.S. Fish and Wildlife Service (USFWS), it is estimated that 117 million wetland acres—or roughly 56 percent of U.S. wetlands—have been lost since colonial times in the lower 48 states. Of the estimated 220 million wetland acres thought to exist then, only 103 million remain. According to the report, 22 of the 50 states have lost at least half of their wetlands.

This translates into sobering statistics. More than 60 acres of wetlands have been lost every hour for the past 200 years in the contiguous United States—more than 500,000 acres a year. Most of this loss has occurred in freshwater drainages in inland areas, where 95 percent of all wetlands exist. The majority of these wetlands were converted by farmers to cropland; others were consumed by urban and suburban development, dredging, or mining, and the discharge of pollution.

Roughly one-third of the nation's total wetlands loss has occurred in the midwestern farm states. Arkansas, Connecticut, Illinois, Indiana, Iowa, Kentucky, Maryland, Missouri, and Ohio have all lost more than 70 percent of their original wetlands. California has lost the most—91 percent; and Alaska the least—less than 1 percent.

By far, one of the worst ongoing wetlands disasters is that occurring in Florida's Everglades. Since 1930, more than 50 percent of this magnificent ecosystem has been lost—along with 95 percent of its wading bird populations. This is due, in part, to a system of drainage canals and dikes built through the Everglades in the 1940s. In addition, a powerful sugar-cane industry operating on 750,000 acres of Everglades marshland is converting the area into phosphorus-choked swamps. Florida now plans to spend $400 million to reduce the ongoing agricultural pollution and water contamination; it will take $270 million just to destroy the largest of the existing canals. Even so, conservation groups and the federal government have filed a lawsuit charging Florida with environmental negligence. Their claim: Not enough is being done fast enough to save this vast wilderness.

According to the USFWS, the disappearance of America's wetlands has enormous implications for wildlife. Wetlands represent some of the planet's most diverse and productive ecosystems, producing abundant plant food that fuels an intricate food web. It is claimed that an estuary produces as much growth as a tropical forest.

From algae, insects, and amphibians to birds, fish, and mammals such as moose, deer, beavers, bear, elk, muskrat, and otters, wetlands provide important habitat for wildlife. Roughly 75 percent of North America's bird species depend upon wetlands for feeding, nesting, and resting. In fact, an estimated one-third of all species listed federally as endangered or threatened—such as whooping cranes, piping plovers, and bald eagles—depend on the relatively undisturbed open space of wetland habitats. Yet populations of many duck and other wetland-dependent migratory bird species are now at historic lows.

An annual midwinter USFWS survey in January 1989 showed wintering duck populations in the Pacific Flyway (California, Oregon, Washington, Nevada, Idaho, Arizona, and parts of Montana, Wyoming, Utah, Colorado, and New Mexico) to be the lowest on record. About 3.4 million ducks were counted—32 percent below the winter count of 1988, and 46 percent below the winter average of the last 33 years. Wintering mallards fell to a record low, down 37 percent from their long-term winter average; pintails were down 70 percent. Statistics from the Central and Mississippi flyways are even more sobering.

In 1977, President Jimmy Carter

The American bittern inhabits marshlands, where it constructs platformlike nests with aquatic vegetation.

officially recognized the value of the country's wetlands. His Executive Order II990 directed federal agencies to avoid the unnecessary alteration of destructions of wetlands. The initial efforts to save wetlands that followed Carter's decree focused primarily on coastal areas. Today, conservation and political efforts are also being directed to control and reverse the attrition of inland wetlands.

By 1989, the accelerating loss of wetlands and a corresponding decline in ducks and migratory birds moved President Bush to issue a national challenge of "no net loss of wetlands." Under this policy, conversion of wetlands would be allowed only if the user restored previously converted wetlands—or created new ones.

Critics remain skeptical, however, claiming that Bush's decree was more of a politically correct action than a truly feasible one. "Environmental engineers must have big egos," says Scheibel, "if they think they can replicate what it took nature hundreds, thousands, or even millions of years to evolve. For

society as a whole, it is a lot cheaper to preserve wetlands than to try to re-create them." Bogs alone take 12,000 years to develop.

While most individuals agree with the principle of preserving wetlands, attempts at putting the no-net-loss policy into practice have caused repercussions from bogs to banks. Enforcement of the no-net-loss policy has discouraged banks from lending money to farmers because of the uncertain status of their lands. If a farmer's land might be designated as wetlands, its value is decreased in the lender's eyes. The policy has also prevented farmers from selling excess land to realtors—assets many had counted on for retirement funds. All this has left farmers fuming. The effort, they say, represents a confiscation of their assets and a federally imposed restriction on their rights as private property owners.

Ironically, between 1940 and 1960, the U.S. Department of Agriculture (USDA) subsidized the drainage of 60 million acres of wetlands for farmland. Now, under the 1985 Food Security Act's "Swampbuster" provision, farmers who

convert wetlands to croplands will be denied eligibility for all USDA farm programs, including subsidized loans, crop insurance, price and income supports, disaster payments, and storage payments.

There appears to be no easy solution to wetlands conservation. According to Senator John Chafee (R-Rhode Island), "when you save wetlands you antagonize somebody." Agricultural analysts have pointed out that it's unrealistic to expect private landowners to do what is best for society out of the kindness of their hearts. Proposals to Congress are emerging that would offer tax incentives for wetlands conservation—to pay farmers to reconvert land that could then become part of a wetlands conservation bank.

Meanwhile, the Army Corps of Engineers and the Environmental Protection Agency recently agreed on a standard definition of wetlands that permits classification as such by an area's soil type alone. Previously, an area was considered wetlands only if it was inundated with water during the growing season (March through October) or if it contained indigenous wetlands-indicator plants such as cattails. "To get around such loose definitions," says Scheibel, "developers would often scrape off the vegetation."

The more stringent definition could have far-reaching effects. One Washington State community has already spent $530,000 putting water and sewer lines into an area destined for commercial and industrial development. Total project costs for the area were expected to reach $9 million. With the new soil type criterion, will the Army Corps of Engineers now designate this area wetlands? That's just what happened in Eugene, Oregon—after the city spent over $12 million preparing a large area for industrial development.

As a result, there are appeals of wetlands designations. In Massachusetts alone, there is a backlog of over 500 cases, which have been waiting as long as two years for review by only four state hearing officers and using up resources needed for other types of environmental law enforcement.

And there are appeals of wetlands appeals. Last year, the New York State Department of Environmental Conservation announced that it would appeal the decisions on previously settled wetlands appeals because the decisions appeared to uphold interests of developers, without ever establishing clear standards for identifying wetlands.

The federal Clean Water Act, the River and Harbor Act, National Environmental Policy Act, the Coastal Zone Management Act, and the "Swampbuster" provision of the 1985 Food Security Act all regulate activities in wetlands. Many state regulations affect development in or near wetlands, too. And local jurisdictions also have regulations that affect projects proposed in or adjacent to wetlands.

"And yet, in spite of all this," says Scheibel, "wetlands continue to be destroyed. More people need to get involved at the community level

to help identify, map, and protect the remaining wetlands."

So just what is being done? Several private conservation groups have long been pushing for change, but now local, state, and federal governments are enacting and updating wetlands-saving measures, too.

In October, Secretary of the Interior Manuel Lujan announced a comprehensive plan aimed at stemming destruction of the nation's vanishing wetlands. The Wetland's Action Plan, developed by the USFWS, outlines strategies for protecting, enhancing, and restoring wetlands with the goal of getting all Interior agencies working together. Under the American/Canadian North American Waterfowl Management Plan, more than 50 wetlands conservation and restoration projects are being initiated. Since this plan was launched in 1986, roughly 600,000 wetland acres have been conserved in North America.

Sales of the Federal Migratory Waterfowl Hunting and Conservation Stamp (Duck Stamp) continue to fund acquisition of critical wetlands for inclusion in the National Wildlife Refuge System. Since 1934, the Duck Stamp program has generated more than $350 million to purchase nearly 4 million wetland acres, about 4 percent of the refuge system.

Nearly 360,000 acres of threatened wetlands and surrounding habitats for wildlife in the United States, Canada, and Mexico will be acquired, improved, and restored in the first projects approved for funding under the 1989 North American Wetlands Conservation Act. Meanwhile, a 60-nation consortium of the best available scientific expertise in wetlands conservation worldwide has been organized as the Convention on Wetlands of International Importance, chaired by the director of the USFWS. Even cartoon icon Garfield has gotten into the act by promoting the values of wetlands along with his creator, Jim Davis, on nationwide public-service announcements.

On the state level, California, Louisiana, Minnesota, New York, North Dakota, and Texas were recently awarded for exemplary wetlands conservation, restoration, and enhancement programs. North Dakota, actively involved in wetlands conservation since 1938, passed landmark legislation in 1987 establishing a statewide policy of no net loss of wetlands. This bill was the first of its kind in the nation and the prototype for the national no-net-loss wetlands decree.

Since 1968, the Louisiana Department of Wildlife and Fisheries has invested $70 million to acquire more than 252,000 acres of forested wetlands and 78,000 acres of coastal marsh to prevent conversion to croplands. The state has also pioneered methods of wetlands restoration in the Mississippi River Delta by installing sediment fences to develop new areas of marsh.

Under the Save Minnesota's Wetlands program, established in 1951, 650,000 acres of wetlands and associated uplands have been acquired. In 1979, the state Legislature passed the Protected Waters Inventory to map all wetlands in the state. Completed in 1984, this mapping effort resulted in nearly 5 million acres of basins being given legal protection.

The passage of the Texas Waterfowl Stamp Act in 1981 marked the beginning of an aggressive campaign to preserve wetlands in the state. Texas Parks and Wildlife Department has now committed more than $13.3 million to wetland conservation, and 88,000 acres have been protected there.

Finally, New York has been actively involved in wetlands preservation and restoration for half a century. During the past two decades, several multimillion-dollar land-acquisition bond programs have been supported with wetlands as a priority. The state currently manages 45,000 acres of wetlands and 7,200 acres of open water. And New York has pioneered an innovative approach to wetlands management—using beavers. By regulating the trapping of these tail-slapping rodents, the state has allowed beavers to create many new acres of wetland.

"Times have changed," concludes Scheibel of Wetlands Watch. "With public education, more people now appreciate wetlands, not as wastelands, but as wonderlands—valuable natural resources that need preserving. However, wetlands continue to be destroyed, not because people don't care, but because they don't know. Public awareness remains critical."

Discussion Question

1. You are assigned the task of explaining the importance of wetlands to local residents. Outline the major points you will make to your audience.

In Search of the Early Everglades

Scientists Struggling to Repair Damage to the River of Grass Must Contend with a Lack of Data and Funds

Bill Sharp and Elaine Appleton

SOURCE
National Parks, January/February 1993, Vol. 6, Nos. 1–2

YOU CROUCH on a small dry hummock in the midst of the seemingly endless wetland grasses of the Everglades. It is 1930, and you are surrounded by an abundance of wildlife beyond the scale of anything you have ever seen. You stand, startling a vast colony of wading birds that takes to the air *en masse,* obliterating the sun. Mammals from mice to manatees abound, along with fish, alligators, and other species.

Fifty miles to the south, in a skiff on the waters of Florida Bay, you can look from horizon to horizon and see nothing but shallow waters dominated by dense seagrasses. The bay is home to fish, shrimp, sea turtles, and birds. Like all early visitors to these places, you are astounded by the numbers of plants, animals, and insects. In 17 years, a portion of this massive ecosystem will become Everglades National Park.

Today, 46 years after formation of the park, both deliberate and inadvertent modifications have severely depleted this abundance. The massive nesting colonies of Everglades' wading birds are gone. Those missing birds have plenty of company. Only 50 Florida panthers exist today, and only two are believed to live within the park; seagrass, mangroves, and sponges periodically die in large sections of Florida Bay at the south end of the park; and economically important populations of fish and shrimp perish in their nursery grounds as rotting vegetation robs the bay's water of life-giving oxygen.

Everglades National Park in Florida is not the same park it was before farming, along with commercial and residential development, exploded in the southern end of the state. In fact, at least one scientist says the Everglades began sliding downhill before becoming a national park in 1947. And the damage continues. Each busload of European tourists and carload of vacationing Americans has a bit less of the Everglades to enjoy than the visitors who came before them. The traditional role of NPS [National Park Service] dictates that the agency ensure the public's enjoyment of the parks while preserving the resources. In the face of limited funding and public demand for access, preservation and the research necessary to sustain it have become secondary pursuits.

The Everglades is far from the only national park facing serious problems; however, its issues are among the most urgent because the park's water supply—its lifeblood—has been diverted, dammed, poisoned, and otherwise damaged. In the vast majority of cases where parks are threatened, it is not the inability to take action that prevents implementing remedies to save them but the lack of clear knowledge of the mechanisms causing the damage.

"We have a mandate to protect

Changes to the Everglades' eco-system have depleted the number of wading birds, such as the roseate spoonbill.

the Everglades," says Mike Soukup, research director for the National Park Service's South Florida Research Center. "We need to voice the needs of the park [in the interest of] its long-term preservation. The issues are so pressing and the threats so immediate for the Everglades that unless research information is available at the appropriate time, our efforts here will fail."

A limnologist, Soukup specializes in freshwater ecosystems. He directs research to help understand the Everglades and works to share that information effectively. Among his goals are preserving the park and returning waters to conditions prevalent before humans intervened.

Changes made to the Everglades' natural systems in the late 1940s and after were made with little or no thought to environmental consequences. And now that the changes

have been in place for some time, each alteration has its proponents, says Soukup, and many of these have political or economic influence. NPS and conservationists working to support the park must deal with the entrenched interests of developers and the conflicting interests of flood control, demands for a water supply, and agricultural interests who want water but oppose regulation of their runoff.

Some opposition to Everglades restoration comes from unlikely corners. Bass fishermen oppose filling canals to restore water flow patterns, because the canals provide easy access to fishing areas. The U.S. Fish and Wildlife Service has opposed water-level changes that might increase risks to the endangered Everglades kite, which have become dependent on habitat in certain areas and may not adapt

quickly enough to survive. The most effective answers to these dilemmas, more often than not, involve better data, ecosystem perspective, and careful approach.

Research in the National Park Service has a fundamental purpose, according to Dominic Dottavio, NPS southeast region chief scientist and deputy associate regional director for science and natural resources in Atlanta. "Research is a tool to help us understand what we have, how it is changing, and what we have to do to protect it in perpetuity," he says. "In order to manage the resources effectively, you have to have scientific research and resource management research. How do we get rid of exotic plant species, and what is the impact of activities in areas adjacent to our parks? We can answer these questions only through good-quality research."

This map compares early conditions of the historic Everglades with existing conditions of the Everglades today.

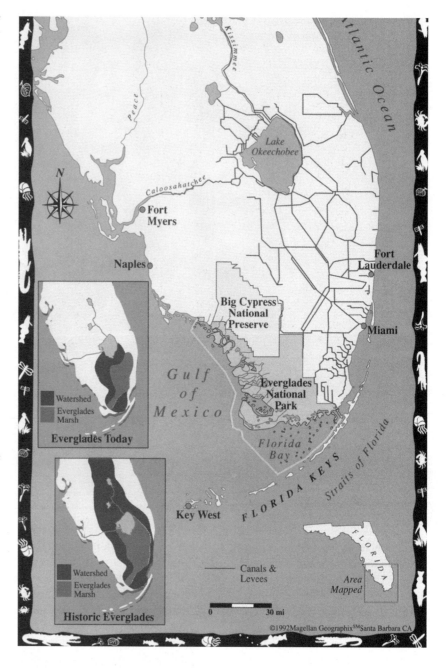

The Park Service's attitude has not always been so positive. A National Research Council report published in August, *Science and the National Parks,* notes that a dozen major reviews over the past 30 years—including at least two issued by NPCA—strongly recommended, to little avail, that NPS improve its research program. The chairman of the committee that wrote the report, Paul G. Risser, ecologist and provost at the University of New Mexico, Albuquerque, uses Sequoia National Park in California to illustrate the need for greater study.

"We thought without research that maintaining the sequoias meant preventing fires among the trees," he says. "Because of research, we now know that if there are no fires, the trees cannot reproduce. Without periodic fires, a layer of soil litter builds up that seeds cannot penetrate [and as a result they cannot] compete and grow. Occasional burns open the understory and provide a good seedbed."

What the sequoias are to Sequoia National Park, water is to the Everglades. Yet, detailed research about the role of water in the health of the park is new. The South Florida Research Center was founded in 1978 in response to rapid degradation of the Everglades. By that time, a network of huge water diversion projects was robbing the park of a sizable percentage of its water as well as disrupting distribution and affecting quality.

"Vast changes to the water management . . . brought us to the point where the Everglades doesn't really exist as a sustainable ecosystem," says Soukup. "It needs to be thoroughly restored to bring it back to the characteristics for which the national park was established."

Understanding water in the Everglades requires mastering complex relationships of quantity, quality, distribution, and timing across a vast region, nearly all of it disrupted. It is an extremely complex ecosystem. Water flows slowly but steadily along the "river of grass" leading southward from Lake Okeechobee. The lake and nearly every water body and waterway south of it to the Everglades has been changed. Water is diverted for human use, much of it never to return to the system. The water that is returned is sometimes full of nutrients that affect the natural development of Everglades' plant communities. And any water releases are irregular and in locations convenient to people, not the Everglades.

Soukup's team aims to identify the historical characteristics of water that typified the Everglades. The team also will look at the relationship of these factors with the wading bird colonies and the biological activity that promoted the establishment of the park.

"The holes in our knowledge center on what the system was like in quantitative terms," says Soukup. "How much water was in what areas, for what time periods, and with what variability? How much water flowed into Florida Bay in an average year, and what were the extremes? Those baseline data for the original ecosystem don't exist."

In the absence of data, Everglades researchers turn to computer modeling to simulate early conditions in the park. Using the models, researchers test hypothetical water delivery systems to determine how best to ensure that water reaches the right portions of the park.

Extending the use of modeling allows the park's researchers to determine how variations of water delivery affect wildlife populations such as wood storks, alligators, and panthers. Other work predicts how changes in the water supply in the northern Everglades will affect quality when that water reaches Florida Bay and, in turn, the coral reefs in the Florida Keys.

But even as research and new modeling tools begin to provide a pathway for recovery in the Everglades, the effort is starving for lack of funds. When it was founded, the South Florida Research Center was provided with $1.6 million in funding. In the 14 succeeding years, the center has taken on additional park responsibilities, while funding has remained the same. For the past two years, the center had been scheduled for a significant increase, only to see it cut late in the budget process. "In real dollars, the amount of money available for Everglades research has decreased substantially since 1978," says the Park Service's Dottavio.

As a result, even the most dedicated researchers experience severe frustration. "Funding for research in the Park Service is absolutely abysmal," says one researcher. "We are facing monumental problems ecologically, and these funding problems make it worse. I feel like a runner coming up on the finish line, and I realize that I am about to fall. It appears that the federal government and the Park Service are not capable of identifying the true needs and getting the dollars there."

To do the job right, Soukup

wants the center's budget increased from the current $1.6 million to $5 million. And to include the entire larger ecosystem in the effort, he says, an additional $3 million should be made available to Big Cypress and Biscayne national parks, which are adjacent to the Everglades. This includes both research and resource management efforts.

And Hurricane Andrew worsened the situation. Homestead, Florida, which suffered the worst damage from the storm, is headquarters for the Everglades and home to most of its employees. Simply placing a phone call to or from the center, difficult in the best of times because of inadequate phone systems, was nearly impossible for weeks following the storm. Of 250 people employed at the park, the homes of 175 were seriously damaged, according to NPS. The research center lost much of its roof, and water damaged many records and papers. NPS officials estimate that cleanup efforts will take several years and put the initial recovery needs for Big Cypress, Biscayne, and Everglades at $52 million. Most of this money will go toward repairing infrastructure and not for research.

Hurricane damage will undoubtedly draw attention from the Everglade's other problems; however, it is not likely to still the requests for more research funding. The intensity of the cries for money for the Everglades might seem enough to elicit quick responses from Washington, but the problems are more complex than that, notes John Dennis, chief of the science branch of the Park Service's Wildlife and Vegetation Division. "Because of the way the federal budget is set, each bureau has to weigh all its competing needs," he says. "NPS distributes funds internally based on priority, and it is fairly easy to place the human needs of park visitors at a higher level than those of research."

Throughout NPS, much of the research requested by professionals goes unfunded. In its 1988 report

Natural Resources Assessment and Action Program, NPS identified 2,500 natural resource and research projects that would go unfunded from 1988 to 1992. Anne Frondorf, chief of planning and information in the Park Service's Wildlife and Vegetation Division, estimates that at least 750 of the unfunded projects were research. And while she reports that some progress in funding has been made, it is likely that similar numbers of proposed NPS research projects [will] continue to go begging.

Discussion Question

1. Why is research so important to the restoration of the Everglades?

Saving Disney's Land

Jeff Porro

■

SOURCE
Nature Conservancy,
March/April 1993, Vol. 43, No. 2

FOR YEARS, developers gazed longingly at the 8,500-acre Walker Ranch near Walt Disney World in central Florida. The property is ideally located for houses and shopping centers—but its woods, prairie and wetlands are the home of 15 endangered and threatened species, including the greatest concentration of active bald eagle nests in the southeastern states.

Now, thanks to a creative effort

by The Walt Disney Corporation, federal and state regulatory agencies and The Nature Conservancy, the ranch will become a preserve and an environmental learning center.

The effort began late in 1991, when Disney decided to try a new approach to developing 10,000 acres of its property close to Disney World. Instead of seeking government approval parcel by parcel, Disney asked regulatory agencies for permission to build on the whole tract during the next 20 years. To compensate, Disney officials offered to help preserve an entire ecosystem—the Walker Ranch—and to set aside an additional 8,900 acres on its property through a conservation easement.

Government officials agreed to the plan, provided that a conservation group manage the ranch. "The name that continued to surface as the best in the business was The Nature Conservancy," says Don Killoren, a vice president of Disney Development Company who was actively involved in the negotiations.

Conservancy staffers prepared a management plan for the property that was accepted by all of the parties. Disney then bought the ranch and donated it to the Conservancy, which will restore damaged parts of the tract and preserve the entire ecosystem.

Disney will pay for restoration and management of the preserve for 20 years and will establish a $1 million endowment for stewardship after 2012. They will also build an environmental learning center that will educate the public about restoring and safeguarding ecosystems. Conservancy officials estimate the total value of the Disney project at between $35 million and $45 million.

Jora Young of the Conservancy's Florida regional office gives much credit for the plan's success to Disney, and also to the Conservancy's growing reputation as a land manager.

"We've had a reputation for being good at acquisition," says Young. "Now our reputation as managers is starting to be one of the tools we can use to achieve our conservation mission."

Discussion Questions

1. Critically analyze and discuss the land preservation deal described in this article.

2. What information would you need to know in order to determine if this really was a good deal for wildlife and habitat?

Finding Refuge in the Wetlands

Jeff Porro

■

SOURCE
Nature Conservancy,
March/April 1993, Vol. 43, No. 2

TWENTY-FOUR hours before the end of the 1992 presidential campaign, George Bush did a big favor for the wetlands in Bill Clinton's home state. He signed into law the Arkansas-Idaho Land Exchange Act,

Conservancy efforts helped to move this Arkansas wetland from the hands of a forest products company to federal protection as a wildlife refuge.

which added to the federal wildlife refuge system 41,000 acres of bottomland hardwoods along Arkansas' White River.

The land—which provides habitat for mallards, bald eagles and black bears—had been owned by the Potlatch Corporation, a forest products firm. In exchange for the property, Potlatch received 17,625 acres of federal land in Idaho.

The swap was an important victory for The Nature Conservancy's Arkansas field office, which has spent a decade protecting wetlands in eastern Arkansas, according to state director Nancy DeLamar. When Potlatch expressed an interest in a land exchange, Conservancy staffers provided background information and organized meetings between the company, the Arkansas congressional delegation and the U.S. Fish and Wildlife Service to nudge the deal to completion.

"Nancy DeLamar and her colleagues consistently demonstrated the flexibility and dedication necessary to make this land exchange a success," said Senator Dale Bumpers (D-Arkansas), who played a leading role in the negotiations.

Conservationists are particularly happy that the swap will connect two existing protected areas, the 110,000-acre White River National Wildlife Refuge and the 57,000-acre Cache River National Wildlife Refuge. "It is a wetlands resource comparable to some of the biggest in the country, such as the Everglades and the Okefenokee swamp," says DeLamar.

Discussion Question

1. This very brief article explains a way nonprofit groups can help promote habitat protection. What other ways can groups like The Nature Conservancy help set aside vital habitat?

'Good Times' Are Killing the Keys

From Coral Reefs to Mangrove Forests, a Fragile Tropical World Is Endangered

George Laycock

■
SOURCE
Audubon, September/October 1991, Vol. 93, No. 5

George Laycock, a longtime *Audubon* contributor, has traveled to the Florida Keys frequently since the 1950s, both as a writer and a fisherman. He is the author of 52 books on the outdoors and wildlife.

FLORIDA is subdivided into sixty-seven counties—Monroe County and sixty-six normal ones. Monroe County will always be unique because it is a collection of little tropical islands sprinkled across the glistening waters off the southern tip of the peninsula.

Here the tides wash coconuts ashore on warm sand beaches. Tropical butterflies pay silent visits to wild orchids and drift among trees named gumbo-limbo, lignum vitae, and poisonwood. Hidden in the deep shadows of the palmettos live tiny Key deer and native raccoons, rabbits, and rodents. Pelicans patrol their fishing waters, while graceful, long-legged wading birds stalk the

tidal flats. White-crowned pigeons arrive from Caribbean islands to raise their young in the mangrove forests that fringe the Keys. In the sea-grass meadows, myriad larvae grow into new generations of shrimp, lobsters, and marine fish. Brilliantly colored tropical fish cruise the coral reefs.

This is a fragile environment where, until the early days of this century, the pressures within the ecosystem were limited largely to those that the wild species exerted on each other. But their world was destined for dramatic and often devastating changes.

These islands in the sun, scarcely rising above mean high tide, hold a magnetic attraction for people from distant, colder climates. And the tide of sun worshipers washing over the Florida Keys continues to grow. On weekends long lines of automobiles towing boats and packed to the brim with flippers, masks, coolers, folding chairs, suntan lotion, and diapers, choke the two-lane U.S. Highway 1, the only road into the Keys. This may be good news to those who sell T-shirts, but to other residents it simply means more noise, crowds, and traffic gridlock.

And who can blame them for coming? Nowhere else in North America can motorists find a highway that carries them from island to island for a hundred miles out into a world of Caribbean seascapes, coral reefs, and little green islands nestled in cobalt waters.

Beyond the people developments, the region is predominantly in public holdings. There are marine sanctuaries, four national wildlife refuges, state parks, and a portion of Everglades National Park.

Yet everywhere, even in the protected wilderness areas out in Florida Bay, where jet skiers zoom across bird-rich tidal flats, the overcrowding continues out of control. Some 80,000 people live in the Keys, and land there is still being subdivided and homes built. On top of

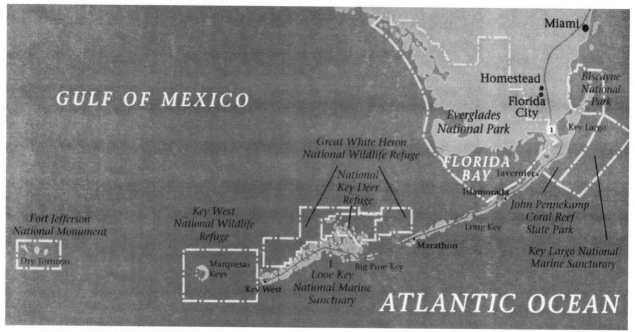

GULF OF MEXICO

ATLANTIC OCEAN

Miami

Homestead

Biscayne National Park

Florida City

Everglades National Park

Key Largo

FLORIDA BAY   Tavernier

Great White Heron National Wildlife Refuge

National Key Deer Refuge

Islamorada

John Pennekamp Coral Reef State Park

Fort Jefferson National Monument

Key West National Wildlife Refuge

Long Key

Marathon

Key Largo National Marine Sancturary

Dry Tortugas

Marquesas Keys

Big Pine Key

Key West

Looe Key National Marine Sanctuary

Human impact threatens wildlife and natural habitat in the Florida Keys.

this tourists by the hundreds of thousands arrive by boat, automobile, aircraft, and even by bicycle. This growing population pressure keeps both people and wildlife under siege.

There appears to be no element of this once natural world that escapes the impact of the growing human invasion and the steady loss of wildlife habitat. Beneath the blue-green waters on the ocean side of the Keys, the spectacular coral reefs—the only such living reefs in North America—are threatened. Marine biologists are concerned about the dwindling fish populations in Florida Bay, as well as the impacts of chemical pollution and floating sewage. The little Key deer die at abnormally high rates, while other animals ranging from crocodiles to ospreys face serious trouble.

Mass tourism was slow in coming. As long as the Keys could be reached only by boat, and for some decades afterward, they retained much of their old character as isolated islands with quiet fishing villages. The major population center was Key West, rich in its history of pirates, wreckers,

spongers, and cigar makers, and its lore of Ernest Hemingway, John James Audubon, shrimp boats, and turtle kraals. The Keys needed two major engineering developments if they were ever to shake their tranquil image and begin showing signs of progress.

Railway magnate Henry Flagler had the answer to the first of these, and visitors driving in the Keys, especially over the bridges that link these islands, may sometimes recall his dream. His Florida East Coast Railroad reached Miami in 1896, but instead of sensibly viewing this as the terminus of his line, Flagler sent his engineers and laborers island hopping out into the Atlantic, laying the rails that forged the main islands into a single chain all the way to Key West.

Then, on a January morning in 1912, Flagler himself arrived in a blaze of red-white-and-blue glory on the first train ever to chuff into Key West. Politicians and newsmen could scarcely find words to express their admiration and wonder. Historian Jefferson B. Browne wrote of the price that men had paid: Workmen were lost to drowning, crushed

beneath concrete, "blown to atoms by dynamite," and swept away by hurricanes, but the end seemed to justify the sacrifice. Flagler, he added, "was humanity crystallized, patriotism embodied! The work is done. Let it speak for itself now and forever!"

But forever is a long time. Only twenty-three years later, in the stormy season of 1935, Flagler's railroad was demolished by what Florida Keys tourism promoters now refer to as "wind and erosion" because they can't bring themselves to say "hurricane." With winds of more than 200 miles per hour (a storm is classed a hurricane when its winds exceed 74 mph) and waves estimated at twenty-six feet, enough "wind and erosion" hit Lower Metacumbe Key to blow railroad cars off the tracks, destroy the roadbed, and kill at least 409 people.

Although this hurricane terminated rail service into the Keys, Flagler's visionary project did pioneer the route for an overseas highway on the railroad bed and bridges. By 1938, Hudsons, Hupmobiles, Studebakers, Buicks, and Ford sedans rumbled all the way out to Key West, crossing forty-two overseas bridges, one of them seven miles long. This "113-mile-long cul-de-sac," as one resident calls it, stitched the islands together. At last the Florida Keys were delivered from isolation. They would never be the same again.

But there was one more natural barrier blocking development in the Keys: As long as people had lived here, their freshwater supply was limited largely to what the forty-inch annual rainfall could wash into their cisterns from the rooftops. Visitors were expected to conserve. Into the 1940s, a sign in Captain Lowe's hotel in Tavernier on Key Largo asked, "Why take three baths a week when one will do?" (Some fifty years later a small sign in a local motel bathroom still asks visitors to "Please take short showers.")

As World War II approached, the U.S. Navy came to Key West in force and, in 1940, in cooperation with the newly formed Florida Keys Aqueduct Commission, built a pipeline to carry water 130 miles from wells at Florida City on the mainland. Now, with freshwater and a highway reaching the Keys, the game board was set up for the developers. With the end of the war, they swooped in, aided by jubilant politicians. Prosperity was coming to Paradise at last.

During the 1950s, developers began to plat the Keys, undeterred by the fact that acre after acre of the proposed home lots was covered by a foot or two of tidal waters where mangroves still grew and wild birds nested. Monroe County's five commissioners, exercising their own home-rule brand of government, were allowed by the state to exploit their island environment in their own fashion. They welcomed dredgers and builders. Said one conservationist of that period, "Fast-buck opportunists swarmed over the Keys like locusts." Dredgers dug up the coral, which belonged to the people of Florida, and sold it to developers who used it as fill to

convert more swamps, mangrove forests, and shallow bays to dry land. Whole new communities rose beneath the tropical sun. Leading right to residents' yards were newly dredged pleasure-boat canals whose deep, quiet waters became sumps for raw sewage.

By the 1980s infusions of state and federal money for improving the highway and waterline had shifted tourism into high gear. Sleepy fishing villages awakened. Neon-lighted tourist facilities blossomed. Dreamers bought building lots, dry or submerged, often without seeing them. Real estate became big business. It still is. Today, roughly 1,700 people are licensed to sell real estate in Monroe County.

Over the years environmental problems have mounted. Because of the extensive dredging and filling, prime wildlife habitat has been destroyed. National Audubon's Robert Porter Allen was among the earliest scientists to note the changes. He came to the Upper Keys in 1939 to study roseate spoonbills and stayed on to establish Audubon's research headquarters at Tavernier. Audubon Vice-President Alexander "Sandy" Sprunt IV took over there in 1960.

Sixty-one percent of wading-bird feeding grounds disappeared from Lower Metacumbe Key north through Key Largo, and of this forty-four percent was due to dredging and filling. "Spoonbills moved off their feeding grounds on Plantation Key and Key Largo, crossing the Florida Bay to the Everglades National Park," says Sprunt. In the warm midsummer days, algae bloomed in the boat harbors and canals. Divers on the coral reefs began noticing greater turbidity in the once crystalline waters.

In addition, the world-famous fishing in Florida Bay was deteriorating. And it is still going downhill. Some fishing bird populations continue to fall. Broad expanses of sea grass, nursery habitat for countless marine organisms, are dying in the bay. There are frequent fish kills.

While biologists may still be reluctant to say outright what causes the massive kills of fish and sea grass, there are prominent suspects. Audubon scientists are finding salt concentrations in parts of Florida Bay at twice the level of seawater, due perhaps to increased municipal and agricultural use of Everglades water, which reduces the flow of freshwater into the bay. In addition, there are abnormally high levels of nutrients (some say from sewage from the Keys and agricultural runoff from the Homestead area), and this means low oxygen levels, especially in recent hot summers.

Shrimpers have watched their catch plummet. In Key West, Sergeant Jim Brown of the Florida Marine Patrol says that while ten years ago there would be 500 shrimp boats for his officers to check in the peak shrimping season, the fleet last winter was down to fewer than 200 boats. In 1990 the shrimp catch was measured at 2.4 million pounds in Monroe County, down from an annual average of 6.4 million pounds for the years 1968 to 1988. Imports of pond-grown shrimp are considered a major factor in the decline of the Florida shrimp industry. But marine biologists also suspect that habitat degradation has caused shrimp populations to decrease over the last couple of years.

Even in the wilderness areas, humans are a growing threat to wildlife. One biologist who hesitates to be identified because of the political clout of the recreational watercraft lobby says, "It makes me want to cry. We first started to see Jet Skis in 1988, and now they are in all the refuge areas. They're going into the most pristine areas, places that have never before been disturbed. They seek out the flat, calm tidal creeks too shallow for boat traffic. They scare eagles off the nest. Ospreys don't even lay eggs in these areas anymore. They chase fish and shorebirds off the flats."

A November 1990 report prepared by a work group that included

Scars caused by boat propellers and jet skiers are clearly visible on the ocean bottom.

representatives of various environmental agencies and organizations, including the National Audubon Society, the Florida Keys Audubon Society, and the Wilderness Society, finds that "virtually no corner of the Keys seascape is unaffected by boating impacts." This group reports that between 5,000 and 10,000 acres of sea-grass meadows, important to a wide range of wildlife, have been "severely impacted" by propellers scraping the bottom and by turbidity and siltation. At least four endangered species—the American crocodile, bald eagle, West Indian manatee, and Atlantic green turtle—are among those suffering from this habitat destruction. Even if given immediate protection, the sea-grass meadows may not recover for many decades.

In a September 20, 1990, letter to Florida Attorney General Bob Butterworth, the Wilderness Society's Florida Keys representative, Ross S. Burnaman, wrote, "Sadly, little is done to stop the widespread destruction of sea-grass beds and disturbance of shallow water habitat."

In Marathon, Peter I. Kalla, nongame biologist for the Florida Game and Fresh Water Fish Commission, keeps track of more than thirty species on Florida's list of threatened and endangered terrestrial animals in the Keys, several of them found nowhere else. The endangered species in the Keys include the American crocodile, Key Largo wood rat, Lower Keys marsh rabbit, and Schaus' swallowtail butterfly. But the best known of these troubled animals is North America's smallest white-tailed deer. Isolated in its saltwater environment long ago, the Key deer adapted to life among the palmettos and mangroves and evolved into an animal different in many ways from its larger northern cousins.

In 1954, after many people thought it was already too late to save the tiny deer, the federal government designated the National Key Deer Refuge on Big Pine Key. The first refuge manager there was the late Jack Watson—burly, gruff, profane, and incorruptible. An Audubon writer visiting Watson after he had been on the job some ten years reported that the cigar-smoking, gun-toting warden had taught the poachers some respect.

"You have to deal with these damned outlaws in a way they understand," Watson said. In those days he knew which of his neighbors were poaching, and when he couldn't collect evidence that would stand up in court, he intimidated them. On Big Pine Key the disappearance of any dog, for any reason, was blamed on "that damned federal warden"—usually with justification because Old Blue and Old Chaser risked their lives if they chose to run on Watson's refuge. Key deer began a slow increase in numbers. Current estimates tell us they number between 250 and 300, a gene pool that geneticists believe still places them at extremely high risk.

Highway mortality is a major threat to these little deer. The number of deer killed by highway vehicles has climbed to more than forty each year since 1985, and this does not include those that make it off the road to die unseen in the palmettos.

Meanwhile, the deer suffer from poor nutrition because people feed them illegally. They even feed them carrots, bag and all, and several deer have died with plastic bags in their guts. According to biologist Willard D. Klimstra, a leading authority on the species, people have converted 25 to 30 percent of the deer on Big Pine Key into semidomesticated panhandlers. Klimstra, who has studied these little deer for nearly a quarter of a century, is convinced that the number of "yard deer" is growing.

"This is disastrous," he says. "It makes them zombies. They lose their natural fear of dogs and people." Tourists hand-feed them, trying to hug them and be photographed with them. This summer two residents were convicted in federal court for coaxing a deer close enough to beat it to death with a ball bat.

The Endangered Species Act makes it a crime to act in any way that alters the behavior of an endangered animal, and artificial feeding clearly alters behavior. Says Klimstra, "We could do more than we have toward enforcing the laws that are on the books."

But loss of habitat poses the most severe threat to the Key deer over the long haul. "We can get a handle on the road kill business," says one biologist, "but there is no way to counter this loss of habitat."

"I can't see anything ahead but a declining population," says Kalla. Most biologists agree. And the Key deer is symbolic of several species of wildlife caught in the Florida Keys people crush.

One note of hope was struck in 1975, when Florida designated the Keys an "Area of Critical State Concern" to save resources that belonged to the state. To the dismay of frustrated promoters and developers in the Keys, the Florida Department of Community Affairs could now limit building in Monroe County. Sandy Sprunt credits this designation with saving North Key Largo, the largest block of undeveloped land in the Keys.

But how does this meddling by outsiders sit with the local drumbeaters? Gene Lytton, a member of one of the oldest families in the Keys, a former Monroe County Commissioner, and long known as part of the "concrete coalition," echoes the developers' sentiment that there can be no economic health without growth. After identifying himself as an "environmental realist," he notes that he is a strong advocate of private property rights. "The state emasculates local government," he says. "They tell you what you can do on your own property.

They say, 'Oh, you can't build here because you've got this pretty little wang poo plant.'

"It's the 'got miners' causing the trouble. They come down here with their General Motors pensions and build a house, and they say, 'I got mine,' and they don't want anybody else to get what they've got."

Florida has searched, without much success, for ways to slow development on Big Pine Key. One conservationist, who recently counted thirty-one houses under construction in prime deer habitat, called the official effort to slow building there a "farce." Big Pine Key, with some 3,800 building lots remaining, has perhaps the greatest development potential left in the Keys. Inevitably, the deer are losing the contest for living space. Builders are still granted more than one hundred building permits a year on Big Pine Key.

Throughout the Keys the human buildup has brought problems that seem to defy solution. Monroe County, for example, seems unable to cope with what may be this country's most primitive sewage disposal arrangement. "Fecal contamination" is a given. Cesspools were outlawed with the advent of zoning some thirty years ago, but R. J. Helbling, of the Florida Department of Environmental Regulation, says the Keys still have 5,000 to 8,000 active cesspools.

"Where else in the United States can you get away with running raw sewage right into the groundwater?" asks marine biologist Ron Jones of Florida International University in Miami. "In most places, they would tar and feather you."

Meanwhile, the Key's 15,000 or more septic tanks are not much of an improvement, because porous soils and high groundwater levels offer inadequate leach beds.

In addition there are, according to Helbling, thousands of live-aboard boats that are supposed to be equipped with holding tanks, but that illegally flush directly into the

harbors. Low cost. Low upkeep. And don't tell anybody where they can do it. As the mayor of Key West once said in a televised interview, "Whales do it in the water, why can't we?"

But after years of resistance, Key West recently brought on line a new sewage-disposal plant that meets EPA standards. Until the summer of 1990 this oldest city in south Florida, population 30,000, collected its sewage and piped it untreated into the water less than a mile offshore.

In addition the Keys have run out of space for disposing of solid waste. Residents pay heavily for the fleets of trucks shuttling throwaway plastic, garbage, and diapers to landfills on the mainland. And because they are charged extra for added weight, the old refrigerator or mattress often goes into the mangroves in the dark of night.

Meanwhile, there is growing concern because disturbing levels of toxic chemicals are being found in hard and soft corals in the reefs. Peter W. Glynn of the University of Miami's Rosenstiel School of Marine and Atmospheric Science reported in *Marine Pollution Bulletin* in 1989 that the toxics include heavy metals and such organochlorine pesticides as lindane, heptachlor, chlordane, and DDT. These are also found in fish and shellfish that people eat. Nobody can be sure where these chemicals are coming from. Glynn explains that his group's findings are not final, and additional laboratory tests are still under way to verify the levels of contamination.

Protecting the fragile coral reefs is an overwhelming problem. The touch of a foreign object—an anchor chain or even the foot or hand of a resting diver—can kill some corals. In 1957 a gathering of scientists in Everglades National Park recommended formation of a new underwater state park to protect the coral reef off Key Largo. This was the beginning of the famous John Pennekamp Coral Reef State Park.

Beyond these state waters the federal government protects the reef in the Key Largo National Marine Sanctuary. Congress recently passed a Florida Keys National Marine Sanctuary bill that extends the federal sanctuary all the way from Key Biscayne in the Upper Keys to the Dry Tortugas.

Boaters and divers flock to the coral reefs. Visitors to the Key Largo National Marine Sanctuary have increased to more than a million a year, nearly double the number when the sanctuary opened in 1975. Boaters are supposed to tie to buoys provided for the purpose instead of dropping anchor, but in spite of enforcement efforts and heavy fines, anchor chains still drag occasionally across the coral, boat hulls scrape bottom, and human hands touch the delicate formations. The growing numbers of visitors to the coral reefs call for a constant program of educating people on how to protect these sensitive areas. Billy Causey, manager of the Looe Key National Marine Sanctuary, believes that ongoing educational efforts— including brochures, slide shows, and underwater signs for divers—are working and that most people want to protect the reefs.

Oil could become another threat to the reefs and the Keys. The petroleum industry wants to move into offshore areas leased in the early 1980s. President Bush, however, issued a statement saying there will be no new leasing or drilling until at least the year 2000. "But this could change anytime, like 'no new taxes,'" says Chris Robertson of the Florida Public Interest Research Group. "It's not law."

Yet there are moments when the traveler hears that perhaps not all is lost. "I tend to be an optimist," says Audubon wading-bird biologist George Powell. "If you're talking about the water around the Keys, the future does not look good. But if they would develop adequate sewage systems—and it's still possible— water quality could improve. Terres-

trial habitat is a somewhat brighter picture. There are some fairly sizable chunks of habitat being preserved on Key Largo and Big Pine Key."

Organizations working to conserve lands in the Keys include The Nature Conservancy, the South Florida Water Management District, the U.S. Fish and Wildlife Service, the State of Florida, and the Florida Keys Land and Sea Trust. The South Florida Water Management District recently made $2 million available for the Nature Conservancy to purchase a 200-acre wetlands addition to the Key deer refuge. The Nature Conservancy also recently purchased 1,000 acres on North Key Largo, as well as land on Long Key for a new marine-science research and education center.

Some residents see a glimmer of hope in the results of the November 1990 election. Environmentalist Jack London, who has a background in college-level teaching, commercial fishing, house restoration, and writing, upset Gene Lytton in the biggest election landslide the Keys has ever seen. Says London, "We have to think of the quality of life. We have to curb the proliferating urban sprawl." The Monroe County Commission is at work on tests of sewage-treatment plans that might take some of the biological pressures off the marine ecosystem. The commission also is considering a new comprehensive county plan that would reduce the number of building permits for the Keys from about 550 a year to perhaps 125. Seasoned environmentalists in the Keys wonder, however, if it is already too late to see the marine ecosystem recover its health.

For the traveler Key West is the end of the road, and to those who remember the Key West of yesterday, the visit does little to lift the spirits. Even development booster Gene Lytton admits, "Key West is maxed out."

Historian Jefferson B. Browne would be shocked. Writing in 1912, he declared Key West "unchanging

and unchangeable" and predicted great prosperity. But prosperity came as part of a deal. The dredging and filling ended, but developers still force back the frontiers. Quaint, old conch houses—single-story, white-frame structures standing shaded and flower-bordered along narrow alleys—share space with high-rise condo complexes, where people live antlike and with resort hotels, where rooms go for $250 or $300 a night. Currently, there are some 600 single-family dwellings for sale in Key West. Northerners, thinking they are buying a little bit of the Old Key West, pay $200,000 or more for one of these small houses, while old-timers, forced out by high taxes, move off to Ocala. These "conchs in exile," as they call themselves, reminisce with other displaced Key Westers about ocean sunsets, Key lime pie, and the quiet island life, while longing for what will never be—a return to the Keys.

They leave behind the tourism promoters proud of the 2 million visitors a year coming to the Florida Keys. And still, the tourism people spend $3 million of hotel-bed-tax money annually, advertising around the world for yet more people to come.

The drumbeaters favor adding two lanes to the existing highway and bridges to better accommodate the people and promote growth. The four-lane highway is part of Monroe County's comprehensive plan due late this year. However, the plan, which the state requires of all Florida counties and cities, will limit the highway across the Keys to four lanes. "Please view this as progress," says one county official, "because there were plans to make part of it six lanes." Indeed, another county official told *Audubon*, "The four-lane highway is needed if our growth is to continue at our historic rate"— a view with which there is widespread agreement.

"Have you seen our sunsets?" the locals ask. Evening brings Key West visitors crowding onto the public dock. There are musicians and painters, and the scene here at the end of the road takes on a carnival atmosphere. People toss coins to a performing juggler. A thin dark-haired woman with a broad smile sells brownies from her bicycle, and craftsmen peddle wire jewelry.

Then the disappearing sun washes its reds and yellows across the evening sky and reflects its warm colors in the calm waters. And as it slides from view below the horizon, a thousand worshipers give the setting sun another standing ovation.

Discussion Question

1. List all of the problems facing the Keys and the surrounding waters. How can they be curtailed or prevented?

Return of the Red Wolf

In the Advancing Science of Captive Breeding and Species Reintroduction, the Red Wolf Leads the Pack

Chris Kidder

■

SOURCE
Nature Conservancy,
September/October 1992, Vol. 42, No. 5

Chris Kidder is a freelance journalist living on North Carolina's Outer Banks.

IT'S BARELY DAWN when the pickup truck turns off the highway and bumps down a sand road, through acres of coastal field and marsh of the Alligator River National Wildlife Refuge in northeastern North Carolina. Miles later, crossing blackwater creeks to forest thick with loblolly pine and bald cypress, the truck pulls up near a chain link fence.

Behind the fence are wolves. Red wolves, *Canis rufus*. Today, a family of six—constituting nearly 4 percent of the world's population of this species—is being moved to a new wilderness home in the Smoky Mountains of Tennessee to start a new population.

The wolves have done well here at Alligator River. They have fed themselves, multiplied and passed their initial test of wilderness fitness.

Now five years into a precedent-setting recovery experiment, the red wolf has become a showcase carnivore in a contemporary approach to biological conservation—endangered species reintroduction. However, their handling on this May morning suggests anything but that of a rare gem.

"Let's get in there and get them out," says Mike Phillips, donning coveralls. Phillips, the U.S. Fish and Wildlife Service (FWS) biologist who heads the red wolf project at the refuge, lifts the roof from one of the wooden wolf houses, slips a muzzle on wolf #357, an adult male cowering in the corner, and stuffs the animal, limp with fright, into one of the transport kennels.

The wolf's mate, #337, has taken refuge in another house with her four pups. Phillips gets a loop around her neck and holds her while Mike Morse, a FWS field technician, gathers the pups and puts them in another kennel. Finally, mother wolf is crated and the kennels loaded on the truck.

"Contacts with humans should not be pleasant," explains Phillips, who has spent the last five years trying to establish these wolves in the wild. He's seen lack of fear sentence wolves to lives in captivity or death on the highways. If the red wolf is to survive beyond chain link pens, their handling must be strictly business. Such are the complexities of the uncertain science of captive breeding and reintroduction.

Five years ago, a pair of captive-raised red wolves was brought here to the refuge and experimentally released, becoming the first wild-extinct species of carnivore to be returned to its native range. With additional releases, there are now perhaps two dozen roaming the refuge—enough that surplus wolves are now booked for new homes to start new families. Because of the red wolves' success, managers of other critically endangered wildlife, from California condors to black-footed ferrets, are taking lessons from Alli-

At Alligator River National Wildlife Refuge, government biologists are reintroducing the red wolf to its former habitat.

gator River on the experimental craft of endangered species reintroduction. Says Phillips, boldly, "Our management program portends the future for conservation biology around the world."

Thousands of red wolves once roamed throughout the Southeast. But invading humans, falling forests, disease and predator eradication campaigns helped drive them toward extinction. As they retreated, coyotes invaded from the west, filling niches left by the wolves and interbreeding with lonely survivors. By 1973, biologists determined that the once widespread red wolf had been pushed into just 1,700 square miles of coastal swamp straddling the Louisiana and Texas border. By 1980, the red wolf was thought extinct in the wild.

Breeding in captivity "was the red wolf's only chance for survival," says Roland Smith, who now directs the red wolf captive breeding program through the Point Defiance Zoo in Tacoma, Washington. Red wolves and any look-alikes were rounded up. By 1980, FWS biologists had captured more than 400 wild canids. The canids were evaluated for physical characteristics unique to red wolves: large ears, long, spindly legs, and overall lanky appearance. True red wolves weighed between 40 and 80 pounds—heavier than the average coyote, lighter than the average gray wolf—and their short, sparse coats were mottled brown, black, gray, yellow or—not realized by the biologist who named the species—rarely red. After genetic studies further separated out the purest from the inbreeds and hybrids, 14 red wolves were chosen to serve as founders for the species.

Under Smith's direction, wolves were strategically moved from one facility to another in a game of musical mates. And from those 14 original red wolves, nearly 200 descendants are known to be alive today.

The captive breeding program proved successful from the start. But transforming captive-bred wolves into wild wolves would involve much more than opening cage doors and waving goodbye. Wolves born in captivity needed time to acclimatize to their new surroundings, to adjust from a regular captive diet to one of sporadic road-kill carcasses and live prey. When a cage was finally opened, it was left that way, and food provided periodically should the wolves need help in the early and frightening transition.

Even this canine boot camp could not guarantee a completely wild product. The first male released in the Smokies was 10 years old and accustomed to people. He kept daytime hours and raided nearby poultry farms. He was recaptured and returned to captivity this spring.

"You have a hard time teaching old dogs new tricks," says Chris Lucash, Phillips' counterpart at the Great Smoky Mountains project. "He was all we had available. That's one of the problems you face with an endangered species. You can't always take and release the best."

Should the red wolves learn to avoid people, there is no assurance they will avoid their smaller cousin, the coyote. Once an embodiment of the Old West, the modern, adaptable coyote now occupies every state in the nation, including North Carolina. They have been known to interbreed with red wolves. And hybrids do not bode well for the future of the red wolf.

Although recognized now as a separate species under the Endangered Species Act, "the taxonomic identity of the red wolf has been in question ever since the animal was first discovered 200 years ago," says a recent FWS report. Recent studies raise new questions about the red wolf's genetic integrity, suggesting former trysts with gray wolves and coyotes. Because of genetic ties to other canids, the experts may decide red wolves aren't unique enough to justify the expense of the recovery program, says Phillips. The Endangered Species Act does not protect hybrids.

To some, however, the genetic debate overlooks the wolf's broader ecological profile. "Whatever the red wolf is," says Curtis Carley, who established the FWS red wolf recovery program in 1973, "the wolves we have in captivity seem to breed true and represent the southeastern canine that has been recorded as part of our heritage."

Genetic confusion aside, there remains yet another crucial question bearing on the red wolf's future: Where to put them? For the reintroduction specialists, it has come down to the infinitely complex task of finding habitat acceptable both to wolves and humans.

The first proposed reintroduction site, located on the Kentucky-Tennessee border, was scrapped in 1984 after three years of planning when factions from all demographic corners rallied against the plan. Ranchers feared for their livestock, hunters for their game. Even conservationists opposed the plan, fearing the wolves themselves would be left defenseless at the hands of humans. By some accounts, the proposal failed on the grounds of poor public relations.

The same mistake would not be repeated at Alligator River. When Warren Parker, FWS red wolf coordinator from 1985 through 1990, proposed reintroducing red wolves at Alligator River, he aimed to win approval from neighboring citizens in Dare, Hyde and Tyrrell counties. He made use of a 1982 amendment to the Endangered Species Act, allowing for an experimental population of wolves, made up of surplus individuals deemed unessential for the species' survival. (In this case, these were the wolves whose genes were no longer needed in the captive breeding program.) These surplus wolves could be introduced and—more importantly to the public—removed if they got into trouble.

Parker's bargain for public support ensured hunters and trappers continued access to the refuge, a privilege not as easily allowed under the original Endangered Species Act. The experimental classification allowed more aggressive control when the wolves ventured onto private land and leniency in prosecuting the public when wolves were accidentally killed. The new classification worked so well at Alligator River that it has been applied to another endangered carnivore, the black-footed ferret.

Shortly before Parker began introducing locals to their potential new neighbor, The Nature Conservancy completed negotiations for the donation of 118,000 acres of North Carolina wetlands to the Fish and Wildlife Service for creation of the Alligator River National Wildlife Refuge.

"Nobody in their wildest dreams ever thought about the red wolf at that point," says Merrill Lynch, ecologist for the Conservancy's North Carolina field office. "We were looking at large landscape protection." The forested coastal swamp and 100-mile estuarine shoreline comprised the greater part of a peninsula bounded by the Alligator River and the Albemarle and Pamlico sounds. The refuge harbored alligators, bears, deer and more than 40 species of breeding birds. It also fostered multitudes of marsh rabbits and small rodents, which, coincidentally, nicely suited the tastes of wild red wolves. With Parker's diplomacy and the Conservancy's real estate wizardry, the red wolf finally had a home.

While many conservationists dislike the species-specific approach to preserving biological diversity, most admit that programs to save species like the red wolf have benefits beyond the main attraction. "You're dealing with a top predator with large habitat requirements," explains Bob Jenkins, The Nature Conservancy's science director. "If they can be met, it will provide habitat for thousands of other species that might not otherwise get attention."

Most of the Alligator River National Wildlife Refuge is in Dare County, where tourism is the major industry. In Dare County, enthusiasm for the red wolf program is high. The Dare County-based Coastal Wildlife Refuge Society has announced plans to raise $4.5 million for a visitors' center on Roanoke Island. The red wolf will be the hook for fund raising, the centerpiece of proposed exhibits.

Yet while Dare County citizens plan to promote red wolves, their neighbors in Hyde and Tyrrell counties take a more guarded view of the reintroduction program. "I don't think people care as long as the wolf stays out of their way." says Lindsay Mooney, a Hyde County hunter. A wolf left the refuge recently and locals claim it killed at least one domestic cat and a gaggle of geese before being recaptured. Now, says a local trapper, "people in Hyde County, they're dead set against the wolf."

Public opposition in Hyde and Tyrrell counties would be a serious setback. "Without the support of the public, we're dead in the water," says FWS red wolf coordinator Gary Henry. "We're aiming at allowing the wolves to disperse naturally westward."

"Westward" lies the 110,310-acre Pocosin Lakes refuge, an area identified by Conservancy ecologists as prime habitat for red wolves, purchased by the Conservation Fund and recently donated to FWS. For the program to be cost effective, "we've got to get 50 wolves out there in the next couple years," says Phillips. "That probably can't be done without more land."

The red wolf and other "glamour" species can educate people about the need to preserve biological diversity, says Phillips. "People get excited about wolves and that gives us a chance to point out other important things that people don't get excited about."

On the other hand, glamour species preservation is almost always large and very expensive. "Of course it's necessary to educate the public about endangered species," says Jenkins. "Whether the red wolf program is a cost-effective device for that aspect I can't say." If the goal is biological diversity, focusing solely on large species is a dangerous policy, says Lynch. "We can't overlook protecting the small things."

"The wolf is the acid test for how willing we are to allow predators to survive," says Jan DeBlieu, author of *Meant to Be Wild,* a book about endangered species programs. "What are we willing to do to restore the natural landscape?"

People can provide the red wolves with wilderness homes, but the business of basic survival will be up to the wolves themselves. And the natural hazards are many. Female wolf #337, now headed for the Smokies, has been a survivor. At Alligator River, her previous litter of pups was killed and her mate run down on the highway.

There will also be some mandatory intrusions on her freedom. If she and her new mate, #357, are released in the Smokies, they—and any pups they may sire—like all wolves in the program, will be medicated, tracked and managed to fit promises made to an apprehensive public.

The goal to restore the red wolf in its native range and preserve the species for generations to come may be greater than the sum of its parts. It may not be enough that experimental reintroduction and captive breeding have been successful. Most of the Alligator River wolves that live in the wild are radio-collared for surveillance. Their range, their health and even their sex lives are intensely managed.

"For full recovery to have taken place for the red wolf," says Jenkins, echoing a concern other conservationists have already voiced, "that cannot mean every red wolf is radio-collared and recaptured at will. It has to be a living population going where, and doing what, it wants."

Discussion Questions

1. Make a list of concerns regarding the reintroduction of red wolves that various interest groups, including livestock growers, pet owners, parents, and poultry farmers, might have.

2. How could these concerns be addressed?

The Beetles and the Wily Weed

A Natural Weapon Will Join Wetlands War

Gwen Florio

■

SOURCE
The Philadelphia Inquirer,
March 15, 1992

FIRST, The Villain: Like all the best bad guys, his appearance is deceiving. People like his looks. "How gorgeous!" they exclaim upon seeing his lovely lavender cloak. Little do they know the evil that lurks within.

Call him The Purple Peril.

Our Hero is just right, too: Small, nondescript—a little grubby, actually—the kind of fellow nobody notices as he scurries about his business.

Call him *Hylobius transversovittatus*— a.k.a. "Bug."

He's the U.S. Department of Agriculture's secret weapon in a decades-long war against purple loosestrife, a

tall, handsome wildflower with the dismaying habit of elbowing other plants out of their wetlands homes.

The towering purple plant and the tiny, long-nosed bug have been waging an epic battle for centuries in Europe.

Soon, the war will commence on America soil, with the first volleys being fired at the John Heinz National Wildlife Refuge at Tinicum, just south of Philadelphia in Delaware County.

If all goes well, thousands of Hylobius beetles—eventually joined by backup contingents of beetles known as *Galerucella pusilla* and *Galerucella calmarientsis*—will begin munching their way across the refuge this summer, nibbling at the Purple Peril's roots, lunching on its leaves.

It's enough to make a loosestrife turn pale.

Hold your sympathy, says Jackie Burns, an outdoor recreation planner at the refuge.

The loosestrife, which in the summertime covers acres of the refuge with its dramatic blossoms, is a real people-pleaser, she said. But pleasing people is not the nature preserve's primary mission.

USDA entomologist Stephen Hight called the plant "a problem for wetlands and wetlands wildlife, especially waterfowl." Ducks and geese find loosestrife inedible, and it crowds out the native plants they like to eat.

"It has kind of a negative impact for waterfowl," said Hight.

The loosestrife is a European import that made its way to this country in the 19th century, he said. The plant quickly spread across the northern half of the continent, ranging from New Brunswick, Canada, to North Carolina, and all the way out to the Pacific Northwest.

People tried uprooting it, cutting it, burning it, and blasting it with herbicides—only to see the loosestrife pop back up again every spring.

Meanwhile, its only natural enemies, the beetles, remained in Europe.

Until now, that is. After 10 years of research, the USDA is bringing them to Tinicum and a refuge near Buffalo. If they prove effective, they'll be sicked on loosestrife across the country, Hight said.

The project has cost about $500,000—cheap compared with herbicides, he said. "They can cost millions . . . and that doesn't even include negative environmental impacts," he said.

The money came from both the USDA and the U.S. Fish and Wildlife Service, which funded testing in Europe to ensure that the beetles would stick to loosestrife, and not develop a palate for tasty American morsels.

As larvae, Hylobius chew on the loosestrife's roots; as adults, they eat the leaves. Nearly 4,000 Hylobius eggs wintered at Tinicum in two big, screen tents in the middle of a huge patch of 6-foot-tall loosestrife.

USDA permission to release them should come by the end of the month, Hight said. The Galerucella are still in Europe, and could be released in this country by late July if they pass quarantine and other regulations, he said. Like the adult Hylobius, they feed on the loose-strife's leaves.

"Our hope is that this will be kind of a one-two punch—hitting it above ground and below," he said.

Despite the size of the project, visitors to the refuge shouldn't fear millions of crunchy beetles under-foot. "It's the classic thing—the plant population is very high, so the insect population would build through a few years. Then the plant population would start dropping off; so, too would the insects," he said.

Nor should die-hard loosestrife fans worry that the plant will disappear entirely. Hight said the USDA is hoping to kill off 80 percent of the loosestrife—in the next *decade.*

"All we want to do is reduce the population," he said." . . . It's a great looking plant."

Discussion Question

1. Critically analyze what this article suggests about introducing nonnative plant species into foreign environments.

Air Pollution

Although non-Southerners may equate air pollution with Southern California, the industrial Midwest, or perhaps urban centers in the Northeast, the states in the southern region of the United States have significant problems with air quality. For example, the annual emissions estimates of toxic pollutants required by the Emergency Planning and Community Right to Know Act of 1986 show that in 1987 Texas, Louisiana, Tennessee, Virginia, Georgia, and North Carolina were among the top 10 states with respect to toxic air emissions. Facilities located in the 14 states in the South reported air toxics emissions of over 1 billion pounds, just under half of the total emissions nationwide. Although improvements have been made, the southern states continue to be a leader in toxic air pollution emissions.

Moreover, 11 states in the South contain urban areas that fail to meet the EPA standard for ozone. The city of Houston's ozone problem is classified as severe. The cities of Atlanta, Georgia; Baton Rouge, Louisiana; Beaumont-Port Arthur, Texas; El Paso, Texas; and Washington, D.C., are in the serious category. Twenty-four areas in West Virginia, North Carolina, Florida, Kentucky, Tennessee, Alabama, South Carolina, Louisiana,and Texas have moderate to marginal ozone pollution.

Background Reading: Chapters 15 and 16.

Back on Track

Earth Day Success Story: The Chattanooga Choo-Choo No Longer Spews Foul Air

William Oscar Johnson

■

SOURCE
Sports Illustrated,
April 30, 1990, Vol. 72

HERBERT O. FRY has fished the Tennessee River around Chattanooga since he was a boy, and he's 73 now. He sums up his life this way: "I sold yarn, millions of dollars worth to Jews in Brooklyn, to Swedes in Minnesota, to points west, points east, all points of the compass. I traveled everywhere selling yarn, but no matter where I was I always was just *achingly* interested in catching fish."

Fry retired from the yarn game 11 years ago and now stays home in Chattanooga soothing his ache for fish and raising hell about the state of the environment—mainly about the state of the river. Last weekend he attended some local Earth Day activities in the city's new Tennessee Riverpark, a lovely two-mile greenbelt that stretches along a previously cluttered and inaccessible section of the riverbank. The $4.8 million park is beautifully landscaped, equipped with walking paths, picnic tables, state-of-the-art playground appa-

ratus, five sturdy fishermen's piers and at its center, an attractive building that includes a combination delicatessen bait shop, where you can buy a submarine sandwich and a cup of earthworms in a single transaction. The building was named after Fry because this park was his vision—and he applied his considerable salesman's charm and fisherman's tenacity to the project until the damned thing got done.

Since he is obviously something of an expert on citizen action in the interest of the environment, Fry was asked if he thought Earth Day 1990, which engaged the attention of the world on Sunday, would really make much difference in the grand scheme of things. He said, "It could change the world if it got embedded in the mind of ordinary folks. But I think there is very little chance of that."

The last time there was an Earth Day, the ordinary folks of Chattanooga scarcely noticed. The year was 1970, and the city was under siege from the sky.

Ironically, the town's nicest features were the cause of its worst problems. Chattanooga is gently cupped in one of the most beautiful river valleys on earth within a ring of some of the most beautiful mountains you ever saw. Strong winds rarely blow and the sun shines 213 days an average year.

Unfortunately, the sun hits the rim of the mountaintops before it hits the valley floor. This causes a near-perpetual inversion, which with the lack of wind, holds the lower layers of air close to the ground. That air was laden with tons of pollutants in 1970. It had been growing increasingly dirty since before the Civil War as Chattanooga, with its river, its railroads and its coal mines attracted more and more industry—from foundries to coke plants to the noxious Volunteer Army Ammunition Plant (VAAP), which produced more TNT than any other plant on earth.

In the decade before the first Earth Day, which was mainly an American event, the federal government studied air pollution in U.S. cities. Its tests for particulate pollution covered two types: TSP, total suspense particulate, such as bits of floating soot and dirt, and BSO, benzene-soluble organic particles— same stuff, smaller particles. In 1969 the government announced that among the 60 cities tested during the period of 1961–65, Chattanooga was worst in TSP and No. 2, or runner up worst, to Los Angeles, in BSO.

So Earth Day 1970 was just one of the many dark and dirty days in Chattanooga, a city in which the mid-'60s death rate from tuberculosis was double that of the rest of Tennessee and triple that of the rest of the U.S., a city in which the filth of the air was so bad it melted nylon stockings off women's legs, in which executives kept supplies of clean white shirts in their offices so they could change when a shirt became too gray to be presentable, in which headlights were turned on at high noon because the sun was eclipsed by the gunk in the sky. Some citizens could actually identify different sections of town by nose alone—a stink of rotten eggs in one place, acrid metal in another, coal smoke in another, the pungent smell (and orangish haze) of nitrogen dioxide near the VAAP. One part of town, site of a city dump, was known simply as Onion Bottom.

People joked, "We like to look at what we're breathing before we inhale it." A billboard appeared bearing this alarming question: DEEP DOWN INSIDE . . . WOULDN'T YOU RATHER BREATHE CLEAN AIR? Anyone who has lived in Chattanooga for a while has memories of what the old air did. Linda Harris, local chairman of Earth Day 1990 and acting director of the Chattanooga Nature Center, recalls her childhood: "Our eyes stung, and our noses itched. The milkman left milk bottles at dawn, and when we brought them in a couple of hours later, we could write our names in the dirt that had collected in the moisture on the bottles." Wayne Cropp, director of

the Chattanooga-Hamilton County Air Pollution Control Bureau, says, "You could always tell which way the ammo plant was in town by looking at the bushes in that part of town: The side away from the plant grew green, and the side toward it was brown." Fry recalls, "As a boy, I had a morning paper route. I delivered before daybreak. If I had a cold, my nose would be running black by the time I got home."

The established federal standard for acceptable TSP is 75 micrograms per cubic meter of air; Chattanooga averaged 214 micrograms in 1969. There were days when the TSP count climbed into the thousands, and sometimes filters in the monitoring equipment got so clogged it broke down. Nevertheless, apathy reigned among Chattanoogans until that federal report suddenly identified their city as being the foulest air polluter of them all. "That was our warning heart attack," says William Sudderth, a second-generation Chattanoogan and a riverfront developer. The whole town suddenly gave its full attention to the crisis. There was no single power source behind the clean-air crusade—no benevolent despot, no politician, no environmental activists, no enlightened aristocracy, no Earth Day sponsors. "Everyone kind of rose up together and responded," Harris says.

And what followed was not a miracle but a nuts-and-bolts model of how tough government, cooperative businessmen and a very alarmed public can make a dirty world clean again. With a variety of local and federal sticks and carrots to motivate them, some of the worst corporate polluters installed filtering and scrubbing devices that, in some cases, reduced smokestack pollution by 99%. Polluters who threatened to move out of town if they were forced to meet federal standards were told to go ahead and get out. One of the greatest contributions to a cleaner atmosphere occurred in 1977 when the VAAP was put into mothballs by the Department of Defense.

It didn't happen right away, but day by day the Chattanooga choo-choo blew out less and less corruption. In 1981 the TSP level dipped below the 75 mark for the first time—and stayed there. In '83 the chamber of commerce launched a campaign to boost the town with the previously laughable slogan "Chattanooga Shines!" And in '89 the annual TSP reading was the lowest it had ever been—just 49 micrograms per square meter; and BSO, now superseded by a category called PM_{10} (Particulate Matter 10 microns or less in diameter) is well within the government's standard.

Well, that's a marvelous Earth Day bedtime story—but Chattanooga didn't stop with its air. Next came the river. The downtown banks of the Tennessee had been so cluttered with ancient factories, warehouses, and shipping piers that there was no way for ordinary citizens to get near the water. That is changing as we speak. Fry's previously unreachable section of the river will soon be only part of Riverpark, a mammoth $750 million, 20-year undertaking that will clean up 22 miles of river shore. The centerpiece will be in the middle of the city, at Ross' Landing (named after the celebrated Cherokee chief John Ross), where a state-of-the-art freshwater aquarium will be completed by the summer of 1992. Tall as a 12-story building, it will have a re-created Southern Appalachian mountain forest under a glass pyramid, complete with waterfall, trout stream and otter pool, as well as living, swimming examples of practically every freshwater fish and reptile known to man.

Discussion Questions

1. Many people are pessimistic about our ability to respond to environmental problems. Does this article make you more or less hopeful? Why?

2. Explain why the Chattanooga valley is so vulnerable to air pollution.

Houston Citizens Negotiate with Giant Oil Company and Win Toxic Reductions

The Environmental Exchange

■

SOURCE
What Works: Air Pollution Solutions,
Report No. 1, 1992

ENVIRONMENTAL activists in the Lone Star State have their work cut out for them. Industries in Texas report releases of more toxics into the air than in any other state—close to 800 million pounds in 1989. That's 67% more than Louisiana, the number two state. And petrochemical companies have long been a powerful force in the Texas economy.

But George Smith and his colleagues in GHASP (Galveston-Houston Association for Smog Prevention) have found a way to fight back.

GHASP is a coalition of clean air activists in the Houston area. It includes both individual members and representatives of groups like Sierra Club, Audubon Society, and the Houston Outdoor Nature Club. GHASP focuses its efforts on the permitting process and has been involved in dozens of permitting battles, intervening when it believes a company poses an unreasonable risk to a community in the area. "We try to pick the fruits that look juiciest," Smith explains.

For example, when an ARCO petrochemical plant announced a major expansion that would dramatically increase benzene vapor emissions in the Channelview neighborhood, GHASP swung into action. Benzene is highly carcinogenic, causes leukemia, and short-term high doses can kill you.

Examining the permit application, GHASP questioned whether ARCO was meeting state regulations, which required the use of the best pollution control technology available. Information from an anonymous consultant provided evidence for GHASP's claim that ARCO had underestimated emission levels and failed to include sufficient pollution reduction measures in its permit application.

When it appeared that GHASP's complaints about the application might force the Texas Air Control Board to hold a public hearing, ARCO decided to avoid further delays in the permitting process and began direct negotiations with GHASP that produced dramatic changes in three important areas:

- *Pollution Reduction.* ARCO agreed to significant pollution control measures: installing hydrocarbon scrubbers, using vapor recovery for barge loading, improving benzene tank seals, and implementing a better wastewater discharge process. None of these steps had been included in the original permit. The result? Benzene emissions of more than 20 tons per year were avoided. That's a lot of carcinogenic pollution people in the area won't have to breathe.

- *Citizen Involvement.* Channelview residents are more involved in air toxic issues in the wake of GHASP's success. Residents were not actively involved in GHASP's negotiations with ARCO, in part because they did not want to be viewed as anticompany, and in part because some citizens were preoccupied with battling an incinerator in the area. But after negotiating with GHASP, ARCO decided to encourage more citizen input. ARCO and another company in the area asked GHASP to help set up a citizens' advisory panel involving local residents; residents have responded enthusiastically to the opportunity to keep formal tabs on the companies in their area.

- *Stricter Standards.* GHASP convinced the Air Control Board to use a more accurate way to estimate hydrocarbon evaporation from petrochemical plants—a step Smith believes will be indispensable to similar fights in the future. "When other plants of this type are applying for a permit, it will be much easier now to convince them to install leakless valves," says Smith.

"There are two ways for citizens to make a difference in Texas," Smith continues. "First, they should contest permits and negotiate with companies to provide for reductions. Second, they should complain to the Air Control Board. Our involvement in the permitting process puts the Texas Air Control Board and companies out there on notice: they can't slide by with incomplete permits or insufficient control technology."

Though not infallible and hardly easy, the strategy of intervening in the permit process can work—and is spreading. GHASP frequently gets calls from other citizens' groups around the country. Smith notes, "We can't fight their battles for them, but we can provide advice on effectively intervening in the permitting process."

Discussion Questions

1. After reading this article, what advice would you give to companies interested in locating in or near a community?

2. How do the efforts of GHASP contribute to building a sustainable society?

Reducing Toxic Emissions

Low-Income Neighborhood Blocks Houston Ammonia Facility

The Environmental Exchange

■

SOURCE
What Works: Air Pollution Solutions,
Report No. 1, 1992

WHEN LaRoche Industries decided in 1990 to put a 105,000-gallon ammonia storage facility in Cloverleaf, Texas, its permit application to the Texas Air Control Board let the cat out of the bag. The company argued that the Cloverleaf neighborhood was "appropriate" for the facility because it consisted mainly of "small poorly maintained houses . . . small junky businesses . . . and very low quality housing."

This reasoning did not sit well with the blue-collar, multiethnic residents of Cloverleaf, which lies just east of Houston. Residents wanted to know why LaRoche considered it more "appropriate" for them to bear the risks of housing huge quantities of flammable ammonia than their

higher-salaried neighbors. Their worries about the storage tanks were heightened by an accident that had taken place a few months earlier: a 3,000-gallon ammonia tanker had crashed and ruptured on a Houston freeway, killing seven people, hospitalizing 50, and requiring the evacuation of more than 1,000 others.

At the first community meeting attended by the Texas Air Control Board, residents and activists from the environmental group called Texans United hung posters showing enlarged reproductions of the offensive language in the permit application. "The Air Control Board officials flew into our first meeting expecting to smooth things over with a few fearful homeowners," remembers one of the community's leaders, Karla Land. "But when they looked around the hall and saw those posters, they knew they'd walked into a hornet's nest of organized opposition. The Board seemed a little stunned, promised to consider our objections, then headed back to Austin."

The community held a second meeting, inviting LaRoche and the Air Board to meet with residents, school board members, and state and county officials. The residents received a welcome surprise when Land stood up to read a letter to LaRoche from the Air Board's permitting engineer, which stated: "I have determined that the facility has an unreasonable potential to emit large quantities of ammonia. A large number of individuals in the immediate area would be adversely affected by accidental release. Based on the information submitted to date, my recommendation would be that the application be denied." The Air Board accepted its engineer's recommendation, and a Board official later admitted that this was the first time the Air Board had ever denied a permit application.

LaRoche also sent a letter to community leaders in Cloverleaf, informing them of its decision to look elsewhere for their storage tank site. While residents were relieved, many

expressed concern about the risks the tank would eventually pose in some other neighborhood. Some Cloverleaf residents decided to act on their concern, and they now volunteer for Texans United to help provide resources and organizing assistance to other low-income communities in the state.

Discussion Question

1. Some people say that environmental issues are of no concern to low-income communities. Do you agree?

His Family Ravaged by Cancer, an Angry Louisiana Man Wages War on the Very Air That He Breathes

Susan Reed

■

SOURCE
People, May 25, 1991, Vol. 35, No. 20

AMOS FAVORITE is 68, and he has a story to tell. "I lost my uncle James Barber and his son Israel," he says, ticking off the names on his fingers. "I lost another uncle, Jorden Favorite, his wife, Pipine, and their son Isaac. Then there was another uncle, Herbert Favorite. My cousins

"I'll fight till I die to get zero discharges from those places," says Amos Favorite (outside Union Texas petroleum's plant in Geismar, La.).

Lizzie and Emily died in their 40s. Then there was my aunt's husband, Josene LeBlanc." Favorite shakes his head—nine members of his family gone in 13 years. "Our people used to live to be 95, 100 years old," he says. "Now even the young people are dyin'."

Favorite thinks he knows the cause of those deaths, and if you visit him in his hometown of Geismar, in Ascension Parish, La., he might climb into his secondhand blue 1984 Lincoln Continental and take you on what he calls a "toxic tour." Located 10 miles south of Baton Rouge, Ascension is a poor and rural area, a place of pastures, live oaks and small brick houses. At the intersection of highways 30 and 73, however, the bucolic scene changes abruptly. Across the road sits a massive, Oz-like city of steel pipe, cylinders and tanks. Steam billows from the cooling towers of plants with names like Rubicon, Liquid Carbonic and Borden Chemical, and flames from burning chemical waste leap from 200-foot-high stacks. Strange odors and a fine mist fill the air.

Inhabited mostly by blacks, Ascension Parish lies in the heart of a notorious 85-mile stretch of the

Mississippi Valley that has come to be known as Cancer Alley. Between Baton Rouge and New Orleans, 125 companies produce 20 percent of America's petrochemicals for such products as fertilizers, herbicide, gasoline, paint and plastic. In the process, more toxic chemicals are spewed into local air, land and water than in any other state in America— more than 2 billion pounds from 1987 to 1989. Dr. David Ozonoff, an environmental epidemiologist at Boston University's School of Public Health, calls the chemical corridor "a public-health catastrophe." Dr. Velma Campbell, who was a physician at New Orleans's Ochsner Clinic, says, "The area is like a massive human experiment conducted without the consent of the experimental subjects."

No one is angrier about the situation than Favorite, a retired aluminum-plant worker who is president of a fledgling environmental group called Ascension Parish Residents Against Toxic Pollution. "We got some deadly poisons out here," he says, as he drives past BASF, Vulcan, Triad, CF Industries, Shell Oil and Air Products & Chemicals, some of the 18 chemical companies that crowd the 296-square-mile

parish. In 1989 these companies discharged 124.4 million pounds of toxic chemicals, including the carcinogens vinyl chloride and benzene, as well as ammonia and mercury, which affect the nervous system, and chloroform, toluene and carbon tetrachloride, which can deform fetuses. "We call it toxic gumbo," says Ramona Stevens, another local activist.

Yet there is little agreement about the severity of the problem—or even whether there *is* a problem. According to the National Cancer Institute, Louisiana has the highest lung cancer mortality rate in the country among white males, and Ascension Parish ranks in the top 10 percent nationwide for pancreatic cancer. So far, however, scientists have failed to make a direct link between those numbers and chemical pollution. "Carcinogens in Louisiana have been well regulated," insists Richard Kleiner of the Louisiana Chemical Association, an industry lobby group. Kleiner argues that smoking and diet contribute more to the state's high cancer rates than chemical emissions. "Studies don't suggest that the environment plays a major role," says Kleiner.

If that is so, critics ask, then why

have three major chemical companies—Dow, Georgia Gulf and Exxon—spent millions of dollars recently to relocate people living near their plants in the Mississippi Valley? "Companies are moving people instead of reducing pollution," says Marylee Orr, executive director of the Louisiana Environmental Action Network. "They're dealing with the effects, not the causes."

One difficulty in establishing a cause-and-effect link, says Dr. Linda Pickle, bio-statistics director at the Lombardy Cancer Research Center at Georgetown University in Washington, D.C., is that "environmental pollution is very difficult to assess. There are so many chemicals in the air down there, and no one really knows the effects of long-term, low-level exposure. Cancer takes at least 10 to 20 years to develop, so the current mortality rates in the state probably reflect conditions two decades ago." Still, she cautions, "Studies on asbestos and cancer have suggested that toxic chemicals may very well have a synergistic effect. People who smoke *and* breathe chemicals are likely to be at higher risk."

The absence of a direct link, argues Lois Gibbs, executive director of a Virginia-based group called Citizens' Clearinghouse for Hazardous Waste, allows chemical producers too much leeway. "Chemicals should be guilty until proven innocent," says Gibbs, who notes that the Environmental Protection Agency regulates only eight airborne pollutants. "In this country, they're innocent until proven guilty. And the public is guilty of hysteria until proven right."

Amos Favorite and his neighbors say they don't need statistics to prove that something is terribly wrong. They talk about the former pharmacist in nearby St. Gabriel, Kay Gaudet, who counted 643 miscarriages among local women in 1987. They talk about mustard greens that turn black overnight in

their gardens, pecan trees that have stopped bearing nuts, new aluminum window screens that disintegrate in a month, and a white powder that settles on cars and eats away paint. Residents talk about pervasive asthma and their children's chronically runny eyes and noses. Recently, Favorite noticed that huge branches from the parish's blackberry trees had started falling to the ground. "The chemical plant people tell me it's the frost," he says. "I been here 68 years and never seen anything like this."

Favorite and his neighbors breathed, bathed in and drank toxic chemicals for decades without knowing it. That changed on June 15, 1984, when BASF, located in Geismar, locked out 370 local members of the Oil, Chemical and Atomic Workers union. Richard Miller, an organizer for the union, traveled to Louisiana to support the workers. "When I got here, I looked around for allies," he says. "I thought 'Who are the other victims'? The residents of Geismar were the first ones I saw."

Amos Favorite attended a meeting Miller helped convene at the Geismar firehouse and volunteered to become the first president of Ascension Parish Residents Against Toxic Pollution. "Organizing was frightening for many of the people," says Miller. "This industry is economically very powerful in Louisiana. The communities' needs are great, and their resources are few. Often it forces them to accept the companies' money and sing to their tune."

Still, Favorite's group has managed to score significant victories. In 1986 it filed suit to force the state Department of Environmental Quality [DEQ] to increase a $66,700 fine levied against BASF for violations, including the excess release of phosgene and toluene into the air. (The fine was eventually raised to $150,000.) In 1988 the group sued the DEQ again, this time to compel it to draw up regulations to phase

out underground disposal of hazardous wastes. Favorite's group won the suit, which is now under appeal by the industry. Ascension's environmental activists have blocked a proposed asbestos dump in Geismar and succeeded in rerouting chemical trucks out of their neighborhoods. Indeed, there is some indication that progress is being made. A new state law will mandate a drop of 50 percent in toxic air pollution from 1987 levels by the end of 1994. Says Dr. Paul Templet, head of the Department of Environmental Quality: "Small environmental groups like Ascension Parish influence this agency, and we influence industry."

Much of the credit for those changes belongs to Favorite, an intriguing character who has been an activist for one cause or another most of his life. "I came to Geismar in 1925, when I was 2 years old," he says. "My mother was a single parent. She gave me to my grandparents to raise when I was 6 weeks old." Favorite grew up on the Waterloo sugarcane plantation in Geismar. "It was educated slavery," he says. "Us colored children were only allowed to go to school three months a year until seventh grade. It cost too much to go see the doctor in Gonzales. The plantation vet would look at us when he came to check the animals." Favorite left school to cut cane for 20 cents a ton at the age of 9, the same year his mother died.

World War II offered Favorite a chance to escape. Trained in chemical-warfare defense—the scientific background is useful to him still—he served with the 503rd Antiaircraft Artillery Battalion on Saipan and Okinawa. After the war, he married Rosemary James (now his wife of 45 years) and studied electronics under the GI Bill. Hired in 1958 by the Ormet aluminum plant to clean storage tanks, he later filed a class-action suit on behalf of 37 blacks that charged discrimination in the labor contract. The group won $37,000 in an out-of-court settlement, and Favorite returned to Ormet victorious.

At home, Favorite's children started following their father's example. With the coming of public-school desegregation in the 1960s, Amos's 16-year-old daughter, Barbara—one of nine children—told him she wanted to attend the all-white high school in Geismar. Favorite agreed. "She's stubborn, just like me," he says. "She used to come home with busted eggs all over her, spit on her, strawberries on her clothes. The Klan burned crosses on my lawn. I called up the Justice Department in Washington, D.C., and they put me in touch with an NAACP lawyer down here. He sued the hell out of that school. When the school told us they couldn't guarantee her safety, federal marshals came in to protect her."

Favorite doesn't always have the support of his family. "Some of my kin get angry with me," he says softly. "After I testified at the State Capitol in Baton Rouge about my relatives who died of cancer, a TV crew came to film their graves. A few relatives threatened to sue me. They don't understand that it's silence that hurts us most."

So Favorite continues spreading the word. On a recent Friday evening, he gathered a group of black deacons in Geismar's local recreation hall. There, a representative from the Louisiana Coalition for Tax Justice explained that Ascension Parish had lost $94 million over the last 10 years because of tax breaks granted to the chemical companies by the state. "We are subsidizing them for poisoning us," says Favorite, amazed. "Discrimination wasn't nothin' compared to what we live with now. The war I fought in was nothin' compared to the one we're fightin' now. As long as these chemical plants keep pollutin', we are livin' in the shadow of death. All of us."

Discussion Questions

1. According to this article, what is one reason the poor are disproportionately affected by air pollution?

2. In this article, Lois Gibbs notes that chemicals should be proven guilty until proven innocent. What does she mean and do you agree with this idea? Why or why not?

3. Critics are quoted as saying that three major companies are paying to relocate people living near their plants in the Mississippi Valley, which is proof that environmental pollutants are causing cancer. Do you agree with this statement? What other interpretations could one derive from the actions of these companies?

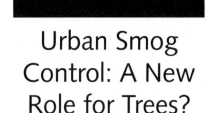

Urban Smog Control: A New Role for Trees?

J. Raloff

■
SOURCE
Science News, July 5, 1990, Vol. 138

TREES CAN serve as a major source of the hydrocarbons contributing to smog ozones. But trees can also help cool the air, thus slowing the heat-driven photochemical reactions that brew ozone from hydrocarbons and nitrogen oxides. Indeed, because smog production is so temperature-sensitive, trees that cool cities may do more to limit ozone than to foster it.

This conclusion, drawn from new computer simulations, suggests that sprawling urban growth may take an unnecessarily large toll on air quality if planners don't spare as many trees as possible, says study coauthor William L. Chameides, an atmospheric chemist at the Georgia Institute of Technology in Atlanta. He believes city planners and air-pollution-control managers should join forces to "think about how they want their [region] to grow"—and especially "where they want to leave green spaces."

Noting that most U.S. ozone-mitigation strategies focus on limiting hydrocarbons, Chameides says the new findings also reinforce the importance of shifting to approaches that place at least as much emphasis on nitrogen oxides.

Growing cities sacrifice many trees to development. Atlanta, whose population has increased 30 percent each decade since 1970, has lost about 20 percent of its trees over the past 15 years, Chameides says. During that same period, average summer temperatures have climbed steadily in both the city and its surroundings. But Atlanta's increase, totalling almost 4°F, dwarfs those of its rural neighbors—in one case by a factor of about 5. In a paper to appear in the *Journal of Geophysical Research* later this summer, Chameides and Georgia Tech colleague Carlos A. Cardelino argue that the urban temperature increase probably stems from a deforestation-fostered "heat-island effect," in which asphalt and dark-roofed structures become massive heat reservoirs.

On the basis of Atlanta's 1985 tree cover of about 57 percent, Chameides and Cardelino computed likely ozone concentrations for a typical summer day in Atlanta under a range of scenarios. Each scenario explored some facet of the city's changes over the past 15 years, such as tree loss or an estimated reduction of as much as 50 percent in hydrocarbon emissions from human activities, including traffic and industrial facilities.

Comparisons of these simulations suggest that the sharp, steady rise in Atlanta's summer temperatures over the past 15 years has resulted in a large net increase in hydrocarbon emissions from vegetation—despite the loss of one-fifth of the city's tree cover. This apparent contradiction reflects the fact that a tree's hydrocarbon-emission rate increases dramatically as temperatures climb. Indeed, Chameides and Cardelino say the estimated vegetative-hydrocarbon increase from Atlanta's remaining trees would have canceled the city's decrease in automobile and industrial hydrocarbon emissions.

Arthur H. Rosenfeld, a leading analyst of urban heat-island effects, has long advocated tree planting to reduce urban heating and to sequester the carbon dioxide emissions that threaten to initiate a global warning. "But I hadn't even thought about that [vegetative-hydrocarbon feedback on ozone from tree cutting], so I'm happy to have [the Georgia Tech team] publish it," says Rosenfeld, who directs the Center for Building Science at Lawrence Berkeley (California) Laboratory.

The new simulations make "an important contribution," adds John R. Holmes, director of research at California's Air Resources Board (CARB) in Sacramento. The results also dovetail with observations made by CARB—a state agency, responsible for some of the toughest auto-emissions regulations in the nation. During the 1970s, when CARB's vehicle-emissions controls focused on hydrocarbons, Los Angeles experienced a reduction in ozone, but only downtown, where traffic was highest and vegetation was lowest. Ozone levels continued to increase in the areas where they had always been highest—downwind. It wasn't until "we had large reductions in both hydrocarbons and nitrogen oxides that ozone levels went down, rapidly all over," Holmes says.

Gary Z. Whitten, who models urban ozone at Systems Applications, Inc., in San Rafael, California, says the Atlanta report makes a good argument for controlling nitrogen oxides, but he points out

that the computer model failed to account for several factors that could greatly affect ozone production, such as a heat-driven increase in atmospheric mixing. Whitten views the analysis as "interesting" but only a "first start" at defining the complex interactions among trees, temperature and smog.

Discussion Questions

1. Explain the central thesis of this article. What evidence is used to support it?

2. Has the author provided a balanced view?

The Case of the Disappearing National Park

Aaron Sugarman

■

SOURCE
Condé Nast Traveler, May 1992

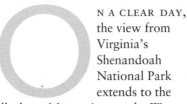N A CLEAR DAY, the view from Virginia's Shenandoah National Park extends to the Allegheny Mountains, on the West Virginia border some 30 miles away. The trouble is, clear days are rare, particularly in summer, when a haze of smudgey sulfur dioxide particles often blankets the park. Much of the time it's hard to make out the Massanutten Range just seven miles away.

Last year at the park, 35 percent of the days from May to October were classified by the National Park Service as having visibility of zero to ten miles. Only 16 percent of the days had good visibility. July had no good days at all.

And the problem seems likely to grow worse. In the last six years, the Virginia Air Pollution Control Board has approved permits to build 23 power plants in the state and is currently considering another ten applications. Park Service projections indicate that the proposed plants would emit tons of sulfur dioxide, further reducing visibility at Shenandoah.

"Anyone who knows the park knows that visibility is terrible in the summer," says John Christiano, chief of the Park Service's Air Quality Division. "The state needs to do something with existing sources of pollution before approving more."

Virginia officials argue that more than 90 percent of the pollutants afflicting Shenandoah come from outside Virginia. Elizabeth Haskell, Virginia's Secretary of Natural Resources, says visibility will not improve unless neighboring states cut emissions. "We care about the park as much as the environmentalists do," she adds. "We just can't solve the problem alone."

Virginia officials also say they are carefully following the strict pollution-control standards of the nation's Clean Air Act.

One of the goals of this act, passed in 1970 and amended several times since, is to reduce the pollution-induced haze that hangs over many national parks. In fact, to benefit Shenandoah, it specifically mandates sizable cuts in emissions from a big coal-burning unit in West Virginia and other facilities bordering Virginia by 1995.

What worries environmentalists is that the proliferation of electric plants within Virginia could more than offset any gains. The Environmental Protection Agency predicts that sulfur dioxide emissions in the state will more than double by the year 2010.

Officials at Virginia's air pollution agency say the Park Service has yet to furnish sufficient proof that any of the proposed projects will reduce the visibility at Shenandoah. Unless the state agency is able to quantify the effect a plant will have on the park, it has no legal grounds to reject a permit, says spokeswoman Mary E. Major.

That creates a bind, because existing pollution-forecasting systems are fairly limited. The Park Service says the evidence is clear enough, though: Visibility is most impaired in the summer, when air currents often flow from the south, southwest, and east—where several proposed plants would be located. "Common sense suggests that emissions from these areas will indeed reach the park a substantial percentage of the time," concludes a Park Service report.

The Environmental Protection Agency, the ultimate arbiter in this conflict, has only added to the regulatory muddle. Based on a forecasting program that it acknowledges is insufficient, the agency issued a statement saying that it "cannot conclude that the emissions from the proposed facilities would cause adverse impact."

David Carr, Jr., a staff attorney for the Southern Environmental Law Center, finds that contradictory. "The EPA says the plants won't cause adverse impact, but it also says sulfur dioxide levels in Virginia will increase. Where do they think the stuff comes from?"

While the technicalities are thrashed out, the staff at Shenandoah expect another hazy summer. "The hard part is that the view is the attraction here," says David Haskell, Chief of Natural Resources for Shenandoah. "We don't have caves or redwoods. People come for the Skyline Drive, the Blue Ridge Mountains, and the Appalachian Trail. It certainly defeats the purpose if they can't see anything."

Discussion Question

1. Explain why environmentalists and Park Service officials are concerned about the proposed power plants.

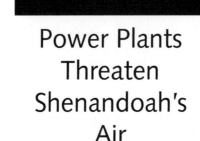

Power Plants Threaten Shenandoah's Air

Heather Swain

■

SOURCE
National Parks, March/April 1991, Vol. 65

BETWEEN 15 AND 20 proposed major power plants in Virginia threaten Shenandoah National Park with deteriorating air quality and other damage.

Nineteen plants have recently sought air pollution permits from the state, and four of these have been granted. The Department of Interior, which must ensure that requirements of the Clean Air Act are met, has issued a preliminary determination that one plant, operated by Multitrade Limited and located about 65 miles from Shenandoah, will adversely affect the park. Interior also warned it may issue similar determinations in regard to other proposed plants.

Shenandoah has suggested a halt in permitting until further determi-

nations can be made. Virginia admits concern over air quality, but the state claims it cannot delay permitting under current regulations. Elizabeth Haskell, state secretary of natural resources, said, "Providing all . . . regulatory requirements are met, there is no basis for delay or denial of the application, and a permit must be issued."

Shenandoah has compiled evidence of damage caused by air pollution over the past decade. Human-created pollution substantially decreases visibility within the park 90 percent of the time, and park visibility has dropped 50 percent over the last 40 years. "Poor visibility is the single most frequent complaint we hear from visitors," said superintendent William Wade.

Although progressive damage to visibility is the most obvious problem resulting from pollutants the power plants would emit, they also would contribute to acid rain, harming the park's streams, and elevated ozone concentrations, damaging sensitive vegetation and posing a potential threat to human health.

Given the severity of current problems, the Interior Department holds that any significant increase in emissions would be damaging. Interior has proposed that no permits be issued to any new facilities without an offsetting agreement, whereby a new source's emissions would be matched by a decrease of at least the same amount elsewhere in the state. Interior also recommends that Virginia develop an emissions control strategy, setting a total emissions cap for the entire state.

The park is particularly concerned about the cumulative effects of air pollution from multiple sources. Although most of the plants seeking permits are relatively small, their combined impact could be significant. Plans for many of them do not incorporate state-of-the-art pollution controls. The state, however, says it must consider each facility separately. Virginia claims it may consider facilities with emissions under a certain level non-impacting by looking at the effects individually rather than collectively.

David Haskell, chief of natural resources and science at Shenandoah, has indicated that Interior will probably appeal to the Environmental Protection Agency [EPA] if the state continues to permit facilities that the park considers threatening. EPA could dismiss the appeal or could move to overturn the permits.

He said the park hopes private companies will make offsetting offers on their own, and that Virginia will "take a more aggressive stand in setting regulations for the whole state" so that the matter will come to an EPA appeal.

Elizabeth Fayad, NPCA staff counsel, said, "NPCA is considering alternatives, including litigation, to assure that Shenandoah's natural beauty and resources will not be further adversely impacted by these sources."

Discussion Questions

1. The two articles on Shenandoah National Park give slightly different views on the problems and effectiveness of the solutions. Did you detect any major differences?

2. Which article seems more objective? Which one seems more comprehensive?

Water Pollution

The 14 states in the southern United States have approximately 560,000 kilometers (350,000 miles) of rivers and streams, 3.6 million hectares (9 million acres) of lakes, and 60,000 square kilometers (23,000 square miles) of estuaries. The region contains about 30% of the nation's river and stream miles, 40% of the nation's lake acreage, and 65% of U.S. estuaries.

Under the Clean Water Act, individual states designate lakes and segments of rivers by end use, for example, as a supply of drinking water, contact recreation, or aquatic habitat. The water quality for each lake or segment of river varies with its designated use. For example, water for drinking must contain a lower concentration of suspended solids than water for agricultural uses.

Recently, all states reported to the EPA on the quality of water in their lakes and rivers. The southern states assessed 46% of their rivers and streams. According to this survey, 70% of the river and stream miles fully supported their designated uses, 21% supported them partially, and 9% did not. The major causes of river or stream impairment were (in decreasing order) siltation, nutrient pollution, pollution with infectious agents, organic pollutants, and metal pollution.

Of the 66% of the South's lake acreage, 79% fully supported designated uses, 14% partially supported designated use, and 7% did not. The major pollutants in decreasing order were salinity, organic enrichment, nutrient pollution, and siltation.

Success in meeting these goals stems in large part from controls on point sources. The remaining problem areas are largely the result of nonpoint water pollution, runoff from farms and the urban landscape, for example. Controlling nonpoint water pollution is proving to be difficult and is intimately linked with the way people treat the land.

Background Reading: Chapter 17.

Strip-Mine Shell Game

Despite a Landmark Law, the Appalachian Landscape Remains Unreclaimed

Ted Williams

■

SOURCE
Audubon, November/December 1992, Vol. 94, No. 6

NEVER IS Appalachia more beautiful than when her hollows are bandaged in ground fog as they were on the morning in August when photographer Bill Campbell and I cruised north from Charleston, West Virginia, in Mike Mallory's Cessna 182. By the time we reached Fellowsville the sun, hot and bright over our starboard wing, had exposed most of the strip-mine wounds along the higher coal seams.

There on the lushly forested slope of Laurel Mountain was F&M Coal company's unreclaimed pit. Bulldozer cuts, like mouse bites in yellow soap, converged from all compass points. Brown, green turquoise, and orange wastewater festered in scum-ringed impoundments of sundry shapes and sizes before bleeding out of the mountain in a thousand seeps.

As strip mines go, the F&M pit is not especially large. But along with its two partially revegetated sister mines, it is a world-class producer of sulfuric acid and toxic metals—800,000 gallons of runoff per day right into aquifers, wells, and the left fork of Sandy Creek, now fishless for at least six miles. Then, after killing off about three miles of Sandy's main stem, F&M poison pours into Tygart Lake—a 1,750-acre jewel that supplies drinking water to Taylor County, plus smallmouth and largemouth bass, walleye, channel cats, and panfish to anglers from all over America. "The buffering capacity of the lake is being used up, so it can't neutralize other acid-mine sources," declares Frank Jernejcic, a fisheries biologist with the West Virginia Division of Natural Resources. "If we had the wrong combination of rainfall, low water conditions, and everything else, we could conceivably kill Tygart Lake. We need people to know how they're being screwed. Everyone here just accepted it as the cost of doing business."

None of this was supposed to happen under the Surface Mining Control and Reclamation Act (SMCRA), signed into law in 1977, seven years before the F&M mines got going. The statute was bitterly contested by the coal industry because it empowered citizens to control strip miners as well as deep miners who damage surface features. Over its six-year gestation it survived 2 vetoes by Gerald Ford, 183 days of hearings and legislative consideration, 18 days of House action, 3 House-Senate conferences and reports, 11 committee reports, and 52 recorded votes in the House and Senate. So when Jimmy Carter finally signed it, environmentalists caroused wildly; then most of them rode off on the spoor of different dragons.

Under SMCRA, reclamation of abandoned mines was to be funded by a tax on coal. Reclamation of new mines was to be required and guaranteed by bonds that would be forfeited by the permittee in cases of noncompliance or bankruptcy. A new bureaucracy—the Office of Surface Mining Reclamation and Enforcement (OSM)—was to police all coal extraction. Government or industry (or both) was to pay expenses, including attorney's fees, when citizens brought successful administrative or judicial complaints for noncompliance or nonenforcement.

One of the environmentalists who didn't ride off is Cindy Rank, president of the West Virginia Highlands Conservancy, which regularly sues the state and the OSM, thereby facilitating sporadic enforcement of SMCRA. She's found that getting the states and the OSM to implement the law has been harder than getting Congress to pass it. "Unfortunately, the public thinks we have a wonderful remedy," she lamented as we strolled along a bright orange creek in Dola, 35 miles west of the F&M site.

"Why?" I inquired. "Can't they see?"

They could if they were here, she explained. But they're not; there's been a mass exodus of people from rural Appalachia.

"Nobody lives with strip mines anymore. They've moved to the megalopolises where there are more glitzy issues—toxic-waste dumps, garbage washing into the ocean. . . ."

For a while we stood there—Rank staring at the creek with angry blue eyes, me staring at where my loafers had been because I was sinking past my ankles into slimy orange mine waste. We could see the

"yellow boy," as the iron is called when it oxidizes and precipitates out of solution, clotting like stockyard offal 30 feet after it left the deep mine, whose 1,000 subsurface acres are now flooded with groundwater. From the plane I'd traced the creek to its confluence with a larger, coffee-colored river even more fouled by mine drainage.

This mine, along with its toxic spoil banks and stripped access areas, was abandoned by Glory Coal Company in 1984. Glory just took the coal, declared bankruptcy, and walked, owing the federal government $387,000 in abandoned-mine-reclamation fees. Despite legal obligations, neither the state nor the OSM has moved a muscle to treat the water or reclaim the site. How was such a mess possible 15 years after passage of one of the toughest and most enlightened environmental statutes Americans ever demanded from their lawmakers?

Basically, it took Tom Galloway three days to make me understand the answer. Galloway, 46, is the attorney who helped write SMCRA and who represents Cindy Rank's Highland Conservancy, the National Wildlife Federation, and other environmental groups in their efforts to get the law working. I had met him at his office on the District of Columbia's "opulent K-Street Corridor . . . scorned as a special interest vice by those who hold themselves to serve only the public interest," as former National Coal Association president Carl Bagge described it. Actually, I'd found the place spare and unpretentious—just like the trim man with the natural smile who greeted me at the door and who had discarded a career in corporate law in order to represent environmentalists, often charging them virtually nothing save the attorney's fees recovered from government and/or industry if he wins. This system of private, public-interest lawyering is about as lucrative as, say, freelance writing; but it has paid the bills for a decade.

Coal moguls and OSM bureaucrats, who for the past 12 years have been essentially the same people, revile Galloway with a dedication that suggests his effectiveness. Galloway's office is decorated with cartoons, cut from industry publications, that depict him as a vulture hovering over dying coalmen or a Rambo-like figure stomping on the nation's energy reserves. One caption reads: "With the overwhelming zeal of a Khomeini commando environmental champion L. Thomas Galloway has penetrated West Virginia's borders intent on spilling the blood of its coal operators and enforcement officials."

No coal-industry apologist is more galled by Galloway than OSM director Harry Snyder, who used to be a lobbyist for CSX Railroad, a heavy investor in coal. It's as if Galloway were a white whale who'd made off with the director's leg. Snyder has pushed legislation that would do away with SMCRA's attorney's-fee provisions on grounds that the agency has become a "welfare system for a small coterie of attorneys." According to government officials, the director has (1) told members of Congress that Galloway held up a substantive agreement to address deficiencies in the Applicant Violator System in exchange for attorney's fees; (2) tried to have the federal government investigate him; and (3) asked the solicitor's office to get him disbarred.

Snyder denies that he ever sought to have Galloway investigated or disbarred and says he harbors no animosity toward him. "Tom Galloway and I often disagree," Snyder says, "but lawyers are paid to disagree and to represent their clients."

No one intimidates Tom Galloway. Not the Kentucky strip miners who impaneled a grand jury, purportedly to investigate interference with business but really to try to get him held as a subpoenaed witness in the infamous Clay County jail, where he might not have survived the night had he not left the state. Not the Virginia coal worker with the long record who promised to kill him. And certainly not the bloated coal barons who puff and blow from both sides of OSM walls.

An only child, Galloway was six when his father was murdered by a mental patient brandishing a Luger. Raised by his mother, he grew up in a house with no running water in a dour Kentucky locale called Tick Ridge, near the Ohio River and a coalfield. Mine blight was always present in his youth, but it didn't traumatize him. As a means of escaping poverty he won scholarships to Florida State and the University of Virginia Law School, where he became a star. Commitment to the earth and the public good came later.

"The [Interior] Department has done its best to gut the statute," Galloway told me at six in the morning as he aimed our rental car toward West Virginia. "You bring a case, you better be prepared to litigate for ten years. No side gives up. The war continues."

The strain of getting SMCRA on the books exhausted environmentalists, he explained. "We passed a big federal statute. Everybody thinks it's great, and they move away. It becomes so detailed and technocratic that no one pays attention to it." Except the coal industry, which has mastered the art of manipulating the act for its own purposes.

Still, SMCRA has been better than nothing. The OSM predicts that by the turn of the century 5 to 10 percent of the mines abandoned before 1977 will be reclaimed, using coal-tax money generated by the act. Of the 24,000 operations—including surface and deep mines, refuse piles, and prep plants—that come under the act and that were started after 1977, about 17,000 have been "reclaimed." While many of these are grassed-over deserts useless for wildlife, regulatory authorities are so elated to see any sort of vegetation that they release the reclamation

bonds posted by the coal companies (which would have proved thoroughly inadequate had they been forfeited). On many of these sites the post-mining land-use designation is "pastureland"; but given their remoteness, they are immune from even bovine perusal. "You'd think there was a cattle boom in Appalachia," remarks Galloway.

During the seven-hour drive from Washington to Mallory's airfield in Charleston, Galloway regaled me with stories about strip miners. There was, for example, the adventure that began five years ago in Lee County, Virginia, and is unfolding as I write.

Galloway told it this way: State coal regulators—responding to Galloway's complaints about wildcatting (illegal mining) in Virginia—told reporters he was wrong, that they'd stopped it all. Galloway got mad and took reporters, OSM inspectors, and congressional staff investigators on a state-police-escorted tour of Lee County, finding a wildcat mine not in days or hours, but in minutes. The congressional staffers then asked to stop by Mine 44-05668, shut down by the state, so the reporters might see for themselves how well the act was working. The vehicles rounded a corner on a high "bench" road between stripped gullies, and there was the shut-down mine, swarming with wildcatters, who promptly whipped out guns. The trooper drew his own pistol and, when this failed to impress, a shotgun. Meanwhile, a mining truck had pulled up behind the entourage, sealing escape. As the trooper walked back to deal with the trucker—whom he recognized as a parolee and to whom he promised permanent lodging in federal prison if he didn't remove his buns and wheels in, say 30 seconds—a front-end loader lowered its bucket and started for the cruiser, perched precariously on the bench, from which Galloway was taking in the show. Just before impact, the front-end loader stopped and idled. As the

mining truck withdrew, the trooper leapt back into the cruiser and executed a squealing 180 degree turn.

"Wait!" cried an OSM inspector who had jumped into the backseat. "I've got to issue a closure order here."

"Are you out of your—mind?" inquired the trooper. "We'll send in a—SWAT team tomorrow!" And with that, he fishtailed off the site.

According to a previous finding by the state Division of Mine Land Reclamation, the mine was controlled by Aubra Dean of Waynesboro, Virginia—despite the fact that it was officially listed under the ownership of Garry L. Williams, who had been Dean's foreman on the site. State and federal documents show that Dean has a long record of running afoul of mining laws. For instance, in November 1979 he was "forever barred" (but temporarily, as it turned out) from mining in Kentucky because, to use the words of the hearing officer, "the existence for two years of two violations on each of four permits and three violations on a fifth permit would seem to make them [Dean and his company, Dean Trucking] repeatedly in violation of the statutes and regulations of the Commonwealth."

Operations in Kentucky and Virginia controlled by Aubra Dean and Carl McAfee, a lawyer, have collected citations from the Mine Safety and Health Administration as well. According to information officer Kathy Snyder, the tally over the past five years has been 820. Among the fines imposed was one for $298,182 (which the men have not paid as of this writing) for a 1991 cave-in that killed four workers. The accident happened at Mine 44-05668—where Galloway and the state trooper had their set-to with the illegal miners and which since 1990 has been officially listed under ownership of Dean and McAfee, with Williams now listed as superintendent.

Neither the states nor the OSM have been able or willing to seriously discipline such coal operators. But

Dean and McAfee had come to the attention of coal authorities in Kentucky, thanks in part to a software package that Galloway and the National Wildlife Federation put together on a PC for $450,000 because the OSM's $20 million system didn't work. If the Department for Surface Mining Reclamation and Enforcement documents alleged links between the men and unreclaimed mines in Harlan County, and if the messes aren't cleaned up, the department can invoke SMCRA to bar Dean and McAfee from mining in the United States. Catalogued in the National Wildlife Federation-Applicant Violator (NWF-AV) system, as the software package is called, is information on cited mining permittees, including several hundred thousand ownership-and-control relationships and 28,500 uncorrected violations involving more than $300 million in unpaid penalties.

Strip miners can get around SMCRA by setting up shell companies and skimming off profits in the form of royalties, coal-washing fees, etc. Then when it comes time to clean up the stripped land and treat the acid drainage, the shells declare bankruptcy and the affluent perpetrators can walk away, assets insulated. No one effectively matched the shells to the strippers until the NWF, Galloway, and his various citizen clients started tracking them on their computer.

While I floundered in the orange ooze of the Stygian river that surged from the abandoned Glory mine in Dola, Galloway held court 100 yards downstream with three inspectors from the West Virginia Division of Environmental Protection. With the NWF-AV system he had tied the bankrupt shell of Glory Coal to Pittston Coal Company—a national giant that for years has gotten around SMCRA via such legal ploys as festooning mountains with adjacent two-acre mines. (Before environmentalists got this loophole plugged, the act had exempted mines

of fewer than two acres so as not to inconvenience mom-and-pop, pick-and-shovel operations.) Pittston officials had departed the site minutes before we arrived—this at the suggestion of the state inspectors, who feared an unpleasant confrontation.

Mine regulators had known about the Pittston connection for the better part of a decade, but enforcement is not something that coal states or the OSM like to get dirtied up with. West Virginia, Kentucky, and most other mining states have been granted "primacy," which means that the OSM lets them deal with reclamation, supposedly stepping in only if they fail to enforce SMCRA as, invariably, they do. Yet in these states the OSM issues fewer than 25 citations a year, most of which it "vacates" by decreeing that the offense never existed or "terminates" by cutting an easy deal with the company. Now the agency is in the process of emasculating or throwing out scores of SMCRA regulations that it deems burdensome to industry, including the one banning deep miners from national parks and wildlife refuges.

"So it's a matter of punishing the coal company?" a mine inspector is saying to Cindy Rank. He is arguing that it would be better to "mitigate," to take the money Pittston will pay to clean up this mess and use it to clean up a worse one somewhere else. Rank is having none of it.

Nor is Galloway. "For an enforcement system to work," he says as soon as the inspectors are out of earshot, "it has to be credible and rational. Once you get into political horse trading, which mitigation is, the political system takes over, not the regulations. And the political system in West Virginia [as elsewhere in Appalachia] is completely devoid of environmental ethics. . . . Their attitude should not be to write off this stream and fix a better one. It should be to go downstream and look for the people responsible for other pollution. They have these brand-new legal tools, which should

prevail. It's not just that they don't do it; they don't even *think* about doing it."

Certainly they can't be accused of thinking at Laurel Mountain. When F&M Coal Company went belly-up, after stripping the mountain of its coal and poisoning Sandy Creek and Tygart Lake in the process, West Virginia's mine regulators cashed $268,000 worth of reclamation bonds it had been required to post. About this time they discovered that reclamation and acid neutralization was going to cost more than *$4 million*. Thomas Rodd and Galloway, as attorneys for the Laurel Mountain/Fellowsville Area Clean Watershed Association, asked the state Division of Environmental Protection to make up the balance from the coal-tax-generated reclamation fund, but it refused, arguing there'd be nothing left for other cleanups.

Now there may be a new source of revenue, provided by the perpetrators themselves. The NWF-AV computer system has sniffed out the Oz figures who allegedly operated behind F&M. And last April Rodd and Galloway formally requested that the OSM include in its enforcement action Donald Frazee, Edward Frazee, Inter-State Lumber Company (formed by the Frazees), and Jno. McCall Coal Company. (The *F* in F&M stands for Frazee, the *M* for McCall.) "Individuals cannot evade enforcement by hiding behind a corporate shell," wrote Rodd and Galloway.

Is effective regulation of strip-mining possible with more citizen participation in SMCRA? I put the question to one of the leading experts on the subject, Richard diPretoro, a geologist and environmental consultant based in Morgantown, West Virginia. He explained that I was missing the point, that there is no such thing as effective reclamation. "If you assume Laurel Mountain was a ten on a zero-to-ten scale of natural health and beauty," he said, "and if you assume the old 'shoot-and-shove' mining, where they'd blast and push the spoil over the side with a dozer, is a zero, then I think full-scale, modern reclamation will rate a two or a three."

All methods of coal removal and burning are inherently destructive, and the United States is glutted with coal—far more than it can use. Our refusal to internalize the real costs of coal, our tradition of hacking it out and burning it up without any genuine effort to protect the earth, explains why electric rates in Appalachia are among the cheapest in the nation and why this country can afford to export coal to countries that take better care of their land than we do.

Maybe the answer isn't no more coal mining, which diPretoro has suggested, thereby getting himself rejected by the West Virginia senate as a governor-picked nominee to the Public Energy Authority. But probably the answer is no more strip-mining and better-regulated deep mining.

SMCRA was designed to deal with coal-extraction methods now considered dainty and obsolete. These days, machines the size of 10-story buildings eat whole mountains and spit the spoil into hollows, where it hangs menacingly over people who bet their lives on strip-mining technology. "Mountaintop removal," as the method is called, is good for the locals, profess coal operators and regulators. How so? Why, because Appalachia is too hilly. Snip off a few mountains and you can build schools, malls, and hospitals on the "reclaimed" stumps. To hear them talk, mountaintop removal rates a 10 on the natural-health-and-beauty scale.

Buried in the congressional testimony that led to SMCRA's passage is the story of Mrs. Bige Ritchie, an impoverished octogenarian who frantically and vainly tried to stop the bulldozers that came to her family cemetery after coal. "I thought my heart would bust in my breast," she said, "when I saw the

coffins of my children come out of the ground and go over the hill."

One of the "burdensome" regulations currently being trashed by the OSM protects cemeteries from cave-ins caused by deep mining. "Subsidence," as cave-ins are collectively termed in coal industrialese, is more common these days because companies no longer are required to support the earth, or "overburden," on top of a mine. And now companies speak of "planned subsidence," as if it were just part of the cost of doing business. Planned or otherwise, subsidence causes things to pop out of the earth at odd angles. Coffins, perhaps? In this, SMCRA's 15th year, we appear to have gnawed around the coal seam and met our own dragline tracks on the far side of the mountain.

Discussion Questions

1. This article paints a fairly pessimistic view of coal mining in the Appalachian region. Do you think it is accurate?

2. Do the assertions made in this article represent most mining operations or only a select few?

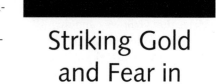

Striking Gold and Fear in South Carolina

Peter Applebome

■

SOURCE
New York Times National, June 5, 1991

PEOPLE LAUGHED a decade ago when geologists started poking holes in the pine forest at the edges of the South Carolina Piedmont looking for gold.

But they stopped laughing when mining companies started buying up huge chunks of land, when a mining giant began gouging out cavernous pits for one of the nation's largest gold-processing operations, and when South Carolina suddenly became the only non-Western state among the 10 largest gold producers.

Now the question is how far South Carolina's gold rush will go, and whether the state fully understands the environmental aspects of a mining revolution that allowed the United States to increase gold production more than ninefold in the 1980's.

At issue are mining operations that use tons of cyanide to leach gold out of vast quantities of rock—up to 30 tons of ore to get one ounce of gold in South Carolina. The new technologies, which allow mining of less-rich ores than were commonly mined in the past, are part of a little-noticed explosion in gold mining that could soon put other states in the same unexpected situation as South Carolina.

'It Can't Happen Here'

"Communities that are concerned about the impact that mineral development might have on other things they care about, like water quality, need to be aware this is something that could happen to them, even if it has not in the past," said Philip M. Hocker, president of the Mineral Policy Center, a non-profit organization concerned with environmental issues in mining. "Many places have thought, 'Well, it can't happen here,' and they found out it could."

The mining industry says mines have stringent safety standards and an estimable safety record. But for now, the economy boom in gold mining is more certain than the environmental costs.

In the 1980's, the United States went from an importer of gold to an exporter, passing the Soviet Union to become second only to South Africa in production. United States gold production grew from one million ounces in 1980 to nearly 10 million ounces a year now.

"I don't know of another industry that's gone from a $6.7 billion deficit to an $8 billion surplus in that short a time," said John Lutley, president of the Gold Institute, a trade association in Washington.

A Question of Control

But many residents here, 15 miles north of the state capital in Columbia, worry whether mining technology widely used in the remote and arid West is appropriate in South Carolina. And they wonder how aggressively a state with a reputation for lax environmental standards will monitor a process that uses tons of cyanide and produces other dangerous metals as a by-product of mining.

"We have a little chemical company near here the state has never been able to control," said Daisy Hollis, an outspoken critic of the mining who lives on 10 acres near the largest mine. "When I heard the people who couldn't control that were going to control these mines, it sent chills up my spine."

South Carolina is hardly synonymous with gold mining, but it does have a long mining history. Gold was a part of Indian lore, and the first official discovery was at the Haile Mine in 1827. Before gold was discovered in California in 1848, South Carolina was the nation's largest gold-mining state.

After many lean decades, mining resumed in 1985. Now, thanks largely to the Kennecott Ridgeway mine that opened here in 1989, South Carolina is the eighth-largest gold-mining state, producing 220,000 ounces in 1990. The mine, run by the Kennecott Corporation, which is owned by Standard Oil of Ohio, cost $100 million to start up. It is expected to produce 1.4 million ounces over its 10-to-12-year life span.

Local 'Mountains' Are Born

The mine, spread out over 2,400 acres, consists of two pits, which will both be 400 feet deep and stretch over 85 acres. Its "carbon-in-pulp" technology allows the gold to be extracted from the ore using sodium cyanide in large leach tanks. The residue and cyanide solution is then deposited in a huge 381-acre tailing pond. The discarded ore is deposited in two vast heaps the locals call Mount Cyanide and Mount Overburden.

To many people, the mine, which employs 200 people, is a valuable economic asset in the depressed north-central part of the state.

"They said they wanted to be a good neighbor, and I think they are," said Ronnie Murphy, who sold *four* acres now at the center of a mining pit. "I really don't see where

the mine has done any damage. There's two things Fairfield County needs: more jobs and more tax money.

And plant officials say the mine has taken extensive steps to make sure that no chemicals leach into the ground, and that air and water quality are carefully monitored. Industry officials say that while cyanide is deadly, it breaks down quickly when exposed to sunlight, does not build up in animal tissues and is not believed to cause birth defects or cancer.

Even industry critics say the new cyanide processes are more economical and more environmentally sound than older processes, which used chemicals like mercury. But environmental concerns persist, intensified by indications that the mine's success is likely to breed others.

Questions have been raised elsewhere about the long-term effects of cyanide in mining. Cyanide residue from mining has occasionally killed migratory birds and caused foul-smelling tap water in the West, raising concerns in Montana, Nevada and California and leading to new precautions by mining companies.

But the questions are magnified here because of South Carolina's rainy climate, which could enhance the chances of leaks of cyanide and other poison into the underground water supply; its temperature inversions, which could trap gases in the atmosphere; and its attention-getting location, near the state capital.

Glenn Miller, a professor of biochemistry at the University of Nevada who is an expert on cyanide in mining, said a key difference was that solutions which tend to evaporate in arid Nevada can overflow in rainy South Carolina.

"The risks to surface water and ground water, I think, are substantially greater in South Carolina than they are in Nevada," he said.

Residents' fears were heightened last November when a serious leak at the Brewer Gold Mine in Jefferson, S.C., occurred after a dam broke, causing 10 million gallons of cyanide-containing storm water to spill into nearby creeks and rivers.

Nor are environmental groups satisfied with the company's system—primarily fireworks and sound effects—for keeping birds out of the lethal water of the tailing pond at the mine here. A legal challenge to the mine led to the formation of a citizens' monitoring group, some of whose members remain convinced that the mine is a bad idea.

"We don't want this kind of industry that's ruining what we do love about South Carolina, the wildlife and the beautiful countryside and the clean air and the good drinking water," said Joyce Brown, chairman of the citizens' panel.

Others say their problem is with the degree of supervision and regulation.

"I think mining is a necessary thing," said Carl Schulz, a committee member who is a professor of environmental health at the University of South Carolina School of Public Health. "I would just feel a lot more comfortable if I felt there was a less permissive regulatory atmosphere than exists in the state of South Carolina."

Gold booms have a history of going bust, but mining opponents worry that the current operations are just the start. Dr. Schulz said he could foresee seven mines operating near here by 1993.

To some, the real worry is that the effects of the mine may not be clear for years. "The thing we're wary about is we've got a major program in place, moving very, very large quantities of material and creating a very new physical condition, which has only been going on for six years," said Mr. Hocker. "It's well known in mining that the environmental problems from a mine frequently won't show up for 20 or 30 years after the mining is under way.

Discussion Questions

1. Describe the main points of this article.

2. Do you think the widespread use of cyanide leaching to extract gold is environmentally sound? Why or why not?

The Green Revolution in Wastewater Treatment

More Than 150 Municipal and Industrial Artificial Marshlands Now Successfully Treat Wastewater Across the United States

Becky Gillette

■

SOURCE
Biocycle, December 1992, Vol. 33, No. 12

SEWAGE TREATMENT plants aren't normally the kind of place that draw admiring visitors . . . unless you're talking about the kind of systems where the "plant" is actually made up of plants of the green, growing, flowering variety. At these sites, artificial marshland wastewater treatment systems, not only is water being purified in a natural, very economical manner, but the resulting marshlands are so attractive that

they could do double duty as nature preserves.

City officials in Crowley, Louisiana considered their system so attractive that they actually built a conference center in the middle of the marshland! Crowley, one of the larger systems with a capacity of four million gallons per day, has been having BOD [biological oxygen demand] levels as low as two or three.

It has been about seven years now since the nation's first municipal artificial marshland wastewater treatment systems came on line. More than 150 municipal and industrial projects are now in operation across the United States.

Artificial marshlands (also known as constructed wetlands) work by funneling wastewater through aquatic plant systems. Organics in the wastewater are absorbed and biodegraded by plants and—the real workhorses in these natural systems—the microorganisms that thrive on plant roots and stems. The marsh systems are estimated to cost less than half as much to construct as conventional mechanical treatment systems (see sidebar on systems in Union and Picayune, Mississippi). The marsh systems are normally designed to be gravity flow.

The systems are less susceptible to mechanical breakdown and shock-loading, and more flexible dealing with stormwater intrusion. On the debit side, the marshlands require more land than mechanical treatment plants, and the technology for maximum treatment results is still being defined.

In essence, artificial marsh systems use aquatic plants to "farm" human wastewater. In fact, at a marsh treatment facility operated by the Mississippi Gulf Coast Regional Wastewater Authority near Ocean Springs, a farmer, Linwood Tanner, was hired as chief operator.

"What we needed most was a farmer, not a wastewater plant operator," says Donald E. Scharr, P.E.,

Economics of Artificial Marshland Wastewater Treatment

If artificial marshlands were just better for the environment but cost the same as conventional mechanical treatment plants, it is doubtful that the technology would now be utilized. But the systems are also extremely economical.

One example is Union, Mississippi, which must meet advanced secondary treatment levels because the stream it discharges into dries up during part of the year. In the late 1980s, Union faced a critical problem. It still owed $334,000 for a conventional sewage treatment facility that was obsolete. This city of less than 2,000 residents would have had to pay an additional $660,000 to upgrade the system, plus a $64,000 annual power bill. A new mechanical treatment system would have cost an estimated $1.2 million with annual maintenance costs of $55,000 and energy costs of $60,000.

Bill C. Wolverton, an environmental scientist with Wolverton Environmental Services, proposed an alternative 14-acre artificial marshland treatment system designed to treat 500,000 gallons of sewage per day. The cost was only $450,000, with annual main-tenance of about $2,000 and annual energy costs of $300 for aerators in the primary lagoon.

Union's marsh alternative saved them an estimated $750,000 in initial construction costs, plus continues to save an estimated $110,000 in operating costs.

A case with even greater economy is Picayune, Mississippi. Picayune had serious trouble with ground and stormwater intrusion into their sewage collection system. The amount of wastewater fluctuated from one to three million gallons per day depending on rainfall. An engineering group estimated that it would cost $11 million to correct the collection system, and another $4–$8 million for a new mechanical treatment plant.

Instead, Wolverton designed a duckweed marsh system that has the potential for actually making the city some money. The system, which went on line in October 1992, cost $300,000. The 25-acre marshland is designed to treat one to two million gallons of wastewater per day. Duckweed will be harvested and sold as animal feed. Part of the economy was possible because Picayune must meet only secondary treatment standards of 30 BOD. Land costs were also low in Picayune.

opments," says Woody Reed, a private consultant in Vermont who has studied constructed wetlands across the country. "There seems to be a tremendous interest in the use of the systems. I think the potential is greater than we've taken advantage of so far. But people are aware of the concept, and they are pushing to apply it."

Reed, one of the authors of the book *Natural Systems for Waste Management and Treatment,* says much knowledge has been gained in recent years. "We have learned that the early designs were somewhat deficient in hydraulics," Reed says. "They weren't properly designed to allow the water to flow through the gravel beds. We've learned that the systems originally designed have too short a detention time, and provided insufficient oxygen for effective removal of ammonia nitrogen. We're working on ways to increase detention time by increasing the size of the marsh. You can do that if you're building a brand-new marsh. If you have an existing marsh, you can find some other method to provide the oxygen.

"Probably the most important thing is that three years ago, there was really no consensus about how these things should be designed. Now there is more agreement, more of an approach to a consensus about how these things should be designed."

Modifications in Mississippi

Barry Royals, Chief of the Surface Water Program for the Mississippi Department of Environmental Quality (DEQ), says that while the 15 municipal wetlands now in use in Mississippi are working well overall, modifications would make them work even better.

According to Royals, a major difficulty has been that when artificial marshlands were initially installed, there was not enough testing and data gathered to allow someone to design the system with any degree of

Operations & Maintenance Manager for the authority. "We needed someone who knew how to grow plants, and figured we could teach him about the wastewater. But I'm sure Linwood never expected to end up farming cattails!"

Farming cattails and other aquatic plants to purify wastewater is a fast-growing new treatment technology. In most cases, primary treatment is with a sewage lagoon, with secondary treatment accomplished by the marsh systems. Treatment results have been positive for the most part, although authorities agree refinements are needed to get the best results with natural systems.

"Everything I've done for the past three years convinces me these (wetland systems) are an ideal concept for small communities, schools, parks, office buildings, individual homes and apartment devel-

An artificial marsh system in Union, Mississippi, uses a lagoon and shallow channels filled with aquatic plants for treatment.

competence to guarantee the quality of the effluent.

"The artificial wetlands are a new system and are not yet refined to the point where, when they are installed, we know with a high degree of certainty what the quality of water is going to be coming out the other end," Royals says. "We're increasing the size of these systems to make them perform to a level that will meet the permit requirements."

Overall, though, Royals and other pollution officials in Mississippi have been pleased with the marsh treatment alternative.

"We think the wastewater produced by wetlands systems is very good as a whole," says Glen Odom, Chief of Municipal National Pollutant Discharge Elimination System (NPDES) for the Mississippi DEQ. "It is an excellent technology for small towns. As a whole, the systems are meeting 15 BOD limits."

Some, like the Mississippi Coast facility, are regularly logging in under 10 BOD. One of the first municipal wetland systems put in operation, located in Collins, Mississippi, has only logged two 11 BODs since 1989; the rest of the time the BODs have been 10 or below, according to Odom.

Royals explains that system performance depends on the type of wastewater being treated. A case in point is a system in Pelahatchie, located near Jackson, Mississippi. Due to a combination of design problems and more industrial flow, the system hasn't been performing up to standards and has to be modified.

"These systems are biological systems, natural systems," Royals says. "And like any biological system, it is subject to being upset."

He adds that conventional mechanical treatment systems also suffer when contaminated with incompatible chemicals.

In general, Mississippi's experience has been that the artificial marshlands are meeting secondary type permit requirements. But when permit limits are very stringent, the systems are having difficulty meeting those advanced secondary limits. "What we're finding is that you have to have more acreage out there in order to achieve the very stringent permit requirements," adds Royals. Mississippi DEQ officials now believe 30 to 40 acres of marshland are needed for every one million gallons of wastewater treated per day.

Refinements at TVA

Researchers with the Tennessee Valley Authority (TVA) have come to similar conclusions that systems need refinement to obtain the best results. "Basically what we have found is that wetland systems work real well for common parameters such as BOD and suspended solids," says James Watson, Senior Environment Engineer, Water Quality Dept., Tennessee Valley Authority (TVA). "But as far as more advanced parameters, ammonia nitrogen and phosphorus, results were not quite what we had hoped for. That's where the technology is now."

TVA has been conducting a number of different studies to evaluate constructed wetland technology for discharge from municipalities, industries, individual homes and agricultural operations. "There's still a lot to be done, but resources (for studying the systems) are very scarce," Watson says.

At TVA, the plants used most often in artificial marshlands have been cattails, soft-stem bulrush and reeds. TVA has also experimented with some decorative species such as water iris, arrowhead and sweetflag.

Watson believes the constructed wetlands have good potential for solving wastewater problems at an economical cost, especially if some simple adaptations compatible with the wetlands' technology prove effective in improving treatment results. Sand or gravel filters added to wetlands show promise for improving treatment results, while keeping the system inexpensive to operate.

TVA has constructed three municipal demonstration projects in western Kentucky. The largest in Benton treats waste from 5,000 people; a system in Pembroke treats waste from 1,000; and another in Hardin serves 500–600 people. Several different concepts were incorporated into the designs to evaluate variables.

In Benton, a two-cell lagoon system has a 16-acre primary lagoon and a 10-acre secondary lagoon. The secondary lagoon was converted into a three-cell wetland. One cell has a two-foot-deep gravel marsh with a subsurface flow. The other two cells have surface flow with an average depth of 12 to 18 inches.

According to Watson, the Benton system was meeting ammonia nitrogen limits in winter, but not in summer. "We also were occasionally violating fecal coliform and dissolved oxygen limits." He added that while there was good reduction in fecal coliform—90 percent plus—permit limits were very low, "so a good percentage of time, we weren't meeting fecal coliform limits."

He says fecal coliform is a minor problem that can be easily resolved with disinfection. Another problem common to wetlands, relatively low dissolved oxygen, also can be solved without great expense or trouble by adding natural or mechanical aeration at the end.

For compliance with ammonia nitrogen limits at Benton, TVA is in the process of adding recirculation (running it through wetlands twice). Wastewater will trickle through a pea gravel vertical flow filter before it goes into the wetland the second time. Dr. Bill C. Wolverton, an environmental scientist with Wolverton Environmental Services, Incorporated, has designed many of the municipal artificial marsh systems now in operation in Mississippi, Louisiana and Alabama. Wolverton says ammonia nitrogen is nontoxic

to fish and other aquatic life as long as pH is in a neutral range.

Most states still have fairly stringent ammonia-nitrogen permit limits. Any possible exemptions for natural systems would have to be worked out on a state-by-state basis.

Management Techniques

There are still many questions to be answered about how to best manage the natural systems. For example, experts have differing views concerning the need for harvesting. Some experts believe the artificial marsh systems can be managed naturally without periodic harvesting. Wolverton believes harvesting is necessary for two reasons: (1) Nutrient removal will be greatest when plants are still growing and not yet mature; (2) when plants die back in winter, they can decompose and release BOD minerals into the water. That would add to—rather than help—water quality problems.

"If you only have to meet secondary treatment levels, harvesting may not be necessary," Wolverton says. "But if you have to meet advanced secondary levels, you will need some type of harvesting. You might not need it every year, but every two or three years."

The new Micro/Agro™ systems Wolverton has been designing are dual systems constructed so that half the system can be taken our of operation when maintenance or harvesting of plants is necessary.

"What we've found is that you must have a management scheme," Wolverton says. "You can't just have a plant that is good at removing waste. You have to have a way to harvest those plants when they reach maximum size."

Wolverton says there have been major advances in understanding the natural waste treatment systems. "After 20 years of studies, a better understanding of the biological processes involved in the plant/microbial wastewater treatment system has been achieved," he

says. "What we are doing now is different than what we were doing five years ago. And in five years, we will be doing it differently again. But we're on the right track. That's what is important."

One major change in Wolverton's designs is eliminating the use of rock/reed filters in larger municipal systems. Although rock/reed filters can treat wastewater in a smaller area than surface marsh filters, the rocks are expensive, particularly in areas such as Louisiana and Mississippi where natural rock is scarce. There also have been clogging problems with some rock filters.

"Due to earlier hydraulic flow and filter-clogging problems, large volume aquatic plant wastewater treatment systems in the southern United States are now designed without rock filters," Wolverton says.

He has another reason for dispensing with the rock filters, and that is a desire to recycle the wastewater and produce not just clean water but also valuable feed products. Wolverton's newest, large municipal system, which is located in his hometown of Picayune, Mississippi, uses duckweed exclusively as the plant to filter the waste.

"Because of their food value, cold tolerant properties and ease of harvesting, duckweed (*Lemna* sp.) is becoming the favorite aquatic plant used in large open-channel wastewater treatment systems," Wolverton says.

He adds that studies have shown the duckweed channels 1.2 feet deep with hydraulic retention times of seven to eight days can reduce BOD levels of domestic sewage from an average of 35.5 mg/l to an average of 3 mg/l while maintaining a pH effluent level of about 7.24. "This data is being used successfully in designing cost-effective duckweed/microbial wastewater treatment for upgrading anaerobic and facultative lagoon effluent to secondary and advanced secondary levels," Wolverton says.

Duckweed has the protein equivalent of soybean meal and requires only drying—no other processing—before being fed to animals. The small, floating aquatic plants also grow prolifically.

"Approximately 15,000 pounds of high-quality protein can be expected per acre per year from duckweed grown in nutrient enriched wastewater," Wolverton says. "One acre of duckweed has the potential to remove approximately 2,351 pounds of nitrogen, 588 pounds of phosphorus, 784 pounds of potassium, 784 pounds of calcium, 235 pounds of sulfur, 313 pounds of magnesium and 392 pounds of chloride annually in the southern United States."

Most natural systems for wastewater treatment thus far have been installed in the southern U.S., but experts believe there is potential for designing systems to work well in northern areas, as well.

James Watson, the TVA researcher, points out that natural systems have been in use throughout cold climates in Europe for a number of years. "I believe we'll be able to modify the systems to come up with appropriate designs for cold areas," Watson predicts. He believes that subsurface gravel marshes have the best potential in colder climates.

There are still other issues unresolved concerning natural systems. Other factors include the amount of rainfall and sunlight, temperature fluctuations, aging of the system, animal activity, plant blight and insect problems.

At the Mississippi Gulf Coast marsh facility, operators have had to turn into trappers to deal with the nutria problem. Nutria, an aquatic animal in the rodent family, cut down cattails for food.

In 1991 the marsh had an invasion of army worms that caused great concern when operators were seeing an entire acre of cattails being mowed down each day. Pesticides would have only added to the water quality problem, so there was little to be done. Fortunately, the army worms didn't kill the cattails, but just cut off the stems at the water line. The worm invasion soon went away, and treatment was never adversely affected.

Part of the reason artificial marsh systems have been such an easy "sell" is that the technology involved is simple to understand. But, as evident from the issues facing those designing, managing and studying these systems, it isn't simple at all. There are many complex factors that go into creating the optimum natural wastewater treatment system. Like nature itself, apparent simplicity masks an underlying complex chain of interactions that are still only dimly understood.

Discussion Question

1. With your knowledge of sewage treatment and ecology, especially food chains and nutrient cycles, describe the process that occurs in wetland wastewater treatment plants.

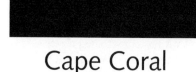

Cape Coral Reclaims Its Water Supply

Ellen Underwood

■

SOURCE
WaterLines, Winter 1992

ANYONE who has lived in Florida for any length of time knows water can be a sensitive subject. It's either restricted or flowing too freely. It has been called "a limiting factor to growth in the state," and has been the subject of many heated debates in government chambers.

But nowhere has it been a more sensitive subject than in the city of Cape Coral on the lower west coast of Florida where in the past decade, the hottest topic around town has been water. More specifically, it has been the planned "dual water system" that has been the center of controversy. The system is called "dual water" because two pipes are run to each household—one for high quality drinking water and a second for non-potable irrigation water, consisting of treated canal water and reclaimed wastewater

No one is more familiar with the issue than Joseph Mazurkiewicz, the mayor of the city for eight years. At 39 and slightly graying, Mazurkiewicz, known as "Mayor Joe" to many, was elected at age 31 in his first attempt at running for office.

"It's been a long haul," Mazurkiewicz said in December. In the past eight years, he has seen fellow councilmen both win and lose seats depending on their position on the dual water issue. Two special referendums have been held in which issues relating to the adoption of a dual water system won by a landslide majority both times. And more ink has been used in newspapers covering this issue than any other story in recent years.

Today, after much public debate and education regarding its health and cost aspects, the dual water system is well underway, with the first phase scheduled to be completed this January. Homes are to be connected as soon as February. The plan calls for 26 square miles of the city to be connected to the system in three years: eight square miles now and the remaining 18 in two additional phases.

Cape Coral is unique because it is the only city in Florida which has plans today to eventually connect its entire utility system to non-potable irrigation water. "The city has planned for its water supply needs through the year 2010," said Chip Merriam, the District's intergovernmental representative in Fort Myers.

Cape Coral has grown very quickly and has been forced to look at its long-term water supply needs. Cape Coral presently is the 14th largest city in Florida in population, and is second largest in land area—114 square miles. Today there are 78,000 residents, but at build-out the population will grow to around 380,000. These figures are even more astonishing considering the city was just a dream in 1957.

Cape Coral is a typical land development story in Florida. Developers saw the raw coastal land in the late 1950s as a gold mine. They moved in with dredges and bulldozers, built 400 miles of canals, and used the fill-dirt to build lots. Very quickly, lots were sold to

mostly out-of-staters for very little money—often for several thousand dollars or less.

People liked the coastal location, with access to the Gulf of Mexico and estuaries. The original developers went bankrupt and successive developers carried on the dream of Cape Coral. Soon, all lots were sold. In 1971, the city was officially incorporated. Today, Cape Coral continues to grow by 5,000 to 8,000 people annually. The lifestyle is almost exclusively residential, with little business or agricultural development. However, some critical infrastructure needs, such as water, were not addressed during the initial development.

As with many Florida cities, Cape Coral relies on groundwater as its primary source of drinking water. The city's potable water comes from a deep aquifer which requires treatment at a reverse osmosis plant. However, this aquifer is limited and projections show it will not be able to meet the demand of the city at build-out. In fact, the maximum amount of water available from the aquifer without harmful effects to the system, such as additional saltwater intrusion and impacts to other aquifers, is only enough for potable use—leaving none available for irrigation.

In 1981, Don Kuyk, the city utilities director, originated the idea to use canal water for irrigation, saving the city's underground water resources for potable, indoor use. Strongly supporting the program and helping to develop it was Robert Godman, a retired NASA engineer who lives in Cape Coral. Soon after, reclaimed water was included in the plan as a way to dispose of wastewater and as a back-up for the canal water during times of low flow.

"It was a stroke of genius," said Steve Lamb, director of the District's Regulation Department, which issues the city's water use permit. "The city planned this system long before the District began requiring the use of reclaimed water through-

out our 16-county area," he added. "They are way ahead of us in that area."

District studies support the fact that Cape Coral does not have a safe, reliable and cost-effective water supply for future needs without a secondary system. The District's draft water supply plan for the lower west coast identifies areas where demand exceeds supply. When the dual water system is not a part of the modeling process, "Cape Coral is a problem area," said Terry Clark, supervising professional in the Planning Department. "When it is in there, Cape Coral is not a problem."

"The residents of Cape Coral should be commended for their foresight," Clark said, echoing the sentiments of many District officials. "This is a very positive example of a community coming together to solve a problem."

Steve Kiss, Assistant City Utilities Director, recently dropped a copper penny into the final holding tank where reclaimed water is treated with chlorine before being stored in a 5 million gallon storage tank. The penny reflected the sunlight on its journey down, turning over and over in the water as it sank to the bottom of the cement tank. The water was so clear, one could still see the penny lying on the bottom of the tank even though the floor and walls were painted black.

During a short tour of the facility, Kiss and utilities director Kuyk pointed out that every conceivable detail has been considered in the building of the dual water system. "In the mid-1980s, it was a model in the country," Kuyk said. Today, as other municipalities are looking for ways to take advantage of reclaimed water, the technology is becoming more popular.

At one edge of the property sits hundreds of cement lock boxes. The South Florida Water Management District has committed $500,000 worth of funding for the installation of the lock boxes for homeowners in phase one. Similar to water meter

Cape Coral's Dual Water System: Two pipes are run to each household—one for high-quality drinking water and a second for non-potable irrigation water, consisting of treated canal water and reclaimed wastewater.

boxes, these boxes allow the key-holder access to the dual water system. The boxes will include a one-inch PVC hook-up for connection to residential irrigation systems, and a 3/4-inch hose bib for those homeowners who wish to irrigate with a garden hose.

• Lamb feels that Cape Coral is on the cutting edge of reuse technology. "Wastewater is a commodity that is very important . . . it can solve long-term problems," he said. "A lot of utilities are beginning to look at it with new eyes, as a resource for the future."

It used to be that wastewater was primarily reserved for golf course irrigation. When Lamb looks at what Cape Coral and other cities are doing, he wonders if there are better

uses for the resource. The city of Hollywood in Broward County is looking at it for use as a saltwater barrier. Palm Beach County is considering recharging its wetlands with reclaimed water. And some utilities are examining reuse as a method of recharging diminishing groundwater levels.

But in Cape Coral, the mayor sees it as the capital project which will not only provide water for the city's growth into the 21st century, but one which will make the city more beautiful. "The hope is that once it is up and fully operational, there will be a great greening of the city . . . it will enhance the value of the community as a whole."

Discussion Questions

1. Describe the steps being taken in Cape Coral, Florida, to reduce demand on its aquifer.

2. The plan outlined in this article may suffice to supply water to Cape Coral, but at what cost? Estimate the costs and projected growth.

3. If you were a citizen, would you promote other measures to stop or slow growth or lessen its impact? What measures would you promote?

Citrus Trees Blossom with Reclaimed Water

John L. Jackson, Jr. and Phil Cross

■

SOURCE
Water Environment & Technology,
February 1993, Vol. 5, No. 2

John L. Jackson, Jr., Florida Cooperative Extension Service, Tavares, Fla., and *Phil Cross*, Metcalf & Eddy Services, Inc., Winter Garden, Fla.

THE Water Conserv II project in Orlando, Florida, is the world's largest water reuse system of its kind and the first in Florida to use reclaimed water for the irrigation of crops intended for human consumption. The system combines irrigation of citrus groves and rapid infiltration basins (RIBs).

Currently, 27 growers use the free, reclaimed water on approximately 4040 ha (10,000 ac) of citrus groves. When the citrus growers do not irrigate, or when the available reclaimed water exceeds demand, the excess is discharged into RIBs. About two-thirds of the area's discharge is used for citrus irrigation and one-third for RIBs.

Growers have the benefit of a free source of water with adequate pressure that is not subject to drought restrictions and that has been shown to increase crop yield. The city of Orlando and Orange County, suppliers of the reclaimed water, have reduced demand on the Florida aquifer, saved water disposal costs, and met the conditions of a recent zero discharge mandate.

The Water Conserv II system has a flow of approximately 1096 L/s (25 mgd). Reclamation facilities are designed to meet EPA Class I reliability requirements to ensure a clear, odorless, virus-free product that is safe for human contact and that will not contaminate groundwater quality.

The system's current permitted capacity is 1928 L/s (44 mgd), with a design capacity of 2192 L/s (50 mgd), and a peak capacity of up to 3287 L/s (75 mgd). The reclaimed water is pumped through 8.8 km (5.5 mi) of pressurized pipeline to a junction point on the Florida Turnpike, where the flows are combined in a common 1.37-m-dia (54-in.-dia) pipeline that runs 25 km (15.5 mi) to the distribution center. This distribution center is the focal point for operation and maintenance of a 34-km (21-mi) transmission pipeline, a 50-km (31-mi) distribution network, 6464 ha (1600 ac) of RIBs, 23 supplemental wells, 24 surge protection sites, and associated facilities.

The distribution center consists of four 19×10^6-L (5×10^6-gal) storage reservoirs, a distribution pump station, and operations and maintenance facilities. The storage reservoirs are used for flow equalization, while the distribution center pumps reclaimed water to the citrus groves for irrigation, or to the RIBs for disposal. The system is computer controlled, and chlorination facilities provide disinfection.

Grower Agreements

The Water Conserv II project has a grower agreement, which is a 20-year contract between the citrus grower and the city and county that is binding to the landowners. Under the agreement, the reclamation facilities guarantee a weekly amount of free, reclaimed water to each grower at 275.8 kPa (40 lb/in.²) under normal operating conditions. They also guarantee that the reclaimed water is free of virus and fecal coliform, and that it meets standards agreed on by the growers and the city and county. In many cases, these standards are more stringent than official limits.

Growers have the right to terminate the agreement if it is determined that the reclaimed water is detrimental to crop productivity or quality. Growers can increase or refuse their weekly allotment of reclaimed water for 4 weeks per year, but for no more than 2 weeks consecutively.

A buy-out clause is provided whereby growers can end participation in the 20-year agreement by repaying the city and county at the rate of $890/ha ($3600/ac) after the first year, with the repayment rate decreasing by 5% each year thereafter.

Freeze Protection: A Key Element

When the system was being designed, participating citrus growers agreed that they needed reclaimed water on cold nights to spray trees to protect against freeze damage. When water is sprayed on the trees, ice forms, maintaining a temperature of 0°C (32°F). Citrus does not sustain freeze damage until the temperature reaches −2.2°C (8°F) for 4 or more hours.

However, after estimates showed that there would not be enough reclaimed water available for all growers to have freeze protection, 23 wells were installed to supplement the reclaimed water during freeze events and during peak irrigation periods. The wells are collectively capable of supplying an additional 3217 L/s (91 mgd) of water.

Citrus Crop Thrives

Studies on the effect of reclaimed water on citrus production over the last 6 years indicate no obvious problems with the citrus trees, and

Reclaimed water provides freeze protection for young trees in Florida citrus groves.

the water has the potential of providing nutrients to the trees.

Growers have reported that the use of reclaimed water has increased yields substantially (estimates are from 10% to 30%). Although fruit production data have not been collected over a wide area, one study shows that Hamlin orange fruit production increased 23% from trees using reclaimed water.

Studies show that the growth rate of young Hamlin and navel orange trees increased 225% and 443%, respectively, from 1987 to 1988 in areas using reclaimed water. Even though there was a varietal difference, the reclaimed water appears to be promoting greater canopy growth.

Nitrogen and phosphorus levels on citrus leaves have been higher in those areas that receive reclaimed water. There has also been an increase in sodium and iron levels, while the manganese and zinc content have been lower than from trees receiving well water. No consistent trends were observed for leaf

potassium, calcium, magnesium, and copper levels. Although the leaf nitrogen and phosphorus levels are higher in areas that use reclaimed water, the amount is well within the acceptable range.

Sodium content in citrus leaves in areas using reclaimed water was almost three times that of the well water locations; however, the amount of sodium in the Water Conserv II blocks has begun to stabilize and is still well within the optimum range for citrus.

Soil Analysis

Although surface soil analysis does not show any consistent trends in the accumulation of nutrients, an examination of the entire soil profile reveals a different picture. During 6 years of operation, the soil profile in the reclaimed water area has exhibited a trend toward higher levels of nitrogen and phosphorus, while the potassium, calcium, magnesium, and sodium levels have varied from year to year.

The soil moisture level in the Water Conserv II blocks is higher than in the well blocks. During the early years of the project, when growers were using larger amounts of reclaimed water, it became apparent that the fruit had been adversely affected. Levels of sugar per box of fruit were 0.3 kg (0.75 lb) less in the Water Conserv II blocks.

As growers have improved water management techniques, however, there has been no significant deference in the sugar content of the fruit from the Water Conserv II blocks and that from the well blocks. Researchers have concluded that the reduced sugar content resulted from excessive irrigation, not from the reclaimed water. Soil moisture levels in the Water Conserv II blocks are still elevated.

Discussion Questions

1. Outline the program described in this article.

2. Do you think it might be useful in your region?

Hazardous Waste

A ccording to the
United States EPA,
the nation's indus-
tries produce about
270 million metric
tons of hazardous
waste each year. Nearly all of this
waste is in a liquid form and 90% to
95% is treated or disposed of on site
in wastewater treatment facilities, in
surface impoundments, or in deep
injection wells.

According to a report entitled
the Green Index, written by Bob Hall
and Mary Lee Kerr at the Southern
Institute, 9 of the top 13 states in
hazardous waste production are in
the southern region. Together, the
southern states generate 66% of the
nation's hazardous waste.

Current production poses threats
to people and the environment. Moreover, previous improper disposal has
created a legacy of toxic hot spots in
desperate need of cleanup. This cleanup effort could cost many billions of
dollars, underscoring the importance
of prevention.

Background Reading: Chapter 19
(sections 19.1 and 19.2).

Southern Exposure

*Decades of Pollution Have Left
Their Mark on Dixie—And a
New Breed of Activist Vows
That Old Times There Won't
Be Forgotten*

Donald G. Schueler

SOURCE
Sierra, November/December 1992,
Vol. 77, No. 6

Donald G. Schueler is the author of
Incident at Eagle Ranch (University of
Arizona Press, 1980). He is working on a
book about the Yucatan, *Temple of the
Jaguars*, to be published in the spring of
1993 by Sierra Club Books.

F YOU happen to live in the
Deep South, as I do, the
Green Index makes for a punishing read. Published by the
Institute for Southern Studies,
it ranks the 50 states on the
basis of 256 environmental indicators—everything from air pollution
to waste disposal. No state rates a
gold star in the final tally, but that
doesn't make it any less humiliating
to find every one of the Deep South
states—Louisiana, Mississippi,
Alabama, South Carolina, Georgia,
Arkansas, and Texas—clustered at
the bottom of the list, with my own
home state, Louisiana, coming in
next-to-last. True, this miserable
showing is not really unexpected;

nevertheless, it hurts to see the bad
news laid out like that when I didn't
even know anyone was keeping
score.

I'm not surprised by the survey
results, because during my adult
years I have watched the environmental pillaging of the Deep South
from a front-row seat. Thirty years
ago I was a very-wet-behind-the-ears
environmental activist—there was
no other kind down here back
then—and I still remember what it
felt like, coming away from one
hearing after another conducted by
the U.S. Army Corps of Engineers or
this or that congressional committee,
knowing that my comrades and I
had been outgunned again—that one
more river or bottomland forest was
doomed, and that another carpet-
bagger industry had won every
extortionate concession it demanded
from state officials. I saw it all
happen—and yet the scale of the
destruction still seems unreal, as
though I had read it in a novel.
Sometimes I feel the same profound
sadness Augustus McCrae felt in
Larry McMurtry's *Lonesome Dove*,
when he stared at empty, bone-
littered prairies where, just a little
while back, he had watched millions
of grazing buffalo.

Not so long ago—the late 1950s,
early '60s—most of the Deep South
was still a backwater, impoverished,
insular, racially divided. But it was
also a region of hauntingly beautiful
natural landscapes. The air and
water were clean and clear.
Although the Big Woods of
Faulkner's stories were coming
down, there was still available to
everyone a seemingly limitless
domain of piney forests, languid
creeks and bayous, teeming marshes,
and fertile bottomlands—places
where, no matter how dirt-poor or
oppressed you were, it was possible
to feel free and happy for hours or
days on end.

Yet most all of that has been
destroyed. Not just abused and over-
exploited, mind you, but *erased*. In
order to reap the benefits of govern-

This "Keep Out" sign is an environmentally ineffective means of dealing with the problems of hazardous waste.

ment agricultural subsidies, farmers cleared and drained the second-largest forested wetlands in the world and converted them into an immense soybean field. The Corps of Engineers entombed the region's rivers. At the behest of giant paper mills, loggers transformed millions of acres of upland forests into biologically sterile pine plantations. Most dismaying of all, because the straitjacketed Mississippi River no longer delivers the rich sediments it once spread across a vast delta and coastal plain, the nation's richest coastal wetlands were, and still are, sinking into the Gulf of Mexico at rates as high as 50 square miles a year.

Adding injustice to these indignities are the poisons. Toxics have left the South's air and water the most industry-befouled in the entire country. According to the *Green Index,* the region leads the nation in overall per-capita exposure to industrial poisons in the air and water, and it produces a disproportionate share of the most dangerous chemicals, those that cause cancer, birth defects, or nerve damage.

When it comes to the treatment

and disposal of hazardous wastes, the Deep South is right near the bottom. Two of the nation's largest dumps for toxic and radioactive garbage (much of which is imported from other regions) are sited in impoverished, mostly black counties in South Carolina and Alabama, while a mostly white Louisiana parish harbors the country's largest (and most dangerous) incinerator of hazardous waste.

At first glance it might seem hard to understand how the South, during the very years when the environmental movement was coming of age, could suffer this ecological equivalent of Sherman's March to the Sea. Part of the explanation lies in the fact that the civil-rights movement was gaining significant national attention during the same period; Dixie, always regarded as benighted and backward, was now shunned and boycotted by public and private organizations, including environmental ones. This pariah status also made it easier for politicians to agree tacitly to make the Deep South a "sacrifice zone," a sump for the rest of the nation's toxic detritus.

"What you've got operating down here is a colonial attitude," says Jim Price, the Sierra Club's Southeast staff director, "a conscious decision by local government and big business to prey on folks who are politically and economically weak."

Of such vulnerable people the Deep South has always had more than its share. Southerners, white and black, are even less well educated than the deplorable national average. They have lower incomes and a higher rate of infant mortality, and in their adult years they die earlier than Americans elsewhere. It is not a coincidence that they should also be exploited to an unusual degree—by their own government, and by corporations that poison their air and water, devastate their forests and wetlands, and bury hazardous wastes in their backyards.

Patsy Ruth Oliver, of Texarkana, Texas, can give you an earful about how that exploitation works out in human terms—and she will, if you give her half a chance. She can tell you about the succession of wood-treatment companies that dumped creosote wastes on what was later to become Carver Terrace, the subdivision in which she lives, lacing its soils and water with a staggering array of deadly chemicals. And about the real-estate developer who deemed the site "good enough for niggers to live in." But it is when she starts talking about Big Government, as represented by the regional office of the Environmental Protection Agency, that she really gets worked up.

"There was this one time," she recalls, "when this little EPA white man was sitting behind his desk telling me I had to be patient about living with poisons, like he was saying 'Take two aspirin.' I don't know, it just went through me like an electric shock, him saying that. So I told him, 'Listen, the only people in Carver Terrace who are patient are the ones lying underground!' I just didn't know I had it in me to say that."

About 75 families live in Oliver's modest subdivision. "We had nice people moving in here," she says. "We thought we were getting out of the ghetto, moving into middle-class society, trying to teach our children what the American dream is all about. But what we got was a nightmare, a toxic prison." She points to her gray cat, crouched listlessly beside the door of her small brick home. "The vet can't do nothing for her. She's covered with tumors. I mean, it gets to be ridiculous. Your neighbors are sick, your dogs are sick, even the cat's sick! It's so awful, you almost have to laugh, even though *that's* sick, too."

According to Oliver's count, at least 25 people in Carver Terrace, including her own mother, have died of cancer during the last ten years: "We got the hearse coming in here more than the Yellow Cab." So far no statistical evidence has been gathered to support her claim that the cancer death rate is abnormally high, but there is no doubt that the subdivision deserves its present designation as a Superfund site. Tarry globs of creosote poke out of lawns, turn

the soil pitchblack, stink up the neighborhood, and leave greasy, evil-smelling residues in sinks and toilets—a bad-enough situation made worse by the floods that regularly inundate the area. Oliver and many of her neighbors suffer from eye irritations, headaches, skin rashes, nausea, and respiratory problems, all of which can be caused by exposure to creosote's toxic ingredients.

As early as 1980, investigators for the EPA and the Texas Department of Water Resources unofficially warned residents that it might be dangerous for their children to play out of doors. But the EPA's official follow-up has not endeared the agency to Patsy Oliver. "They did test after test here," she fumes, "but they never studied *us*. They put a fence around the gravel pit next door and a sign saying it was contaminated, yet they'd tell us we didn't have a thing to worry about. Then they put in some grass sod to cover up the creosote, which lasted as long as a snowball in an oven, and they said everything was fine!"

Oliver says that her experience with the EPA turned her from "one hysterical housewife" into a "true believer in the environmental movement." With the help of a local group, Friends United for a Safe Environment, she became a leader ("I guess I had the big mouth") in a successful effort to publicize her neighborhood's plight. Other groups joined the fight against an obdurate EPA, which still insisted, in spite of much evidence to the contrary, that the site posed "no immediate threat." Finally, in 1991, the rallies and marches paid off. Although it took an act of Congress to accomplish it, the Army Corps of Engineers must now do what the EPA probably should have done five years ago—namely, buy out the residents of Carver Terrace so they can move to safer homes. About half the community has left, but Oliver and others are still waiting to receive settlements from the Corps. "I'm

hoping God has finally put some truth into those people," says Oliver, frowning. "But I'll believe that when I'm out of here!"

Understandably, Patsy Ruth Oliver is not likely to feel all choked up with gratitude when she finally does escape the toxic prison on Carver Terrace. Yet in a way she is lucky. For countless others living in the Deep South, escape is not an option.

Ironically, the proliferation of environmental regulations elsewhere in the United States has played an important role in inducing many of the country's most pollution-prone industries to come to roost in Dixie. Bob Hall of the Institute for Southern Studies hardly exaggerates when he laments that "everything in the South is available for the taking." Nowhere else are regulatory agencies more reluctant to crack down on polluting industries, or governments more eager to grant them tax breaks. Georgia and Alabama, for example, allow influential pulp and paper companies to dump dioxin into state waters in quantities of dozens of times greater than those recommended by the EPA. And in Louisiana, again just for example, 30 giant corporations, including many of the country's worst polluters, received $2.5 billion in property-tax exemptions during the 1980s, though they created few permanent new jobs.

One explanation for these look-the-other-way environmental policies and giveaway tax breaks is that so many communities in the Deep South are company towns, relying in an almost feudal way on the patronage of one or two major industries. In most of the Deep South's state capitals, the prevailing political attitude resembles that of a debt-ridden developing nation: Political bosses encourage outsiders to buy the region's human and natural resources at bargain prices. Inevitably, skeptics suggest that some of the politicians themselves are among the resources for sale.

Consider the relationship of Louisiana's current governor, Edwin Edwards, with businessman Jack Kent. In the last election, Kent spent hundreds of thousands of dollars helping his buddy Edwards regain the office he had held some years earlier. Kent could easily afford the expense: He is the sole stockholder of Marine Shale Processors, a company in southwest Louisiana that for more than a decade—most of it coinciding with Edward's previous tenure in office—has successfully operated the nation's largest hazardous-waste incinerator, without a license and in flagrant violation of a swarm of EPA regulations. Not surprisingly, Kent resents meddlesome environmentalists, publicly advocating the use of "green-colored ax handles" on them. (His workers, taking the hint, once gave a Greenpeace demonstrator at his plant a good pummeling.) Yet Governor Edwards seems unruffled by his friend's excesses. He goes about his business nonchalantly, happily dismantling much of the environmental legislation enacted during the administration of his predecessor, Buddy Roemer.

If most politicians in the Deep South's state capitals are environmental Neanderthals, it must be said in fairness that those the South sends to Washington are proto-hominids. According to the League of Conservation Voters, the overall environmental voting records of the region's congressional delegations range rather narrowly from pretty sorry to downright shameful. Four southern senators—Trent Lott (R) of Mississippi, Jesse Helms (R) of North Carolina, Phil Gramm (R) of Texas, and Howell Heflin (D) of Alabama—have the distinction of possessing some of the worst environmental voting records in Congress.

No one better epitomizes the anti-ecological attitudes of southern politicians than my home state's senior senator, J. Bennett Johnston (D), whose recent machinations provide a sterling example of what

can happen when an influential member of Congress and the executives of a powerful corporation start scratching each other's backs.

The citizens of Claiborne Parish, Louisiana, possess a good deal of insight into this symbiosis. An international consortium, Louisiana Energy Services (LES), has been planning to build the nation's first private uranium-enrichment plant in Forest Grove, an impoverished rural black community just a few miles outside Homer, the parish seat. There LES would produce enriched uranium for nuclear-energy utilities. Unfortunately, the LES plant would also produce thousands of tons of radioactive waste that would be stored on-site for at least 30 years, or maybe forever.

No one is more eager to see LES settle in Claiborne Parish than Bennett Johnston; and since Johnston chairs the Senate Committee on Energy and Natural Resources, the consortium could hardly have asked for a more valuable ally. He joined with LES in assuring Claiborne Parish residents that the proposed plant would be a bonanza for the whole area, citing economic benefits that, according to critics, were grossly exaggerated.

Johnston sponsored an amendment, inconspicuously attached to an unrelated Senate bill, that would have speeded up the licensing of privately owned uranium-enrichment plants, LES being the only available beneficiary. When the bill moved to the House, the senator made a rare appearance before the House Interior Subcommittee on Energy and Environment to push his amendment. He arrived brandishing a geiger counter and various props, most notably a uranium fuel pellet and a dinner plate he said one "can buy in a supermarket." Obligingly, the fake pellet did not kill anyone or even burn a hole in the senator's hand. And, as intended, the plate set the geiger counter popping like mad—Johnston's point being, of course, that the uranium plant posed

no more risk of radiation than a common, easily purchased household item. Only later would opponents of Johnston's amendment learn that the plate he had brandished was a piece of uranium-coated orange Fiestaware, last manufactured in the 1940s.

Johnston's efforts on behalf of LES have set many citizens of Claiborne Parish on edge. They don't want a radioactive neighbor, and Ronnie Anderson, for one, is puzzled about why LES or anyone else would want to build a uranium-enrichment plant in the first place. Anderson, a devoted family man, oil-company technologist, and staunch member of Homer's First Baptist Church, questions the need for producing more of a substance that is already in abundant supply. But Anderson's main concern is what the company plans to do with all those tons of radioactive waste the plant will generate. Officials at LES insist that the material will be safely contained in a closed system and that, years hence, it will be disposed of somewhere else. The trouble is, "somewhere else" does not exist. At present, no dumpsite is available that could accommodate LES's "low-level" radioactive wastes.

Ronnie Anderson's common sense led him to become a vocal anti-LES activist, a fact that has not exactly endeared him to parish leaders. Some of his neighbors won't give him the time of day, and Chamber of Commerce types in Homer have gone so far as to take out an ad in the local paper denouncing "a small handful of people"—of whom Anderson is considered a ringleader—for spreading "innuendoes, half truths, facts taken out of context, and misrepresentation" thereby "irresponsibly frightening many citizens."

Such bashing of upstart members of the community is common in the South, where industry chieftains are used to getting their way. When business moguls and their political allies are crossed, as happens with more and more frequency of late, they can be quite touchy. Take, for instance, the display put on for us not long ago by Jim Bob Moffett, CEO of Louisiana's huge Freeport McMoran Corporation, with petrochemical facilities located in the industrial corridor between Baton Rouge and New Orleans known as "Cancer Alley." Since the company had contributed to an endangered-species exhibit at the New Orleans zoo, Moffett was asked to give a polite little speech at the dedication ceremony. But evidently he was still brooding about the way Buddy Roemer, our maverick governor at the time, had refused his company permission to dump huge quantities of radioactive gypsum into the Mississippi. He chose the occasion to stun his distinguished, ecologically minded audience by telling them that, rather than put up with pestering environmentalists, Louisiana businesses would do well to "kick their butts and send them home."

Jim Bob's intemperate outburst can be justly construed as an expression of corporate arrogance. Yet behind the blustering words it is also possible to detect a note of self-pity. The fact is that Moffett and other industry bigwigs have been getting worried lately. In the last few years they have had to contend with a growing number of people from all walks of life who, like Oliver and Anderson, are becoming "true believers" in the environmental movement. Out of sheer desperation, these homegrown activists are learning how to kick too.

A case in point is J. Wesley Cooper of Natchez, Mississippi, an antiques dealer and authority on antebellum houses (he lives in one himself). Given his patrician ways, he might seem an unlikely candidate for the role of fiery-eyed environmental activist; yet that is what he has become. In the mid-1980s, Cooper suspected that there might be a connection between the hundreds of chemical-filled drums he

knew to be rusting at a large Armstong Tire and Rubber Company dump near his estate and the fact that he had just undergone a serious operation for cancer. When he demanded that the site be tested, the state Bureau of Pollution Control had Armstrong hire a firm to examine it. "At first," Cooper rumbles, "they said nothing was dumped there, which was a barefaced lie. Then they said there were only drums on top of the ground, with nothing buried underneath, even though you could see the damn things sticking right out of the dirt."

Barred from the site, Cooper rented a backhoe and invited the news media and everybody else he could think of to convene at another Armstrong dump close by. Then, while the television cameras whirred, he uncovered drum after leaking drum, their oily contents spilling out as they were brought to light.

Most environmental protest in the South is aimed at industries whose practices directly threaten human health. But a growing number of people also dream of restoring habitat that has been lost.

Such a one is Michael Caire, who, in addition to practicing obstetrics and gynecology in West Monroe, Louisiana, has appointed himself a champion of the south-central subspecies of black bear, *Ursus americanus luteolus*. The bruin still survives, though just barely, in remnants of Louisiana's bottomlands, an area of wetlands forested with tupelo, gum, oak, bald cypress, and other hardwoods. Caire is quick to tell you that *luteolus* and its habitat are as inseparable as red beans and rice: "The bear is *the* indicator species. So restoring healthy populations of the critters means restoring viable parts of the ecosystem itself."

The ambitious young doctor envisions putting back together the Humpty Dumpty fragments of bottomland forests that still survive amid soybean fields. They can be linked, he argues, by using easements to expand riparian corridors

and by allowing unproductive agricultural tracts to revert to woodlands. "The way I see it, restoration is going to be the rallying cry of the environmental movement in the future," says Caire. "If we can get this program started, it will be the ideal model for other restoration projects. But what's needed is a symbol, the Louisiana black bear, which is a lot more charismatic than, shall we say, the spotted owl. It's Faulkner's bear, it's Teddy Roosevelt's teddy bear."

Caire and others pestered the U.S. Fish and Wildlife Service for years to add the bear to the endangered species list. But in doing so they found themselves bucking the Deep South's timber industry, which does not want interference from an animal requiring even more protected forest acreage than the spotted owl. To forestall a federal listing, the state's timber companies suddenly developed a great interest in the bear's welfare, and sponsored a Black Bear Conservation Committee to work up a management plan. Fish and Wildlife, which had postponed action on the bear for years, obligingly went along, suggesting that, after all, *luteolus* might not be a genuine subspecies requiring protection.

In response, Defenders of Wildlife, represented by the Sierra Club Legal Defense fund, filed suit against the foot-dragging federal agency, finally forcing it to list the bear as a threatened species. (Fewer than 100 individuals may survive.) Caire, meanwhile, is pressing the Black Bear Committee to take its task seriously. He went so far as to become a committee member in hopes of voting in favor of the bears. "Whatever it takes to help them recover," says Caire, "that's what we're going to do."

Whatever it takes. That imperative, compounded of equal parts of frustration, defiance, and hope, sums up the attitude of thousands of do-it-yourself activists who are now taking a stand in this most environ-

---

Signs of Trouble Down South

Some of the states that produce the largest volume of hazardous wastes spend the least to manage them. A state's spending to regulate solid and hazardous wastes to some extent indicates its commitment to tackling environmental problems.

HAZARDOUS WASTE GENERATED IN STATE *(Pounds per capita/rank among all states*)* *(* 1 = least waste generated; 50 = most waste generated)*		STATE SPENDING ON WASTE MANAGEMENT *($ per ton of waste/rank among all states*)* *(* 1 = most spending; 50 = least spending)*	
DEEP SOUTH STATES		**DEEP SOUTH STATES**	
Georgia	12,498/49	Georgia	$.08/50
Louisiana	6,097/46	Louisiana	.30/44
Texas	4,731/44	Texas	.16/49
Alabama	3,685/42	Alabama	.18/48
South Carolina	3,180/41	South Carolina	.28/46
Mississippi	1,919/39	Mississippi	.26/47
Arkansas	49/14	Arkansas	.80/30
OTHER STATES (for comparison)		**OTHER STATES** (for comparison)	
Tennessee	13,932/50	Tennessee	$.32/40
West Virginia	12,476/48	West Virginia	.40/38
New Jersey	2,378/40	New Jersey	1.57/22
New York	1,798/38	New York	1.65/21
California	733/32	California	3.54/15
Ohio	556/30	Ohio	1.32/24
Maine	12/6	Maine	10.70/1
South Dakota	3/1	South Dakota	.32/42

Source: The Institute for Southern Studies, 1991-1992 Green Index (Island Press, 1991)

mentally oppressed of all the nation's regions. True, it often seems that only a miracle will let them succeed. The corporate and governmental policymakers who control the South from within and without remain as closely allied as ever. In these days of widespread economic hardship, the demand for jobs-at-any-price often seems more insistent than at any time in the last decade.

Considering the odds against them, southern grassroots activists are already accomplishing what could pass for small miracles. On their account, politicians lose votes, bureaucrats lose face, and industrialists lose their tempers and their nerve. In at least some southern

states, toxics in the air and water are thinning out a little, and patches of wildlands are getting more protection.

True, the day when southerners can proudly point to favorable ratings in the *Green Index* is a long way off. But old-time environmental activists like myself can take heart knowing that the Olivers, Andersons, Coopers, and Caires of the region's counties and parishes persist in the fight. They—we—are determined to force change. After all, we may be poorer on average than other Americans, but given the facts about our situations, we are no more willing to be taken advantage of than anyone else.

Discussion Questions

1. Do you agree with the author's characterization of the South in this article? Why or why not?

2. If so, what steps must be taken to prevent future disasters and clean up past mistakes?

Savannah River: The Deadly Cover-Up

Tom Horton

■

SOURCE
Outdoor Life, April 1992,
Vol. 189, No. 4

T'S A RAINY autumn day in the Southeast. The old Savannah River slides lazily toward the Atlantic between the dense, yellowing foliage of hardwoods that line both Georgia and South Carolina shores. Outboard roaring, Zan Bunch deftly punches the bow of the johnboat into a stumpy nook of the riverbank. He hits a button on the electro-shockers and a fat largemouth of around three pounds rolls to the surface where Jeff Jones nets it.

In little more than an hour, the two state of Georgia biologists will capture more than a dozen each of bass, redbreast sunfish, bowfin, catfish, crappies, and other species proving you might think what thousands of fishermen already know: that the river here, 40 miles south of Augusta, supports an abundant fresh water fishery.

But the fish they're collecting are proof of something ominous. In the last few years the Georgia Department of Natural Resources has become increasingly alarmed by levels of radioactive cestum, strontium, and tritium in Savannah River fish that range from dozens to hundreds of times higher than normal.

The source of the problem lies behind serene-looking wooded swamps on the South Carolina side of the river, where for more than 20 river miles only an occasional yellow "No Trespassing" sign hints at anything out of the ordinary. Sprawling across a federal reservation of 300 square miles in rural Barnwell and Aiken counties are clusters of nuclear reactors and associated industrial facilities that for four decades have produced the deadly radioactive plutonium and other fissionable materials needed to arm America's atomic arsenal.

What is happening to the fish is part of the increasingly bitter aftertaste of a single-minded devotion to weapons building that allowed federal agencies to operate in secrecy, virtually without being overseen in regard to impacts on the environment and public health. In the interest of building weapons for national security, the fox was allowed to guard the henhouse: The Department of Energy set its own guidelines and regulated itself.

The Savannah River Site, as it is called (you follow Atomic Road south out of Augusta), is one of 15 major facilities nationwide in the U.S. Department of Energy's nuclear weapons production complex. The public, until recently, only knew of the end products of the weapons production—the sleek missiles and awesome thermonuclear bombs that lent world-shattering potency to our gleaming fleets of submarines, bombers and fighter jets. Now, amid admissions from the Department of Energy that its environmental oversight has been a miserable failure, the facade is parting to reveal the guts of the bomb-making process. It

is a process that has carried "a price few who promoted the enterprise could have anticipated . . . [resulting in] the release of vast quantities of hazardous chemicals and radionuclides to the environment," according to a recent analysis by Congress' Office of Technology Assessment.

So extensive and hazardous are the waste problems at Savannah River and at other parts of the nuclear weapons production complex that the Office of Technology Assessment and other government oversight agencies say cleanup is liable to take many decades. The costs may end up dwarfing the bailout of the nation's Savings and Loans. Cleanup estimates for America's nuclear weapons complex range as high as $200 billion just for Savannah River.

Scattered across Savannah River Site's vast reservation are more than 200 areas where the environmental legacy of the Cold War must now be dealt with. In one area, partially buried, sit rows of steel tanks, some of them corroding. Each holds, in temporary storage, mixtures that include some of the deadliest, high-level radioactive wastes ever produced. There are 51 of these "tank farms," and each is larger than the U.S. Capitol dome. An estimated total of 35 million gallons of high-level radioactive wastes are stored at the site.

In another area, millions of cubic feet of lower-level radioactive wastes have been buried in the soil, some from the earlier years of operation contained by nothing more than cardboard boxes. These unlined seepage basins, only now being closed, have allowed radioactive and hazardous chemical wastes to contaminate soil and ground water at levels ranging up to hundreds of thousands of times greater than normal.

Even if the money and the technology to clean it all up is forthcoming, the process will take decades. and some wastes, like

radioactive tritium, are virtually impossible to remove from ground water because tritium is itself a form of water. Meanwhile, the contamination keeps seeping out of the swamps and creeks at the site into the Savannah River, contaminating the entire food web.

The fish are absorbing radioactive materials both from river water and from sediments where radioactive cesium was accidentally discharged years ago. The radioactivity is passed upward through successively higher and higher levels of the food chain, becoming more concentrated at each level. Ecologists say that the chemistry of Southeastern soft-water rivers like the Savannah is such that fish there tend to concentrate toxic materials twice as highly as in many other parts of the country. Largemouth bass sampled in the river seem to have concentrated cesium to an especially high degree.

What does it all mean to the outdoorsman who fishes this stretch of the river? The levels of radiation are millions of times less than anything that could cause acute poisoning. But even very low exposures to radiation have the potential to cause cancer and to cause changes in human genetic material.

The Department of Energy says that there is no threat to human health from any pollution at the weapons complex. Though levels of radiation in some fish may be hundreds of times above normal, they are still very, very small, the Department of Energy says—far below what every human is normally exposed to from sunlight, X-rays at the doctor's office and worldwide "background" levels in the atmosphere from nuclear testing of past years.

But many experts disagree in varying degrees with this line of thought. Jim Setser, chief of the Georgia Department of Natural Resource's radiation program, says that his experts are convinced the Department of Energy badly underestimates the degree of the threat;

but he stops short of calling for a ban, or health advisory, on eating Savannah River fish, stating, "We don't have the evidence yet to back it up."

Dr. Karl Z. Morgan, a health physicist who pioneered in the field of radiation's effects on health, has been critical of the Savannah River Site. He says, "There is no such thing as a safe level of exposure, any more than it is safe to drive without a seat belt. I would say, 'No, it is not safe to eat those fish.'"

Morgan and Setser both point out that the fish are only part of the many ways in which the weapons complex exposes people to radiation. Georgia has identified dairy operations 16 miles away from the weapons complex that routinely show tritium in milk about 20 times greater than normal background levels. Rainwater several miles from the site has even higher levels.

Potentially a much greater concern to the state, Setser says, is that ground water in the heavily contaminated aquifers lying beneath the weapons site will someday make its way beneath the Savannah River and into the wells that are the major source of water for southeastern Georgia.

Even downstream as far as Savannah, 100 river miles from where the complex borders the river, the Savannah's water continues to show low, but unmistakable, evidence of radioactive contamination.

Officials at the Savannah River Site counter that the elevated levels of radioactivity Georgia has been monitoring in the environment are "nothing new," and are no cause for alarm. When placed in perspective, the worst-case human exposures to all radiation from the weapons complex have amounted to less than 1 percent of what people get from normal background exposures, says Steve Wright, director of environmental operations for the Department of Energy at the Savannah River Site.

Wright agrees that levels of radioactive materials in milk, rainwater and river water may be many times greater than normal, but he notes that they are still well below health standards set by the U.S. Environmental Protection Agency and other agencies. For example, though levels of tritium found in milk and in drinking water supplies downstream as far as Savannah and Beaufort, South Carolina, are 10 to 100 times above normal, they are still far less than half the amount allowed by federal health standards.

But Setser's department feels that the Department of Energy and its contractor, Westinghouse, which operates the weapons site, underestimate the public's potential for radiation exposure by as much as 50 times. One reason, they say, is that site officials fail to take into account people who eat large amounts of fish caught in the river near the site. Morgan's health studies further assert that current federal safety guidelines underestimate the toxicity of radioactive tritium by as much as five times.

Another group that thinks the government is overlooking the health risks associated with living around the site is the Energy Research Foundation, a Columbia, South Carolina, environmental group that specializes in analyzing nuclear weapons issues. Tim Connor, a journalist and researcher with the foundation, says that no assessment of radiation risks has ever taken into account the decades of past exposure.

"An example would be the unlucky fisherman who is fond of largemouth bass he happens to catch [near the plant]," Connor said while testifying recently at a hearing on the complex's operations. "The cesium he ingests through eating the flesh of a fish he catches next year will be added to the cesium retained in his body from past consumption of fish . . . what science tells us about low-level radiation is that it has a cumulative effect."

Government agencies like the Office of Technology Assessment that have scrutinized the weapons complex simply say that far too little is known about health hazards from radiation releases for the Department of Energy to dismiss risks.

In the last year or so, prodded by U.S. Rep. Lindsay Thomas (D-GA) Congress has told the Department of Energy to begin more extensive studies of both ground water and fish contamination along the Savannah. And prodded by the Energy Research Foundation and others, the federal Centers for Disease Control is studying possible health risks from living near the weapons site.

In addition to apparently lenient concern for the human side of the equation, the government's weapons making has also proceeded without regard for laws and regulations protecting wetlands and streams at the Savannah River Site. In producing the nuclear material for America's weapons, the site has devastated hundreds of acres of forested swamps and rendered miles of streams devoid of life for decades.

The Savannah River Site used five nuclear reactors to make the tritium and plutonium that are the basic fuels for atomic weapons. These reactors required continuous flows of river water—tens of millions of gallons a day—to prevent a disastrous meltdown of the reactor cores. In many cases this water was discharged at scalding temperatures into streams and swamps feeding the Savannah.

Although the federal Clean Water Act banned such practices, the Environmental Protection Agency effectively gave the Department of Energy a loophole by letting it meet temperature standards at the boundary of the site, miles away from the reactors, rather than at a the point of their discharge into streams.

A typical result was Four Mile Creek, a sizable stream into which the site's C Reactor discharged. From about 1955 until 1985, when the reactor was shut down, as much as six miles of the creek leading to the Savannah River were devoid of life most of the time because of the high temperatures, according to studies conducted by University of Georgia ecologists.

In addition, the elevated water levels from the discharge damaged or destroyed wooded wetlands along the creek at the rate of five acres per year. Recent studies, the ecologists say, show that the wetlands are coming back, and that the fish communities are rebounding. But full recovery, they say, "will likely take decades."

Other reactors caused similar or greater damage to other tributaries. Until the last decade or so, site officials kept even the temperature of the water discharged "classified" for national security reasons, according to scientists who worked at the site then.

The Department of Energy's secrecy and self-regulation that has resulted in the contamination and radioactive hazards at the Savannah River Site is a situation that only began changing in recent years. Only in 1988 did the Department of Energy begin revealing to Congress the records of dozens of serious mishaps that occurred at the site's nuclear reactors, many of which released radiation into the air or water. One close miss in 1965—covered up for 23 years—probably came within minutes of a meltdown that would have destroyed the reactor, releasing huge quantities of high-level radioactivity.

"The history of Savannah River has been one of complete, criminal disregard of the environment . . . the bills are now coming due," says Jim Simon, an attorney for the Natural Resources Defense Council, a private, national environmental group that has filed more than a dozen lawsuits during the last two decades to force the federal government to protect the environment at the site.

You might think that there would be widespread alarm in Georgia and South Carolina over the mounting evidence of mismanagement and environmental problems at the weapons complex, all of which have been reported by the local and national news media. But more than 20,000 jobs depend directly on the Savannah River Site. The economic implications of shutting it down seem a lot more real than threats of cancers or chromosomal damage that might take decades to become evident, assuming the impact of radioactivity could even be sorted out from effects of diets, smoking and other environmental health factors.

National security remains a powerful force in these parts, despite the rapid winding down of the Cold War. Presently the Department of Energy's nuclear weapons production is in tatters nationwide. The aging reactors that made bomb-grade plutonium and tritium at Savannah River have been shut down since 1988 because of safety and environmental concerns. Built with 1950's technology, the reactors are nearing the end of the 40-year life span for which they were designed; and during the 1980s, vital maintenance budgets and money for waste handling consistently were cut by the Reagan administration and Congress, according to sources at the Du Pont Company, which ran the site until Westinghouse took over in 1989.

Two other critical weapons assembly facilities, Hanford in Washington state and Rocky Flats near Denver, Colorado, are shut down for extended periods for similar problems. In addition, a total of nearly 2,000 Department of Energy and Department of Defense facilities have been targeted for cleanups, an estimated $400 billion project that will affect many states throughout the nation.

Rick Ford, a Department of Energy spokesman at the Savannah River Site, says, "The fact that we currently have no capability to make more [nuclear weapons fuel] makes people in high places very uncomfortable."

A glaring illustration of this is the Department of Energy's continued push to quickly resume tritium production at its so-called K Reactor at Savannah River, despite several unresolved questions about its environmental impacts or whether there is any need for it.

For decades the K Reactor poured millions of gallons of scalding water daily from its cooling system, seriously damaging hundreds of acres of wooded wetlands and miles of streams feeding the Savannah River. Pushed by the Natural Resources Defense Council and other concerned groups, site officials agreed to construct, at the cost of more than $100 million, a dramatically less-harmful cooling system for the old reactor. It will be completed by the end of this year; but the Department of Energy contended it could not wait for it.

The rush to get K Reactor up and running again confirmed the worst fears of Setser and others last December 22 when an old heat exchanger ruptured and released 150 gallons of tritium-contaminated water. The malfunction spilled as much radioactivity into Steele Creek, a Savannah River tributary, as the reactor normally released in a year.

So much tritium hit the river that 80 miles downstream at Savannah, Georgia, radiation crews measured levels as high as 60,000 picocuries per liter—violating federal drinking water standards threefold. On the South Carolina side, a metropolitan water supply that pumps from the river had to be temporarily shut down. Two industrial water users in Georgia, the Savannah Foods and Industries sugar refinery and Fuji Vegetable Oil, were temporarily shut down but have since reopened. Oyster beds near the mouth also were closed to harvesting.

In official statements, Westinghouse described the release of tritium as "small," having "no impact on human health," and a spokesman said it was "nothing to write home about." However, angry government officials in both South Carolina and Georgia said they would ask the Department of Energy to indefinitely delay the reactor's restart until an investigation was made.

Setser said he had seen an internal memo showing that the equipment was suspect. "They [Savannah River complex officials] knew that heat exchanger was bad, but they decided not to replace it anyhow," he said. "It is just one more in a long series of times they didn't have their ducks in a row. It has made us all aware of just how vulnerable everything downstream from the place is."

A great debate is waging over the long-range future of nuclear weapons making along the Savannah River. Georgia Gov. Zell Miller recently opposed a new, $5.6 billion reactor at the weapons complex, which is one of the two finalists for the reactor among the nationwide nuclear weapons complex.

In a letter dated June 17, 1991, to federal energy secretary Admiral James D. Watkins, the governor accused the department of continuing to downplay concerns about contaminated fish and ground water. He said that an intensive cleanup of existing problems at the weapons facility would generate more jobs and economic benefits for the region than a return to production of radioactive materials.

Another argument against resuming production at places like Savannah River is that many arms experts say that enough tritium for new nuclear warheads can either be recycled from existing, outdated weapons, or produced more safely and cleanly by other processes than those at Savannah River. As for plutonium, the other main ingredient of warheads, even the Department of Energy says that it has plenty on hand to meet any foreseeable needs.

A long-term concern of environmental officials in both South Car-

olina and Georgia is that the federal government's stated commitment to clean up the Savannah River Site may not prove equal to the huge task. David Wilson, an official in the South Carolina Bureau of Solid and Hazardous Waste, says that Georgia's environmental concerns "are reasonable, given the history of the place." He says, "Our mind-set is the government will spend whatever it takes to clean up, but will the federal budget run dry? We just don't know."

Tim Connor of the Energy Research Foundation says that he has long-term concerns about the potential for fishermen to begin exploiting the abundant populations of bass and other species in contaminated lakes and ponds on the sprawling Savannah River Site, as pullback from weapons production occurs, which he feels is inevitable.

The ponds at the site, closed to most fishing pressure for several decades, harbor what one site official calls "the best bass fishing this side of Cuba."

"I don't think it would be smart for anyone to be eating fish from some of those places on a regular basis for many, many decades, maybe centuries," Connor said.

Westinghouse and the Department of Energy are proud of the cleanup progress they have already made, and point to it as proof that they will not stick the states with a toxic nightmare. On the standard tour of the big weapons site, a group of earnest young engineers and scientists take you to a grassy knoll that was once a highly contaminated settling basin for wastes. Beneath it remain billions, perhaps trillions of gallons of polluted ground water; but by literally inventing new cleanup technologies they have succeeded in pumping it up and removing, so far, nearly half of an estimated 450,000 pounds of toxic waste from the water. At another location, a billion-dollar plant is nearly complete that will soon begin the elaborate entombment of mil-

lions of gallons of the nastiest wastes stored at Savannah River. But it will take the plant 15 years to process just what is already there, if no new weapons production is ever begun; and the federal government still has not agreed upon a final, safe resting place for the waste-filled canisters. They will be stored temporarily at Savannah River Site in massive pads of concrete set in the ground.

It is all impressive to the layman, and part of what the congressional Office of Technology Assessment report on cleanup of the nationwide nuclear weapons complex calls "laudable efforts," but the efforts are not enough to keep the report from concluding that "the prospects for effective cleanup . . . in the next several decades are poor." The report calls for "significant changes" that include a virtual end to self-regulation by the Department of Energy, substituting independent oversight of its activities at local, state and national levels.

Back in his office in Atlanta, Setser stops poring over the years of statistical data his department has been gathering on radioactivity in fish and milk, rainwater, drinking water and collard greens and turns philosophical for a moment: "In too many ways, the [Savannah River Site] still is not responsible to anyone but itself, and that just has to change. There was a time when Americans trusted institutions more—schools, government, doctors, or at least they didn't question them—but that has all changed now. We have become less trusting and there is hell to pay."

Discussion Questions

1. Describe the environmental problems outlined in this article.

2. How could problems such as the ones described in this article be prevented in the future?

Toxic Buyouts: You Can't Go Home Again

Jim Schwab

∎

SOURCE
The Neighborhood Works,
August/September, 1992, Vol. 15, No. 4

Jim Schwab, a Chicago writer, is working on a book about blue-collar environmentalism.

JANICE DICKERSON grew up in the African-American community of Reveilletown, La., learning that her people's heritage was embodied in its survival. The small, unincorporated settlement, just two miles south of Plaquemine, was founded by freed slaves after the Civil War. The church celebrated its centennial just a few years ago. Dickerson was the fourth generation of her family to live on her property. The community's continuity allowed her to trace her family back to her great-great-grandmother, who was born in slavery.

But because of pollution problems that followed the arrival of a chemical manufacturer, the historic town today no longer exists. Its residents have scattered throughout southern Louisiana; Dickerson is part of a core of six families that moved to the same Baton Rouge neighborhood. The town's ruin taught Dickerson some vital lessons about negotiating with corporate polluters when life nearby is no longer tenable.

For the last three years, she has applied those lessons as an organizer for the Gulf Coast Tenants Organization, which works with other communities based near large polluters. Several million people live in the industrial corridor from Baton Rouge to New Orleans where one-fifth of the nation's petrochemicals are produced.

Increasingly, communities which have organized against companies because of toxic emissions are instead turning to a last resort solution—buyouts—where the company pays the costs of relocating residents of an entire community. Less drastic alternatives, say residents, may produce a generation of sick children and shortened lifespans for their parents because of health hazards associated with living near petrochemical plants. These communities have concluded that when it is clear a neighboring facility will not or cannot clean up its act, it's time to get out.

Getting Information

In 1969 when Georgia-Pacific built its plant within 300 feet of some Reveilletown homes, Dickerson says, the common perception was that "black folks aren't concerned about the environment." That perception did not sit well with Dickerson, who first noticed problems with lawns and gardens when the company later added a vinyl chloride unit. She recalls thinking, "If this stuff is killing grass and trees and vegetables, it's got to be doing something to people." Dickerson then organized Victims of a Toxic Environment United (VOTE United) to "raise hell with everybody we could."

Often residents have no way of linking individual symptoms, such as the nausea, headaches, and respiratory problems experienced by Reveilletown residents, to an area-wide problem, and relating that to plant emissions. For that reason, Dickerson says, "We decided we wanted health testing done." But "nobody had any vision of leaving Reveilletown."

Instead, VOTE United focused on enlisting plaintiffs for a class action lawsuit it filed in 1984 against Georgia-Gulf Corp., a Georgia-Pacific subsidiary that bought out the parent company. The suit, according to Ramona Stevens, an environmental consultant to the group, charged that the firm's toxic air emissions were damaging resident's health and sought monetary damages and relocation. VOTE United's struggle dragged until the 1986 Superfund Amendments and Reauthorization Act created the Toxic Release Inventory [TRI], which provided the group with an effective new tool. The TRI is a mandatory annual disclosure by companies of the nature and volume of toxic substances they release into the environment. Though imperfect, it has given communities vast amounts of new information about company operations.

Stevens cautions that TRI data are often rough estimates and not totally reliable. Nonetheless, as Dickerson notes, the data have provided a far better window into company operations than activists previously had. "Before 1986, we didn't know what the hell was going on over there," Dickerson says. The group had gotten some help from her brother and father, who had worked in the chemical industry. But after 1986, she says, "we started getting information on what was being produced over there and sharing it with people involved in the lawsuit."

The information produced a dramatic change of heart among the residents, who began for the first time to feel that moving was the only long-term way to protect their families' health. Residents eventually settled out-of-court, but a court order has sealed the records so plain-

Examples of Emissions from Petrochemical Companies in Louisiana and Their Potential Health Effects

Benzene: carcinogen; can cause birth defects and damage to reproductive organs

Ethylene family: (e.g., dichlorethylene) possible carcinogens; environmentally harmful to wildlife; may harm reproductive ability; continued exposure causes damage over a period of years; may cause damage from short exposure (burns, breathing problems)

Hydrochloric acid: short exposure may cause problems such as breathing, burns, etc.; may cause damage over period of years

Methyl chloride: carcinogen; may harm reproductive ability; causes harm through short exposure (breathing, burns, etc.)

Phthalic anhydride: may harm the nervous system—brain, nerves, spinal cord

Sulfuric acid: short exposure can cause burns, breathing problems; causes damage from long-term exposure; environmentally damaging to wildlife

Vinyl chloride: carcinogen; linked to miscarriages; birth defects; infertility; damages genes and chromosomes

Source: Data from the Toxic Release Inventory, Office of Toxic Substances, 1991, and the Right-to-Know Network, OMB Watch.

---

tiffs cannot discuss the settlement of health findings, according to Stevens.

"After we started doing research on the consequences of keeping our children in that kind of environment, one of our attorneys said Georgia-Gulf was willing to buy out the community," Dickerson says. "We resisted, but as the data started surfacing, we were placed in almost a no-win situation." With Ashland Chemical just three-fourths of a mile to the south, the village of 150 people "was surrounded." Noting that the residents expected little help at the time from Louisiana's notoriously lax regulatory agencies, Dickerson adds that "we accepted that offer for our children."

But even after reconciling themselves to abandoning their historic community, Reveilletown residents learned how to hang tough. "We made a pact," Dickerson says. "If everybody didn't get what they wanted, nobody would move. No

one would end up with a mortgage, and people would get some change for the kids."

Housing costs, however, were the easy part of the negotiations. Health issues proved stickier. "They were saying you could not in all honesty say the exposure to those chemicals caused that illness. Our biggest problem is getting health officials to link health conditions with specific exposures," Dickerson says.

But she is convinced life has improved. "My mother took Benadryl before and doesn't now," she says. "The kids don't have colds as frequently as they did. My general assessment is that if we had stayed, by now it may have been too late."

Tracking Health Effects

Health effects are also an issue for Janis Terrell, who lives in the largely white, blue-collar Fairlea Subdivision in Port Arthur, Texas. Unlike Reveilletown, Fairlea grew up

with the nearby Fina oil refinery, built in the 1930s. Few people in the strongly pro-union neighborhood ever questioned the value of having a refinery for a neighbor until early 1991, when Fina started up a new sulphur recovery unit just 1,500 feet from the nearest house. Throughout the spring and summer, sporadic releases of hydrogen sulfide, which smells like rotten eggs, sent numerous residents to the hospital. According to Terrell, a spokesman for Fairlea Residents Opposed to Noise and Toxins (FRONT), the emissions aggravated existing lung problems for some residents.

Residents soon began keeping logs of air pollution incidents, phoning them in to the Texas Air Control Board office in Beaumont and pleading with city officials for a crackdown on Fina. According to Dale L. Watson, assistant city planner, Port Arthur has an industrial performance standards agreement with Fina, which sits just beyond the city limits but within the city's zoning authority.

As pollution episodes continued, residents' patience wore thin. Their interest in staying in the area also waned. For one thing, the persistent problem had made people aware that the L-shaped refinery property left the neighborhood, bounded on a third side by Sabine Lake, with only one route of escape in an emergency. That route involved three parallel roads which cross a railroad track. Residents realized, says Terrell, that "should we have to evacuate while a train is coming, we would be in a lot of trouble."

One man instigated a letter-writing campaign by distributing the address of the president of the Belgian company's U.S. subsidiary. When new problems arose in July of 1991 with flaring and emissions of black smoke, residents showered letters on the executive at his firm's headquarters in Dallas, demanding a buyout of the neighborhood and relocation of its residents.

Solidarity Pays Off

The local plant manager began to research the possibilities, and the company hired a real estate group to negotiate an agreement. But Terrell notes that, at one point, the company attempted to bypass FRONT by approaching residents individually. The organization's firm stance in insisting on representing residents prevailed, however, and by February, Fina was offering $4,000 in relocation assistance to each family. In addition, FRONT negotiated a valuation procedure in which each party could choose one appraiser from a list of 10 that Fina and FRONT jointly compiled. The total buyout cost for Fairlea's 211 households is now expected to top $10 million and may take up to eight years to complete. The neighborhood voted last year to let those who were sick move first.

Terrell also has plans for after the settlement. Learning from others' experiences, she wants FRONT to maintain an ongoing directory of former residents in order to monitor long-term health effects. The group has consulted with Dr. Marvin Legator, a toxicologist at the University of Texas at Galveston, who according to Terrell, told the group that proper testing on the effects, particularly on humans, of many industrial chemicals has never been done or is woefully inadequate.

Terrell says that one sticking point with Fina has been the company's insistence that residents sign a waiver releasing Fina from any responsibility for future health problems. Terrell has other complaints about Fina's handling of the buyout. She says Fina should beef up its buyout budget to speed up the process, and complains that it is still being less generous than Conoco in Ponca City, Okla., and Dow Chemical in Morrisonville, La., on the north end of Plaquemine.

Environmental Health of Louisiana

Toxic chemical releases, lbs. per capita. Ranking among 50 states (least amount ranked #1)

Total Toxic Release to The Environment

67.7

Rank: 50

Cancer-Causing Chemicals Released

5.41

Rank: 48

Birth Defect Toxins Released

0.17

Rank: 47

Nerve Damaging Toxins Released

8.67

Rank: 48

Source: 1991-1992 Green Index, a State-by-State Guide to the Nation's Environmental Health, by Bob Hall and Mary Lee Kerr. Island Press, 1991. Institute for Southern Studies.

Voluntary Buyout

Dow Chemical has set the standard for corporate behavior in a buyout situation. Small wonder— Dow didn't wait for the community to file a lawsuit or bang on its door demanding such a solution. It hired the same Prudential relocation team later employed by Fina to put together a package to initiate the proposal. For that, at least, it has

won some praise for its proactive stance. Its $10 million acquisition of Morrisonville, home to 250 residents, was a model of generosity and neighborly responsibility, at least in the eyes of chemical industry observers. Whenever it announced a policy change that would result in a more generous offer to those who had not yet sold out, the company retroactively provided previous sellers the same benefits. Families living in small wooden shacks often ended up with more than $50,000.

TRI data published by the Louisiana Chemical Association also show that Dow has been making substantial progress in pollution prevention, reducing the plant's air emissions by 40 percent from 1989 to 1990. Dow appears to have made substantial progress in reducing one major known carcinogen, benzene— cutting emissions by about 70 percent to 64,000 pounds.

The company had experienced no major accidents at Morrisonville and no one was specifically organizing against them, but it did reduce its potential liability by millions of dollars. With no neighbors left, there would be far fewer victims to compensate in the aftermath of an accident.

Still, the move out of Morrisonville ended the heritage of one more historic African-American community, founded in the 1870s by freed slaves from a nearby plantation. Just north of Plaquemine, Morrisonville had survived even when the government in 1932 order a relocation to high ground for flood control along the Mississippi River. But as Dow expanded after opening its plant in 1958, the steady encroachment of noise and other impacts foreclosed the village's future. A few of the residents relocated to Morrisonville Estates, a cluster of homes and trailers that Dow built for them several miles down the road. Others, like many of Reveilletown's residents, simply dispersed to nearby cities. In the process, they lost forever their vegetable gardens, their fishing, their trees, and the landmarks that shaped a lifetime.

Even Dow USA spokesman Guy Barone bemoans the loss of heritage that came with buying out the historic African-American community, noting only that "we tried to take a bad situation and make it good as could be." Referring to the original siting decision more than 30 years ago, he says that today's company officials "had to work with the hand that history dealt us."

Elizabeth Avants, a member of Alliance With and Action to Restore the Environment (AWARE), a Plaquemine environmental group that worked with Morrisonville residents, sums up the skepticism many of them still feel. "No amount of compensation can repay you for having lived next to one of these plants," she says. "They are cutting people's lives short for money."

Prove It!

Health effects remain a bone of contention in almost every buyout. Dale Emmanuel is the plant manager for Placid Oil Refining Co., which this year bought out the Sunrise community just north of Port Allen, across the Mississippi River from the state capital in Baton Rouge. When Sunrise residents sued Placid to force a buyout based on health problems residents say are associated with living near the plant, Emmanuel says, his firm asked for proof of health impacts from the plaintiffs and never got any.

Ada Mae Gaines, the president of Sunrise's VOTE United, organized by Janice Dickerson and named after her Reveilletown organization, sees it differently. Gaines says that her 12-year-old grandson has leukemia and another, 20 years old, must have his sinuses drained periodically.

The reality is that medical science often cannot pinpoint the cause of an illness, says Tom Callender, a Lafayette, La., doctor who handles a large volume of environmental

health cases. He suggests pointedly that the burden of proof should be on the company to show that it is producing a safe product in the safe fashion. The alternative, he suggests, is an unwieldy and hopelessly expensive system in which victims and government agencies must test literally thousands of chemicals that often interact to create compound effects whose causes are virtually untraceable and unmeasurable.

TRI data for 1989-1990 showed mixes results in pollution reduction at Exxon's Baton Rouge chemical plant. Although benzene emissions were down by 60 percent to 165,000 pounds, methyl chloride, which has produced birth defects in animal tests, rose by 45 percent to 680,000 pounds.

Exxon is in the midst of a buyout controversy with Baton Rouge's Garden City Community Alliance (GCCA). Ledell J.R. Hannon, vice-president of GCCA, calls Exxon's dealings with the neighborhood "one of the worst buyout situations in the nation. We're ready to move."

Not Interested

But Exxon community affairs coordinator David Gardner says that the company is not interested in buying the neighborhood. Exxon's property purchases are for the purpose of improving the area's safety and appearance, according to Gardner, who says that GCCA is at odds with Exxon because it wants the company to pursue a goal the company does not share.

Hannon contends, however, that Exxon is undermining the area's stability. "When businesses move out," Hannon notes, "no one wants to move in. Vacancy rates rise, crime goes up, and drug pushers arrive."

The bad blood also extends to the treatment of low-income homeowners desperate to move out of an area in which Exxon may be the only available buyer. Court records show that one homeowner received only $6,000 for his house, far short

of GCCA's modest demand for $37 per square foot, roughly in line with Dow's purchases. Hannon says that Exxon will not even discuss a relocation fee. Gardner, for his part, suggests that the sellers have been happy with their transactions and that GCCA is "not really representative" of the neighborhood. In any event, he says, Exxon is dealing strictly with individuals and not engaging in collective negotiations.

To Hannon and the Louisiana Coalition for Tax Justice (a group he chairs), the Exxon situation is a

---

Health Statistics for Louisiana

Cancer Deaths

    9,100 per 100,000

    Rank: 46

Workers in High Risk
   Jobs (in state)

    6.61%

    Rank: 48

Workers in Toxic
   Industries (in state)

    45.9%

    Rank: 46

Premature Deaths
   (before age 65)

    1,558 per 100,000

    Rank: 47

Source: 1991-1992 Green Index, a State-by-State Guide to the Nation's Environmental Health, by Bob Hall and Mary Lee Kerr. Island Press, 1991. Institute for Southern Studies.

failure in corporate ethics. "Any-body should be able to buy property cheaply if they can," he says, "but we're talking about a problem they created. They're throwing small change in the faces of poor people."

Basic Questions

Neighborhood buyouts by pol-luting industries are a relatively recent phenomenon, largely trace-able to a growing awareness among adjacent communities of the hazards and long-term health effects of toxic exposure. Neighbors are learning hard lessons about the value of soli-darity and good information in pur-suing an equitable settlement, especially when they are being forced to surrender a heritage and camaraderie that can never be replaced.

But Janice Dickerson, in partic-ular, is impatient with some environ-mentalists who, she says, have dis-played little sensitivity to residents by suggesting that "buyouts are not the answer." Some, she says, suggest that people should stay to monitor the companies' performance. She asks, "Why don't they move into my house? But don't ask me to expose my kids."

A better question is why so many plants, particularly in the rural South, were located so close to resi-dential areas, particularly those that preexisted the plants. The answer, all to often, is that local zoning was either nonexistent or ineffective. Almost no Louisiana parishes have any industrial performance stan-dards, which would regulate such nuisance factors as noise, glare, odors, and vibrations. State agencies have been much weaker than their counterparts elsewhere.

Finally, racism, though not always a factor, has surely played a part. In that sense, at least, neigh-borhoods demanding fair buyout settlements are helping to ensure that the piper of environmental justice will get his due, however much he may have been delayed by a system that was geared to serve busi-ness first. There is even serious hope for more enlightened business prac-tices. Dow's Barone suggests that, "though I may catch flak for saying it," state and local land-use planning must play a stronger role in helping to prevent such situations in the future.

Discussion Questions

1. Explain why it has become nec-essary for some companies to buy homes of and relocate local residents.

2. Could these costly moves have been prevented? How?

Solid Waste

According to the U.S. Environmental Protection Agency, Americans produced about 175 million metric tons of municipal solid waste in 1990. Although regional production values are not available, the 14 southern states contain about one-third of the U.S. population and, therefore, produce about 60 million metric tons of garbage. Nationwide, the two largest components of the solid waste stream are paper waste and yard waste.

One fact, often overlooked in the debate over solid waste, is that industrial facilities in the South and elsewhere produce far greater quantities of waste—nationwide, about 7000 million metric tons of nonhazardous solid waste. This amount dwarfs the measly 175 million metric tons produced by households and businesses.

Solid waste has become a major headache for urban centers in the South and elsewhere largely because of disposal difficulties. Although there generally is no shortage of landfill space, obtaining approval for new landfills from environmental administrators and local populations has proved extremely difficult. Strict regulations to prevent groundwater contamination and other problems and the widespread NIMBY syndrome combine with the political force dubbed the NIMTOO syndrome (not in my term of office). These forces are being met by a new movement in some locales, a phenomenon called the YIMBYFAP (yes in my backyard, for a price).

All the attention paid to disposal ignores a fundamental reality of our time: We need alternatives, such as source reduction, composting, and recycling, not just because of land-filling problems, but because they make good sense from an ecological and economic standpoint. Recycling, for example, saves considerable amounts of energy, water, and valuable resources. It also produces far fewer pollutants than making goods from virgin materials. Although it is not 100% clean, it is much better than the conventional strategy. It's important to keep the bigger picture in mind as the debates over solid waste flare up.

Background Reading: Chapter 19 (sections 19.3 and 19.4).

A Tale of One City

Making Recycling a Top Priority in Charlotte, North Carolina

Liz Chandler

■

SOURCE
EPA Journal, July/August 1992,
Vol. 18, No. 3

Chandler is a reporter for the
Charlotte Observer.

ON A FALL night in 1989, Hurricane Hugo roared into Charlotte, North Carolina, turning this so-called "City of Trees" into a city of trash. The 90 mile per hour winds snapped trees and downed power lines to create more waste overnight than all of Mecklenburg County's 511,000 people generate in a year. By dawn, the city and county had launched a cleanup that would last 18 months, cost more than $27 million, and become the biggest test ever for the community's commitment to recycling.

To preserve landfill space, government officials decided to collect and grind 300,000 tons of trees into mulch. "The hurricane inundated us with waste—particularly yard waste," says Charlie Willis, chairman of the county's solid waste advisory board. "We couldn't bury it. We didn't have a place for it all. And we thought it would be a step backward from our commitment to recycling. So we had to go out and find places to stack thousands of tons of trees until we converted them to mulch."

The nation's 35th most populous city, Charlotte has become accustomed to adversity in waste disposal. The city, which spreads over 80 percent of Mecklenburg County, boomed during the 1980s, growing to the nation's third largest banking center, reaching the final cut for a new National Football League team, and landing on *Newsweek*'s Top 10 "hottest" cities list. And along with a 10-year growth spurt that added 25 percent more people, came tons more trash. Today, residents and businesses generate about 600,000 tons of garbage a year, and now the community is without a public landfill.

The 300-acre dump that served for 20 years hit capacity in April. Citizen lawsuits have blocked attempts to open a new one. Neighboring South Carolina, which borders Mecklenburg County, recently joined the fight to stop a planned 574-acre landfill on the South Carolina state line. The South Carolina legislature passed a law in June that would make that state's low-level radioactive waste dump off-limits to North Carolina if any North Carolina county builds a landfill within a mile of the South Carolina border. Lacking its own

1989's Hurricane Hugo created more waste in one night than Mecklenburg County, North Carolina, normally generates in one year.

landfill, Mecklenburg County has contracted for space in a private dump—a move that has pushed up dump fees and sparked a new round of legal battles.

It is the constant struggle of siting new landfills that has propelled recycling from a modest experiment to a top priority in this community. Citizens have embraced recycling with 75 percent of eligible households participating at least once a month in Charlotte's voluntary curbside program. Seven local governments in Mecklenburg County, including Charlotte, have joined in a waste management plan that makes recycling the most favored disposal method.

The plan calls for reducing the amount of trash buried or burned by 25 percent per capita by next July. The goal jumps to 40 percent reduction by 2001. So far, the community has reduced disposal by 13 percent since 1990—a long way from next year's target.

To meet the goals, the city and county have set up aggressive recycling programs. This year, the two governments will spend $6.2 million on recycling collection and processing, up from $1.8 million three years ago. They also plan a $10 million recycling center for business waste. In January, curbside collection will be extended to 84,000 apartments and condominiums. And

a new law requires all private haulers who serve businesses and citizens outside Charlotte to provide recycling to residential customers.

There's already an array of recycling programs in Charlotte: residential curbside pickup, yard waste pickup, dropoff centers, and a remove-and-resell operation for bulky appliances left at the landfill. The county mines and sells metals from ash at its incinerator. It also runs small recycling programs for cardboard, car batteries and motor oil.

Curbside collection in Charlotte reaches 110,000 single family homes and is nationally recognized for its high participation. Residents toss glass bottles, aluminum and steel cans, newspapers, and milk containers and other plastics into bright red "Curb it" bins and place the bins at the curb. City collectors carry the material to recycling headquarters, where glass is smashed, cans are crushed, and plastic and newspaper are baled for sale.

Recycling headquarters is a converted warehouse about twice the size of a high school gym. Recyclables are dumped on the floor, pushed by tractors onto conveyor belts, and separated by workers into bins. Loads are then baled for sale to buyers. A private company runs the processing center at no cost to the county. In return, the company

keeps whatever profits it makes.

Charlotte took a big step in recycling in 1991 when it cranked up yard waste collection. Residents place tree limbs and brush at the curb. Grass clippings and leaves must go in clear plastic bags, so workers can see what's inside. City haulers drop the materials at two sites, where county workers take over.

With a shredder and three giant machines called tub grinders, county workers grind scrap wood and limbs into mulch. Other yard waste is composted: It's laid out in long rows, and sometimes has nitrogen added to it. Within six months, the waste becomes a black humus, which is used to enhance soil.

Both products are sold to residents and landscapers. The county last year made $100,000, an amount expected to jump this year because of a new state law that prohibits putting yard waste in landfills. Mulch goes for about $1 for a 20-pound bag; compost sells for $3.50 for a 40-pound bag. The county has a revenue sharing deal with a private company to sell compost at local retail stores.

Recycling coordinators won't accept new materials until they are certain they have buyers. Charlotte, for example, claims to be the first city to accept spiral paper cans— from products like frozen juice—for recycling. Officials added the cylindrical cardboard cans only because they struck a deal with a nearby company that converts the cans into low-grade paper for cones that are used for textile yarn, carpet, and tape. "We would never have picked up spiral cans unless the market had approached us," says Cary Saul, the county's deputy engineering director. "We know markets have been a problem. We're cautious. You have to know your market before you get into something."

As competition for buyers heats up across the country, Charlotte recycling organizers say they've maintained strong markets because they supply a clean, thoroughly sorted product. "We're going after the best quality paper we can get to supply our plants," says Kenny King, a buyer who has a contract for Charlotte's newspaper. "We've got to have a large quantity. We want it source-separated . . . They do an excellent job at it."

Says the county's Saul: "When things get tight, buyers are going to want the best material. They don't want a bunch of newspaper with broken glass mixed in."

Despite everything that Charlotte is doing, city and county leaders admit they probably won't make the 25-percent reduction goal next July. More likely, it will be 1994, after a new commercial recycling center opens. The center is the city's first significant venture into attacking the business waste stream, which accounts for 55 percent of all trash.

The success of the center is not guaranteed. County leaders last year took steps to ensure a steady supply of recyclables with a "flow control" ordinance. It required that all waste generated in Mecklenburg County be dumped at a county disposal facility, and it gave county officials power to dictate which facility haulers used. The plan was to direct commercial waste rich with paper, cardboard, and other recyclables to the new recycling center. But one

Mecklenburg County company, wanting to use its own landfill outside the county, sued and blocked the new law.

Now the county has turned to old-fashioned competition to lure haulers. Officials are devising strategies, such as cut-rate dump fees and tax incentives, to lure haulers to use county facilities. "We can't go too far, because it could backfire," says Saul. "If we make disposal too cheap, there's no incentive for companies to cut the amount they produce."

Despite recycling's popularity, nobody believes it's a cure-all. The county's waste management strategy also relies heavily on incineration. Landfilling is the least-favored option.

"Recycling is one of the options we've chosen to manage a portion of our waste," says Saul. "It's part of the answer. . . . We want to recycle everything we can, then incinerate the rest. We only want to put things that won't burn in the landfill."

Mecklenburg County's incinerator opened in 1989; it burns 200 tons of refuse daily. Steam generated from incineration heats buildings at the University of North Carolina-Charlotte in the winter. And the electricity produced is sold to the Duke Power Company. The plant burns about 11 percent of the county's waste; energy sales help defray about half the $2 million annual operating costs. The rest of the money comes from dump fees charged at county disposal sites.

The county plans a second incinerator that will burn 600 tons daily. Set to open in 1996, the $90 million burner is to be paid for through revenue bonds and energy sales. All told, the two incinerators are projected to burn 26 percent of the county's waste in 1996.

So far, the incinerators have dodged the kind of citizen's legal challenges that stalled a new landfill. Officials have calmed neighbors' fears by monitoring air quality

around its current burner and publicizing the results. The county pledges similar tests around the new incinerator and will build in anti-pollution devices, such as stack scrubbers and a bag house.

Mecklenburg County has spent more than $100,000 to study a myriad of waste disposal options, including such obscure methods as refuse-derived fuel and bioconversion. Officials say they're confident the recycle-burn-bury combination is best. Still, it's going to take more than a disposal plan to reach the 40-percent reduction goal by 2001. "The real key is reducing waste at the source," says Saul. "That takes time. That means plants have to change processes, and people have to change habits."

Board chairman Willis says people are beginning to do that. "There's a tremendous intangible benefit derived from curbside recycling. Those red boxes are like advertisements. Public awareness is so great that it filters into everything people do. They buy smarter. They think greener. They are more aware of their impact on the environment.

Discussion Question

1. What factors contribute to the success of Charlotte's recycling and composting programs?

North Carolina County Institutes Sticker System

New Program Serves 30,000 Households and Already Has Resulted in a Doubling of Recycling Participation Rates

Robert Bracken

■

SOURCE
BioCycle, February 1992,
Vol. 33, No. 2

N NOVEMBER 1991, Craven County, North Carolina became one of the first communities in the Southeast to introduce volume based garbage collection. The program requires all participating households to attach a special sticker to each 33-gallon bag of garbage. Priced at $1.25 each, the stickers provide an ongoing financial incentive to cut back on waste. Since the program started, the county has reported a doubling of recycling participation rates and record tonnages of recyclables collected.

In conjunction with implementing volume based pricing, Craven County also extended curbside garbage and recyclables collection to the rural areas. A tax-funded network of roadside dumpsters (greenboxes) was dismantled when curbside service was introduced. Altogether, the program delivers volume based garbage collection and recycling to about 30,000 households, including six municipalities plus the unincorporated areas.

The major change for municipal residents accustomed to curbside pickup for monthly flat fee was adjusting to the sticker routine. Two program features, however, make this job easier: (1) A minimum supply of stickers is mailed to households each month; and (2) if they choose, residents can place garbage loose into cans and attach the appropriate number of stickers (in roughly 33-gallon increments) to the outside of the can. A standard 90-gallon trash cart, for example, would need three stickers.

For larger households, and those not recycling, the volume based program probably means higher trash collection charges. But for smaller households and those that recycle or practice source reduction, it may translate into lower monthly trash bills.

Pressures to Reduce Waste

The volume based program came in response to a broad range of pressures. Like every community in North Carolina, Craven County faces a number of new state laws dealing with solid waste, including a waste reduction goal of 25 percent by 1993 and 40 percent by 2001. State law also bans several types of waste from sanitary landfills, including large appliances (white goods), whole tires, lead acid batteries and, in 1993, yard waste. In addition, the state encourages local governments to set up enterprise funds and rely on fees rather than taxes to fund solid waste programs.

A year before implementation, county planners began developing a collection program that could meet state mandates and provide financial incentives for waste reduction. Variable rate and volume based program options were developed based on conversations with managers of similar programs and a review of publications. While program details varied widely, variable rate programs were grouped into four major categories; bags, stickers, variable can size and weight based systems.

Programs using subscription to varying sizes of garbage cans had the advantage of providing permanent disposal outlets to all residents— thereby reducing the potential for illegal dumping and ensuring a predictable revenue base. The logistics, however, of getting the can capacity and rate increments just right, combined with the need to maintain a large can inventory to cover switches to a different size, made this option seem less attractive.

Truck-mounted weighing systems were viewed as too experimental and capital intensive, although paying for garbage disposal by the pound has advantages. It should provide the convenience and permanence of rigid containers and at the same time send very accurate price signals to customers.

The sticker system was chosen because of its low setup costs, proven track record, and the fact that stickers could be mailed to customers and, unlike bags, would not take up valuable shelf space at retail stores. Sheets of stickers are used instead of rolls for ease of storage under cash registers and for quick sale in multiple sets.

Since program revenues are tied to sticker usage, accurate estimates of how many stickers households use—or bags of garbage they generate—are critical. Managers of other volume based programs told the county to expect a significant shrinkage of waste volume at the curb, due to increased recycling and source reduction as well as other dollar stretching consumer behavior such as overstuffing bags. Therefore, the sticker price had to be adjusted accordingly.

The $1.25 sticker price was set at a level to cover the costs of collection and disposal. It is estimated that

Households must attach stickers to each 33-gallon bag of garbage. Sticker sales cover the costs of collection and disposal.

$.30 of the sticker price goes to cover the $25 per ton tipping fee.

Initial plans called for a system with a flat base fee which covered only the costs of curbside recycling—leaving sticker revenue to cover all garbage related collection and disposal costs. However, the risks of loading too much cost onto the sticker had to be taken into account. As the sticker price rises, the potential for illegal dumping and revenue shortfalls (due to nonparticipation) increases. To reduce these risks, the county decided to provide, in addition to recycling, a base garbage collection service combined with a higher base fee.

Monthly Allotment

Under the adopted program, the county sends all households a monthly allotment of four stickers, enough to meet the needs of a waste conscious or small household. If a household uses up the minimum supply, additional stickers must be purchased at various retail outlets—mainly grocery stores—or government offices throughout the county. Unused stickers can be carried over from month to month, but are not currently redeemable for cash. Although an informal market prob-

ably exists between residents with excess stickers and those in need of more, these exchanges are not seen as detrimental to program success. The county is studying ways to credit customer accounts for unused stickers, thereby providing small waste generators with fee increments below the four sticker base.

By providing weekly curbside recycling to all participating households, the county offers a clear alternative to the costly garbage bag. In the county's view, not offering this alternative would have defeated the purpose of variable rates, which is to promote recycling and waste reduction.

In addition to curbside recycling service, seven staffed dropoff centers are open to receive the following items free of charge: white goods and bulky items, yard waste, mixed paper, corrugated cardboard, lead acid batteries, used oil and rinsed agrichemical containers. These centers are especially important as outlets for materials that are either too bulky for placement in garbage containers, banned from the landfill, or not easily collected by curbside recycling trucks.

Thus far, multifamily dwellings with dumpster service are not included in the sticker program.

Metering trash deposited in common dumpsters poses technical problems that would seem to preclude volume and weight incentives in this sector.

Roadside Greenboxes Eliminated

Replacing roadside dumpsters in the rural areas with curbside collection service enabled the county to consolidate the unincorporated areas and municipalities into a single solid waste collection and funding system. The County wide system was formally established through a series of interlocal agreements between the county and its municipalities. Two small towns opted not to participate but may join the compact at a later date. American Refuse Systems (ARS), a North Carolina based hauler and limited partner with Waste Management Inc., was chosen as the lead contractor to manage the garbage collection and recycling operations. Several other private haulers also collect waste under a franchise agreement with the county. Revenues to these haulers are based on the number of stickers collected on their routes.

Under the old system, municipal residents complained that their county property taxes were used to fund a dumpster network that they never used, while at the same time, they were charged monthly fees for city garbage pickup. Moreover, the greenboxes were an eyesore, frequently overflowing with garbage, and could not allow for monitoring of incoming waste, e.g. to screen banned items. And, of course, no financial incentive for waste reduction was possible under this system. The greenbox, in effect, was a bottomless garbage pit, the cost of which was buried in the county property tax bill.

The county sought to address these problems by establishing a county wide curbside service based on variable user fees. Importantly, the estimated cost to rural households for the curbside system was comparable to the cost of establishing a network of staffed convenience centers, a common collection alternative in rural areas. The favorable rural curbside costs could be achieved by integrating the rural routes with the high density municipal routes in the context of a county wide program.

Public Acceptance Challenge

Program implementation, however, was far from trouble free. The biggest problem was, and remains, public acceptance. Despite public meetings, press conferences and meetings, the perception of many residents was simply that the program was being forced on them. Many could not understand why the old system had to be changed. The program, they argued, should have been put to a vote.

The fact that the sticker system was paired with the extension of curbside garbage collection county wide has generated multiple reactions. Rural residents, especially those who were comfortable with hauling their own garbage to roadside dumpsters at no charge, were upset by now having to pay for garbage collection. They argued that

taxes should have at least been reduced correspondingly. Other rural residents were concerned about increased truck traffic or about dogs and other animals tearing apart garbage bags left at the curb. Many feared that illegal dumping would run rampant, that stickers would be stolen or counterfeited, or that people would not understand program details.

At the height of public reaction to the plan, shortly after kickoff, dissenters marched through downtown New Bern, the county seat, vowing to boycott the program. A lawsuit was filed by a local activist (the case is pending) and the newspaper started a daily column called "Trash Watch" to track program developments.

The solid waste advisory committee has taken an active and useful role in transferring public reaction into a managed forum that reports to the Board of Commissioners. Both before and after program implementation, it met to consider modifications. One proposed change would have eliminated the monthly four sticker allotment and associated fee, requiring residents to purchase each sticker as needed. The recycling fee would be billed annually to residences as a special charge on their property tax bill. For now, however, the county intends to stay with the monthly billing routine.

Aside from the high profile protests by some residents, operational problems so far have been minimal. Few, if any, illegal dumping incidents, including raids on commercial dumpsters, have been reported. The base monthly supply of stickers, along with a tough new litter ordinance, may be contributing factors.

Recycling Rate Doubles

Advocates of the program point to signs of an emerging success story: recycling participation and tonnages of recyclables collected

have increased dramatically. In communities that already had curbside recycling, participation (measured by setout rates) more than doubled immediately after the sticker system went into effect—from about 30 percent to more than 70 percent. Since the program extended curbside recycling to new areas of the county—more than doubling the recycling customer base—an increase in tonnage of recyclables was expected. However, the per capita poundage increased as well, from six pounds per setout to seven pounds, which is impressive given the much broader customer base.

According to Tyler Harris, county manager, successful implementation of a volume-based program requires a comprehensive educational and promotional campaign well in advance of program startup. "Convincing people of the need to change is the hardest part," he says.

Craven County's transition to volume based garbage collection appears to be working, as more people grow accustomed to the new routine. As one county official says, "People are just saying 'we're going to get by with one bag of garbage per week'—and they're doing it."

Discussion Questions

1. Describe all of the features of Craven County's effort to reduce waste and promote recycling.

2. Why does a volume-based system work so well? What advantages does it offer over a flat monthly fee?

Drop-Off Recycling in Tampa: Maximizing Efficiency and Quality

**Barbara Kropf and
Caroline Mixon**

■

SOURCE
Resource Recycling, January 1993,
Vol. 12, No. 1

O NE of Florida's largest cities has shown that drop-off recycling service can be a successful way to make sure all residents can take part in material recovery. The City of Tampa has learned a number of lessons in how to develop and implement a cost-effective drop-off recycling program.

In 1988, the City of Tampa began its recycling program, shortly after the State of Florida passed its comprehensive solid waste management act. The act requires all counties to recycle 30 percent of their waste by January 1995. Two pilot programs were implemented in September 1989, one for curbside recycling collection and the other for curbside yard waste recycling collection.

Both pilot projects proved to be successful, though neither has been expanded citywide. The recycling program currently services 26,000 single-family homes and 3,000 multi-family units, and the curbside yard waste program services approximately 30,000 homes. Tampa has 278,000 residents and 80,000 homes. Residential response has been very positive.

Tampa is committed to providing every citizen the opportunity to recycle, so the city looked at implementing a drop-off recycling program to fill the gap created by the decision not to go citywide with curbside collection. In addition to servicing the entire community, the city was seeking a way to recover and recycle materials not included in the curbside program. Other nearby communities in Florida had implemented drop-off programs and were having tremendous success, at much lower costs per ton than curbside recycling collection.

Program Development and Implementation

In putting together Tampa's drop-off program, the city first had to decide on the number and location of sites, the types of collection containers, and the type of truck needed to collect the recyclables. At the time the program was developing, Durbin Paper Company was collecting newspaper in six-cubic-yard drop-off containers throughout the county. The city considered piggybacking on these existing sites, thereby avoiding initiating its own newspaper collection program.

Eight-cubic-yard dumpsters were chosen for plastics, and igloos (dome-shaped collection containers) for aluminum and glass containers. An 18-cubic-yard open-body truck was retrofitted with a crane to collect the material from the igloos. The city's solid waste department already had several front-end loader trucks to collect plastics. Although roll-off containers might have been the best choice to collect the newspaper and plastics, front-end loader containers were chosen because of an existing fleet of front-end loader containers and trucks.

Before implementing the program, city staff checked with the local end-use markets concerning material specifications and requirements. The local market/processor provided signage for some of the materials to be collected. This proved important because problems developed quickly with plastic contamination.

Site Development

City recycling staff worked with Kash n' Karry, a local supermarket chain, to establish drop-off sites at some of its stores. Four of its locations were chosen as drop-off sites. Other city sites include parks, schools, shopping centers, a museum, and the local Pepsi bottling plant.

The program began in November 1990 with 10 sites for newspaper, aluminum, glass and plastics. An additional 12 sites were for newspaper and plastics only. Today this has been expanded to a total of 15 sites for all materials, and 10 sites for newspaper and plastics only.

Site location was only the first step. Agreements were negotiated with each site owner/manager. Signage was developed for the sites including trailblazer signs, general signs for each site, and signage for each recyclable material. Signage is a critical element for drop-off sites. People may not see the sites themselves from the road, so the trailblazer signs pointing to the sites are extremely important. Additional signage was then developed to instruct people about acceptable and unacceptable materials for each of the drop-off containers.

A major glitch in the plans occurred when Durbin Paper was bought by Waste Management, Inc. which informed the city a few days before the opening of the drop-off sites that it planned to remove the newspaper containers permanently and suggested that the city pay for servicing the drop-off centers.

In response, city staff scrambled to locate every six-cubic-yard container in storage, fixing and painting those that needed it and getting them placed in time for the grand opening of the drop-off sites. Due to the time constraints, the containers for collecting newspaper started out the same color as the garbage containers, but eventually they were all replaced with ones in light blue with white lids, the color designated for recovered newspapers.

Once the program was underway, however, it was immediately successful, with costs now averaging just $27 per ton.

The Problem with Plastics

Due to contamination problems, collecting plastics through drop-off programs has proven to be difficult, sometimes even ineffective. The multitude of different resin types, lack of public knowledge and education, and changing handler or reclaimer specifications have made it extremely difficult to market recovered plastics from drop-off programs without first removing contaminants.

Several months into Tampa's drop-off recycling program, problems arose with plastics collection. The drop-off sites initially accepted all plastics marked with a 1 (polyethylene terephalate, PET) or a 2 (high density polyethylene, HDPE). The local processor, which had been paying $20 per ton of plastics collected, began complaining of contamination problems and first refused to pay for the plastics and later to accept them at all. For several weeks, the city continued to collect the plastics, taking them to the waste-to-energy plant where they were burned. Then, another local processor agreed to accept the material at $0 value.

Based on actual sorts of the material, contamination in the Tampa drop-off program plastics stream averaged 29.5 percent. As the cost of upgrading the material became exorbitant and end-use market specifications became tighter, all local markets cut off the city. The material could not be sold, or even marketed at negative value, so alternative options were examined. Recycling planners chose a comprehensive strategy of public education, one that would rely on modifying the behavior of participants.

Dealing with Contamination

Educating the public to achieve behavior modification required an understanding of the existing situation; therefore, several analytical tools were used before initiating a change in existing education components.

One such tool was a participant survey. The survey was conducted to obtain input and to serve as a benchmark of the public's knowledge. Time was spent surveying users of the drop-off containers to obtain information ranging from general demographics to individual opinions. Program participants were surveyed twice, before and after program changes. A brochure was designed to answer the questions most frequently asked by the survey participants. After finding out which elements confused the public, a plan was formed to deal with these problems, and a pilot program was ready for implementation.

The goal of the implementation phase of this pilot project was to develop a new, more cost-effective plastics recycling system using the necessary tools to control contamination, while at the same time, incorporating maximum efficiency in drop-off collection and municipal participation in creating markets.

As a first step, the existing collection system was evaluated. Particular attention was paid to maximum utilization of available resources. Eliminating, moving or adding drop-off sites was considered in order to capture more recyclable material in the most efficient manner. Criteria used for consideration of site changes were:

- current effectiveness of existing sites
- distance between sites
- number of collection containers at each location
- equipment requirements
- efficiency of collection routes
- population served, convenience of access and public visibility.

Four of the existing sites were selected to serve as a pilot program for the analysis. The survey found that two of the primary complaints of participants were the awkwardness of the drop-off container lids and the length of time to read the signage and difficulty in interpreting it. Therefore, new signage and lids for the containers were developed.

The new signage alone was found to decrease sorting time by 50 percent (3.5 minutes compared to 7 minutes). The new signs identified plastic containers by product type as well as resin content, i.e., soft drink, water and milk, laundry detergent, fabric softener, dishwashing liquid and bleach.

The new lids, which were actually the existing lids with holes cut in them to allow recyclable materials to be deposited without having to lift the entire lid, cut bulk unloading and contamination by film bags.

After these modifications took place, a composition analysis showed that contamination decreased from nearly 30 percent to less than 7 percent.

Collection Efficiency

Plastic is collected by the city on each Sunday, Tuesday, Wednesday and Friday. Three 40-cubic-yard front-loading trucks are used on two routes. Time and volume data were collected for each store on the route. Nine of the 20 sites in the drop-off

Drop-Off Program Statistics

Program Costs

Labor	(5,192 hours)	$61,721.62
Vehicles	(3,459 hours)	41,463.60
Total costs		$103,185.22

Program Revenues (1)	**Tonnage**	**Revenues**
Glass	493.33	$1,447.75
Aluminum	12.90	7,404.68
Plastic	207.90	(2,340.70)
Newspaper	2,867.10	675.94
Total	3,581.23	$7,187.67

Total program cost	$103,185.22
(Less) city revenues (2)	(7,187.67)
Net program cost per ton	$26.81
Disposal cost savings (3)	$86,415.08

(1) Average revenue per ton:

Glass	
Clear	$10
Colored	0
Aluminum	780
Plastic	(25)
Newspaper	0

(2) City revenues represent the funds received from the sale of recovered materials.

(3) Disposal cost savings reflect the weighted average cost to burn and landfill one ton of processible refuse ($24.13 per ton).

program collected less than four cubic yards of plastics per week. These were categorized as unproductive and were targeted to be moved or eliminated.

Because the compaction ratio of the truck on its fullest route was 1.7, truck capacity could be better utilized by the city. The actual achievable compaction ratio depends on many factors, such as hydraulic type ("full pack" and "half pack") and the condition of equipment, whether lids are left on the plastics that are collected, and truck packing techniques.

Site maintenance is critical to program success. A site with litter and blowing debris does not attract citizens to recycling. Nor is it a sign of a "good neighbor." To address this issue, the city requires staff to maintain each of the sites—collecting litter, checking signage and controlling vandalism.

Although vandalism has not been a major problem, it is one that city staff had not anticipated. There is the obvious sabotage: Sometimes garbage has been placed in the recycling containers and, on occasion, containers have been burned. Igloos for aluminum cans have required locks to prevent theft; once, an entire igloo was stolen (and later recovered, although without the aluminum cans it had contained).

The city has had to move several drop-off sites as well. One shopping center requested that the sites be moved because the owner felt that the presence of the recycling center was a deterrent to attracting new tenants. Another landlord limited the number of containers that could be placed on-site, forcing the city to move the center to a location that could handle the volume of material. When sites are relocated, citizens must be informed of the new location and the continued importance of their participation in the program.

Lessons Learned

The city's drop-off recycling program has provided high visibility for recycling in the city with much lower costs than curbside recycling collection. Not having to contract for recycling collection service has provided incredible flexibility in managing the program and implementing changes.

If you manage the basics well and keep citizens involved in the program, the centers are a natural for recycling events. Minimum route time, maximum truck fullness, pleasing aesthetics and clear recycling instructions were all achieved in Tampa.

Discussion Questions

1. Explain the main features of Tampa's drop-off recycling program and why it is successful.

2. What problems were encountered and how were they addressed?

3. The economics of this program (see the table Drop-Off Program Statistics) suggest that it costs the city about $27 per ton. However, disposal cost savings were not determined in this calculation. Using disposal cost savings, what is the real cost of this program?

Responding to Andrew: Recycling After the Storm

Robert Steuteville

■

SOURCE
BioCycle, October 1992,
Vol. 33, No. 10

UT OF 1.9 million Dade County, Florida, residents, an estimated 250,000 were left homeless by Hurricane Andrew. Two hundred and thirty-five of those homeless people were employees of the Metro Dade County Department of Solid Waste Management (DSWM). Yet virtually all of the DSWM employees were back to work—home or no home—putting in 12 hour days, seven days a week, immediately after the storm.

For the solid waste crews, each homeless person represents large amounts of construction and demolition (C&D) debris that has to be collected and recycled, landfilled or burned. Up to 75,000 housing units were destroyed. Trees were leveled—in some cases entire forests—all over the county, leaving massive amounts of vegetative waste.

Preliminary estimates put the total amount of waste produced by Andrew in Florida at 18 million tons. The closest comparison is Hurricane Hugo, less damaging than Andrew, which ripped into the Carolinas three years ago, generating 300,000 tons of wood and yard waste in the Charlotte, North Carolina, area alone. In the Charleston, South Carolina, region, 10 to 14 years of landfill space were used up following Hugo.

During the first few days after Andrew, DSWM's first priority was to restore garbage collection to head off public health problems, according to Debbie Higer, chief of the service development department. The hurricane occurred on Monday, August 24, and by Thursday regular trash collection was restored to all but the most severely impacted areas of the county.

Community Recycling, Dade County's contractor for curbside service in 11 municipalities and unincorporated areas, began regular collection of recyclables in some areas two days after the storm. A week later, the contractor was collecting everywhere but Florida City, one of the most devastated areas, according to Jean Marie Massa, Community Recycling's manager for marketing. The collections were yielding unusually large amounts of plastic bottles and cans—the remnants of bottled water and food consumed during extended electrical outages.

Dealing with Storm Debris

The long-term goal for recycling goes well beyond restoration of normal services. County officials want to recover as much of the debris from the storm as possible. Since large portions of the waste are branches, stumps, tree trunks and other yard debris, the county's mulching program will help to accomplish that goal. "We already have one of the most aggressive yard trash mulching programs in the country," says Higer. "We processed 80,000 tons of material in the last year."

The mulching program has been underway for a year and a half. Yard trash curbside pickup normally occurs twice a month for 240,000 single family households. Also there are 20 sites where the public can drop off yard trash and pick up finished mulch for their gardens or landscape projects.

Systematic Collection

The hurricane will require a tremendous expansion of the program. DSWM and contractors will systematically, over an extended period of time, go through the county three times to pick up yard trash and wood waste, demolition debris and other bulky waste. Residents are being asked to take their waste to the curb. "There are some areas that are completely devastated—there is simply nothing there. They are being treated differently in terms of collection." In the southern sections of the county with the worst damage, contractors hired by the Army are doing most of the work.

In the northern sections, there are opportunities for separation of vegetative waste—although much of the material was commingled, Higer says. "If it is an area with a lot of downed tress, we try to separate it—use separate trucks—so that we get the clean yard trash, and put it aside for mulching." The county expects a strong demand for mulch, boosted by all of the rebuilding that will be going on in the next year or two.

The material is mulched at the South Dade County Landfill, where there is a reduced tipping fee for yard trash. The county has a contract with NR Associates of Boynton Beach, Florida, which uses three tub grinders to prepare vegetative waste for mulching. The largest grinder stayed at the South Dade mulching site, according to Harry Elrod, NR general manager. The other two were moved around to where Army engineers were piling waste.

By taking the grinders to the waste, rather than the reverse, transportation is cut down dramatically,

Toppled Trees in Louisiana

The devastation was not nearly as bad in Louisiana as it was in southern Florida, but still there is massive amounts of damage and debris. Lafayette, Louisiana, a city of 82,000 people, was better prepared to deal with toppled trees than any other city or parish in the state. With the state's only yard trash composting operation, Lafayette was processing about 21 tons daily prior to the storm. The city took in 1,400 tons of material in 12 days after the storm, according to Morgan White, Lafayette recycling coordinator.

At that point, the city had to renegotiate with the contractor and find additional firms to supplement hauling. Branches and leaves covered the ground nearly everywhere, even though the city was not the worst hit in the state. "In the parks you couldn't even see the ground," White says. "To the greatest extent possible we are taking the stuff in for composting. The contractor has made a commitment to cut as much firewood as possible. What can't be recovered is probably going to a construction landfill."

Lafayette's composting operation occupies 10 acres in the middle of a 163 acre site. One tub grinder is used to reduce branches and stumps into mulchable material. The tub grinder was operating non-stop, and attempts to windrow were put on hold. Space

could be a problem, White says, because the site was already approaching capacity before the storm. The state gave permission to use an additional three acres for storage.

The city is trying to avoid a major expansion of the facility so neighbors are not bothered. Despite the problem of too much material, the composting operation provides "a definite advantage," White feels. "It is fortunate that we have a facility that can accommodate a portion of the waste, so it is not just landfilled."

In Baton Rouge, like most of the state, yard debris was being burned to keep it out of the landfill, according to Susan Hamilton, the city recycling coordinator. However, there is some recycling of hurricane debris going on in the city, which was hit by winds of 80 miles per hour. In order to meet a 25 percent state recycling goal, the city is seeking a contractor to build a composting facility next year on a new landfill site. It looks like the city will have a head start on collections of organic material. "We're stockpiling a portion of the waste and we're going to get a chipper out there as fast as we can," says Hamilton.

Some of the waste initially went to Total Woods Waste of Baton Rouge, a company that charges landscapers to drop off branches and brush. The firm recycles some material as firewood, and burns other material—

incorporating the ashes in a fertilizer product. Because of the cost, however, the city discontinued deliveries to Total Wood Waste until federal emergency funds are available.

Louisiana State University in Baton Rouge was lent a grinder by the James River Corp., a paper mill. Material from the extensively wooded campus was piled over a three-acre parking lot, and about 50 people spent a week grinding and hauling. The mulch was used to cover tree roots all over campus.

Particularly in south Louisiana, there were reports of entrepreneurs with pickup trucks recycling pieces of scrap metal and other materials left on the side of the road. "I would imagine that a scrap dealer, or someone who wanted to sell to a scrap dealer, could make some money," says Hamilton.

In terms of waste management and recycling, neither Florida nor Louisiana were entirely prepared for a storm like Andrew. The response, particularly in Florida, is evolving as funds become available, officials refine plans and contractors come on board. Higer explains: "We have been inundated with potential contractors. People have been on the phone with ideas. We have been looking from an internal perspective at what is feasible. Then we get back in touch with the contractors."

Elrod says. "Crews take it to one centrally located area, so they won't have to be traveling all of the main streets, hauling it to the landfill. The grinding reduces the size of the material to the point where you can haul 100 yards at a time," says Elrod. Although NR was the only yard waste processing contractor for

the county prior to the storm, there is no way that the company can keep up with the current volume, Elrod explains.

The county is looking to hire more contractors for grinding and chipping. The Federal Emergency Management Association approved the purchase of grinding equipment

for the 20 neighborhood dropoff sites, as well as nine mobile units. Five large dropoff centers have been added in parks and other public properties. As these fill up, the county will select some more holding sites for the material.

The ultimate destination of the processed waste depends on its level of contamination. The material that is contaminated with C&D wood and other items is being burned. In areas of the county, like Homestead and Florida City, which were nearly destroyed, the waste is so mixed and diverse that source separation is often out of the question. Higer describes the typical waste mound: "It looks [like] a huge creature stepped on a house and dumped it in a pile."

Officials realize that it is important to move large piles of debris away from homes as quickly as possible, because of the potential fire hazard. "It is stacked so high," says Elrod. "The houses are surrounded by trees that are dying. They've got to get them away from the houses, because if they did catch fire, the people would be caught in an oven."

Other Materials

Wood is the largest percentage of the waste, but huge quantities of other materials are being collected, including white goods, tires, glass, metal, concrete and hazardous or special wastes. "As we get into it, we are either going to be contracting (for disposal or recycling) or recycling this material ourselves," Higer says. "We are already recycling white goods. Hazardous waste will be handled in the proper manner."

Solvents and paints will become a greater concern as rebuilding hits full stride. "We want to make sure these do not get into the landfill," she says. The county plans to set up special collection sites for household chemicals.

Special contracts are being explored for individual materials. The government of Haiti, a country that is completely deforested and in desperate need of fuel, is interested in receiving large amounts of wood waste for firewood or to make charcoal. A vendor is being sought to separate and recycle water jugs and beverage containers from the emergency food distribution depots. An estimated 6,000 cubic yards of

HDPE, 1,000 cubic yards of PET, and another 1,000 cubic yards of commingled glass and aluminum have been collected.

Dade County has no program for recycling C&D debris and doesn't have the necessary separation equipment. Officials have plans to catch up in the area of recycling C&D debris, especially because it will make up more and more of the hurricane waste once the rebuilding hits full stride. They are considering contracting with companies for separation and recycling of many kinds of commingled debris. For example, Higer says, nails could be removed from C&D wood waste, or metal springs from mattresses. "As time moves on, because we are not going to run out of material, we're looking at setting up more sophisticated separation measures."

Collecting and processing of material will be going on for months, and maybe longer. "It's not going anywhere," Higer says. "We've been told by the people who had to deal with Hugo that it will be years. I don't even want to think about that now." She adds, "We will be using every mechanism available. The department has a strong bias in favor of recycling. We would love to be able to recycle it all. The reality is that we must use a variety of means to deal with the waste, including disposal and burning."

Discussion Questions

1. Explain steps being taken in southern Florida to recycle and compost waste in the wake of Hurricane Andrew.

2. Many scientists believe that rising global temperatures, which are believed to be caused by greenhouse gases emitted from human activities, will continue to warm the world's oceans. This, in turn, adds energy to storms such as Hurricane Andrew. If this scientific prediction is true, what implications will this have for your community? What steps should be taken?

Florida's New Market Strategy

Kathleen Meade

■

SOURCE
Waste Age, July 1992,
Vol. 23, No. 7

FLORIDA GOVERNOR Lawton Chiles has introduced a strategic plan to increase market development for recycling in the state. The plan combines the bid process for waste removal and government procurement in a unique "recycling business ventures" concept.

Spearheaded by former state legislator Robert McKnight, and supported by such legislative leaders as George Kirkpatrick, the plan grew out of recommendations from the governor's Recycled Markets Advisory Committee. According to the plan, Florida would designate recycling business ventures that would competitively bid to both remove specific recyclable materials from state institutions and sell the recycled materials back to the state.

"Under the plan, anyone who wants to sell the state certain products must become a part of a venture," McKnight explains. "That could include a hauler, an end-user, and maybe an intermediate processor. We will buy those products that are competitively bid, provided the waste is collected and separated first."

Florida is leaving it up to recycled

product manufacturers, processors, and haulers to join and bid together on these state contracts. By doing so, officials hope to rely on economic incentives—rather than state mandates—to spur market development.

McKnight envisions competitive bids on a commodity-by-commodity basis. The state will first take an inventory of each of the different types of recyclable wastes in state agencies, institutions, universities, schools, facilities, etc. Since this is the first such plan of its kind, Florida plans to implement it in pieces, taking into account which commodities best lend themselves to ventures, he adds.

"We are going with those ventures that make the most sense," McKnight says. "Paper is a venture that is obvious."

Here's how the competitive bid might work for paper: A state office generates such recyclable paper as newsprint, coated stock, computer print-out, card stock, high grades, mixes paper, corrugated cardboard, phone directories, and other books. A hauler that collects that paper could sell it to a mill (or through an intermediate processor). That mill, in turn, could sell back to the state such items as copier paper, towels, tissues, forms, napkins, FAX paper, corrugated cardboard, or gypsum dry-wall partitions. That mill would have to certify, however, that it used the state's waste paper to make its recycled products. All of the players involved in this transaction would make up the recycling venture, and there would be more than one venture for each commodity the state buys.

Each competitive bid, McKnight explains, would have a list of qualified Florida recycling business ventures attached to it. A number of different haulers can be involved, but they must be tied to end-users that will make recycled products from Florida's recyclables.

To start with an easy product, the first venture in this plan will be spent fuels, McKnight says. First, the state will conduct an inventory of its waste fuels. Then, the state will issue a request for proposals (RFP) for the hauling of spent fuels and the sale of recycled fuels back to the state. This RFP could be out by the summer, he predicts, but first, the state plans to hold workshops to explain the process to all the interested parties.

Government procurement of this nature alone could spur markets in the state, McKnight notes. "Between the products we're buying now and the DOT [Florida Department of Transportation] road-based recycling program, we're talking about potentially $200 to $300 million per year," he says.

But the program could go well beyond state wastes and state purchases, McKnight envisions. Cities and counties, which collect recyclables from residents across the state, could also participate in the program with the wastes over which they have jurisdiction, he says.

Retailers, too, could become an integral part of the program, by selling products made with Florida's recyclables on their shelves. Ventures will be required to show that their recycled products are being sold, McKnight notes. "We're insisting that the venture be substantial enough to stand on its own two feet," he adds.

Without a law, the state cannot force companies to put labels on their products stipulating recycled content from Florida's wastes. But McKnight is predicting that retailers will recognize items that are part of a venture with materials such as shelf tags. "We think the retail community will cooperate," he says. "The retailers have not decided exactly how they're going to do it. We won't impose on them exactly how they must do it."

With a first-time, closed-loop program such as this, a number of controversial issues have already arisen. Topics such as anti-trust, interstate commerce violations, and flow control have already been discussed, McKnight notes. Another issue is whether such a program would push small haulers, without the national connections, out of business; McKnight claims it won't.

"We are competitively bidding the removal of the waste," McKnight says. "But this is an RFP, not just an invitation to bid where price is best." Crafted into the RFPs, he adds, will be priorities for ventures that hire minorities, use Florida-based companies and include independent haulers.

Discussion Questions

1. Explain how Florida's marketing strategy works. What is the rationale behind this program? How does it contribute to sustainability?

2. What laws and regulations may impede such programs?

Composting in Texas

Kathleen Sheehan

■

SOURCE
Waste Age, August 1992,
Vol. 23, No. 8

ACCORDING TO Gordon Pierce, city manager for the city of Nacogdoches, Texas, composting was popular in Texas long before the June 1991 signing of Texas Senate Bill 1340 (SB 1340) by Texas Governor Ann Richards. SB 1340 is the state's comprehensive recycling legislation which has set a 40% recycling goal by 1994. It is perhaps the nation's most aggressive recycling goal, compared, for example, to California's mandated AB 939 which stipulates a 25% goal by 1995, and a 50% goal by 2000. Pierce says SB 1340 simply stimulated the already popular issue. But, whether or not Texas was interested in composting before SB 1340's passage, there are those like Horace Sullivan, Houston's solid waste planning manager, who feel you simply can't be successful at recycling 40% of the waste stream without a composting component in your waste management plan.

SB 1340 directs Texas state governmental agencies to give preference to products made from recyclable materials, thereby helping to create a market for them. Since composting has been classified by the U.S. EPA as recycling, composting's popularity could only increase, Pierce says.

SB 1340 also directs the Texas General Land Office to conduct a statewide recycling marketing study and promotion campaign. Garry Mauro, Texas land commissioner, has taken the directives one step further and has started corporate recycling councils designed to inspire private industry to become just as involved in the issue as the Texas government.

At the first meeting of the Texas Recycling Industry Advisory Committee last February, Mauro told attendees that "Recycling is on the threshold of becoming a full-fledged industry here in Texas—a profitable, job-creating industry." Members of the committee include representatives of environmental groups, the state's largest waste management companies, and experts in the individual components of recyclables such as plastics, paper, and glass.

"We're interested in composting here in Nacogdoches, not to save money—we don't think we will," Pierce says. "In fact, we think it will cost us money. We still have a lot of land available at a low price here in Texas, so the cost of landfilling is not as much an issue here as in other parts of the country. While other areas of the U.S. talk about disposal costs of $95 a ton, I'm disposing at $10 a ton, and meeting all the standards. With such low landfill disposal costs, it's clear that cities in Texas that are involved in composting are not involved so much from a cost savings standpoint, but more from an environmental standpoint. Composting is simply the right thing to do."

There are no big municipal solid waste (MSW) composting facilities in Nacogdoches; their involvement is limited to composting yard debris and forest scrap. But considering the fact that yard debris accounts for 31 million tons (18% by weight, or 10.3% by volume) of the nation's MSW, the involvement in composting of Nacogdoches, and count-less other cities of its size across the state, is consequential.

The city, in conjunction with Texas A&M University, presents composting seminars geared to city managers from some of the smaller cities across Texas, and brings information to those who otherwise might have difficulty obtaining it.

On a Larger Scale

While there are small-scale composting projects springing up throughout Texas, there is a large-scale MSW composting project in the northeast portion of the state that has been operational since 1972. Located in Big Sandy, the facility utilized garbage and sewage sludge exclusively for the first eight years of its existence. Today, the facility composts a variety of organic materials including MSW, sewage sludge, hardwood sawdust, brewery sludges, and chicken manure.

"Since the early 1980s, people have been coming from all over the world to see demonstration projects at Big Sandy. Our last demonstration, a few months ago, included people from as far away as South Africa," says Billy Toups, a regional director for Bedminster Bioconversion Corp. (Cherry Hill, N.J.). Bedminster technology is used exclusively at Big Sandy. "We now have a facility in Arizona and one under construction in Tennessee." Toups adds, "So we expect Texas will become less of a destination for people across the country. West Coast people are now beginning to go to our Arizona facility, and obviously, we anticipate people going to Tennessee once that facility is completed."

Vital Earth, Inc., a sister company to Bedminster, owns and operates the Big Sandy facility which comfortably processes 25 tpd of solid wastes and about 15 tpd of liquid wastes. The compartmented rotary kiln at Big Sandy accelerates the composting process, and a plant with several like kilns is capable of processing several hundred tons a day. There are now three plants in the U.S. that utilize the Bedminster technology, and a fourth is under construction. By volume, the compost produced in a Bedminster plant is about 33% of the input solid waste, Toups says, and the compacted volume of the non-degradable residue from the process is about 15% of input solid waste.

"Ironically, the Bedminster process was not developed as a way to recycle or reduce waste," Toups adds. "The focus was returning to the soils the nutrient and organic value contained in what the inventor, Eric Eweson, called 'urban by-products.' We are not really a garbage company, but a soils company. Rather than 'getting rid of garbage' by composting it, our focus is on producing soil products that are useful and needed in horticultural and agricultural applications."

Facility in Transition

"There's got to be a better way than just burying all this stuff," says Leonard Camarillo, commissioner of Precinct #4, Hidalgo County, Texas. When he took office in 1983, trash disposal problems, such as an overburdened landfill, led him to find an alternative plan to deal with the county's waste. After careful study of the problems, he found that what would work best for the area was a shredding/composting facility.

Opened in February 1991, and operated by the county since October last year, the facility is the county's first major recycling/composting effort. The plant itself covers the front six acres of an 18-acre site, with the back 12 to be used as compost settling beds. Currently, it is set up only to shred the incoming waste, and it produces 100 tons of shredded material a day. Bids are out for the composting equipment necessary to bring the facility up to its full potential.

The facility's biggest problem,

Camarillo says, is the tremendous amount of plastics in the waste stream. In addition to removing the plastics to improve the quality of the eventual compost, the county is also involved in research projects with Texas A&M University.

Right now, the facility receives refuse from approximately 90,000 area residents. It shreds approximately 70 tons of waste per day from the cities of San Juan and Mercedes, and approximately 30 tpd from surrounding rural areas—only a portion of the 500 tons generated by the county each day. But as soon as the market is firmly established for the finished product, Camarillo expects to see other cities sending their wastes to the Hidalgo County facility.

Combined Process Facility

Clay Roming, the engineer that designed a sewage sludge plant and compost system for the Brazos River Authority, talked to *Waste Age* about the facility, which had just received an engineering excellence award by the Consulting Engineers Council of Texas. Built in 1975, the facility treats all wastewater generated by the city of Belton and approximately 50% of the wastewater generated by the city of Temple. It serves the wastewater needs of approximately 39,000 citizens and has been producing compost for about two years now.

Temple-Belton, as the facility is called, is an advanced secondary plant that produces a very high quality effluent. The average flow through the facility is six million gallons a day, which produces about 6,000 pounds of solids a day after processing. Dewatered sludge is mixed with wood chips as a bulking agent in preparation for composting, but when a bulking agent is not available, Roming says, "We recycle the compost itself by using it as a bulking agent.

"We actually fell into composting

here as a means of solving existing odor problems in our sludge treatment process. Subsequently, we find we not only solved our odor problems, but have substantially reduced the burden on our landfills, due to our large usage of woodwaste products in the process."

Houston Plans a Composting Facility

Ulysses Ford, III, director of the department of public works for the city of Houston, says he's glad to see composting move from an "interesting little backyard project" to its rightful place as an economically, as well as environmentally sound solid waste handling method. He believes that composting can divert almost as much waste from landfills as an incinerator, and at a more economical cost—a belief that is shared by Houston's solid waste planning manager, Horace Sullivan. According to Sullivan, a recent study shows composting to be less costly than incineration. He warns, though, that before the study, many people had considered composting a much simpler process than it turned out to be.

Ford talks about a potential landfill diversion of approximately 50-60% of the total residential waste stream when composting is used. If other recycling efforts such as dropoff and buy-back centers are included, he estimates an optimum achievable city-wide level of 75% recycling.

Houston has been a proponent of recycling by composting since at least the mid-1960s to the early 1970s, when the city had two facilities up and running. One facility was unable to control odors, and the other was unable to find solid markets for the compost; both eventually closed.

Ford says odor problems can be overcome by selecting a good design and developing an odor management plan before, during, and after processing; by implementing good operating procedures; and by providing for the appropriate treatment of process gases.

He addresses the issue of markets by noting that a host of uses for compost has been established, and adds that assessing potential health risks could aid in the selection of markets.

Houston has an MSW composting plan in the works with WPF, Inc., an Ohio-based consulting and project management company that specializes in solid waste composting facilities. The agreement calls for a 1,000-tpd enclosed plant to be built in Houston. The city has guaranteed 750 tons of refuse per day based on a 4-day work week. That amount represents only one-third of what the city generates per day—about 2,500 tpd. Since there are no waste-to-energy facilities in Houston, the other two-thirds of the waste stream will go to recycling facilities and landfills.

Although the proposed facility will certainly help Houston reach the 40% recycled goal called for in SB 1340, work had begun on the project long before anyone was aware of SB 1340, says Edward T. Chen, Houston's deputy assistant director of recycling, Public Works Department.

Chen relayed information about another, little-known project underway in Houston—a 6,000-sq.-ft. facility the likes of which, he says, would be "unique in the nation." Current plans call for a three-component facility: a multimedia recycling educational center, a retail store (with all recycled products including compost), and a recycling industry showplace where companies can display their expertise and technologies. Funding for this unique undertaking is coming from the city, private industry, and trade associations, according to Chen. He looks for plans to come to fruition this fall.

Establishing Standards

Ford calls for nationwide uniform product standards, a view shared by many others. Rep. Al Swift (D-Wash.), sponsored H.R. 3865, entitled the National Waste Reduction, Recycling, and Management Act. The proposed federal legislation includes several provisions that would help advance composting as a key component of integrated waste management plans. The legislation would, among other things, require that EPA establish standards for all compost products, and it strongly suggests that 75% of yard waste be diverted from landfilling or combustion by 1995. It also outlines federal procurement guidelines for compost, and would exempt organic materials destined for composting from fees on interstate transportation of MSW.

Even the President of the United States has addressed composting. Late last year, President Bush issued Executive Order 12780, directing federal agencies to initiate recycling programs that foster waste reduction and the recycling of materials, including "the composting of organic materials." Congressman George Hochbrueckner (D-N.Y.), who calls composting the "sleeping giant of waste management," followed up on Bush's order with detailed questionnaires to at least 28 different federal agencies about their plans to implement the order.

Hochbrueckner's office has issued a report based on replies received from federal agencies, Congress, and the federal judiciary. The report, entitled "A Survey of Recycling Activity Within the Federal Government," reveals some successes such as the recycling of 89,000 tons of asphalt mix by the Commerce Department in 1990, and the fact that the House of Representatives has been composting yard waste from the Capitol grounds for years. The report also reveals some areas in which the government needs to learn more from private sector experts.

For example, several agencies said they would consider implementing on-site composting of biodegradable wastes from office buildings, and using finished compost in landscaping, but that they needed to have a plan presented to them in order to do so. While agencies are interested in the implications of increased recycling activities, the report points out that few activities, if any, have taken place to implement the order. Copies of the report are available through Rep. Hochbrueckner's office. Contact Mary Anne Weber at 515/689-6767.

Private industry is also interested in defining composting industry standards. Novon Products Group (Morris Plains, N.J.), which is a commercial division of Warner-Lambert Co., is active in the Institute for Standards Research (ISR), a subsidiary of the American Society for Testing and Materials. ISR's program advisory committee chose composting as the first of eight waste disposal systems to investigate, and one of their first goals is to develop and evaluate a series of standard laboratory tests that will help predict the behavior of biodegradable products in real disposal systems.

Novon researches and develops its own family of biodegradable, starch-based products and provides for their commercial use. Made up primarily of annually renewable sources of starch—such as potatoes or corn, plus other degradable or natural additives—their products decompose completely when disposed of in biologically active environments, such as compost facilities or wastewater treatment systems, according to Dr. Kenneth D. Tracy, Novon's vice president, environmental technology.

The company is currently testing materials in a compost-type environment in Belgium, and has a large-scale demonstration program underway in Maine. The Maine demonstration is being called a closed-loop project because it handles the compost from cradle to grave. Compostables are collected from super markets and restaurants, delivered to a farm where they are composted, and then applied directly to the farm land. In addition, Novon is pursuing partnerships and trials with composters in other areas of Europe and the U.S.

Tracy sees the pre-product testing standards Novon is using and the post-product testing standards proposed by the Solid Waste Composting Council (SWCC, Washington, D.C.), as vital to the composting industry.

SWCC's function is to promote composting and its role in integrated waste management, and toward that end, its members are hard at work on the issue of creating standards for the industry. Members include composters, compost users, consumer product companies, academic institutions, consultants, public officials, non-profit organizations, and environmentalists. Its charter calls for nationally and internationally acceptable compost product standards, and for the provision of product classifications and quality controls.

A draft of the SWCC's proposed standards is available as a background aid to people in state and other government agencies to assist them in their efforts to implement standards within the governmental regulatory framework—legislation like Texas Senate Bill 1340, the House of Representatives Bill 3685, and executive orders from President Bush.

Discussion Questions

1. Why is composting generally not considered economical in Texas?

2. Would efforts to include external costs of landfilling and incineration change the economics of composting? What economic factors should be included in this determination?

Urbanization

n 1990, the 14 states in the southern region contained approximately 80 million people, nearly one-third of the U.S. population. The region also contains 38 of the nation's largest 100 metropolitan areas. Moreover, population growth from 1980 to 1990 increased by 13.4%, second only to that occurring in the West, which grew by 22%. The fastest population growth occurred in Florida, Texas, Georgia, Virginia, and North Carolina. The slowest growth occurred in Kentucky, Louisiana, and Tennessee. West Virginia's population actually shrank during that decade by 8%.

Population growth brings with it many problems collectively called urbanization issues. When improperly regulated, it usurps valuable farmland and puts pressure on forests, wetlands, and other important wildlife habitat. It increases congestion on highways and streets and causes deterioration in air and water quality. It places extraordinary demands on government to provide services.

Fortunately, many locales have begun to try to reduce, even prevent, the damaging effects of population growth. Despite this progress, much more is needed to protect the valuable resources that we share with many species on this planet.

Background Reading: Chapter 23 (see discussion entitled "Sustainability Through Land-Use Plannings" in section 23.3).

Growthbusters Versus Growth Management: The Sarasota Saga

Marie L. York

■
SOURCE
Environmental and Urban Issues,
Fall 1990, Vol. 18, No. 1

N THE FACE of the onslaught of 1000 new Floridians every day, it is a predictable reaction that current residents want to halt growth. Sarasota County, with the third highest per capita income of any county in Florida, was the latest to face the citizen outcry to stop the immigration. Nearly 10,000 citizens, frustrated by a delayed water project and drought-induced water restrictions, clogged roads and a near-capacity landfill, signed a petition, which placed onto the September 1990 ballot the opportunity to stop all development for two years.

The most controversial aspect of the three-part referendum to amend the county charter was a proposed two-year moratorium on the issuance of all construction permits in the county, excluding remodeling and expansions. The second part required that after the moratorium was over, it would take a unanimous vote of the county commissioners to approve any rezoning that would

result in an increase in density, except in affordable housing. The third part called for the levy of the maximum amount of impact fees legally allowed.

The proposed moratorium failed by a three-to-one margin. The three no-growth candidates running for county commission seats did not win, nor did they qualify for runoff elections. No incumbent avoided a runoff election, however, partly due to an incredibly large number of candidates—six in one district alone. Interestingly, a large number of voters only cast one ballot regarding the referendum, and did not vote on any other issue or candidate.

In the post-referendum wake, the county is left with nearly 25 percent of voters who distrust the effectiveness of the growth management process, or who do not trust their leaders and administrators to carry out the intention of the comprehensive plan. Also left is a highly organized and cohesive business and development constituency, galvanized into existence by the no-growth threat. The economic studies also revealed the county's dependence upon the construction industry and the local government's increasing reliance upon impact and development fees, rather than ad valorem taxation as a source of revenue.

The proposed moratorium was construed as an insult to the comprehensive planning process because neither the recently adopted comprehensive plan nor the concurrency management system and land development regulations were being given an opportunity for implementation. Nathaniel Reed, president of 1000 Friends of Florida, wrote: "The desire to control the negative effects of growth is understandable; in fact, it is the basis for growth management. But the decision to allow no growth at all for two years is at best premature, at worst a recipe for disaster."

Several ironies have emerged from this saga. Over the last decade

Sarasota County's planning efforts have won national and state planning awards. 1000 Friends of Florida has recognized their current comprehensive plan as one of the better plans in the state. Another irony is that the structure of growth management enforcement in Florida is largely dependent upon the citizens who have the right to challenge in the event of noncompliance. The moratorium would have rendered the plan nonfunctional, and would have thus precluded that right for lack of implementation. A third irony is that much of the need for revenues for services and infrastructure was not coming from the demands of new growth but from past growth that had not been assessed its fair share when approved for development.

Florida's growth is often compared to that of California. Florida's professional planning and development communities have been frightened by the California example. California, unlike Florida, has not passed statewide comprehensive planning legislation to deal effectively with its growth. Citizen initiatives have instead forced growth moratoria and revenue limiting measures. The Florida growth management legislation is an effort to minimize the negative effects of growth, while attempting to avoid the California reaction of moratoria and restrictive zoning. The California tax revolt, which gave birth to Proposition 13, froze assessments for property owners, which in turn translated to higher taxes for newcomers. Sarasota residents were accused of trying to duplicate this effect, in addition to limiting competition from new businesses.

Many accusations flew. Taxicabs sprouted signs opposing the moratorium. A local car wash bypassed its usual television promotion to sponsor anti-moratorium advice. T-shirts became walking propaganda. The Growth-restraint and Environmental Organization (GEO) became known as the growthbusters;

the opponents coalesced into the Citizens for Responsible Solutions (CRS). By June 30, 1990, GEO had 425 members and $7000. The CRS quickly garnered $272,000 and claimed 1200 members. A highly organized political machine evolved with voter registration drives, phone banks, targeted mailings and sample ballots.

The CRS hired a Pennsylvania firm to analyze the economic impact of the moratorium on building permits. The results of this study indicated that: (1) if the moratorium was passed the county could anticipate a deep recession; (2) the local government would experience a drastic shortfall in revenue; (3) the construction industry would be decimated with a loss of construction jobs totaling 8000; and (4) the ripple effect would result in a total job loss of 12,000, with the unemployment rate rising to the 12 to 15 percent range. Wages and salaries loss over the two-year period was predicted to be $560 million, with a total personal income loss of $1.6 billion. Population was predicted to fall by 4 percent, with sales of existing single-family homes dropping 85 percent, while home prices increased 12 to 13 percent.

The Sarasota County Commission had a firm in Cambridge, Massachusetts, conduct a more comprehensive analysis to study the possible effects of the proposal. This report was equally gloomy, with 12,400 to 15,000 jobs predicted to be lost in the first year alone, with losses to the county coffers of $6.8 to $9.9 million in 1991 and $9.6 to $12.5 in 1992. After that year, the net effect on county revenues was predicted to be positive due to the lagged increase in property tax values.

The anticipated change in housing values was not an across-the-board increase. Lower income houses were expected to decrease in value as a result of increased supply due to higher unemployment. The higher priced housing stock was

expected to jump dramatically with the increase offsetting the value decline in lower income level homes and vacant lands.

It is commonly understood that moratoria create winners and losers. The winners are usually existing residents and property owners, given that the supply of housing, commercial and industrial space becomes limited, and the price of existing development is driven upward. The losers are potential new residents and businesses unable to make a free choice in the search for housing and space. Moratoria change the population mix of a community, favoring the wealthy and discriminating against persons of lower income.

Most supporters of Florida's growth management system feel that the Sarasota saga ending in a clear and pronounced rejection of the moratorium was a good thing. Fortunately for Florida, growth management legislation is in place throughout the state. The challenge, which the state, its local governments and the private sector face, is to stand shoulder-to-shoulder and implement the system, in a full and fair fashion. Only then will the threat of no-growth local wildfires be squelched—through the responsible management of growth.

Discussion Questions

1. This article illustrates the strong pro-growth forces in our society. It also illustrates how public opinion is shaped by economic forces. Explain what is meant by these comments.

2. How could the growthbuster movement have been prevented?

Planning for Growth in Arlington County, Virginia

Richard Ward

■

SOURCE
Urban Land, January 1991

This article was adapted by permission of *Urban Land*.

ARLINGTON County, Virginia, is a densely developed inner-ring suburb in the rapidly growing Washington, D.C. area. Occupying about 26 square miles and located across the Potomac River from Washington, D.C., Arlington County has some 170,000 residents in 85,000 households. Originally part of the 10-mile-sided square surveyed in 1791 to become the nation's capital, it was returned to the state of Virginia by Congress in 1846. By the end of World War II, the county was largely built out as a bedroom community for federal workers in downtown Washington. It also contained a number of well-known federal establishments including Arlington National Cemetery, Fort Myer, the Pentagon, and Washington National Airport.

The establishment of the Pentagon as headquarters of the War Department (later the Department of Defense) in 1941 set the stage for the dramatic dispersal of federal office operations from downtown Washington into the surrounding Maryland and Virginia suburbs after the war. By the early 1960s, capacity limits in Washington led the General Services Administration (GSA), the government's landlord, to focus its search for space for a mushrooming federal bureaucracy on Arlington County and to switch from a build-and-own to a lease-space approach. GSA's new policies gave impetus to a frenzied wave of speculative office development in the 1960s and 1970s.

The locational advantage of Arlington's centrality and adjacency to downtown Washington grew in importance as the federal government continued to expand into the suburbs. The county has experienced almost 30 years of steady office growth, fueled first by the growth of federal agencies and then by the arrival of private firms doing business (directly or indirectly) with the government.

The completion in 1979 of the two Metro rapid-transit system lines that cross Arlington County enhanced further the county's attraction as an office location.

Planning for Growth

In the 1960s, the derelict industrial/warehouse area known as Rosslyn was ripe for redevelopment. But the electorate hesitated to adopt traditional urban renewal techniques, out of fear that public housing would accompany the county's acceptance of federal grants and urban renewal powers. Thus, the county came up with its own market-driven, carrot-and-stick redevelopment procedures.

In 1962, it formulated a plan for a Rosslyn office district. The plan established a system of above-grade pedestrian walkways and a high-capacity street system to feed traffic into the Potomac River bridge system. In return for each prospective developer's agreement to submit to a rigorous site-plan review procedure and to contribute a share of the street and pedestrian circulation systems, the developer became eligible to receive an increase in building height from four to 10 stories and in floor/area ratio (FAR) from 1.0 to 3.25, and for a decrease in the amount of on-site parking required.

In the late 1960s, Arlington undertook a similar planning effort on a larger scale for the Jefferson Davis Highway corridor near Washington National Airport. Two large mixed-use complexes emerged that are known today as Crystal City and Pentagon City. Distinct from Rosslyn's pattern of individual building projects, these incorporated retail, hotels, offices, and apartments within environments under common ownership and development.

The county's carrot-and-stick approach succeeded in achieving some important project objectives. For example, Rosslyn the construction of a high-capacity road network and a system of grade-separated pedestrian walkways both were funded in large part by exactions from private developers. The Rosslyn area was quickly redeveloped with six- to 12-story office towers at an FAR of 2.5 to 3.5. In anticipation of the arrival of the regional rapid rail system, on-site parking was required at the low ratio of one space per 500 square feet.

However, the overall design quality of Rosslyn's redevelopment proved to be disappointing. In planning for other mixed-use districts centered on Metro stations, the county would seek to build on and improve the development incentives process initiated in the Rosslyn plan.

Well before Metro construction commenced, the county, anticipating both positive and negative economic and environmental impacts, began a process of sector planning for station areas to take maximum advantage of the opportunities the system offered and mitigate is undesirable effects.

In general, this sector planning

process involved a clear commitment to accommodate growth stimulated by the transit stations and a corresponding commitment to contain development within designated transit impact zones, avoid undesirable spillover effects beyond these zones, and preclude the kinds of urban design shortcomings evident in Rosslyn's redevelopment. Sector plans would encourage mixed-use development, focus on good architecture and urban design, reinforce the unique character of each station area, and clearly demarcate the high-density transit zones from surrounding low-density residential areas.

The planning effort for the five stations on the Rosslyn/Ballston corridor involved extensive public participation. It produced detailed sector plans for each station affected. If the five plans are realized, the number of residential units located within one-quarter mile of each of the stations in the corridor will double from the present 12,100 to about 25,000; the total commercial area will more than double to 25 million square feet.

- The Rosslyn plan (completed in 1977) refines the 1962 plan. It envisions an expansion in office space from a late-1970s base of 4.5 million square feet to 7.5 million square fee ultimately. It also encourages residential and retail development to give the area a better mix of land uses.

- The Courthouse plan (1981) looks for a growth in office space from the current 2.4 million square feet to 3.4 million, and the development of almost 4,000 units of high-density housing.

- The plan for the Clarendon area (1984) envisages growth from a base of about 1 million square feet of office and retail space to 4 million square feet, and the addition of about 1,000 residential units.

- The Virginia Square plan (1983) adds about 2.5 million square feet currently in place and doubles the supply of housing to about 3,200 units.

- Ballston (1981) is projected to expand to 8 million square feet of retail and office space from a current base of 2.8 million. With 1,300 transit-related residential units already in place and 1,200 under construction, the Ballston area is planned to have 8,200 residential units at buildout.

The Plans Work Out

Arlington's success in growth management can be measured in several ways. First, a substantial amount of growth has occurred over the past 25 years. This has not taken place in new subdivisions spreading across vacant or farm lands at the urban fringe. Rather, it has come in the form of a major office, retail, hotel, and high-density apartment infill and redevelopment. In fact, the county has seen its office inventory grow from less than 1 million square feet in the early 1960s to over 29 million square feet today, with equally dramatic expansion in its inventories of apartment units, hotel rooms, and retail floor space. At the same time, the number of single-family residences has changed very little.

A second indicator of success, and the most important in the planning context, is the fact that this process of internal growth has been *managed*. Instead of lurching from crisis to crisis, as is too often the result when local governments confront rapid growth and change, Arlington has planned for, anticipated, and accommodated growth in a systematic and orderly manner. Capital investment in necessary supporting public infrastructure has occurred in pace with private investment. The physical boundaries of areas of change have been defined clearly vis-à-vis those neighborhoods

to be conserved and buffered from growth. The county has maintained a dynamic planning process with opportunity for public participation. The specific content of Arlington's countywide comprehensive and sector plans has been adjusted over the course of two decades, but its vision from the 1960s is well on the way to fulfillment in the 1990s.

The expansion of retail in the Rosslyn/Ballston rail corridor and the Jefferson Davis Highway corridor is especially noteworthy. In 1986, Ballston Common replaced a 1950s-era retail center at Ballston, providing a net increase of 720,000 square feet of commercial space. This development clearly has become the focus of a new downtown in central Arlington. In 1989, the Fashion Centre at Pentagon City on Metro's blue/yellow line in the Jefferson Davis corridor opened, adding more than 1 million square feet of upscale retail space, a Ritz-Carlton hotel (opened in 1990), and a 172,000-square-foot office tower.

Much of the credit for the county's successful growth management must go to the general atmosphere of "good government" that derives from the high quality and consistency of its political, administrative, and civic leadership. Ellen Bozman, a member of the county board for the past 17 years, typifies this leadership, as do William Hughes and Thomas Parker, who have served in and skillfully led the planning and economic development divisions of county government each for more than 20 years. Arlington has a county manager form of government, which has worked well in responding to citizens' needs and wishes while keeping the business of government moving forward efficiently and effectively.

In the words of William Barnes Lawson, long-term legal counsel to local developers, "The key to Arlington's success has been the integrity of the local elected officials. This is helped by the fact that all five county board members are elected at large and thereby represent community interests rather than narrow parochial concerns."

Lawson also emphasizes the high quality and long-term consistency of the county's plans and planning processes. The existence of consistent plans lets developers know what the rules of the game are and what is expected of them. Developers react positively to this certainty.

The combination of the explicit sector plans and incentive zoning tools, argues Lawson, requires developers to "put their programs where their mouths are"—in other words, to deliver on the promises of their plans for development in return for a supportive and reasonably predictable public planning and approval process. The relative certainty of the outcome of the plan-approval process within the framework of the county's long-established land use plan, in turn, translates into relatively predictable land values. Incentives allow developers to "earn their way to the top," to develop at higher densities in exchange for providing public infrastructure, design improvements, and housing in mixed-use projects.

While she recognizes the importance of the county's incentive zoning and sector planning, county board member Ellen Boxman credits much of Arlington's success in managing growth to the citizen participation process leading up to the county's 1975 Long Range Community Plan and Program. Over a three-year period, more than 60 public meetings took place to consider three growth scenarios: no growth, high growth, and growth concentrated in the Rosslyn/Ballston and Jefferson Davis corridors. Citizens, rejecting the extreme visions of remaining an aging, mostly residential enclave or of massively changing Arlington's character and scale, widely accepted public policy based on the third scenario. When a wave of new growth hit the county, Arlington was prepared to handle it

within the framework of its 1975 plan. In Bozman's words, the county "was prepared to successfully effect decisions at the cutting edge of development" without interminable debate and controversy.

Robert Buchanon, a developer with the Radnor Buchanon organization, provides another perspective. He notes that "Arlington had the vision to understand the enviable position it occupied in the Washington metropolitan region and then acted incisively to secure that position for future generations. The county planned for growth before it happened and then, when the market came of age, it was able to steer growth in the direction of the plan.

An Effective System

According to Thomas Parker, director of economic development in Arlington County, a strong economy has given the county a second wave of development that will help it overcome the planning and development shortcomings in its original growth sector, Rosslyn. Arlington has learned much through its experience. It has learned to harness the market resource represented by a diverse population to create a series of mixed-use environments. Its areas of dense, mixed-use development take full advantage of the county's two transit-served corridors and preserve surrounding, low-density residential neighborhoods. In purchasing and upgrading pre-1940s housing at an unprecedented level, a new generation of residents is showing its enthusiasm for the easy access to jobs, shopping, and transit brought about by the county's planned development policies.

Furthermore, as Lawson notes, "Arlington's success with incentive zoning as a growth management tool is now being emulated by many surrounding suburban communities." However, he says, in his work representing real estate interests throughout the region he has found

no community with a system as effective as Arlington's.

The Arlington experience demonstrates that in a demand-driven economy, public planning can shape the pattern of development to achieve clearly stated public objectives. Planning with the community for growth, rather than waiting to react to growth pressures, can pay handsome dividends for all. Without the availability of condemnation and other traditional incentives for redevelopment, a carefully focused, zoning-based redevelopment strategy can work providing it incorporates realistic incentives for development to take place within public planning guidelines.

Discussion Questions

1. Most urban/suburban regions illustrate one of the key problems with modern society: linearity. In many cities, good usable land and building sites are located within the confines of the city, but rather than make use of them, new developments spring up on the periphery. What is the danger of this approach?

2. How have the planners and officials of Arlington County sought to reduce sprawl? Given the material presented here, how could they have further promoted a sustainable future?

Multiparty Development and Conservation Agreements

Douglas R. Porter

■

S O U R C E
Urban Land, February 1992

Douglas R. Porter is a growth management consultant and is founder and
president of The Growth Management
Institute, a nonprofit research and educational organization in Washington, D.C.

HEN developers run into regulatory obstacles regarding preservation of environmental fea-
tures such as wetlands and wildlife, they are turning increasingly to special planning processes to help them over the hurdles. Such plan-
ning efforts often bring together a variety of governmental entities, one or more property owners, and a host of special interests in a cooperative planning exercise that sorts out environmental issues.

Sometimes authorized in federal law, sometimes in state legislation, and sometimes simply organized as ad hoc ventures by developers and public officials, areawide planning (also termed "special area planning" or "focal-point planning") typically attempts to reach negotiated agreements that reconcile environmental and development concerns. In general, however, the participants in this process operate in a never-never land outside the planning frame-
works of local governments and the regulatory thickets of federal, state, and local agencies.

Practitioners in the development field should understand how these special planning processes may help them. Also, while laudable in most respects, collaborative planning approaches could be better served by the governments and regulatory agencies that ultimately benefit from them.

Planning Gaps and Glitches

Developers are discovering that soggy ground or the presence of certain creatures, some quite diminu-
tive, can disrupt and even bring to a halt the most splendid proposals. Conflicts with environmental goals appear to be on the increase as urban development pushes into sensitive environments, and as govern-
mental requirements for the preservation of endangered species, wetlands, and other environmental features have both broadened and intensified.

All of this means that developers are encountering more circumstances in which they must somehow come to terms with restrictions on development to save important environ-
mental features. Often they are required to make major contributions of land, time, and cash.

Such restrictions, although bur-
densome, frequently are less onerous than the regulatory maze itself. Developers ordinarily confront a confusion of rules, regulations, and procedures laid down by a multitude of federal and state agencies and one or more local jurisdictions, each operating independently of the others. Various special-interest groups add their concerns to the brew. Developers usually must resort to hiring well-paid attorneys and specialists to find a path for them through the permitting maze.

Two problems seem uppermost. One is overlapping jurisdictions among a multiplicity of governmental agencies, each with its own rules and its own ways of reaching decisions. The second problem—
probably the more serious one—is the general absence of any public decision-making structure allowing environment/development conflicts to be considered and reconciled in a rational, creative manner.

If a wetland or habitat of an endangered species lies wholly within a single property in a single jurisdiction, the local planning and zoning apparatus, if coordinated with state and federal requirements, may offer a suitable venue for reaching agreements about preservation and development. Too often, however, one or more of the fol-
lowing conditions exists:

- The community's comprehensive plan and zoning ordinance have not identified specific environ-
mental features worth preserving and have not defined ways in which competing developmental and environmental goals might be reconciled, thereby providing little guidance to property owners.

- The features in question cross property lines or jurisdictional boundaries. Attempts to coordinate development decisions among multiple property owners and communities, however, usually are frustrated by the case-
by-case procedures established by permitting agencies.

- The regional agencies that might provide guidance for multiparty or interjurisdictional agreements lack plans with sufficient force and definition to assist in such decisions.

In short, public planning mecha-
nisms in most communities and regions fail to link satisfactorily with the environmental regulatory

machinery set up by federal, state, and local governments.

Federal agencies have taken some cautious steps to bridge this procedural gap. The U.S. Army Corps of Engineers (Corps) and the U.S. Environmental Protection Agency (EPA), charged with issuance of wetlands permits, have encouraged advance identification programs for wetlands that serve to notify property owners of potential development problems and to establish some wetlands preservation priorities.

The Coastal Zone Management Act of 1972 approaches the problem by encouraging coastal states to designate critical areas and prepare areawide plans for them. The Endangered Species Act of 1973, as amended, represents a third approach. The law authorizes limited invasions of endangered species habitats if a regional conservation plan is in place and other conditions are met. (See Timothy Beatley, "Regional Approaches to Wildlife Habitat Conservation." *Urban Land*, August 1991.)

A number of states have established special procedures for dealing with areas of critical concern. New Jersey's Hackensack Meadowlands Development Commission focuses planning and regulatory activities across a number of jurisdictions on the goal of achieving a proper balance between environmental preservation and development. Florida's Chapter 380 process establishes multijurisdictional planning procedures for designated critical areas and has been applied in several areas, including the East Everglades. Maryland's Chesapeake Bay Critical Area Commission devises and implements criteria to guide development and preservation in 61 jurisdictions on the bay and its tributaries.

Experience with these approaches has been studied by an informal working group whose meetings were cosponsored by ULI and the Environmental Law Institute. The meetings were initiated by Lindell Marsh, a member of ULI's Federal Policy Council, and chaired by John M. DeGrove, director of the FAU/FIU Joint Center for Environmental and Urban Problems at Florida Atlantic University in Fort Lauderdale. The group also sponsored case studies describing specific experiences with resolving environment/development conflicts (see "A Quartet of Wetlands Plans," *Urban Land*, April 1990). Some lessons and conclusions drawn from this wealth of information may help others resolve similar conflicts.

How Areawide Planning Should Work

Rarely are areawide planning exercises undertaken unless a long history of clashes among developers, public officials, and environmental groups or agencies has convinced all parties that some extraordinary means of accommodating the various interests will be required. Initiated either by developers, special interest groups, or public agencies, a planning group is formed to represent the variety of stockholders involved. The group may have official status, as in the state-appointed Chesapeake Bay Commission or the Resource Planning and Management Committee for the East Everglades, or it may be convened as an ad hoc group such as the committees typically organized to formulate habitat conservation plans.

Funding for staffing and research comes from developer and public agency contributions, from federal and state grants, and sometimes from foundation or corporate donations. Normally, little money is available to pay for the expenses of individuals to attend meetings.

Like its counterparts everywhere, a planning group obtains and discusses information, appoints subcommittees to follow up specific issues, holds public hearings or workshops, adapts to constant member turnover, and reports its conclusions to one or more official agencies. The findings and conclu-sions of some groups flow directly into official acts, such as adoption of habitat conservation plans by the U.S. Fish and Wildlife Service (FWS), or multijurisdiction agreements among local governments. In other cases, results of group efforts serve simply to advise an array of agencies and interests.

This approach to preserving important environmental features in developing areas is useful in at least three important ways. First, it views environmental concerns broadly over a large area containing numerous property holdings, a way of looking at issues that heartens environmentalists who decry the cumulative effects of incremental permit approvals and the ecological discontinuities that often result from narrowly defined plans. Developers avoid the "death by a thousand cuts" that individual permitting efforts seem to entail. Public officials find useful the consensus-building forum that mutes discord.

Second, discussions and negotiations often yield creative solutions that satisfy the objectives of all interests. Planning group participants tend to focus on reconciling differences and finding solutions. They avoid stonewalling. Tradeoffs and mitigations become possible that otherwise might be difficult with single parcels or with unilateral decisions. The costs of preservation can be equitably distributed among the various interests.

Third, the involvement of many interests in public discussion builds community confidence and trust in the outcomes, encouraging long-term adherence to agreements and plans. This provides greater predictability and certainty for advocates of both environmental and development values.

For all these reasons, special areawide planning may prove an invaluable aid to developers seeking approvals and environmentalists seeking preservation of important resources.

The Ways It Doesn't Work

For all its promise, collaborative planning is time-consuming and resource-intensive if only because of the diversity of participants and the range of issues. Yet these investments would be eminently worthwhile if the consensus reached were equitable, implementable, and enforceable. Unfortunately, such is often not the case.

The process can go wrong for many reasons, among them:

Constraints on Federal Agencies. Especially when federal agencies are not lead agencies in the process, getting their representatives to the table is difficult. Regional offices of agencies such as EPA and the Corps plead a lack of staff and travel budgets. In the planning for the East Everglades, federal agencies were only sporadically represented. As a result, EPA basically disagreed with the committee's final consensus on tradeoffs among agricultural and preservation uses.

Another problem is that federal agencies find it difficult to agree to the classification of some otherwise protected areas as expendable. The EPA/Corps advance wetlands identification process, for example, can easily identify wetland areas that are unsuitable for development but finds it hard to acknowledge that some wetlands may be fillable. David Davis, EPA's deputy director of wetlands, oceans, and watersheds, says that the agencies are working on overcoming that psychological barrier. "We find ourselves needing to be able to make that kind of judgment," he says, especially when wetlands identification occurs within an overall planning framework.

The wholehearted participation of federal agencies in group planning endeavors may be stymied by their wariness about regulatory takings. The Bush Administration's insistence on enforcing Executive Order 12630—which directs federal agencies to identify actions that might stir takings claims and make those agencies potentially liable for compensation to landowners—may be keeping EPA and other agencies from joining these planning efforts.

Lack of Funding for Planning and Implementation. With their less than official status, areawide planning efforts must contend with funding uncertainties and shortages that limit staffing, research, and follow-up management. In a few cases, a lead local or state agency may provide basic staff resources, and federal and state agencies sometimes provide special grant funds. Most often, however, it falls to developers to foot much of the bill. The lack of any central funding source leads to limitations on the comprehensiveness of planning and dampens the entire effort.

And planning costs are negligible compared to the costs of implementing plans, which often call for the acquisition of hundreds or thousands of acres to preserve unique habitats or expansive wetlands. Compensation for landowners forced to reserve large acreages is often critical to obtaining their support for preservation plans and to avoid taking claims.

Acquisition and compensation costs can be enormous. In Riverside County, California, for example, protection of the Stephen's kangaroo rat will require land acquisition costing from $50 million to $250 million. The proposed Austin, Texas, preserve for 15 species will cost over $80 million. Beyond acquisition costs, management of preserves and other elements of plans require continuing funding.

Local public officials are inclined to turn to developers and landowners for funds. Many of the plans to date have been funded primarily by developer fees. Riverside County, for example, has raised almost $25 million from a $1,950-per-acre development fee for its kangaroo rat preserve.

But state and federal funds should also be part of the financing packages, since preservation efforts are state and national priorities and their benefits spread well beyond the boundaries of the immediately affected properties. In addition, developers may understandably not agree to plans that depend solely on private funding.

The federal laws that require preservation of wetlands and wildlife habitats have not been accompanied by suitable funding to compensate landowners. The federal Land and Water Conservation Fund has provided some acquisition funds. In some cases, such as Florida's East Everglades, the state has stepped in to raise the necessary funds. The Nature Conservancy has funded some planning and acquisition in several habitat conservation plans. But these essentially ad hoc financing efforts do not provide a predictable mechanism for implementing plans. There is a clear need for state and federal programs to assisting in funding areawide planning efforts.

Lack of Enforceable Assurances. Since many special collaborative planning efforts depend on ad hoc, voluntary planning groups, it is not surprising to find problems in follow-up implementation and management.

Except for FWS official commitments to habitat conservation plans, federal agencies by and large are bound only by the moral force of adopted areawide plans as they review individual permit applications. This is a frail foundation, as was proved by the fate of an elaborately negotiated plan for Grays Harbor, Washington, an early prototype of special area management plans. The plan came to naught when new personnel in the Corps, a key player, reneged on its staff's previous consent to plan implementation.

State and local governments, as well, are subject to the vagaries of political trends that may cause them to withdraw support from adopted plans. In addition, areawide plans usually depend heavily on actions by assorted agencies and departments, actions that are easily unhinged by budget and staff changes.

Experience demonstrates the critical necessity of assigning follow-up responsibilities to specific organizations or agencies. Follow-up problems on plans forged by critical area committees in Florida, which are disbanded when plans are adopted, are instructive. Also, agreements that involve several property owners may go adrift if one or more of them run into economic or other problems.

Areawide plans, then, are only as good as the long-term viability and consistency of outlook of the public and private interests with a stake in the outcome. The legislation and regulations governing areawide planning efforts should be augmented to provide assurances to all interests that the plans, once formulated, will be honored.

The Complexity of Environmental Values. It should come as no surprise that areawide planning often runs afoul of disagreements over the definition of critical environmental features, the determination of viable preservation areas, and other technical matters. Many environmental interests find it difficult to accept any diminution of resource lands, much less the use of created wetlands or habitats in new locations to replace lost lands.

Questions of adequate biodiversity and the long-term viability of preserved habitats are difficult to resolve. Many of these disagreements stem from insufficient understanding of the particular wetlands and species involved, but research to provide an adequate understanding would take years to carry out.

These uncertainties will make for continuing controversy during and after the planning process. To have any hope of long-term implementation, plans must build in ongoing research activities and latitude for revisions based on research findings. To do so, of course, will add more uncertainty to the implementation process.

Future Prospects

With all these drawbacks, is there any hope for the concept of collaborative planning? The answer has to be yes, if only because environmental and development conflicts undoubtedly will continue to increase, and the collaborative planning approach represents one of the few means of reconciling the host of issues raised by such conflicts. As it now works, it is a necessary but costly and complex procedure, and certainly less than satisfactory in its long-term reliability.

Two lines of action would significantly help to improve the process. First, federal and state agencies should construct a regulatory home for such efforts, including incentives for undertaking them and mechanisms for recognizing their results in later agency actions. The incentives should include a funding element and the recognition should include a form of binding agreement that developers can "take to the bank" and that environmental interests can bank on as well.

Second, efforts should be made at all levels to weave a more consistent and coherent relationship among planning and regulatory entities involved in development and conservation. Local governments, for example, should identify wetlands areas and habitats of endangered species, and should reflect the development restrictions on these lands in their comprehensive plans, zoning, and subdivision requirements and procedures. Those plans and ordinances could also make room for the multiparty agreements resulting from areawide planning.

State and regional agencies concerned with growth management and environmental preservation should establish guidelines for meshing environmental permitting requirements with state, regional, and local planning processes. State agencies, in particular, could assist in smoothing relationships between federal regional offices and local

governments, including authorization, for example, of multiparty development and conservation agreements.

In short, collaborative, areawide planning needs help and support throughout the various levels of government if the concept's advantages are to be truly realized.

Discussion Questions

1. This article explains ways that conflicts could have been avoided in Sarasota. Summarize the key strategies to involve multiple stakeholders in urban growth management and planning.

2. List and describe the pros and cons of this approach.

Small Town Thinking

Erik Hagerman

■

SOURCE
World Watch, July/August 1991, Vol. 4, No. 4

FEW AMERICAN ideals have promised so much and delivered so little as the suburb. More than 40 years have passed since Levittown, New York, inaugurated the age of large-scale suburban housing developments, and what once epitomized the American dream is looking more like a nightmare.

Mile after mile of countryside has been overrun by faceless strip development festooned in neon and plastic. A once-liberating road and highway network has become a mire of congestion. Trimmed hedges and ornamental front lawns now create secluded refuges rather than extensions of community.

Finally, in its shortsighted attempts to bring people to nature, the suburb has ended up defiling the environment. No aspect of modern life has generated more environmental damage than our car-dependent transportation system—from the alarming amounts of raw materials it consumes, to the pollutants it puts out, to the very roads themselves—and sprawling suburban development has played a primary role in shaping this system.

This may be about to change. A small but increasingly influential group of architects and town planners, known as "neo-traditionalists," are turning disappointment with suburbs into blueprints for a new kind of community. Their inspiration is none other than the American small town.

For this new generation of town-builders, the appeal of the small town lies not in nostalgia, but in functionality. "The problem with current suburbs is not that they are ugly," says Andres Duany, who, with his wife Elizabeth Plater-Zyberk, runs an architecture and planning firm based in Miami, Florida. "The problem is they don't work."

Small towns do work, largely because they were designed for people, not cars. According to Duany, "Most of the needs of daily life can be met within a three-to-four-acre area, and generally within a five-minute walk of a person's home." Just about any town of less than 50,000 people built before the turn of the century can serve as an example, as can the older sections of cities such as Annapolis, Maryland, and Charleston, South Carolina.

The ill-fated love affair with the auto changed this time-honored design. As planners switched their focus from walking to driving, they unintentionally dismantled many of the basic physical features that made American communities such pleasant places to live. Roads were widened and parking lots expanded; sidewalks and tree-lined streets were eliminated.

The neo-traditionalists want to return these features to the American townscape. In the process, they hope to bring people out of their cars, and perhaps return some vitality to what they see as an ailing sense of community.

The cornerstone of the new approach is the concept of mixed use, which brings homes near offices, and both near shopping. Add to this low buildings, streets that encourage walking and downplay cars, and nearby parks and town squares, and you have the basic framework of a viable small town.

Real, operating examples of the new traditionalist approach are few because of years of lag time between design and development. The furthest along of these projects are just beginning to open, but they are easy to distinguish from most of the suburban development that is now taking place.

At Kentlands, a neo-traditional community being built in Gaithersburg, Maryland, the sections of neighborhoods that have been built—with their narrow, grid-like streets, compactly sited houses, and back alleys lined with garages—bring to mind not suburbia but nearby Annapolis, Maryland, and Georgetown in Washington, D.C. It will be 10 years before the development fills out its design of a town of close-to-home neighborhood parks, corner stores, public lakes, elementary and pre-schools, and a downtown center. But Kentlands already evokes the sense of place that most suburbs noticeably lack.

It's this ineffable quality that lies at the heart of neo-traditionalism's rapidly growing popularity. "[Neo-traditionalism] is unique in the history of modern architecture in that it has been as much a popular as a professional phenomenon," says Vincent Scully, art professor at Yale University and one of the country's most influential architecture critics. Duany and Plater-Zyberk, he says, are "bringing to fruition the most important contemporary movement in architecture."

Scully's comments are borne out by a recent Gallup poll. Asked where they would prefer to live, more people chose a small town than suburbs, farms, or cities.

These desires have not been lost on developers. Many have gotten no farther than the nostalgic image of neo-traditionalism and continue to peddle what are essentially the same old subdivison, dressed up with a sidewalk here and a gazebo there. "Like a frog turning into a prince," Philip Langdon recently wrote in *The Atlantic,* "the [conventional] pod becomes a 'village' with a kiss from the marketing staff."

But an increasing number of developers are attempting to build the real thing. Over the past eight years, Duany and Plater-Zyberk have designed more than 30 new towns and urban retrofits, ranging from Birmingham, Alabama, to Los Angeles, and from 60 acres and about 100 people to 10,000 acres and more than 25,000 people.

Peter Calthorpe, an architect and town planner based in San Francisco and another pioneer in the neo-traditional approach to community design, is finding himself in similar demand. His list of current projects includes Sutter Bay, one of California's largest residential development projects. The project will consist of 14 villages and house 175,000 people on a 25,000-acre site 10 miles north of Sacramento. What makes Sutter Bay particularly unique is that Sacramento has already agreed to extend is new light-rail transit system to link the community with the city center.

Unfortunately, in most regions of the United States, actually building a small town might get you arrested. Many of the distinguishing characteristics of small towns—narrow streets, on-street parking, shops near residences—are forbidden by codes written when auto dependence was thought to be a sign of progress.

As a result, many neo-traditionalists are concentrating their attention on persuading local zoning boards to revise these codes. Says Duany: "Planning codes and zoning ordinances are the genetic codes that determine what communities will look like in the future. It's unrealistic to think that we can retrain 35,000 planners. The most effective route is to get this [kind of town design] in the codes, and then let them follow it."

In each of their many projects, Duany and Plater-Zyberk helped communities push through revised sets of zoning codes to re-allow small-town features. To spread the word further, they recently completed a set of neo-traditional ordinances that virtually any community can add directly to its existing zoning codes. Calthorpe has put together a similar package for the county of Sacramento, which asked him in 1990 to design a plan that would make transit and pedestrian orientation a part of all new development in the county.

Where the bureaucratic barriers to neo-traditional communities have come down, the public has responded. At Duany and Plater-Zyberk's first project, Seaside, a mixed-use resort community on the coast of the Florida panhandle, lot prices have increased 500 percent in the past eight years. Seaside is still the only fully completed example of the neo-traditional approach, and it is somewhat unrepresentative because of its resort status, but 12 other projects are now under construction, with homes virtually all selling at or faster than the market pace.

Americans finally seem to be recognizing something Europeans have known for quite some time: people can guide and control the shape of their communities. The recognition comes none too soon. As Duany noted at the end of a recent lecture, "[Unless things change], all the energy that we put into all this growth is going to be the heritage of misery. If we're not careful, we are going to be remembered as the generation that destroyed America."

Discussion Questions

1. Explain the main development concept described in this article and ways to implement it.

2. How will this idea help promote sustainability?

Acknowledgments

Text Credits

Barbara Sleeper, "Wetlands, Wonderlands," *Animals,* Jan/Feb 91, Vol. 24, No. 1. Reprinted with permission from *Animals* magazine published by the Massachusetts Society for the Prevention of Cruelty to Animals, 350 S. Huntington Avenue, Boston, MA 02130.

Bill Sharp/Elaine Appleton, "In Search of the Early Everglades." Reprinted by permission from *National Parks* magazine, Jan/Feb 93, Vol. 6, Nos. 1–2. Copyright ©1993 by National Parks and Conservation Association.

Jeff Porro, "Saving Disney's Land." Published originally in *Nature Conservancy,* Mar/Apr 93, Vol. 43, No. 2. Reprinted by permission of the author.

Jeff Porro, "Finding Refuge in the Wetlands." Published originally in *Nature Conservancy,* Mar/Apr 93, Vol. 43, No. 2. Reprinted by permission of the author.

George Laycock, "'Good Times' Are Killing the Keys," *Audubon* magazine, Sept/Oct 91, Vol. 93, No. 5. Reprinted with permission of author and National Audobon Society.

Chris Kidder, "Return of the Red Wolf," published originally in *Nature Conservancy,* Sept/Oct 92, Vol. 42, No. 5. Reprinted by permission of the author.

Gwen Florio, "The Beetles and the Wily Weed," *The Philadelphia Inquirer,* Mar 15, 92. Reprinted with permission from *The Philadelphia Inquirer.*

William Oscar Johnson, "Back on Track," *Sports Illustrated,* Apr 30, 90, Vol. 72, ©1990 Time, Inc., all rights reserved.

The Environmental Exchange, "Houston Citizens Negotiate with Giant Oil Company and Win Toxic Reductions," *What Works: Air Pollution Solutions,* Report No.1, 1992. Reprinted with permission of The Environmental Exchange.

The Environmental Exchange, "Reducing Toxic Emissions: Low-Income Neighborhood Blocks Houston Ammonia Facility," *What Works: Air Pollution Solutions,* Report No.1, 1992. Reprinted with permission of The Environmental Exchange.

Susan Reed, "His Family Ravaged by Cancer, an Angry Louisiana Man Wages War on the Very Air That He Breathes," *People.* May 25, 91, Vol. 35, No. 20, copyright ©1991.

Janet Raloff, "Urban Smog Control: A New Role for Trees?," *Science News,* July 5, 90, Vol. 138. Reprinted with permission from *Science News,* the weekly news magazine of science, copyright 1990 by Science Service, Inc.

Aaron Sugarman, "The Case of the Disappearing National Park," May 92. Courtesy *Condé Nast Traveler.* Copyright ©1992 by The Condé Nast Publications Inc.

Heather Swain, "Power Plants Threaten Shenandoah's Air." Reprinted by permission from *National Parks* magazine, Mar/Apr 91, Vol. 65. Copyright ©1991 by National Parks and Conservation Association.

Ted Williams, "Strip-Mine Shell Game," *Audubon* magazine, Nov/Dec 92, Vol. 94, No. 6. Reprinted with permission from the National Audubon Society.

Peter Applebome, "Striking Gold and Fear in South Carolina," *NY Times National,* June 5, 91. Copyright ©1991 by The New York Times Company. Reprinted by permission.

Becky Gillette, "The Green Revolution in Wastewater Treatment," *BioCycle,* Dec 92, Vol. 33, No. 12. Reprinted with permission from J.G. Press, Inc.

Ellen Underwood, "Cape Coral Reclaims Its Water Supply," *WaterLines,* Winter 1992. Reprinted with permission of the South Florida Water Management District.

John L. Jackson, Jr./Phil Cross, "Citrus Trees Blossom with Reclaimed Water," *Water Environment & Technology,* Feb 93, Vol. 5, No. 2, ©1993, Water Environment Federation, used with permission.

Donald G. Schueler, "Southern Exposure." Published originally in *Sierra,* Nov/Dec 92, Vol. 77, No. 6. Copyright ©1992 by the Sierra Club. Reprinted by permission of the author.

The table, "Signs of Trouble Down South," reprinted with permission from the Nov/Dec 92 issue of *Sierra* magazine.

Tom Horton, "Savannah River: The Deadly Cover-Up," *Outdoor Life,* Apr 92, Vol. 189, No. 4. Copyright ©1992 by Tom Horton. Reprinted by permission of Sterling Lord Literistic, Inc.

Jim Schwab, "Toxic Buyouts: You Can't Go Home Again," *The Neighborhood Works* magazine, Aug/Sept 92, Vol. 15, No. 4, 2125 W. North Ave, Chicago, IL 60647. Subscriptions $30/year.

Liz Chandler, "A Tale of One City," *EPA Journal,* July/Aug 92, Vol. 18, No. 3. Reprinted with permission.

Robert Bracken, "North Carolina County Institutes Sticker System," *BioCycle,* Feb 92, Vol. 33, No. 2. Reprinted with permission of author and J.G. Press, Inc.

Barbara Kropf/Caroline Mixon, "Drop-Off Recycling in Tampa: Maximizing Efficiency and Quality," *Resource Recycling,* Jan 93, Vol. 12, No. 1. Reprinted with permission of *Resource Recycling* magazine, P.O. Box 10540, Portland, OR 97210, (503) 227-1319.

Robert Steuteville, "Responding to Andrew: Recycling After the Storm," *Bio-Cycle,* Oct 92, Vol. 33, No. 10. Reprinted with permission from J.G. Press, Inc.

Kathleen Meade, "Florida's New Market Strategy," *Waste Age,* July 92, Vol. 23, No. 7. Copyright ©1992 by National Solid Waste Management Association.

Kathleen Sheehan, "Composting in Texas," *Waste Age,* Aug 92, Vol. 23, No. 8. Copyright ©1992 by National Solid Waste Management Association.

Marie L. York, "Growthbusters Versus Growth Management: The Sarasota Saga," *Environmental and Urban Issues,* Fall 1990, Vol. 18, No. 1. Reprinted with permission from Florida Atlantic University/ Florida International University (FAU/FIU) Joint Center for Environmental and Urban Problems.

Richard Ward, "Planning for Growth in Arlington County, Virginia." Reprinted with permission from *Urban Land,* Jan 91, by ULI–Urban Land Institute, 625 Indiana Ave NW, Washington, D.C. 20004.

Douglas R. Porter, "Multiparty Development and Conservation Agreements." Reprinted with permission from *Urban Land,* Feb 92, by ULI–Urban Land Institute, 625 Indiana Ave NW, Washington, D.C. 20004.

Erik Hagerman, "Small Town Thinking," *World Watch,* July/Aug 91, Vol. 4, No. 4. Reprinted with permission from Worldwatch Institute.

Photo and Map Credits

P. vi: © W. Cody/Westlight; p. 4: © Gerry Ellis; p. 7: © John Shaw/Tom Stack & Associates; p. 8: © National Parks and Conservation Association; p. 10: © Nancy Webb; p. 12: © Gary Cox; p. 14: © Catherine Karnow/Woodfin Camp & Assoc.; p. 18: © Grady Allen; p. 26: *People Weekly* © Taro Yamasaki; p. 40: © Becky Gillette; p. 46: © Phil Cross; p. 48: © Tracy Glantz; p. 63: Courtesy of Mecklenburg County Public Service; p. 66: American Refuse System/Greg Peverall.

Principles of Sustainability

A sustainable society is one in which all human activity takes place and is maintained over time within the limits set by the environment—most importantly, the capacity of the environment to assimilate waste, to provide food, and to supply a host of other resources, such as water and building materials. A sustainable society is one that meets its needs in ways that permit other species and future generations to meet their needs. A sustainable society is based on two sets of principles: ethical and operational.

Ethical Principles

1. All life depends on a healthy, well-functioning ecosystem. (Planet care is the ultimate form of self-care.)
2. The Earth has a limited supply of resources, which must be shared by all living things. ("There's not always more, and it's not all for us.")
3. Humans are a part of nature, subject to its rules. There are no exceptions. We violate the laws of nature at our own risk.
4. Human success results from cooperation with nature, fitting into the web of life, rather than domination and control.

Operational Principles

A sustainable society is built on six principles: (1) conservation (frugality and efficiency), (2) recycling, (3) renewable resources, (4) restoration, (5) population control, and (6) adaptability.

1. Conservation   Conservation means using only what you need. It requires a cutback on unnecessary consumption. Be a conscientious consumer: Buy only what you need and purchase durable goods.

Conservation also means using resources much more efficiently. Currently, U.S. citizens and businesses waste one-half, perhaps three-fourths, of the energy they consume. Every day, huge amounts of water and other resources are wasted. Although often viewed as one of society's greatest faults, this wastefulness also offers one of our greatest opportunities for improvement. By becoming more efficient, we can significantly reduce energy demand, habitat damage, pollution, and ensure a steady supply of resources for future generations.

2. Recycling   To recycle means to use materials again and again. Recycling saves energy and reduces all forms of pollution. It stretches limited supplies of finite resources and protects wildlife habitat. It also creates unparalleled employment and business opportunities. Currently, the United States recycles only about 14% of its municipal garbage. It could, however, recycle and compost as much as 50% to 90%.

3. Renewable Resources   Renewable resources, like forests or wind energy, are regenerated by natural processes. Americans could conceivably acquire ten times as much energy—each year—from renewable resources as from all of the fossil fuel remaining in the Earth. Several renewable energy supplies are now cost competitive with traditional sources, and they provide electricity without the severe environmental impacts of conventional sources.

Whenever possible, choose renewable resources (wood, paper, cotton, or wool) over nonrenewable resources (plastics and synthetic cloth) and support government programs that tap into economical renewable supplies.